Organic Reactions

Organic Reactions

VOLUME 65

WILEY

JOHN WILEY & SONS, INC., PUBLICATION

Published by John Wiley & Sons, Inc., Hoboken, New Jersey

Published simultaneously in Canada.

Library of Congress Catalog Card Number 42-20265

ISBN 0-471-68260-8

Printed in the United States of America

10 9 8 7 6 5 4 3 2 1

PREFACE TO THE SERIES

In the course of nearly every program of research in organic chemistry the investigator finds it necessary to use several of the better-known synthetic reactions. To discover the optimum conditions for the application of even the most familiar one to a compound not previously subjected to the reaction often requires an extensive search of the literature; even then a series of experiments may be necessary. When the results of the investigation are published, the synthesis, which may have required months of work, is usually described without comment. The background of knowledge and experience gained in the literature search and experimentation is thus lost to those who subsequently have occasion to apply the general method. The student of preparative organic chemistry faces similar difficulties. The textbooks and laboratory manuals furnish numerous examples of the application of various syntheses, but only rarely do they convey an accurate conception of the scope and usefulness of the processes.

For many years American organic chemists have discussed these problems. The plan of compiling critical discussions of the more important reactions thus was evolved. The volumes of *Organic Reactions* are collections of chapters each devoted to a single reaction, or a definite phase of a reaction, of wide applicability. The authors have had experience with the processes surveyed. The subjects are presented from the preparative viewpoint, and particular attention is given to limitations, interfering influences, effects of structure, and the selection of experimental techniques. Each chapter includes several detailed procedures illustrating the significant modifications of the method. Most of these procedures have been found satisfactory by the author or one of the editors, but unlike those in *Organic Syntheses* they have not been subjected to careful testing in two or more laboratories.

Each chapter contains tables that include all the examples of the reaction under consideration that the author has been able to find. It is inevitable, however, that in the search of the literature some examples will be missed, especially when the reaction is used as one step in an extended synthesis. Nevertheless, the investigator will be able to use the tables and their accompanying bibliographies in place of most or all of the literature search so often required.

Because of the systematic arrangement of the material in the chapters and the entries in the tables, users of the books will be able to find information desired by reference to the table of contents of the appropriate chapter. In the interest of economy the entries in the indices have been kept to a minimum, and, in particular, the compounds listed in the tables are not repeated in the indices.

The success of this publication, which will appear periodically, depends upon the cooperation of organic chemists and their willingness to devote time and effort to the preparation of the chapters. They have manifested their interest already by the almost unanimous acceptance of invitations to contribute to the work. The editors will welcome their continued interest and their suggestions for improvements in *Organic Reactions.*

Chemists who are considering the preparation of a manuscript for submission to *Organic Reactions* are urged to write either secretary before they begin work.

CONTENTS

CHAPTER 1

THE PASSERINI REACTION

Luca Banfi and Renata Riva

Department of Chemistry and Industrial Chemistry, University of Genoa, Genoa, Italy

CONTENTS

banfi@chimica.unige.it
Organic Reactions, Vol. 65, Edited by Larry E. Overman et al.
ISBN 0-471-68260-8

INTRODUCTION

Isocyanides are a very special class of organic compounds which may behave as acyl anion equivalents. However, with very few exceptions,[1–6] isocyanides do not react with carbonyl compounds in the absence of an acid. In this chapter, we consider all reactions involving the interaction of a carbonyl compound or an acetal with an isocyanide and an acid, processes that frequently but not invariably result in the incorporation of the acid residue in the final product.

In the "classic" Passerini reaction, discovered in 1921 by Mario Passerini[7] in Florence, Italy, the acid is always a carboxylic acid and the products are α-acyloxy amides (Eq. 1).

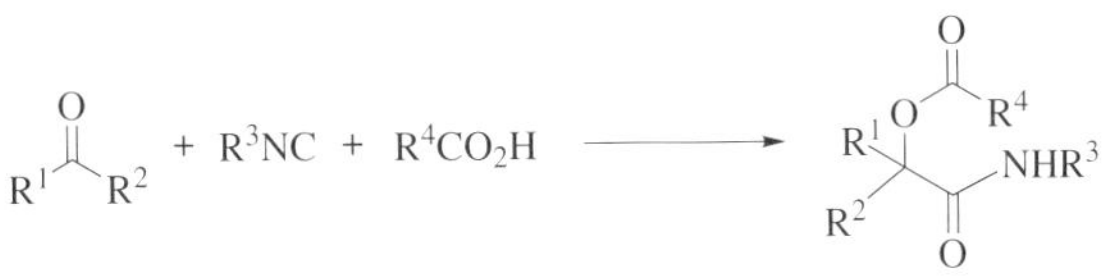
O
R1 R2
+ R3NC + R4CO2H
R1 O R4
NHR3
R2
O

(Eq. 1)

A more recent variation, which is often referred to as the Passerini reaction as well, employs a variety of mineral (H+) or Lewis acids (LA) to afford α-hydroxy amides (Eq. 2). These compounds may also be formed in the presence of carboxylic acids such as formic acid or trifluoroacetic acid.

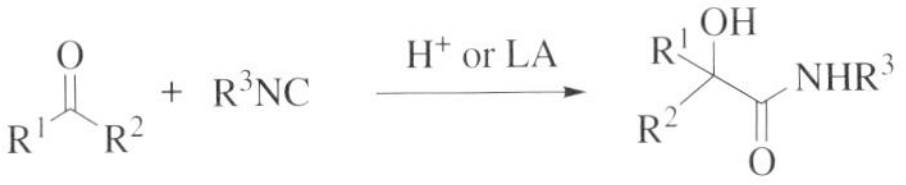
O
R1 R2
+ R3NC
H+ or LA
R1 OH
NHR3
R2
O

(Eq. 2)

Interactions of isocyanides with carbonyl compounds and some protic or Lewis acids can also lead to a large variety of products other than α-hydroxy amides, including several heterocyclic systems. The type of compounds that can be obtained by these "Passerini-type" reactions will be discussed later.

Acetals can also react with isocyanides in the presence of Lewis acids. The usual products are α-alkoxy amides (Eq. 3).

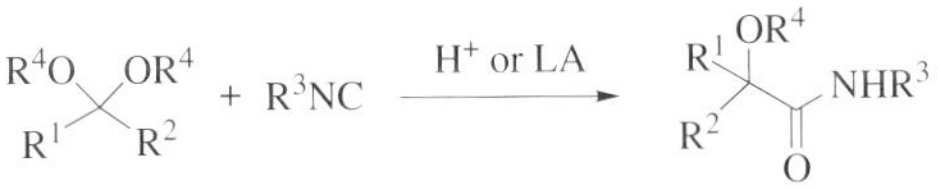
R4O OR4
R1 R2
+ R3NC
H+ or LA
R1 OR4
NHR3
R2
O

(Eq. 3)

The "classic" Passerini reaction is one of the oldest multicomponent reactions and is the first based on isocyanides to be discovered. This methodology is experiencing a growing interest in recent years, because of the usefulness of multicomponent reactions in combinatorial synthesis. While it is still probably the best method for producing α-acyloxy amides in a highly convergent manner, its synthetic scope has been increased recently by employing bifunctional substrates, which are able to undergo secondary reactions. In this way, a larger variety of products can be synthesized and complex biologically active substances can be accessed quickly. In addition, the variation leading to α-hydroxy amides often represents the method of choice for the formation of these types of compounds. Finally, modifications that form products different from α-acyloxy amides or α-hydroxy amides, especially those leading to heterocyclic systems, may find useful applications in synthesis.

Early results in this field were reviewed by Passerini.[8] More recently, several reviews of isocyanide-based multicomponent reactions, including the Passerini reaction, have appeared.[9–17] A closely related reaction is the Ugi four-component condensation.[11] In addition to the three components of the classic Passerini reaction, a primary amine is also involved. The Ugi condensation is believed to proceed through the formation of an imine, which then undergoes a three-component reaction with the isocyanide and the carboxylic acid to afford a *β*-acylamino amide (Eq. 4). This reaction is not covered in this chapter.

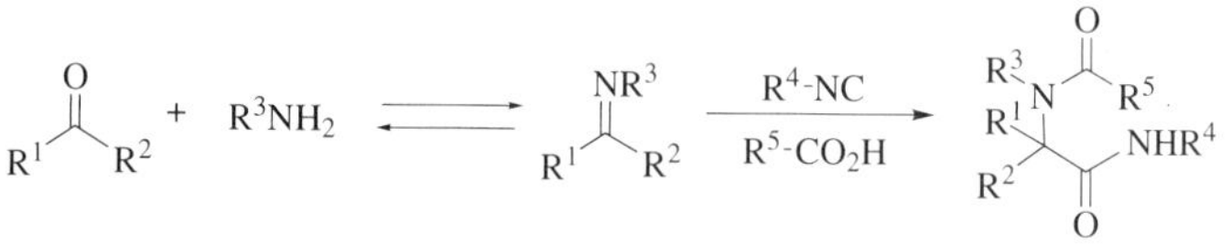
R1
R2
O
+
R3NH2
NR3
R1
R2
R4-NC
R5-CO2H
R3
N
R1
R2
O
R5
NHR4
O

(Eq. 4)

MECHANISM

Although several different mechanisms have been proposed,[7,18–22] the one most generally accepted for the classical Passerini reaction was proposed by Ugi (Eq. 5).[9,23] According to this mechanism, the final product arises from a rearrangement of the intermediate **1**.

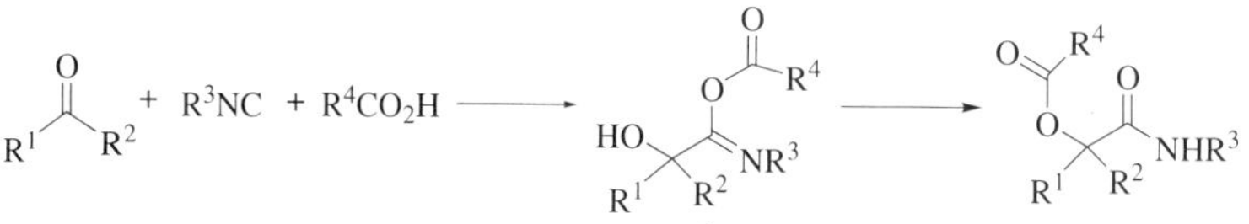
O
R1
R2
+ R3NC + R4CO2H
O
O
R4
HO
NR3
R1
R2
O
R4
O
O
NHR3
R1
R2

1

(Eq. 5)

There is some experimental evidence for the involvement of **1** as intermediate. In the specific example of Eq. 6,[9] compound **2**, corresponding to **1** in Eq. 5, could be isolated owing to its cyclic nature, which prevents the facile O → O migration of the acyl group.

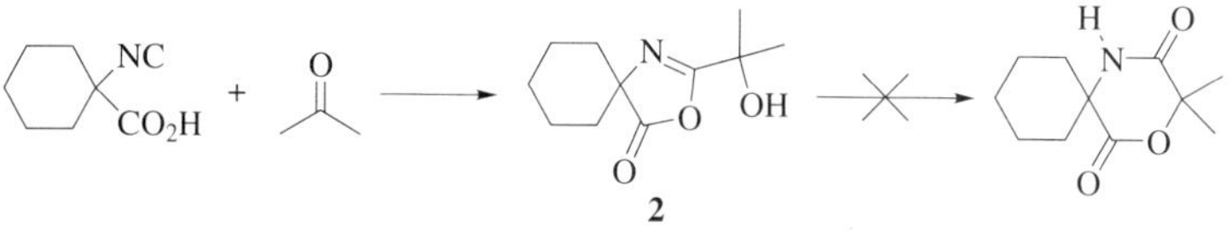
NC
CO2H
+
O
N
O
O
OH
H
N
O
O
O

2

(Eq. 6)

It is not clear whether the acyl migration is the rate-limiting step or, as it has been proposed for the related Ugi condensation,[11] a fast process. Formation of **1** is most likely irreversible, as suggested by the high asymmetric induction achieved with bulky chiral isocyanides.[24]

There is more uncertainty on how intermediate **1** is formed. One possibility involves the simultaneous reaction of an isocyanide with a carbonyl compound and a carboxylic acid giving intermediate **1**. In this hypothesis, the reaction proceeds through a relatively non-polar, cyclic transition state **3**. This can be either 5-membered or 7-membered, depending on which carboxylate oxygen participates. Eq. 7 depicts both transition states. This mechanism is in agreement with the fact that the reaction rate depends on all three components[18] and that the reaction is faster in non-polar solvents.[25]

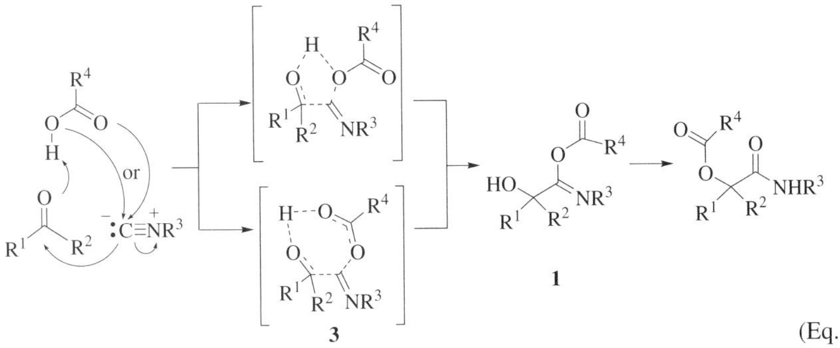

(Eq. 7)

Since the simultaneous union of three molecules is a very rare process in organic chemistry, an alternative possibility is shown in Eq. 8. Three separate steps are involved: protonation of the carbonyl, nucleophilic attack of the isocyanide onto the protonated carbonyl giving nitrilium ion **4**, and final reaction of the latter with the carboxylate to afford intermediate **1**. This three-step mechanism would, however, proceed through a charged transition state, which is not consistent with the higher rate of the reaction in non-polar solvents.

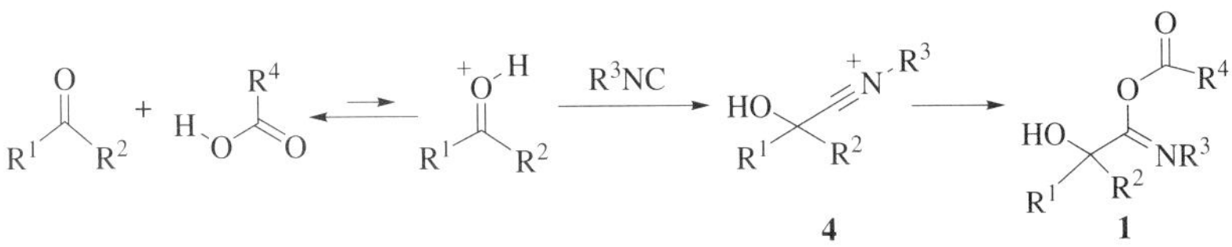

(Eq. 8)

A third, more appealing, possibility is the two-step mechanism shown in Eq. 9,[11] which involves the reaction of the isocyanide with a loosely bound adduct **5**, formed by reaction of the carboxylic acid with the carbonyl compound. This adduct may also be seen as a tight ion pair resulting from protonation of the carbonyl by the carboxylic acid. In this scenario, the rate-limiting step would entail the union of two species (the isocyanide and the adduct formed by the protonated carbonyl and the carboxylate) giving transition states **3**.

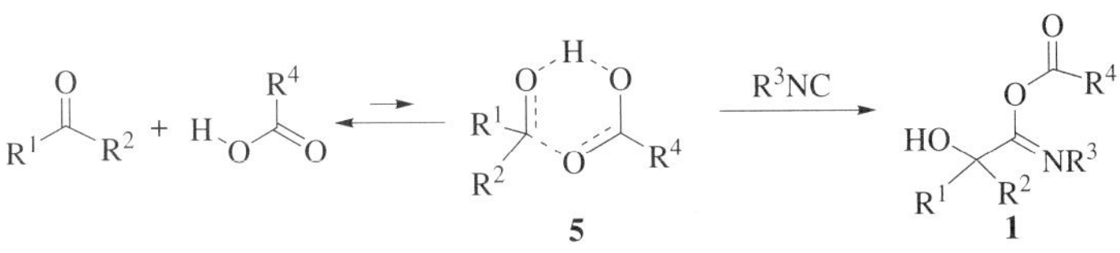

(Eq. 9)

Isocyanides are known to react with carboxylic acids giving formamides and anhydrides, probably by the mechanism shown in Eq. 10.[26] In the presence of a carbonyl compound, however, α-acyloxy amides are the major products. Anhydride formation as a side reaction becomes significant only when the carbonyl compound is a bulky ketone, suggesting that the reaction of the isocyanide with the protonated carbonyl compound, as compared with protonation of the isocyanide, is preferred. The reasons for this preference are not very clear at present, but stabilization of the transition state by simultaneous reaction with the two components of the adduct **5**, according to the mechanism of Eq. 9, is a conceivable explanation.

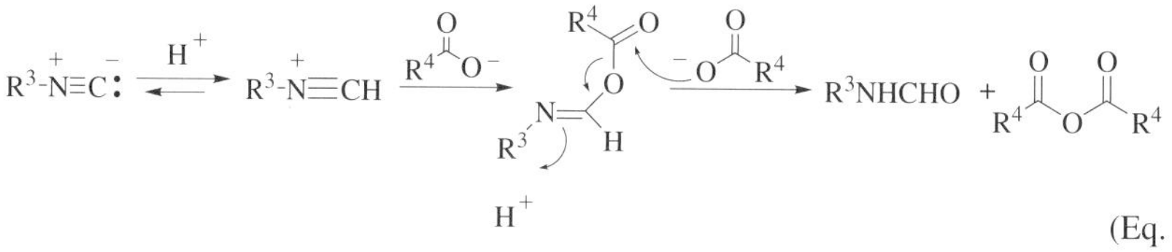

(Eq. 10)

Isocyanides also react with carbonyl compounds when an acidic species other than a carboxylic acid is employed. These condensations can be considered variations of the Passerini reaction and are covered in this chapter. As discussed in detail in the "Scope and Limitations" section, the type of products formed depends mainly on the nature of the acidic species. For example, use of strong mineral acids, pyridinium salts, or Lewis acids such as titanium (IV) chloride results in the formation of α-hydroxy amides. Their formation can be rationalized by a general mechanism similar to that of the first part of the classical Passerini reaction (Eq. 11). In this case, however, the nucleophile (A^-) that attacks the isocyanide carbon is different, and intermediate **6** is simply hydrolyzed during work-up. In the case of titanium (IV) chloride this mechanism was deduced by a thorough study, which excluded the intermediacy of organotitanium species.[27]

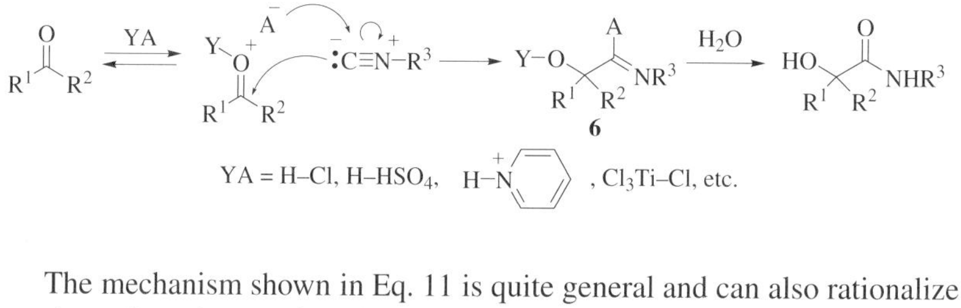

(Eq. 11)

The mechanism shown in Eq. 11 is quite general and can also rationalize the formation of products other than α-hydroxy amides. For example, when the acid species is HN_3 (Eq. 12) an intermediate analogous to **6** is formed, but it undergoes a cyclization process giving a tetrazole.[23]

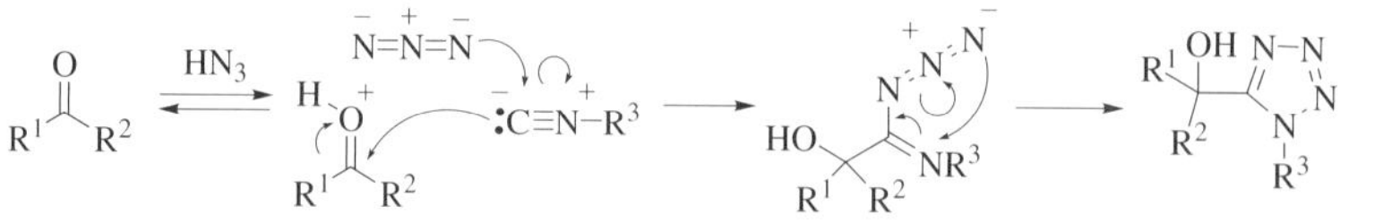

(Eq. 12)

When a good nucleophile is already present in the structure of the carbonyl component, it can react intramolecularly and outcompete A^- to afford special products.[28] When the Lewis acid, such as BF_3, cannot release a nucleophile A^-, the only available nucleophile is another molecule of isocyanide, and iminooxetanes are formed (Eq. 13).[29,30] The fate of these intermediates will be discussed later.

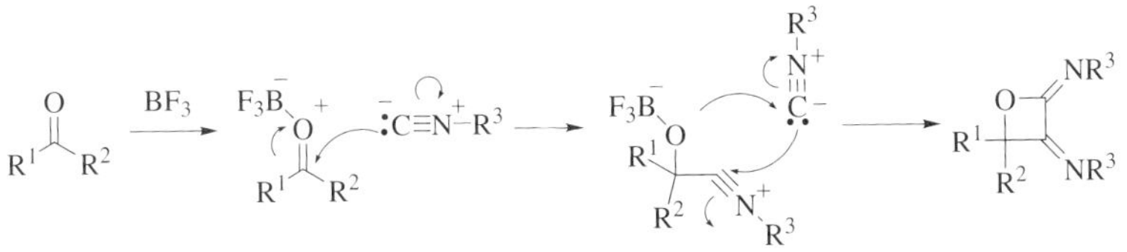
BF3
F3B
R1
R2
C≡N–R3
R3
NR3

(Eq. 13)

SCOPE AND LIMITATIONS

Classical Passerini Reaction

Functional Group Compatibility. The classic Passerini reaction takes place under mild conditions in a slightly acidic medium and thus tolerates a variety of different functional groups. As shown in the Tabular Survey, esters, nitriles, amides, urethanes, imides, β-lactams, sulfonamides, sulfoxides, sulfones, enamines, alkyl and aryl chlorides, acetals, epoxides, phosphonates, azides, aromatic nitro groups, *N*-nitroguanidines, and azo groups are all compatible with this reaction. Primary aliphatic nitro compounds are known to react with isocyanides giving nitrile oxides and are probably incompatible with the Passerini reaction.[31,32]

There are no examples of the Passerini reaction involving substrates containing a primary amine. This functional group appears to be incompatible, probably because substrates containing this functionality typically undergo the related Ugi condensation (Eq. 4).[11] There may be competition between these two reactions, with the Passerini condensation favored in non-polar solvents and in the absence of a catalyst. For example, Eq. 14 shows a reaction where the Passerini product was isolated during an attempted double Ugi reaction involving lysine.[33]

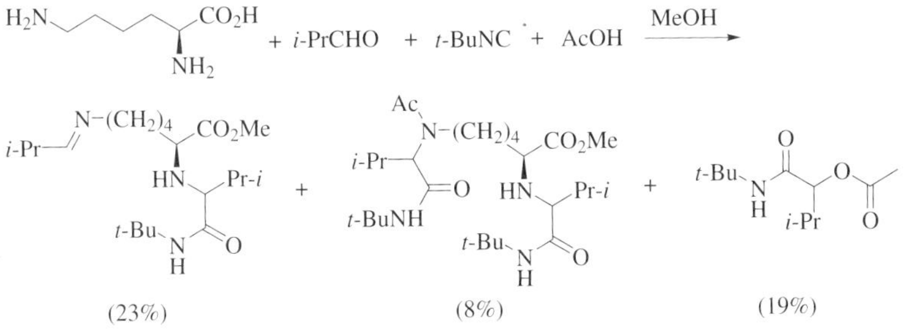
H2N
CO2H
NH2
+ i-PrCHO + t-BuNC + AcOH
MeOH
N–(CH2)4
CO2Me
i-Pr
HN
Pr-i
t-Bu
N
H
O
+
Ac
t-BuNH
O
(23%)
(8%)
(19%)

(Eq. 14)

However, normally unprotected α-amino acids cannot be used as an acid component for the Passerini reaction, since they prefer to react with carbonyl compounds and isocyanides according to the Ugi Five-Center-Four-Component Reaction (U-5C-4CR) (Eq. 15).[33,34]

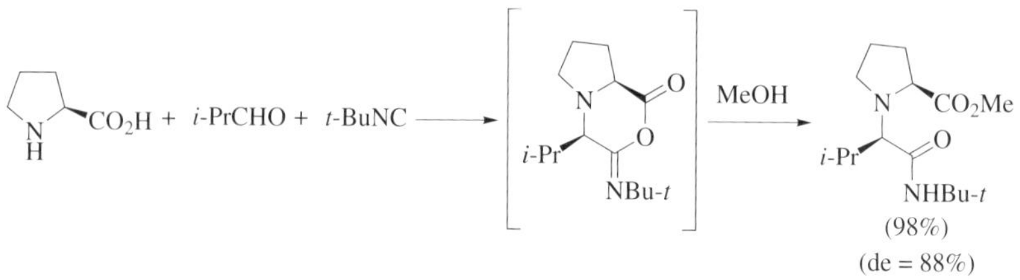
CO2H + i-PrCHO + t-BuNC
N
H
O
N
O
i-Pr
NBu-t
MeOH
N
CO2Me
O
i-Pr
NHBu-t
(98%)
(de = 88%)

(Eq. 15)

When two equivalents of the isocyanide and the carbonyl compound are reacted with one equivalent of anthranilic acid in the presence of HN_3, both Passerini and Ugi reactions can take place in one pot, giving a complex product derived from the union of six molecules (Eq. 16).[33]

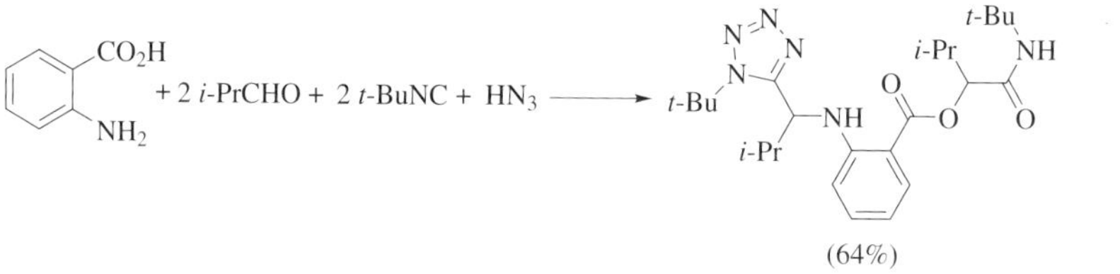
CO2H
NH2
+ 2 i-PrCHO + 2 t-BuNC + HN3
N=N
N
N
t-Bu
i-Pr
NH
O
O
t-Bu
i-Pr
NH
O
(64%)

(Eq. 16)

Secondary amines also participate in Ugi condensations[33,35] and could react with intermediate **1** as well (Eq. 5), making them incompatible with the Passerini reaction.

Tertiary amines are not able to undergo the Ugi reaction, and would be expected not to interfere. However, the failure to condense 2-(morpholinoethyl)isonitrile with benzaldehyde and acetic acid has been reported recently.[36] The buffering effect of the strongly basic amine may explain this result, since the Passerini reaction may be performed on weakly basic pyridinecarboxy aldehydes.[37,38] Isocyanides containing an enamine group can also be used.[39–41]

Examples with hydroxy-containing substrates are rare, possibly reflecting interference by the extra hydroxy group during the intramolecular acyl transfer step. However, appropriately positioned phenols are tolerated in both the carboxylic acid[3,42,43] and in the aldehyde component.[44] An example is shown in Eq. 17.[43]

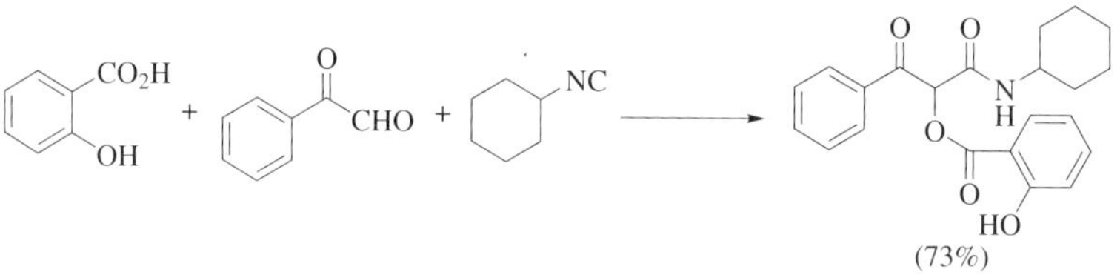
CO2H
OH
+
O
CHO
+
NC
O
O
N
H
O
O
HO
(73%)

(Eq. 17)

Carbonyl Compounds. Both aromatic and aliphatic aldehydes are usually good substrates for Passerini reactions. Protected α-amino,[45–51] α-alkoxy,[40] α-oxo aldehydes,[43,44,52–55] and β-oxo aldehydes[56] may also be used. In the latter two examples, the aldehyde reacts selectively over the keto functionality, as shown in Eq. 17. In addition, there are examples of the Passerini reaction utilizing sugar-derived aldehydes, in both monosaccharides[57,58] and polysaccharides.[59] An interesting example involves the use of three sugar-derived components to afford a complex pseudo-oligosaccharide, as shown in Eq. 18.[58]

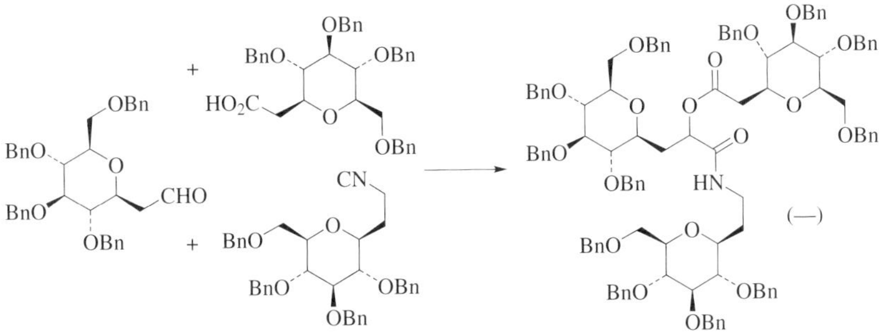

(Eq. 18)

Formaldehyde as a reactant represents a special example. In the classical Passerini reaction formaldehyde has been used successfully in the anhydrous, monomeric form (Eq. 19).[60] On the other hand, in the variations that employ HN_3[23] or H_2SO_4[61] as acidic components, an aqueous solution can be employed (see p. 22).

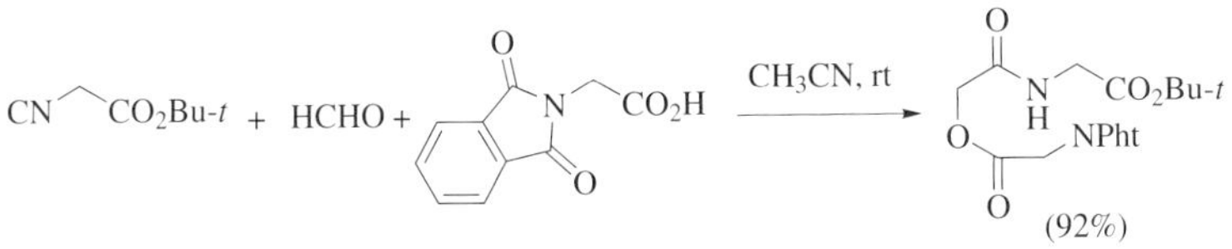

(Eq. 19)

With ketones, the reaction is generally slower. The bulkiness of the carbonyl substrate, especially when combined with a bulky isocyanide, may completely prevent the desired reaction.[3,62,63] The use of high pressure has been shown to dramatically increase the yield in reactions involving bulky reactants such as methyl isopropyl ketone and *tert*-butyl isocyanide (Eq. 20).[63]

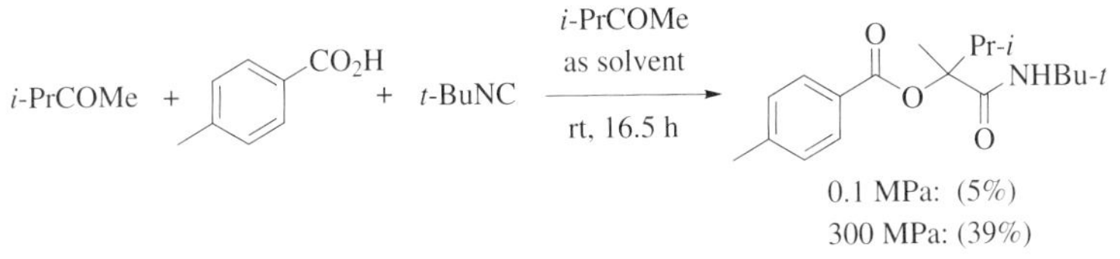

(Eq. 20)

β-Oxo esters,[64] α-oxo nitriles,[65] and α-diketones[3,63,66,67] are also useful substrates; in the latter reactions, only one of the carbonyl groups undergoes the condensation.

α,β-Unsaturated ketones and aldehydes, such as benzalacetone, cholestenone, and crotonaldehyde, are not reactive enough to undergo the Passerini condensation.[62]

Perhalogenated aldehydes and ketones show high reactivity in the classical Passerini reaction[6] and can even react with isocyanides in the absence of the carboxylic component. This behavior will be discussed later.

Isocyanides. Simple, unfunctionalized isocyanides react well in the Passerini reaction. They are usually prepared and purified prior to the condensation, one exception being the in situ preparation of methyl isocyanide through photochemical degradation of methyl isothiocyanate.[68] Methyl isocyanide can also be prepared prior to use.[69,70]

In recent years, polyfunctionalized, non-commercially available isocyanides have seen more use, particularly those derived from α-amino esters. Apart from the commercially available α-isocyano acetates, a variety of α-substituted[40,47,48] or α,α'-disubstituted isocyano acetates[71–74] have been prepared and successfully employed. With the latter reagents, the Passerini products have been obtained only with aliphatic aldehydes; ketones and aromatic aldehydes are not reactive enough.[73]

α-Substituted-α-isocyano amides (corresponding to dipeptides) have also been prepared in enantiomerically pure form and employed in the Passerini reaction.[45,49] Other examples of polyfunctionalized isocyanides include sugar-derived,[50,51,57] porphyrin-derived,[75] and phosphonylmethyl-derived compounds (used both in classical[40,76] and in Lewis acid catalyzed Passerini reactions[27,77,78]). Finally, isocyanides containing a boronic ester have also been employed in the Lewis acid catalyzed Passerini condensation.[79]

On the other hand, isocyano-1,3,5-triazines are inert in the classical Passerini reaction.[80] However, they do react in the BF_3-catalyzed variation (see p. 28).

Isocyanides are usually prepared by dehydration of the corresponding formamides. An alternative method involves functionalization of a simpler isocyanide. In this manner, a large number of different substrates may be accessed in a convergent fashion. This approach is particularly interesting in light of diversity-oriented synthesis because it increases the number of different starting materials from 3 to 4. This strategy, known as "reagent explosion," can take advantage of the relative acidity of the α-hydrogen atoms in α-isocyanoacetates. As shown in Eq. 21, the reaction of methyl isocyanoacetate with a secondary amine and the acetal of 1-(diethoxymethyl)imidazole, in the presence of camphorsulfonic acid (CSA), affords functionalized isocyanides **7**, which may be employed in subsequent Passerini reactions.[39] This strategy has been utilized for preparing a library of 4,620 different adducts in solution.

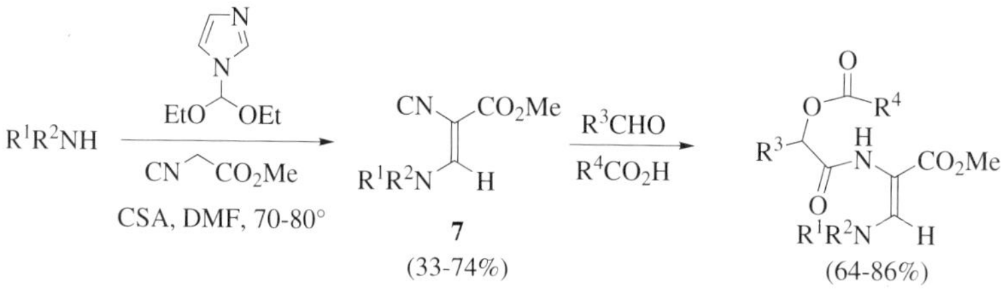
R¹R²NH
EtO OEt
CN CO₂Me
CSA, DMF, 70-80°
CN CO₂Me
R¹R²N H
7
(33-74%)
R³CHO
R⁴CO₂H
R³
R⁴
CO₂Me
R¹R²N H
(64-86%)

(Eq. 21)

Unsaturated isocyanides of general formula **7**, in addition to other types of α-alkylydene isocyano acetates, have also been transformed through the Passerini reaction into a library of azinomycin analogs.[41] The alkylation of α-lithio methyl isocyanide is another method to obtain diverse isocyanides from a single precursor.[81,82]

A variety of α-isocyano acetamides may be prepared from α-isocyano acetic esters by reaction with amines under acidic catalysis. However, when the α-isocyano acetic esters are monosubstituted at the α-position and are optically active, this methodology leads to racemization.[78]

Carboxylic Acids. While good results are typically achieved with a variety of aliphatic or aromatic acids, lower yields have been observed with bulky carboxylic acids, for example, in the Passerini reaction involving menthoxyacetic acid with ketones.[83] A significant decrease in yield with increased branching is observed in the reaction of cyclohexyl isocyanide with acetone and valeric, isovaleric, and pivalic acids.[63] This problem can be overcome in part by carrying out the reaction at $-20°$[18] or at high pressure.[63]

With formic acid, the formation of the α-hydroxy amide as the main product along with the normal Passerini adduct has been observed from the sugar-derived isocyanide **8** (Eq. 22).[50] In another example, a lower yield (compared to that observed with acetic acid) of the usual Passerini adduct was reported, although the by-products were not identified.[84]

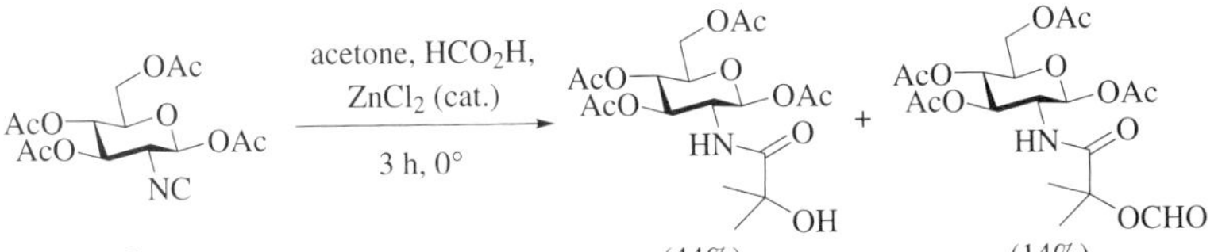

OAc
AcO
AcO
O
OAc
NC
acetone, HCO2H,
ZnCl2 (cat.)
3 h, 0°
HN
OH
OCHO

8 (44%) (14%) (Eq. 22)

The formation of α-hydroxy amides as by-products is also observed when *N*-Boc-α-amino acids are employed as carboxylic acid components in reactions with protected α-amino aldehydes.[45] In the example shown in Eq. 23, the desired Passerini adduct **9** is contaminated with the corresponding hydroxy amide **10**. Interestingly, the ratio of the two products does not seem to depend on the dryness of the reaction medium or on the reaction time. The reasons for this behavior are not clear and surely deserve a more thorough examination. A possible hypothesis is that intermediate **1** (Eq. 5) undergoes deacylation promoted by an external nucleophile (e.g., another molecule of the carboxylic acid giving an anhydride) or by an internal nucleophile (e.g., the NHBoc group). There may be a relationship between the extent of this side reaction and the acid strength of the carboxylic acid, since both formic acid and *N*-Boc-α-amino acids are more acidic than normal aliphatic and aromatic acids.

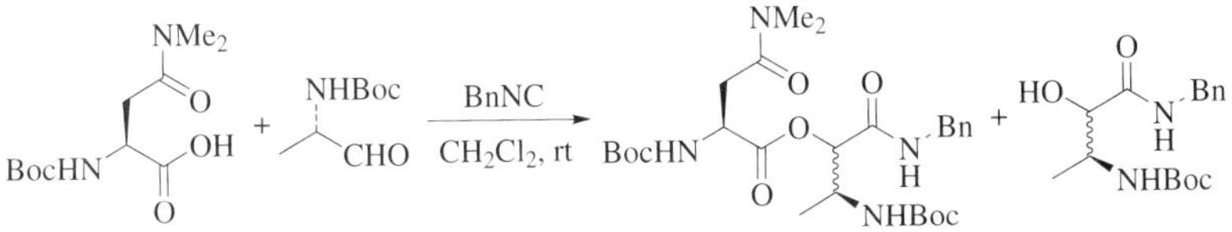

NMe2
BocHN
OH
NHBoc
CHO
BnNC
CH2Cl2, rt
Bn
HO
NHBoc

9 (60%) **10** (32%) (Eq. 23)

However, other acids with electron-withdrawing groups, such as α-cyanoacetic acid,[52] sulfonylacetic acids,[54] chloroacetic acid,[60,71] trichloroacetic acid,[74] azidoacetic acid,[60] fluoroacetylaminoacetic acid,[76] and phosphonylacetic acids,[44] give high yields in the Passerini reaction. α,β-Unsaturated acids[66] and alk-2-ynoic acids[85] also work well in this reaction.

As noted earlier, unprotected α-amino acids are typically poor substrates in this reaction, since they preferentially undergo the Ugi 5C-4CR reaction.[86] However, aspartic and glutamic acids are exceptions, since each bears an additional carboxylic acid moiety. In the presence of two equivalents of an aldehyde and an isocyanide, both the Ugi 5C-4CR and the Passerini reactions occur, giving rise to adducts **11** and **12**, respectively, in a single step (Eqs. 24, 25). The high chemoselectivity of these reactions is worth noting; the Passerini condensation occurs exclusively at the carboxylic acid of the side chain. Interestingly, with glutamic acid, adduct **12** is quantitatively converted in situ into the corresponding piperazinedione **13**.

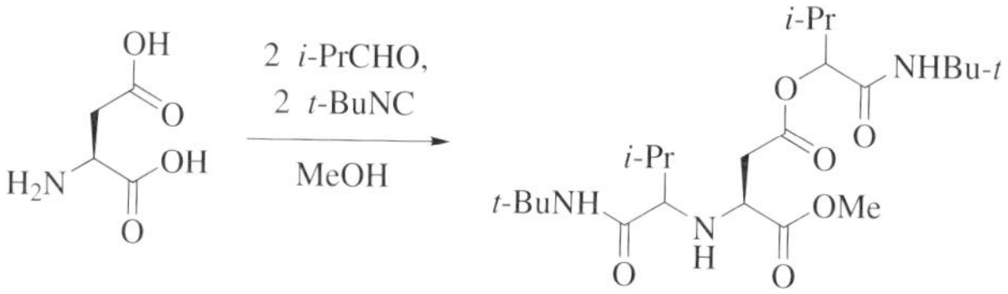
OH
O
H2N
OH
O
2 i-PrCHO,
2 t-BuNC
MeOH
i-Pr
NHBu-t
O
O
O
i-Pr
O
t-BuNH
N
H
OMe
O
O

11 (85%)

(Eq. 24)

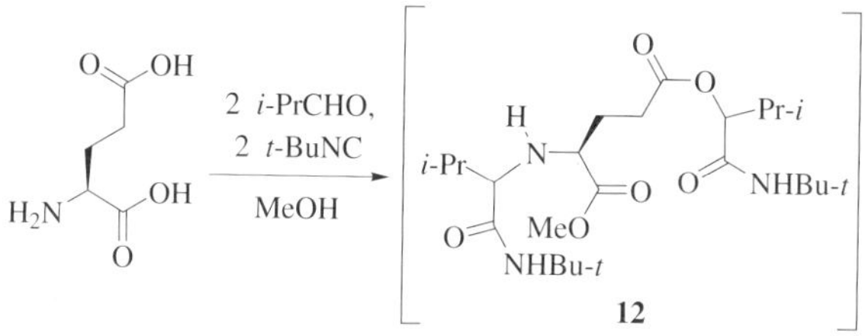
O
OH
H2N
OH
O
2 i-PrCHO,
2 t-BuNC
MeOH
O
O
Pr-i
H
N
i-Pr
O
O
NHBu-t
O
MeO
NHBu-t

12

(Eq. 25)

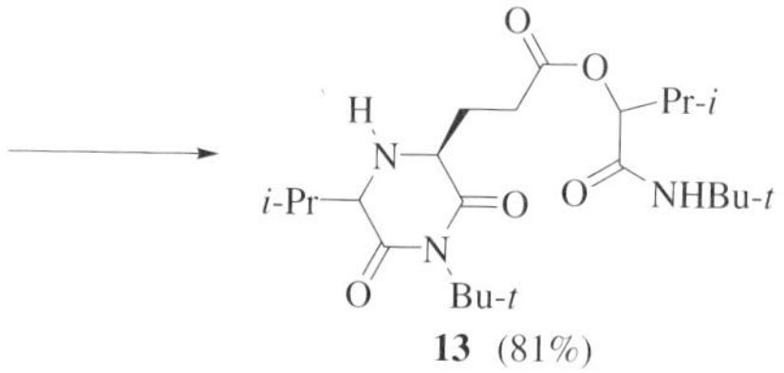
O
O
Pr-i
H
N
i-Pr
O
O
NHBu-t
N
O
Bu-t

13 (81%)

Intramolecular Reactions. An intramolecular variant of the Passerini reaction requires that two of the three reacting functionalities be present in the same molecule. Of the three possible combinations, the one that has been most studied involves bifunctional substrates containing both the carbonyl and carboxylic acid functionalities, which leads to lactones.

For example, salicylaldehydes react with isocyanides to afford isobenzofuranones such as **14** (Eq. 26).[81,87] In an analogous manner, levulinic acid furnishes γ-lactones (Eq. 27).[88] The orthoamide **17** is obtained in excellent yield and with high stereoselectivity starting from oxo acid **15**, most likely through cyclization of the 7-membered lactone **16** (Eq. 28).[89] Finally, an 8-membered lactone is also obtained through an intramolecular Passerini reaction.[90] Although these are the only examples found in the literature, it is conceivable that other lactones of different ring sizes might be obtained by choosing the appropriate oxo acid.

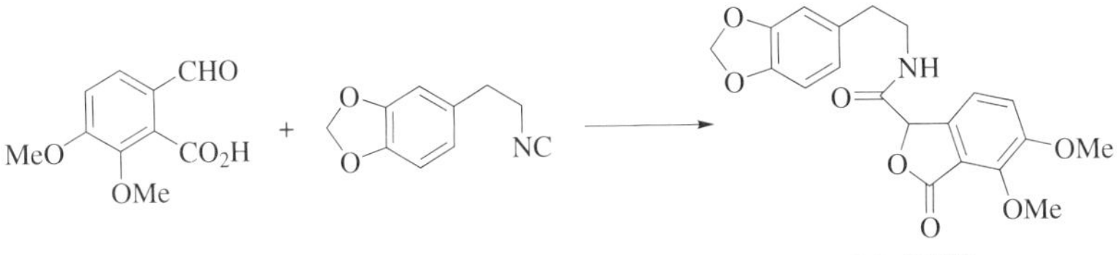

14 (71%)

(Eq. 26)

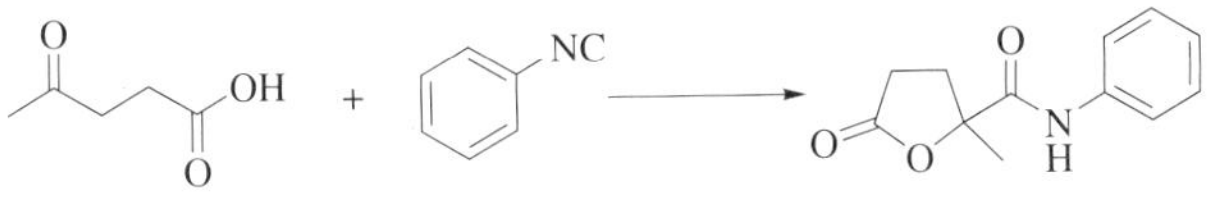

(66%)

(Eq. 27)

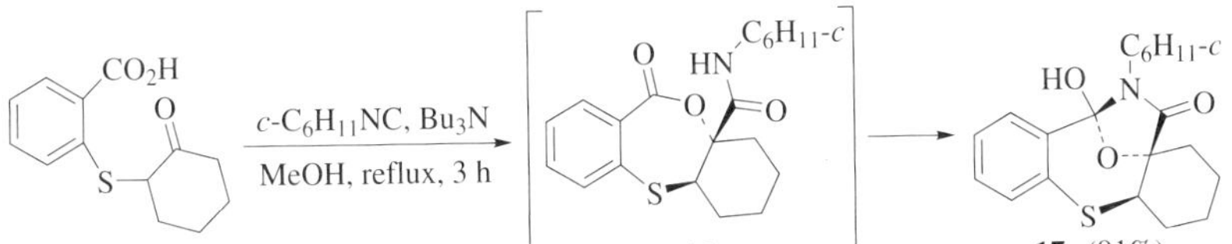

15 **16** **17** (91%)

(Eq. 28)

The Passerini reaction of carboxymethylcellulose or hyaluronic acid **18** with an isocyanide and glutaraldehyde is a method for generating cross-links in the polymer (Eq. 29).[59] After the first intermolecular reaction with one of the two carbonyl groups, a second intramolecular reaction involves the polysaccharide carboxy group and the second carbonyl of glutaraldehyde.

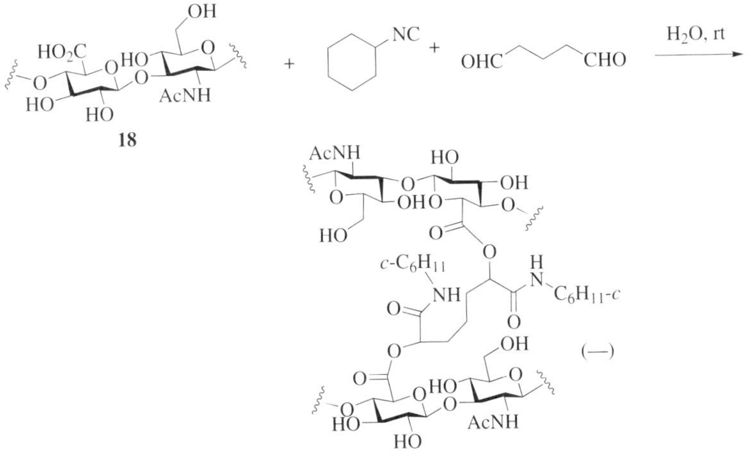

18

(Eq. 29)

Cross-links can also be obtained using partially oxidized[91] pullulane and scleroglucan, which contain both carboxy and aldehyde groups.[59] Ester bridges obtained by this procedure are partially labile and, therefore, cross-linking using the related Ugi condensation seems more promising.[92]

The only example in the literature of the intramolecular condensation of an isocyano acid with a ketone was discussed earlier (Eq. 6). In that reaction, the intermediate oxazolinone does not isomerize into the α-acyloxy amide. It would be interesting to explore the behavior of other isocyano acids where the two functional groups are more distant. However, because isocyanides are known to react with carboxylic acids, increasing the distance may lead to unstable species.

Although there are no reported examples of the Passerini reaction involving an isocyano aldehyde or an isocyano ketone, the reaction of isocyano ketone **19** with a carboxylic acid was attempted, but it did not provide the expected Passerini adduct **20** (Eq. 30).[93] The two main products, **21** and **22**, appear to arise from an intermolecular reaction of the isocyanide with the carboxylic acid and are analogous to the products obtained by reacting phenylacetic acid with methyl isocyanoacetate in the absence of any carbonyl compound (Eq. 31).[93] It should be noted that formamides and anhydrides are the usual products of the reaction of isocyanides with carboxylic acids, and that compound **21** is the product of cyclization-dehydration of an acyclic formamide. The fact that the intermolecular reaction of the isocyanide with the acid is favored over intramolecular attack at the carbonyl may reflect a geometric constraint that prevents the approach of the isocyanide carbon to the carbonyl. Increasing the distance between the two functional groups may remove this constraint, and it is conceivable that an intramolecular Passerini reaction of an isocyano ketone or isocyano aldehyde is possible provided that the ring in the transition state is sufficiently large.

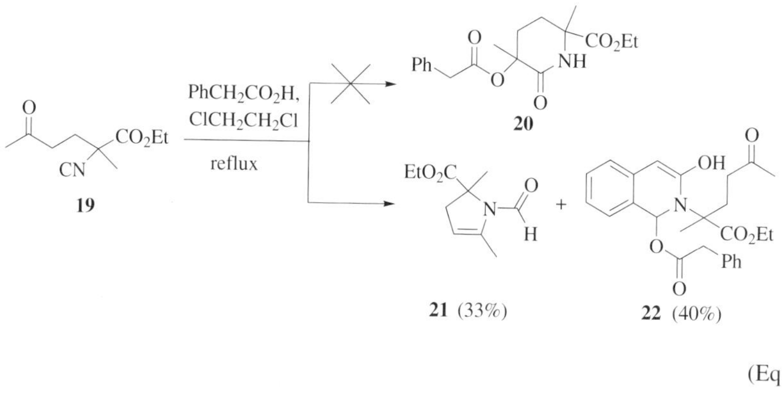
O
CO2Et
CN
19
PhCH2CO2H,
ClCH2CH2Cl
reflux
Ph
O
O
CO2Et
NH
O
20
EtO2C
N
O
H
+
OH
N
O
CO2Et
O
Ph
O
21 (33%)
22 (40%)

(Eq. 30)

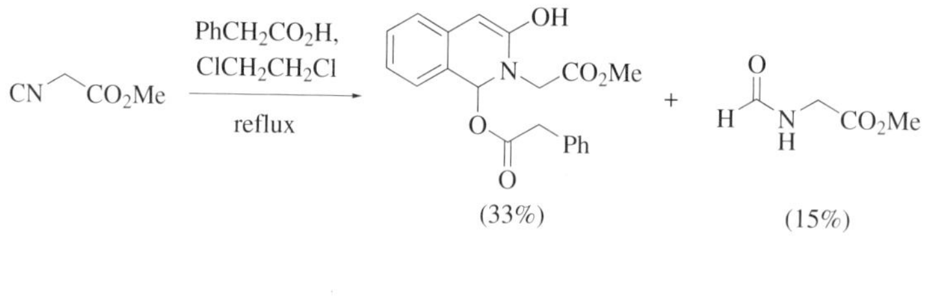
CN
CO2Me
PhCH2CO2H,
ClCH2CH2Cl
reflux
OH
N
CO2Me
O
Ph
O
+
O
H
N
H
CO2Me
(33%)
(15%)

(Eq. 31)

Passerini Reactions Followed by Secondary Transformations. The scope of these reactions can be broadened by incorporating one or more functional groups within the reacting substrates to allow secondary transformations to take place after the initial condensation. This secondary transformation may occur spontaneously (domino reaction) or by the subsequent exposure to additional reactants. This sequence of reactions may be performed in one pot or following the isolation of the Passerini adducts. Thus, a short, convergent access to a large variety of products can be achieved.

For example, when the condensation is carried out starting from arylglyoxals, the resulting α-acyloxy ketones **23** are cyclized with ammonium formate to afford oxazoles **24** in moderate overall yields (Eq. 32).[43,53]

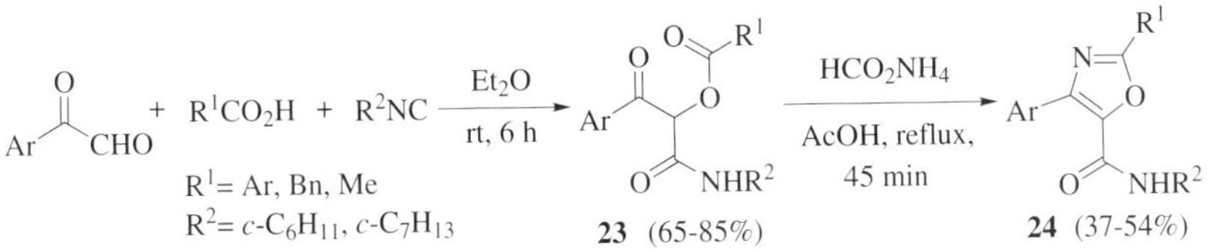
Ar
CHO
O
+ R¹CO₂H + R²NC
Et₂O
rt, 6 h
R¹= Ar, Bn, Me
R²= c-C₆H₁₁, c-C₇H₁₃
23 (65-85%)
HCO₂NH₄
AcOH, reflux,
45 min
24 (37-54%)
NHR²

(Eq. 32)

Employing carboxylic acids bearing electron-withdrawing groups (E) at the α-position provides adducts that can undergo Knoevenagel-type reactions giving substituted 2(*5H*)-furanones (butenolides) **25** (Eq. 33). This strategy has been successfully applied using α-cyanoacetic acid,[52] α-sulfonyl acetic acids,[54] and α-(2-nitrophenyl)acetic acid.[55] The presence of the electron-withdrawing group makes compounds **25** quite acidic and, therefore, the base employed for the Knoevenagel cyclization (triethylamine or piperidine) must be used in stoichiometric amounts. Butenolides **26** are isolated after acidification and may be converted into 5-methoxyfurans **27** by treatment with diazomethane (for E = CN or $ArSO_2$).[52,54]

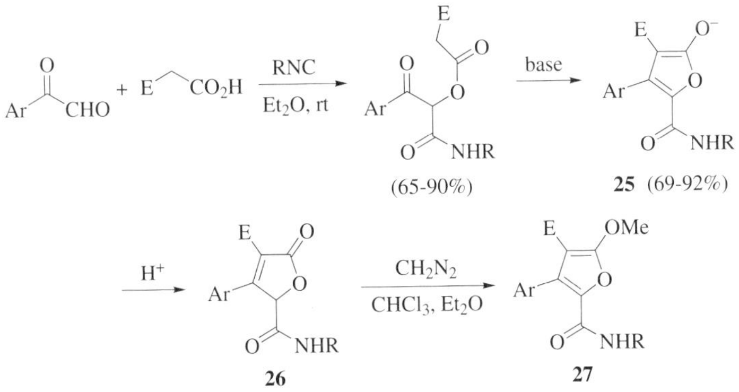
Ar
CHO
+ E
CO₂H
RNC
Et₂O, rt
NHR
(65-90%)
base
25 (69-92%)
H⁺
26
CH₂N₂
CHCl₃, Et₂O
OMe
27

(Eq. 33)

For E = o-$O_2NC_6H_4$, reduction of the nitro group followed by transacylation affords 2-oxoindole derivatives **28** in excellent yields (Eq. 34).[55]

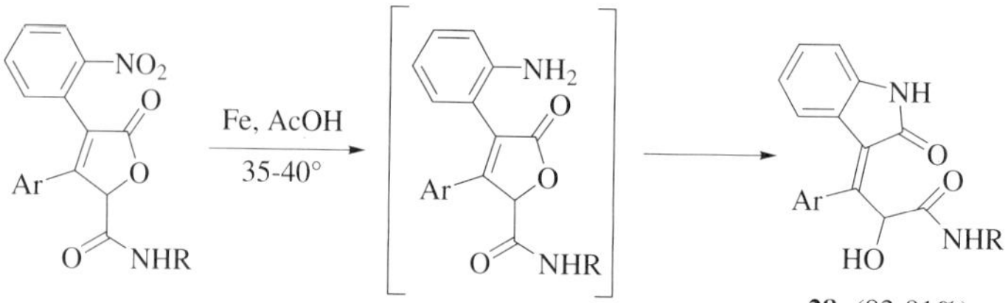

28 (82-91%)

(Eq. 34)

Similar compounds have been obtained in a one-pot sequence through a tandem Passerini/Horner-Wadsworth-Emmons reaction. The 3-position of the butenolide is either unsubstituted or substituted with an aryl group (Eq. 35).[44] Unsubstituted phosphono acids (R^1 = H) tend to give better yields (52–87%) than aryl substituted ones (R^1 = Ar; 13–47%).

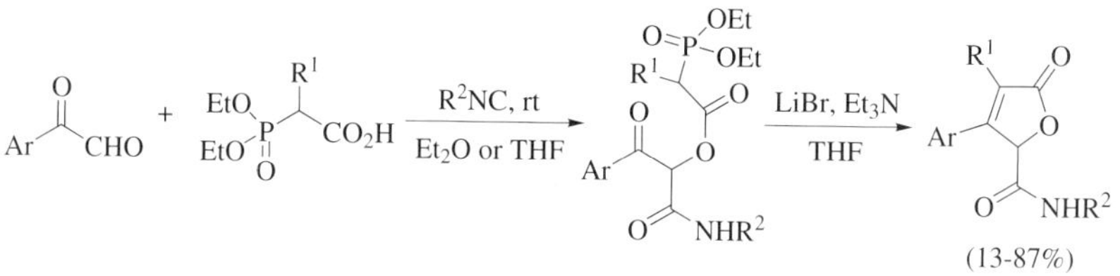

(Eq. 35)

When a related reaction is attempted starting from β-oxo aldehydes, the elimination products **29** are formed instead (Eq. 36).[56]

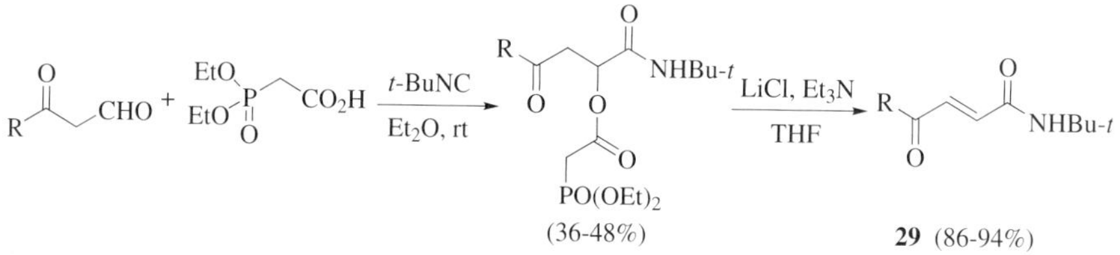

29 (86-94%)

(Eq. 36)

2(*5H*)-Furanones may also be obtained via a Knoevenagel cyclization, utilizing other types of dicarbonyl compounds (e.g. diacetyl) and unactivated carboxylic acids. Eq. 37 shows an example where cyclization is mediated by cesium fluoride and benzyltriethylammonium chloride.[66] In this process ester hydrolysis is a significant side reaction, thus limiting the scope.

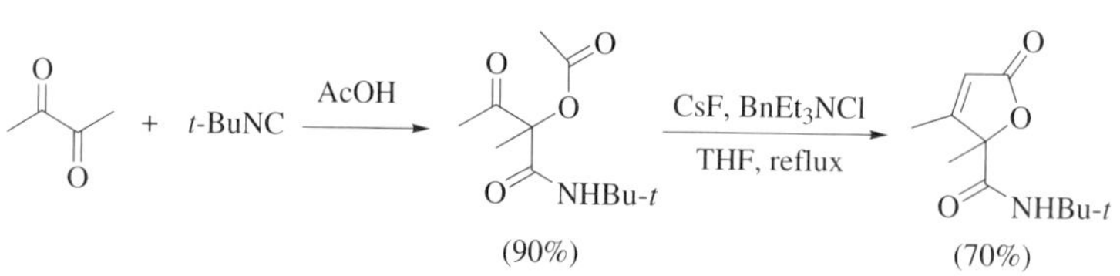

(Eq. 37)

The Passerini products from α-halo ketones may be further transformed by subsequent intramolecular displacement processes. Treating the condensation products from α-chloro ketones with cesium fluoride and benzyltriethylammonium chloride produces β-lactams (Eq. 38),[94] whereas treatment with potassium hydroxide gives epoxides. This second transformation is less useful for diversity-oriented synthesis, since one of the three initial Passerini components (the acid) is lost. Starting from α,α'-dichloroacetone, both substitution pathways can be performed in sequence, affording 1-oxa-5-azaspiro[2,3]hexan-4-ones (Eq. 39).[95]

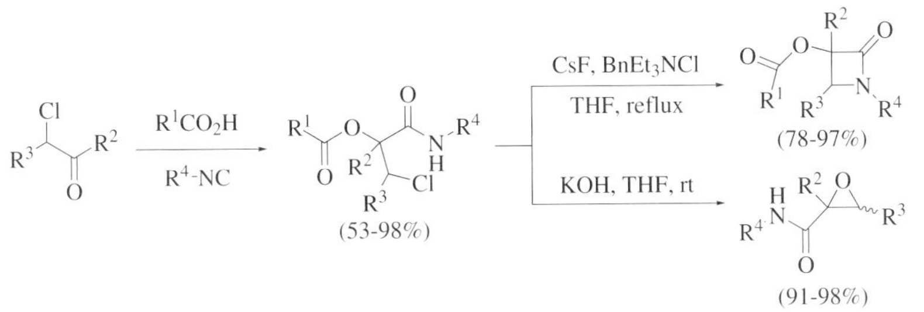
Cl
R3
R2
O
R1CO2H
R4-NC
R1
O
O
R2
R3
N
H
R4
Cl
(53-98%)
CsF, BnEt3NCl
THF, reflux
KOH, THF, rt
R1
R2
R3
R4
N
(78-97%)
H
N
R4
R2
R3
(91-98%)

(Eq. 38)

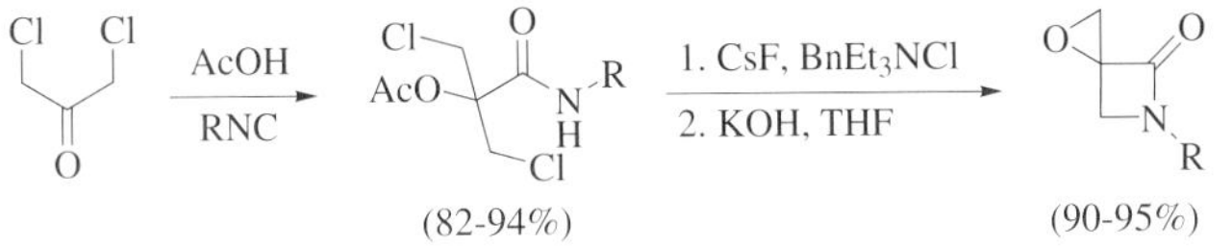
Cl
Cl
O
AcOH
RNC
AcO
N
H
R
(82-94%)
1. CsF, BnEt3NCl
2. KOH, THF
N
R
(90-95%)

(Eq. 39)

Larger rings are obtained when the halogen is more distant to the carbonyl involved in the Passerini reaction. With α-bromoketone **30**, a five-membered ring is formed on treatment with cesium fluoride (Eq. 40).[67]

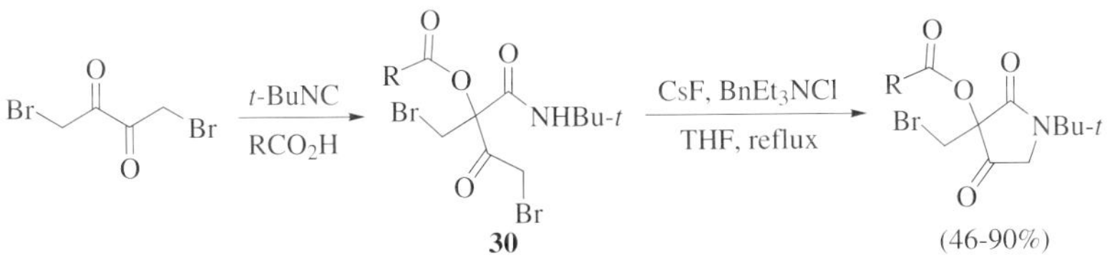
Br
Br
t-BuNC
RCO2H
NHBu-t
CsF, BnEt3NCl
THF, reflux
NBu-t
(46-90%)

30

(Eq. 40)

Ester hydrolysis following a classic Passerini condensation may be used to prepare α-hydroxy amides. This approach is complementary to the direct methods that will be described later.[47] This method involves the loss of the carboxylic component, which is a disadvantage from the point of view of atom-economy and/or diversity-oriented synthesis. In contrast, deprotection-transacylation reactions allow the retention of the diversity element inherent in the original condensation. A representative example involving the Passerini adducts **31** is shown in Eq. 41. This protocol leads to good overall yields starting from either *tert*-butoxycarbonyl (Boc)-protected[45,46,49] or fluorenylmethyloxycarbonyl (Fmoc)-protected[48,96] amino aldehydes. In both examples, the deprotection/transacylation protocol can be carried out in one pot, affording the rearranged α-hydroxy-β-acylamino amides **32**, which have been used as peptidomimetics and transition state analog (TSA) protease inhibitors. Moreover, they can be easily oxidized to the even more interesting α-oxo-β-acylamino amide scaffolds.

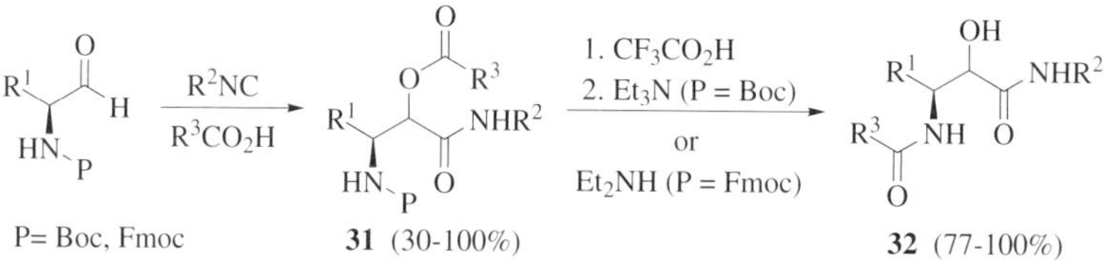

(Eq. 41)

This two-step methodology has been applied to combinatorial solution-phase synthesis (for P = Boc), using solid-phase scavengers to remove excess reagents or side products. In this way, a library of 9,600 compounds has been prepared.[97] Finally, this reaction sequence is applicable to solid-phase syntheses (later in the text).[96,98]

When R^3CO_2H is a protected α-amino acid, the two *N*-protecting groups can be either orthogonal[48,49] or identical.[45] An example of the latter arrangement is shown in Eq. 42. Application of the deprotection/transacylation protocol to the Passerini adduct **33** affords β-acylamino amide **34**, endowed with a free amino group, which can be further acylated in order to introduce a fourth diversity element. In this way a four-unit peptidomimetic can be assembled in only two discrete steps (the conversion of **33** into **35** is accomplished in a one-pot fashion).[45]

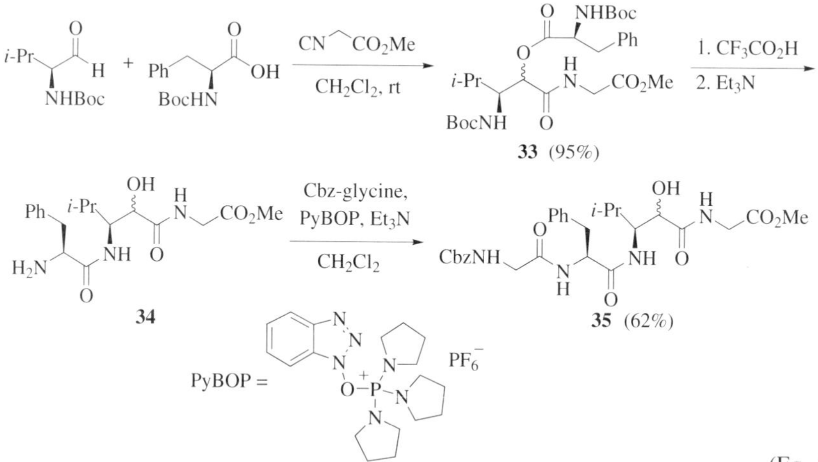

(Eq. 42)

Another interesting secondary reaction that can be applied to a Passerini adduct is the intramolecular Diels-Alder reaction (IMDA), leading to complex polycyclic structures. In order to exploit this methodology, the condensation must be carried out with one component containing a diene and another containing a dienophile. In the related Ugi reaction, this strategy has been applied to various dienophiles,[99–101] but there is only a single application reported thus far for the Passerini reaction that employs 2-furaldehyde and a variety of acetylenic acids as substrates (Eq. 43).[85] This sequence of reactions could not be performed in one pot, since the thermal Diels-Alder reaction, while successful with the corresponding Ugi adducts, is not feasible in this case. However, the Lewis acid (Me_2AlCl)-catalyzed IMDA affords the desired oxabicyclic derivatives **36** in good yields, except when R^1 is hydrogen. The overall yields of the two steps range from moderate to good, and the diastereoselectivity in the IMDA step is high. Only one diastereoisomer was isolated.

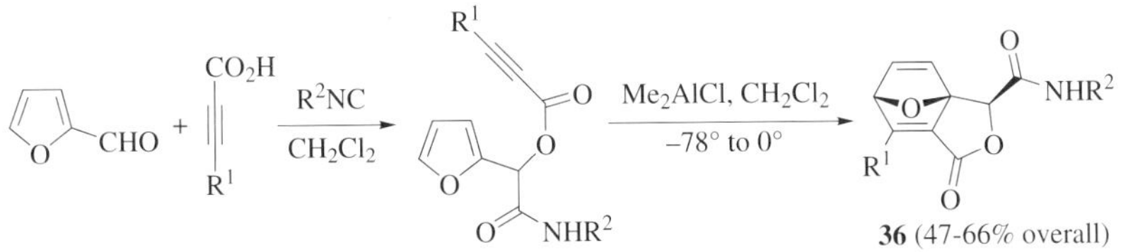
CO2H
CHO
+
R1
R2NC
CH2Cl2
R1
O
O
O
O
NHR2
Me2AlCl, CH2Cl2
–78° to 0°
O
NHR2
O
R1
O
O

36 (47-66% overall)

(Eq. 43)

Ring-closing metathesis (RCM) has also been applied to appropriate products of the Passerini reaction giving complex, natural product-like macrocycles.[102] The example shown in Eq. 44 is particularly interesting, since it illustrates the sequential application of two secondary transformations, namely the already mentioned oxazole synthesis and the RCM.

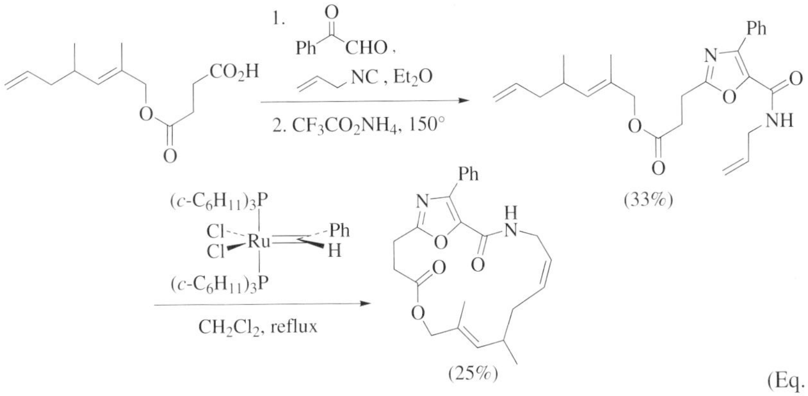
CO2H
O
O
1. Ph CHO,
NC, Et2O
2. CF3CO2NH4, 150°
Ph
N
O
O
NH
O
O
(33%)
(c-C6H11)3P
Cl
Cl
Ru
Ph
H
(c-C6H11)3P
CH2Cl2, reflux
Ph
N
O
H
N
O
O
O
(25%)

(Eq. 44)

Passerini-Type Reactions That Form Products Other Than α-Acyloxy Amides

The reaction between an aldehyde or a ketone with an isocyanide in the presence of a protic (either mineral or organic) or Lewis acid can afford several different products including α-hydroxyamides, α-oxo-β,γ-unsaturated amides, pyrrolo[1,2-*a*]quinoxaline derivatives, diimino thioanhydrides, 2,3-bis(alkylimino) oxetanes, indoles, indolenines, tetrazoles, cyclic imines, or enamines. The formation of these products, which is strongly influenced by the structure of the reactants and by the reaction conditions, is discussed in the following sections. Isocyanides do not usually react with carbonyl compounds in the absence of an acid species; the few exceptions are presented below.

Reactions without Acid. Only some perhalogenated aldehydes and ketones react with isocyanides without the need of an acid catalyst. This, of course, gives rise to different types of products. Under anhydrous conditions, imino dioxolanes **37** are formed (Eq. 45).[5,6]

2
O
F3C
CF3
+
R−NC
CF3
F3C
O
CF3
RN
O
CF3

37 (75-90%)

(Eq. 45)

The formation of these products is reasonable on the basis of the mechanism shown in Eq. 46, involving attack of the isocyanide carbon on the carbonyl. The resulting ionic intermediates **38** can only react with another ketone molecule, affording the imino dioxolanes **37**. The high reactivity of the ketone allows reaction with the isocyanide even in the absence of acid catalysis.

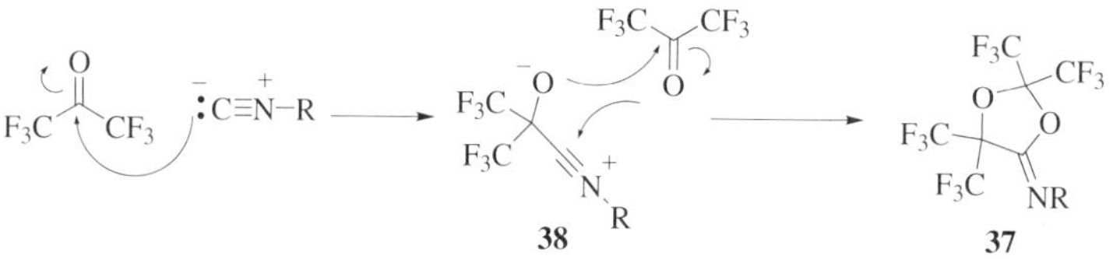
O
F3C
CF3
:C≡N−R
F3C
CF3
O
F3C
O
F3C
N
R
F3C
CF3
O
O
F3C
F3C
NR

38 **37**

(Eq. 46)

The reaction can take a different course when an additional oxo functionality is present in the trifluoro ketone, as in the case of hexafluoroacetylacetone.[103] In this reaction the stoichiometry is 1:1 and the isolated product is the pyrrolin-2-one **40**, possibly arising from rearrangement of the dihydrofuran **39**. A possible mechanism for the formation of **39** is shown in Eq. 47. This intermediate may also arise from conjugate addition of the isocyanide to the mono-enol form of the starting diketone.

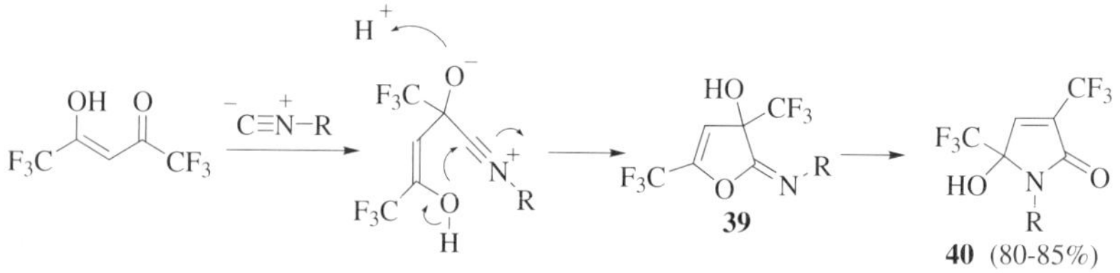
OH O
F3C
CF3
C≡N−R
H
F3C
O
N
R
F3C
O
H
HO
CF3
F3C
O
N
R
CF3
F3C
HO
N
O
R

39 **40** (80-85%)

(Eq. 47)

In contrast, when the perhalogenated carbonyl compound is used in its hydrate form, the carbonyl hydrate can play the role of acid Y-A in the general mechanism shown in Eq. 11 to promote the normal reaction pathway. Thus, hydrates of some perhalogenated aldehydes and ketones have been reported to react (in Et_2O or in water) with isocyanides in the absence of any acid species to give α-hydroxy-amides.[1-3,104] An example is shown in Eq. 48.[1] The same behavior is observed with hexafluoroacetylacetone when the reaction is carried out in water (Eq. 49).[104] The α-hydroxy amide can also be formed by hydrolytic opening of imino dihydrofuran **39**.

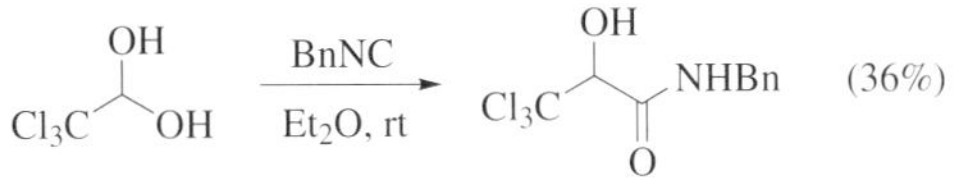
OH
Cl3C OH
BnNC
Et2O, rt
OH
Cl3C
NHBn
O
(36%)

(Eq. 48)

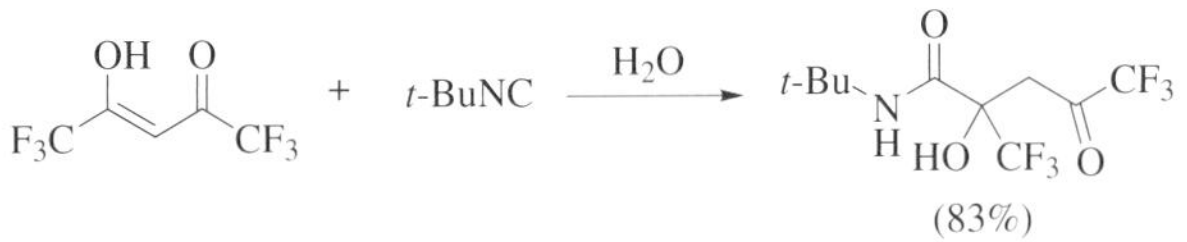
OH O
F3C CF3
+ t-BuNC
H2O
O
t-Bu N
H
CF3
HO CF3 O
(83%)

(Eq. 49)

The reaction of trichloroacetaldehyde hydrate with isodiazomethane, which has an isocyanide-like structure, is noteworthy (Eq. 50). Reaction of the amine with the first equivalent of chloral is followed by a Passerini-type condensation with a second equivalent giving the observed α-hydroxy hydrazide.[105]

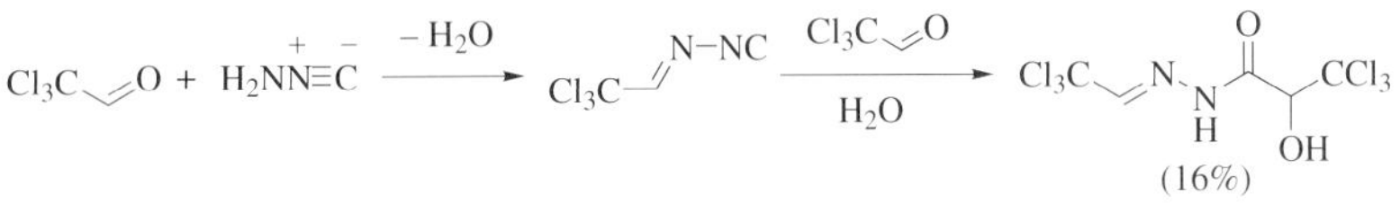
Cl3C O + H2NN≡C
– H2O
Cl3C N–NC
Cl3C O
H2O
Cl3C N N
H
O
CCl3
OH
(16%)

(Eq. 50)

Reactions with Protic Acids. *Mineral Acids.* Aqueous hydrochloric,[19,20,106-109] hydrobromic,[110] sulfuric,[20,61,111-114] nitric,[20] and phosphoric[20] acids can be used in the Passerini reaction to afford α-hydroxy amides **41** (Eq. 51). The formation of these products can be explained most likely by the intervention of intermediate **6** (see Eq. 11), as previously described in the Mechanism section. However, a detrimental side reaction involving solvolysis of the isocyanide may occur under these conditions.[20]

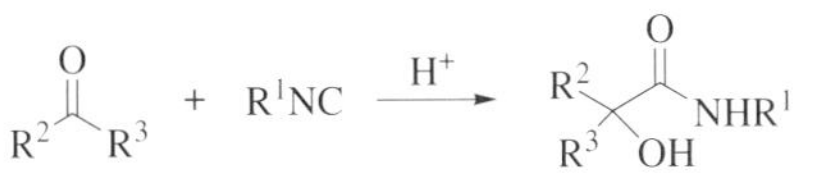
O
R2 R3
+ R1NC
H+
O
R2
R3 OH
NHR1

41

(Eq. 51)

The best yields (up to 89%) and the higher reaction rates are obtained using a stoichiometric amount of acid,[20] whereas catalytic amounts of the acid give lower yields (30–40% at 0° and up to 60% at 35°). A representative example is shown in Eq. 52.[20] The reaction can accommodate simple aldehydes[19,20,106,111,112] or ketones[19,20,106-109,111-114] which are typically employed as solvents in the reaction. For this reason, the scope of the reaction is restricted to carbonyl compounds with low molecular weights. In contrast, the Passerini reaction in the presence of hydrochloric acid is less accommodating when a stoichiometric amount of the carbonyl compound is used. Many unidentified products are obtained after a rapid exothermic reaction.[84]

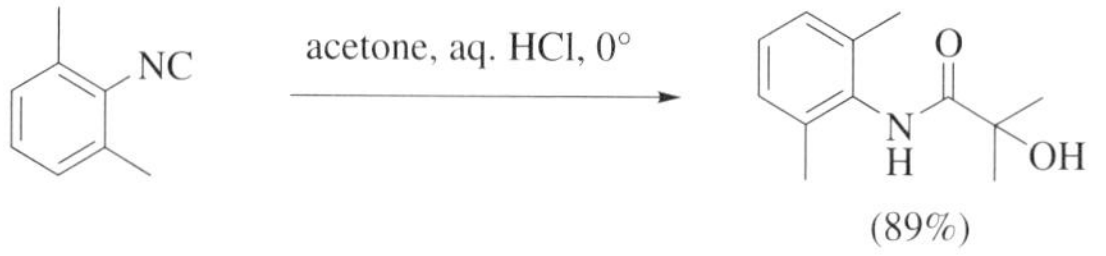
NC
acetone, aq. HCl, 0°
O
N
H
OH
(89%)

(Eq. 52)

Paraformaldehyde reacts with various isocyanides in the presence of dilute sulfuric acid, affording the corresponding α-hydroxy amides in quite good yields.[10] Moreover, aqueous formaldehyde[61] or glyoxal[61] reacts with α-trisubstituted isocyanides to give the corresponding α-hydroxy amides in moderate yields (Eq. 53). This result is in contrast to the classical Passerini reaction, which requires anhydrous monomeric aldehydes in order to give the desired α-acyloxy acetamides.[60] However, the reaction can be performed with stoichiometric amounts of the desired aldehyde.

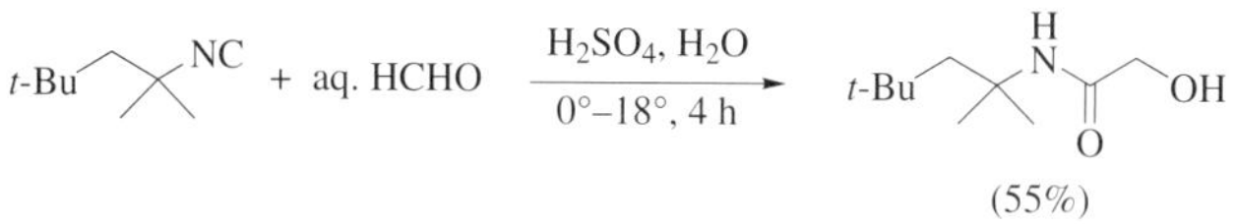
t-Bu
NC
+ aq. HCHO
H2SO4, H2O
0°–18°, 4 h
t-Bu
H
N
OH
O
(55%)

(Eq. 53)

Trifluoroacetic Acid-Pyridine and Trimethylammonium Hydrochloride. The use of a large excess of carbonyl compound may be avoided by promoting the reaction with a mixture of trifluoroacetic acid and pyridine.[115–117] Typical conditions involve the use of 2 equivalents of trifluoroacetic acid and 4 equivalents of pyridine. The use of trifluoroacetic acid was first reported in the reaction with 2-picolinaldehyde and *tert*-butyl isocyanide to give α-hydroxy amide **42** (Eq. 54).[115] In this case, the aldehyde bearing a pyridine nucleus behaves as a pyridine equivalent. The same reaction performed in the presence of hydrochloric acid[20] gives a complex mixture, in which only **43**, arising from the condensation with an additional molecule of aldehyde, could be identified.[115]

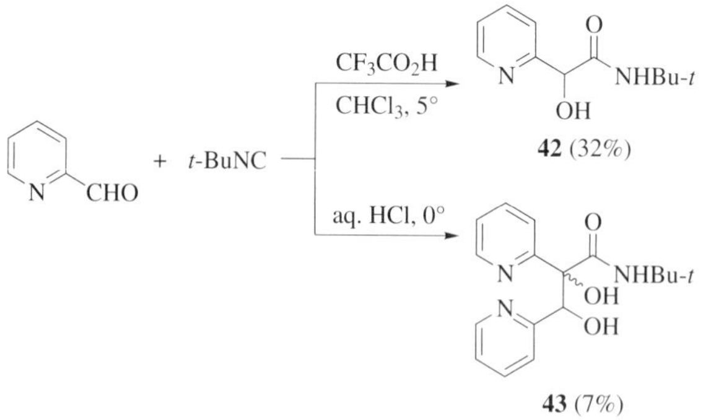
N
CHO
+ t-BuNC
CF3CO2H
CHCl3, 5°
N
O
NHBu-t
OH
42 (32%)
aq. HCl, 0°
N
O
NHBu-t
OH
N
OH
43 (7%)

(Eq. 54)

Trifluoroacetic acid-pyridine promoted reactions do not afford α-trifluoroacetoxy amides, but rather the corresponding α-hydroxy amides. The reaction using 3,4-dichlorobenzaldehyde gives the best results with an equimolecular amount of trifluoroacetic acid and pyridine, whereas the yield suffers when trifluoroacetic acid alone or a mineral acid is used.[115] The rate has been reported to be influenced by the order of addition of the reagents.[115] The yields range from moderate to good when aromatic or aliphatic aldehydes or aliphatic cyclic or acyclic ketones and *tert*-butyl isocyanide are used. The reaction does not work with aromatic ketones.[115]

A possible explanation for α-hydroxy amide formation is that the intermediate α-trifluoroacetamides are hydrolyzed during the work-up (Eq. 55).[116] Although trifluoroacetates of secondary alcohols are reasonably stable under acidic conditions,[118] neighboring group assistance by the amide carbonyl may facilitate the cleavage via intermediate **44**.

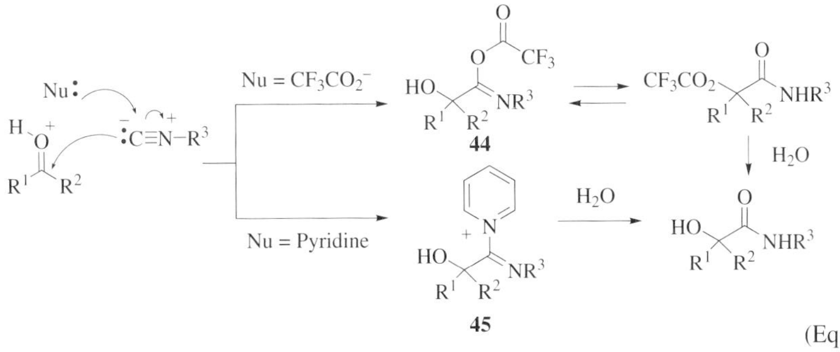

(Eq. 55)

An alternative possibility is that pyridine acts as a nucleophile, instead of trifluoroacetate, adding onto the isocyanide carbon to form intermediate **45**, which is easily hydrolyzed on work-up. Pyridine is used in excess[116,117] or in equimolar amounts.[115] This hypothesis is corroborated by the fact that preformed trimethylammonium hydrochloride catalyzes the Passerini reaction between isobutyraldehyde and cyclohexyl isocyanide, affording the corresponding α-hydroxy amide.[84]

The same reaction can be applied to *N*-protected optically active α-amino aldehydes, utilizing different achiral or chiral isocyanides,[116,117] giving the corresponding α-hydroxy amides in moderate to high yield. One of the most stereochemically complex examples of this series is shown in Eq. 56.[116]

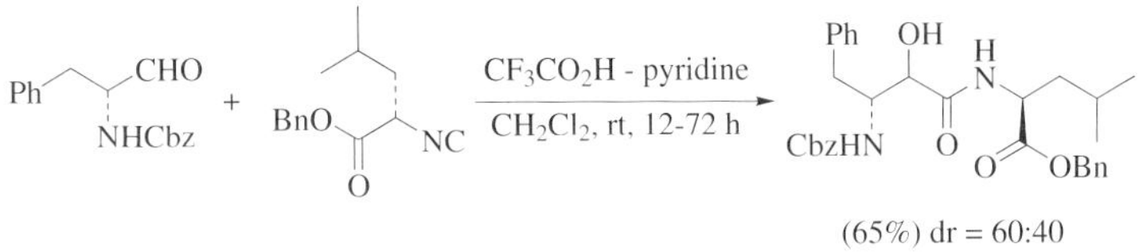

(Eq. 56)

The salts of other tertiary amines can also be used; tertiary alkyl amines give inferior results, whereas 2,6-lutidine, 2,4,6-collidine, 2,6-di-*tert*-butylpyridine (pK_a in the range 5.2–7.4) are optimal. In the reaction between (*S*)-2-*N*-Boc-3-phenylpropanal and *tert*-butyl isocyanide, replacement of pyridine with 2,4,6-collidine increases the yield from 24 to 78%.[116]

Salicylic Acid. Salicylic acid typically affords the expected α-acyloxy amides.[3] However, when the carbonyl compounds are 2-(arylaminothiocarbonyl)cyclohexanones, 1,3,4,5,6,7-hexahydrobenzo[*c*]thiophenes **46** are obtained in good yields using stoichiometric amounts of salicylic acid (Eq. 57).[28] Other acids, such as catalytic *p*-toluenesulfonic acid or stoichiometric benzoic acid, give only fair to moderate yields. A variety of aromatic and aliphatic isocyanides can be used.

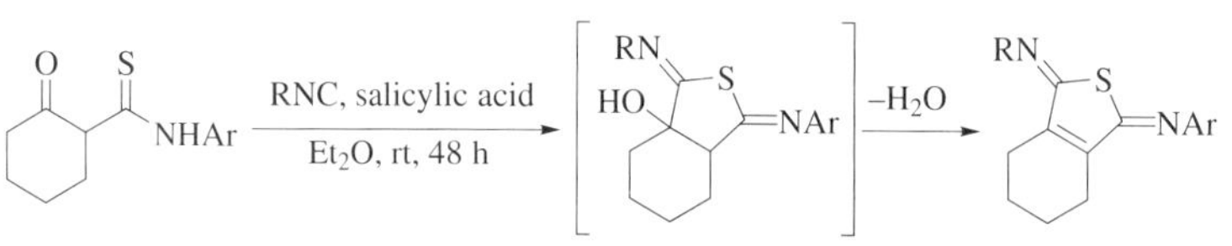

O
S
NHAr
RNC, salicylic acid
Et_2O, rt, 48 h
RN
HO
S
=NAr
$-H_2O$
RN
S
=NAr

46 (65-83%)

(Eq. 57)

The formation of compounds of this type can be explained by nucleophilic trapping of the isocyanide-derived carbocation by the thioamide sulfur, followed by dehydration.

Sulfonic Acids. A single example exists in which a stoichiometric amount of a sulfonic acid is used in a Passerini condensation involving acetone and 2,4-bis(dimethylamino)-6-isocyano-1,3,5-triazine, giving the corresponding α-hydroxyamide in moderate yield.[80]

m-Chloroperbenzoic Acid. The reaction between aliphatic aldehydes and two equivalents of *tert*-butyl isocyanide catalyzed by *m*-chloroperbenzoic acid does not give the usual Passerini-type products. Instead, 2,3-bis[alkylimino]oxetanes of the general formula **47** are isolated (Eq. 58).[119] The use of catalytic BF_3 gives the same types of products (see p. 28).[22,29,120] These reaction conditions are claimed to be unsuitable for aromatic aldehydes or for ketones, although no explanation for this behavior was provided.[119]

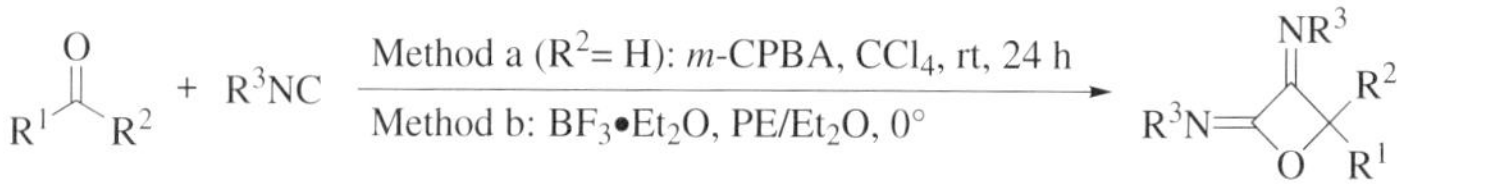

O
R^1 R^2
+ R^3NC
Method a (R^2= H): *m*-CPBA, CCl_4, rt, 24 h
Method b: BF_3•Et_2O, PE/Et_2O, 0°
NR^3
R^2
R^3N=
O R^1

47 (a: 58-94%; b: 40-96%)

(Eq. 58)

Hydrazoic Acid, Aluminum Azide, and Trimethylsilyl Azide. As previously noted (Eq. 12), hydrazoic acid can behave as the acidic component in the Passerini reaction to afford tetrazole derivatives **48** (Eq. 59). The formation of these compounds can be rationalized by the standard mechanism of attack of protonated carbonyl ion on the isocyanide, followed by imidoyl azide formation and cyclization through N–N bond formation.[23]

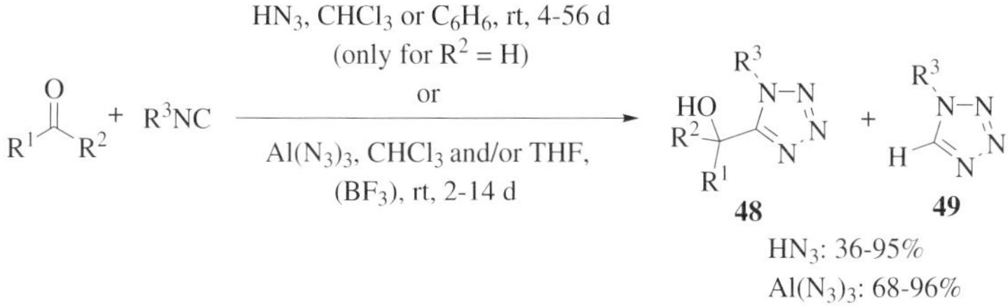
(Eq. 59)

Aliphatic aldehydes react smoothly, but aromatic aldehydes and ketones give the corresponding tetrazoles in very low yields because of the competitive formation of unsubstituted tetrazoles **49**, which arises from the reaction between the isocyanide and hydrazoic acid. Hydrazoic acid may be successfully replaced by aluminum azide in this reaction. This reagent does not react with isocyanides in the absence of carbonyl compounds; therefore, substituted tetrazoles **48** are obtained in good yields with aliphatic and aromatic aldehydes and ketones. Under some conditions, a catalytic amount of BF_3 may accelerate the condensation.[23]

Other derivatives of hydrazoic acid can also be used for the synthesis of tetrazoles. The best yields have been achieved with trimethylstannyl azide or trimethylsilyl azide.[121] The latter is the reagent of choice because it is not explosive (as is hydrazoic acid) and is less toxic than the tin reagent. This method has been applied to a series of *N*-Boc protected α-amino aldehydes (Eq. 60).[122] Trimethylsilyl ethers **51** that may be formed (up to 40%) as by-products can be easily hydrolyzed to alcohols **50** by fluoride treatment. Tetrazoles **50** are converted into cis-constrained norstatine analogs after deprotection of the amine and acylation with polymer-bound tetrafluorophenol esters or sulfonates.

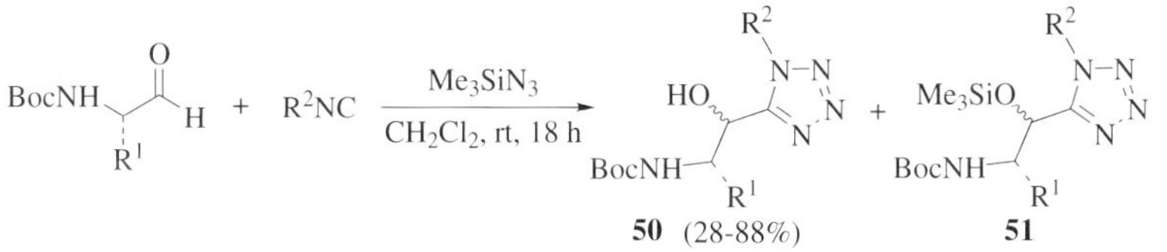
(Eq. 60)

Replacing hydrazoic acid with hydrogen cyanide, thioacetic acid, monophenylphosphoric acid, or picric acid results in a complete lack of reaction.[23]

Thiocarboxylic Acids. Thiocarboxylic acids react with methyl 3-(*N,N*-dimethylamino)-2-isocyanoacrylate **52** and aldehydes in the presence of a catalytic amount of $BF_3{\cdot}OEt_2$ to afford 2-(1-acyloxyalkyl)thiazoles **53** (Eq. 61). The yields are low to moderate. The reaction does not take place without BF_3, which is the best among a series of Lewis acids tested. The mechanism probably entails a standard Passerini reaction in which the thiocarboxylic acid plays the role of the carboxylic acid, followed by cyclization and elimination of dimethylamine.[123]

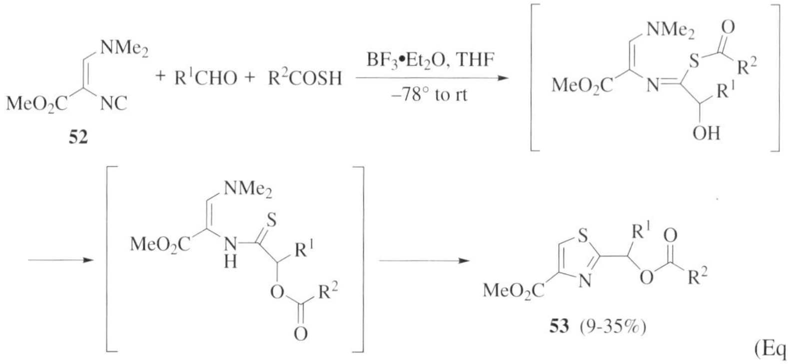

(Eq. 61)

Reactions with Lewis Acids. *Titanium Tetrachloride.* The modified Passerini reaction in the presence of $TiCl_4$ is an excellent method for obtaining α-hydroxy amides. The first proposed mechanism, involving the insertion of the isocyanide carbon into the Ti–Cl bond giving intermediate **54** (Eq. 62)[70] followed by reaction with the carbonyl moiety, was later ruled out.[27,124,125] The currently accepted mechanism, supported by X-ray data, involves the formation of hexacoordinated titanium complexes instead of organometallic adducts.

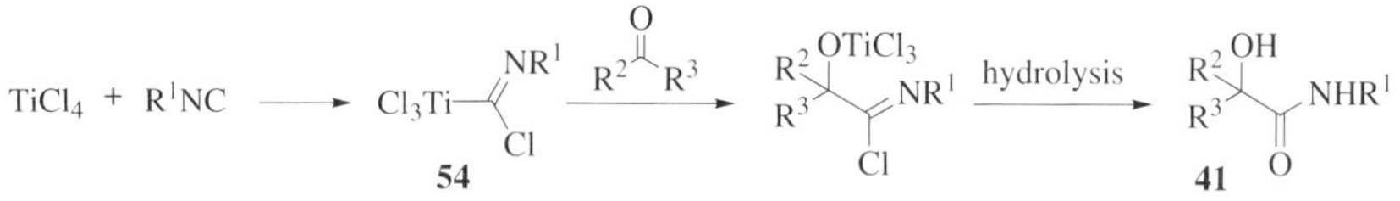

(Eq. 62)

The outcome of the reaction can, therefore, be rationalized by the mechanism shown in Eq. 63, with the intervention of several titanium complexes. On work-up, intermediate **55** is hydrolyzed giving α-hydroxy amide **41**. By choosing an appropriate isocyanide bearing an organic fragment that can act as a tridentate ligand, it is possible to fill all the empty coordination sites on titanium. In this fashion, the crystalline compound **56,** which is a specific example of the putative intermediate **55**, was isolated and characterized by X-ray analysis.

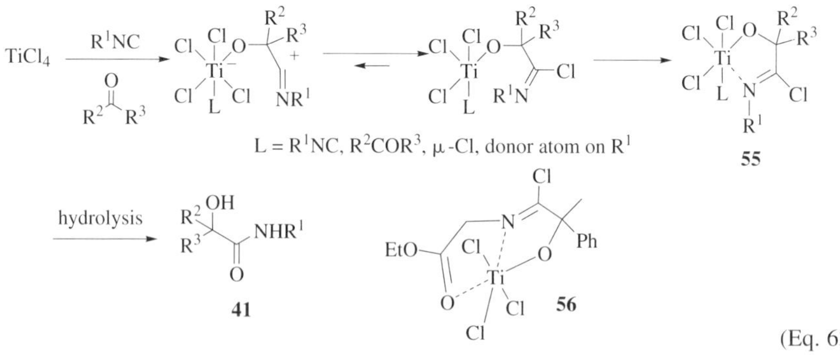

(Eq. 63)

This reaction can be applied to a large variety of carbonyl compounds: aliphatic[70,78,126] or aromatic[27,70,78,125,126] aldehydes, acyclic[70,126] and cyclic[70,78,126] ketones, aromatic ketones,[27,70,78,125,126] fluorinated ketones,[127,128] and even diketones; in the latter only one carbonyl group reacts.[27,125] Diverse isocyanide types that react under these conditions range from simple aliphatic (acyclic[70,126] or cyclic[127]) or aromatic[78,128] isocyanides to isocyano acetates,[27,78,125] isocyano acetamides,[78] isocyano phosphonates,[27,78,125] and isocyano methyl boronates.[79] In some examples, optically active isocyanides derived from chiral α-amino acids are used.[78]

With simple isocyanides (for example methyl isocyanide) the yields are excellent even when readily enolizable ketones, such as acetone or acetophenone, are used.[70,126] On the other hand, the very low yields obtained with *tert*-butyl methyl ketone[70] and adamantanone[70] reflect the steric sensitivity of the reaction.

Low yields of α-hydroxy amides are obtained with tertiary or benzylic isocyanides.[78] The main products are cyanohydrins, presumably resulting from *N*-dealkylation of intermediates such as **57**, facilitated by the stability of the resulting carbocation (Eq. 64)

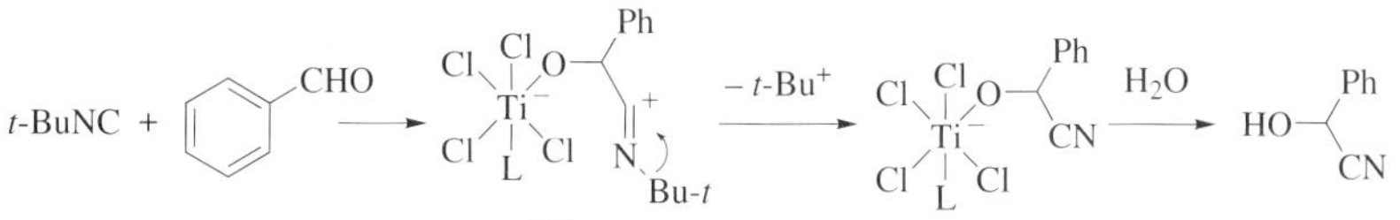
t-BuNC +
CHO
Cl
Ti
O
Ph
N
Bu-t
L
− t-Bu⁺
CN
H₂O
HO

57

(Eq. 64)

Another exception is represented by the reaction of cinnamaldehyde with methyl isocyanide: the corresponding α-hydroxy aldehyde is isolated in only 36% yield, owing to the competing Michael addition of the isocyanide.[70] This is not unexpected, since it is known that α,β-unsaturated compounds may undergo 1,4-addition with isocyanides.[119,129–133] The final products depend upon the nature of both the isocyanide and the Lewis acid (Eq. 65). For example, when aryl or methyl isocyanides are used with diethylaluminum chloride, unsaturated *N*-substituted imino dihydrofurans **58**, readily convertible into the corresponding lactones, are obtained in good yields, provided that the α,β-unsaturated ketone can assume a *s*-cis conformation.[132] However, with *tert*-butyl isocyanide and titanium tetrachloride, the main product is always the hydrocyanation derivative **59**.[133] In both examples, the isocyanide attacks the β-carbon, but the subsequent course of the reaction is different. In the first instance, the oxygen atom of the intermediate enolate completes the addition by intramolecular attack on the isocyanide carbon atom. In the second example, the *N*-*tert*-butylimidoyl cation intermediate rapidly loses a *tert*-butyl cation, giving the titanium enolate of the β-cyano ketone **59**.

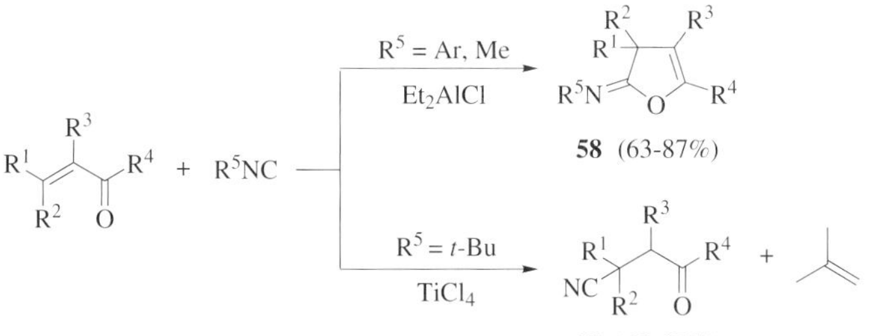
R¹
R²
R³
R⁴
O
+ R⁵NC
R⁵ = Ar, Me
Et₂AlCl
R⁵N
R⁵ = t-Bu
TiCl₄
NC
+

58 (63-87%)

59 (63-87%)

(Eq. 65)

With the exceptions noted above, the yields range from moderate to excellent. For reasons that are not apparent, yields appear to be lower when more highly functionalized isocyanides are used.[78] Furthermore, the conditions of the $TiCl_4$-mediated reaction are compatible with a variety of functional groups such as aromatic and heterocyclic rings, ethers, esters, amines, amides, halides, phosphonates, and boronates.

An additional example is represented by the reaction involving acetals, silyl enol ethers, and isocyanides to afford, at low temperatures and in one pot, γ-alkoxy-α-hydroxy amides through a domino aldol condensation-Passerini reaction (Eq. 66).[134] The more nucleophilic character of the silyl enol ethers relative to the isocyanide makes the aldol condensation faster. Thus, a β-alkoxy aldehyde is formed initially, which, after being activated by titanium tetrachloride, reacts with the isocyanide in the same fashion as just described. Only silyl enol ethers derived from aldehydes react in this fashion; those derived from ketones do not react even at room temperature.

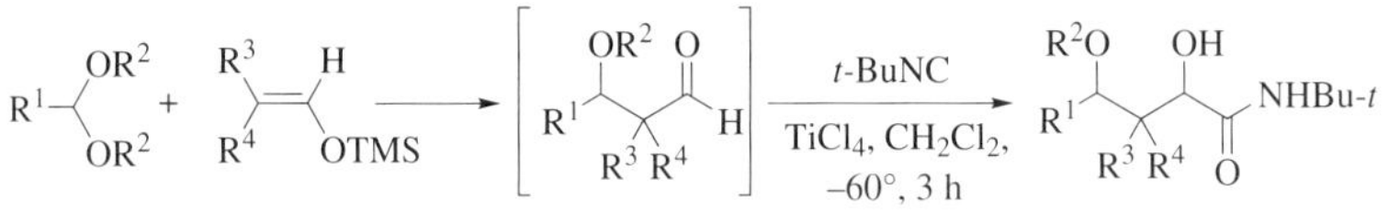

OR2
R1
OR2
R3
H
R4
OTMS
OR2 O
R1
H
R3 R4
t-BuNC
TiCl4, CH2Cl2,
–60°, 3 h
R2O
OH
R1
NHBu-t
R3 R4
O

(Eq. 66)

Boron Trifluoride and Aluminum Trichloride. Although BF_3 has been widely used as a Lewis acid, either in stoichiometric or catalytic amounts, the reaction between isocyanides and carbonyl compounds gives α-hydroxy amides **41** as the main products in only a few examples. Among these, the reaction between cyclohexanone and an isocyano triazine[80] and the reaction between paraformaldehyde and isocyanooctaethylporphyrin are worth mentioning.[75] The difference between $TiCl_4$ and BF_3 may be explained by the absence of an accessible nucleophile (Cl^- in the case of $TiCl_4$) in the latter case.

When BF_3 and one[111,112] or two[113,135] molar equivalents of isocyanide are used in the reaction with carbonyl compounds,[111,135] the corresponding α-hydroxy amides are typically formed only in small amounts. Under these conditions, α-oxo-β,γ-unsaturated amides,[111,112] 2,3-bis(alkylimino)oxetanes,[29] indolenines,[113] or indoles[135] are usually the main products.

The nature of the products obtained in the BF_3-mediated Passerini reaction is strongly dependent upon the quantity of Lewis acid used and on the work-up conditions. As noted earlier (Eq. 58), when a catalytic amount of BF_3 is added to the reaction medium and a neutral or slightly basic work-up is performed, the products are 2,3-bis(alkylimino)oxetanes, such as **60** (Eq. 67).[22,29,120] Two equivalents of isocyanide are required in order for the reaction to go to completion. In one specific example employing acetaldehyde, the 1,4-dioxepane **61**, derived from participation of three isocyanide and two aldehyde molecules, was isolated as a by-product.[22,120]

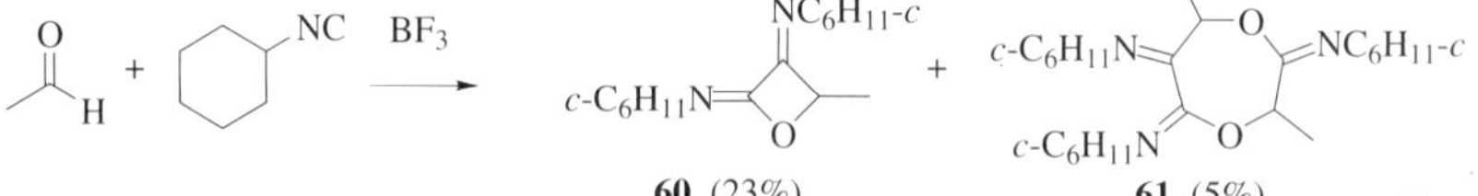

O
H
NC
BF3
NC6H11-c
c-C6H11N
O
c-C6H11N
O
NC6H11-c
c-C6H11N
O

60 (23%) **61** (5%)

(Eq. 67)

The formation of these compounds can be explained as a variation of the standard Passerini mechanism in which the second equivalent of isocyanide is the only nucleophile able to react with the carbon atom of the initially incorporated isocyanide molecule (see Eq. 13). Several aldehydes or ketones, including chlorinated examples were successfully employed, with yields ranging from 40% to 93%.[29] Formaldehyde must be used in the anhydrous monomeric form.[29] Diverse isocyanides can be used, the optimal results being obtained from α-substituted derivatives.[29] The best is *tert*-butyl isocyanide, since it is the most resistant to polymerization.[30,136] The diimino oxetanes can be transformed into other compounds through reactions involving aqueous or gaseous hydrochloric acid (Eq. 68).[22,137]

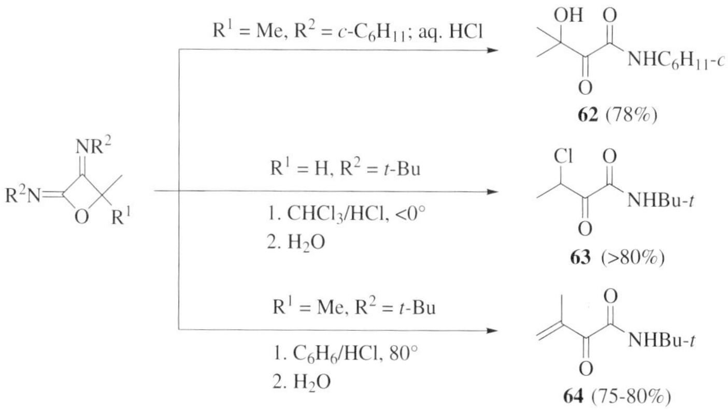
NR^2
R^{2}N
O
R^1
R^1 = Me, R^2 = *c*-C_6H_{11}; aq. HCl
OH
O
NHC_6H_{11}-*c*
62 (78%)
R^1 = H, R^2 = *t*-Bu
1. $CHCl_3$/HCl, <0°
2. H_2O
Cl
NHBu-*t*
63 (>80%)
R^1 = Me, R^2 = *t*-Bu
1. C_6H_6/HCl, 80°
2. H_2O
64 (75-80%)

(Eq. 68)

These transformations involve ring opening of the protonated oxetane to afford an α-*N*-alkylimino amide, which is readily hydrolyzed during aqueous work-up. When R^1 = methyl, this opening reaction presumably involves a carbocation intermediate, whereas for R^1 = H an S_N2 nucleophilic substitution at the sp^3 oxetane carbon is more likely. When aqueous HCl is used, the carbocation is trapped by water and β-hydroxy-α-oxo amides such as **62** are formed.[22] When R^1 = H and chloroform saturated with gaseous hydrogen chloride is used at low temperature, the oxetane oxygen is displaced by the chloride anion to provide β-chloro-α-oxo amides **63**.[137] On the other hand, when R^1 = methyl, and a slow continuous addition of hydrogen chloride is carried out at higher temperatures, the higher stability of the intermediate carbocation and the lower concentration of the chloride nucleophile favors an elimination process giving β,γ-unsaturated α-oxo amides **64**.[137]

From carboxylic acids and oxetanes derived from aldehydes, α-acylamino-β-hydroxy acrylamides **65** are obtained via an O $\rightarrow$ N acyl migration (Eq. 69).[137] When α- or β-halo carboxylic acids are used, the enol intermediate **65** cyclizes through alkylation of the enol by the alkyl halide giving β- or γ-lactams in moderate yield.[138]

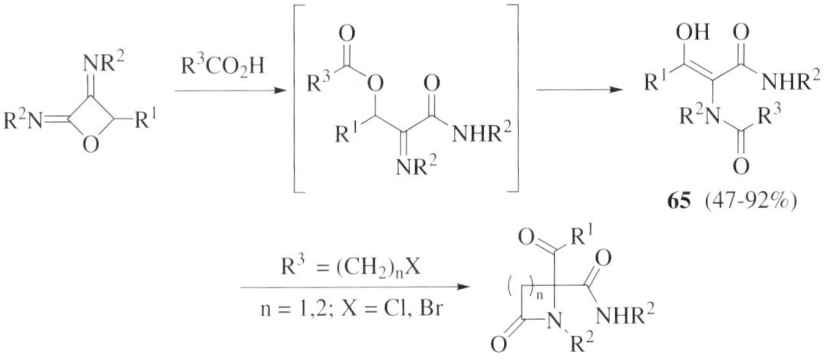

65 (47-92%) (34-55%)

(Eq. 69)

The diiminooxetanes are quite stable up to 200°. However, if 2,3-bis(secondary alkylimino)oxetanes are heated at 200°, they undergo a thermal rearrangement leading to 3-imidazolin-5-ones **66** (Eq. 70). On the other hand, 2,3-bis(tertiary-alkylimino)oxetanes are stable under these conditions, but when heated to 300°, they rearrange to afford cyclobutanone derivatives **67**. In both of these thermal rearrangements, several byproducts are also formed.[139]

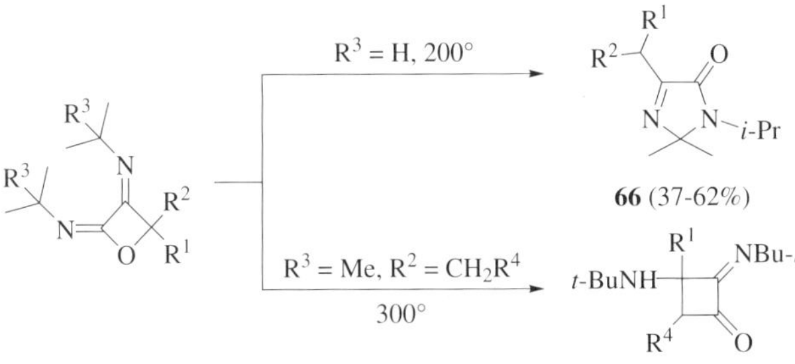

66 (37-62%)

67 (30-64%)

(Eq. 70)

The condensation of *tert*-butyl isocyanide with various acyclic[112] or cyclic[111] ketones using stoichiometric instead of catalytic amounts of $BF_3{\cdot}OEt_2$ affords α-oxo-β,γ-unsaturated amides **69**, albeit in low yields (Eq. 71). These compounds are presumably derived from the ring opening of oxetanes **68**, followed by hydrolysis of the resulting α-alkylimino-β,γ-unsaturated amides, as previously described in Eq. 68.

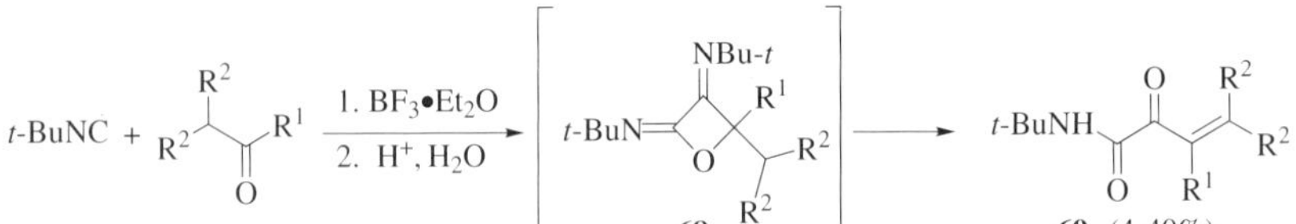

68 **69** (4-40%)

(Eq. 71)

When two equivalents of an aryl isocyanide are employed, the intermediate oxetanes **70** lead to indolenine derivatives **71** (Eq. 72).[30,113,114,140,141] The reaction proceeds via a different rearrangement. However, the use of phenyl isocyanides with, at most, one additional substituent on the aromatic ring results in poor yields with respect to acyclic ketones and only moderate yields for cyclic derivatives.[113,140,141] In particular, phenolic isocyanides are unsuitable for the preparation of indolenines **71**, giving only the α-hydroxy amide, albeit in low yield. Free amino groups must be protected because they can act as Lewis bases binding to BF_3.[113]

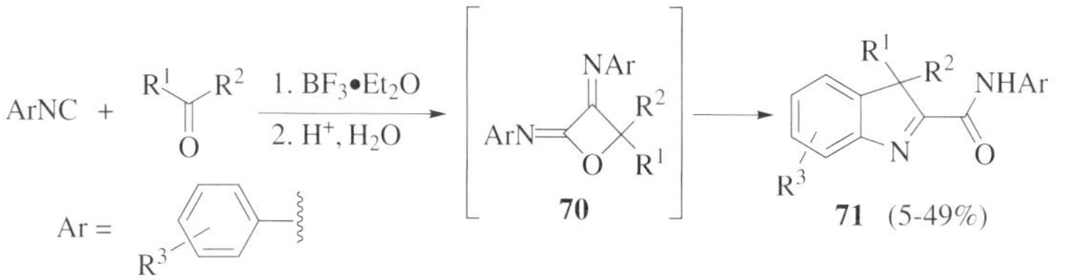
ArNC +
R1
R2
O
1. BF3•Et2O
2. H+, H2O
NAr
ArN
R2
O
R1
70
R1
R2
NHAr
N
O
R3
71 (5-49%)
Ar =
R3

(Eq. 72)

On the other hand, better results are reported with both acyclic and cyclic sterically unhindered ketones in reactions with α-naphthyl isocyanide. Although the second cyclization giving the final product could, in principle, involve either C-2 or C-8 (naphthalene ring numbering), the reaction is regioselective for C-2 attack (Eq. 73).[30,114,141]

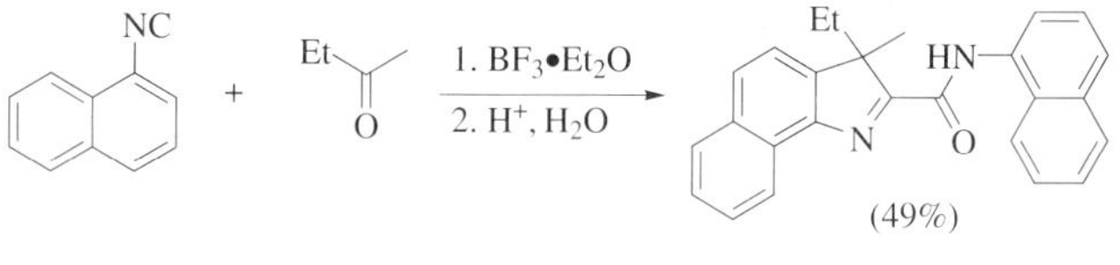
NC
+
Et
O
1. BF3•Et2O
2. H+, H2O
Et
HN
N
O
(49%)

(Eq. 73)

When o,o′-disubstituted aromatic isocyanides are used, the normal formation of indolenines is prevented. Only α-*N*-aryl-β,γ-unsaturated amides such as **72** (the imines of α-oxo amides **69**) are isolated (Eq. 74).[141] Evidently, the imine is stabilized against hydrolysis by the bulkiness of the aryl residue.

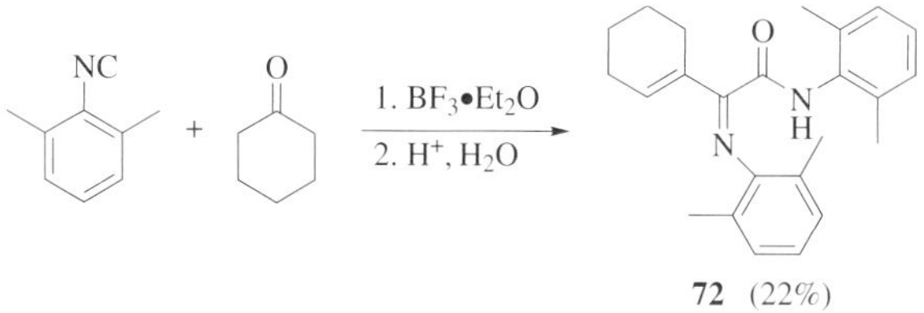
NC
+
O
1. BF3•Et2O
2. H+, H2O
O
N
H
N
72 (22%)

(Eq. 74)

When aromatic ketones are reacted with *tert*-butyl isocyanide and stoichiometric amounts of BF_3, 2-aminoacyl indoles **73** are the major products (Eq. 75).[135,142] When the aromatic ketone contains an acetoxy substituent it is usually cleaved to the corresponding phenol.[135] There is only one example of an aliphatic ketone, 1-acetylcyclohexene, that gives the same reaction; however, the yield of isolated indole is very low.[30] Other α,β-unsaturated ketones usually react with isocyanides and Lewis acids by 1,4-additions, as described previously.[132,133]

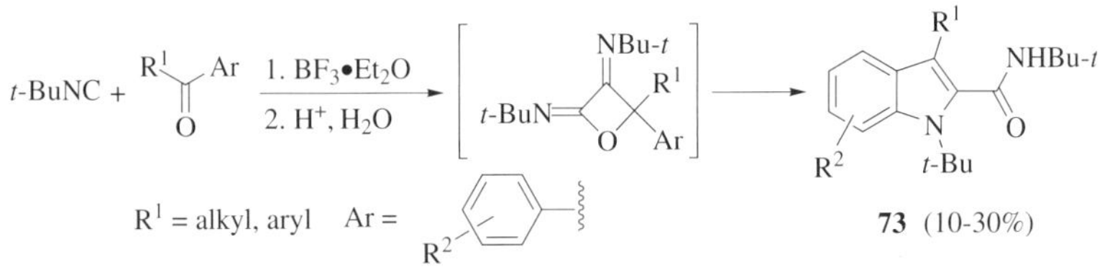
t-BuNC +
R1
Ar
O
1. BF3•Et2O
2. H+, H2O
NBu-t
t-BuN
R1
O
Ar
R1
NHBu-t
N
O
R2
t-Bu
R1 = alkyl, aryl
Ar =
R2

73 (10-30%)

(Eq. 75)

The formation of the indolenines **71** and indoles **73** can be rationalized by the mechanisms shown in Eqs. 76 and 77, involving BF_3-mediated opening of the oxetane ring to afford a carbocation, whose fate depends on the structure of the starting material. When the starting ketone is aromatic, the cyclization depicted in Eq. 76 affords indoles.[30] On the other hand, when the ketone is aliphatic and the isocyanide is aromatic, the Friedel-Crafts alkylation shown in Eq. 77 occurs, leading to 3-*H*-indolenines. When no aryl groups are present in any of the substrates, elimination leads to β,γ-unsaturated-α-oxo amides.

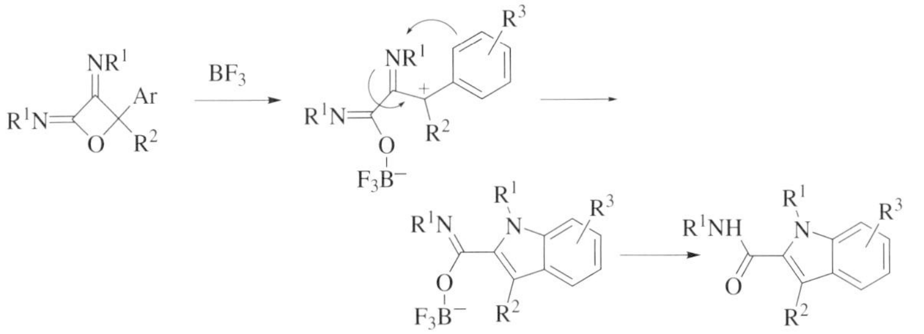
NR1
R1N
Ar
O
R2
BF3
NR1
R3
R1N
R2
O
F3B−
R1
R1N
N
R3
O
F3B−
R2
R1NH
R1
N
R3
O
R2

(Eq. 76)

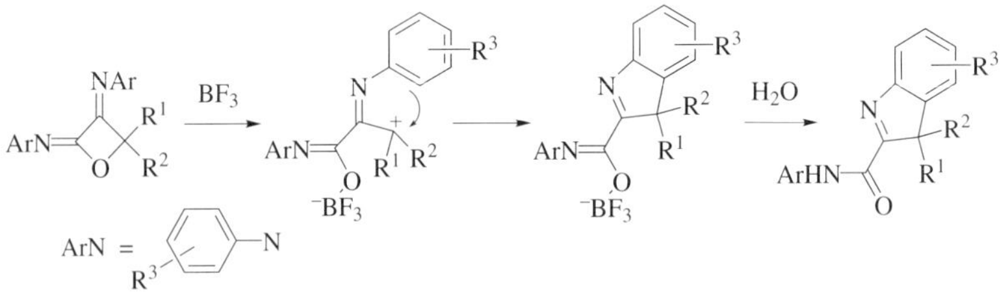
NAr
ArN
R1
O
R2
BF3
N
R3
ArN
O
−BF3
R1
R2
N
R3
R2
ArN
R1
O
−BF3
H2O
N
R3
R2
ArHN
R1
O
ArN =
R3
N

(Eq. 77)

When the isocyanide or the carbonyl compound contains nucleophilic moieties that can attack the isocyanide carbon, the condensation does not lead to diimino oxetanes and only one equivalent of isocyanide is consumed. The use of substrates containing a pyrrole ring presents such an example. The reaction of 1-(2-isocyanophenyl)pyrroles with different aldehydes or ketones (cyclic, acyclic, aromatic, aliphatic) gives 4-(1-hydroxyalkyl)pyrrolo[1,2-*a*]quinoxalines **74** in moderate to excellent yields (Eq. 78).[143,144] The best results are obtained when catalytic amounts of BF_3 are added in two portions at 20-minute intervals. Titanium tetrachloride, tin tetrachloride, zinc chloride, and aluminum chloride are unsatisfactory for this condensation. When the same reaction is performed without any added carbonyl compound, pyrrolo[1,2-*a*]quinoxalines are isolated in nearly quantitative yield. A similar reaction can be performed using electrophiles other than a carbonyl compound, such as epoxides and acetals. The reaction of the latter, discussed in a forthcoming section, furnishes compounds **75** (Eq. 78).[144]

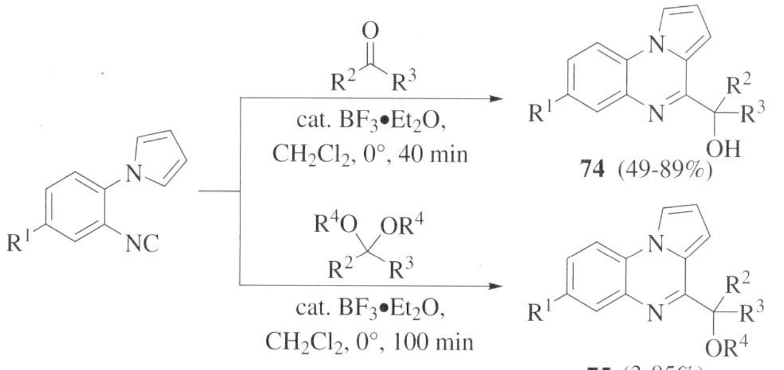
R2 R3
O
cat. BF3•Et2O,
CH2Cl2, 0°, 40 min
R4O OR4
cat. BF3•Et2O,
CH2Cl2, 0°, 100 min
R1
NC
N
OH
OR4

74 (49-89%)

75 (3-85%)

(Eq. 78)

Starting from 2-(1-pyrrolyl)benzaldehyde, α-addition of the carbonyl and the pyrrole onto the isocyanide leads to pyrrolo[1,2-*a*]quinolin-5-ols **76** (Eq. 79). These adducts are fairly unstable; therefore, they should be isolated as the acetates **77**.[145]

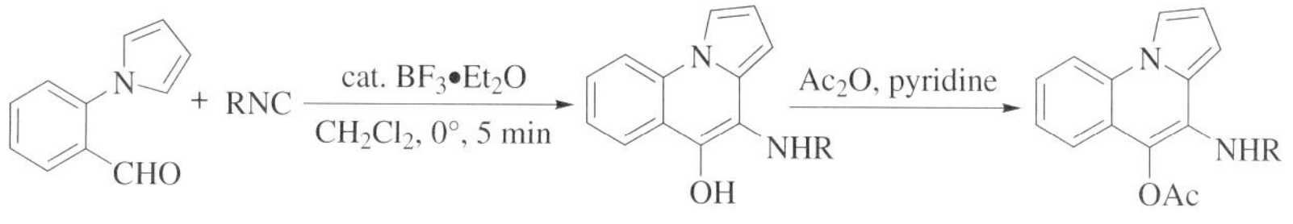
+ RNC
cat. BF3•Et2O
CH2Cl2, 0°, 5 min
Ac2O, pyridine
CHO
NHR
OH
OAc

76

77 (45-59%)

(Eq. 79)

An allenic moiety is also nucleophilic enough to attack the isocyanide carbon. Thus, the reaction of *tert*-butyl isocyanide with β-allenic aldehydes or ketones in the presence of catalytic aluminum trichloride affords five-membered cyclic compounds **78** and **79** (Eq. 80).[146] When R^3 = H, the enamino ketone tautomer **78** predominates.

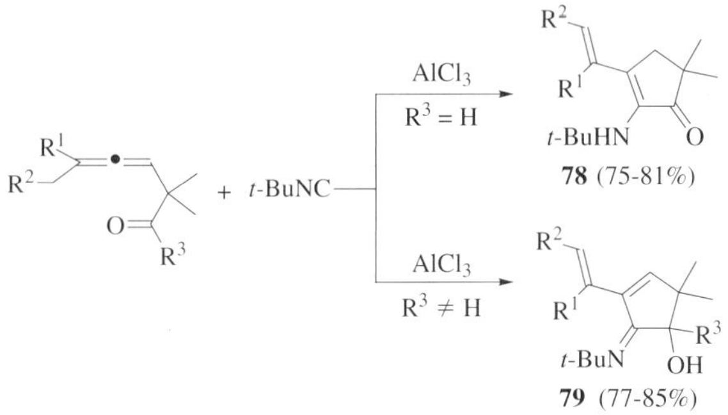

(Eq. 80)

Other Lewis Acids. In the one example of the use of phosphorus oxychloride as the acid component in the Passerini reaction,[141] the α-hydroxy amide is obtained in very low yield together with small amounts of a β,γ-unsaturated-α-oxo amide (the product typically formed in the presence of BF_3).

The combination of catalytic zinc (II) trifluoromethanesulfonate and stoichiometric trimethylsilyl chloride is able to promote the Passerini reaction only in a few specific examples. The efficiency and the outcome of the reaction depend on the nature of the isocyanide. The initial steps are presumed to be silylation of the carbonyl compound followed by nucleophilic attack of the isocyanide to afford nitrilium ion **80** (Eq. 81). The addition of a second molecule of isocyanide can then form another nitrilium ion **81**.

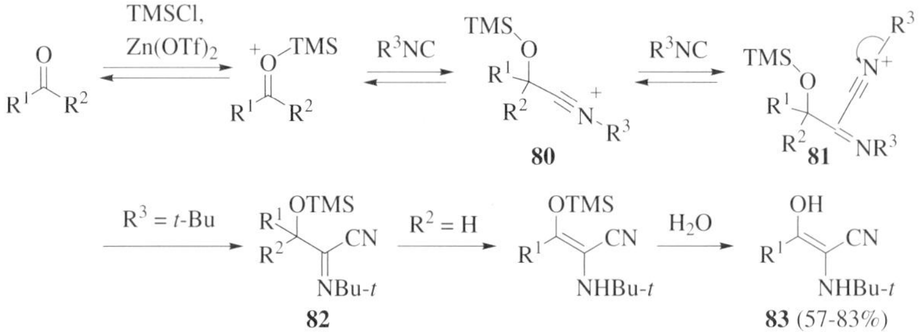

(Eq. 81)

These steps are similar to those involved in the previously described BF_3-catalyzed Passerini reaction, but here cyclization of **81** to a diimino oxetane is not possible since the oxygen atom is silylated. Only when the isocyanide structure allows a suitable alternative transformation of **80** or **81** into a more stable species does the reaction occur. For example, when *tert*-butyl isocyanide is used, nitrilium ion **81** undergoes a dealkylation reaction, giving imino nitriles **82** (Eq. 81).[147] If the initial substrate is an aldehyde, the imino nitriles **82** are further converted, via tautomerization and hydrolytic work-up, into α-cyano enamines **83**, which are useful intermediates for the synthesis of 4-cyano-oxazoles. Starting with ketones, the final products are **82**, which resist desilylation during work-up. In contrast, with other simple isonitriles, such as cyclohexyl isocyanide, no reaction takes place.[36] Interestingly, although dealkylation could in principle occur at the level of nitrilium ion **80** to afford a cyanohydrin, no such type of product is detected. Moreover, in contrast with other Lewis acid mediated condensations of *tert*-butyl isocyanide (see Eq. 67), 1,2-addition predominates over 1,4-addition[147] with α,β-unsaturated aldehydes.

Zinc triflate and trimethylsilyl chloride are also able to promote the Passerini condensation when the isocyanide bears a donor group, such as the morpholine in derivative **84**, that can attack the intermediate nitrilium ions (Eq. 82). Lone pair donation by the morpholine nitrogen converts **85** into a more stable cation, which affords α-hydroxy amides in moderate to good yields after hydrolytic work-up.[36]

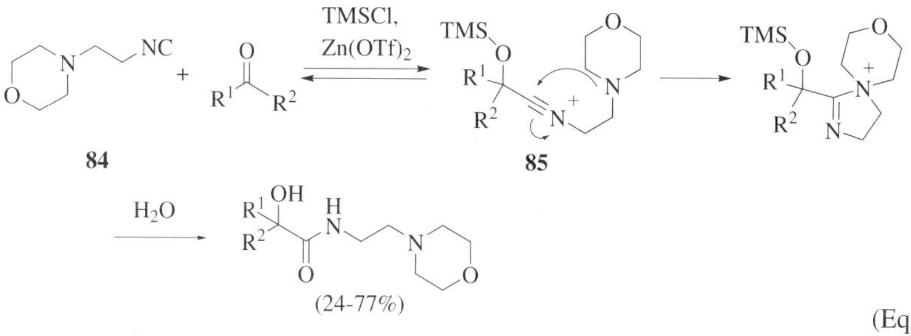
TMSCl,
Zn(OTf)2
84
85
H2O
(24-77%)

(Eq. 82)

A similar neighboring effect is observed with ethyl isocyanoacetate (Eq. 83). The carbonyl oxygen may either trap nitrilium ion **86**, derived from a 1 : 1 adduct between the carbonyl component and the isocyanide, or a nitrilium ion similar to **81** (Eq. 81), representing a 2 : 1 isocyanide/carbonyl adduct. In the first scenario, the final products are the 5-ethoxyoxazoles **87**, whereas in the second, 1,4-oxazinones **88** are formed. With aldehydes, the first pathway predominates. With cyclohexanone, oxazole **87** is accompanied by oxazinone **88**. By using three equivalents of isocyanide, **88** becomes the main product and is isolated in 45% yield.[36]

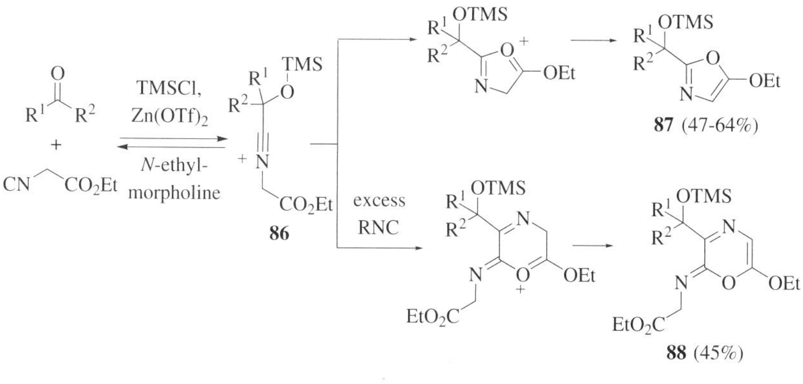
TMSCl,
Zn(OTf)2
N-ethyl-
morpholine
86
excess
RNC
87 (47-64%)
88 (45%)

(Eq. 83)

Recently, this method was extended using α-isocyano acetamides, yielding 5-amino-oxazoles as the main products.[36a]

Passerini Reactions Between Acetals, Isocyanides, and Various Acid Species Affording α-Alkoxy Amides or Other Products

Acetals can replace carbonyl compounds in the Passerini reaction. The typical products are α-alkoxy amides, but under specific conditions other types of compounds may be obtained.

Reactions with Protic Acids. *Trifluoroacetic Acid.* The Passerini reaction between veratraldehyde dimethyl acetal and three isocyanides in the presence of trifluoroacetic acid furnishes the corresponding α-alkoxy amides in high yields (Eq. 84), whereas the reaction of the corresponding aldehyde does not work.[148]

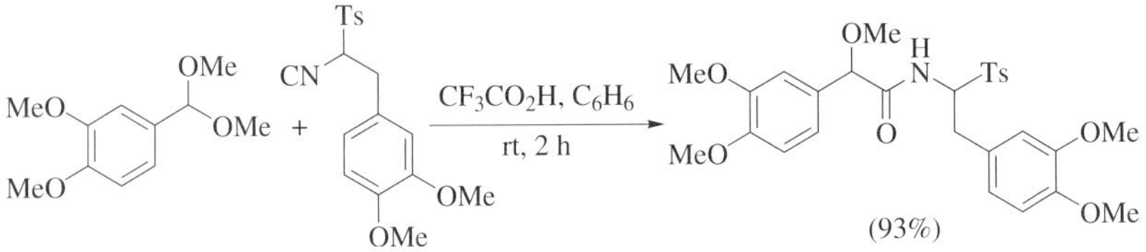

Ts
OMe
CN
MeO
OMe +
MeO
OMe
OMe
CF3CO2H, C6H6
rt, 2 h
OMe
H
N
Ts
MeO
MeO
O
OMe
OMe
(93%)

(Eq. 84)

Reactions with Lewis Acids. *Titanium Tetrachloride and Diethylaluminum Chloride.* The reaction between acetals and isocyanides in the presence of stoichiometric amounts of $TiCl_4$ affords different products depending on the structural features of the organic reagents and the amount of isocyanide employed.

For example, various acetals derived from aliphatic and aromatic aldehydes react with an equimolar amount of cyclohexyl isocyanide in the presence of titanium tetrachloride, at temperatures ranging from −70° to −30°, giving the corresponding α-alkoxy amides **90** in 66–90% yield (Eq. 85).[149] Similar results are obtained using an aryl isocyanide. Ketals, however, produce the products in much lower yield. An α-alkoxyimidoyl chloride species is most likely the intermediate obtained after the nucleophilic attack of the isocyanide onto the α-alkoxy carbenium ion formed between the starting acetal and $TiCl_4$. When the same reaction is performed with *tert*-butyl isocyanide or β-(trimethylsilyl)ethyl isocyanide and the temperature is allowed to reach 20°, the products formed are *O*-alkyl cyanohydrins **91**. The intermediate nitrilium ion **89** loses a stabilized carbocation.[150]

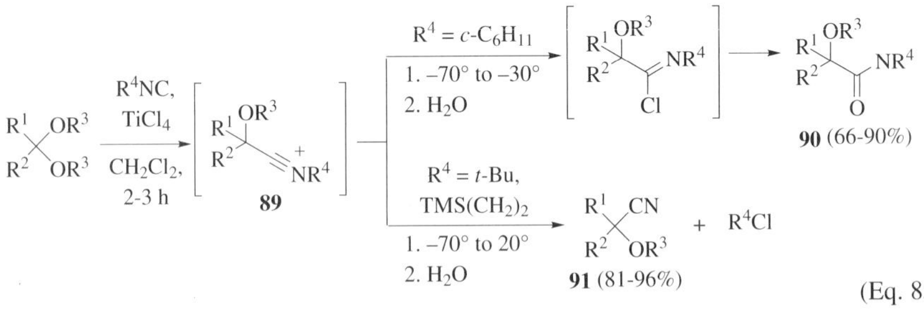

R1 OR3
R2 OR3
R4NC,
TiCl4
CH2Cl2,
2-3 h
R1 OR3
R2
+
NR4
89
R4 = c-C6H11
1. –70° to –30°
2. H2O
R1 OR3
R2
NR4
Cl
R1 OR3
R2
NR4
O
90 (66-90%)
R4 = t-Bu,
TMS(CH2)2
1. –70° to 20°
2. H2O
R1 CN
R2 OR3
+ R4Cl
91 (81-96%)

(Eq. 85)

If two equivalents of *tert*-butyl isocyanide are used and the temperature is allowed to reach room temperature, different products are obtained from aldehyde-derived (either aromatic or aliphatic) acetals (Eq. 86). In this case, the second molecule of isocyanide attacks the imidoyl cation species, giving intermediate **92**, which on loss of *tert*-butyl cation affords a new class of compounds, namely the β-alkoxy-α-cyano enamines **93**, along with variable amounts of the normal α-alkoxy amide products **94** and α-chloro ethers **95** (derived from the Lewis acid mediated replacement of one alkoxy group of the acetal by a chlorine atom).[151]

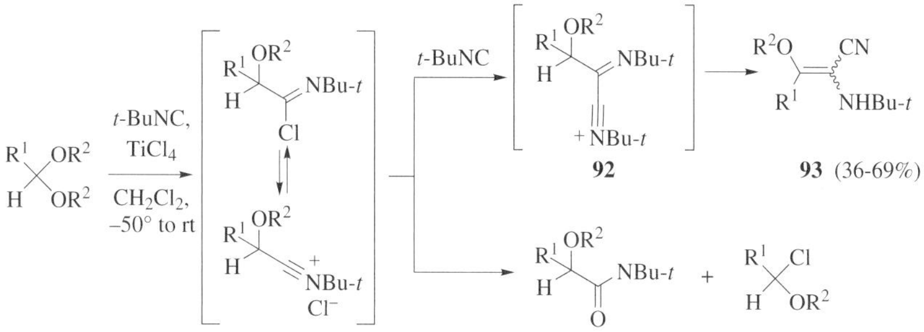
t-BuNC,
TiCl4
CH2Cl2,
–50° to rt
t-BuNC
NBu-t
Cl
Cl⁻
+NBu-t

92

93 (36-69%)

94 (0-30%)

95 (0-12%)

(Eq. 86)

The reaction of acetals with *tert*-butyl isocyanide can also be performed using a stoichiometric amount of diethylaluminum chloride as the Lewis acid.[134] This feature allows for the use of either an acetal or a ketal. Presumably, the reaction follows the same pathway as the $TiCl_4$-mediated transformation; however, α-imino nitriles **96** are isolated instead (Eq. 87). It should be noted that when R^2 = H, the enamines **93** are the tautomers of **96**. Treatment of **96** (R^2 = H) with a protic acid leads to enamines **93**. Apparently, diethylaluminum chloride is unable to promote this tautomerization. On the contrary, $TiCl_4$ is able to promote this transformation (probably because of the presence of some HCl), and only enamines **93** are isolated.

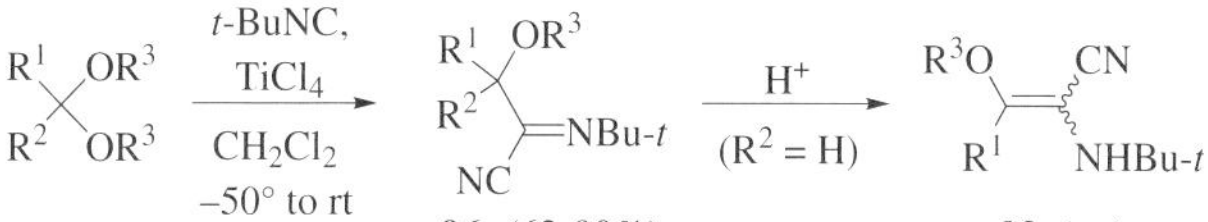
t-BuNC,
TiCl4
CH2Cl2
–50° to rt
H+
(R2 = H)

96 (63-90%)

93 (—)

(Eq. 87)

Finally, there is a single example in which an α-chloro ether is subjected to nucleophilic displacement by a variety of different nucleophiles including cyclohexyl isocyanide in the presence of titanium chloride. This compound shows a reactivity similar to that observed with acetals, and α-alkoxy amides are produced in good yields.[152]

Additionally, *O*-alkylated-*N*-carbomethoxy hemiaminals, when treated with phenyl isocyanide and $TiCl_4$, afford the corresponding secondary α-*N*-carbomethoxyamides. This transformation is more closely related to the Ugi reaction rather than the Passerini condensation.[153,154]

Boron Trifluoride. Following the protocol discussed previously (Eq. 78), 4-(1-alkoxyalkyl)pyrrolo[1,2-*a*]quinoxalines **75** can be obtained by reaction of acetals with 1-(2-isocyanophenyl)pyrroles in the presence of catalytic amounts of BF_3.[144] These reactions are slower with respect to the analogous carbonyl compounds, but the yields are usually good, except when acetaldehyde or bromoacetaldehyde dimethyl acetals are used.

Stereochemical Aspects

Racemization Issues. There are two main stereochemical aspects of the Passerini reaction: the potential racemization of preexisting stereocenters, and the stereoselectivity in the formation of the new stereogenic center.

With respect to the first point, the mild, nearly neutral conditions of the reaction are expected to avoid racemization processes in most cases. For example, non-racemic protected α-amino aldehydes were demonstrated to be configurationally stable in the classical Passerini reaction[45–51] as well as in the pyridinium trifluoroacetate-[116] or $TiCl_4$[78]-mediated reactions. However, other α-chiral aldehydes may not be configurationally stable under Passerini conditions. For example, it has been shown that aldehydes having an α-alkyl substituent undergo racemization during the related Ugi condensation.[155]

The use of enantiomerically pure α-substituted isocyano acetates or isocyano acetamides may pose racemization problems, although there is experimental evidence that demonstrates that the use of both esters[47,48,78] and amides[45] occurs without loss of stereochemical integrity (Eq. 88). Care should be taken during the preparation of chiral α-isocyano esters from the corresponding formamides: whereas the use of diphosgene or triphosgene under controlled temperatures (especially with *N*-methylmorpholine as the base) seems to afford products endowed with high optical purity,[48,78,156–159] the combination of other dehydrating agents and bases, such as phosphorus oxychloride and diisopropylamine, leads to various degrees of racemization.[160–162] Racemization during formamide dehydration is less problematic with α-isocyano amides.[163–165]

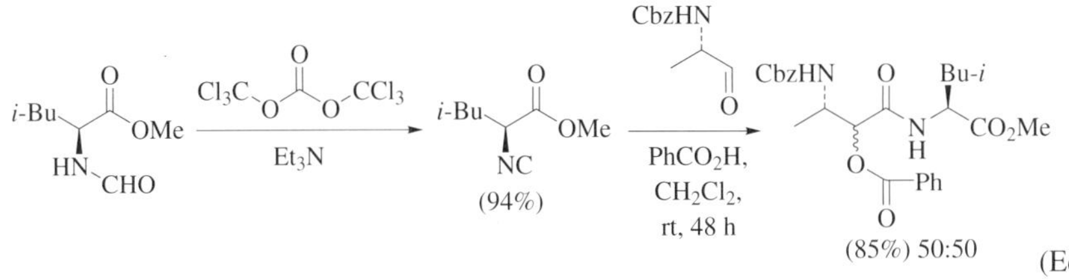
i-Bu
OMe
HN
CHO
Cl3C
CCl3
Et3N
i-Bu
OMe
NC
(94%)
CbzHN
PhCO2H,
CH2Cl2,
rt, 48 h
CbzHN
Bu-i
CO2Me
Ph
(85%) 50:50

(Eq. 88)

Several reports indicate that chiral, non-racemic carboxylic acids do not racemize during the Passerini condensation.[18,45,46,48,49,83,166–168]

Diastereoselective Passerini Reactions. The Passerini condensation of unsymmetrically substituted ketones or aldehydes generates a new stereogenic center. When at least one of the three components is chiral, two different diastereoisomers are formed. Most often, the diastereoselectivity of the Passerini reaction, whether employing carboxylic acids or other acid species, is moderate, ranging from 1 : 1 to

4:1. This relatively low diastereoselectivity is somewhat surprising for the reactions involving aldehydes with an α-stereogenic center (see Tables 3 and 7), which often proceed with high stereoselectivity in other types of nucleophilic additions. The low steric requirement of the isocyano group may account for the low stereoselection, which is a significant limitation of the Passerini reaction. However, there are a few notable exceptions to this general behavior. One of them is the intramolecular reaction of keto acid **15** (Eq. 28),[89] which possesses an α-stereogenic center and affords only one of the two possible diastereoisomeric products.

Another variation that has shown significant promise in the development of a stereoselective Passerini reaction involves the use of a camphor-derived auxiliary (Eq. 89).[24] Chiral camphor-derived isocyanide **97** gives high asymmetric induction in reactions with some aliphatic aldehydes.[24] The chiral auxiliary may then be removed following the condensation reaction.[169]

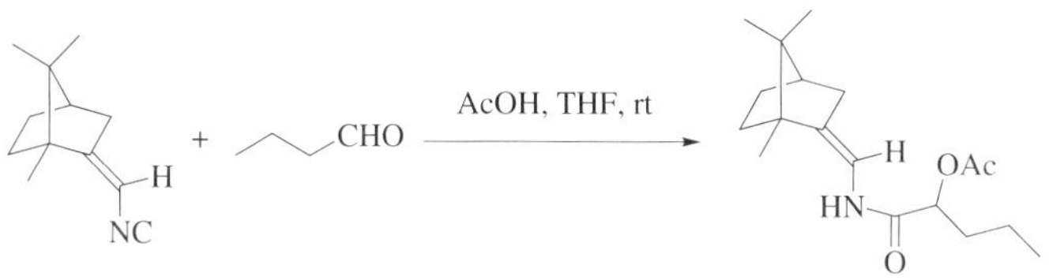
+
CHO
AcOH, THF, rt
H
NC
H
OAc
HN
O

97 (94%) % de = 93 (Eq. 89)

The galacturonic acid derivative **98** is another excellent chiral inducer for the classic Passerini reaction (Eq. 90).[168] However, the diastereoselectivity is not always high. For example, by using *o*-tolyl isocyanide it decreases to 56:44.

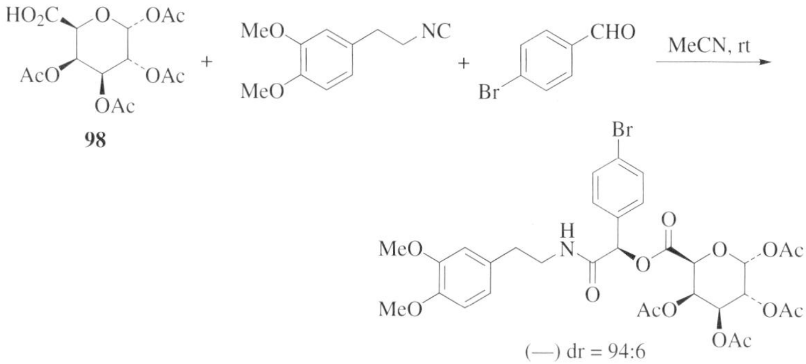
HO2C
O
OAc
AcO
OAc
OAc
98
+
MeO
MeO
NC
+
Br
CHO
MeCN, rt
Br
MeO
MeO
H
N
O
O
O
O
OAc
AcO
OAc
OAc

(—) dr = 94:6 (Eq. 90)

The pyridinium trifluoroacetate mediated Passerini reaction of chiral racemic cyclic ketone **99** with *tert*-butyl isocyanide is reported to be highly stereoselective, affording a single isomer (Eq. 91).[115]

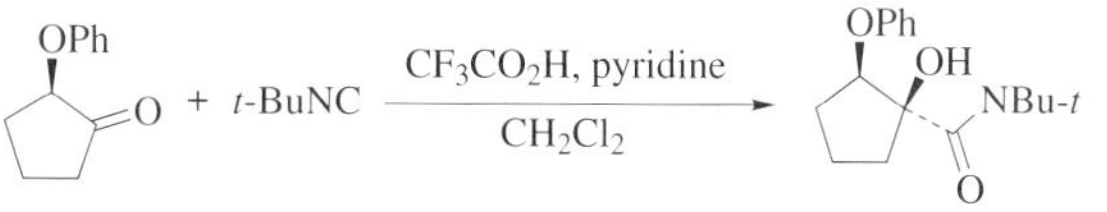
OPh
O
+ t-BuNC
CF3CO2H, pyridine
CH2Cl2
OPh
OH
NBu-t
O

99 (33%) (Eq. 91)

These four examples suggest that high induction may be obtained in suitable sterically constrained systems.

The reaction of acetaldehyde diethyl acetal, trimethyl 1-propenyloxysilane, and *tert*-butyl isocyanide in the presence of $TiCl_4$ (following the general reaction scheme depicted in Eq. 66), which affords a γ-alkoxy-β-hydroxyamide with three stereocenters, generates predominantly one of the four possible diastereoisomers.[134]

It should be stressed that this general lack of stereoselectivity is not always a problem. In several applications, α-hydroxy amides are prepared only as intermediates for the synthesis of α-oxo amide enzyme inhibitors, in which the secondary hydroxy function is oxidized to give the corresponding ketone, thus losing the stereochemical information generated in the Passerini condensation.[116,117]

Asymmetric Catalytic Passerini Reactions. The Lewis acid mediated Passerini reactions seem well suited for the development of a chiral catalyst. However, preliminary attempts that used chiral alkoxy titanates gave unsatisfactory results, affording α-hydroxy amides with no enantioselection at all.[78] Moreover, a problem associated with the use of a chiral Lewis acid is the poor catalytic turnovers that necessitate the use of stoichiometric quantities of the chiral mediator. These problems have been solved by using stoichiometric amounts of a mild achiral Lewis acid (tetrachlorosilane) together with catalytic amounts of a chiral Lewis base activator, such as phosphoramide **100** (Eq. 92).[170]

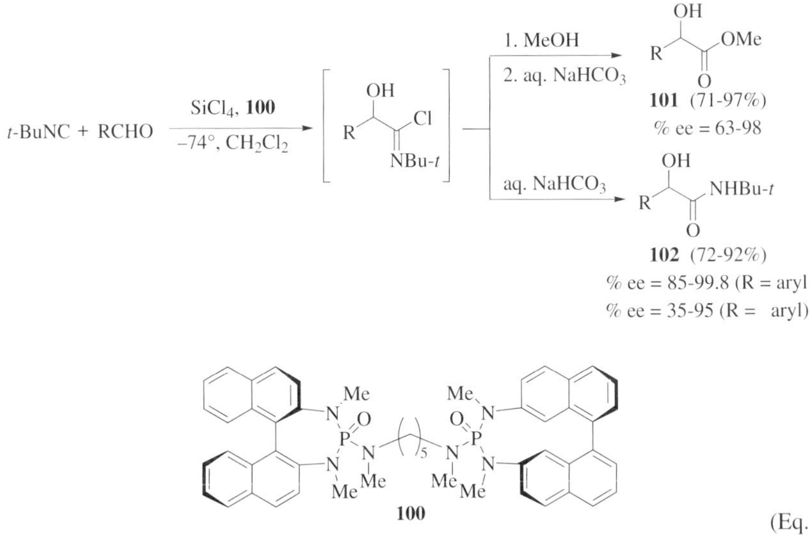

t-BuNC + RCHO
SiCl4, 100
–74°, CH2Cl2
OH
Cl
R
NBu-t
1. MeOH
2. aq. NaHCO3
OH
R
OMe
O
101 (71-97%)
% ee = 63-98
aq. NaHCO3
OH
R
NHBu-t
O
102 (72-92%)
% ee = 85-99.8 (R = aryl
% ee = 35-95 (R = aryl)
Me
N
O
P
N
N
Me
Me
5
Me
O
N
P
N
N
Me
Me
Me
100

(Eq. 92)

Depending on the work-up conditions, either the methyl esters **101** or the α-hydroxy amides **102** can be isolated. Using *tert*-butyl isocyanide and aromatic aldehydes the enantiomeric excesses are consistently high, whereas with aliphatic aldehydes they are lower and greatly depend on the steric bulk. A decrease of selectivity is also observed on using less encumbered isonitriles. Therefore, for the preparation of methyl esters **101**, *tert*-butyl isocyanide is the reagent of choice.

Development of a chiral catalyst for the classical Passerini reaction is even more difficult since the Lewis acid usually replaces the carboxylic acid as the third component, leading to α-hydroxy amides or to other kinds of products, as described above. Nevertheless, after a thorough screening of combinations of Lewis acids/chiral ligands, the couple formed by diol **103** and Ti(OPr-i)$_4$ was found to afford moderate yields and enantiomeric excesses with a series of substrates (Eq. 93).[171] It should be noted that a stoichiometric quantity of the chiral inducer was needed in this screening experiment. This study represents the first example of an asymmetric classical Passerini among three achiral components and opens the way to further improvements.

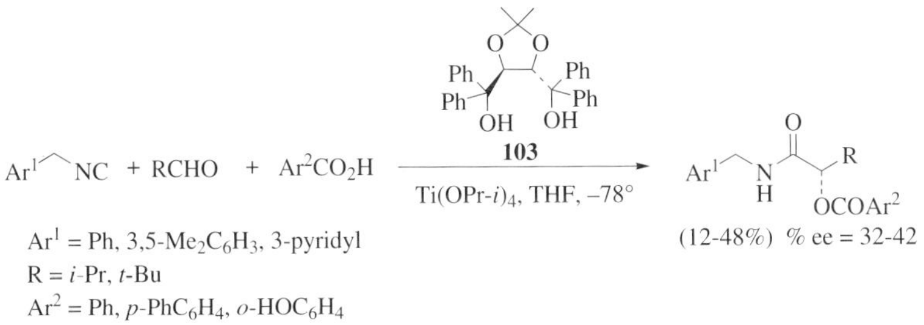
Ph
Ph
Ph
Ph
OH
OH
103
Ar1
NC
+ RCHO
+ Ar2CO2H
Ti(OPr-i)4, THF, –78°
Ar1
N
H
R
OCOAr2
Ar1 = Ph, 3,5-Me2C6H3, 3-pyridyl
R = i-Pr, t-Bu
Ar2 = Ph, p-PhC6H4, o-HOC6H4

(Eq. 93)

(12-48%) % ee = 32-42

Solid-Phase Synthesis

At present, the solid-phase Passerini reaction has been reported using only supported carboxylic acids or isocyanides. Compound **104**, a glycine bound to a resin through a photocleavable carbamate, reacts with a variety of isocyanides and aldehydes to afford adducts that are converted into the corresponding acetamides by photochemical cleavage in the presence of Ac$_2$O (Eq. 94).[12] The yields are not always good. Difficult conversions include aromatic aldehydes (especially β-naphthaldehyde) in combination with allyl or butyl isocyanides.

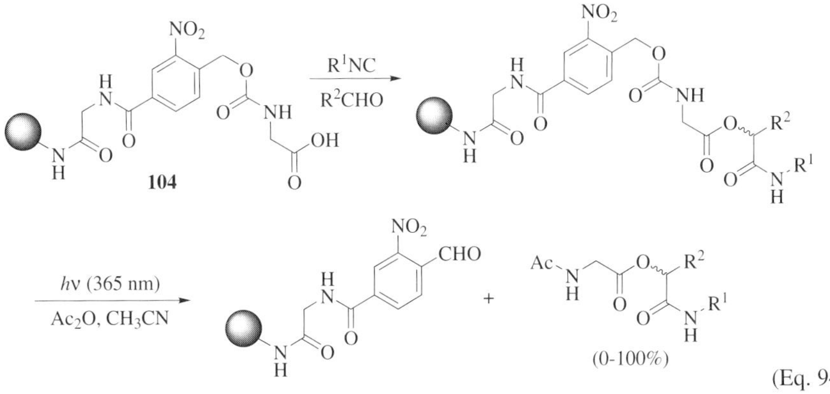
NO2
NH
OH
104
R1NC
R2CHO
NO2
NH
R2
R1
hν (365 nm)
Ac2O, CH3CN
NO2
CHO
+
Ac
R2
R1
(0-100%)

(Eq. 94)

β-Isocyano propionates, such as **105** or **106**, linked to aminomethyl polystyrene resin through acid[96] or photochemically[98] cleavable linkers, react well with *N*-(Fmoc)-phenylalaninal or *N*-(Boc)-phenylalaninal, respectively, and phenylacetic acid (Eq. 95 and 96). After a deprotection/transacylation step, the adducts are removed from the resin. The overall yields of the final *α*-hydroxy-*β*-acylamino amides are comparable to those obtained in the solution phase, indicating that the combinatorial synthesis of libraries of this class of compounds is feasible. Using the acid-cleavable linker and three different solid-supported isocyanides (also derived from *α*-aminoacids), a mini-library of 12 adducts has been synthesized.[96]

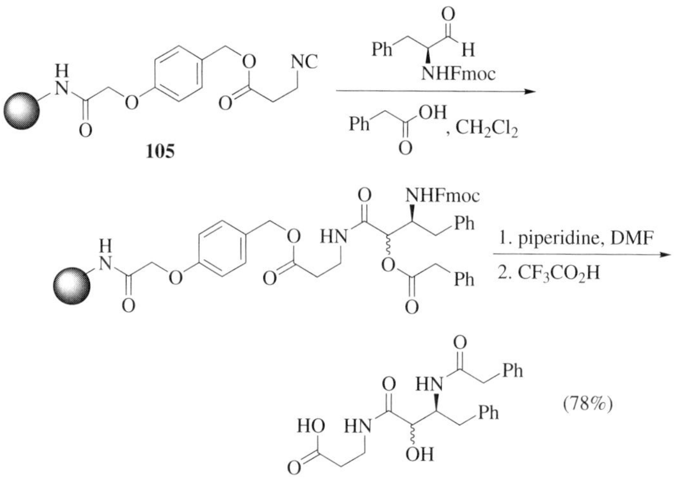

105
NHFmoc
, CH_2Cl_2
1. piperidine, DMF
2. CF_3CO_2H
(78%)

(Eq. 95)

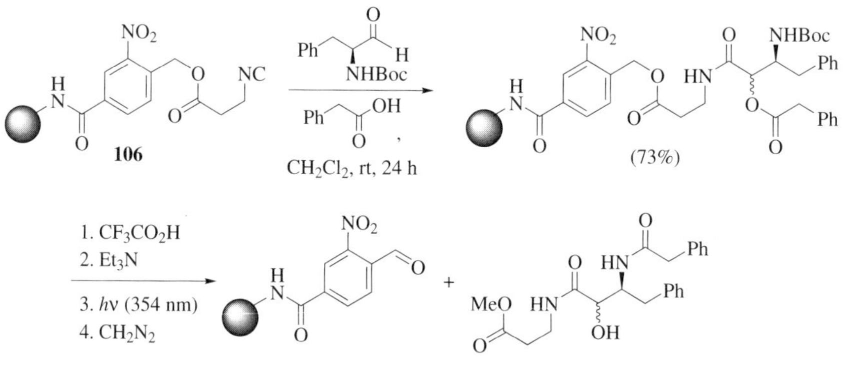

106
NHBoc
CH_2Cl_2, rt, 24 h
(73%)
1. CF_3CO_2H
2. Et_3N
3. *hν* (354 nm)
4. CH_2N_2
(55%)

(Eq. 96)

Isocyanides supported through a Wang linker, such as **107**, have also been used for the synthesis of *α*-hydroxy amides through the pyridinium trifluoroacetate-mediated Passerini reaction. Eq. 97 shows the application of this methodology to the synthesis of the enzyme inhibitor poststatin (**108**).[162]

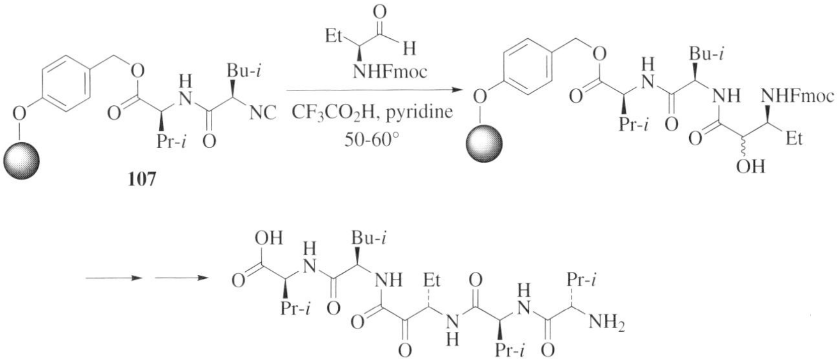

107

108 (58%)

(Eq. 97)

Isocyanide **109,** supported through a Rink linker, reacts with Fmoc-protected α-amino aldehydes affording the α-hydroxy amides, such as **110,** in excellent yield (Eq. 98).[162] After Fmoc deprotection, acylation of the resulting amine, and oxidation, the linker is cleaved affording the primary amide **111**. This is a convenient method for the preparation of primary α-oxo amides, which are not directly accessible through the Passerini reaction.

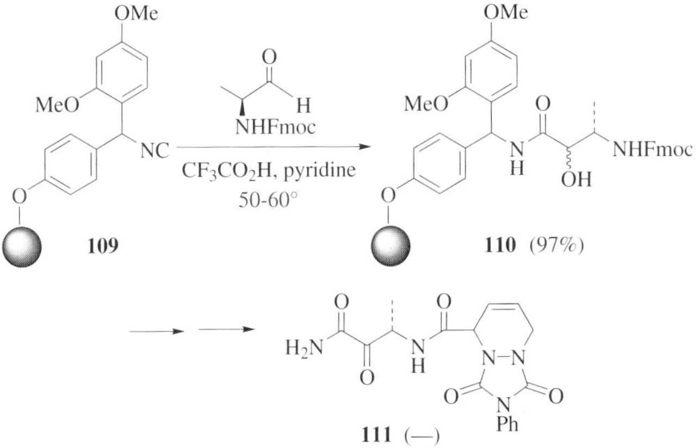

109

110 (97%)

(Eq. 98)

111 (—)

APPLICATIONS TO SYNTHESIS

The Passerini reaction has been demonstrated to be the most convergent approach to depsipeptides, an important family of antitumor agents characterized by a peptide structure incorporating an ester or a lactone functionality.[60] Chloroacetic acid, azidoacetic acid, or glycine derivatives are treated with simple aldehydes and *tert*-butyl isocyanoacetate giving, after suitable functional group manipulation, depsipeptide precursors, such as **112** (Eq. 99). After *tert*-butyl ester cleavage, the resulting depsipeptides have been used for further Passerini reactions, allowing the preparation of pentadepsipeptides, such as **113** (Eq. 100).[60]

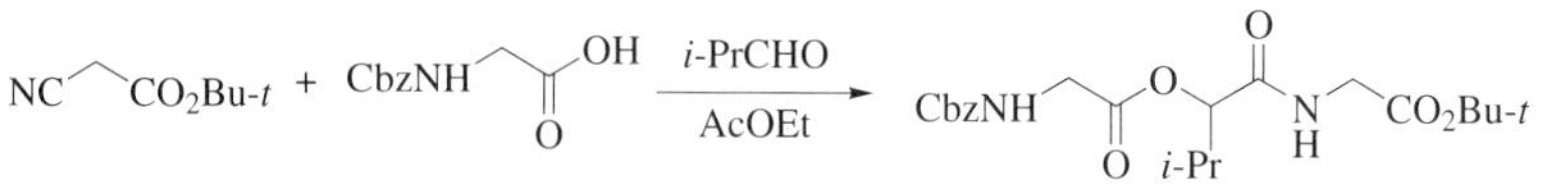
NC
CO2Bu-t
+ CbzNH
OH
O
i-PrCHO
AcOEt
CbzNH
O
O
i-Pr
O
N
H
CO2Bu-t

112 (80%)

(Eq. 99)

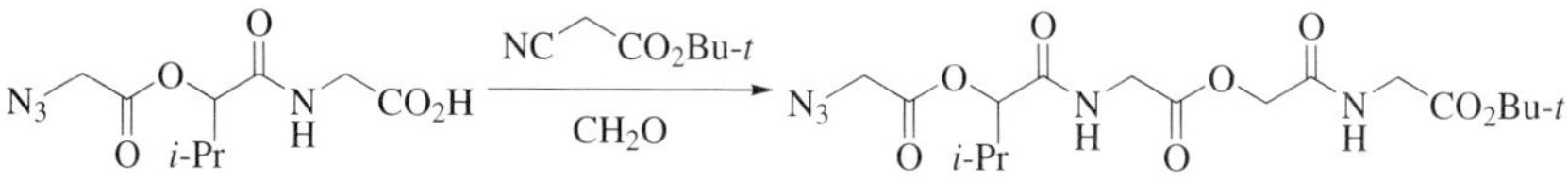
N3
O
O
i-Pr
O
N
H
CO2H
NC
CO2Bu-t
CH2O
N3
O
O
i-Pr
O
N
H
O
O
O
N
H
CO2Bu-t

113 (56%)

(Eq. 100)

In a synthetic approach toward analogs of the azinomycins **116**, the α-acyloxy amide moiety is assembled using a Passerini reaction starting from epoxy aldehyde **114**. In order to obtain the acyl enamine present in the natural compounds, two alternative routes were followed. The first makes use of α-phosphono isocyano acetate (**115**) in order to provide a functionality able to undergo subsequent Horner-Wadsworth-Emmons reactions (Eq. 101).[40]

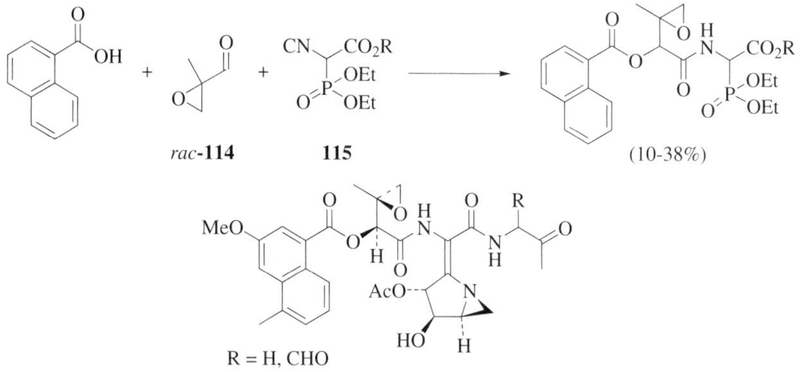
O
OH
+
O
O
+
CN
CO2R
OEt
O=P
OEt
O
O
O
O
H
N
CO2R
O
OEt
P
O
OEt
MeO
O
O
O
H
H
N
O
O
R
N
H
O
AcO
N
HO
H

rac-**114** **115** (10-38%)

R = H, CHO

116

(Eq. 101)

Alternatively, an isocyano acetate already containing the double bond is used (Eq. 102).[40,41] Of particular interest are the examples involving functionalized isocyanides containing the aziridine nucleus, such as **117**. With these aziridinyl isocyanides, the addition of pyridine is essential to obtain the desired products. However, under these conditions, the geometry of the starting alkene is not fully conserved.

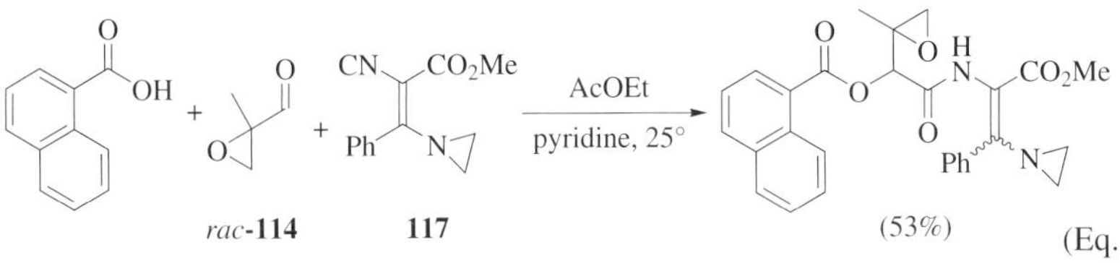
O
OH
+
O
O
rac-114
+
CN
CO2Me
Ph
N
117
AcOEt
pyridine, 25°
O
O
O
H
N
CO2Me
O
Ph
N
(53%)

(Eq. 102)

The above-mentioned synthetic applications are characterized by the fact that the acyl group is retained in the exact position where it was introduced during the Passerini reaction. On the other hand, in the concise synthesis of the prolyl endopeptidase inhibitor eurystatin A (**122**) shown in Eq. 103,[48] the acyl group is again retained, but shifted to another position. The condensation of three amino acid derived components, an Fmoc-protected alaninal, a leucine- derived isocyanide, and a protected ornithine, allow the one-pot preparation of methyl and benzyl esters **118** and **119**. Removal of the Fmoc group under mild basic conditions causes an in situ *O*- to *N*-acyl migration to afford α-hydroxy amides **120** and **121**, bearing the entire acyclic skeleton of eurystatin. A sequence of deprotection steps, macrocyclization, and final oxidation of the secondary alcohol give the target compound in good overall yield. The negligible diastereoselection in the Passerini step is unimportant because the stereogenic center created is eventually lost during the final oxidation.

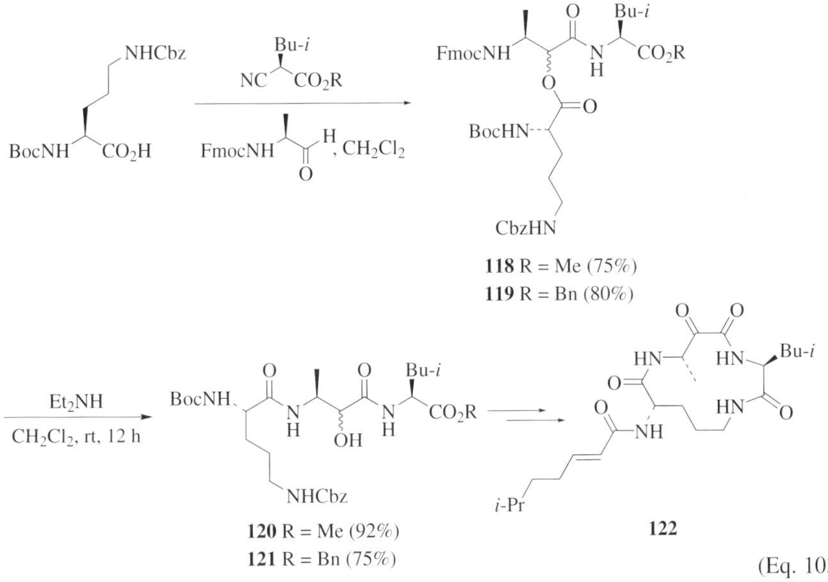
NHCbz
BocNH
CO2H
Bu-i
NC
CO2R
FmocNH
H
O
, CH2Cl2
FmocNH
O
Bu-i
N
H
CO2R
O
O
BocHN
CbzHN
118 R = Me (75%)
119 R = Bn (80%)
Et2NH
CH2Cl2, rt, 12 h
BocNH
O
N
H
OH
O
Bu-i
N
H
CO2R
NHCbz
120 R = Me (92%)
121 R = Bn (75%)
O
O
HN
HN
Bu-i
O
HN
O
O
NH
i-Pr
122

(Eq. 103)

An earlier preparation of eurystatin A (**122**) also takes advantage of the Passerini reaction in a key step (Eq. 104).[47] This synthesis is, however, less efficient in terms of atom economy, since the acyl group (benzoyl) is not retained, but is removed soon after the Passerini reaction. The third amino acid fragment (ornithine) is joined only in a later step.

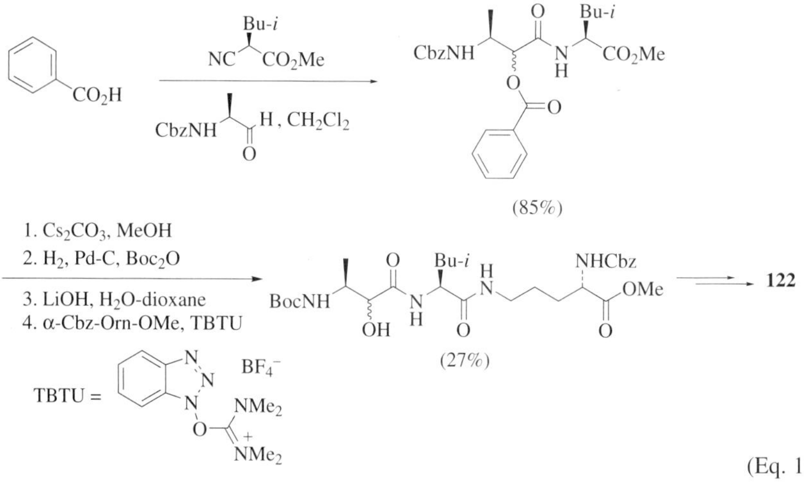

(Eq. 104)

An interesting intramolecular Passerini reaction was used for a convergent racemic synthesis of hydrastine (**126**), a phthalidyl isoquinoline alkaloid (Eq. 105).[81] The key step exploits opianic acid (**124**) as a bifunctional synthon, which reacts with isocyanide **123** to afford lactone **125**. The resulting amide group is then employed for the cyclization under Bischler-Napieralski conditions giving, after catalytic reduction of the intermediate iminium salt and reductive *N*-methylation, the final target **126** as an approximately equimolar diastereomeric mixture.

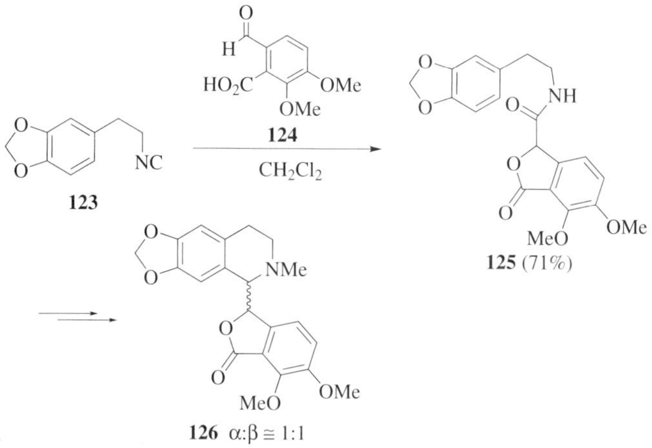

(Eq. 105)

Whereas the isocyanide typically acts in the Passerini reaction as an equivalent of the [CONHR] acyl anion, in an approach to amphimedine (**129**) and related marine alkaloids,[37] methyl isocyanide is employed as an equivalent of a carbalkoxy group (Eq. 106). Conversion of the *N*-methyl amide **127** into the methyl ester **128** is carried out through an unusual sequence involving *N*-nitrosation followed by rearrangement. Finally, deacetylation affords intermediate **128**, which could not be converted into amphimedine (**129**). In this application, neither the acyl group from the carboxylic component nor the NHMe group from the isocyanide component is retained in the target.

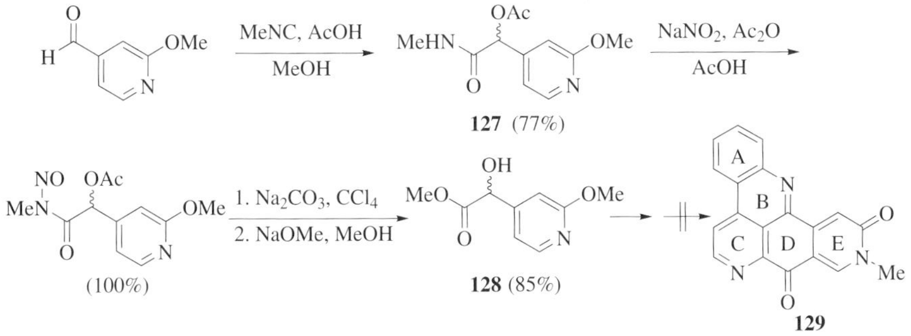
O
H
OMe
N
MeNC, AcOH
MeOH
OAc
MeHN
OMe
O
N
127 (77%)
NaNO2, Ac2O
AcOH
NO
OAc
MeN
OMe
O
N
(100%)
1. Na2CO3, CCl4
2. NaOMe, MeOH
OH
MeO
OMe
O
N
128 (85%)
A
B
N
C
D
E
O
N
Me
N
O
129

(Eq. 106)

Another example where the α-acyloxy amide moiety introduced by a Passerini reaction is not retained is represented by the synthesis of a series of 2-pyridylethanolamines **132**, endowed with β-adrenergic agonist properties and structurally related to the well-known bronchodilator salbutamol (Eq. 107).[38] For this synthesis, the isocyanide is employed as an aminomethyl anion equivalent. Thus, after the Passerini reaction, the products **130** are deacetylated and submitted to reductive treatment with borane to provide secondary amines **131**.

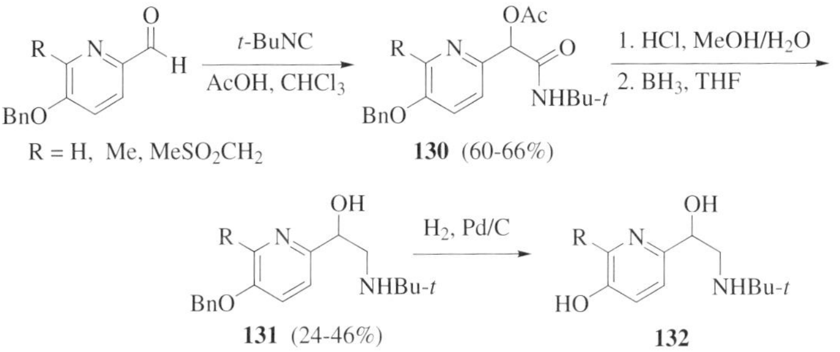
R
N
O
H
BnO
R = H, Me, MeSO2CH2
t-BuNC
AcOH, CHCl3
OAc
R
N
O
NHBu-t
BnO
130 (60-66%)
1. HCl, MeOH/H2O
2. BH3, THF
OH
R
N
NHBu-t
BnO
131 (24-46%)
H2, Pd/C
OH
R
N
NHBu-t
HO
132

(Eq. 107)

A family of ten α-keto amide inhibitors of factor Xa, a trypsin-like serine protease involved in blood coagulation, was prepared by taking advantage of the pyridinium trifluoroacetate-mediated Passerini reaction.[117] Eq. 108 shows a representative example. Several α-amino aldehydes are reacted with *tert*-butyl isocyanide under these conditions (also on a 70-g scale). With all of these aldehydes, the resulting *tert*-butyl amide is not maintained, but cleaved in order to couple the resulting acid with a suitable amine. On the other hand, the amino group arising from the starting amino aldehyde is coupled with an arginine derivative. Final oxidation of the secondary alcohol furnishes the desired α-keto amide targets as single stereoisomers. The overall sequence is, therefore, demonstrated to be stereoconservative.

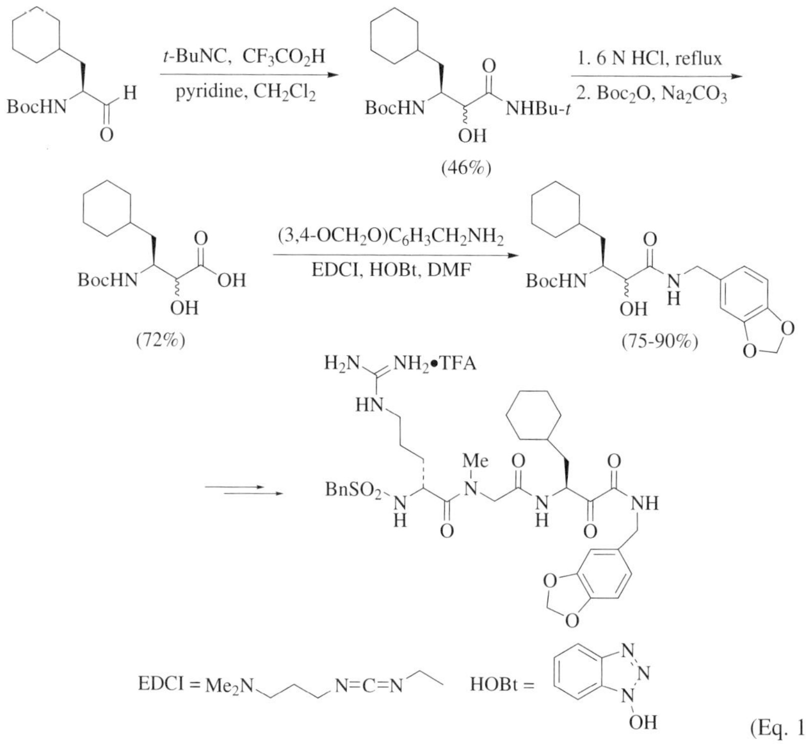

t-BuNC, CF3CO2H
pyridine, CH2Cl2
BocHN
H
O
NHBu-t
OH
(46%)
1. 6 N HCl, reflux
2. Boc2O, Na2CO3
OH
(72%)
(3,4-OCH2O)C6H3CH2NH2
EDCI, HOBt, DMF
N
H
(75-90%)
H2N
NH2•TFA
HN
BnSO2
Me
NH
EDCI = Me2N
N=C=N
HOBt =

(Eq. 108)

In an attempted synthetic approach to benzylisoquinoline alkaloids, such as papaverine (**135**), the Passerini reaction between isocyanide **133** and veratraldehyde dimethyl acetal in the presence of trifluoroacetic acid was investigated (Eq. 109).[148] This represents one of the very few synthetic applications of the Passerini reaction of acetals. Interestingly, the related reactions with the corresponding aldehyde failed. Removal of the sulfonyl group, followed by Bischler-Napieralski cyclization, affords the dihydroisoquinoline **134**. Unfortunately, the final elimination to give papaverine (**135**) failed.

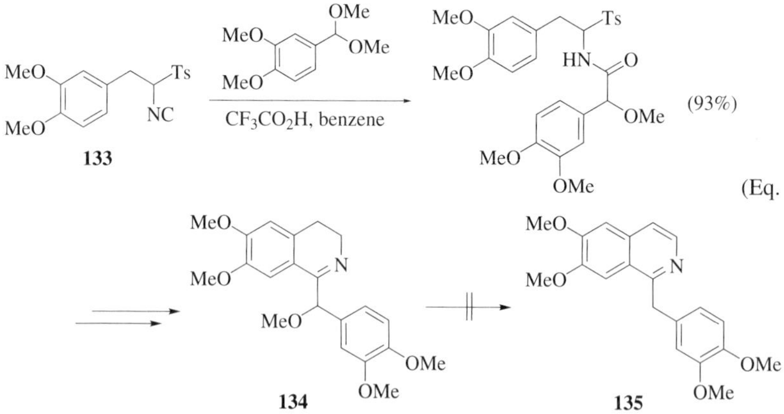

MeO
Ts
NC
OMe
CF3CO2H, benzene
HN
O
(93%)
N

133 **134** **135**

(Eq. 109)

COMPARISON WITH OTHER METHODS

The synthesis of secondary α-acyloxy amides through the Passerini reaction is unique, since there are no other ways to assemble this type of compound in a single synthetic step. Additionally, the synthesis of secondary α-hydroxy amides by the use of protic or Lewis acids is probably the most direct and general route to these substances. All of the alternative preparations of these two classes of compounds require at least two synthetic steps. From the point of view of briefness, therefore, both types of Passerini condensations (the "classic" and the one leading to α-hydroxy amides) are superior to other methods for most target molecules.

However, there may be reasons that make the alternative methods comparable or even preferred in some cases. Some drawbacks of the Passerini reaction are the incompatibility of certain carbonyl compounds as well as the difficulty in preparing certain isocyanides (not to mention the notorious stench of compounds containing this functionality). However, the most important limitations of the Passerini reaction are related to stereochemical issues, since there are no current methods to make the process enantioselective. In addition, when one of the substrates is chiral and enantiomerically pure, Passerini reactions tend to be only moderately stereoselective.

Among the many possible methods of synthesizing α-acyloxy amides and α-hydroxy amides, those involving a limited number of steps or presenting clear advantages from a stereochemical point of view have been selected for comparison with the Passerini reaction.

α-Acyloxy amides may be prepared by the Ag_2O-mediated reaction of carboxylic acids with α-halo amides (typically α-bromo amides) (Eq. 110).[172–175] These compounds may, in turn, be obtained by the reaction of α-halo carboxylic acid halides with a primary amine.[174]

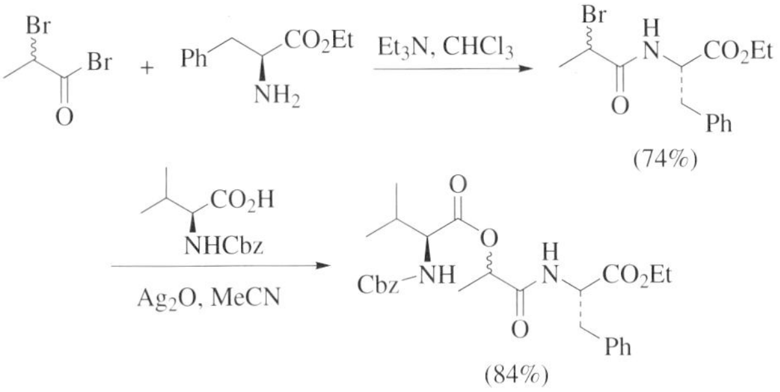
Br
Br
O
Ph
CO_2Et
NH_2
Et_3N, $CHCl_3$
Br
H
N
CO_2Et
O
Ph
(74%)
CO_2H
NHCbz
Ag_2O, MeCN
O
O
H
N
CO_2Et
Cbz
NH
O
Ph
(84%)

(Eq. 110)

This method is convergent and rather concise: it joins in two steps a haloacyl halide, an amine, and a carboxylic acid. However, this procedure, which requires more steps than the corresponding Passerini reaction, has no advantages from a stereochemical point of view, and may be preferred only when the required amines are more easily available than the corresponding isocyanides. An alternative method for preparing secondary α-bromo amides is the Ritter reaction of α-bromo nitriles. This reaction, however, requires harsh reaction conditions and is limited to amides *N*-substituted with tertiary groups.[110]

One route to α-acyloxy amides is based on the rearrangement of *O*-acyl hydroxamates (Eq. 111).[176] This methodology is also highly convergent, since *O*-acylhydroxamates are prepared in two steps from three different components (two acyl chlorides and an *N*-substituted hydroxylamine). The overall sequence (3 steps) is rather concise, but still longer than the Passerini condensation. This rearrangement seems limited to hydroxamates having a β,γ-double bond or an α-aryl substituent, making it complementary to the Passerini reaction. Products such as **136** cannot be obtained by the latter process, owing to the reported unreactivity of α,β-unsaturated aldehydes under classical Passerini conditions,[62] and to the complex reaction that takes place under Lewis acid catalysis. Another advantage of this strategy is the use of *O*-methylhydroxylamine instead of the highly volatile methyl isocyanide when the desired products are *N*-methyl amides.

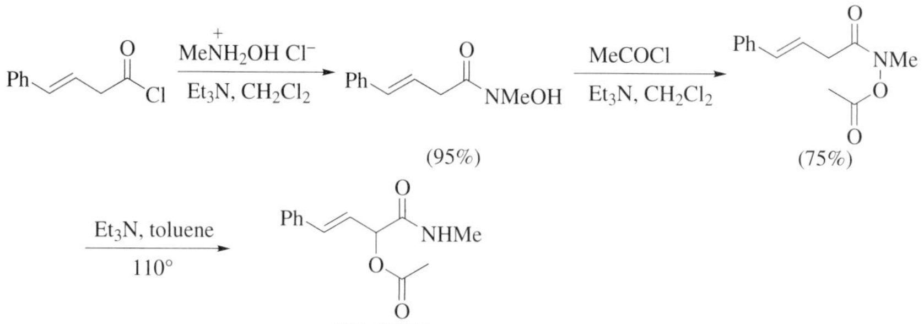

Ph
O
Cl
$MeNH_2OH$ Cl⁻
Et_3N, CH_2Cl_2
Ph
O
NMeOH
(95%)
MeCOCl
Et_3N, CH_2Cl_2
Ph
O
NMe
O
O
(75%)
Et_3N, toluene
110°
Ph
O
NHMe
O
O

136 (99%)

(Eq. 111)

Apart from these two methodologies, α-acyloxy amides are commonly prepared through acylation of α-hydroxy amides.[177] The two most direct methods to prepare α-hydroxy amides (other than the Passerini reaction) are the α-oxidation of amide enolates[178–182] and the condensation of carbonyl compounds with carbamoyl-metal species.[183–185] In both instances, however, these methods work well only for tertiary amides, whereas the products of the Passerini reaction are secondary amides. There is only one report concerning the α-oxidation of a secondary amide; however, the yields are low and the method has a limited scope.[186]

α-Hydroxy amides can also be prepared by the reduction of α-oxo amides. This method may be particularly useful from a stereochemical point of view, since asymmetric reduction of α-oxoamides in the presence of chiral catalysts is known.[187–192] The problem is the preparation of α-oxo amides. Obviously, they can be obtained from (racemic) α-hydroxy amides through oxidation.

A useful alternative is to react an α-oxo ester with a primary amine in the presence of Me_3Al.[188,193] The most general way to prepare α-oxo esters involves reaction of an organometallic compound with a dialkyl oxalate[194] or with a cyano formate such as **137**.[195] The overall three-step sequence is shown in Eq. 112. The only advantage over the Passerini reaction lies in the asymmetric reduction step.

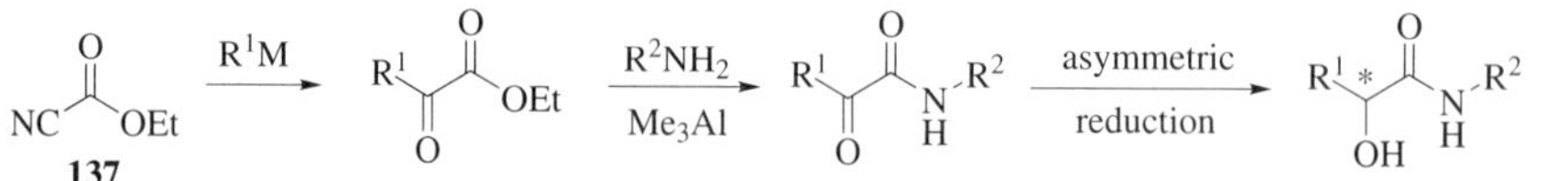

O
NC
OEt
R^1M
R^1
O
OEt
O
R^2NH_2
Me_3Al
R^1
O
N
H
R^2
O
asymmetric
reduction
R^1
*
O
N
H
R^2
OH

137

(Eq. 112)

Thus far, the routes that have been used most frequently for the preparation of α-hydroxy amides begin with α-hydroxy acids or α-hydroxy esters.

From the acids, amide formation has been carried out in several ways. The most useful procedures are those that do not require previous hydroxyl protection. Coupling may be achieved with *N,N'*-dicyclohexylcarbodiimide (DCC)/*N*-hydroxybenzotriazole (HOBT),[196] DCC/*N*-hydroxysuccinimide (HOSU),[197] *N*-(dimethylaminopropyl)-*N'*-ethylcarbodiimide (EDCI)/HOBt,[198–203] carbonyldiimidazole (CDI),[204] or benzotriazolyloxy-tris(dimethylamino)phosphonium hexafluorophosphate (BOP).[205,206] A method that seems well suited for bulk synthesis involves simple heating of the sodium carboxylate with the hydrochloride of a primary amine (Eq. 113).[207] Although protection-deprotection of the hydroxyl is expected to lengthen the synthetic route, an efficient one-pot procedure has been reported.[208]

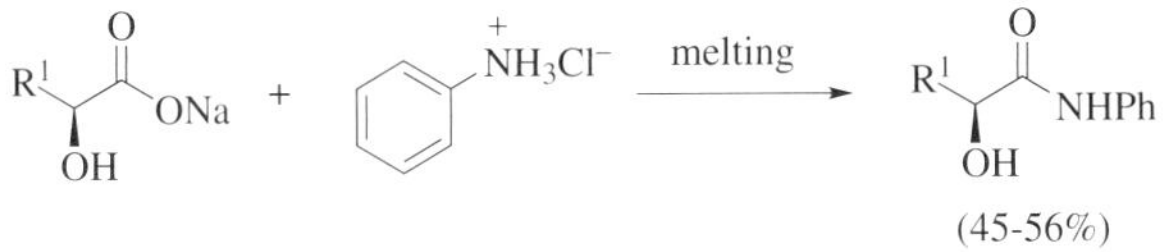

O
R1
ONa
OH
+
NH3Cl−
melting
O
R1
NHPh
OH
(45-56%)

(Eq. 113)

α-Hydroxy amides can also be prepared from α-hydroxy esters. Traditional methods involve simple heating of the latter with a primary amine.[187,209–213] Milder methods to achieve this transformation employ aluminum amides obtained by treatment of a primary amine with Me_3Al[193] or with $LiAlH_4$,[214] or take advantage of enzymatic catalysis.[215]

The synthesis of α-hydroxy amides from α-hydroxy acids or α-hydroxy esters has advantages over the use of the Passerini reaction, particularly when the starting materials are readily available in enantiomerically pure form. For example, a variety of α-hydroxy acids can be obtained from the pool of chiral compounds, and can be obtained in enantiomerically pure form by the nitrosation of α-amino acids.[216] There are also several efficient asymmetric methods for the preparation of α-hydroxy esters. In addition to chemical[217,218] or enzymatic[219] asymmetric reduction of α-oxo esters, they can be synthesized by oxidation of chiral *N*-acyl oxazolidinones,[220] asymmetric substitution of a chiral α-halo ester,[221] or hydrolysis of 1,1,1-trichloro-2-alkanols, which are produced in high enantiomeric excess by asymmetric reduction of the corresponding ketones.[222]

As stated previously, the direct homologation of aldehydes or ketones to α-hydroxy amides by means of amidoyl anion equivalents different from isocyanides is possible only when the final targets are tertiary amides. It is possible, however, to homologate carbonyl compounds to α-hydroxy acids and then transform them into α-hydroxy amides using the various coupling methodologies already cited above. This strategy has been frequently used beginning with protected α-amino aldehydes and continuing in two main ways. The first route (Eq. 114) involves the synthesis of a cyanohydrin,[201–203,205,223] which may be synthesized by traditional methods (using NaCN/HCl or NaCN on the bisulfite adduct, or NaCN/Ac_2O or acetone cyanohydrin), or by the addition of trimethylsilyl cyanide,[224,225] tributyltin cyanide,[226] or diethylaluminum cyanide.[223] Generally, hydrolysis of the nitrile brings about the removal of the amine protecting group, which then needs to be reintroduced before transformation into the secondary amide **138**. Therefore, the overall number of steps from the protected α-amino aldehydes to **138** is either three or four.

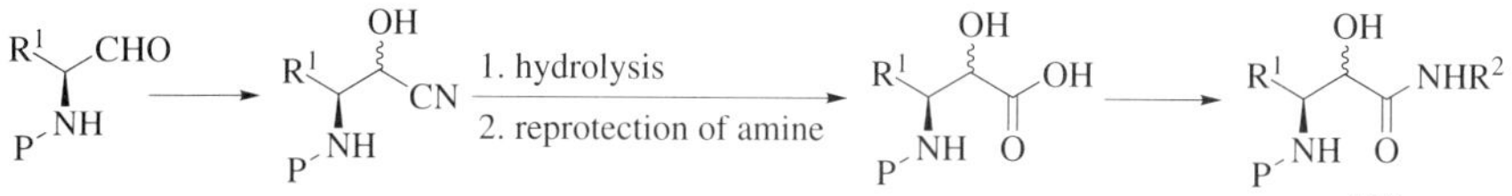

OH
OH
OH
1. hydrolysis
2. reprotection of amine

138

(Eq. 114)

The second route, which is accomplished in three synthetic steps, makes use of tris(alkylthio)methane as the acyl anion equivalent (Eq. 115).[200,227–229] After condensation, the orthothioformate is converted into the carboxylic acid by means of mercury(II) compounds.

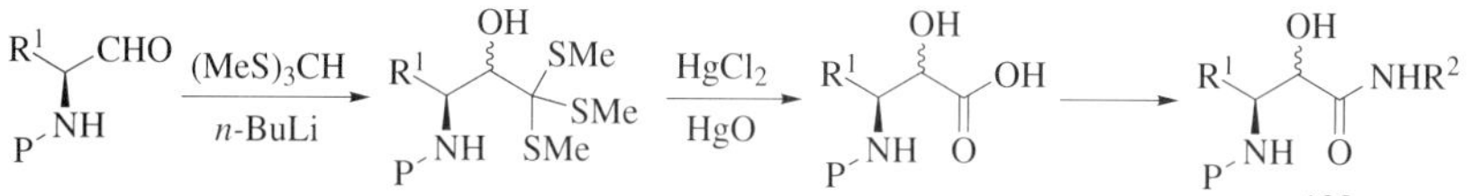

(MeS)3CH
n-BuLi
HgCl2
HgO

138

(Eq. 115)

Both methodologies appear to be less efficient for the generation of compounds such as **138** than the previously described Passerini routes.

The route described in Eq. 114 may be more synthetically attractive only if a diastereoselective preparation of amides **138** is desired. While the Passerini condensations involving protected α-amino aldehydes are generally poorly stereoselective, better diastereoselectivities, although not dramatically high, are sometimes achieved in the conversion of the same carbonyl substrates into cyanohydrins.[223]

Another route to α-hydroxy amides is via nucleophilic opening of α,β-epoxy amides, which are, in turn, prepared by the coupling of 2,3-epoxy acids with primary amines or by the oxidation of α,β-unsaturated amides (Eq. 116). Unfortunately, there is no example of the reductive opening of secondary α,β-epoxyamides, nor of their opening by carbon nucleophiles. Therefore, this methodology is limited to the synthesis of β-heteroatom substituted α-hydroxy amides, such has β-alkoxy,[230] alkylthio,[231] or halo[230] derivatives.

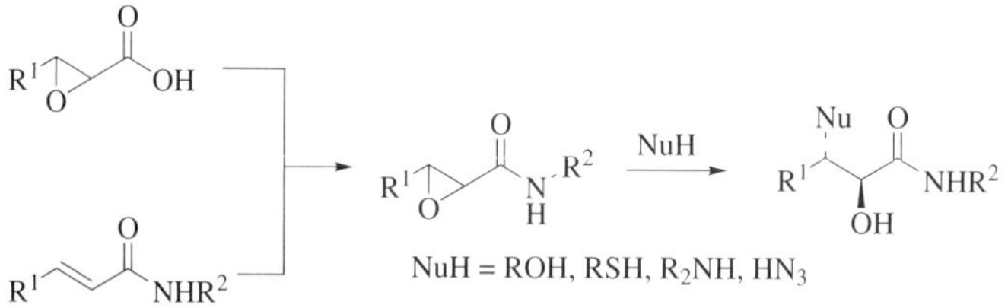

NuH
NuH = ROH, RSH, R2NH, HN3

(Eq. 116)

The most interesting products from a synthetic point of view are those resulting from epoxide opening with nitrogen nucleophiles. This opening may be performed either with primary or secondary amines[232–235] or with magnesium azide (Eq. 117).[236–240]

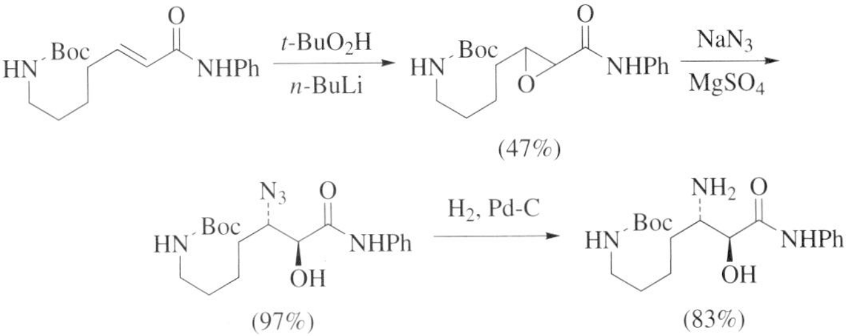

(Eq. 117)

The overall synthesis of α-hydroxy-β-amino amides from unsaturated amides through the magnesium azide route requires only three steps. Therefore, it compares well with the homologations of α-amino aldehydes described in Eqs. 114–115 and it is only slightly longer than the pyridinium trifluoroacetate-mediated Passerini reaction. Moreover, the diastereoselectivity is very high since it is controlled by the configuration of the starting double bond. On the other hand, the main drawback of this method is the fact that the generation of optically active compounds by this strategy is not as easy as in the syntheses starting from protected α-amino aldehydes, which take advantage of the large number of available enantiomerically pure α-amino acids (the precursors of α-amino aldehydes). While there are presently no direct methods for the asymmetric epoxidation of unsaturated amides, α,β-epoxy amides can be generated from optically active α,β-epoxy acids, which are, in turn, enantioselectively synthesized via asymmetric epoxidation of allylic alcohols,[231,236] nucleophilic asymmetric epoxidation,[241] asymmetric dihydroxylation,[242] or asymmetric Darzens reactions.[243]

Finally, an interesting new methodology to prepare α-hydroxy amides involves the hydrogenolytic cleavage of 3-benzyloxy-4-aryl-2-azetidinones **139** (Eq. 118).[244,245] The method is highly convergent and concise, since the azetidinones **139** are assembled in one step by the Staudinger condensation of activated benzyloxyacetic acid derivatives (e.g., the acyl chloride) with imines. However, it is limited to azetidinones having an aryl or heteroaryl group at position 4. For the generation of α-hydroxy-β-arylpropionamides, such as **140**, this method may be an attractive alternative to the Passerini reaction, since it starts from aromatic aldehydes, which are more readily available than the arylacetaldehydes needed for the synthesis of **140** through the Passerini condensation.

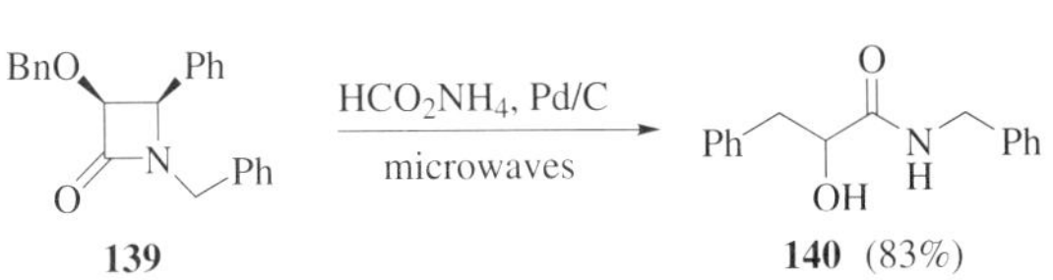

(Eq. 118)

EXPERIMENTAL CONDITIONS

The classical Passerini reaction is reported to proceed faster in a low polarity medium. Therefore, it is typically carried out in solvents such as dichloromethane, ethyl acetate, diethyl ether, or tetrahydrofuran. When the carbonyl component is volatile and cheap (e.g., acetone), it may be employed in excess as the solvent. Only in a limited number of reports has the use of more polar solvents, such as dimethylformamide or dimethyl sulfoxide, been described.[39] Alcoholic solvents, such as methanol or ethanol, are not well suited for this reaction.[246] Recently, however, a remarkable rate acceleration has been reported for some Passerini reactions carried out in water, compared with dichloromethane.[246] This acceleration does not seem to be related to the polarity of the solvent, but to the cohesive energy density or hydrophobic effects instead. This method is probably limited to substrates that are at least partially soluble in water. The Passerini reaction is also accelerated by high pressure.[63]

The classical Passerini reaction is typically carried out at room temperature, with reaction times varying from a few hours to several days (in the case of ketones). There are only two reports of the use of Lewis acids as additives.[50,171]

The Passerini-type reactions mediated by protic or Lewis acids are faster and, especially with strong Lewis acids, are often carried out at 0° or even lower temperatures. Dichloromethane is the solvent of choice for these reactions.

EXPERIMENTAL PROCEDURES

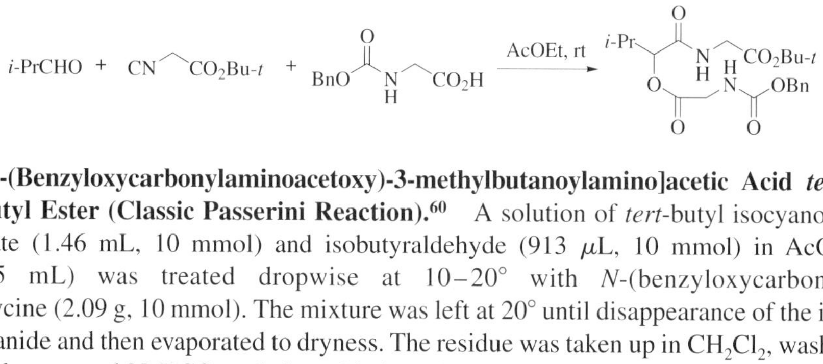

i-PrCHO + CN CO2Bu-*t* +
BnO N H CO2H
AcOEt, rt
i-Pr
O
N H
CO2Bu-*t*
O
H N
OBn
O O

[2-(Benzyloxycarbonylaminoacetoxy)-3-methylbutanoylamino]acetic Acid *tert*-Butyl Ester (Classic Passerini Reaction).[60] A solution of *tert*-butyl isocyanoacetate (1.46 mL, 10 mmol) and isobutyraldehyde (913 μL, 10 mmol) in AcOEt (15 mL) was treated dropwise at 10–20° with *N*-(benzyloxycarbonyl) glycine (2.09 g, 10 mmol). The mixture was left at 20° until disappearance of the isocyanide and then evaporated to dryness. The residue was taken up in CH_2Cl_2, washed with saturated $NaHCO_3$ solution, dried, and evaporated to dryness. Crystallization from toluene gave the pure product (3.38 g, 80%), mp 68–70°. Anal. Calcd for $C_{21}H_{30}N_2O_7$: C, 59.70; H, 7.16; N, 6.63. Found: C, 59.20; H, 7.27; N, 6.60.

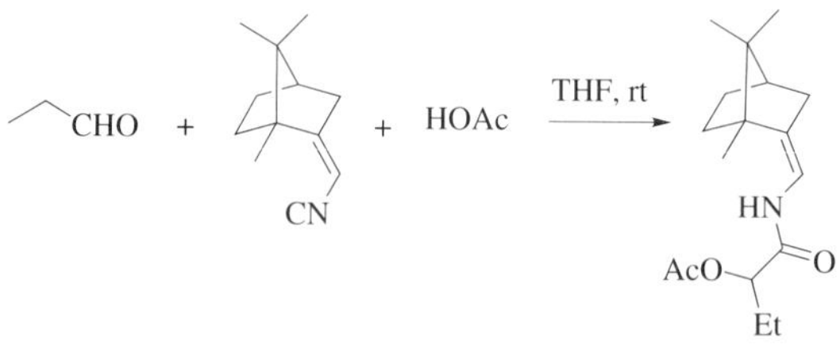

CHO +
CN
+ HOAc
THF, rt
HN
AcO
O
Et

1-[(1,7,7-Trimethylbicyclo[2.2.1]hept-2-ylidenemethyl)carbamoyl]propyl Acetate (Classic Passerini Reaction of a Chiral Isocyanide).[24] A solution of propanal (289 μL, 4.00 mmol), acetic acid (227 μL, 4.00 mmol), and 2-isocyanomethylene-1,7,7-trimethylbicyclo[2.2.1]heptane (400 mg, 2.28 mmol) in THF (15 mL) was stirred for 40 hours at room temperature. After removal of the solvent, the residue was taken up in Et_2O, washed with water and brine, and dried over Na_2SO_4. Evaporation afforded the title product (630 mg, 94%) as a yellow, sticky solid. The diastereomeric excess, determined by GC-MS, was 93%. ^{1}H NMR ($CDCl_3$) (in the NMR spectrum some peaks appeared doubled because of the presence of rotamers: the average ppm value is indicated) δ 8.03 (broad s, 1H, N*H*), 6.48 (m, 1H, C=C*H*), 5.23 (m, 1H, C*H*–O), 2.44 (broad d, J = 15.5 Hz, 1H, CH_2C=C), 2.17 (s, 3H, CH_3C=O), 1.91 (m, 2H, CH_2), 1.88 (d, J = 15.5 Hz, CH_2C=C), 1.84–1.71 (m, 2H, C*H*–CH_2C=C, CH_2), 1.65 (m, 1H, CH_2), 1.47 (m, 1H, CH_2), 1.26 (s, 3H, CH_3), 1.21 (m, 1H, CH_2), 0.93 (m, 3H, CH_2CH_3), 0.87 (s, 3H, CH_3), 0.84 (s, 3H, CH_3). ^{13}C NMR ($CDCl_3$) δ 169.1, 165.9 (*C*=O), 132.0 (*C*=CH), 112.6 (C=*C*H), 74.7 (*C*H–O), 51.4, 48.9 (quaternary *C*), 44.5 (*C*H), 35.7, 35.1, 27.8, 25.0 (CH_2), 20.8, 19.9, 18.3, 14.7, 8.8 (CH_3). EIMS (70 eV) m/z M^+ 293 (10), 165 (14), 148 (16), 133 (12), 122 (15), 105 (21), 101 (19), 43 (100).

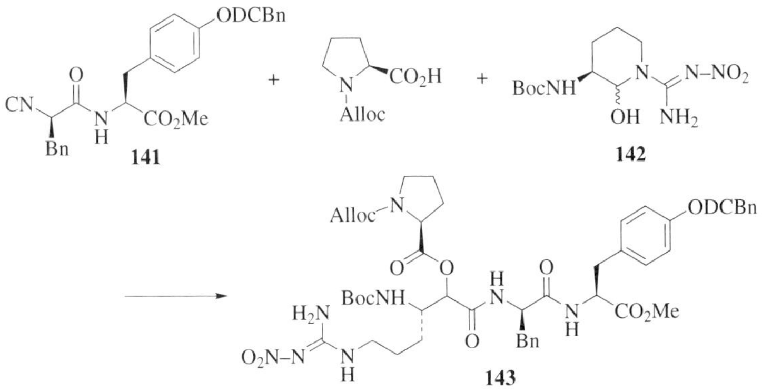
ODCBn
O
CN
N
H
CO2Me
Bn
141
+
N
CO2H
Alloc
+
BocNH
N
N–NO2
OH
NH2
142
Alloc
N
ODCBn
O
O
H
N
O
BocNH
H2N
N
H
CO2Me
O
Bn
O2N–N
N
H
143

Compound 143 (Classic Passerini Reaction of a Protected α-Amino Aldehyde).[49] A solution of isonitrile **141** (DCBn = 2,6-dichlorobenzyl) (590 mg, 1.15 mmol), *N*-(allyloxycarbonyl)-L-proline (298.0 mg, 1.50 mmol), and protected L-argininal **142** (382.2 mg, 1.26 mmol) in anhydrous CH_2Cl_2 (4.6 mL) was stirred at room temperature for 16 hours. The solvent was slowly removed in vacuo and the resultant thick residue was stirred for 1 day. Standard extractive work-up in EtOAc gave a crude product that was purified by silica gel chromatography (CH_2Cl_2/2-PrOH 98:2) to afford compound **143** as an amorphous, colorless solid (59%). The diastereoisomeric ratio (1.6:1) was determined by reverse-phase HPLC [C18, (H_2O + 0.1% CF_3CO_2H)/CH_3CN 95:5 to 25:75]; ^{1}H NMR (CD_3OD) δ 7.44 (m, 2H), 7.35 (m, 1H), 7.16–7.29 (m, 3H), 7.03–7.14 (m, 4H), 6.95 (m, 2H), 5.85 + 5.97 (minor + major, m, 1H), 5.25 + 5.27 (2 s, 2H), 5.16–5.37 (m, 2H), 5.03 + 5.11 (minor + major, m, 1H, C*H*OC=O), 4.57–4.76 (m, 2H), 4.50 (m, 2H), 4.40 (m, 1H), 3.88 + 4.01 (2 m, 1H), 3.68 + 3.72 (major + minor, s, 3H), 3.39–3.59 (m, 2H), 3.02–3.25 (m, 4H), 2.95 (m, 1H), 2.70–2.89 (m, 1H), 2.30 (m, 1H), 1.90–2.12 (m, 3H), 1.49–1.70 (m, 2H), 1.40 + 1.45 (2 s, 9H), 1.32–1.47 (m, 2H). MS: $[MH]^+$ 1013.19.

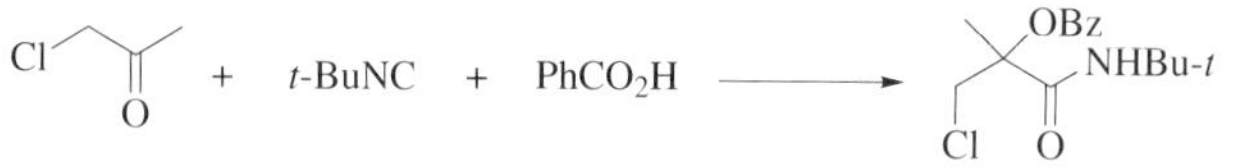

2-Benzoyloxy-*N*-*tert*-butyl-3-chloro-2-methylpropionamide (Classic Passerini Reaction of a Ketone).[94] Chloroacetone (796 μL, 10 mmol) was added dropwise to an ice-cooled mixture of benzoic acid (1.22 g, 10 mmol) and *tert*-butyl isocyanide (1.13 mL, 10 mmol). The mixture was stirred for 15 hours at room temperature. Trituration from petroleum ether afforded the acyloxy amide (2.92 g, 98%), mp 76°; IR (Nujol) 1674, 1715, 3400 cm^{-1}; ^1H NMR ($CDCl_3$) δ 7.90 (m, 2H), 7.50 (m, 3H), 6.25 (broad s, 1H, N*H*), 4.34, 4.20 (AB syst., $J = 12$ Hz, 2H, CH_2Cl), 1.82 (s, 3H, CH_3), 1.42 (s, 9H, $(CH_3)_3C$).

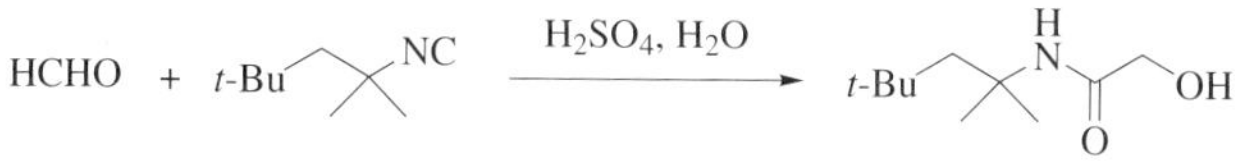

2-Hydroxy-*N*-(1,1,3,3-tetramethylbutyl)acetamide (Mineral Acid Mediated Passerini Reaction).[61] A solution of concentrated H_2SO_4 (3.5 mL, 66 mmol) in H_2O (25 mL) was added dropwise into an ice-cooled, stirred mixture of 1,1,3,3-tetramethylbutyl isocyanide (6.96 g, 50 mmol) and 35% aqueous formaldehyde (20 mL, 252 mmol). After being stirred for 4 hours at room temperature, the mixture was extracted with Et_2O and the organic extracts were washed with saturated aqueous $NaHCO_3$, and dried ($MgSO_4$). Evaporation of the solvent followed by fractional distillation at 123° (0.01 Torr) afforded the pure α-hydroxy amide (5.1 g, 55%); ^1H NMR ($CDCl_3$) δ 6.77 (s, 1H, N*H*), 4.93 (t, 1H, O*H*), 3.93 (d, $J = 4.7$ Hz, 2H, CH_2OH), 1.77 (s, 2H, CH_2), 1.48 (s, 6H, CH_3), 1.05 (s, 9H, CH_3).

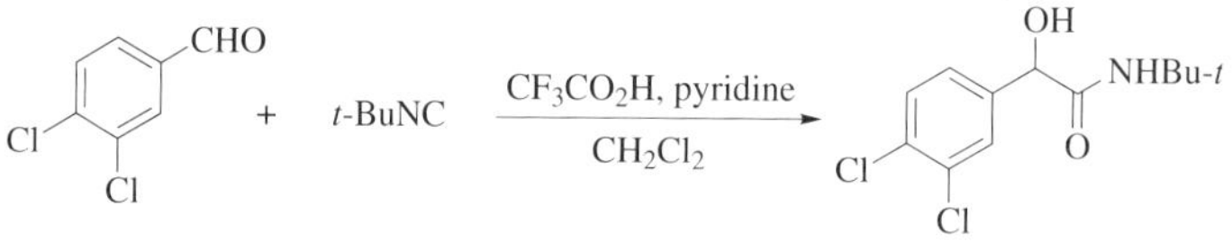

***N*-*tert*-Butyl-2-(3,4-dichlorophenyl)-2-hydroxyacetamide (Passerini Reaction Mediated by Trifluoroacetic Acid-Pyridine).**[115] A mixture of purified 3,4-dichlorobenzaldehyde (8.75 g, 50 mmol), *tert*-butyl isocyanide (2.08 g, 35 mmol), and pyridine (3.96 g, 50 mmol) in CH_2Cl_2 (25 mL) was cooled to −5° under N_2 and treated dropwise with stirring, at −5° to + 5°, with CF_3CO_2H (2.90 g, 25 mmol). The mixture was warmed to room temperature for 1 hour and then treated (with ice cooling in order to maintain the temperature between 20° and 30°) with additional CF_3CO_2H (2.90 g). The mixture was stirred for 2 hours at room temperature and then treated for 2 hours at room temperature with a solution of $NaHSO_3$ (15 g) in H_2O (100 mL). Filtration gave the bisulfite adduct of the starting aldehyde (mp 164–165°) (9.0 g, 32.2 mmol). The CH_2Cl_2 layer of the filtrate was dried (Na_2SO_4) and concentrated, giving the crude product. Crystallization from chlorobutane gave the pure α-hydroxy amide, mp 116–117° (2.90 g, 30%, 60% based on recovered $NaHSO_3$ adduct); ^1H NMR ($CDCl_3$) δ 7.42 (d, J = 1.0 Hz, 1H), 7.35 (d, J = 7.0 Hz, 1H), 7.13 (dd, J = 1, 7 Hz, 1H), 4.82 (d, J = 4.0 Hz, 1H, O*H*), 4.72 (d, J = 4.0 Hz, 1H, C*H*–OH), 1.28 (s, 9H).

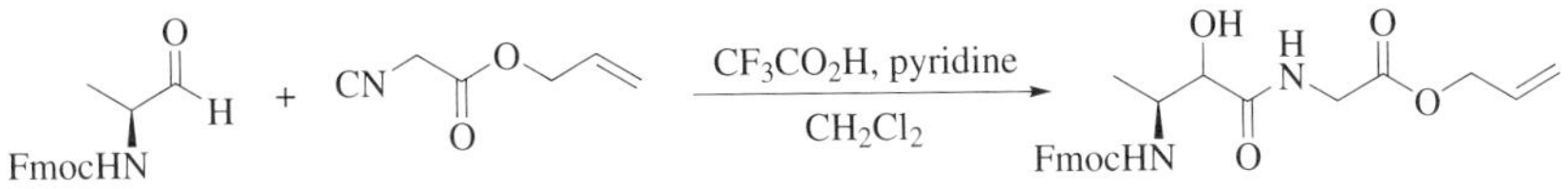
FmocHN
H
CN
CF3CO2H, pyridine
CH2Cl2
OH
FmocHN

[3-(9-Fluorenylmethoxycarbonyl)amino-2-hydroxybutanoylamino]acetic Acid Allyl Ester (Passerini Reaction Mediated by Trifluoroacetic Acid-Pyridine).[116] Trifluoroacetic acid (770 μL, 10.0 mmol) was added dropwise to a cooled solution (−10°) of freshly prepared *N*-(9-fluorenylmethoxycarbonyl)-L-alaninal (1.477 g, 5.00 mmol), allyl α-isocyanoacetate (845 mg, 6.75 mmol), and pyridine (1.62 mL, 20.0 mmol) in CH_2Cl_2 (5 mL), while maintaining the temperature below 0°. After 1 hour at 0°, the bath was removed and the reaction mixture was stirred at room temperature for 72 hours. After concentration, the resulting slurry was taken up in AcOEt and washed successively with three portions each of 1 N HCl, saturated $NaHCO_3$ solution, and brine. The organic layer was dried over Na_2SO_4, filtered, and concentrated. The crude product was purified by flash column chromatography on silica gel, giving the α-hydroxy amide (1.82 g, 83%).

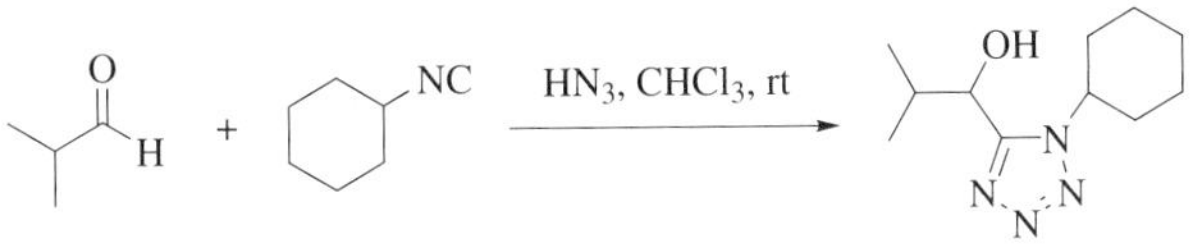
H
NC
HN3, CHCl3, rt
OH

1-(1-Cyclohexyl-1*H*-tetrazol-5-yl)-2-methylpropan-1-ol (HN$_3$-Mediated Passerini Reaction).[23] A 7.3% solution of HN_3 in $CHCl_3$ (40 mL, 68 mmol) was treated under ice-cooling with freshly distilled isobutyraldehyde (4.33 g, 60 mmol) and dropwise with cyclohexyl isocyanide (5.45 g, 50 mmol). The mixture was stirred at 20° for 3 days. The resulting colorless crystals were collected and crystallized from cyclohexane, giving pure tetrazole, mp 97–99° (9.54 g, 85%). Anal. Calcd for $C_{11}H_{20}N_4O$: C, 58.90; H, 8.99; N, 24.98. Found: C, 58.88; H, 8.85; N, 25.16.

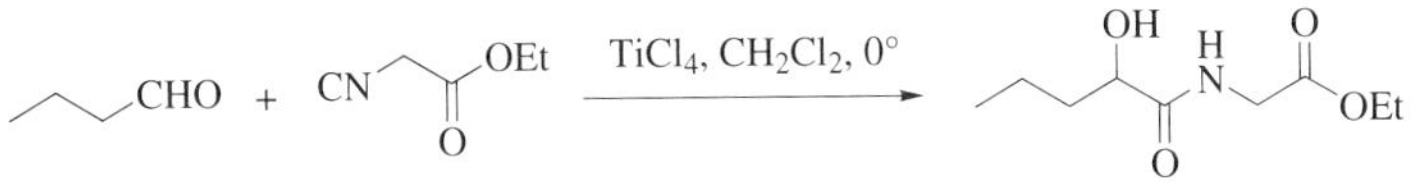
CHO
CN
OEt
TiCl4, CH2Cl2, 0°
OH
OEt

***N*-(2-Hydroxypentanoyl)glycine Ethyl Ester (TiCl$_4$-Mediated Passerini Reaction).**[78] To a solution of ethyl α-isocyanoacetate (546 μL, 5.0 mmol) in CH_2Cl_2 (25 mL) was added a 2 M solution of $TiCl_4$ in CH_2Cl_2 (2.75 mL, 5.5 mmol) at 0° under argon. The color of the clear solution changed from yellow to dark brown.

After a few minutes a pale yellow precipitate was formed. About 60 minutes later the mixture was treated with butyraldehyde (0.45 mL, 5.0 mmol), whereby the precipitate disappeared within a few minutes. The clear solution was stirred until TLC showed the disappearance of starting material, then it was hydrolyzed, and 20 minutes later the two layers were separated. The aqueous layer was extracted twice with CH_2Cl_2, and the combined organic layers were washed with saturated $NaHCO_3$ solution, water, brine, water, and finally dried over Na_2SO_4. Removal of the solvent afforded the crude product, which was purified by crystallization (Et_2O/*n*-hexane) giving the α-hydroxy amide as white crystals, mp 84.4–85.5° (0.95 g, 96%); IR (KBr) 3295, 3245, 1749, 1737, 1639, 1625, 1542 cm^{-1}; ^1H NMR ($CDCl_3$) δ 7.11 (broad s, 1H, N*H*), 4.21 (q, J = 7.3 Hz, 2H, OCH_2), 4.18 (dd, J = 3.8, 7.9 Hz, 1H, C*H*–O), 3.97–4.13 (m, 2H, NCH_2), 3.04 (broad s, 1H, O*H*), 1.40–1.87 (m, 4H, CH_2CH_2), 1.29 (t, J = 7.3 Hz, 3H, CH_3CH_2O), 0.95 (t, J = 7.3 Hz, 3H, CH_3CH_2C). Anal. Calcd for $C_9H_{17}NO_4$: C, 53.19; H, 8.43; N, 6.89. Found: C, 52.98; H, 8.43; N, 6.75.

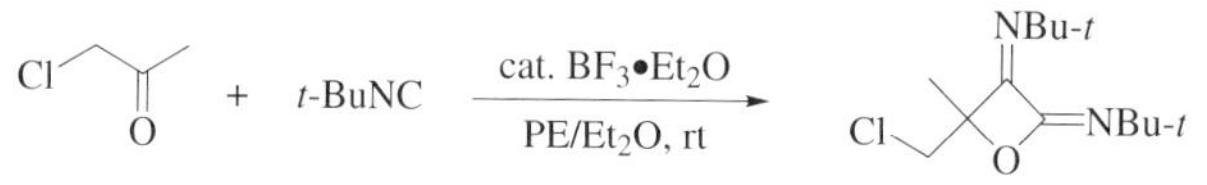
Cl
O
+ t-BuNC
cat. BF3•Et2O
PE/Et2O, rt
NBu-t
Cl
NBu-t
O

2,3-bis(*tert*-Butylimino)-4-(chloromethyl)-4-methyloxetane (Passerini Reaction Mediated by CatalyticAmounts of BF_3).[29] A solution of $BF_3 \cdot Et_2O$ (1.0 mL, 10.8 mmol) in Et_2O (40 mL) was added slowly dropwise during 40 minutes to an ice-cooled solution of chloroacetone (18.4 g, 200 mmol) and *tert*-butyl isocyanide (33.2 g, 400 mmol) in petroleum ether (40 mL). During the addition the temperature was not allowed to rise above 12°. After being stirred for 4 hours at room temperature, the mixture was treated with saturated $NaHCO_3$ solution (400 mL) and extracted with CH_2Cl_2. The organic phase was dried (Na_2SO_4) and evaporated. The residue was distilled (bp 60–65° at 0.06 Torr), giving the diimino oxetane as colorless crystals mp 48–50° (47.5 g, 92%). Anal. Calcd for $C_{13}H_{23}ClN_2O$: Cl, 13.70; N, 10.83. Found: Cl, 14.10; N, 10.90.

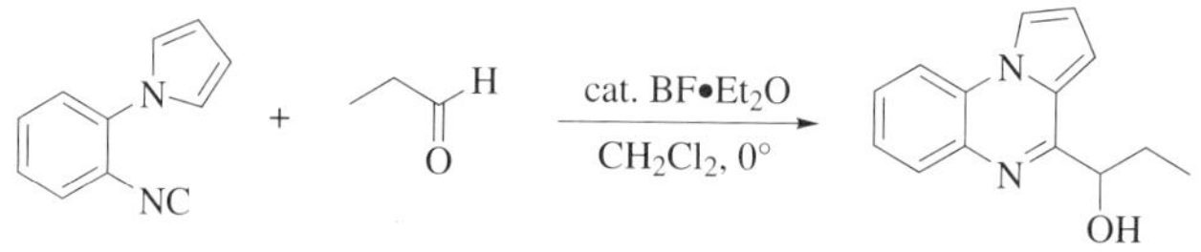
N
NC
+
H
O
cat. BF•Et2O
CH2Cl2, 0°
N
N
OH

4-(1-Hydroxypropyl)pyrrolo[1,2-*a*]quinoxaline (BF_3-Mediated Passerini Reaction of 1-(2-Isocyanophenyl)pyrroles).[144] To a magnetically stirred solution of 1-(2-isocyanophenyl)pyrrole (170 mg, 1.00 mmol) and propanal (58 mg, 1.00 mmol) in dry CH_2Cl_2 (20 mL) was added $BF_3 \cdot Et_2O$ (14 mg, 0.10 mmol) at 0° under argon. After 20 minutes, more $BF_3 \cdot Et_2O$ (14 mg, 0.10 mmol) was added to the reaction mixture and stirring was continued for an additional 20 minutes. After addition of a saturated $NaHCO_3$ solution and extraction with CH_2Cl_2, the organic extracts were dried (Na_2SO_4), evaporated to dryness, and chromatographed to afford the pyrroloquinoxaline (200 mg, 89%); R_f 0.68 (*n*-hexane/AcOEt 1:1); IR (neat) 3395, 3158, 1613, 1365, 756 cm^{-1}; ^1H NMR ($CDCl_3$) δ 7.90–8.00 (m, 2H), 7.87 (dd, J = 1.6, 8.4 Hz, 1H), 7.40–7.60 (m, 2H), 6.85–6.90 (m, 2H), 5.00–5.10 (m, 1H), 4.82 (d, J = 6.3 Hz, 1H), 2.10–2.20 (m, 1H), 1.75–1.90 (m, 1H), 1.03 (t, J = 7.4 Hz, 3H); HRMS (m/z): calcd for $C_{14}H_{14}N_2O$: 226.1107; found, 226.1124.

TABULAR SURVEY

We have attempted to cover the literature thoroughly through the end of 2003. When protective groups were present (including esters), they have not been considered in the carbon count.

Tables 1–2 and 7–9 have been ordered according to isocyanides.

Tables 3–5 refer to classic Passerini reactions that can lead to diastereoisomeric mixtures. They are ordered according to the chiral component. When two or three chiral components are employed, the reaction is listed only in the Table related to the first occurring chiral compound. When a chiral isocyanide or carboxylic acid was used, but no new stereogenic centers were generated (reactions involving formaldehyde or symmetric ketones), the reactions are listed in Tables 1 and 2. In nearly all cases, the relative configuration of the major diastereoisomer was not determined. When it was established, the structure of the major isomer is depicted.

Table 6 refers to intramolecular Passerini reactions and has been ordered according to the bifunctional component.

Table 10 refers to enantioselective Passerini reactions and has been ordered according to the chiral catalyst.

A dash (—) indicates that no yield is given in the reference.

The following abbreviations have been used in the tables:

All	allyl
Alloc	allyloxycarbonyl
Bn	benzyl
Bz	benzoyl
dr	diastereoisomeric ratio
DCBn	2,6-dichlorobenzyl
Fmoc	9-fluorenylmethoxycarbonyl
MPa	mega pascal
NEM	*N*-ethylmorpholine
P	pressure
PE	petroleum ether
Pht	phthaloyl
Pmc	2,2,5,7,8-pentamethylchroman-6-sulfonyl
py	pyridine
TMS	trimethylsilyl
Ts	*p*-toluenesulfonyl
Tr	triphenylmethyl

TABLE 1. α-ACYLOXY AMIDES FROM ACHIRAL ISOCYANIDES, ALDEHYDES, AND CARBOXYLIC ACIDS

Isocyanide	Aldehyde and Carboxylic Acid		Conditions	Product(s) and Yield(s) (%)	Refs.
C_2					
MeNC	t-BuCHO	$PhCO_2H$	Et_2O, rt, 10 h	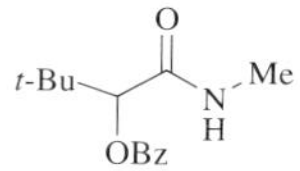 (64)	69
	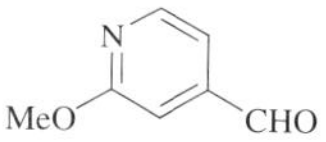	AcOH	MeOH, reflux, 2.5 h	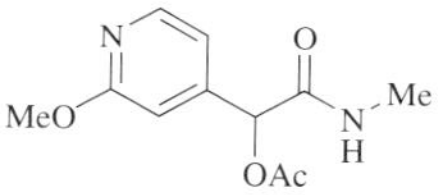 (77)	37
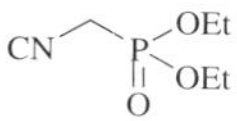	R^1CHO	$R^2CH_2CO_2H$	CH_3CN, 80°, 8 h	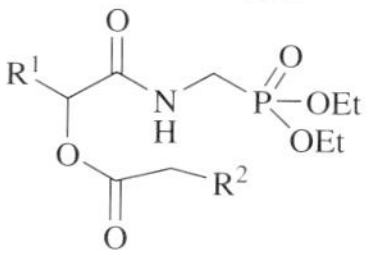 R^1 / R^2 i-Pr / AcNH (74) i-Pr / F_3CCONH (75) i-Pr / $CbzNHCH_2CONH$ (89) i-Pr / $TrNHCH_2CONH$ (66) i-Pr / PhtN (92) Ph / PhtN (87)	76
C_3					
	RCHO	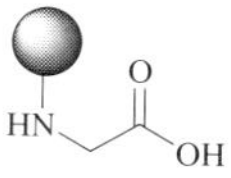	rt, 2 d	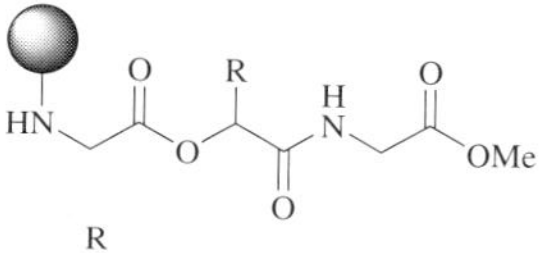 R n-Pr (>70) $n-C_5H_{11}$ (>70) Ph (>70) 2-naphthyl (>70)	12

	i-PrCHO	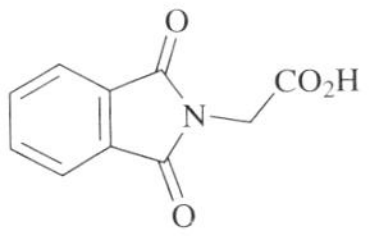	MeOH, rt	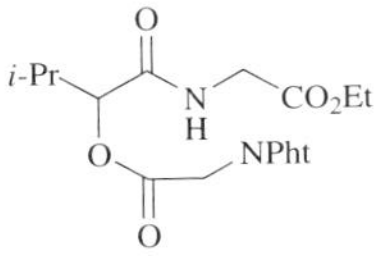 (79)	60
	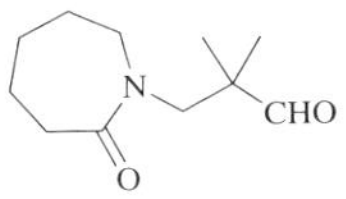	$PhCO_2H$	Et_2O, rt, 5 d	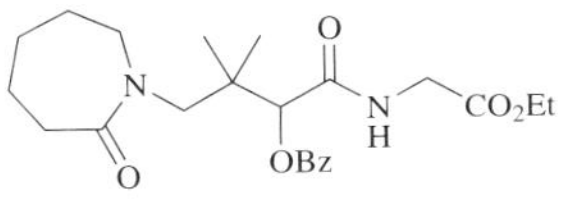 (75)	247
	R^1CHO	R^2CO_2H	rt	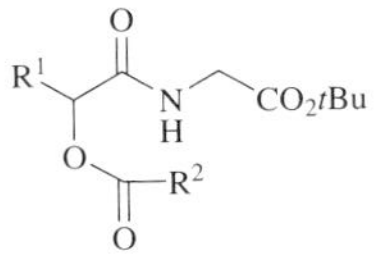	60

Solvent	R^1	R^2	
MeCN	H	L-AcNHCH(i-Pr)	(18)
MeCN	H	$PhtNCH_2$	(92)
AcOEt	H	$N_3CH_2CO_2CH(i\text{-}Pr)CONHCH_2$	(56)
MeCN	Me	N_3CH_2	(76)
MeCN	$MeSCH_2CH_2$	$PhtNCH_2$	(56)
AcOEt	i-Pr	$ClCH_2$	(73)
AcOEt	i-Pr	N_3CH_2	(88)
AcOEt	i-Pr	$ClCH_2CONHCH_2$	(29)
AcOEt	i-Pr	$CbzNHCH_2$	(80)
MeCN	i-Pr	$PhtNCH_2$	(85)

C_4

n-PrNC	PhCHO	$PhCO_2H$	Et_2O, rt, 4 d	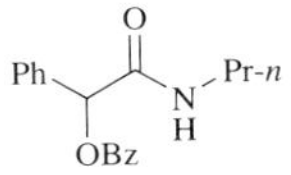 (62)	2

TABLE 1. α-ACYLOXY AMIDES FROM ACHIRAL ISOCYANIDES, ALDEHYDES, AND CARBOXYLIC ACIDS (*Continued*)

Isocyanide	Aldehyde and Carboxylic Acid		Conditions	Product(s) and Yield(s) (%)		Refs.
C_4						
	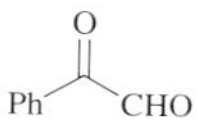	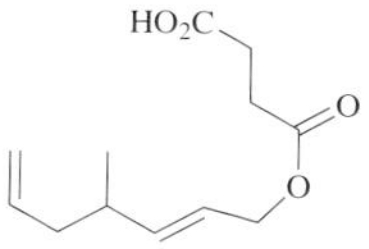	Et_2O, rt	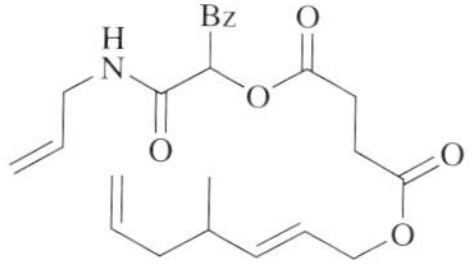	(88)	102
	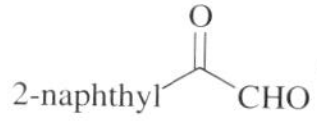		1. Et_2O, rt, 18 h 2. THF, LiCl, Et_3N, rt, 4 h	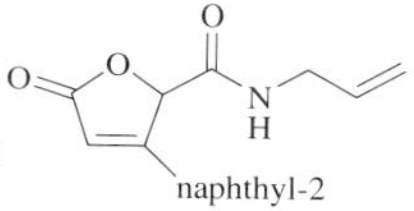	(75)	44
	RCHO	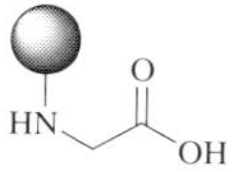	rt, 2 d	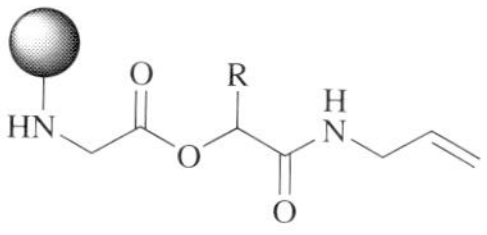 R — *n*-Pr (30-70); *n*-C_5H_{11} (>70); Ph (<30); 2-naphthyl (<30)		12
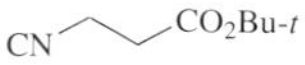	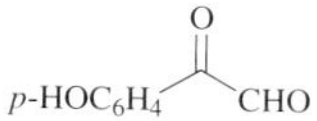		1. Et_2O, rt, 18 h 2. THF, LiCl, Et_3N, rt, 4 h	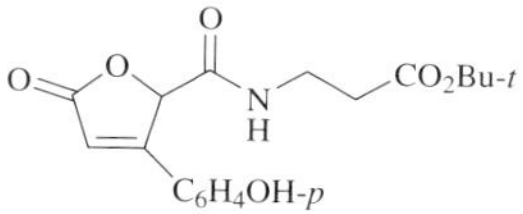	(81)	44

C_5					
n-BuNC	RCHO	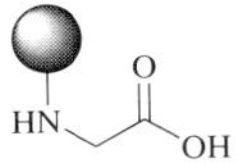	rt, 2 d	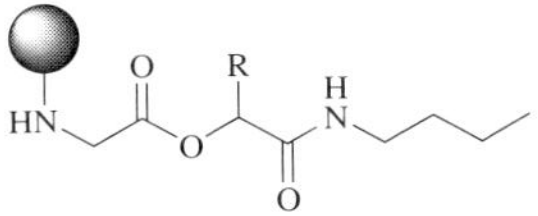 R / n-Pr (>70) / n-C_5H_{11} (>70) / Ph (30-70) / 2-naphthyl (<30)	12
	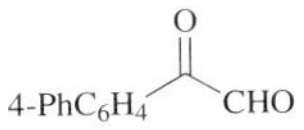		1. Et_2O, rt, 18 h 2. THF, LiCl, Et_3N, rt, 4 h	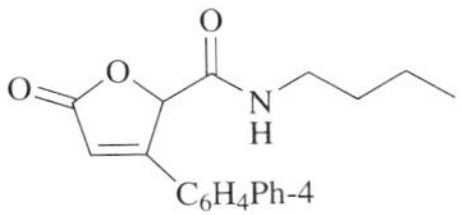 (85)	44
t-BuNC	*i*-PrCHO	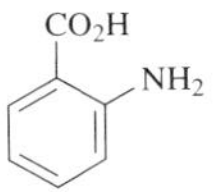	HN_3, MeOH, rt, 24 h	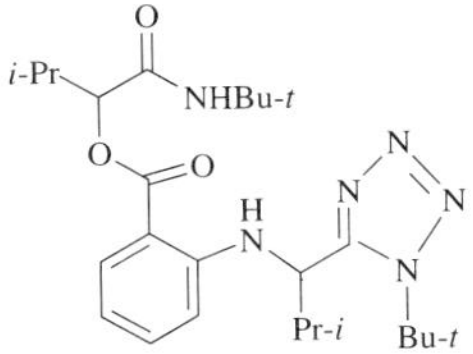 (64)	33
	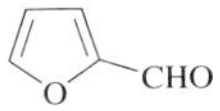		CH_2Cl_2, rt	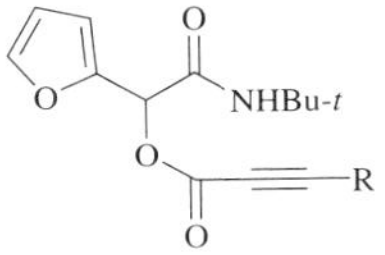 R / Me (82) / Ph (69)	85

TABLE 1. α-ACYLOXY AMIDES FROM ACHIRAL ISOCYANIDES, ALDEHYDES, AND CARBOXYLIC ACIDS (*Continued*)

Isocyanide	Aldehyde and Carboxylic Acid		Conditions	Product(s) and Yield(s) (%)			Refs.
C_5							
t-BuNC	R^1CHO	R^2CO_2H	H_2O, rt, 3-6 h	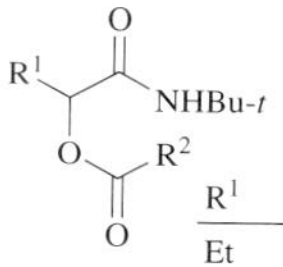			246
				R^1	R^2		
				Et	$Me_2C{=}CH$	(89)	
				Et	*c*-$C_5H_9CH_2$	(72)	
				Et	*p*-MeC_6H_4	(87)	
				Et	3,5-$(MeO)_2C_6H_3$	(82)	
				i-Pr	$Me_2C{=}CH$	(86)	
				i-Pr	*c*-$C_5H_9CH_2$	(87)	
				i-Pr	*p*-MeC_6H_4	(91)	
				i-Pr	3,5-$(MeO)_2C_6H_3$	(82)	
				i-$PrCH_2$	$Me_2C{=}CH$	(88)	
				i-$PrCH_2$	*c*-$C_5H_9CH_2$	(85)	
				i-$PrCH_2$	*p*-MeC_6H_4	(87)	
				i-$PrCH_2$	3,5-$(MeO)_2C_6H_3$	(85)	
				Bn	$Me_2C{=}CH$	(88)	
				Bn	*c*-$C_5H_9CH_2$	(88)	
				Bn	*p*-MeC_6H_4	(93)	
				Bn	3,5-$(MeO)_2C_6H_3$	(87)	
	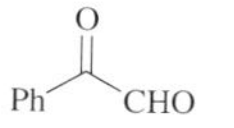	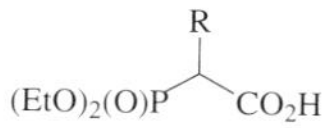	1. Et_2O, rt, 18 h 2. THF, LiCl, Et_3N, rt, 4 h	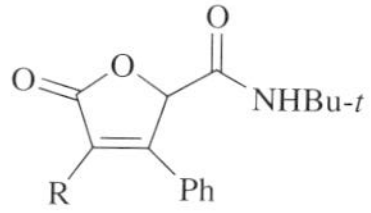	R		44
					H	(87)	
					Ph	(36)	
	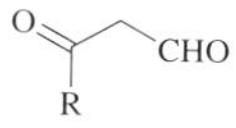		Et_2O, rt, 14 h	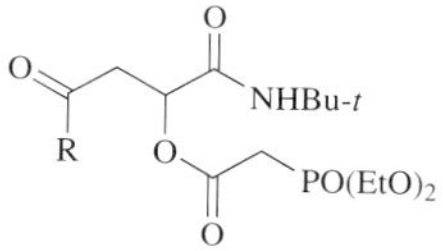	R		56
					t-Bu	(48)	
					p-PhC_6H_4	(36)	

	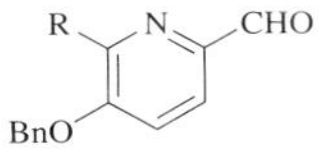	AcOH	$CHCl_3$, reflux, 5 h	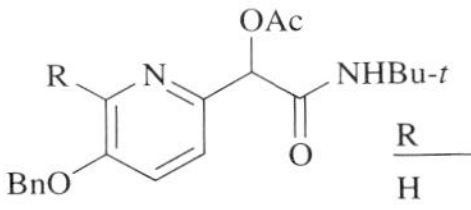 R: H (66); $MeS(O_2)CH_2$ (—)	38
C_6 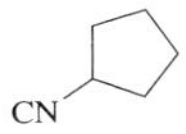	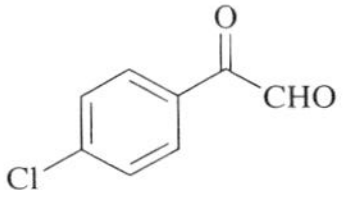	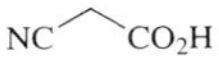	Et_2O, rt, 6 h	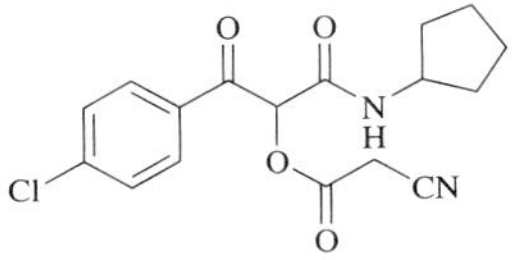 (80)	52
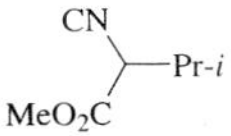	HCHO		CH_3CN, rt	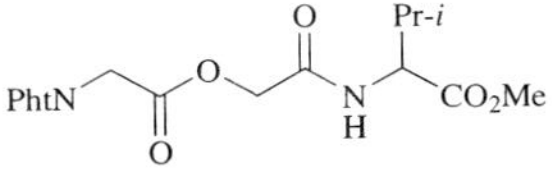 (94)	60
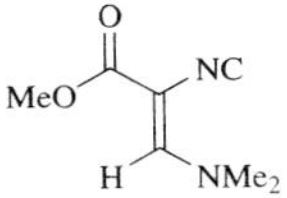	c-$C_6H_{11}CHO$	$PhCO_2H$	AcOEt, rt	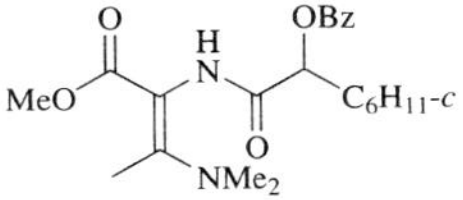 (64)	39
C_7 n-$C_6H_{13}NC$	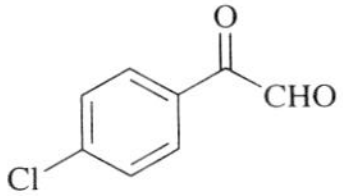		Et_2O, rt, 6 h	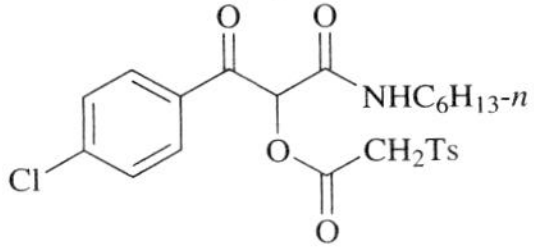 (65)	54
c-$C_6H_{11}NC$	i-PrCHO	RCO_2H	MeOH, rt	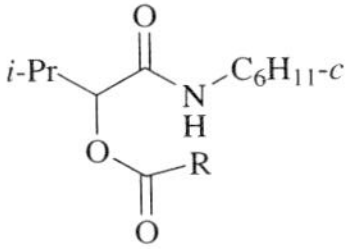 R: H (20); Me (69)	84
	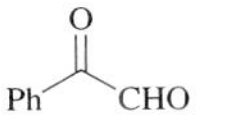	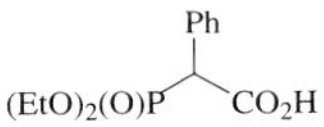	1. Et_2O, rt, 18 h 2. THF, LiCl, Et_3N, rt, 4 h	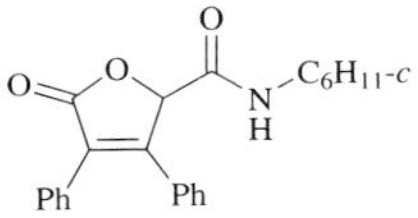 (42)	44

TABLE 1. α-ACYLOXY AMIDES FROM ACHIRAL ISOCYANIDES, ALDEHYDES, AND CARBOXYLIC ACIDS (*Continued*)

Isocyanide	Aldehyde and Carboxylic Acid		Conditions	Product(s) and Yield(s) (%)			Refs.
C_7							
c-$C_6H_{11}NC$	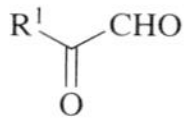	R^2CO_2H	Et_2O, rt	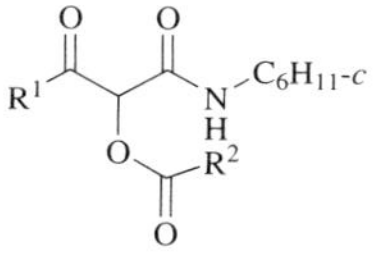			
			Time	R^1	R^2		
			6 h	2-thienyl	$NCCH_2$	(78)	52
			6 h	Ph	p-$MeC_6H_4S(O_2)CH_2$	(77)	54
			6 h	Ph	Ph	(75)	43
			6 h	Ph	o-ClC_6H_4	(82)	43
			6 h	Ph	o-HOC_6H_4	(73)	43
			24 h	Ph	o-$O_2NC_6H_4CH_2$	(77)	55
			24 h	p-$MeOC_6H_4$	o-$O_2NC_6H_4CH_2$	(85)	55
			24 h	p-ClC_6H_4	o-$O_2NC_6H_4CH_2$	(90)	55
			6 h	p-ClC_6H_4	o-HOC_6H_4	(73)	43
			6 h	p-ClC_6H_4	o-ClC_6H_4	(68)	43
			6 h	p-ClC_6H_4	Ph	(77)	43
			6 h	p-ClC_6H_4	p-$MeC_6H_4S(O_2)CH_2$	(80)	54
			6 h	p-ClC_6H_4	$PhS(O_2)CH_2$	(71)	54
			6 h	p-ClC_6H_4	$NCCH_2$	(80)	52
			6 h	p-MeC_6H_4	$NCCH_2$	(80)	52
			6 h	p-MeC_6H_4	Me	(70)	53
			6 h	p-MeC_6H_4	p-$MeC_6H_4CH_2$	(77)	53
			24 h	p-MeC_6H_4	o-$O_2NC_6H_4CH_2$	(84)	55

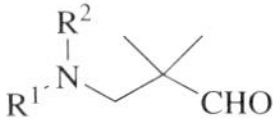 R^3CO_2H — Et_2O, rt

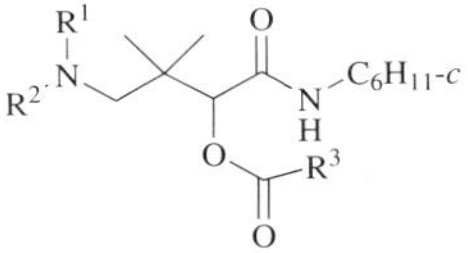

Time	R^1	R^2	R^3		Refs.
10 d	$MeS(O_2)$	Me	Me	(91)	152
24 h	—$(CH_2)_3CO$—		Ph	(84)	247
1 d	—$(CH_2)_5CO$—		Ph	(83)	247
5 d	—$CH_2CH(Ph)CH_2CO$—		Ph	(80)	247

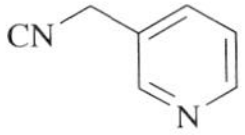 *i*-PrCHO — $PhCO_2H$ — Et_2O, rt, 14 h

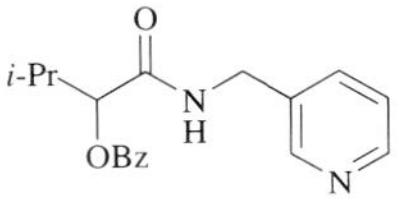 (76) 171

PhNC — R^1CHO — R^2CO_2H — rt

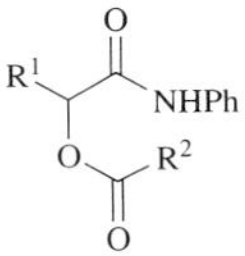

Solvent	Time	R^1	R^2		Refs.
none	—	Me	Me	(75)	3
Et_2O	5 d	2-furyl	Ph	(—)	42
Et_2O	6 d	2-furyl	*o*-HOC_6H_4	(—)	42
none	4 d	Ph	Ph	(35)	64
Et_2O	5 d	*m*-$O_2NC_6H_4$	Ph	(82)	3
Et_2O	27 d	3,4-$(OCH_2O)C_6H_3$	Ph	(34)	3

TABLE 1. α-ACYLOXY AMIDES FROM ACHIRAL ISOCYANIDES, ALDEHYDES, AND CARBOXYLIC ACIDS (*Continued*)

Isocyanide	Aldehyde and Carboxylic Acid		Conditions	Product(s) and Yield(s) (%)	Refs.
C_7					
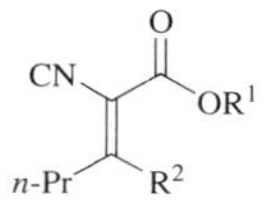	*n*-PrCHO	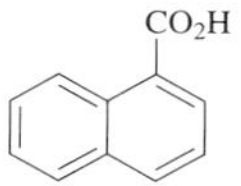	CH_2Cl_2, rt, 8-12 h	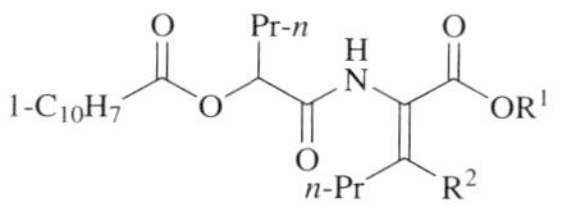 R^1 R^2 Me H (54) Et H (55) Me Br (69) Et Br (61)	41
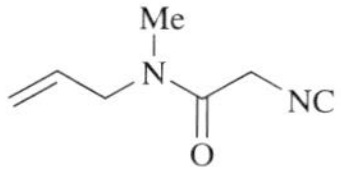	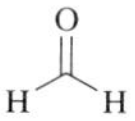	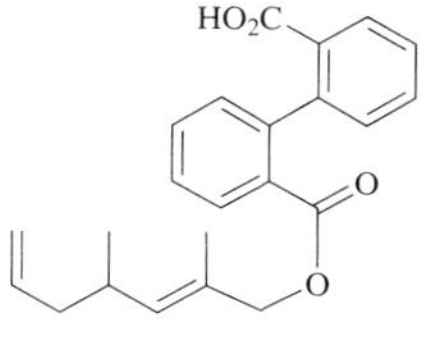	Et_2O, rt, 3 d	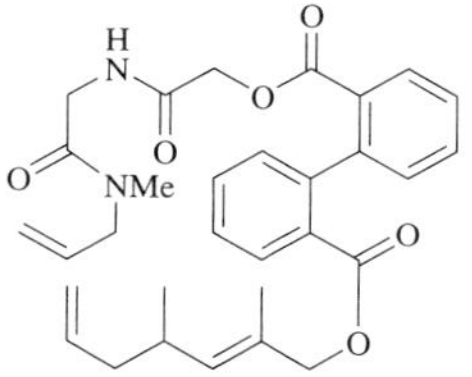 (67)	102
C_8					
p-MeC_6H_4NC	PhCHO	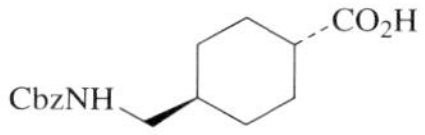	Neat, 5°, 3 d	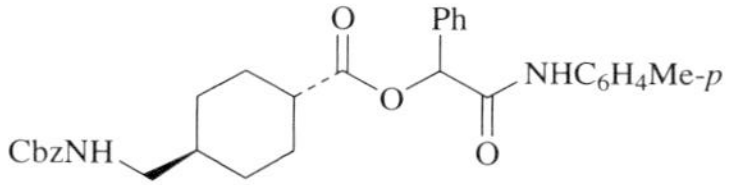 (30)	248

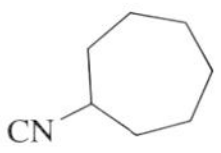 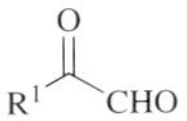  Et$_2$O, rt, 6 h

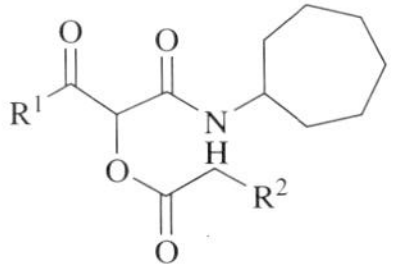

R^1	R^2		
2-thienyl	NC	(80)	52
Ph	NC	(82)	52
Ph	H	(65)	53
Ph	Ph	(78)	53
Ph	*p*-MeC_6H_4	(80)	53
Ph	*p*-$O_2NC_6H_4$	(85)	53
p-ClC_6H_4	NC	(82)	52
p-ClC_6H_4	H	(73)	53
p-ClC_6H_4	Ph	(84)	53

BnNC 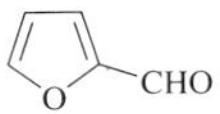  CH_2Cl_2, rt

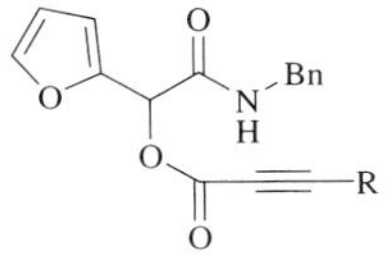

85

R	
H	(79)
Me	(86)
Ph	(79)

RCHO 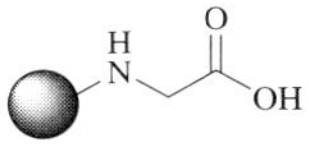 rt, 2 d

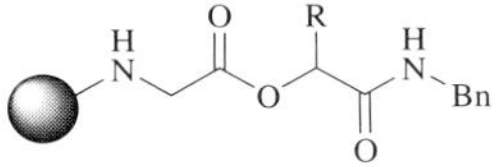

12

R	
n-Pr	(>70)
n-C_5H_{11}	(>70)
Ph	(30-70)
2-naphthyl	(30-70)

TABLE 1. α-ACYLOXY AMIDES FROM ACHIRAL ISOCYANIDES, ALDEHYDES, AND CARBOXYLIC ACIDS (*Continued*)

Isocyanide	Aldehyde and Carboxylic Acid	Conditions	Product(s) and Yield(s) (%)	Refs.
C_8				
BnNC	R^1CHO R^2CO_2H	H_2O, rt, 3-6 h	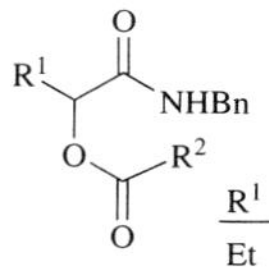	246
	R^1CHO R^2CO_2H	Et_2O, rt, 14 h	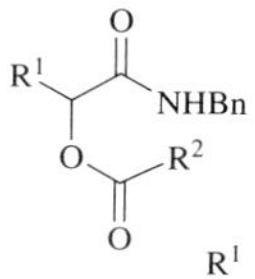	171

Products for ref. 246:

R^1	R^2	
Et	$Me_2C{=}CH$	(89)
Et	c-$C_5H_9CH_2$	(86)
Et	p-MeC_6H_4	(92)
Et	3,5-$(MeO)_2C_6H_3$	(85)
i-Pr	$Me_2C{=}CH$	(89)
i-Pr	c-$C_5H_9CH_2$	(85)
i-Pr	p-MeC_6H_4	(93)
i-Pr	3,5-$(MeO)_2C_6H_3$	(89)
i-$PrCH_2$	$Me_2C{=}CH$	(90)
i-$PrCH_2$	c-$C_5H_9CH_2$	(87)
i-$PrCH_2$	p-MeC_6H_4	(85)
i-$PrCH_2$	3,5-$(MeO)_2C_6H_3$	(94)
Bn	$Me_2C{=}CH$	(86)
Bn	c-$C_5H_9CH_2$	(81)
Bn	p-MeC_6H_4	(85)
Bn	3,5-$(MeO)_2C_6H_3$	(90)

Products for ref. 171:

R^1	R^2	
i-Pr	Ph	(86)
i-Pr	o-HOC_6H_4	(93)
t-Bu	Ph	(81)
t-Bu	p-PhC_6H_4	(95)

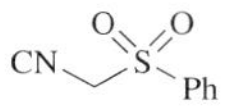	RCHO	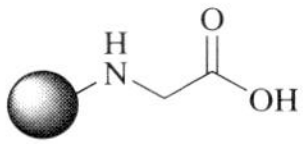	rt, 2 d	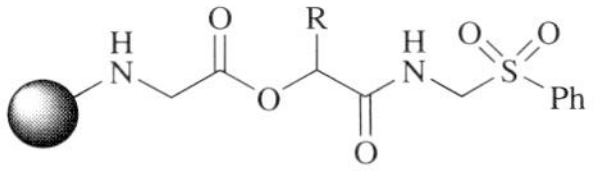	12
				R: n-Pr (>70); n-C_5H_{11} (>70); Ph (>70); 2-naphthyl (30-70)	
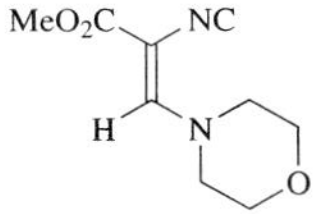	c-$C_6H_{11}CHO$	$PhCO_2H$	DMF, rt	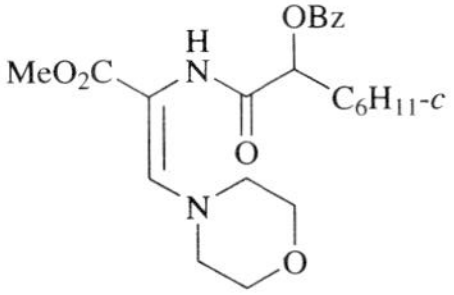 (86)	39
C_9 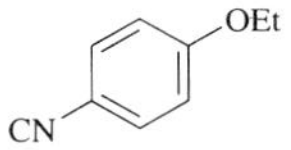	R^1CHO	R^2CO_2H		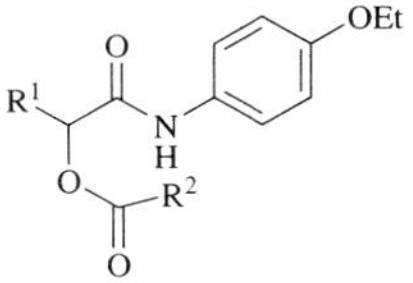	

Conditions	R^1	R^2	Yield	Refs.
Et_2O, rt, 4 d	2-furyl	Ph	(—)	42
Et_2O, rt, 4 d	2-furyl	Me	(—)	42
Et_2O, rt, 4 d	2-furyl	o-HOC_6H_4	(—)	42
Neat, rt	p-$O_2NC_6H_4$	Ph	(70)	249
Neat, rt	p-$O_2NC_6H_4$	o-HOC_6H_4	(46)	249

TABLE 1. α-ACYLOXY AMIDES FROM ACHIRAL ISOCYANIDES, ALDEHYDES, AND CARBOXYLIC ACIDS (*Continued*)

Isocyanide	Aldehyde and Carboxylic Acid		Conditions	Product(s) and Yield(s) (%)	Refs.
C_{10}					
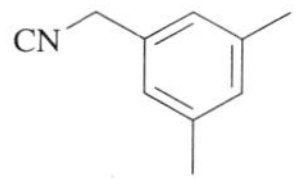CN	*i*-PrCHO	$PhCO_2H$	Et_2O, rt, 14 h	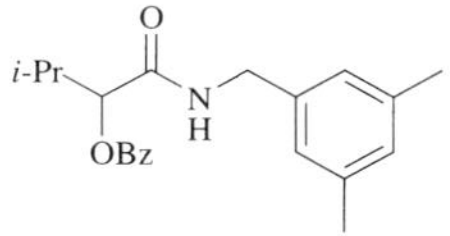*i*-Pr O N H OBz (47)	171
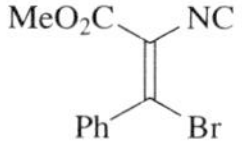MeO_2C NC Ph Br	RCHO	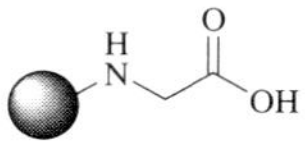H N O OH	rt, 2 d	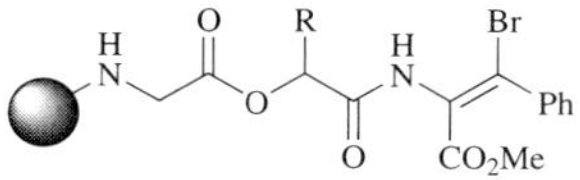H N O R O O H N Br Ph CO_2Me R: *n*-Pr (>70); *n*-C_5H_{11} (>70); Ph (>70); 2-naphthyl (<30)	12
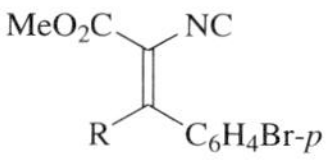MeO_2C NC R C_6H_4Br-*p*	*i*-PrCHO	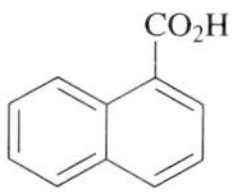CO_2H	CH_2Cl_2, rt, 8-12 h	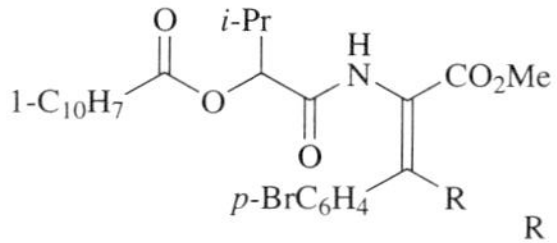O *i*-Pr 1-$C_{10}H_7$ O O H N CO_2Me *p*-BrC_6H_4 R R: H (84); Br (81)	41
C_{12}					
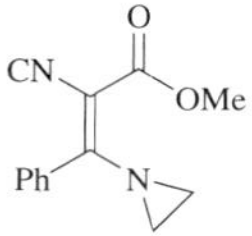O CN OMe Ph N	RCHO	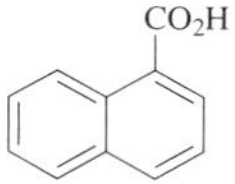CO_2H	CH_2Cl_2, rt, 8-12 h	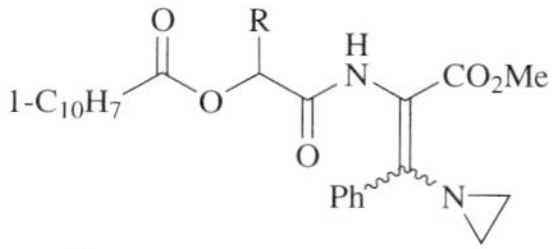O R 1-$C_{10}H_7$ O O H N CO_2Me Ph N R: *n*-Pr (60); 1-naphthyl (52)	41

C_{13}	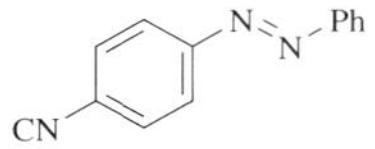	PhCHO	$PhCO_2H$	Neat, rt, 2d	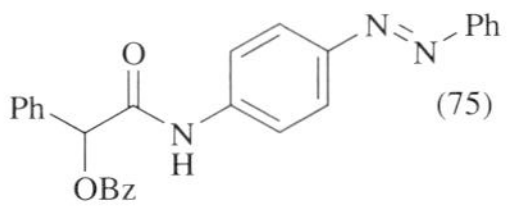	(75)	64
C_{13}	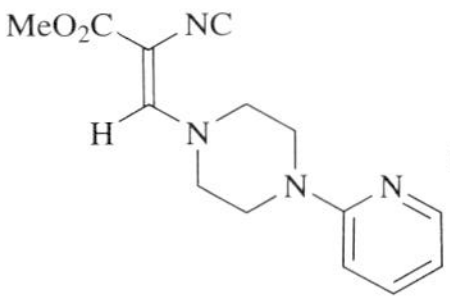	c-$C_6H_{11}CHO$	$PhCO_2H$	DMSO, rt	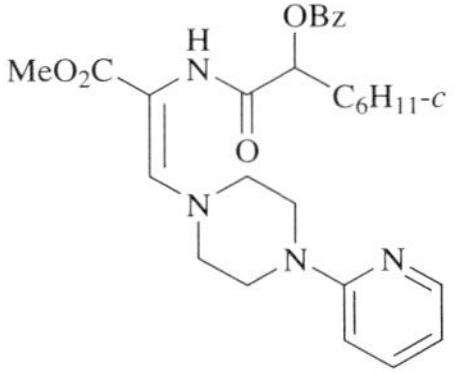	(68)	39
C_{16}	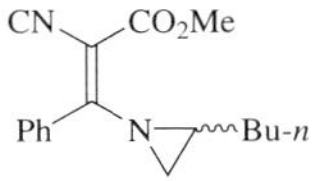	n-PrCHO	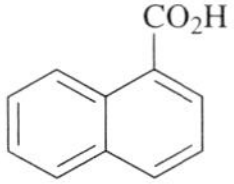	CH_2Cl_2, rt, 8-12 h	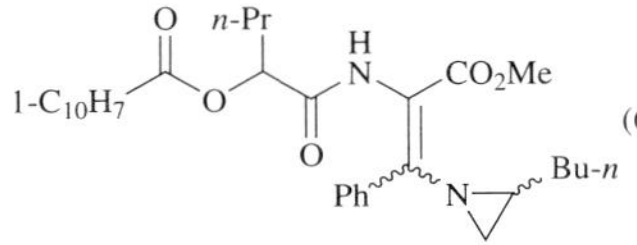	(68)	41
C_{29}	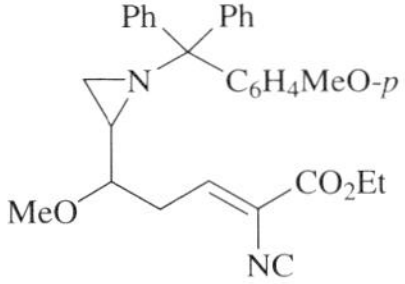	i-PrCHO	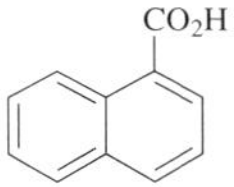	CH_2Cl_2, rt, 8-12 h	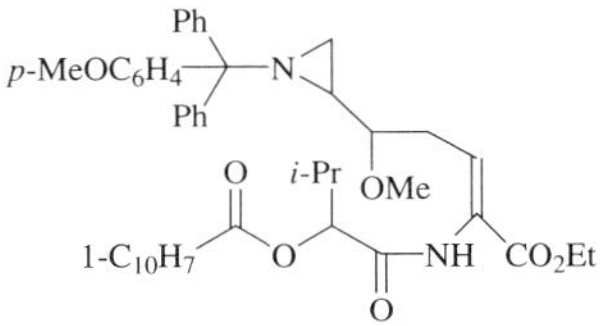	(90)	41

TABLE 2. α-ACYLOXY AMIDES FROM ACHIRAL ISOCYANIDES, KETONES, AND CARBOXYLIC ACIDS

Isocyanide	Ketone and Carboxylic Acid		Conditions	Product(s) and Yield(s) (%)	Refs.
C_2					
MeNCS[a]		$PhCO_2H$	Acetone, $h\nu$ (254 nm), rt, 24 h	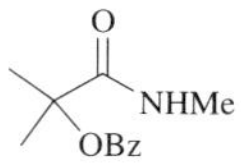 (—)	68
C_3					
EtNC	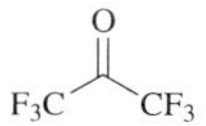	AcOH	Et_2O, rt	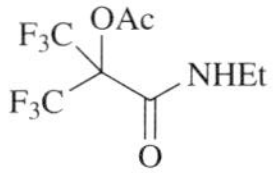 (71)	6
CN⌒CO_2Et	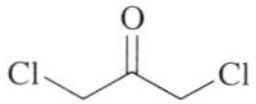	RCO_2H	CH_2Cl_2, rt	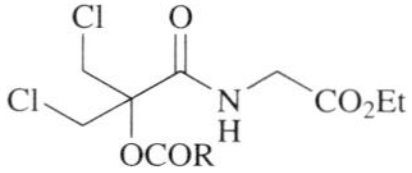 R: Me (90); Ph (90)	95
CN⌒CO_2Bu-*t*		$PhtNCH_2CO_2H$	Acetone, rt	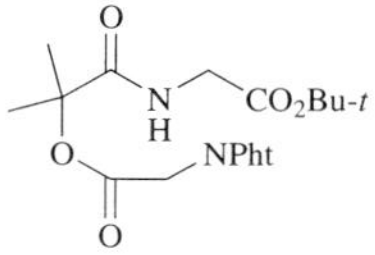 (51)	60
C_4					
i-PrNC	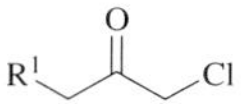	R^2CO_2H		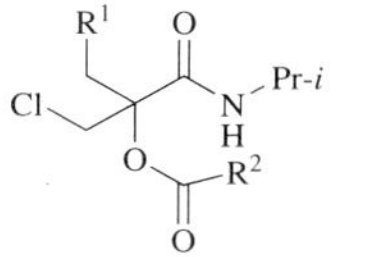	
				R^1 R^2	
			Neat, rt, 15 h	H Ph (82)	94
			CH_2Cl_2, rt	Cl Me (89)	94, 95
			CH_2Cl_2, rt	Cl Ph (72)	95

C$_5$

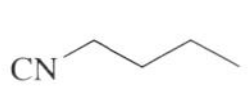 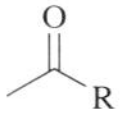 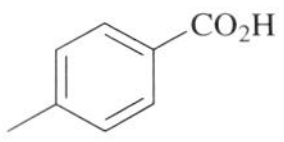 rt, 16.5 h 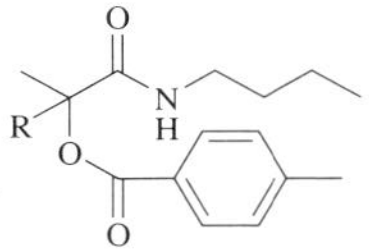 63

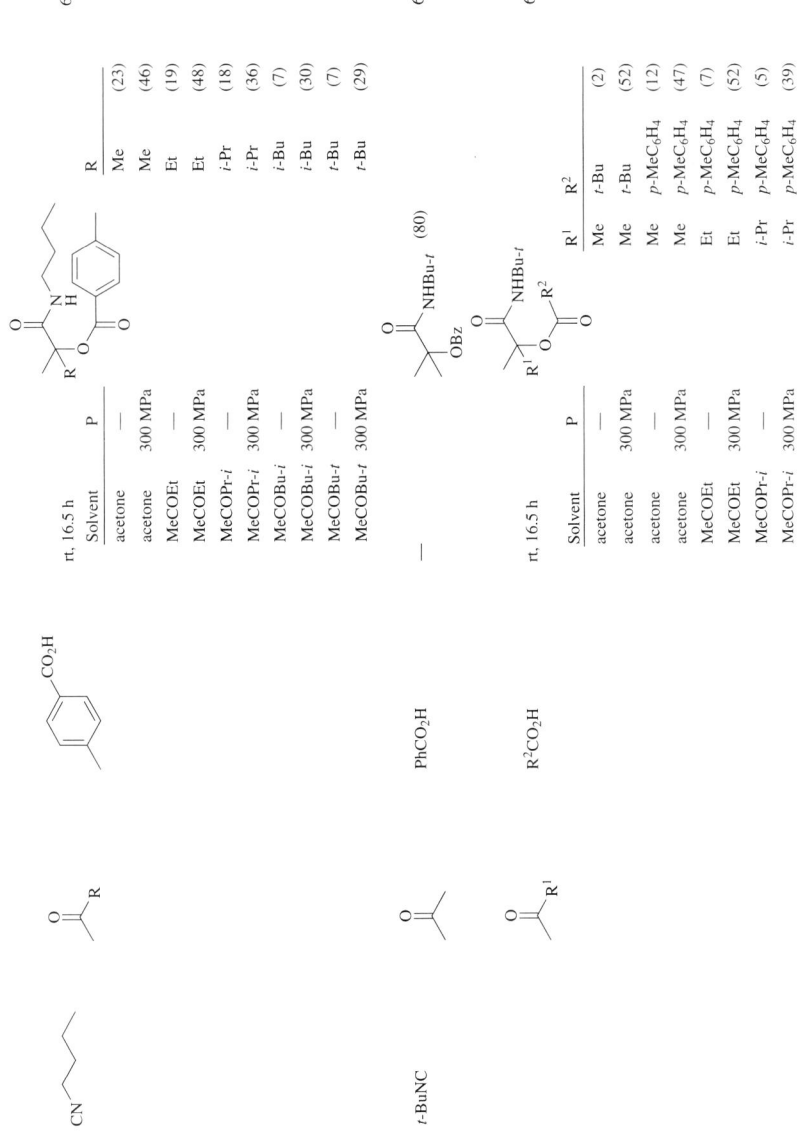

Solvent	P	R	
acetone	—	Me	(23)
acetone	300 MPa	Me	(46)
MeCOEt	—	Et	(19)
MeCOEt	300 MPa	Et	(48)
MeCOPr-*i*	—	*i*-Pr	(18)
MeCOPr-*i*	300 MPa	*i*-Pr	(36)
MeCOBu-*i*	—	*i*-Bu	(7)
MeCOBu-*i*	300 MPa	*i*-Bu	(30)
MeCOBu-*t*	—	*t*-Bu	(7)
MeCOBu-*t*	300 MPa	*t*-Bu	(29)

t-BuNC  $PhCO_2H$ — 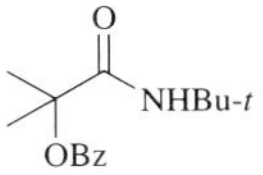 (80) 66

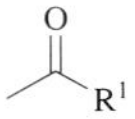 R^2CO_2H rt, 16.5 h 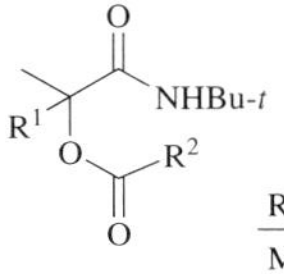 63

Solvent	P	R^1	R^2	
acetone	—	Me	*t*-Bu	(2)
acetone	300 MPa	Me	*t*-Bu	(52)
acetone	—	Me	*p*-MeC_6H_4	(12)
acetone	300 MPa	Me	*p*-MeC_6H_4	(47)
MeCOEt	—	Et	*p*-MeC_6H_4	(7)
MeCOEt	300 MPa	Et	*p*-MeC_6H_4	(52)
MeCOPr-*i*	—	*i*-Pr	*p*-MeC_6H_4	(5)
MeCOPr-*i*	300 MPa	*i*-Pr	*p*-MeC_6H_4	(39)
MeCOBu-*i*	—	*i*-Bu	*p*-MeC_6H_4	(1)
MeCOBu-*i*	300 MPa	*i*-Bu	*p*-MeC_6H_4	(12)
MeCOBu-*t*	600 MPa	*t*-Bu	*p*-MeC_6H_4	(73)

TABLE 2. α-ACYLOXY AMIDES FROM ACHIRAL ISOCYANIDES, KETONES, AND CARBOXYLIC ACIDS (*Continued*)

Isocyanide	Ketone and Carboxylic Acid		Conditions	Product(s) and Yield(s) (%)			Refs.
C_5							
t-BuNC	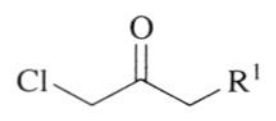	R^2CO_2H		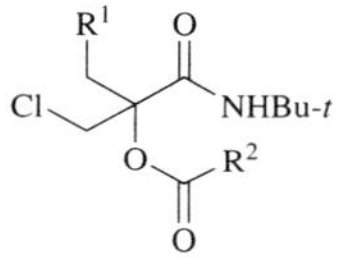			
				R^1	R^2		
			Neat, rt, 15 h	H	Me	(78)	94
			Neat, rt, 15 h	H	*n*-Pr	(73)	94
			Neat, rt, 15 h	H	Ph	(98)	94
			Neat, rt, 15 h	H	(*E*)-PhCH=CH	(73)	94
			CH_2Cl_2, rt	Cl	Me	(61)	94, 95
			CH_2Cl_2, rt	Cl	Ph	(68)	94, 95
	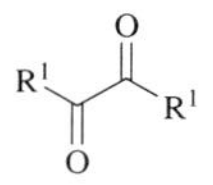	R^2CO_2H	—	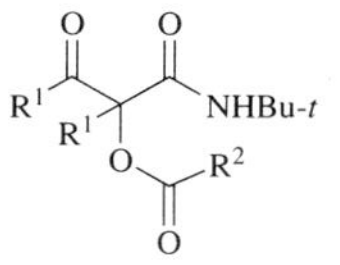			
				R^1	R^2		
				Me	Me	(90)	66
				Me	Ph	(93)	66
				Me	(*E*)-PhCH=CH	(85)	66
				CH_2Br	Me	(93)	67
				CH_2Br	Ph	(84)	67
				CH_2Br	(*E*)-PhCH=CH	(66)	67
				Ph	Me	(85)	66
				Ph	Ph	(94)	66

C_7					
c-C_6H_{11}NC	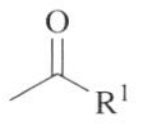	R^2CO_2H	rt, 16.5 h	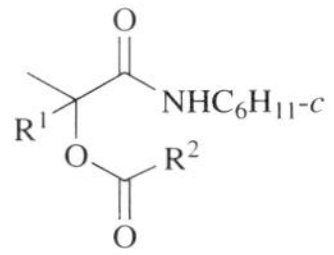	63

Solvent	P	R^1	R^2	
acetone	—	Me	Me	(41)
acetone	300 MPa	Me	Me	(89)
acetone	—	Me	*n*-Bu	(16)
acetone	300 MPa	Me	*n*-Bu	(51)
acetone	—	Me	*i*-Bu	(9)
acetone	300 MPa	Me	*i*-Bu	(38)
acetone	—	Me	*t*-Bu	(6)
acetone	300 MPa	Me	*t*-Bu	(28)
acetone	—	Me	*p*-MeC_6H_4	(17)
acetone	300 MPa	Me	*p*-MeC_6H_4	(40)
MeCOEt	—	Et	*p*-MeC_6H_4	(15)
MeCOEt	300 MPa	Et	*p*-MeC_6H_4	(36)
MeCOPr-*i*	—	*i*-Pr	*p*-MeC_6H_4	(20)
MeCOPr-*i*	300 MPa	*i*-Pr	*p*-MeC_6H_4	(39)

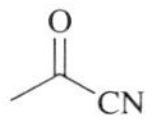	HCO_2H	Neat	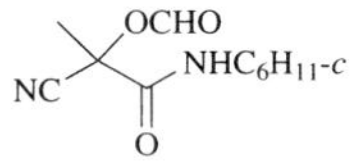	(58)	65
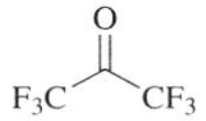	AcOH	Et_2O, rt	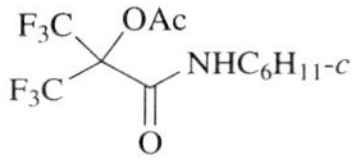	(75)	6

TABLE 2. α-ACYLOXY AMIDES FROM ACHIRAL ISOCYANIDES, KETONES, AND CARBOXYLIC ACIDS (*Continued*)

Isocyanide	Ketone and Carboxylic Acid		Conditions	Product(s) and Yield(s) (%)	Refs.
C_7					
c-$C_6H_{11}NC$	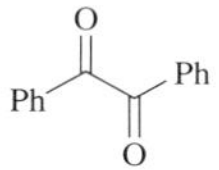	RCO_2H	Et_2O, rt, 16.5 h	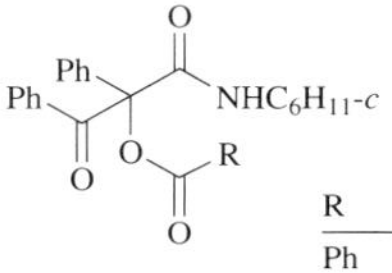	63
			P	R	
			—	Ph (20)	
			300 MPa	Ph (33)	
			—	*o*-MeC_6H_4 (18)	
			300 MPa	*o*-MeC_6H_4 (28)	
			—	*m*-MeC_6H_4 (13)	
			300 MPa	*m*-MeC_6H_4 (24)	
			—	*p*-MeC_6H_4 (10)	
			300 MPa	*p*-MeC_6H_4 (20)	
PhNC		AcOH	Neat, rt, 15 d	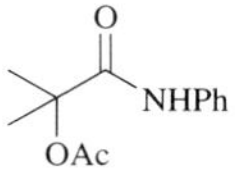 (30)	64
		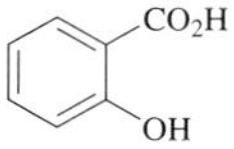	Acetone, rt, 7 d	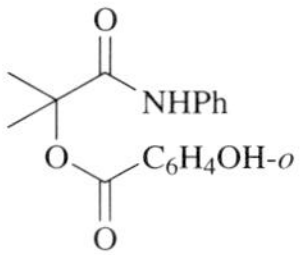 (20)	3
	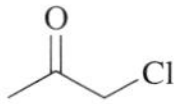	RCO_2H	Et_2O, rt, 7 d	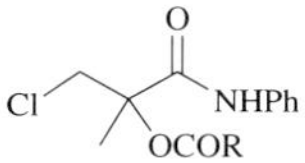 R: Me (—), Ph (—)	3
	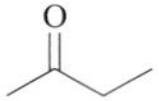	AcOH	Neat, rt, 7 d	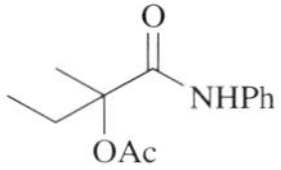 (78)	3

	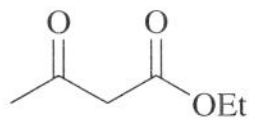	AcOH	Neat, rt, 10 d	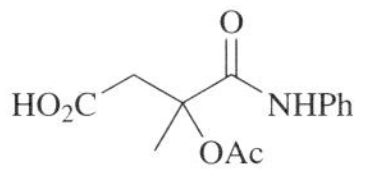	(45)	64
	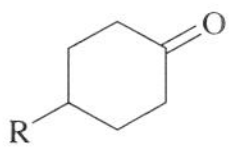	$PhCO_2H$	Et_2O, rt Time: 12 h / 4 d	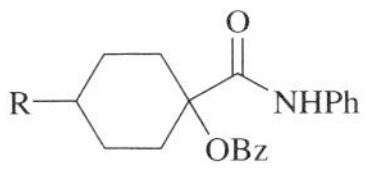 R: H (94) / Me (85)		250
	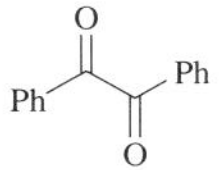	$PhCO_2H$	Et_2O, rt, 10 d	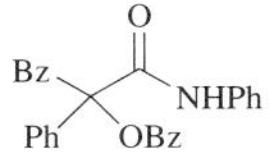	(100)	3
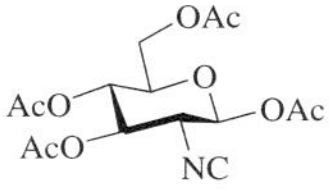	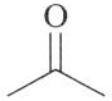	AcOH	$ZnCl_2$ (cat), CH_2Cl_2, 0°, 5 h	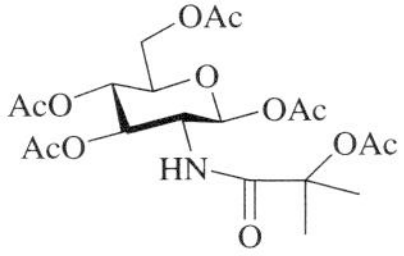	(15)	50
C_9 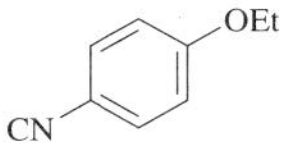		AcOH	Neat, rt, 3 d	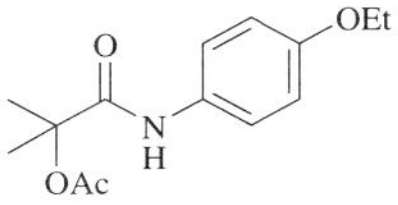	(—)	249
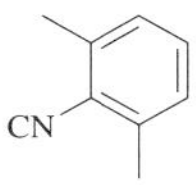	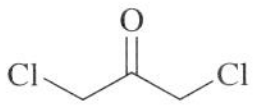	AcOH	CH_2Cl_2, rt	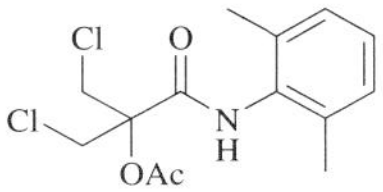	(71)	95

TABLE 2. α-ACYLOXY AMIDES FROM ACHIRAL ISOCYANIDES, KETONES, AND CARBOXYLIC ACIDS (*Continued*)

	Isocyanide	Ketone and Carboxylic Acid		Conditions	Product(s) and Yield(s) (%)	Refs.
C_9	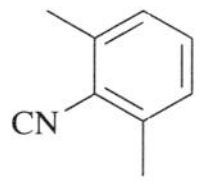	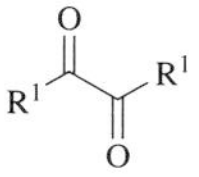	R^2CO_2H	—	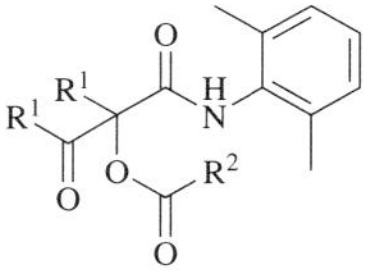	66
					R^1 / R^2	
					Me / Me (88)	
					Me / Ph (75)	
					Me / (*E*)-PhCH=CH (79)	
					Ph / Me (82)	
					Ph / Ph (90)	
					Ph / (*E*)-PhCH=CH (89)	
C_{13}	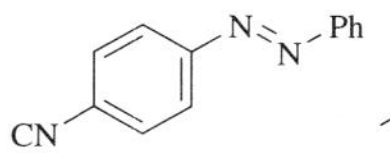	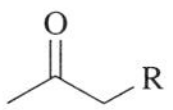	AcOH		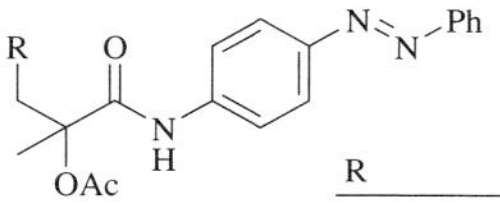	
					R	
				Acetone, rt, 15 d	H (55)	7
				Neat, rt, 10 d	CO_2Et (73)	64

[a] The substrate is converted in situ into methyl isocyanide.

TABLE 3. α-ACYLOXY AMIDES FROM CHIRAL CARBONYL COMPOUNDS, ISOCYANIDES, AND CARBOXYLIC ACIDS

Carbonyl Ccompound	Isocyanide and Carboxylic Acid		Conditions	Product(s), Yield(s) (%), dr	Refs.
C_3 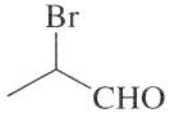	t-BuNC	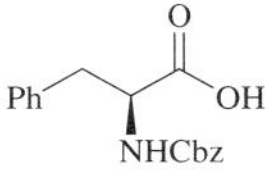	CH_2Cl_2, rt, 4 h	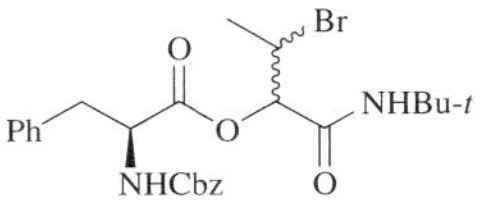 (76), —	166
	t-BuNC	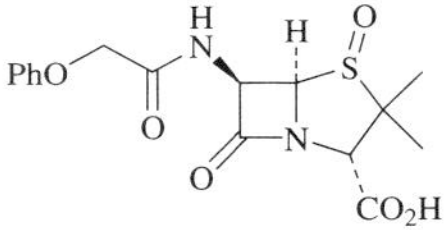	CH_2Cl_2, THF, rt, 4 h	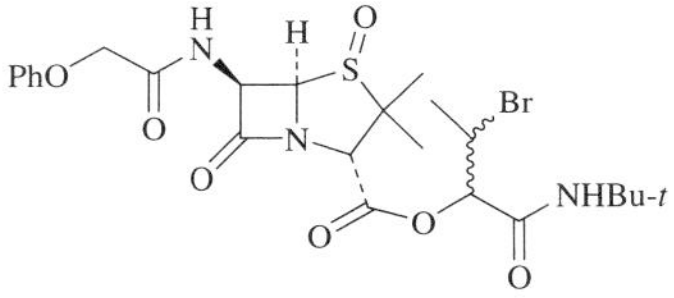 (72), —	167
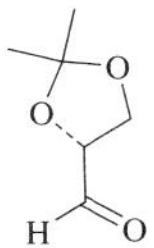		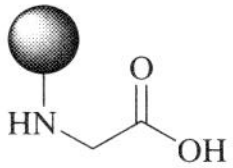	rt, 2 d	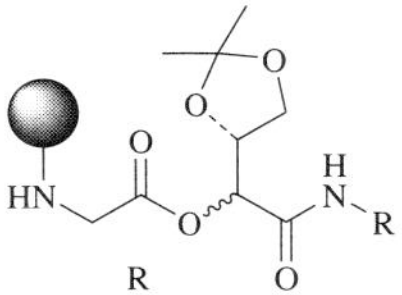	12
				R: CO_2Me (>70), —	
				R: CH_2=CH (30-70), —	
				R: n-Pr (>70), —	
				R: Ph (30-70), —	
				R: $PhS(O)_2$ (>70), —	
	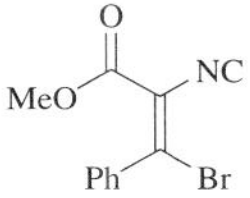	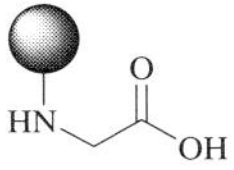	rt, 2 d	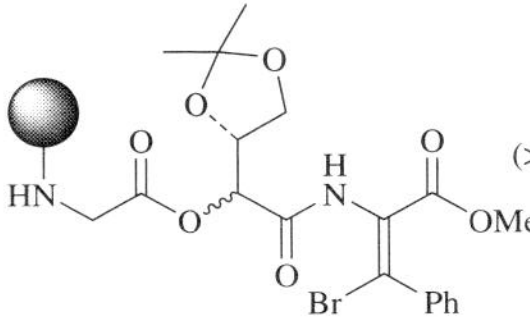 (>70), —	12

TABLE 3. α-ACYLOXY AMIDES FROM CHIRAL CARBONYL COMPOUNDS, ISOCYANIDES, AND CARBOXYLIC ACIDS (*Continued*)

Carbonyl Ccompound	Isocyanide and Carboxylic Acid		Conditions	Product(s), Yield(s) (%), dr	Refs.
C_3 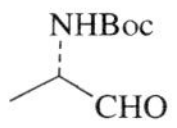	R^1NC	R^2CO_2H	CH_2Cl_2, rt, 20-40 h	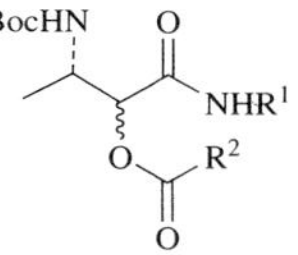 R^1 / R^2 CH_2CO_2Me / Bn — (71), 63:37 c-C_6H_{11} / p-MeC_6H_4 — (80), 67:33 Bn / n-Pr — (85), 65:35	46
	BnNC	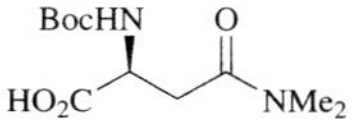	CH_2Cl_2, rt, 24 h	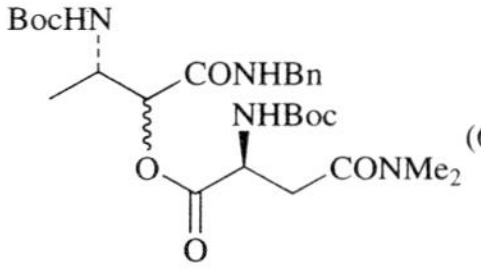 (60), 64:36	45
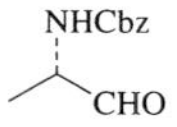	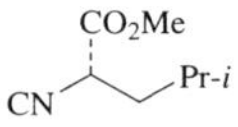	$PhCO_2H$	CH_2Cl_2, rt, 48 h	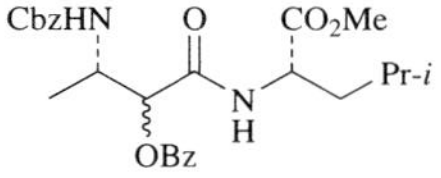 (85), 50:50	47
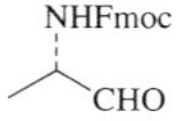	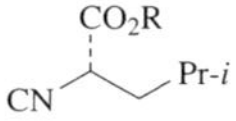	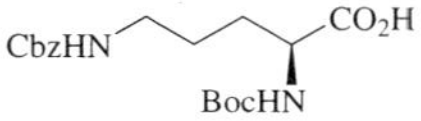	CH_2Cl_2, rt, 4 d	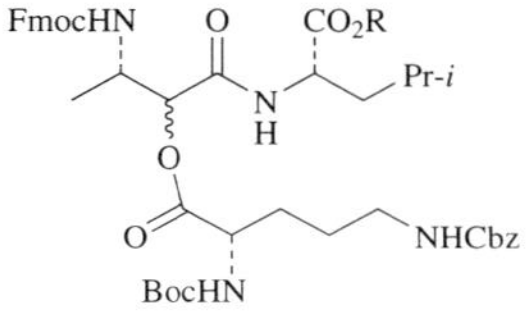 R Me — (75), 55:45 Bn — (80), 55:45	48

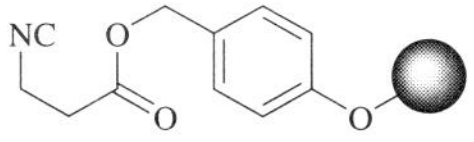

RCO_2H | CH_2Cl_2, rt, 24 h

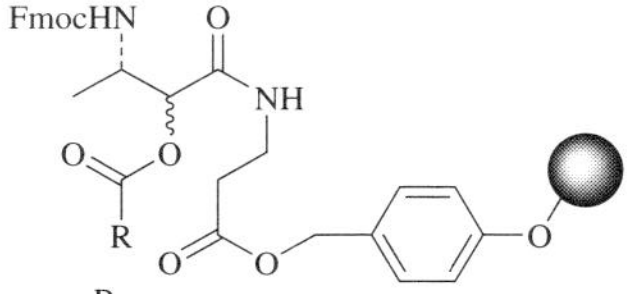

R	
$BzNHCH_2$	(100), —
Bn	(100), —

96

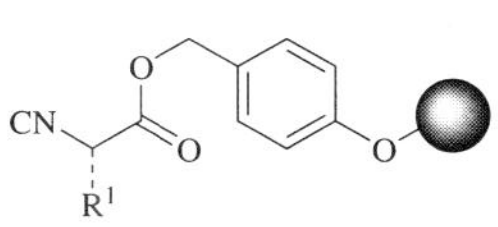

R^2CO_2H | CH_2Cl_2, rt, 24 h

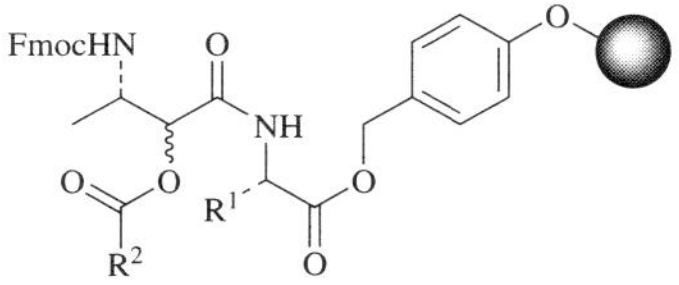

R^1	R^2	
Me	$BzNHCH_2$	(>72), —
Me	Bn	(>67), —
i-Bu	$BzNHCH_2$	(>87), —
i-Bu	Bn	(>93), —

96

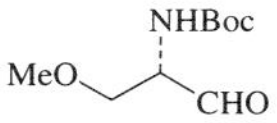

R^1NC | R^2CO_2H | CH_2Cl_2, rt, 20-40 h

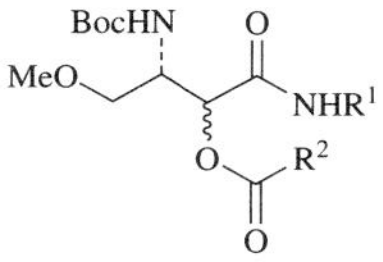

R^1	R^2	
c-C_6H_{11}	Bn	(95), 61:59
Bn	*n*-Pr	(95), 57:43

46

TABLE 3. α-ACYLOXY AMIDES FROM CHIRAL CARBONYL COMPOUNDS, ISOCYANIDES, AND CARBOXYLIC ACIDS (*Continued*)

Carbonyl Compound	Isocyanide	and Carboxylic Acid	Conditions	Product(s), Yield(s) (%), dr	Refs.
C_4 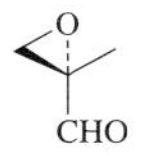	RNC	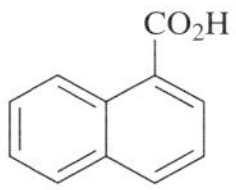	AcOEt, rt	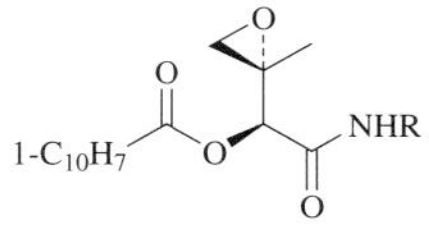	40
				R: $(EtO)_2P(O)CH_2$ (75), S:R = 78:22	
				R: EtO_2CCH_2 (73), S:R = 78:22	
				R: $(EtO)_2P(O)CH(CO_2Bu\text{-}t)$ (20), S:R = 78:22	
				R: $(EtO)_2P(O)CH(CO_2CH_2CH_2SiMe_3)$ (38), S:R = 78:22	
				R: $(EtO)_2P(O)CH(CO_2Ph)$ (10), S:R = 78:22	
	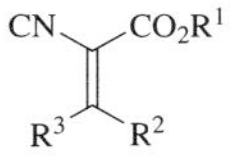	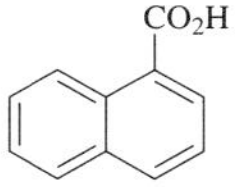	AcOEt, rt	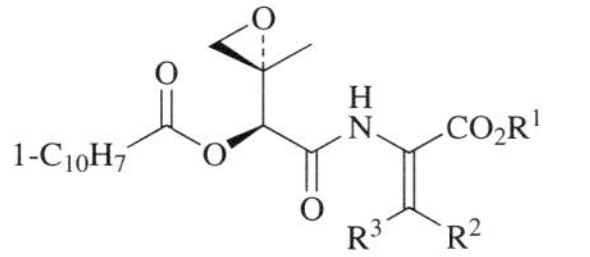	
				R^1 = Me, R^2 = Br, R^3 = *n*-Pr (39), —	41
				R^1 = Et, R^2 = H, R^3 = *n*-C_6H_{13} (50), S:R = 79:21	40
	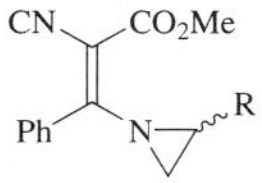	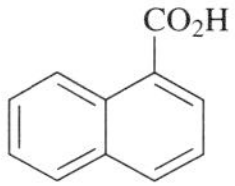	CH_2Cl_2, rt, 8-12 h	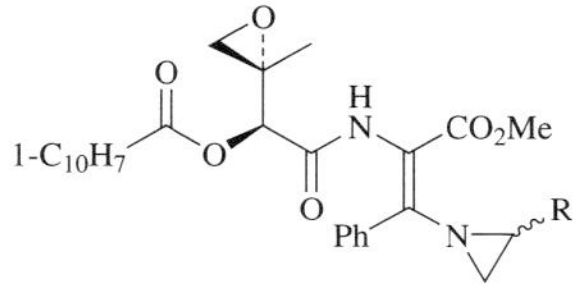	41
				R: H (53), —	
				R: *n*-Bu (56), —	

Reactant	Isocyanide	Acid	Conditions	Product(s) and Yield(s) (%), Ratio	Refs.
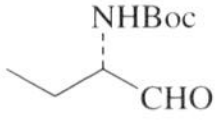	R^1NC	R^2CO_2H	CH_2Cl_2, rt, 20-40 h	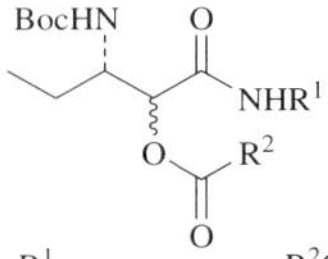 R^1 — R^2CO_2H CH_2CO_2Me — L-Boc-Leu — (59), 66:34 $CH_2CH_2CO_2Bn$ — L-Cbz-Leu — (69), 66:34 n-Bu — L-Cbz-Leu — (65), 61:39 t-Bu — $PhCH_2CO_2H$ — (83), 69:31	46
	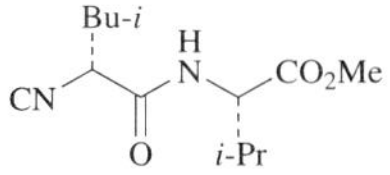	RCO_2H	CH_2Cl_2, rt, 24 h	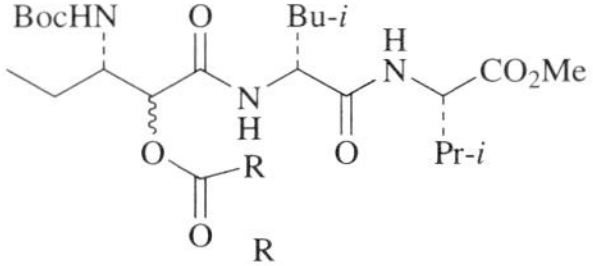 R L-Cbz-Phe — (48), — L-Boc-Val — (39), —	45
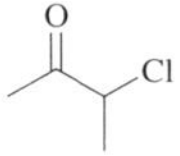	t-BuNC	RCO_2H	Neat, rt, 15 h	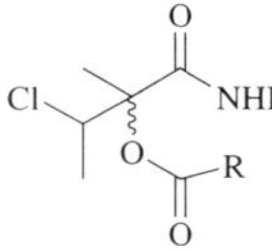 R Me — (84), — n-Pr — (69), — Ph — (90), —	94
C_5 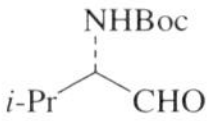	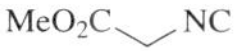	RCO_2H	CH_2Cl_2, rt, 20-40 h	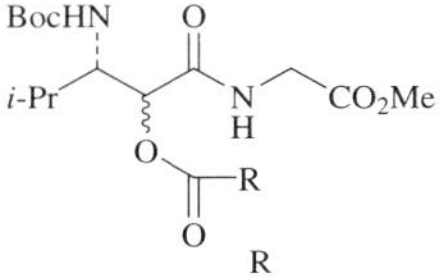 R Cbz-Gly — (77), 60:40 L-Boc-Phe — (95), 52:48 D-Boc-Phe — (89), 63:37	46 45, 46 46

TABLE 3. α-ACYLOXY AMIDES FROM CHIRAL CARBONYL COMPOUNDS, ISOCYANIDES, AND CARBOXYLIC ACIDS (*Continued*)

Carbonyl Compound	Isocyanide and Carboxylic Acid		Conditions	Product(s), Yield(s) (%), dr	Refs.
C_5 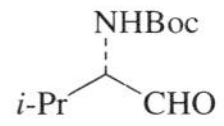 NHBoc i-Pr CHO	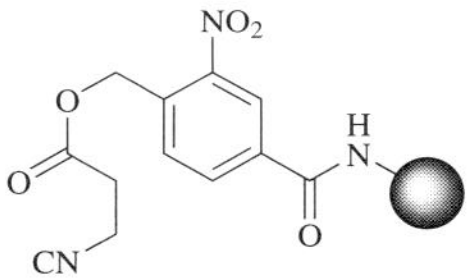 NO2 O O CN O H N	$BnCO_2H$	CH_2Cl_2, rt, 24 h	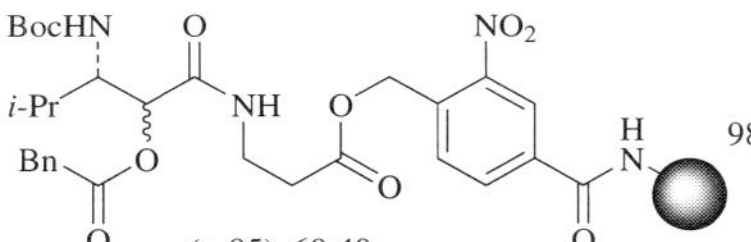 BocHN O NO2 i-Pr NH O Bn O O O H N O (> 95), 60:40	98
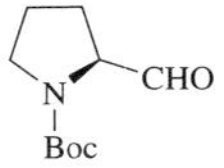 CHO N Boc	R^1NC	R^2CO_2H	CH_2Cl_2, rt, 20-40 h	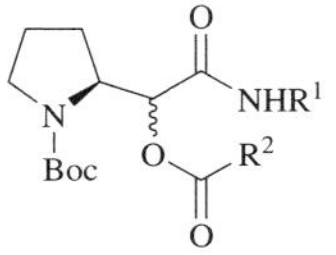 O NHR1 N Boc O R2 O	
				R^1 / R^2CO_2H	
				t-Bu / $PhCH_2CO_2H$ (94), 65:35	46
				c-C_6H_{11} / $PhCH_2CO_2H$ (85), 69:31	46
				c-C_6H_{11} / L-Boc-Phe (83), 75:25	45
				Bn / *n*-$PrCO_2H$ (93), 62:38	46
	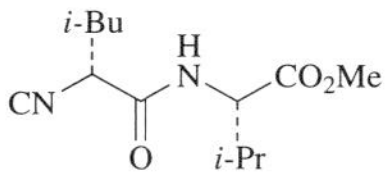 i-Bu H CN N CO2Me O i-Pr	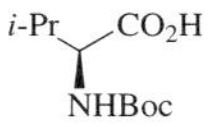 i-Pr CO2H NHBoc	CH_2Cl_2, rt, 48 h	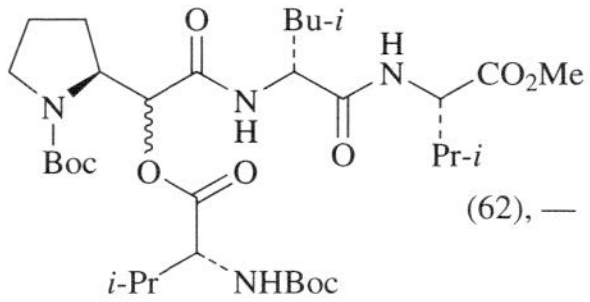 O Bu-i H N CO2Me N H Boc O O O Pr-i i-Pr NHBoc (62), —	45
	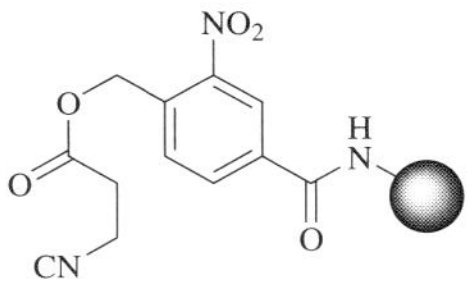 NO2 O O CN O H N	$BnCO_2H$	CH_2Cl_2, rt, 24 h	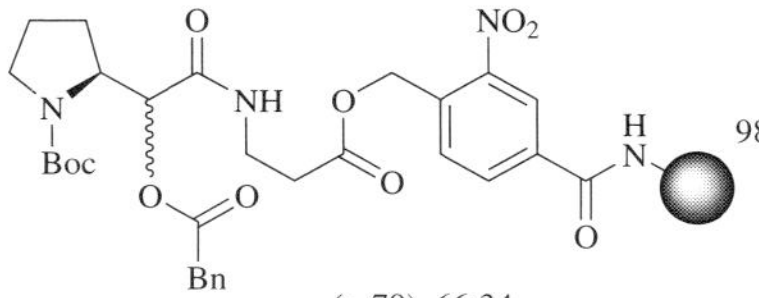 O NO2 N NH O Boc O O O Bn H N O (> 79), 66:34	98

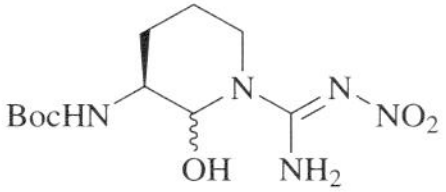	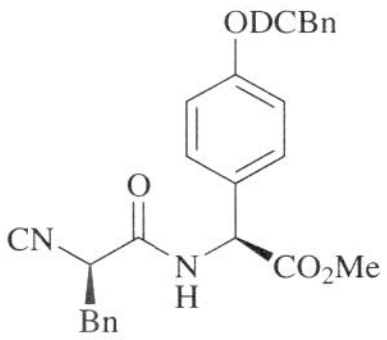	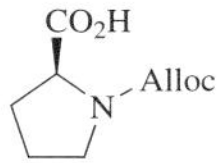	CH_2Cl_2, rt, 16 h	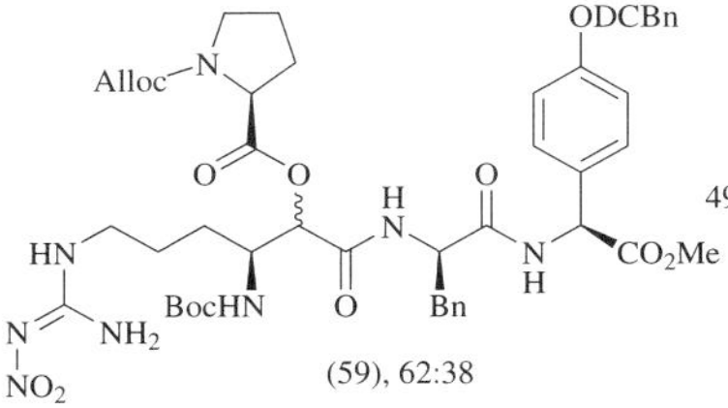 (59), 62:38	49
C_7					
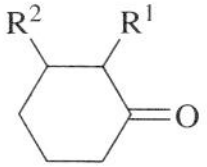	PhNC	$PhCO_2H$	Et_2O, rt	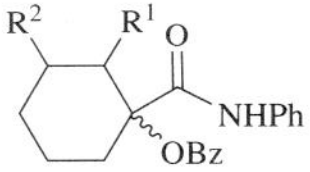 R^1 / R^2 Me / H (76), — H / Me (93), —	250
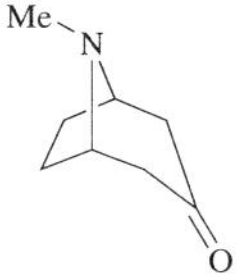	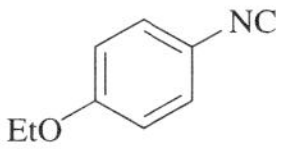	$PhCO_2H$	$CHCl_3$, rt, 10 d	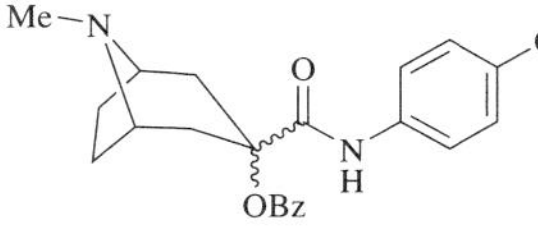 (34), —	251
C_8					
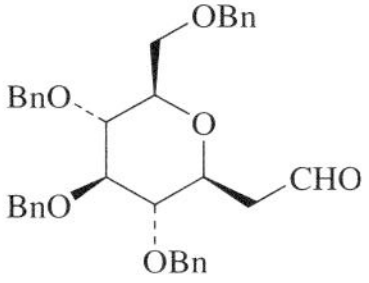	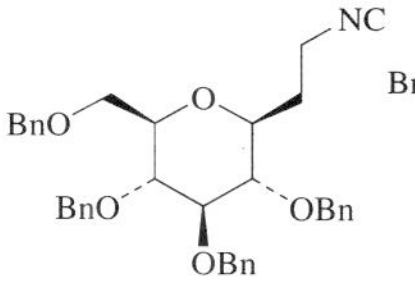	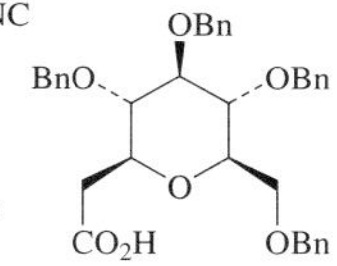	—	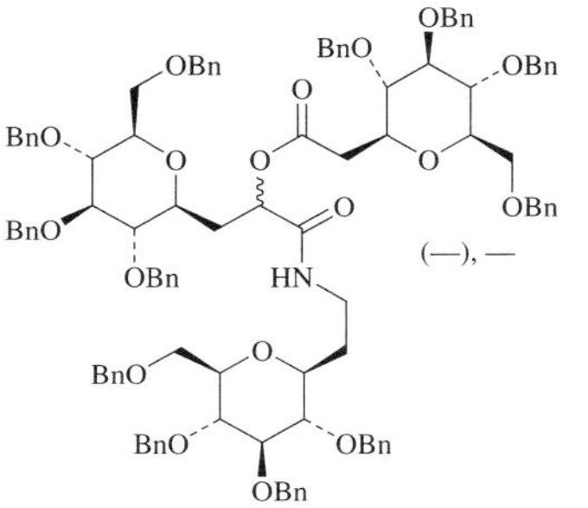 (—), —	58

TABLE 3. α-ACYLOXY AMIDES FROM CHIRAL CARBONYL COMPOUNDS, ISOCYANIDES, AND CARBOXYLIC ACIDS (*Continued*)

Carbonyl Compound	Isocyanide	and Carboxylic Acid	Conditions	Product(s), Yield(s) (%), dr	Refs.
C_8					
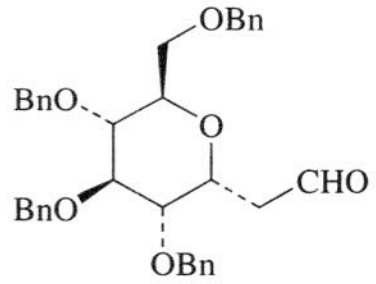	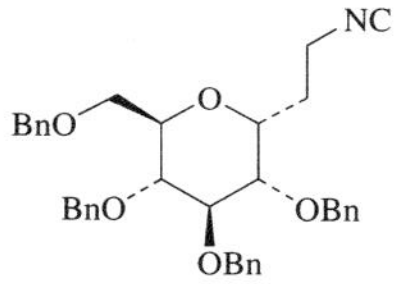	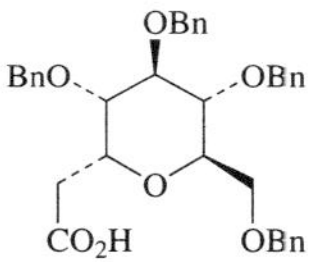	—	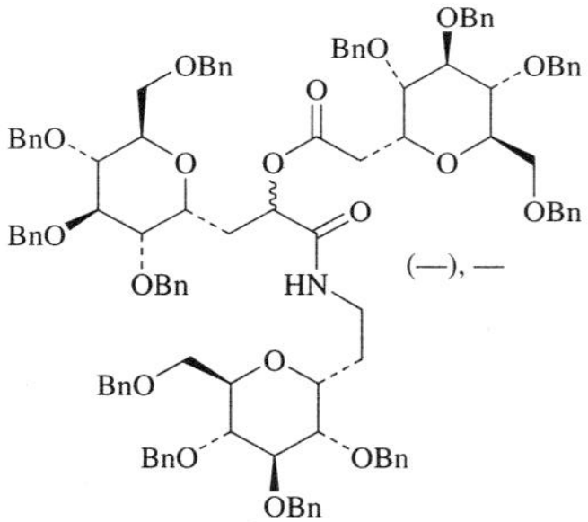 (—), —	58
C_9					
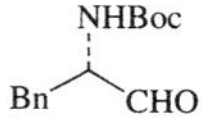	R^1NC	R^2CO_2H	CH_2Cl_2, rt, 20-40 h	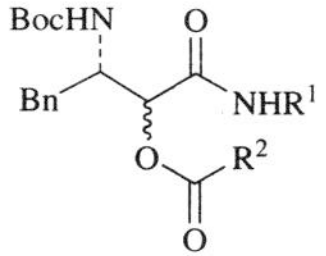	46
				R^1 / R^2	
				CH_2CO_2Me / $PhCH_2$ (91), 64:36	
				CH_2CO_2Me / *s*-Bu (73), 63:37	
				t-Bu / $PhCH_2$ (82), 71:29	
				Bn / Ph (67), 62:38	
				Bn / $PhCH_2$ (81), 67:33	
	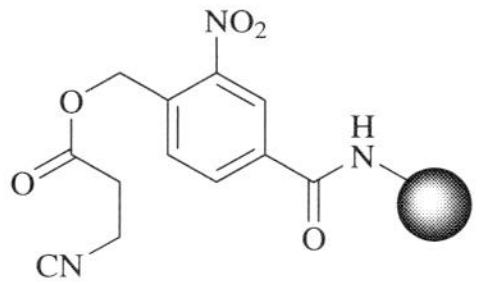	$BnCO_2H$	CH_2Cl_2, rt, 24 h	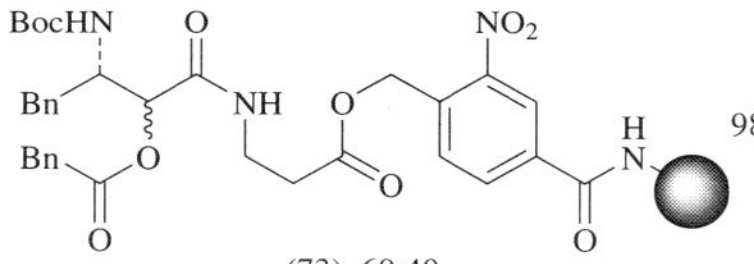 (73), 60:40	98

Aldehyde	Isocyanide	Acid	Conditions	Product(s) and Yield(s) (%), ratio	Refs.
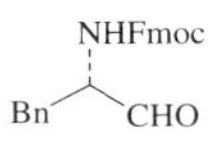	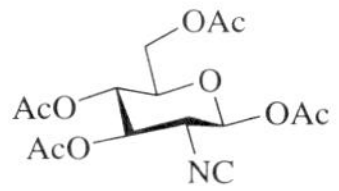	AcOH	CH_2Cl_2, rt, 4 d	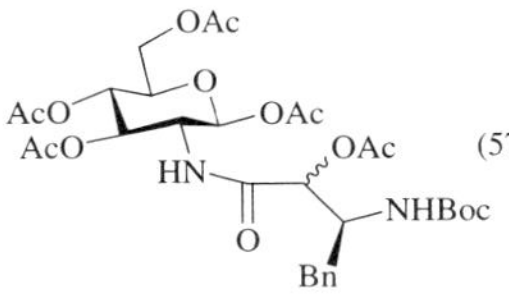 (57), 60:40	50
	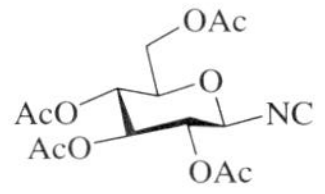	AcOH	CH_2Cl_2, rt, 3 d	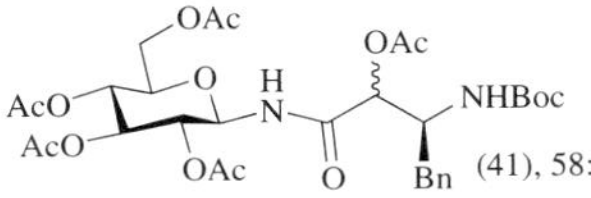 (41), 58:42	50
	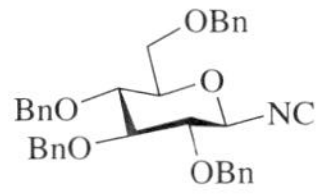	AcOH	CH_2Cl_2, rt, 8 d	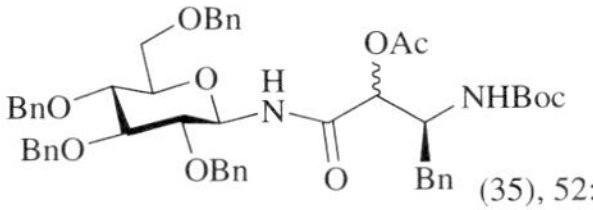 (35), 52:48	50
	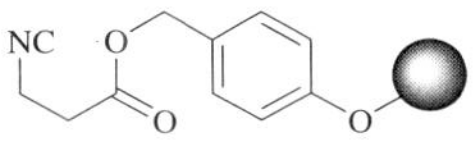	RCO_2H	CH_2Cl_2, rt, 24 h	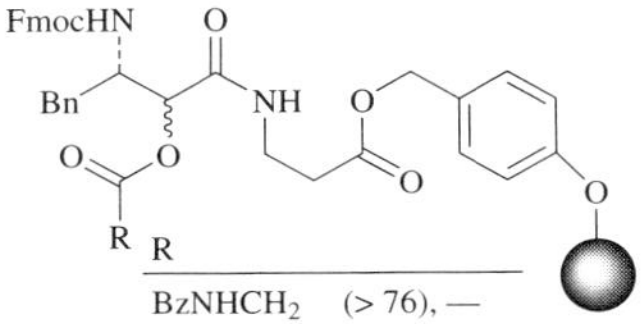	96
				R: $BzNHCH_2$ (> 76), —	
				R: $PhCH_2$ (> 78), 60:40	
	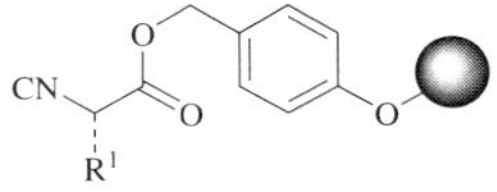	R^2CO_2H	CH_2Cl_2, rt, 24 h	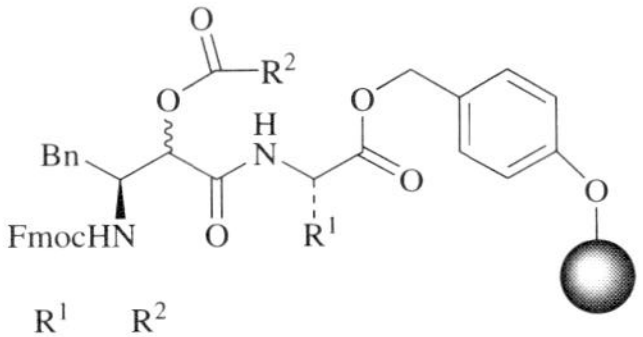	96
				R^1 = Me, R^2 = $BzNHCH_2$ (> 64), —	
				R^1 = Me, R^2 = $PhCH_2$ (> 62), —	
				R^1 = *i*-Bu, R^2 = $BzNHCH_2$ (> 63), —	
				R^1 = *i*-Bu, R^2 = $PhCH_2$ (> 83), —	

TABLE 3. α-ACYLOXY AMIDES FROM CHIRAL CARBONYL COMPOUNDS, ISOCYANIDES, AND CARBOXYLIC ACIDS (*Continued*)

Carbonyl Compound	Isocyanide and Carboxylic Acid	Conditions	Product(s), Yield(s) (%), dr	Refs.
C_{27}				
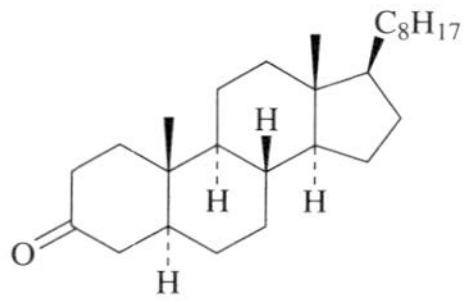	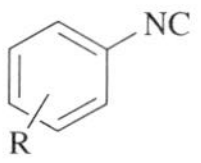 $PhCO_2H$	Et_2O, rt	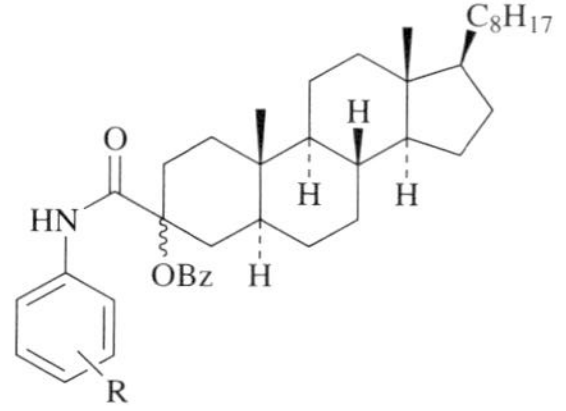 R / Time: H / 7 d (30), —; *o*-OMe / 6 d (53), —; *p*-Me / 5 d (43), —	62

TABLE 4. α-ACYLOXY AMIDES FROM PROCHIRAL CARBONYL COMPOUNDS, CHIRAL ISOCYANIDES, AND CARBOXYLIC ACIDS

Isocyanide	Carbonyl Compound and Carboxylic Acid		Conditions	Product(s), Yield(s) (%), dr			Refs.
C_5							
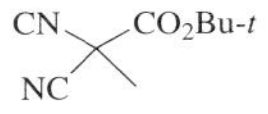	i-PrCHO	$ClCH_2CO_2H$	AcOEt, rt, 5 d	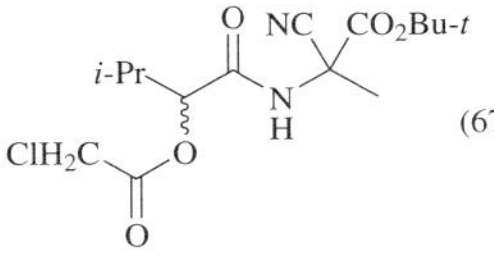		(67), 75:25	74
C_6							
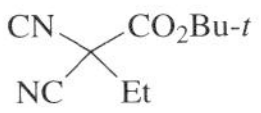	R^1CHO	R^2CO_2H		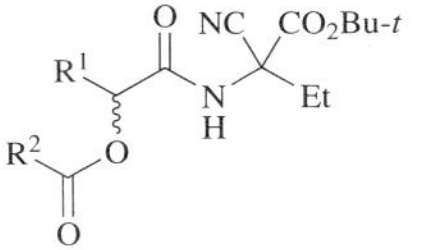			
				R^1	R^2		
			CH_2Cl_2, rt, 2 d	Et	p-$FC_6H_4CH_2$	(17), —	73
			CH_2Cl_2, rt, 2 d	i-Pr	$BrCH_2$	(30), —	73
			CH_2Cl_2, rt, 2 d	i-Pr	Cl_2CH	(45), —	73
			AcOEt, rt, 5 d	i-Pr	$ClCH_2$	(57), 60:40	74
			CH_2Cl_2, rt, 2 d	i-Pr	p-$O_2NC_6H_4CH_2$	(11), —	73
C_7							
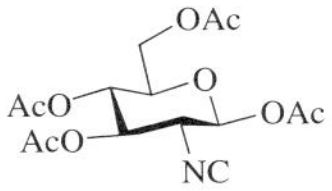	R^1CHO	R^2CO_2H	CH_2Cl_2, rt	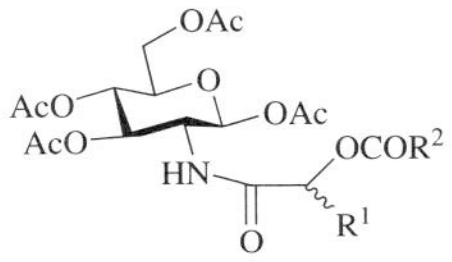			50
			Time	R^1	R^2CO_2H		
			3 d	$BocNHCH_2$	AcOH	(53), 57:43	
			8 d	i-Pr	AcOH	(23), 53:47	
			28 h	i-Pr	Boc-Gly	(90), 53:47	
			3 d	i-Pr	L-Boc-Phe	(35), 58:42	
			12 d	p-$MeOC_6H_4$	AcOH	(17), 56:44	

TABLE 4. α-ACYLOXY AMIDES FROM PROCHIRAL CARBONYL COMPOUNDS, CHIRAL ISOCYANIDES, AND CARBOXYLIC ACIDS (*Continued*)

	Isocyanide	Carbonyl Compound	Carboxylic Acid	Conditions	Product(s), Yield(s) (%), dr	Refs.
C_7	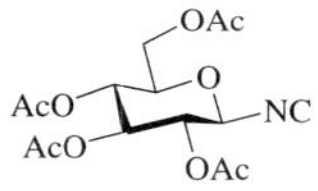	RCHO	AcOH	CH_2Cl_2, rt Time: 24 h; 2 d	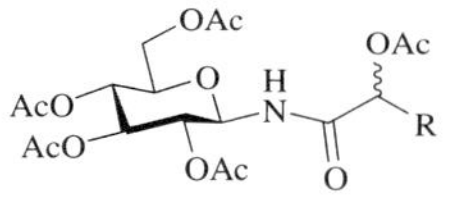 R = Et (80), 50:50 R = $BocNHCH_2$ (23), 55:45	50
	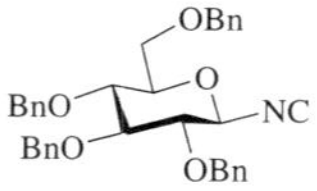		AcOH	CH_2Cl_2, rt, 6 d	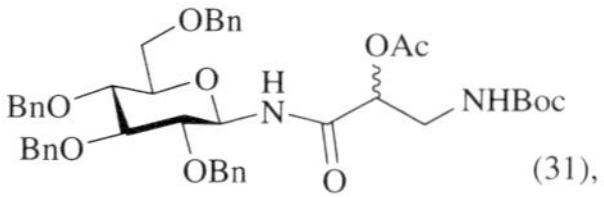 (31), 57:43	50
C_8	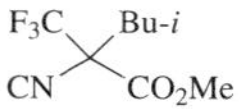	BnCHO	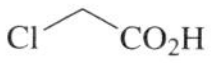	AcOEt or CH_3CN, rt	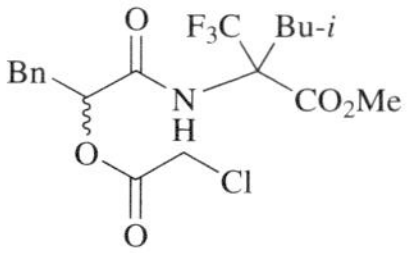 (47), —	71
C_{10}	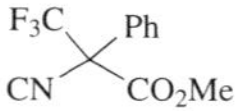	*i*-PrCHO	RCO_2H	AcOEt or CH_3CN, rt	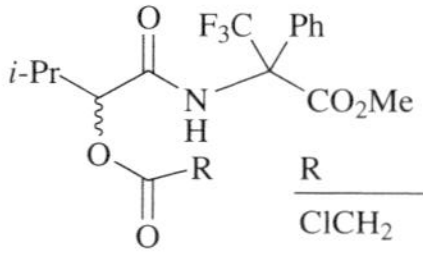 R = $ClCH_2$ (73), 50:50 R = $PhtNCH_2$ (73), 50:50	71
C_{11}	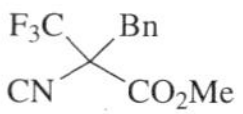	*i*-PrCHO	$PhtNCH_2CO_2H$	AcOEt or CH_3CN, rt	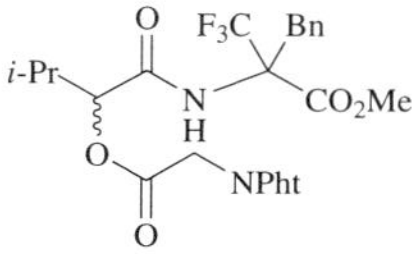 (72), 50:50	71

 Ar = p-$FC_6H_4CH_2$	i-PrCHO	Cl_2CHCO_2H	CH_2Cl_2, rt, 2 d	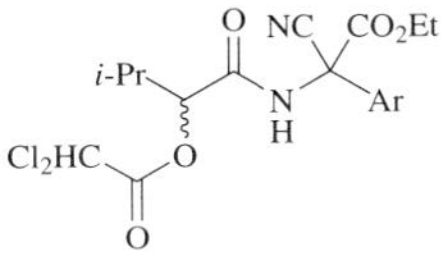	(27), —	73
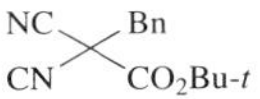	i-PrCHO	ClH_2CCO_2H	CH_2Cl_2, rt, 2 d	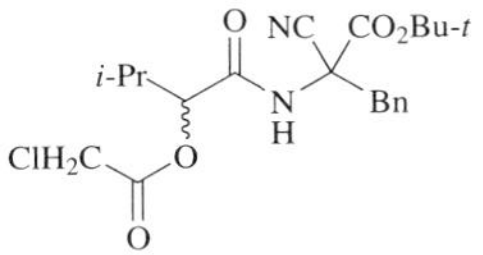	(63), 52:48	74
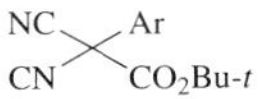 Ar = p-$R^1C_6H_4CH_2$	R^2CHO	R^3CO_2H	CH_2Cl_2, rt, 2 d	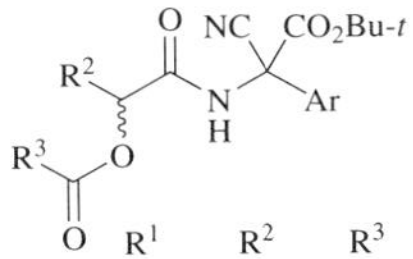		73

R^1	R^2	R^3	
H	Me	ICH_2	(36), —
H	Et	ICH_2	(46), —
H	i-Pr	ICH_2	(53), —
H	i-Pr	F_3C	(33), —
H	i-Pr	$NCCH_2$	(27), —
H	i-Pr	p-$O_2NC_6H_4CH_2$	(35), —
F	n-Pr	Cl_2CH	(53), —
F	i-Pr	$CH_3CH(OH)$	(7), —
F	i-Pr	Cl_2CH	(60), —
Cl	i-Pr	$ClCH_2$	(61), —
Cl	i-Pr	$BrCH_2$	(51), —
Cl	i-Pr	ICH_2	(57), —
NO_2	Et	Cl_2CH	(55), —
NO_2	i-Pr	Cl_2CH	(58), —
NO_2	i-Pr	Cl_3C	(38), —
NO_2	i-Pr	p-$FC_6H_4CH_2$	(28), —

TABLE 4. α-ACYLOXY AMIDES FROM PROCHIRAL CARBONYL COMPOUNDS, CHIRAL ISOCYANIDES, AND CARBOXYLIC ACIDS (*Continued*)

Isocyanide	Carbonyl Compound and Carboxylic Acid		Conditions	Product(s), Yield(s) (%), dr	Refs.
C_{12}					
 Ar = p-$R^2C_6H_4CH_2$	R^3CHO	R^4CO_2H	CH_2Cl_2, rt, 2 d	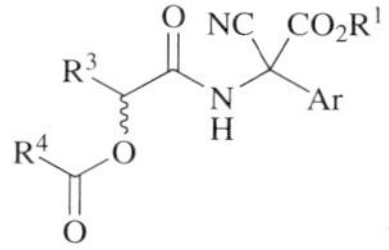	73

R^1	R^2	R^3	R^4	
Et	Me	Et	$ClCH_2$	(27), —
Et	Me	*i*-Pr	$ClCH_2$	(26), —
Et	Me	*i*-Pr	F_3C	(28), —
Et	Me	*i*-Pr	p-$O_2NC_6H_4CH_2$	(23), —
t-Bu	Me	*i*-Pr	$BrCH_2$	(49), —
t-Bu	Me	*i*-Pr	FCH_2	(39), —
t-Bu	Me	*i*-Pr	ICH_2	(63), —
t-Bu	Me	*i*-Pr	Cl_2CH	(58), —
t-Bu	Me	*i*-Pr	F_3C	(38), —
t-Bu	Me	*i*-Pr	p-$FC_6H_4CH_2$	(22), —
t-Bu	Me	*i*-Pr	p-$O_2NC_6H_4CH_2$	(23), —
t-Bu	OMe	*i*-Pr	ICH_2	(50), —
t-Bu	OMe	*i*-Pr	Cl_2CH_2	(53), —

Isocyanide	Carbonyl Compound	Carboxylic Acid	Conditions	Product(s), Yield(s) (%), dr	Refs.
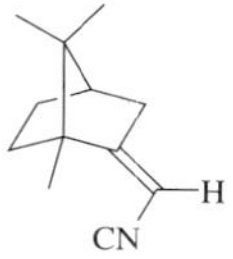	RCHO	AcOH	THF, rt, 40 h	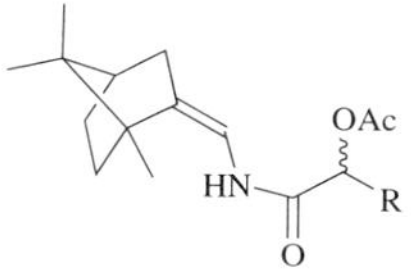	24

R	
Me	(96), 96:4
Et	(94), 96:4
i-Pr	(84), 96:4
t-Bu	(89), 96:4

TABLE 5. α-ACYLOXY AMIDES FROM PROCHIRAL CARBONYL COMPOUNDS, ISOCYANIDES, AND CHIRAL CARBOXYLIC ACIDS

Carboxylic Acid	Isocyanide and Carbonyl Compound		Conditions	Product(s), Yield(s) (%), dr	Refs.
C_3 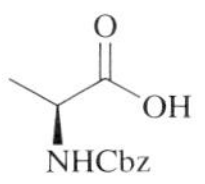O OH NHCbz	t-BuNC	RCHO	THF, rt, 2h CH_2Cl_2, rt, 2 h CH_2Cl_2, rt, 4 h	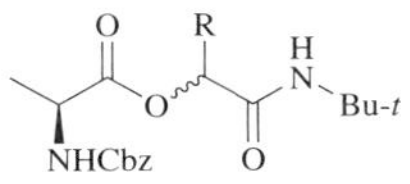O R H N Bu-t NHCbz O R: $ClCH_2$ (75), —; Cl_3C (73), —; Br_3C (95), —	166
C_4 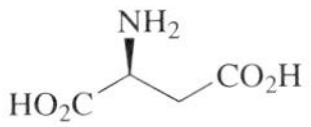NH2 HO2C CO2H	t-BuNC	i-PrCHO	MeOH, rt	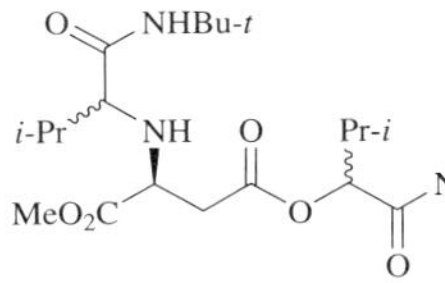O NHBu-t i-Pr NH O Pr-i NHBu-t MeO2C O O (85), 50:27:13:10	86
C_5 CO2H	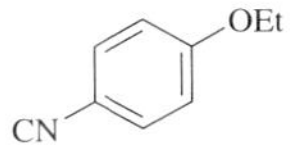OEt CN	PhCHO	Neat, rt, 42 h	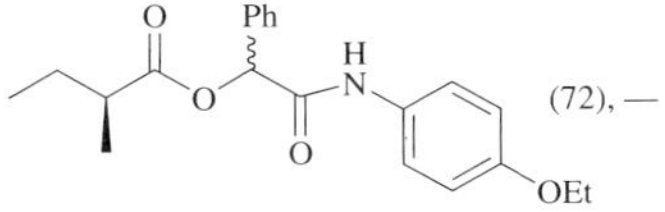O Ph H N O O OEt (72), —	18, 83
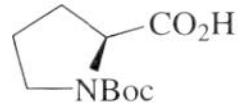CO2H NBoc	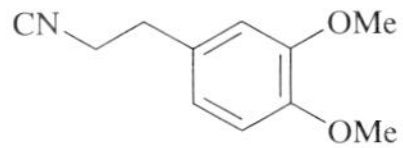CN OMe OMe	PhCHO	MeCN, rt, 16 h	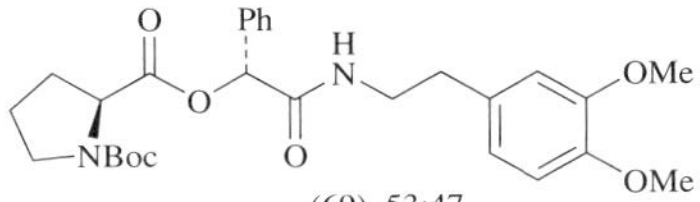O Ph H N OMe O NBoc O OMe (69), 53:47	168
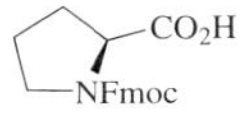CO2H NFmoc	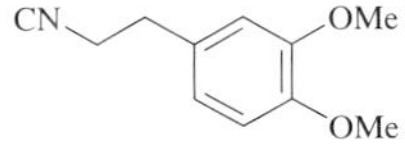CN OMe OMe	PhCHO	MeCN, rt, 16 h	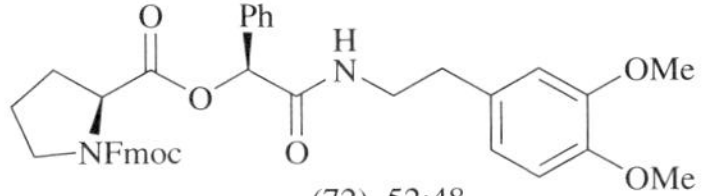O Ph H N OMe O NFmoc O OMe (72), 52:48	168
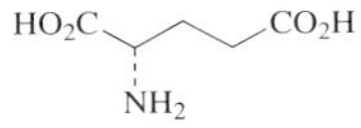HO2C CO2H NH2	t-BuNC	i-PrCHO	MeOH, rt	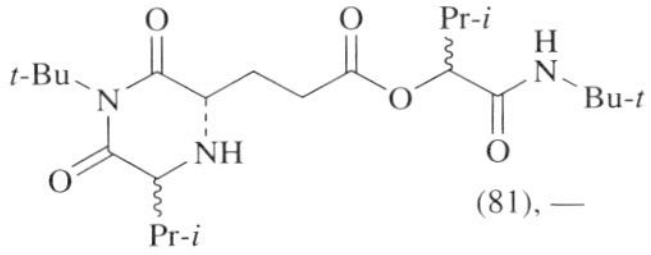O O Pr-i H t-Bu N N Bu-t O NH O O Pr-i (81), —	86

TABLE 5. α-ACYLOXY AMIDES FROM PROCHIRAL CARBONYL COMPOUNDS, ISOCYANIDES, AND CHIRAL CARBOXYLIC ACIDS *(Continued)*

Carboxylic Acid	Isocyanide and Carbonyl Compound		Conditions	Product(s), Yield(s) (%), dr	Refs.
C_6					
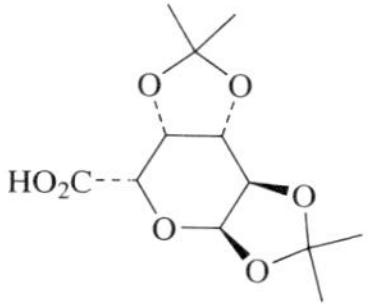	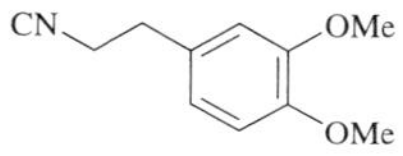	PhCHO	MeCN, rt, 16 h	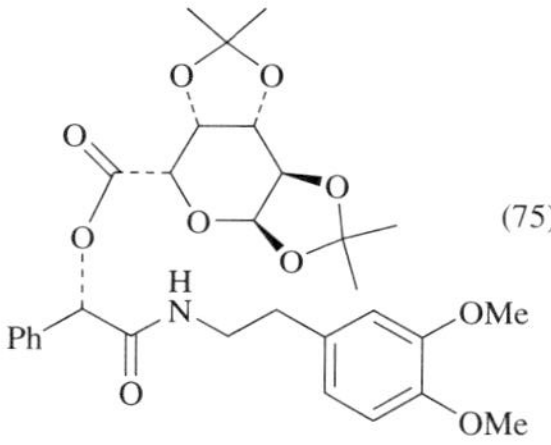 (75), 53:47	168
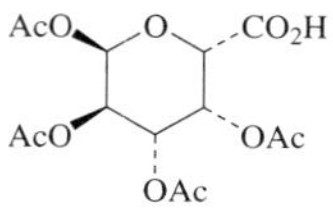	R^1NC	R^2CHO	MeCN, rt, 16 h	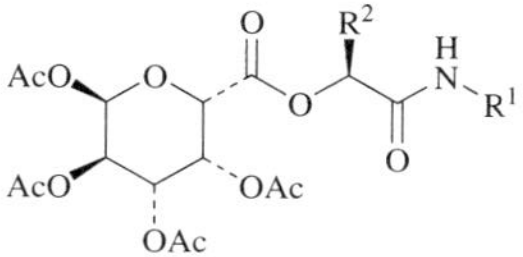	168
	R^1	R^2			
	3-MeO-4-(HC≡$CCH_2O)C_6H_3CH_2CH_2$	*p*-BrC_6H_4		(—), 98:2	
	3,4-$(MeO)_2C_6H_3CH_2CH_2$	*p*-BrC_6H_4		(—), 94:6	
	3-MeO-4-$HOC_6H_3CH_2CH_2$	*p*-ClC_6H_4		(—), 71:29	
	o-MeC_6H_4	*p*-BrC_6H_4		(—), 56:44	
	c-C_6H_{11}	1-naphthyl		(—), 91:9	
	Bn	*p*-(*i*-Pr)C_6H_4		(—), 90:10	
C_9					
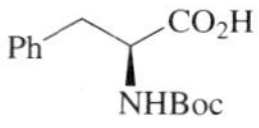	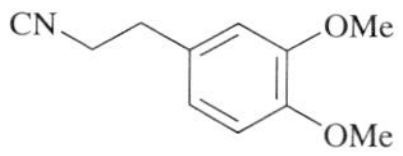	PhCHO	MeCN, rt, 16 h	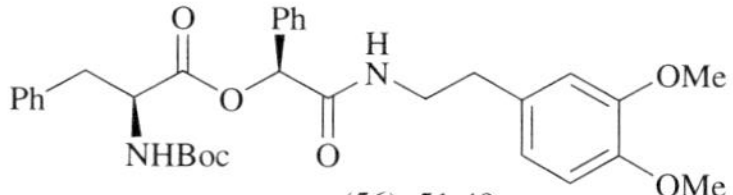 (56), 51:49	168
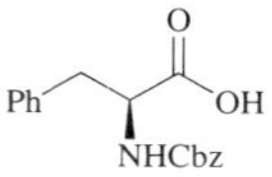	*t*-BuNC	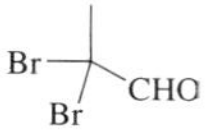	CH_2Cl_2, rt, 4 h	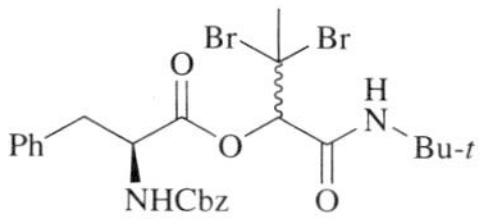 (95), —	166

C_{11}	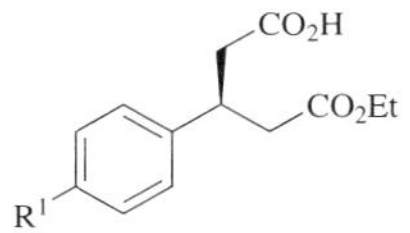	R^2NC	R^3CHO	i-Pr_2O, EtOH	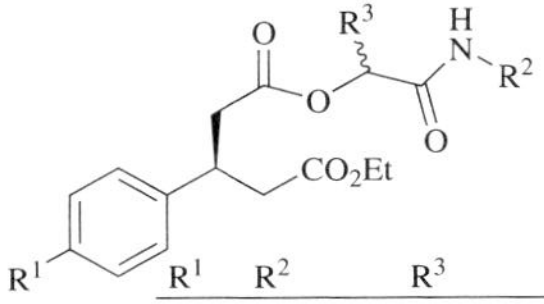	252

R^1	R^2	R^3	
H	Bn	Et	(41), 50:50
H	Bn	i-Bu	(41), 50:50
Cl	Bn	i-Bu	(35), 50:50
H	n-C_6H_{13}	i-Bu	(40), 50:50
H	Bn	n-C_7H_{15}	(36), 50:50
H	Bn	Ph	(18), 50:50

C_{12}	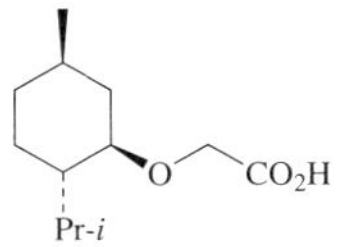	PhNC	R^1COR^2	Neat, –20°, 18 d	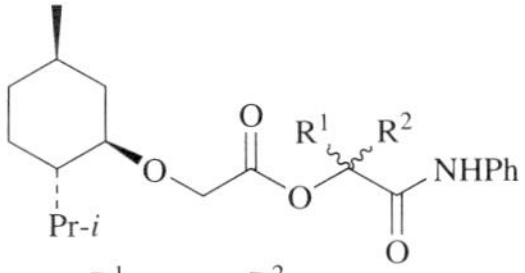	

R^1	R^2		
Et	Me	(>75), —	18
i-Pr	H	(98), —	18
n-Pr	Me	(>26), —	18
n-C_6H_{13}	Me	(19), 80:20	83
Ph	H	(56), —	18,83

C_{16}	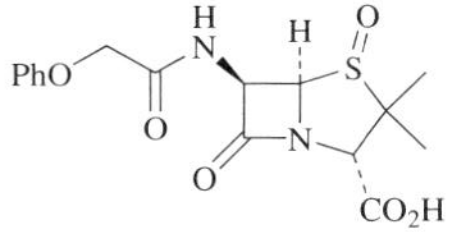	t-BuNC	RCHO		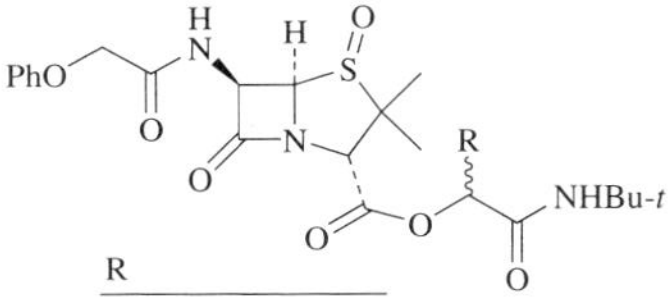	167

	R	
THF, rt, 2 h	$ClCH_2$	(58), —
CH_2Cl_2, THF, rt, 4 h	Cl_3C	(99), —
CH_2Cl_2, THF, rt, 4 h	Br_3C	(85), —
CH_2Cl_2, THF, rt, 4 h	CH_3CBr_2	(95), —

TABLE 5. α-ACYLOXY AMIDES FROM PROCHIRAL CARBONYL COMPOUNDS, ISOCYANIDES, AND CHIRAL CARBOXYLIC ACIDS *(Continued)*

Carboxylic Acid	Isocyanide and Carbonyl Compound		Conditions	Product(s), Yield(s) (%), dr	Refs.
C_n					
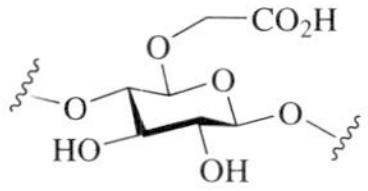	c-$C_6H_{11}NC$		H_2O, rt	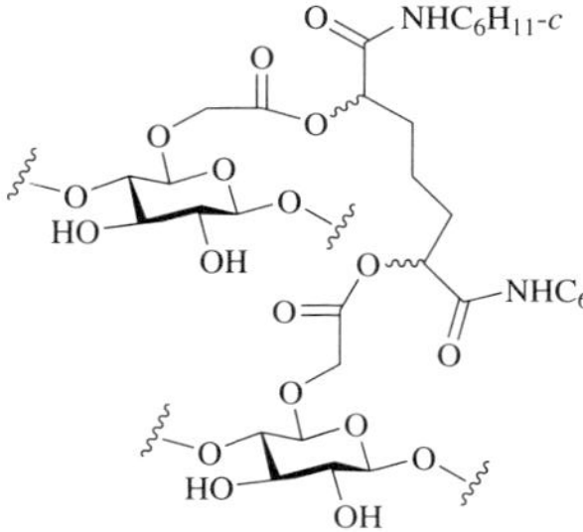 (—), —	59
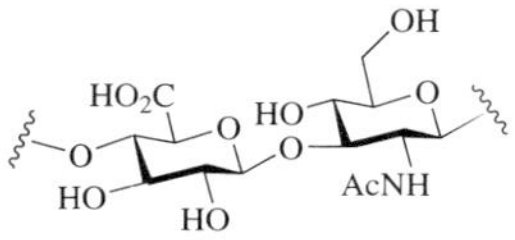	c-$C_6H_{11}NC$		H_2O, rt	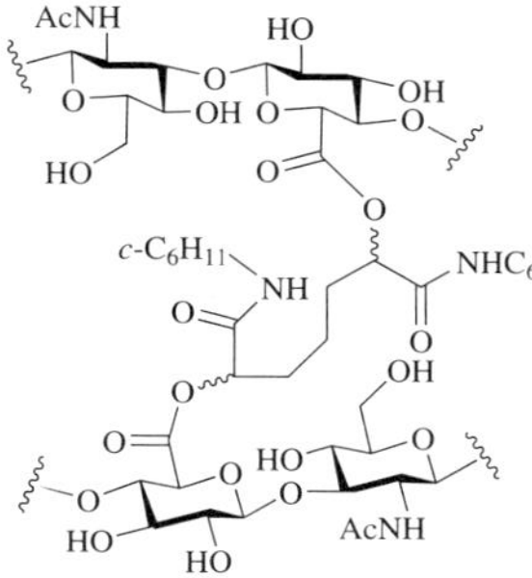 (—), —	59

TABLE 6. α-ACYLOXY AMIDES OR OTHER PRODUCTS FROM INTRAMOLECULAR PASSERINI REACTIONS

	Bifunctional Component	Other Component	Conditions	Product(s) and Yield(s) (%)	Refs.
C_5	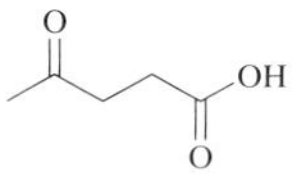O OH O	PhNC	Et_2O, rt, 10 d	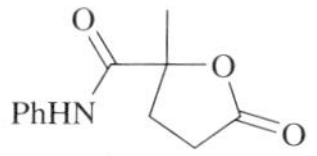O PhHN O O (66)	88
C_7	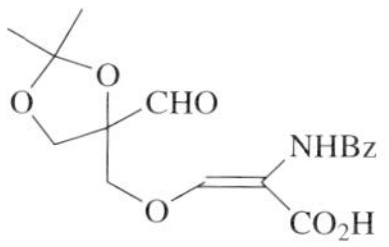O O CHO NHBz O CO2H	c-$C_6H_{11}NC$	MeOH-H_2O, rt, 20 h	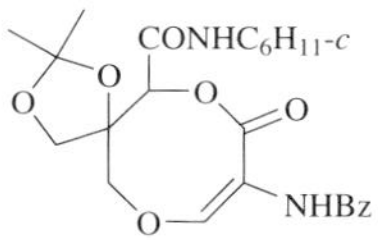CONHC6H11-c O O O O O NHBz (36)	90
C_8	NC CO2H	O	—	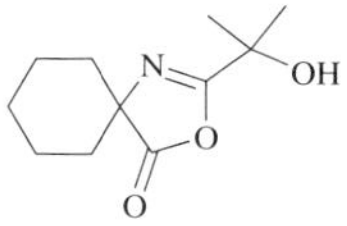N OH O O (—)	9
	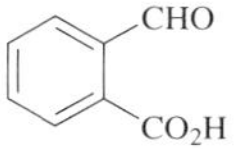CHO CO2H	PhNC	$CHCl_3$, rt, 4 d	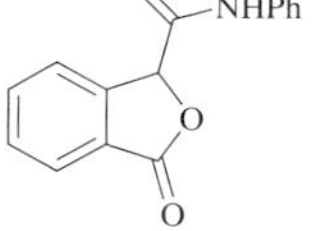O NHPh O O (—)	87
		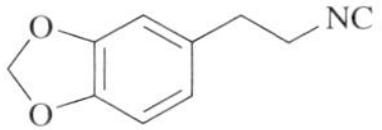O O NC	CH_2Cl_2, 3 h	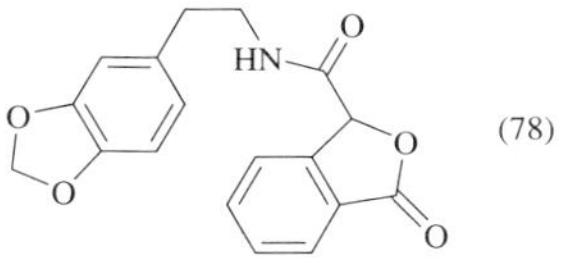HN O O O O O (78)	81
C_9	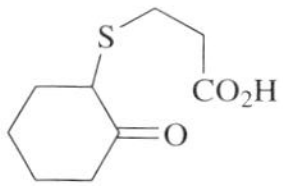S CO2H O	c-$C_6H_{11}NC$	MeOH, n-Bu_3N, rt, 24 h	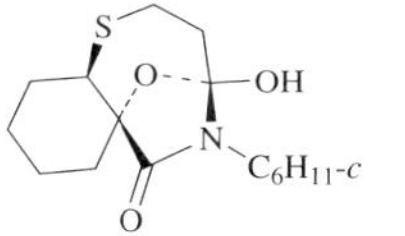S O OH N C6H11-c O (17)	89

TABLE 6. α-ACYLOXY AMIDES OR OTHER PRODUCTS FROM INTRAMOLECULAR PASSERINI REACTIONS (*Continued*)

Bifunctional Component	Other Component	Conditions	Product(s) and Yield(s) (%)	Refs.
C_{10}				
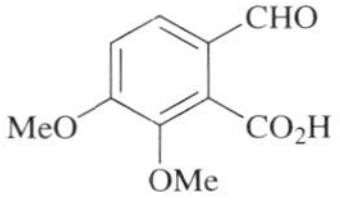	PhNC	1. EtOH, rt 2. KOH (aq)	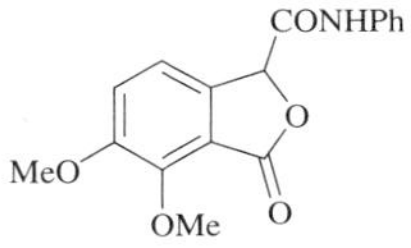 (—)	87
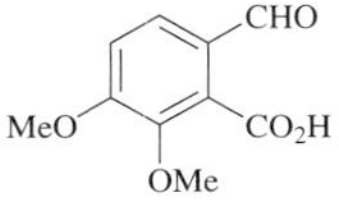	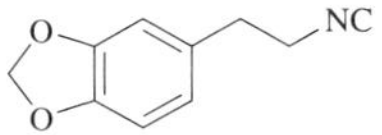	CH_2Cl_2, 3 h	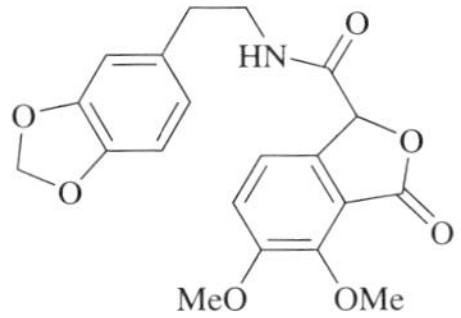 (71)	81
C_{13}				
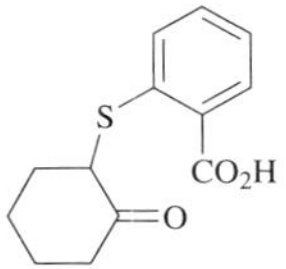	RNC	MeOH, n-Bu_3N, rt, 24 h	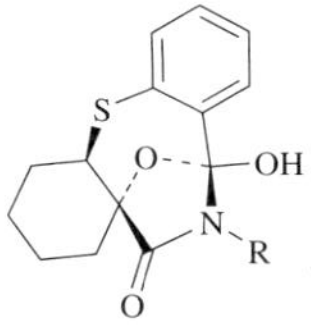 R n-Bu (87) c-C_6H_{11} (91) p-$O_2NC_6H_4CH_2$ (73)	89
Cn				
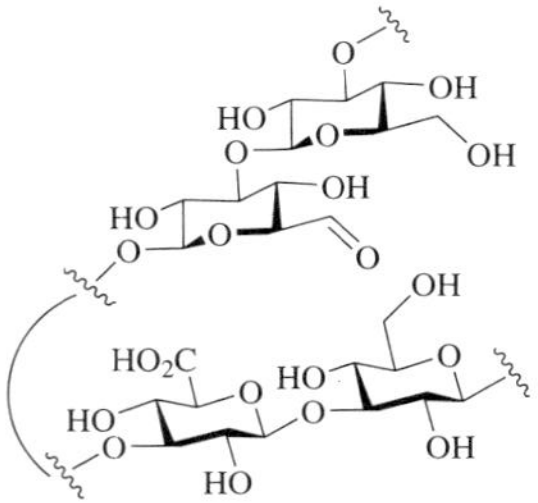	c-$C_6H_{11}NC$	H_2O, rt	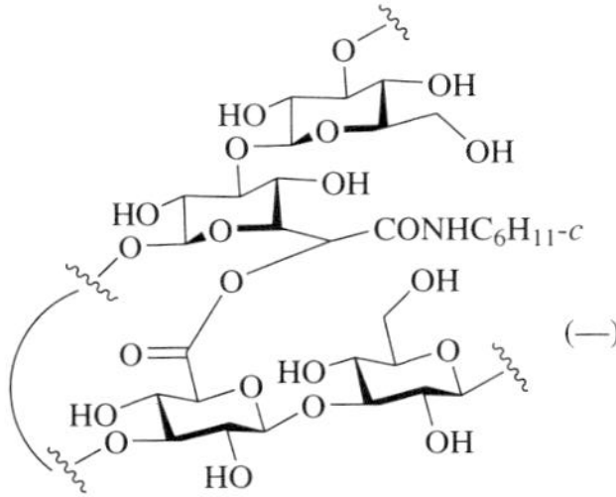 (—)	59

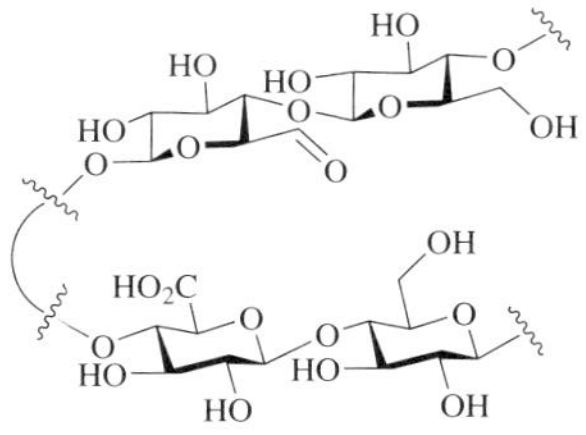	c-$C_6H_{11}NC$	H_2O, rt	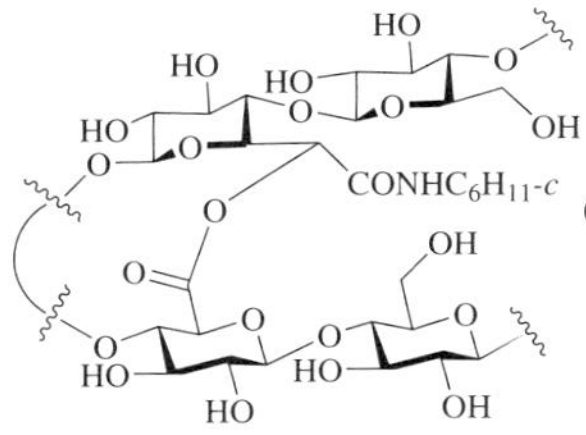	(—)	59

TABLE 7. α-HYDROXY AMIDES FROM ISOCYANIDES, CARBONYL COMPOUNDS, AND ACID SPECIES

Isocyanide	Carbonyl Compound	Conditions	Product(s), Yield(s) (%), dr	Refs.
C_2				
MeNC	RCHO	1. $TiCl_4$, CH_2Cl_2, –60° to T_1 2. HCl (aq), T_2	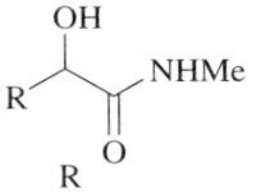	70
			R / T_1 / T_2	
			t-Bu / –20° / –20° / (65)	
			n-C_5H_{11} / 0° / 0° / (96)	
			Ph / 0° / 0° / (90)	
			p-$MeOC_6H_4$ / 0° / 0° / (73)	
			p-BrC_6H_4 / 0° / 0° / (95)	
			$PhCH_2$ / 0° / 0° / (86)	
			PhCH(Me) / 0° / 0° / (96), 50:50	
			(*E*)-PhCH=CH / 5° / 0° / (36)	
	RCOMe	1. $TiCl_4$, CH_2Cl_2, –60° to T_1 2. HCl (aq), 0°	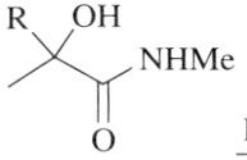	70
			R / T_1	
			Me / 0° / (84)	
			t-Bu / 0° / (17)	
			n-C_6H_{13} / 0° / (82)	
			Ph / rt / (82)	
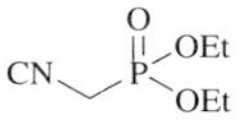	PhCHO	1. $TiCl_4$, CH_2Cl_2, 0° 2. H_2O	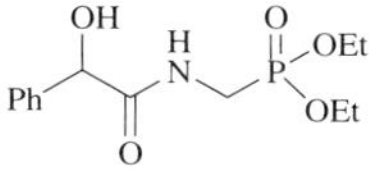 (95)	78
	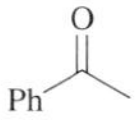	1. $TiCl_4$, toluene, rt 2. CH_2Cl_2, H_2O, rt, 30 min	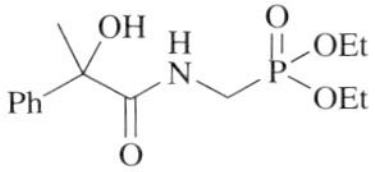 (43)	27
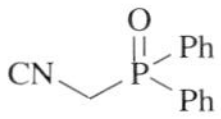	Cl_3CCHO	—	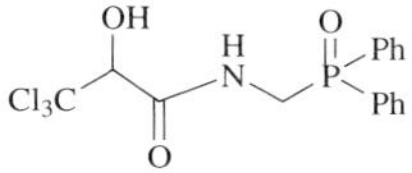 (41)	77

	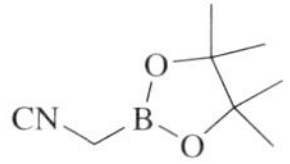	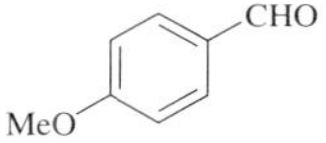	$TiCl_4$, CH_2Cl_2, 0° to rt	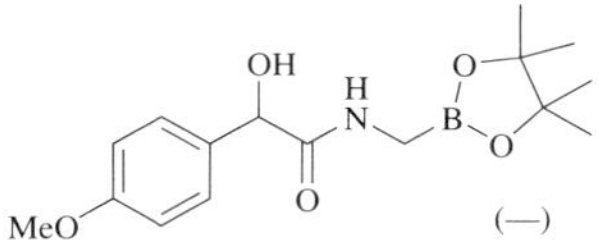 (—)	79
C_3		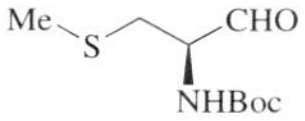	CF_3CO_2H, pyridine, CH_2Cl_2, rt, 12-72 h	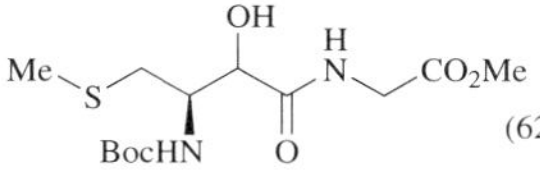 (62), —	116
		RCHO	1. $TiCl_4$, CH_2Cl_2, 0° 2. H_2O	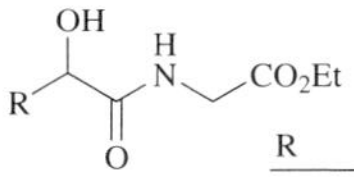 R: *n*-Pr (96); *t*-Bu (76); Ph (44); *p*-BrC_6H_4 (90); *p*-$MeOC_6H_4$ (90)	78
		PhCHO	$SiCl_4$, pyridine *N*-oxide, CH_2Cl_2, –74°	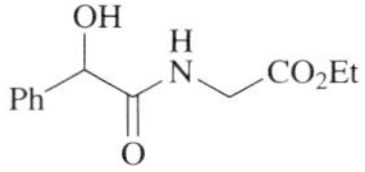 (72)	170
		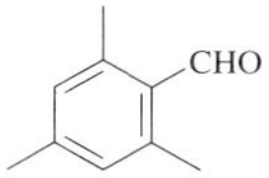	1. $TiCl_4$, toluene, rt, 6 h 2. CH_2Cl_2, H_2O, rt, 20 min	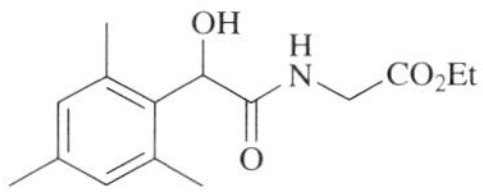 (72)	27
			CF_3CO_2H, pyridine, CH_2Cl_2, rt, 12-72 h	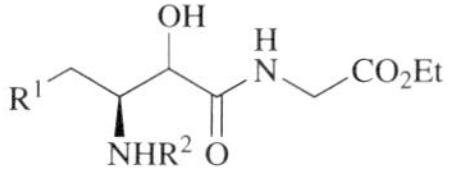	

R^1	R^2		
$O_2NNHC(NH)$-$NHCH_2CH_2$	Boc	(38), 50:50	116
p-(*t*-BuO)C_6H_4	Fmoc	(69), 50:50	116, 117

TABLE 7. α-HYDROXY AMIDES FROM ISOCYANIDES, CARBONYL COMPOUNDS, AND ACID SPECIES (*Continued*)

Isocyanide	Carbonyl Compound	Conditions	Product(s), Yield(s) (%), dr	Refs.
C_3				
	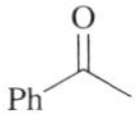	1. $TiCl_4$, CH_2Cl_2, 0°, 17.5 h 2. H_2O	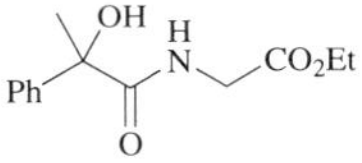 (81)	27, 78
	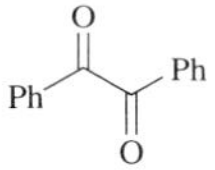	1. $TiCl_4$, toluene, rt 2. CH_2Cl_2, H_2O, rt, 30 min	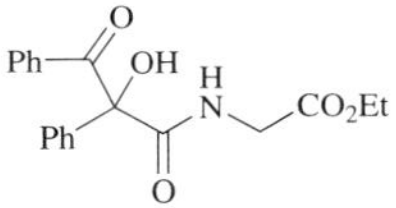 (47)	27
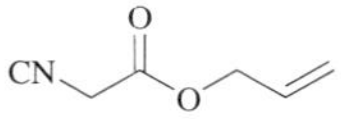	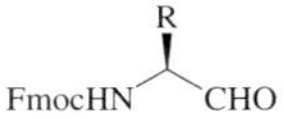	CF_3CO_2H, pyridine, CH_2Cl_2, rt, 12-72 h	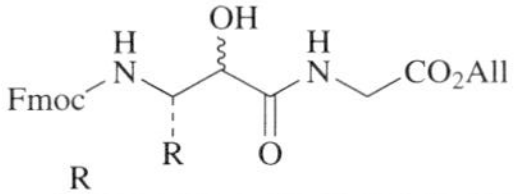	116
			R	
			H (77)	
			Me (83), —	
			t-$BuOCH_2$ (68), —	
			Et (73), —	
			(*R*)-*t*-BuOCH(Me) (62), —	
			(*S*)-*t*-BuOCH(Me) (74), —	
			t-BuO_2CCH_2 (60), —	
			n-Pr (73), —	
			i-Pr (68), —	
			n-Bu (69), —	
			i-Bu (85), —	
			PmcNH-C(=NH)NH$(CH_2)_3$ (76), —	
			BocNH$(CH_2)_4$ (79), —	
			$PhCH_2$ (67), —	
			t-$BuOC_6H_4CH_2$ (66), —	
	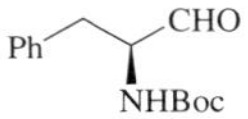	CF_3CO_2H, pyridine, CH_2Cl_2, rt, 12-72 h	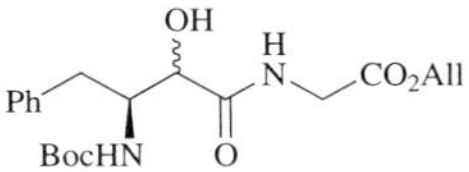 (67), —	116

		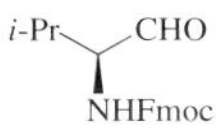	CF_3CO_2H, pyridine, CH_2Cl_2, rt, 12-72 h	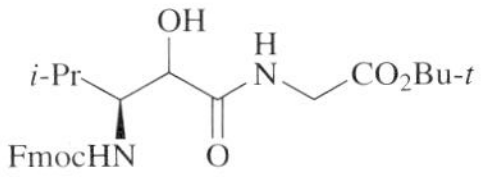 (68), —	116
C_4					
	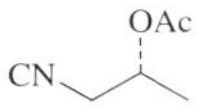	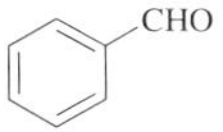	1. $TiCl_4$, CH_2Cl_2, 0° 2. H_2O	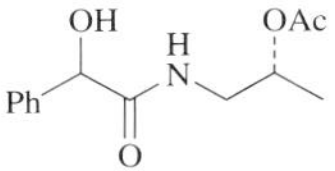 (55), —	78
		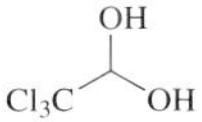	Et_2O, rt, 7 d	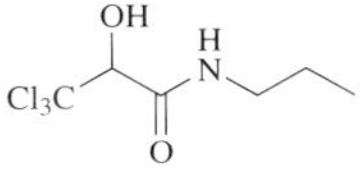 (—)	2
C_5					
	t-BuNC	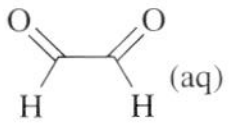	H_2SO_4, 18°, 10 h	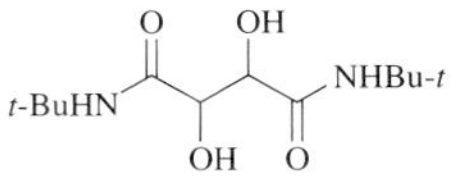 (35), —	61
		RCHO		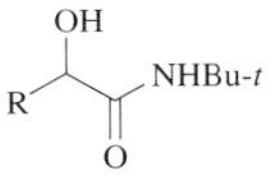	
		R	H_2SO_4, H_2O, 0°, 3 h		
		Et	BF_3•Et_2O, PE - Et_2O, 0°, 1 h	(47)	111
		Et	H_2SO_4, H_2O, 0°, 3 h	(10)	111
		n-Pr	BF_3•Et_2O, PE - Et_2O, 0°, 1 h	(50)	111
		n-Pr	H_2SO_4, H_2O, 0°, 3 h	(10)	111
		i-Pr	H_2SO_4, H_2O	(70-80)	112
		s-Bu	CF_3CO_2H, $CHCl_3$, –5° to rt	(53)	112
		2-pyridyl	$SiCl_4$, CH_2Cl_2, –74°	(32)	115
		Ph	$SiCl_4$, HMPA, CH_2Cl_2, –74°	(79)	170
		Ph	$SiCl_4$, pyridine N-oxide, CH_2Cl_2, –74°	(90)	170
		Ph	CF_3CO_2H, pyridine, CH_2Cl_2, rt, 2 h	(94)	170
		p-$TfOC_6H_4$	CF_3CO_2H, pyridine, CH_2Cl_2, rt, 2 h	(69)	115
		3,4-$Cl_2C_6H_3$	CF_3CO_2H, pyridine, CH_2Cl_2, rt, 2 h	(52)	115
		$PhOCH_2$		(69)	115

TABLE 7. α-HYDROXY AMIDES FROM ISOCYANIDES, CARBONYL COMPOUNDS, AND ACID SPECIES (*Continued*)

Isocyanide	Carbonyl Compound	Conditions	Product(s), Yield(s) (%), dr	Refs.
C_5				
t-BuNC	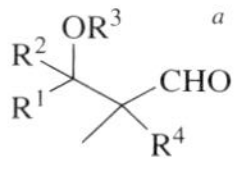 [a]	1. $TiCl_4$, CH_2Cl_2, –60° 2. Na_2CO_3 (aq)	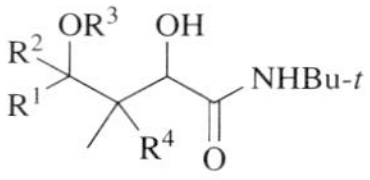	134
			R[1] / R[2] / R[3] / R[4]: H / H / Me / Me (54) Me / H / Et / H (50) Me / H / Et / Me (61) Me / Me / Me / Me (22) *n*-Pr / H / Me / Me (55) C_6H_{13} / H / Me / Me (70) Ph / H / Me / Me (40)	
	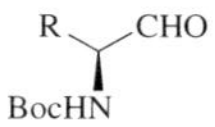	CF_3CO_2H, base, CH_2Cl_2, rt, 12-72 h	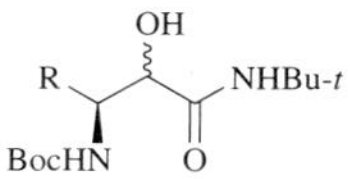	116
	R	**Base**		
	$O_2NNHC(=NH)NH(CH_2)_3$	pyridine	(92), —	
	$PhCH_2$	pyridine	(24), —	
	$PhCH_2$	2,4,6-collidine	(78), —	
	c-$C_6H_{11}CH_2$	pyridine	(46), —	

Isocyanide	Carbonyl compound	Conditions	Product(s) and Yield(s) (%)	Refs.
	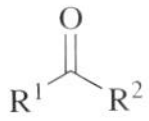		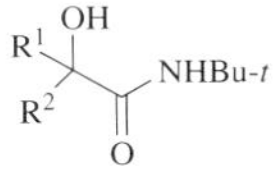	
	R^1 / R^2			
	Me / Me	HCl, 0°	(72)	20
	Me / Et	H_2SO_4, H_2O	(38)	112
	Me / *i*-Pr	$BF_3•Et_2O$ (stoich.), Et_2O, 0°	(5)	112
	Me / *i*-Pr	H_2SO_4, H_2O	(11)	112
	CF_3 / CF_3COCH_2	H_2O, rt, 3 h	(83)	104
	i-Pr / *i*-Pr	$BF_3•Et_2O$ (stoich.), Et_2O, 0°	(<5)	112
	i-Pr / *i*-Pr	H_2SO_4, H_2O	(1)	112
	Me / $PhOCH_2$	CF_3CO_2H, pyridine, CH_2Cl_2, rt, 2 h	(33)	115
	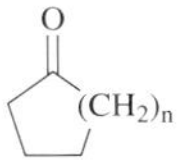	H_2SO_4, H_2O, 0°	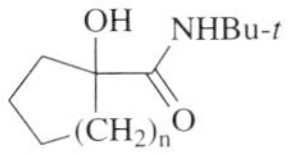	111
			n / Time	
			1 / 3 h (70)	
			2 / 3 h (75)	
			3 / 3 h (20)	
			4 / 3 h (10)	
			8 / 2 d (4)	
	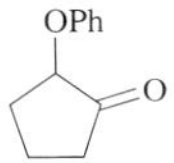	CF_3CO_2H, pyridine, CH_2Cl_2, rt, 2 h	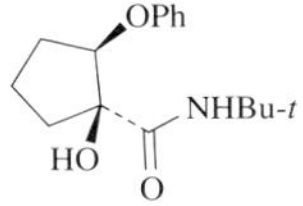 (33), 100:0	115
$(t\text{-BuNC})_2CuBr$		HBr, 30°/40°, 2.5 h	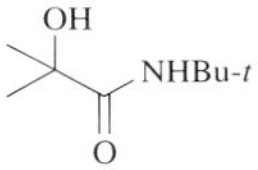 (67)	110

TABLE 7. α-HYDROXY AMIDES FROM ISOCYANIDES, CARBONYL COMPOUNDS, AND ACID SPECIES (*Continued*)

Isocyanide	Carbonyl Compound	Conditions	Product(s), Yield(s) (%), dr	Refs.
C_7				
c-$C_6H_{11}NC$	RCHO		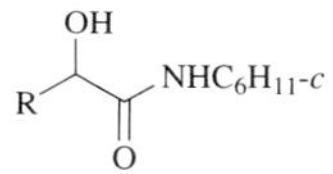	
	R	Acid		
	Et	HCl, Me_2NNH_2 MeOH, rt, 70 h	(22)	106
	n-Pr	HCl MeOH, rt, 48 h	(42)	106, 107
	n-Pr	4-Methylpyrazolidine MeOH, rt, 48 h	(50)	106, 107
	i-Pr	HCl acetone, 0°	(88)	20
	i-Pr	Me_3N•HCl MeOH, rt	(35)	84
	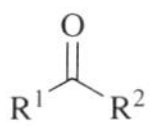		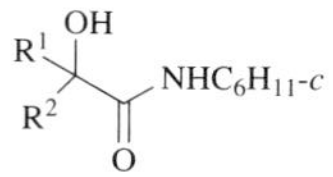	
	R^1 R^2			
	Me Me	HCl, neat, 0°	(71)	20
	CF_3 CF_3COCH_2	H_2O, rt, 3 h	(76)	104
	—$(CH_2)_5$—	H_2SO_4, H_2O, 0°, 3 h	(60)	111
	—$(CH_2)_5$—	HCl, Me_2NNHMe, MeOH-H_2O, rt, 10 d	(40)	107
	CF_3 Ph	1. $TiCl_4$, CH_2Cl_2, 0°, 3 h 2. aq HCl, rt, 1 h	(46)	127
PhNC	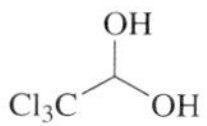	Et_2O, rt, 2 d	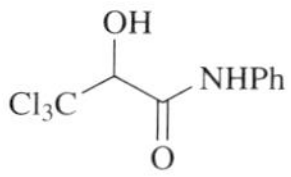 (36)	1
		1. $TiCl_4$, CH_2Cl_2, 0° 2. H_2O	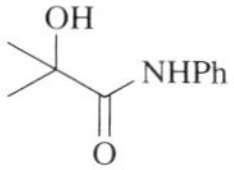 (88)	78
	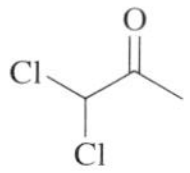	H_2O, rt, 28 d	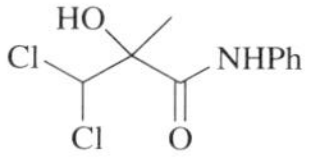 (13)	3

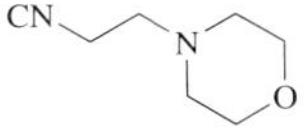	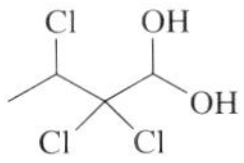	Et_2O, rt, 4-5 d	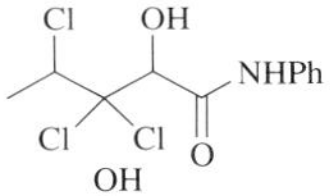 (53)	1
	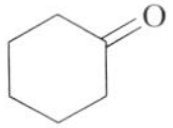	1. $TiCl_4$, CH_2Cl_2, 0° 2. H_2O	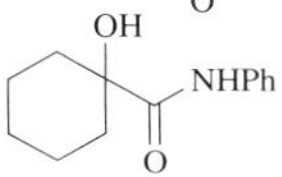 (59)	78
	PhCHO	1. $TiCl_4$, CH_2Cl_2, 0° 2. H_2O	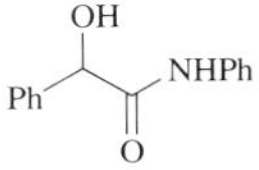 (98) **I**	78
	PhCHO	$SiCl_4$, pyridine *N*-oxide, CH_2Cl_2, –74°	**I** (76)	170
	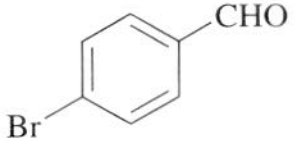	1. $TiCl_4$, CH_2Cl_2, 0° 2. H_2O	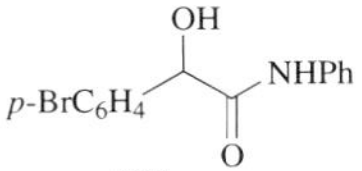 (98)	78
	BnCHO	1. $TiCl_4$, CH_2Cl_2, 0° 2. H_2O	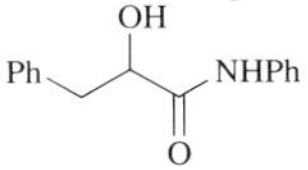 (98)	78
	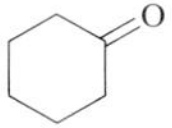	$Zn(OTf)_2$, TMSCl, CH_2Cl_2, rt, 24 h	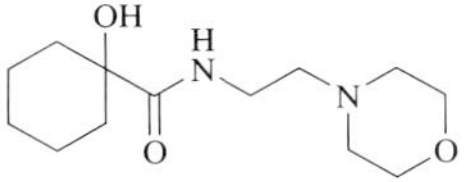 (57)	36
	PhCHO	1. $TiCl_4$, CH_2Cl_2, –70°/rt, 24 h 2. H_2O	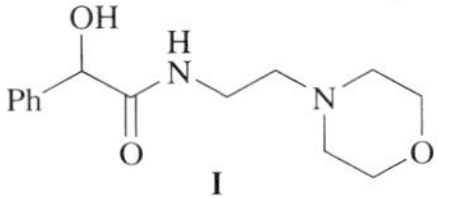 (67) **I**	78
	PhCHO	$Zn(OTf)_2$, TMSCl, CH_2Cl_2, rt, 24 h	**I** (77)	36
		1. $TiCl_4$, CH_2Cl_2, 0°, 24 h 2. H_2O	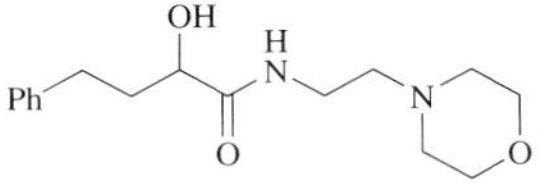 (57) **I**	78
		$Zn(OTf)_2$, TMSCl, CH_2Cl_2, rt, 24 h	**I** (74)	36

TABLE 7. α-HYDROXY AMIDES FROM ISOCYANIDES, CARBONYL COMPOUNDS, AND ACID SPECIES (*Continued*)

Isocyanide	Carbonyl Compound	Conditions	Product(s), Yield(s) (%), dr	Refs.
C_7 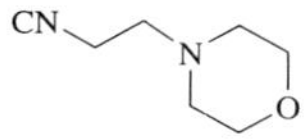	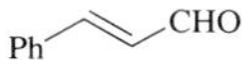	$Zn(OTf)_2$, Me_3SiCl, CH_2Cl_2, rt, 24 h	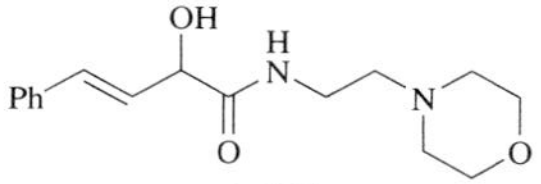 **I** (61)	36
	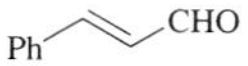	Me_3SiOTf, CH_2Cl_2, rt, 24 h	**I** (73)	36
	n-$C_9H_{19}CHO$	$Zn(OTf)_2$, Me_3SiCl, CH_2Cl_2, rt, 24 h	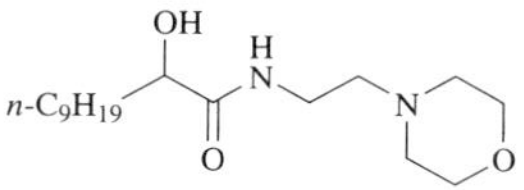 (60)	36
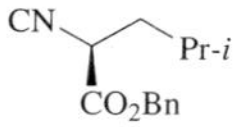	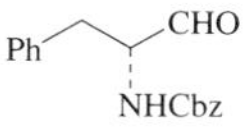	CF_3CO_2H, pyridine, CH_2Cl_2, rt, 12-72 h	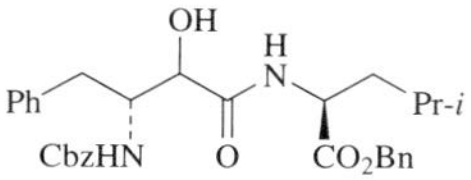 (65), 60:40	116
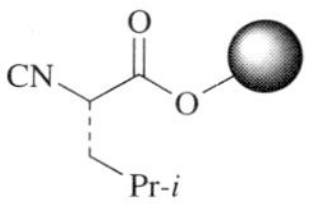		CF_3CO_2H, pyridine, THF 50°, 30 h	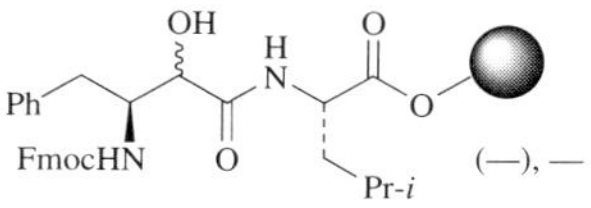 (—), —	162
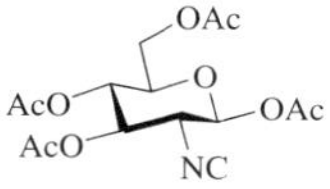		HCO_2H. $ZnCl_2$ (cat.), CH_2Cl_2, 0°, 3 h	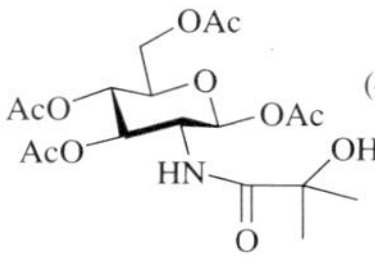 (44) + 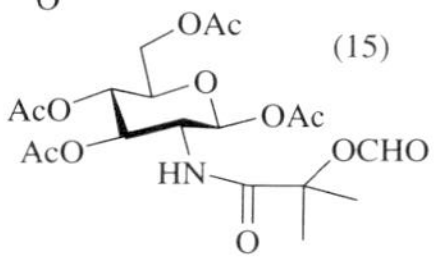 (15)	50

	Nitrile	Carbonyl Compound	Conditions	Product(s) and Yield(s) (%)	Refs.
C_8	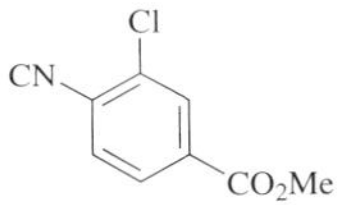	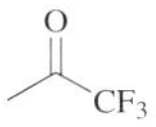	1. $TiCl_4$, CH_2Cl_2, 0°, 8 h 2. HCl (aq), rt, 1 h	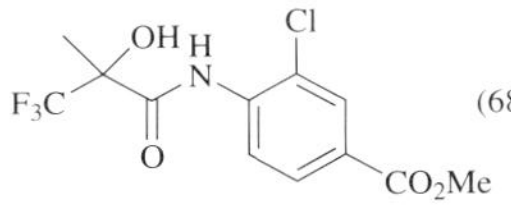 (68)	128
	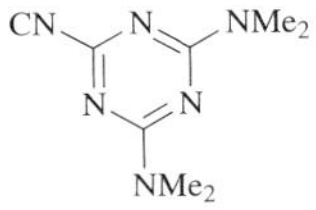		$PhSO_3H$, $CHCl_3$, acetone, rt, 2 h	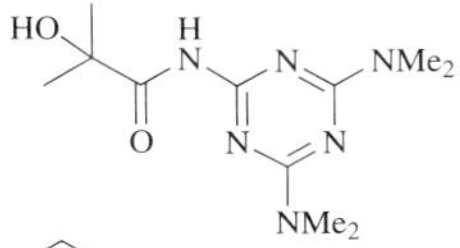 (35)	80
		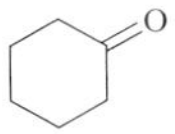	BF_3•Et_2O, $(ClCH_2)_2$, MeOH, –10°, 30 min	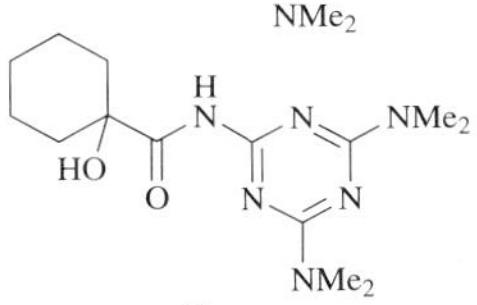 (30)	80
C_9	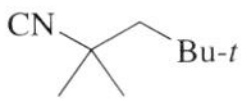	HCHO (aq)	H_2SO_4, 18°, 4 h	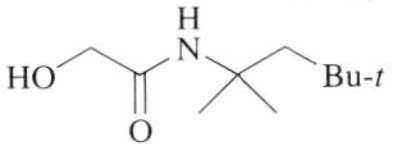 (55)	61
	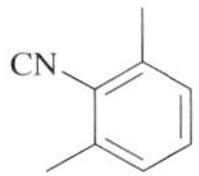	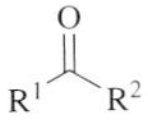	HCl, acetone, 0°	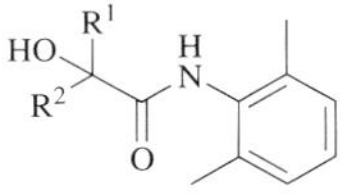 (89)	20
		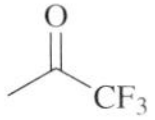	1. $TiCl_4$, CH_2Cl_2, 0°, 8 h 2. HCl (aq), rt, 1 h	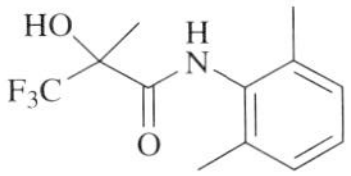 (23)	128
	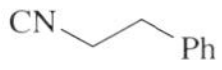	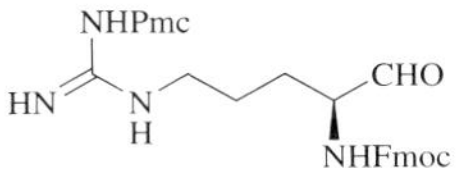	CF_3CO_2H, pyridine, CH_2Cl_2, rt, 12-72 h	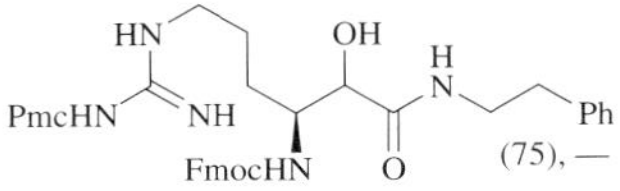 (75), —	116

R^1	R^2	
Me	Me	(89)
Me	Et	(63)
—$(CH_2)_6$—		(85)
Ph	H	(62)

TABLE 7. α-HYDROXY AMIDES FROM ISOCYANIDES, CARBONYL COMPOUNDS, AND ACID SPECIES (*Continued*)

Isocyanide	Carbonyl Compound	Conditions	Product(s), Yield(s) (%), dr	Refs.
C_9				
	PhCHO	$SiCl_4$, pyridine-*N*-oxide, CH_2Cl_2, −74°	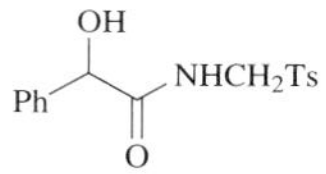 (69)	170
	PhCHO	1. $TiCl_4$, CH_2Cl_2, 0° 2. H_2O	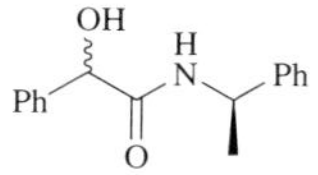 (15), 53:47	78
C_{10}				
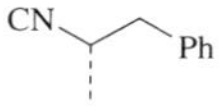	PhCHO	1. $TiCl_4$, CH_2Cl_2, 0° 2. H_2O	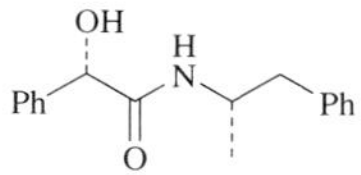 (47), 53:47	78
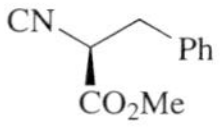	RCHO	1. $TiCl_4$, CH_2Cl_2 2. H_2O	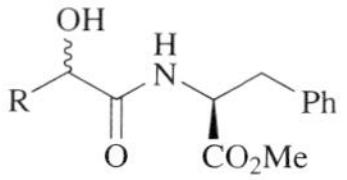	78
			R — Temp — Time	
			n-Pr — −25° — 20 h — (34), 50:50	
			Ph — 0° — 10 h — (66), 50:50	
			p-BrC_6H_4 — 0° — — — (38), 50:50	
			PhCH(Me) — 0° — — — (35), 25:25:25:25	
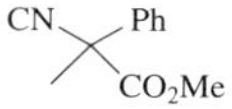	PhCHO	1. $TiCl_4$, CH_2Cl_2, −10° to rt, 24 h 2. H_2O	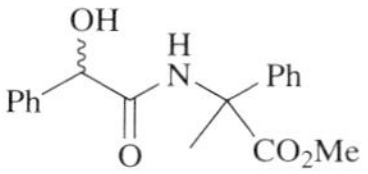 (>34), —	78
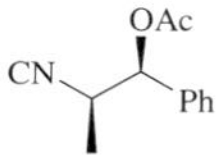	PhCHO	1. $TiCl_4$, CH_2Cl_2, 0° 2. H_2O	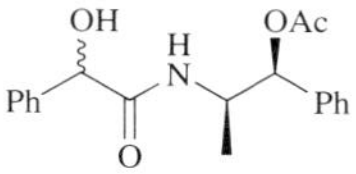 (>33), —	78

C_{11} 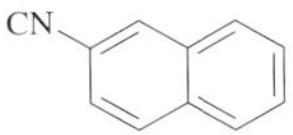CN	O	HCl, acetone, 0°	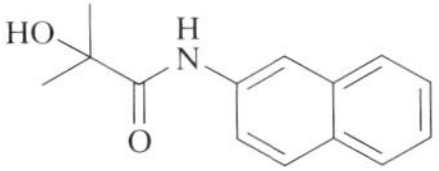HO H N O (84)	20
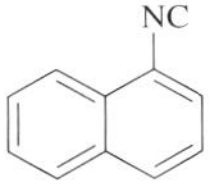NC	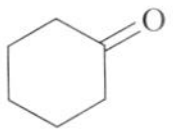O	$POCl_3$, Et_2O, 0°, 2 h	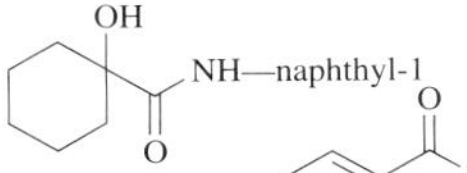OH NH—naphthyl-1 O (9) + 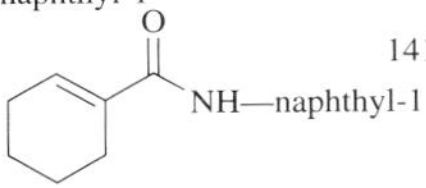O NH—naphthyl-1 (11)	141
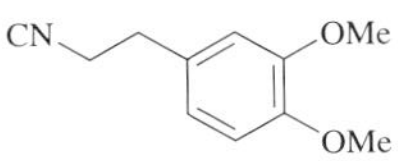CN OMe OMe	PhCHO	1. $TiCl_4$, CH_2Cl_2, –60° to rt, 15 h 2. H_2O	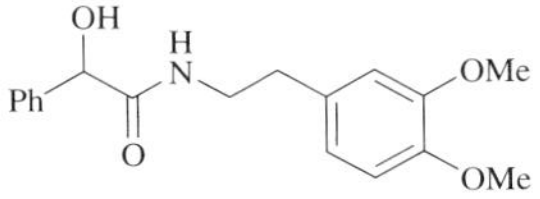OH H N Ph O OMe OMe (23)	78
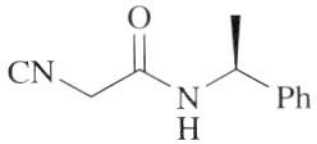O CN N H Ph	CHO	1. $TiCl_4$, CH_2Cl_2 2. H_2O	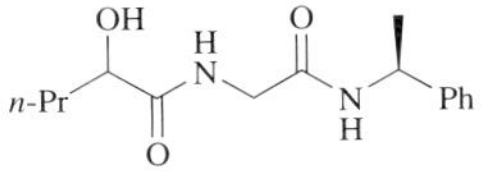OH H N O n-Pr O N H Ph (52), —	78
	PhCHO	1. $TiCl_4$, CH_2Cl_2, 0° 2. H_2O	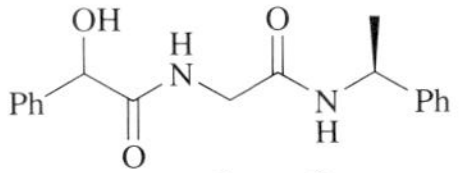OH H N O Ph O N H Ph (53), —	78
C_{12} 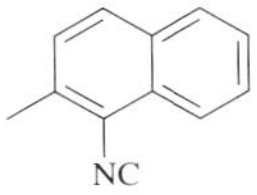NC	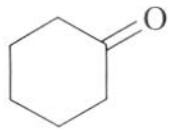O	H_2SO_4, H_2O, 0°	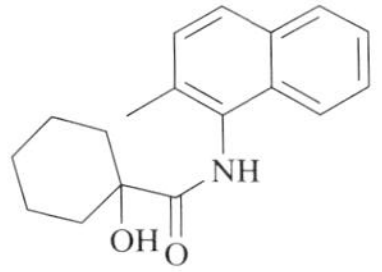NH OH O **I** (24)	114
	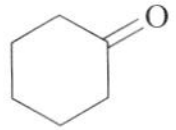O	BF_3•Et_2O, Et_2O, 0°, 1.5 h	**I** (3)	114
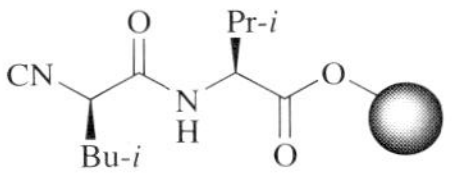O Pr-*i* CN N H O Bu-*i* O	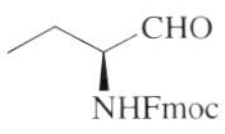CHO NHFmoc	CF_3CO_2H, pyridine, THF 50°, 30 h	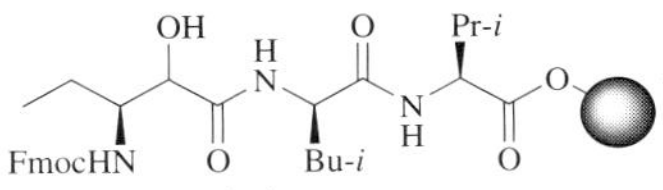OH H N O Pr-*i* O FmocHN O Bu-*i* N H O (—)	162

TABLE 7. α-HYDROXY AMIDES FROM ISOCYANIDES, CARBONYL COMPOUNDS, AND ACID SPECIES (*Continued*)

Isocyanide	Carbonyl Compound	Conditions	Product(s), Yield(s) (%), dr	Refs.
C_{14}				
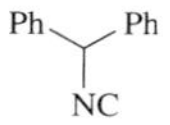		HCl, acetone, rt, 30 min	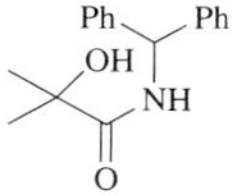 (85)	108
C_{15}				
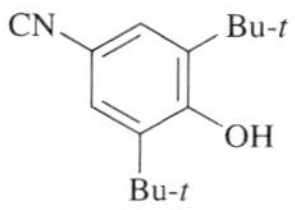	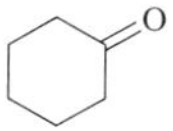	H_2SO_4, C_6H_6, rt, 2 h	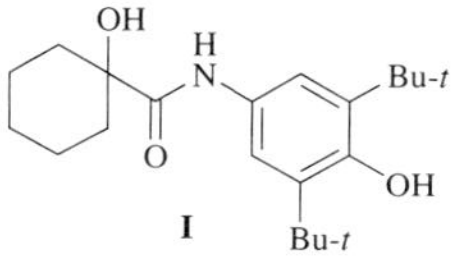 **I** (—)	113
	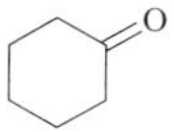	BF_3•Et_2O, Et_2O, 0°, 1.5 h	**I** (—)	113
C_{16}				
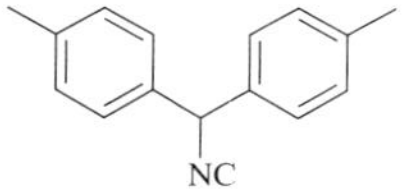		HCl, acetone, rt	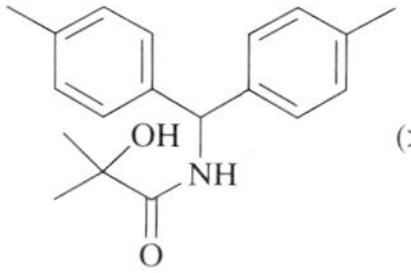 (>90)	109
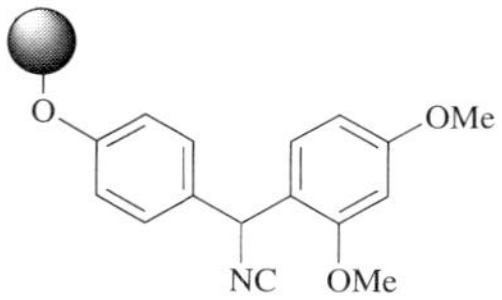		CF_3CO_2H, pyridine, THF 50°, 30 h	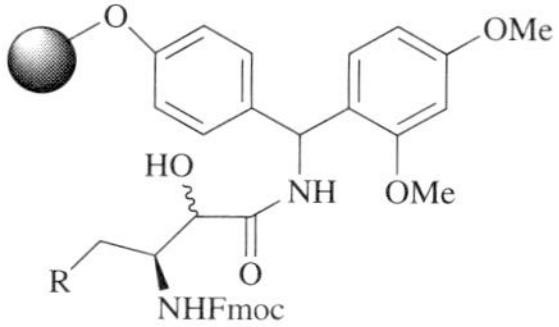	162

R	
H	(97), —
Ph	(—), —
c-C_6H_{11}	(—), —

C_{18}				
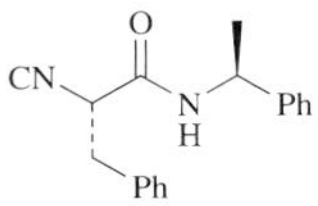	RCHO	1. $TiCl_4$, CH_2Cl_2 2. H_2O	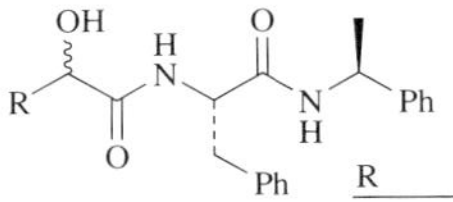 R *n*-Pr (16), — Ph (31), —	78
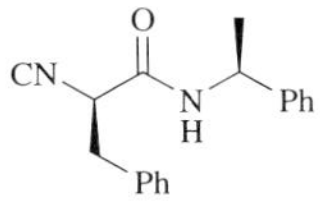	PhCHO	1. $TiCl_4$, CH_2Cl_2, 0° 2. H_2O	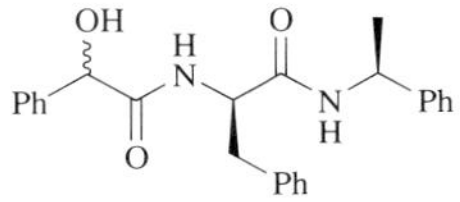 (11), —	78
C_{37}				
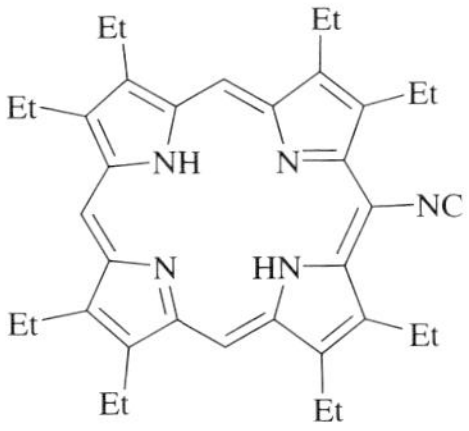	$(CH_2O)_n$	BF_3•Et_2O, toluene	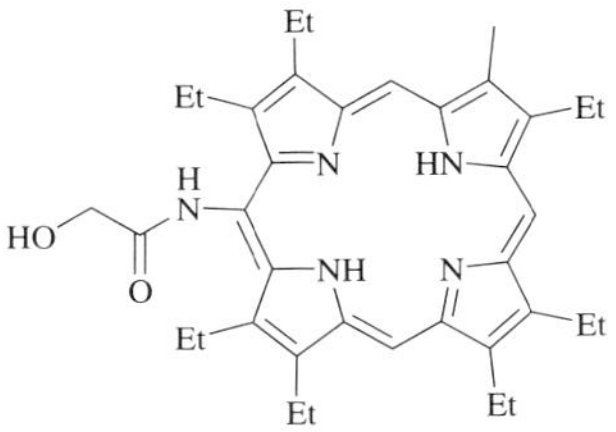 (28)	75

[a] The β-alkoxy aldehydes were formed in situ by reaction of an acetal with an enoxysilane. The Lewis acid was added to the preformed mixture of acetal, enoxysilane, and isocyanide.

TABLE 8. REACTIONS OF ISOCYANIDES WITH CARBONYL COMPOUNDS AND ACID SPECIES GIVING OTHER PRODUCTS

Isocyanide	Carbonyl Compound and Acid Species		Conditions	Product(s) and Yield(s) (%)	Refs.
C_3					
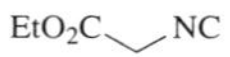	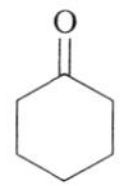	$Zn(OTf)_2$	TMSCl, NEM, CH_2Cl_2, rt	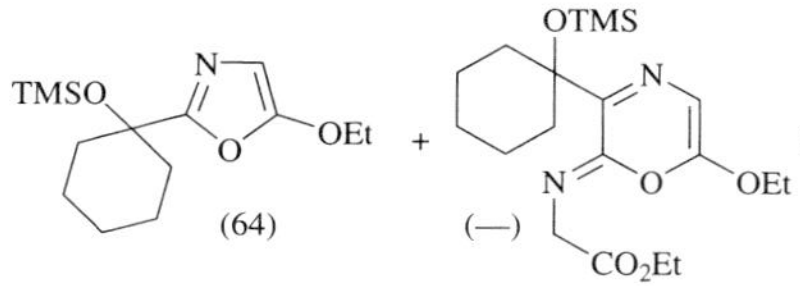 (64) + (—)	36
	RCHO	$Zn(OTf)_2$	TMSCl, NEM, CH_2Cl_2, rt	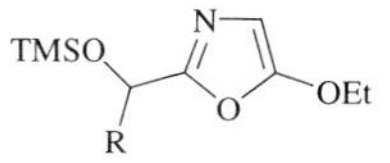 R Ph (53) $PhCH_2CH_2$ (58) (*E*)-PhCH=CH (54) n-C_9H_{19} (47)	36
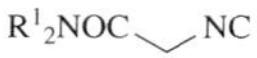	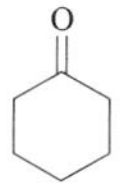	$Zn(OTf)_2$	R^2_3SiCl, NEM, CH_2Cl_2, rt	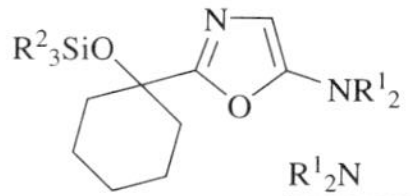 R^1_2N / R^2 Me_2N / Me (79) Me_2N / Et (77) morpholino / Et (80)	36a
	R^2CHO	$Zn(OTf)_2$	R^3_3SiCl, NEM, CH_2Cl_2, rt	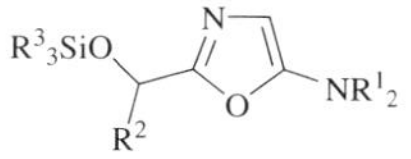 R^1_2N / R^2 / R^3 Me_2N / *t*-Bu / Me (84) Me_2N / Ph / Me (29) Me_2N / $BnCH_2$ / Me (74) Me_2N / $CH_3(CH_2)_8CH_2$ / Me (82) Me_2N / $CH_3(CH_2)_8CH_2$ / Et (72) morpholino / $BnCH_2$ / Et (73)	36a

C_4					
	i-PrCHO	$BF_3 \cdot Et_2O$ (cat.)	PE, Et_2O	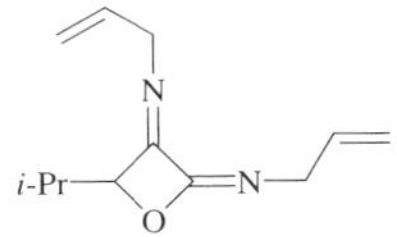 (53)	29
i-PrNC	RCHO	$BF_3 \cdot Et_2O$ (cat.)	PE, Et_2O	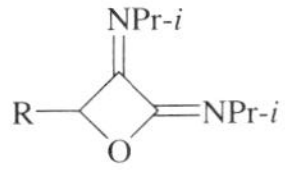 R: Me (73); Ph (63)	29
	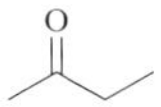	$BF_3 \cdot Et_2O$ (cat.)	Et_2O, –78°	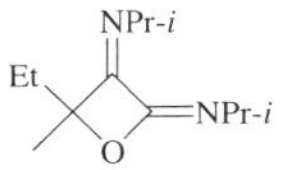 (—)	139
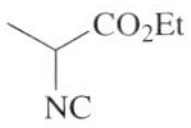	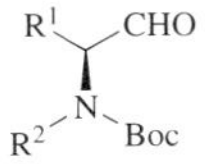	$TMSN_3$	CH_2Cl_2, rt, 18 h	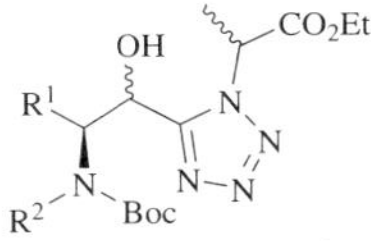	122
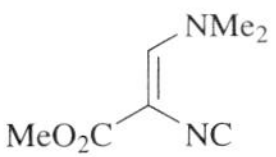	R^1CHO	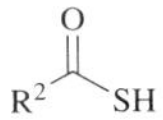	$BF_3 \cdot Et_2O$, THF, –78° to rt	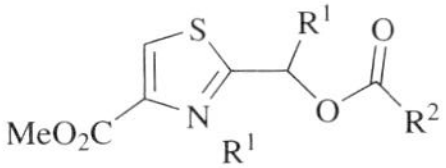	123

R^1	R^2	
Me	H	(77)
—$(CH_2)_3$—		(74)

R^1	R^2	
p-MeC_6H_4	Me	(12)
$BnCH_2$	Me	(32)
t-Bu	Me	(29)
c-C_6H_{11}	Me	(31)
n-C_5H_{11}	Me	(28)
i-Bu	Me	(31)
c-C_3H_5	Me	(35)
CH_2=$CHCH_2(Me_2)C$	Me	(15)
i-Bu	CF_3	(11)
p-$MeOC_6H_4$	Me	(10)
o-BrC_6H_4	Me	(9)
i-Pr	Ph	(19)
$MeSCH_2CH_2$	Me	(23)

TABLE 8. REACTIONS OF ISOCYANIDES WITH CARBONYL COMPOUNDS AND ACID SPECIES GIVING OTHER PRODUCTS (*Continued*)

Isocyanide	Carbonyl Compound and Acid Species			Conditions	Product(s) and Yield(s) (%)	Refs.
C_5						
R^1NC	R^2CHO				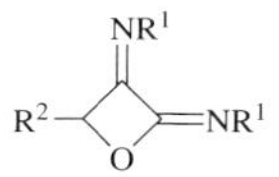	
R^1	R^2					
i-Bu	*i*-Pr		BF_3•Et_2O (cat.)	PE, Et_2O	(70)	29
t-Bu	H		BF_3•Et_2O (cat.)	PE, Et_2O	(40-45)	29
t-Bu	Me		BF_3•Et_2O (cat.)	PE, Et_2O	(91)	29
t-Bu	CCl_3		BF_3•Et_2O (cat.)	PE, Et_2O	(68)	29
t-Bu	Et		*m*-CPBA	CCl_4, rt, 24 h	(58)	119
t-Bu	*n*-Pr		BF_3•Et_2O (cat.)	PE, Et_2O	(73)	29
t-Bu	*i*-Pr		*m*-CPBA	CCl_4, rt, 24 h	(94)	119
t-Bu	*t*-Bu		*m*-CPBA	CCl_4, rt, 24 h	(77)	119
t-Bu	Bn		BF_3•Et_2O (cat.)	PE, Et_2O	(60)	29
t-Bu	Ph		BF_3•Et_2O (cat.)	PE, Et_2O	(74)	29
t-Bu	*p*-ClC_6H_4		BF_3•Et_2O (cat.)	PE, Et_2O	(73)	29
t-Bu	*p*-$AcOC_6H_4$		BF_3•Et_2O (cat.)	PE, Et_2O	(60)	29
t-Bu	4-methyl-3-cyclohexenyl		BF_3•Et_2O (cat.)	PE, Et_2O	(80)	29
t-BuNc	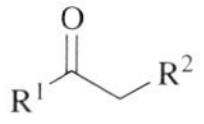		BF_3•Et_2O (cat.)		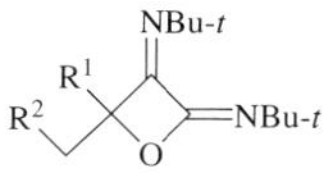	
	R^1	R^2				
	Me	H		PE, Et_2O	(88)	29, 139
	Me	Cl		PE, Et_2O, rt, 4 h	(92)	29
	$ClCH_2$	Cl		PE, Et_2O	(96)	29
	Me	$(MeO)_2CH$		PE, Et_2O	(76)	29
	$AcOCH_2$	$AcOCH_2$		PE, Et_2O	(84)	29
	Me	Et		Et_2O, –78°	(—)	139
	c-C_3H_5	Me		Et_2O, –78°	(—)	139

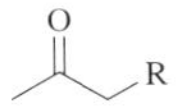	BF_3•Et_2O (stoich.)	PE, Et_2O, 0°	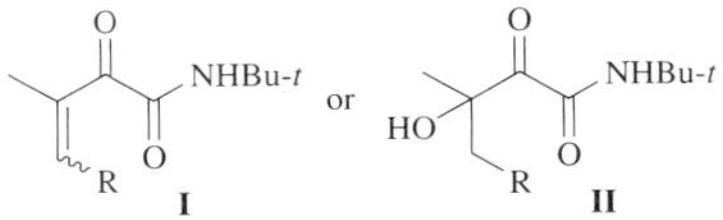	
			R: H; I (15); II (—)	112
			R: Cl; I (—); II (40)	29
			R: Et; I (9); II (—)	112
RCHO	$Zn(OTf)_2$	TMSCl, NEM, CH_2Cl_2, rt, 4 h	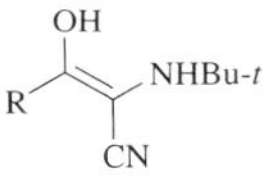	
			R: Ph (75)	147
			R: p-FC_6H_4 (64)	
			R: p-$MeOC_6H_4$ (57)	
			R: $PhCH_2CH_2$ (64)	
			R: (E)-PhCH=CH (83)	
			R: n-$C_9H_{19}CHO$ (63)	
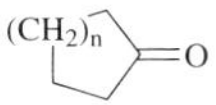			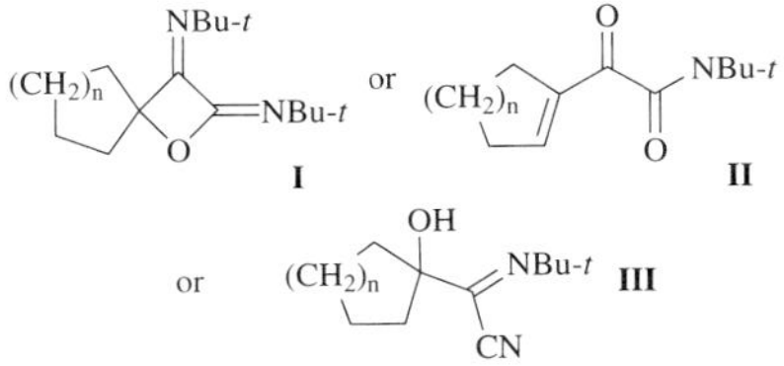	
n			I / II / III	
1	BF_3•Et_2O (stoich.)	PE, Et_2O, 0°	(—) (4) (—)	111
2	BF_3•Et_2O (stoich.)	PE, Et_2O, 0°	(—) (40) (—)	111
2	BF_3•Et_2O (cat.)	PE, Et_2O	(93) (—) (—)	29
2	$Zn(OTf)_2$	TMSCl, NEM, CH_2Cl_2, rt, 4 h	(—) (—) (56)	147
3	BF_3•Et_2O (stoich.)	PE, Et_2O, 0°	(—) (35) (—)	111
4	BF_3•Et_2O (stoich.)	PE, Et_2O, 0°	(—) (20) (—)	111
8	BF_3•Et_2O (stoich.)	PE, Et_2O, 0°	(—) (6) (—)	111

TABLE 8. REACTIONS OF ISOCYANIDES WITH CARBONYL COMPOUNDS AND ACID SPECIES GIVING OTHER PRODUCTS (*Continued*)

Isocyanide	Carbonyl Compound and Acid Species	Conditions	Product(s) and Yield(s) (%)	Refs.
C_5				
t-BuNC	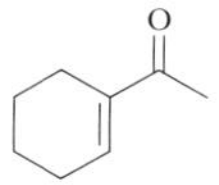 BF_3•Et_2O	—	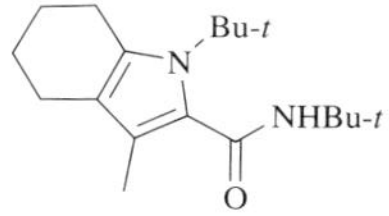 (5)	30
	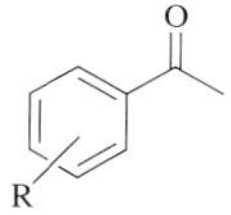 BF_3•Et_2O (stoich.)	PE, 0°, 1-2 h	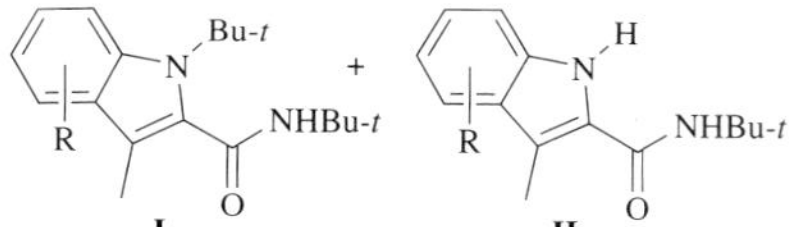 **I** + **II**	
			R: H, **I** (15), **II** (—)	135, 142
			R: *p*-Cl, **I** (18), **II** (—)	135, 142
			R: *p*-MeO, **I** (20), **II** (—)	135
			R: *o*-MeO, **I** (17), **II** (—)	135
			R: *p*-AcO, **I** (10), **II** (—)	135
	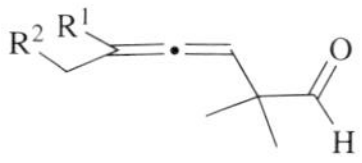 $AlCl_3$	—	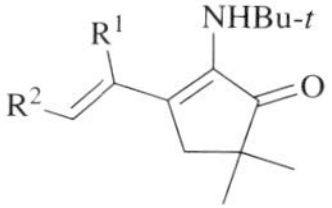 R^1 = Me, R^2 = Me (78); R^1 = H, R^2 = Me (75); R^1R^2 = —$(CH_2)_4$— (81)	146
	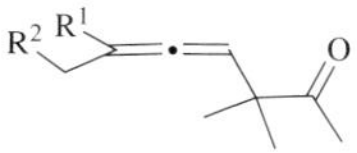 $AlCl_3$	—	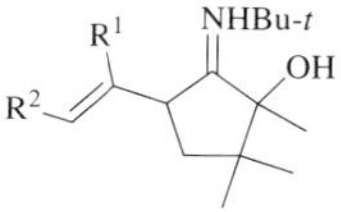 R^1 = Me, R^2 = Me (82); R^1 = H, R^2 = Me (85); R^1R^2 = —$(CH_2)_4$— (77)	146
	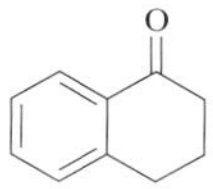 BF_3•Et_2O (stoich.)	PE, 0°, 1-2 h	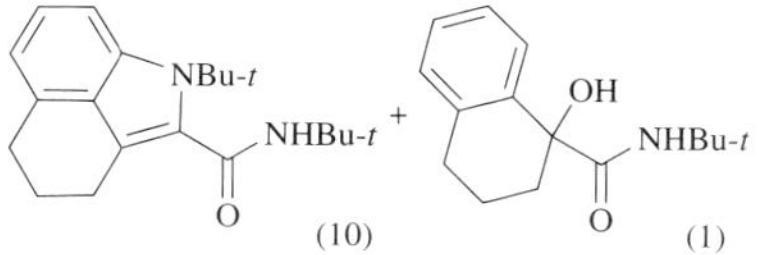 (10) + (1)	135

	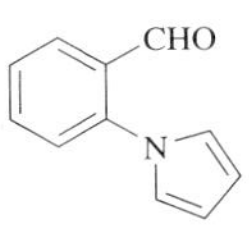	BF_3•Et_2O (cat.)	1. CH_2Cl_2, 0°, 5 min 2. Ac_2O, py, 10 min, rt	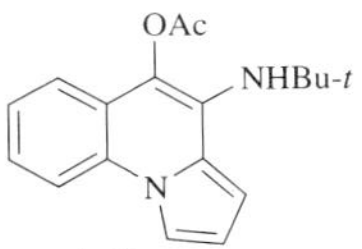 (51)	145
	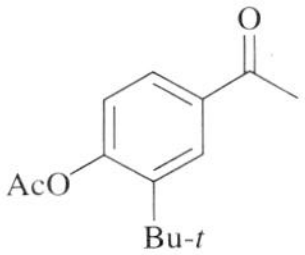	BF_3•Et_2O (stoich.)	PE, 0°, 1-2 h	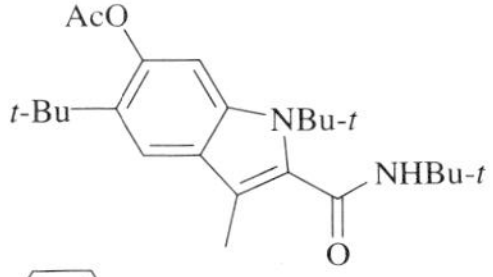 (21)	135
	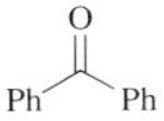	BF_3•Et_2O (stoich.)	PE, 0°, 1-2 h	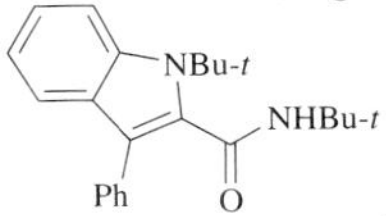 (30)	135, 142
	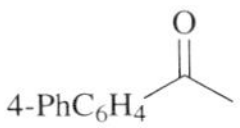	BF_3•Et_2O (stoich.)	PE, 0°, 1-2 h	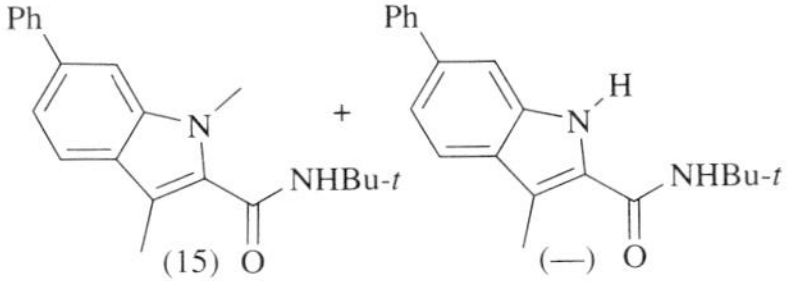 (15) + (—)	135, 142
	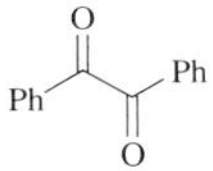	BF_3•Et_2O (stoich.)	PE, C_6H_6, 0°, 1-2 h	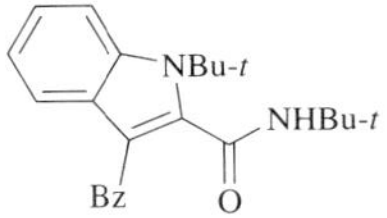 (15)	135, 142
	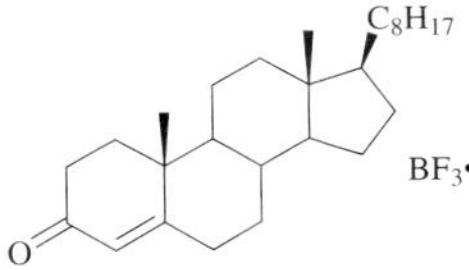	BF_3•Et_2O (stoich.)	—	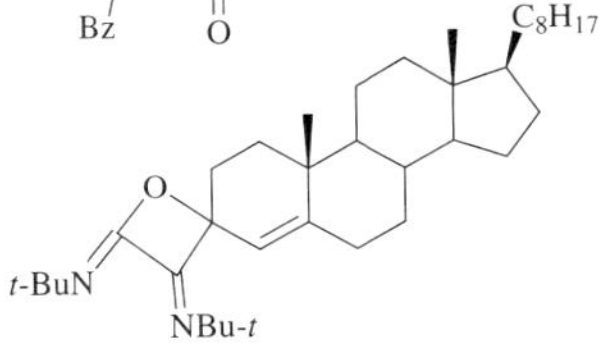 (—)	30
C_6 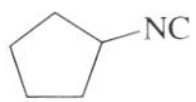	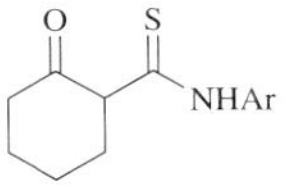	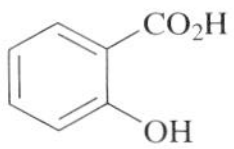	Et_2O, rt, 48 h	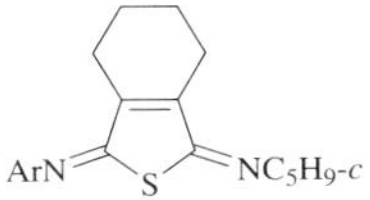 Ar: Ph (65); p-ClC_6H_4 (70)	28

TABLE 8. REACTIONS OF ISOCYANIDES WITH CARBONYL COMPOUNDS AND ACID SPECIES GIVING OTHER PRODUCTS (*Continued*)

Isocyanide	Carbonyl Compound and Acid Species		Conditions	Product(s) and Yield(s) (%)	Refs.
C_7					
c-$C_6H_{11}NC$	CH_3CHO	$BF_3{\bullet}Et_2O$ (stoich.)	Et_2O, −78°, 2 d	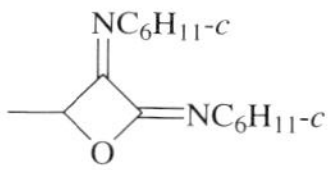 (23) + 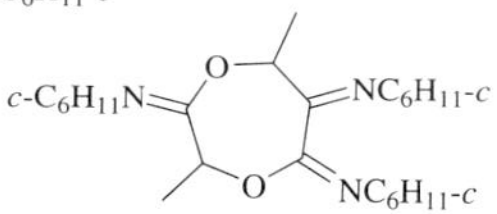 (5)	120
	R^1CHO			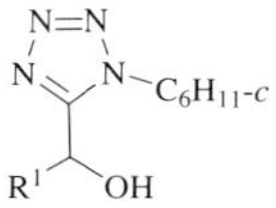	
	R^1				
	Cl_3C	HN_3	$CHCl_3$, rt, 1 h	(91)	23
	i-Pr	HN_3	$CHCl_3$, rt, 3 d	(85)	23
	Ph	$Al(N_3)_3$	$BF_3{\bullet}Et_2O$, THF, rt, 3 d	(94)	23
	$PhCH_2CH_2$	$TMSN_3$	THF, rt	(80^a, 40^b)	121
	$PhCH_2CH_2$	$TMSN_3$	MeOH, rt	(37^a, 33^b)	121
	$PhCH_2CH_2$	$TMSN_3$	CH_2Cl_2, rt	(95^a, 95^b)	121
	$PhCH_2CH_2$	Me_3SnN_3	THF, rt	(100^a, 100^b)	121
	$PhCH_2CH_2$	Me_3SnN_3	MeOH, rt	(100^a, 95^b)	121
	$PhCH_2CH_2$	Me_3SnN_3	CH_2Cl_2, rt	(100^a, 100^b)	121
	$PhCH_2CH_2$	$(PhO)_2PON_3$	THF, rt	(2^a, 2^b)	121
	$PhCH_2CH_2$	$(PhO)_2PON_3$	MeOH, rt	(0^a, 0^b)	121
	$PhCH_2CH_2$	$(PhO)_2PON_3$	CH_2Cl_2, rt	(6^a, 8^b)	121
	$PhCH_2CH_2$	NaN_3	THF, rt	(25^a, 5^b)	121
	$PhCH_2CH_2$	NaN_3	MeOH, rt	(0^a, 0^b)	121
	$PhCH_2CH_2$	NaN_3	CH_2Cl_2, rt	(59^a, 51^b)	121
	$PhCH_2CH_2$	$TMSN_3$	$BF_3{\bullet}Et_2O$, THF, rt	(15^a, 18^b)	121
	$PhCH_2CH_2$	Me_3SnN_3	$BF_3{\bullet}Et_2O$, THF, rt	(49^a, 36^b)	121
	$PhCH_2CH_2$	$(PhO)_2PON_3$	$BF_3{\bullet}Et_2O$, THF, rt	(0^a, 0^b)	121
	(S)-$PhCH_2CH(NHBoc)$	$TMSN_3$	CH_2Cl_2, rt, 18 h	(61)	122

	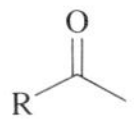	BF_3•Et_2O (cat.)		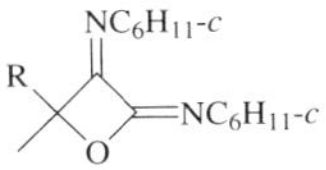	
	R				
	Me		PE, Et_2O	(60)	29
	Me		Et_2O, –78°, 1 d	(64)	22
	Et		Et_2O, –78°	(—)	139
	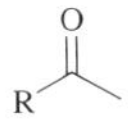			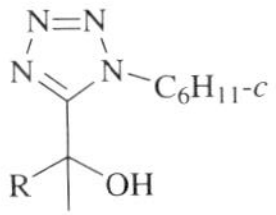	23
	R				
	Me	$Al(N_3)_3$	THF, rt, 4 d	(96)	
	Me	HN_3	$CHCl_3$, rt	(32)	
	Ph	$Al(N_3)_3$	BF_3•Et_2O, THF, rt, 14 d	(68)	
	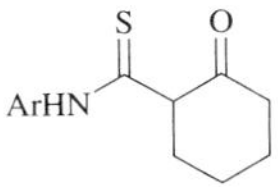	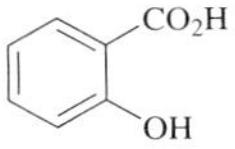	Et_2O, rt, 48 h	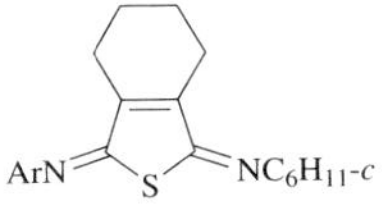 Ar: Ph (70); p-ClC_6H_4 (75)	28
PhNC	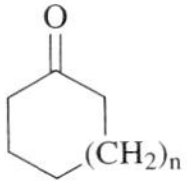	BF_3•Et_2O (stoich.)	Et_2O, 0°, 1 h	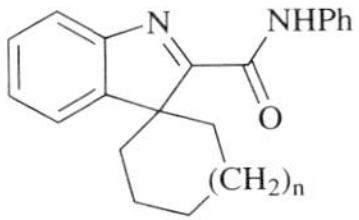 n: 1 (40); 2 (30)	113, 140
	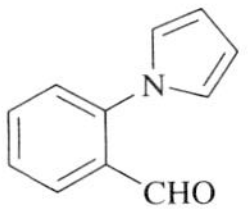	BF_3•Et_2O (cat.)	1. CH_2Cl_2, 0°, 5 min 2. Ac_2O, py, 10 min, rt	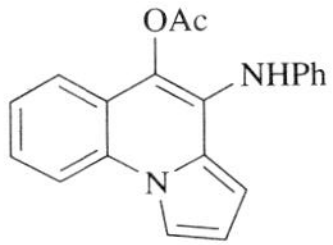 (54)	145

TABLE 8. REACTIONS OF ISOCYANIDES WITH CARBONYL COMPOUNDS AND ACID SPECIES GIVING OTHER PRODUCTS (*Continued*)

Isocyanide	Carbonyl Compound and Acid Species		Conditions	Product(s) and Yield(s) (%)		Refs.
C_8						
BnNC		$Al(N_3)_3$	$CHCl_3$, rt, 2 d	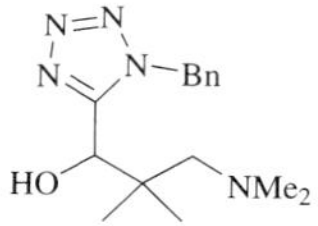	(82)	23
	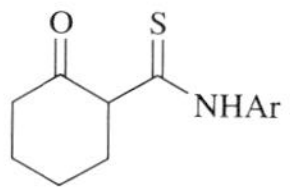	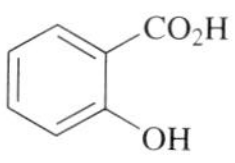	Et_2O, rt, 48 h	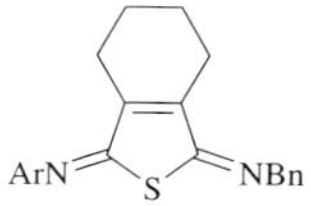	Ar: Ph (66); 4-ClC_6H_4 (68)	28
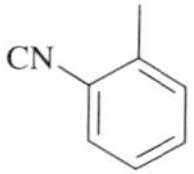	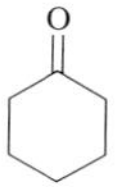	BF_3•Et_2O (stoich.)	Et_2O, 0°, 1 h	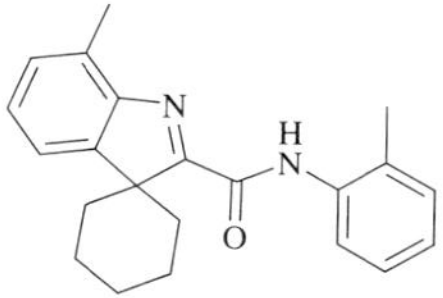	(37)	113
	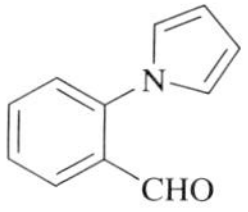	BF_3•Et_2O (cat.)	1. CH_2Cl_2, 0°, 5 min 2. Ac_2O, py, rt, 10 min	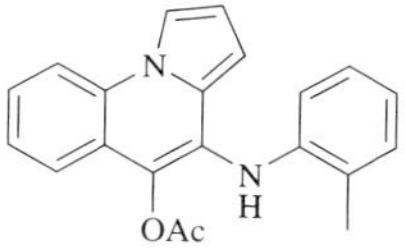	(59)	145
	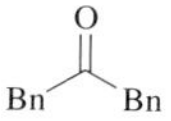	BF_3•Et_2O (stoich.)	Et_2O, 0°, 1 h	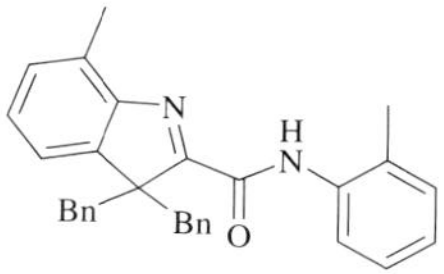	(5)	113
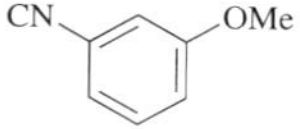	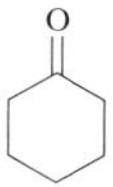	BF_3•Et_2O (stoich.)	Et_2O, 0°, 1 h	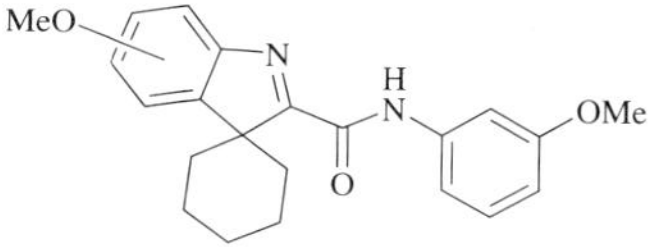	(25)	113

	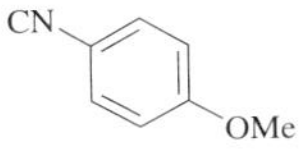	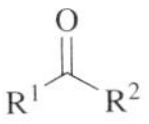	$BF_3 \cdot Et_2O$ (stoich.)	Et_2O, 0°, 2 h	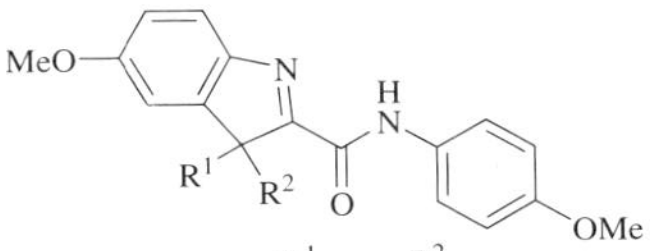	
					R^1 / R^2: Me / Et (26)	141
					Me / *i*-Pr (5)	113
					Et / Et (23)	141
					—$(CH_2)_5$— (33)	113
					Bn / Bn (14)	141
C_9	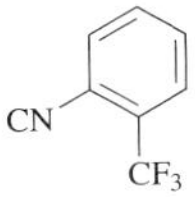	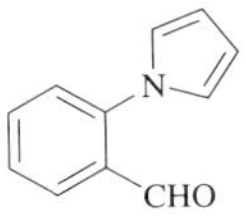	$BF_3 \cdot Et_2O$ (cat.)	1. CH_2Cl_2, 0°, 5 min 2. Ac_2O, py, rt, 10 min	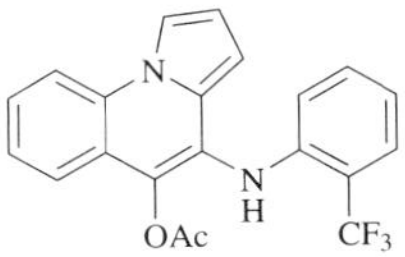 (45)	145
	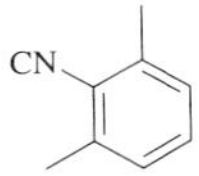	RCHO			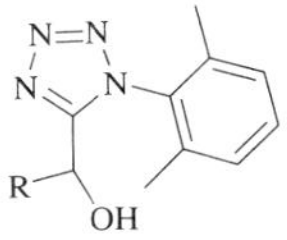	
		R: H (aq)	HN_3	THF, $CHCl_3$, rt, 56 d	(36)	23
		Cl_3C	HN_3	C_6H_6, rt, 6 d	(71)	23
		BocNH	$TMSN_3$	CH_2Cl_2, rt, 18 h	(28)	122
		$Me_2NCH_2C(CH_3)_2$	$Al(N_3)_3$	THF, $CHCl_3$, rt, 10 d	(96)	23
		(*S*)-$PhCH_2CH(NHBoc)$	$TMSN_3$	CH_2Cl_2, rt	(82)	122
		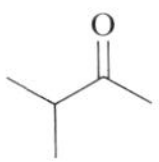	$BF_3 \cdot Et_2O$ (stoich.)	Et_2O, 0°, 2 h	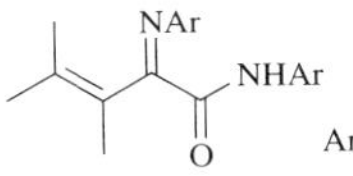 (50) Ar = 2,6-$Me_2C_6H_3$	141

TABLE 8. REACTIONS OF ISOCYANIDES WITH CARBONYL COMPOUNDS AND ACID SPECIES GIVING OTHER PRODUCTS (*Continued*)

	Isocyanide	Carbonyl Compound and Acid Species		Conditions	Product(s) and Yield(s) (%)	Refs.
C_9	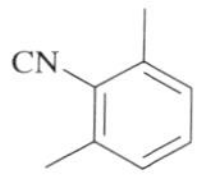	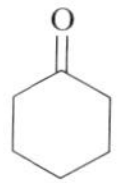	BF_3•Et_2O (stoich.)	Et_2O, 0°, 2 h	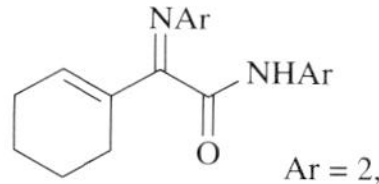 (22) Ar = 2,6-$Me_2C_6H_3$	141
		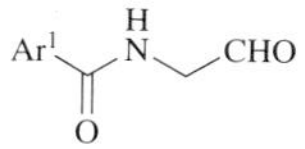	$TMSN_3$	CH_2Cl_2, rt, 18h	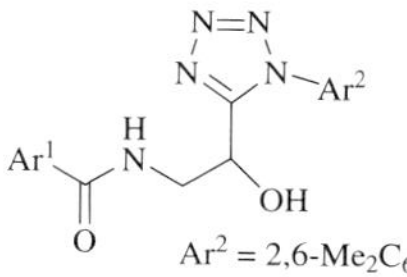 Ar^2 = 2,6-$Me_2C_6H_3$ Ar^1: 2-pyridyl (69); *m*-NCC_6H_4 (70); 3-quinolyl (86)	122
		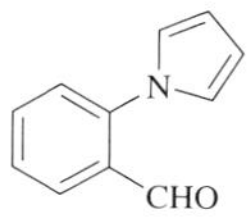	BF_3•Et_2O (cat.)	1. CH_2Cl_2, 0°, 5 min 2. Ac_2O, py, 10 min, rt	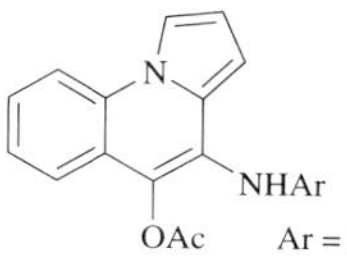 (53) Ar = 2,6-$Me_2C_6H_3$	145
	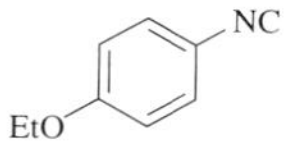	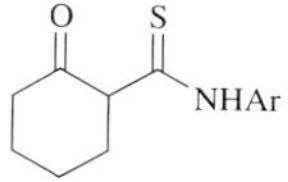	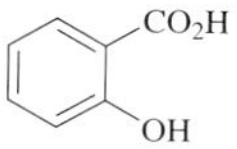	Et_2O, rt, 48 h	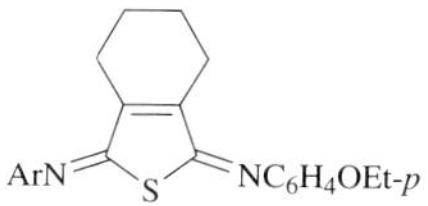 Ar: Ph (78); 4-ClC_6H_4 (83)	28
C_{11}	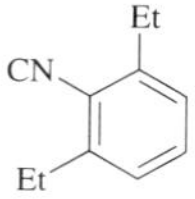	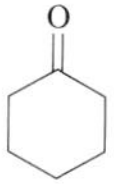	BF_3•Et_2O (stoich.)	Et_2O, 0°, 2 h	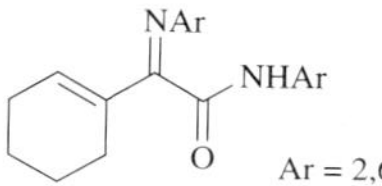 (21) Ar = 2,6-$Et_2C_6H_3$	141
	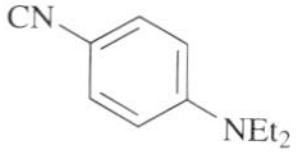	Cl_3CCHO	HN_3	$CHCl_3$, rt	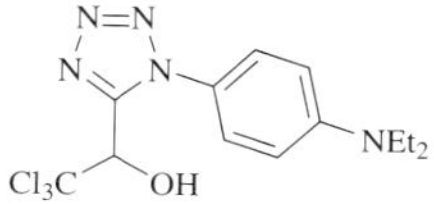 (95)	23

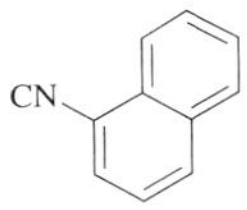 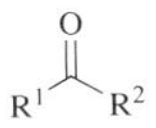 BF_3•Et_2O (stoich.) — Et_2O, 0°, 1.5 h

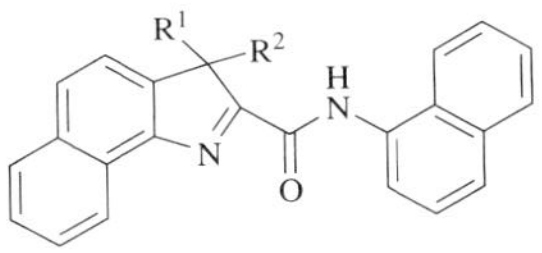

R^1	R^2	
Me	Et	(49)
Et	Et	(39)
—$(CH_2)_3$—		(45)
n-Pr	*n*-Pr	(37)
Me	$PhCH_2$	(43)
Et	$PhCH_2$	(15)
—$CH_2CMe_2CH_2CMe_2CH_2$—		(34)

114

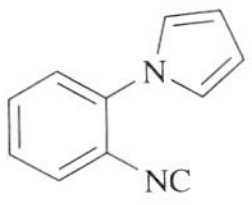 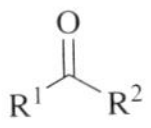 BF_3•Et_2O (cat.) — CH_2Cl_2, 0°, 40 min

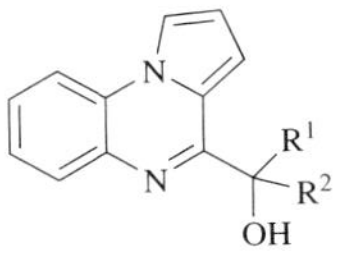

R^1	R^2	
Et	H	(89)
Me	Me	(76)
Me	CO_2Et	(84)
i-Pr	H	(85)
t-Bu	H	(52)
2-furyl	H	(49)
Me	$(CH_2)_2CO_2Et$	(74)
—$(CH_2)_5$—		(78)
Ph	H	(81)
Me	Ph	(59)

144

TABLE 8. REACTIONS OF ISOCYANIDES WITH CARBONYL COMPOUNDS AND ACID SPECIES GIVING OTHER PRODUCTS (*Continued*)

Isocyanide	Carbonyl Compound and Acid Species		Conditions	Product(s) and Yield(s) (%)	Refs.
C_{12}					
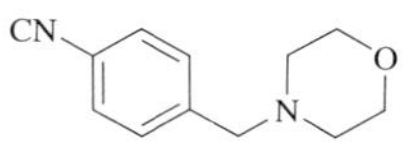	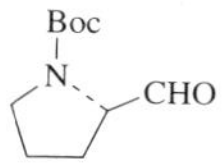	$TMSN_3$	CH_2Cl_2, r.., 18 h	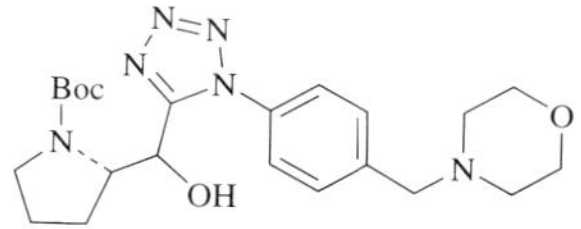 (40)	122
	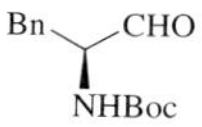	$TMSN_3$	1. CH_2Cl_2, rt 2. Bu_4NF	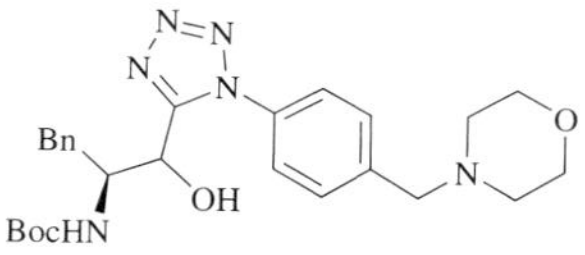 (91)	121
C_{13}					
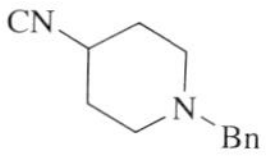	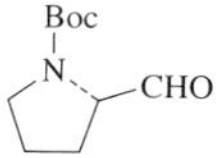	$TMSN_3$	CH_2Cl_2, rt	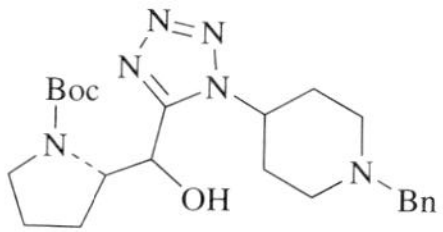 (88)	122
	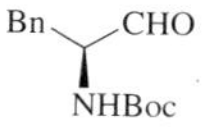	$TMSN_3$	1. CH_2Cl_2, rt 2. Bu_4NF	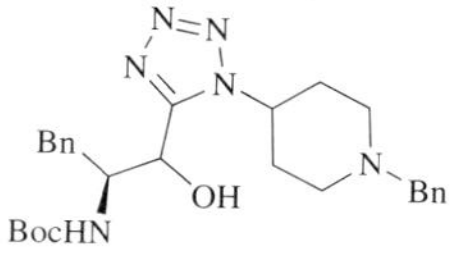 (99)	121

[a] The yields were determined by the electrospray method.

[b] The yields were determined by UV (215 nm).

TABLE 9. α-ALKOXY AMIDES OR OTHER PRODUCTS FROM REACTION OF ISOCYANIDES WITH ACETALS AND ACID SPECIES

Isocyanide	Acetal	Conditions	Product(s) and Yield(s) (%)	Refs.
C_5				
t-BuNC	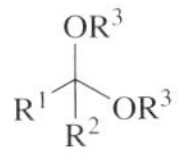	CH_2Cl_2, rt	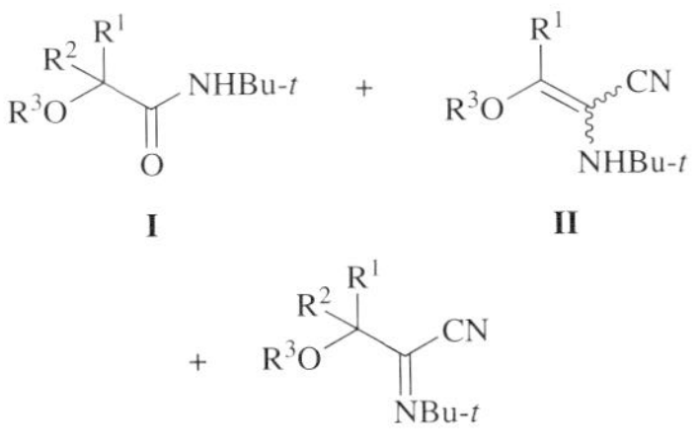 I, II, III	

R^1	R^2	R^3	Acid	Time	I	II	III	Refs.
H	H	Me	$TiCl_4$	3 h	(—)	(50)	(—)	151
Me	H	Et	$TiCl_4$	3 h	(14)	(52)	(—)	151
Me	H	Et	$AlEt_2Cl$	12 h	(—)	(—)	(90)	253
Et	H	Me	$TiCl_4$	3 h	(8)	(65)	(—)	151
Et	H	Me	$AlEt_2Cl$	12 h	(—)	(—)	(64)	253
Me	Me	Me	$AlEt_2Cl$	12 h	(—)	(—)	(90)	253
n-Pr	H	Me	$TiCl_4$	3 h	(7)	(57)	(—)	151
n-Pr	H	Me	$AlEt_2Cl$	12 h	(—)	(—)	(63)	253
i-Pr	H	Me	$TiCl_4$	3 h	(30)	(36)	(—)	151
i-Pr	H	Me	$AlEt_2Cl$	12 h	(—)	(—)	(78)	253
t-Bu	H	Me	$AlEt_2Cl$	12 h	(—)	(—)	(75)	253
n-C_6H_{13}	H	Me	$TiCl_4$	3 h	(1)	(57)	(—)	151
n-C_6H_{13}	H	Me	$AlEt_2Cl$	12 h	(—)	(—)	(90)	253
Ph	H	Me	$TiCl_4$	3 h	(—)	(69)	(—)	151
Ph	H	Me	$AlEt_2Cl$	12 h	(—)	(—)	(65)	253

TABLE 9. α-ALKOXY AMIDES OR OTHER PRODUCTS FROM REACTION OF ISOCYANIDES WITH ACETALS AND ACID SPECIES (*Continued*)

Isocyanide	Acetal	Conditions	Product(s) and Yield(s) (%)	Refs.
C_7				
$c\text{-}C_6H_{11}NC$	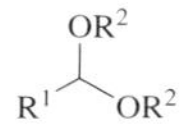	$TiCl_4$, CH_2Cl_2	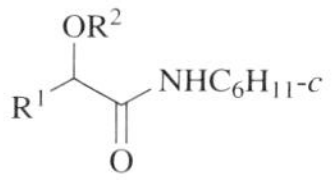	149
	R^1 / R^2	Temp. / Time		
	Me / Et	–35° / 2 h	(90)	
	(*E*)-MeCH=CH / Me	–45° / 3 h	(80)	
	Ph / Me	–70° / 3 h	(90)	
	Ph / Et	–45° / 3 h	(89)	
	$PhCH_2CH_2$ / Et	–45° / 3 h	(82)	
	(*E*)-PhCH=CH / Et	–35° / 2.5 h	(85)	
	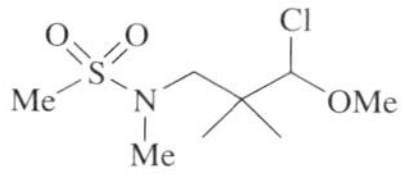	$TiCl_4$, $CHCl_3$, –78 to rt	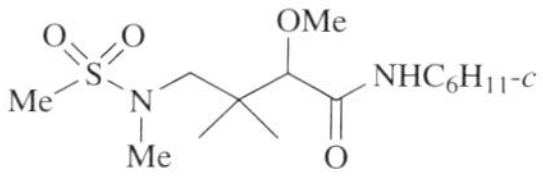 (69)	152
	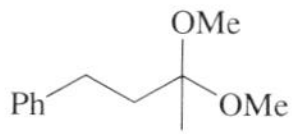	$TiCl_4$, CH_2Cl_2, –45°, 3 h	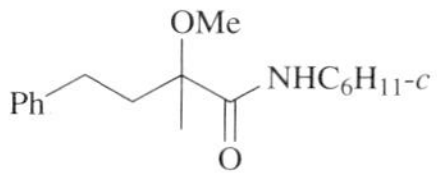 (66)	149
	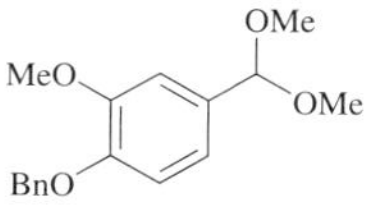	CF_3CO_2H, C_6H_6, rt, 1 d	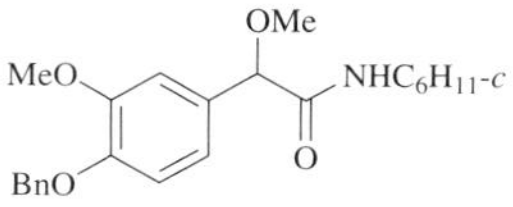 (76)	148
C_9				
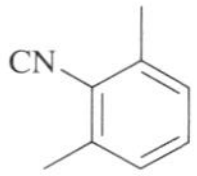	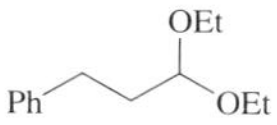	$TiCl_4$, CH_2Cl_2, –45°, 3 h	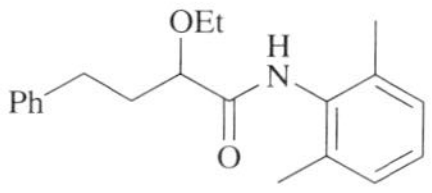 (81)	149

C_{11} 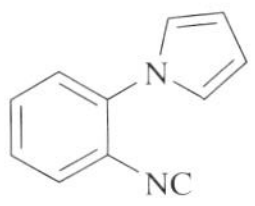	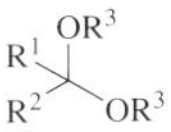	BF_3•Et_2O, CH_2Cl_2, 0°, 100 min	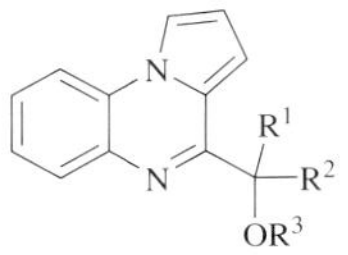		144
	R^1 / R^2 / R^3				
	Me / H / Me		(3)		
	Me / H / Et		(85)		
	CH_2=CH / H / Et		(42)		
	Me / Me / Et		(53)		
	Ph / H / Me		(82)		
C_{12} 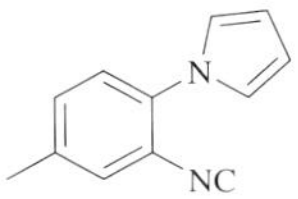	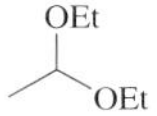	BF_3•Et_2O, CH_2Cl_2, 0°, 100 min	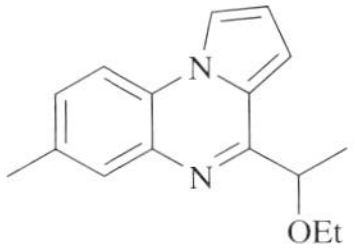	(82)	144
C_{18} 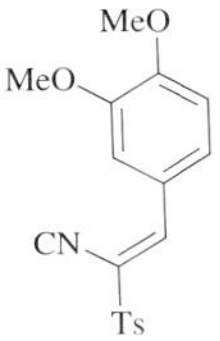	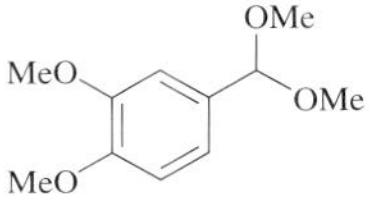	CF_3CO_2H, C_6H_6, rt, 4 h	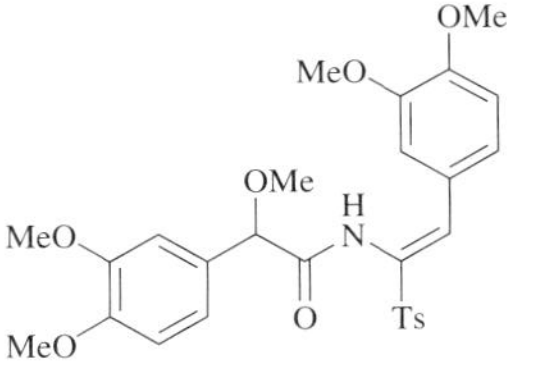	(70)	148
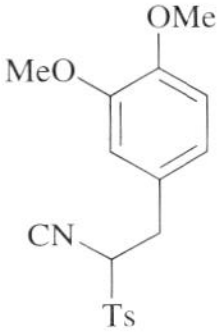	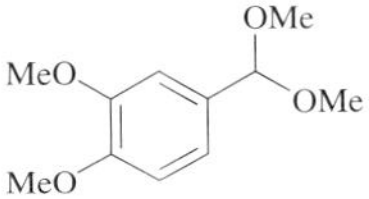	CF_3CO_2H, C_6H_6, rt, 2 h	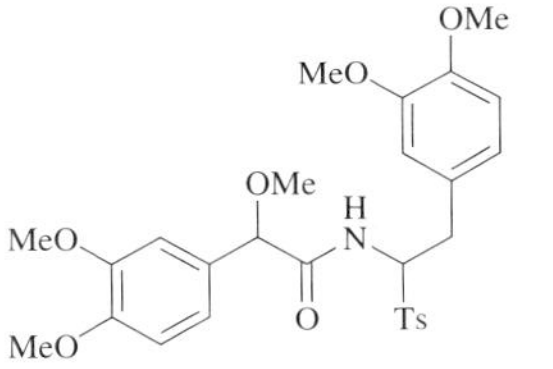	(93)	148

TABLE 10. ENANTIOSELECTIVE PASSERINI REACTIONS

Isocyanide	Catalyst	Aldehyde and Acid Species		Conditions	Product(s), Yield(s) (%), ee (%)	Refs.
C_3						
$CN\diagdown\!\diagup CO_2Et$	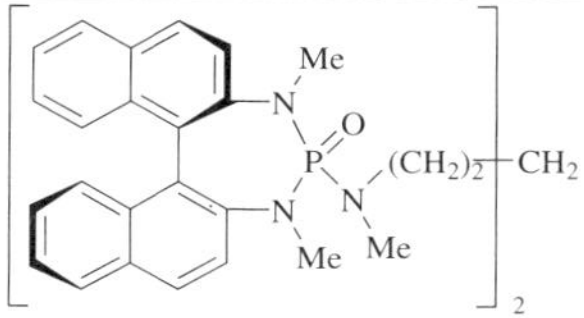	PhCHO	$SiCl_4$	1. CH_2Cl_2, –74° 2. $NaHCO_3$ (aq)	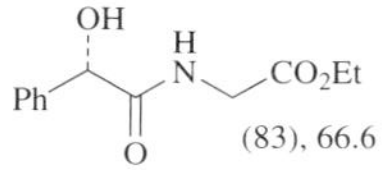 (83), 66.6	170
C_5						
t-BuNC	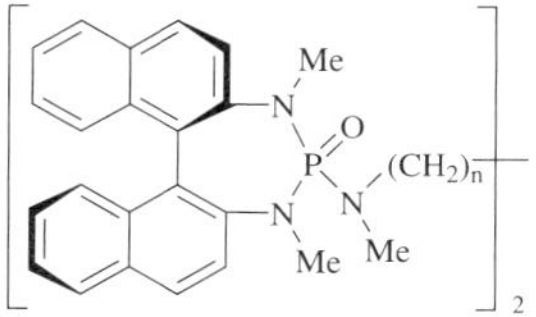	PhCHO	$SiCl_4$	1. CH_2Cl_2, –74° 2. $NaHCO_3$ (aq)	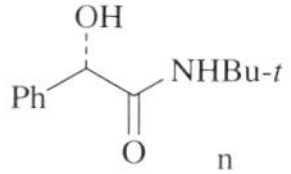 n 2 (93), 92.8 3 (86), 91.2	170
	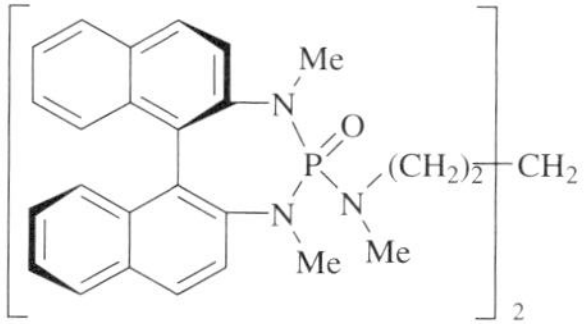	RCHO	$SiCl_4$	1. CH_2Cl_2, –74° 2. $NaHCO_3$ (aq)	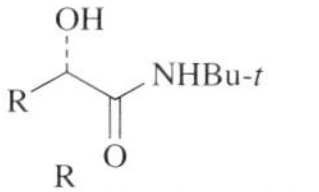 R 2-furyl (83), 91.8 Ph (96), >98 *c*-C_6H_{11} (53), 74.2 *p*-MeC_6H_4 (91), 99.8 *p*-$F_3CC_6H_4$ (89), 93.0 *p*-$MeOC_6H_4$ (89), 96.6 $PhCH_2CH_2$ (92), 63.8 (*E*)-PhCH=CH (81), 95.6 $PhC\equiv CCH_2$ (86), 34.8 (*E*)PhCH=CMe (76), 54.0 1-naphthyl (92), 84.4 2-naphthyl (93), 99.6	170

	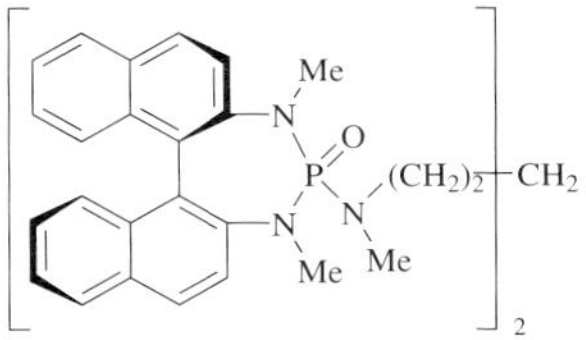	RCHO	$SiCl_4$	1. CH_2Cl_2, –74° 2. MeOH 3. $NaHCO_3$ (aq)	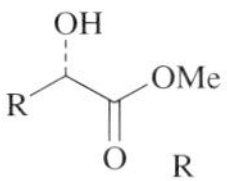 R — Ph (97), >98; $PhCH_2CH_2$ (71), 95.8; (*E*)-PhCH=CH (88), 63.6	170
C_7						
PhNC	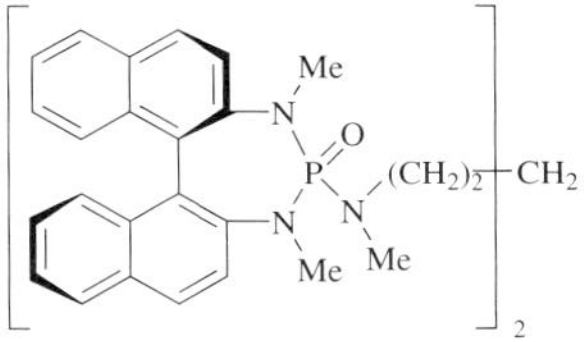	PhCHO	$SiCl_4$	1. CH_2Cl_2, –74° 2. $NaHCO_3$ (aq)	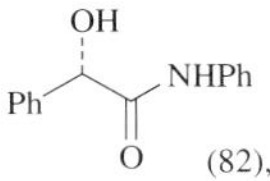 (82), 46.4	170
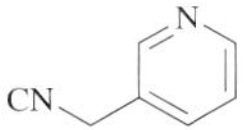	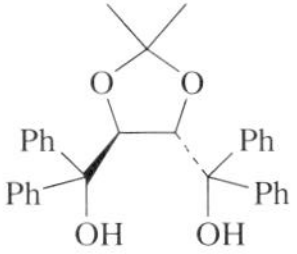 + $Ti(OPr\text{-}i)_4$	*i*-PrCHO	$PhCO_2H$	THF, –78°	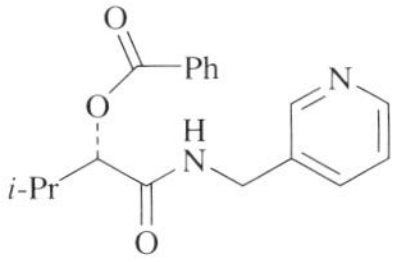 (28), 42	171
C_8						
BnNC	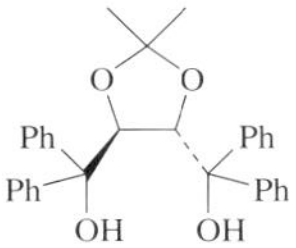 + $Ti(OPr\text{-}i)_4$	R^1CHO	R^2CO_2H	THF, –78°	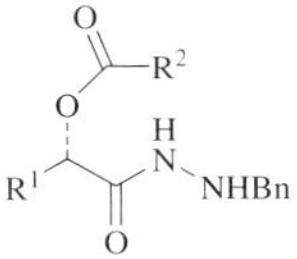	171

R^1	R^2CO_2H	
i-Pr	$PhCO_2H$	(46), 36
i-Pr	*o*-$HOC_6H_4CO_2H$	(12), 32
t-Bu	$PhCO_2H$	(48), 34
t-Bu	*p*-$PhC_6H_4CO_2H$	(46), 36
i-Pr	(*S*)-Boc-Phe	(56), 12

TABLE 10. ENANTIOSELECTIVE PASSERINI REACTIONS (*Continued*)

	Isocyanide	Catalyst	Aldehyde and Acid Species		Conditions	Product(s), Yield(s) (%), ee (%)	Refs.
C_9		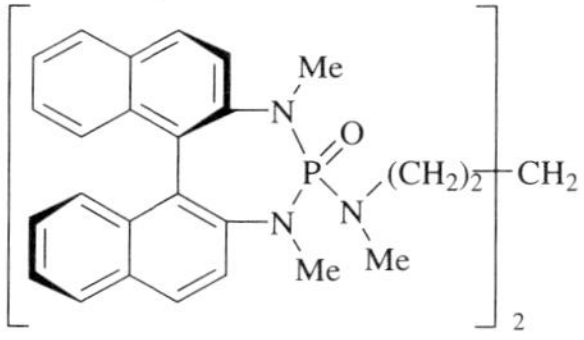		$SiCl_4$	1. CH_2Cl_2, –74° 2. $NaHCO_3$ (aq)	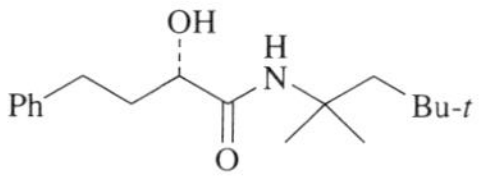 (87), 40.0	170
	$TsCH_2NC$	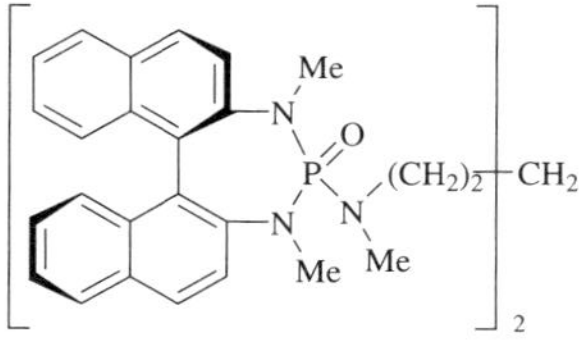	PhCHO	$SiCl_4$	1. CH_2Cl_2, –74° 2. $NaHCO_3$ (aq)	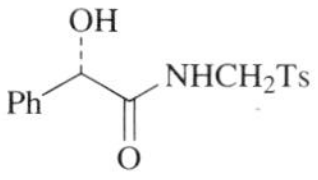 (80), 77.0	170
C_{10}	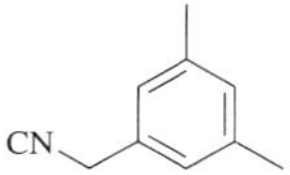	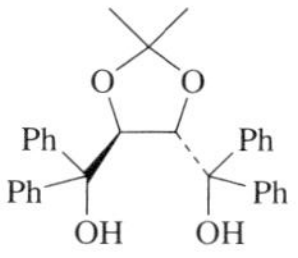 + Ti(OPr-*i*)$_4$	*i*-PrCHO	$PhCO_2H$	THF, –78°	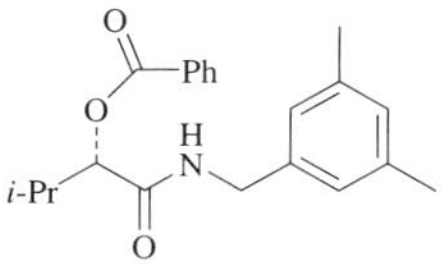 (31), 32	171

REFERENCES

[1] Passerini, M. *Gazz. Chim. Ital.* **1922**, *52–1*, 432.
[2] Passerini, M. *Gazz. Chim. Ital.* **1926**, *56*, 826.
[3] Passerini, M. *Gazz. Chim. Ital.* **1924**, *54*, 529.
[4] Passerini, M. *Gazz. Chim. Ital.* **1925**, *55*, 721.
[5] Gambaryan, N. P.; Rokhlin, E. M.; Zeifman, Y. V.; Chen, C.-Y.; Knunyants, I. L. *Angew. Chem., Int. Ed. Engl.* **1966**, *5*, 947.
[6] Middleton, W. J.; England, D. C.; Krespan, C. G. *J. Org. Chem.* **1967**, *32*, 948.
[7] Passerini, M. *Gazz. Chim. Ital.* **1921**, *51–2*, 126.
[8] Passerini, M. *Accad. Naz. Lincei Memorie* **1927**, 378.
[9] Marquarding, D.; Gokel, G.; Hoffmann, P.; Ugi, I. In *Isonitrile Chemistry*; Ugi, I., Ed.; Academic Press: New York, 1971, p. 133.
[10] Ugi, I.; Lohberger, S.; Karl, R. In *Comprehensive Organic Synthesis*; Trost, B. M., Heathcock, C. H., Eds.; Pergamon: Oxford, 1991; Vol. 2, p. 1083.
[11] Dömling, A.; Ugi, I. *Angew. Chem., Int. Ed. Engl.* **2000**, *39*, 3169.
[12] Armstrong, R. W.; Combs, A. P.; Tempest, P. A.; Brown, S. D.; Keating, T. A. *Acc. Chem. Res.* **1996**, *29*, 123.
[13] Bienaymé, H.; Hulme, C.; Oddon, G.; Schmitt, P. *Chem. Eur. J.* **2000**, *6*, 3321.
[14] Ugi, I. *Angew. Chem., Int. Ed. Engl.* **1982**, *21*, 810.
[15] Ugi, I.; Dömling, A.; Horl, W. *Endeavour* **1994**, *18*, 115.
[16] Ugi, I.; Heck, S. *Comb. Chem. High Throughput Screen.* **2001**, *4*, 1.
[17] Ugi, I.; Dömling, A. In *Combinatorial Chemistry-A Practical Approach*; Fenniri, H. Ed.; Oxford Univ. Press: Oxford, 2000, p. 287.
[18] Baker, R. H.; Stanonis, D. *J. Am. Chem. Soc.* **1951**, *73*, 699.
[19] Hagedorn, I.; Eholzer, U.; Winkelmann, H. D. *Angew. Chem., Int. Ed. Engl.* **1964**, *3*, 647.
[20] Hagedorn, I.; Eholzer, U. *Chem. Ber.* **1965**, *98*, 936.
[21] Kagen, H.; Lillien, I. *J. Org. Chem.* **1966**, *31*, 3728.
[22] Saegusa, T.; Taka-ishi, N.; Fujii, H. *Tetrahedron* **1968**, *24*, 3795.
[23] Ugi, I.; Meyr, R. *Chem. Ber.* **1961**, *94*, 2229.
[24] Bock, H.; Ugi, I. *J. Prakt. Chem.* **1997**, *339*, 385.
[25] Ugi, I. *Angew. Chem., Int. Ed. Engl.* **1962**, *1*, 8.
[26] Gautier, A. *Ann. Chim. (Paris)* **1869**, *17*, 222.
[27] Carofiglio, T.; Cozzi, P. G.; Floriani, C.; Chiesi-Villa, A.; Rizzoli, C. *Organometallics* **1993**, *12*, 2726.
[28] Bossio, R.; Marcaccini, S.; Pepino, R.; Torroba, T. *J. Chem. Soc., Perkin Trans. 1* **1996**, 229.
[29] Kabbe, H.-J. *Chem. Ber.* **1969**, *102*, 1404.
[30] Zeeh, B. *Synthesis* **1969**, 65.
[31] El Kaïm, L.; Gacon, A. *Tetrahedron Lett.* **1997**, *38*, 3391.
[32] Dumestre, P.; El Kaïm, L.; Grégoire, A. *Chem. Commun.* **1999**, 775.
[33] Ugi, I.; Ebert, B.; Hörl, W. *Chemosphere* **2001**, *43*, 75.
[34] Demharter, A.; Hörl, W.; Herdtweck, E.; Ugi, I. *Angew. Chem., Int. Ed. Engl.* **1996**, *35*, 173.
[35] Ugi, I.; Steinbrückner, C. *Chem. Ber.* **1961**, *94*, 2802.
[36] Xia, Q.; Ganem, B. *Org. Lett.* **2002**, *4*, 1631.
[36a] Wang, Q.; Xia, Q.; Ganem, B. *Tetrahedron Lett.* **2003**, *44*, 6825.
[37] Subramanyam, C.; Noguchi, M.; Weinreb, S. M. *J. Org. Chem.* **1989**, *54*, 5580.
[38] Jen, T.; Frazee, J. S.; Schwartz, M. S.; Kaiser, C.; Colella, D. F.; Wardell Jr., J. R. *J. Med. Chem.* **1977**, *20*, 1258.
[39] Bienaymé, H. *Tetrahedron Lett.* **1998**, *39*, 4255.
[40] Moran, E. J.; Armstrong, R. W. *Tetrahedron Lett.* **1991**, *32*, 3807.
[41] Kim, S. W.; Bauer, S. M.; Armstrong, R. W. *Tetrahedron Lett.* **1998**, *39*, 7031.
[42] Ridi, M. *Gazz. Chim. Ital.* **1941**, *71*, 462.
[43] Bossio, R.; Marcaccini, S.; Pepino, R. *Liebigs Ann. Chem.* **1991**, 1107.
[44] Beck, B.; Magnin-Lachaux, M.; Herdtweck, E.; Dömling, A. *Org. Lett.* **2001**, *3*, 2875.
[45] Banfi, L.; Guanti, G.; Riva, R.; Basso, A.; Calcagno, E. *Tetrahedron Lett.* **2002**, *43*, 4067.

[46] Banfi, L.; Guanti, G.; Riva, R. *Chem. Commun.* **2000**, 985.
[47] Schmidt, U.; Weinbrenner, S. *J. Chem. Soc., Chem. Commun.* **1994**, 1003.
[48] Owens, T. D.; Araldi, G.-A.; Nutt, R. F.; Semple, J. E. *Tetrahedron Lett.* **2001**, *42*, 6271.
[49] Owens, T. D.; Semple, J. E. *Org. Lett.* **2001**, *3*, 3301.
[50] Ziegler, T.; Kaisers, H.-J.; Schlömer, R.; Koch, C. *Tetrahedron* **1999**, *55*, 8397.
[51] Ziegler, T.; Schlömer, R.; Koch, C. *Tetrahedron Lett.* **1998**, *39*, 5957.
[52] Bossio, R.; Marcaccini, S.; Pepino, R.; Torroba, T. *Synthesis* **1993**, 783.
[53] Bossio, R.; Marcaccini, S.; Pepino, R.; Polo, C.; Torroba, T. *Org. Prep. Proced. Int.* **1992**, *24*, 188.
[54] Bossio, R.; Marcaccini, S.; Pepino, R. *Liebigs Ann. Chem.* **1994**, 527.
[55] Marcaccini, S.; Pepino, R.; Marcos, C. F.; Polo, C.; Torroba, T. *J. Heterocycl. Chem.* **2000**, *37*, 1501.
[56] Beck, B.; Picard, A.; Herdtweck, E.; Dömling, A. *Org. Lett.* **2004**, *6*, 39.
[57] Lockhoff, O. *Angew. Chem., Int. Ed. Engl.* **1998**, *37*, 3436.
[58] Lockhoff, O.; Frappa, I. *Comb. Chem. High Throughput Screen.* **2002**, *5*, 361.
[59] de Nooy, A. E. J.; Masci, G.; Crescenzi, V. *Macromolecules* **1999**, *32*, 1318.
[60] Fetzer, U.; Ugi, I. *Liebigs Ann. Chem.* **1962**, *659*, 184.
[61] König, S.; Lohberger, S.; Ugi, I. *Synthesis* **1993**, 1233.
[62] Baker, R. H.; Schlesinger, A. H. *J. Am. Chem. Soc.* **1945**, *67*, 1499.
[63] Jenner, G. *Tetrahedron Lett.* **2002**, *43*, 1235.
[64] Passerini, M. *Gazz. Chim. Ital.* **1921**, *51–2*, 181.
[65] Neidlein, R. *Arch. Pharm. (Weinheim, Ger.)* **1966**, *299*, 603.
[66] Sebti, S.; Foucaud, A. *Tetrahedron* **1986**, *42*, 1361.
[67] Sebti, S.; Foucaud, A. *J. Chem. Res. (M)* **1987**, 791.
[68] Schmidt, U.; Kabitzke, K.; Boie, I.; Osterroht, C. *Chem. Ber.* **1965**, *98*, 3819.
[69] Quast, H.; Aldenkortt, S. *Chem. Eur. J.* **1996**, *2*, 462.
[70] Schiess, M.; Seebach, D. *Helv. Chim. Acta* **1983**, *66*, 1618.
[71] Burger, K.; Mütze, K.; Hollweck, W.; Koksch, B. *Tetrahedron* **1998**, *54*, 5915.
[72] Burger, K.; Schierlinger, C.; Mütze, K. *J. Fluorine Chem.* **1993**, *65*, 149.
[73] Müller, S.; Neidlein, R. *Helv. Chim. Acta* **2002**, *85*, 2222.
[74] Bergemann, M.; Neidlein, R. *Helv. Chim. Acta* **1999**, *82*, 909.
[75] Johnson, C. K.; Dolphin, D. *Tetrahedron Lett.* **1998**, *39*, 4753.
[76] Rachon, J. *Chimia* **1983**, *37*, 299.
[77] Kreutzkamp, N.; Lämmerhirt, K. *Angew. Chem., Int. Ed. Engl.* **1968**, *7*, 372.
[78] Seebach, D.; Adam, G.; Gees, T.; Schiess, M.; Weigand, W. *Chem. Ber.* **1988**, *121*, 507.
[79] Versleijen, J. P. G.; Faber, P. M.; Bodewes, H. H.; Braker, A. H.; van Leusen, D.; van Leusen, A. M. *Tetrahedron Lett.* **1995**, *36*, 2109.
[80] Hashida, Y.; Imai, A.; Sekiguchi, S. *J. Heterocycl. Chem.* **1989**, *26*, 901.
[81] Falck, J. R.; Manna, S. *Tetrahedron Lett.* **1981**, *22*, 619.
[82] Schöllkopf, U.; Henneke, K.-W.; Madawinata, K.; Harms, R. *Liebigs Ann. Chem.* **1977**, 40.
[83] Baker, R. H.; Linn, L. E. *J. Am. Chem. Soc.* **1948**, *70*, 3721.
[84] McFarland, J. W. *J. Org. Chem.* **1963**, *28*, 2179.
[85] Wright, D. L.; Robotham, C. V.; Aboud, K. *Tetrahedron Lett.* **2002**, *43*, 943.
[86] Ugi, I.; Demharter, A.; Hörl, W.; Schmid, T. *Tetrahedron* **1996**, *52*, 11657.
[87] Passerini, M.; Ragni, G. *Gazz. Chim. Ital.* **1931**, *61*, 964.
[88] Passerini, M. *Gazz. Chim. Ital.* **1923**, *53*, 331.
[89] Marcaccini, S.; Miguel, D.; Torroba, T.; Garcia Valverde, M. *J. Org. Chem.* **2003**, *68*, 3315.
[90] Just, G.; Chung, B. Y.; Grözinger, K. *Can. J. Chem.* **1977**, *55*, 274.
[91] de Nooy, A. E. J.; Rori, V.; Masci, G.; Dentini, M.; Crescenzi, V. *Carbohydr. Res.* **2000**, *324*, 116.
[92] de Nooy, A. E. J.; Capitani, D.; Masci, G.; Crescenzi, V. *Biomacromolecules* **2000**, *1*, 259.
[93] Banfi, L.; Calcagno, E.; Guanti, G.; Riva, R., unpublished work, University of Genoa.
[94] Sebti, S.; Foucaud, A. *Synthesis* **1983**, 546.
[95] Sebti, S.; Foucaud, A. *Tetrahedron* **1984**, *40*, 3223.
[96] Basso, A.; Banfi, L.; Riva, R.; Piaggio, P.; Guanti, G. *Tetrahedron Lett.* **2003**, *44*, 2367.
[97] Tadesse, S.; Balan, C.; Jones, W.; Viswanadhan, V.; Hulme, C. 223rd ACS National Meeting, Orlando, FL; ORGN 241, 2002.

[98] Banfi, L.; Basso, A.; Guanti, G.; Riva, R. *Molecular Diversity* **2003**, *6*, 227.
[99] Paulvannan, K. *Tetrahedron Lett.* **1999**, *40*, 1851.
[100] Paulvannan, K.; Jacobs, J. W. *Tetrahedron* **1999**, *55*, 7433.
[101] Lee, D. S.; Sello, J. K.; Schreiber, S. L. *Org. Lett.* **2000**, *2*, 709.
[102] Beck, B.; Larbig, G.; Mejat, B.; Magnin Lachaux, M.; Picard, A.; Herdtweck, E.; Dömling, A. *Org. Lett.* **2003**, *5*, 1047.
[103] Yavari, I.; Shaabani, A.; Asghari, S.; Olmstead, M.; Safari, N. *J. Fluorine Chem.* **1997**, *86*, 77.
[104] Shaabani, A.; Bazgir, A.; Soleimani, K.; Bijanzahdeh, H. R. *J. Fluorine Chem.* **2002**, *116*, 93.
[105] Müller, E.; Kästner, P.; Beutler, R.; Rundel, W.; Suhr, H.; Zeeh, B. *Liebigs Ann. Chem.* **1968**, *713*, 87.
[106] Zinner, G.; Kliegel, W. *Arch. Pharm. (Weinheim, Ger.)* **1966**, *299*, 746.
[107] Zinner, G.; Bock, W. *Arch. Pharm. (Weinheim, Ger.)* **1971**, *304*, 933.
[108] Songstad, J.; Stangeland, L. J.; Austad, T. *Acta Chem. Scand.* **1970**, *24*, 355.
[109] Engemyr, L. B.; Martinsen, A.; Songstad, J. *Acta Chem. Scand., Ser. A* **1974**, *28*, 255.
[110] Otsuka, S.; Mori, K.; Yamagami, K. *J. Org. Chem.* **1966**, *31*, 4170.
[111] Müller, E.; Zeeh, B. *Liebigs Ann. Chem.* **1966**, *696*, 72.
[112] Zeeh, B.; Müller, E. *Liebigs Ann. Chem.* **1968**, *715*, 47.
[113] Zeeh, B. *Chem. Ber.* **1968**, *101*, 1753.
[114] Zeeh, B. *Tetrahedron* **1968**, *24*, 6663.
[115] Lumma, W. C. J. *J. Org. Chem.* **1981**, *46*, 3668.
[116] Semple, J. E.; Owens, T. D.; Nguyen, K.; Levy, O. E. *Org. Lett.* **2000**, *2*, 2769.
[117] Semple, J. E.; Levy, O. E.; Minami, N. K.; Owens, T. D.; Siev, D. V. *Bioorg. Med. Chem. Lett.* **2000**, *10*, 2305.
[118] Lansbury, P. T.; Nickson, T. E.; Vacca, J. P.; Sindelar, R. D.; Messinger II, J. M. *Tetrahedron* **1987**, *43*, 5583.
[119] Moderhack, D. *Synthesis* **1985**, 1083.
[120] Saegusa, T.; Taka-ishi, N.; Fujii, H. *Polymer Lett.* **1967**, *5*, 779.
[121] Nixey, T.; Hulme, C. 223rd ACS National Meeting, Orlando, FL; ORGN 244, 2002.
[122] Nixey, T.; Hulme, C. *Tetrahedron Lett.* **2002**, *43*, 6833.
[123] Henkel, B.; Beck, B.; Westner, B.; Mejat, B.; Dömling, A. *Tetrahedron Lett.* **2003**, *44*, 8947.
[124] Carofiglio, T.; Floriani, C.; Chiesi-Villa, A.; Guastini, C. *Inorg. Chem.* **1989**, *28*, 4417.
[125] Carofiglio, T.; Floriani, C.; Chiesi-Villa, A.; Rizzoli, C. *Organometallics* **1991**, *10*, 1659.
[126] Seebach, D.; Beck, A. K.; Schiess, M.; Widler, L.; Wonnacott, A. *Pure Appl. Chem.* **1983**, *55*, 1807.
[127] Aicher, T. D.; Anderson, R. C.; Gao, J.; Shetty, S. S.; Coppola, G. M.; Stanton, J. L.; Knorr, D. C.; Sperbeck, D. M.; Brand, L. J.; Vinluan, C. C.; Kaplan, E. L.; Dragland, C. J.; Tomaselli, H. C.; Islam, A.; Lozito, R. J.; Liu, X.; Maniara, W. M.; Fillers, W. S.; DelGrande, D.; Walter, R. E.; Mann, W. R. *J. Med. Chem.* **2000**, *43*, 236.
[128] Bebernitz, G. R.; Aicher, T. D.; Stanton, J. L.; Gao, J.; Shetty, S. S.; Knorr, D. C.; Strohschein, R. J.; Tan, J.; Brand, L. J.; Liu, C.; Wang, W. H.; Vinluan, C. C.; Kaplan, E. L.; Dragland, C. J.; DelGrande, D.; Islam, A.; Lozito, R. J.; Liu, X.; Maniara, W. M.; Mann, W. R. *J. Med. Chem.* **2000**, *43*, 2248.
[129] Morel, G.; Marchand, E.; Foucaud, A.; Toupet, L. *J. Org. Chem.* **1990**, *55*, 1721.
[130] Yavari, I.; Anary-Abbasinejad, M.; Alizadeh, A. *Monatsh. Chem.* **2002**, *133*, 1221.
[131] Marcaccini, S.; Torroba, T. *Org. Prep. Proc. Int.* **1993**, *25*, 141.
[132] Ito, Y.; Kato, H.; Saegusa, T. *J. Org. Chem.* **1982**, *47*, 741.
[133] Ito, H.; Kato, H.; Imai, H.; Saegusa, T. *J. Am. Chem. Soc.* **1982**, *104*, 6449.
[134] Pellissier, H.; Gil, G. *Tetrahedron* **1989**, *45*, 3415.
[135] Zeeh, B. *Chem. Ber.* **1969**, *102*, 678.
[136] Müller, E.; Zeeh, B. *Tetrahedron Lett.* **1965**, 3951.
[137] Kabbe, H.-J. *Chem. Ber.* **1969**, *102*, 1410.
[138] Kabbe, H.-J.; Joop, N. *Liebigs Ann. Chem.* **1969**, *730*, 151.
[139] Saegusa, T.; Taka-ishi, N.; Ito, H. *Bull. Chem. Soc. Jpn.* **1971**, *44*, 1121.
[140] Zeeh, B. *Angew. Chem., Int. Ed. Engl.* **1967**, *6*, 453.
[141] Zeeh, B. *Chem. Ber.* **1969**, *102*, 1876.
[142] Zeeh, B. *Tetrahedron Lett.* **1967**, 3881.
[143] Kobayashi, K.; Matoba, T.; Irisawa, S.; Matsumoto, T.; Morikawa, O.; Konishi, H. *Chem. Lett.* **1998**, 551.

[144] Kobayashi, K.; Irisawa, S.; Matoba, T.; Matsumoto, T.; Yoneda, K.; Morikawa, O.; Konishi, H. *Bull. Chem. Soc. Jpn.* **2001**, *74*, 1109.
[145] Kobayashi, K.; Nakahashi, R.; Takanohashi, A.; Kitamura, T.; Morikawa, O.; Konishi, H. *Chem. Lett.* **2002**, 624.
[146] Gil, G.; Zahra, J.-P. *Tetrahedron Lett.* **1985**, *26*, 419.
[147] Xia, Q.; Ganem, B. *Synthesis* **2002**, 1969.
[148] Barrett, A. G. M.; Barton, D. H. R.; Falck, J. R.; Papaioannou, D.; Widdowson, D. A. *J. Chem. Soc., Perkin Trans. 1* **1979**, 652.
[149] Mukaiyama, T.; Watanabe, K.; Shiono, M. *Chem. Lett.* **1974**, 1457.
[150] Ito, H.; Imai, H.; Segoe, K.; Saegusa, T. *Chem. Lett.* **1984**, 937.
[151] Pellissier, H.; Meou, A.; Gil, G. *Tetrahedron Lett.* **1986**, *27*, 2979.
[152] Shipov, A. G.; Zheltonogova, E. A.; Baukov, Y. I. *J. Gen. Chem. USSR (Engl. Transl.),* **1992**, 2150.
[153] Shono, T.; Matsumura, Y.; Tsubata, K. *Tetrahedron Lett.* **1981**, *22*, 2411.
[154] Irie, K.; Aoe, K.; Tanaka, T.; Saito, S. *J. Chem. Soc., Chem. Commun.* **1985**, 633.
[155] Kelly, G. L.; Lawrie, K. W. M.; Morgan, P.; Willis, C. L. *Tetrahedron Lett.* **2000**, *41*, 8001.
[156] Skorna, G.; Ugi, I. *Angew. Chem., Int. Ed. Engl.* **1977**, *16*, 259.
[157] Kamer, P. C. J.; Cleij, M. C.; Nolte, R. J. M.; Harada, T.; Hezemans, A. M. F.; Drenth, W. *J. Am. Chem. Soc.* **1988**, *110*, 1581.
[158] Bowers, M. M.; Carroll, P.; Joullié, M. M. *J. Chem. Soc., Perkin Trans. 1* **1989**, 857.
[159] Eckert, H.; Forster, B. *Angew. Chem., Int. Ed. Engl.* **1987**, *26*, 894.
[160] Bayer, T.; Riemer, C.; Kessler, H. *J. Peptide Sci.* **2001**, *7*, 250.
[161] Obrecht, R.; Hermann, R.; Ugi, I. *Synthesis* **1985**, 400.
[162] Huber, V. J.; Nagula, G.; Lum, C.; Farber, K.; Goodman, B. A. 221st ACS National Meeting: San Diego, CA; ORGN 140, 2001.
[163] Urban, R.; Marquarding, D.; Seidel, P.; Ugi, I.; Weinelt, A. *Chem. Ber.* **1977**, *110*, 2012.
[164] Visser, H. G. J.; Nolte, R. J. M.; Zwikker, J. W.; Drenth, W. *J. Org. Chem.* **1985**, *50*, 3133.
[165] Visser, H. G. J.; Nolte, R. J. M.; Zwikker, J. W.; Drenth, W. *J. Org. Chem.* **1985**, *50*, 3138.
[166] Eckert, H. *Synthesis* **1977**, 332.
[167] Eckert, H. *Z. Naturforsch.* **1990**, *45b*, 1715.
[168] Frey, R.; Galbraith, S. G.; Guelfi, S.; Lamberth, C.; Zeller, M. *Synlett* **2003**, 1536.
[169] Keating, T. A.; Armstrong, R. W. *J. Am. Chem. Soc.* **1995**, *117*, 7842.
[170] Denmark, S. E.; Fan, Y. *J. Am. Chem. Soc.* **2003**, *125*, 7825.
[171] Kusebauch, U.; Beck, B.; Messer, K.; Herdtweck, E.; Dömling, A. *Org. Lett.* **2003**, *5*, 4021.
[172] Cavicchioni, G.; D'Angeli, F.; Casolari, A.; Orlandini, P. *Synthesis* **1988**, 947.
[173] Cavicchioni, G. *Tetrahedron Lett.* **1987**, *28*, 2427.
[174] Hussain, R.; Toth, I.; Gibbons, W. A. *Liebigs Ann. Chem.* **1991**, 963.
[175] Ganu, V. S.; Shaw, E. *J. Med. Chem.* **1981**, *24*, 698.
[176] Clark, A. J.; Al-Faiyz, Y. S. S.; Broadhurst, M. J.; Patel, D.; Peacock, J. L. *J. Chem. Soc., Perkin Trans. 1* **2000**, 1117.
[177] Davies, J. S.; Howe, J.; LeBreton, M. *J. Chem. Soc., Perkin Trans. 2* **1995**, 2335.
[178] Hartwig, W.; Born, L. *J. Org. Chem.* **1987**, *52*, 4352.
[179] Wasserman, H. H.; Lipshutz, B. H. *Tetrahedron Lett.* **1975**, *16*, 1731.
[180] Davis, F. A.; Chen, B.-C. *Chem. Rev.* **1992**, *92*, 919.
[181] Guertin, K. R.; Chan, T.-H. *Tetrahedron Lett.* **1991**, *32*, 715.
[182] Chen, B.-C.; Zhou, P.; Davis, F. A.; Ciganek, E. *Org. React.* **2003**, *62*, 1.
[183] Cunico, R. F. *Tetrahedron Lett.* **2002**, *43*, 355.
[184] Screttas, C. G.; Steele, B. R. *Org. Prep. Proced. Int.* **1990**, *22*, 269.
[185] Ramon, D. J.; Yus, M. *Tetrahedron* **1996**, *52*, 13739.
[186] Amadéi, E.; Alilou, E. H.; Eydoux, F.; Pierrot, M.; Réglier, M.; Waegell, B. *J. Chem. Soc., Chem. Commun.* **1992**, 1782.
[187] Wang, G.-Z.; Mallat, T.; Baiker, A. *Tetrahedron: Asymmetry* **1997**, *8*, 2133.
[188] Aldea, R.; Alper, H. *J. Org. Chem.* **1998**, *63*, 9425.
[189] Pasquier, C.; Naili, S.; Pelinski, L.; Brocard, J.; Mortreux, A.; Agbossou, F. *Tetrahedron: Asymmetry* **1998**, *9*, 193.

[190] Pasquier, C.; Eilers, J.; Reiners, I.; Martens, J.; Mortreux, A.; Agbossou, F. *Synlett* **1998**, 1162.
[191] Chiba, T.; Miyashita, A.; Nohira, M.; Takaya, H. *Tetrahedron Lett.* **1993**, *34*, 2351.
[192] Yamamoto, K.; Rehman, S. U. *Chem. Lett.* **1984**, 1603.
[193] Basha, A.; Lipton, M.; Weinreb, S. M. *Tetrahedron Lett.* **1977**, *48*, 4171.
[194] Li, X.; Yeung, C.; Chan, A. S. C.; Lee, D.-S.; Yang, T.-K. *Tetrahedron: Asymmetry* **1999**, *10*, 3863.
[195] Tatlock, J. H. *J. Org. Chem.* **1995**, *60*, 6221.
[196] Snider, B. B.; Song, F.; Foxman, B. M. *J. Org. Chem.* **2000**, *65*, 793.
[197] Ley, J. P.; Bertram, H.-J. *Tetrahedron* **2001**, *57*, 1277.
[198] Kotsovolou, S.; Chiou, A.; Verger, R.; Kokotos, G. *J. Org. Chem.* **2001**, *66*, 962.
[199] Manabe, K. *Tetrahedron* **1998**, *54*, 14465.
[200] Boatman, P. D.; Ogbu, C. O.; Eguchi, M.; Kim, H.-O.; Nakanishi, H.; Cao, B.; Shea, J. P.; Kahn, M. *J. Med. Chem.* **1999**, *42*, 1367.
[201] Bennett, J. M.; Campbell, A. D.; Campbell, A. J.; Carr, M. G.; Dunsdon, R. M.; Greening, J. R.; Hurst, D. N.; Jennings, N. S.; Jones, P. S.; Jordan, S.; Kay, P. B.; O'Brien, M. A.; King-Underwood, J. T.; Raynham, T. M.; Wilkinson, C. S.; Wilkinson, T. C. I.; Wilson, F. X. *Bioorg. Med. Chem. Lett.* **2001**, *11*, 355.
[202] Donkor, I. O.; Zheng, X.; Miller, D. D. *Bioorg. Med. Chem. Lett.* **2000**, *10*, 2497.
[203] Adang, A. E. P.; Peters, C. A. M.; Gerritsma, S.; de Zwart, E.; Veeneman, G. *Bioorg. Med. Chem. Lett.* **1999**, *9*, 1227.
[204] Rickman, B. H.; Matile, S.; Nakanishi, K.; Berova, N. *Tetrahedron* **1998**, *54*, 5041.
[205] Han, W.; Hu, Z.; Jiang, X.; Decicco, C. P. *Bioorg. Med. Chem. Lett.* **2000**, *10*, 711.
[206] Chatterjee, S.; Dunn, D.; Tao, M.; Wells, G.; Gu, Z.-Q.; Bihovsky, R.; Ator, M. A.; Siman, R.; Mallamo, J. P. *Bioorg. Med. Chem. Lett.* **1999**, *9*, 2371.
[207] Girreser, U.; Noe, C. R. *Synthesis* **1995**, 1223.
[208] Kelly, S. E.; LaCour, T. G. *Synth. Commun.* **1992**, *22*, 859.
[209] Frankland, P. F.; Slator, A. *J. Chem. Soc.* **1903**, *83*, 1349.
[210] Choi, D.; Stables, J. P.; Kohn, H. *Bioorg. Med. Chem.* **1996**, *4*, 2105.
[211] Gawronski, J.; Gawronska, K.; Skowronek, P.; Rychlewska, U.; Warzajtis, B.; Rychlewski, J.; Hoffmann, M.; Szarecka, A. *Tetrahedron* **1997**, *53*, 6113.
[212] Ziólkowski, M.; Czarnocki, Z. *Tetrahedron Lett.* **2000**, *41*, 1963.
[213] Bonnet, D.; Joly, P.; Gras-Masse, H.; Melnyk, O. *Tetrahedron Lett.* **2001**, *42*, 1875.
[214] Solladié-Cavallo, A.; Bencheqroun, M. *J. Org. Chem.* **1992**, *57*, 5831.
[215] Khumtaveeporn, K.; Ullmann, A.; Matsumoto, K.; Davis, B. G.; Jones, J. B. *Tetrahedron: Asymmetry* **2001**, *12*, 249.
[216] Degerbeck, F.; Fransson, B.; Grehn, L.; Ragnarsson, U. *J. Chem. Soc., Perkin Trans. 1* **1993**, 11.
[217] Singh, V. K. *Synthesis* **1992**, 607.
[218] Solladié-Cavallo, A.; Bencheqroun, M. *Tetrahedron: Asymmetry* **1991**, *2*, 1165.
[219] Kim, M. J.; Whitesides, G. M. *J. Am. Chem. Soc.* **1988**, *110*, 2959.
[220] Evans, D. A.; Morrissey, M. M.; Dorow, R. L. *J. Am. Chem. Soc.* **1985**, *107*, 4346.
[221] Koh, K.; Durst, T. *J. Org. Chem.* **1994**, *59*, 4683.
[222] Corey, E. J.; Link, J. O. *Tetrahedron Lett.* **1992**, *33*, 3431.
[223] Andrés, J. M.; Martinez, M. M.; Pedrosa, R.; Pèrez-Encabo, A. *Tetrahedron: Asymmetry* **2001**, *12*, 347.
[224] Reetz, M. T.; Drewes, M. W.; Harms, K.; Reif, W. *Tetrahedron Lett.* **1988**, *29*, 3295.
[225] Herranz, R.; Castro-Pichel, J.; Vinuesa, S.; Garcia-Lopez, M. T. *J. Org. Chem.* **1990**, *55*, 2232.
[226] Herranz, R.; Castro-Pichel, J.; Vinuesa, S.; Garcìa-Lopez, T. *Synthesis* **1989**, 703.
[227] Brady, S. F.; Sisko, J. T.; Stauffer, K. J.; Colton, C. D.; Qiu, H.; Lewis, S. D.; Ng, A. S.; Shafer, J. A.; Bogusky, M. J.; Verber, D. F.; Nutt, R. F. *Bioorg. Med. Chem.* **1995**, *3*, 1063.
[228] Burkhart, J. P.; Peet, N. P.; Bey, P. *Tetrahedron Lett.* **1990**, *31*, 1385.
[229] Hengeveld, J. E.; Grief, V.; Tandanier, J.; Lee, C. M.; Riley, D.; Lartey, P. A. *Tetrahedron Lett.* **1984**, *25*, 4075.
[230] Woydowski, K.; Liebscher, J. *Tetrahedron* **1999**, *55*, 9205.
[231] Chong, J. M.; Sharpless, K. B. *J. Org. Chem.* **1985**, *50*, 1560.
[232] Bhatia, B.; Jain, S.; De, A.; Bagchi, I.; Iqbal, J. *Tetrahedron Lett.* **1996**, *37*, 7311.

[233] Das, B. C.; Iqbal, J. *Tetrahedron Lett.* **1997**, *38*, 2903.
[234] De, A.; Ghosh, S.; Iqbal, J. *Tetrahedron Lett.* **1997**, *38*, 8379.
[235] Punniyamaruthy, T.; Iqbal, J. *Tetrahedron Lett.* **1997**, *38*, 4463.
[236] Behrens, C. H.; Sharpless, K. B. *J. Org. Chem.* **1985**, *50*, 5696.
[237] Cacciola, J.; Fevig, J. M.; Stouten, P. F. W.; Alexander, R. S.; Knabb, R. M.; Wexler, R. R. *Bioorg. Med. Chem. Lett.* **2000**, *10*, 1253.
[238] Cacciola, J.; Alexander, R. S.; Fevig, J. M.; Stouten, P. F. W. *Tetrahedron Lett.* **1997**, *38*, 5741.
[239] Khanjin, N. A.; Hesse, M. *Helv. Chim. Acta* **2001**, *84*, 1253.
[240] Nakayama, K.; Kawato, H. C.; Inagaki, H.; Nakajima, R.; Kitamura, A.; Someya, K.; Ohta, T. *Org. Lett.* **2000**, *2*, 977.
[241] Reetz, M. T.; Lauterbach, E. H. *Tetrahedron Lett.* **1991**, *32*, 4477.
[242] Torres-Valencia, J. M.; Cerda-Garcia-Rojas, C. M.; Joseph-Nathan, P. *Tetrahedron: Asymmetry* **1998**, *9*, 757.
[243] Wang, Y.-C.; Li, C.-L.; Tseng, H.-L.; Chuang, S.-C.; Yan, T.-H. *Tetrahedron: Asymmetry* **1999**, *10*, 3249.
[244] Banik, B. K.; Barakat, K. J.; Wagle, D. R.; Manhas, M. S.; Bose, A. K. *J. Org. Chem.* **1999**, *64*, 5746.
[245] Ojima, I. *Acc. Chem. Res.* **1995**, *28*, 383.
[246] Pirrung, M. C.; Das Sarma, K. *J. Am. Chem. Soc.* **2004**, *126*, 444.
[247] Shipov, A. G.; Zheltonogova, E. A.; Oleneva, G. I.; Kobzareva, V. P.; Macharashvili, A. A.; Mozzhukhin, A. O.; Shklover, V. E.; Struchkov, Y. T.; Baukov, Y. I. *J. Gen. Chem. USSR (Engl. Transl.)* **1992**, 2549.
[248] Okano, A.; Inaoka, M.; Funabashi, S.; Iwamoto, M.; Isoda, S.; Moroi, R.; Abiko, Y.; Hirata, M. *J. Med. Chem.* **1972**, *15*, 247.
[249] Passerini, M.; Ragni, G. *Gazz. Chim. Ital.* **1934**, *64*, 909.
[250] Passerini, M. *Gazz. Chim. Ital.* **1923**, *53*, 410.
[251] Passerini, M.; Cima, E. *Ann. Chim. Farm. (Rome)* **1940**, *18*, 5.
[252] Ostaszewski, R.; Portlock, D. E.; Fryszkowska, A.; Jeziorska, K. *Pure Appl. Chem.* **2003**, *75*, 413.
[253] Pellissier, H.; Gil, G. *Tetrahedron Lett.* **1988**, *29*, 6773.

CHAPTER 2

DIELS-ALDER REACTIONS OF IMINO DIENOPHILES

GEOFFREY R. HEINTZELMAN, IVONA R. MEIGH, YOGESH R. MAHAJAN, AND STEVEN M. WEINREB

Department of Chemistry, The Pennsylvania State University, University Park, Pennsylvania 16802
and
Johnson & Johnson Pharmaceutical Research and Development, Raritan, New Jersey 08869

CONTENTS

---

gheintze@prdus.jnj.com; smw@chem.psu.edu

Organic Reactions, Vol. 65, Edited by Larry E. Overman et al.
ISBN 0-471-68260-8

ACKNOWLEDGMENTS

We are grateful to the National Science Foundation and the National Institutes of Health for financial support of the imino Diels-Alder research done at the Pennsylvania State University. We also thank Dr. Magnus W. P. Bebbington and Nancy Weinreb for assistance in preparing this Chapter.

INTRODUCTION

In comparison to all-carbon [4+2]-cycloadditions, imino Diels-Alder (ImDA) reactions are still in their infancy. The last four decades, however, have seen intensive research activity in this area, the result of which is that synthetic organic chemists now have a powerful tool at their disposal for the rapid construction of highly functionalized six-membered nitrogen heterocycles, often in a regio-, diastereo- and enantioselective manner.

This chapter encompasses the topic of imino [4+2]-cycloaddition reactions, of which there are a number of variants, but the coverage of this review has been limited exclusively to imine and iminium ion dienophiles that undergo inter- and intramolecular Diels-Alder reactions with acyclic and cyclic all-carbon 1,3-dienes affording 1,2,5,6-tetrahydropyridines as the initial cycloadducts. There are a number of reports of reactions of imines with oxygenated dienes where it is uncertain whether the cyclic product arises via a concerted Diels-Alder cycloaddition mechanism or alternatively through a two-step process involving an initial Mannich-like reaction followed by a Michael ring closure. In many of these experiments, uncyclized Mannich by-products have been isolated in varying amounts. Situations such as these where there is mechanistic ambiguity, as well as those few instances where the reactions clearly proceed via a Mannich mechanism,[1,2] have been included in the chapter since it is difficult to know where to draw the line on coverage.

A comprehensive discussion of each structural type of imino dienophile covers the literature up to the middle of 2004. A general overview of mechanistic, regiochemical, and stereochemical considerations is also presented. However, specific relevant exo/endo issues, remote diastereoselectivity and the use of chiral auxiliaries are discussed under each particular imine type. Separate sections have also been included that describe intramolecular reactions, as well as more recent advances in enantioselective cycloadditions involving chiral catalysts.

ImDA cycloadditions can be either thermal or acid-catalyzed, and a wide array of structurally diverse imino dienophiles can be utilized. Many of the imino dienophiles discussed are highly reactive and/or hydrolytically labile; therefore, they have often been formed in situ. Thus, the nature of the reactive dienophile species is not always clear, being often dependent upon specific reaction conditions. In general, it is assumed that Lewis or Brønsted acid catalyzed cycloadditions proceed via an iminium ion species rather than a neutral imine. In addition, at times one can only surmise as to the reacting geometry of the imine. Where appropriate, a brief discussion of methods used for imine generation is included.

Electron-deficient imines, such as *N*-sulfonyl-, *N*-acyl-, and *C*-acylimines, are the most reactive dienophiles, forming cycloadducts with a wide variety of 1,3-dienes. Unactivated (*N,C*-alkyl/aryl-substituted) and electron-rich (heteroatom-substituted) imino dienophiles can also undergo ImDA reactions with highly reactive electron-rich dienes or under Lewis or protic acid catalysis. A number of detailed reviews on the topic of ImDA reactions have been published over the last twenty-five years.[3–8]

MECHANISM, INCLUDING GENERAL REGIOCHEMICAL AND STEREOCHEMICAL CONSIDERATIONS

A number of mechanistic and theoretical studies of ImDA reactions have appeared in the literature.[9–16] However, for the purposes of this mechanistic overview, ImDA reactions have been classified into three broad categories: thermal reactions of unactivated neutral imines, Lewis acid catalyzed reactions, and Brønsted acid catalyzed cycloadditions, each of which is considered separately.

Initially, theoretical calculations were conducted on reactions of simple formaldimine systems with unactivated dienes (Eq. 1),[9,10] for which comparable experimental data were lacking. Recently, however, more complex systems, such as those depicted in Eqs. 2, 6, 8 and 9, also have been considered.[11,15,16] From all the computational and experimental evidence available to date, a number of conclusions can be made with respect to the mechanism of the ImDA cycloaddition. In particular, it has been established that the ImDA reaction can proceed in either a concerted or stepwise manner, depending on the nature of the reactants and the reaction conditions. It should be noted, however, that there are a number of reports in which experimental results contradict theoretical predictions. These examples are discussed below.

The reaction of 1,3-butadiene and formaldimine (Eq. 1) has been modeled in detail by ab initio molecular orbital calculations.[9,10,12] At all levels of theory (HF/3-21G*, RHF/6-31G*, MP2/6-31G*, and B3LYP/6-31G*), it was found that the reaction proceeds in a concerted but asynchronous manner; at lower levels of theory, a more synchronous transition state was suggested. Of the two possible diastereomeric reaction channels, the exo-lone pair transition state **1** was calculated to be 4.3–5.5 kcal mol^{-1} more stable than the endo orientation in **2**.

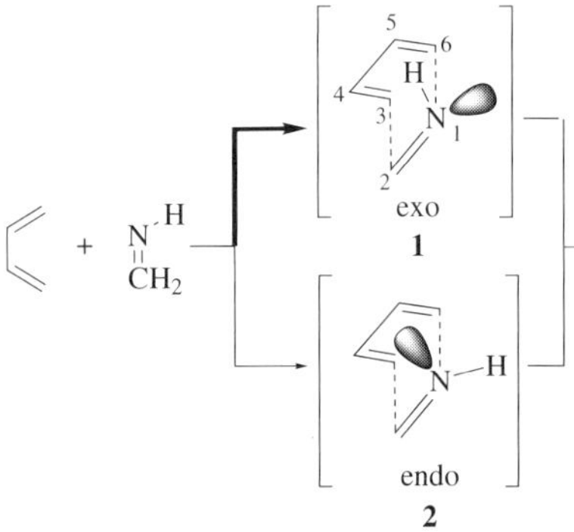
5
6
H
4
3
N
1
2
exo
1
N
H
CH2
+
NH
N
H
endo
2

(Eq. 1)

The exo-lone pair effect also applies to the more highly substituted *N*-methylformaldimine,[9,10] and to a certain extent to more complex systems,[11,15] but as the number of substituents on the imine and the stereoelectronic demands increase, the effect is significantly perturbed.[11] This exo-lone pair preference is thought to arise from electrostatic repulsions between the lone pair on nitrogen and the butadiene π system when the lone pair is endo. Surprisingly, the newly forming N1–C6 bond (exo 1.930 Å; endo 1.982 Å) was calculated to be shorter than the C2–C3 bond (exo

2.356 Å; endo 2.266 Å) in both the exo and endo transition states. This fact appears contradictory from the perspective of frontier molecular orbital (FMO) theory, but has also been found to hold for pyridine-substituted imine **3** (Eq. 2),[15] as well as imines that bear carbonyl groups.[11] Furthermore, a shorter N1–C6 bond would suggest that the direction of electron donation is from the imine to the butadiene, which is somewhat surprising. According to FMO theory, ImDA reactions are predicted to occur through a $HOMO_{diene}$–$LUMO_{dienophile}$ controlled cycloaddition,[17] and thus, the partial C2–C3 bond should be more fully formed than the N1–C6 bond because the carbon of the imine has the larger LUMO coefficient. Apparently dienophile HOMO/diene LUMO interactions are also important in these transition structures. It has also been suggested that the forming bond lengths of the ImDA reaction of formaldimine in Eq. 1 are not as expected because of the presence of unfavorable steric interactions between the two hydrogens on C2 and C3 in transition state **1**, which arise because the system twists in the transition state in order to minimize lone pair-π system repulsion.[12]

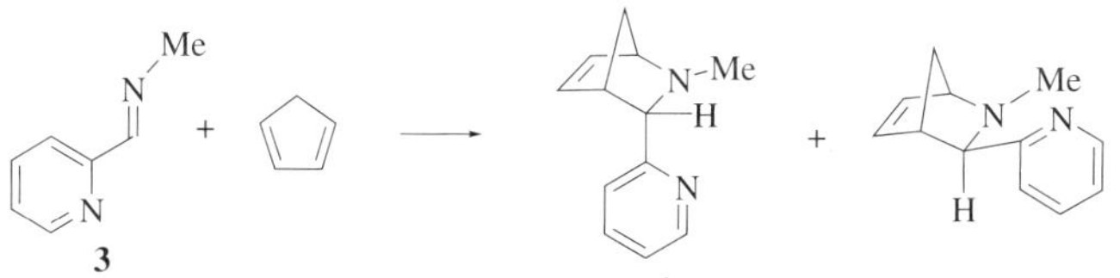
Me
N
N
+
N-Me
H
N
+
Me
N
N
H
3 endo exo (Eq. 2)

Semiempirical (PM3), as well as ab initio (HF/6-31G* and B3LYP/6-31G*) calculations, conducted on the thermal reaction between pyridine-substituted imine **3** and cyclopentadiene (Eq. 2) also found that the reaction proceeds in an asynchronous, concerted manner through a cyclic transition state.[15] The large activation energy value (45 kcal mol^{-1}) predicted for the reaction would indicate that experimentally this reaction should not take place. As yet there is no literature evidence to suggest that such a reaction has ever been attempted.

It has been calculated that Lewis acid coordination to the imine nitrogen lowers the activation energy of the ImDA cycloaddition (ca. 2.5 kcal mol^{-1}, MP2/6-31G*) by lowering the dienophile LUMO energy. Although there is a preference for the coordinating species to occupy the exo position, it is somewhat less than the lone-pair preference in uncomplexed imine systems (ca. 3.6 kcal mol^{-1}).[9,10] In Lewis acid coordinated systems, the partially formed N1–C6 bond (exo 2.310 Å; endo 2.319 Å) is calculated to be longer than the forming C2–C3 bond (exo 2.002 Å; endo 2.025 Å), and the reaction therefore is asynchronous in nature. In these reactions, the Lewis acid catalyst is strongly coordinated to the nitrogen atom of the imine or cycloaddition product. As a consequence, product complexation can deactivate or inhibit the catalyst, and therefore, stoichiometric amounts of the catalyst are often required in order to achieve complete conversion.

Upon Brønsted acid protonation of the nitrogen of formaldimine, the imine carbon atom in cation **4** becomes highly electrophilic. The result of this protonation is that the mechanism for the corresponding ImDA reaction with 1,3-butadiene, which is considered to be concerted for neutral unactivated imines or Lewis acid catalyzed ImDA reactions, is calculated to shift to a stepwise process that involves a tandem Mannich-Michael type reaction (Eq. 3). As depicted, the N1–C6 and C2–C3 forming bond lengths calculated for transition state **5** are 3.058 and 1.919 Å, respectively.[10] From the planar geometry about C5 and C6 it can be concluded that the N1–C6 bond is essentially unformed. The structures of the possible intermediates resemble allylic cations such as **6**, with the C2–C3 bond lengthened from 1.602 to 1.773 Å as a result of hyperconjugation with the allylic system. This reaction is expected to be highly exothermic (-32 kcal mol^{-1}). Solvated systems for both the corresponding Lewis and Brønsted acid catalyzed reactions were also modeled computationally and showed little difference in comparison to the respective gas-phase reaction models.

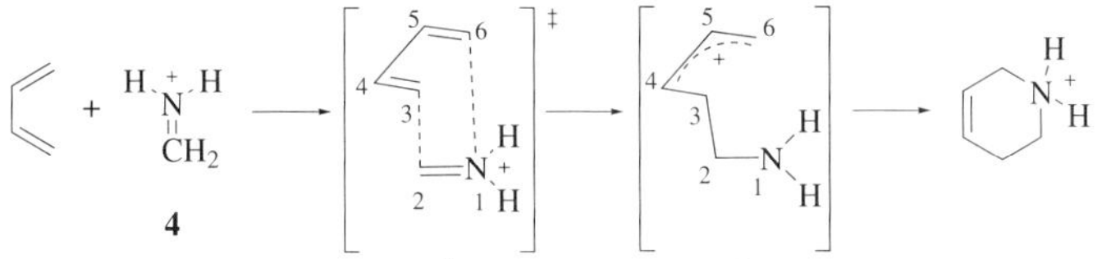
H
N
H
CH2
5
6
4
3
H
N
2
1 H
‡
5
6
4
3
H
N
2
1
H
H
N
H

4 **5** **6** (Eq. 3)

NMR kinetic studies conducted on the ImDA reaction of the *N,N*-dimethylmethyleneammonium ion (**7**) under aprotic conditions with a number of different 1,3-dienes (Eq. 4) also indicate that this reaction proceeds by a stepwise mechanism via allyl cation **8**, and thereby support the results obtained from ab initio calculations.[13]

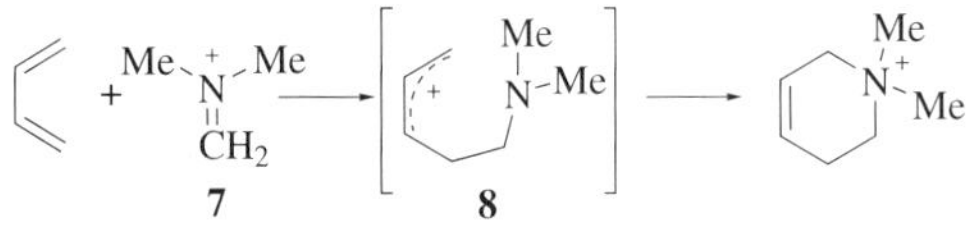
Me
N
Me
CH2
Me
N-Me
Me
N
Me

7 **8** (Eq. 4)

On the other hand, experimental kinetic studies of the ImDA reaction of the iminium cation **7** and cyclopentadiene ruled out a stepwise mechanism. These studies concluded that the transition state for the pericyclic process is 6.5 kcal mol^{-1} lower in energy than that calculated for the transition state in the stepwise process.[13] The preference for a concerted pathway in this example was believed to be a consequence of the s-cis-fused diene system. More recently, density functional theory calculations carried out for the same reaction using the B3LYP/6-31G* basis set have contradicted the conclusions of the experimental kinetic findings.[12] The computations also rule out the formation of diradical species in this reaction. Furthermore, analysis of the geometry of the transition state indicate a highly asynchronous and polar reaction in which the carbon atom of the iminium salt **7** undergoes nucleophilic attack by cyclopentadiene giving an acyclic carbocation intermediate like **9** (Eq. 5). These calculations indicate that the N1–C6 bond is essentially unformed in the transition state (3.135 Å), that the C2–C3 bond has taken on sp^3-sp^3 character, while the C6 carbon remains sp^2 hybridized, in agreement with a C4–C5–C6 allylic cation structure.

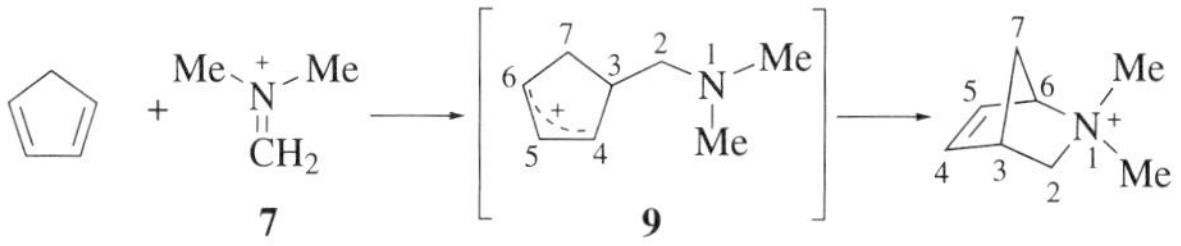
Me
N
Me
CH2
7
Me
N
Me
9
Me
N+
Me

(Eq. 5)

Similarly, cycloaddition of the *C*-pyridine-substituted diprotonated iminium salt **10** and cyclopentadiene (Eq. 6) in strong acid media was calculated to proceed by a stepwise process through either of the two intermediates corresponding to **11** or **12** in which the imine nitrogen in the acyclic intermediate is sp^3 hybridized.[15,16] The exo transition structure **12** was predicted to be only 0.4 kcal mol^{-1} more favorable than the endo transition state, a preference which is significantly diminished in comparison to the totally unprotonated species (3.2 kcal mol^{-1}, Eq. 2). Protonation of the imine nitrogen atom reduces the unfavorable interactions between the N1 lone pair and the π system of the diene, thereby lowering the exo selectivity for the protonated process.

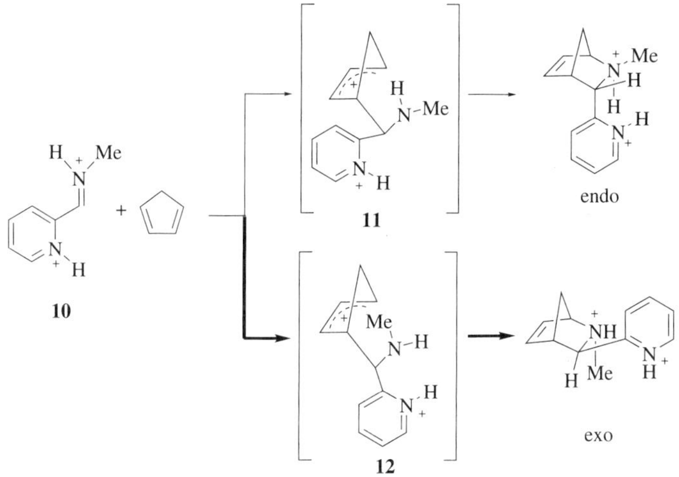
10
11
12
endo
exo

(Eq. 6)

It is well established that electron-withdrawing carbonyl and sulfonyl substituents activate imines toward ImDA reactions. Ab initio calculations at the HF/3-21G* and MP2/6-31G* levels of theory, and FMO theoretical analysis of thermal ImDA reactions that involve electron-deficient imines, closely parallel observed experimental reactivities.[11] Dienophiles become increasingly more reactive as the number of electronegative substituents increases because each substituent lowers the LUMO dienophile energy, and in turn the enthalpy of activation. Protonation of the imino dienophile has a similar effect, also stabilizing the transition states.[15] Calculations indicate that *C*-acyl imines should be more reactive than *N*-acyl imines, which in turn should be more reactive than the corresponding sulfonyl counterparts.[11] This difference in reactivity between imines that bear acyl and sulfonyl substituents can be rationalized on the basis that unlike acyl substituents, sulfonyl groups do not participate in secondary orbital overlap interactions.

In summary, theoretical calculations for ImDA reactions of unactivated neutral imines and Lewis acid catalyzed reactions indicate that both proceed through a concerted but asynchronous mechanism.[9–12,15] On the other hand, it is clear from computational studies and the limited experimental mechanistic evidence available that Brønsted acid catalyzed ImDA reactions proceed in a stepwise manner via intermediates such as **6** in Eq. 3.[10,12,13,16]

Mechanistic and theoretical studies have established that ImDA reactions, whether thermal or acid-catalyzed, have significant dipolar character in the transition state. Thus, the regiochemical outcome for imino dienophile cyclizations with unsymmetrical dienes can often be qualitatively predicted by invoking the simple mechanistic model shown in Eq. 7. Of the four possible dipolar forms **13–16**, intermediates **13** and **14** lead to regioisomeric tetrahydropyridine **17**, whereas intermediates **15** and **16** give the alternate isomer **18**. From the available data, structures **14** and **16** are only applicable in thermal cycloadditions in which the X and Y groups on the imino dienophile carbon are electron-withdrawing, and thereby have good carbanion-stabilizing properties.[4] In all other thermal, as well as acid-catalyzed reactions, only the relative stabilities of intermediates **13** and **15** need be considered.

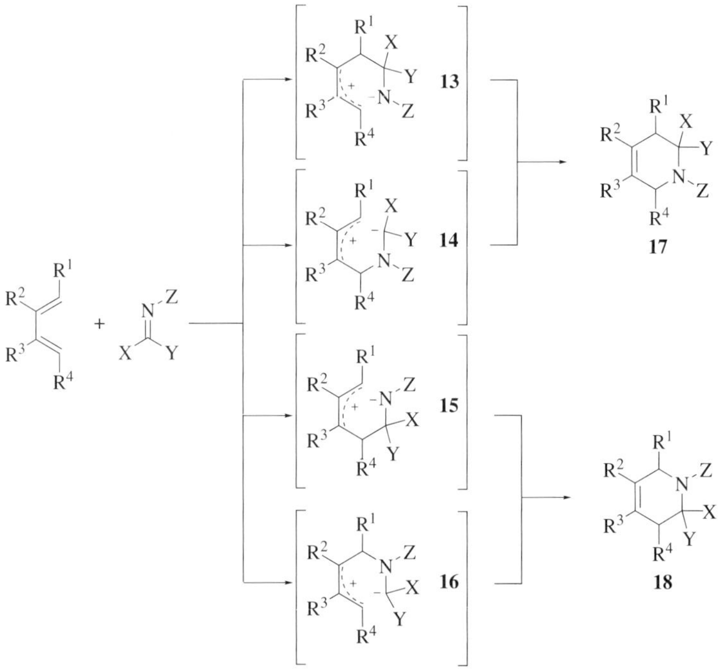
R1
R2
R3
R4
N
Z
X
Y
+
13
14
15
16
17
18

(Eq. 7)

ImDA reactions often show excellent stereoselectivity, but ambiguities can arise because the nitrogen atom in both the reactant imine and ImDA adduct undergoes lone pair inversion.[18] In many instances, when an acyclic imine is used, the dienophile apparently prefers to adopt an E geometry in order to minimize unfavorable steric interactions.[19] However, one cannot be certain whether the actual reacting dienophile has the E or Z geometry because imine isomerization is reasonably facile. Thus, application of the Alder rule of endo addition becomes difficult, if not impossible, and as a result, stereochemical issues usually require analysis on a case by case basis. Of course, these problems do not arise when the imine geometry has been fixed as Z by virtue of ring constraints.

Theoretical calculations[11] and experimental data[18,20–30] indicate that carbonyl functions preferentially adopt an endo orientation in the transition state. In contrast, both computational[11] and experimental[31–39] evidence suggests that sulfonyl substituents generally prefer to be exo disposed. When both carbonyl and sulfonyl functionalities are present, stereochemical selectivities appear to be reversed. For example, cycloaddition of imine **19** with cyclopentadiene gives exclusively product **21** (Eq. 8).[40]

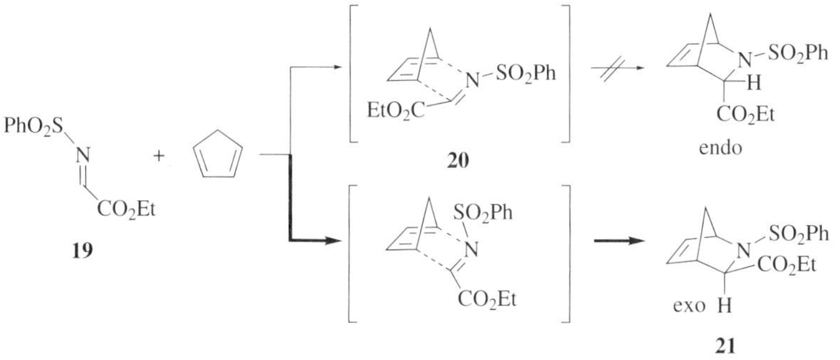
PhO2S
N
CO2Et
19
+
EtO2C
N-SO2Ph
20
N-SO2Ph
H
CO2Et
endo
SO2Ph
N
CO2Et
SO2Ph
N
CO2Et
exo H
21

(Eq. 8)

These results highlight the limitations of applying a molecular orbital interpretation to all ImDA reactions, since from a molecular orbital analysis, the reaction would be expected to proceed through transition state **20**, which combines both an endo *C*-carbonyl and an exo sulfonyl moiety. However, the observed stereocontrol leading exclusively to cycloadduct **21** can be rationalized, if one considers the reaction to be reversible and the outcome to be governed by thermodynamic considerations. In general, *N*-sulfonyl imines are thermodynamically more stable than their carbonyl-substituted counterparts. Thus, the overall transition state energy of the reaction for the former dienophile is lower. This fact supports the hypothesis that reactions involving *N*-sulfonylimines have greater potential for thermodynamic reversibility.[11] Theoretical calculations conducted for thermal ImDA cycloadditions of dienophiles in which both *C*- and *N*-carbonyl functions are present indicate that *C*-carbonyls should show a preference for endo disposition.[11] However, experimental evidence shows that this hypothesis is incorrect, and that, in fact, the major products are usually the exo *C*-acyl cycloadducts.[41–44]

Although a plethora of studies have been conducted into the mechanism of the ImDA reaction, extensive studies concerning the diastereoselectivity and enantioselectivity of reactions involving chiral auxiliaries are lacking. Recently, semiempirical (PM3) calculations were conducted on the reaction between the protonated chiral imine **22** and cyclopentadiene (Eq. 9).[15] The reaction proceeds in a stepwise fashion, whereby the first step involves nucleophilic attack at the electron-deficient carbon atom of the iminium salt **22** by the electron-rich cyclopentadiene; the subsequent step involves ring closure. Of the four possible cycloadducts, the major product was predicted to be **23**, which would arise from exo addition to the Si-face of the dienophile. Indeed, the theoretical calculations generally agree with the observed experimental results, as **23** and **24** are produced in an 87:13 ratio in the reaction.[45] The exo selectivity is presumably because of steric interactions along the endo approach between the pyridinium cation and the cyclopentadiene, while the facial selectivity results from minimization of steric hindrance between the bulky chiral substituent of the imine and the cyclopentadiene.

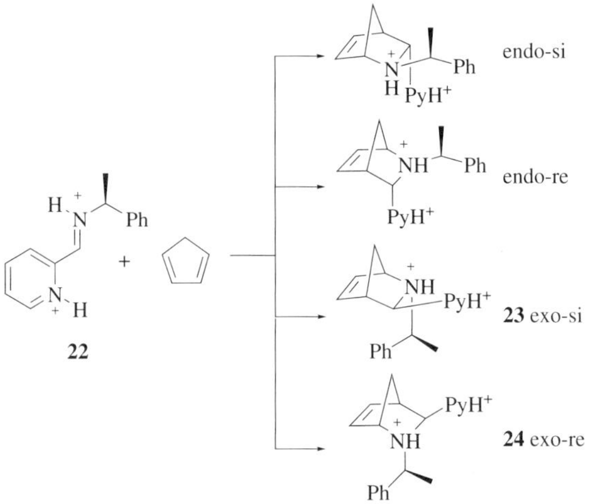
H
N
Ph
N
H
+
22
endo-si
endo-re
23 exo-si
24 exo-re

(Eq. 9)

Finally, a number of reactions of electron-rich oxygenated dienes with various imines under Lewis acid catalysis provide adducts that may arise by a stepwise Mannich-Michael process, rather than via an actual concerted [4+2]-cycloaddition mechanism. Thus, in the example shown in Eq. 9a there is reasonable mechanistic evidence based partly on isolation of acyclic by-products that vinylogous amide **26** is in fact formed via the initial Mannich intermediate **25**, which then undergoes a subsequent Michael-type cyclization to afford the observed product.[46] Situations such as this have generally been noted in the appropriate tables.

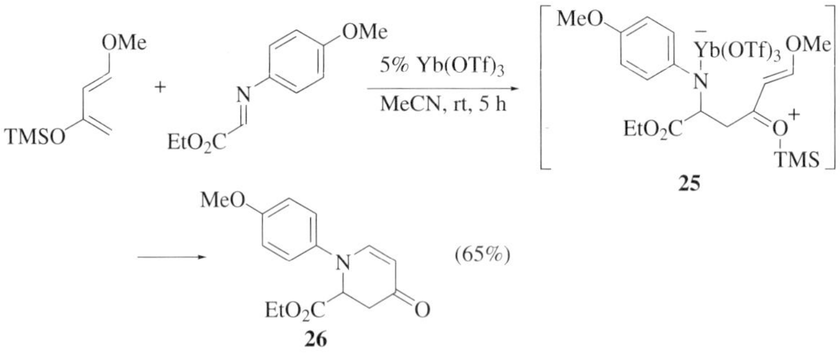
OMe
TMSO
+
OMe
N
EtO2C
5% Yb(OTf)3
MeCN, rt, 5 h
MeO
Yb(OTf)3
OMe
N
EtO2C
O
TMS
25
MeO
N
EtO2C
O
26
(65%)

(Eq. 9a)

SCOPE AND LIMITATIONS

Intermolecular Cycloadditions

Acyclic *N*-Acylimines and *N*-Cyanoimines. The earliest examples of [4+2]-cycloadditions of *N*-acylimines involve transient intermediates generated from the corresponding bis-carbamates in the presence of a catalytic amount of boron trifluoride etherate (Eq. 10).[47] Subsequent ^{1}H NMR investigations indicate that for substituted bis-carbamates a protonated (*E*)-*N*-acyliminium dienophile is probably the reactive species.[31] A number of symmetrical cyclic and acyclic dienes have been utilized for these cycloadditions, affording the expected 1,2,5,6-tetrahydropyridine adducts in variable yields.[47–50]

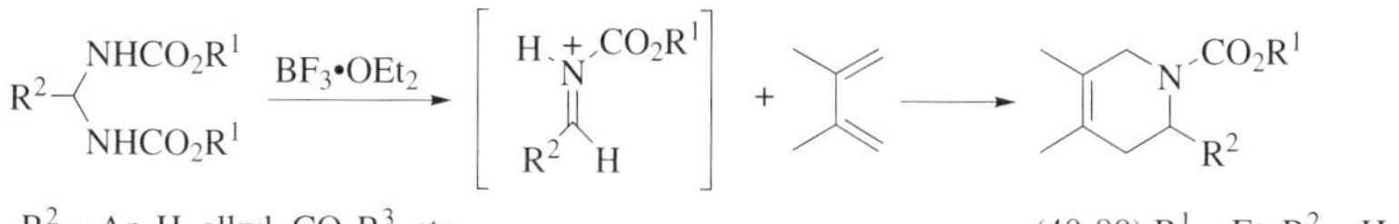

(Eq. 10)

R^2 = Ar, H, alkyl, CO_2R^3, etc. (40-80) R^1 = Et, R^2 = H

Reactions of *C*-unsubstituted,[47,51] *C*-substituted,[24,30,41,47,52–57] or *C*-disubstituted[58–60] *N*-acylimines or iminium ions with unsymmetrical dienes generally proceed with high regioselectivity, giving products in better than 9:1 ratios. For example, imine **27** adds to isoprene giving a mixture of regioisomeric adducts.[47] Formation of the major product can be rationalized on the basis of the mechanistic model illustrated in Eq. 7 via the dipolar forms **13** and **15** (see Mechanism section).

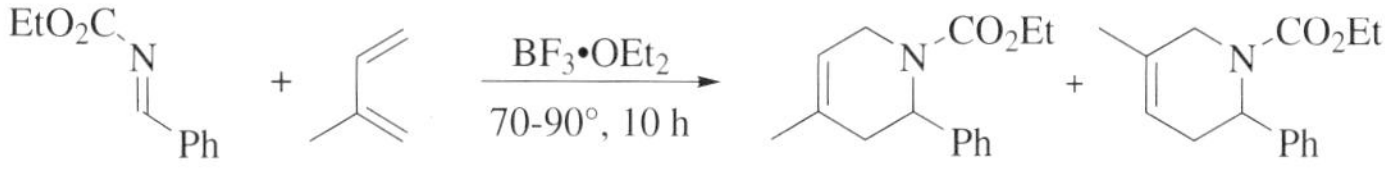

27 >90:10 (40-80%)

Similarly, *C*-phosphonate dienophile **28** reacts with Danishefsky's diene (**29**) to exclusively afford one adduct of dihydropyridinone **30**.[57] Triacylimino dienophiles also usually show good regioselectivity (Eq. 11).[41,59,60] However, predicting the regiochemistry of addition with these triacylated dienophiles is not straightforward, as it can be envisaged that all four dipolar forms **13–16** are of similar energy.

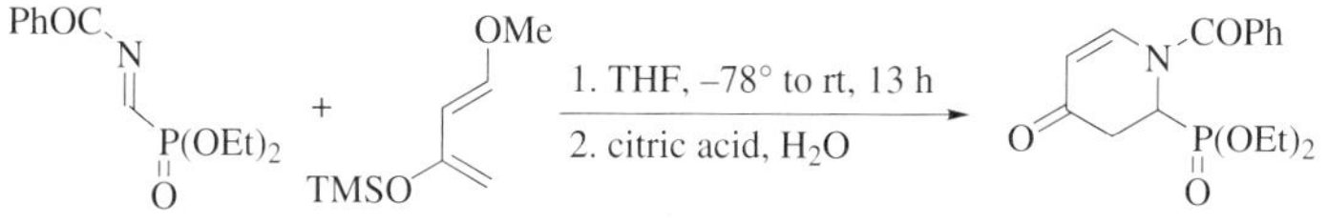

28 **29** **30** (31%)

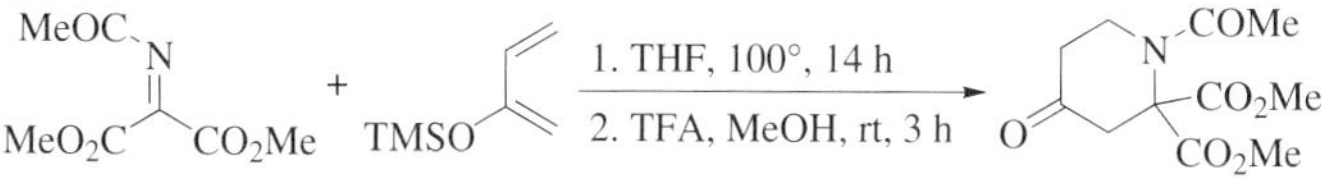

(16%) (Eq. 11)

The stereoselectivity of a large number of thermal or $BF_3{\cdot}OEt_2$-catalyzed ImDA reactions of cyclohexadienes and cyclopentadienes with *C*-substituted *N*-acylimines has been investigated (Eq. 12). Both *C*-alkyl and *C*-aryl substituted *N*-acylimines give predominantly the exo product in about a 4 : 1 or higher ratio, except when R^2 is α-branched or is highly substituted.[18,22,23,25–27,29,31,61] In the latter situations, the reaction generally affords a 1 : 1 mixture of cycloadducts, or provides the endo product in excess.[18,20,21,23,28,31] This result can be rationalized by assuming the existence of a concerted reaction pathway that involves an (*E*)-*N*-acyliminium ion[31] in which the *N*-acyl group prefers to be endo to the diene owing to secondary orbital interactions.[11] However, when the *C*-substituent on the imine is bulky, steric considerations come into play. Interaction of the dienophile substituent with the bridge of the cyclic diene has a destabilizing effect in a transition state in which the *N*-acyl group is endo. Thus, a reversal of stereochemistry is observed. Interestingly, thermal or $AlCl_3$-catalyzed cycloaddition of *N*-(2,2,2-trichloroethylidene)acetamide **31** with cyclopentadiene does not follow this trend, giving instead exclusively the exo product in moderate yields.[20,21] On the other hand, reaction of benzylidenecarbamic acid phenyl ester **32** with 1,3-cyclohexadiene affords the exo adduct in only 38% diastereomer excess.[32]

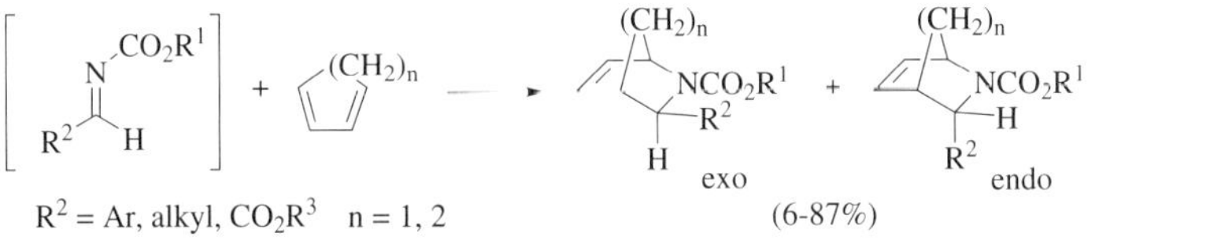
CO2R1
N
R2
H
(CH2)n
(CH2)n
NCO2R1
R2
H
exo
(CH2)n
NCO2R1
H
R2
endo
R2 = Ar, alkyl, CO2R3   n = 1, 2
(6-87%)

(Eq. 12)

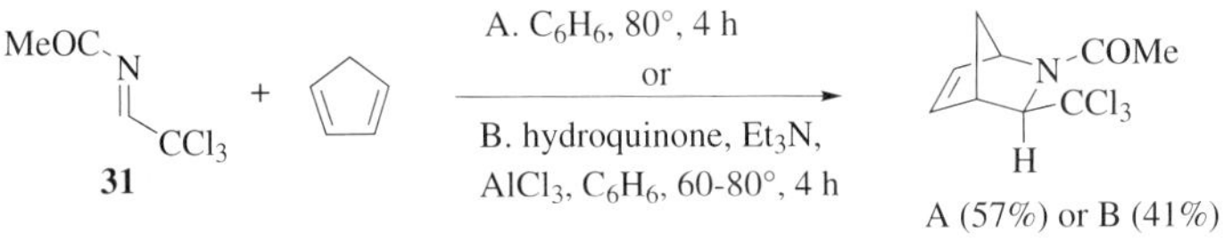
MeOC
N
CCl3
31
A. C6H6, 80°, 4 h
or
B. hydroquinone, Et3N,
AlCl3, C6H6, 60-80°, 4 h
N
COMe
CCl3
H
A (57%) or B (41%)

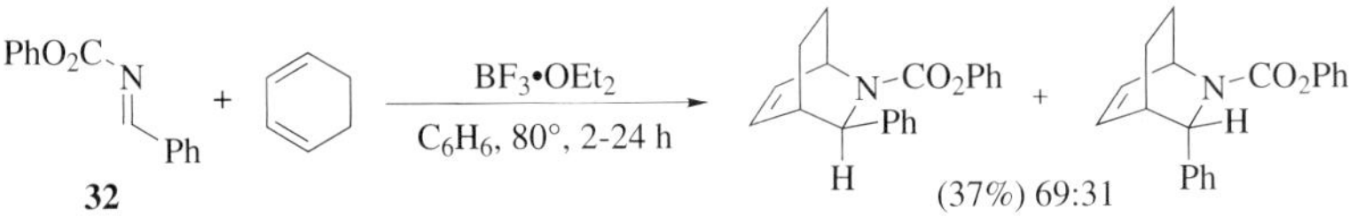
PhO2C
N
Ph
32
BF3•OEt2
C6H6, 80°, 2-24 h
N
CO2Ph
Ph
H
(37%) 69:31
N
CO2Ph
H
Ph

The relative endo-directing ability of an *N*-acyl group has been found to be stronger than that displayed by a competing *C*-acyl substituent in imines that bear both *N*- and *C*-acyl moieties. Therefore, Lewis acid catalyzed cycloadditions of such dienophiles with 1,3-cyclohexadienes typically give exo isomers as the major product (Eq. 13).[23,29,62] Analogous ImDA cyclizations of such dienophiles with cyclopentadiene under thermal conditions are reported to give rise to the exo adducts exclusively.[41–43] Interestingly, these experimental results directly contradict predictions based on theoretical calculations (see Mechanism section).

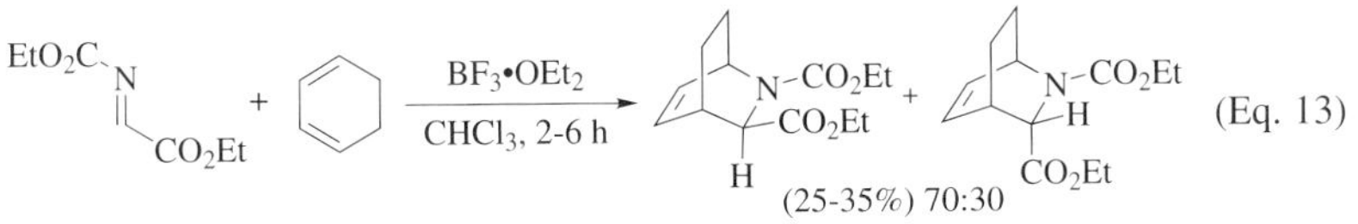

When allowed to react with *C*-aryl substituted imines, medium-ring cyclic dienes such as 1,3-cycloheptadiene or dialkylated 1,3-cyclohexadiene **33** also give the corresponding ImDA products, but with no or considerably lower diastereoselectivity.[28,61]

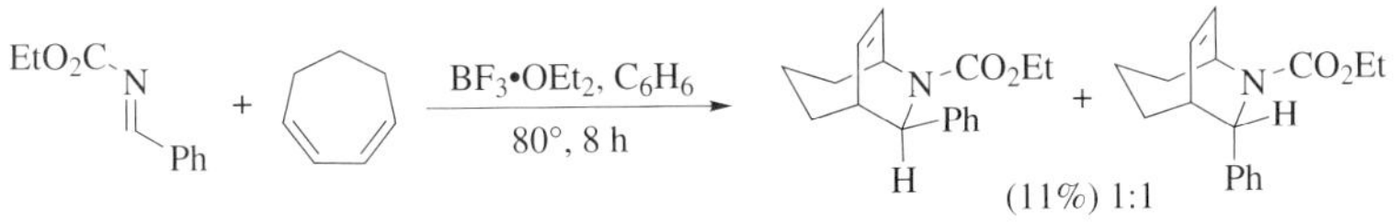

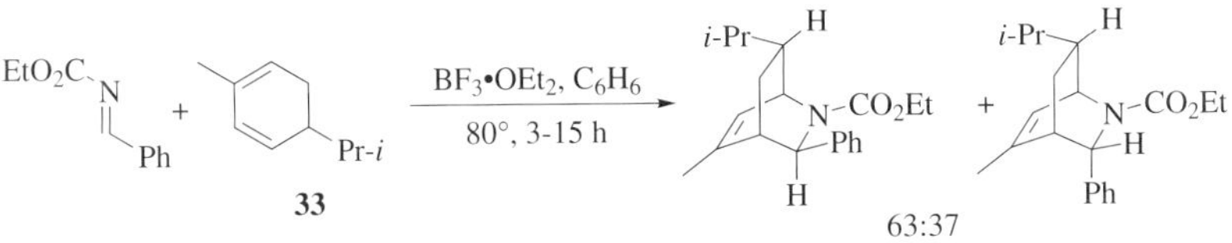

Cycloaddition of *N*-acyliminium ion **34** with the tricyclic tetraene **35** is found to proceed primarily from the face opposite the tetrahydrofuran ring, as would be expected on the basis of steric effects.[63]

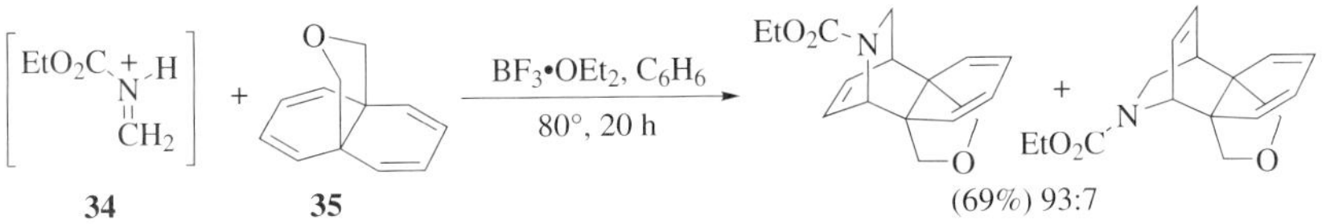

Several dehydroglycyl peptides such as imine **36** have been prepared through the corresponding α-chloroglycyl derivatives. These dienophiles react with cyclopentadiene, and constitute the only examples to date of cycloadditions of chiral *N*-acylimines.[44] The reaction of imine **36** with cyclopentadiene proceeds with high stereoselectivity, giving the expected exo cycloadduct in 80% diastereomer excess. The 3S configuration of the major product was determined by X-ray crystallographic analysis of the dioxopiperazine derivative. The observed 3S stereochemistry can be rationalized by assuming that the *N*-acylimino ester adopts the E configuration, and that the diene approaches endo to the *N*-acyl group from the less hindered α-face, opposite to the sterically demanding alkyl group. Endo addition to the E *N*-acylimine from the sterically more hindered β-face would result in formation of the minor diastereomer.

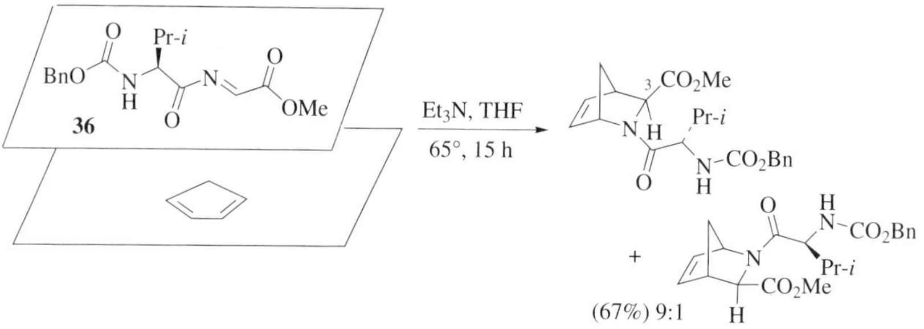

Treatment of *N*-benzylidene methyl carbamate **37** with the 1-aryl-3-alkyldiene **38** in toluene at low temperature employing $BF_3 \cdot OEt_2$ as catalyst affords exclusively cycloadduct **39** in 95% yield. The two phenyl substituents are trans disposed in the product, as is expected on the basis of a transition state with the carboxylic ester substituent endo to the diene.[30]

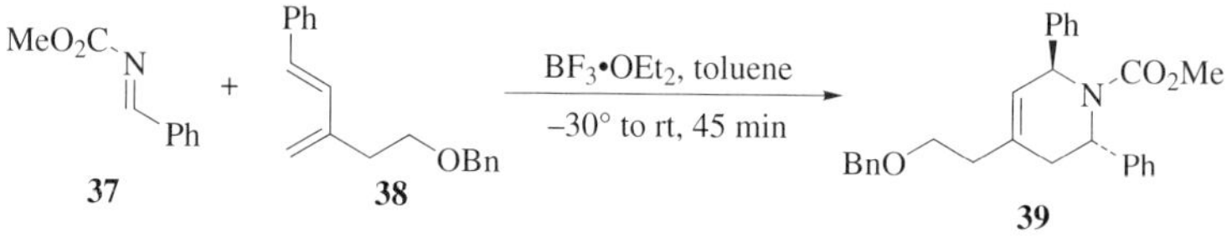

Only a few *N*-cyanoimine Diels-Alder reactions have been reported to date, probably because of the tendency of these imines to polymerize. Tricyanoimine **40** reacts with 2,3-dimethyl-1,3-butadiene and cyclopentadiene in benzene at room temperature, giving the corresponding cycloadducts **41** and **42**, respectively, in moderate yields (Eq. 14).[64,65] Treatment of *C*-acyldicyanoimine **43**, which is prepared in situ by thermal decomposition of methyl 3,3-diazido-2-cyanoacrylate, with 2,3-dimethylbutadiene also gives tetrahydropyridine **44** cleanly as a crystalline solid, whereas reaction with cyclopentadiene affords an inseparable mixture of the endo and exo bicyclic adducts **45** and **46**.[65]

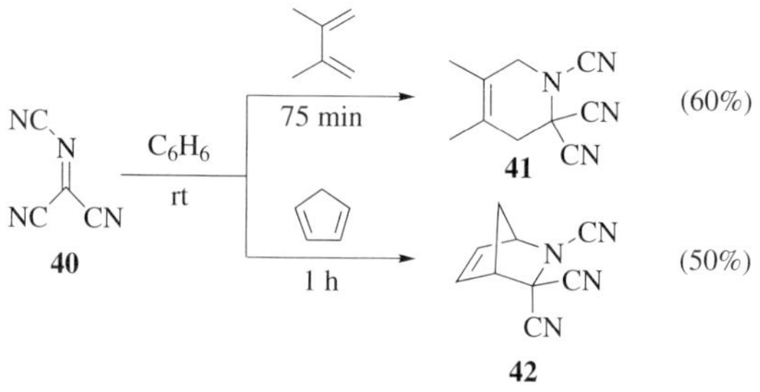

(Eq. 14)

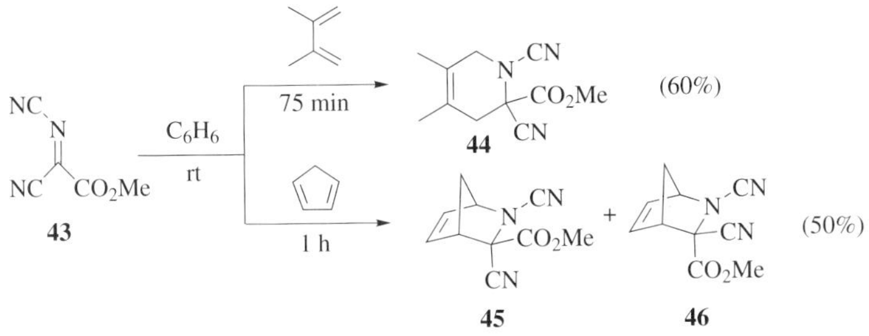

Cyclic *N*-Acylimines. Although a number of structurally diverse cyclic *N*-acylimino dienophiles have been utilized in ImDA reactions, the most common systems are dehydrohydantoins **48**. Two methods have been developed for the in situ generation of such species: thermal or acid-promoted elimination of methanol from α-methoxyhydantoin **47**,[66-70] and *N*-chlorination–dehydrochlorination of a mono-substituted hydantoin **49**.[71]

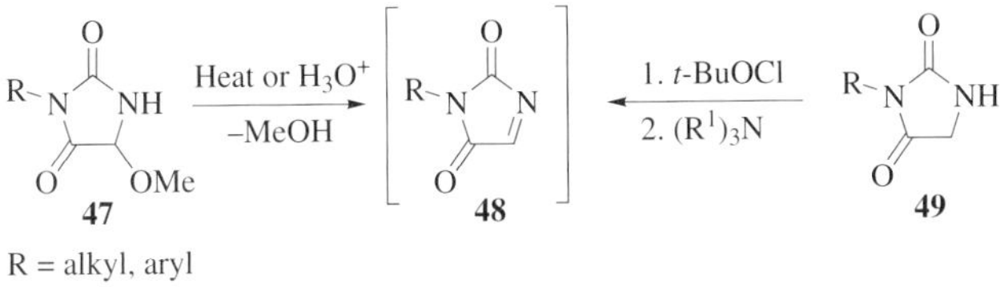

There is no ambiguity about the reacting configuration of these cyclic *N*-acylimines. Cycloadditions of dehydrohydantoins with unsymmetrical dienes proceed both regio- and stereoselectively via a transition state in which the two carbonyl groups are clearly endo to the diene (Eq. 15).[66] The observed regiochemistry of the cycloadducts is again consistent with dipolar forms **13** and **15** in Eq. 7. Of note is the fact that regioselective addition does not occur when isoprene is used as the diene; instead, a 60:40 mixture of regioisomers **50** and **51** is isolated.[66,71]

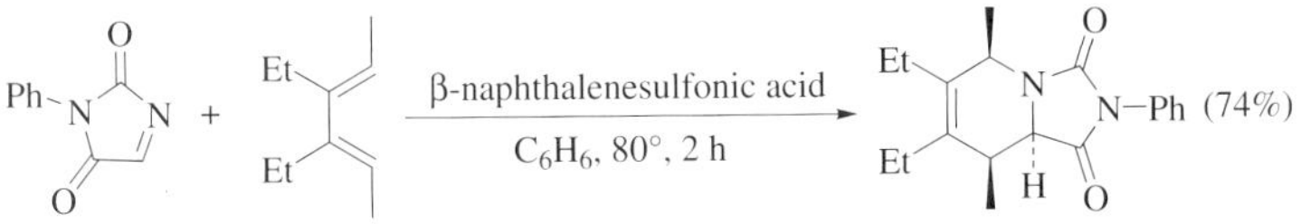

(Eq. 15)

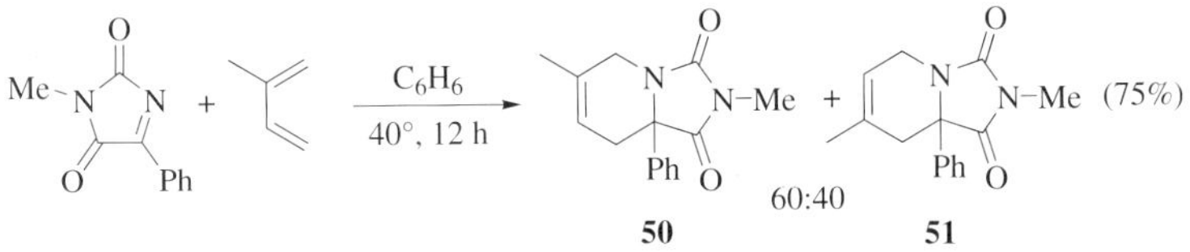

A regiochemical study that was conducted using several trisubstituted dienes with dehydrohydantoin **52**, derived thermally from methoxy compound **47**, indicates that as the R substituent becomes more electron-donating, the ratio of regioisomers changes in accord with the qualitative mechanistic model proposed in Eq. 7.[69] These cycloadducts also tend to epimerize readily at C4 during chromatographic purification.

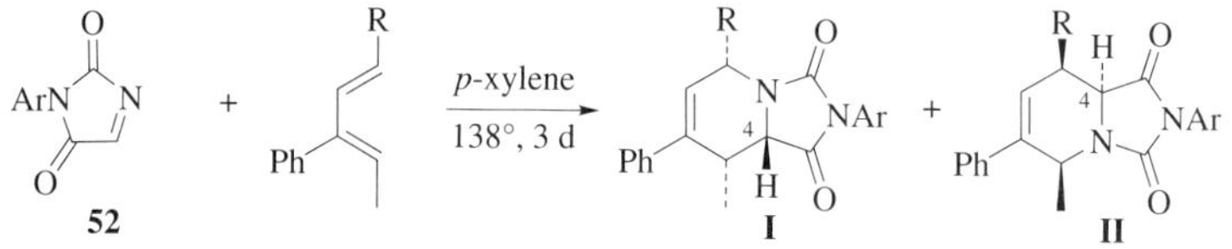

52 Ar = 4-ClC_6H_4 **I** **II**

R	I	II	I:II
CO_2Me		(13%)	0:100
H	(2%)	(22%)	8:92
CH_2OAc	(34%)	(22%)	55:45
Me	(26%)	(11%)	67:33
Et	(37%)	(9%)	75:25

2,5-Diaza-2,4-cyclopentadienone (**53**), generated from an insoluble polymer-bound precursor, undergoes cycloaddition with the polymer-bound ester **54** of 2-furoic acid giving cycloadduct **55**, which on hydrolysis affords the bicyclic adduct **56**, albeit in poor yield.[72]

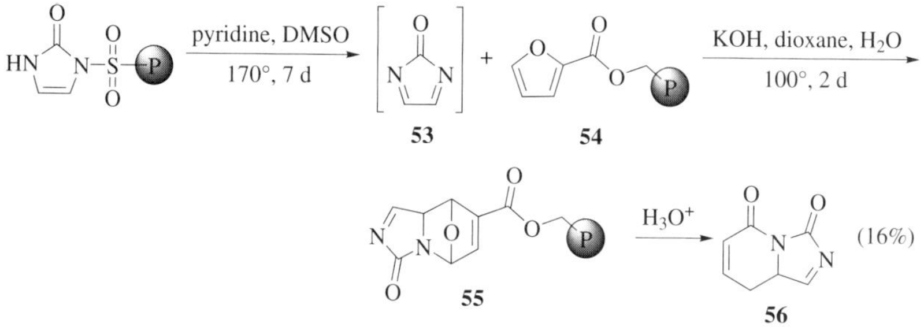

53 **54** **55** **56** (16%)

Reaction of dehydropyrrolidinone **57** under acidic conditions with a number of cyclic and acyclic 1,3-dienes gives exclusively endo products such as that shown in Eq. 16. However, a 1 : 1 mixture of stereoisomers **58** and **59** is obtained when imine **57** and 2-methyl-1,3-pentadiene are stirred with formic acid at room temperature for 5 hours. This result suggests that steric repulsion between the C3 methyl group of the diene and a methylene group of the imine, as depicted in Figure 1, competes with the electronic preference for endo products, and results in a lack of stereoselectivity.

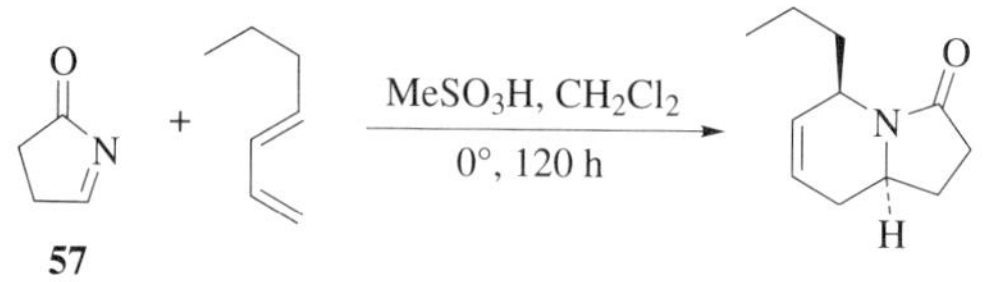

57 (21%) (Eq. 16)

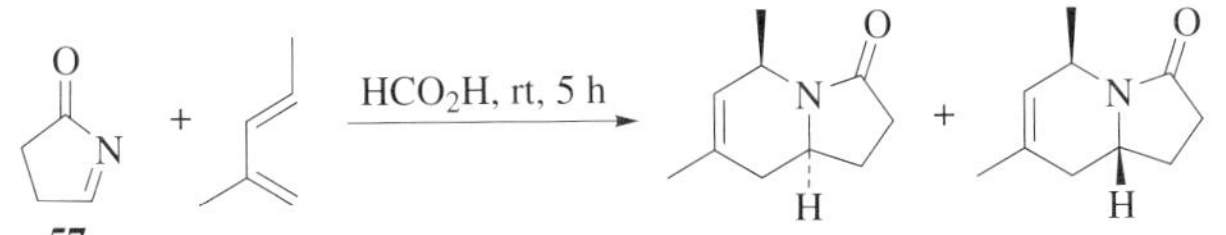
HCO2H, rt, 5 h

57 58 (70%) 1:1 59

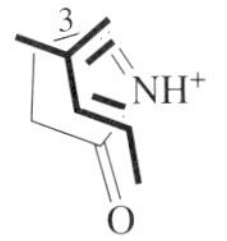

Figure 1 Endo transition state showing the steric repulsion between C3 methyl of the diene and the methylene of the cyclic iminium ion.

Interestingly, titanium tetrachloride catalyzed ImDA reactions of *N*-Boc iminium ion **60** with Danishefsky's diene has been shown to provide either the 5β-isomer **61** or 5α-isomer **62**, simply by changing the order of addition of reagents.[73] Addition of the diene to a mixture of the Lewis acid and the iminium ion precursor gives exclusively tetracycle **61**, whereas addition of the Lewis acid to a mixture of the diene and the iminium ion precursor affords adduct **62**. It was suggested that the former pathway proceeds via initial cleavage of the Boc group forming a Lewis acid complexed imine, whereas the latter reaction proceeds via a metal complexed *N*-acyliminium ion species.

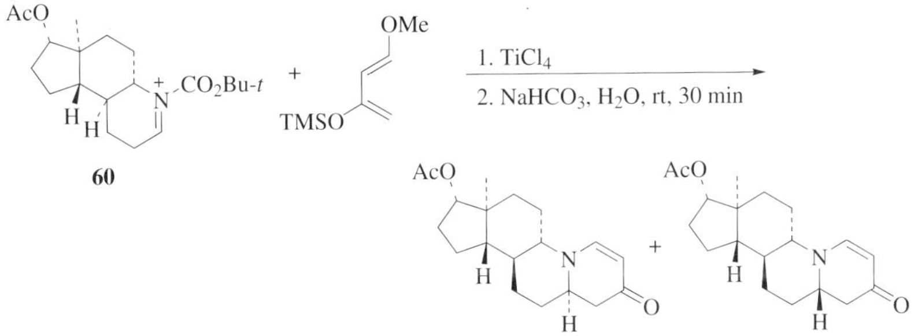
1. TiCl4
2. NaHCO3, H2O, rt, 30 min

60

61 (3-32%) 62

Both benzoxazinones and benzothiazinones have been found to be highly reactive imino dienophiles, which afford the expected endo [4+2]-cycloadducts in good to excellent yields (Eq. 17).[74,75]

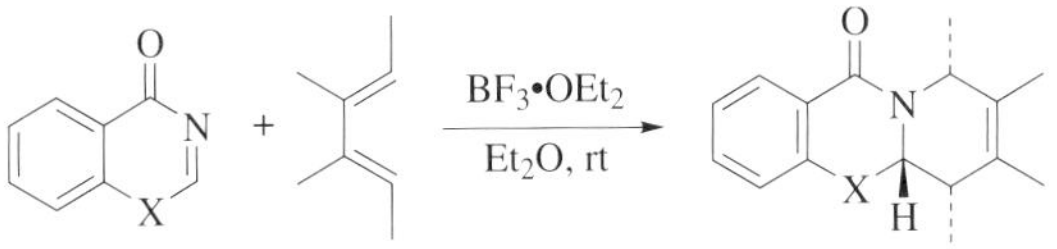
BF3•OEt2
Et2O, rt

(Eq. 17)

X = O (86%)
X = S (50%)

Azacyanoquinones such as **64**, which can be generated in situ by thermolysis of the corresponding 2,3-diazido-1,4-benzoquinone **63**, undergo cycloadditions with both cyclic and acyclic dienes giving adducts such as **65** and **66**, respectively.[76] Configurations of these products have not been determined, but the reactions proceed about ten times faster than the analogous cycloadditions of dehydrohydantoins.

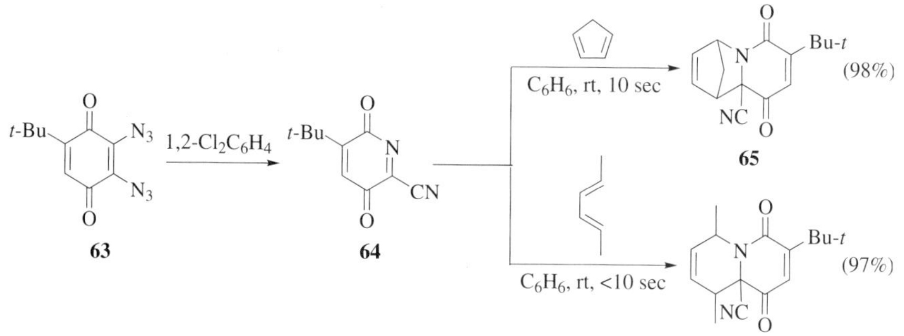
t-Bu
N3
N3
O
O
1,2-Cl2C6H4
t-Bu
N
CN
O
O
C6H6, rt, 10 sec
N
Bu-t
(98%)
NC
O
O
C6H6, rt, <10 sec
N
Bu-t
(97%)
NC
O
O

63 **64** **65** **66**

Azetinone **68**, which is postulated to be the reactive intermediate generated when 4-acetoxyazetidinone **67** is heated at reflux in acetonitrile with zinc chloride, has been trapped successfully with a series of siloxydienes.[77-79] The cycloadducts always arise from endo addition of the acylimine to the diene. Surprisingly, mixtures of diastereomers in undisclosed yield are isolated when pure Z diene **69** is employed, even though one would expect that the reaction should provide only the 5α product **70**. This outcome is rationalized on the basis that the diene is undergoing Z/E isomerization under the Lewis acidic reaction conditions.

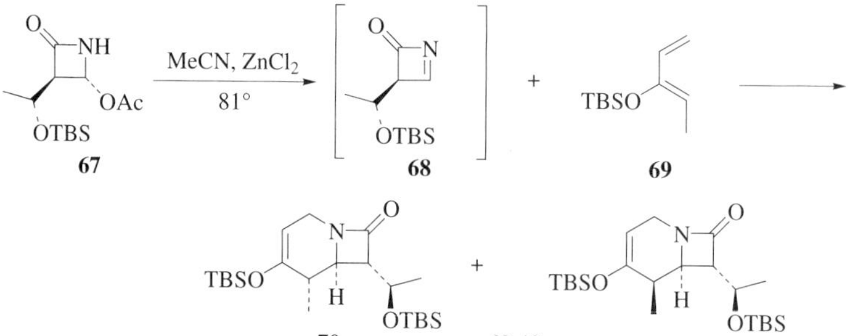
O
NH
OAc
OTBS
MeCN, ZnCl2
81°
O
N
OTBS
+
TBSO
TBSO
N
O
H
OTBS
+
TBSO
N
O
H
OTBS
60:40

67 **68** **69** **70**

Acyclic *C*-Acylimines. [4+2]-Cycloadditions of various acyclic *C*-acylimines with 1,3-dienes under thermal conditions, or using acid catalysis, proceed with complete regioselectivity giving single adducts in good yields. A number of examples of this process are depicted below.[19,53,80,81] Although the observed regiochemistry can be rationalized on the basis of the mechanistic model shown in Eq. 7, via the dipolar forms **13** and **15**, the adducts obtained when electron-rich Danishefsky-type dienes are utilized may arise by a stepwise Mannich-Michael process rather than by concerted cycloaddition.

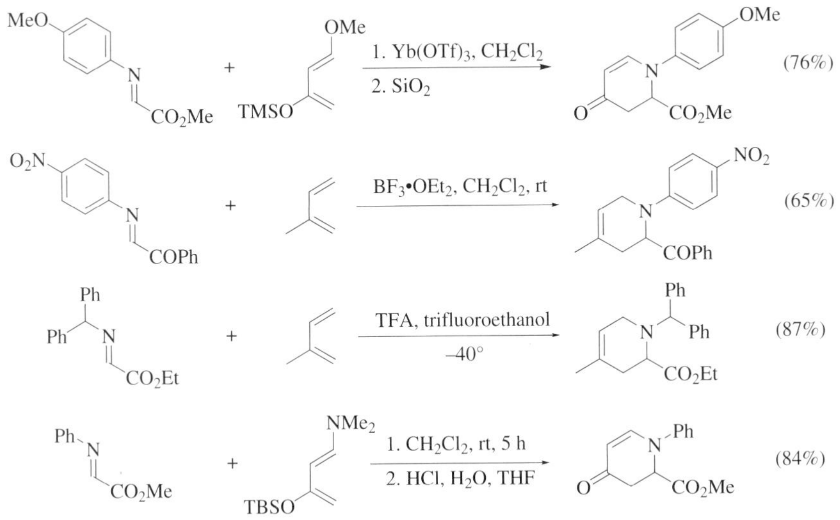
MeO
N
CO2Me
+
OMe
TMSO
1. Yb(OTf)3, CH2Cl2
2. SiO2
OMe
N
O
CO2Me
(76%)
O2N
N
COPh
+
BF3•OEt2, CH2Cl2, rt
NO2
N
COPh
(65%)
Ph
Ph
N
CO2Et
+
TFA, trifluoroethanol
–40°
Ph
N
Ph
CO2Et
(87%)
Ph
N
CO2Me
+
NMe2
TBSO
1. CH2Cl2, rt, 5 h
2. HCl, H2O, THF
N
Ph
O
CO2Me
(84%)

In one report the presence of a small amount of a second regioisomer **72** is noted in three similar cycloaddition reactions of dienophile **71** (Eq. 18).[82] However, in an almost identical reaction that is conducted in the absence of a Lewis acid catalyst at room temperature in dimethylformamide using either (*S*)- or (*R*)-**71** as the dienophile, a mixture of regioisomers is not observed.[81,83–85]

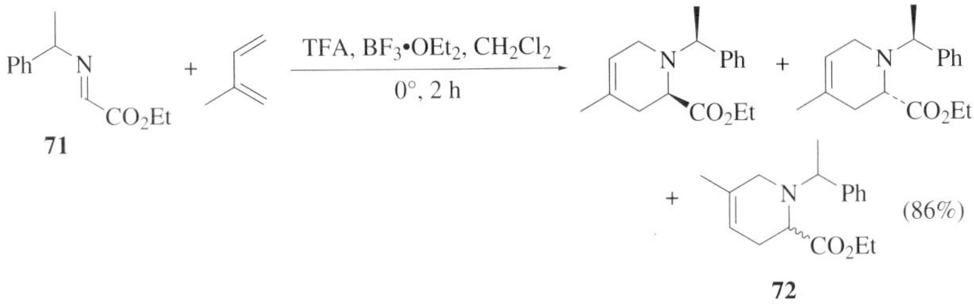
Ph
N
CO2Et
71
+
TFA, BF3•OEt2, CH2Cl2
0°, 2 h
N
Ph
CO2Et
+
N
Ph
CO2Et
+
N
Ph
CO2Et
72
(86%)

(Eq. 18)

Each of the reactions of acyclic dienes 1,3-pentadiene, 3-methyl-1,3-pentadiene, and (*E*,*E*)-hexa-2,4-diene with the imine **73** derived from benzhydrylamine and ethyl glyoxylate affords a single adduct. These products presumably arise via attack of an E imine on the diene via a transition state with the acyl group endo.

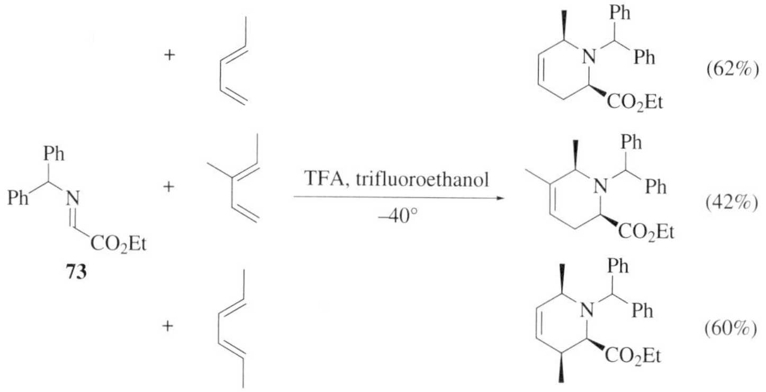

On the other hand, simple acyclic *N*-substituted *C*-acylimines undergo ImDA reactions with cyclic dienes to almost always give the exo acyl cycloadducts as the major products. The ratio of exo to endo products, however, varies significantly. For example, both benzylimino acetic acid **(74)** and the benzyl ester derivative **75** react with cyclopentadiene, giving the adducts in an identical 1.4 : 1 exo to endo ratio (Eq. 19).[86,87] On the other hand, in the presence of CuCl or the aluminum dichloride alkoxide of menthol as catalyst, *N-p*-nitrophenyl- and *N-p*-chlorophenyl-substituted imines **76** and **77** afford solely the exo product, albeit in undisclosed or poor yields (Eq. 20).[19] Similarly, in the reaction of benzylimine ketone **78** and cyclopentadiene under aqueous acidic conditions a 10 : 1 mixture of exo/endo cycloadducts is isolated (Eq. 21).[86]

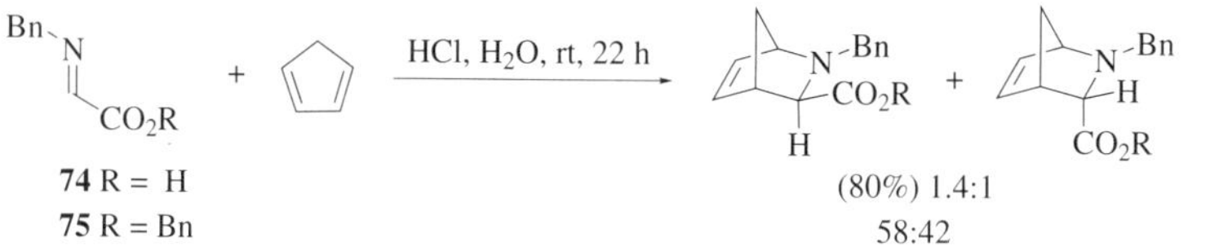

(Eq. 19)

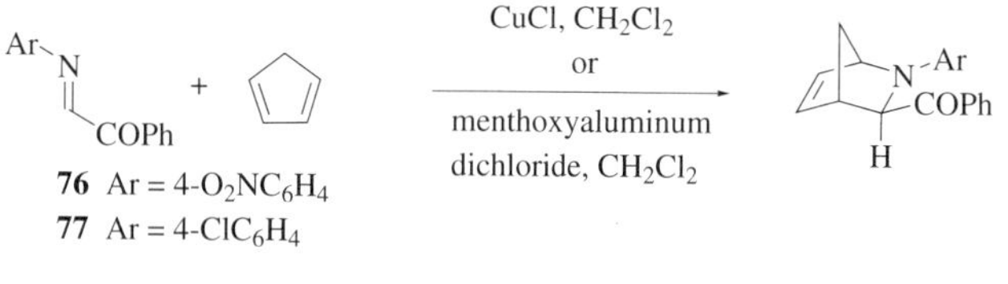

(Eq. 20)

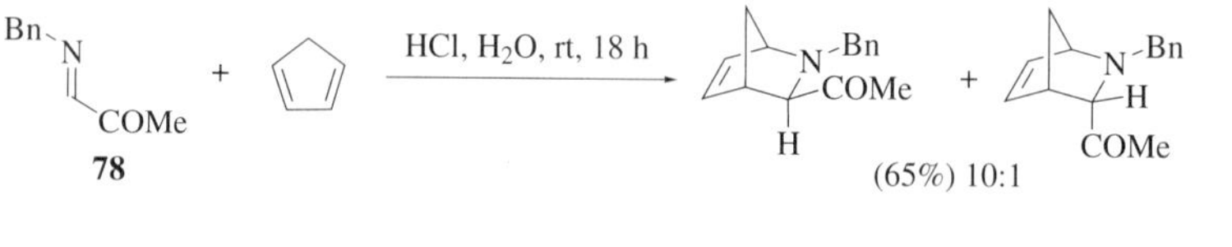

(Eq. 21)

In contrast to the results just discussed, some reactions utilizing aqueous Diels-Alder methodology involving an imine unsubstituted on nitrogen and cyclopentadiene afford the endo cycloadduct as the major, or only product isolated (Eq. 22).[86,88]

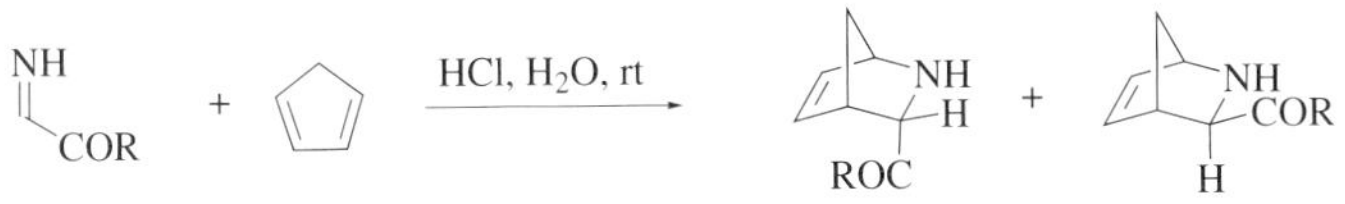
NH
COR
HCl, H_2O, rt
NH
H
ROC
NH
COR
H

(Eq. 22)

R = Me (84%) 2:1
R = Ph (89%) 2:1
R = OMe endo product only

Numerous examples in which a chiral auxiliary has been tethered to either, or both, ends of a *C*-acylimine can be found in the literature. The ImDA reactions with such imines are usually regioselective, and often give excellent diastereoselectivities. Such ImDA reactions have synthetic utility in the preparation of enantiomerically pure compounds because, in principle, it is possible to separate the diastereomers produced in the cycloaddition and then to remove the chiral auxiliary. For example, (1-phenylethylimino)acetic acid ethyl ester (**71**) or the methyl ester analog **79** react with several acyclic dienes in the presence of a Lewis acid catalyst affording a mixture of diastereomeric adducts. Diastereoselectivities could be increased by decreasing the reaction temperature, albeit sometimes at the expense of the product yield.[82,89]

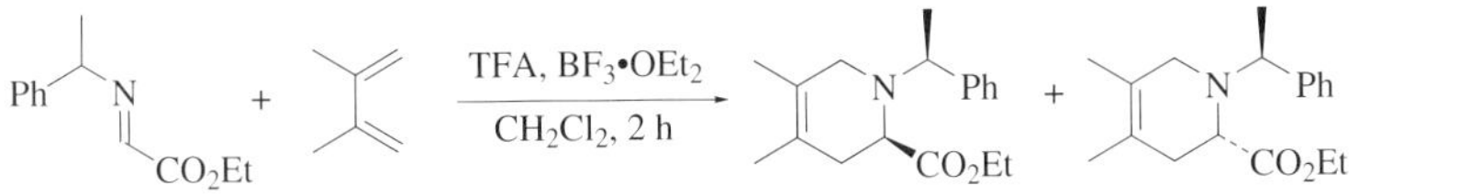
Ph
N
CO_2Et
TFA, BF_3•OEt_2
CH_2Cl_2, 2 h
N
Ph
CO_2Et
N
Ph
CO_2Et

71 (racemic imine used)

0°	(89%) 65:35
−80°	(35%) 80:20

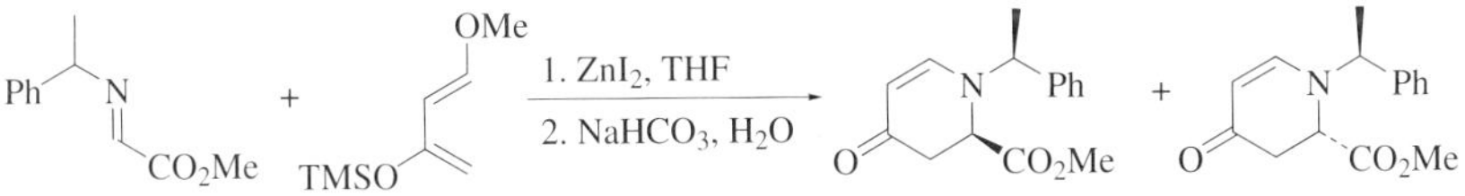
Ph
N
CO_2Me
OMe
TMSO
1. ZnI_2, THF
2. $NaHCO_3$, H_2O
O
N
Ph
CO_2Me
O
N
Ph
CO_2Me

79 (racemic imine used)

0 to 20°, 3 h	(85%) 62:38
−80° to rt, 15 h	(85%) 70:30

Reactions of chiral *C*-acylimines with cyclic dienes in which the chiral moiety is connected, either directly to the imine nitrogen atom or to the acyl moiety, often give rise to high diastereoselectivities.[83,84,90–99] For example, imine **80** derived from (*S*)-1-phenylethylamine and ethyl glyoxylate undergo reaction with cyclopentadiene in the presence of boron trifluoride etherate to afford a mixture of four diastereomers with an exo/endo ratio of >98 : 2; the selectivity between the two exo isomers is also high (90 : 10).[91]

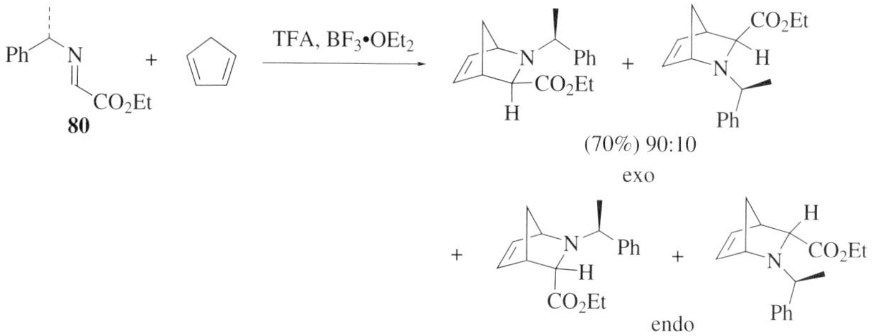
Ph
N
CO2Et
80
+
TFA, BF3•OEt2
N
Ph
+
CO2Et
H
CO2Et
H
N
Ph
(70%) 90:10
exo
+
N
Ph
+
H
CO2Et
H
CO2Et
N
Ph
endo

With sterically encumbered auxiliaries such as the isoborneol–dicyclohexyl sulfonamide derivative shown in Eq. 23, only the exo diastereomers are isolated.[92,98] A similar preference for exo addition is observed with spiro[2.4]hepta-1,3-diene (Eq. 24).[91] The pattern that emerges in the above reactions of *N*-α-phenethylamine-derived imines is that R chiral auxiliaries preferentially promote Si face attack, whereas those with an S absolute configuration give predominantly products from Re face attack.

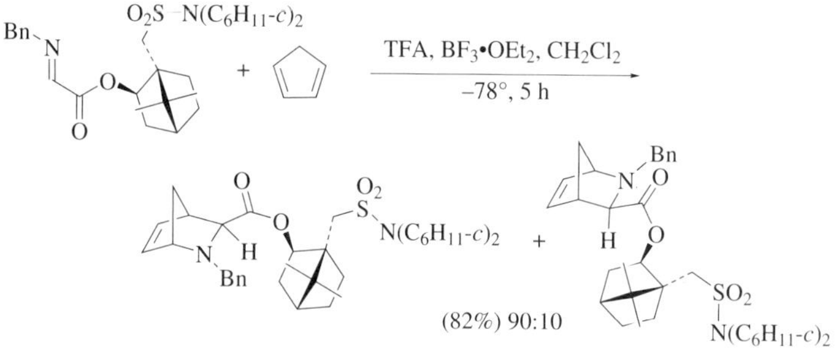
O2S N(C6H11-c)2
Bn
N
O
O
+
TFA, BF3•OEt2, CH2Cl2
−78°, 5 h
O
O
O2
S
N(C6H11-c)2
N
H
Bn
+
Bn
N
O
H
O
SO2
N(C6H11-c)2
(82%) 90:10

(Eq. 23)

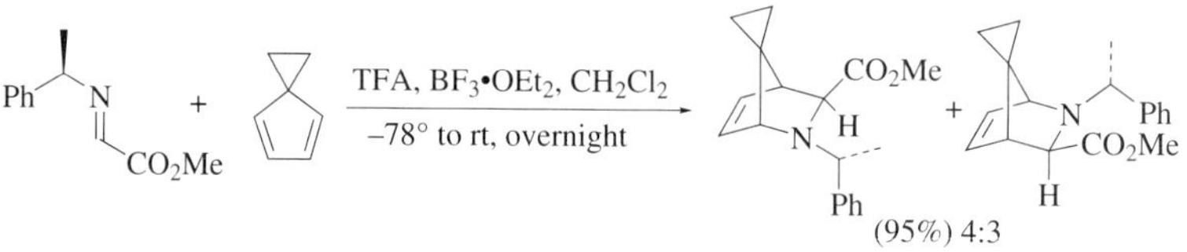
Ph
N
CO2Me
+
TFA, BF3•OEt2, CH2Cl2
−78° to rt, overnight
CO2Me
N
H
Ph
+
N
Ph
CO2Me
H
(95%) 4:3

(Eq. 24)

Finally, some investigations with imine substrates bearing two chiral auxiliaries have been reported. The use of glyoxylate ester imines with pantolactone or borneol as the chiral auxiliary on oxygen and either a (*S*)- or (*R*)-1-phenylethyl moiety on nitrogen leads to modest but unspecified diastereoselectivity in reactions with 2,3-dimethylbutadiene (Eq. 25).[100]

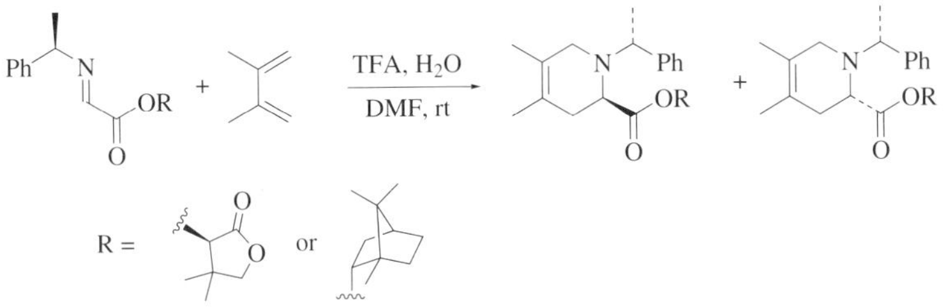
Ph
N
OR
O
+
TFA, H_2O
DMF, rt
N
Ph
OR
O
+
N
Ph
OR
O
R =
O
O
or

(Eq. 25)

Similarly, reactions involving various cyclic and acyclic dienes with imines bearing the *N*-(*S*)-1-phenylethyl and 8-phenylmenthyl ester auxiliaries in trifluoroethanol as solvent also give modest, if somewhat higher, diastereoselectivities (Eq. 26).[100] However, when the diastereomeric imine derived from (*R*)-1-phenethylamine is employed, only single diastereomeric ImDA adducts in about 50% yield and >95% de are isolated with both cyclic and acyclic dienes (Eq. 27). As expected, the major products isolated from reaction with cyclic dienes are the exo adducts derived from Si face attack of the imine, whereas in the acyclic diene examples all cycloadditions proceed through a *C*-acyl endo transition state.

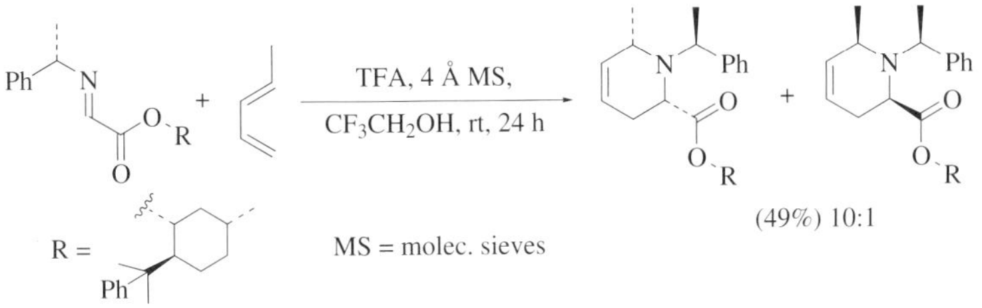
Ph
N
O
R
O
+
TFA, 4 Å MS,
CF_3CH_2OH, rt, 24 h
N
Ph
O
O
R
+
N
Ph
O
O
R
(49%) 10:1
R =
Ph
MS = molec. sieves

(Eq. 26)

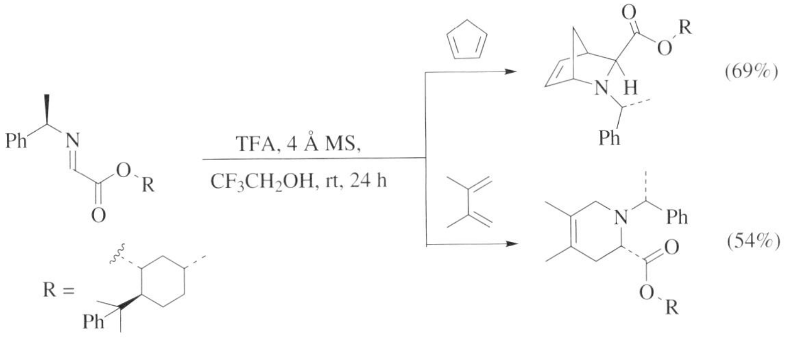
Ph
N
O
R
O
R =
Ph
TFA, 4 Å MS,
CF_3CH_2OH, rt, 24 h
O
R
O
N H
Ph
(69%)
N
Ph
O
O
R
(54%)

(Eq. 27)

Cyclic *C*-Acylimines. A relatively small number of examples of cyclic *C*-acylimines participating in ImDA reactions have been reported. For instance, 2-phenylindolo-3-one (**81**) is found to react sluggishly with acyclic dienes in the absence of a catalyst; alternatively, in the presence of $AlCl_3$ extensive formation of by-products is observed. However, using *p*-toluenesulfonic acid in benzene at room temperature single regioisomeric ImDA cycloadducts are obtained in moderate to good yield.[101,102] The only exception arises when isoprene is utilized as the diene; here an inseparable mixture of regioisomers **82** and **83** is isolated.[101–103] Interestingly, the reactions of imine **81** with (*E,E*)- or (*E,Z*)-hexa-2,4-diene proceed stereospecifically, giving the corresponding adducts **84** and **85**, respectively, although the relative stereochemistry has not been established. [4+2]-Cycloadditions of the more reactive 2-ethoxycarbonylindol-3-one with unsymmetrical dienes are less regioselective.

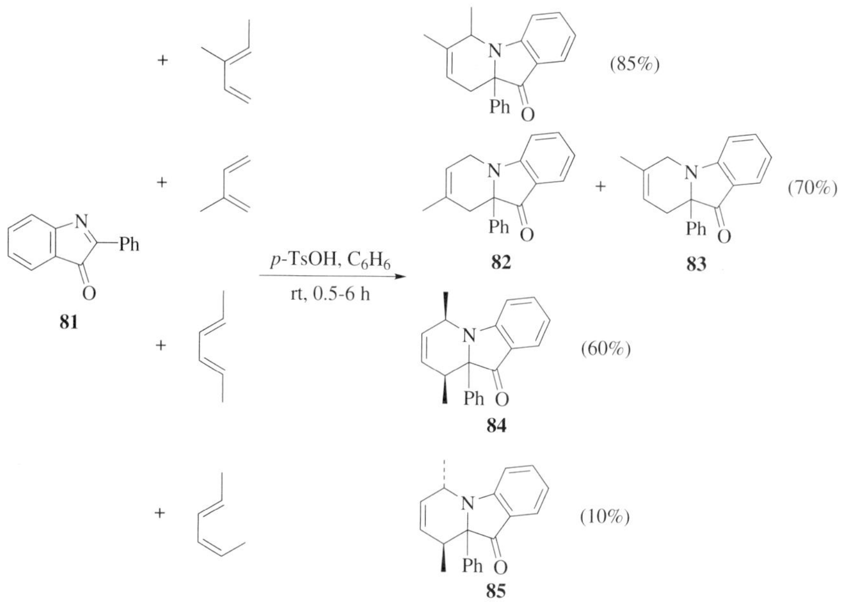
+
(85%)
+
+
(70%)
p-TsOH, C6H6
rt, 0.5-6 h
+
(60%)
+
(10%)
N
Ph
O
81
82
83
84
85

2,5-Aryl-substituted 1,3,4-oxadiazin-6-ones **86**,[104] (5*S*)-5-phenyl-3,4-dehydromorpholin-2-one (**87**),[105] as well as 3-azacyclopentadienone (**88**)[72] and 3,4-diaza-2,5-diphenylcyclopentadienone (**89**),[72,106–108] have also been reported to undergo ImDA reactions to afford cycloadducts in varying yields.

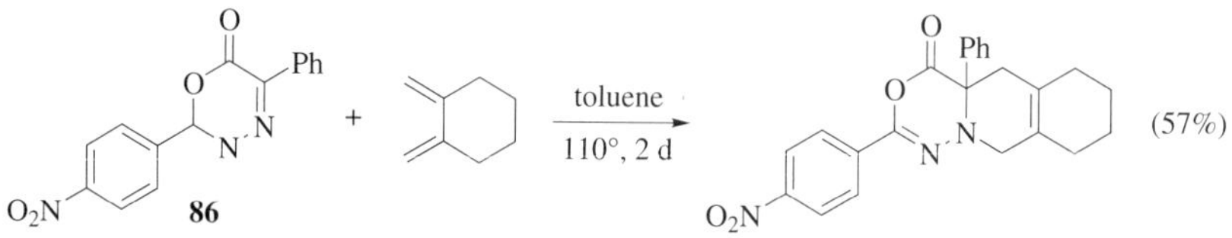
toluene
110°, 2 d
(57%)
86
O2N
Ph

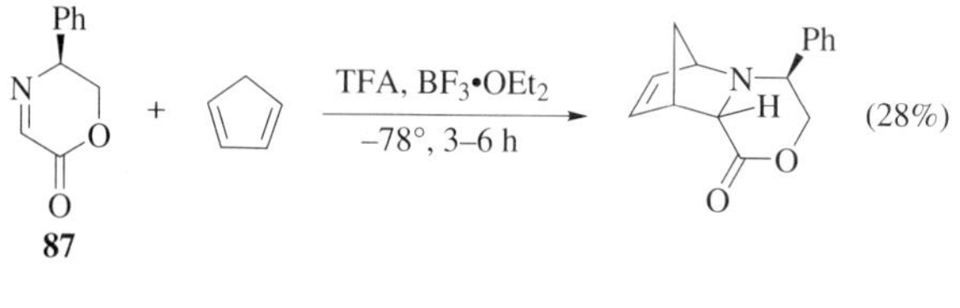
TFA, BF3•OEt2
–78°, 3–6 h
(28%)
87
Ph

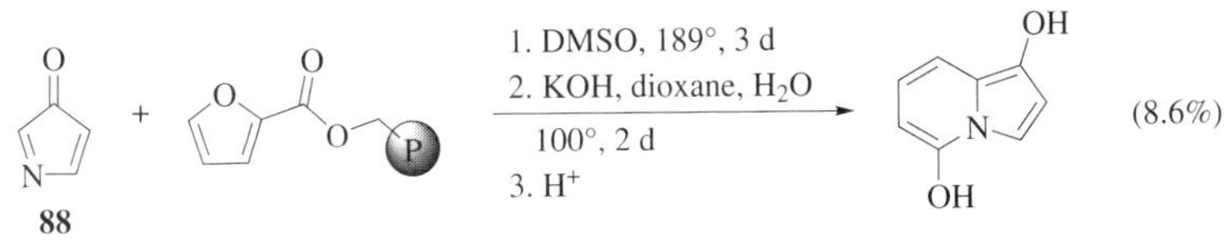

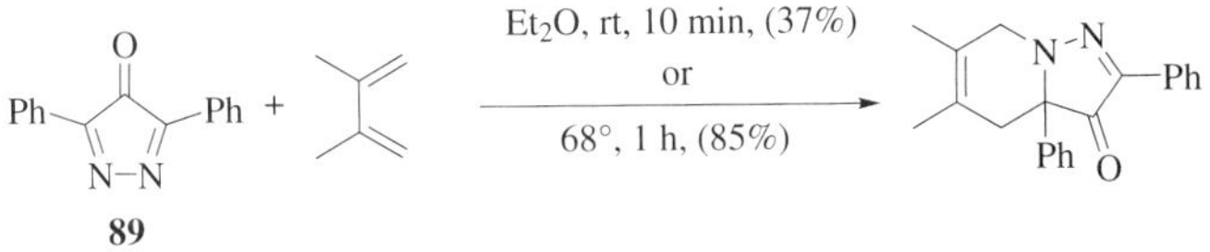

***N*-Sulfonylimines and *N*-Phosphorylimines.** In early work, *N*-tosylimines were generated in situ for use in ImDA reactions from non-enolizable aldehydes such as chloral by reaction with *N*-sulfinyl-*p*-toluenesulfonamide in the presence of a Lewis acid.[109] Similarly, glyoxylate-derived *N*-sulfonylimines were prepared in the same fashion.[31,32,37] It was discovered later that *N*-sulfonylimines could also be obtained by the thermal reaction of glyoxylate esters with *p*-toluenesulfonyliso-cyanate (Eq. 28).[35] Subsequently, *N*-sulfinylsulfonamide-based methodology was extended to the in situ generation of *N*-tosylimines from a variety of enolizable aldehydes such as propionaldehyde (Eq. 29).[33] It is presumed that imine formation involves initial [2+2]-cycloaddition of the *N*-sulfinyl compound with the aldehyde affording intermediate **90**, which upon elimination of sulfur dioxide (or carbon dioxide in isocyanate reactions) provides a Lewis acid coordinated reactive species **91**. *N*-Sulfonylimines have also been generated from *N*-*p*-tosylsulfilimines of napththol[1,8-*de*]dithiane in the presence of $BF_3{\cdot}OEt_2$ (Eq. 30)[34], as well as by bromination of glycinates with subsequent base-mediated dehydrobromination (Eq. 31).[110,111]

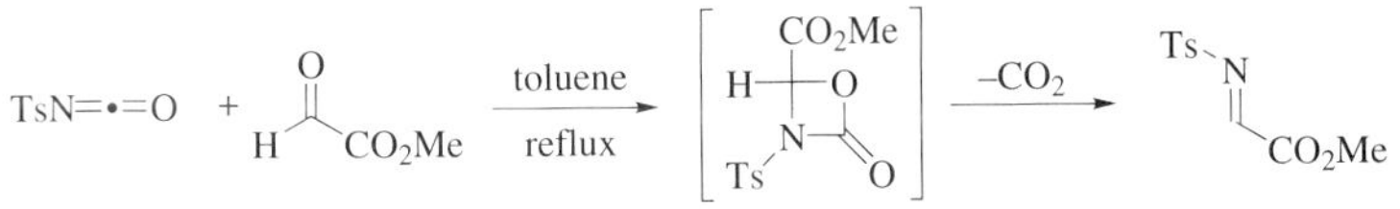

(Eq. 28)

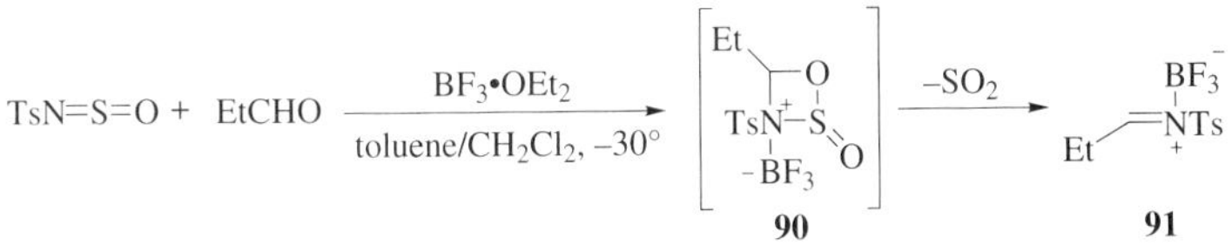

(Eq. 29)

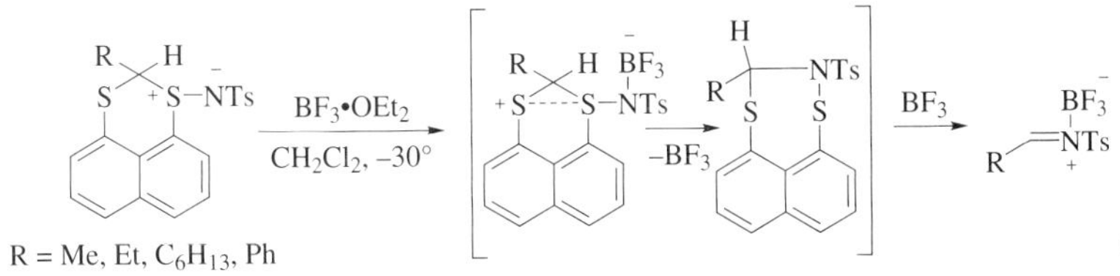

(Eq. 30)

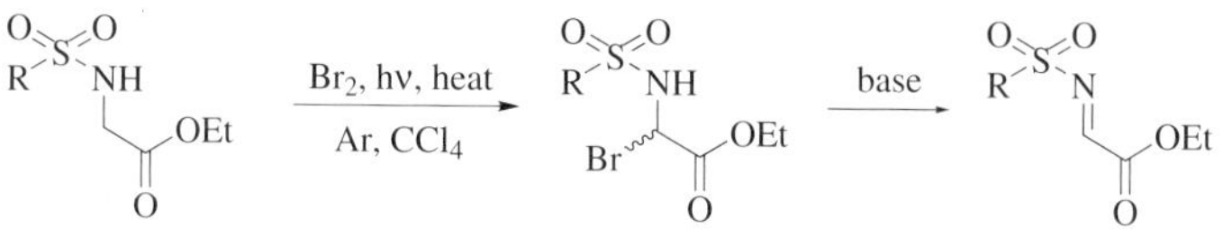

(Eq. 31)

Cycloaddition reactions of *N*-sulfonylimines and unsymmetrical dienes typically afford single regioisomers; a number of representative examples of this process are shown in Eqs. 32–35. These results can be nicely rationalized on the basis of the dipolar mechanistic model in Eq. 7.[35,57,59,112]

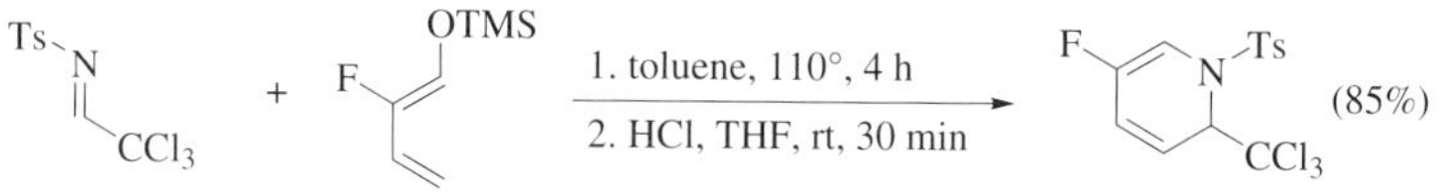

(Eq. 32)

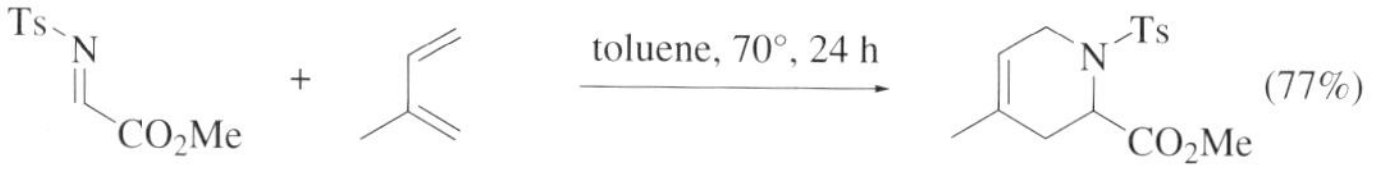

(Eq. 33)

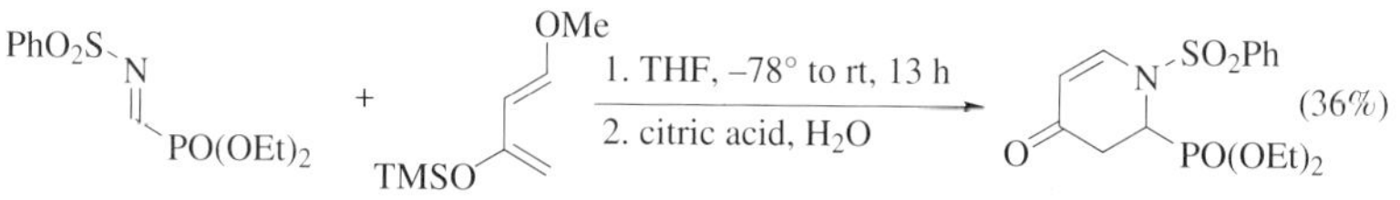

(Eq. 34)

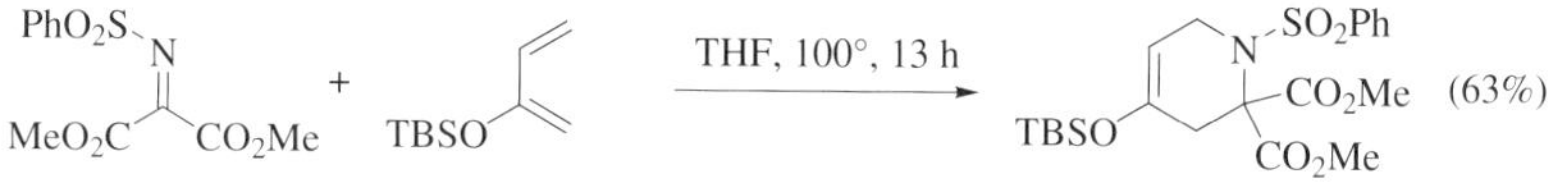

(Eq. 35)

Curiously, the thermal cycloaddition of *N*-sulfonylimine **92** with 1-vinyl-6-methoxy-3,4-dihydronaphthalene (**93**) gives rise to a 3:1 mixture of regioisomeric adducts **94** and **95** in almost quantitative yield.[37] The major adduct **94** is in accord with predictions based on the dipolar mechanistic model. Alternatively, steric hindrance arising between the C5 allylic methylene group of the diene and the ester function may account for formation of the minor adduct **95**.

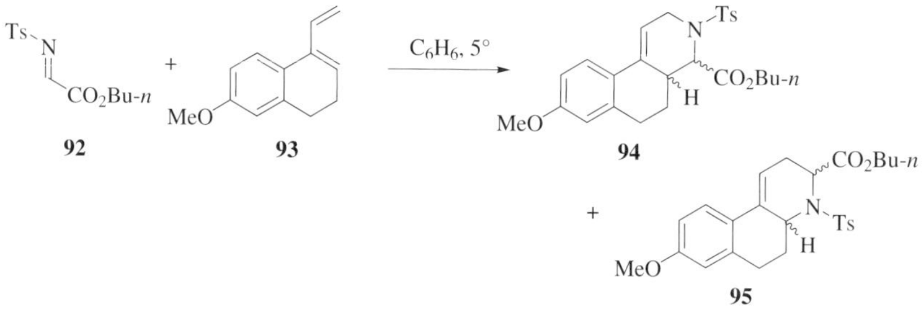

92 **93** **94** **95**

In general, [4+2]-cycloadditions of *N*-sulfonylimines and acyclic dienes show only modest stereoselectivity, as illustrated by the examples presented in Eqs. 36–38.[33–35,113] However, E diene ether **97** reacts with ethyl (*N*-tosylimino) acetate (**96**), affording a single stereoisomeric ImDA adduct **98**, which is assigned the 2,6-cis configuration.[114] Similarly, in the presence of zinc chloride, the ImDA reaction of a 4:1 mixture of Z/E isomers of [1-(2-methoxyvinyl)propenyl-oxy]trimethylsilane (**99**) with dienophile **96** affords a 22:1 mixture of enones **100** and **101**. It should be noted that the yields and ratios of cycloadducts **100** and **101** are highly dependent on the specific reaction conditions employed. For example, the thermal condensation of diene **99** with imine **96** (toluene, room temperature, 3 hours) affords a 4.7:1 mixture of **100**:**101** in 51% yield; alternatively, if $AlCl_3$ is used as the catalyst (toluene, −78°, 3 hours), a 53% yield of a 7:1 mixture of **100**:**101** is obtained. Although the major cis product **100** may arise from addition of the Z diene via a transition state in which the acyl group of the imine is endo, and the minor trans enone **101** could be derived from the E diene via an endo transition state, it is possible that the observed mixtures are not the kinetic distributions, but the result of enone or diene isomerization. Indeed, in acid-catalyzed isomerizations, both the pure major (**100**) and minor (**101**) enone ester isomers provide the same 4:1 equilibrium mixture of **101**:**100**.

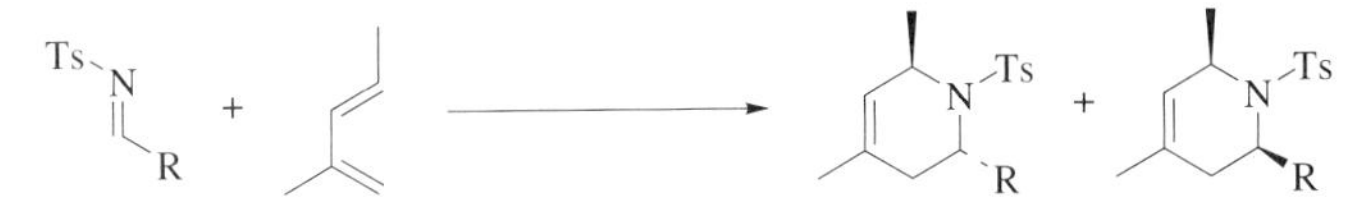
Ts
N
R
+
N
Ts
R
+
N
Ts
R

R = Et — BF_3•OEt_2, $MgSO_4$, toluene, CH_2Cl_2, −30°, 3 h — (68%) 2:1 (Eq. 36)

R = CO_2Me — toluene, rt, 24 h — (83%) 2:1

Ts
N
R
+
N
Ts
R
+
N
Ts
R

R = Et — BF_3•OEt_2, $MgSO_4$, toluene, CH_2Cl_2, −30°, 3 h — (61%) 6:1 (Eq. 37)

R = Ph — BF_3•OEt_2, CH_2Cl_2, −20° — (52%) 3:1

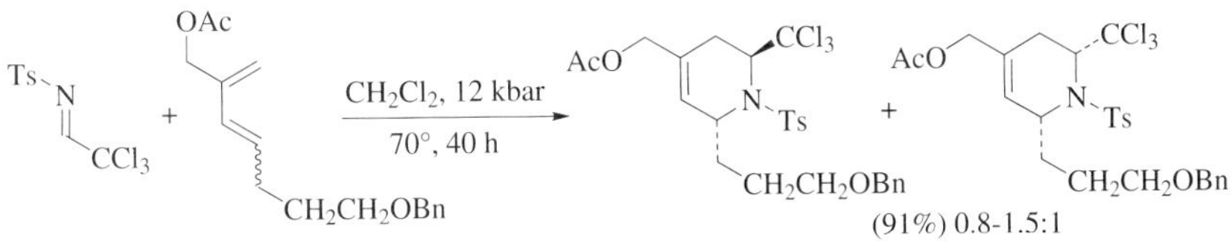
Ts
N
CCl3
+
OAc
CH2CH2OBn
CH2Cl2, 12 kbar
70°, 40 h
AcO
CCl3
N
Ts
CH2CH2OBn
+
AcO
CCl3
N
Ts
CH2CH2OBn
(91%) 0.8-1.5:1

(Eq. 38)

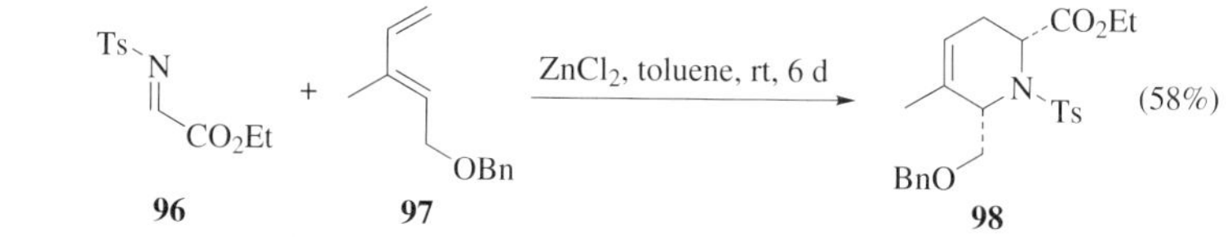
ZnCl2, toluene, rt, 6 d
(58%)
96
97
98

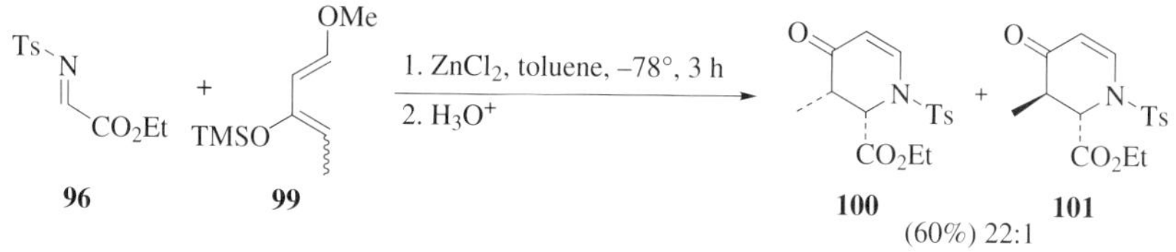
1. ZnCl2, toluene, –78°, 3 h
2. H3O+
96
99
100
101
(60%) 22:1

Interestingly, only one regio- and stereoisomer **105** is obtained when 7-methoxy-4-vinyl-1,2-dihydronaphthalene (**103**) and trichloromethyl *N*-sulfonylimine **102** are stirred in benzene at room temperature for 16 hours, but a 1.5:1 mixture of stereoisomers **105** and **106** is obtained when the reaction is conducted using the 4-(1-methoxyvinyl) derivate **104**.[36,38] Similarly, the thermal cycloadditions of siloxydienes **108** and **109** with *N*-benzylidene benzenesulfonamide (**107**) proceed stereoselectively, affording cycloadducts **110** and **111**, respectively, in moderate yields.[115]

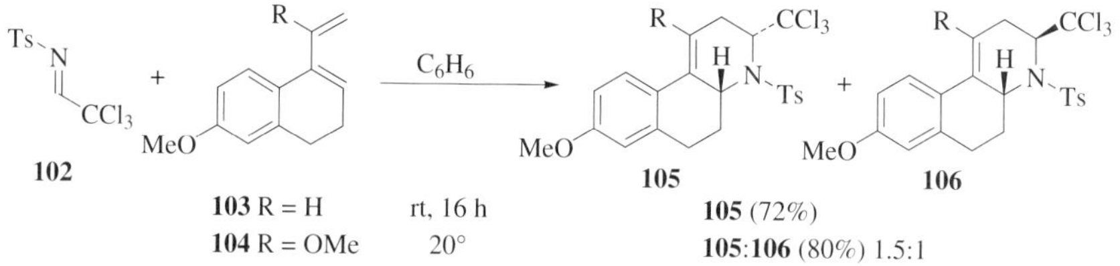
C6H6
102
105
106
103 R = H rt, 16 h 105 (72%)
104 R = OMe 20° 105:106 (80%) 1.5:1

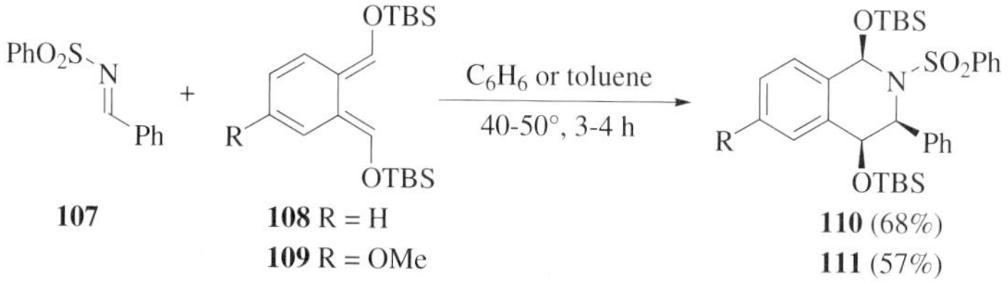
C6H6 or toluene
40-50°, 3-4 h
107
108 R = H
109 R = OMe
110 (68%)
111 (57%)

Thermal ImDA reactions of disubstituted *C*-phosphonate or *C*-acyl *N*-sulfonylimines with cyclopentadiene or cyclohexadiene afford exclusively exo-bicyclic adducts (Eqs. 39 and 40).[35,57,110,116–118] When siloxycyclohexadienes **112** are utilized, both the exo ketones **113** and **115**, and endo ketones **114** and **116** are isolated, with the exo products predominating (Eq. 41). With the TBS ether, the major exo adduct is the unhydrolyzed silyl enol ether **117**; the corresponding endo isomer **118** is also observed.[119–122]

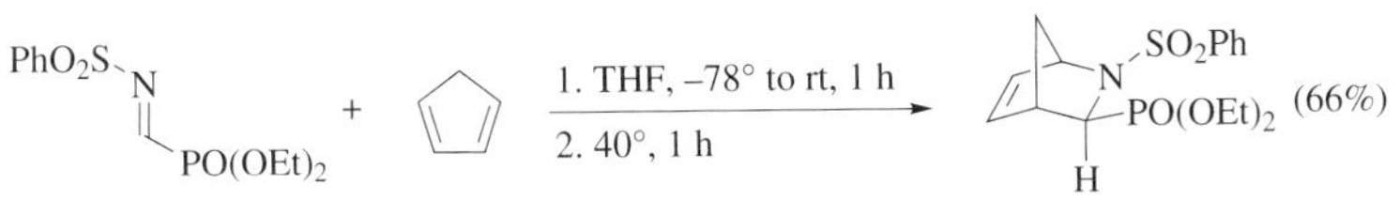
PhO2S
N
PO(OEt)2
+
1. THF, –78° to rt, 1 h
2. 40°, 1 h
N
SO2Ph
PO(OEt)2
H
(66%)
(Eq. 39)

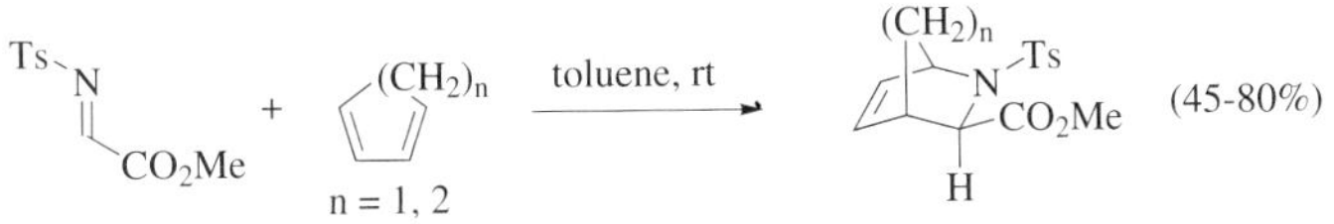
Ts
N
CO2Me
+
(CH2)n
n = 1, 2
toluene, rt
(CH2)n
N
Ts
CO2Me
H
(45-80%)
(Eq. 40)

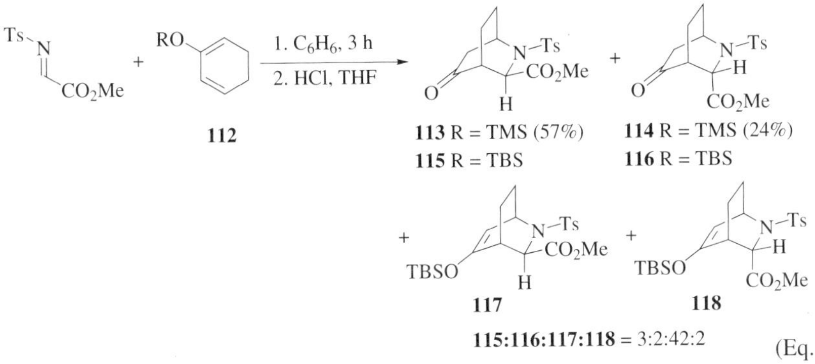
Ts
N
CO2Me
+
RO
112
1. C6H6, 3 h
2. HCl, THF
N–Ts
CO2Me
O
H
113 R = TMS (57%)
115 R = TBS
+
N–Ts
H
O
CO2Me
114 R = TMS (24%)
116 R = TBS
+
N–Ts
CO2Me
TBSO
H
117
+
N–Ts
H
TBSO
CO2Me
118
115:116:117:118 = 3:2:42:2
(Eq. 41)

Lewis acid catalyzed [4+2]-cycloadditions of 1,3-cyclohexadiene with unbranched *C*-alkyl- or *C*-aryl-substituted *N*-sulfonylimines proceed with modest stereoselectivity usually yielding the exo adducts as the major products (Eqs. 42 and 43).[31,32]

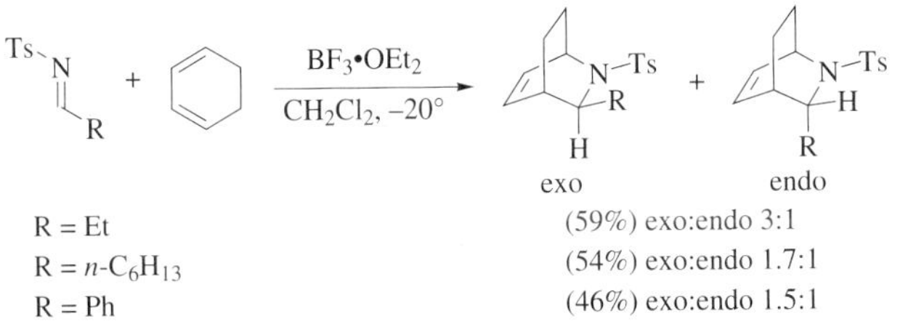
Ts
N
R
+
BF3•OEt2
CH2Cl2, –20°
N–Ts
R
H
exo
+
N–Ts
H
R
endo
R = Et
R = n-C6H13
R = Ph
(59%) exo:endo 3:1
(54%) exo:endo 1.7:1
(46%) exo:endo 1.5:1
(Eq. 42)

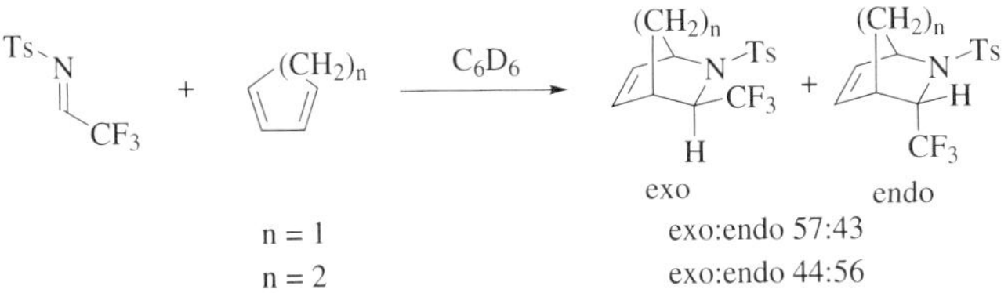
Ts
N
CF3
+
(CH2)n
C6D6
(CH2)n
N
Ts
CF3
H
exo
+
(CH2)n
N
Ts
H
CF3
endo
n = 1
n = 2
exo:endo 57:43
exo:endo 44:56
(Eq. 43)

The chiral allylic ether dienes **121** and **122** have been combined with (toluene-4-sulfonylimino)acetic acid methyl ester (**119**) and (toluene-4-sulfonylimino)acetic acid *n*-butyl ester (**120**). These reactions represent the only examples to date of

N-sulfonylimines undergoing ImDA reactions with dienes bearing chiral centers.[39,123] In each example, the reaction gives a single regio- and diastereoisomer in good yield. The stereochemical outcome can be rationalized by the E dienophile, with the *C*-acyl substituent endo, adding to the Si face of the double bond with the adjacent chiral center.

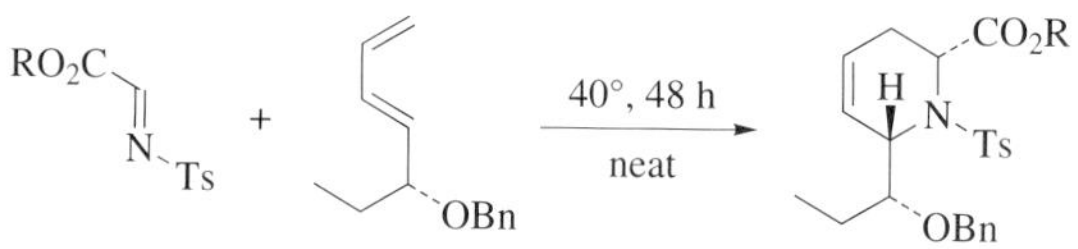
RO2C
N
Ts
+
OBn
40°, 48 h
neat
CO2R
H
N
Ts
OBn

119 R = Me **121** (74%)
120 R = *n*-Bu (83%)

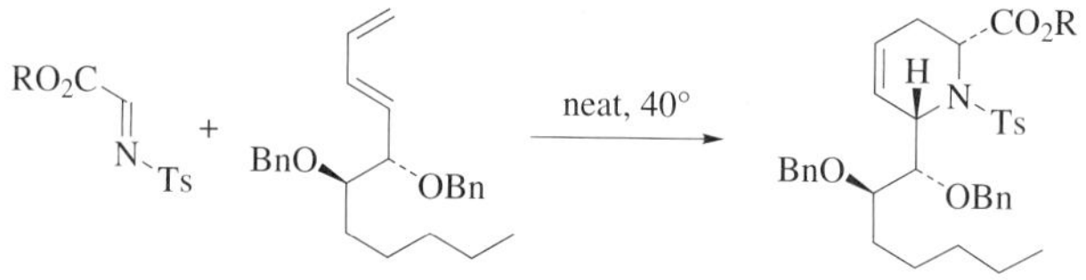
RO2C
N
Ts
+
BnO
OBn
neat, 40°
CO2R
H
N
Ts
BnO
OBn

119 R = Me **122** 2 d (67%)
120 R = *n*-Bu 6 d (81%)

There are examples of asymmetric ImDA reactions in which the imine bears a chiral auxiliary, but there is only one reported example in which the chiral auxiliary is on the imine nitrogen. All other cycloadditions involve *C*-carboxylate-substituted auxiliaries.

Camphorsulfonamide-derived imine **123** reacts with Danishefsky's diene, both in the presence and absence of Lewis acids, and in a number of different solvents, giving mixtures of diastereomeric adducts **124** and **125**,[111,124] which result from the cycloaddition proceeding through two possible diastereomeric exo transition states. Diastereoselectivities are only moderate, and varying the polarity of the reaction solvent has little effect. However, slightly better diastereoselectivities are achieved in carbon tetrachloride (**124**:**125**, 67:33) than in acetonitrile, ether, or dichloromethane (**124**:**125**, 58:42, 60:40, and 54:46, respectively). Of the Lewis acids screened, titanium tetraisopropoxide (25 mol %) gives rise to the highest diastereomer excess observed (**124**:**125**, 70:30), but with only 25% conversion after 6 hours, although yields of the ImDA adducts actually decrease as the amount of catalyst is increased. Boron, zinc, and other titanium-based Lewis acids afford cycloadducts in diastereomer excesses comparable to those obtained under thermal conditions. Interestingly, diethylaluminum chloride shows a modest reversal in selectivity, giving a diastereomeric mixture of 41:59 (**124**:**125**).

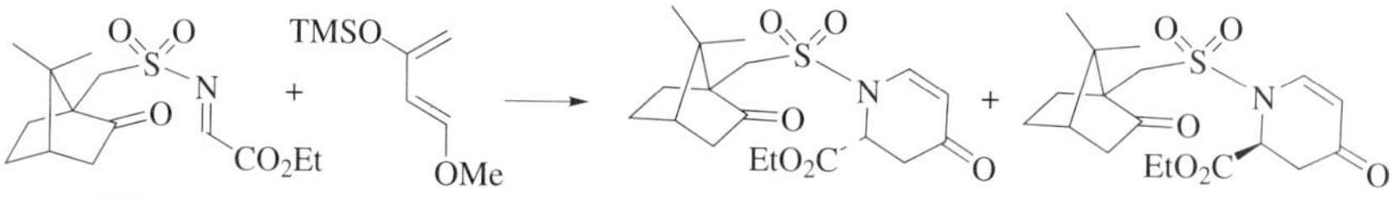
O
S
O
N
O
CO2Et
+
TMSO
OMe
O
S
O
N
O
EtO2C
O
+
O
S
O
N
O
EtO2C
O

123 **124** **125**

Reaction of (−)-menthyl, (−)-bornyl, and (−)-8-phenylmenthyl *C*-carboxylate *N*-tosylimines (**126–128**) with cyclopentadiene gives the corresponding exo Diels-Alder adducts in excellent yields, but with poor diastereoselectivities (**129**, 56:44; **130**, 53:47; **131**, 60:40).[110]

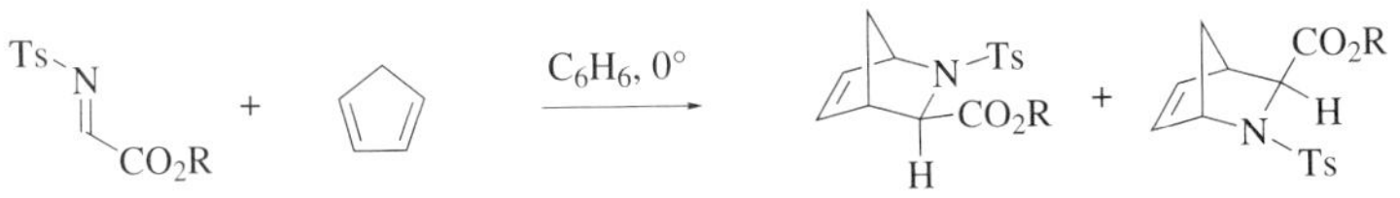
Ts
N
CO2R
+
C6H6, 0°
N
Ts
CO2R
H
+
CO2R
H
N
Ts

126 R = (–)-menthyl
127 R = (–)-bornyl
128 R = (–)-8-phenylmenthyl

129 56:44
130 53:47
131 60:40

Synthetically more useful diastereoselectivities are obtained for the cycloaddition of chiral *N*-sulfonylimines **132** and **133** with cyclopentadiene. Ethyl (*S*)-lactate **132** and (*R*)-pantolactone **133**-derived glyoxylate imino esters afford, after optimization of reaction conditions, 12:88 and 85:15 mixtures of diastereomeric adducts, respectively, in 50–60% yield.[125] The stereochemical outcome for the ethyl (*S*)-lactate example is rationalized by the transition state depicted in Fig. 2, in which the Lewis acid complexed imine adopts an s-trans conformation and the imine double bond is attacked from the Si face so that steric interactions between the diene and dienophile are minimized.

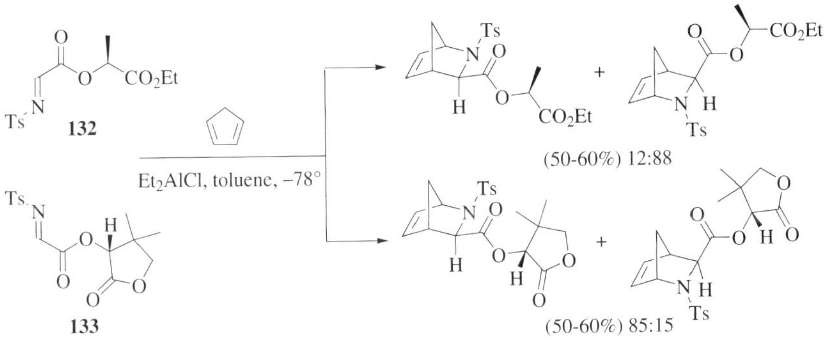
132
133
Et2AlCl, toluene, –78°
(50-60%) 12:88
(50-60%) 85:15

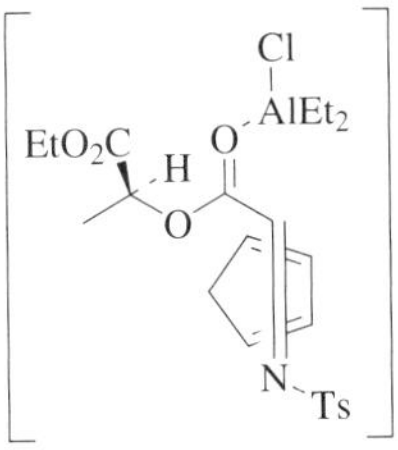
Cl
AlEt2
EtO2C
H
O
O
N
Ts

Figure 2 Proposed transition state depicting the approach of the diene to the Lewis acid-imine complex from the sterically less hindered Si-face

Of the four possible diastereomers, [4+2]-cycloaddition of cyclopentadiene with *N*-glyoxyloyl-(2*R*)-bornane-10,2-sultam heterodienophile **134** gives rise only to the (6*R*)-**135** and (6*S*)-**136** cycloadducts.[126] The exact diastereomer ratios of the two products are dependent on the specific reaction conditions used, and with appropriate choice of conditions the diastereoselectivity of the reaction can be reversed. For example, under high pressure at ambient temperature, and in the presence of a weak Lewis acid (2 mol% $Eu(fod)_3$), a 25:75 mixture of diastereomers **135** and **136** is formed in moderate yield (64%). Alternatively, at low temperatures (−78°) and at atmospheric pressure with titanium tetrachloride (50 mol%) as catalyst, the same products are obtained in an 80:20 ratio, albeit in poorer overall yield (38%).

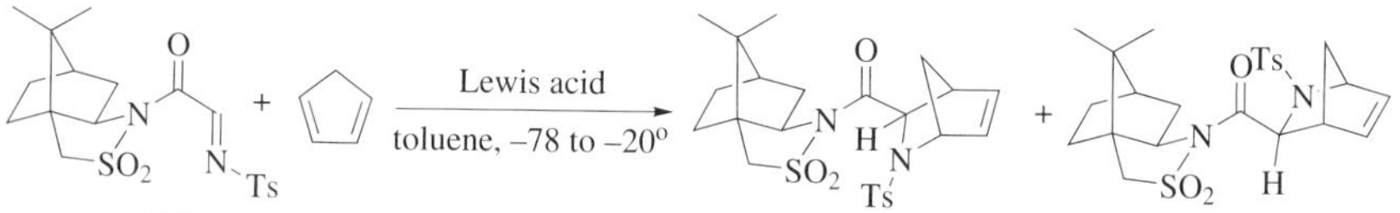
O
N
SO2
N
Ts
+
Lewis acid
toluene, –78 to –20°
O
N
H
N
SO2
Ts
+
O
Ts
N
N
SO2
H
134 **135** **136**

There are only a few reported examples of cyclic *N*-sulfonylimines participating in ImDA reactions. In each one, Danishefsky's diene reacts thermally with either 3-methyl- (**137**) or 3-chloro-1,2-benzisothiazole-1,1-dioxide (**138**) to yield dihydropyridones **139** and **140**, respectively; the latter is obtained after HCl elimination.[127,128] The related imines 3-methoxy- and 3-phenylthio-1,2-benzisothiazole do not react with Danishefsky's diene, possibly because of the deactivation of the imine by the electron-donating 3-alkoxy or 3-thio substituents.

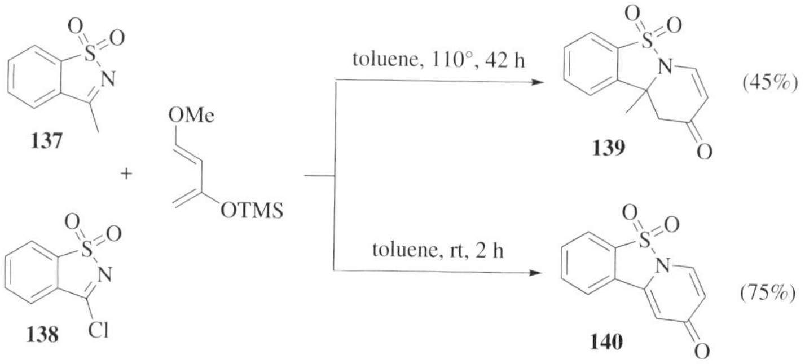
O
S
O
N
OMe
+
OTMS
O
S
O
N
Cl
toluene, 110°, 42 h
toluene, rt, 2 h
O
S
O
N
O
(45%)
O
S
O
N
O
(75%)
137 **138** **139** **140**

To date there is only a single report of the use of *N*-phosphorylimines in Diels-Alder processes.[129] It was found that aryl-substituted *N*-phosphorylimines **141** react with Danishefsky's diene in the presence of a catalytic amount of cupric triflate, followed by exposure to trifluoroacetic acid, affording vinylogous amides **142**. However, it is unclear if these reactions proceed by initial Mannich addition followed by Michael cyclization to produce enone **142** during the subsequent treatment with acid as shown in Eq. 9, or by a concerted Diels-Alder mechanism.

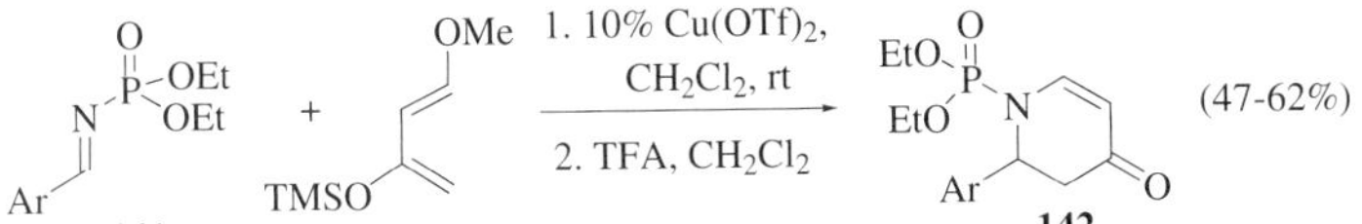

141 **142**

Ar = 2-furyl, Ph, 1-naphthyl, 4-MeOC$_6$H$_4$, 4-MeC$_6$H$_4$

Unactivated Alkyl/Arylimines Including Iminium Salts. In early work, *N,C*-alkyl/arylimino dienophiles were used in the form of their dialkylmethyleneammonium salts for ImDA reactions with simple dienes such as 2,3-dimethylbutadiene and isoprene because neutral imines tend to be unreactive as dienophiles.[130,131] However, it was later demonstrated[132] that alkyl/arylimines react with electron-rich oxygenated dienes in the presence of $ZnCl_2$, giving ImDA cycloadducts under mild conditions. Another important discovery entailed the preparation of [4+2]-cycloaddition adducts from simple unactivated imines and non-functionalized dienes under mildly acidic aqueous conditions.[133] Subsequently, several Lewis acids such as $TiCl_4$,[134] lanthanide triflates,[135] In(OTf)$_3$,[136] $BiCl_3$ and Bi(OTf)$_3$,[137] Nafion®-H,[138] SmI_2,[139] and montmorillonite K-10[140] were shown to promote various cycloaddition reactions of alkyl/arylimines. As a result of these advances, alkyl/arylimino Diels-Alder reactions have become the most widely utilized type of cycloadditions during the past two decades.

Reactions of simple alkyl- or aryl-substituted imines and iminium ions with unsymmetrical dienes have generally been found to proceed with high regioselectivity, in many instances leading to only one regioisomeric product. For example, *N*-benzylmethyleneamine, generated in situ from the corresponding amine and formaldehyde, reacts with isoprene in the presence of hydrochloric acid under aqueous reaction conditions to afford only one regioisomeric adduct **143**.[133] Formation of this product can be rationalized on the basis of the mechanistic model in Eq. 7 via the dipolar forms **13** and **15**.

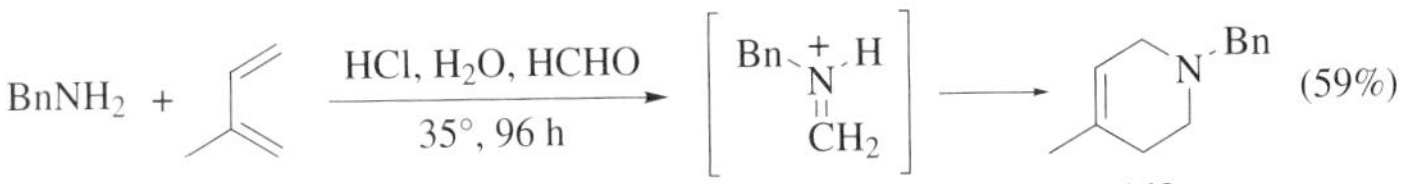

143

Irrespective of the nature of the substituents (alkyl or aryl), imines react with electron-rich dienes such as Danishefsky's diene, affording exclusively one regioisomeric dihydropyridinone adduct **144**.[141]

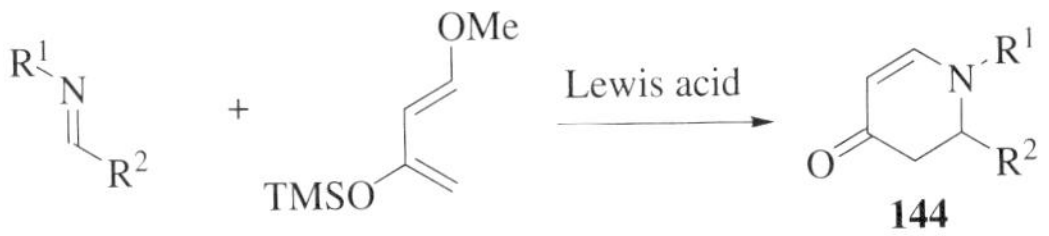

144

Vinylketenes, generated either by photochemical Wolff rearrangement of α'-silyl-α'-diazo-α,β-unsaturated ketones or by 4π-electrocyclic ring opening of 2-silylcyclobutenones, react regioselectively with *N*-silylimines affording a single dihydropyridinone.[142]

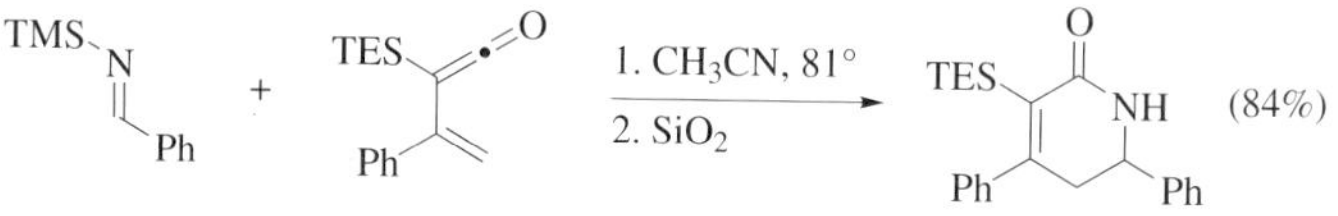
TMS
N
Ph
+
TES
O
Ph
1. CH3CN, 81°
2. SiO2
O
TES
NH
Ph
Ph
(84%)

Vinylallenes with an electron-donating substituent such as an alkyl group attached to C1 also participate in ImDA reactions with alkyl/arylimines under mild conditions.[143] Thus, in the presence of a Lewis acid catalyst vinyl allene **145** reacts with an alkyl/arylimine, affording only one regio- and stereoisomeric octahydroquinoline **146** in good yield. This product arises via a transition state having the *C*-aryl substituent of the imine endo to the diene.

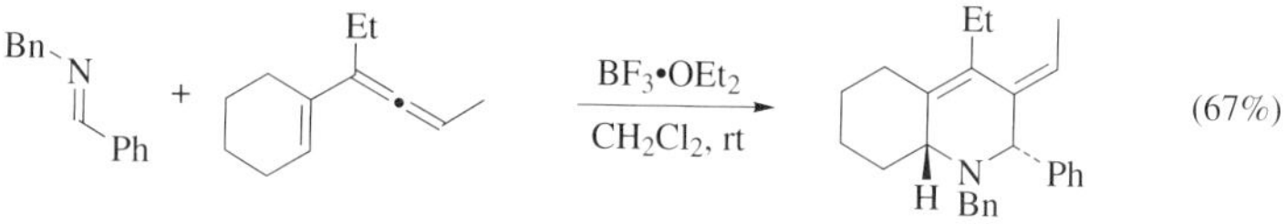
Bn
N
Ph
+
Et
BF3•OEt2
CH2Cl2, rt
Et
N
H Bn
Ph
(67%)

145 **146**

Imines bearing alkyl/aryl substituents also react with dienes possessing electron-withdrawing groups, probably via an inverse electron demand ImDA process. When *N*-benzyl-*p*-methoxybenzylideneamine is heated with the quinodimethane diene **147**, generated thermally in situ from the corresponding benzocyclobutene derivative, isoquinoline derivative **148** is obtained as the only regioisomeric product in moderate yield.[144] The formation of this product can again be rationalized on the basis of the mechanistic model in Eq. 7 via the dipolar forms **13** and **15**.

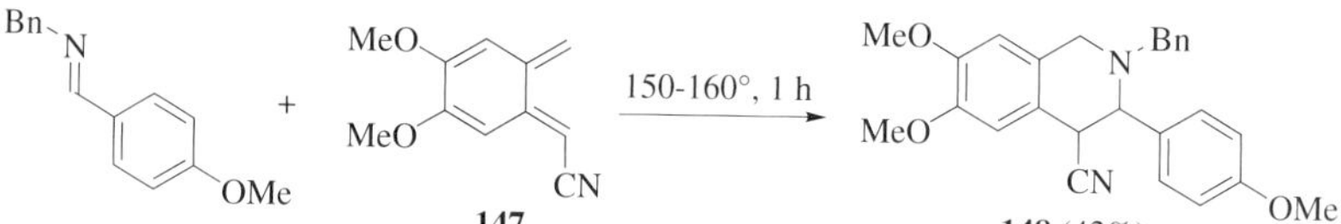
Bn
N
OMe
+
MeO
MeO
CN
150-160°, 1 h
MeO
MeO
N
Bn
CN
OMe

147 **148** (42%)

N-Methylbenzylideneamine adds to 1,3-di(phenylsulfonyl)-1,3-butadiene, an electron-deficient diene, in a regioselective manner, affording tetrahydropyridine **149** as the only product in excellent yield.[145,146]

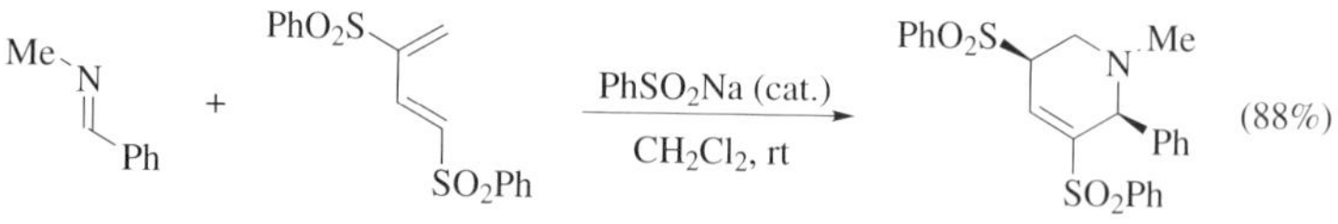
Me
N
Ph
+
PhO2S
SO2Ph
PhSO2Na (cat.)
CH2Cl2, rt
PhO2S
N
Me
Ph
SO2Ph
(88%)

149

The exo/endo stereoselectivity of thermal, Lewis or Brønsted acid catalyzed ImDA reactions of dienes with alkyl/aryl-substituted imines depends on several factors such as temperature, the nature of the Lewis acid catalyst, the substituents on both the imine nitrogen and the carbon, and on the diene employed. Various Lewis acid catalyzed Diels-Alder reactions of cyclic dienes with alkyl/aryl-substituted imines have been reported. In the majority of these reactions the formation of the exo product is favored over the endo isomer, albeit only to a small extent. For example, the reaction of *N*-benzylbenzylideneamine with cyclopentadiene in the presence of $La(OTf)_3$ leads to the exo cycloadduct as the major product, but in only 60% diastereomer excess (Eq. 44).[147]

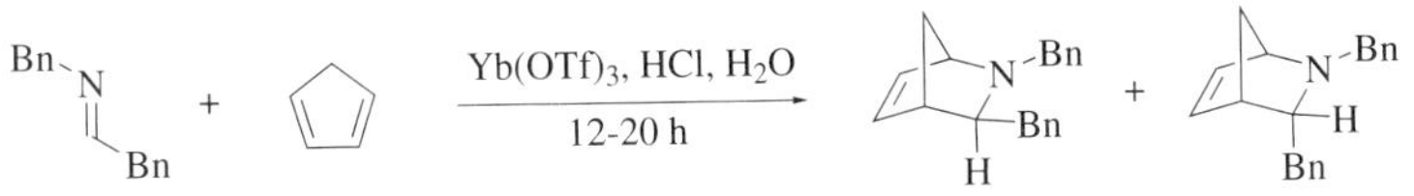
Bn
N
Bn
+
Yb(OTf)3, HCl, H2O
12-20 h
N
Bn
Bn
H
+
N
Bn
H
Bn

(72%) 4:1

(Eq. 44)

However, in some examples the formation of the endo isomer is favored over the exo isomer. For instance, when imine **150** adds to cyclohexenone in the presence of $InCl_3$, the exo isomer is obtained in 44% diastereomer excess (Eq. 45).[148] On the other hand, when imine **151** is subjected to the Diels-Alder reaction conditions, the endo isomer is obtained as the major product, albeit again with very low diastereoselectivity.

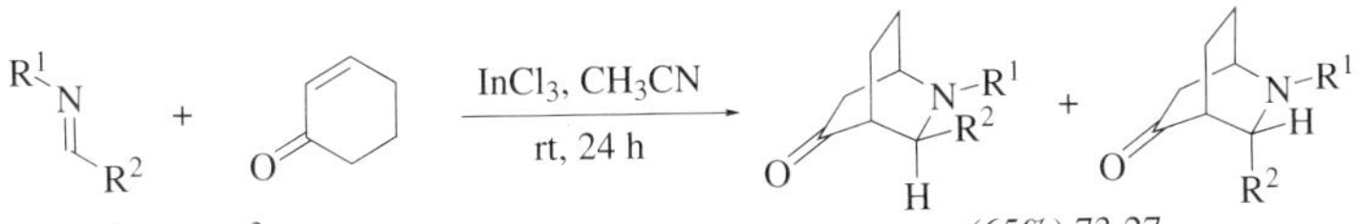
R1
N
R2
+
O
InCl3, CH3CN
rt, 24 h
N
R1
R2
H
O
+
N
R1
H
R2
O

150 R^1 = Ph, R^2 = 4-ClC_6H_4
151 R^1 = R^2 = 4-ClC_6H_4

(65%) 73:27
(74%) 47:53

(Eq. 45)

Acyclic dienes, on the other hand, often participate in ImDA reactions to form cycloadducts with good exo/endo selectivities. When *N*-benzylpropylideneamine is combined with the 2-siloxy-1,3-butadiene **152** in the presence of TMSOTf, two diastereomeric products are obtained in an 89:11 ratio with the exo isomer as the major product.[149,150]

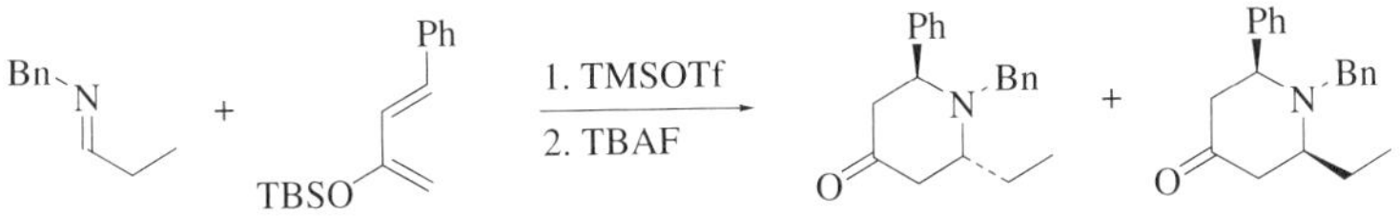
Bn
N
+
Ph
TBSO
1. TMSOTf
2. TBAF
Ph
N
Bn
O
+
Ph
N
Bn
O

152

(49%) 89:11

When *N*-methylbenzylideneamine reacts with bis(phenylsulfonyl)-substituted diene **153**, the intermediate endo cycloadduct **154** is formed, which rapidly undergoes a formal 1,3-hydrogen shift affording tetrahydropyridine **155** in 80% yield.[151]

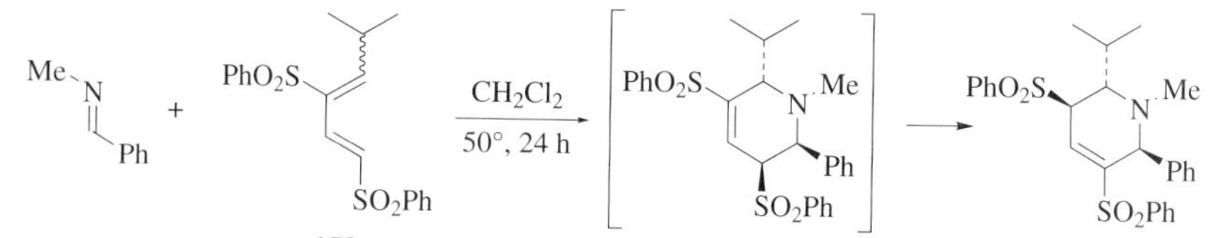
Me
N
Ph
+
PhO2S
SO2Ph
CH2Cl2
50°, 24 h
PhO2S
N
Me
Ph
SO2Ph
PhO2S
N
Me
Ph
SO2Ph

153 **154** **155** (80%)

Addition of vinylallenes to imines employing Lewis acids as catalysts is highly diastereoselective, leading to the formation of only the endo isomer in moderate yields.[143] For example, vinylallene **156** adds to *N*-benzylbenzylideneamine, affording the endo adduct **157** in 54% yield. A small amount of the triene **158** is also formed because of a competing ene reaction.

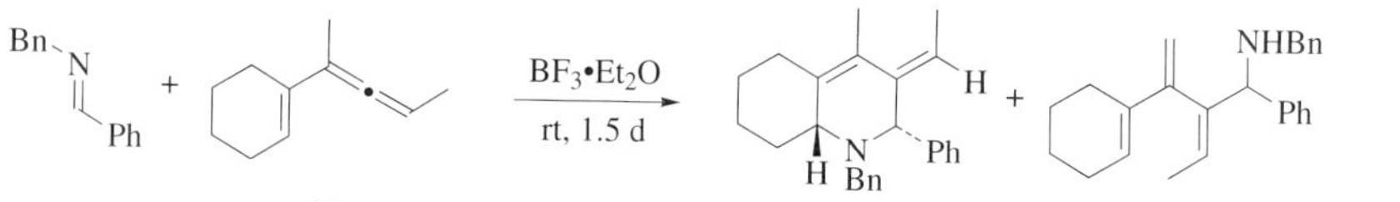
Bn
N
Ph
+
BF3•Et2O
rt, 1.5 d
H
N
Ph
H Bn
+
NHBn
Ph

156 **157** (54%) **158** (11%)

The use of chiral imines derived from amino acids in cycloadditions with cyclic dienes enhances the diastereoselectivity of the reaction.[152,153] When formaldiminium ions **159** react with cyclopentadiene, the cycloadducts are obtained with good diastereoselectivities with **160** as the major product and **161** as the minor. Similar cycloadditions with acyclic dienes, however, proceed with only moderate diastereoselectivities.

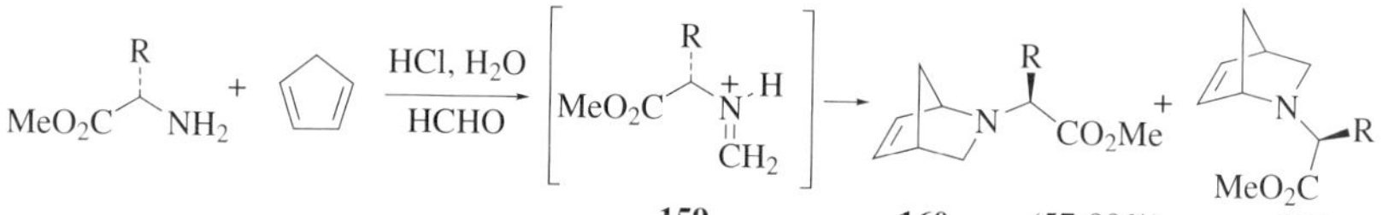
R
MeO2C
NH2
+
HCl, H2O
HCHO
R
MeO2C
N
H
CH2
R
N
CO2Me
+
N
R
MeO2C

159 **160** (57-99%) **161**

The observed stereoselectivity with dienophile **159** is explained by invoking an interaction between the oxygen atom of the ester carbonyl group and the electrophilic imine carbon. Secondary orbital interactions between the carbonyl group and diene, and the steric hindrance of the R group favors attack from the Re face in an exo transition state leading to the observed adduct **160** (Fig. 3). The minor diastereomer **161** is formed via the endo transition state shown.

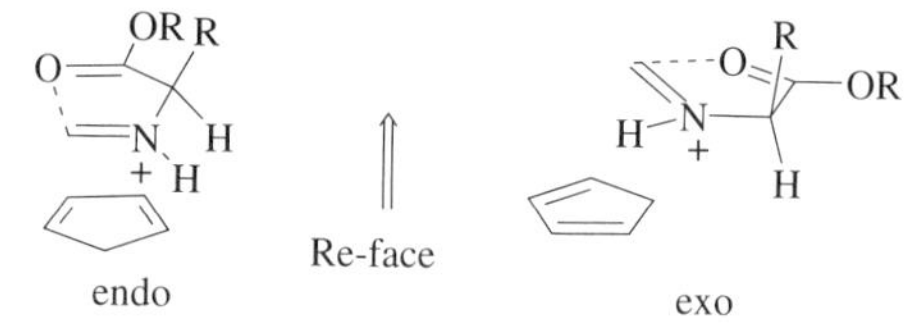
OR R
O
N
H
+ H
endo
Re-face
R
O
OR
H
N
+
H
exo

Figure 3 Transition states proposed to account for the formation of diastereisomeric adducts **160** and **161**

As mentioned previously, the stereoselectivity of ImDA reactions of alkyl/arylimines depends on a variety of factors. By manipulating these variables, it is possible to tilt the exo/endo ratio in favor of one isomer. Some extreme examples demonstrate the possibility of obtaining only the exo adduct under one particular set of conditions and the endo adduct under a different set of conditions.

When silyloxydiene **162** is exposed to *N*-phenylbenzylideneamine in the presence of $AlCl_3$ at $-40°$ for 15 minutes, cycloadducts **163** and **164** are obtained, where the exo product **163** (cis relation between the new stereocenters) is favored (exo/endo = 87/13).[154] However, when the reaction is conducted at room temperature and the mixture is stirred for 2 hours, the endo cycloadduct **164** predominates (exo/endo = 2/98). This thermodynamic control of the cycloaddition leading to the more stable endo adduct is presumably a result of a retro-ImDA reaction that can occur at room temperature.[155]

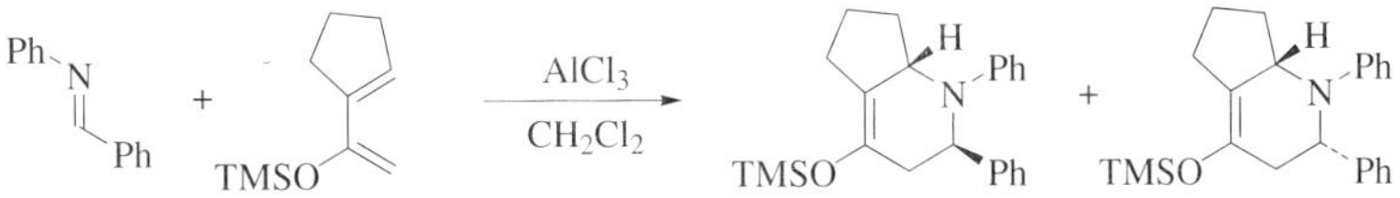
Ph
N
Ph
+
TMSO
AlCl3
CH2Cl2
H
N
Ph
TMSO
Ph
+
H
N
Ph
TMSO
Ph

162 **163** **164**

–40°, 15 min (75%) 87:13
rt, 2 h (75%) <2:98

Similarly, *N*-phenyl-*p*-methoxybenzylideneamine adds to silyloxydiene **165** in the presence of $AlCl_3$ at 20°, affording predominantly the endo cycloadduct **167** (thermodynamic control).[156] Surprisingly, when the Lewis acid is changed from $AlCl_3$ (1 equiv) to TBSOTf (0.1 equiv), the exo isomer **166** (i.e., the "kinetic" product) is obtained as the major adduct. When the exo adduct is resubjected to the ImDA reaction conditions, it is readily converted into the endo isomer in the presence of $AlCl_3$, whereas only 30% conversion is observed in the presence of TBSOTf or TMSOTf. Furthermore, TMSOTf does not afford the same selectivity as TBSOTf (**166**/**167** = 70/30). Thus, the exo stereocontrol appears to be attributable to the nature of the Lewis acid employed and the structure of the Lewis acid-imine complex.

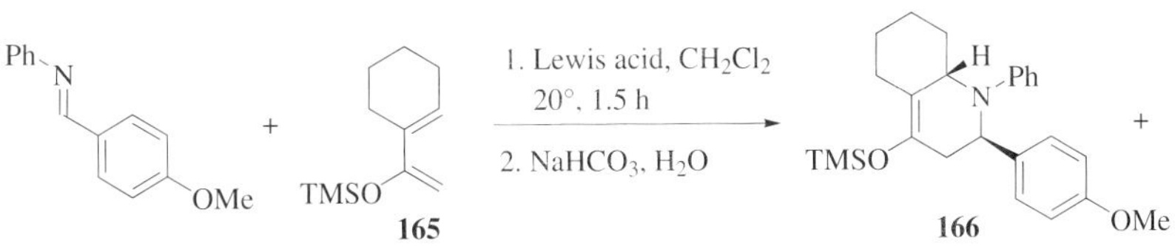
Ph
N
OMe
+
TMSO
1. Lewis acid, CH2Cl2
20°, 1.5 h
2. NaHCO3, H2O
H
N
Ph
TMSO
+
OMe

165 **166**

Lewis Acid	**166**:**167**	**166**+**167**
$AlCl_3$ (1.0)	<2:98	(71%)
TBSOTf (0.1)	>98:2	(84%)

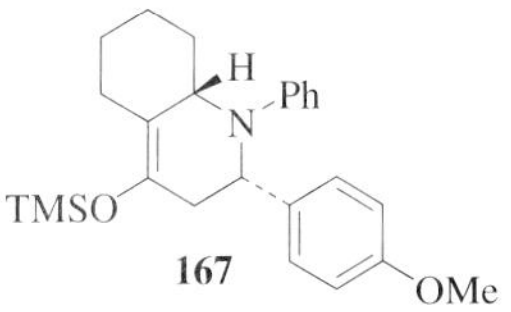
H
N
Ph
TMSO
OMe

167

The ImDA reactions of vinylketenes bearing silyl substituents are not only highly regioselective but also exhibit high endo/exo selectivity. Thus, when vinylketene **170** is treated with *N*-silylimine **168** at room temperature, the endo isomer **171** is obtained in >98% diastereomer excess.[142] However, when the silyl moiety on the imine nitrogen is replaced by a methyl group (e.g., **169**), the exo adduct **172** becomes the major product, albeit with only moderate diastereoselectivity.

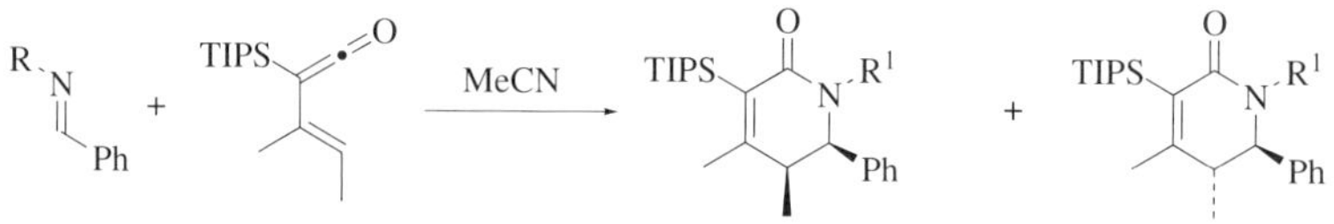
R
N
Ph
+
TIPS
O
MeCN
TIPS
O
N
R¹
Ph
+
TIPS
O
N
R¹
Ph

168 R = TMS **170** 81°, 1.5 h **171** R^1 = H (79-83%) >99:1 **172**
169 R = Me 120°, 42 h R^1 = Me (71%) 1:3

Treatment of diene **174**, possessing an electron-donating nitrogen substituent, with *N*-silyl- or *N*-aryl-substituted benzylideneamines **173** in the presence of $ZnCl_2$ affords a drastic difference in the product diastereoselectivity that could be directly linked to the nature of the substituent on the imine nitrogen.[157] When R^1 = TMS, the cycloaddition proceeds diastereoselectively to afford the endo adduct **175** as the major product, whereas with R^1 = PMP (p-$MeOC_6H_4$) the exo adduct **176** is obtained predominantly. This observation suggests that the reversal of stereoselectivity arises from a difference in approach of the imines to the dienophile during the cycloaddition process (Fig. 4). With *N*-silylimines, the addition proceeds via an endo transition state to afford adduct **176**. However, *N*-arylimines approach the diene via an exo transition state in such a way that the aromatic substituent on the imine nitrogen has maximum overlap with the diene π-electron system, affording **176**.

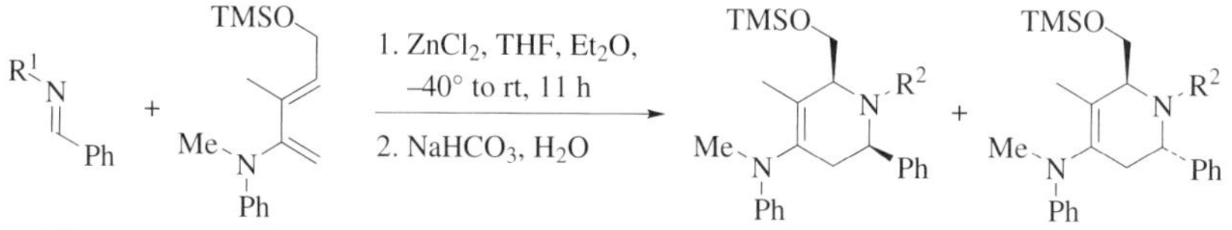
R¹
N
Ph
+
TMSO
Me
N
Ph
1. ZnCl₂, THF, Et₂O,
–40° to rt, 11 h
2. NaHCO₃, H₂O
TMSO
N
R²
Me
N
Ph
Ph
+
TMSO
N
R²
Me
N
Ph
Ph

173 **174** **175** **176**

R^1	R^2	**175:176**	**175+176**
TMS	H	>95:5	(78%)
PMP	PMP	<5:95	(91%)

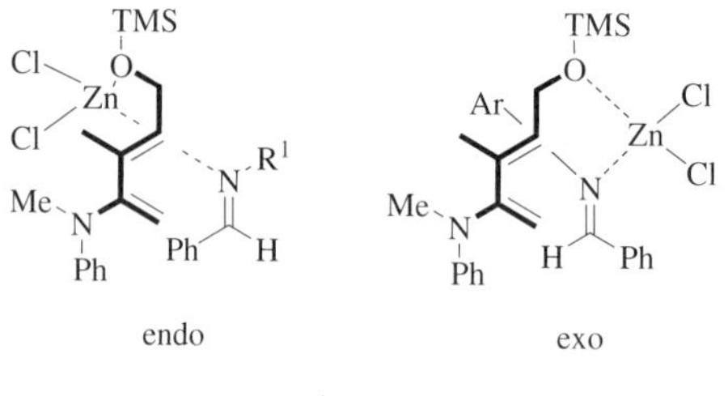
TMS
O
Cl
Zn
Cl
N
R¹
Me
N
Ph
Ph
H
endo
TMS
O
Ar
Cl
Zn
Cl
N
Me
N
Ph
H
Ph
exo

Figure 4 Transition states proposed to rationalize the observed endo and exo selectivity for *N*-silyl- and *N*-aryl-substituted imines, respectively.

The diastereoselectivity of ImDA reactions of silyl vinylketenes is not only affected by the nature of the substituent on the imine nitrogen, but also by the nature of substituents on the diene component.[142] Vinylketene **179** ($R^1 = R^2 =$ Me) bearing acyclic alkyl substituents undergoes an ImDA reaction with *N*-silylimine **178**, affording exclusively the endo cycloadduct **180**. However, vinylketene **179** (R^1, $R^2 = -(CH_2)_4-$), possessing a cyclohexyl ring, yields exclusively the exo adduct **177**, because of lower steric hindrance in the exo approach to the diene.

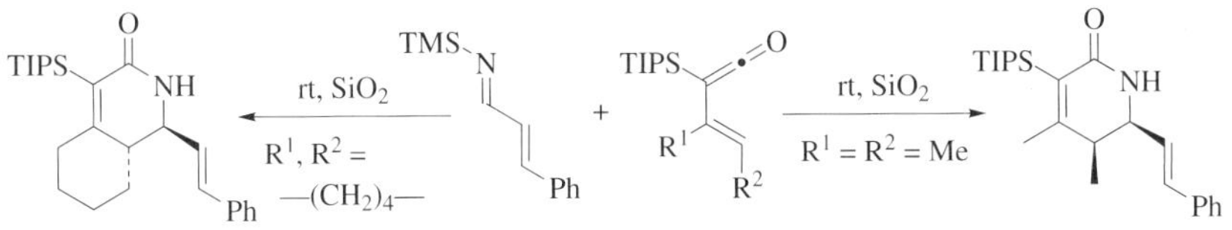
TIPS
O
NH
Ph
rt, SiO2
R1, R2 =
—(CH2)4—
TMS
N
Ph
+
TIPS
O
R1
R2
rt, SiO2
R1 = R2 = Me
TIPS
O
NH
Ph

177 (73%) **178** **179** **180** (78%)

Asymmetric alkyl/arylimino Diels-Alder reactions have been explored using either imines bearing auxiliaries derived from chiral amines, or imines prepared from chiral aldehydes or ketones. Usually the inducing stereocenter(s) of imines, derived from chiral aldehydes or ketones, remains an integral part of the product that is not easily removed. Exceptions include imines having chiral arene-chromium, olefin-iron complex fragments.[158] In most of the reactions involving chiral amines, the inducing stereocenter can be removed under mild conditions after the cycloaddition.

One of the first examples of an asymmetric cycloaddition of an unactivated imine[159] includes cyclohexylidene-protected imine **181**, prepared from L-threonine. Reaction of **181** with (1,3-dimethoxybuta-1,3-dienyloxy)trimethylsilane (Brassard's diene) in the presence of a Lewis acid such as Et_2AlCl affords a single stereoisomer of lactam **182** in 82% yield. The syn configuration of cycloadduct **182** is proposed to derive from a chelation-controlled addition process.

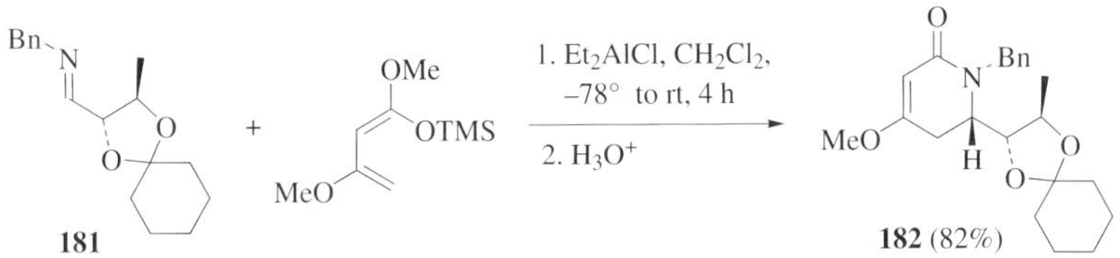
Bn
N
O
O
+
OMe
OTMS
MeO
1. Et2AlCl, CH2Cl2,
–78° to rt, 4 h
2. H3O+
O
N
Bn
MeO
H
O
O

181 **182** (82%)

In order to investigate the effects of the substituents and Lewis acids on the diastereoselectivity of this type of cycloaddition, a series of α-alkoxy imines **183** was prepared and their reactions with Brassard's diene was studied.[160] With $SnCl_4$, the chelation-controlled diastereomers syn-**184** are obtained in low yields for all three substrates, with a higher syn selectivity being observed when the steric bulk of the side chain is increased. With Et_2AlCl as the catalyst, both the imine substrates **183** bearing either a small substituent (R = *n*-C_5H_{11}) or a large substituent (R = *t*-Bu) lead to a high degree of chelation-controlled products, whereas with the dienophile containing a medium-sized side chain (R = *i*-Pr), only poor diastereoselectivity is observed. Furthermore, as the amount of Et_2AlCl is increased from less than one equivalent to two equivalents, syn selectivity is also increased. Based on this result and the fact that the medium-sized substrate **183** (R = *i*-Pr) gives only low diastereoselectivity, it is concluded that when Et_2AlCl is the catalyst, a more complicated explanation than either a chelation- or nonchelation-controlled mechanism is required.

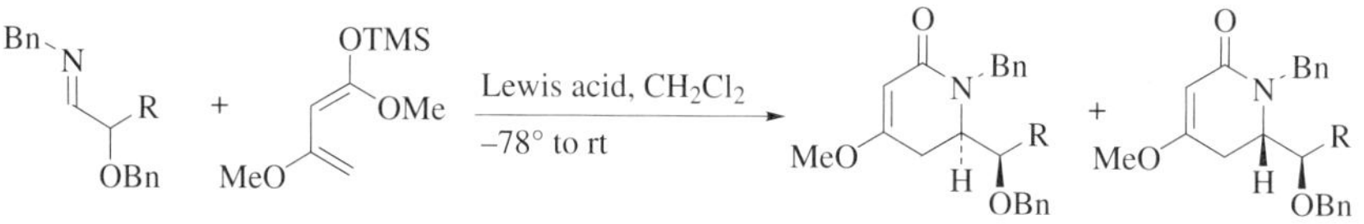
Bn
N
R
OBn
OTMS
OMe
MeO
Lewis acid, CH2Cl2
–78° to rt
O
N
Bn
R
MeO
H
OBn
O
N
Bn
R
MeO
H
OBn

183 R = n-C_5H_{11}, *i*-Pr, *t*-Bu | $SnCl_4$ (10-20%) | Et_2AlCl (55-75%) | syn-**184** | anti-**184**

The ImDA reaction of chiral α-silyloxy aldimine **185** with activated 2-siloxy-1,3-diene **186** affords only two diastereomers, ketopiperidines syn-**187** and anti-**187**.[150,161] The diastereoselectivity here is dependent on the Lewis acid, solvent, and temperature. When the reaction is performed in hexane with $Zn(OTf)_2$ as the catalyst, ketopiperidine syn-**187** is obtained as the major product. However, with $TiCl_4$ as the catalyst and isobutyronitrile as the solvent, ketopiperidine anti-**187** is obtained as the major product. These results can be rationalized using the chelation model shown in Fig. 5. Although $Zn(OTf)_2$ can in principle participate in chelation, it is insoluble in hexane, and thus a non-chelation model is proposed to explain the anti-stereoselectivity with this catalyst.

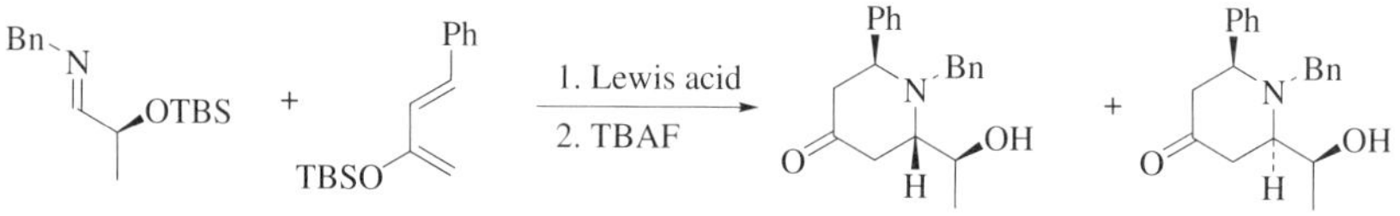
Bn
N
OTBS
Ph
TBSO
1. Lewis acid
2. TBAF
Ph
N
Bn
O
OH
H
Ph
N
Bn
O
OH
H

185 | **186** | syn-**187** | anti-**187**

$Zn(OTf)_2$, n-C_6H_{13} (93%) 89:11

$TiCl_4$, *i*-PrCN (87%) <2:98

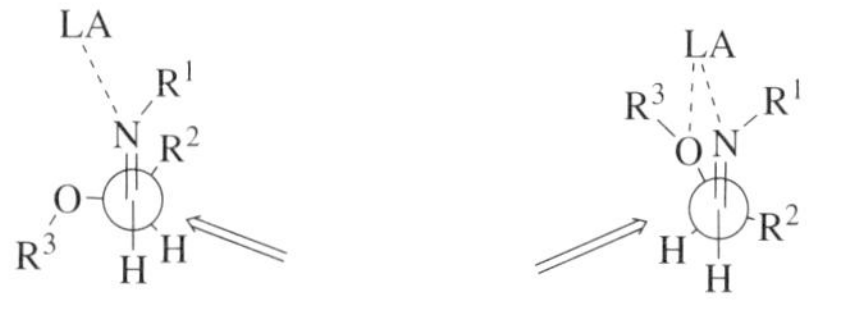
LA
R1
N
R2
O
R3
H
H
LA
R3
R1
O
N
R2
H
H

non-chelation model chelation mode

Figure 5 Chelation and non-chelation models proposed to account for the opposite stereoinduction in the presence of $Zn(OTf)_2$ and $TiCl_4$

Recently, an asymmetric reaction of imine **188**, derived from (*R*)-2,3-di-*O*-benzylglyceraldehyde and benzylamine, with Danishefsky's diene has been reported.[162–164] Among the various Lewis acids tested, ZnI_2 proved best, affording syn-**189** in 88% yield and 90% diastereomer excess. Surprisingly, irrespective of the chelating ability of the Lewis acid, the same syn diastereomer is obtained as the major product with both ZnI_2 and Et_2AlCl. To explain these results, a chelation model (for ZnI_2) and a non-chelation model (for Et_2AlCl and BF_3) similar to that in Fig. 5 is proposed.

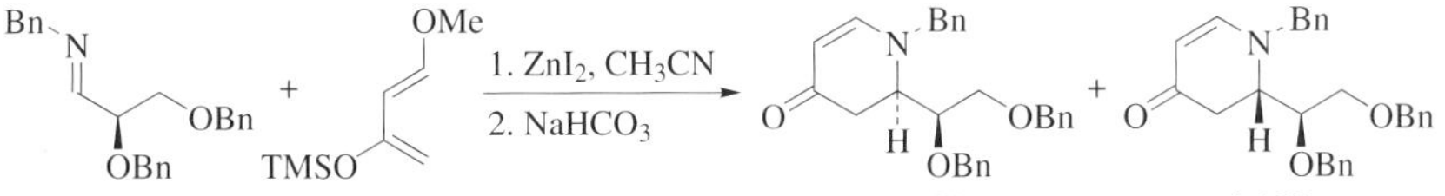
Bn
N
OBn
OBn
OMe
TMSO
1. ZnI2, CH3CN
2. NaHCO3
N
Bn
O
H
OBn
OBn
Bn
O
H
OBn
OBn

188 syn-**189** anti-**189**

(88%) 95:5

ImDA reactions of imines derived from 4-oxoazetidine-2-carboxaldehydes have been investigated with the intention of using a chiral *β*-lactam moiety to obtain stereochemical induction.[165] Addition of imines **190** to Danishefsky's diene in the presence of a Lewis acid affords the cycloadducts **191** with moderate selectivities. Further experimentation using ZnI_2 as the catalyst shows that the diastereoselectivity depends to some extent on the nature of the substituent on the imine-nitrogen (R^3), while substituents on the *β*-lactam ring (R^1 and R^2) have little or no effect.

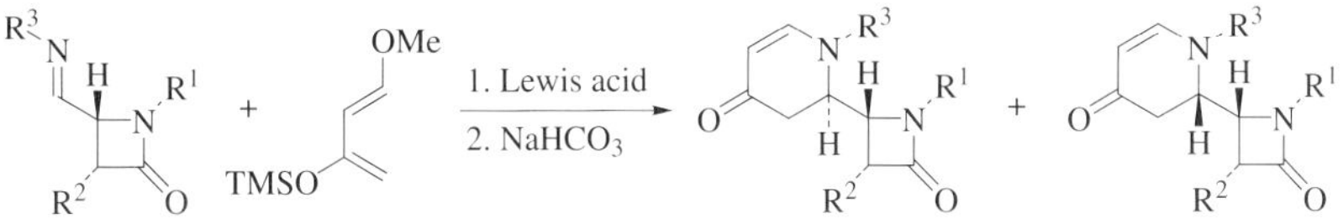
R3
N
H
R1
N
R2
O
OMe
TMSO
1. Lewis acid
2. NaHCO3
R3
N
H
R1
O
H
R2
O
R3
N
H
R1
O
H
R2
O

190 **191** (57-87%) 20-64% de

R^1 = 4-$MeOC_6H_4$ or 4-MeC_6H_4
R^2 = PhO, MeO, or BnO
R^3 = 4-$MeOC_6H_4$ or Bn

Interestingly, when imine **192** is treated with simple alkyl-substituted dienes such as 2,3-dimethylbutadiene, the products obtained correspond to the participation of *N*-arylimines **193** as azadienes rather than as heterodienophiles.

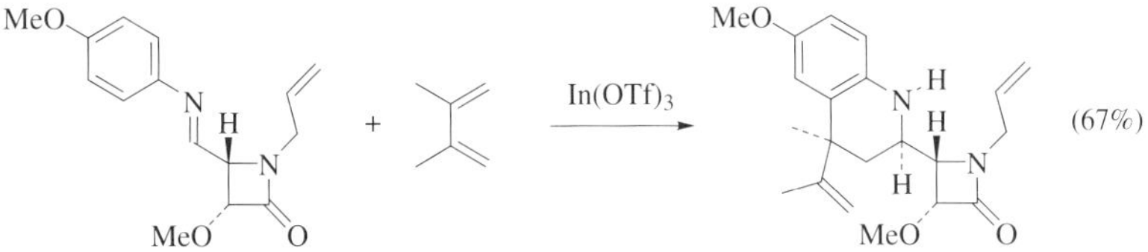
MeO
N
H
N
MeO
O
In(OTf)3
MeO
N
H
H
N
H
MeO
O
(67%)

192 **193**

Recently, tricarbonylchromium complexes have been introduced as novel chiral auxiliaries for alkyl/arylimino Diels-Alder reactions.[158,166-169] When metal-complexed chiral ortho-substituted benzaldehyde imines, e.g., **194**, are treated with Danishefsky's diene in the presence of 1.2 equivalents of a Lewis acid, only a single diastereomer such as **195** is usually obtained. The major diastereomer presumably forms by the approach of a diene on an E imine from the side opposite the large tricarbonylchromium group. Furthermore, placing a methoxy group in the ortho position of the aryl ring seems to lower the selectivity slightly.[158] Extending the distance between the arene-chromium group and imine by a methylene or ethenyl group significantly lowers the stereoselectivity.[166]

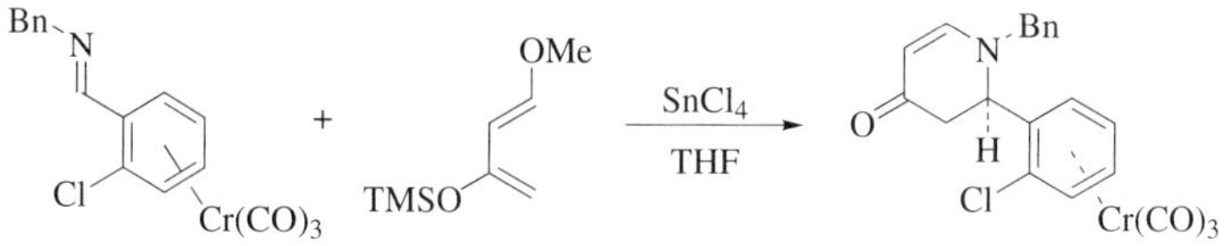
Bn
N
OMe
+
Cl
Cr(CO)3
TMSO
SnCl4
THF
Bn
N
O
H
Cl
Cr(CO)3

194 **195** (98%) >98:2

Use of imines derived from chiral amino alcohols as the dienophiles for asymmetric ImDA reactions have also been explored. This type of chiral auxiliary was chosen because of its ability to form metal chelates.[170] The imines **196** combine with Danishefsky's diene and these reactions show some distinctive and instructive trends.

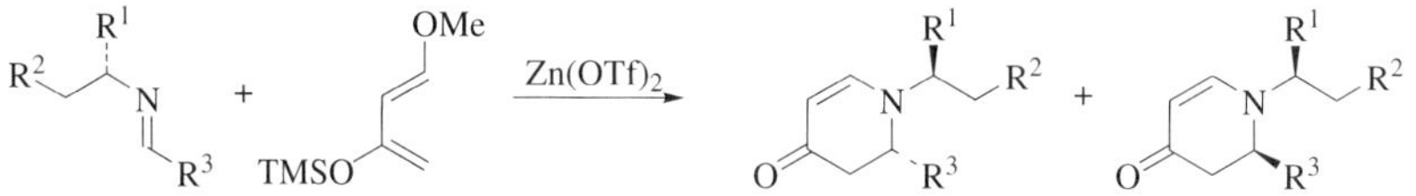
R1
R2
N
R3
+
OMe
TMSO
Zn(OTf)2
R1
N
R2
O
R3
+
R1
N
R2
O
R3

196 **197** **198**

R^1	R^2	R^3	**197+198**	**197:198**
Me	OMe	Ph	(63%)	59:41
i-Pr	OMe	Ph	(67%)	87:13
Ph	OTMS	*n*-Pr	(45%)	58:42
Ph	OH	Ph	(60%)	>95:5
Ph	H	Ph	(69%)	80:20

The size of the group R^1 directly influences the stereoselectivity in these cycloadditions. As the bulk of R^1 is increased, the diastereoselectivity is enhanced. In contrast, the substituent R^3 on the imine carbon has little influence on the process, although no reaction occurs when R^3 is a bulky *tert*-butyl group. Imines that can form a cyclic chelate give similar diastereoselectivities for the ImDA adduct **197** as those that did not have a second chelating site. Furthermore, the imines with a more Lewis basic second chelating group ($R^2 = CO_2Me$, OH) give higher diastereoselectivities than imines with a less Lewis basic site (R^2 = OTMS). To account for the stereoselectivity, the chelate model is invoked for chelating substrates; for those substrates that can only form monodentate coordination to the Lewis acid (Fig. 6) a modified Felkin-Anh model is applied.

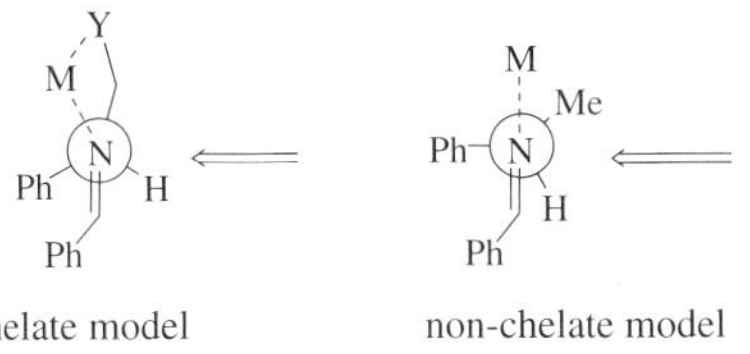
chelate model
non-chelate model

Figure 6 Chelate and non-chelate models proposed to account for the different stereoselectivities exhibited by substrates with or without a second chelating group

The use of imines with a chiral carbohydrate template have been investigated as dienophiles in diastereoselective ImDA cycloadditions.[171–175] The reaction of galactosyl-derived imines **199** with 2,3-dimethyl-1,3-butadiene or isoprene in the presence of $ZnCl_2$ as the catalyst proceeds in good yields, but affords mixtures of three diastereomeric products (one from the α-anomer) with moderate selectivities.[171]

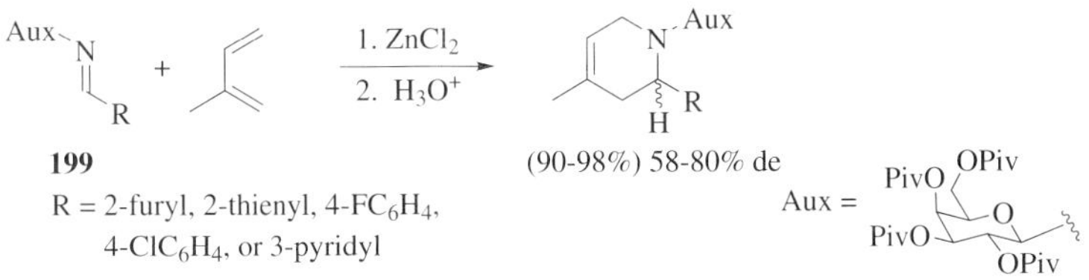
1. ZnCl2
2. H3O+
199
R = 2-furyl, 2-thienyl, 4-FC6H4, 4-ClC6H4, or 3-pyridyl
(90-98%) 58-80% de
Aux =

However, excellent selectivities are achieved by reaction of imine **199** with Danishefsky's diene, affording dehydropiperidinone derivatives **200** in high yield and diastereoselectivity. The reaction of imine **199** with isoprene is believed to be a concerted ImDA process, whereas reaction with Danishefsky's diene most likely proceeds by a stepwise Mannich/Michael-type process as the corresponding uncyclized Mannich products can be isolated if the reaction is quenched with aqueous NH_4Cl solution.[173] It was proposed that the diene approaches the sterically less shielded Si-face of the galactosylimine-Lewis acid complex as outlined in Fig 7.[171,173]

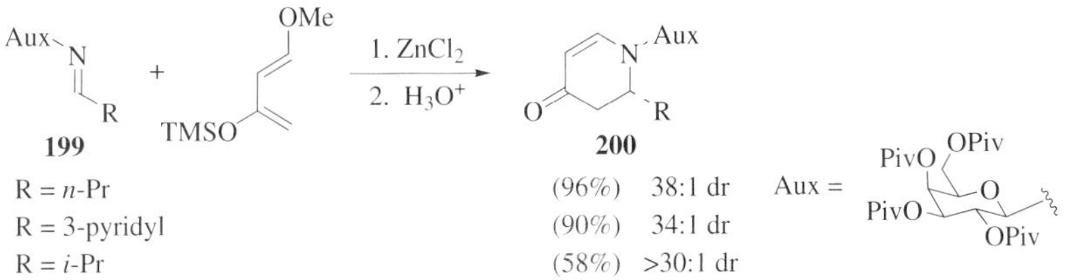
1. ZnCl2
2. H3O+
199
R = n-Pr
R = 3-pyridyl
R = i-Pr
200
(96%) 38:1 dr
(90%) 34:1 dr
(58%) >30:1 dr
Aux =

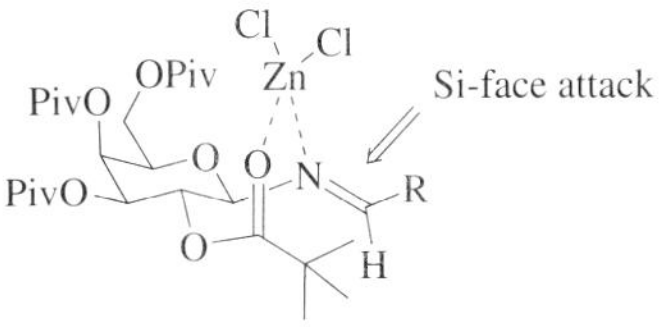
Si-face attack

Figure 7 $ZnCl_2$-imine **199** complex showing attack at the sterically less hindered Si-face

Asymmetric ImDA reactions of chiral imines **201** derived from α-amino acid esters with activated dienes leading to cycloadducts **202** have been studied extensively. Excellent diastereoselectivities are obtained when valine and isoleucine esters are employed as the auxiliaries. Isolation of a Mannich addition product (26% yield) after aqueous workup of the reactions of imines **201** with Danishefsky's diene again suggests that the reaction probably proceeds via a stepwise Mannich/Michael sequence. Irrespective of the use of either chelating Lewis acids ($ZnCl_2$ and $TiCl_4$), or nonchelating Lewis acids (BF_3 and Et_2AlCl), the same major stereoisomeric adduct **202** is obtained.

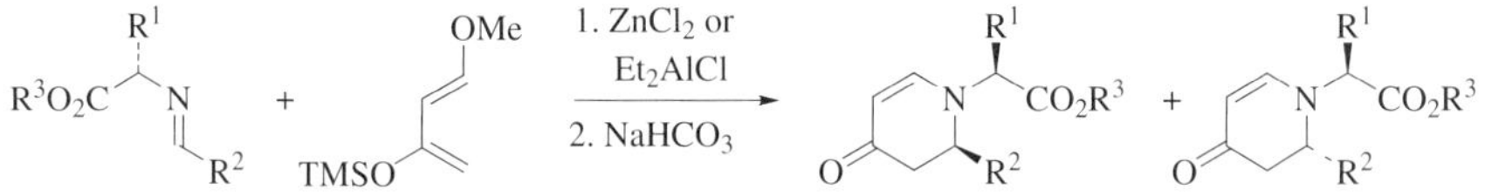

1. ZnCl2 or Et2AlCl
2. NaHCO3

201 **202** (45-80%)

R^1 = *i*-Pr, *i*-Bu
R^2 = alkyl, aryl
R^3 = Me, Et, Bn

The stereochemical outcome of the reaction with non-chelating Lewis acids can be rationalized by assuming a modified Felkin-Anh type of transition state **A** (Fig. 8), in which the Lewis acid coordinates to the nitrogen atom of the imine, and where the —CCO_2R substituent is oriented perpendicular to the imine. To account for the lack of reversal of the stereochemical outcome by changing from a non-chelating to a chelating Lewis acid, it is suggested that an equilibrium exists between the cis and trans imine in the presence of $ZnCl_2$, and that the cis imine probably reacts faster. This explanation is based on the observation of different ^1H NMR chemical shifts of the aldimine proton and the amino acid α-H with and without the presence of one equivalent of $ZnCl_2$. The existence of the chelated intermediate **B** using one equivalent of $ZnCl_2$ is supported by the observation that the sense of asymmetric induction is reversed by using two equivalents of $ZnCl_2$. This result can be rationalized by the coordination of one equivalent of $ZnCl_2$ to the imine nitrogen atom and the second equivalent of $ZnCl_2$ to the ester carbonyl oxygen as shown in **C**.

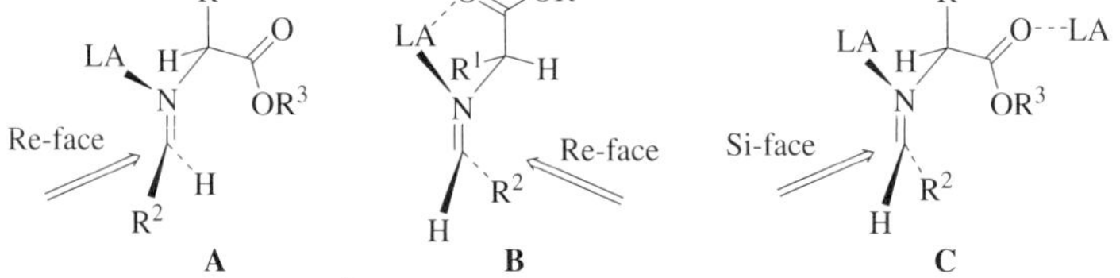

Re-face
Re-face
Si-face

A **B** **C**

Figure 8 Models for coordination of imine **201** to chelating and non-chelating Lewis acids

The effect of double asymmetric induction with imines derived from either enantiomer of *N*-α-methylbenzylamine and *O,O*-dibenzyl-protected glyceraldehyde has been studied.[163] When imine **203** is treated with Danishefsky's diene in the presence of ZnI_2, only one stereoisomeric product is detected. Imine **204**, on the other hand, affords a mixture of cycloadducts with low selectivity. The diastereoselectivities of these processes reflect the net directing effect of both stereogenic units present in the starting imines **203** and **204** (i.e., 1,2-induction from the group at C2, plus or minus 1,3-induction from the *N*-α-methylbenzyl moiety).

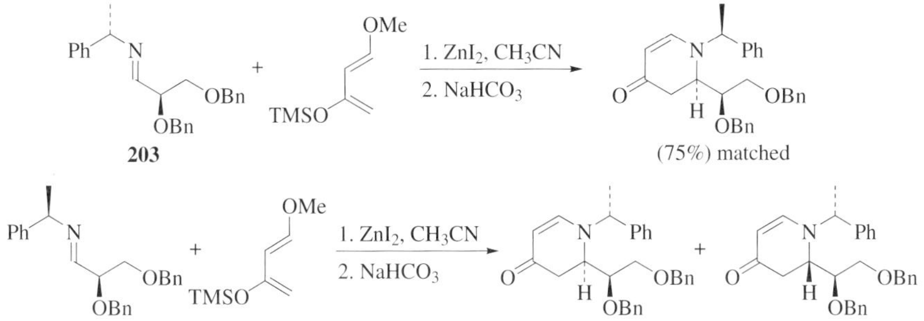
Ph
N
OBn
OBn
+
OMe
TMSO
1. ZnI2, CH3CN
2. NaHCO3
O
N
Ph
H
OBn
OBn
Ph
N
OBn
OBn
+
OMe
TMSO
1. ZnI2, CH3CN
2. NaHCO3
O
N
Ph
H
OBn
OBn
+
O
N
Ph
H
OBn
OBn

203

(75%) matched

204

(73%) 64:36
mismatched

Diene **206** bearing a chiral pyrrolidine auxiliary undergoes $ZnCl_2$-catalyzed cyclodditions with *N*-silylimines **205** to afford the piperidones **207** in moderate to good yields and with good to excellent enantioselectivities.[176–179] The hydroxyl substituent R^2 seems to have more influence on the enantioselectivity of the cycloadditions than would be expected, given that it is relatively distant from the reacting atoms and that rotation can occur about the C–O bond. The effect of R^2 according to the enantiomer excesses found for compounds **207** is as follows: TMS>Me>MOM>TBS.

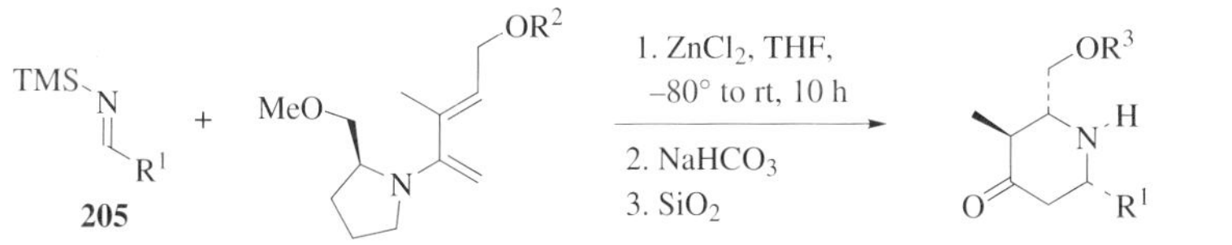
TMS
N
R1
+
MeO
N
OR2
1. ZnCl2, THF,
−80° to rt, 10 h
2. NaHCO3
3. SiO2
OR3
N
H
O
R1

205

206

207

R^1 = 3-furyl, 4-MeOC$_6$H$_4$, Ph
R^2 = TMS, Me, MOM, TBS

R^3 = H, Me, MOM, TBS
(28-65%)

Azirines. A number of reports have appeared over the last three decades that describe the involvement of azirines in hetero Diels-Alder processes.[180] The large majority of these reactions utilize the neutral azirines in cycloadditions with various dienes under thermal conditions, although several recent examples exist of Lewis acid promoted reactions. As a result of ring strain, azirines generally tend to be more reactive in ImDA cycloadditions than acyclic and larger ring analogs. Azirines bearing a variety of alkyl-, aryl-, and acyl-substituent groups can be effective partners in such reactions. These reactions are usually both regio- and stereoselective.

The early work on azirine Diels-Alder reactions involved thermal cycloadditions of a number of reactive cyclopentadienones with alkyl- and aryl-substituted azirines ultimately affording azepine derivatives.[181–185] For example, azirine **208**, produced in situ by thermolysis of the corresponding vinyl azide, reacts with cyclopentadienone **209** in toluene at reflux to form a transient [4+2]-cycloadduct **210** (Eq. 46). The stereochemistry of the initial adduct **210** was suggested to be endo, although this supposition was not proven. This intermediate then loses carbon monoxide, eventually yielding azepine **211**.[182–185] Cycloadditions of an unsymmetrical cyclopentadienone with some substituted azirines reveal low regioselectivity, resulting in the production of mixtures of isomeric azepines. Similar types of azirines, such as **212**, also react thermally with 1,3-diphenylisobenzofuran, affording isolable Diels-Alder adducts like **213**, shown to have the exo stereochemistry (Eq. 47).[186,187]

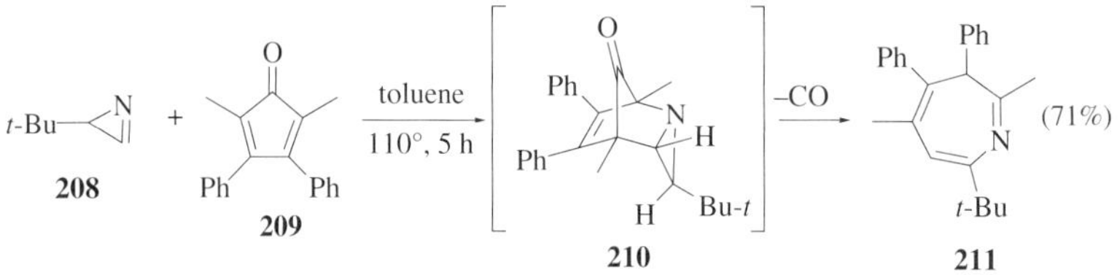
t-Bu
N
O
Ph
Ph
toluene
110°, 5 h
O
Ph
Ph
N
H
H
Bu-t
–CO
Ph
Ph
N
t-Bu
(71%)

208 **209** **210** **211** (Eq. 46)

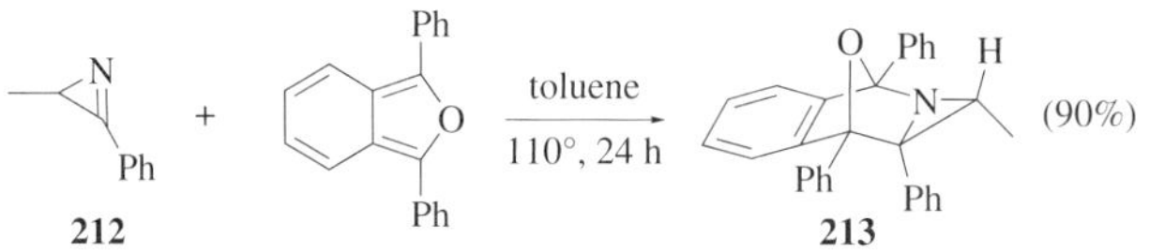
N
Ph
Ph
O
Ph
toluene
110°, 24 h
O
Ph
H
N
Ph
Ph
(90%)

212 **213** (Eq. 47)

Thermal cycloaddition reactions of simple unactivated alkyl- and aryl-substituted azirines with acyclic 1,3-dienes have not been reported. However, this type of cycloaddition can be effected with both cyclic and acyclic dienes if a Lewis acid catalyst is employed. Thus, 2-phenylazirine reacts regio- and stereoselectively with Danishefsky's diene using a variety of catalysts affording adduct **214** in moderate yields.[188,189] Similarly, this azirine adds to 2-trimethylsiloxy-1,3-cyclohexadiene both thermally and under Lewis acid catalysis affording adduct **215**.[189] In both of these examples, the azirine ring is endo in the transition state.

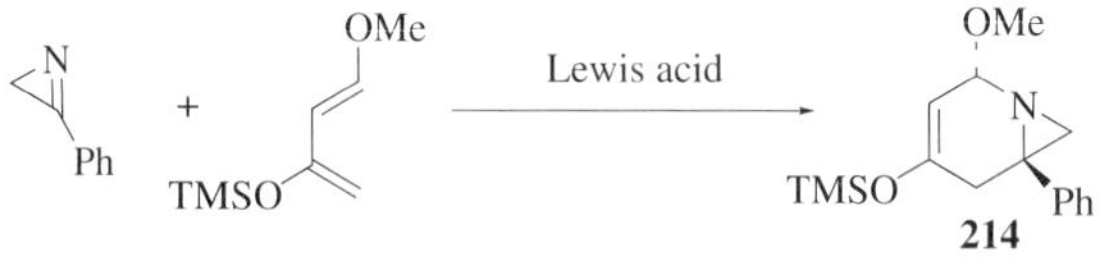
N
Ph
OMe
TMSO
Lewis acid
OMe
N
TMSO
Ph

214

$YbCl_3$, toluene, 75°, 6 h	(55%)
$ZnCl_2$, toluene, 75°, 8 h	(40%)
$CuCl_2$, toluene, 75°, 12 h	(35%)
$BF_3•Et_2O$, CH_2Cl_2, −78°, 0.3 h	(45%)

215

$YbCl_3$, toluene, 90°, 6 h	(50%)
toluene, 90°, 12 h	(48%)

In general, azirines activated with an acyl group at C2 react thermally with a variety of cyclic and acyclic 1,3-dienes, affording Diels-Alder adducts as shown in Eqs. 48–50.[190,191] The regioselectivity in these examples is the same as is seen with simple acyclic *C*-acylimines. Interestingly, these reactions also tend to be highly stereoselective, but with the three-membered ring of the azirine, rather than the *C*-acyl group, being endo to the diene in the transition state. The reasons for this preference are not clear, but may arise at least partly from the calculated preference for an exo nitrogen lone pair in ImDA reactions.[9,10] On the other hand, in similar reactions of these azirines with furans, the adducts with the acyl group endo are the exclusive products (Eq. 51).[192] There is some evidence, however, that these furan Diels-Alder reactions are reversible, and that the initially formed exo acyl isomers are converted into the more stable endo products. In addition, this type of reversibility may account for the observed stereochemical results in the type of isobenzofuran reactions shown above in Eq. 47.

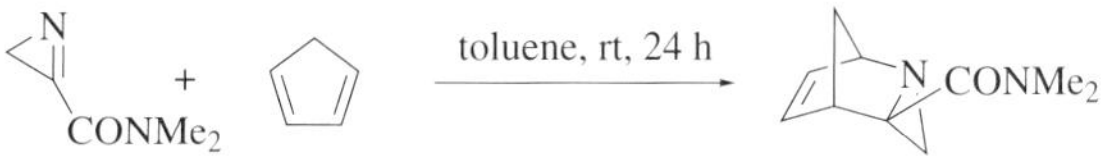

toluene, rt, 24 h

(42%) (Eq. 48)

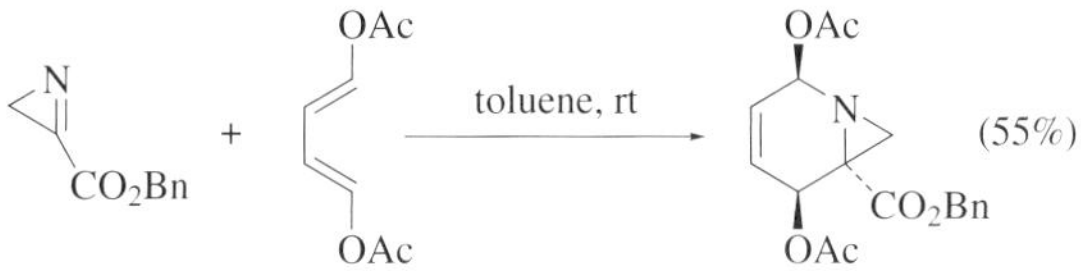

toluene, rt

(55%) (Eq. 49)

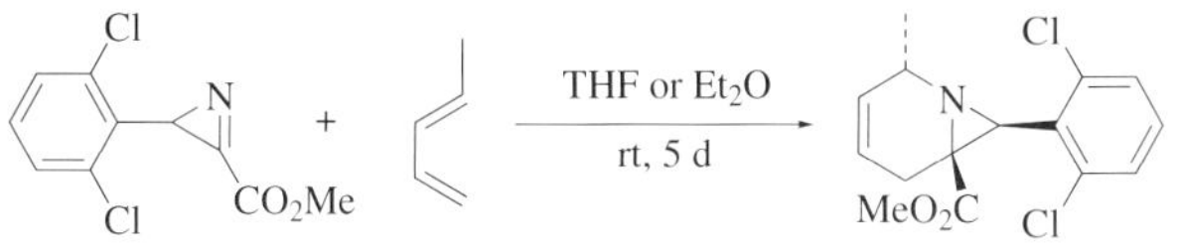

THF or Et_2O
rt, 5 d

(78%) (Eq. 50)

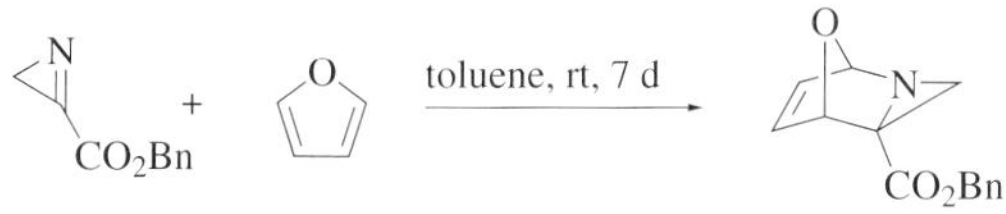

toluene, rt, 7 d

(100%) (Eq. 51)

C-Phosphoryl-activated azirines have recently been used as heterodienophiles.[193] Thus, thermal cycloaddition of azirine **216** with Danishefsky's diene is regio- and stereoselective, giving a single product **217**. As is usually observed for *C*-acyl substituted azirines, the products of these reactions are derived from the three-membered ring being endo in the transition state.

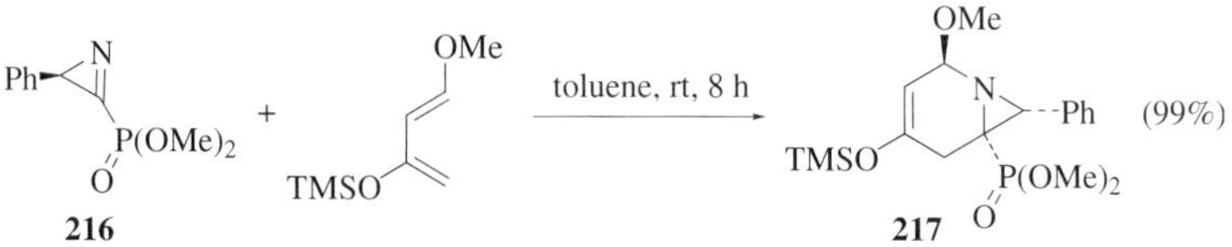
Ph
N
P(OMe)$_2$
O
216
+
OMe
TMSO
toluene, rt, 8 h
OMe
N
Ph
(99%)
TMSO
P(OMe)$_2$
O
217

Two recent reports describe diastereoselective ImDA reactions of a *C*-acyl azirine wherein a sugar-derived chiral auxiliary is attached to the 1,3-diene.[194,195] Thus, diene **219** reacts thermally with azirine ester **218**, affording a single cycloadduct assigned stereostructure **220**. It is suggested that this product arises via endo attack of the azirine ester on the preferred diene conformer (Fig. 9).

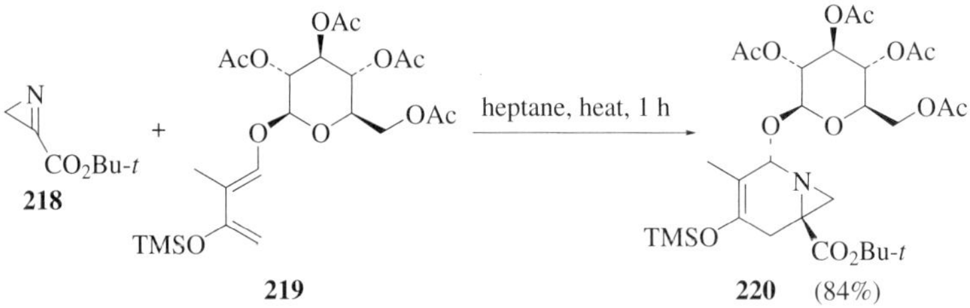
N
CO_2Bu-*t*
218
+
OAc
AcO
OAc
OAc
O
O
TMSO
219
heptane, heat, 1 h
OAc
AcO
OAc
OAc
O
O
N
TMSO
CO_2Bu-*t*
220
(84%)

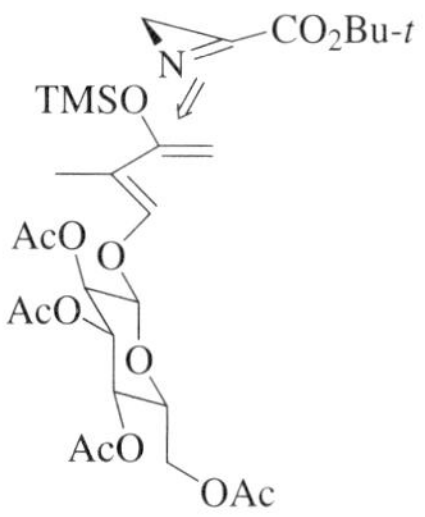
CO_2Bu-*t*
N
TMSO
AcO
O
AcO
O
AcO
OAc

Figure 9 Transition state showing the attack of the azirine ester **218** on the diene **219**

Several attempts to effect diastereoselective ImDA reactions of *C*-acyl azirines bearing chiral auxiliaries have been reported in recent years. These approaches have had varying degrees of success. For example, azirine ester **221** bearing the isobornyl sulfonamide chiral auxiliary reacts thermally with dienes to afford cycloadducts, at best with only modest levels of diastereoselectivity.[196,197] Thus, addition of dienophile **221** to 1-methoxybutadiene leads to a 2:1 diastereomeric mixture of adducts **222** and **223**, although it is not established which isomer is the major product. Similarly, the thermal cycloaddition of azirine amide **224** with cyclopentadiene shows no diastereoselectivity at all.[191]

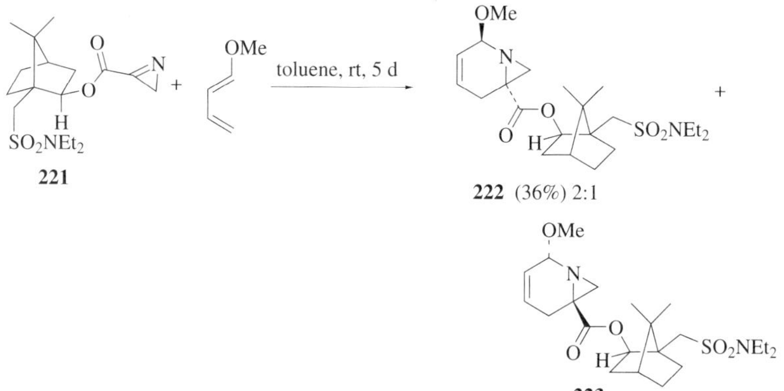
O
N
OMe
toluene, rt, 5 d
O
H
SO2NEt2
N
O
O
H
SO2NEt2
+
+
OMe
N
O
O
H
SO2NEt2

221 **222** (36%) 2:1 **223**

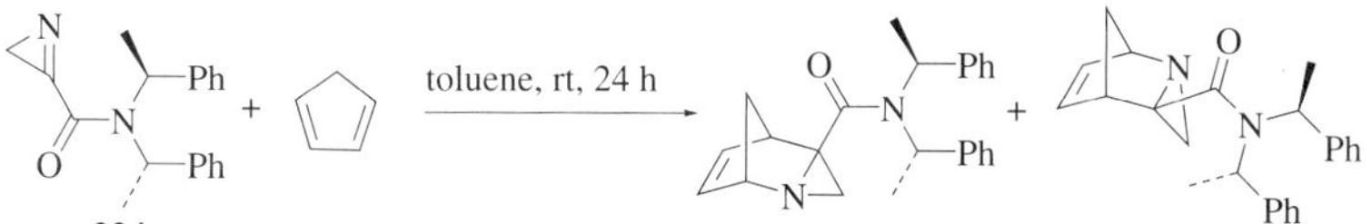
N
Ph
N
O
Ph
+
toluene, rt, 24 h
O
N
Ph
Ph
N
+
N
O
N
Ph
Ph

224 (53%) 1:1

However, much more promising results have recently been obtained in ImDA reactions with some other azirine derivatives using Lewis acid catalysis.[198,199] When azirine **225** attached to a phenylmenthol auxiliary is used as the dienophile, no diastereoselectivity is observed in the thermal cycloaddition with 1-methoxybutadiene, and adducts **226** and **227** are obtained in equal amounts (Eq. 52). However, if a Lewis acid is used to promote the reaction, the rate of cycloaddition is significantly increased and synthetically useful levels of diastereoselectivity are obtained in favor of adduct **226**. Magnesium bromide in particular is an effective catalyst with this system. One can rationalize formation of adduct **226** as the major product by assuming endo attack by the diene on the least congested face of the chiral azirine in the stacked parallel conformation shown in Fig. 10. Similar results are obtained with some other cyclic and acyclic dienes.

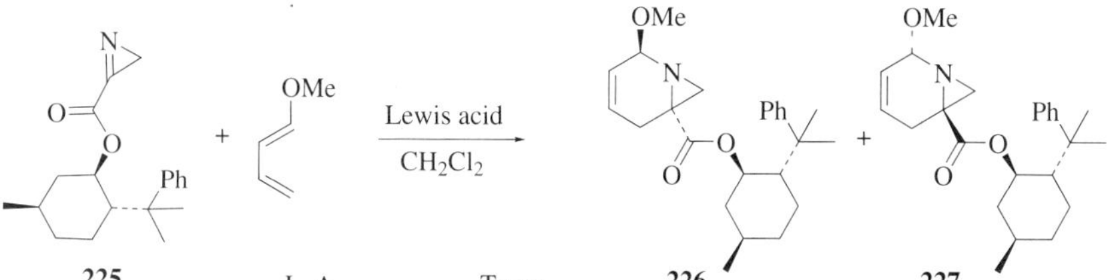
N
O
O
Ph
+
OMe
Lewis acid
CH2Cl2
OMe
N
Ph
O
O
+
OMe
N
Ph
O
O

225 **226** **227**

L. A.	Temp	
none	rt	(100%) 50:50
$MgBr_2 \cdot OEt_2$	−100 to −75°	(82%) 93.5:6.5
$ZnCl_2 \cdot OEt_2$	−100°	(62%) 90:10
$YbCl_3$	−78 to −20°	(43%) 63.5:36.5

(Eq. 52)

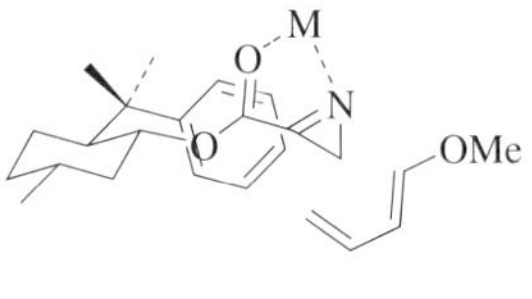

Figure 10 Transition state depicting the attack of the diene from the less congested face of the chiral azirine **225**-Lewis acid complex

Another dienophile that has been investigated is acyl azirine **228** containing Oppolzer's chiral sultam auxiliary. Only moderate levels of diastereoselectivity are observed; curiously, there is little selectivity difference between thermal reactions and those promoted by Lewis acids (Eq. 53).[198] The configuration of the major isomer is not specified.

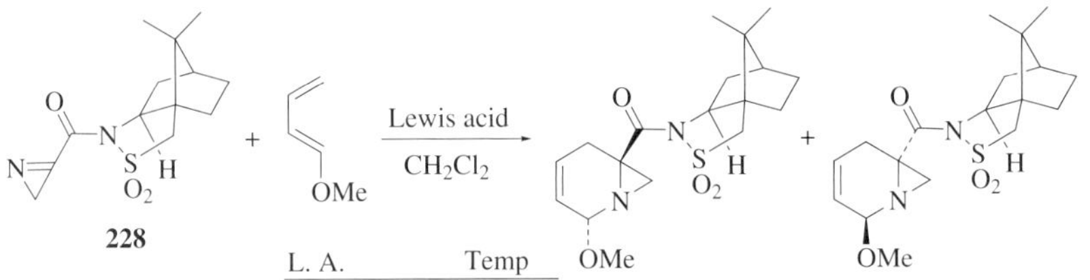

228

L. A.	Temp	
none	rt	(95%) 60.5:39.5
$MgBr_2$•OEt_2	–78°	(38%) 70:30
$ZnCl_2$•OEt_2	–78°	(69%) 36.5:63.5
$YbCl_3$	–78 to –60°	(100%) 62.5:37.5

(Eq. 53)

Oxime Derivatives. Oximes have not been widely utilized as dienophiles in ImDA reactions. In general, it appears that activation is needed for this functional type to participate in cycloadditions. However, there are brief reports that describe the reactions of some simple alkyl-substituted *O*-silyl oximes such as **229** with cyclopentadiene and furan to give ImDA adducts (Eq. 54).[200,201]

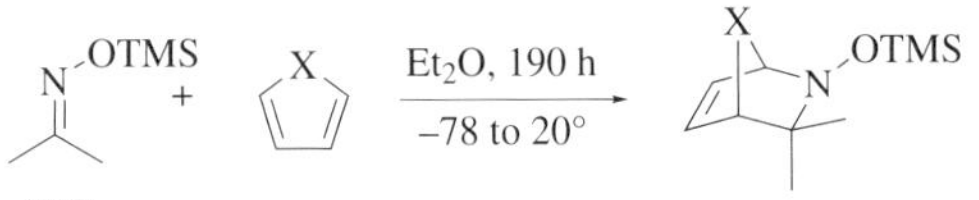

229

X = CH_2 (50%)
X = O (32%)

(Eq. 54)

Extensive investigation on the thermal cycloadditions of various types of substituted oximino malonate and malononitrile derivatives **230** with cyclopentadiene have appeared. Imino Diels-Alder adducts such as **231** are obtained (Eq. 55).[202,203] In general, it was found that as the substituent R on oxygen is changed, oxime reactivity decreases in the order: Ts > Ms > $COC_6H_4NO_2$-*p* > COPh. Similarly, as the carbon substituents on oxime **230** (R^1/R^2) are varied, the Diels-Alder reactivity decreases in the order: CN ≫ CO_2R > $CONH_2$. It was also found that R^1/R^2 carbonyl substituents are better endo directors than is a cyano group, although exo/endo equilibration in the products complicates this issue.

230 **231** (20-90%) (Eq. 55)

R = Ts, $COC_6H_4NO_2$-4, COPh, Ms
R^1, R^2 = CN, CO_2Et, CO_2Me, $CONH_2$

A few examples of cycloadditions of one of the most reactive oximino dienophile, **232**, with some acyclic dienes have also been reported.[204] Thus, unsymmetrical diene **233** reacts regioselectively with dienophile **232**, affording adduct **234** in good yield. One can rationalize this selectivity by considering the dipolar transition states **14** and **16** in Eq. 7. With isoprene, the regioselectivity of addition to **232** is low, affording a mixture of isomeric adducts **235** and **236**. These reactions are not general, and some dienes such as butadiene and 1,4-diphenylbutadiene do not react with **232**. However, electron-rich oxygenated dienes such as **237** do react with oxime **232** regioselectively at low temperature affording adducts **238**.[205]

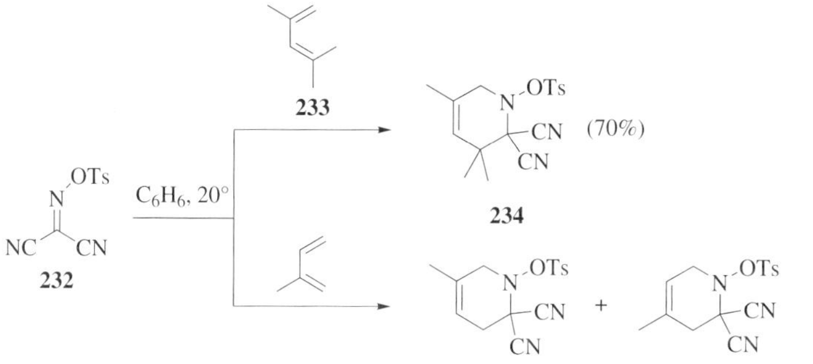

C_6H_6, 20°
(70%)
(77%) 58:42

232 **233** **234** **235** **236**

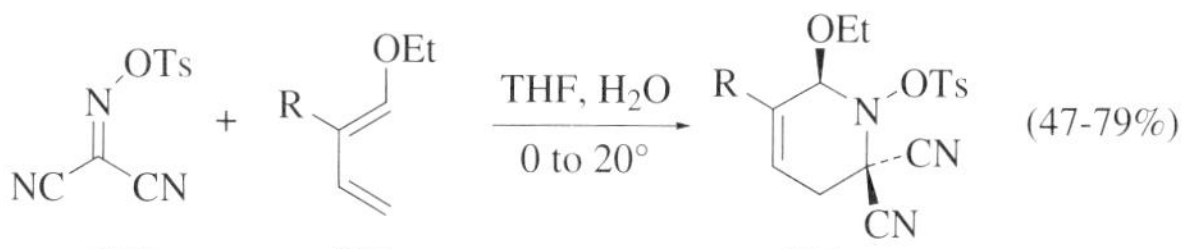

THF, H_2O
0 to 20°
(47-79%)

232 **237** **238**

R = Me, Et, *i*-Pr, Pr, Bu, *n*-C_5H_{11}

Although oximino malonate systems **230** where $R^1 = R^2 = CO_2R$ are totally unreactive with cyclopentadiene under thermal conditions, it was subsequently discovered that this type of oximino compound will act as a dienophile either when lithium perchlorate is used as a catalyst or under high pressure.[206,207] For example, cycloaddition of acyloxime **239** can be effected producing adduct **240** in moderate yields.

OAc
N
MeO2C
CO2Me
+
OAc
N
CO2Me
CO2Me

239 **240**

$LiClO_4$ (5 M), Et_2O, rt, 24 h	(49%)
11 kbar, toluene, 60°, 48 h	(25%)

These studies were extended to the Meldrum's acid derived oximino compound **241**, which is found to have greater reactivity as a dienophile towards cyclopentadiene than the corresponding acyclic systems. However, in order to effect reactions with acyclic dienes high pressure is required (Eq. 56).[208] No investigations with unsymmetrical dienes are included.

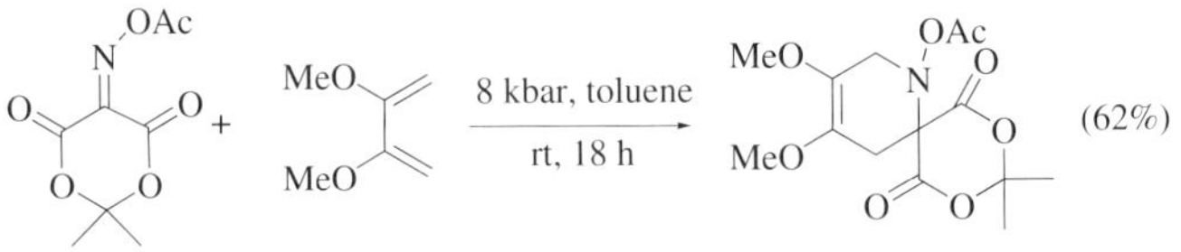
OAc
MeO
MeO
8 kbar, toluene
rt, 18 h
MeO
MeO
OAc
(62%)

(Eq. 56)

241

In an extension of this work, simply replacing the acetoxy group of **241** with a tosyloxy function increases the ImDA reactivity of the oximino compound significantly.[209,210] Tosyloxy compound **242** is found to react regioselectively with a wide array of dienes. Cycloadditions with **242** can be effected thermally in benzene at reflux affording moderate yields of products. The regioselectivity in these examples is the opposite of that observed with simpler imines. However, with two equivalents of dimethylaluminum chloride as catalyst, cycloaddition occurs in high yields under very mild conditions (Eq. 57). Other catalysts such as $TiCl_4$, $ZnCl_2$, $AlCl_3$, and protic acids are less effective. It is suggested that the reactive species in these cycloadditions is the ionic aluminum complex shown in Fig. 11.

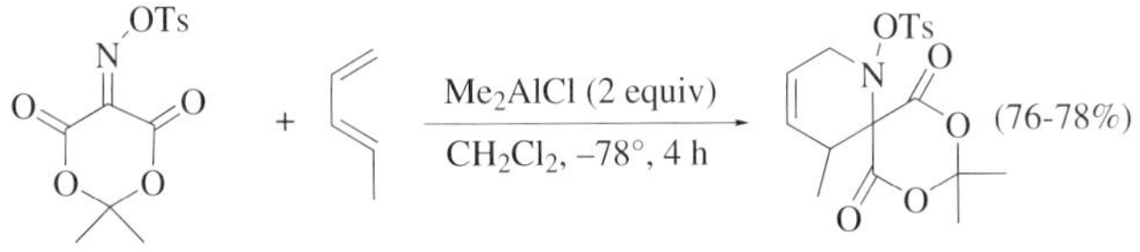
OTs
Me2AlCl (2 equiv)
CH2Cl2, –78°, 4 h
OTs
(76-78%)

(Eq. 57)

242

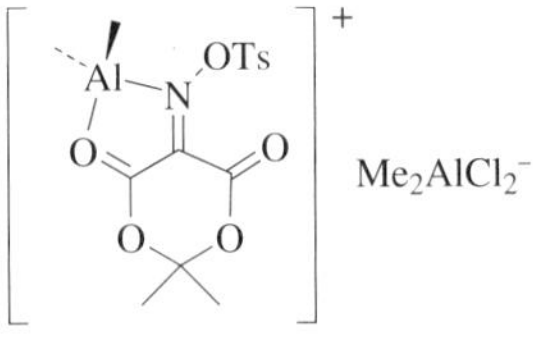
Al
OTs
N
Me2AlCl2–

Figure 11 Proposed structure of the reactive species in the cycloadditions of compound **242** in the presence of Me_2AlCl

Electron-Rich (*C*-Heteroatom-Substituted) Imines. Research on the applications of electron-rich imines as dienophiles in cycloaddition reactions with the aim of synthesizing nitrogen heterocycles has been reported.[211] Imino chlorides such as **243** are used in these studies. These unstable intermediates are generated in situ by reacting an amide[211] or an oxime[212] with $POCl_3$. Aromatic dienes such as isosafrole (**244a**) and methyl isoeugenol (**244b**) condense with imino chlorides **243** at high temperature, affording ImDA cycloadducts **245**.

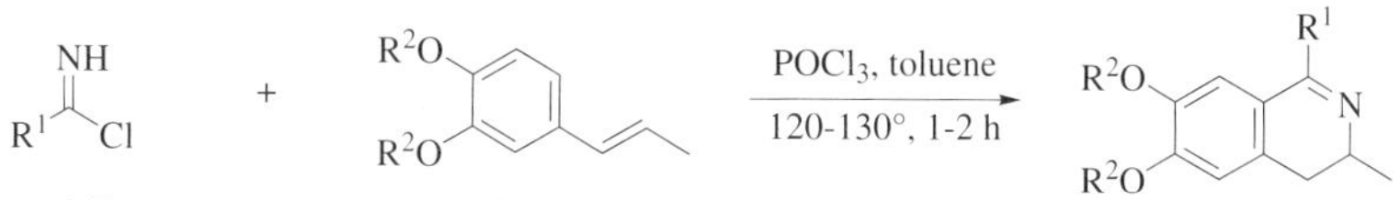
NH
R1 Cl
+
R2O
R2O
POCl3, toluene
120-130°, 1-2 h
R1
R2O
N
R2O

243 **244** **245**

R^1 = H, Ph, Bn, 4-$O_2NC_6H_4$, 4-$MeOC_6H_4$, PhCH=CH, 4-$MeOC_6H_4(CH_2)_2$

a R^2 = —CH_2—
b R^2 = Me

Due to the polymerization of the dienes under the strongly acidic conditions used to generate the imino chlorides in situ, the yields of these reactions are not always reproducible. The nature of the imine substituent R^1 significantly affects the reactivity of imino chlorides. For example, condensation between dienophile **243** (R^1 = *p*-$NO_2C_6H_4$) and **244a** is difficult to effect whereas **243** (R^1 = *p*-$MeOC_6H_4$) successfully condenses with **244a**, affording the cycloadduct **245**.[213] Imino hydrosulfates **246**, generated in situ from the corresponding nitriles and sulfuric acid, also react with aromatic dienes **244a/b** to give ImDA adducts **247**.[214] Here, too, the yields are unfavorably affected due to polymerization of the diene components under the reaction conditions.

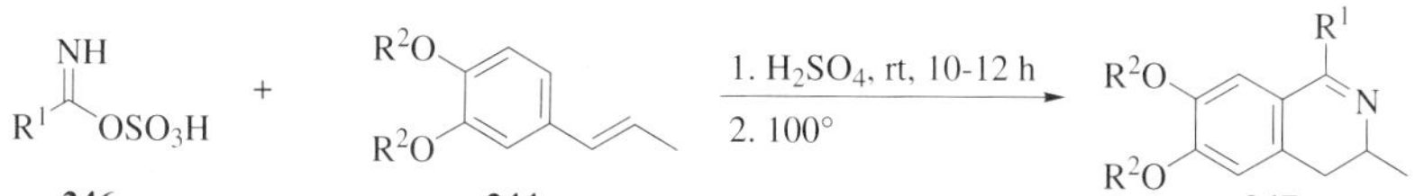
NH
R1 OSO3H
+
R2O
R2O
1. H2SO4, rt, 10-12 h
2. 100°
R1
R2O
N
R2O

246 **244** **247**

R^1 = Me, Ph, Bn

a R^2 = —CH_2—
b R^2 = Me

Fluoroimines **248**, when combined with dienes **249** at low temperatures, lead to the formation of ImDA adducts **250** after warming to 0° or higher.[215] Alternative nitrogen-substituted imines such as **251** can undergo cycloaddition reactions with highly electron-deficient dienes **252**, affording dihydropyridine derivatives **253**.[146] Introduction of a substituent at C4 of diene **252** drastically reduces the rate of the reaction, and thus a longer reaction time is required [7 days for **252** (R = Ph) compared to 12 hours for **252** (R = H)].

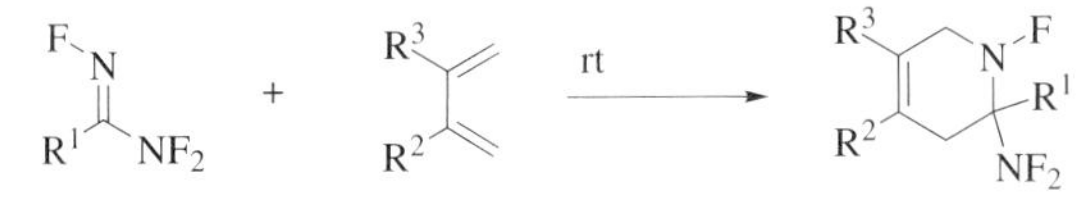

248 **249** **250**

$R^1 = F, NF_2$

a $R^2 = R^3 = H$
b $R^2 = R^3 = Me$
c $R^2 = H, R^3 = Me$

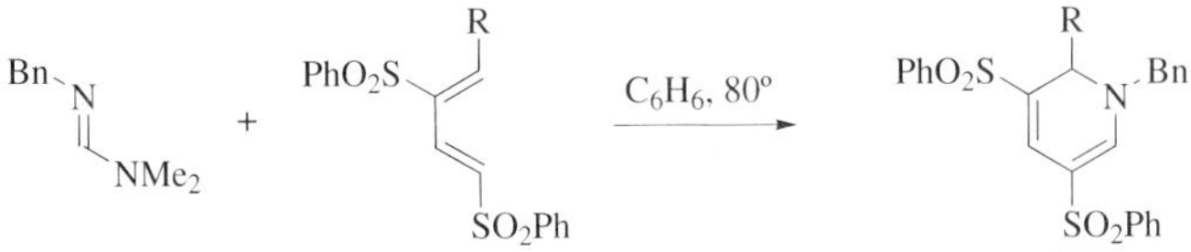

251 **252** **253**

R = H (52%)
R = Ph (83%)

Miscellaneous Imines. *Isocyanates and Isothiocyanates.* Very few examples of [4+2]-cycloadditions of 1,3-dienes with isocyanates have been reported that appear to proceed via concerted Diels-Alder-like mechanisms. In a single example, the electron-rich diene **254** reacts with *p*-tolyl isocyanate affording adduct **255**.[216]

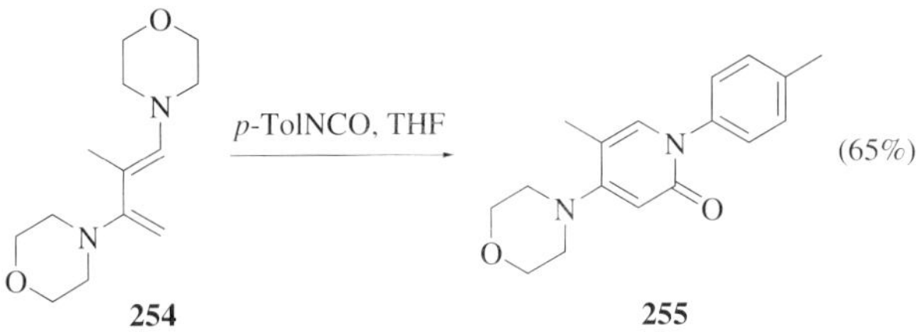

254 **255**

Thiophenyl-substituted dienes, generated by thermal extrusion of SO_2 from various 3-sulfolenes **256**, react with arylsulfonyl isocyanates at 110–130°, affording adducts **257**.[217,218]

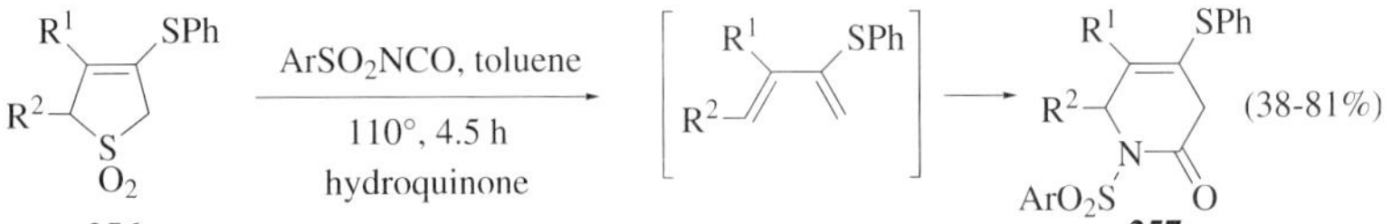

256 **257**

$R^1 = H, SPh$
$R^2 = H, Me$, 4-pentenyl

Only a single report of some simple cycloadditions with a few *N*-acylisothiocyanates has appeared to date.[219] These reactions appear to take several months at room temperature to go to completion and therefore are of little practical value.

Ketenimines and 2-Azaallenes. Only a single report exists of cycloadditions of ketene immonium salt **258** with a few simple dienes.[220] Cyclic dienes react with **258** to give adducts such as **259** in good yields. With acyclic dienes, however, [2+2]-cycloadducts are the primary products. An exception is 2,3-dimethylbutadiene, which prefers to exist in the s-cis conformation, and therefore gives a moderate yield of adduct **260**.

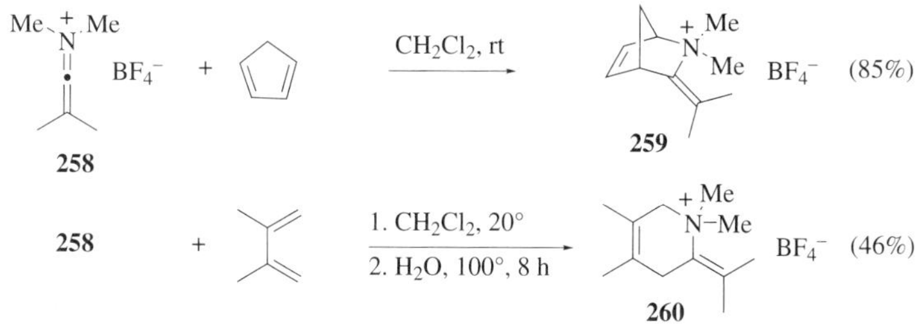
Me
Me
N
BF4⁻
+
CH2Cl2, rt
Me
N
Me
BF4⁻
(85%)
258
259
258
+
1. CH2Cl2, 20°
2. H2O, 100°, 8 h
Me
N–Me
BF4⁻
(46%)
260

There is one publication describing azaallenium salts reacting in ImDA reactions.[221] For example, salt **261** reacts with isoprene affording [4+2]-cycloadducts **262** and **263** in moderate yield in a 7.5:1 ratio. Only one regioisomeric series of compounds is formed in this process. An ene product derived from azaallenium salt **261** and isoprene is also isolated in low yield.

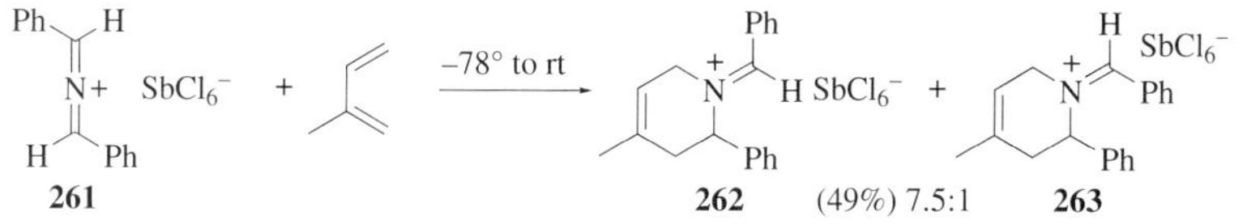
Ph
H
N+
SbCl6⁻
+
H
Ph
261
−78° to rt
Ph
N
H SbCl6⁻
+
Ph
262
(49%) 7.5:1
H
SbCl6⁻
N
Ph
Ph
263

Intramolecular Cycloadditions

Many of the types of imines which have been used in intermolecular Diels-Alder reactions have been applied to intramolecular cycloadditions. However, much of this work is fragmentary and systematic studies of intramolecular imino [4+2]-cycloaddition reactions are lacking. The most widely utilized type of dienophiles in these processes have been *N*-acylimines and simple alkylimines. Only a few scattered applications of other types of imines have been reported.

***N*-Acylimines.** Several examples of intramolecular reactions of *N*-acylimines have appeared over the past thirty years. Since *N*-acylimines are usually unstable, these dienophiles are commonly generated in situ, most often by thermolysis of *N*-acetoxymethyl amides like **264**.[222,223] Thus, heating substrate **264** through a column of glass helices at 370–390° leads to the transient *N*-acylimine **265**, which cyclizes in good yield affording tricycle **266**. Gas-phase pyrolysis has been used to convert benzocyclobutene **267** via *N*-acylimine quinodimethane **268** into the Diels-Alder adduct **269**.[224]

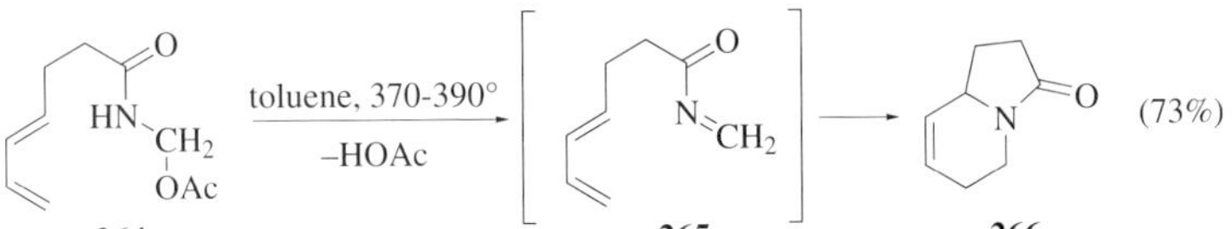
toluene, 370-390°
–HOAc
(73%)

264 265 266

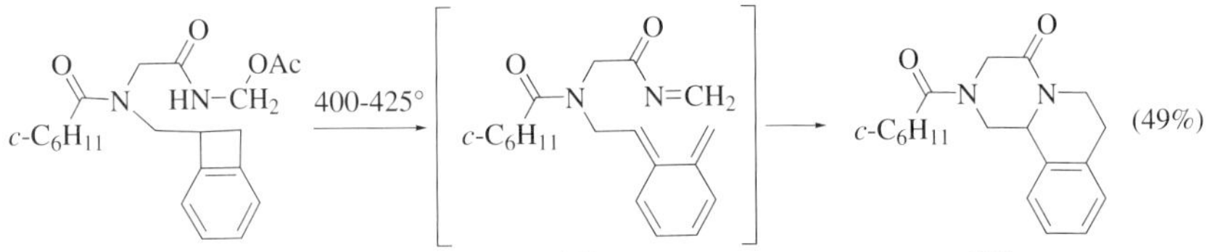
400-425°
(49%)

267 268 269

So-called type 2 Diels-Alder reactions of *N*-acylimines have been described as methods to construct a variety of bridgehead olefins.[225,226] For example, thermolysis of diene substrate **270** leads to the strained bridged heterocycle **271** in modest yield. However, superior yields of products can be obtained if the tether length is increased by one or two carbons. Increasing the tether by three carbons above that in acylamine **270** leads to a sharp reduction in the yield of cycloadduct.

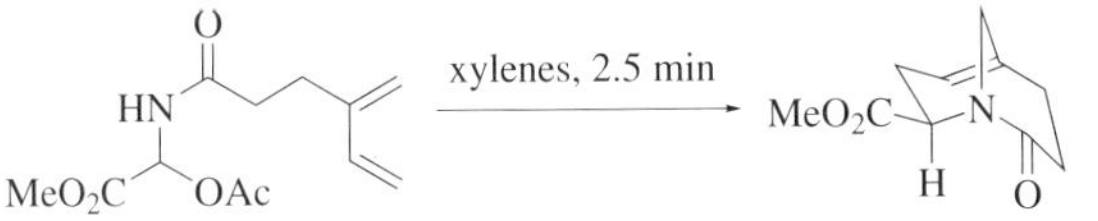
xylenes, 2.5 min

270 271

250° (35%)
252° (29%)

In some instances, remote stereochemistry can be controlled in intramolecular *N*-acylimine cycloadditions.[227,228] Heating substrate **272** in *o*-dichlorobenzene affords a single stereoisomeric quinolizidone adduct **274** in high yield. This cyclization apparently proceeds via a transition state as shown in structure **273** where the acyl group is in a preferred endo position and the benzyloxymethyl group is pseudo-equatorial. However, in some related cycloadditions leading to indolizidone ring systems, the stereoselectivity drops significantly with respect to the remote allylic center.[229,230]

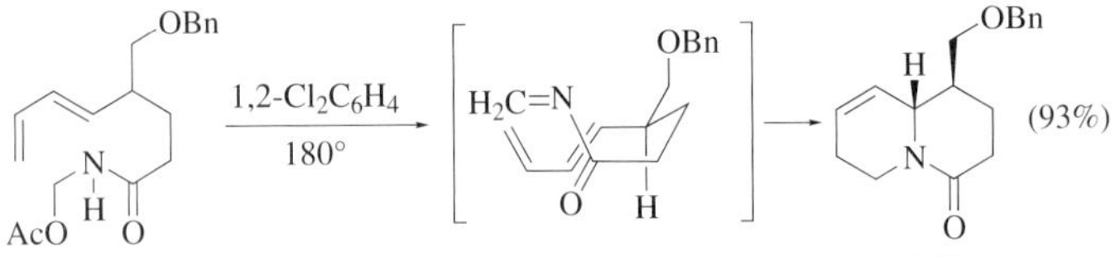
1,2-$Cl_2C_6H_4$
180°
(93%)

272 273 274

Some interesting stereochemical features have been observed in a study of the intramolecular ImDA cycloadditions of a few *N,C*-diacylimines.[229,231] Thermolysis of substrate **275** produced a single stereoisomeric adduct **277** in good yield. It was suggested that the intermediate diacylimine **276** cyclizes via a transition state in which the *N*-acyl group, rather than the *C*-acyl moiety, occupies an endo position with the remote alkyl substituent in a pseudo-equatorial position, leading to the observed relative stereochemistry. An imine with the E geometry is most likely the reacting species here, although one cannot rule out intervention of the Z isomer. Similar results are obtained in closely related intramolecular cycloadditions that produce 6,5-fused systems.

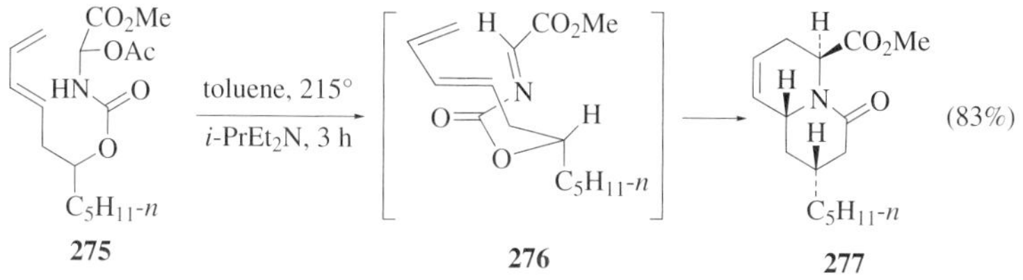

toluene, 215°
i-PrEt2N, 3 h
(83%)
275
276
277

***C*-Cyanoimines.** Rather surprisingly, there appear to be no examples in the literature of intramolecular Diels-Alder reactions of simple *C*-acylimines despite the fact that intermolecular reactions of this type of dienophile have been widely used. However, the use of *C*-cyanoimines in intramolecular [4+2]-cycloadditions has been reported.[232] For example, exposure of triflamide **278** to cesium carbonate leads to a 80:20 E/Z mixture of isolable *C*-cyanoimines **279**, which upon heating in toluene at 120° affords a single Diels-Alder adduct **280** in good yield. Interestingly, in most of the reported reactions both geometric imine isomers lead to the same stereoisomeric product, formed via a transition state having the *N*-alkyl chain endo to the diene.

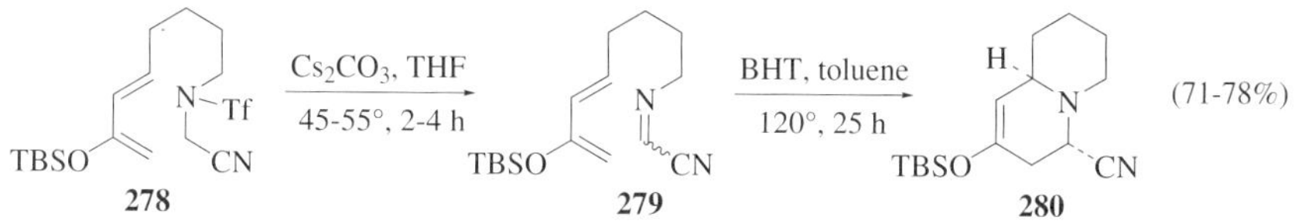

Cs2CO3, THF
45-55°, 2-4 h
BHT, toluene
120°, 25 h
(71-78%)
278
279
280

***N*-Sulfonylimines.** To date only a single example of an intramolecular Diels-Alder reaction of an *N*-sulfonylimine has been reported.[33] Aldehyde diene **281** is treated with *N*-sulfinyl-*p*-toluenesulfonamide in the presence of boron trifluoride etherate to generate *N*-sulfonylimine **282**, presumably as a Lewis acid complex, which undergoes cycloaddition in situ to afford the cis-fused bicyclic sulfonamide **283**.

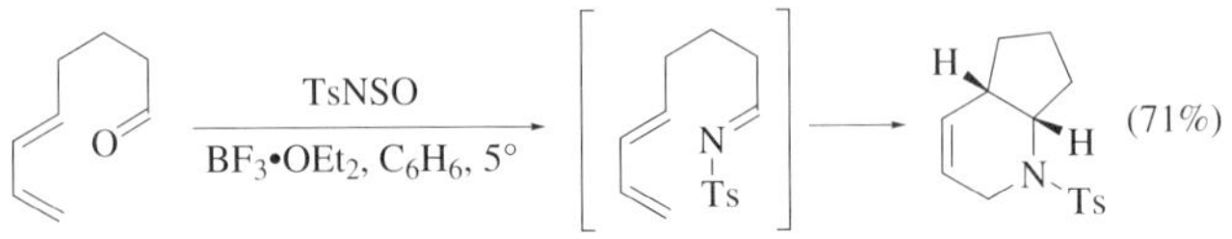

TsNSO
BF3•OEt2, C6H6, 5°
N
Ts
H
H
N
Ts
(71%)

281 282 283

Unactivated Alkylimines. Intermolecular cycloadditions of simple alkyliminium dienophiles in water have been extended to intramolecular reactions, which have been used in alkaloid total synthesis.[233] For example, slow addition of diene aldehyde **284** to an aqueous solution of methylamine hydrochloride leads to formation of the bridged Diels-Alder adduct **285** in good yield.[234] Similarly, exposure of diene amine **286** to formaldehyde under aqueous conditions leads to cycloadduct **287** in high yield.[133]

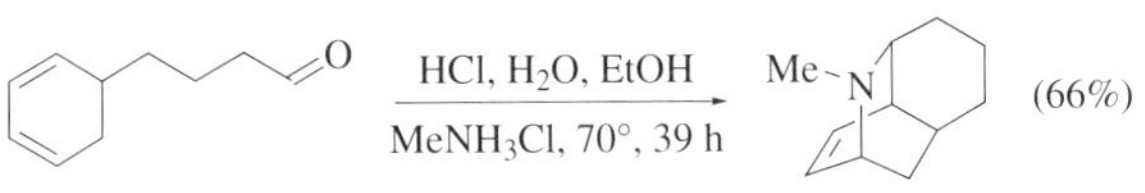

O
HCl, H2O, EtOH
MeNH3Cl, 70°, 39 h
Me
N
(66%)

284 285

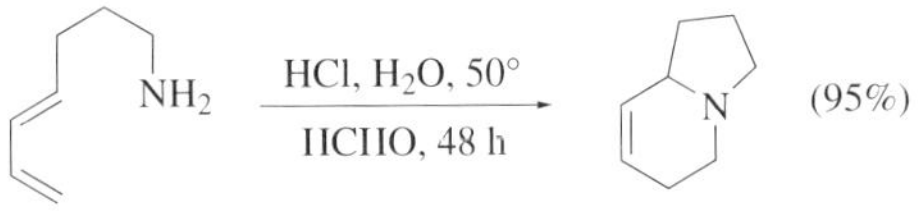

NH2
HCl, H2O, 50°
HCHO, 48 h
N
(95%)

286 287

Although these Diels-Alder reactions are stereospecific with respect to the diene geometry, modest endo/exo and remote stereoselectivity has often been observed. Treatment of enantiomerically pure aldehyde **288** with ammonium chloride under aqueous conditions leads to a 2.2 : 1 mixture of diastereomeric adducts **290** and **292**.[235] It is suggested that this process involves an iminium ion intermediate that cyclizes via transition state conformation **289** affording **290** as the major product. Transition state **291** would afford the minor cycloadduct **292**. The latter conformation is presumably destabilized by an eclipsing interaction between H_a and H_b.

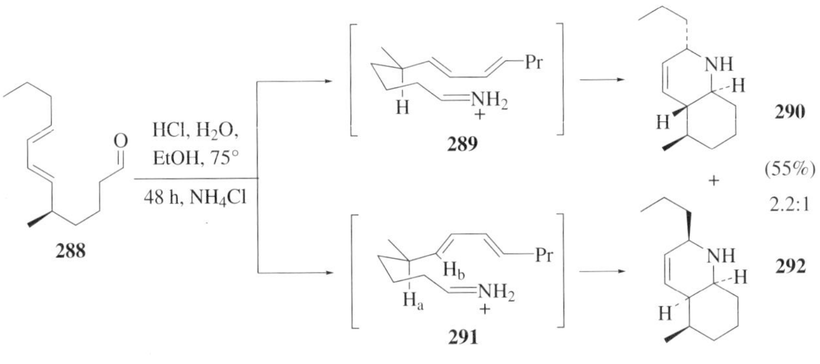

O
HCl, H2O,
EtOH, 75°
48 h, NH4Cl
288
Pr
H
NH2
+
289
NH
H
H
290
+
(55%)
2.2:1
Pr
Hb
Ha
NH2
+
291
NH
H
H
292

A comparison study probed the effects of aqueous reactions versus those promoted by other polar media such lithium perchlorate/ether.[236] Using endocyclic imine substrates like **293**, it was found that the aqueous Diels-Alder reaction of the corresponding TFA salt affords adduct **294** in high yield. On the other hand, if the reaction is run in 5 M lithium perchlorate/diethyl ether, tricycle **294** is formed in poor yield. Similar results are obtained with related systems.

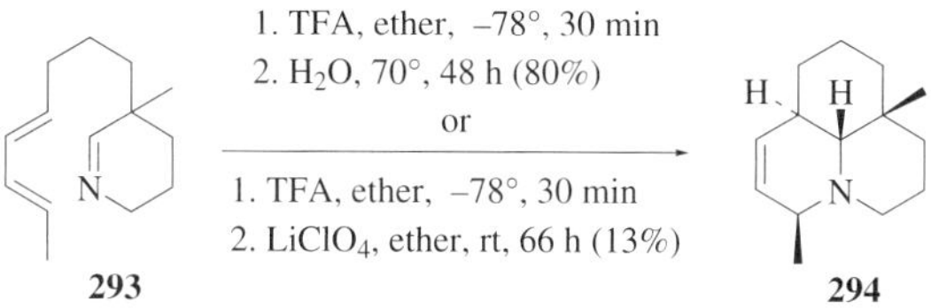
1. TFA, ether, –78°, 30 min
2. H_2O, 70°, 48 h (80%)
or
1. TFA, ether, –78°, 30 min
2. $LiClO_4$, ether, rt, 66 h (13%)
H
H
N
N
293
294

A recent report has described examples of intramolecular ImDA reactions that utilize a vinylallene moiety as the diene component.[237] Cyclization of the *N*-benzylimine derived from aldehyde **295** at room temperature affords a single stereoisomeric tricycle **296** in good yield. With some related substrates subjected to $BF_3 \cdot Et_2O$ catalysis at −78°, the yields of adducts vary depending on the tether length and the nature of the substituent on the allene.

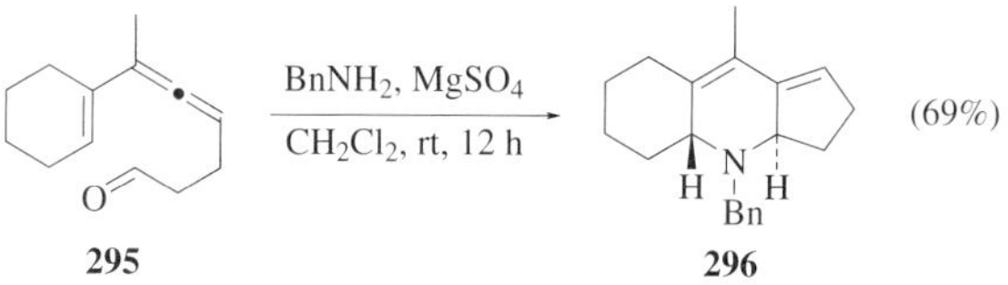
$BnNH_2$, $MgSO_4$
CH_2Cl_2, rt, 12 h
(69%)
O
N
H
H
Bn
295
296

Oxime Derivatives. In an early example of a thermal intramolecular Diels-Alder reaction of an oxime methyl ether with a quinodimethane diene, heating benzocyclobutene substrate **297** affords a mixture of epimeric Diels-Alder adducts **298** and **299** in unspecified ratio and yield.[238]

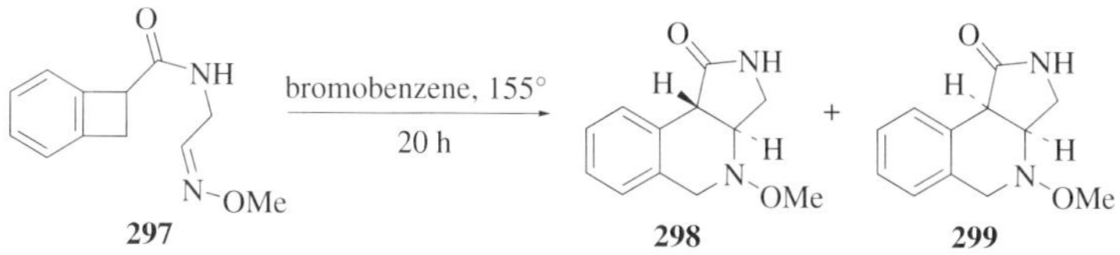
O
NH
N-OMe
bromobenzene, 155°
20 h
O
NH
H
H
N
OMe
+
O
NH
H
H
N
OMe
297
298
299

In a more recent study involving intramolecular cycloadditions of *O*-acyloxime derivatives, readily prepared dicyano or cyano ester substrates **300** are found to cyclize on heating in toluene at reflux or under high pressure at room temperature, affording cycloadducts **301**.[239] Interesting stereoselectivity is observed with cyano ester substrate **302**, which leads exclusively to Diels-Alder product **304**. This cyclization presumably occurs via transition state **303** where the cyano group is endo, a result in contrast to the intermolecular reactions with cyclopentadiene studied earlier where it was found that the ester moiety prefers the endo orientation.[202,203]

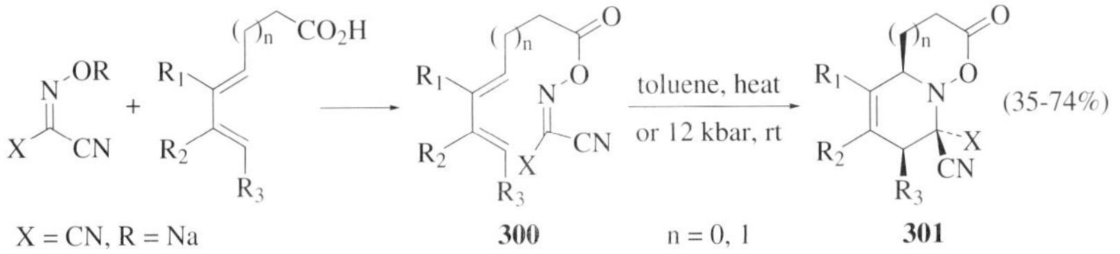
toluene, heat
or 12 kbar, rt
(35-74%)
300
n = 0, 1
301

X = CN, R = Na
X = CO_2Et, R = H
R_1 = H, Me; R_2 = H, Me; R_3 = H, Ph

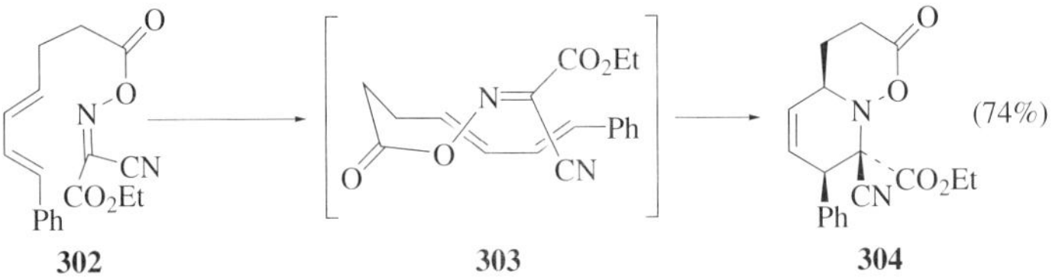
(74%)
302
303
304

Cycloadditions Using Asymmetric Catalysis

Although a great deal of progress has been made over the past decade in enantioselective hetero Diels-Alder reactions of various carbonyl compounds using chiral Lewis acid catalysts, the analogous enantioselective ImDA methodology remained unavailable until quite recently.[7,240] This fact may be the result of some common problems associated with the use of imines as substrates in catalytic enantioselective reactions, such as: 1) The nitrogen atom of imines tends to be more Lewis basic than the oxygen atom of carbonyl compounds. As a result, the coordination of the imine, or the product, to the chiral Lewis acid catalyst is stronger than for a carbonyl compound, leading to deactivation or inhibition of the catalyst. Therefore, stoichiometric amounts of chiral Lewis acids are often needed to achieve conversion and high asymmetric induction. 2) The readily interconvertible E/Z configurations of imines allow several possible metal-complexed structures to exist in solution. 3) The imine double bond has a relatively low reactivity and poor electrophilicity. 4) There is a tendency toward deprotonation of the α-acidic proton of enolizable imines to form enamines. Thus, only very recently have reports appeared concerning catalytic enantioselective Diels-Alder reactions of imines.

Chiral boron(III) reagents have been shown to catalyze the ImDA reactions of aldimines with oxygen-substituted highly electron-rich dienes.[241] When imines **305** are treated with Danishefsky-type dienes **306** in the presence of stoichiometric amounts of chiral boron (III) reagent **307**, ImDA adducts **308** are obtained in good yields with up to 90% enantiomer excess. Addition of 4Å molecular sieves to these reactions was found to be necessary for optimal yields.

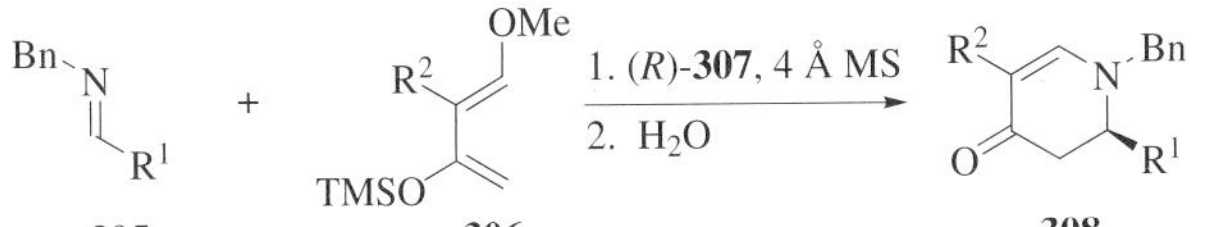

Bn
N
R¹
+
OMe
R²
TMSO
1. (R)-307, 4 Å MS
2. H₂O
R²
N
Bn
O
R¹

305 **306** **308**

R^1 = Ph, *n*-Pr, 2-naphthyl, *c*-C_6H_{11}, 3-pyridyl

a R^2 = H
b R^2 = Me

R^1 = Ph, R^2 = H (75%) 82% ee

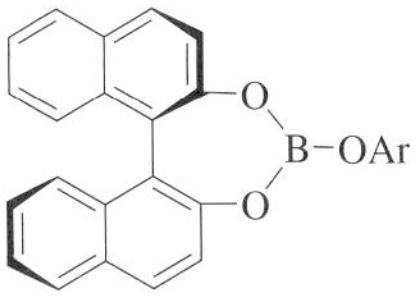

O
B−OAr
O

(*R*)-**307**

Investigations on the effect of the nature of the Lewis acid on the diastereoselectivity of ImDA reactions of chiral imines derived from (*S*)-α-methylbenzylamine and benzaldehyde with Danishefsky's diene reveal that $B(OPh)_3$ can catalyze these reactions effectively. The effect of chiral Lewis acids such as **307** on the diastereoselectivities of this ImDA reaction for double asymmetric induction has been explored.[242,243] When imines **309**, derived from (*S*)-α-methylbenzylamine, are treated with Danishefsky's diene **306a**, the reaction rate increases in the presence of (*R*)-**307** and nearly complete diastereoselectivity is observed (matched case). On the other hand, in the presence of (*S*)-**307**, the same diastereomer is formed, albeit in both lower yield and diastereoselectivity (mismatched case).

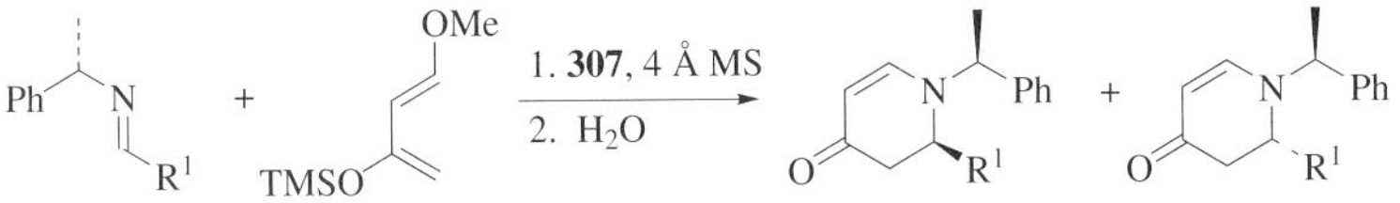

Ph
N
R¹
+
OMe
TMSO
1. 307, 4 Å MS
2. H₂O
N
Ph
O
R¹
+
N
Ph
O
R¹

309 **306a** **310a** R^1 = Ph **310b**

R^1 = Ph, *n*-Pr, *c*-C_6H_{11}, 3-pyridyl

(*R*)-**307**	(61%) 99:1
(*S*)-**307**	(30%) 97:3

In a more recent investigation, the use of another chiral boron catalyst, the Brønsted acid-assisted Lewis acid **311**, for the ImDA reactions was studied.[244] In the presence of stoichiometric amounts of **311**, adduct **310a** (R^1 = Ph) is obtained in 64% yield with 99% de.

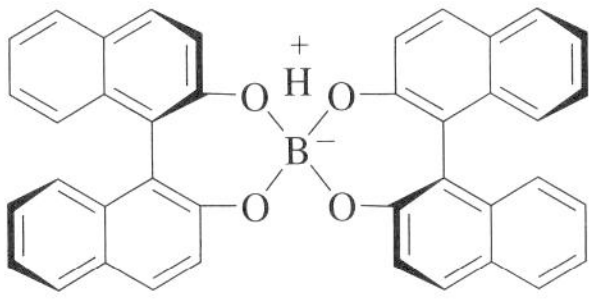

+
O H O
B⁻
O O

311

Although chiral catalysts are successfully used for diastereoselective asymmetric ImDA reactions,[242,243,245,246] there are only a few accounts of their use in enantioselective reactions prior to 1998. Moreover, the early reported diastereoselective and enantioselective ImDA reactions require stoichiometric amounts of the chiral catalyst.

The first catalytic enantioselective ImDA reactions using imines **312** and Danishefsky-type dienes **306**, promoted by a chiral zirconium (IV) complex **313**, was reported in 1998.[247]

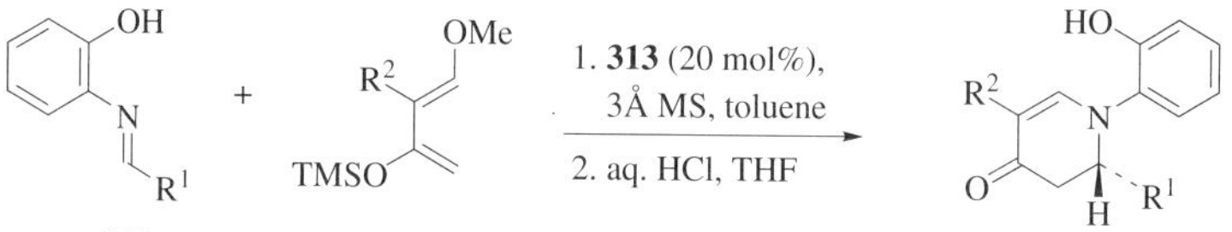
OH
N
R1
+
OMe
R2
TMSO
1. 313 (20 mol%),
3Å MS, toluene
2. aq. HCl, THF
HO
R2
N
O
H
R1

312 **306a, b** **314**

R^1 = Ph, 1-naphthyl, 2-MeC$_6$H$_4$, c-C$_6$H$_{11}$

R^2 = H, Me

R^1 = 2-MeC$_6$H$_4$, R^2 = H; (83%) 82% ee

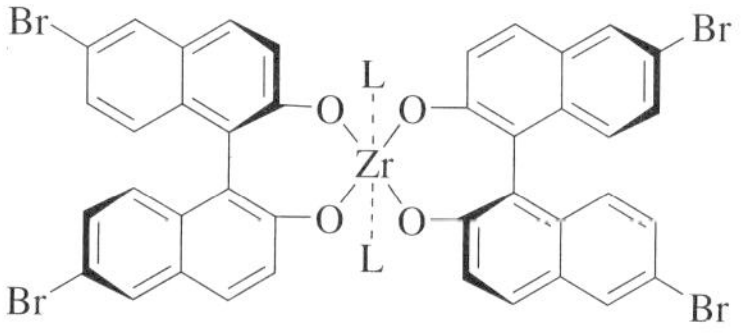
Br
L
O
O
Zr
O
O
L
Br
Br
Br

313

Catalyst **313** is prepared from Zr(OBu-t)$_4$, (R)-Br-BINOL, and a heterocyclic ligand such as 1-methylimidazole, in a ratio of 1:2:2-3. The reaction between imine **312** and Danishefsky-type dienes is strongly influenced by both the choice of ligands and of solvents. The best result (93% yield and 93% enantiomer excess) is obtained from the reaction of **312** (R^1 = 1-naphthyl) and diene **306b** in toluene with 1-methylimidazole as the ligand in the presence of 10 mol % of catalyst **313**. In general, good enantioselectivities are achieved for imines derived from ortho-substituted benzaldehydes. Furthermore, a free ortho-hydroxy group on the N-aryl group is necessary to achieve a high level of enantioselectivity. The absolute configuration of the products is found to be S using catalyst **313** derived from the (R)-Br-BINOL ligand.

Further investigations to examine the catalyst-substrate interaction and to improve selectivities led to the preparation of the chiral Zr (IV) complex **315**.[248] ImDA reaction of imine **312** (R^1 = 2-MeC$_6$H$_4$) with Danishefsky's diene **306a** catalyzed by **315** (20 mol %) gives the corresponding piperidinone derivative in a slightly lower yield of 66% than with catalyst **313**, but with a similarly high enantiomer excess of 84%. The cycloaddition is affected by the solvent, the reaction temperature, and the type of molecular sieves used. The best results are obtained when the reaction is carried out in benzene at room temperature and in the presence of 3Å molecular sieves [(93% yield and 91% enantiomer excess for **314** (R^1 = 2-MeC_6H_4, R^2 = H)].

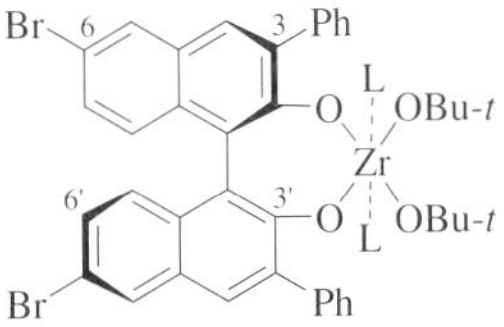
Br
6
3
Ph
L
O
OBu-t
Zr
6'
3'
O
OBu-t
L
Br
Ph

315

L = 1-methylimidazole

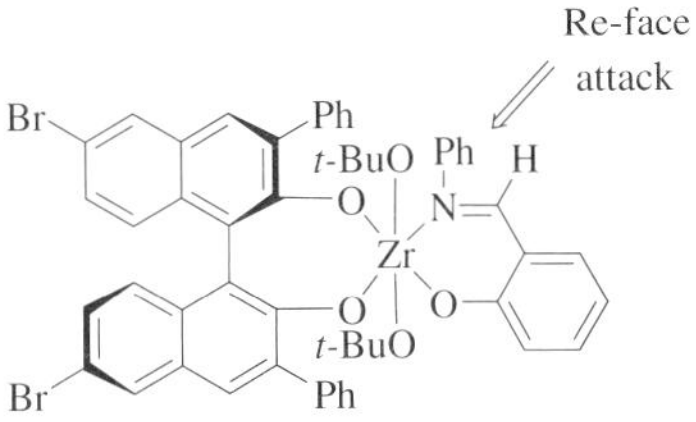
Re-face
attack
Br
Ph
t-BuO
Ph
H
O
N
Zr
O
O
t-BuO
Br
Ph

316

Surprisingly, the absolute configuration of the product is found to be R, opposite to the configuration obtained with catalyst **313**. A mechanistic model involving a chelated imine as shown in **316** is proposed to account for the observed enantioselectivity. The two bulky *tert*-butoxy groups are expected to occupy the axial positions. One of the 3,3′-phenyl groups would then effectively shield one face of the imine, and consequently the diene attacks from the Re face on the opposite side of the imine.

Optimizations with respect to the substituents on the meta- and para-positions of the 3,3′-phenyl rings, on the 6,6′-positions, and for the ligands on zirconium of the catalyst using solid- and liquid-phase methods led to new catalyst **317**.[249] When imine **312** (R^1 = 2-MeC_6H_4) is reacted with Danishefsky's diene **306a** under the optimized conditions, 2 mol % of catalyst **317** is sufficient to produce the cycloadduct **314** (R^1 = 2-MeC_6H_4, R^2 = H) in 94% enantiomer excess, albeit in only 68% yield. However, at 1 mol % of the catalyst, the enantiomer excess drops to 83%. When attached to a solid support, catalyst **317** (20 mol %) affords adduct **314** (R^1 = 2-MeC_6H_4, R^2 = H) in 87% yield and 80% enantiomer excess.

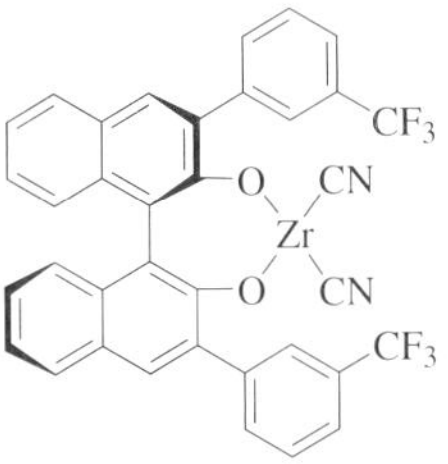
CF3
O
CN
Zr
O
CN
CF3

317

Recently, silver (I)-catalyzed enantioselective ImDA reactions of Danishefsky's diene **306a** with aryl imines **318** were introduced affording cycloaddition products **320** in >77% yield and >88% enantiomer excess.[250] These reactions are effected at 4° in the presence of ≤1 mol % of the catalyst derived from **319**.

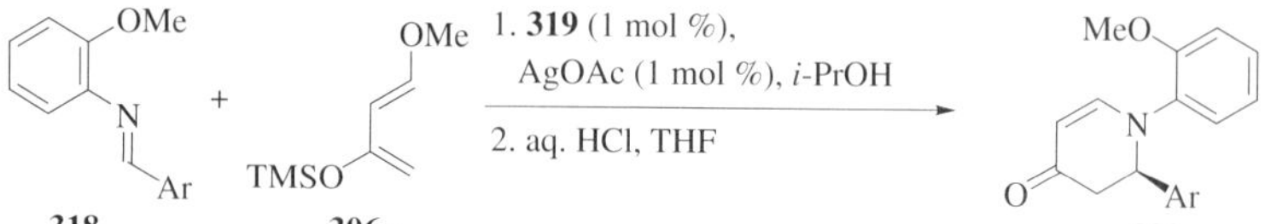
1. 319 (1 mol %), AgOAc (1 mol %), i-PrOH
2. aq. HCl, THF

318 **306a** **320**

Ar = Ph, 4-$MeOC_6H_4$, 4-ClC_6H_4, 2-BrC_6H_4, 3-$O_2NC_6H_4$, 4-$O_2NC_6H_4$, 1-naphthyl, 2-naphthyl, 2-furyl

(86-98%) 88-95% ee

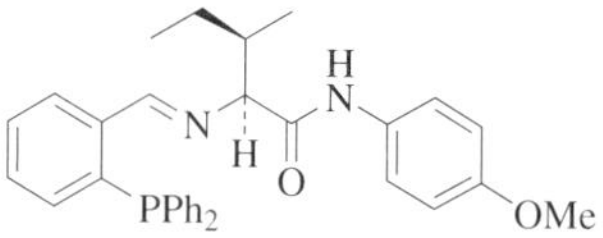

319

The ortho-methoxy group in the *N*-arylimine substrates is necessary to obtain high enantioselectivity. In the absence of 2-PrOH, the yields as well as enantioselectivities are lowered. Furthermore, changing the para-methoxyphenyl substituent on ligand **319** decreases the turnover and enantioselectivity of the cycloaddition reaction.

Catalytic enantioselective cycloaddition reactions of electron-deficient imines, especially of *N*-tosyl α-imino ester **321**, with the Danishefsky-type dienes **306a** and **306c** in the presence of the copper(I) complexes of chiral BINAP derivatives **323**, have been described.[251,252] ImDA reaction of imine **321** with diene **306c** catalyzed by BINAP derivative **323b** (1 mol %) affords the endo diastereomer of the ImDA adduct **322** in 70% yield, with up to 90% diastereomer excess and 98% enantiomer excess.

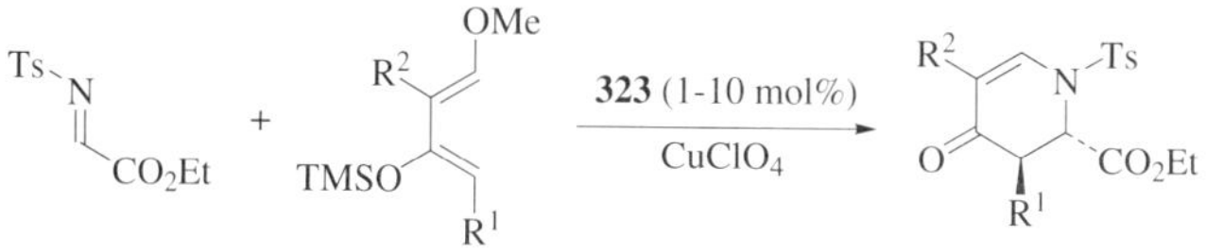
323 (1-10 mol%)
$CuClO_4$

321 **306a** $R^1 = R^2$ = H; **306c** $R^1 = R^2$ = Me **322**

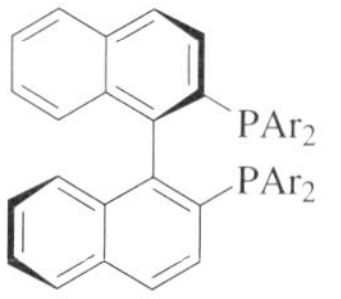

323a Ar = Ph
323b Ar = 4-MeC_6H_4

Metal complexes of several chiral phosphine-substituted oxazoline ligands (Fig. 12) have also been used for the cycloaddition reaction of *N*-tosyl α-imino ester **321** with Danishefsky's diene **306a** and afford the cycloadduct **322** in high yields (up to 97%) and with up to 86% enantiomer excess.

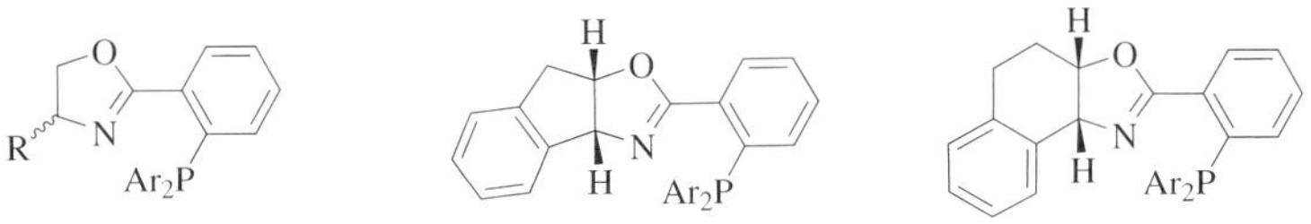
Ar$_2$P
Ar$_2$P
Ar$_2$P

Ar	R
Ph	Ph (R)
Ph	*i*-Pr (S)
Ph	*t*-Bu (S)
4-MeC_6H_4	*t*-Bu (S)

Ar
Ph
4-MeC_6H_4
2,4-$Me_2C_6H_3$

Ar
Ph
4-MeC_6H_4
2,4-$Me_2C_6H_3$

Figure 12 Structures of chiral phosphine-substituted oxazoline ligands tested for the cycloaddition reactions of imine **321** and Danishefsky's diene **306a**

Notably, the copper (I) complex of chiral BINAP derivative **323b** is also capable of catalyzing the ImDA reaction of *N*-tosyl α-imino ester **321** with unactivated dienes. Cyclic dienes such as cyclopentadiene and 1,3-cyclohexadiene react with imine **321** in the presence of **323b**, affording cycloadduct exo-**324a** in 85% yield and 83% enantiomer excess, whereas exo-**324b** is obtained in only 52% yield, but with up to 95% enantiomer excess. The endo isomers **325a** and **325b** are minor products.

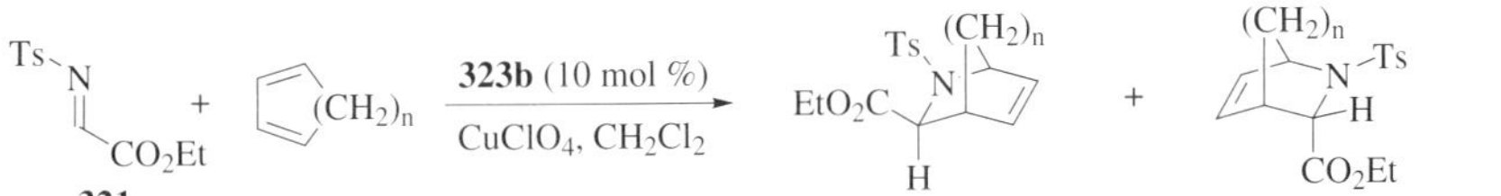
Ts
N
CO_2Et
+
$(CH_2)_n$
323b (10 mol %)
$CuClO_4$, CH_2Cl_2
Ts
$(CH_2)_n$
N
EtO_2C
H
+
$(CH_2)_n$
N
Ts
H
CO_2Et

321

n = 1
n = 2

exo **324a** (85%) 83% ee   *endo* **325a** (7%) 83% ee
exo **324b** (52%) 95% ee   *endo* **325b** (7%) 37% ee

In order to obtain more mechanistic information and to broaden the synthetic scope of the reaction, different ethyl glyoxylate derived imines **326** were treated with Danishefsky's diene **306a** in the presence of BINAP **323b**-$CuClO_4$ complex, leading to the cycloadducts **327** in moderate to good yields and enantioselectivities. The isolation of a Mannich-type product in some of the reactions indicates that these reactions may in fact proceed via the Mannich-Michael pathway.

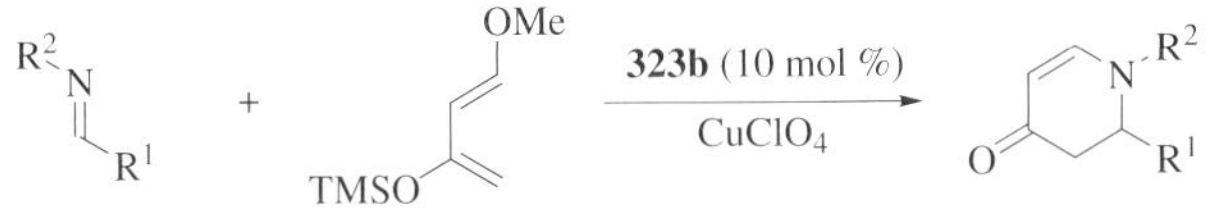
R^2
N
R^1
+
OMe
TMSO
323b (10 mol %)
$CuClO_4$
N
R^2
O
R^1

326   **306a**   **327**

a $R^1 = R^2 = CO_2Et$ (23%) 77% ee
b $R^1 = CO_2Et$, R^2 = 2-MeC_6H_4 (23%) 58% ee
c $R^1 = CO_2Et$, R^2 = Ph (78%) 91% ee
d $R^1 = C_2OEt$, R^2 = 4-MeC_6H_4 (75%) 78% ee

Surprisingly, the ImDA reactions of imines **321** and **326d** with Danishefsky's diene catalyzed by the BINAP-**323b**-$CuClO_4$ complex gives with imine **321** the S enantiomer and with imine **326d** the R enantiomer. These results have been proposed to result from differences in the binding modes of these imines to the catalyst. The *N*-tosyl group of imine **321** can participate in the coordination to the chiral Lewis acid, but the *N*-*p*-methoxyphenyl substituent of imine **326d** cannot.

It has also been demonstrated using different combinations of chiral ligands (BINAP, a bisoxazoline ligand, and 1,2-diphenylethylenediamine) and Lewis acids [$Yb(OTf)_3$, $Cu(OTf)_2$, MgI_2, and $FeCl_3$] that the ImDA adduct **329** can be obtained enantioselectively from imine **328** and Danishefsky's diene **306a**.[80] The highest enantiomer excess of 97% was claimed to result from using a combination of (*1S,2S*)-1,2-diphenylethylenediamine and MgI_2 as the catalyst in acetonitrile. However, a later corrigendum[253] indicated that these results could not be reproduced and rather enantiomer excesses in the range 0–55% are obtained depending upon reaction conditions.

PMP
N
CO2Me
+
OMe
TMSO
N
PMP
O
CO2Me

328 **306a** **329**

The enantioselective ImDA reaction of imine **330** derived from *p*-methoxyaniline and ethyl glyoxylate with Danishefsky's diene in the presence of different metal-BINOL complexes has been investigated. A diethylzinc-(*S*)-BINOL complex gives the best yield and enantioselectivity but only when used in a stoichiometric amount. The enantioselectivity and yield of cycloadduct **331** is severely affected by changes in solvent, temperature, catalyst loading, and reaction time. Surprisingly, the enantioselectivity declines with reaction time and decrease in reaction temperature.

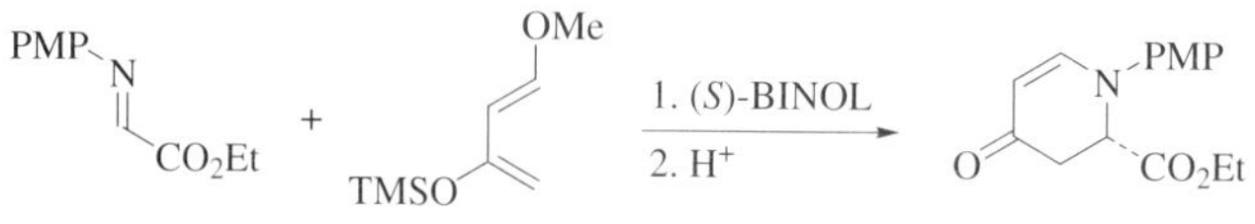
PMP
N
CO2Et
+
OMe
TMSO
1. (S)-BINOL
2. H+
N
PMP
O
CO2Et

330 **306a** **331** (78%) 93% ee

The use of chiral copper complexes of phosphinosulfenyl ferrocenes, such as **332**, as catalysts for enantioselective formal ImDA reactions of some *C*-aryl-*N*-sulfonylimines **333** has been reported.[2] Although ImDA-type cycloadducts are obtained in this process, the reaction clearly proceeds via the formation of an initial Mannich product that subsequently cyclizes on exposure to TFA through an intramolecular conjugate addition process. In general, good yields and high enantioselectivities are obtained in this reaction except for the *C*-arylimine bearing a para-*N,N*-dimethylamino group, where the cycloadduct is obtained in only 39% yield. In general, these reactions are conducted at room temperature, but in a few examples it was found that lowering the temperature to −20° increases the enantioselectivity.

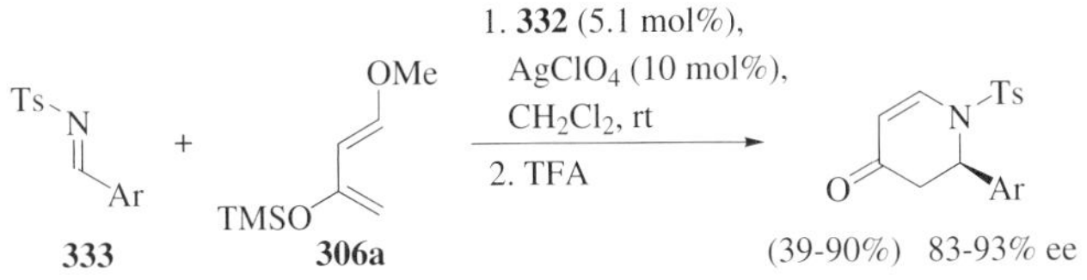

1. 332 (5.1 mol%),
AgClO4 (10 mol%),
CH2Cl2, rt
2. TFA
333
306a
(39-90%) 83-93% ee

Ar = Ph, 2-MeC_6H_4, 4-$MeOC_6H_4$, 4-FC_6H_4, 4-$Me_2NC_6H_4$, 2-naphthyl, PhCH=CH

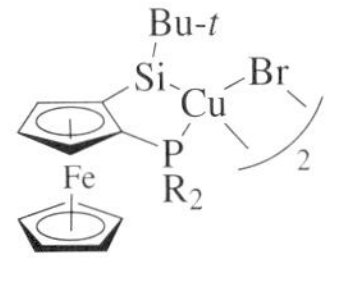

332 R = 1-naphthyl

APPLICATIONS TO SYNTHESIS

Both inter- and intramolecular ImDA reactions have been employed in a number of alkaloid total syntheses over the past twenty-five years.[233,254] In an early example of the use of an intermolecular cyclic *N*-acylimine Diels-Alder reaction in complex molecule synthesis, a thermal cycloaddition of trisubstituted diene **334** with the imine derived in situ from α-methoxyhydantoin **335** affords a 1 : 3 mixture of adducts **336** and **337** in moderate yield.[70,255] Inexplicably, the ImDA reaction does not go to completion, but the overall yield of products can be increased by recycling diene **334**. The major cycloadduct **337** is then utilized to construct the C/D ring system of the fungal metabolite streptonigrin.

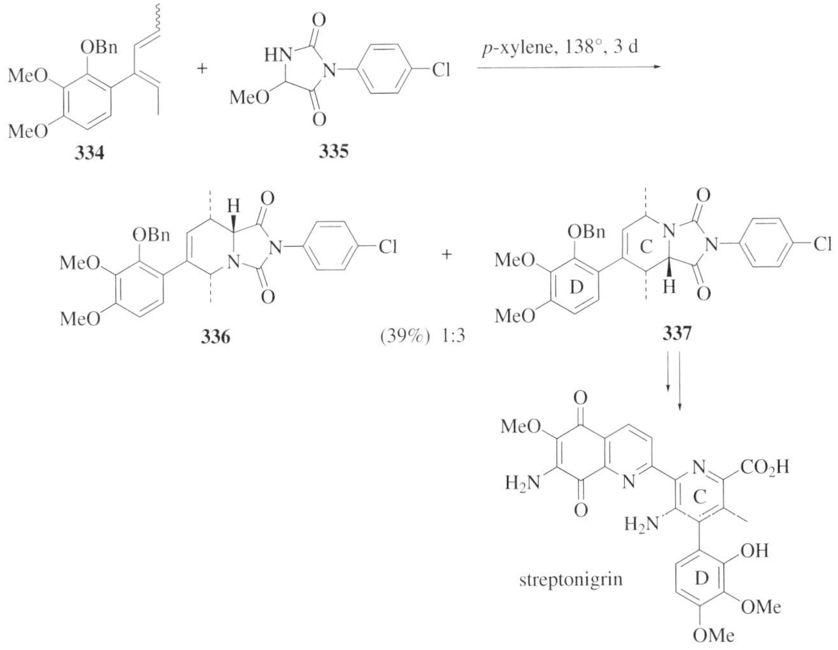

p-xylene, 138°, 3 d
334
335
336
(39%) 1:3
337
streptonigrin

In a more recent alkaloid synthesis involving an intermolecular ImDA reaction, enantiomerically pure cyclic alkylimine **338** reacts with Danishefsky's diene in the presence of ytterbium triflate catalyst to give exo adduct **339** in good yield.[256,257] This adduct is converted in a few subsequent steps into the *Securinega* alkaloid phyllanthine.

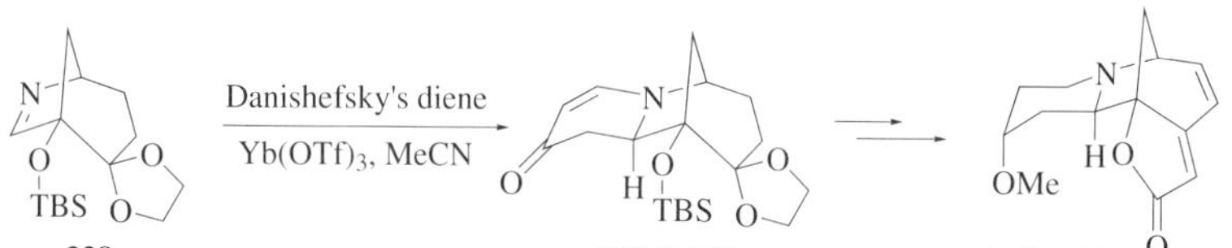
Danishefsky's diene
Yb(OTf)3, MeCN
N
O
TBS
O
O
H
OMe
HO
O

338 **339** (84%) phyllanthine

Several different types of indolizidine alkaloids have been synthesized utilizing intramolecular ImDA reactions of *N*-acylimines. For example, cyclization of the *N*-acylimine diene **341** derived in situ by simultaneous thermal extrusion of SO_2 and acetic acid from 3-sulfolene methylol acetate substrate **340** produces a 5:4 stereoisomeric mixture of bicyclic adducts **342** in good yield.[223,258] This mixture could then be converted into the *Elaeocarpus* alkaloid elaeokanine A.

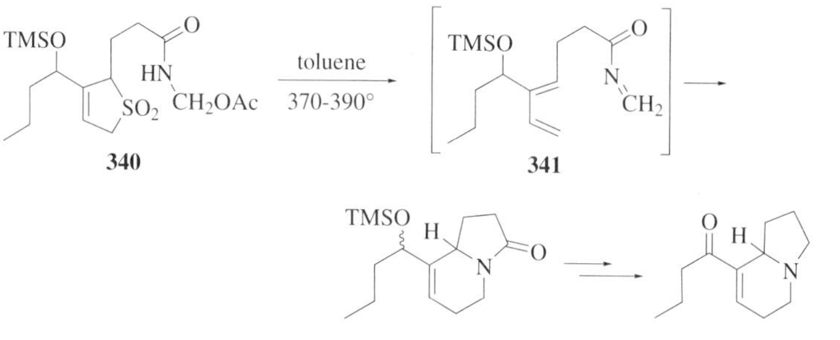
TMSO
O
HN
SO2
CH2OAc
toluene
370-390°
N
CH2
H
O

340 **341**

342 (68%) (+/–)-elaeokanine A

In another indolizidine alkaloid synthesis, substrate **343** cyclizes thermally producing a 1:1.8 mixture of Diels-Alder adducts **344** and **345**.[230] Minor product **344** is transformed in a few steps into the neurotoxic fungal metabolite slaframine. The major cycloadduct **345** could also be converted into the alkaloid via an epimerization sequence.

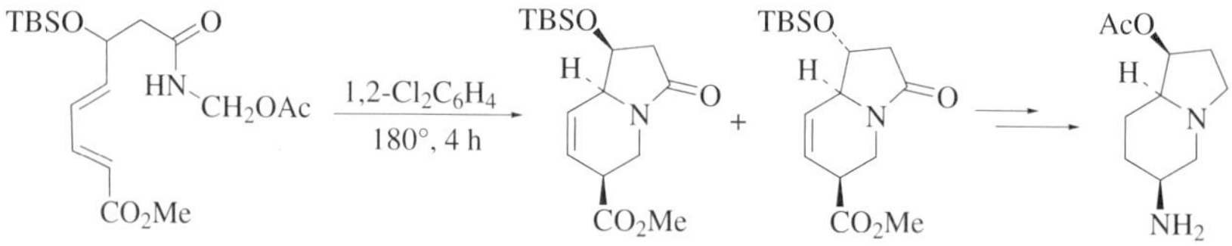
TBSO
O
HN
CH2OAc
CO2Me
1,2-Cl2C6H4
180°, 4 h
H
N
+
AcO
NH2

343 **344** (33%) **345** (57%) (+/–)-slaframine

Intramolecular ImDA methodology has been extended to synthesis of some quinolizidine alkaloids.[227,228] For example, thermolysis of Diels-Alder precursor **346** in *o*-dichlorobenzene affords lactam **347** in 66% yield. This compound is then reduced, affording the quinolizidine alkaloid cryptopleurine. In subsequent related work, it was found that amine **348** reacts with formaldehyde under aqueous conditions at 180° to directly produce racemic cryptopleurine in good yield.[259]

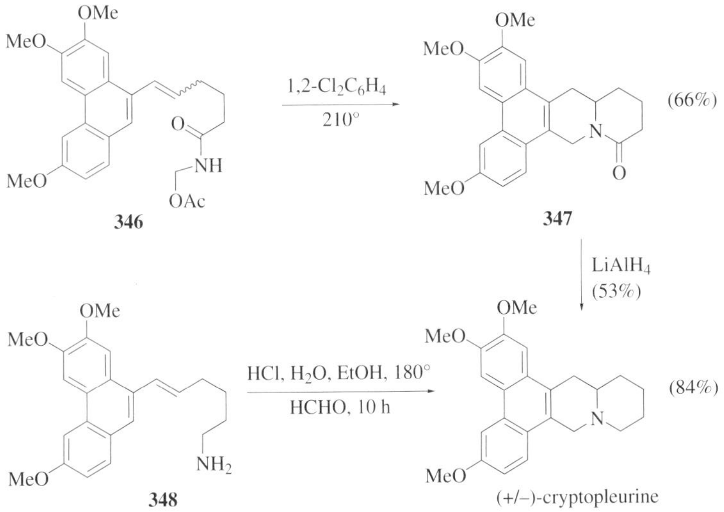
OMe
MeO
MeO
O
NH
OAc
346
1,2-$Cl_2C_6H_4$
210°
OMe
MeO
N
O
MeO
347
(66%)
$LiAlH_4$
(53%)
OMe
MeO
MeO
NH_2
348
HCl, H_2O, EtOH, 180°
HCHO, 10 h
OMe
MeO
N
MeO
(+/–)-cryptopleurine
(84%)

An elegant synthetic approach to the indole alkaloid (+/−)-eburnamonine via an intramolecular Diels-Alder reaction of an endocyclic alkyl-substituted imine with a 3-vinyl indole has been reported.[260,261] Thus, imine substrate **349** cyclizes if exposed to acids such as TFA or CSA in hot benzene, affording pentacycle **350** along with varying amounts of eburnamonine in good overall yield. The initial adduct **350** is isomerized to eburnamonine in high yield using sulfuric acid in hot ethanol. Interestingly, if imine **349** is treated with a lithium salt in benzene at reflux, high yields of adduct **350** are produced uncontaminated with eburnamonine.

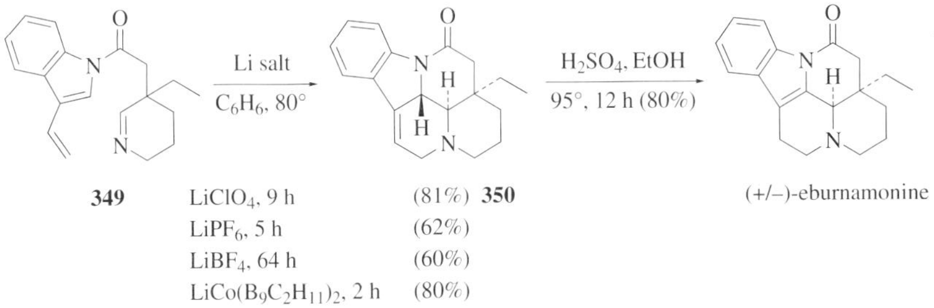
O
N
N
349
Li salt
C_6H_6, 80°
O
N
H
H
N
350
H_2SO_4, EtOH
95°, 12 h (80%)
O
N
H
N
(+/–)-eburnamonine
$LiClO_4$, 9 h (81%)
$LiPF_6$, 5 h (62%)
$LiBF_4$, 64 h (60%)
$LiCo(B_9C_2H_{11})_2$, 2 h (80%)

A clever example of the use of an intramolecular Diels-Alder reaction of an oxime derivative in a total synthesis of lysergic acid was reported several years ago.[262] Heating substrate **351** at 200° in trichlorobenzene promotes a retro-Diels-Alder reaction affording diene **352**, which then undergoes a [4+2]-cycloaddition with the methyl oxime to afford a 3:2 mixture of cycloadducts **353** and **354**. These compounds are then converted into racemic lysergic acid by a short sequence of steps.

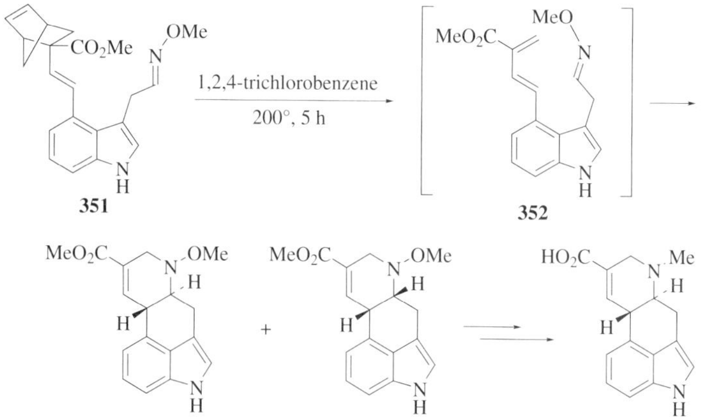
CO2Me
OMe
N
N
H
1,2,4-trichlorobenzene
200°, 5 h
MeO
MeO2C
N
N
H
MeO2C
N
OMe
H
H
N
H
+
MeO2C
N
OMe
H
H
N
H
HO2C
N
Me
H
H
N
H

351 **352**

353 (67%) 3:2 **354** (+/–)-lysergic acid

A one-pot synthesis of 2-carboxyl-substituted pyridines utilizing Diels-Alder reactions of tosyloxy compound **355** has been devised.[209,210] For example, [4+2]-cycloaddition of dienophile **355** with 2,4-dimethylbutadiene yields a single regioisomeric adduct **356**, which without purification is treated with sodium methoxide and *N*-chlorosuccinimide, affording pyridine **357** in good overall yield.

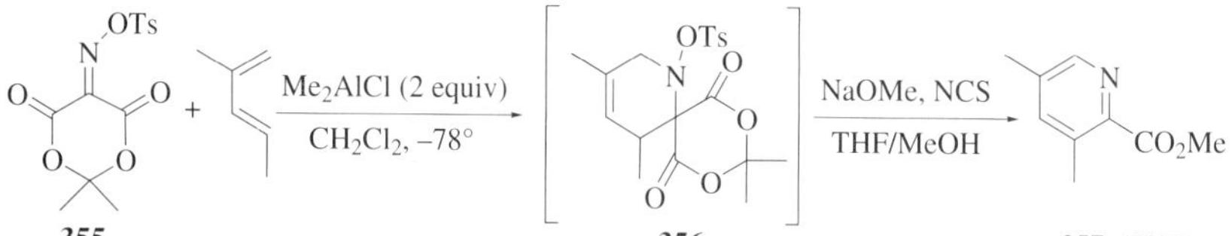
OTs
N
O
O
O
O
+
Me2AlCl (2 equiv)
CH2Cl2, –78°
OTs
N
O
O
O
O
NaOMe, NCS
THF/MeOH
N
CO2Me

355 **356** **357** (70%)

Intramolecular oximino Diels-Alder methodology has recently been used to generate various substituted pyridine derivatives.[239] Thermal cycloaddition of diene acyloxime **358** in refluxing toluene leads to cycloadduct **359**, which upon treatment with cesium carbonate in DMF aromatizes, affording pyridine ester acid **360**.

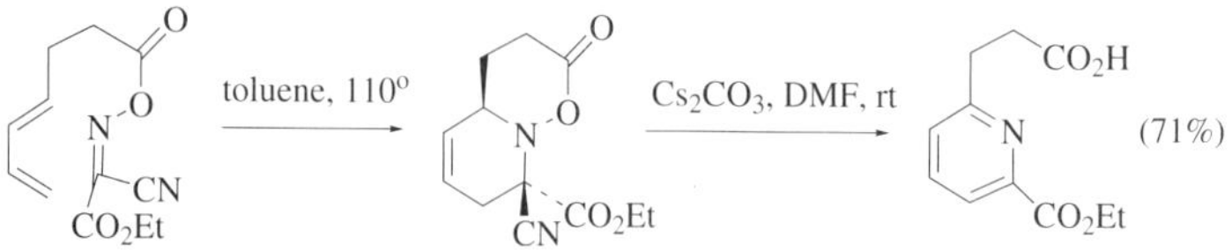
O
O
N
CN
CO2Et
toluene, 110°
O
O
N
CN
CO2Et
Cs2CO3, DMF, rt
CO2H
N
CO2Et
(71%)

358 **359** (70%) **360**

COMPARISON WITH OTHER METHODS

A number of other procedures have been used to construct 1,2,5,6-tetrahydropyridines of the type one obtains via ImDA reactions. This general area has recently been reviewed;[263,264] therefore, only a brief account of the most widely utilized alternative synthetic methods is presented here.

A classical synthesis of 1,2,5,6-tetrahydropyridine systems, which has been used for decades, involves the partial reduction of pyridinium salts.[265] An example of this process is shown in Eq. 58. A problem that can arise with this methodology is the formation of regioisomeric reduction products depending on the substitution pattern of the pyridinium salt.

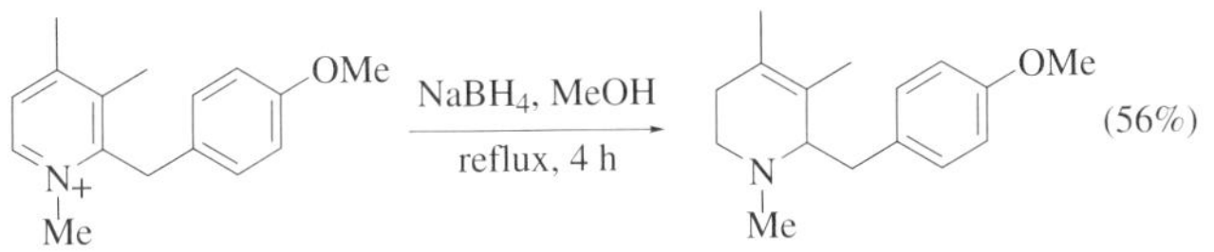
OMe
NaBH4, MeOH
reflux, 4 h
N+
Me
OMe
(56%)
N
Me

(Eq. 58)

In related chemistry, a short enantioselective route to 2-substituted and 2,6-disubstituted 1,2,5,6-tetrahydropyridines has been described starting from chiral pyridinium salt **361**.[266] Addition of a Grignard reagent to **361** leads to intermediate **362**, which is reduced stereoselectively, affording 1,2,5,6-tetrahydropyridines **363** and **364**. Addition of a second Grignard reagent to amino acetal **362** affords a 2,6-disubstituted product.

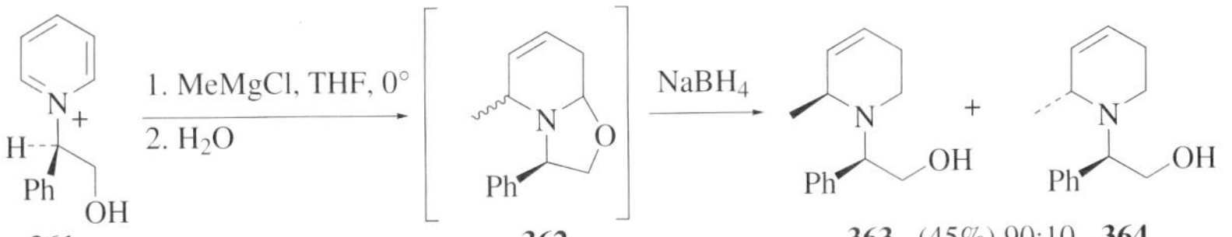
N+
H
Ph
OH
1. MeMgCl, THF, 0°
2. H2O
N
O
Ph
NaBH4
N
Ph
OH
+
N
Ph
OH

361 **362** **363** (45%) 90:10 **364**

A nice modification of the Ireland-Claisen rearrangement has been developed for stereoselective construction of tetrahydropyridines.[267] Thus, an amino lactone like **365** can be converted into silylketene acetal **366**, which subsequently rearranges affording the cis-2,6-disubstituted 1,2,5,6-tetrahydropyridine **367**.

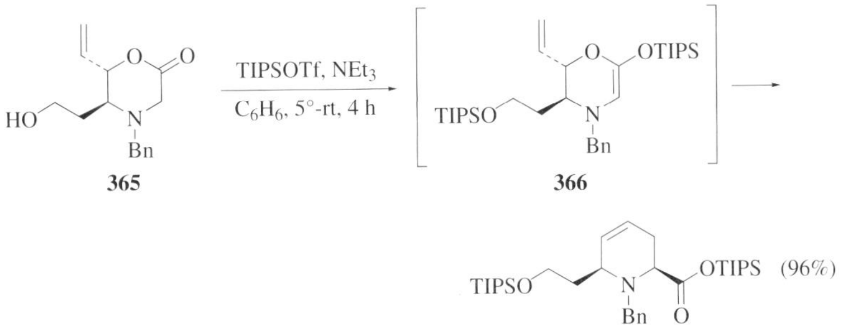
O
O
HO
N
Bn
TIPSOTf, NEt3
C6H6, 5°-rt, 4 h
O
OTIPS
TIPSO
N
Bn
TIPSO
N
Bn
O
OTIPS
(96%)

365 **366** **367**

With the advent of stable, well-behaved transition metal catalysts, ring-closing olefin metathesis has recently become a popular method for constructing nitrogen heterocycles such as 1,2,5,6-tetrahydropyridines.[268,269] For example, ring-closing metathesis of diene **368** using Grubb's ruthenium catalyst affords bicyclic tetrahydropyridine **369** in good yield.[270]

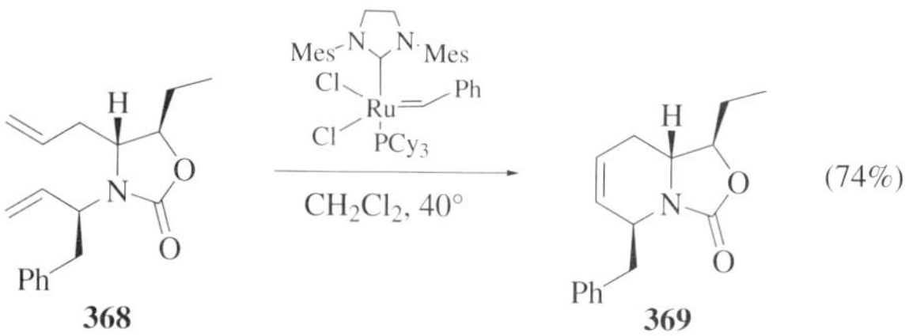
Mes
N
N
Mes
Cl
Ru
Ph
Cl
PCy3
CH2Cl2, 40°
H
N
O
O
Ph
(74%)

368 **369**

Another method that has been used for stereoselective construction of 1,2,5,6-tetrahydropyridines involves an aza-[2,3]-Wittig rearrangement of a vinylaziridine. An example of this transformation is shown in Eq. 59.[271]

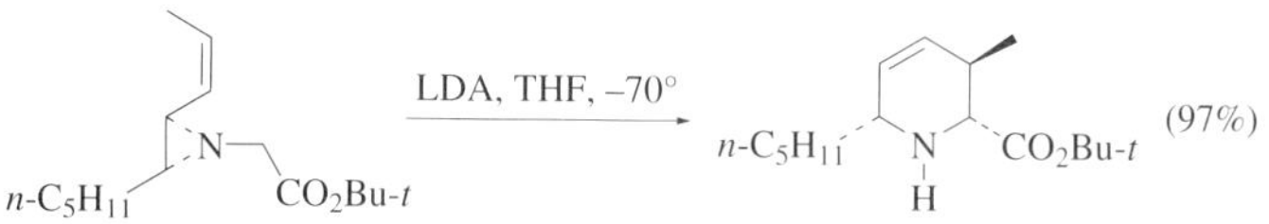
LDA, THF, –70°
n-C5H11
N
CO2Bu-t
H
(97%)

(Eq. 59)

An interesting approach to tetrahydropyridines has been devised that involves an internal cyclization of allylsilane nitrones.[272] Thus, exposure of a nitrone substrate such as **370** to trimethylsilyl triflate probably generates intermediates **371** and **372**, leading to a 1:1 mixture of *cis*- and *trans*-*N*-hydroxy-1,2,5,6-tetrahydropyridines **373** and **374** in high overall yield.

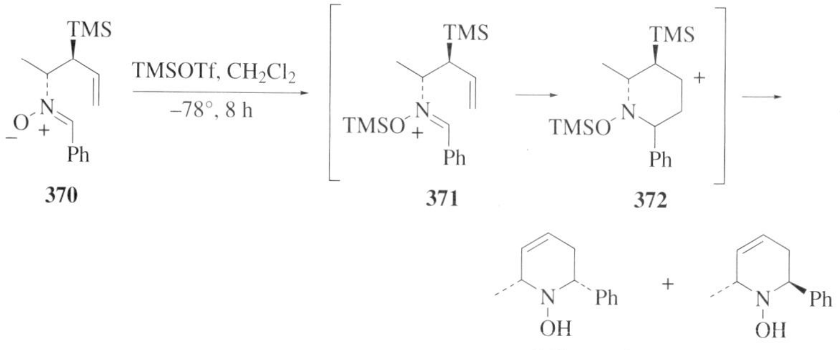
TMS
TMSOTf, CH2Cl2
–78°, 8 h
TMSO
N
Ph
OH
(91%) 1:1

370 **371** **372**

373 **374**

Iminium ion vinylsilane cyclizations have been extensively utilized as a route to 1,2,5,6-tetrahydropyridines. The mechanism of this transformation has also been examined in detail.[273] An example of the process is shown in Eq. 60.

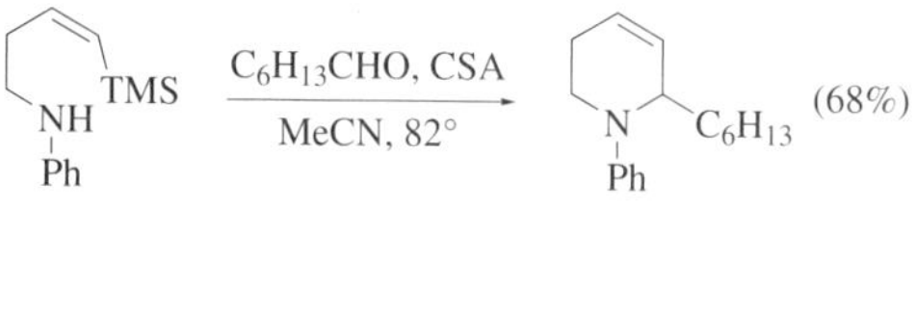
TMS
NH
Ph
C6H13CHO, CSA
MeCN, 82°
N
C6H13
(68%)

(Eq. 60)

Iminium ion allylsilane chemistry has also been used to prepare 1,2,5,6-tetrahydropyridines.[274] The cyclization shown in Eq. 61 is highly stereoselective, leading to the cis-2,6-disubstituted system.

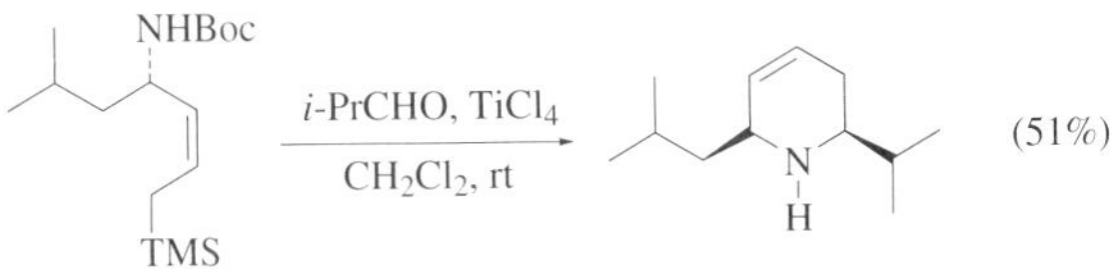
NHBoc
TMS
i-PrCHO, TiCl4
CH2Cl2, rt
N
H
(51%)

(Eq. 61)

Finally, efficient methodology has been developed for synthesizing dihydropyridones of the type that can be obtained via ImDA cycloadditions of highly oxygenated Danishefsky-type dienes.[275] Thus, treatment of readily available 4-methoxypyridine with an acylating agent, followed by a Grignard reagent, affords a 6-substituted 2,3-dihydro-4-pyridone derivative. An example of this transformation is presented in Eq. 62.[276]

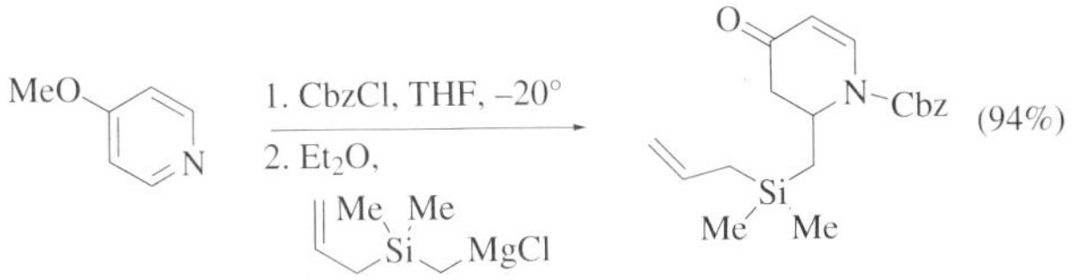
MeO
N
1. CbzCl, THF, –20°
2. Et2O,
Me Me
Si
MgCl
O
N
Cbz
Si
Me Me
(94%)

(Eq. 62)

EXPERIMENTAL CONDITIONS

The diversity of the imino Diels-Alder cycloadditions described in this chapter makes it difficult to generalize regarding reaction conditions. As noted above, some of these reactions can be effected under various thermal conditions. In addition, other processes are promoted by either protic or Lewis acids. A wide range of solvents has also been used in these cycloadditions depending on the particular substrates.

Reactions with *N*-acylimines are generally performed in benzene at reflux in the presence of a Lewis or protic acid, such as boron trifluoride etherate, trifluoroacetic acid, or naphthalenesulfonic acid. In some cases, these reactions can also be performed thermally depending on the reactivity of the imine and the diene.

Most commonly, reactions of *C*-acylimines are carried out with TFA and boron trifluoride etherate in dichloromethane, although other acids and solvents have been used.

Reactions of *N*-sulfonylimines are most often performed under thermal conditions, usually in benzene at reflux. These cycloadditions have also occurred at high pressure. Acidic conditions have also been used, most often with boron trifluoride in dichloromethane at $-20°$.

A wide variety of conditions have been utilized for the cycloadditions of alkyl or arylimines, most often involving an acid catalyst. A widely used procedure forms the imine in situ. An amine hydrochloride is combined with an aldehyde and the diene in water at room temperature. A large number of Lewis acids have been utilized in solvents such as THF or dichloromethane, and the reactions usually take place at or below room temperature. Cycloadditions of alkylimines have also been performed under thermal conditions, but temperatures depend upon the reactivity of the substrates.

EXPERIMENTAL PROCEDURES

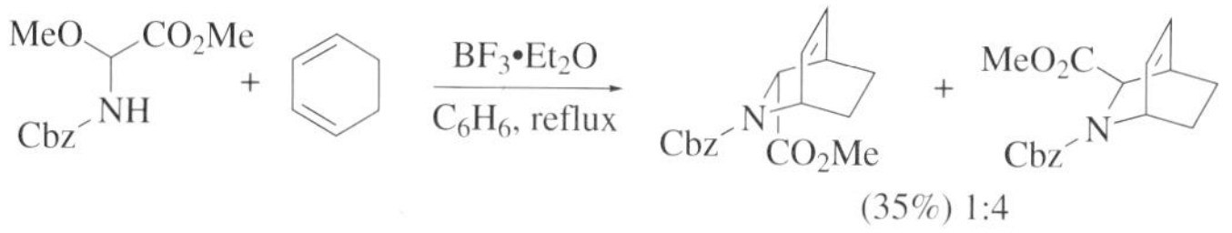

***endo* and *exo* Methyl 2-Benzyloxycarbonyl-2-azabicyclo[2.2.2]oct-5-ene-3-carboxylate (BF_3-Catalyzed Acyclic *N*-Acylimine Diels-Alder Reaction).**[62] $BF_3 \cdot Et_2O$ (4 mL, 0.032 mol) was added to a stirred solution of methyl *N*-benzyloxycarbonyl-2-methoxyglycinate (20.24 g, 0.08 mol) in dry benzene (60 mL) under nitrogen. The mixture was heated to reflux, and cyclohexa-1,3-diene (8 mL, 0.085 mol) in dry benzene (16 mL) was added dropwise over 3 minutes. The mixture was heated under reflux for 1.5 hours, then cooled and poured into saturated aqueous $NaHCO_3$ (100 mL). The organic layer was washed with saturated aqueous $NaHCO_3$ (100 mL), dried ($MgSO_4$), and concentrated to give a yellow oil. The oil was chromatographed on a silica gel column (600 g). Elution with petroleum ether (bp 40–60°)-Et_2O (2:1) gave the exo adduct as an oil (8.5 g, 28%); IR ($CHCl_3$) 1750, 1695 cm^{-1}; ^1H NMR (d_8-toluene, 100°) δ 7.16 (m, 5H), 6.17 (m, 2H), 5.15, 4.95 (ABq, J = 13 Hz, 2H), 4.84 (m, 1H), 3.94 (dd, J = 3, 8 Hz, 1H), 3.44 (s, 3H), 2.68 (m, 1H), 2.55–0.95 (m, 4H); MS *m/z*: M^+ 301 (5), 242 (95), 195 (40), 170 (100), 150 (29), 107 (100). Anal. Calcd for $C_{17}H_{19}NO_4$: C, 67.8; H, 6.3; N, 4.6. Found: C, 68.3; H, 6.6; N, 4.45.

Further elution with the same solvent gave the endo adduct (2.1 g, 7%) as a solid, mp 73.5–75.5°; IR ($CHCl_3$) 1760, 1690 cm^{-1}; ^1H NMR (d_8-toluene, 100°) δ 7.2 (m, 5H), 6.16 (ddd, J = 8, 7, 1 Hz, 1H), 5.93 (ddd, J = 8, 6, 2 Hz, 1H), 5.2, 5.0 (ABq, J = 13 Hz, 2H), 4.84 (m, 1H), 4.21 (d, J = 2.5 Hz, 1H), 3.37 (s, 3H), 2.83 (m, 1H), 1.8–1.0 (m, 4H); MS *m/z*: M^+ 301 (6), 242 (83), 170 (100), 107 (17). Anal. Calcd for $C_{17}H_{19}NO_4$: C, 67.8; H, 6.3; N, 4.6. Found: C, 68.1; H, 6.5; N, 4.45.

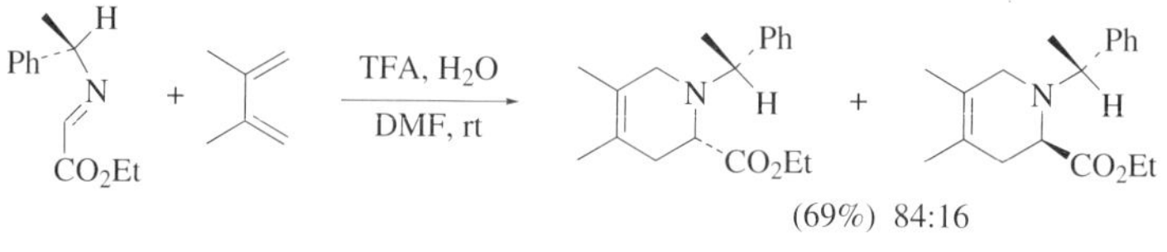

Ethyl (2*S*/*R*)-1-[(*R*)-1-Phenylethyl]-4,5-dimethyl-1,2,3,6-tetrahydropyridine-2-carboxylate (*C*-Acylimine Diels-Alder Reaction with a Chiral Auxiliary).[83] (*R*)-1-Phenylethylamine (10.0 g, 82.6 mmol) and freshly prepared ethyl glyoxylate (8.43 g, 82.6 mmol) were dissolved in dry toluene (30 mL), and the water was removed using a Dean-Stark apparatus by heating the mixture at reflux for 20 minutes. Removal of the solvent under reduced pressure gave the chiral imine as an orange oil (16.93 g, 100%). A portion of this imine (3.75 g, 18.3 mmol) was dissolved in

DMF (12 mL), and then TFA (2.1 g, 18.3 mmol), 2,3-dimethylbutadiene (3.0 g, 36.6 mmol), and water (10 mL) were added. The mixture was stirred at room temperature for 24 hours under argon, and the solvent was then removed under reduced pressure. A solution of the residue in $CHCl_3$ (20 mL) was washed with aqueous $NaHCO_3$ and brine, dried (K_2CO_3), and evaporated under reduced pressure. The resulting residue was purified by flash chromatography (hexane-EtOAc, 98:2), giving the title compound (3.6 g, 69%) as a 84:16 mixture of diastereomers. Crystallization as the hydrochloride salt from EtOAc-hexane gave the pure major (6*S*)-product as white crystals, mp 162–164°; $[\alpha]^{20}_{D}$ −7.5° (*c* 1.00, MeOH); IR (thin film) 3060, 3020, 2980, 2910, 1730 cm^{-1}; ^{1}H NMR (300 MHz, $CDCl_3$) δ 1.25 (t, J = 7.0 Hz, 3H), 1.32 (d, J = 6.8 Hz, 3H), 1.43 (s, 3H), 1.62 (s, 3H), 2.29–2.59 (m, 2H), 2.75–3.17 (ABq, J = 16.4 Hz, 2H), 3.94–4.04 (m, 1H), 3.99 (q, J = 7.0 Hz, 1H), 7.17–7.38 (m, 5H); ^{13}C NMR (75.5 MHz) δ 14.4 (q), 16.3 (q), 18.4 (q), 21.0 (q), 34.8 (t), 52.2 (t), 55.1 (d), 59.8 (t), 61.8 (d), 121.4 (s), 124.4 (s), 126.7 (d), 127.2 (d), 128.3 (d), 146.1 (s), 173.6 (s); HRMS (m/z): calcd for $C_{18}H_{25}NO_2$, 287.1860; found, 287.1860.

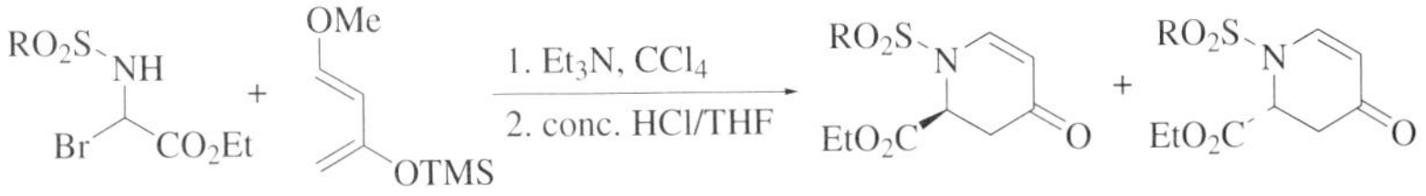

RO2S
NH
Br
CO2Et
OMe
OTMS
1. Et3N, CCl4
2. conc. HCl/THF
RO2S
N
EtO2C
O
R = (1R)-(−)-camphor-10-yl
(58%) 35:65

Ethyl 1-[(1*R*)-Camphor-10-ylsulfonyl]-4-oxo-1,2,3,4-tetrahydropyridine-2(*RS*)-carboxylate (*N*-Sulfonylimine Diels-Alder Reaction with a Chiral Auxiliary).[124] To a solution of ethyl 2-bromo-*N*-[(1*R*)-camphor-10-ylsulfonyl]glycinate (320 mg, 0.81 mmol) in CCl_4 (20 mL) under an argon atmosphere at room temperature was added Et_3N (115 μL, 0.83 mmol) followed by 1-methoxy-3-trimethylsilyloxybuta-1,3-diene (400 μL, 1.85 mmol). The reaction mixture was stirred at room temperature for 12 hours and then concentrated. HCl-THF (1:19, 20 mL) was added and the mixture was stirred at room temperature for 1 hour. The mixture was diluted with water (20 mL) and the organic layer was dried and evaporated. Purification by silica gel chromatography (EtOAc–hexane, 2:1) of the crude oil gave a 35:65 diastereomeric mixture of the title compounds (178 mg, 58%) as a pale brown oil which was not separated. IR (neat) 2967, 1746, 1674, 1601 cm^{-1}; UV (EtOH) 280 nm (ϵ 10,923); ^{1}H NMR (400 MHz; $CDCl_3$) δ 0.9 (s, 3H), 1.10 (s, 1.8H, major diastereomer), 1.11 (s, 1.2H, minor diastereomer), 1.28 (t, 3H), 1.50 (m, 1H), 1.72 (ddd, J = 4.7, 9.3, 14.0 Hz, 0.4H, minor diastereomer), 1.85 (ddd, J = 4.7, 9.3, 14.0 Hz, 0.6H, major diastereomer), 1.96 (asymmetric d, J = 6.0 Hz, 0.4H, minor diastereomer), 2.00 (asymmetric d, J = 6.0 Hz, 0.6H, major diastereomer), 2.09 (m, 1H), 2.16 (t, J = 4.5 Hz, 1H), 2.39 (m, 2H), 3.06 (m, 2H), 3.50 (ABq, J = 15 Hz, sep. 277.0 Hz, 0.8H, minor diastereomer), 3.55 (ABq, J = 14.9 Hz, sep. 160.0 Hz, 1.2H, major diastereomer), 4.24 (m, 2H), 5.17 (m, 1H), 5.39 (dd, J = 1.1 and 8.4 Hz, 0.4H, minor diastereomer), 5.43 (d, J = 8.4 Hz, 0.6H, major diastereomer), 7.61 (dd, J = 1.5, 8.4 Hz, 0.4H, minor diastereomer), 7.72 (dd, J = 1.5, 8.4 Hz, 0.6H, major diastereomer); ^{13}C NMR (100 MHz, $CDCl_3$) δ major diastereomer: 14.0 (CH_3), 19.2 (2 CH_3), 25.4 (CH_2), 27.0 (CH_2), 38.4 (CH_2), 42.5 (CH_2), 42.7 (CH), 48.3 (C), 51.4 (CH_2), 57.0 (CH), 58.6 (C), 62.8 (CH_2), 107.7 (CH), 142.9 (CH), 169.0 (C), 189.5

(C), 214.1 (C); minor diastereomer: 14.0 (CH_3), 19.7 (2 CH_3), 25.4 (CH_2), 26.9 (CH_2), 38.4 (CH_2), 42.4 (CH_2), 42.8 (CH), 48.3 (C), 51.4 (CH_2), 57.0 (CH), 58.5 (C), 62.7 (CH_2), 107.5 (CH), 142.7 (CH), 168.8 (C), 189.6 (C), 214.0 (C); MS-FAB (m/z): 384 (M^+ + H), 215 (M^+ + H − $C_8H_{10}NO_3$), 170 (M^+ + 2H − $C_{10}H_{12}SO_3$, base peak). Anal. Calcd for $C_{18}H_{25}NO_6S$: C, 56.38; H, 6.57; N, 3.65. Found: C, 56.1; H, 6.5; N, 3.6.

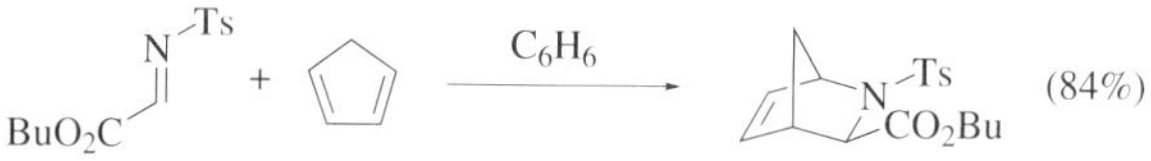

Ts
N
BuO2C
+
C6H6
N
Ts
CO2Bu
(84%)

Butyl 2-(*p*-Tolylsulfonyl)-2-azabicyclo[2.2.1]hept-5-ene-*exo*-3-carboxylate (Thermal *N*-Sulfonylimine Diels-Alder Reaction).[277] To an ice-cooled solution of butyl (*p*-tolylsulfonylimino)acetate (22.2 g, 78.4 mmol) in dry benzene (36 mL) was added freshly distilled and dried ($CaCl_2$) cyclopentadiene (5.18 g, 78.5 mmol). When the exothermic reaction began to subside, the reaction mixture was kept at room temperature for 12 hours and was then concentrated in vacuo. The oily residue was taken up in Et_2O (50 mL) and washed with 5% $NaHCO_3$ solution, dried ($MgSO_4$), and the solvent was removed under reduced pressure. The residue, which solidified upon standing, was crystallized from Et_2O-hexane (1:5), yielding 23.0 g (84%) of butyl 2-(*p*-tolylsulfonyl)-2-azabicyclo[2.2.1]hept-5-ene-*exo*-3-carboxylate as a colorless solid, mp 53–55°; IR (nujol) 1740, 1600 cm^{-1}; Anal. Calcd for $C_{18}H_{23}NO_4S$: C, 61.88; H, 6.64; N, 4.01. Found: C, 61.97; H, 6.59; N, 3.83. [^1H NMR data was reported for the methyl ester, but not for the title compound. Methyl carboxylate analog: ^1H NMR ($CDCl_3$) δ 0.93 (t, J = 5 Hz, 3H), 2.5 (s, 3H), 3.33 (m, 1H), 3.53 (s, 1H), 4.13 (t, J = 6 Hz, 2H), 6.23 (m, 2H), 7.56 (m, 4H)].

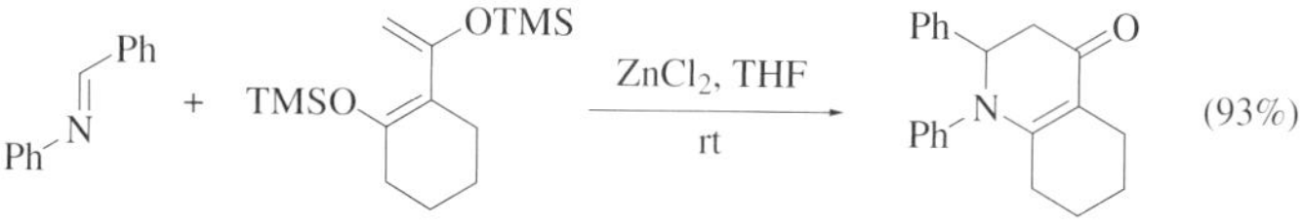

Ph
N
Ph
+
TMSO
OTMS
ZnCl2, THF
rt
Ph
O
N
Ph
(93%)

***N*,2-Diphenyl-2,3,5,6,7,8-hexahydro-1*H*-quinolin-4-one ($ZnCl_2$-Catalyzed Diels-Alder Reaction of an Unactivated Imine).[278]** To 1-trimethylsilyloxy-2-[1-(trimethylsilyloxy)ethylidenyl]-1-cyclohexene (380 mg, 1.34 mmol) were added at room temperature *N*-benzylideneaniline (136 mg, 0.75 mmol) and a catalytic amount of $ZnCl_2$ in THF (2 mL). After being stirred for 20 hours, the reaction mixture was diluted with EtOAc (40 mL) and washed with H_2O (20 mL). The organic layer was dried ($MgSO_4$) and concentrated in vacuo. The residue was purified by silica gel flash column chromatography (30% EtOAc-hexanes), affording 215 mg of the title ketone [93%, R_f 0.28 (50% EtOAc/hexanes)] as a white solid, mp 160–162° (CH_2Cl_2/hexanes); IR (neat) 1630, 1560, 1280 cm^{-1}; ^1H NMR ($CDCl_3$) δ 1.63 (m, 4H), 2.07 (m, 2H), 2.37 (m, 2H), 2.84 (dd, J = 16.4, 7.1 Hz, 1H), 3.12 (dd, J = 16.4, 6.1 Hz, 1H), 4.91 (t, J = 6.6 Hz, 1H), 7.03 (m, 2H), 7.21 (m, 8H); ^{13}C NMR ($CDCl_3$) δ 21.7 (CH_2), 22.1 (CH_2), 22.7 (CH_2), 30.0 (CH_2), 43.3 (CH_2), 65.4 (CH), 110.0 (C), 126.7 (CH), 127.1 (2 CH), 127.5 (CH), 127.6 (2 CH), 128.5 (2 CH), 129.0 (2 CH), 139.9 (C), 144.1 (C), 158.4 (C), 189.9 (C); EIMS (70 eV) m/z: M^+ + 1 304

(14), M^+ 303 (44), 226 (15), 77 (100). Anal. Calcd for $C_{21}H_{21}NO$: C, 83.13; H, 6.98; N, 4.62. Found: C, 83.04; H, 7.13; N, 4.73.

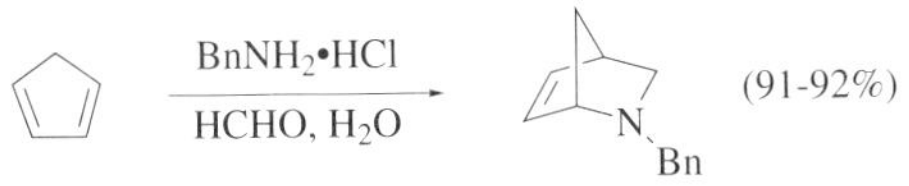
BnNH2•HCl
HCHO, H2O
N
Bn
(91-92%)

N-Benzyl-2-azanorbornene (Aqueous Immonium Diels-Alder Reaction).[279] A 100-mL round-bottomed flask equipped with a Teflon-coated magnetic stirring bar was charged with 24 mL of deionized H_2O and 8.6 g (60.0 mmol) of benzylamine hydrochloride. To the above homogeneous solution was added 6.3 mL (84 mmol) of 37% aqueous formaldehyde solution followed by 9.9 mL (120 mmol) of freshly prepared cyclopentadiene. The flask was stoppered tightly and stirred vigorously at ambient temperature. After 4 hours, the reaction mixture was poured into 50 mL of H_2O and washed with 1 : 1 Et_2O-hexane. The aqueous phase was made basic by the addition of 4.0 g of solid KOH and extracted with Et_2O. The combined Et_2O extracts were dried over anhydrous $MgSO_4$. The solvent was removed under reduced pressure to afford 11.2 g (100%) of *N*-benzyl-2-azanorbornene as a pale yellow oil. Distillation of the crude product at 80–85° (0.05 mm) through a short-path apparatus provided 10.1–10.2 g (91–92%) of pure product as a colorless oil. ^1H NMR (300 MHz, $CDCl_3$) δ 1.42 (dm, $J = 8$ Hz, 1H), 1.52 (dd, $J = 2, 8.5$ Hz, 1H), 1.64 (dm, $J = 8$ Hz, 1H), 2.94 (br s, 1H), 3.18 (dd, $J = 3, 8.5$ Hz, 2H), 3.34, 3.58 (ABq, $J = 13$ Hz, 2H), 3.83 (m, 1H), 6.09 (dd, $J = 2, 6$ Hz, 1H), 6.38 (ddd, $J = 2, 3, 6$ Hz, 1H), 7.2–7.4 (m, 5H). Anal. Calcd for $C_{13}H_{15}N$: C, 84.28; H, 8.16; N, 7.56. Found: C, 84.68; H, 8.36; N, 7.59.

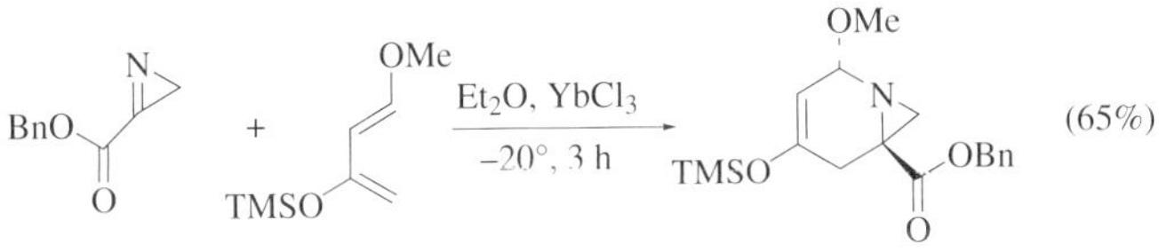
N
BnO
O
+
OMe
TMSO
Et2O, YbCl3
−20°, 3 h
OMe
N
TMSO
OBn
O
(65%)

Benzyl 2-Methoxy-4-trimethylsilyloxy-1-azabicyclo[4.1.0]hept-3-ene-6-carboxylate (Lewis Acid Catalyzed Azirine Cycloaddition).[189] To a solution of azirinecarboxylic acid benzyl ester (28 mg, 0.16 mmol) in Et_2O (6 mL) at −20° was added $YbCl_3$ (11 mg, 0.04 mmol) and after 5 minutes Danishefsky's diene (27 mg, 0.16 mmol) in Et_2O (1 mL). The reaction temperature was maintained at −20°. After 3 hours at this temperature TLC indicated the absence of the starting materials. The reaction mixture was washed with saturated aqueous $NaHCO_3$, dried ($MgSO_4$), and evaporated. Dissolution of the crude product in CH_2Cl_2 (1 mL) and filtration through a Pasteur pipette filled with alumina (pentane-EtOAc, 2 : 1), followed by evaporation provided the title compound as an orange oil (35 mg, 65%). R_f 0.82 (pentane : EtOAc, 2 : 1); IR (neat) 2107, 1737, 1456, 1255, 1191 cm^{-1}; ^1H NMR (400 MHz, $CDCl_3$) δ 7.50–7.25 (m, 5H), 5.28 (d, $J = 10.6$ Hz, 1H), 5.18 (d, $J = 10.6$ Hz, 1H), 5.01 (br s, 1H), 4.59 (br s, 1H), 3.67 (s, 3H), 2.86 (d, $J = 18.1$ Hz, 1H), 2.61 (d, $J = 18.1$ Hz, 1H), 2.17 (s, 1H), 2.12 (s, 1H), 0.20 (s, 9H); ^{13}C NMR (100 MHz, $CDCl_3$) δ 172.5, 148.7, 136.4, 128.9, 128.8, 99.9, 88.4, 67.3, 39.1, 29.3, 27.3, 0.0; CIMS *m/z*: 348 (100); HRMS (*m/z*): calcd for $C_{18}H_{25}NO_4Si$, 348.1631; found, 348.1634.

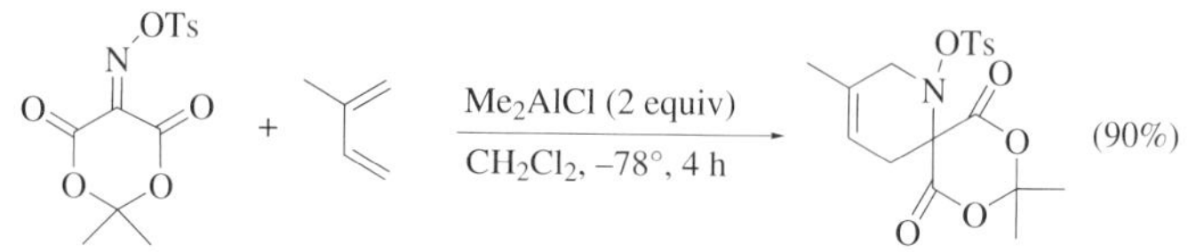

3,3,9-Trimethyl-1.5-dioxo-7-(tosyloxy)-7-aza-2,4-dioxaspiro[5.5]undec-9-ene (Me_2AlCl-Catalyzed Oximino Diels-Alder Reaction).[209,210] A 50-mL, three-necked, round-bottomed flask equipped with an argon inlet adapter, glass stopper, and rubber septum was charged with a solution of 5-(tosyloxyimino)-2,2-dimethyl-1,3-dioxane-4,6-dione (0.327 g, 1.00 mmol) and isoprene (0.204 g, 3.00 mmol) in 14 mL of dichloromethane. The solution was cooled at −78° while Me_2AlCl solution (1.0 M in hexane, 2.0 mL, 2.0 mmol) was added dropwise via syringe over 4 minutes. The resulting orange solution was stirred for 4 hours at −78°, giving a yellow solution that was quenched by addition of 3 mL of saturated sodium potassium tartrate solution. The resulting mixture was allowed to warm to 0°, 15 mL of CH_2Cl_2 and 15 mL of water were added, and the aqueous phase was separated and extracted with three 20-mL portions of CH_2Cl_2. The combined organic phases were washed with 30 mL of saturated aqueous NaCl solution, dried over $MgSO_4$, and concentrated to provide an orange oil. Column chromatography of this material on silica gel (1% methanol-CH_2Cl_2) provided 0.354 g (90%) of the title compound as a white foam; IR ($CHCl_3$) 3020, 1780, 1750, 1385, 1300 cm^{-1}; ^1H NMR (300 MHz, $CDCl_3$) δ 1.67 (s, 3H), 1.69 (s, 3H), 1.88 (s, 3H), 2.48 (s, 3H), 2.72 (br dd, J = 1.2, 3.3 Hz, 2H), 3.93 (s, 2H), 5.33 (br dd, J = 1.2, 3.6 Hz, 1H), 7.36 (d, J = 8.7 Hz, 2H), 7.81 (d, J = 8.4 Hz, 2H); ^{13}C NMR (75 MHz, $CDCl_3$) δ 20.3, 21.7, 28.5, 29.3, 32.8, 57.4, 66.3, 106.2, 113.9, 129.2, 129.55, 129.62, 131.2, 145.9, 164.0.

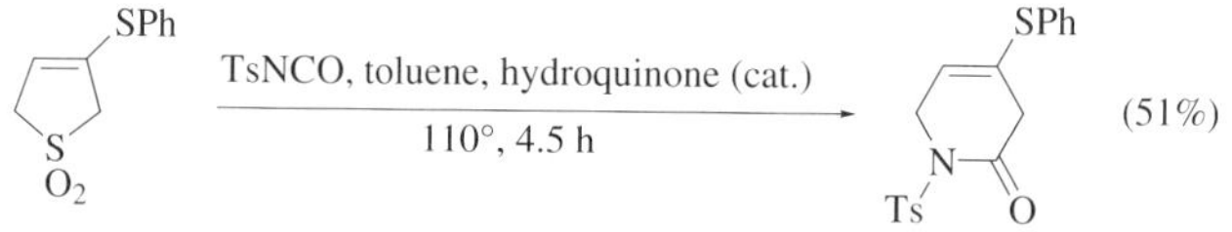

1-[(4-Methylphenyl)sulfonyl]-4-(phenylthio)-3,6-dihydro-2-pyridinone (Diels-Alder Reaction of a 2-Thiosubstituted 1,3-Diene with an Arylsulfonyl Isocyanate).[218] A mixture of 3-thiophenyl-3-sulfolene (226 mg, 1 mmol), *p*-toluenesulfonyl isocyanate (985 mg, 5 mmol), and a catalytic amount of hydroquinone (10 mg, 0.1 mmol) was heated under N_2 in toluene (4 mL) at 110° for 4.5 hours. The solvent was removed under vacuum and to the reaction mixture was added 5% aqueous NaOH solution (50 mL). The mixture was extracted with CH_2Cl_2, and the organic solution was dried (Na_2SO_4) and evaporated. The crude product was purified by silica gel column chromatography (hexane-EtOAc, 8 : 1–4 : 1) to afford the title compound as a white solid (185 mg, 51%), mp 66–68°; IR (film) 3058, 2922, 1691, 1595, 1474, 1461, 1440, 1386, 1356, 1293, 1257, 1187, 1167, 1146, 1088, 853, 705, 690 cm^{-1}; ^1H NMR (300 MHz, $CDCl_3$) δ 2.41 (s, 3H), 3.02 (d, J = 3.4 Hz, 2H), 4.50 (dd, J = 7.3, 3.4 Hz. 2H), 5.80 (br s, 1H), 7.29–7.37 (m, 7H), 7.92 (d, J = 8.2 Hz, 2 H); ^{13}C NMR (75 MHz, $CDCl_3$) δ 21.5, 37.8, 47.5, 118.8, 128.6, 128.8, 129.1, 129.3, 129.4, 130.4, 133.1, 135.5, 145.1, 166.1; EIMS m/z: M^+ 359 (15), 295 (28),

250 (36), 205 (17), 204 (94), 187 (13), 186 (93), 176 (18), 161 (34), 155 (36), 149 (11), 147 (24), 135 (15), 110 (12), 109 (34), 94 (10), 91 (100), 77 (10), 67 (11), 65 (34), 39 (13); EI-HRMS (*m/z*): calcd for $C_{18}H_{17}NO_3S_2$, 359.0650; found, 359.0650.

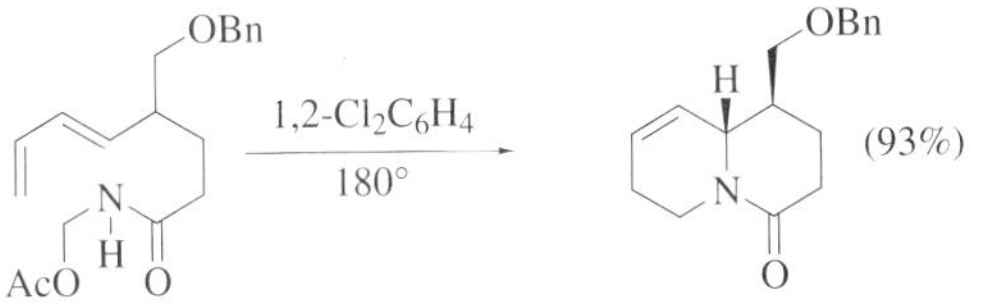

OBn
1,2-Cl2C6H4
180°
N
H
AcO
O
OBn
H
(93%)
N
O

(1*S,9a*R**)-1-(Benzyloxymethyl)-2,3,6,7-tetrahydro-1*H*-quinolizin-4(9a*H*)-one (Intramolecular Thermal *N*-Acylimino Cycloaddition).**[227] (*E*)-(4-(Benzyloxymethyl)octa-5,7-dienamido)methyl acetate (102 mg, 0.31 mmol) was dissolved in *o*-dichlorobenzene (25 mL), and the solution was heated at reflux under nitrogen for 2 hours. The solvent was removed by vacuum distillation, and the residue was purified on silica gel (5 g, EtOAc-hexanes, 1 : 1) to give the title compound as an oil (78 mg, 93%). IR ($CHCl_3$) 3005, 2940, 2880, 1955, 1880, 1820, 1630, 1470, 1420, 1370, 1280, 1240–1200, 1120, 1100, 700–660 cm^{-1}; ^1H NMR (360 MHz, $CDCl_3$) δ 7.38–7.29 (5H), 5.85 (m, 1H), 5.75 (m, 1H), 480 (m, 1H), 4.52 (d, J = 3 Hz, 2H), 3.93 (m, 1H), 3.56 (d, J = 5 Hz, 2H) 2.64–1.65 (m, 8H); MS *m/z*: M^+ 271 (14.3), 180 (100), 150 (14.6), 108 (0.1), 91 (85.9), 82 (37.8); HRMS (*m/z*): calcd for $C_{17}H_{21}NO_2$, 271.1573; found, 271.1576.

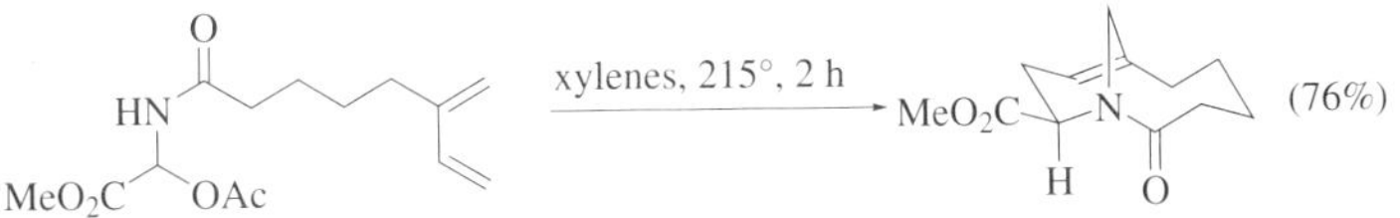

O
HN
MeO2C
OAc
xylenes, 215°, 2 h
MeO2C
N
H
O
(76%)

10-Carbomethoxy-2-oxo-1-azabicyclo[5.3.1]undec-7-ene (Type 2 *N*-Acylimine Diels-Alder Reaction).[226] A solution of methyl (6-methyleneoct-7-enamido) (acetyloxy)acetate (100 mg, 353 μmol) in xylenes (35 mL) was degassed by successive freeze-pump-thaw cycles using a medium-vacuum oil diffusion pump and the mixture was heated in a sealed tube at 215° for 2 hours. The crude product mixture was loaded directly onto a silica gel column and the column was washed with three volumes of hexane prior to elution with Et_2O, affording 60 mg (76%) of the title compound as a colorless crystalline solid. IR (KBr) 3038, 2965, 2936, 2898, 2853, 1747, 1645, 1406, 1364, 1307, 1244, 1208, 1166, 1150, 1056, 1021, 801, 793, 733 cm^{-1}; ^1H NMR (500 MHz, C_6D_6) δ 5.37 (ddd, J = 9.1, 7.7, 1.7 Hz, 1H), 5.10 (ddd, J = 9.3, 4.6, 2.3 Hz, 1H), 3.41 (ddd, J = 15.3,1.6, 1.6 Hz, 1H) 3.29 (s, 3H), 3.03 (ddd, J = 15.4, 3.3,1.6 Hz, 1H) 2.48 (dd, J = 13.3, 9.3 Hz, 1H), 2.39 (ddd, J = 16.1, 8.9, 7.2 Hz) and 2.15 (dddddd, J = 16.0, 7.9, 4.8, 3.1, 1.6, 0.6 Hz, 4H), 1.95 (m, 2H), 1.66 (ddd, J = 14.8, 11.9, 6.2 Hz, 1H), 1.51 (m, 1H), 1.43 (dddd, J = 14.8, 7.6, 5.9, 1.7 Hz, 1H), 1.13 (dddd, J = 14.5, 11.7, 11.7, 6.3 Hz, 1H), 1.05 (ddd, J = 14.6, 10.9, 10.9 Hz, 1H); ^{13}C NMR (125 MHz, C_6D_6) δ 177.4, 173.1, 145.1, 122.4, 52.0, 51.5, 42.8, 37.9, 33.4, 27.7, 26.7, 24.0; EIMS (*m/z*): calcd for $C_{12}H_{17}NO_3$, 223.1208; found, 223.1208.

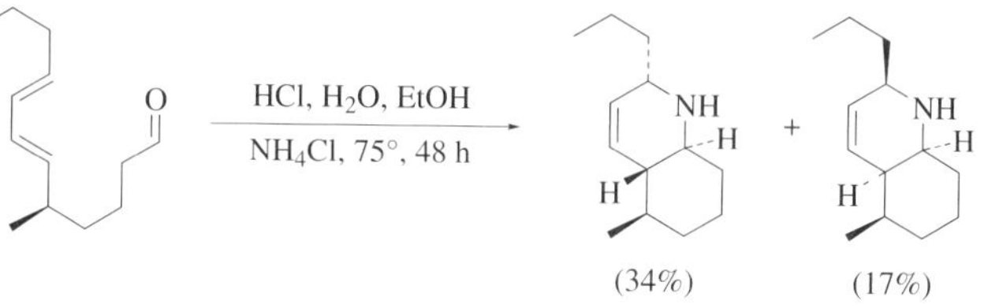

O
HCl, H2O, EtOH
NH4Cl, 75°, 48 h
NH
H
H
+
NH
H
H
(34%)
(17%)

(2*S*,4a*R*,5*R*,8a*S*)-5-Methyl-2-propyl-1,2,4a,5,6,7,8,8a-octahydroquinoline and (2*R*,4a*S*,5*R*,8a*S*)-5-Methyl-2-propyl-1,2,4a,5,6,7,8,8a-octahydroquinoline (Aqueous Intramolecular Immonium Diels-Alder Reaction).[235] To a solution of 254 mg (1.31 mmol) of (*R*,6*E*,8*E*)-5-methyldodeca-6,8-dienal in 100 mL of EtOH was added 100 mL of saturated aqueous ammonium chloride solution. The reaction mixture was heated at 75°. After 48 hours, the mixture was cooled to room temperature and diluted with 160 mL of water. The mixture was washed with hexane, basified with solid KOH, and extracted with Et_2O. The organic extracts were concentrated under reduced pressure to 50 mL and washed with water. The organic layer was dried over anhydrous Na_2SO_4 and concentrated in vacuo, leaving 195 mg of crude product. ^1H NMR ($CDCl_3$) analysis of the crude product indicated a 2.2:1 ratio of Diels-Alder adducts. The isomeric octahydroquinolines were separated by flash chromatography on 60 g of silica gel. Elution with 2% MeOH-$CHCl_3$ provided 87 mg (34%) of the major octahydroquinoline isomer [R_f 0.49 ($CHCl_3$:CH_3OH:concentrated NH_4OH, 85:15:1)]; octahydroquinoline-HCl: $[\alpha]^{23}_D$ + 37.3° (*c* 2.92, MeOH); IR (hydrochloride, CCl_4) 3030, 2960, 2930, 2870, 2860, 2810, 2730, 2690, 2520, 1615, 1605, 1585, 1455, 1450, 1375, 1330, 1295, 1290, 1235, 1165, 1135, 1100, 710 cm^{-1}; ^1H NMR (free amine, $CDCl_3$) δ 5.87 (d, J = 10.4 Hz, 1H), 5.68 (dt, J = 10.4, 3.2 Hz, 1H), 3.33–3.24 (m, 1H), 2.37 (ddd, J = 11.9, 9.4, 3.6 Hz, 1H), 2.33–2.20 (m, 1H), 1.80–1.65 (m, 3H), 1.51–0.90 (m, 15H); ^1H NMR (hydrochloride, $CDCl_3$) δ 6.04–5.78 (m, 3H), 5.67 (dt, J = 10.4, 3.1 Hz, 1H), 3.56–3.48 (m, 1H), 2.54 (ddd, J = 12.5, 9.4, 3.3 Hz, 1H), 1.95 (br d, J = 7.2 Hz, 1H), 1.80–1.00 (m, 10H), 0.96 (d, J = 6.5 Hz, 3H), 0.92 (t, J = 7.2 Hz, 3H); HRMS (m/z): calcd for $C_{13}H_{23}N$, 193.1830; found, 193.1816.

Similarly, 42 mg (17%) of the minor octahydroquinoline was isolated: R_f 0.43; $[\alpha]^{23}_D$ −7.5° (*c* 0.80, MeOH); IR (hydrochloride, CCl_4) 3030, 2970, 2940, 2880, 2800, 2790–2200, 1585, 1460, 1390, 1330, 1280, 1130, 1110, 970, 825, 710 cm^{-1}; ^1H NMR ($CDCl_3$) δ 5.79–5.69 (m, 1H), 3.67–3.56 (m, 1H), 3.28 (ddd, J = 8.6, 4.4, 4.4 Hz, 1H), 2.51 (br s, 1H), 1.86–1.08 (m, 12H), 1.02 (d, J = 7.2 Hz, 3H), 0.95 (t, J = 7.2 Hz, 3H); HRMS (m/z): calcd for $C_{13}H_{23}N$, 193.1830; found, 193.1836.

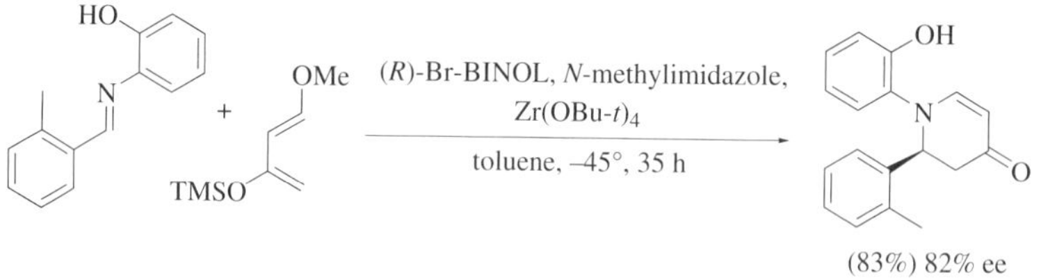

HO
N
+
OMe
TMSO
(R)-Br-BINOL, N-methylimidazole,
Zr(OBu-t)4
toluene, –45°, 35 h
OH
N
O
(83%) 82% ee

(*S*)-1-(2-Hydroxyphenyl)-2-*o*-tolyl-2,3-dihydropyridin-4(1*H*)-one (Catalytic Asymmetric Diels-Alder Reaction of an *N*-Arylimine).[247] 6,6′-Dibromo-1,1′-

bi-2-naphthol (0.088 mmol) in toluene (0.5 mL) and *N*-methylimidazole (0.12 mmol) in toluene (0.25 mL) were added to $Zr(OBu\text{-}t)_4$ (0.04 mmol) in toluene (0.25 mL) at room temperature. The mixture was stirred for 1 hour at the same temperature, and then cooled to −45°. Solutions of (*E*)-*N*-2-methylbenzylidene-1-(2-hydroxyphenyl) amine (0.4 mmol) and Danishefsky's diene (0.6 mmol) in toluene (0.75 mL each) were added successively. The mixture was stirred for 35 hours at the same temperature, and saturated aqueous $NaHCO_3$ was added. The aqueous layer was extracted with CH_2Cl_2, and the crude adduct was treated with THF-1 N HCl (20:1) at 0° for 30 minutes. After a usual work-up, the crude product was purified by chromatography on silica gel, giving the title compound (0.073 mmol, 83%). ^1H NMR ($CDCl_3$) δ 2.10 (s, 3H), 2.68 (dd, *J* = 7.6, 16.6 Hz, 1H), 3.06 (dd, *J* = 7.6, 16.6 Hz, 1H), 5.15 (d, *J* = 7.5 Hz, 1H), 5.52 (t, *J* = 7.5 Hz, 1H), 6.57 (t, *J* = 7.0 Hz, 1H), 6.76–7.00 (m, 6H), 7.38–7.43 (m, 2H), 9.57 (br s, 1H); ^{13}C NMR ($CDCl_3$) δ 19.0, 42.7, 58.6, 98.2, 117.1, 119.6, 126.1, 126.2, 126.6, 127.7, 128.3, 130.9, 131.5, 134.8, 136.6, 151.8, 156.8, 192.0; HRMS (*m*/*z*): calcd for $C_{18}H_{17}NO_2$, 279.1259; found, 279.1271. The enantiomer excess was determined to be 82% after methylation (MeI, K_2CO_3/acetone) by HPLC analysis on a chiral column.

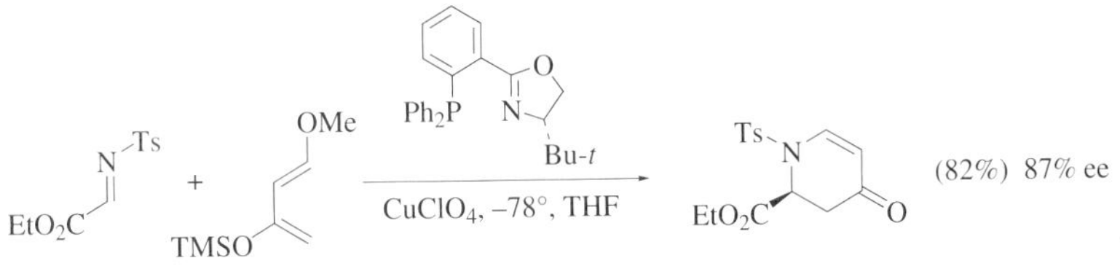
Ts
N
EtO2C
+
OMe
TMSO
Ph2P
O
N
Bu-t
CuClO4, –78°, THF
Ts
N
EtO2C
O
(82%) 87% ee

***N*-Tosyl-4-oxo-1,2,3,4-tetrahydropyridine-2-carboxylic Acid Ethyl Ester (Catalytic Asymmetric Diels-Alder Reaction of an *N*-Sulfonylimine).**[252] $CuClO_4 \cdot$ 4MeCN (13 mg, 0.04 mmol) and (*S*)-4-*tert*-butyl-2-[2-(diphenylphosphanyl)-phenyl]-4,5-dihydrooxazole (30 mg, 0.044 mmol) were added under N_2 to a flame-dried Schlenk tube. The mixture was dried for 1 hour under vacuum and freshly distilled anhydrous THF (1.5 mL) was added with a syringe under N_2, and the light yellow solution was stirred for 0.5 hour. Ethyl 2-(4-methylphenylsulfonamido) acetate (104 mg, 0.4 mmol) was added at room temperature and the mixture was stirred for 3–5 minutes. The solution was cooled to −78° and Danishefsky's diene (1.1–2.0 equiv) was added. The reaction mixture was stirred at that temperature for 20 hours. The reaction was quenched by addition of TFA (0.1 mL) in CH_2Cl_2 (10 mL) at −78° and the mixture was stirred at room temperature for 20 minutes. Evaporation of the solvent gave the crude product, which was purified by flash chromatography (30% EtOAc in pentane) as a light yellow oil (12 mg, 82%). The enantiomer excess was found to be 87% by HPLC using a Chiralpak AD column (2-PrOH-hexane, 15:85, 0.5 mL/min). Crystallization of the product from 2-PrOH-hexane gave colorless crystals with an enantiomer excess of >98.5%: mp 70–71°; $[\alpha]_D^{rt}$ −96.0° (*c* 0.50, $CHCl_3$); ^1H NMR δ 7.75 (d, *J* = 8.2 Hz, 2H), 7.72 (d, *J* = 9.4 Hz, 1H), 7.37 (d, *J* = 8.2 Hz, 2H), 5.37 (d, *J* = 9.4 Hz, 1H), 4.99–4.95 (m, 1H), 4.11–3.95 (m, 2H), 2.83–2.71 (m, 2H), 2.45 (s, 3H), 1.15 (t, *J* = 7.1 Hz, 3H); ^{13}C NMR δ 189.1, 167.9, 145.4, 142.5, 134.9, 130.2, 127.4, 107.7, 62.4, 56.2, 38.1, 21.7, 13.9; HRMS (*m*/*z*): $[M + Na]^+$ calcd for $C_{15}H_{17}NO_5S$, 346.0725; found, 346.0741.

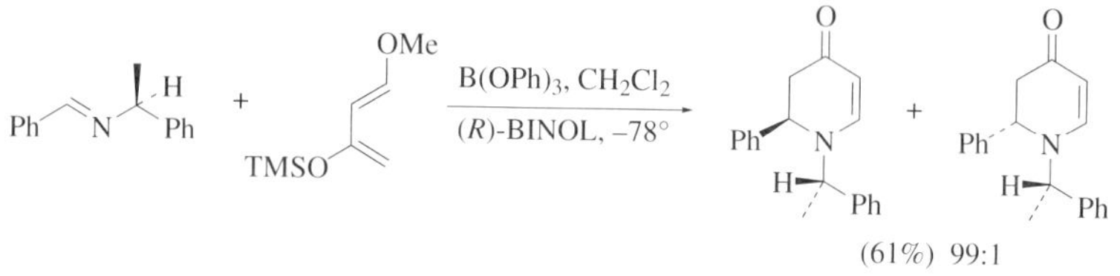
OMe
H
Ph
N
Ph
TMSO
B(OPh)3, CH2Cl2
(R)-BINOL, –78°
O
Ph
N
H
Ph
(61%) 99:1

(2*R*)-2,3-Dihydro-*N*-(*S*)-α-methylbenzyl-2-phenyl-4-pyridone (Catalytic Double Asymmetric Imino Diels-Alder Reaction).[243] To a suspension of powdered 4 Å molecular sieves (1.0 g) in CH_2Cl_2 (10 mL) were added *R*-BINOL (0.35 mmol) and $B(OPh)_3$ (101 mg, 0.35 mmol) at room temperature under argon. After being stirred for 1 hour, the mixture was cooled to 0° and then a solution of (*S*)-*N*-benzylidene-1-phenylethanamine (73 mg, 0.35 mmol) in CH_2Cl_2 (1 mL) was added. After being stirred for 10 minutes at the same temperature, the mixture was cooled to −78°, and a solution of Danishefsky's diene (0.084 mL, 0.42 mmol) in CH_2Cl_2 (1 mL) was added dropwise, followed by stirring at the same temperature for several hours. The solution was washed with water and saturated aqueous $NaHCO_3$, and then dried over $MgSO_4$. Evaporation of the solvent and purification of the residue by column chromatograpy on silica gel gave a 99 : 1 mixture of dihydropyridones in a combined yield of 61%. Major isomer: mp 76°; $[\alpha]_D^{24}$ −181.7° (*c* 1.7, $CHCl_3$); IR (neat) 1650 cm^{-1}; ^1H NMR ($CDCl_3$) δ 1.46 (d, *J* = 7.0 Hz, 3H), 2.63–2.88 (m, 2H), 4.43 (q, *J* = 7.0 Hz, 1H), 4.70 (dd, *J* = 6.6, 8.8 Hz, 1H), 5.04 (d, *J* = 6.0 Hz, 1H), 7.06 (d, *J* = 6.0 Hz, 1H), 7.23–7.55 (m, 10H); MS-FAB (*m*/*z*): 278 (M^+ + 1). Anal. Calcd for $C_{19}H_{19}NO$: C, 82.28; H, 6.90; N, 5.05. Found: C, 81.98; H, 7.11; N, 4.98.

TABULAR SURVEY

The tables contain a comprehensive list of imino Diels-Alder reactions until mid-2004. The examples are arranged by imine type following the organization of the Scope and Limitations section. The imines are organized in order of increasing carbon number of the reactive imine, rather than the precursor. Conditions for the generation of the imine and the structure of its precursor are generally not included.

Reactions of each imine are generally arranged by increasing carbon number of the diene. Again, only the reactive diene species is shown in the tables; footnotes indicate if the diene is formed in situ.

The yields for the reactions are given in parentheses, followed by a ratio of products if applicable. A dash (—) indicates that no conditions or yields were provided. Some details about conditions, stereochemistry of the products, and the identity of side products are often included in the footnotes.

The following abbreviations are used in the tables:

Ac	acetyl
atm	atmosphere(s)
B. A.	Brønsted acid
BHT	2,6-di-*tert*-butyl-4-methylphenol

Bn	benzyl
Boc	*tert*-butoxycarbonyl
Cbz	benzyloxycarbonyl
Cp	cyclopentadienyl
CSA	camphorsulfonic acid
d.e.	diastereomer excess
DBU	1,8-diazabicyclo[5.4.0]undec-7-ene
DIBALH	diisobutylaluminum hydride
DME	1,2-dimethoxyethane
DMF	dimethylformamide
DMSO	dimethyl sulfoxide
ee	enantiomer excess
eq	equivalents
fod	6,6,7,7,8,8,8-heptafluoro-2,2-dimethyl-3,5-octanedionate
HMPT	hexamethylphosphorous triamide
kbar	kilobar
L. A.	Lewis acid
MS	molecular sieves
MOM	methoxymethyl
P	polymeric support
PBS	phosphate-buffered saline
Piv	trimethylacetyl (pivaloyl)
sc	supercritical
SDS	sodium dodecyl sulfate
TBAF	tetrabutylammonium fluoride
TBS	*tert*-butyldimethylsilyl
Temp	temperature
TES	triethylsilyl
Tf	trifluoromethanesulfonyl
TFA	trifluoroacetic acid
THF	tetrahydrofuran
TIPS	triisopropylsilyl
TMEDA	*N,N,N′,N′*-tetramethyl-1,2-ethylenediamine
TMS	trimethylsilyl
Tr	triphenylmethyl
Ts	*p*-toluenesulfonyl
))))	sonication

CHART 1. ASYMMETRIC CATALYSTS USED IN TABLE 12

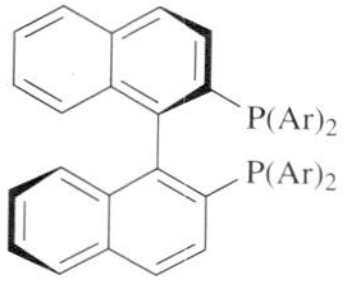

	Ar
1a	Ph
1b	4-MeC_6H_4

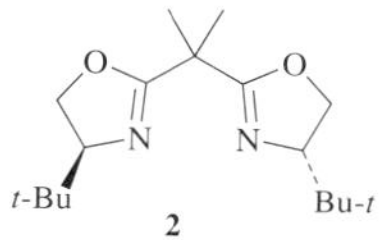

2

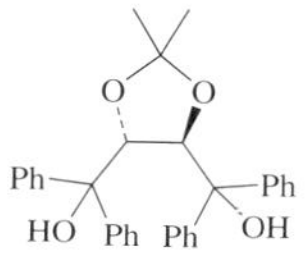

3

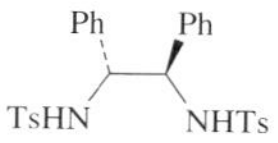

4

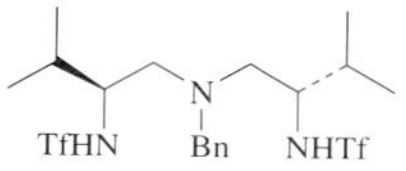

5

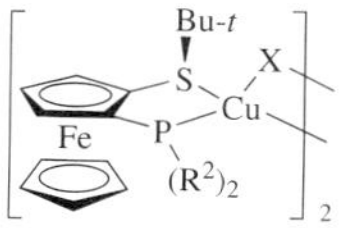

	R^2	X
6a	Ph	Cl
6b	4-FC_6H_4	Cl
6c	2-furyl	Cl
6d	c-C_6H_{11}	Cl
6e	2-MeC_6H_4	Cl
6f	1-naphthyl	Cl
6g	Ph	Br
6h	1-naphthyl	Br

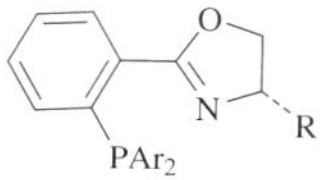

	R	Ar
7a	*i*-Pr	Ph
7b	*t*-Bu	Ph
7c	*t*-Bu	4-MeC_6H_4

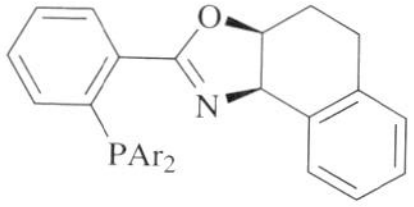

	Ar
8a	Ph
8b	4-MeC_6H_4
8c	2,4-$Me_2C_6H_3$

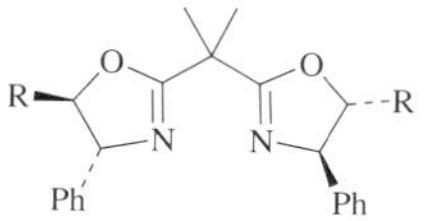

	R
9a	H
9b	Ph

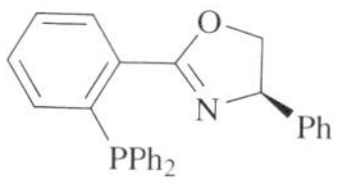

10

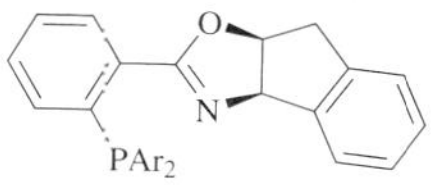

	Ar
11a	Ph
11b	4-MeC_6H_4
11c	2,4-$Me_2C_6H_3$

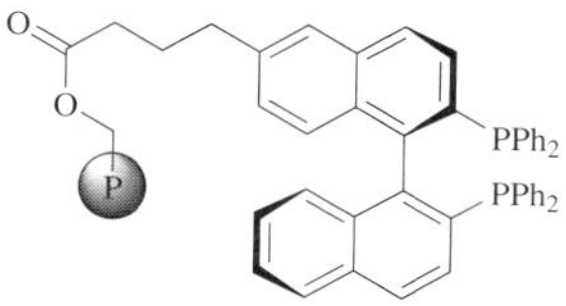

12

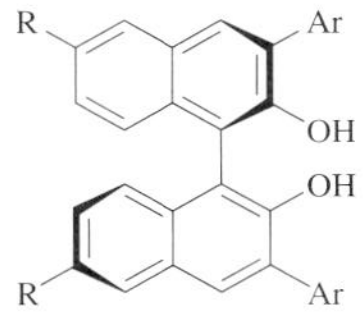

	R	Ar
13a	Br	H
13b	Br	Ph
13c	H	Ph
13d	H	4-FC_6H_4
13e	Br	4-FC_6H_4
13f	H	3-$CF_3C_6H_4$
13g	Br	3-$CF_3C_6H_4$
13h	Br	3-$O_2NC_6H_4$

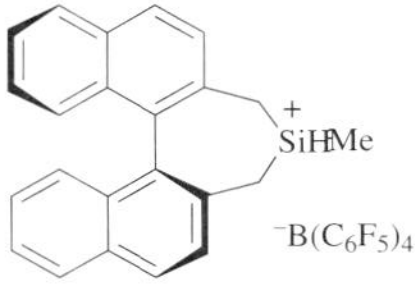

14

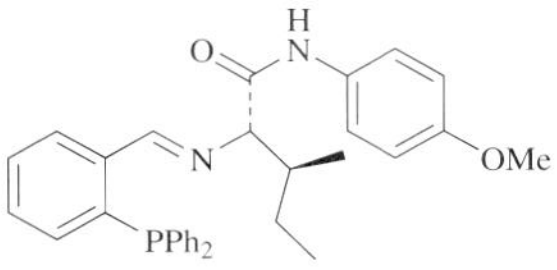

15

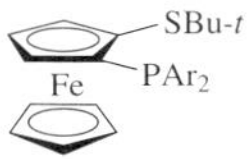

	Ar
16a	Ph
16b	1-naphthyl

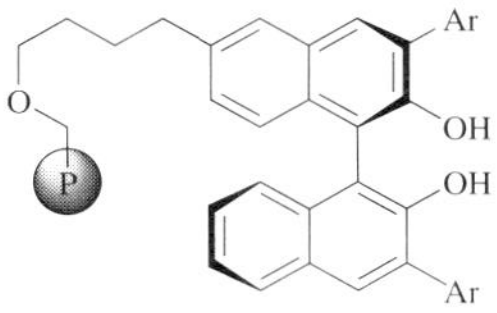

	Ar
17a	Ph
17b	3-MeC_6H_4
17c	2-MeC_6H_4
17d	4-*t*-BuC_6H_4
17e	3,5-$Me_2C_6H_3$
17f	4-PhC_6H_4
17g	2-naphthyl
17h	6-MeO-2-naphthyl
17i	4-FC_6H_4
17j	3,4-$F_2C_6H_3$
17k	3-$CF_3C_6H_4$
17l	3,5-$(CF_3)_2C_6H_3$
17m	3-$MeOC_6H_4$
17n	4-$MeOC_6H_4$
17o	3,4-$(MeO)_2C_6H_3$
17p	4-$EtOC_6H_4$
17q	2-thienyl

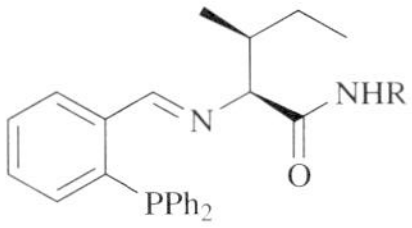

	R
18a	4-$MeOC_6H_4$
18b	4-$CF_3C_6H_4$
18c	2,6-$Me_2C_6H_3$
18d	Bu
18e	Bn
18f	OMe

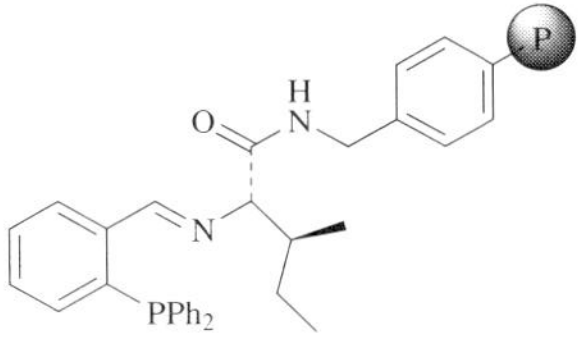

19

TABLE 1. CYCLOADDITIONS OF ACYCLIC *N*-ACYLIMINES AND *N*-CYANOIMINES

	Imine	Diene	Conditions	Product(s) and Yield(s) (%)	Refs.
C_4	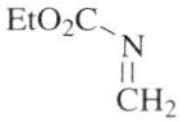		BF_3•OEt_2, *t*-butylcatechol, benzene, 60°, 10 h	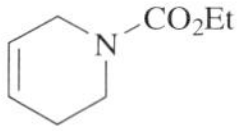 (5)	281, 47
		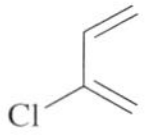	BF_3•OEt_2, benzene, 150°, 7 h	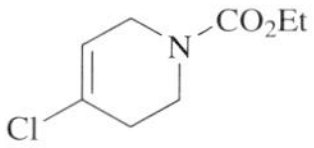 (70)	281, 47
		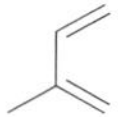	BF_3•OEt_2, benzene, 150°, 6-9 h	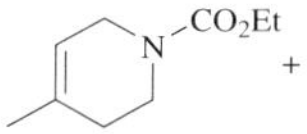 + 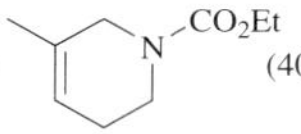 (40-80) >90:10	281, 282, 47
		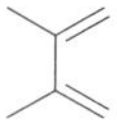	BF_3•OEt_2, *t*-butylcatechol, benzene, 150°, 4 h	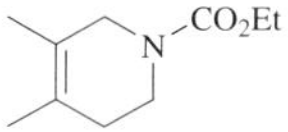 (70)	281, 282, 47
			BF_3•OEt_2	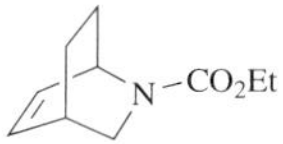	
			Solvent / Temp / Time		
			benzene / 80° / 1 h	(27)	48, 283
			benzene / 80° / 2 h	(39)	49
			benzene / 80° / 1 h	(34)	50
			benzene / 80° / —	(43)	284
			toluene, dioxane / — / —	(25)	280
				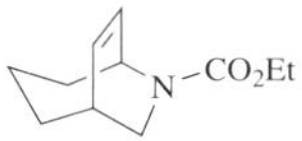	
			BF_3	(—)[a]	285
			BF_3•OEt_2, benzene, 80°, 18 h	(13)[a]	61

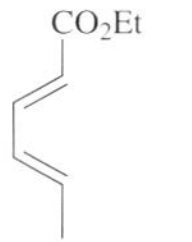	BF_3•OEt_2, benzene, 80°, 19 h	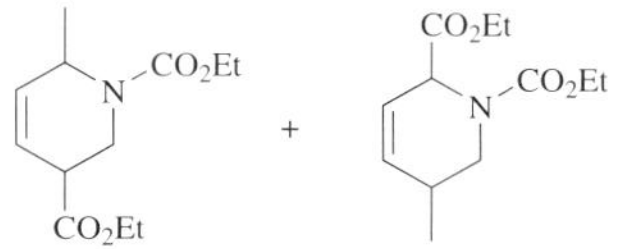 (—) >90:10	281, 47
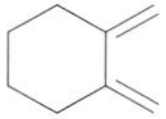	BF_3•OEt_2, benzene, 80°, 14 h	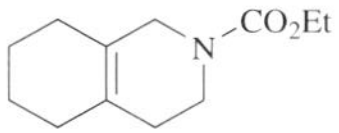 (>60)	281, 282, 47
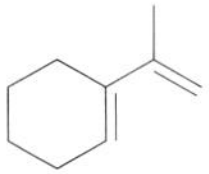	BF_3•OEt_2, benzene, 80°, 14 h	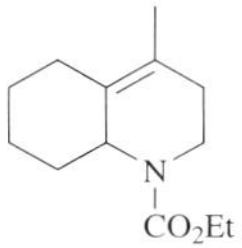 (—)	281, 47
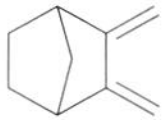	BF_3•OEt_2, benzene, 80°, 14 h	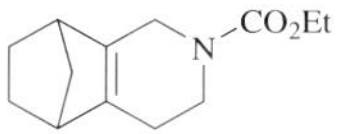 (—)	281
	BF_3•OEt_2, benzene, 80°, 14 h	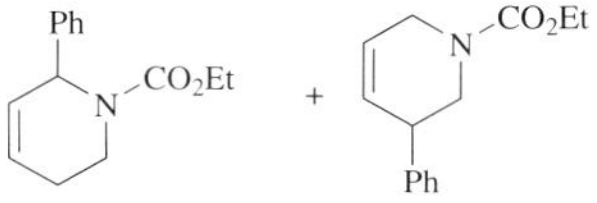 (45) >90:10	281, 282, 47
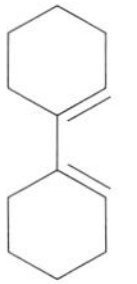	BF_3•OEt_2, benzene, 70-80°, 14 h	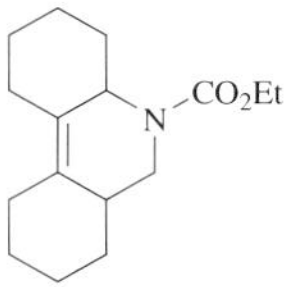 (>80)	281, 282, 47
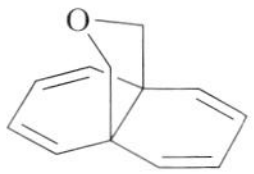	BF_3•OEt_2, benzene, 80°, 20 h	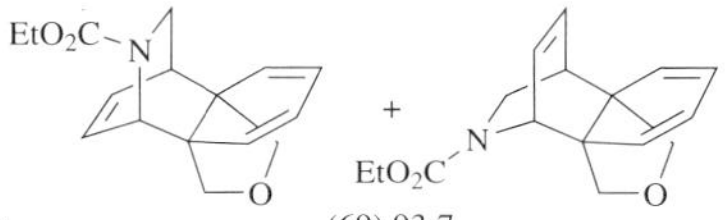 (69) 93:7	63

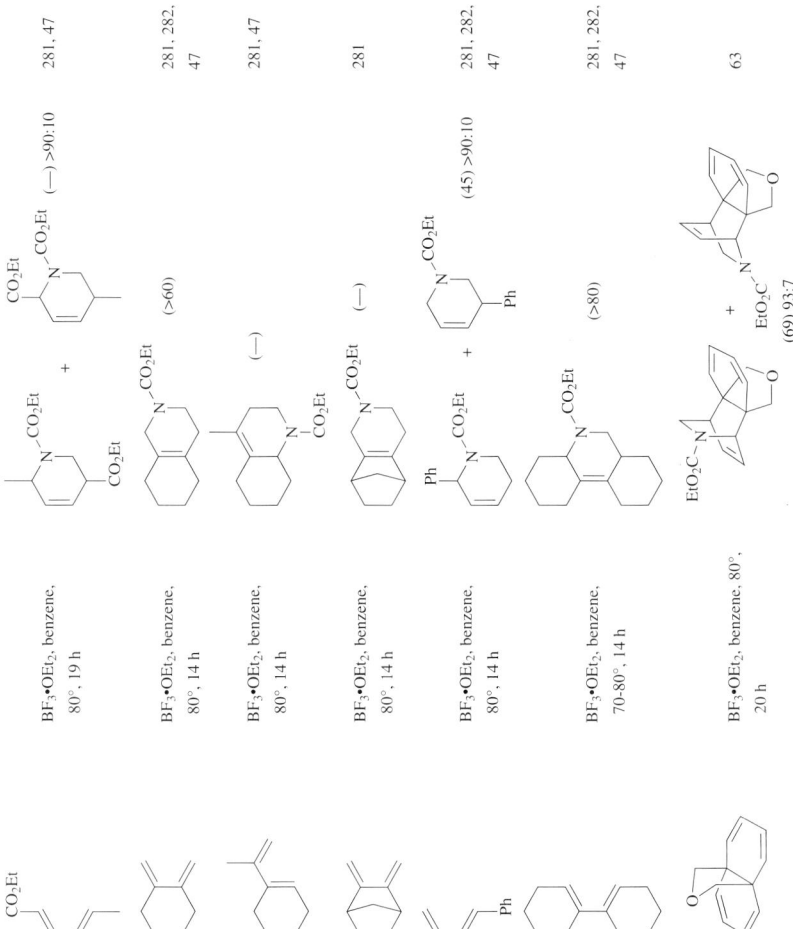

TABLE 1. CYCLOADDITIONS OF ACYCLIC *N*-ACYLIMINES AND *N*-CYANOIMINES (*Continued*)

Imine	Diene	Conditions	Product(s) and Yield(s) (%)	Refs.
C_4				
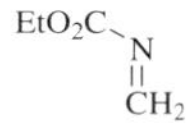EtO2C N CH2	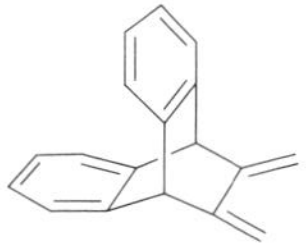	BF_3•OEt_2, benzene, 80°, 10 h	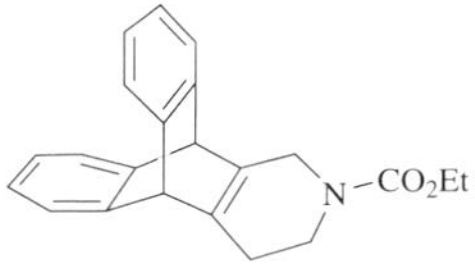N–CO2Et (65)	286
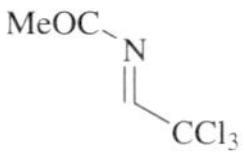MeOC N CCl3			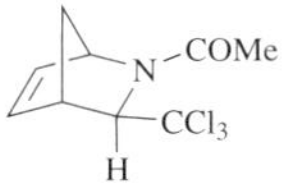N COMe CCl3 H	
		Benzene, 80°, 4 h	(57)	20
		Hydroquinone, Et_3N, $AlCl_3$, benzene, 60-80°, 4 h	(40.9)	21
		Hydroquinone, Et_3N, $AlCl_3$, benzene, 80°, 4 h	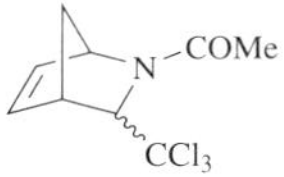N COMe CCl3 (30)[b]	287
		98°, 5 h	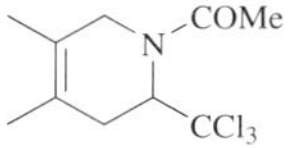N COMe CCl3 (20)[c]	288
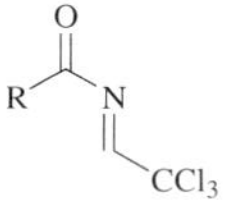O R N CCl3 R = CH_2Cl or CCl_3		Hydroquinone, Et_3N, $AlCl_3$, benzene, 80°, 4 h	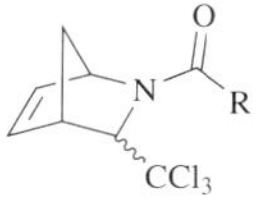O N R CCl3 (55-95)[d]	287
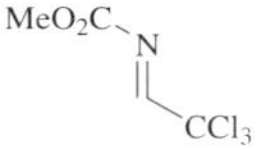MeO2C N CCl3		Benzene, 80°, 5 h	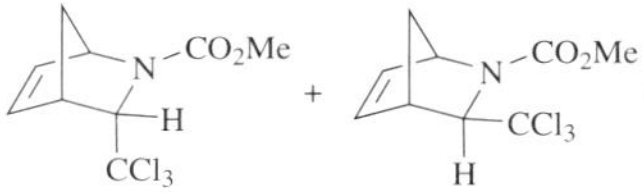N CO2Me H CCl3 + N CO2Me CCl3 H (79) 56:44	20

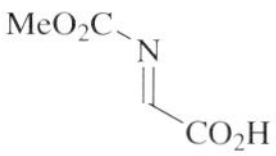	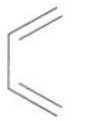	MeSO$_3$H, CH$_2$Cl$_2$, rt	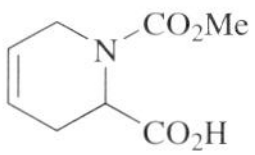	(7-9)[e]	289
		"Acid catalyzed"	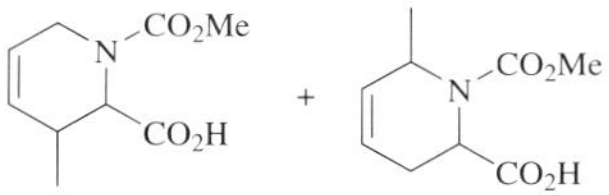	(<20)[e,f]	289
	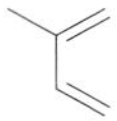	"Acid catalyzed"	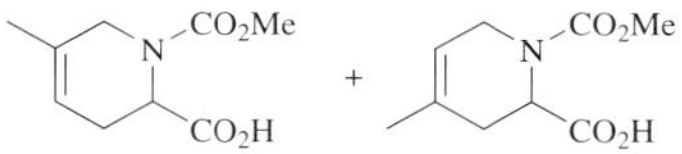	(<20)[e,g]	289
		"Acid catalyzed"	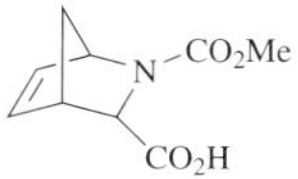	(<20)[d,e]	289
	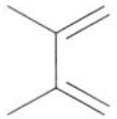	"Acidic conditions"	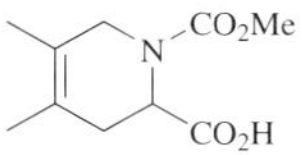	(<20)[e]	289
	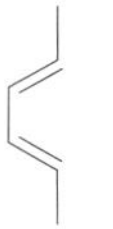	"Acidic conditions"	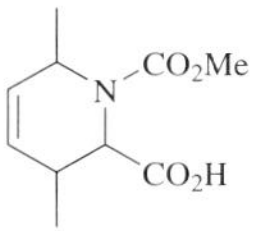	(<20)[d,e]	289
		MeSO$_3$H, THF	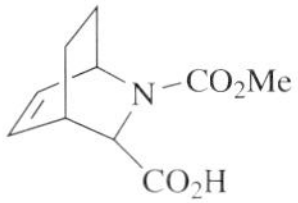	(6)[d,e]	289
	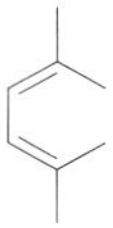	"Acidic conditions"	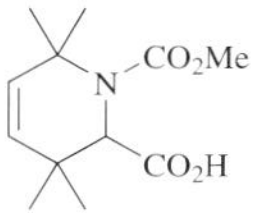	(<20)[e]	289

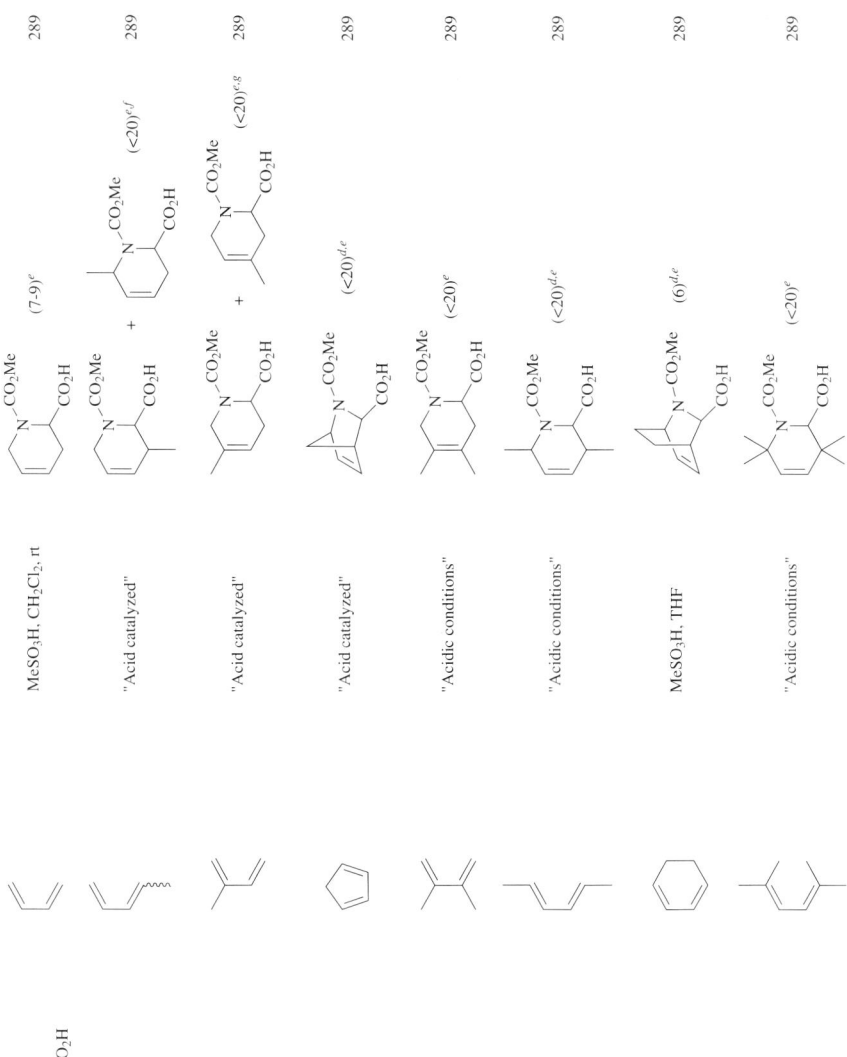

TABLE 1. CYCLOADDITIONS OF ACYCLIC *N*-ACYLIMINES AND *N*-CYANOIMINES (*Continued*)

Imine	Diene	Conditions	Product(s) and Yield(s) (%)	Refs.
C_4 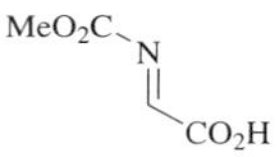		"Acidic conditions"	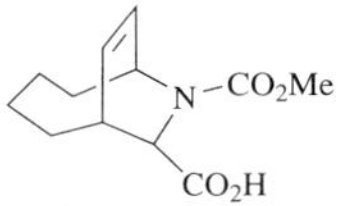 (<20)[d,e]	289
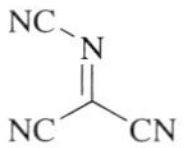		Benzene, rt, 1 h	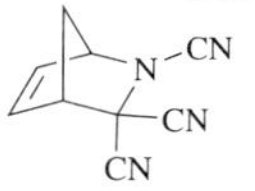 (~50)	65, 64
	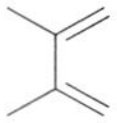		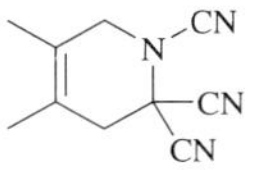	
		—	(48)	64
		Benzene, rt, 75 min	(~60)	65
C_5 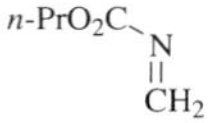	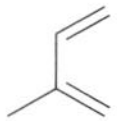	BF_3•OEt_2, 70-90°, 10 h	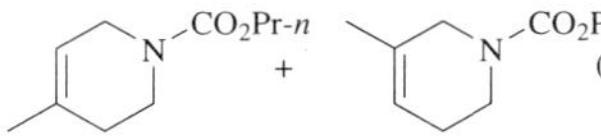 (40-80) >90:10	47
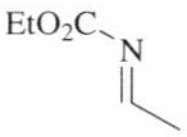		BF_3•OEt_2, 3 h	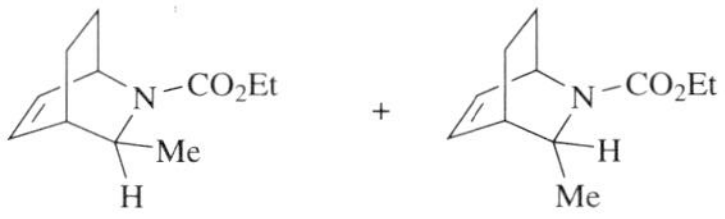	
		Solvent / Temp		
		CH_2Cl_2 40°	(43) 80:20	22, 23
		$CHCl_3$ 61°	(36-41) 82±6:18±6	23
		CH_2Cl_2 40°	(—) 84±9:16±9	23
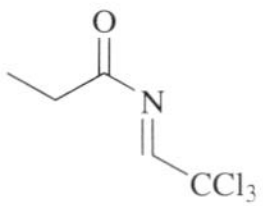		Hydroquinone, Et_3N, $AlCl_3$, benzene, 80°, 4 h	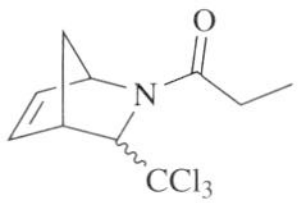 (55-95)[d]	287

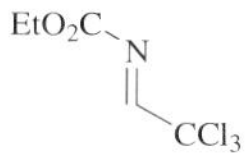		Benzene, 115°, 100 h	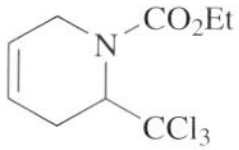	(55)	24
	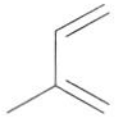	Benzene, 100°, 80 h	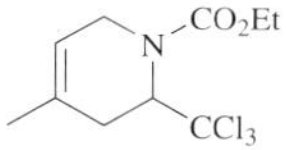	(75)	24
		Benzene	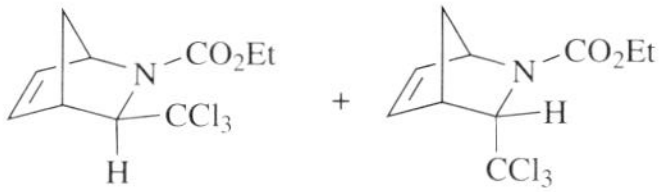 I + II		

Temp	Time	I	II	I + II	I :II	
80°	3 h	(59)	(30)	(—)	(—)	24
80°	3 h	(—)	(—)	(—)	50±3:50±3	31
30°	2 d	(—)	(—)	(87)	50:50	31
145°	3 d	(—)	(—)	(—)	50±3:50±3	31

	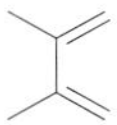	Benzene, 100°, 48 h	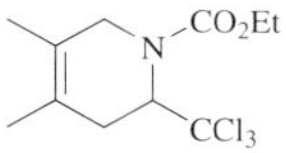	(78)	24

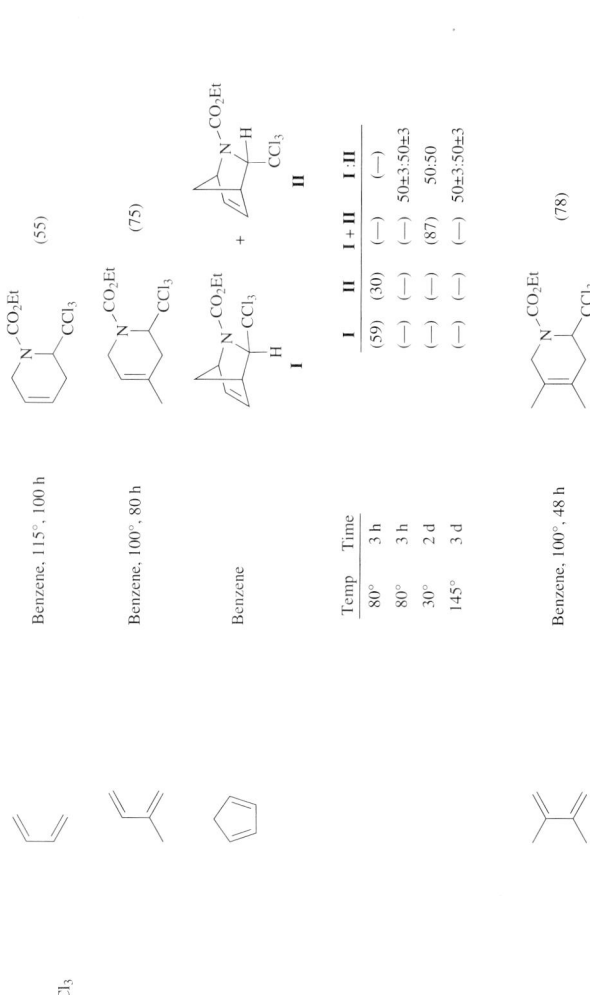

TABLE 1. CYCLOADDITIONS OF ACYCLIC *N*-ACYLIMINES AND *N*-CYANOIMINES (*Continued*)

Imine	Diene	Conditions	Product(s) and Yield(s) (%)	Refs.
C_5				
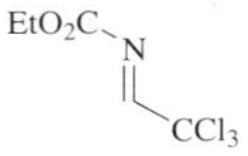			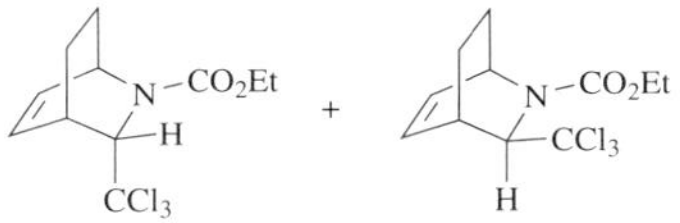	
		BF_3•OEt_2, benzene, 80°[c], 24 h	(—) 80:20	18, 61
		BF_3•OEt_2, benzene, 80°[c]	(—) 75±5:25±5	31
		BF_3, benzene, 30°, 72 h	(—) 75±3:25±3	31, 29
		BF_3•OEt_2, CCl_4, 30°, 72 h	(71) 75±3:25±3	31
		BF_3, $CDCl_3$, 30°, 72 h	(—) 75±3:25±3	31
		Benzene, 80°, 21 d	(trace)	31
		Benzene, 150°, 24 h	(—) 62±3:38±3	31, 29
		Benzene, 140-145°, 64 h	(28) 62±3:38±3	31
		"Without acid catalysis'	(—) 62:38	33
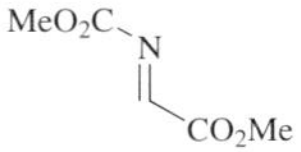		BF_3•OEt_2, $CHCl_3$, 2-6 h	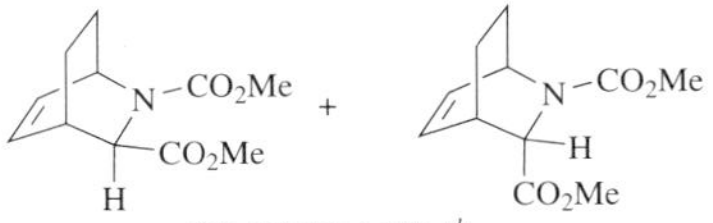 (25-35) 73±4:27±4[h]	29
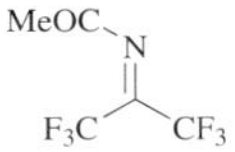		Hydroquinone, 20°, 3 d	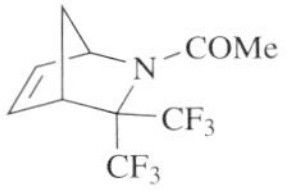 (93.3)	58

	Imine	Diene	Conditions	Product(s) and Yield(s) (%)	Refs.
	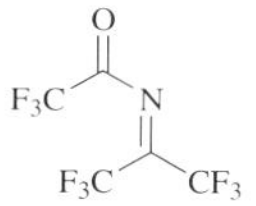		Hydroquinone, ether, −78 to 20°, 3 d	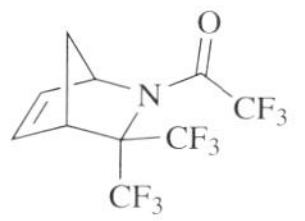 (91.1)	58
	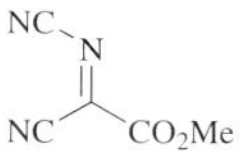		Benzene, rt, 1 h	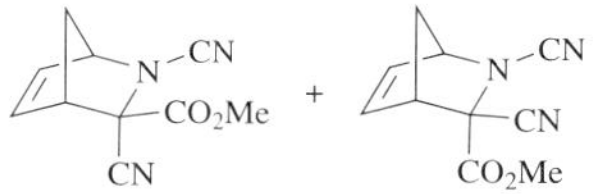 (~50)	65
		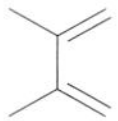		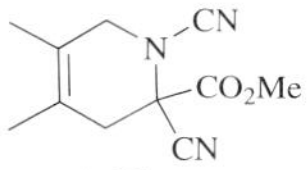	65
			Benzene, rt, 75 min	(~60)	
			MeOH, 60°	(—)	
C_6	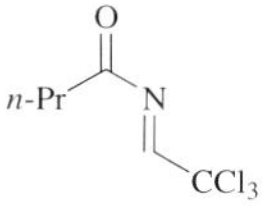		Hydroquinone, Et_3N, $AlCl_3$, benzene, 80°, 4 h	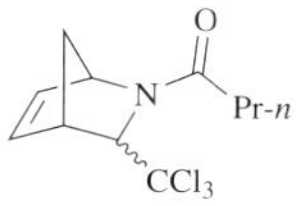 (55-95)[d]	287
	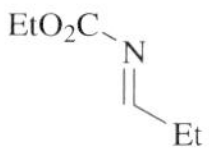		BF_3•OEt_2, $CHCl_3$, 61°, 2 h	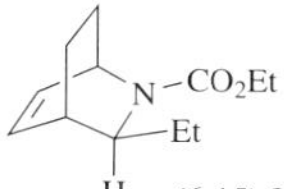 + 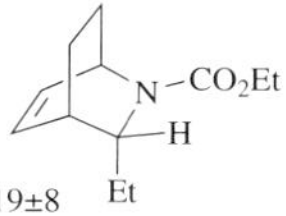 (6-15) 81±8:19±8	23
	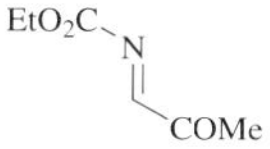		BF_3•OEt_2	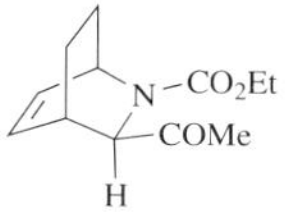 + 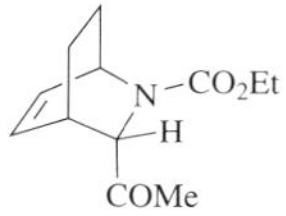	
			Solvent / Temp / Time		
			benzene / 80° / 1 h	(35)[i]	290
			benzene / 80° / 1 h	(—) 67:33	27
			benzene / 80° / 2-24 h	(40-50) 67:33	18, 61, 31, 23
			$CHCl_3$ / — / 2-6 h	(25-35) 67±2:33±2[h]	29

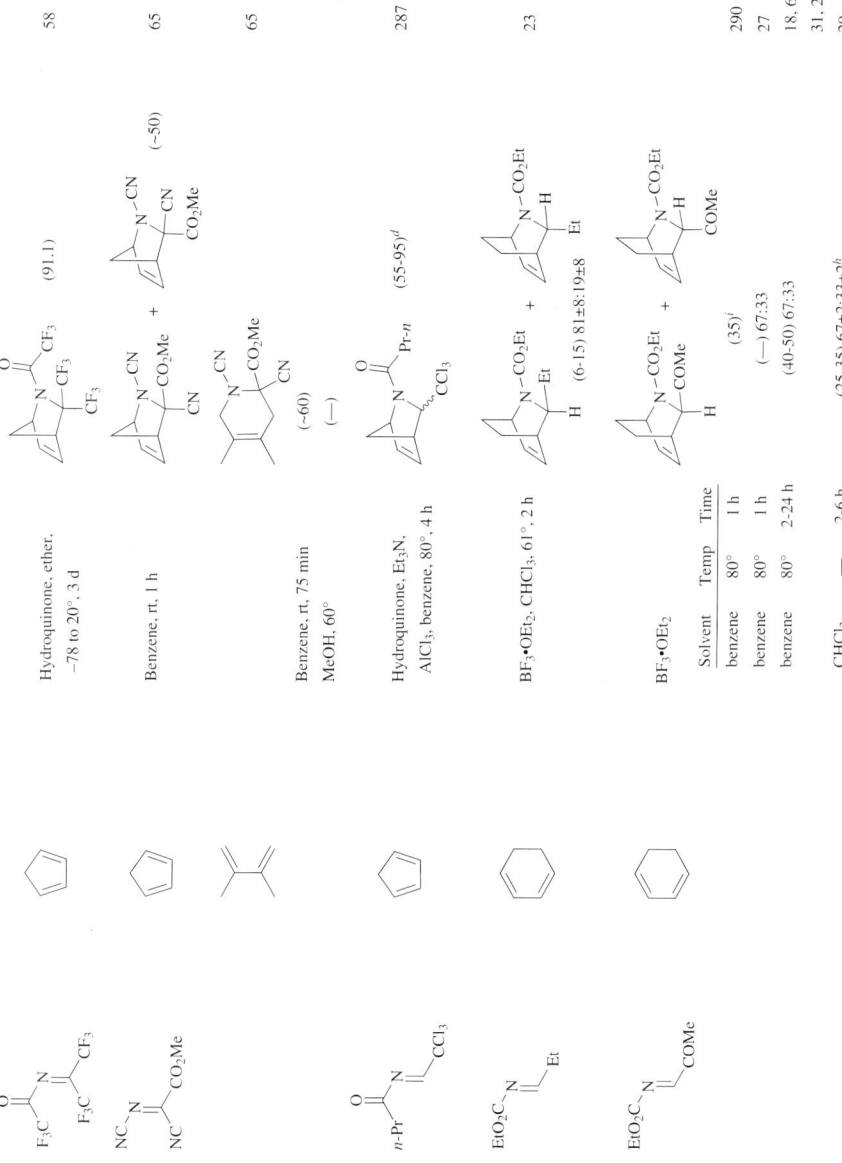

TABLE 1. CYCLOADDITIONS OF ACYCLIC *N*-ACYLIMINES AND *N*-CYANOIMINES (*Continued*)

Imine	Diene	Conditions	Product(s) and Yield(s) (%)	Refs.
C_6 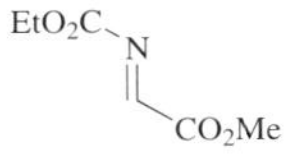			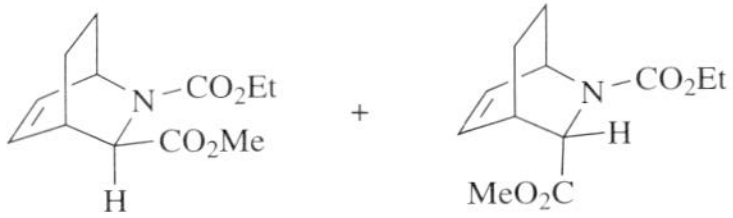	
		BF_3•OEt_2, $CHCl_3$, 2-6 h	(25-35) 65±4:35±4[h]	29
		—	(—) 80:20	23
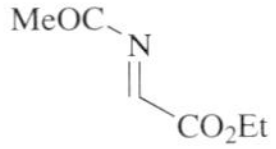			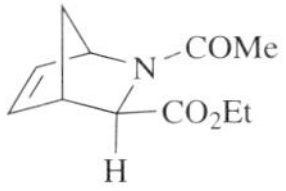	
		DME, 85°, 15 h	(81)	41
		Toluene, 130°, 20 h	(29)	42
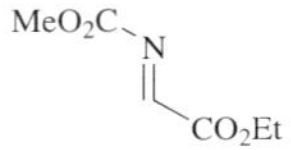		BF_3•OEt_2, $CHCl_3$, 2-6 h	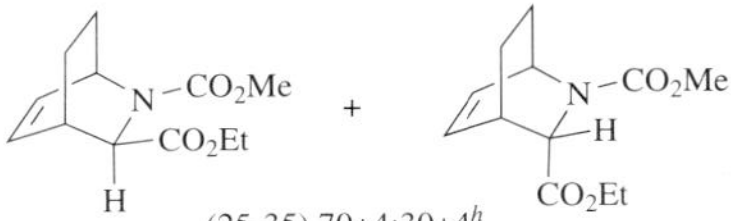 (25-35) 70±4:30±4[h]	29
C_7 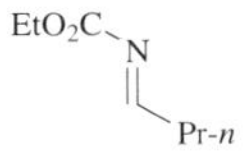		BF_3•OEt_2, $CHCl_3$, 61°, 2 h	 (32) 80±3:20±3	23
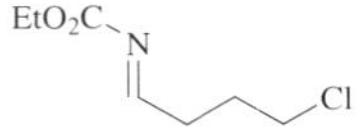	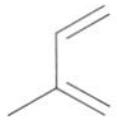	BF_3•OEt_2, benzene, 80°, 16 h	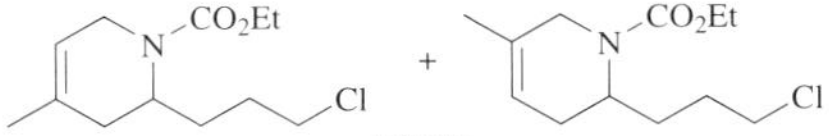 (40-75)	281

		BF_3•OEt_2 Solvent / Temp / Time benzene / 80° / — $CHCl_3$ / 61° / 1 h benzene / 80° / 1 h	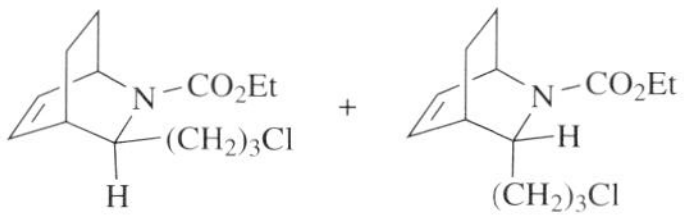 (22) 4:1[j] (20-35) 81±8:19±8 (21) 81±6:19±6	 25 23 23
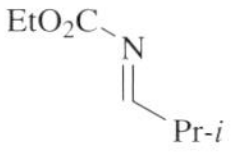	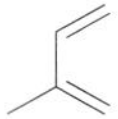	BF_3•OEt_2, benzene, 80°, 16 h	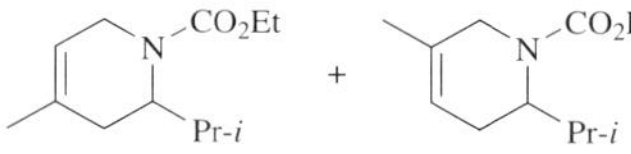 (40-80) >90:10	281, 47
		BF_3•OEt_2, $CHCl_3$, 61°, 1.2 h	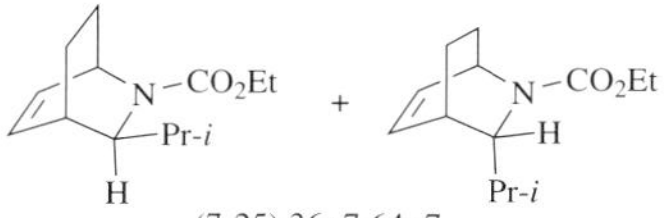 (7-25) 36±7:64±7	23
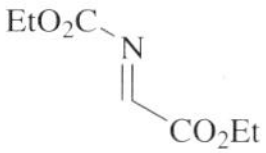		 BF_3•OEt_2, $CHCl_3$, 2-6 h —	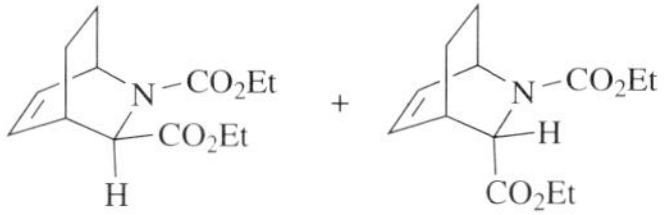 (25-35) 70±4:30±4[h] (—) 70:30	 29 23
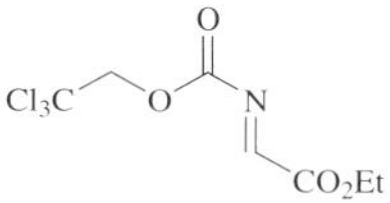		DME, 85°, 15 h	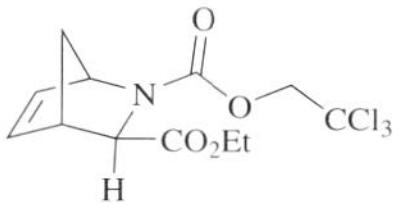 (61)	41

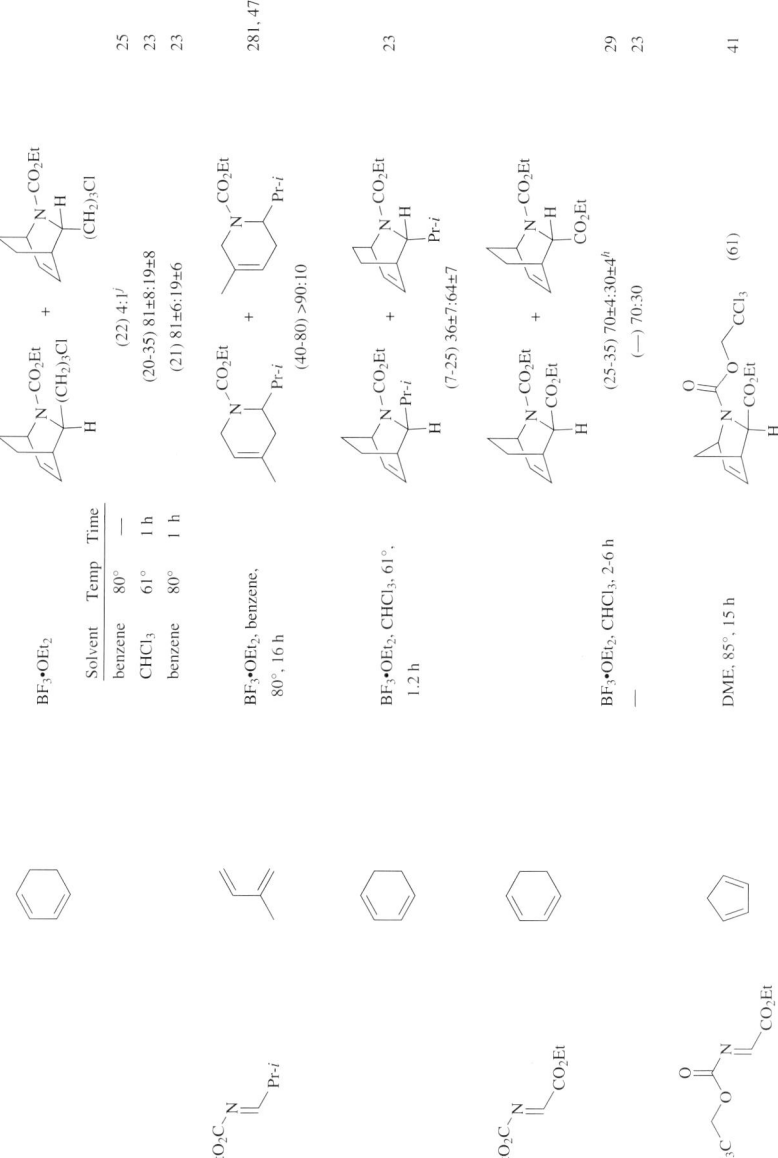

TABLE 1. CYCLOADDITIONS OF ACYCLIC *N*-ACYLIMINES AND *N*-CYANOIMINES (*Continued*)

Imine	Diene	Conditions	Product(s) and Yield(s) (%)	Refs.
C_7				
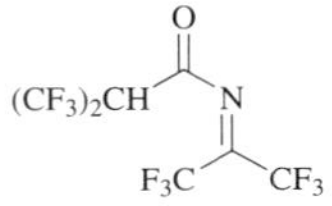	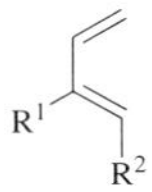	Hydroquinone, 20°, 3 d	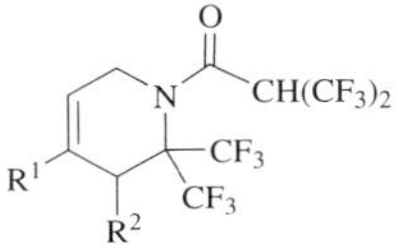 R^1 R^2: H H (63.5); Me H (30.4); H Me (68.6)	58
		Hydroquinone, ether, –78 to 20°, 3 d	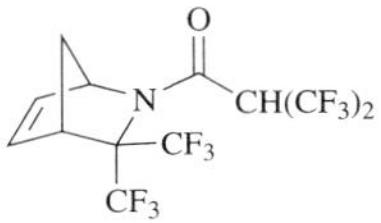 (90.6)	58
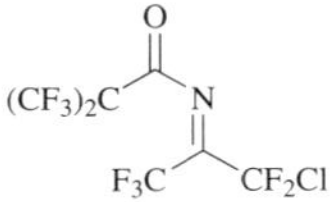		Hydroquinone, ether, –78 to 20°, 3 d	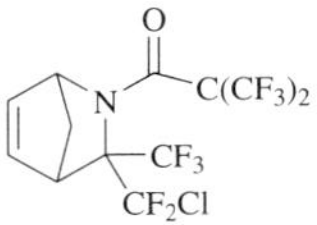 (93.4)[k]	58
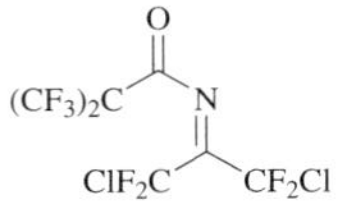		Hydroquinone, ether, –78 to 20°, 3 d	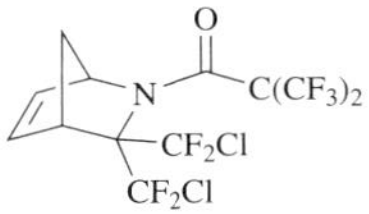 (91.2)	58
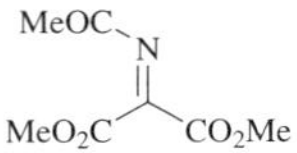	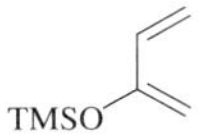	1. THF, 100°, 14 h 2. TFA, MeOH, rt, 3 h	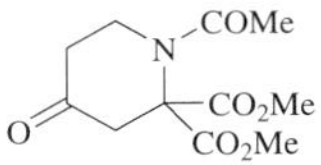 (16)[l]	59
	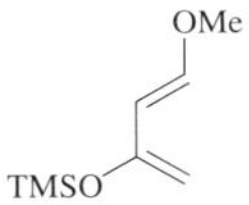	1. THF, 100°, 10 h 2. TFA, MeOH, rt, 3 h	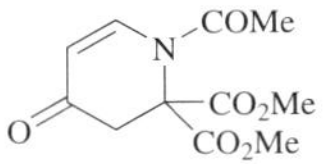 (67)[l]	59

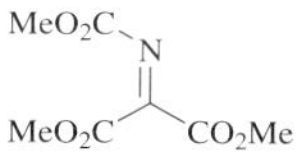	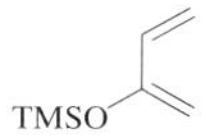	1. Heat 2. TFA, MeOH, rt, 3 h Temp / Time 120° / 8 h 70° / 22 h	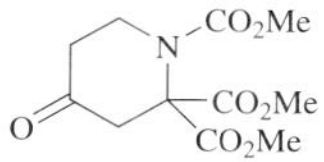 (49)[l] (15)[l]	59
	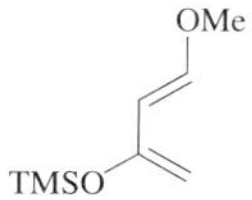	1. Heat 2. TFA, MeOH, rt, 3 h Temp / Time 100° / 5 h 70° / 15 h	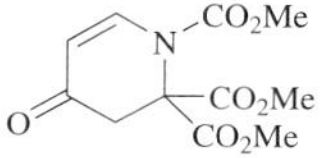 (86)[l] (75)[l]	59
	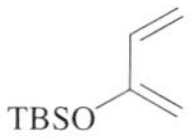	120°, 15 h	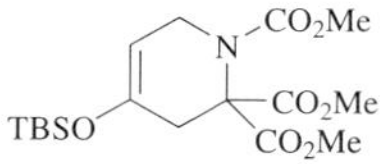 (74)[m]	59
	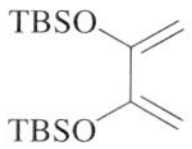	130°, 13 h	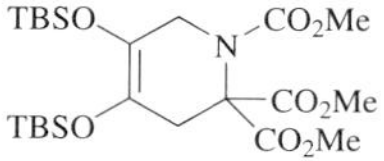 (41)[m]	59
C_8 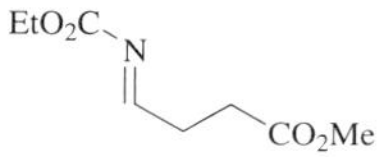	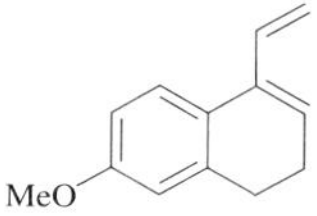	BF_3•OEt_2, benzene, 70°	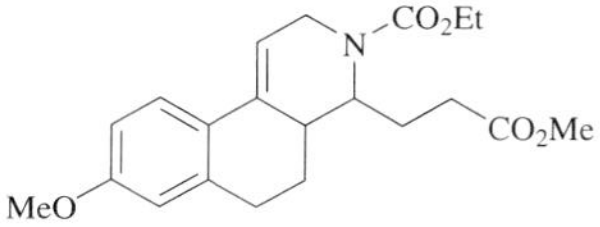 (30-40)[k]	52
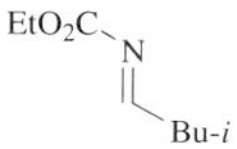		BF_3•OEt_2, $CHCl_3$, 61°, 2 h	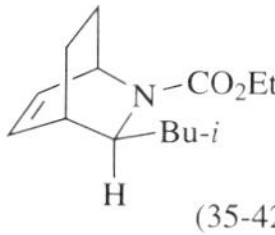 + 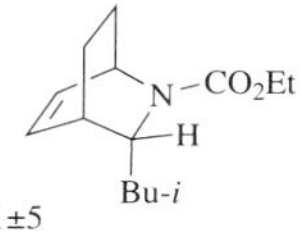 (35-42) 79±5:21±5	23

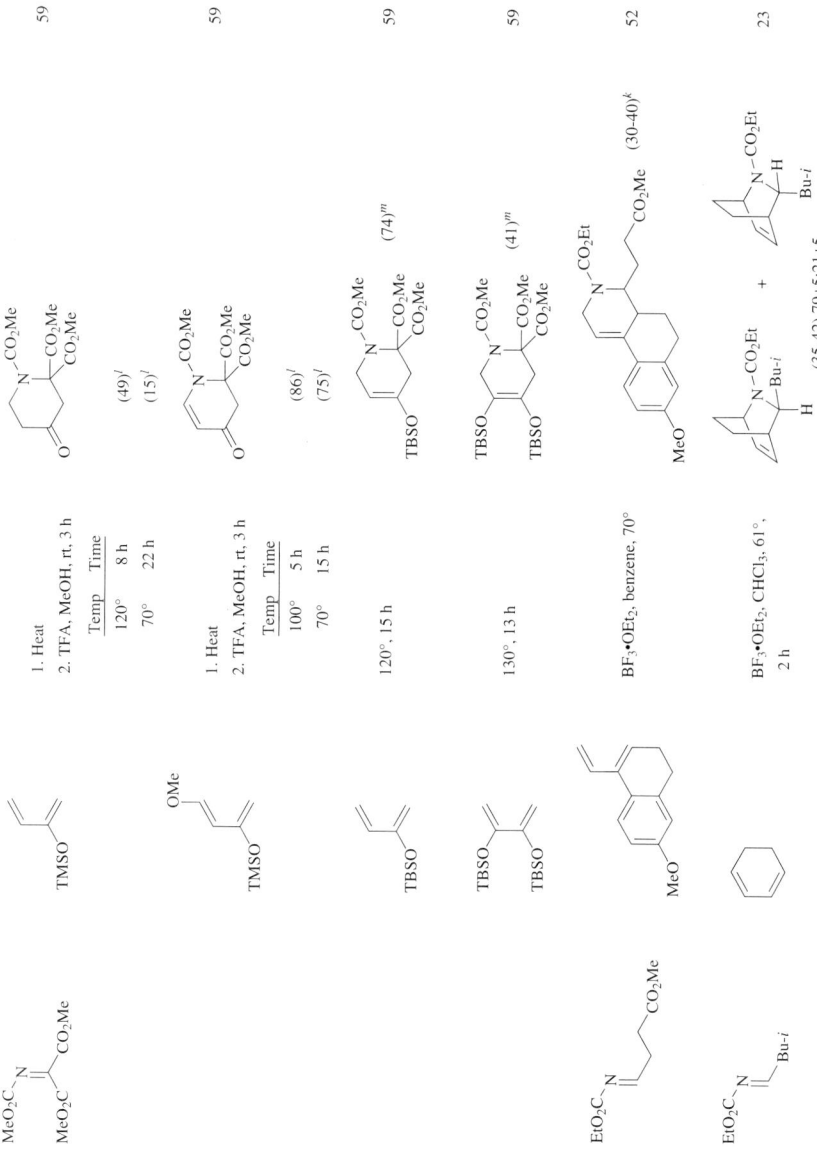

TABLE 1. CYCLOADDITIONS OF ACYCLIC *N*-ACYLIMINES AND *N*-CYANOIMINES (*Continued*)

	Imine	Diene	Conditions	Product(s) and Yield(s) (%)	Refs.
C_8	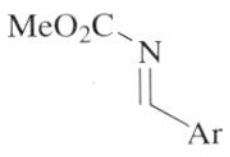 Ar = 3-pyridinyl, 2-Cl-5-pyridinyl		BF_3•OEt_2, $CHCl_3$, 61°, 5 h	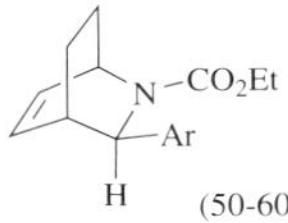 + 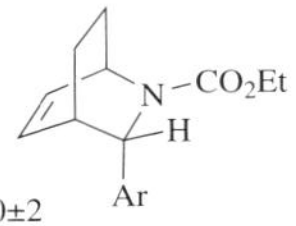 (50-60) 80±2:20±2	26
	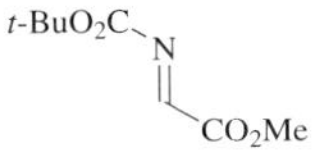	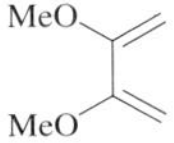	Benzene, 80°, 15 h	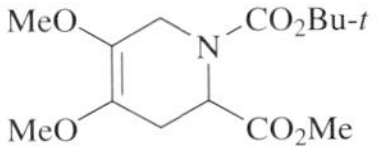 (16)	41
		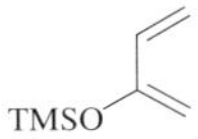	1. Benzene, 80°, 15 h 2. K_2CO_3, MeOH	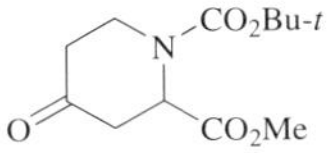 (30)	41
C_9	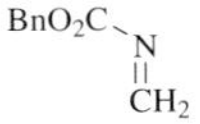	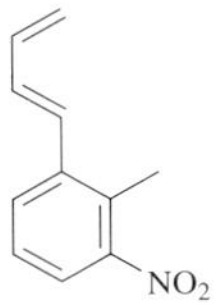	BF_3•OEt_2, benzene, 80°, 25 min	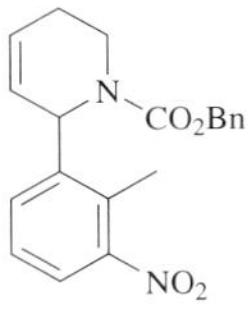 (52)	51
	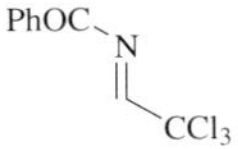		Hydroquinone, Et_3N, $AlCl_3$, benzene, 60-80°, 4 h	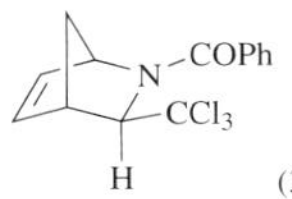 + 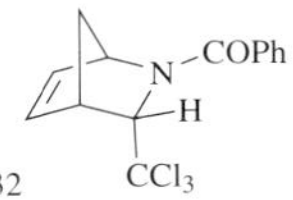 (32.5) 68:32	21
			Hydroquinone, Et_3N, $AlCl_3$, benzene, 80°, 4 h	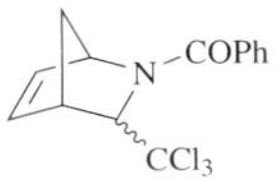 (32.5)[d]	287

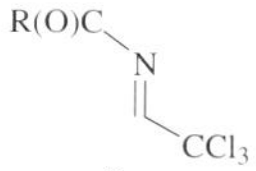		Hydroquinone, Et_3N, $AlCl_3$, benzene, 80°, 4 h	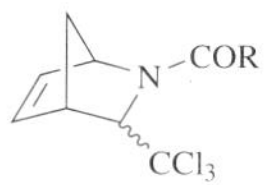	287
R				
4-$O_2NC_6H_4$			(48.1)[d]	
3,5-$(O_2N)_2C_6H_3$			(55-95)[d]	
2-ClC_6H_4			(55-95)[d]	
4-ClC_6H_4			(55-95)[d]	
2-BrC_6H_4			(55-95)[d]	
4-BrC_6H_4			(55-95)[d]	
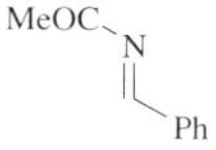	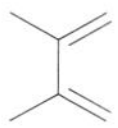	BF_3	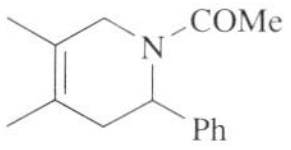 ("lower")	47
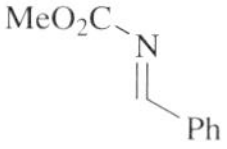		BF_3•OEt_2, benzene, 80°		
		Time		
		2-24 h	(23) 82:18	18
		1 h	(23) 80:20	27
	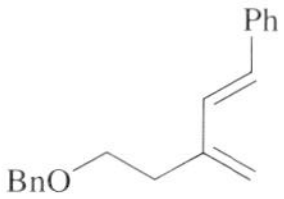	BF_3•OEt_2, toluene, −30° to rt, 45 min	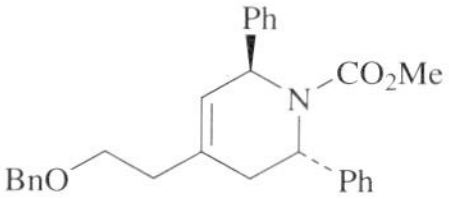 (95)	30
	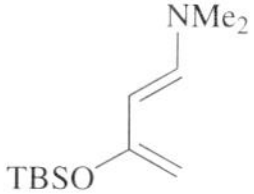	1. CH_2Cl_2, −78°, 3 h 2. HCl, H_2O, THF	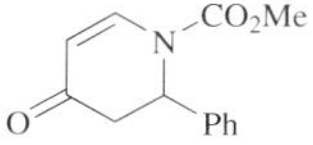 (77)	53
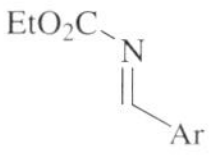 Ar = 3-pyridinyl			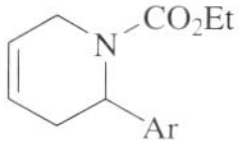	291, 54
		BF_3•OEt_2, DME, 60°	(8)	
		BF_3•2AcOH, AcOH, 80-90°, 15 h	(20)	

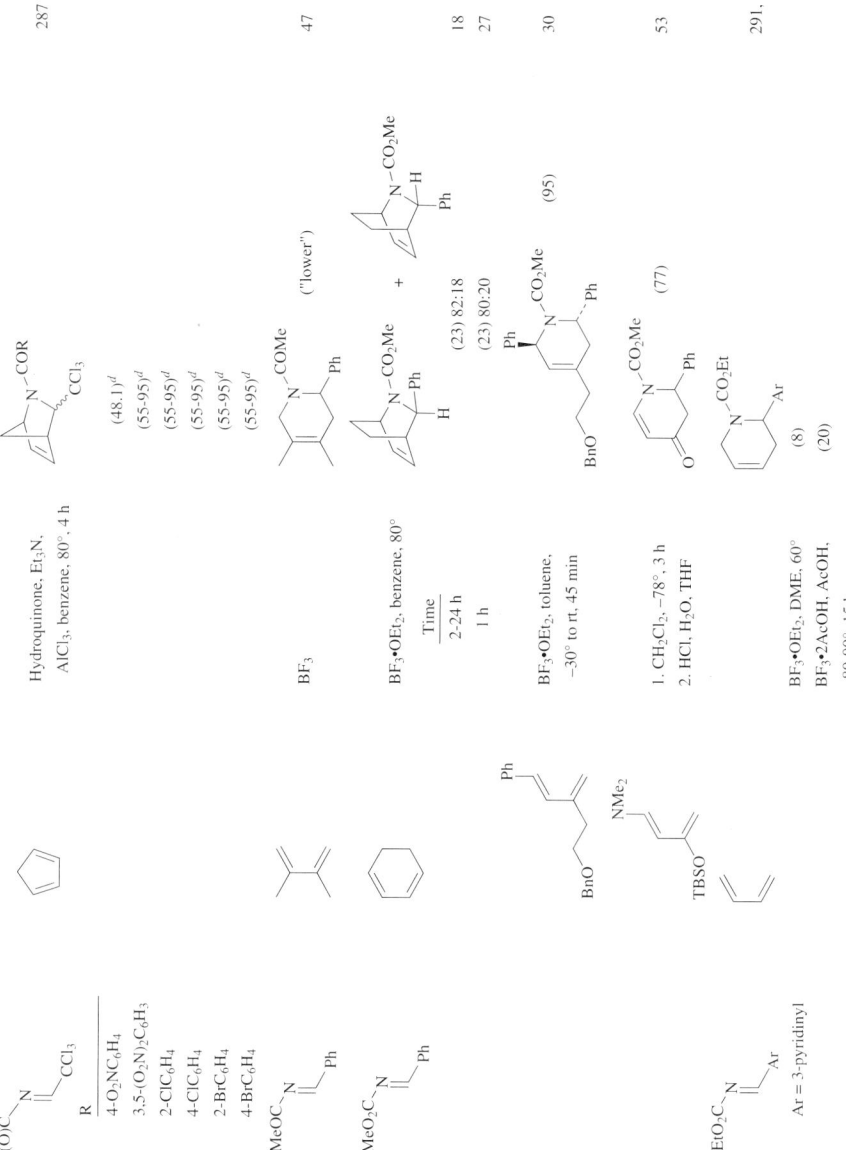

TABLE 1. CYCLOADDITIONS OF ACYCLIC *N*-ACYLIMINES AND *N*-CYANOIMINES (*Continued*)

	Imine	Diene	Conditions	Product(s) and Yield(s) (%)	Refs.
C_9	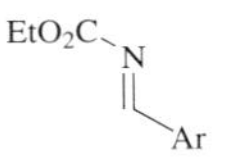 Ar = 3-pyridinyl	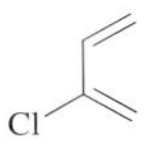	BF_3•2AcOH, AcOH, 60°, 13.5 h	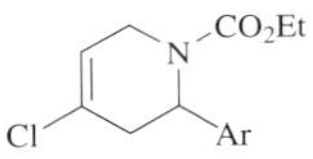 (28)	54
		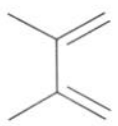	BF_3•OEt_2, DME, 60°	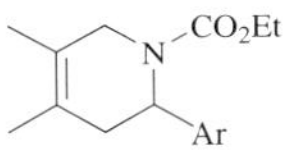	
			Time		
			—	(43)	291
			8 h	(55)[n]	54
	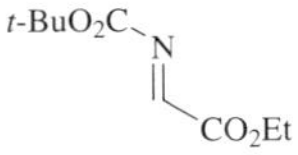			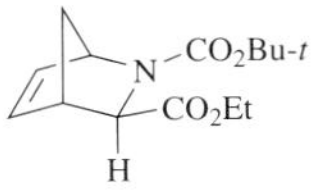	
			DME, 85°, 15 h	(55)	41
			Toluene, rt, overnight	(40-60)	292
		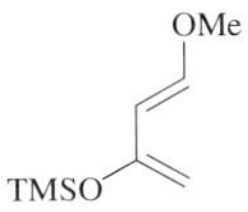	1. Benzene, 80°, 15 h 2. H_3O^+	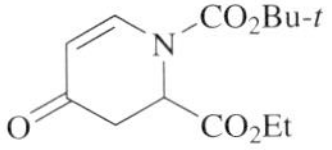 (84)	41
		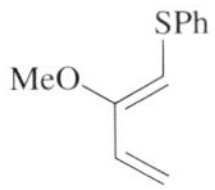	Benzene, 80°, 12 h	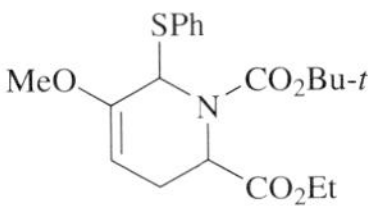 (56)[d]	41
	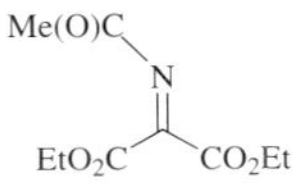			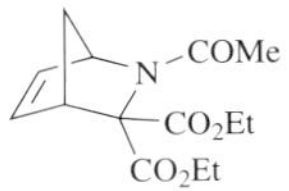	207
			Benzene, 80°, 5 h	(69)	
			Toluene, 20°, 10 h, 11 kbar	(90)	

Imine	Diene	Conditions	Product(s) and Yield(s) (%)	Refs.
	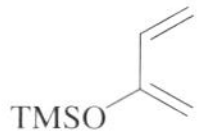	1. Benzene, 60°, 72 h 2. TFA, MeOH, rt, 3 h	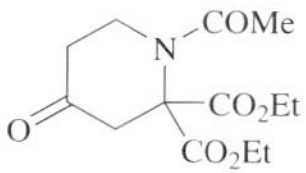 (8)[l]	59
	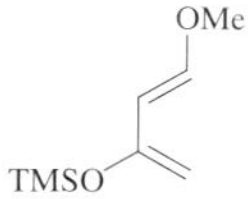		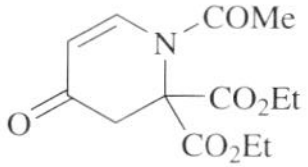 I + 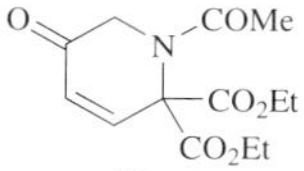 II	
		1. 140°, 2 h 2. H_3O^+	**I** (45) + **II** (15)	41
		1. Benzene, 120°, 2 h 2. TFA, MeOH, rt, 3 h	**I** (53)[l]	59
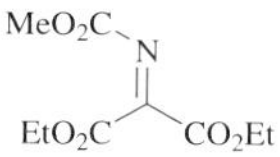	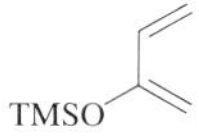	1. Heat 2. TFA, MeOH, rt, 3 h	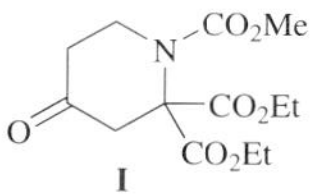 I + 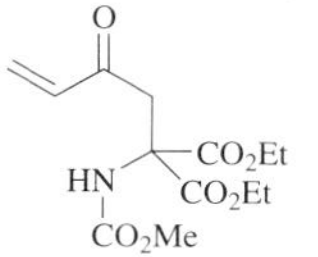 II	59

Temp	Time	I	II	I:II
120°	8 h	(70)	(8)	—
105°	12 h	(—)	(—)	4.3:1
70°	65 h	(—)	(—)	1:1.3

Imine	Diene	Conditions	Product(s) and Yield(s) (%)	Refs.
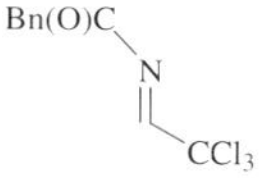		Hydroquinone, Et_3N, $AlCl_3$, benzene, 80°, 4 h	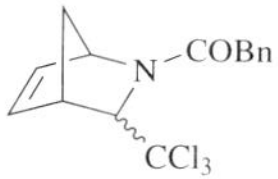 (55-95)[d]	287
C_{10} 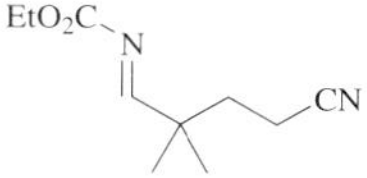	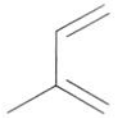	BF_3•OEt_2, benzene, 80°, 16 h	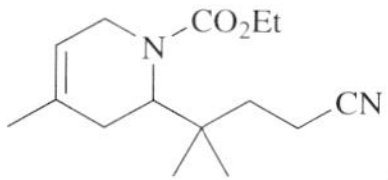 + 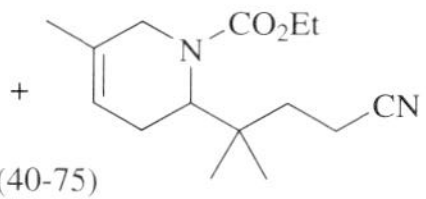 (40-75)	281
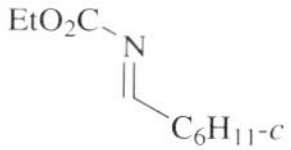		BF_3•OEt_2, $CHCl_3$, 61°, 2 h	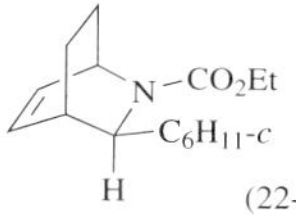 + 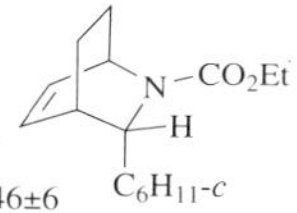 (22-34) 54±6:46±6	23

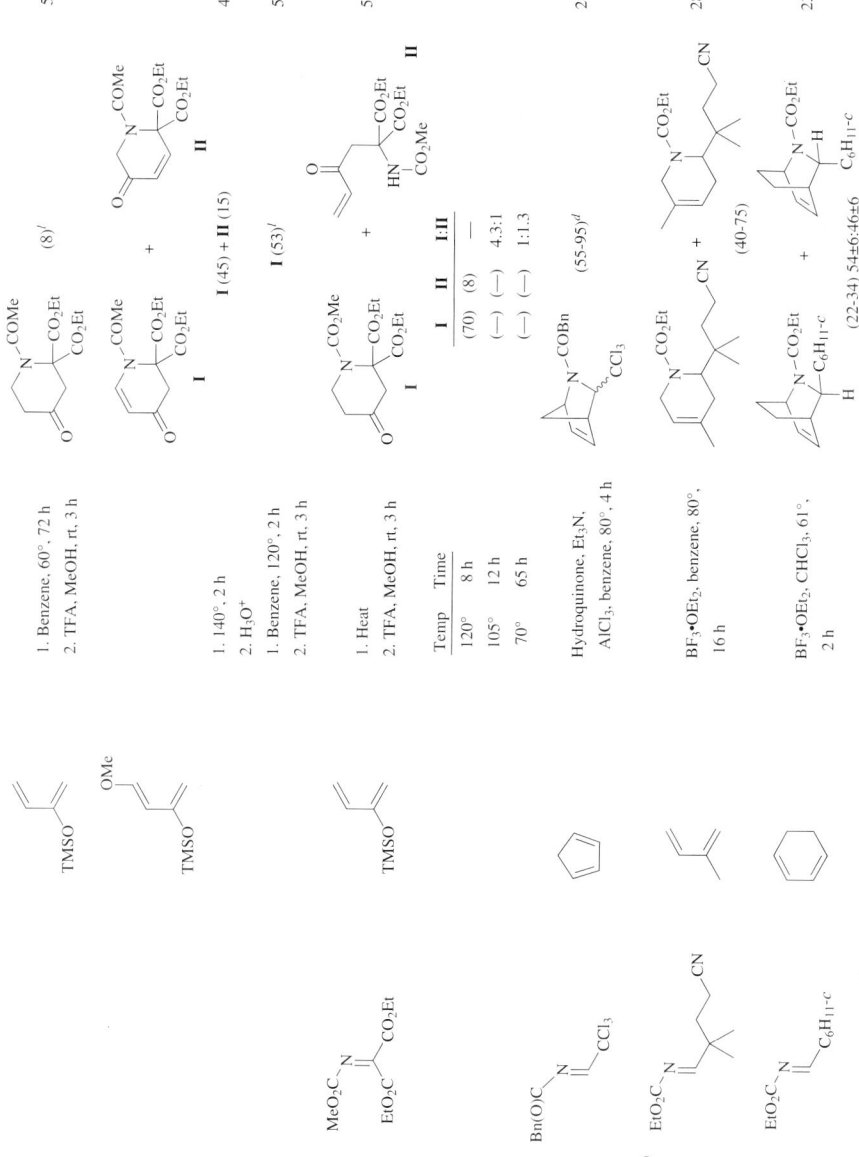

TABLE 1. CYCLOADDITIONS OF ACYCLIC *N*-ACYLIMINES AND *N*-CYANOIMINES (*Continued*)

Imine	Diene	Conditions	Product(s) and Yield(s) (%)	Refs.
C_{10} 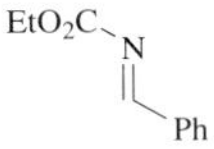	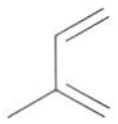	BF_3•OEt_2, benzene, 150°, 4 h	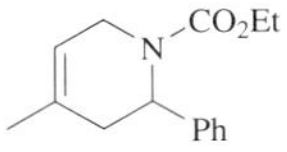 + 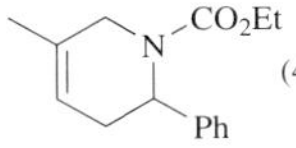 (40-80) >90:10	281, 282, 47
		BF_3•OEt_2, benzene, 80°, 1 h	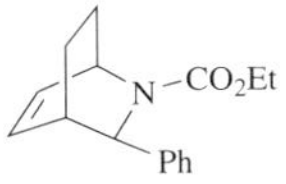 (50)[d]	48
		BF_3•OEt_2, benzene, 80°, 7 h	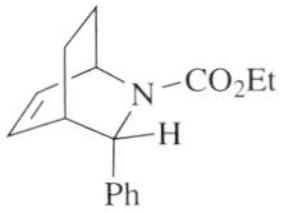 (45)[o]	290
			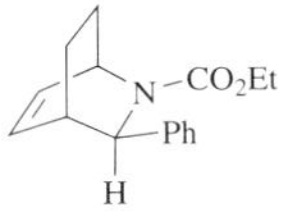 + 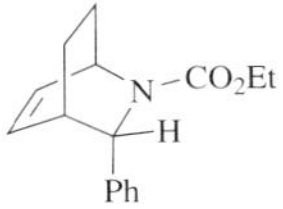	
		Catalyst / Solvent / Temp / Time		
		BF_3•OEt_2 benzene 80° 2-24 h	(40-50) 80:20	18, 61, 31, 23
		BF_3•OEt_2 $CHCl_3$ 61° 2-24 h	(30-40) 87:13	18
		BF_3•OEt_2 $CHCl_3$ — 2-6 h	(25-35) 80±2:20±2[h]	29
		BF_3•OEt_2 CCl_4 76° 2-24 h	(30-40) 86:14	18
		BF_3•OEt_2 CH_2Cl_2 40° 2-24 h	(30-40) 91:9	18
		BF_3•$CuBr_2$ benzene 80° 2-24 h	(60) 81:19	18
		H_2SO_4 benzene 80° 2-24 h	(10-15) 84:16	18
		$AlCl_3$ benzene 80° 2-24 h	(5) 74:26	18
		$SnCl_4$ benzene 80° 2-24 h	(30) 73:27	18
		BF_3•OEt_2 benzene 80° 4 h	(48)	280

		BF_3•OEt_2, benzene, 80°, 8 h	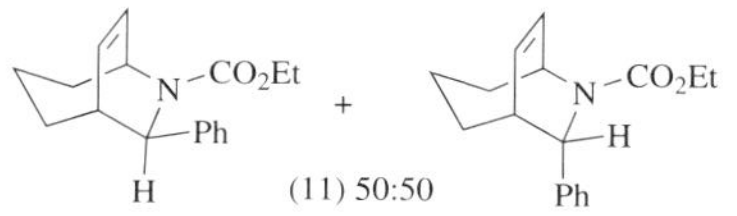 (11) 50:50	61
	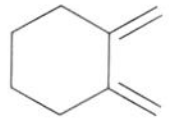	BF_3•OEt_2, benzene, 80°	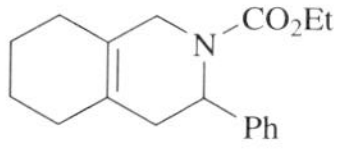 (>60)	281, 282, 47
	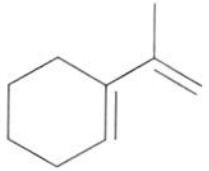	BF_3•OEt_2, benzene, 80°, 14 h	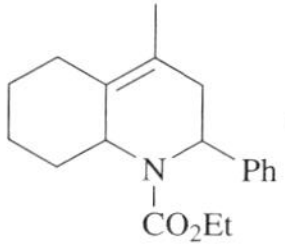 (—)	281
	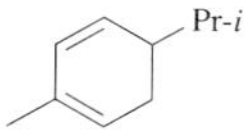	BF_3•OEt_2, benzene, 80°, 15 h	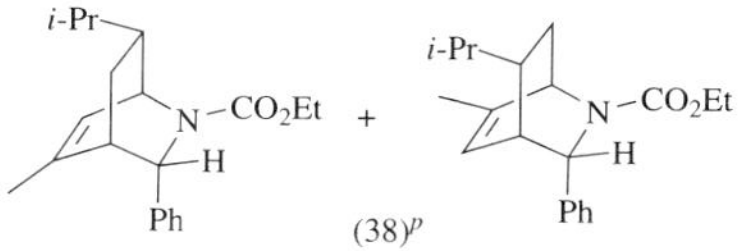 (38)[p]	290
	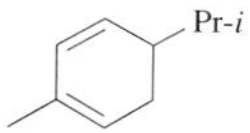	BF_3•OEt_2, benzene, 80°, 3-15 h	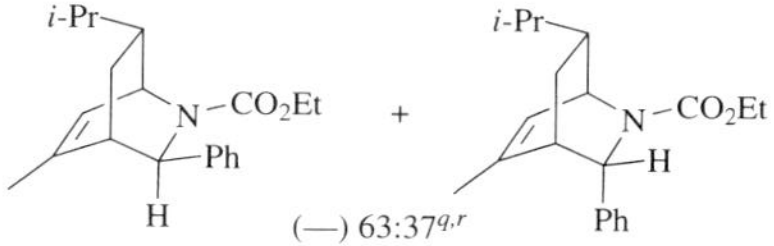 (—) 63:37[q,r]	28
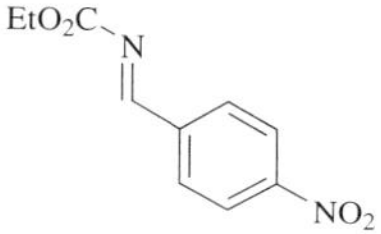	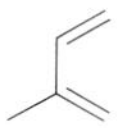	BF_3•OEt_2, benzene, 80°, 16 h	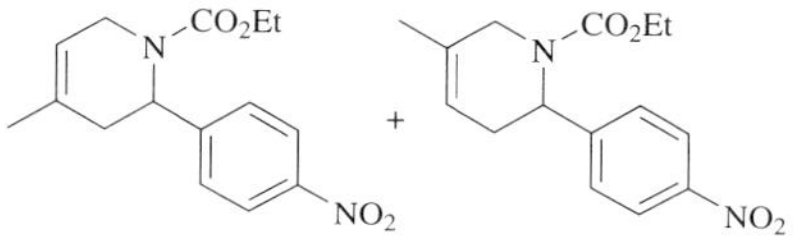 (40-80) >90:10	281, 282, 47

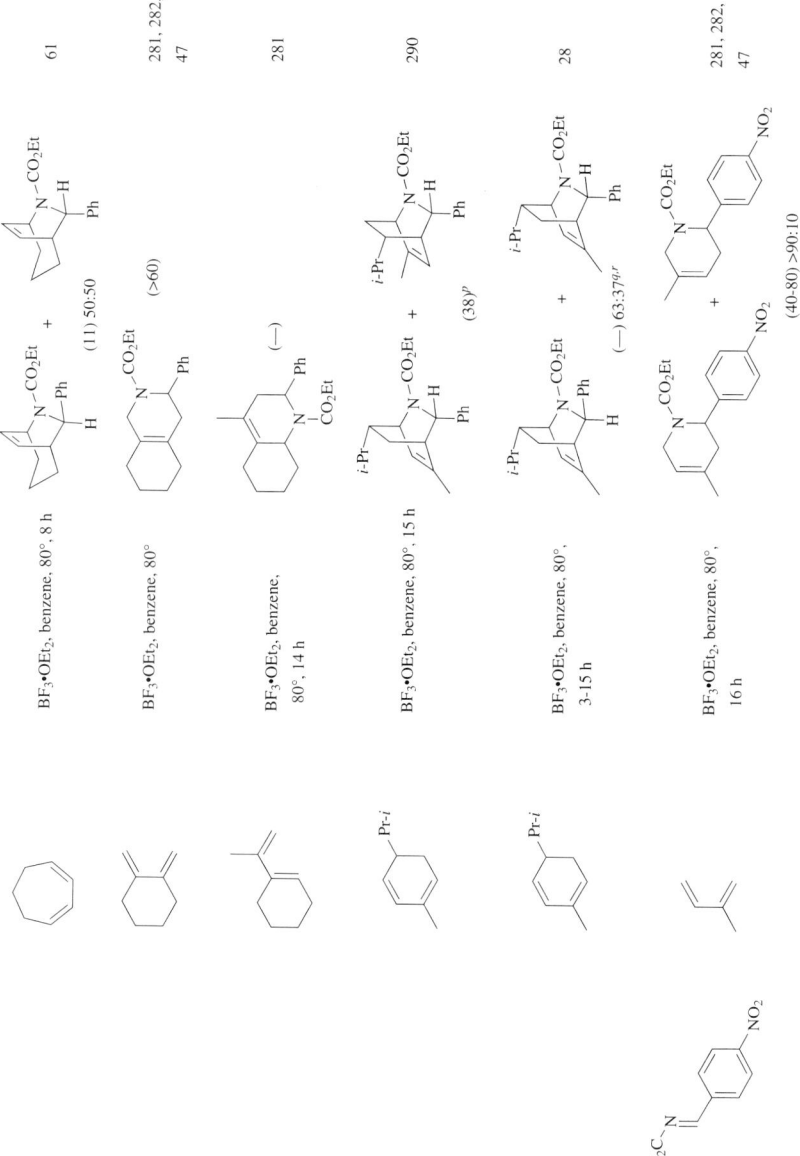

TABLE 1. CYCLOADDITIONS OF ACYCLIC *N*-ACYLIMINES AND *N*-CYANOIMINES (*Continued*)

Imine	Diene	Conditions	Product(s) and Yield(s) (%)	Refs.
C_{10} 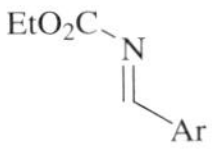 Ar = 4-$O_2NC_6H_4$		BF_3•OEt_2		
		Solvent / Temp / Time		
		benzene / 80° / 2-24 h	(71) 80:20	18, 23
		$CHCl_3$ / — / 2-6 h	(25-35) 80±2:20±2	29
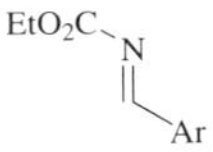	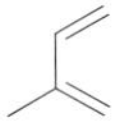	BF_3•OEt_2, benzene, 80°, 16 h	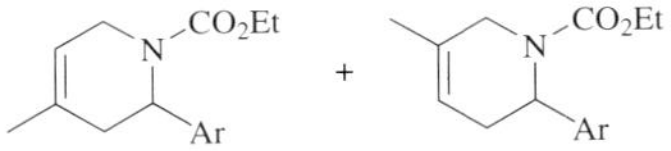	281, 282, 47
Ar				
2-$O_2NC_6H_4$			(40-80) >90:10	
2-ClC_6H_4			(40-80) >90:10	
4-ClC_6H_4			(40-80) >90:10	
3,4-$Cl_2C_6H_3$			(40-80) >90:10	
2,6-$Cl_2C_6H_3$			(40-80) >90:10	
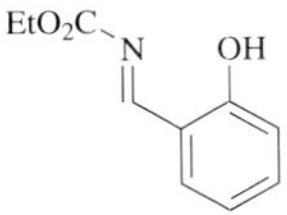		BF_3•OEt_2, benzene, 60-80°, 20 h	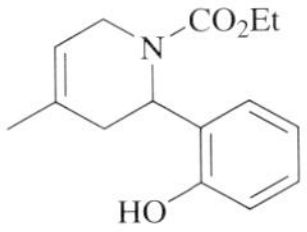 (—)	55
	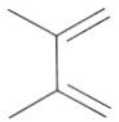	BF_3•OEt_2, benzene, 60 to 80°, 20.5 h	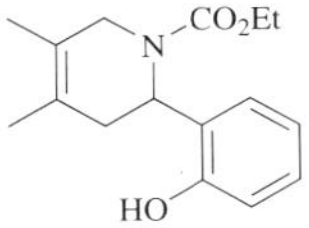 (—)	55

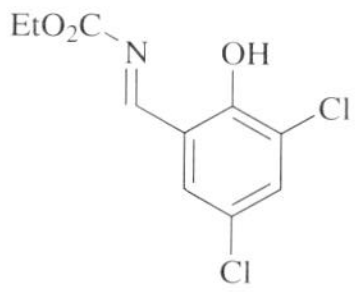	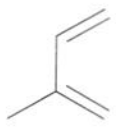	BF_3•OEt_2, benzene, 60 to 70-78°, 20 h	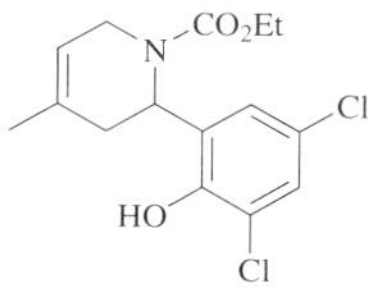	(—)	55
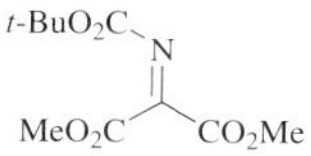	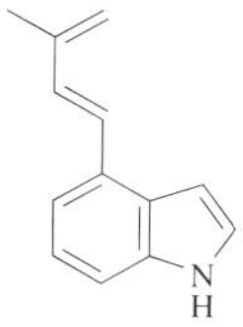	THF, 100°, 36 h, 10^4 atm	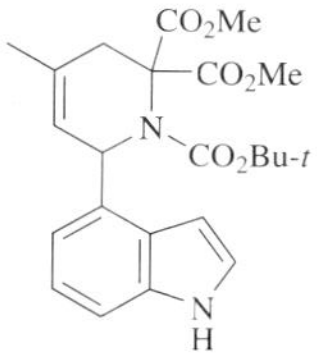	(—)	60
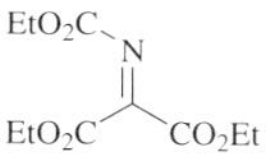		THF, 110°, 10^4 atm	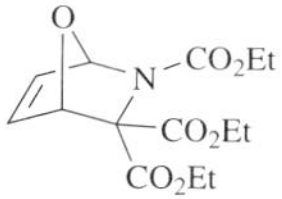	(72)	60
	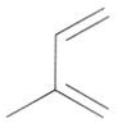	THF, 110°, 30 h, 10^4 atm	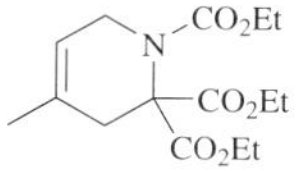	(60)	60
		THF, 65°, 5 h	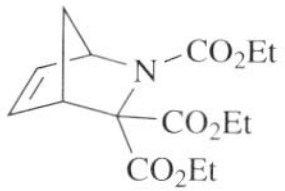	(62)	60
	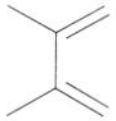	THF, 100°, 24 h, 10^4 atm	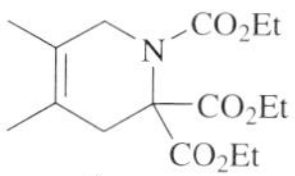	(60)	60
		THF, 110°, 12 h, 10^4 atm	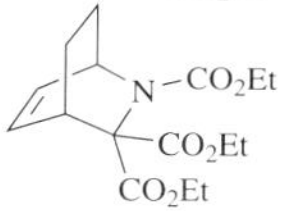	(39)	60

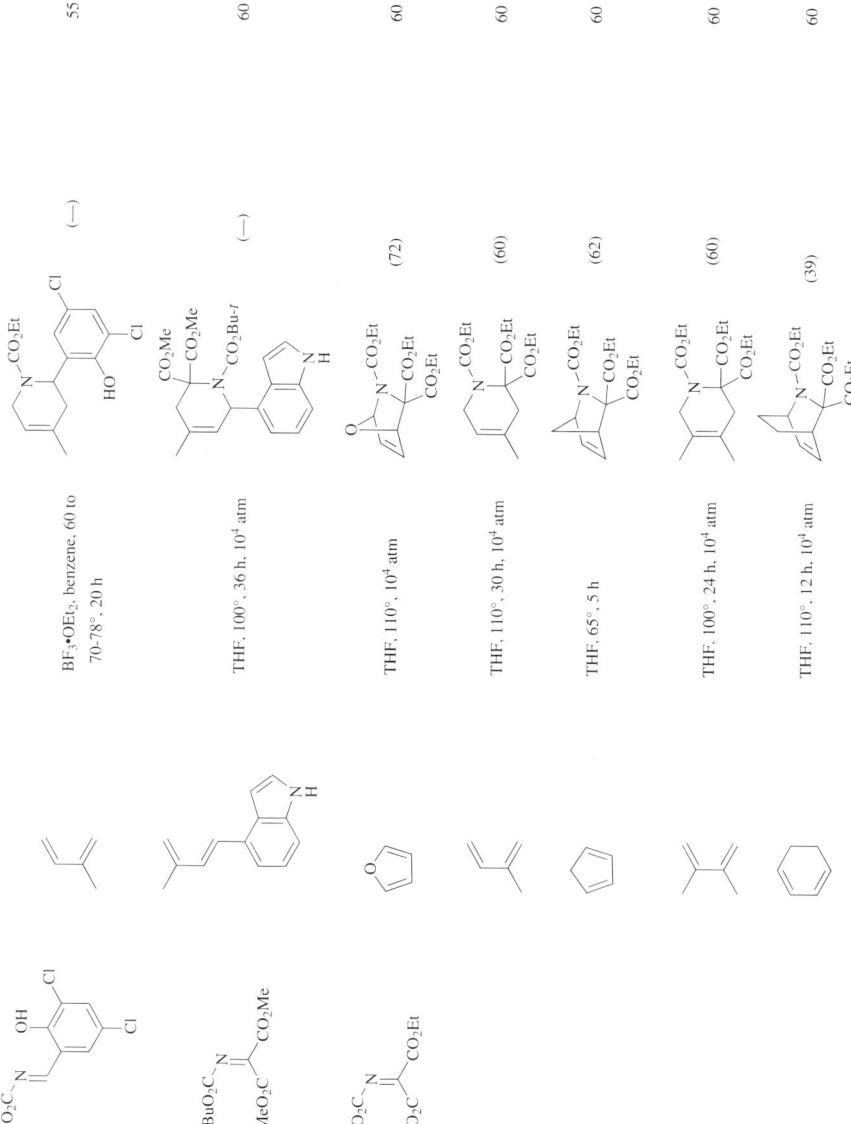

TABLE 1. CYCLOADDITIONS OF ACYCLIC *N*-ACYLIMINES AND *N*-CYANOIMINES (*Continued*)

Imine	Diene	Conditions	Product(s) and Yield(s) (%)	Refs.
C_{10} 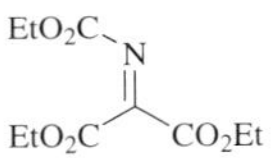		THF, 110°, 36 h, 10^4 atm	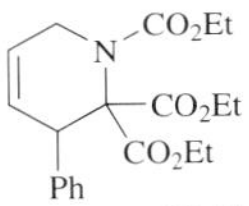 (66)	60
	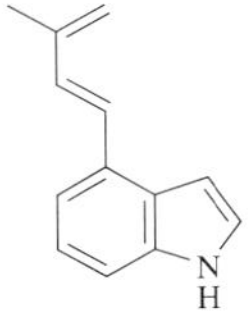	THF, 100°, 36 h, 10^4 atm	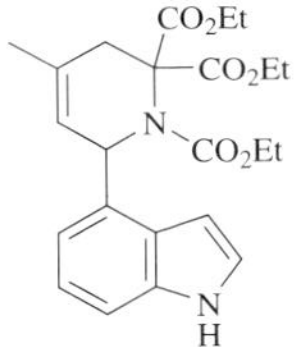 (—)	60
	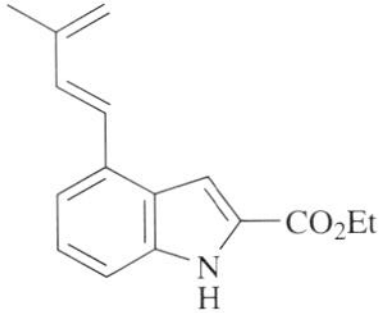	THF, 100°, 36 h, 10^4 atm	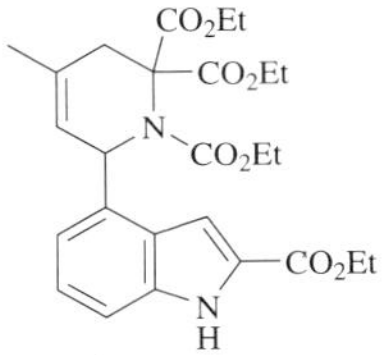 (23)	60
C_{11} 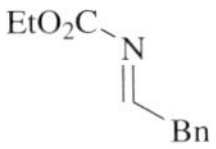		BF_3•OEt_2, $CHCl_3$, 61°, 2 h	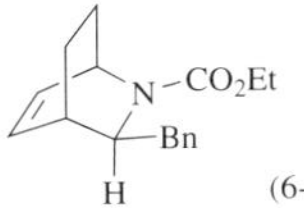 + 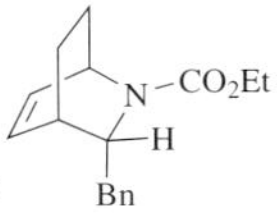 (6-10) >95:5	23
C_{11-13} 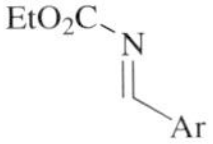		BF_3•OEt_2, benzene, 80°, 1 h	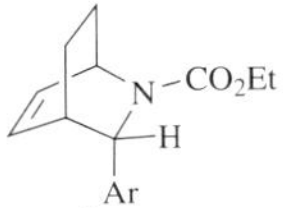 **I** + 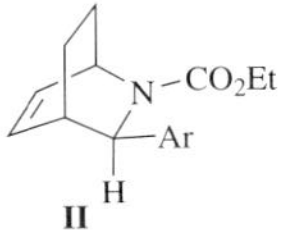 **II**	290
Ar				
2-MeC_6H_4			**I** (44)[o]	
4-$MeOC_6H_4$			**I** (40)[o]	
Bz			**I + II** (—)[i]	
Ar = $C(O)C_6H_3(OMe)_2$-3,4			**I + II** (50)[i]	

C_{11-12} 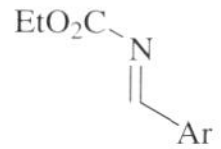		$BF_3 \bullet OEt_2$, benzene, 80°, 2-24 h	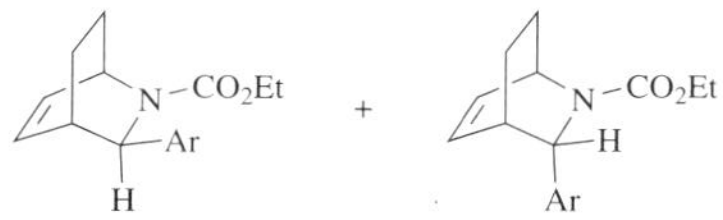	18
Ar: 2-MeC_6H_4; 4-MeC_6H_4; 3,4-$(MeO)_2C_6H_4$			(—) 83:17; (61) 85:15; (64) 79:21	
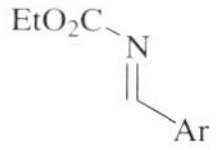 Ar = 4-MeC_6H_4, 2-$MeOC_6H_4$, 3-MeO-4-HOC_6H_3, Bz, 2-$AcOC_6H_4$	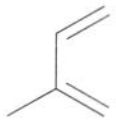	$BF_3 \bullet OEt_2$, benzene, 80°, 16 h	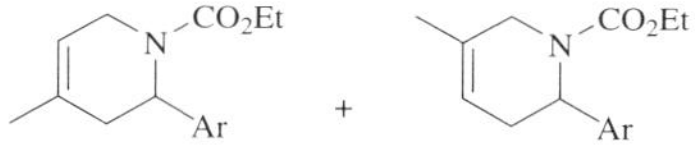 (40-80) >90:10	281, 282, 47
C_{11} 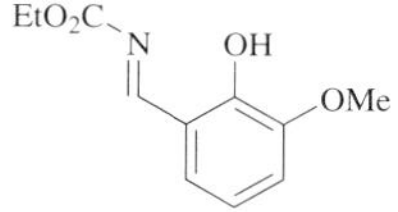	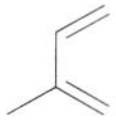	$BF_3 \bullet OEt_2$, benzene, 60-70°, 19-20 h	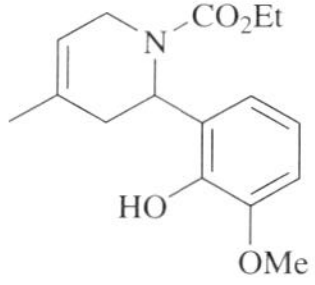 (—)	55
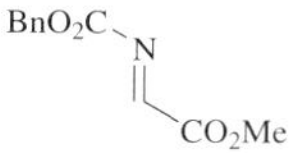		TEA, CH_2Cl_2, rt, 18 h	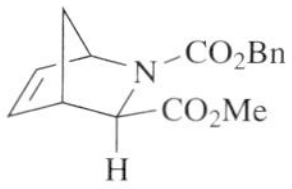 (56)	43
		$BF_3 \bullet OEt_2$	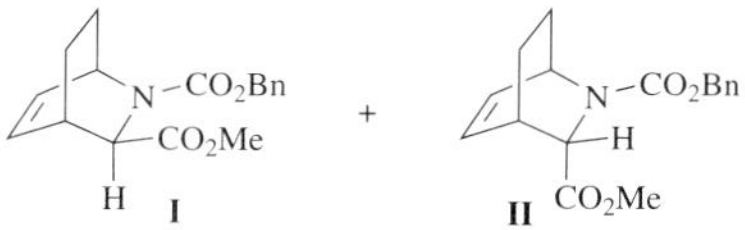 I + II	
		Solvent / Temp / Time: benzene / 80° / 1.5 h	**I** (28)[s] + **II** (7)	62
		Solvent / Temp / Time: $CHCl_3$ / — / 2-6 h	**I** + **II** (25-35), **I**:**II** 80:20[h]	29

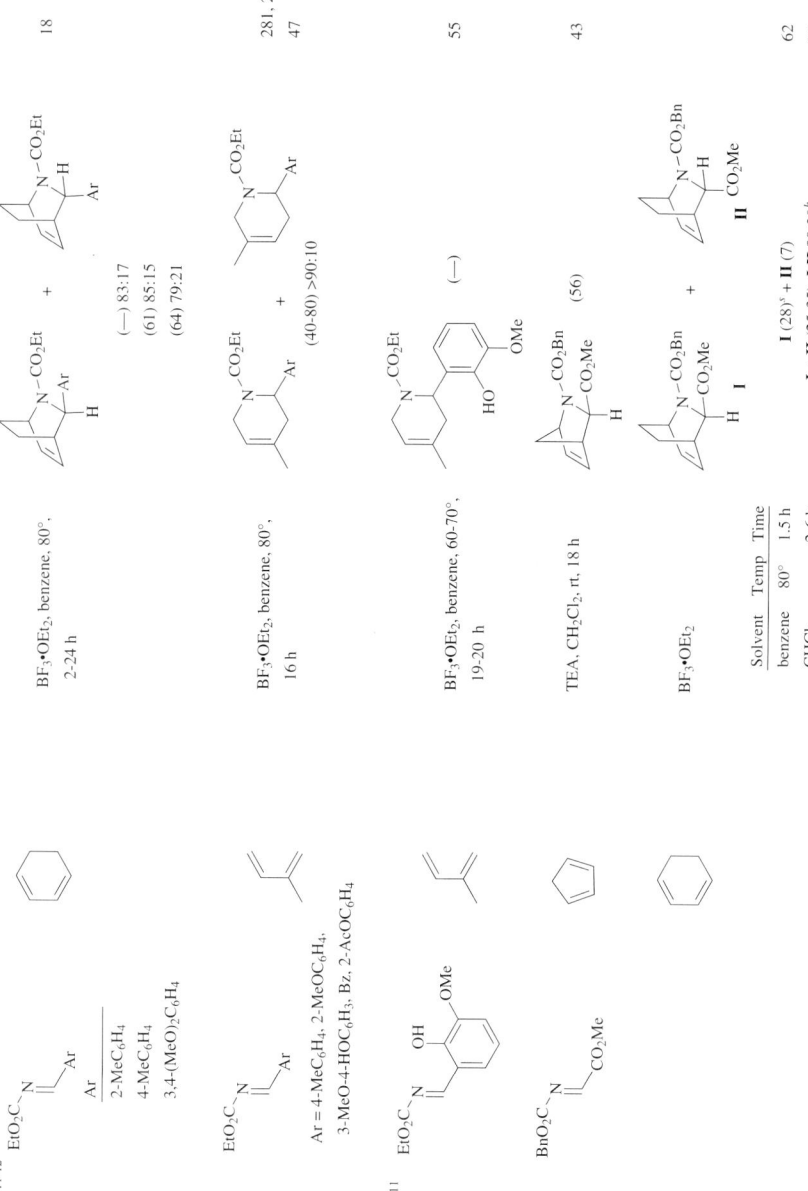

TABLE 1. CYCLOADDITIONS OF ACYCLIC *N*-ACYLIMINES AND *N*-CYANOIMINES (*Continued*)

	Imine	Diene	Conditions	Product(s) and Yield(s) (%)	Refs.
C_{11}	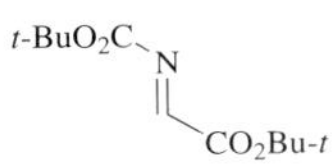	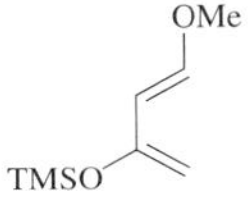		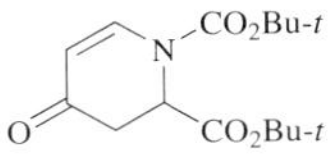	56
			Ti(OEt)$_2$Cl$_2$, THF, −78° to rt, overnight	(72)	
			"Basic", rt	("high")	
			"Thermal"	("high")	
C_{12}	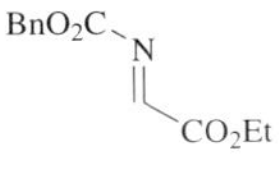			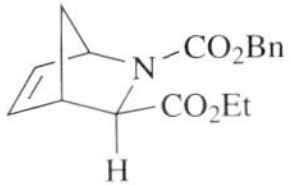	
			DME, 85°, 15 h	(57)	41
			Toluene, 130°, 20 h	(31)	42
	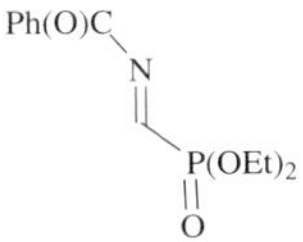		1. THF, −78° to rt, 1 h 2. 40°, 1 h	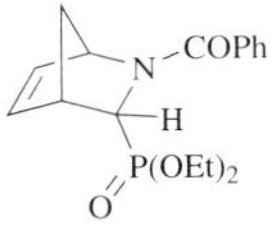 (64)[l]	57
		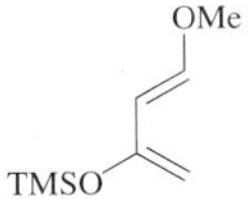	1. THF, −78° to rt, 13 h 2. Citric acid, H_2O	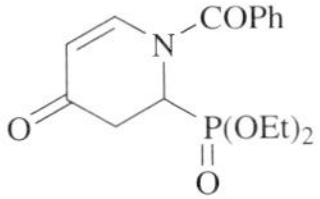 (31)[l]	57
		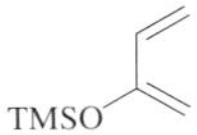	1. THF, −78 to 0°, 13 h 2. Citric acid, H_2O	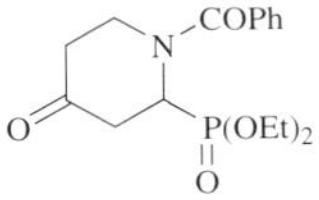 (12)[l]	57

C_{13}				
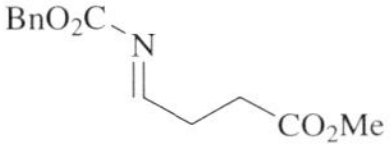BnO2C
N
CO2Me	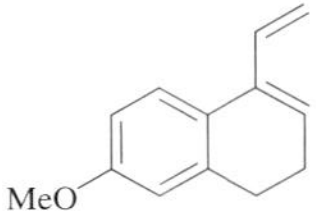MeO	—	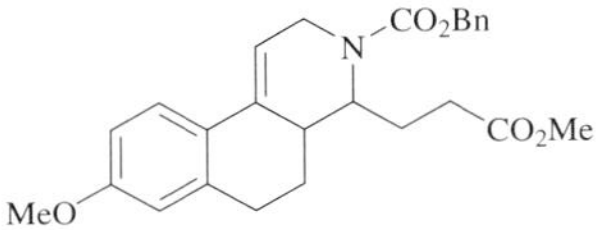N
CO2Bn
CO2Me
MeO (47)[k]	52
	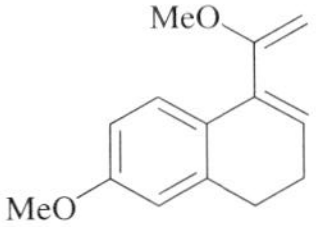MeO
MeO	—	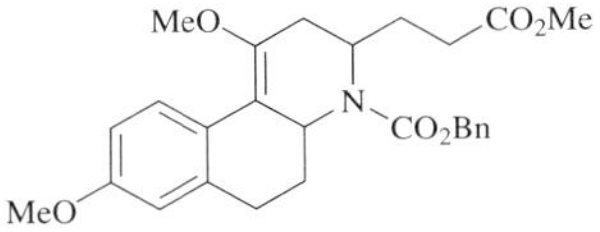MeO
CO2Me
N
CO2Bn
MeO (—)[k]	52
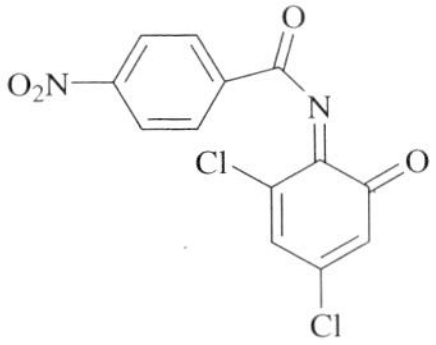O
O2N
N
Cl
O
Cl	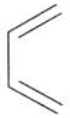	CH_2Cl_2, rt, 1 h	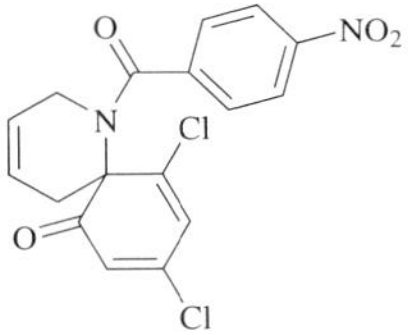O
NO2
N
Cl
O
Cl (93)	293
	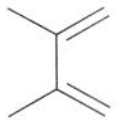	CH_2Cl_2, 30 min	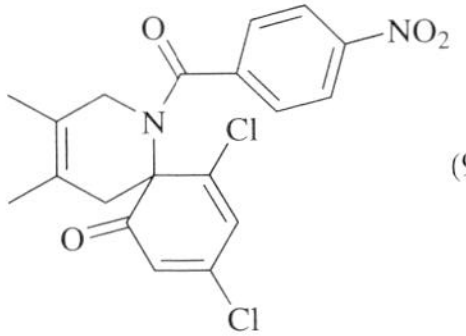O
NO2
N
Cl
O
Cl (92)	294
C_{14-15}				
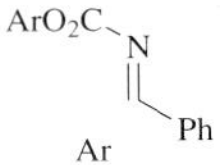ArO2C
N
Ph Ar: Ph, Bn		BF_3•OEt_2, benzene, 80°, 2-24 h	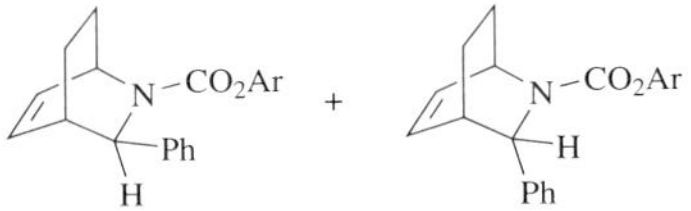N
CO2Ar
Ph
H
+
N
CO2Ar
H
Ph (37) 69:31; (50) 83:17	18

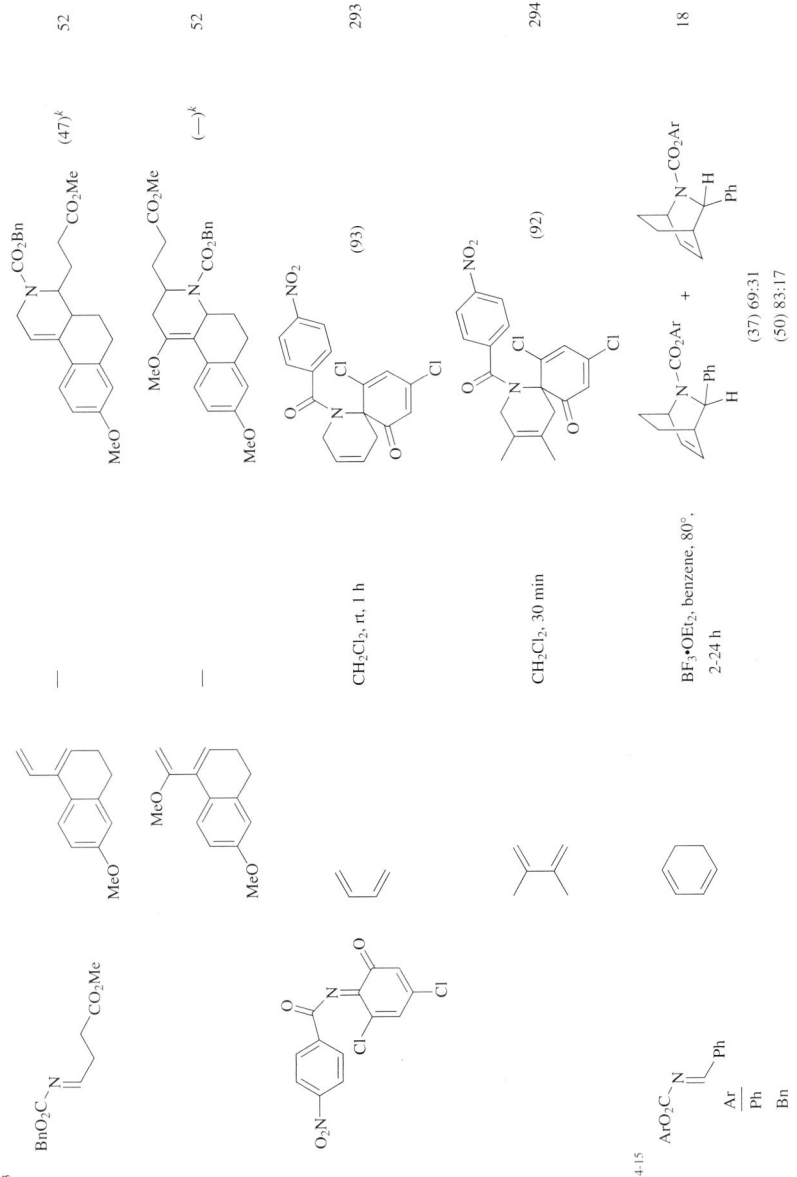

TABLE 1. CYCLOADDITIONS OF ACYCLIC *N*-ACYLIMINES AND *N*-CYANOIMINES (*Continued*)

Imine	Diene	Conditions	Product(s) and Yield(s) (%)	Refs.
C_{16-20} 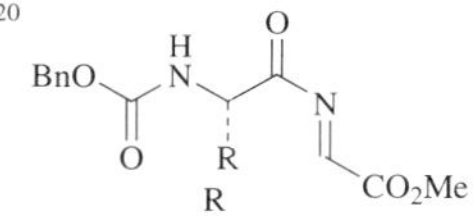 R: *i*-Pr; Bn		Et_3N, THF, 65°, 15 h	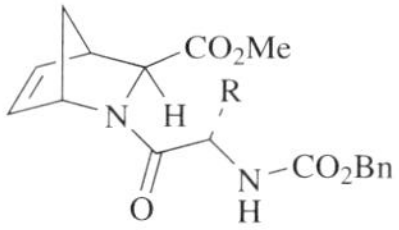 + 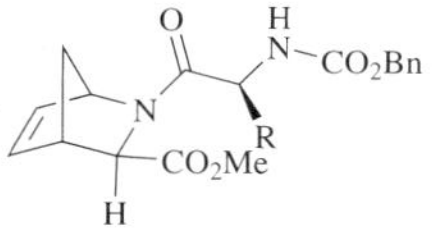 *i*-Pr: (67) 9:1; Bn: (60) 9:1	44
C_{27} 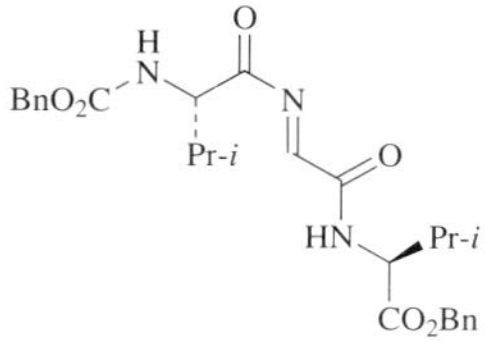		Et_3N, THF, 65°, 15 h	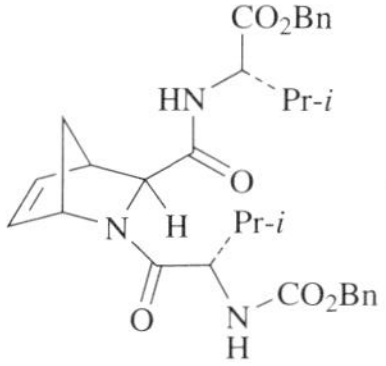 + 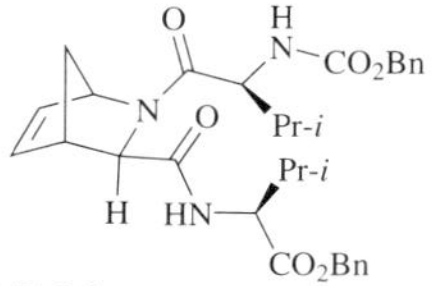 (41) 9:1	44

[a] A small amount of cyclohepta-1,3-dienylmethylcarbamic acid ethyl ester was obtained as a side product.

[b] The endo:exo ratio and stereochemistry of the product is not specified. However, based on the reported melting points,[20, 21] this product was probably obtained as the *exo*-adduct exclusively.

[c] A 55% yield of 6-isopropenyl-2,6-dimethyl-4-trichloromethyl-5,6-dihydro-4*H*-1,3-oxazine was obtained, resulting from a hetero-Diels-Alder reaction in which the imine acts as the diene.

[d] The stereochemistry and ratio of products are not specified.

[e] The major product is the corresponding α-methoxycarbonylamino-γ-alkenyl-γ-butyrolactone, which was obtained in 40-80% yield.

[f] The regioselectivity and stereoselectivity of the reaction are not specified.

[g] The regioselectivity of the reaction is not specified.

[h] Changing the solvent to benzene had little effect on the stereochemistry of the reaction.

[i] A mixture of endo and exo isomers was obtained, but no other stereochemical information is provided.

[j] Use of $AlCl_3$ instead of BF_3 did not affect the yield or stereoselectivity of the reaction.

[k] No stereochemical information is provided.

[l] The acyclic enone resulting from silyl enol ether addition to the imine was also isolated.

[m] The acyclic silyl enol ether resulting from addition to the imine was also isolated.

[n] The corresponding *N*-acetylimine was found to be less reactive than the *N*-ethoxycarbonylimine.

[o] The authors assign the endo-structure to the product, but this may be incorrect based on the results of similar reactions in other references.[18, 29]

[p] The structures of the products are contradicted in a later report.[28]

[q] The reported product structures contradict an earlier study of this reaction.[290]

[r] The reaction could also be run in refluxing chloroform as the solvent or with copper bromide added as a catalyst.

[s] The overall yield of the reaction was not improved by varying the solvent (CCl_4, $CHCl_3$, CH_2Cl_2), the amount of catalyst (10-100 mol% $BF_3 \bullet OEt_2$), the type of catalyst (TFA), or the reaction time (15 min to 2 h).

[t] The hetero-Diels-Alder adduct resulting from cyclopentadiene acting as the dienophile was isolated in 11% yield.

TABLE 2. CYCLOADDITIONS OF CYCLIC *N*-ACYLIMINES

Imine	Diene	Conditions	Product(s) and Yield(s) (%)	Refs.
C_3 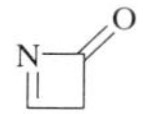	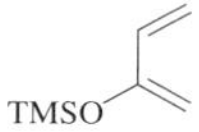	$ZnCl_2$	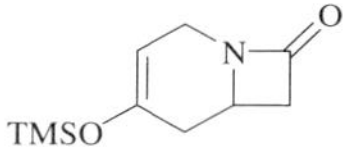 (—)	77
	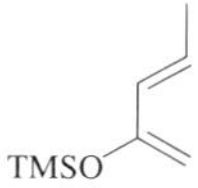	$ZnCl_2$	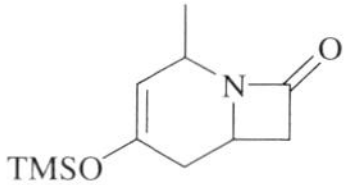 (—)[a]	77
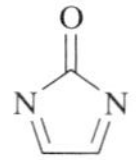	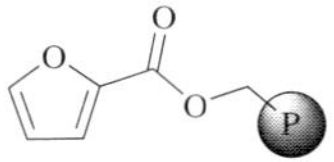	1. Pyridine, DMSO, 170°, 7 d 2. KOH, dioxane, H_2O, 100°, 2 d 3. H^+	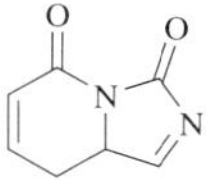 (15.5)	72
C_4 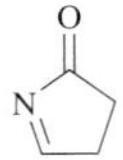	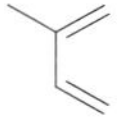		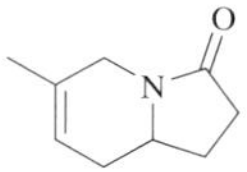	295
		HCO_2H, rt, 2 h	(trace)[b]	
		MeCN, rt, 4 d	(9)[c]	
		$HClO_4$, $CHCl_3$, −10°, 7 d	(10)	
			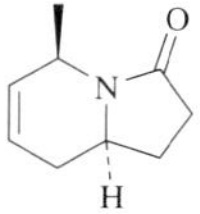	295
		HCO_2H, rt, 2 h	(11)[b]	
		$MeSO_3H$, CH_2Cl_2, 0°, 64 h	(14)	
			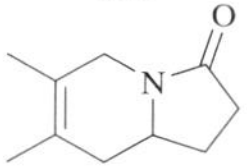	295
		HCO_2H, CH_2Cl_2, rt, 3 h	(16)[b]	
		TsOH, benzene	(19)	

	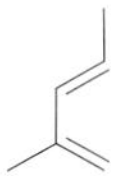	HCO_2H, rt, 5 h	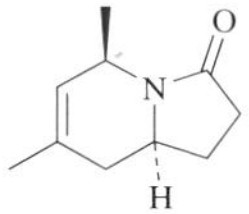 + 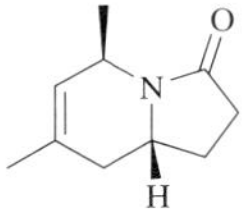 (70) 1:1	295
	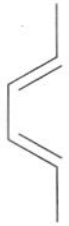	HCO_2H, rt, 2 h $HClO_4$, $CHCl_3$, –10°, 7 d	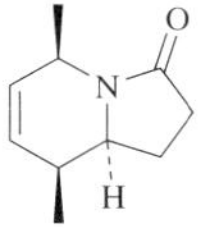 (trace)^b (20)	295
		HCO_2H, rt, 4 h	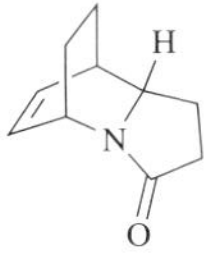 (40)^b	295
	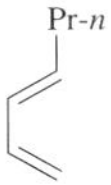	HCO_2H, rt $MeSO_3H$, CH_2Cl_2, 0°, 120 h	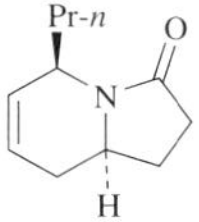 (15-20)^b (21)^c	295
	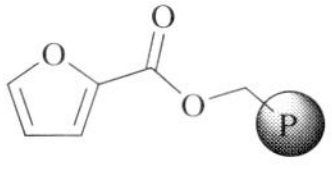	1. $LiNH_2$, HMPT, dioxane, rt, 18 h 2. KOH, dioxane, H_2O, 100°, 2 d 3. H^+	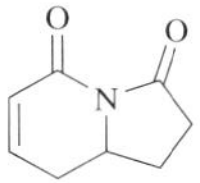 (—)	296
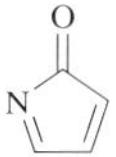	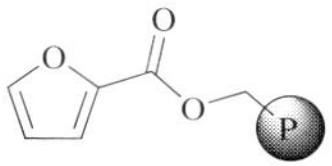	1. DMSO, 100°, 48 h 2. NaOH, dioxane, H_2O, 100°, 2 d 3. H^+	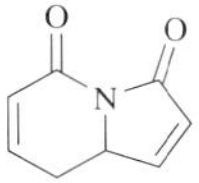 (—)	297

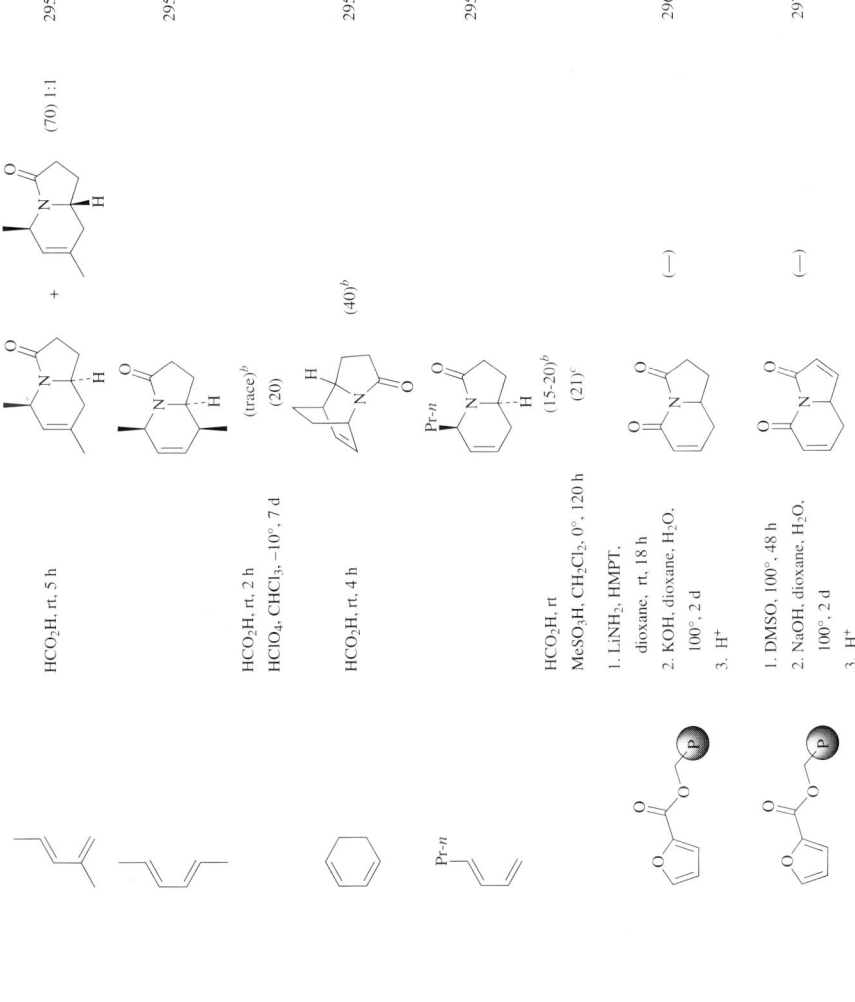

TABLE 2. CYCLOADDITIONS OF CYCLIC *N*-ACYLIMINES (*Continued*)

Imine	Diene	Conditions	Product(s) and Yield(s) (%)	Refs.
C_4 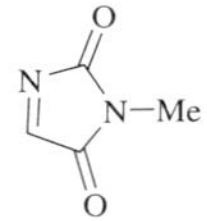		Et_3N, hydroquinone, benzene, rt, 2 h	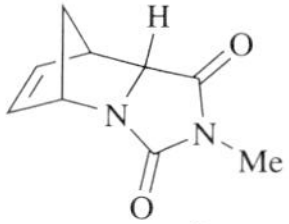 (—)	71
$C_{5\text{-}7}$ 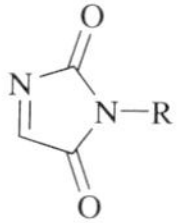	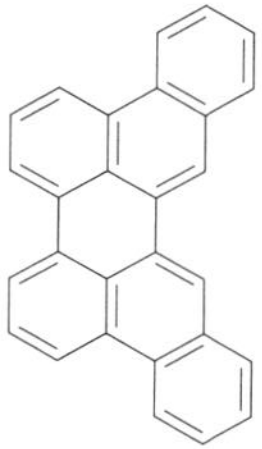		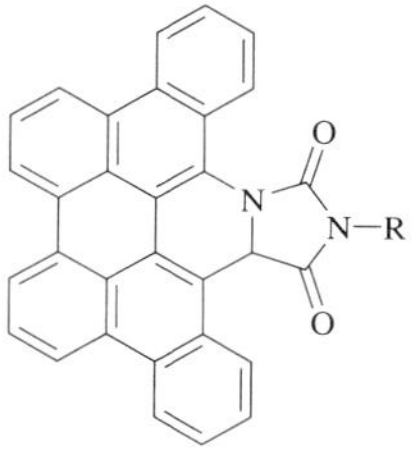	68
R				
Et		*p*-Xylene, 160°, 72 h	(13)	
Et		TFA, benzene, 80°, 24 h	(48)	
n-Pr		*p*-Xylene, 160°, 72 h	(47)	
n-Pr		TFA, benzene, 80°, 24 h	(51)	
n-Bu		*p*-Xylene, 160°, 72 h	(62)	
n-Bu		TFA, benzene, 80°, 24 h	(74)	
C_8 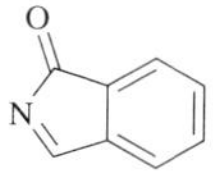	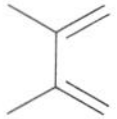	TFA, benzene, 80°, 24 h	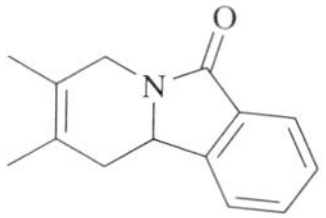 (33)	298
	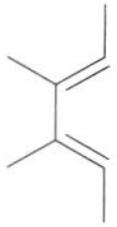	β-Naphthalenesulfonic acid, benzene, 80°, 24 h	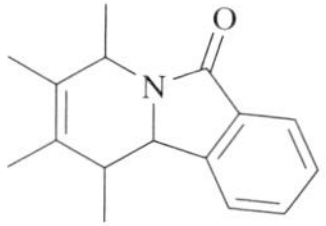 (41)[a]	298

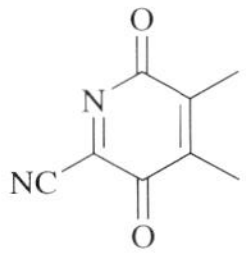		PhCl, MeOH, 130°, 4 h	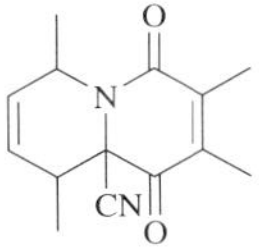 (14)[a]	76
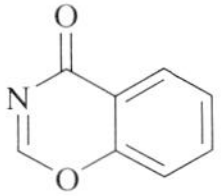	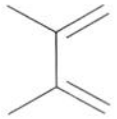	BF_3•OEt_2, rt, 48 h	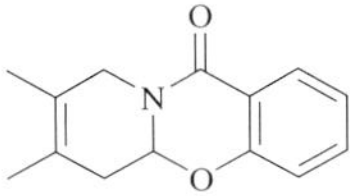 (42)	74
	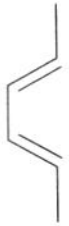	BF_3•OEt_2, ether, rt, 48 h	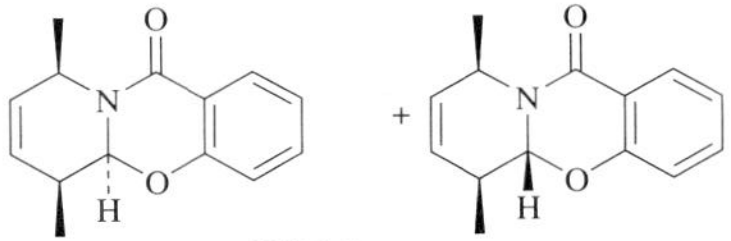 (18) 3:1	74
		BF_3•OEt_2, ether, rt, 48 h	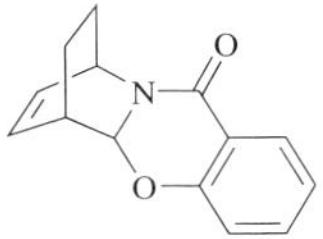 (34)[a]	74
	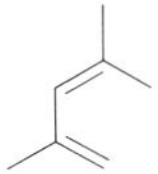	BF_3•OEt_2, ether, rt, 16 h	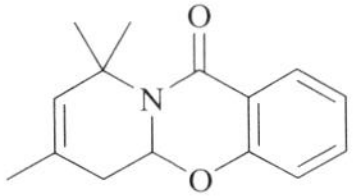 (44)	74
	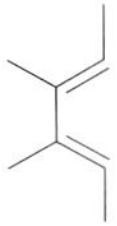	BF_3•OEt_2, ether, rt, 16 h	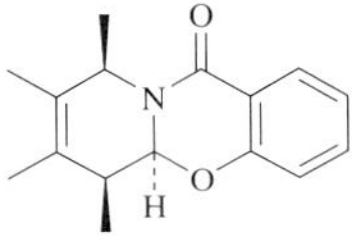 (86)	74

TABLE 2. CYCLOADDITIONS OF CYCLIC *N*-ACYLIMINES (*Continued*)

Imine	Diene	Conditions	Product(s) and Yield(s) (%)	Refs.
C_8				
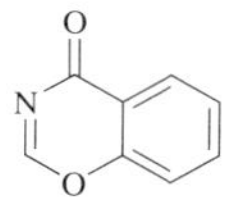	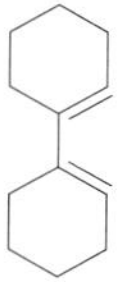	$BF_3{\bullet}OEt_2$, ether, rt	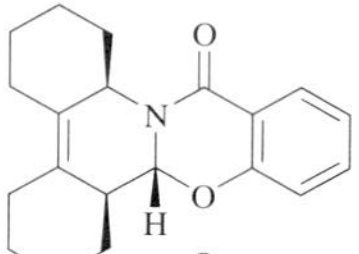 **I** + 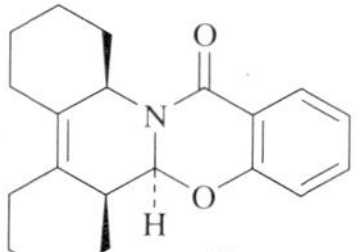 **II**	74
		Time		
		1 h	**I** + **II** (78), **I**:**II** 3:2	
		16 h	**I** (90)	
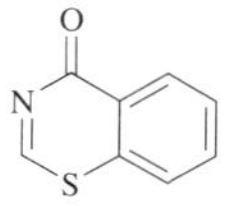	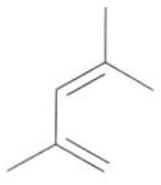	$BF_3{\bullet}OEt_2$, ether, 35°, 3 d	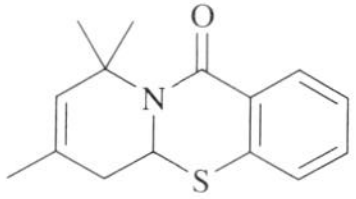 (42)[d]	75
	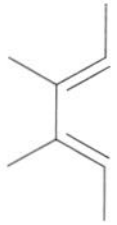	$BF_3{\bullet}OEt_2$, ether, rt, 4 d	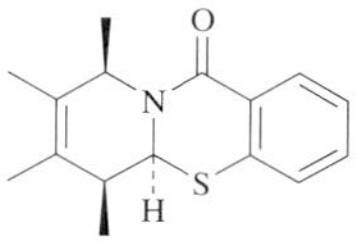 (50)[d]	75
			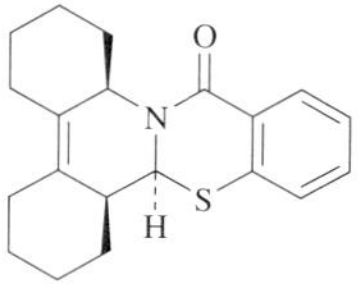	75
		$BF_3{\bullet}OEt_2$, ether, rt, 4 d	(51)	
		p-Xylene, 138°, 24 h	(24)	
		β-Naphthalenesulfonic acid, $CHCl_3$, 61°, 3 h	(40)	

C_9 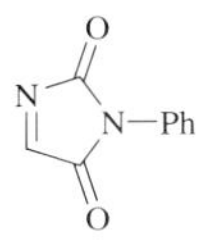	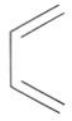		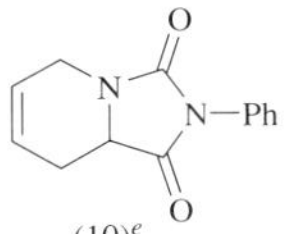	
		BF_3•OEt_2, benzene, rt, 24 h	(10)[e]	67
		β-Naphthalenesulfonic acid, benzene, 80°, 2-4 h	(10)	66
		Toluene, 170°, 3 d	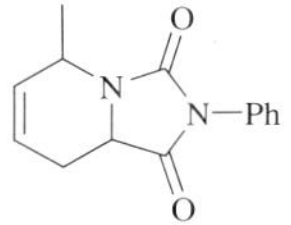 (16-80)[a,f]	66
	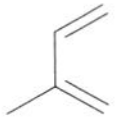	Toluene, 170°, 3 d	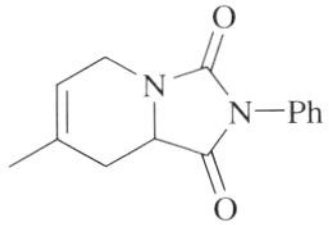 + 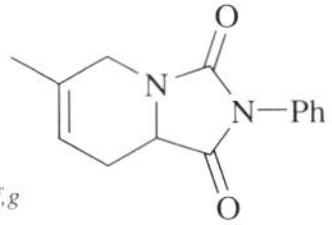 (42)[f,g]	66
	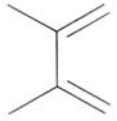	Toluene, 170°, 3 d	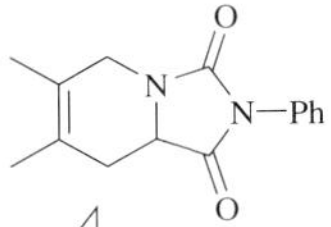 (23)[f]	66
			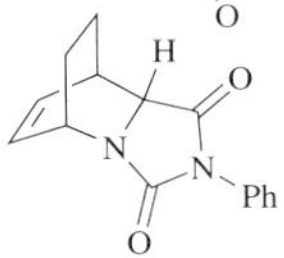	
		Naphthalenesulfonic acid, benzene, 80°, 2 h	(35)	67
		TFA, benzene, 80°, 24 h	(31)	66
		Toluene, 170°, 3 d	(64)	66
		BF_3•OEt_2	(—)	18, 61
		Benzene, 145°	(—)	31
		TsOH, benzene, 80°	(—)	31

TABLE 2. CYCLOADDITIONS OF CYCLIC *N*-ACYLIMINES (*Continued*)

Imine	Diene	Conditions	Product(s) and Yield(s) (%)	Refs.
C_9 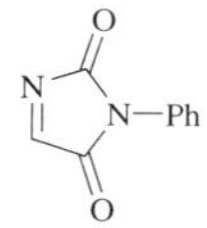	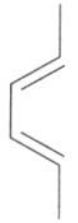	Toluene, 170°, 3 d	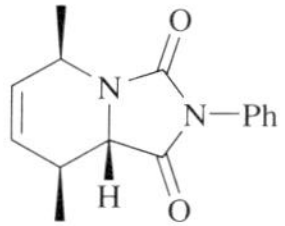 (46)	66
	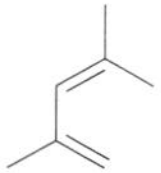	Toluene, 170°, 3 d	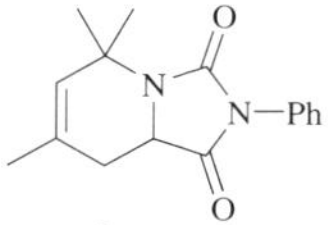 (70)	66
	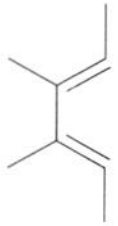		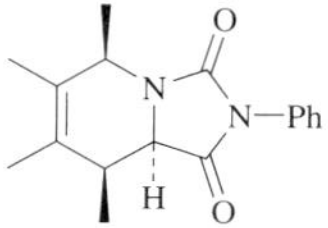 **I** + 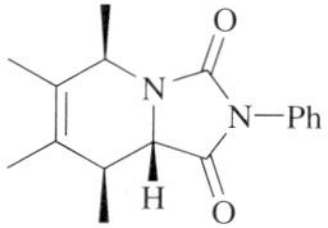 **II**	
		Xylene, 138°	**I** (—)[h]	67
		Toluene, 170°, 3 d	**I** + **II** (80)[i]	66
		Naphthalenesulfonic acid, benzene, 80°, 2 h	**I** (74)[h]	67
		β-Naphthalenesulfonic acid, benzene, 80°, 2 h	**I** (75)[i]	66
	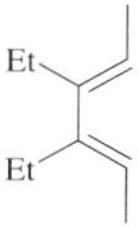		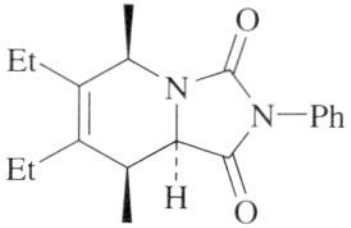	
		Toluene, 170°, 3 d	(16-80)[i]	66
		Naphthalenesulfonic acid, benzene, 80°, 2 h	(70)[h]	67
		β-Naphthalenesulfonic acid, benzene, 80°, 2 h	(74)	66

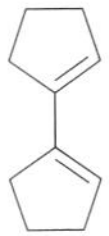	TFA, benzene, 80°, 16 h	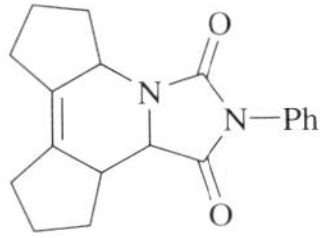 (60-70)[j]	67
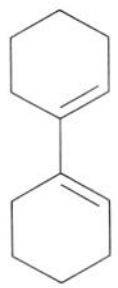		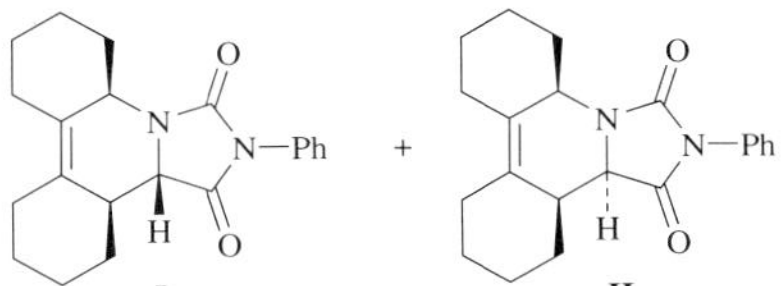 I + II	
	Xylene, 138°	(—)[k]	67
	Toluene, 170°, 3 d	**I** + **II** (70)[i]	66
	TFA, benzene, 80°, 16 h	(60-70)[k]	67
	β-Naphthalenesulfonic acid, benzene, 80°, 2 h	**I** + **II** (76), **I**:**II** 2:1	66
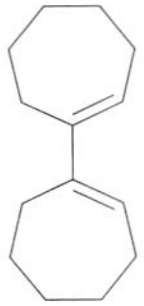	TFA, benzene, 80°, 16 h	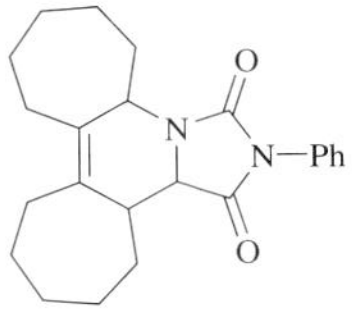 (60-70)[j]	67
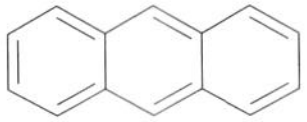	Xylene, 138°, 72 h	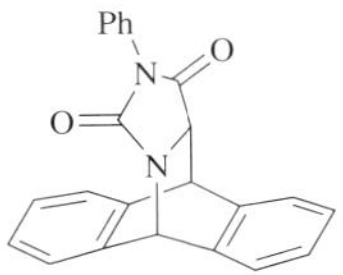 (70-80)	67

TABLE 2. CYCLOADDITIONS OF CYCLIC *N*-ACYLIMINES (*Continued*)

Imine	Diene	Conditions	Product(s) and Yield(s) (%)	Refs.
C_9				
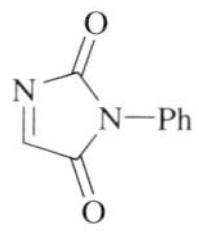	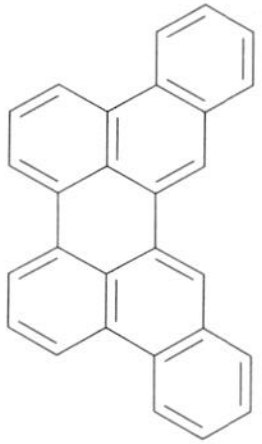	*p*-Xylene, 160°, 72 h TFA, benzene, 80°, 24 h	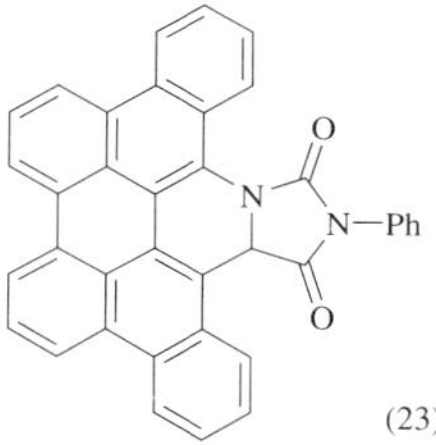 (23) (95)	68
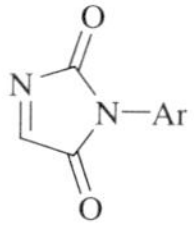 Ar = 4-ClC_6H_4		Benzene, 170°, 3 d β-Naphthalenesulfonic acid, benzene, 80°, 2-4 h	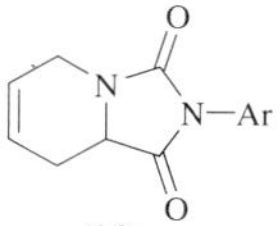 (16) (15)	66
		Toluene, 170°, 3 d	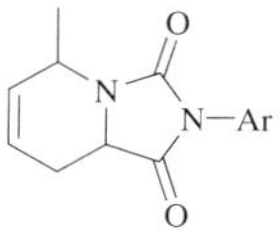 (61)[a,f]	66
	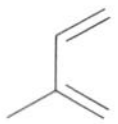	Toluene, 170°, 3 d	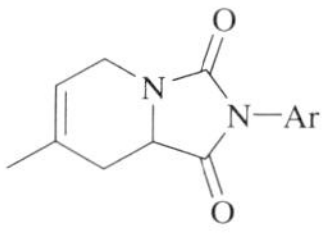 + (46)[f,g] 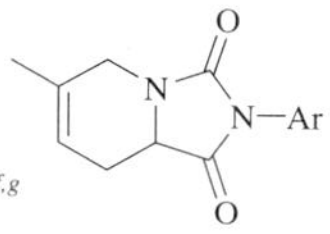	66
	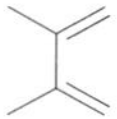	Toluene, 170°, 3 d	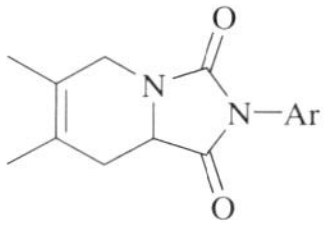 (19)[f]	66

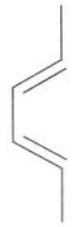	Toluene, 170°, 3 d	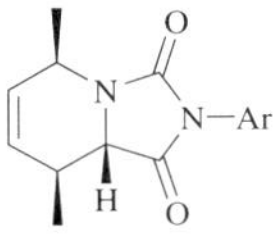	(53)[f]	66
	TFA, benzene, 80°, 18 h	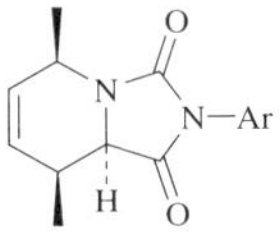	(25)	66
		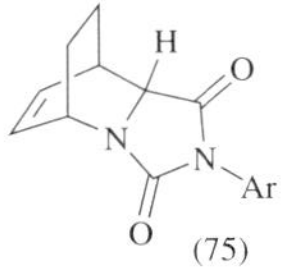		66
	Toluene, 170°, 3 d	(75)		
	TFA, benzene, 80°, 24 h	(33)		
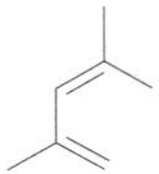		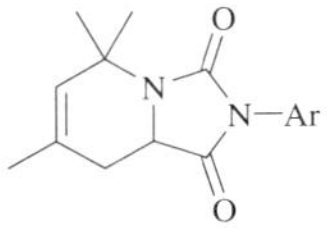		66
	Toluene, 170°, 3 d	(79)		
	β-Naphthalenesulfonic acid, benzene, 80°, 2-4 h	(41)		
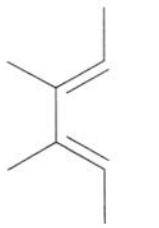		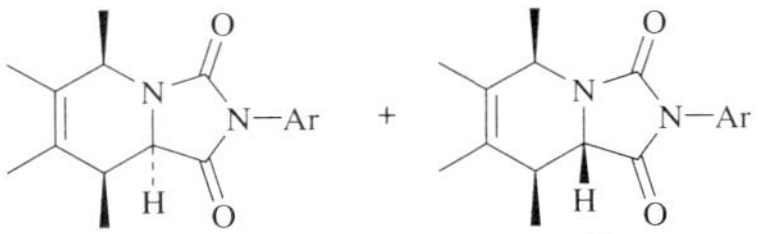 **I** **II**		66
	Toluene, 170°, 3 d	**I** + **II** (75), **I**:**II** 3:1		
	p-Xylene, 138°, 3 d	**I** (46)		
	β-Naphthalenesulfonic acid, benzene, 80°, 2-4 h	**I** (72)		

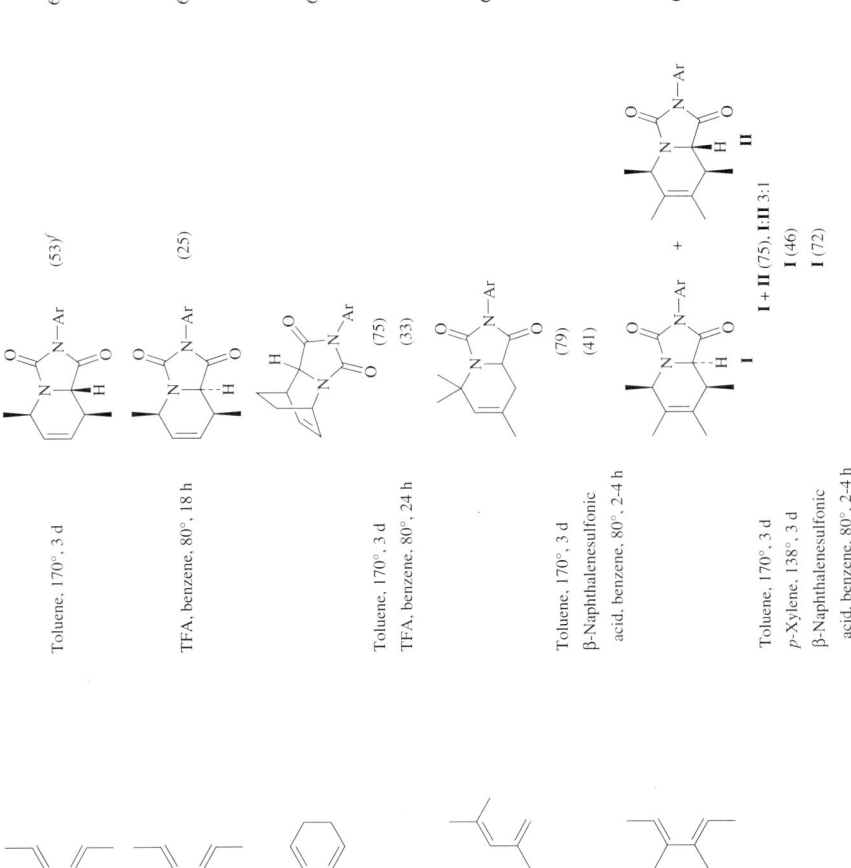

TABLE 2. CYCLOADDITIONS OF CYCLIC *N*-ACYLIMINES (*Continued*)

Imine	Diene	Conditions	Product(s) and Yield(s) (%)	Refs.
C_9				
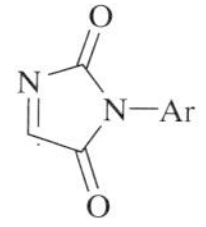 Ar = 4-ClC_6H_4	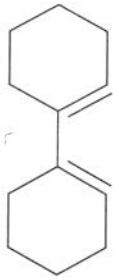	Toluene, 170°, 3 d	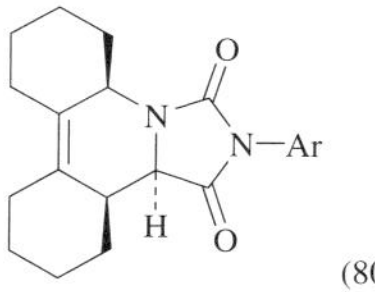 + 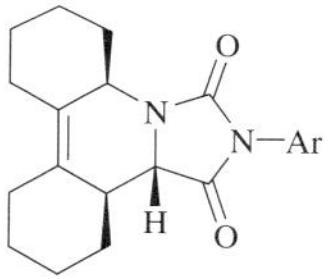 (80)[i]	66
	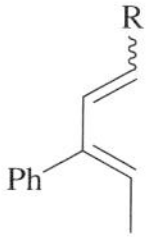	*p*-Xylene, 138°, 3 d	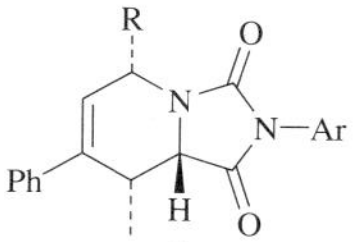 **I** + 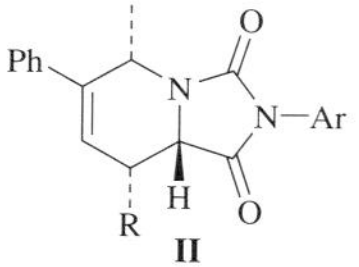 **II**	69
	R = H		**I** (2) + **II** (22)	
	R = Me		**I** (26) + **II** (11)[l]	
	R = Et		**I** (37) + **II** (9)[l]	
	R = CH_2OAc[m]		**I** (34) + **II** (22)[l]	
	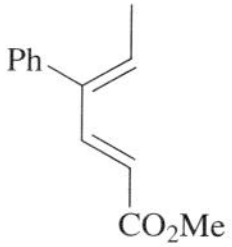	*p*-Xylene, 138°, 3 d	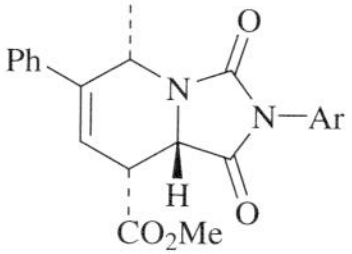 **I** + 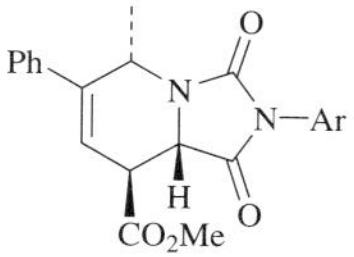 **II** **I** + **II** (13), **I**:**II** 1:1[n]	69
	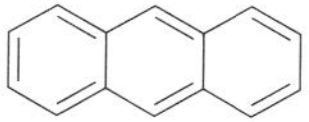	Xylene, 138°, 72 h	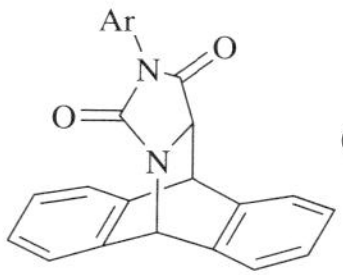 (73)	69

C_{10} 	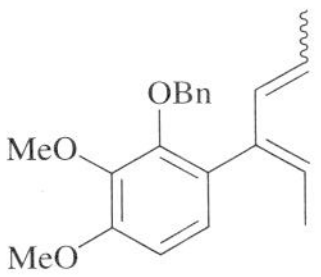	*p*-Xylene, 138°, 3 d	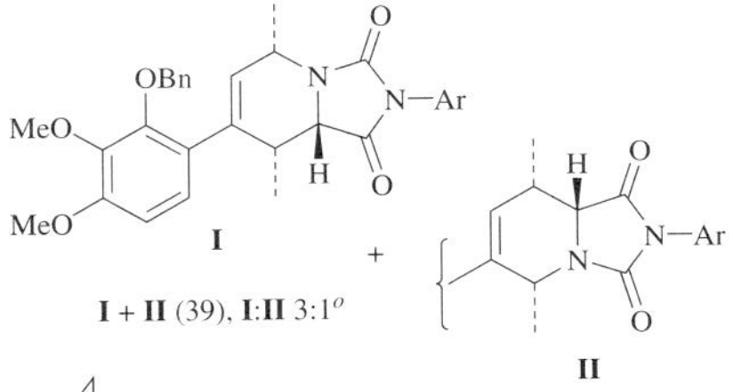 **I** + **II** (39), **I**:**II** 3:1[o]	70, 255
			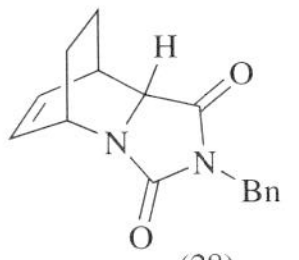	
		"Acid catalyst"	(28)	67
		TFA, benzene, 80°, 24 h	(30)	66
	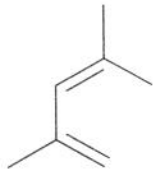	Toluene, 170°, 3 d	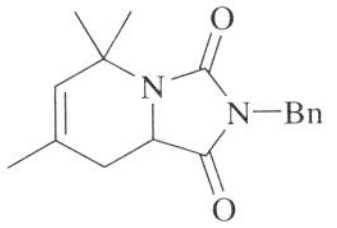 (63)	66
	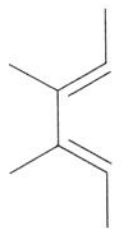		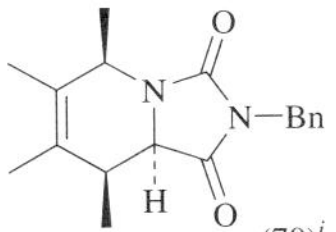	
		Toluene, 170°, 3 d	(70)[i]	66
		"Acid catalyst"	(70)[h]	67
		Xylene, 138°	(—)[h]	67
	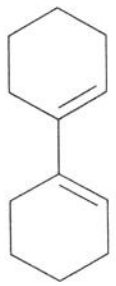		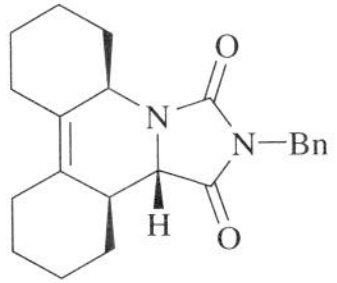 + 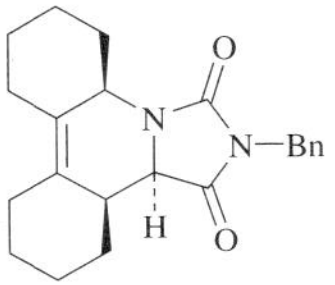	
		Xylene, 138°	(—)[k]	67
		Toluene, 170°, 3 d	(—)[i]	66

TABLE 2. CYCLOADDITIONS OF CYCLIC *N*-ACYLIMINES (*Continued*)

Imine	Diene	Conditions	Product(s) and Yield(s) (%)	Refs.
C_{10} 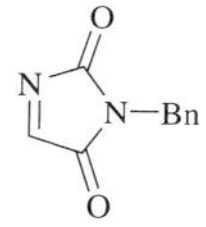	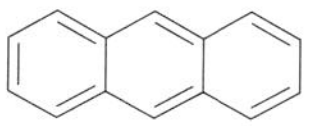		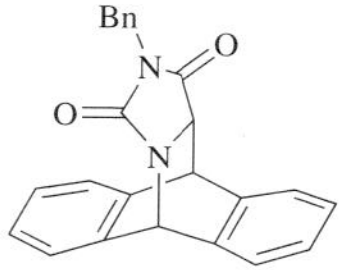	
		Xylene, 138°, 72 h	(70-80)	67
		p-Xylene, 138°, 3 d	(52)	66
	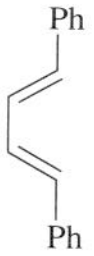		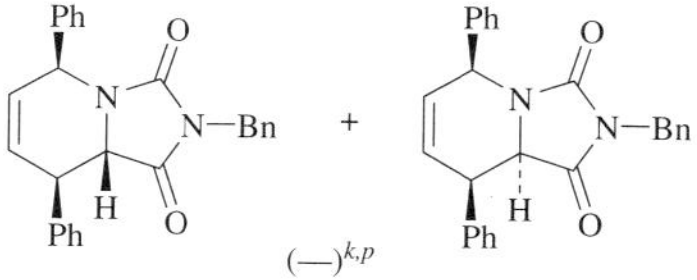	
		TFA, benzene, 80°	(—)[k,p]	67
		Toluene, 170°, 3 d	(28)[i]	66
		β-Naphthalenesulfonic acid, benzene, 80°, 2-4 h	(31)	66
	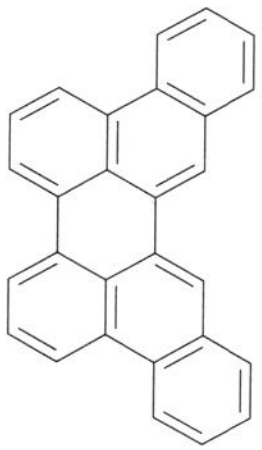		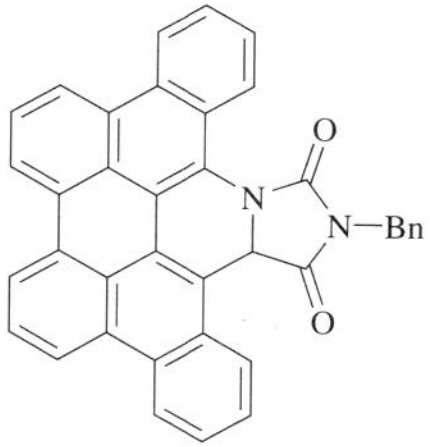	68
		p-Xylene, 160°, 72 h	(60)	
		TFA, benzene, 80°, 24 h	(96)	

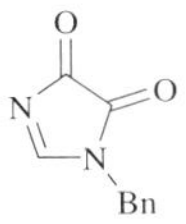	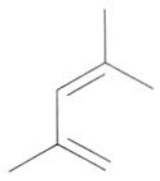	β-Naphthalenesulfonic acid, benzene, 80°, 5 h	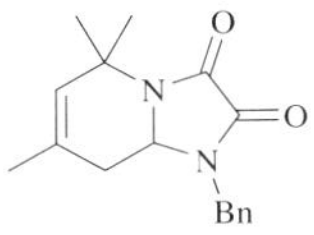 (38)[d]	75
	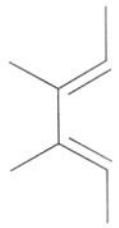	β-Naphthalenesulfonic acid, benzene, 80°, 24 h	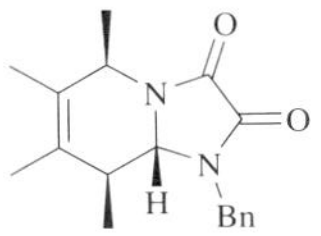 + 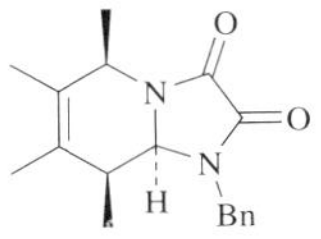 (50)[d]	75
	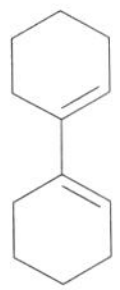	β-Naphthalenesulfonic acid, benzene, 80°, 24 h	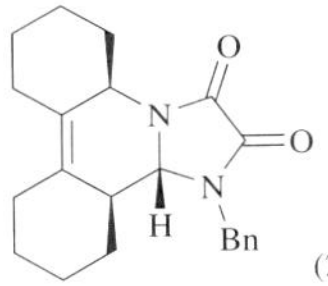 + 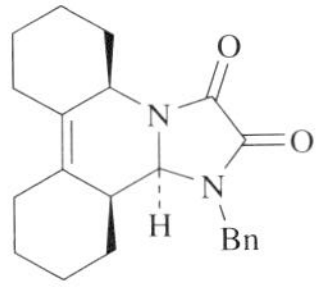 (21)[d]	75
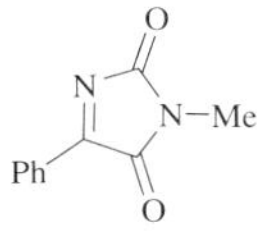	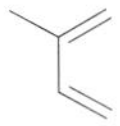	Benzene, 40°, 12 h	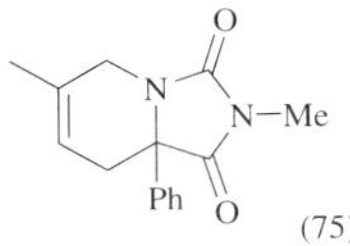 + 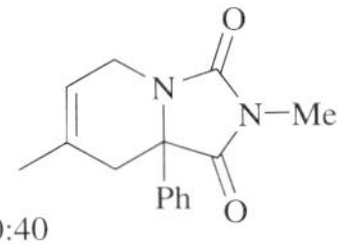 (75) 60:40	71
	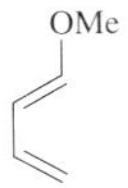	Hydroquinone, benzene, rt, 90 min	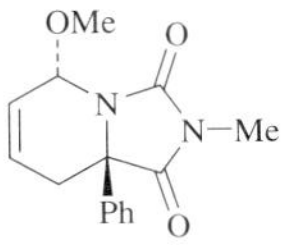 (86)	71
		TMEDA, hydroquinone, benzene, rt, overnight	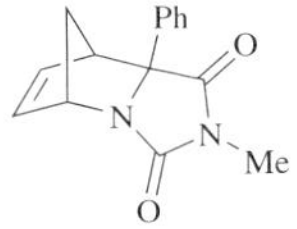 (68)	71

TABLE 2. CYCLOADDITIONS OF CYCLIC *N*-ACYLIMINES (*Continued*)

Imine	Diene	Conditions	Product(s) and Yield(s) (%)	Refs.
C_{10} 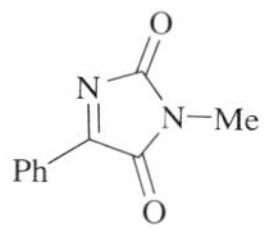	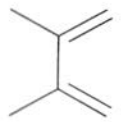	TMEDA, hydroquinone, benzene, 60°, overnight	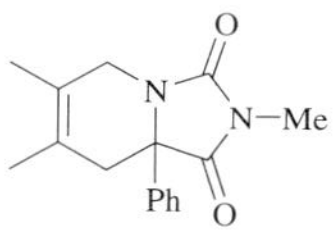 (60)	71
		40°, 10 h	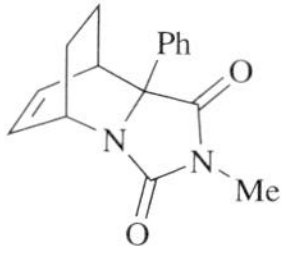 (53)[q]	71
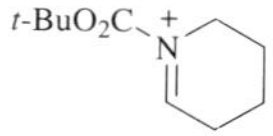	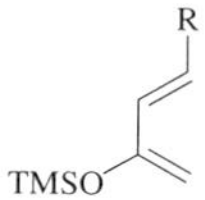	1. TMSOTf, Et_3N, CH_2Cl_2 2. $NaHCO_3$, H_2O, rt, 36 h	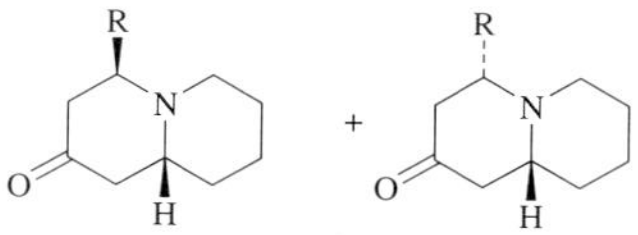	
	R	**Temp / Time**		
	Me	0° / 30 min	(67) 5.5:1.0	299
	Me	rt / 30 min	(67) 5.5:1.0	300
	3,4-$(MeO)_2C_6H_4$	0° / 1 h	(90) 3:2	299
	3,4-$(MeO)_2C_6H_4$	rt / 30 min	(90) 3:2	300
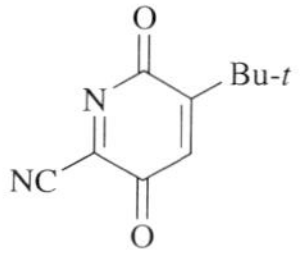		Benzene, rt, 10 sec	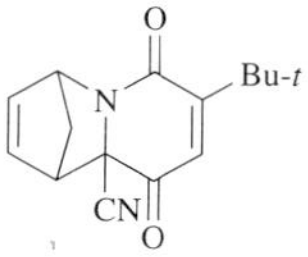 (98)[r]	76
		Benzene, rt, <10 sec	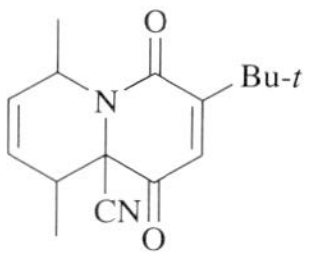 (97)[a]	76

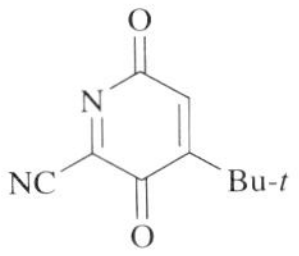		Benzene, rt, <10 sec	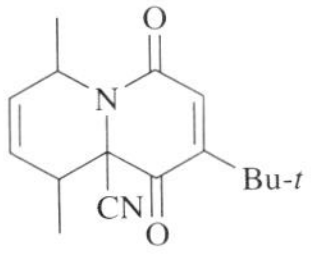 (98)[a]	76
		Benzene, rt, 10 sec	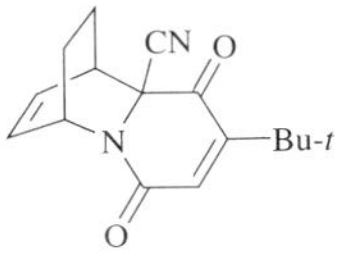 (82)[r]	76
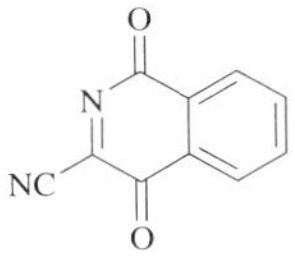		o-Dichlorobenzene, rt, few min	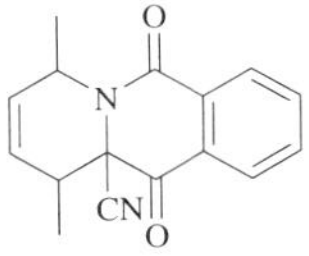 (38)[a]	76
C_{11} 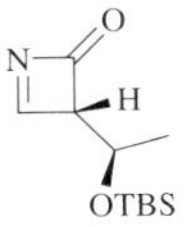	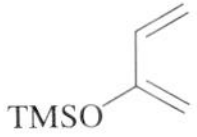	rt, <10 sec	(>95)[a] 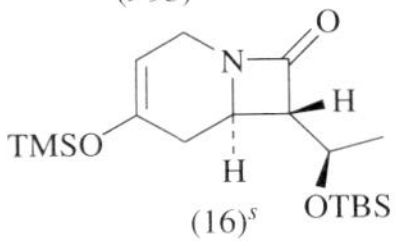	77
		$ZnCl_2$, CH_2Cl_2, rt	(16)[s]	
		$ZnCl_2$, MeCN, 81°	(18)[s]	
	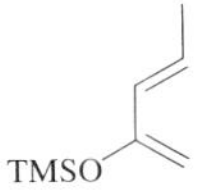	$ZnCl_2$, MeCN, 81°	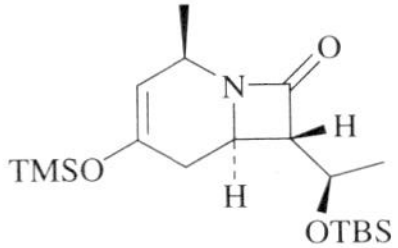 (20)[s,t]	77
		$ZnCl_2$, MeCN, 81°, 3 h	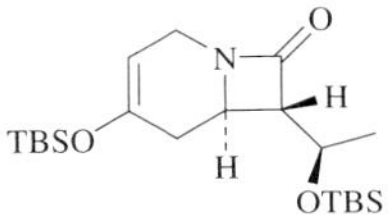 (54-65)	78

TABLE 2. CYCLOADDITIONS OF CYCLIC *N*-ACYLIMINES (*Continued*)

Imine	Diene	Conditions	Product(s) and Yield(s) (%)	Refs.
C_{11} 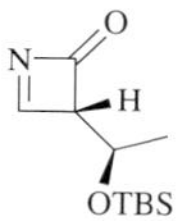	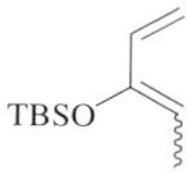		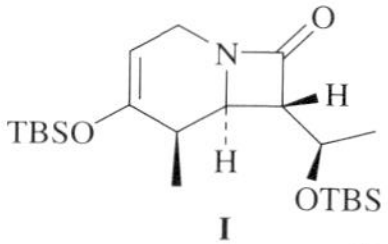 I + 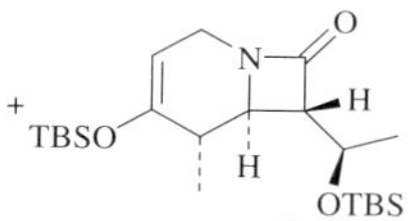 II	79
			I + II I:II	
	Z	$ZnCl_2$	(—) 40:60	
	E	$ZnCl_2$	(—) 80:20	
	Z,E	$ZnCl_2$, MeCN, 81°	(20) 70:30[s]	
	Z,E	$ZnCl_2$, toluene	(65) 70:30[u]	
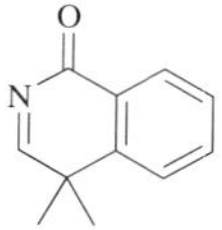	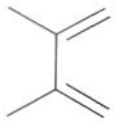	TFA, benzene, 80°, 60 h	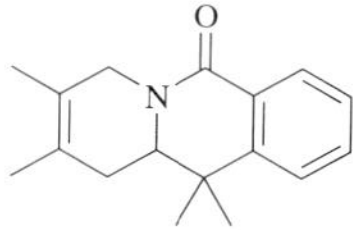 (53)	301
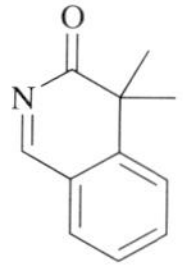	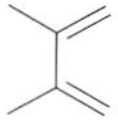	—	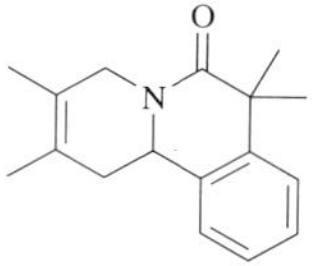 (—)	301
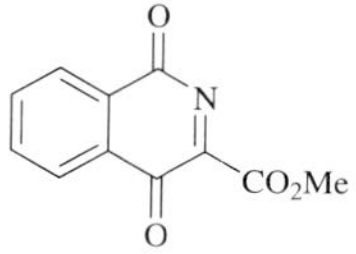		Et_3N, benzene, 80°, 24 h	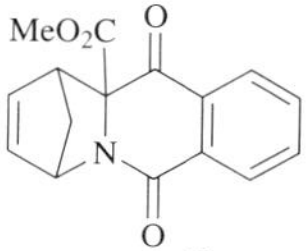 (20)[r]	302
	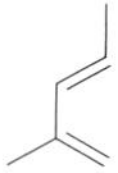	Et_3N, benzene, 80°, 24 h	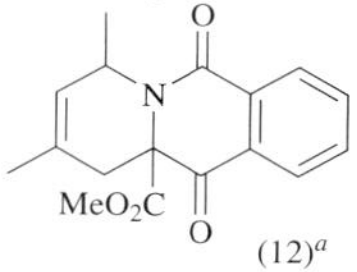 (12)[a] + 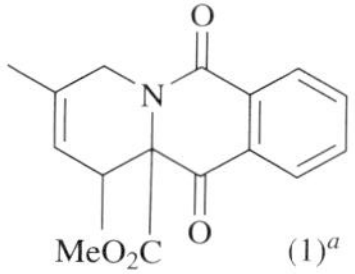 (1)[a]	302

	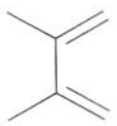		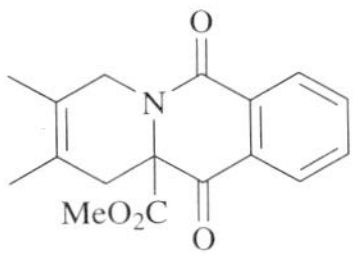	302
		Et_3N, benzene, 80°, 24 h	(15)	
		TsOH, benzene, 80°, 24 h	(23)	
	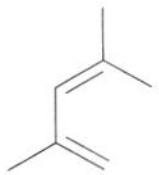	Et_3N, benzene, 80°, 24 h	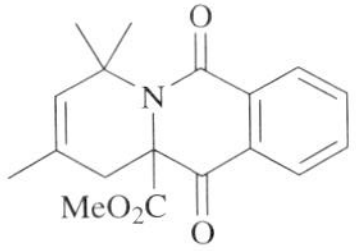 (6)	302
C_{12} 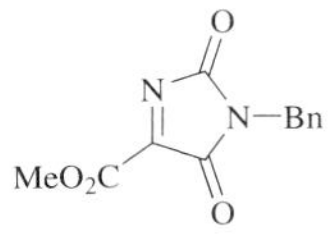	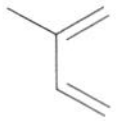	TMEDA, hydroquinone, benzene, rt, 1 h	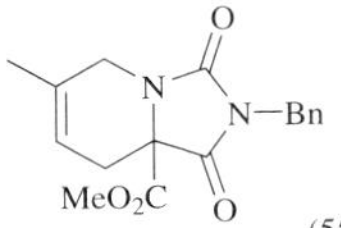 + 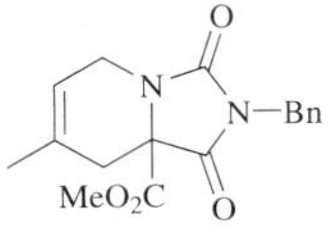 (55) 1:1	71
	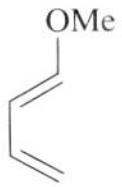	TMEDA, hydroquinone, benzene, rt, 10 min	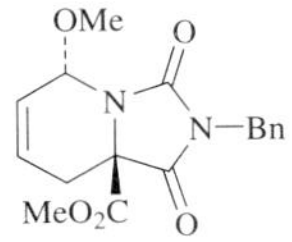 (72)	71
		Et_3N, hydroquinone, benzene, rt, 2 h	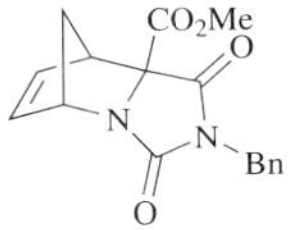 (79)	71
	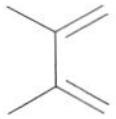	—	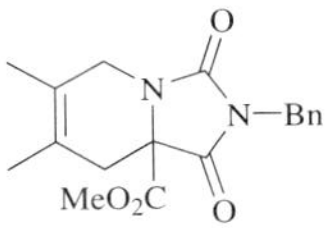 (—)	71

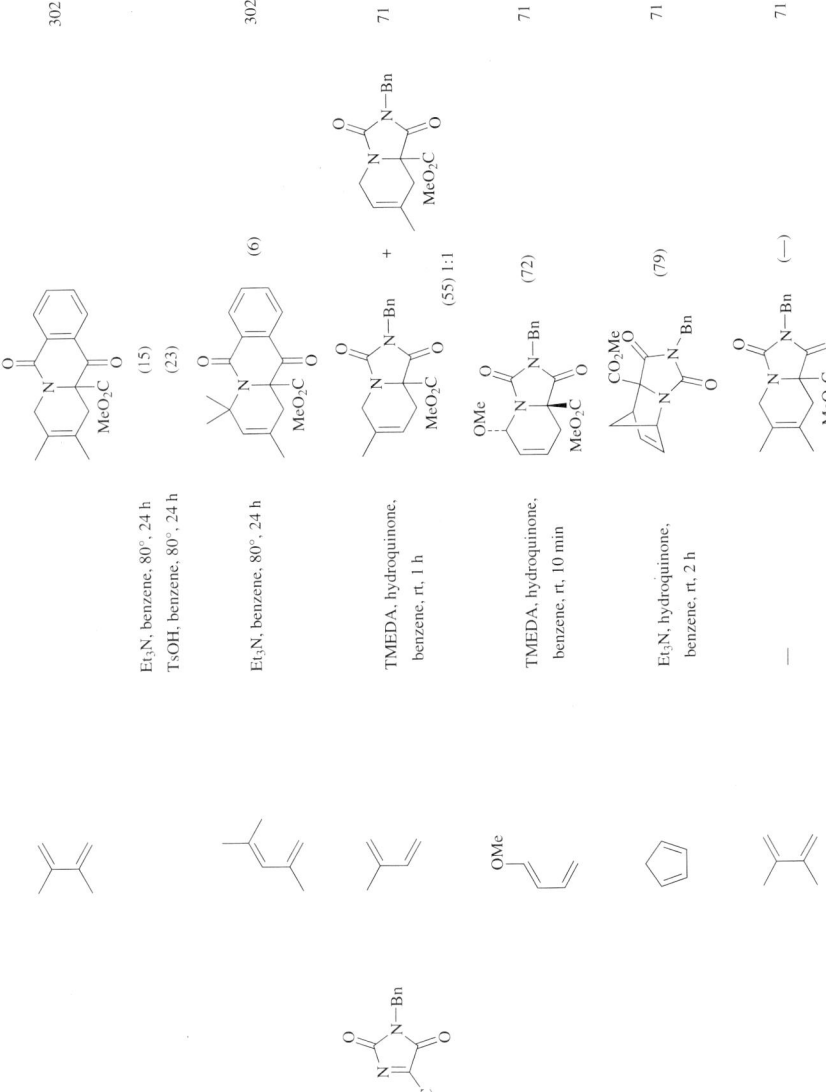

TABLE 2. CYCLOADDITIONS OF CYCLIC N-ACYLIMINES (Continued)

Imine	Diene	Conditions	Product(s) and Yield(s) (%)	Refs.
C_{12}				
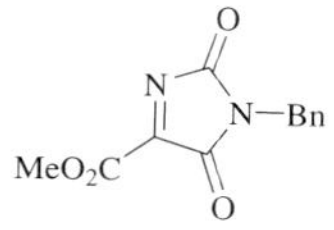		Et_3N, hydroquinone, benzene, rt, overnight	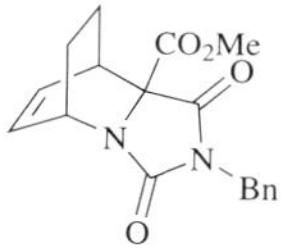 (60)	71
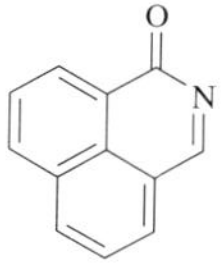	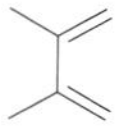	Benzene, rt, 5 d	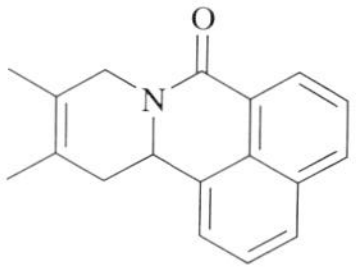 (32)	303
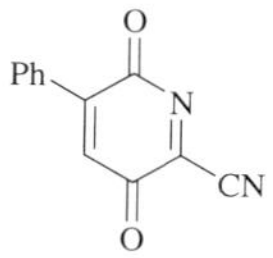		*o*-Dichlorobenzene, rt, 5 min	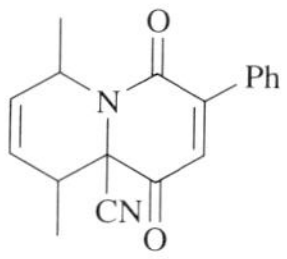 (9)[a]	76
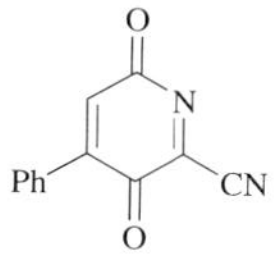		*o*-Dichlorobenzene, rt, 5 min	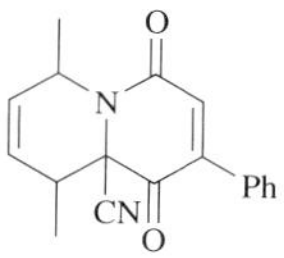 (11)[a]	76
C_{13}				
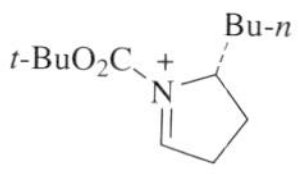	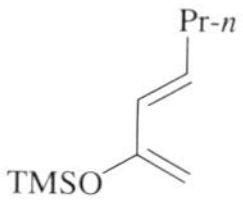	1. Et_3N, TMSOTf, CH_2Cl_2, –78° to rt, 30 min 2. $NaHCO_3$, H_2O, rt, 24 h	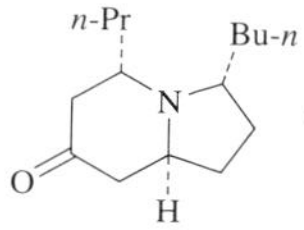 **I** (50)[v] + 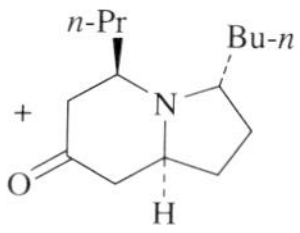 **II** (22)[v]	300
C_{14}				
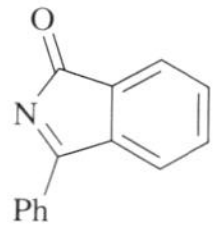	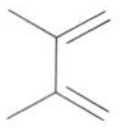	β-Naphthalenesulfonic acid, benzene, 80°, 40 h	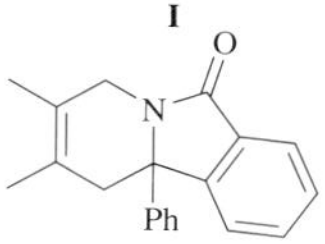 (78)	298

	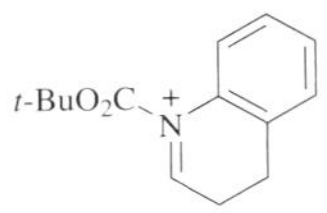	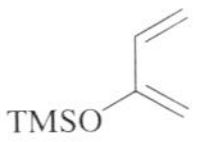	1. Et_3N, TMSOTf, CH_2Cl_2, 0° to rt, 30 min 2. $NaHCO_3$, H_2O, 36 h	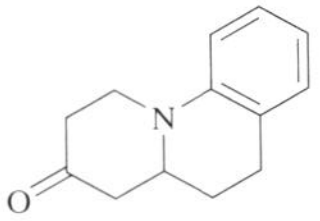 (—)	304
		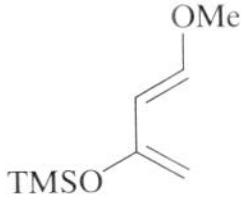	1. Et_3N, TMSOTf, CH_2Cl_2, 0° to rt, 30 min 2. $NaHCO_3$, H_2O, 36 h	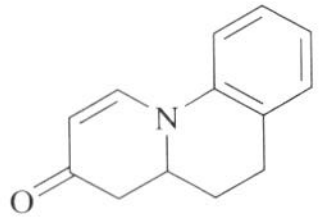 (30)	304
C_{15}	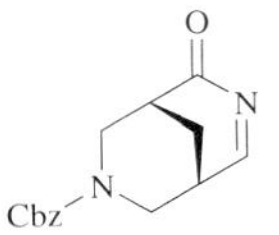	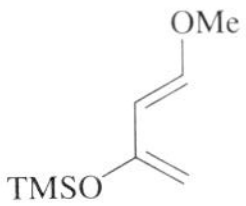	Toluene, 2 d $TiCl_4$	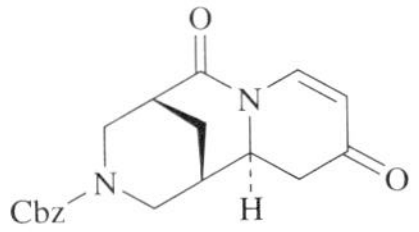 (35)[w] ("very low")	305
	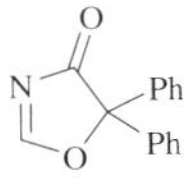	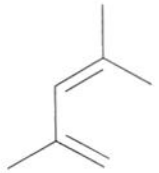	β-Naphthalenesulfonic acid, benzene, 80°, 2 h	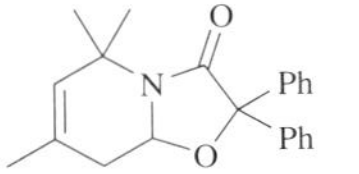 (72)[d]	75
		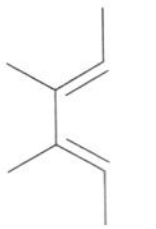	β-Naphthalenesulfonic acid, benzene, 80°, 2 h	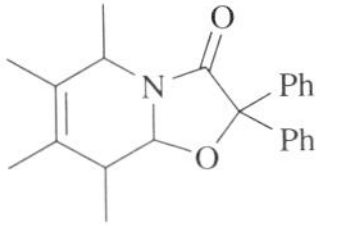 (62)[d,x]	75
		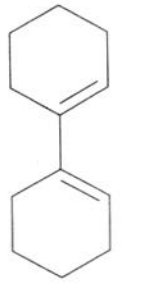	β-Naphthalenesulfonic acid, benzene, 80°, 2 h	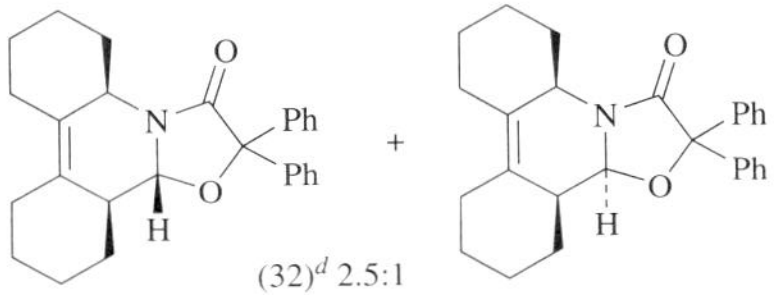 (32)[d] 2.5:1	75

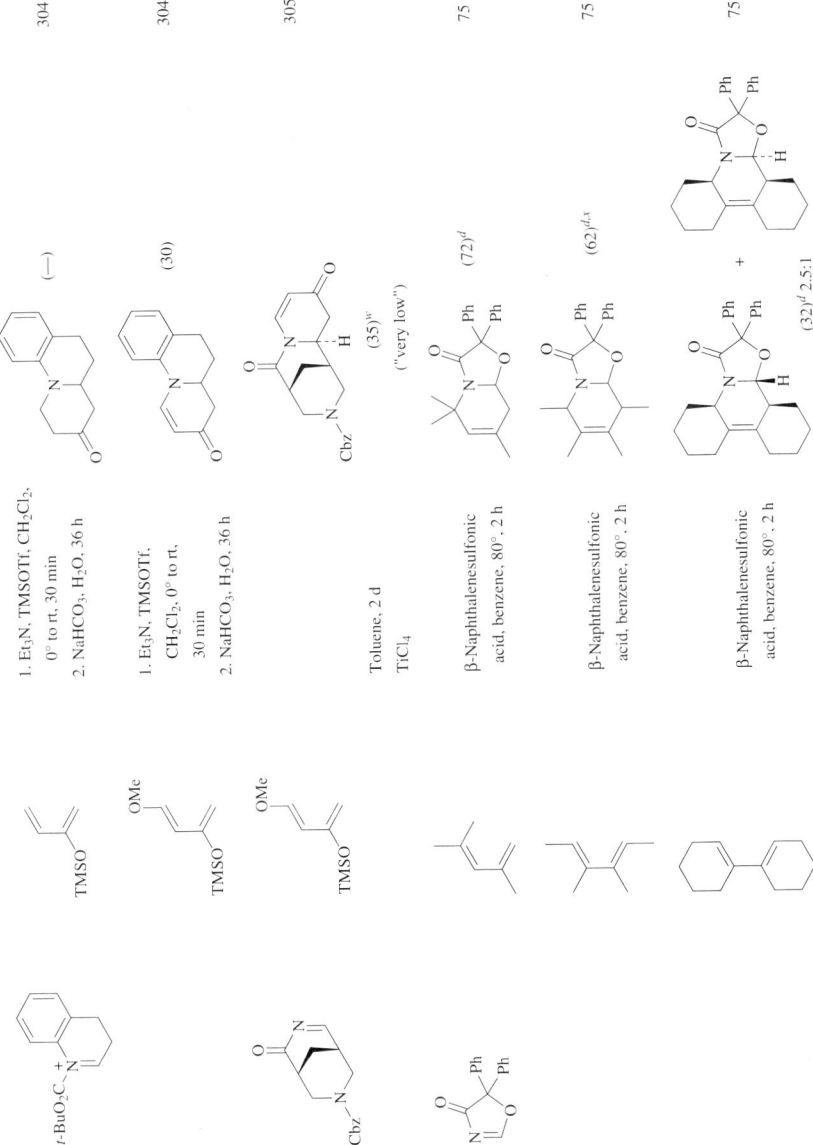

TABLE 2. CYCLOADDITIONS OF CYCLIC *N*-ACYLIMINES (*Continued*)

Imine	Diene	Conditions	Product(s) and Yield(s) (%)	Refs.
C_{15} 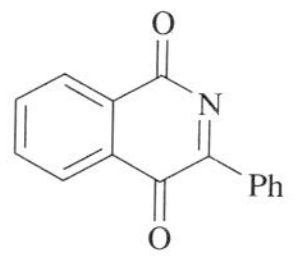	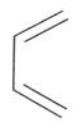	TFA, benzene, rt, 2 h	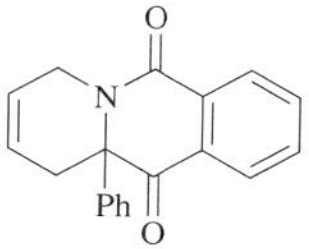 (77)	301
		$CHCl_3$, rt, 4 d	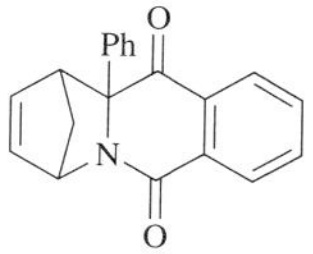 (63)[r]	76
	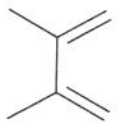	TFA, benzene, rt, 2 h Benzene, rt, 24 h	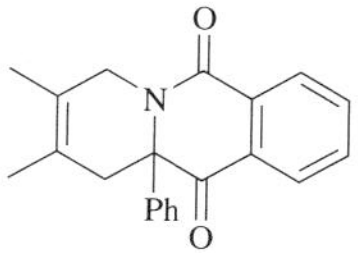 (55) (92)	301
C_{20} 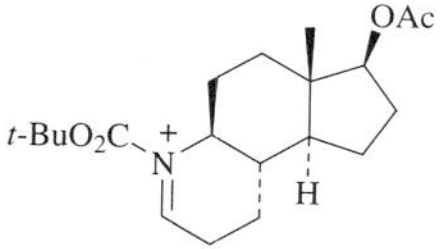	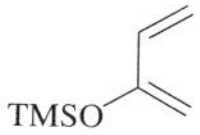	1. TMSOTf, Et_3N, CH_2Cl_2, 0° to rt, 45 min 2. $NaHCO_3$, H_2O, rt, 36 h	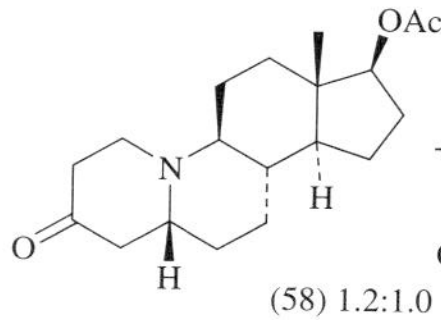 + 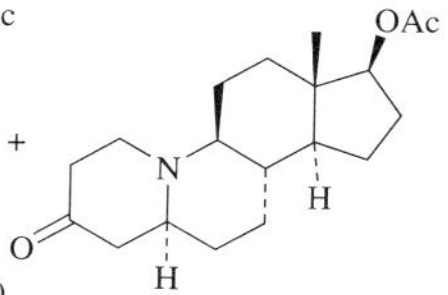 (58) 1.2:1.0	306
	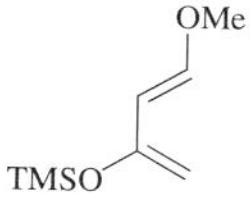	1. TMSOTf, Et_3N, CH_2Cl_2, 0° to rt, 45 min 2. $NaHCO_3$, H_2O, rt, 36 h	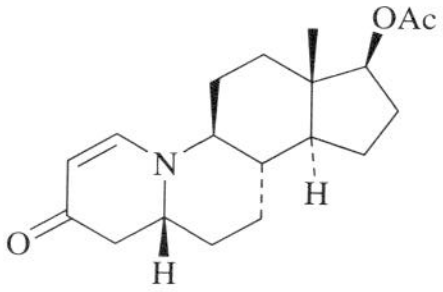 (39)	306

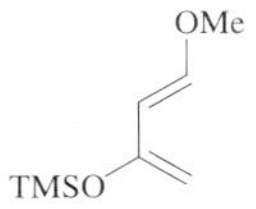

1. Lewis acid
2. $NaHCO_3$, H_2O, rt, 30 min

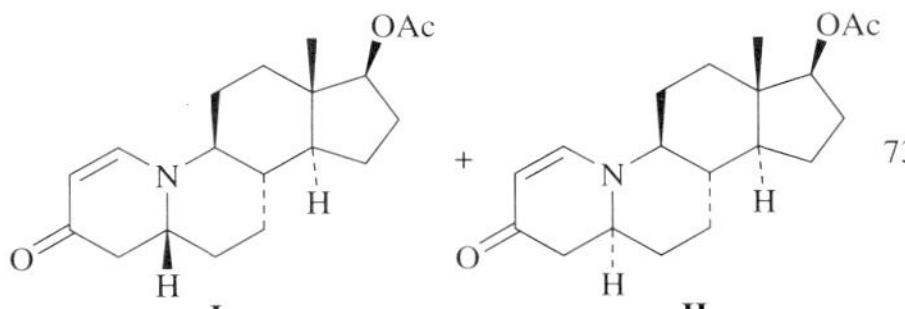

I + II

73

L. A. (eq)	Solvent	Temp	Time	
$TiCl_4$ (1)[y]	CH_2Cl_2	0°	1 h	**I** (15)
$TiCl_4$ (2)[y]	CH_2Cl_2	0°	1 h	**I** (23)
$TiCl_4$ (2)[y]	toluene	0° to rt	1 h	**I** (21)
$TiCl_4$ (2)[z]	CH_2Cl_2	0°	1 h	**II** (18)
$TiCl_4$ (2)[z]	CH_2Cl_2	–30° to rt	1 h	**II** (15)
$TiCl_4$ (2)[z]	CH_2Cl_2	0° to rt	24 h	**II** (15)
$TiCl_4$ (2)[z]	toluene	0° to rt	1 h	**I** (7) + **II** (25)
$AlCl_3$ (2)[z]	CH_2Cl_2	–30° to rt	1 h	**II** (3)
$SnCl_4$ (2)[z]	CH_2Cl_2	–30° to rt	1 h	**II** (3)

C_{23}

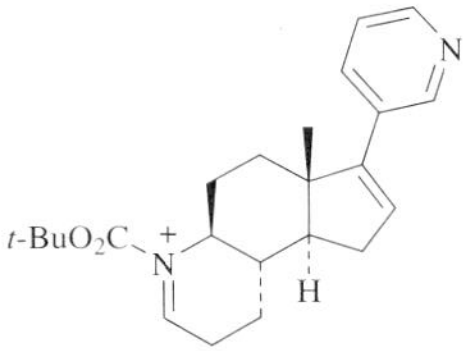

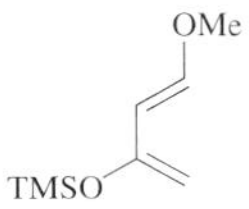

1. Lewis acid, Et_3N, CH_2Cl_2, 0° to rt, 45 min
2. $NaHCO_3$, H_2O, rt, 48 h

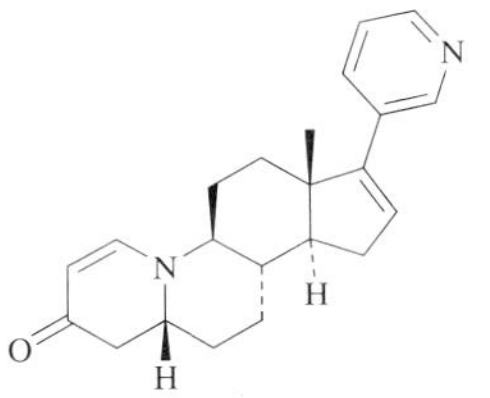

307

L. A.	
TMSOTf	(26)
$TiCl_4$	(trace)

TABLE 2. CYCLOADDITIONS OF CYCLIC *N*-ACYLIMINES (*Continued*)

Imine	Diene	Conditions	Product(s) and Yield(s) (%)	Refs.

[a] No stereochemical information is provided.

[b] The main products of the reaction are the formate esters resulting from formic acid addition to the carbocation formed by *N*-acyliminium ion addition to the diene.

[c] The main products of the reaction are monocyclic lactam dienes.

[d] A lower yield of the cycloadduct(s) was obtained under thermal conditions.

[e] The amidoalkylation product, 3-phenyl-5-(4-phenylbut-2-enyl)imidazolidine-2,4-dione, was obtained in 50% yield.

[f] Acid-catalyzed reactions gave lower yield with di- and mono-substituted dienes.

[g] The authors do not state which regioisomer was obtained or if a mixture was observed.

[h] The authors do not explicitly state the stereochemistry of the product, but based on later results[66] it is most likely the cis-isomer.

[i] High-boiling dienes also reacted in refluxing *p*-xylene, but the yields were generally lower.

[j] Two isomers were obtained, but no stereochemical information is given. They are probably the two isomers corresponding to those obtained with 1,1'-bicyclohexenyl.[66]

[k] Two isomers were obtained, but the stereochemistry of the products is not explicitly stated. Based on later results[66] the products are most likely the ones shown.

[l] The initial ratios of **I**:**II** in the crude mixtures were 67:33 (R = Me), 75:25 (R = Et), and 55:45 (R = CH_2OAc).

[m] This diene was the stereochemically pure trans-diene.

[n] The initial product formed was exclusively adduct **I**, but it epimerized during purification.

[o] Longer reaction times and higher temperatures had little effect on the overall yield. The yield was increased to 56% after one recycling of unreacted starting diene.

[p] An alkylation product was obtained in addition to the two Diels-Alder adducts.

[q] 5-(Cyclohexa-2,5-dienyl)-5-phenyl-3-methylhydantoin was obtained as a minor product.

[r] The authors do not state which isomer (endo or exo adduct) was obtained or if a mixture was observed.

[s] The vinyl ketone resulting from silyl enol ether addition to the imine was obtained as the major product.

[t] The cycloadduct shown is the major stereoisomer obtained. The stereochemistry of the other isomer(s) obtained is not specified.

[u] The vinyl ketone resulting from silyl enol ether addition to the imine was obtained as a minor product.

[v] The ratio of **I**:**II** in the crude mixture was 3:2.

[w] Longer reaction times and higher temperatures had little effect on the yield.

[x] One isomer was obtained, but the stereochemistry was not specified.

[y] The diene was added to the mixture of $TiCl_4$ and the iminium ion precursor.

[z] The Lewis acid was added to a mixture of the diene and the iminium ion precursor.

TABLE 3. CYCLOADDITIONS OF *C*-ACYLIMINES

Imine	Diene	Conditions	Product(s) and Yield(s) (%)	Refs.
C_3				
		HCl, H_2O, rt, 22 h	 + 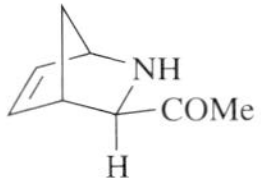 (84) 2:1	86
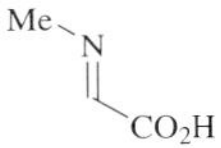		H_2O, rt, 22 h	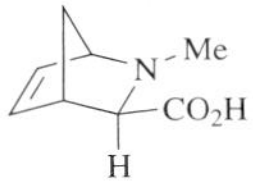 + 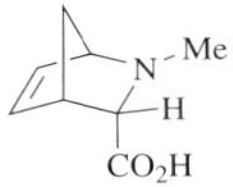 (86) 1.9:1	86
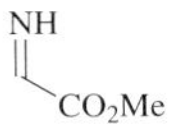		HCl, H_2O	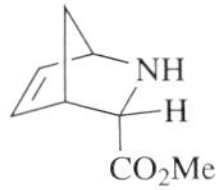 (—)	88
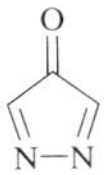	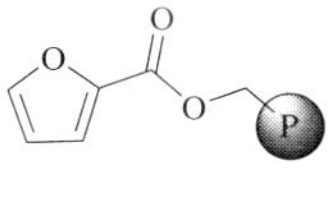	1. TFA, dioxane, 100°, 24 h 2. KOH, dioxane, H_2O, 100°, 2 d 3. H^+	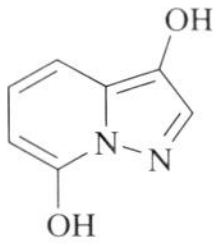 (23.1)	72
C_4				
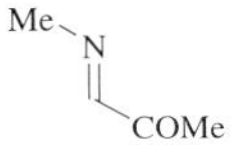		HCl, H_2O, rt, 20 h	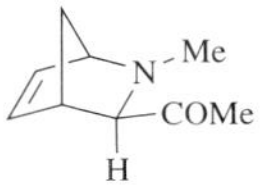 + 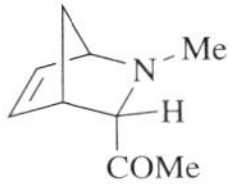 (67) 3.6:1	86
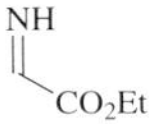		HCl, H_2O, rt, 7 h	 + 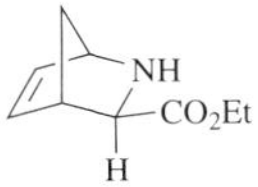 (84)	308
		HCl, H_2O, rt, 7 h	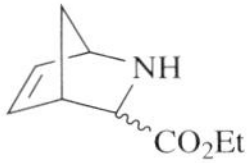 (69)[a]	309

Imine	Diene	Conditions	Product(s) and Yield(s) (%)	Refs.
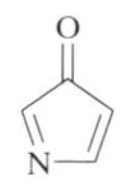	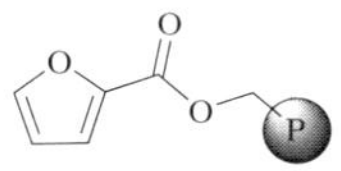	1. DMSO, 189°, 3 d 2. KOH, dioxane, H_2O, 100°, 2 d 3. H^+	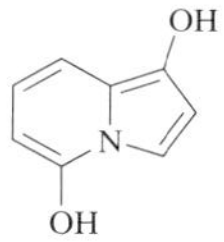 (8.6)	72
C_5 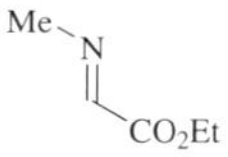	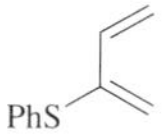	HCl, H_2O, DMF, rt, 24 h	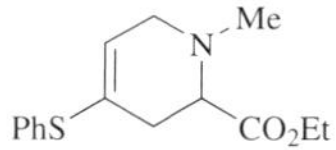 $(68)^b$, $(51)^c$	310
	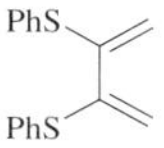	HCl, H_2O, DMF, 24 h	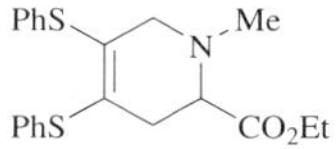	310
		Temp		
		rt	(44)	
		60°	(49)	
C_6 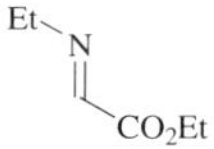	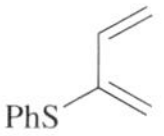	HCl, H_2O, DMF, rt, 24 h	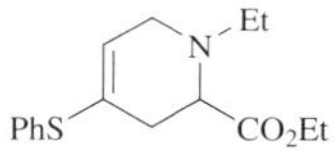 $(64)^b$, $(80)^d$	310
	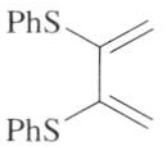	HCl, H_2O, DMF, rt, 24 h	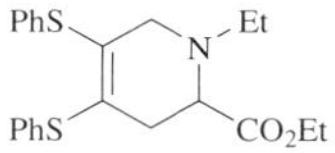 (43)	310
C_7 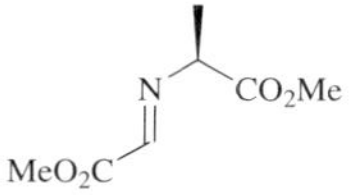		HCl, H_2O, rt, 24 h	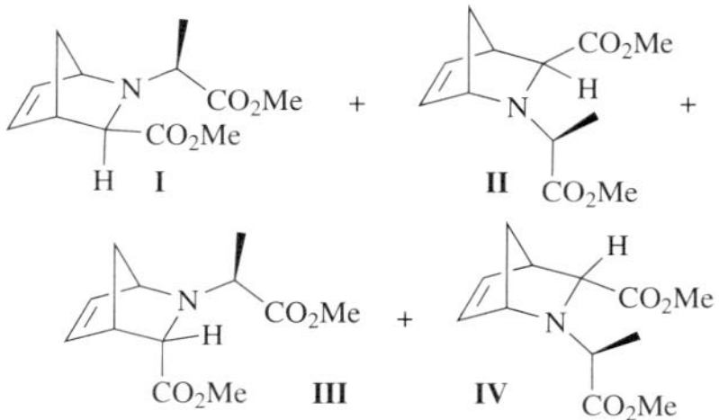 (15) **I**:**II** 83:17, **I** + **II**:**III** + **IV** 7:1	99

Imine	Diene	Conditions	Product(s) and Yield(s) (%), Ratio	Refs.
C_8 		HCl, H_2O, rt	 (89) 2:1	86
C_9 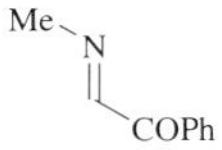		HCl, H_2O, rt, 22 h	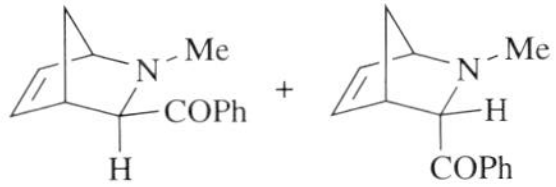 (82) 4.2:1	86
	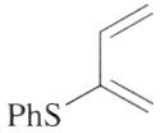	HCl, H_2O, rt, 24 h	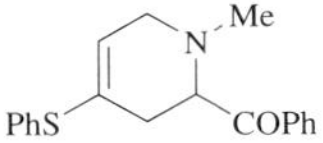	311
		Solvent		
		DMF[e]	(72)	
		DMF[f]	(63)	
		MeCN	(57)	
		EtOH	(39)	
		CH_2Cl_2	(31)	
		THF	(29)	
		H_2O	(trace)	
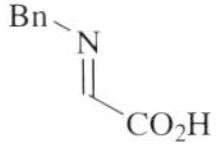		HCl, H_2O, rt, 22 h	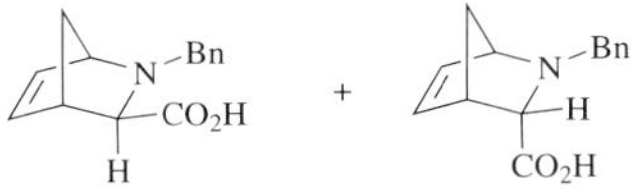 (80) 1.4:1	86
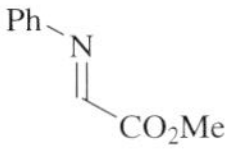	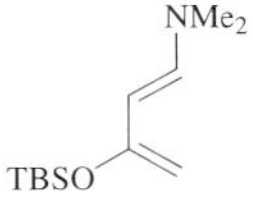	1. CH_2Cl_2, rt, 5 h 2. HCl, H_2O, THF	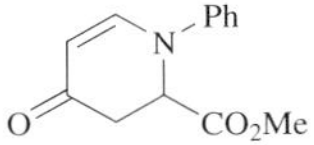 (84)	53
C_{10} 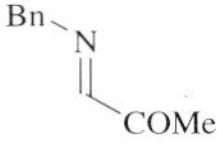		HCl, H_2O, rt, 18 h	 (65) 10:1	86

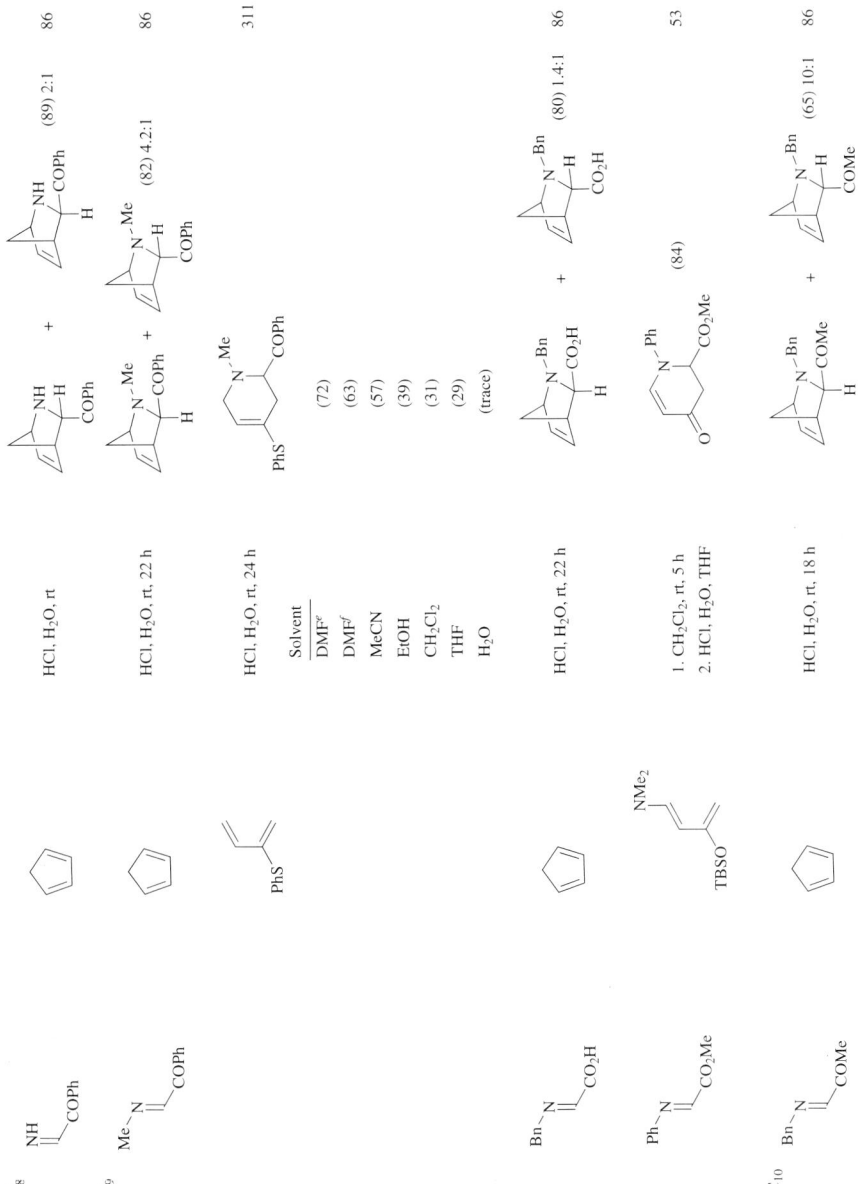

TABLE 3. CYCLOADDITIONS OF *C*-ACYLIMINES (*Continued*)

	Imine	Diene	Conditions	Product(s) and Yield(s) (%)	Refs.
C_{10}	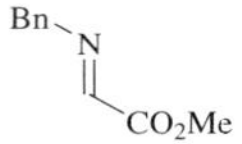		—	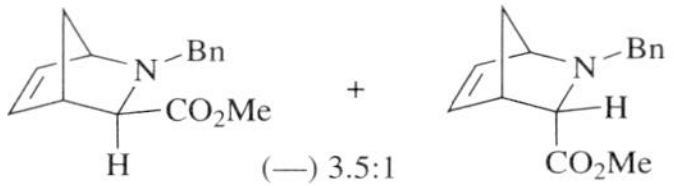 (—) 3.5:1	312
			HCl	(—) 1:2	313
	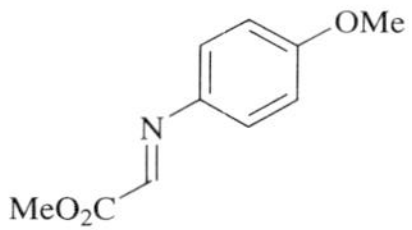	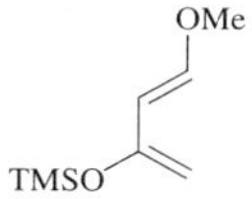	1. $Yb(OTf)_3$, CH_2Cl_2 2. SiO_2	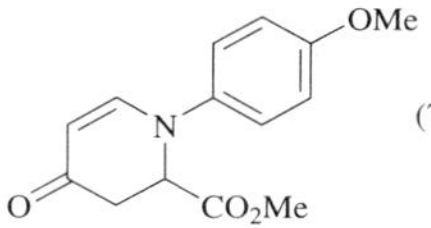 (76)	80
	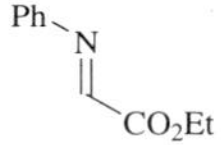	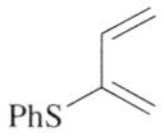	HCl, H_2O, DMF, rt, 24 h	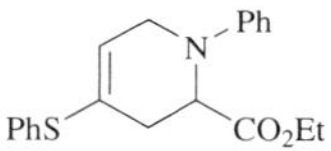 (93)	310
	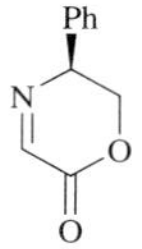	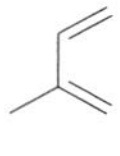	TFA, BF_3•OEt_2, –78°, 3-6 h	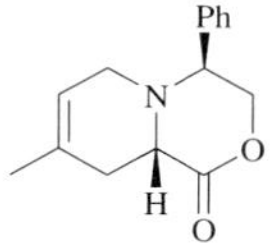 (37)	105
			TFA, BF_3•OEt_2, –78°, 3-6 h	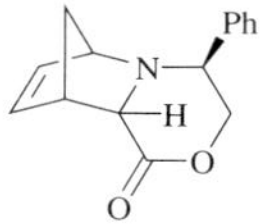 (28)	105
		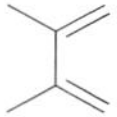	AcOH, BF_3•OEt_2, –78° to rt, 3-6 h	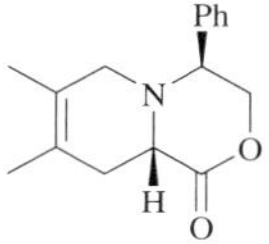 (37)	105
C_{11}	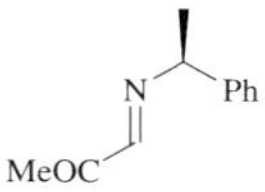		HCl, H_2O, 0°, overnight	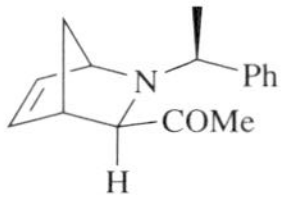 (31)	314

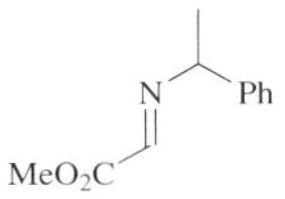		TFA, BF_3•OEt_2, CH_2Cl_2	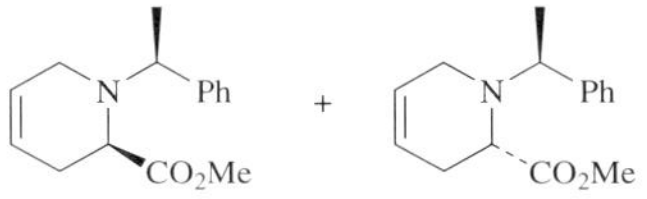	82
		Temp / Time		
		0° / 4 h	(64) 66:34	
		–10° / 2 h	(52) 65:35	
		–10° / 4 h	(60) 65:35	
		–30° / 2 h	(35) 67:33	
	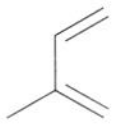	TFA, BF_3•OEt_2, CH_2Cl_2	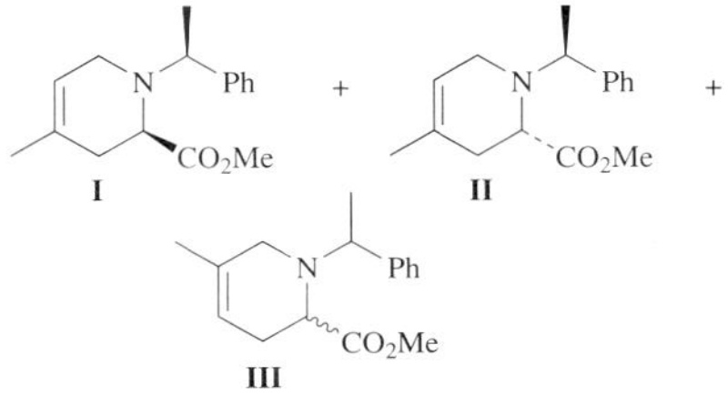I II III	82
		Temp / Time		
		0° / 2 h	**I** + **II** + **III** (72), **I** + **II**:**III** 79:21, **I**:**II** 64:36[g]	
		–80° / 2 h	**I** + **II** + **III** (40), **I** + **II**:**III** 97:3, **I**:**II** 77:23[g]	
		TFA, BF_3•OEt_2, CH_2Cl_2	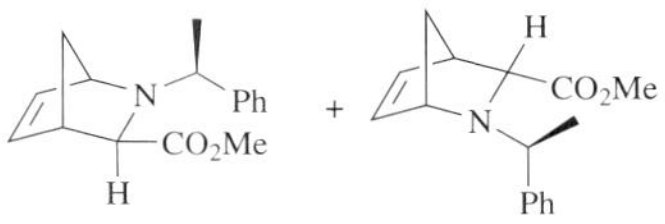	
		Temp / Time		
		–78° / 5 h	(94) 98:2	315
		0° / —	(—) 87:13	315
		–80° / 2 h	(94) 98:2, >98% d.e.[h]	82
		0° / 2 h	(95) 87:13, >98% d.e.[h]	82

TABLE 3. CYCLOADDITIONS OF *C*-ACYLIMINES (*Continued*)

Imine	Diene	Conditions	Product(s) and Yield(s) (%)	Refs.
C_{11} 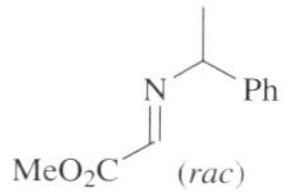 (*rac*)	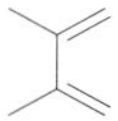	TFA, BF_3•OEt_2, CH_2Cl_2	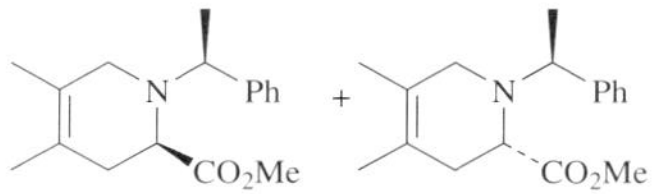	82
		Temp / Time: 20° / 1 h	(74) 62:38	
		0° / 2 h	(87) 64:36	
		–80° / 2 h	(46) 77:23	
		TFA, BF_3•OEt_2, CH_2Cl_2	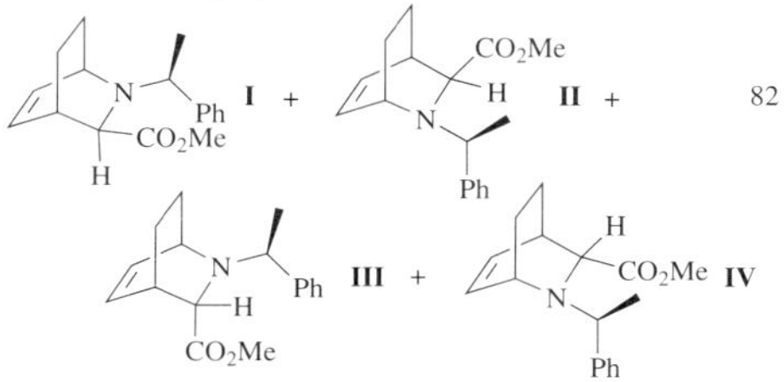 **I** + **II** + **III** + **IV**	82

Temp	Time	I + II + III + IV	I:II	I + II:III + IV
20°	8 h	(95)	77:23	74:26
0°	6 h	(75)	79:21	71:29
–80°	19 h	(50)	88:12	78:22

Imine	Diene	Conditions	Product(s) and Yield(s) (%)	Refs.
	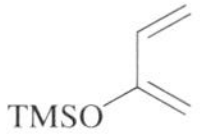	1. TFA, BF_3•OEt_2, CH_2Cl_2 2. H_2O	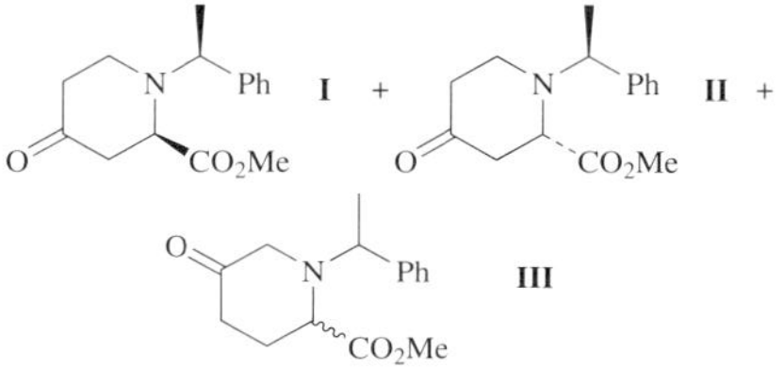 **I** + **II** + **III**	82

Temp	Time	I + II	I:II	I + II:III
0°	2 h	(95)	67:33	>98:2[i]
–80°	2 h	(65)	75:25	>98:2[i]

Reactant	Diene	Conditions	Product(s) and Yield(s) (%)	Refs.
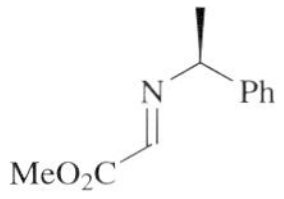	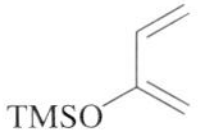	1. ZnI_2, THF, 0-20°, 2 h 2. $NaHCO_3$, H_2O, rt	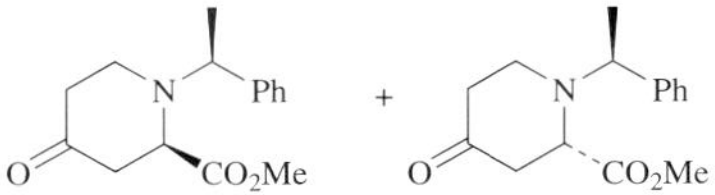 (85) 65:35	89
	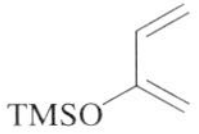	ZnI_2, THF, 0-20°, 2 h	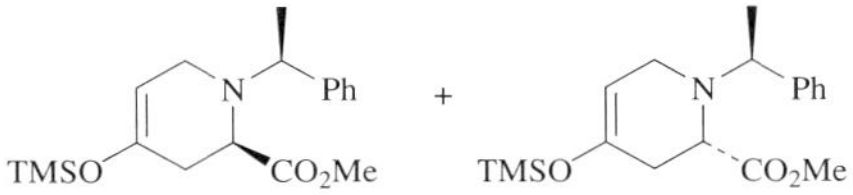 (75) 65:35	89
	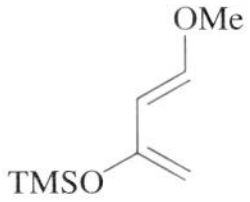	1. ZnI_2, THF 2. $NaHCO_3$, H_2O	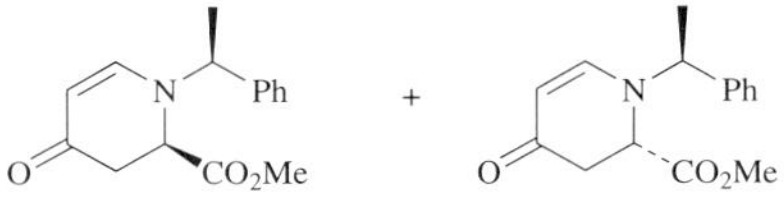	89
		Temp / Time		
		0-20° / 3 h	(85) 62:38	
		–80° to rt / 15 h	(85) 70:30	
		TFA, BF_3•OEt_2, CH_2Cl_2	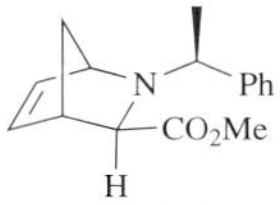	
		Temp / Time		
		–78° / 5 h	(—)	315
		–78° / —	(—)	316
		–75-78° / 20 h	(56)	317

Imine	Diene	Conditions	Product(s) and Yield(s) (%)	Refs.
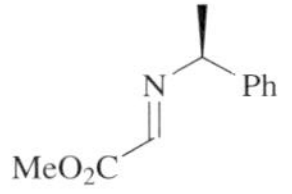	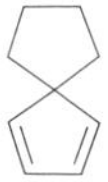	TFA, BF_3•OEt_2, CH_2Cl_2, –78° to rt, overnight	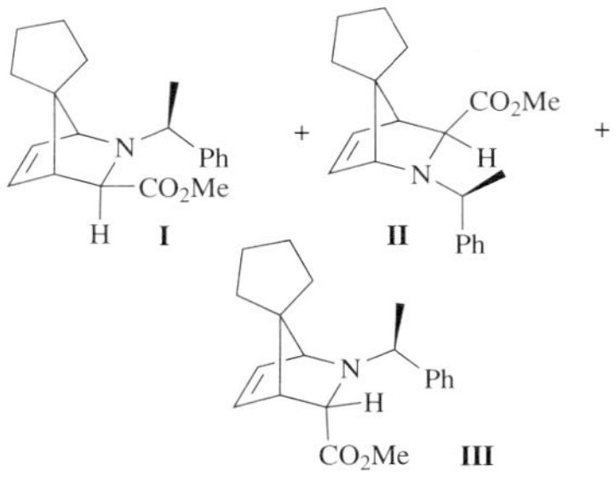 **I** + **II** + **III** **I** (34), **I**:**II**:**III** 70:20:10[j]	91
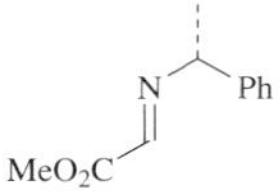		TFA, BF_3•OEt_2, CH_2Cl_2, –78°, 5 h	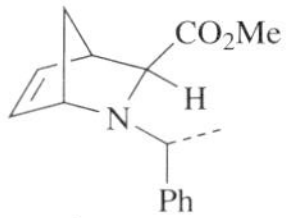 (—)	315
		TFA, BF_3•OEt_2, CH_2Cl_2, –78° to rt, overnight	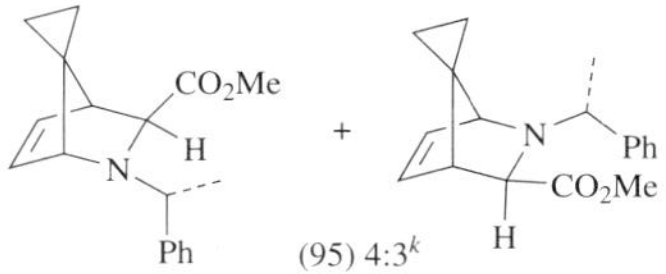 (95) 4:3[k]	91
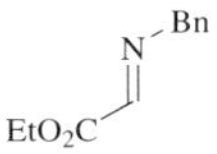	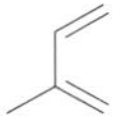	HCl, H_2O, DMF, rt, 35 h	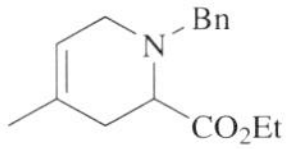 (43)	84, 318
	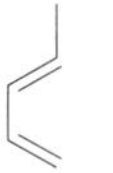	HCl, H_2O, DMF, rt, 45 h	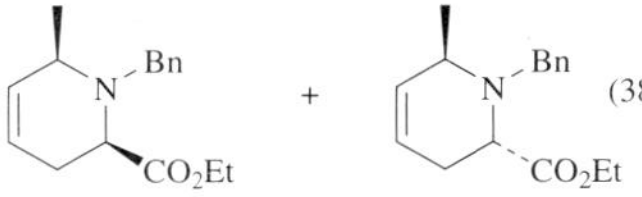 (38) 93:7	84, 318

	HCl, H_2O, rt	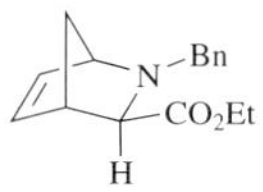 + 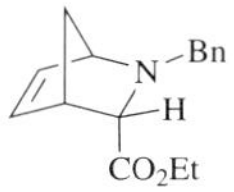		84, 318
	Solvent / Time			
	EtOH / —		(17)	
	H_2O / —		(52)	
	DMF / 15 h		(89) 69:31	
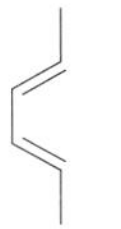	HCl, H_2O, DMF, rt, 48 h	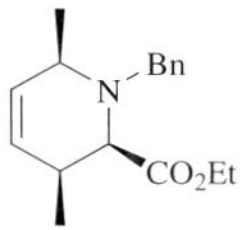	(36)	84, 318
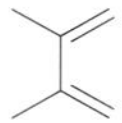	Acid, H_2O, DMF, rt	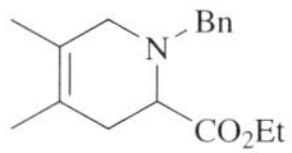		
	Acid / H_2O (eq) / Time			
	TFA / 0.01 / —	(94)[l]		84, 318
	TFA / 0.03 / 35 h	(94)		83
	TFA / 0.1 / —	(94)[l]		84, 318
	TFA / 1.0 / —	(29)		84, 318
	TFA / 10 / —	(9)		84, 318
	HCl / 1.0 / 35 h	(47)		84, 318
	HCl, H_2O, DMF, rt, 32 h	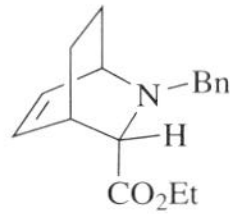 + 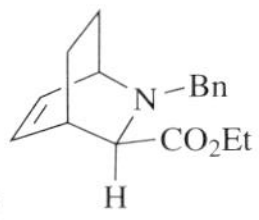	(21) 73:27	84, 318

TABLE 3. CYCLOADDITIONS OF *C*-ACYLIMINES (*Continued*)

Imine	Diene	Conditions	Product(s) and Yield(s) (%)	Refs.
C_{11} 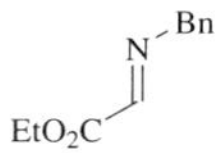	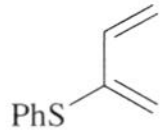	HCl, H_2O, DMF, rt, 24 h	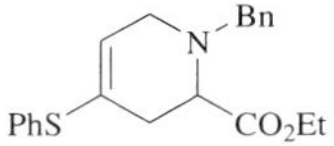 (56)[m], (89)[d]	310
	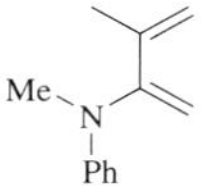	$Yb(OTf)_3$, THF, rt, 12 h	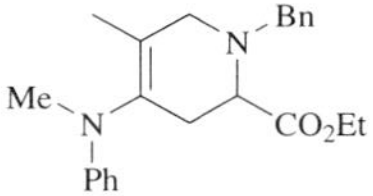 (89)	319
	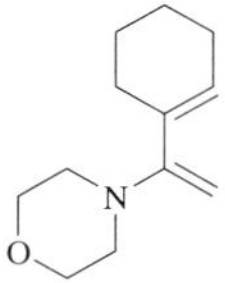	$Yb(OTf)_3$, THF, rt	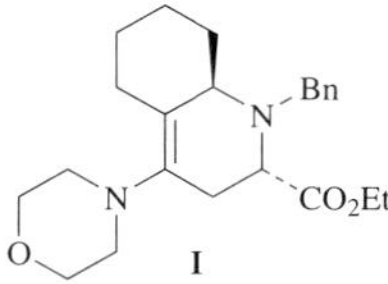 + 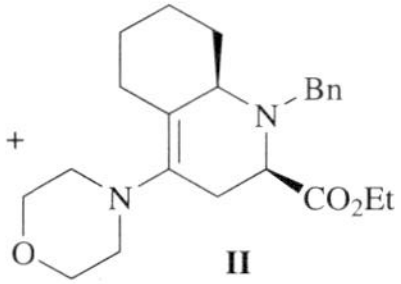 **I + II** (—)[n]	319
	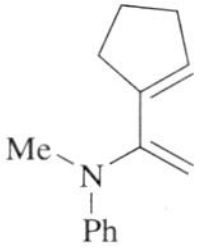	Lewis acid, THF, rt, 12 h	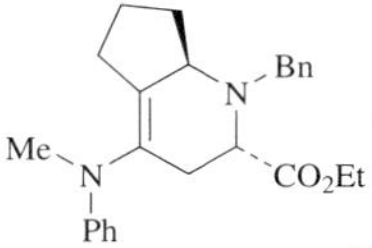 + 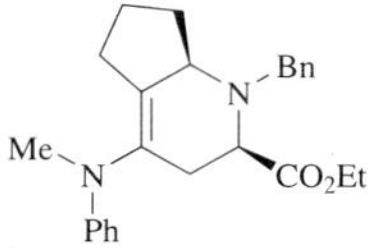	319
		L. A.		
		$CuClO_4$	(40) 3:1	
		$Yb(OTf)_3$	(55) <20:1	
		$Cu(OTf)_2$	(48) <20:1	
	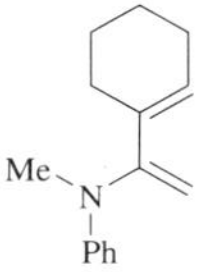	Lewis acid, rt, 12 h	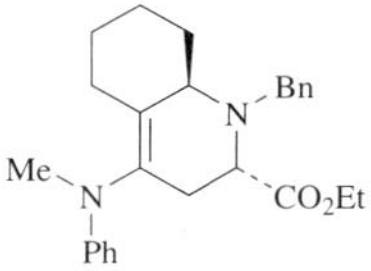 + 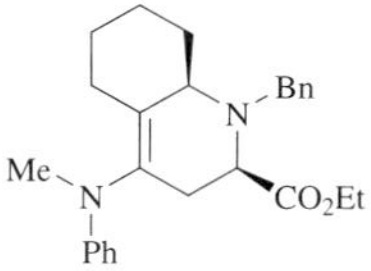	319

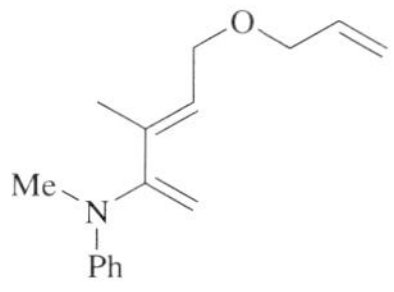

Lewis acid, THF, rt, 12 h

L. A. (eq)	Solvent	
$CuClO_4$ (0.1)	THF	(50) 1:4
$CuClO_4$ (0.1)	CH_2Cl_2	(42) 1:4
$CuClO_4$ (0.2)	CH_2Cl_2	(46) 1:4
$CuClO_4$ (0.2)	THF	(66) 1:7
$Cu(OTf)_2$ (0.2)	THF	(63) 10:1
$Yb(OTf)_3$ (0.2)	THF	(72) 19:1

L. A.	
$CuClO_4$	(50) 3.5:1
$Yb(OTf)_3$	(55) 4:1

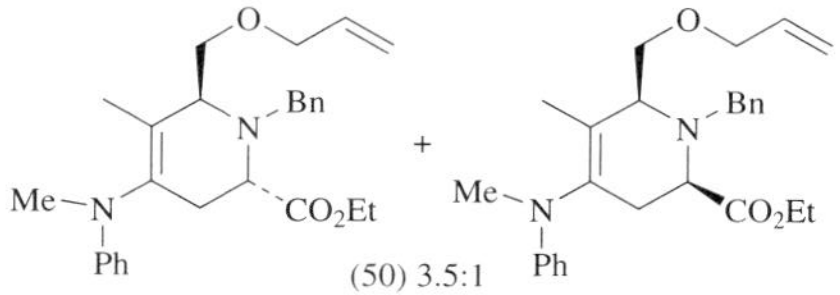

319

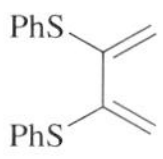

HCl, H_2O, DMF, rt, 24 h

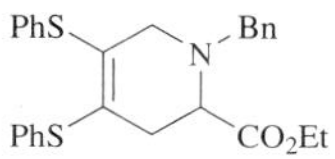

(61)

310

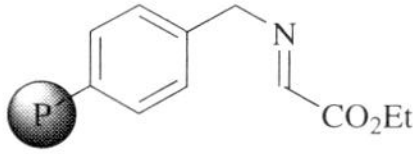

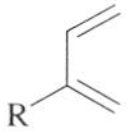

1. $Yb(OTf)_3$, CH_2Cl_2, rt, 12-48 h
2. $ClCO_2CH(Cl)CH_3$, CH_2Cl_2, rt, 3 h
3. MeOH, 65°, 2 h

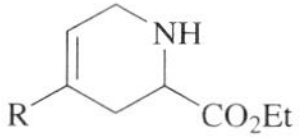

R	
H	(83)
Me	(96)

320

TABLE 3. CYCLOADDITIONS OF *C*-ACYLIMINES (*Continued*)

Imine	Diene	Conditions	Product(s) and Yield(s) (%)	Refs.
C_{11} 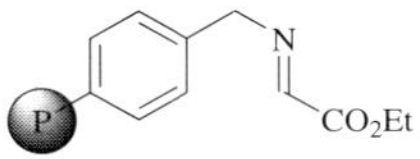	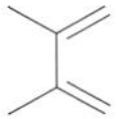	1. Lewis acid, CH_2Cl_2, rt, 12-48 h 2. $ClCO_2CH(Cl)CH_3$, CH_2Cl_2, rt, 3 h 3. MeOH, 65°, 2 h	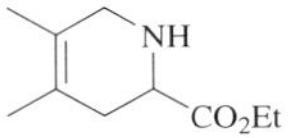	320
		L. A.		
		$La(OTf)_3$•xH_2O	(85)	
		$Nd(OTf)_3$	(92)	
		$Dy(OTf)_3$	(82)	
		$Yb(OTf)_3$	(93)	
		$Yb(OTf)_3$•xH_2O	(91)	
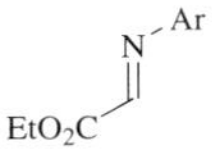 Ar = 4-$MeOC_6H_4$	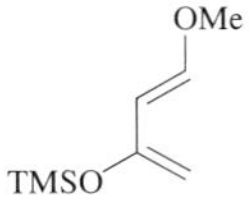	$Yb(OTf)_3$, MeCN, 5 h	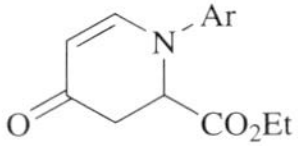 (65)	46
	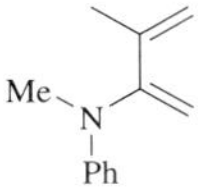	$Yb(OTf)_3$, THF, rt, 12 h	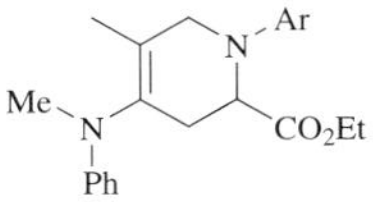 (90)	319
	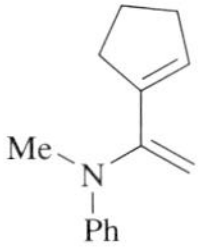	Lewis acid, THF, rt, 12 h	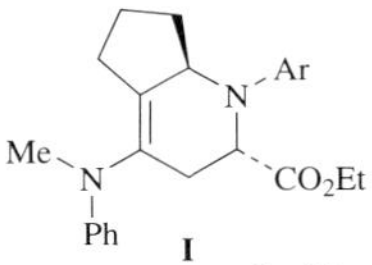 **I** + 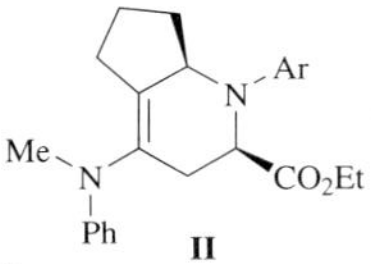 **II**	319
		L. A.	**I + II** / **I:II**	
		$CuClO_4$	(64) 1:1	
		$Yb(OTf)_3$	(64) 1:1	
		$Cu(OTf)_2$	(64) 1:1	

	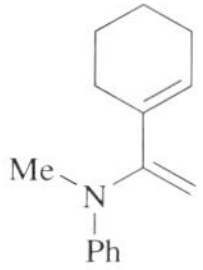	Lewis acid, THF, rt, 12 h	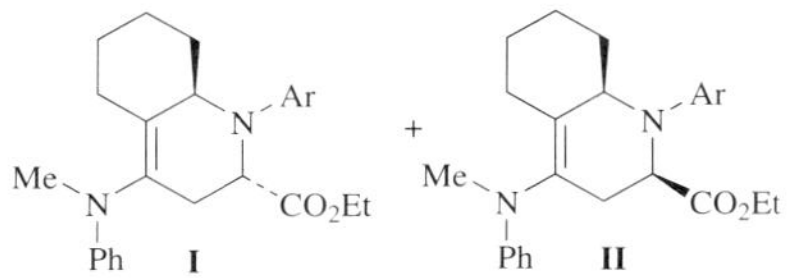 I + II	319
		L. A. / CuClO$_4$ / Cu(OTf)$_2$ / Yb(OTf)$_3$	I + II / (83) / (74) / (77); I:II / 4:1 / 4:1 / 4.5:1	
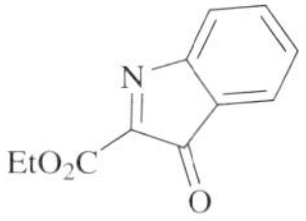	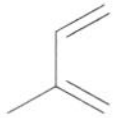	Pb(OAc)$_4$, CH$_2$Cl$_2$, rt, 8-14 h	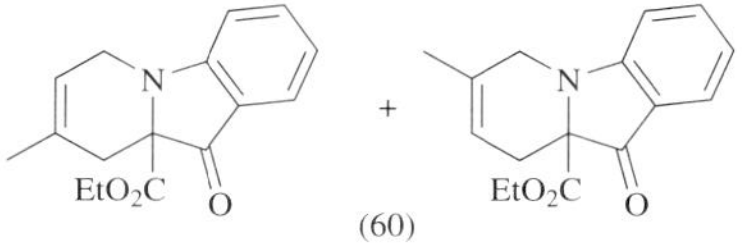 (60)	102, 101
		Pb(OAc)$_4$, CH$_2$Cl$_2$, rt, 8-14 h	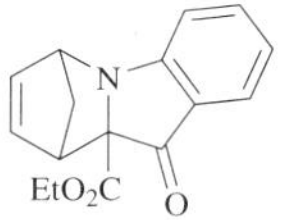 (80)[a]	102, 101
	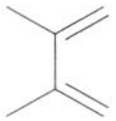	Pb(OAc)$_4$, CH$_2$Cl$_2$, rt, 8-14 h	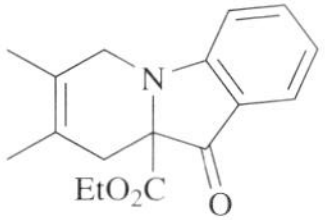 (65)	102, 101
		Pb(OAc)$_4$, CH$_2$Cl$_2$, rt, 8-14 h	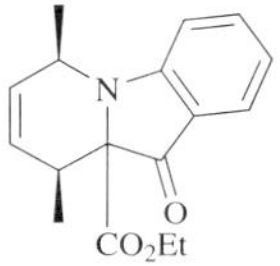 (70)[o]	102, 101

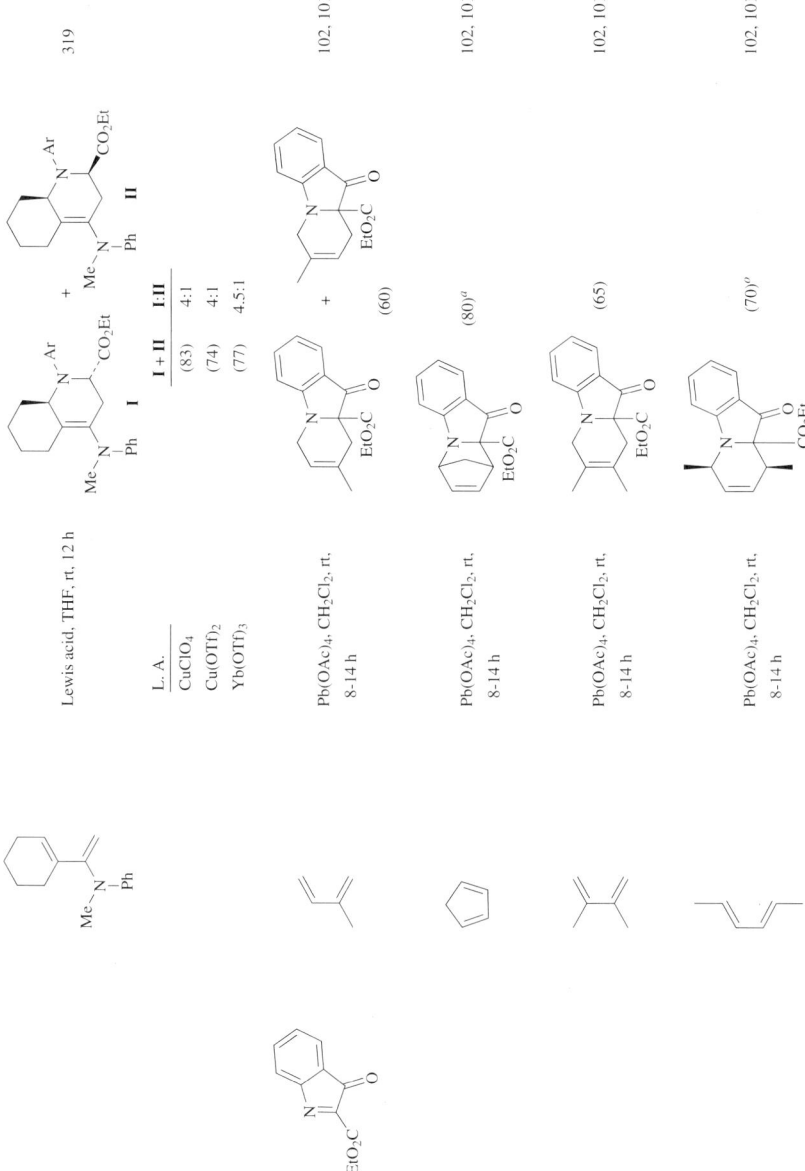

TABLE 3. CYCLOADDITIONS OF *C*-ACYLIMINES (*Continued*)

Imine	Diene	Conditions	Product(s) and Yield(s) (%)	Refs.
C_{11}				
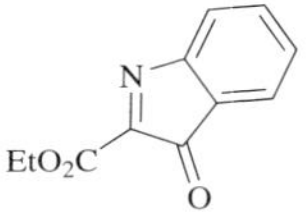	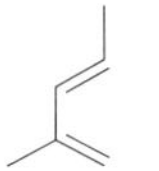	$Pb(OAc)_4$, CH_2Cl_2, rt, 8-14 h	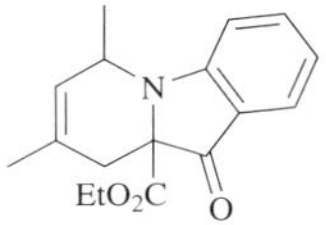 + 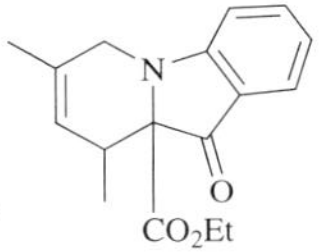 (50)[a]	102, 101
	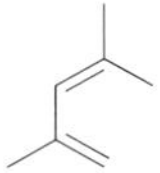	$Pb(OAc)_4$, CH_2Cl_2, rt, 8-14 h	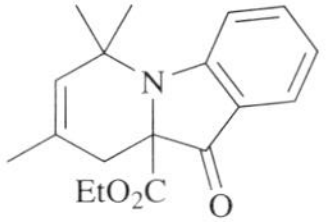 (35)	102, 101
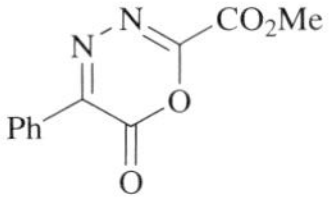	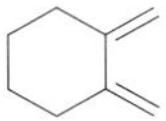	Toluene, 110°, 7 d	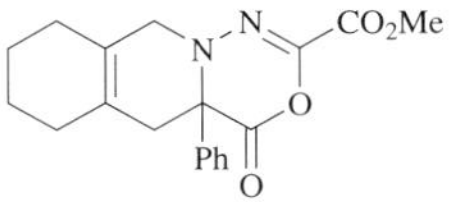 (—)[p]	104
C_{12}				
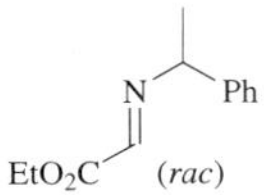 (*rac*)		TFA, BF_3•OEt_2, CH_2Cl_2, 0°, 4 h	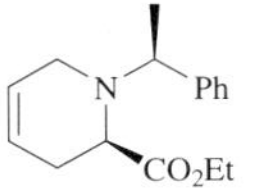 + 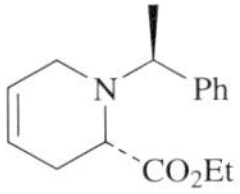 (64) 66:34	82
	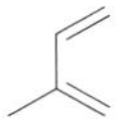	TFA, BF_3•OEt_2, CH_2Cl_2, 0°, 2 h	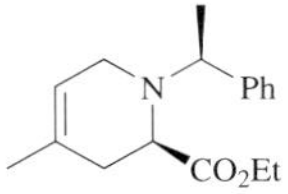 **I** + 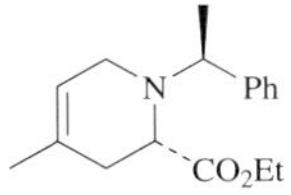 **II** + 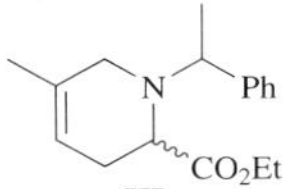 **III** **I** + **II** + **III** (86), **I** + **II**:**III** 79:21, **I**:**II** 68:32[g]	82

		TFA, BF_3•OEt_2, CH_2Cl_2 Temp / Time –80° / 2 h 0° / 2 h	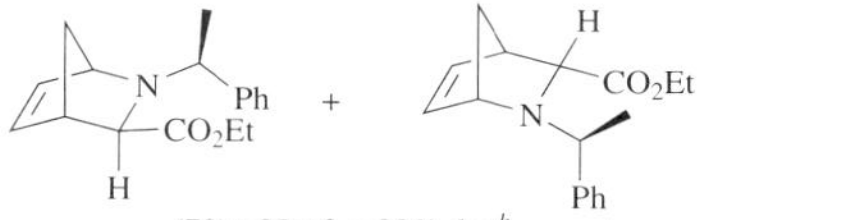 (70) >98:<2, >98% d.e.[h] (75) 87:13, >98% d.e.[h]	82
	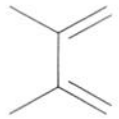	TFA, BF_3•OEt_2, CH_2Cl_2 Temp / Time 0° / 2 h –80° / 2 h	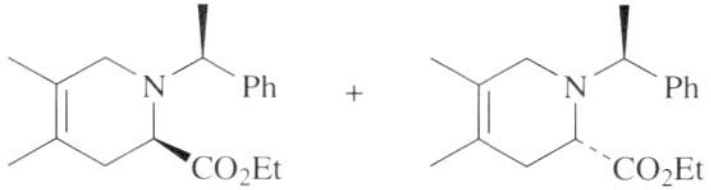 (89) 65:35 (35) 80:20	82
		TFA, BF_3•OEt_2, CH_2Cl_2	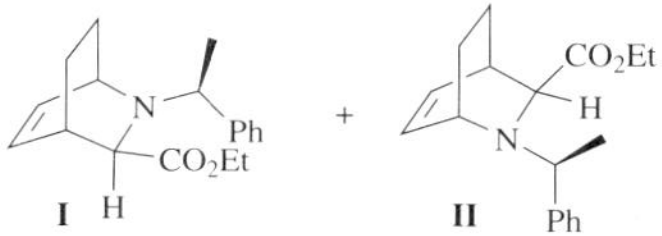 I + II + 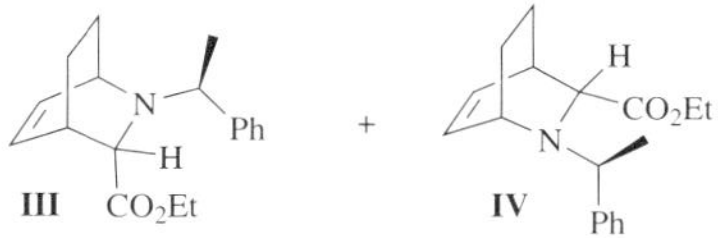 III + IV	82
		Temp / Time 20° / 8 h 0° / 6 h	I + II + III + IV / I:II / I + II:III + IV (90) / 78:22 / 74:26 (60) / 80:20 / 76:24	
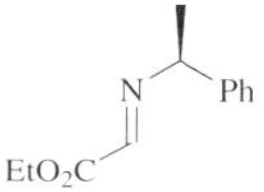	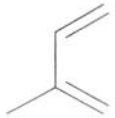	TFA, H_2O, DMF, rt, 30 h	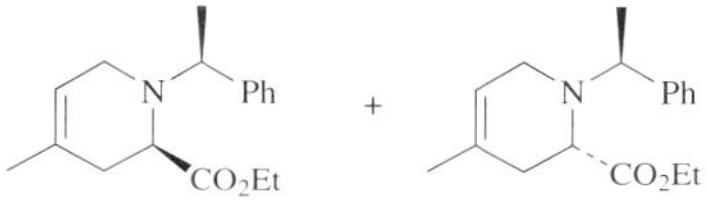 (—)	84

TABLE 3. CYCLOADDITIONS OF *C*-ACYLIMINES (*Continued*)

Imine	Diene	Conditions	Product(s) and Yield(s) (%)	Refs.
C_{12} 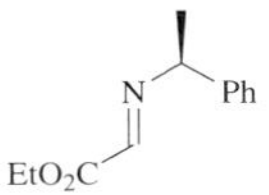			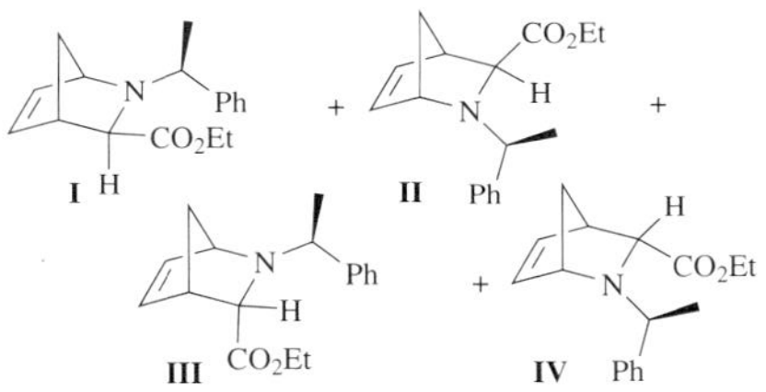	
		TFA, BF_3•OEt_2	**I** (70) **I**:**II** 90:10, **I** + **II**:**III** + **IV** >98:2	91
		TFA, BF_3•OEt_2, CH_2Cl_2	**I** (60)	96, 95, 97, 94
		—	**I** (—)	90
		TFA, BF_3•OEt_2, CH_2Cl_2, –78° to rt, 48 h	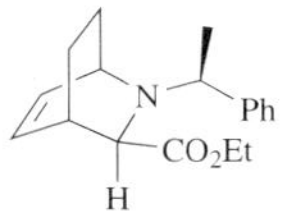 (31)	94, 97
	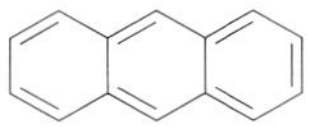	TFA, BF_3•OEt_2, CH_2Cl_2, –78° to rt, overnight	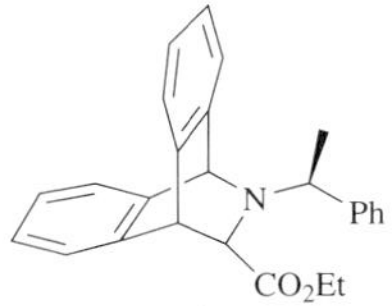 (37)[q]	91
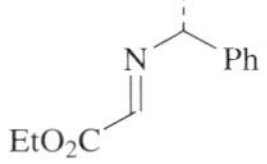	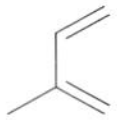	TFA, H_2O, DMF, rt, 30 h	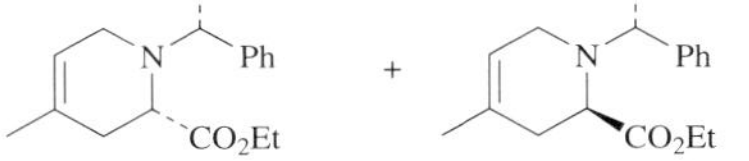 (44) 85:15	83, 84
			(44) 81:19	84

	TFA, H_2O, DMF, rt, 36 h	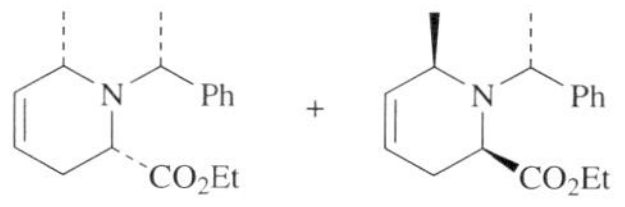	
		(35) 69:31	83
		(35) 68:32	84
		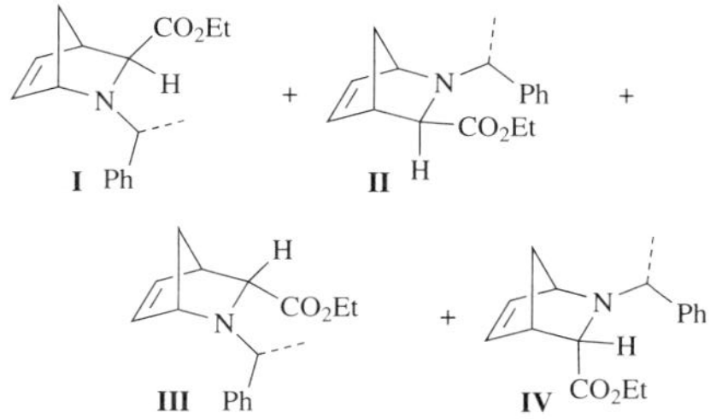	
	HCl, H_2O, rt, 24 h	(52) **I**:**II** 90:10, **I** + **II**:**III** + **IV** 6:1	99
	TFA, H_2O, DMF, rt, 24 h	**I** + **II** + **III** (82), **I** + **II**:**III** 97:3, **I**:**II** 94.5:5.5	84, 83
	—	**I** (—)[r]	321, 322
	—	**I** (—)	94
	TFA, BF_3•OEt_2, CH_2Cl_2, –60° to rt, overnight[s]	**I** (—)	90
	TFA, BF_3•OEt_2, CH_2Cl_2, –60° to rt, overnight[t]	**I** (—)	90
	TFA, DMF, H_2O, rt, overnight	**I** (—)	90

TABLE 3. CYCLOADDITIONS OF *C*-ACYLIMINES (*Continued*)

Imine	Diene	Conditions	Product(s) and Yield(s) (%)	Refs.
C_{12} 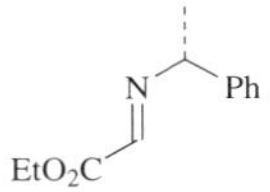	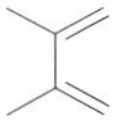	TFA, H_2O, DMF, rt, 24 h	 (69) 86:14	83
			(69) 84:16	84
	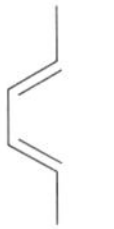	TFA, H_2O, DMF, rt, 36 h	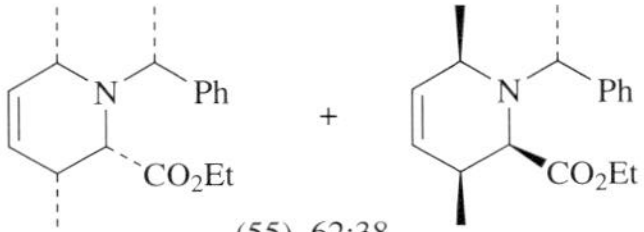 (55) 62:38	83
			(55) 63:37	84
			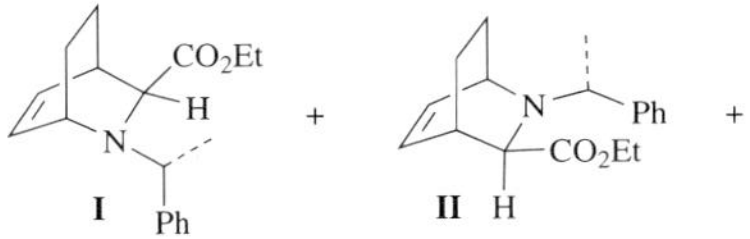 I + II + 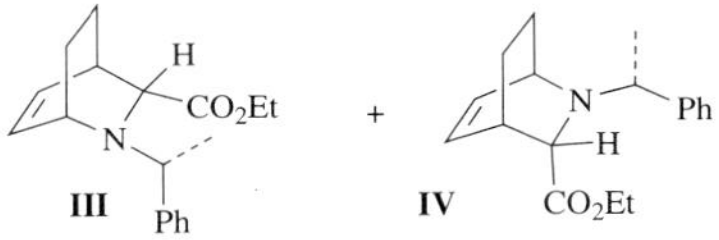 III + IV	
		TFA, H_2O, DMF, rt, 60 h	**I** + **II** + **III** (31), **I** + **II**:**III** 92:8, **I**:**II** 92:8	83, 84
		TFA, BF_3•OEt_2, CH_2Cl_2	**I** (—)	323
		TFA, BF_3•OEt_2, CH_2Cl_2, –80° to rt, 74 h	**I** (65) + **IV** (8)	324, 325
	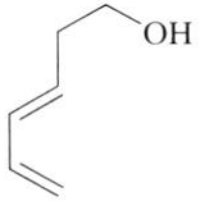	$MeSO_3H$, TFA	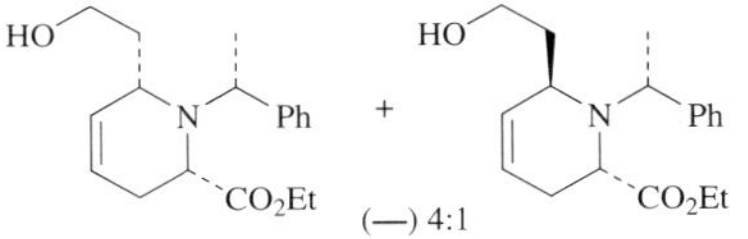 (—) 4:1	326

C_{14}				
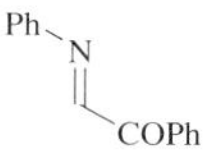	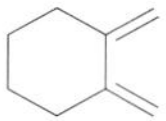	TFA, H_2O, DMF, rt, 84 h	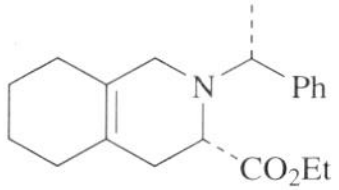 + 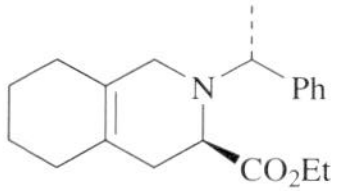 (45) 3:1[u]	327
	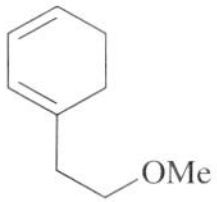	$MeSO_3H$, TFA, CH_2Cl_2, –78° to rt, overnight	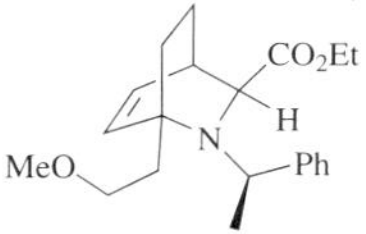 (67)	326
	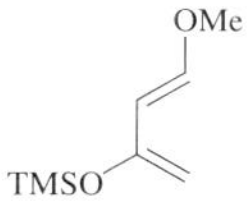	1. $Yb(OTf)_3$, $MgSO_4$, MeCN, rt 2. H_2O	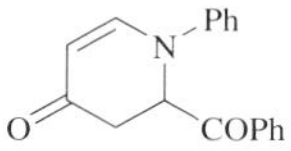 (76)[v]	135
	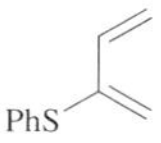	HCl, H_2O, DMF, rt, 24 h	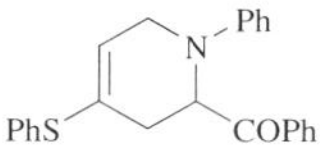 (95)	311
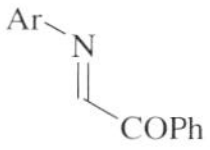 Ar = 4-$O_2NC_6H_4$		BF_3•OEt_2, CH_2Cl_2, rt, few min	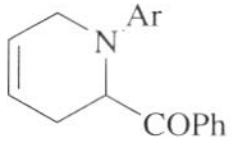 (16)[w]	19
	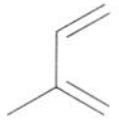	BF_3•OEt_2, CH_2Cl_2, rt, few min	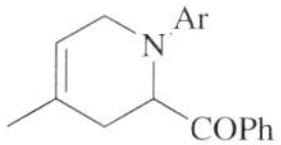 (65)[x]	19
			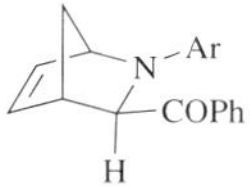	19
		CuCl, CH_2Cl_2	(—)	
		Menthoxyaluminum dichloride, CH_2Cl_2, –20°, 8 h	(8)	

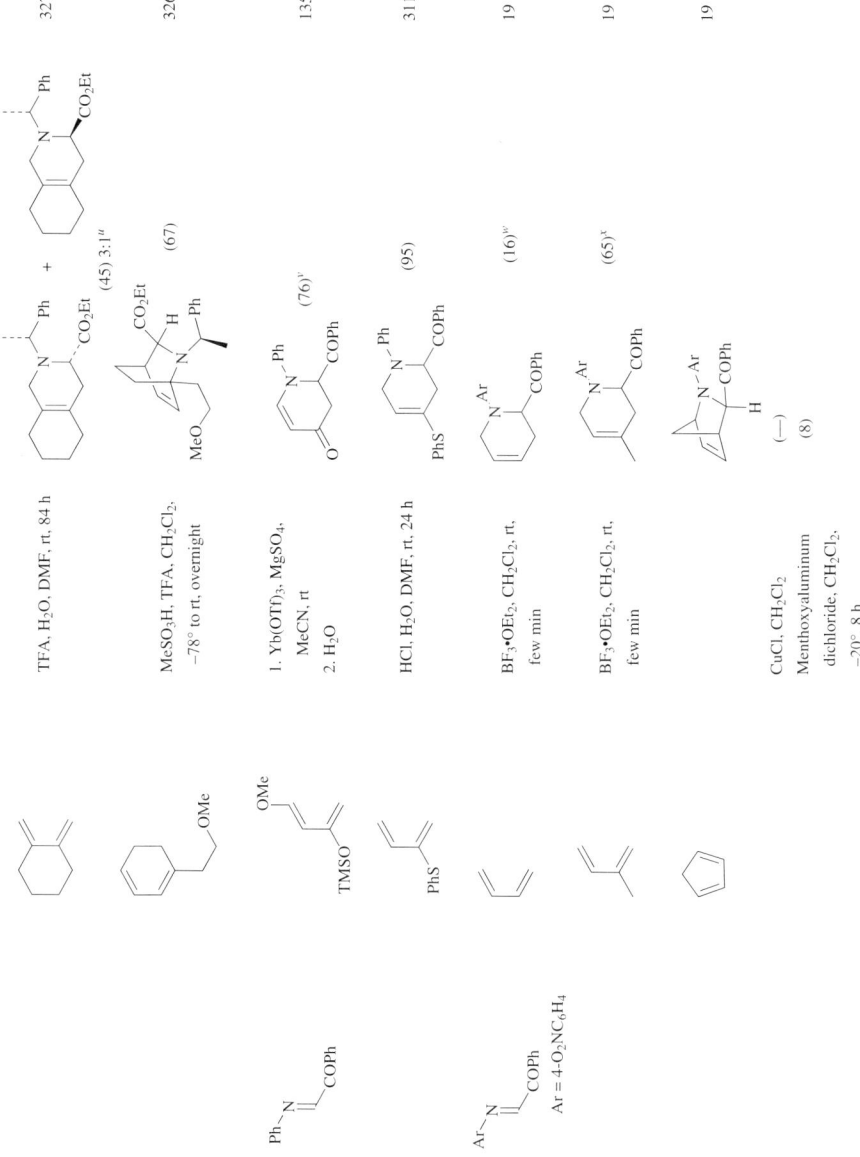

TABLE 3. CYCLOADDITIONS OF *C*-ACYLIMINES (*Continued*)

Imine	Diene	Conditions	Product(s) and Yield(s) (%)	Refs.
C_{14}				
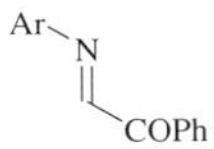 Ar = 4-$O_2NC_6H_4$	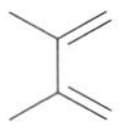	Lewis acid, CH_2Cl_2, rt[y]	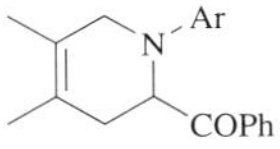 (75)	19
		Lewis acid, $CDCl_3$, rt	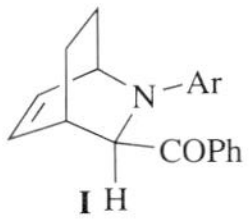 **I** + 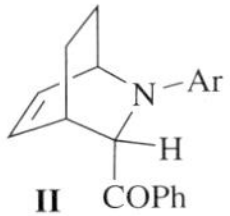 **II** + 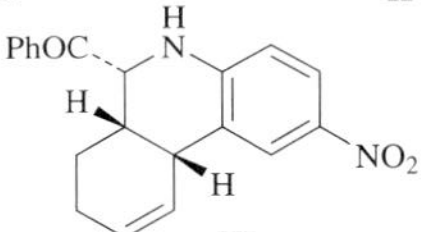 **III**; **I** + **II** + **III** (—)	19
		L. A. — BF₃•OEt₂ / TiCl₄ / SnCl₄ / AlCl₃ / Menthoxyaluminum dichloride	**I**:**II**:**III** — 56:15:29 / 59:18:23 / 44:31:25 / 52:20:28 / 56:17:27	
		SO_2, –78° to rt	**I** + **II** + **III** (—), **I**:**II**:**III** 47:26:27	19
		Lewis acid[z]	**I** (57) + **II** (13) + **III** (28)	
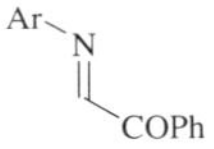 Ar = 4-ClC_6H_4	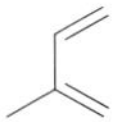	BF_3•OEt_2, CH_2Cl_2, rt, few min	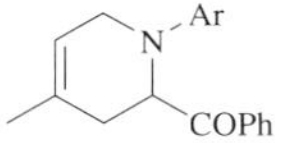 (72)	19

L. A.	**I**:**II**:**III**
BF_3•OEt_2	56:15:29
$TiCl_4$	59:18:23
$SnCl_4$	44:31:25
$AlCl_3$	52:20:28
Menthoxyaluminum dichloride	56:17:27

Imine	Diene	Conditions	Product(s) and Yield(s) (%)		Refs.
		CuCl, CH_2Cl_2 or Menthoxyaluminum dichloride, CH_2Cl_2	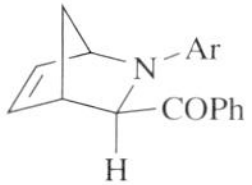N Ar COPh H	(—)	19
	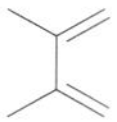	Lewis acid, CH_2Cl_2, rt[y]	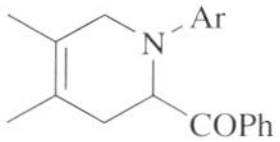N Ar COPh	(88)	19
			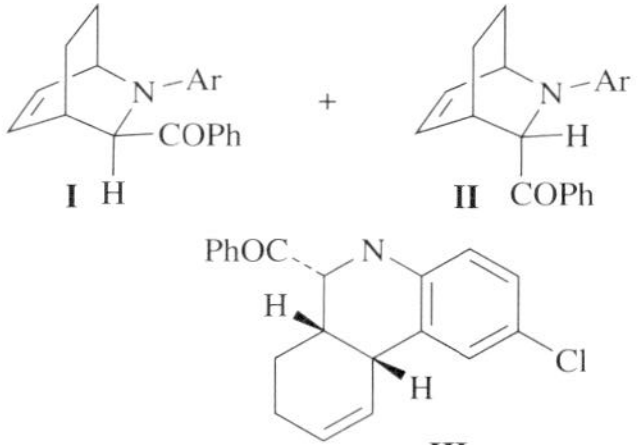N Ar COPh H + PhOC Cl I, II, III		19
		Lewis acid[z]	**I** (44) + **II** (15) + **III** (34)		
		BF_3•OEt_2, CH_2Cl_2, rt[aa]	(—) **I**:**II**:**III** 45:10:45		
		BF_3•OEt_2, CH_2Cl_2, rt[bb]	(—) **I**:**II**:**III** 42:12:46		
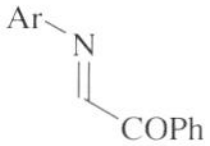Ar N COPh Ar = 3,4-ClC_6H_3		BF_3•OEt_2, CH_2Cl_2, rt, 20 h	N Ar COPh	(52)[a]	328
		BF_3•OEt_2, CH_2Cl_2, rt	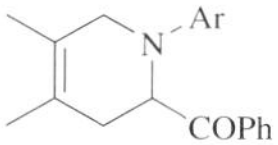N Ar COPh		
		Time			
		4 h	(55)[aa]		328
		16 h	(35)[bb]		328
		4 h	(66)		329

TABLE 3. CYCLOADDITIONS OF *C*-ACYLIMINES (*Continued*)

Imine	Diene	Conditions	Product(s) and Yield(s) (%)	Refs.
C_{14} 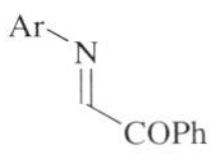 Ar = 3,4-ClC_6H_4		BF_3•OEt_2, CH_2Cl_2, rt, 20 h	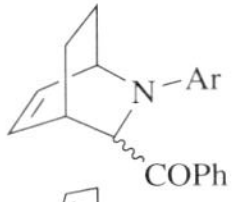 (45)[a]	328
		BF_3•OEt_2, CH_2Cl_2, rt, 20 h	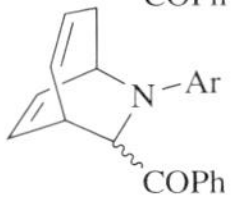 (30)[a]	328
C_{14-15} 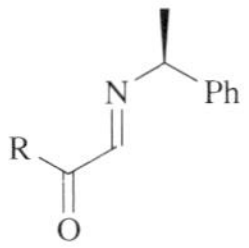			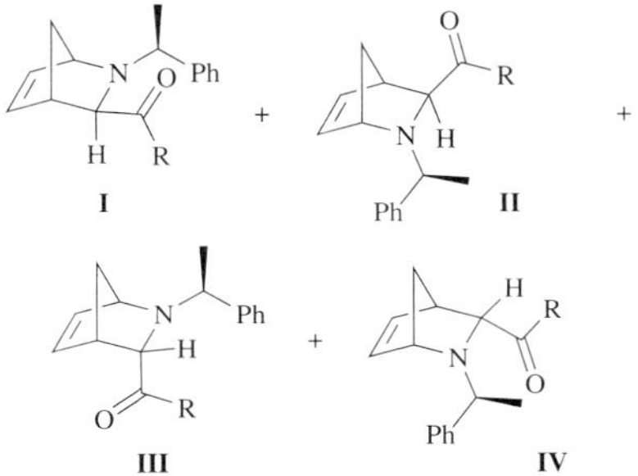 **I** + **II** + **III** + **IV**	
R				
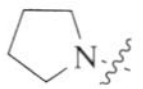		TFA, BF_3•OEt_2, CH_2Cl_2, –78° to rt, overnight	**I** + **II** (72), **I**:**II** 9:1[cc]	93
		TFA, BF_3•OEt_2	**I** (40), **I**:**II** 80:20, **I** + **II**:**III** + **IV** 95:5	91
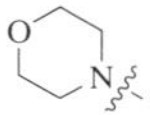		TFA, BF_3•OEt_2, CH_2Cl_2, –78° to rt, overnight	**I** + **II** + **III** + **IV** (29), **I**:**II** 79:21, **I** + **II**:**III** + **IV** 95:5[dd]	93, 91
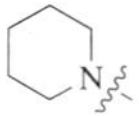		TFA, BF_3•OEt_2, CH_2Cl_2, –78° to rt, overnight	**I** + **II** + **III** + **IV** (80), **I**:**II** 80:20, **I** + **II**:**III** + **IV** 96:4[hh]	93, 91

C_{14} 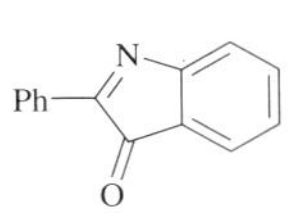	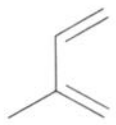		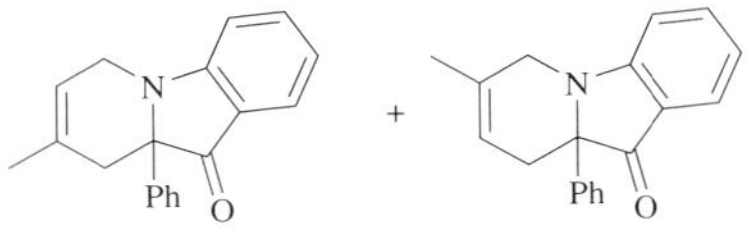 I + II	
		$AlCl_3$ or $HClO_4$, benzene, 80°	(—)[ee]	103
		TsOH, benzene, rt, 0.5-6 h	**I** + **II** (70)	102
		TsOH, benzene, rt, 0.5-6 h	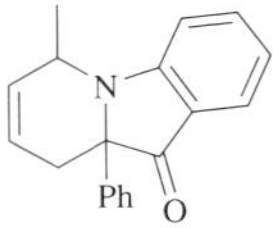 (50)[a]	102
		Benzene, rt	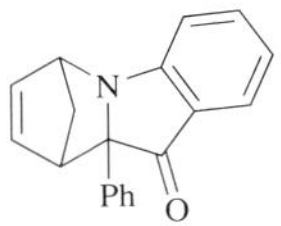 (—)[a]	103
	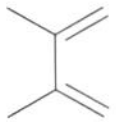	TsOH, benzene, rt, 0.5-6 h	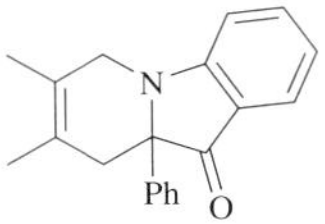 (80)	102
	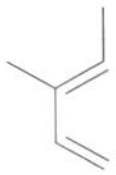	TsOH, benzene, rt, 0.5-6 h	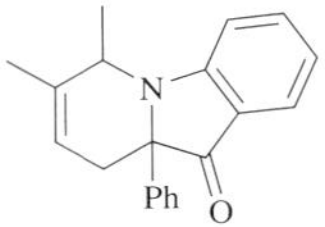 (85)[a]	102
	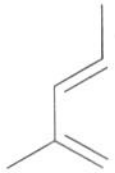	TsOH, benzene, rt, 0.5-6 h	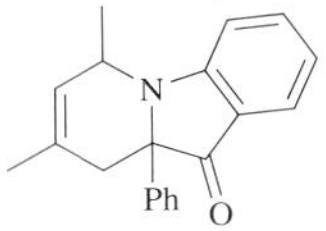 (90)[a]	102

TABLE 3. CYCLOADDITIONS OF C-ACYLIMINES (*Continued*)

Imine	Diene	Conditions	Product(s) and Yield(s) (%)	Refs.
C_{14}				
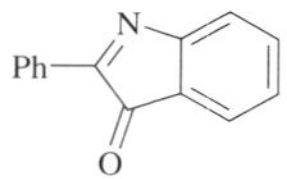		$AlCl_3$ or $HClO_4$, benzene, 80°	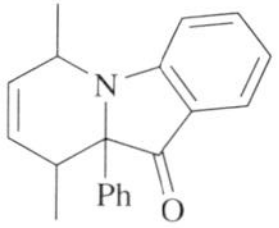 (—)[a]	103
	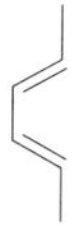	TsOH, benzene, rt, 0.5-6 h	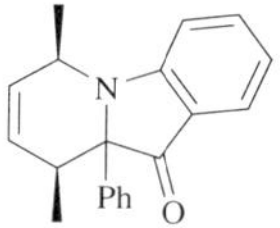 (60)[ff]	102, 101
	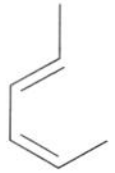	TsOH, benzene, rt, 0.5-6 h	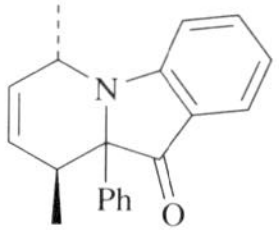 (10)[ff]	102, 101
	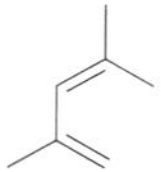	TsOH, benzene, rt, 0.5-6 h	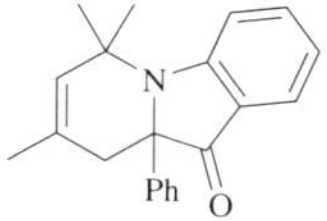 (70)	102, 101
C_{15}				
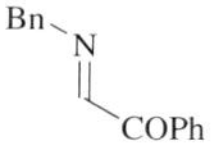		HCl, H_2O, rt, 22 h	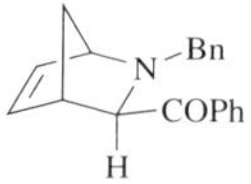 + 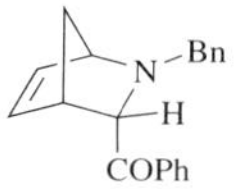 (88) 3:2	86
	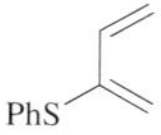	HCl, H_2O, DMF, rt, 24 h	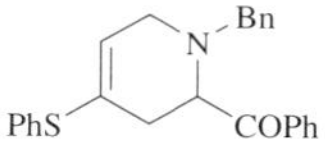 (82)[e], (74)[gg]	311

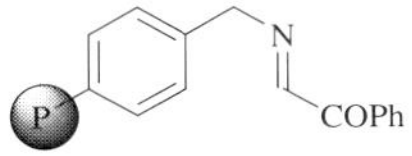

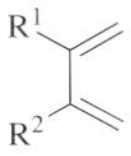

R^1	R^2
H	Me
Me	Me

1. $Yb(OTf)_3$, CH_2Cl_2, rt, 12-48 h
2. $ClCO_2CH(Cl)CH_3$, CH_2Cl_2, rt, 3 h
3. MeOH, 65°, 2 h

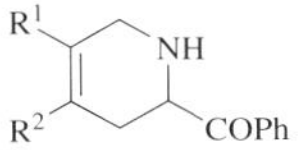

(80)
(82)

320

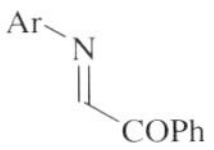

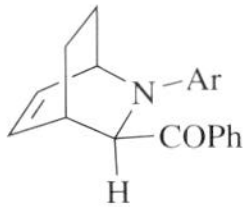

I

+

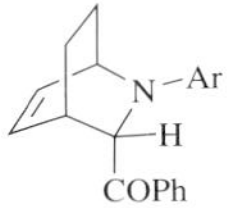

II

+

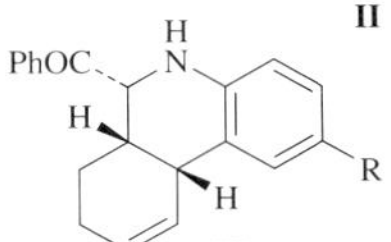

III

19

Ar	Conditions	R	I + II + III	I:II:III
4-MeC_6H_4	BF_3•OEt_2, CH_2Cl_2, rt[aa]	Me	(—)	26:12:62
4-MeC_6H_4	BF_3•OEt_2, CH_2Cl_2, rt[bb]	Me	(—)	29:10:61
4-$MeOC_6H_4$	BF_3•OEt_2, CH_2Cl_2, rt	MeO	(—)	19:8:73

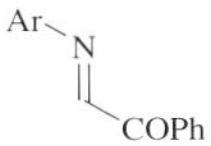

Ar = 4-$MeOC_6H_4$

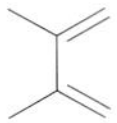

$Sc(OTf)_3$, $MgSO_4$, MeCN, rt

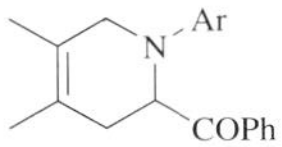

(44)

330

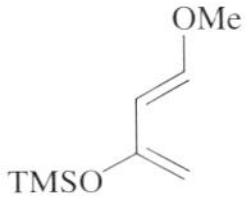

1. $Yb(OTf)_3$, $MgSO_4$, MeCN, rt
2. H_2O

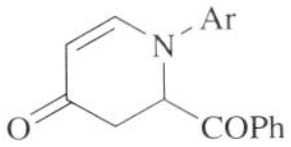

(76)[v]

330

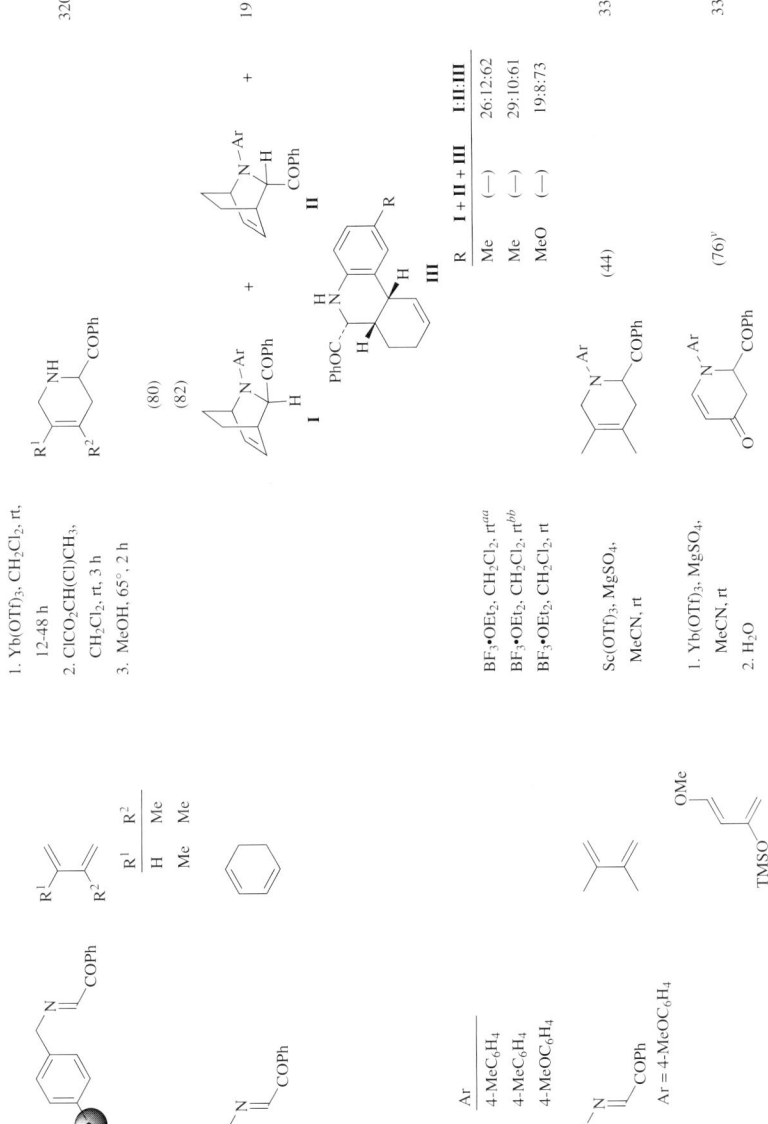

TABLE 3. CYCLOADDITIONS OF *C*-ACYLIMINES (*Continued*)

Imine	Diene	Conditions	Product(s) and Yield(s) (%)	Refs.
C_{15}				
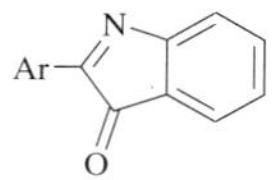 Ar = 4-MeC_6H_4	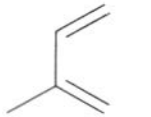	$AlCl_3$ or $HClO_4$, benzene, 80°	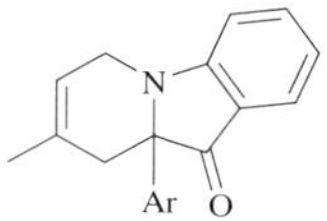 + 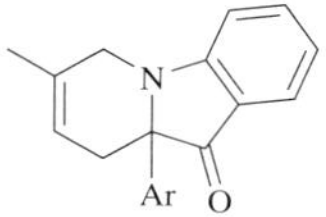 (—)[ii]	103
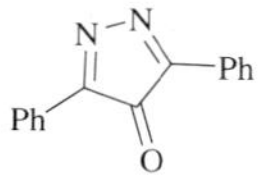		—	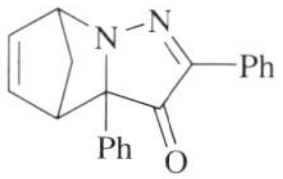 (—)[a]	106, 107
	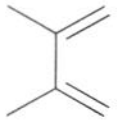		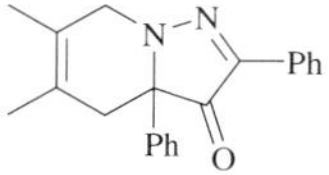	
		Ether, rt, 10 min	(37)	106, 107
		68°, 1 h	(85)	108
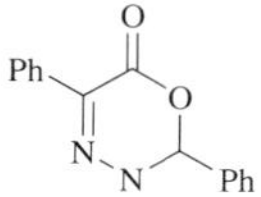	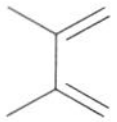	CCl_4, 76°, 10 d	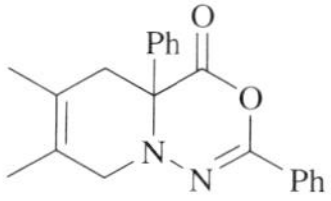 (30)	104
$C_{15\text{-}16}$				
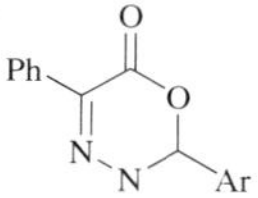	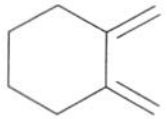	Toluene, 110°	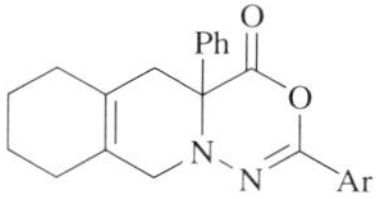	104
Ar		Time		
4-$O_2NC_6H_4$		2 d	(57)	
4-BrC_6H_4		3 d	(71)	
4-MeC_6H_4		3 d	(62)	

C_{16}	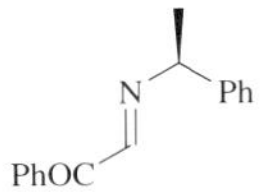			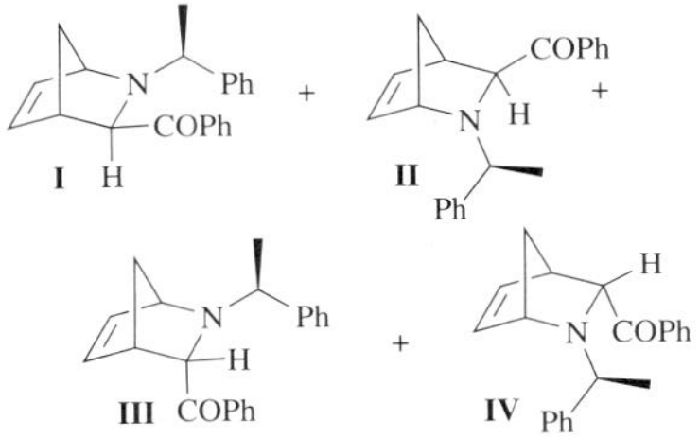	
			TFA, BF_3•OEt_2, CH_2Cl_2, –78°, 5 h	**I** (42)	314
			TFA, BF_3•OEt_2, CH_2Cl_2, –78° to rt, overnight	**I** + **II** (42), **I**:**II** 6:1	94
			TFA, BF_3•OEt_2	**I** (42), **I**:**II** 83:17, **I** + **II**:**III** + **IV** 95:5	91
	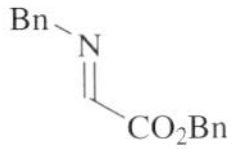		—	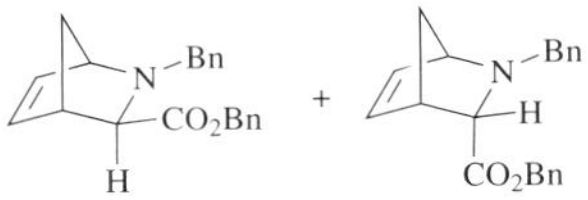 (—) 58:42	87
	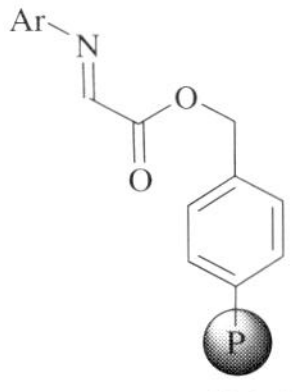 Ar = 4-MeOC_6H_4	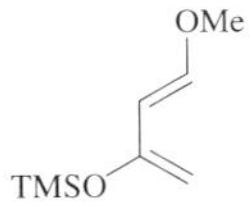	1. $Sc(OTf)_3$, CH_2Cl_2, MeCN, –5°, 20 h 2. NaOMe, THF, MeOH, rt, 1 h 3. HCl, dioxane	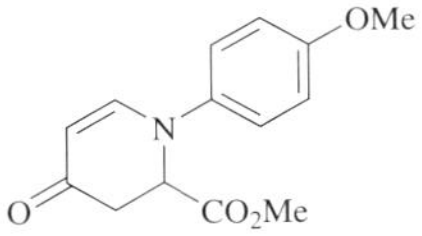 (69)	331, 332
	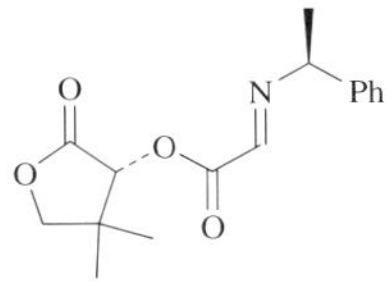	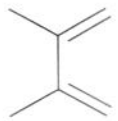	TFA, H_2O, DMF, rt	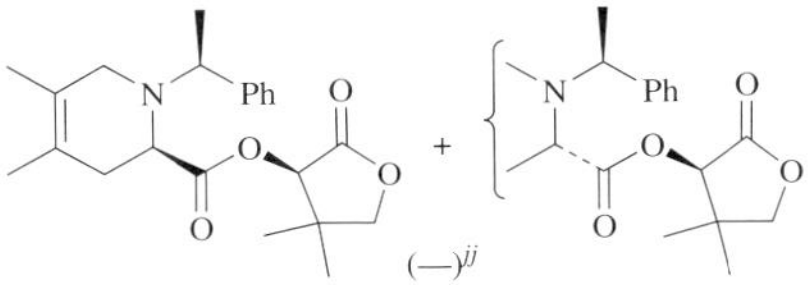 (—)[jj]	100

TABLE 3. CYCLOADDITIONS OF C-ACYLIMINES (*Continued*)

Imine	Diene	Conditions	Product(s) and Yield(s) (%)	Refs.
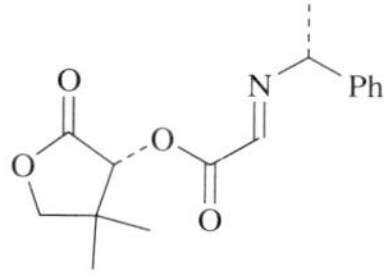	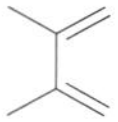	TFA, H_2O, DMF, rt	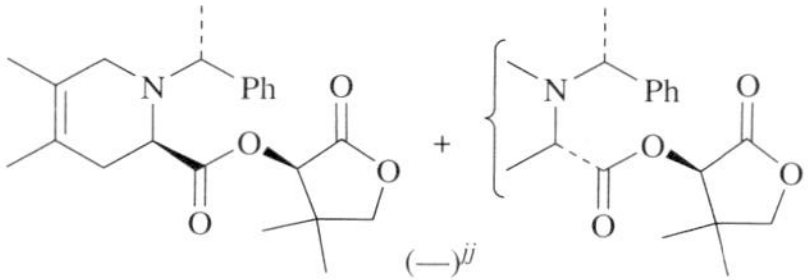 (—)[jj]	100
C_{17} 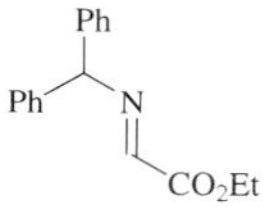	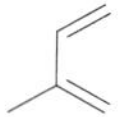	TFA, trifluoroethanol, –40°	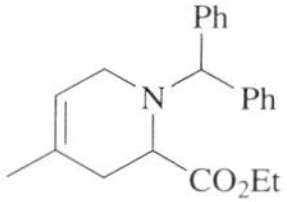 (87)	81
		TFA, trifluoroethanol, –40°	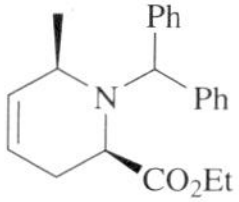 (62)[kk]	81, 85
	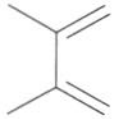	TFA, trifluoroethanol, –40°	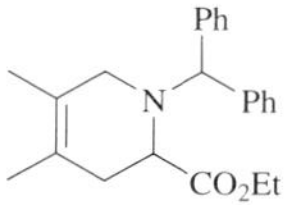 (95)	81
		TFA, trifluoroethanol, –40°	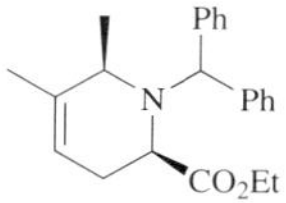 (42)[kk]	81
		TFA, trifluoroethanol, –40°	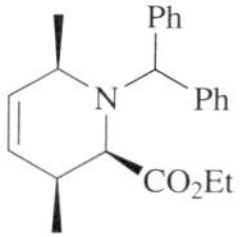 (60)	81

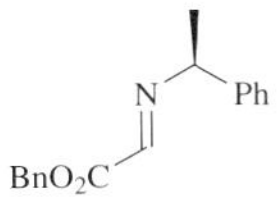			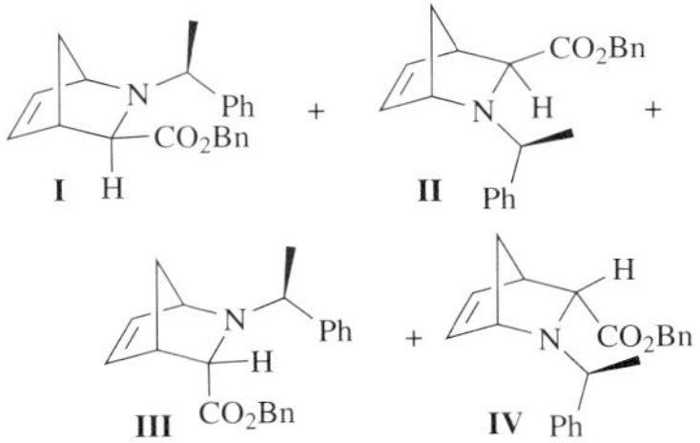	
		TFA, BF_3•OEt_2	**I** (83), **I**:**II** 97:3, **I** + **II**:**III** + **IV** >99:<1	91
		TFA, BF_3•OEt_2, CH_2Cl_2, –78° to rt	**I** (83)	94
		—	**I** (—)[ll]	87
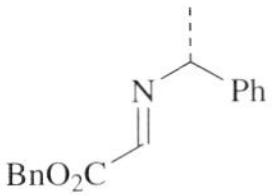			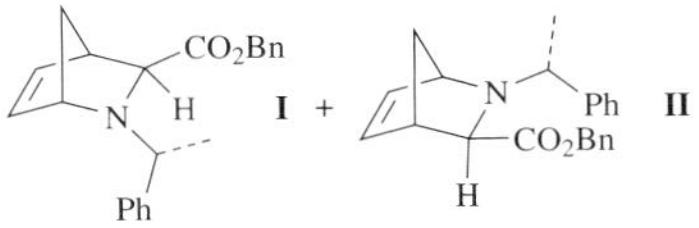	
		TFA, BF_3•OEt_2, CH_2Cl_2, –78°, 5 h	**I** (83), **I**:**II** 98.5:1.5[mm]	333
		—	(—) **I**:**II** 3:1	87
		TFA, BF_3•OEt_2, –78°	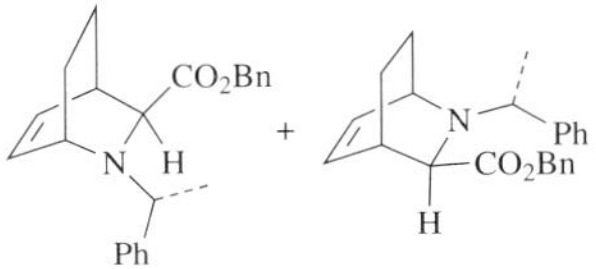 (71) 98.5:1.5	333
C_{18} 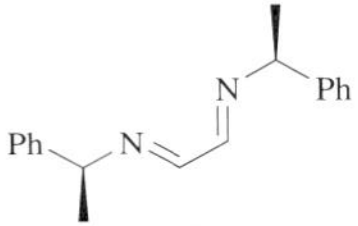		1. TFA, BF_3•OEt_2, CH_2Cl_2, –78°, 7 h 2. $NaHCO_3$, H_2O	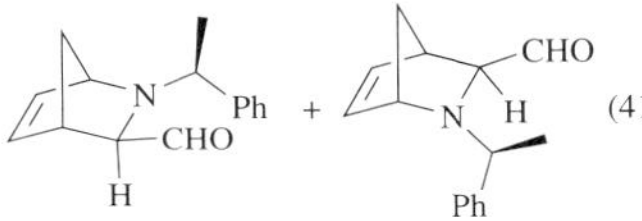 (41) 3:1	334

TABLE 3. CYCLOADDITIONS OF C-ACYLIMINES (*Continued*)

Imine	Diene	Conditions	Product(s) and Yield(s) (%)	Refs.
C_{19}				
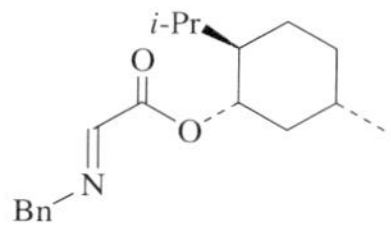		TFA, BF_3•OEt_2, CH_2Cl_2, −78°, 5 h	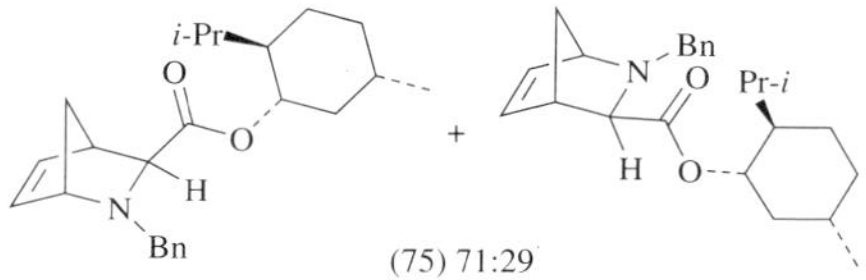 (75) 71:29	92
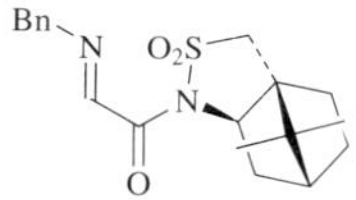		TFA, BF_3•OEt_2, CH_2Cl_2, −78°, 5 h	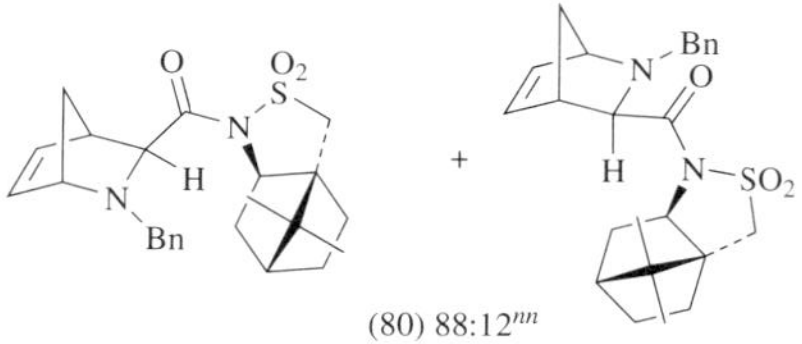 (80) 88:12[nn]	92
C_{20}				
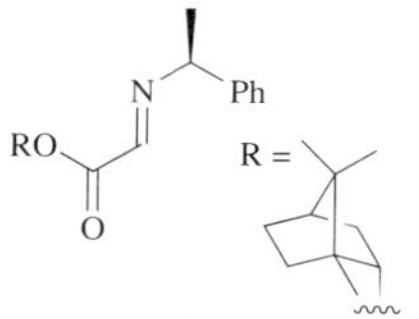	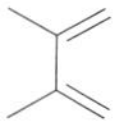	TFA, H_2O, DMF, rt	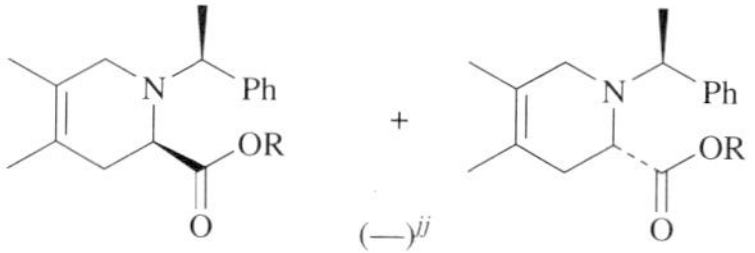 (—)[jj]	100
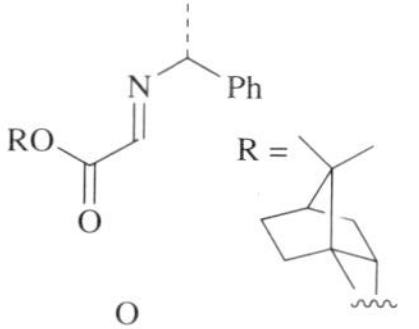	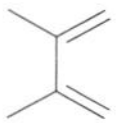	TFA, H_2O, DMF, rt	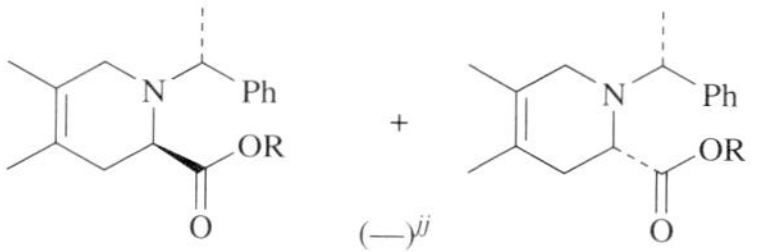 (—)[jj]	100
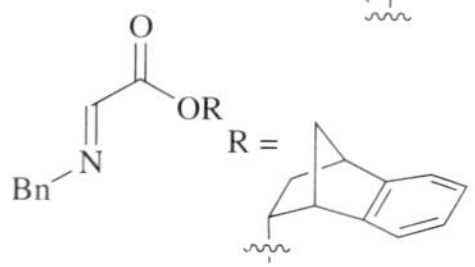		TFA, BF_3•OEt_2, CH_2Cl_2, −78°, 6 h	 (83) 63:37	335

C_{21} 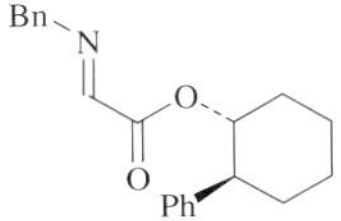		TFA, BF_3•OEt_2, CH_2Cl_2, –78°, 5 h	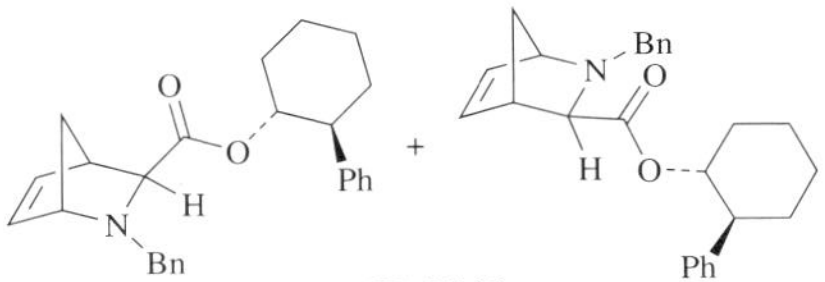 (78) 77:23	92
C_{25} 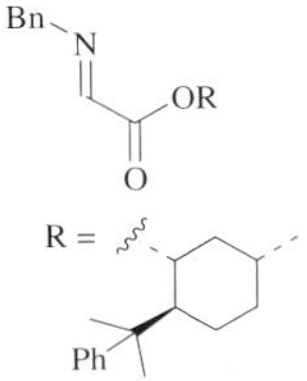		TFA, BF_3•OEt_2, CH_2Cl_2, –78°	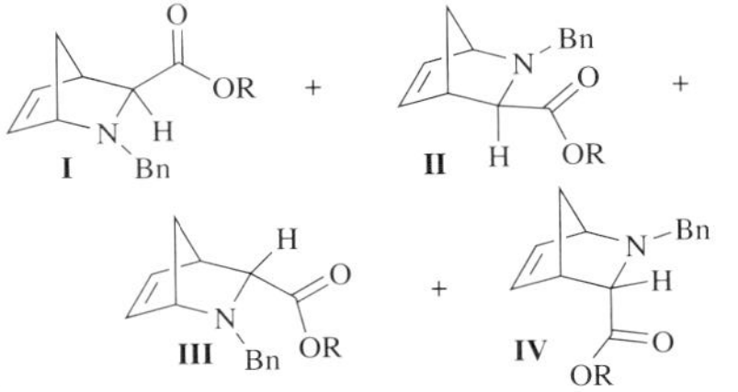I II III IV	
		Time		
		5 h	**I** + **II** (80), **I**:**II** 85:15	92
		—	**I** + **II** + **III** + **IV** (89), **I**:**II** 88:12, **I** + **II**:**III** + **IV** 91:9[oo]	98
C_{26} 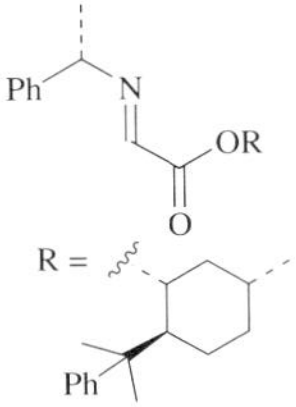	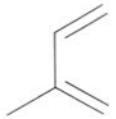	TFA, 4 Å MS, trifluoroethanol, rt, 24 h	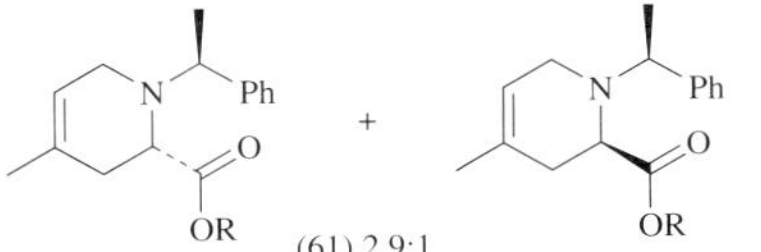 (61) 2.9:1	100
	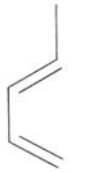	TFA, 4 Å MS, trifluoroethanol, rt, 24 h	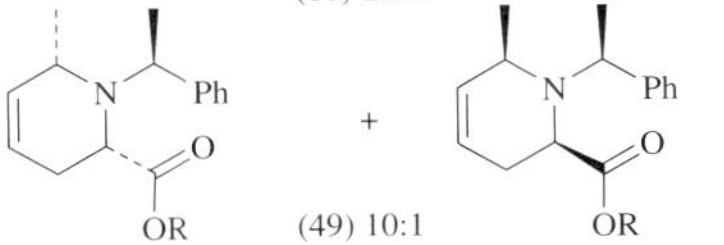 (49) 10:1	100
		TFA, 4 Å MS, trifluoroethanol, rt, 24 h	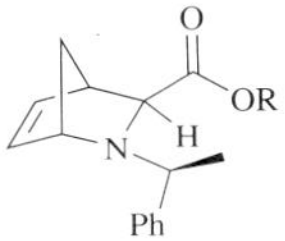 (48) 1.7:1.2:1[pp]	100

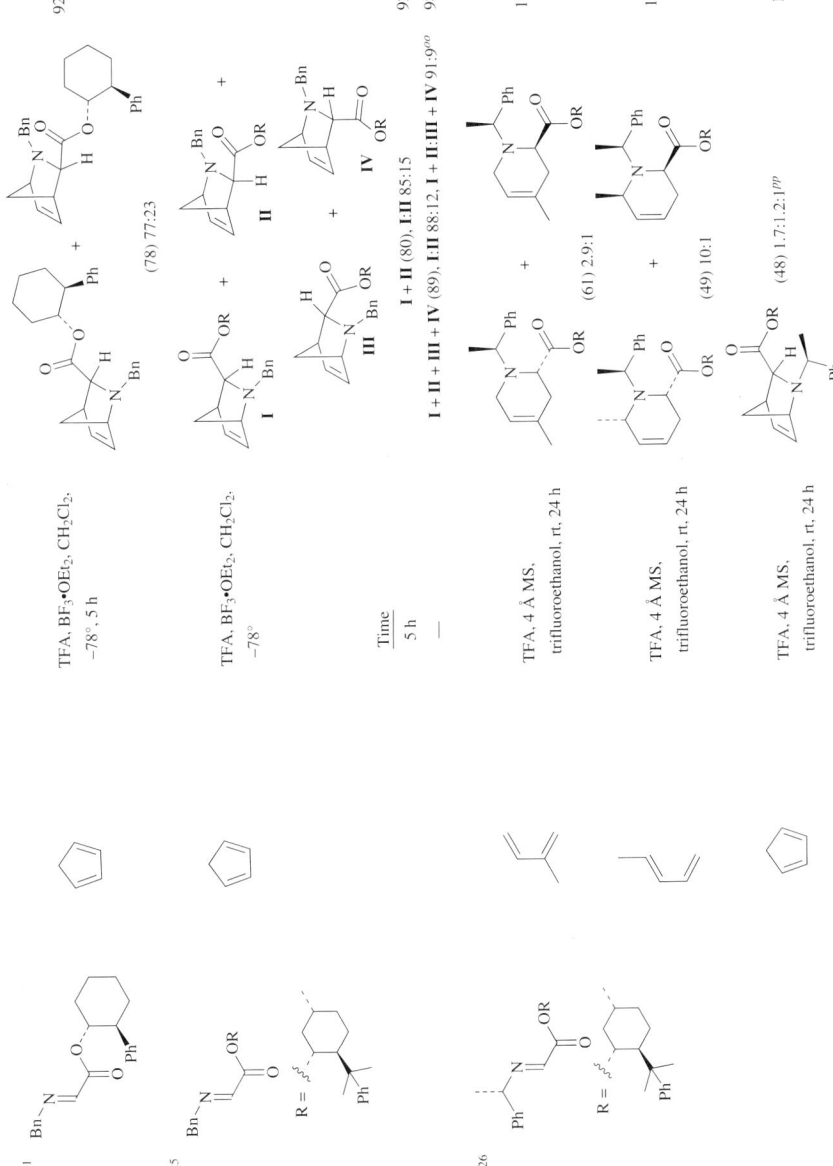

TABLE 3. CYCLOADDITIONS OF *C*-ACYLIMINES (*Continued*)

Imine	Diene	Conditions	Product(s) and Yield(s) (%)	Refs.
C_{26} 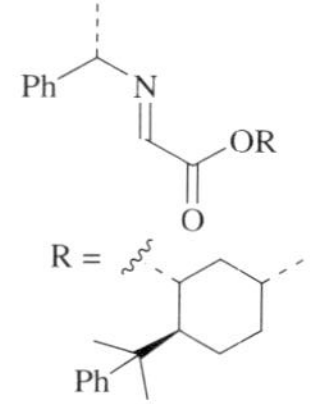	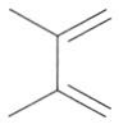	TFA, H_2O, DMF, rt TFA, 4 Å MS, trifluoroethanol, rt, 24 h	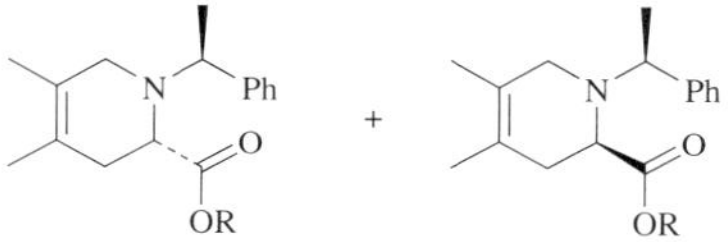 (—)[qq] (54) 1.6:1	100
		TFA, 4 Å MS, trifluoroethanol, rt, 24 h	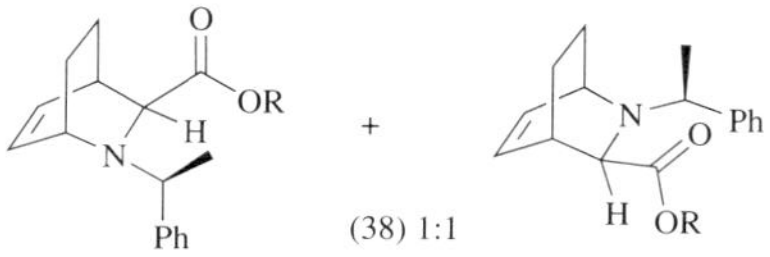 (38) 1:1	100
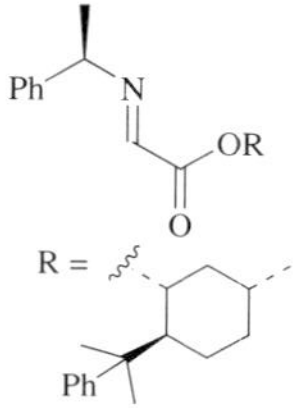	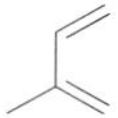	TFA, 4 Å MS, trifluoroethanol, rt, 24 h	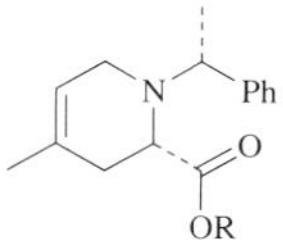 (48)[rr]	100
		TFA, 4 Å MS, trifluoroethanol, rt, 24 h	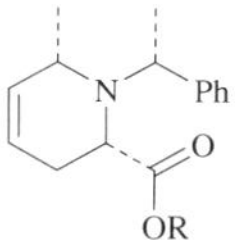 (53)[rr]	100
		TFA, 4 Å MS, trifluoroethanol, rt, 24 h	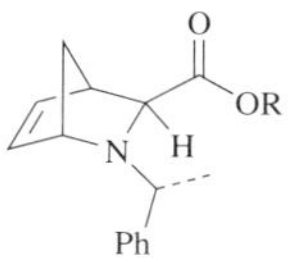 (69)[rr]	100

	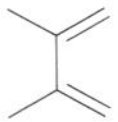	TFA, H_2O, DMF, rt TFA, 4 Å MS, trifluoroethanol, rt, 24 h	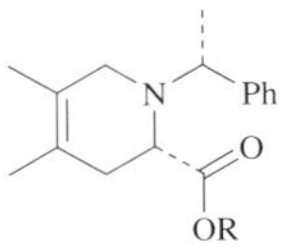 N Ph O OR (—) (54)[rr]	100
		TFA, 4 Å MS, trifluoroethanol, rt, 24 h	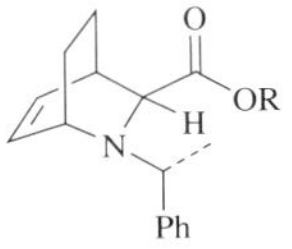 O OR H N Ph (51)[rr]	100
	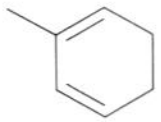	TFA, 4 Å MS, trifluoroethanol, rt, 24 h	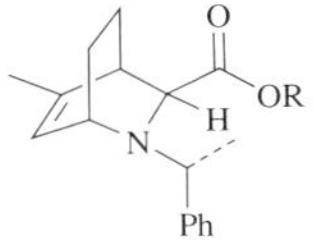 O OR H N Ph (40)[rr]	100
C_{31-33} 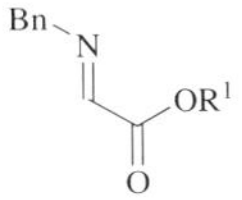 Bn N OR[1] O R^1 = 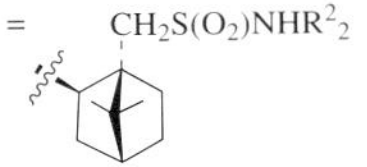 $CH_2S(O_2)NHR^2_2$		TFA, BF_3•OEt_2, CH_2Cl_2, −78°, 5 h	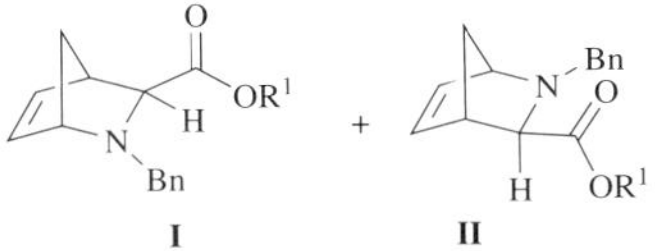 O OR[1] H N Bn + Bn N O H OR[1] I + II	92

R	I + II	I:II
c-C_6H_{11}	(82)	90:10
Bn	(85)	86:14

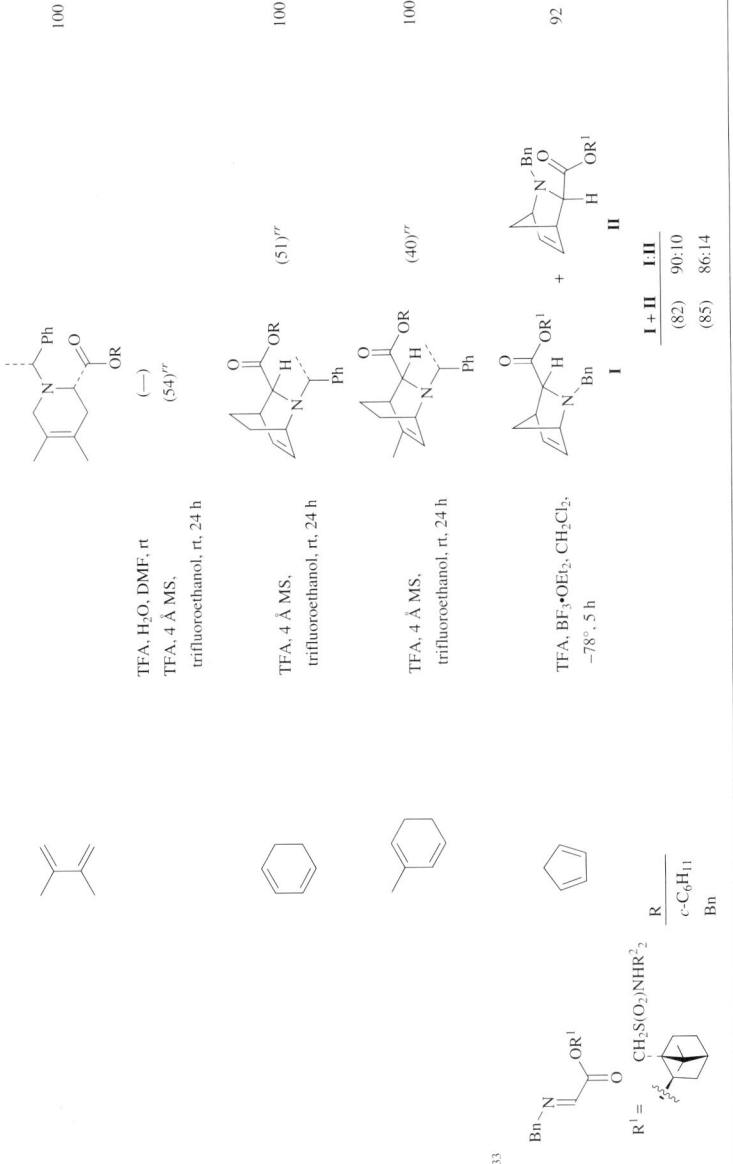

TABLE 3. CYCLOADDITIONS OF *C*-ACYLIMINES (*Continued*)

Imine	Diene	Conditions	Product(s) and Yield(s) (%)	Refs.

[a] No stereochemical information is provided.

[b] The imine was formed in situ with 1 equiv. of the corresponding amine and 10 equiv. of ethyl glyoxylate and then reacted with 1 equiv. of diene.

[c] The imine was formed in situ with 5 equiv. of the corresponding amine and 1 equiv. of ethyl glyoxylate and then reacted with 5 equiv. of diene.

[d] The imine was formed in situ with 1 equiv. of the corresponding amine and 2 equiv. of ethyl glyoxylate and then reacted with 10 equiv. of diene.

[e] The imine was formed in situ with 10 equiv. of both the corresponding amine and phenylglyoxal and then reacted with 1 equiv. of diene.

[f] The imine was formed in situ with 1 equiv. of the corresponding amine and 2 equiv. of phenylglyoxal and then reacted with 6.9 equiv. of diene.

[g] No diastereomeric ratio is provided for the minor regioisomer **III**.

[h] Although the authors report a diastereomeric excess for the major (exo) isomer, only the two isomers shown (endo/exo) were isolated.

[i] Although the authors report the regioselectivity, only two diastereomers (**I** and **II**) were isolated.

[j] The stereochemistry of the major isomer is not indicated, but it is probably the isomer shown based on the results of other reactions with this imine.[315] The stereochemistry of endo-isomer **III** is not indicated; it may be a mixture of the two endo-isomers.

[k] After purification, the major isomer was obtained in 32% yield. The stereochemistry of the major isomer is not indicated, but it is probably the isomer shown based on the results of other reactions with this imine.[315]

[l] Addition of methanol instead of water led to a similar increase in yield.

[m] The imine was formed in situ from 10 equiv. of both the corresponding amine and ethyl glyoxylate and then reacted with 1 equiv. of diene.

[n] No yield or stereochemical ratio is provided, but adduct **I** is the major product observed.

[o] The relative stereochemistry between the carboethoxy and the methyl substituents is not indicated.

[p] The major product, 2-oxo-3-(2-oxo-1-phenyl-2,3,4,5,6,7-hexahydro-1*H*-inden-1-yl)propionic acid methyl ester, was isolated in 7% yield. The crude mixture was a 1.0:2.3 ratio of Diels-Alder cycloadduct to major product.

[q] The stereochemistry of the major isomer is not specified. Although the major isomer was isolated in 37% yield, the crude product showed a diastereomeric ratio of 57:43. The stereochemistry of the two diastereomers observed is not indicated.

[r] The adduct **I** was obtained with "high" diastereoselectivity.

[s] The imine was formed in situ.

[t] The pure imine was isolated prior to performing the hetero Diels-Alder reaction.

[u] A 30% yield of the major product was obtained after purification.

[v] Good yields were also obtained with other lanthanide triflates.

[w] The major product resulted from the hetero Diels-Alder reaction in which the imine acts as the diene and butadiene acts as the dienophile.

[x] Two stereoisomers of (2-methoxy-5-methyl-2-phenyl-5-vinyltetrahydrofuran-3-yl)-(4-nitrophenyl)amine were obtained as side products.

[y] Exact reaction conditions are not provided. The reaction may have also been performed under SO_2 catalysis (–78° to rt, few min).

[z] Exact reaction conditions are not provided. However, the text states that the three products can be obtained under BF_3•OEt_2 (CH_2Cl_2, rt, few min) or SO_2 (–78° to rt, few min) catalysis.

[aa] The reaction was carried out with the free imine.

[bb] The reaction was carried out with the methanol adduct of the imine.

[cc] The major isomer was isolated in 40% yield.

[dd] The major isomer was isolated in 14% yield.

[ee] Only one isomer was obtained, but its structure was not determined. The authors suggest that the adduct is compound **II**.

[ff] The relative stereochemistry between the phenyl and the methyl substituents is not indicated.

[gg] The imine was formed in situ from 1 equiv. of the corresponding amine and 2 equiv. of phenylglyoxal and then reacted with 10 equiv. of diene.

[hh] The major isomer was isolated in 29% yield.

[ii] Only one adduct was obtained, but its structure was not determined.

[jj] Modest asymmetric induction was observed, but the major diastereomer is not specified.

[kk] A small amount of the ene reaction product was also isolated.

[ll] Compound **I** is the major adduct obtained. The identity and ratio of the other products is not specified.

[mm] An endo-isomer was also obtained, but its stereochemistry is not specified.

[nn] The diastereoselectivity is not modified in toluene.

[oo] The major adduct **I** was isolated in 57% yield.

[pp] Three diastereomers were formed. The major adduct is shown, but stereochemical information for the other two isomers is not provided. The yield given is the combined yield for all three isomers.

[qq] Modest asymmetric induction was observed. The major adduct is probably identical to that obtained in trifluoroethanol.

[rr] A single cycloadduct was detected. The diastereomeric excess is given as >95%.

TABLE 4. CYCLOADDITIONS OF *N*-SULFONYLIMINES AND *N*-PHOSPHORYLIMINES

Imine	Diene	Conditions	Product(s) and Yield(s) (%)	Refs.
C_4				
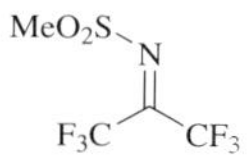		Ether, –78 to 50°, 2 d	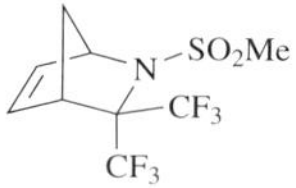 (65)	336
C_5				
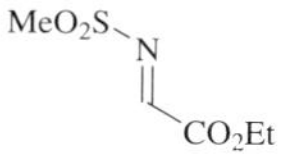		1. Base[a], toluene 2. Cyclopentadiene	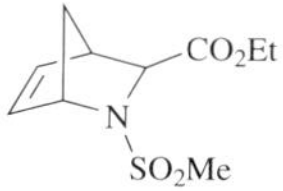	40
		Base / Temp / Time		
		Et_3N / rt / 1 h	(74)[b]	
		i-Pr_2NEt / 0° / 1 h	(98)[b]	
		n-BuLi / 0° / 30 min	(12)[b]	
		n-BuLi / 0° / 1 h	(15)[b]	
		MeLi / –78° / 30 min	(23)[b]	
		MeLi / 0° / 1 h	(29)[b]	
C_{7-8}				
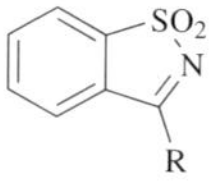	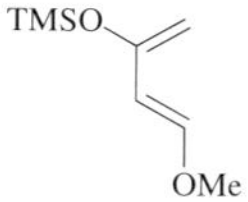	Toluene	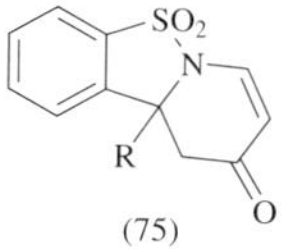	128, 127
R		Temp / Time		
Cl		rt / 2 h	(75)	
Me		110° / 42 h	(45)	
C_{8-9}				
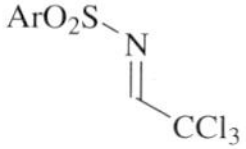		Benzene, heat, 5 h	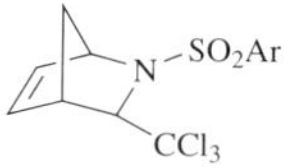	337
Ar				
Ph			(53)[c]	
4-ClC_6H_4			(54)[c]	
4-$MeOC_6H_4$			(60)[c]	

	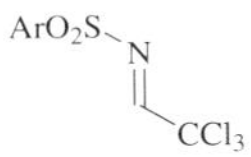	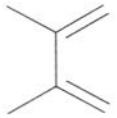	Benzene, heat, 5 h	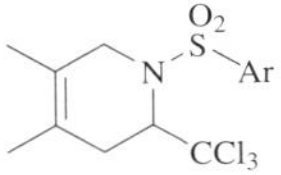	337
	Ar: Ph			(48)	
	4-ClC_6H_4			(68)	
	4-$MeOC_6H_4$			(57)	
C_8	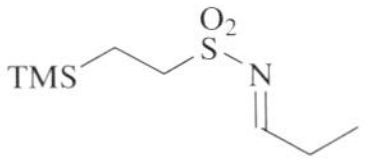	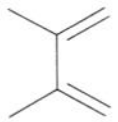	BF_3•OEt_2, $MgSO_4$, toluene, CH_2Cl_2, –30°, 3 h	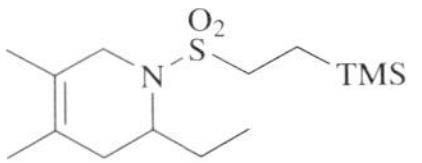 (58)	33
C_9	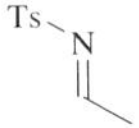	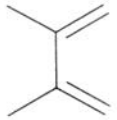	BF_3•OEt_2, CH_2Cl_2, –20°	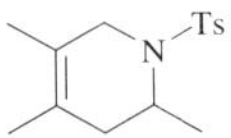 (73)	34
	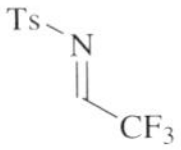		d_6-Benzene, 30°	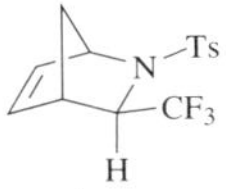 + 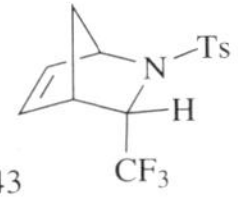 (—) 57:43	31, 32
		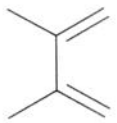	Benzene, 80°, 6 h	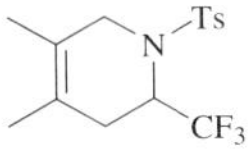 (96)	338
			Benzene, 80°, 7.5 h	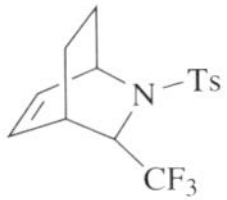 (95)[c]	338
			d_6-Benzene, 80°, 2 h	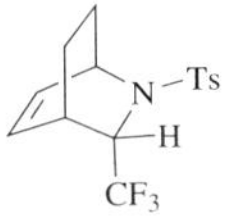 + 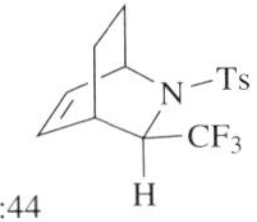 (—) 56:44	31, 32

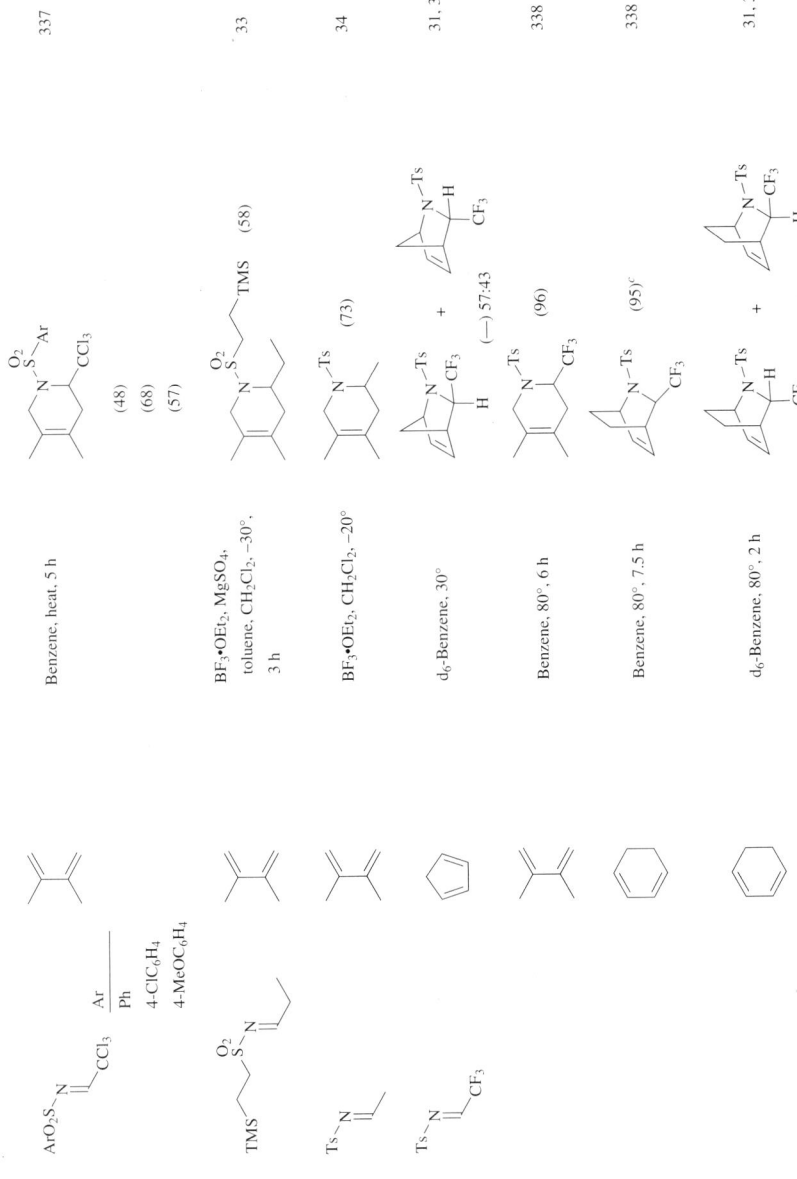

TABLE 4. CYCLOADDITIONS OF *N*-SULFONYLIMINES AND *N*-PHOSPHORYLIMINES (*Continued*)

Imine	Diene			Conditions	Product(s) and Yield(s) (%)	Refs.
C_9 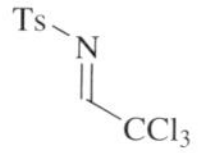	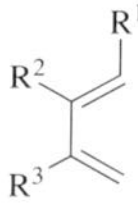				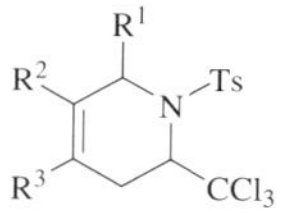	
	R^1	R^2	R^3			
	H	H	H	Benzene, 90°, 20 h	(90)	339
	H	H	H	—	("good")	38
	Me	H	H	Benzene, 80°, 10 h	(70)[e]	339
	H	H	Me	Benzene, 90°, 8 h	(82)	339
	H	H	OMe	—	("good")	38
	H	Me	Me	Benzene, 80°, 6 h	(94)	338, 340
	H	Me	Me	Benzene, heat, 5 h	(85)	337
	Ph	H	H	Benzene, 80°, 10 h	(72) + (7)[f]	339
	H	H	Ph	Benzene, 80°, 5 h	(86)	339
	H	H	4-MeC_6H_4	Benzene, 80°, 2 h	(84)	339
	H	H	4-$MeOC_6H_4$	—	("good")	38
					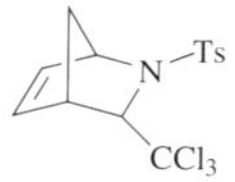	
				Benzene, 80°, 5 h	(93)[c]	338, 340
				Benzene, heat, 5 h	(93)[c]	337

	d_6-Benzene, 30°, 2 h —	 **I + II** (—), **I:II** 78:22 **II** (—)	31, 32 24
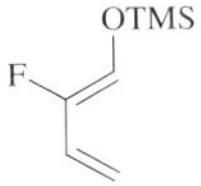	1. Toluene, 110°, 4 h 2. HCl, THF, 25°, 30 min	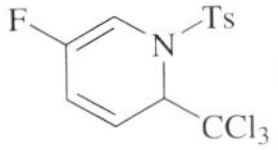 (85)	112
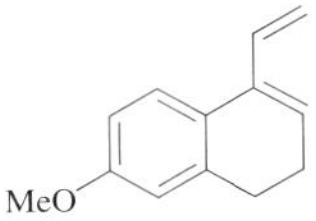	Benzene, rt, 16 h	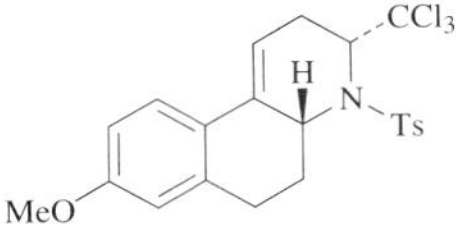 (72)	36, 38
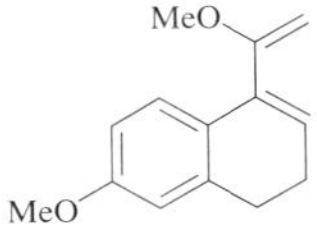		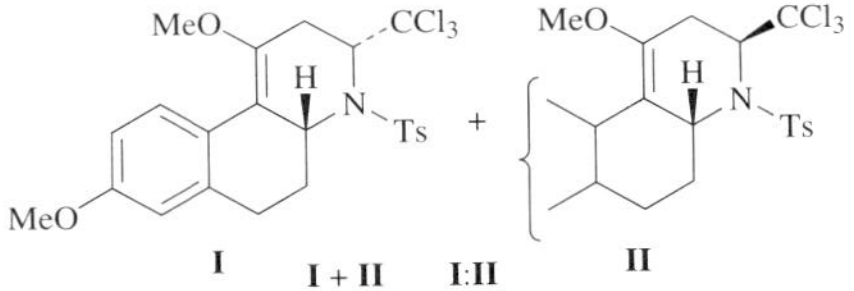	36

Conditions	I + II	I:II
Benzene, 20°, 23 h	(80)	1.5:1[h]
Benzene, 80°	(—)	1.5:1
Ether, 10°	(35)	1.5:1
Ether, 35°	(—)	1.5:1

TABLE 4. CYCLOADDITIONS OF *N*-SULFONYLIMINES AND *N*-PHOSPHORYLIMINES (*Continued*)

Imine	Diene		Conditions	Product(s) and Yield(s) (%)	Refs.
C_9 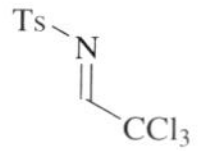	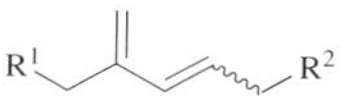		CH_2Cl_2, 12 kbar, 70°	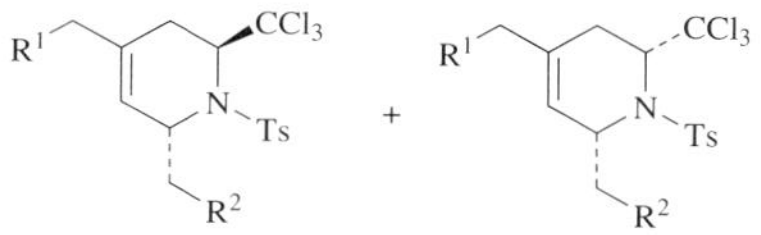	
	R[1]	R[2]	Time		
	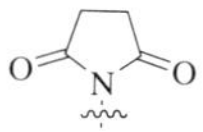	OEt	36-48 h	(77)[g]	113
	n-C_5H_{11}	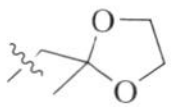	36-48 h	(67)[g]	113
	OAc	CH_2CO_2Bn	36-48 h	(89)[g]	113
	OAc	CH_2CH_2OBn	40 h	(91) 1.3:1[g]	113
	OBn	OTBS	36-48 h	(62)[g]	113
	OTBS	CH_2CH_2OBn	36-48 h	(60)[g]	113
	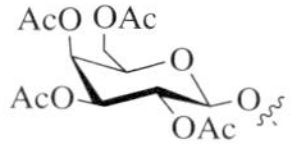	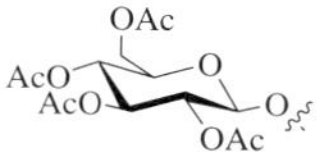	36-48 h	(61)[g]	113, 341
	OTr	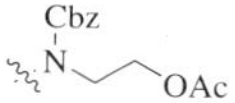	36-48 h	(72)[g]	113
	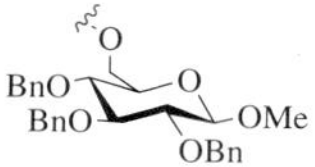	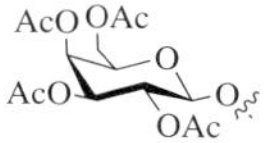	40 h	(57)	341
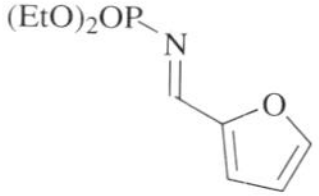	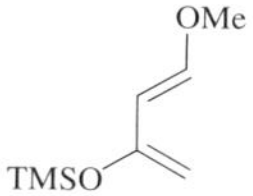		1. $Cu(OTf)_2$, CH_2Cl_2, rt 2. TFA, CH_2Cl_2	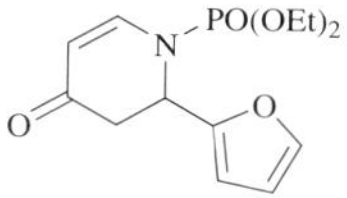 (47)	129

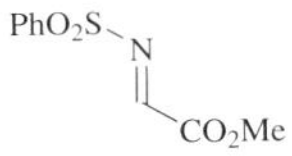
PhO2S
N
CO2Me

1. Base[a], toluene
2. Cyclopentadiene

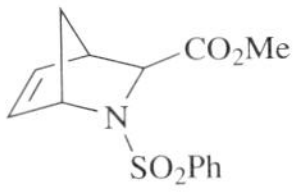
CO2Me
N
SO2Ph

Base	Temp	Time	
Et_3N	rt	1 h	(76)[b]
i-Pr_2NEt	0°	1 h	(92)[b]
n-BuLi	0°	30 min	(15)[b]
n-BuLi	0°	1 h	(18)[b]
MeLi	0°	30 min	(27)[b]
MeLi	0°	1 h	(26)[b]

40

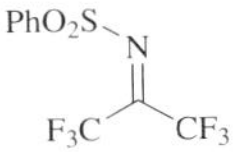
PhO2S
N
F3C
CF3

$CHCl_3$, 60°

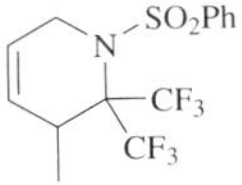
N
SO2Ph
CF3
CF3

(33)

336

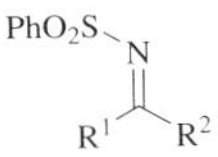
PhO2S
N
R1
R2

R^1	R^2
CF_3	CF_3
CF_2Cl	CF_3
CF_2Cl	CF_2Cl
CF_2NO_2	CF_3

Ether, –78 to 50°, 2 d

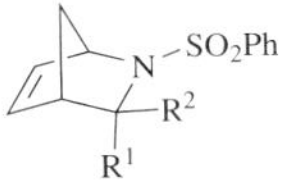
N
SO2Ph
R2
R1

(77)
(68)
(69)
(40)

336

C_{10}

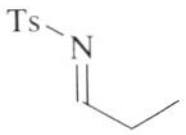
Ts
N

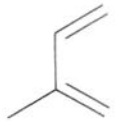

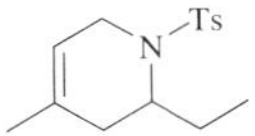
N
Ts

BF_3•OEt_2, $MgSO_4$, toluene, CH_2Cl_2, –30°, 3 h — (63)[d] — 33

BF_3•OEt_2, CH_2Cl_2, –20° — (88) — 34

TABLE 4. CYCLOADDITIONS OF *N*-SULFONYLIMINES AND *N*-PHOSPHORYLIMINES (*Continued*)

Imine	Diene	Conditions	Product(s) and Yield(s) (%)	Refs.
C_{10} 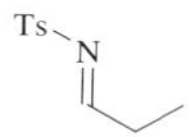				
		BF_3•OEt_2, $MgSO_4$, toluene, CH_2Cl_2, –30°, 3 h	(61)[d] 6:1	33
		BF_3•OEt_2, CH_2Cl_2, –20°	(88) 2:1	34
	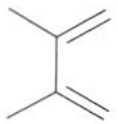		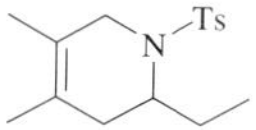	
		BF_3•OEt_2, $MgSO_4$, toluene, CH_2Cl_2, –30°, 3 h	(79)[d]	33
		BF_3•OEt_2, CH_2Cl_2, –20°	(81)	34
	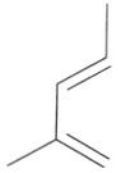	BF_3•OEt_2, $MgSO_4$, toluene, CH_2Cl_2, –30°, 3 h	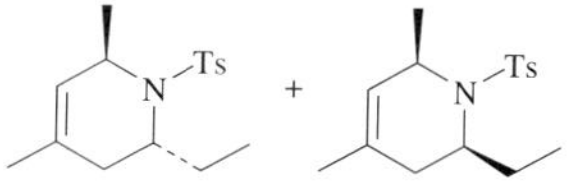 (68)[d] 2:1	33
		BF_3•OEt_2, $MgSO_4$, toluene, CH_2Cl_2, –30°, 3 h	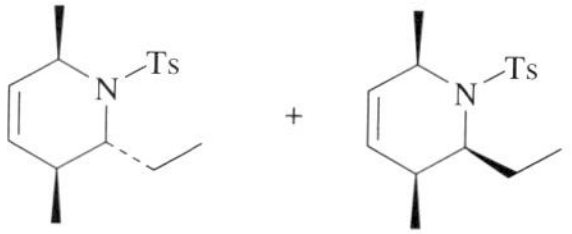 (44)[d] 1:1	33

		BF_3•OEt_2, $MgSO_4$, toluene, CH_2Cl_2, –30°, 3 h	(73)[d] 1.5:1	33
		BF_3•OEt_2, CH_2Cl_2, –20°	(59) 3:1	34

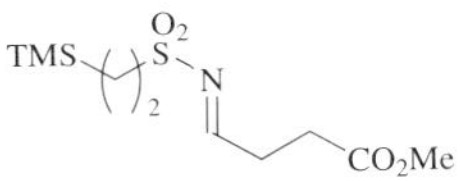	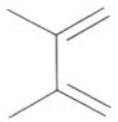	BF_3•OEt_2, $MgSO_4$, toluene, CH_2Cl_2, –30°, 3 h	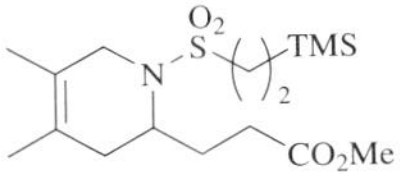 (56)	33
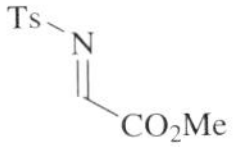	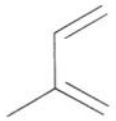	Toluene, 70°, 24 h	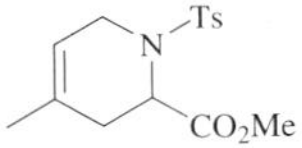 (77)	35
			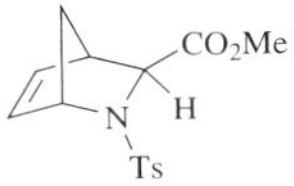	
		Toluene, rt, 24 h	(70)	35
		Toluene, rt, 20 h	(45)	116
	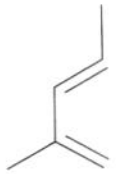	Toluene, rt, 24 h	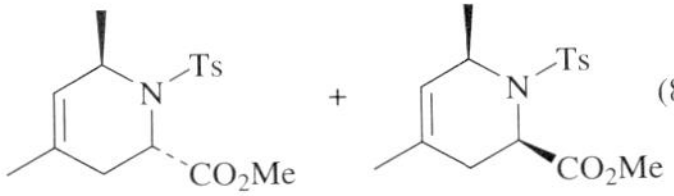 (83) 2:1	35
			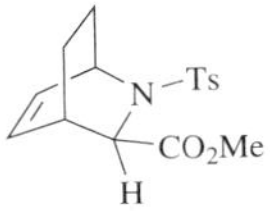	
		Benzene, heat, 16 h	(80)	117
		Toluene, heat	(80)	118
		Toluene, rt, 24 h	(57)	35
	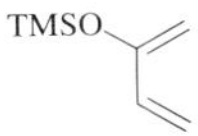	Toluene, 50°, 3 h	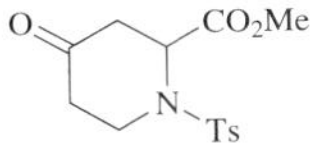 (50)	35
	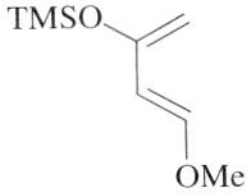	Toluene, rt, 24 h	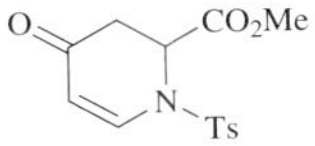 (60)	35

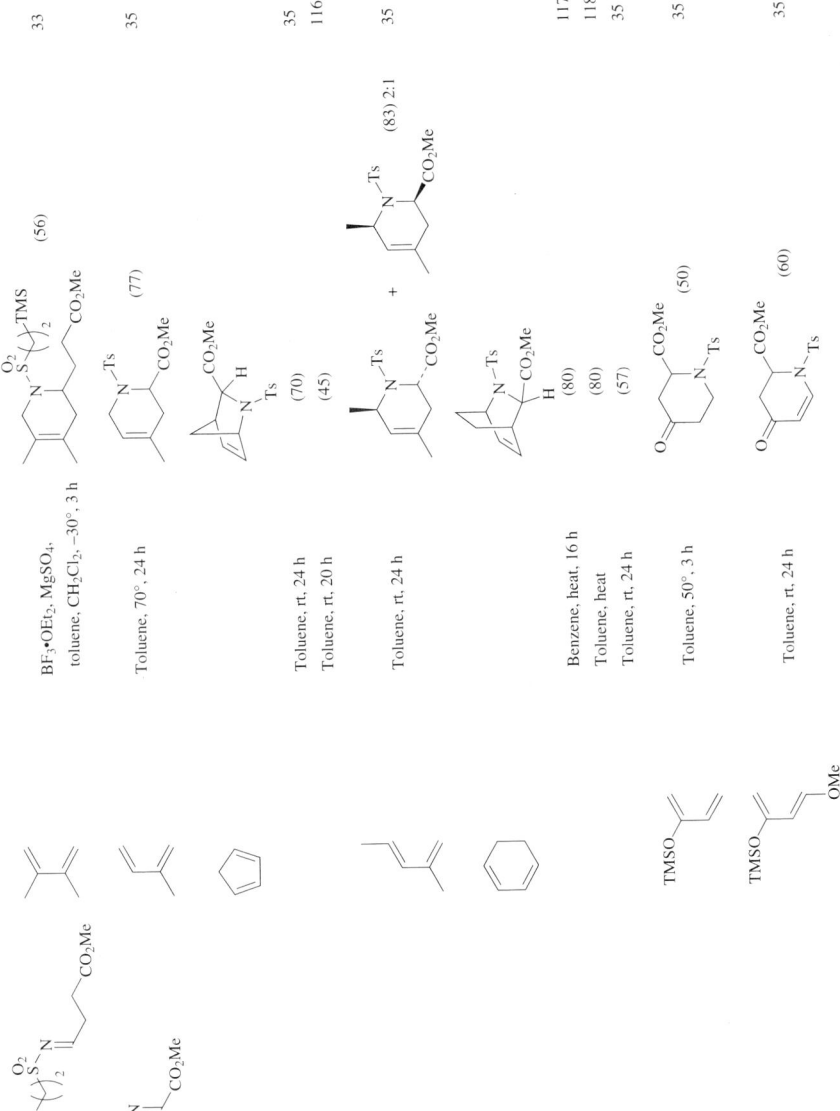

TABLE 4. CYCLOADDITIONS OF *N*-SULFONYLIMINES AND *N*-PHOSPHORYLIMINES (*Continued*)

Imine	Diene	Conditions	Product(s) and Yield(s) (%)	Refs.
C_{10} 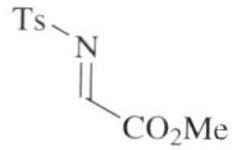	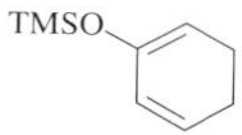	1. Benzene, 5° to rt, 3 h 2. HCl, THF, rt, 1 h	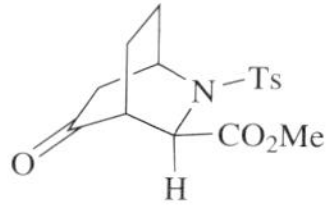 (57) + 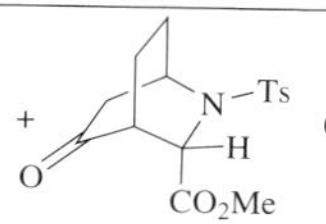 (24)	119, 120
	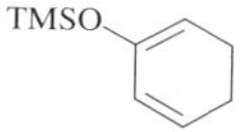		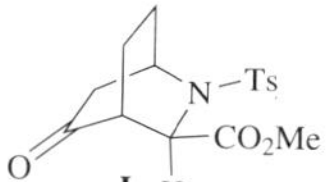 I + 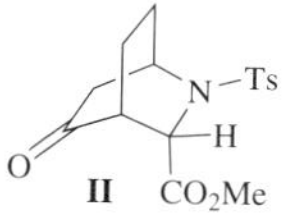 II + 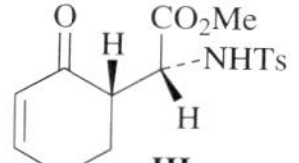 III + 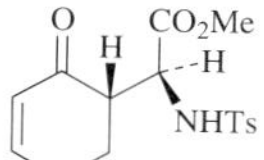 IV	121
		1. Benzene, 5° to rt 2. HCl, THF	**I** (47) + **II** (18.5) + **III** (6) + **IV** (2)	
		1. Benzene, 6° to 20°, 3 h 2. HCl, THF	(74) **I**:**II**:**III**:**IV** 56:10:26:8	
		1. AlCl$_3$, MeCN, –40° to 20°, 2.5 h 2. HCl, THF	(60) **I**:**II**:**III**:**IV** 10:3:77:10	
	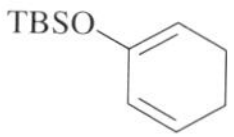	1. Benzene, 6°, 3 h 2. HCl, THF	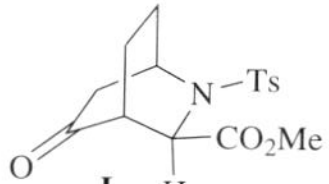 I + 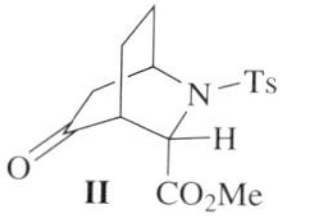 II + 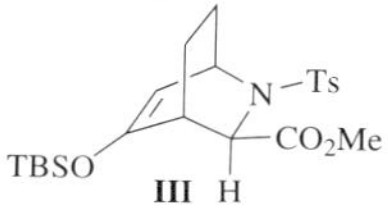 III + 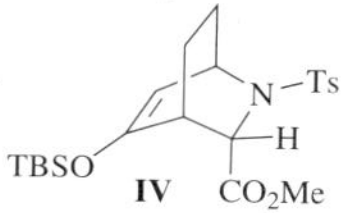 IV + 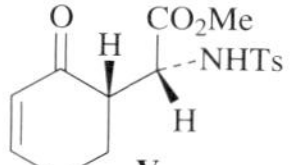 V + 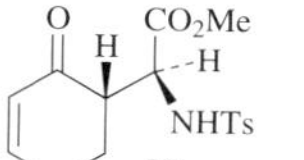 VI (69) **I**:**II**:**III**:**IV**:**V**:**VI** 4:3:61:3:28:2	122

	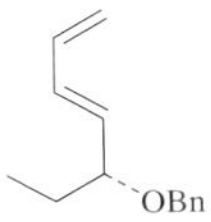	40°, 48 h	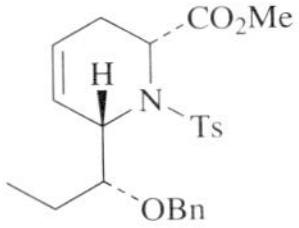 (74)	39
	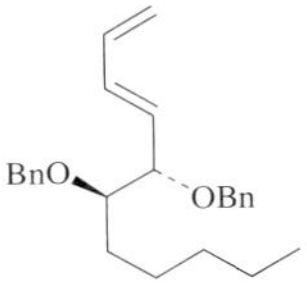	40°, 2 d	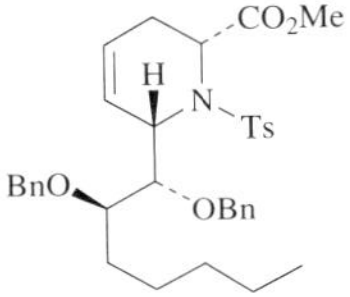 (67)	39
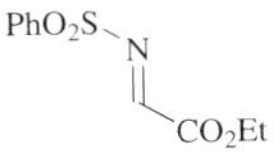		Benzene, 0°	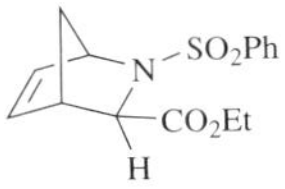 (80)	110
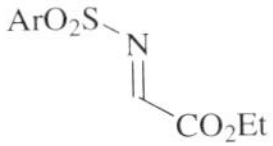			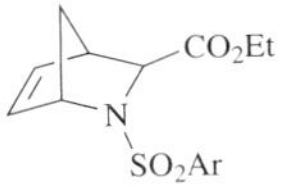	40
Ar				
Ph		—[i]	(<23)[b]	
Ph		—[j]	(5-25)[b]	
4-$O_2NC_6H_4$		—[j]	(5-25)[b]	
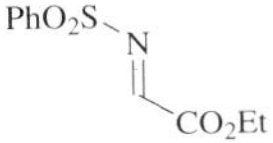		1. NaH[a], CH_2Cl_2, −100°, 30 min 2. Cyclopentadiene, $Ln(OTf)_3$, −100°, 15 min	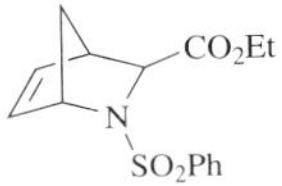 (75)[b]	40

TABLE 4. CYCLOADDITIONS OF *N*-SULFONYLIMINES AND *N*-PHOSPHORYLIMINES (*Continued*)

Imine	Diene	Conditions	Product(s) and Yield(s) (%)	Refs.
C_{10} 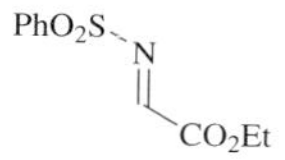		1. Base[a], solvent 2. Cyclopentadiene	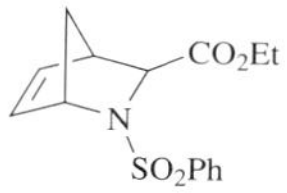	40
	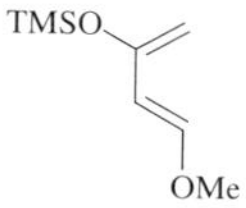	1. DBU[a], CH_2Cl_2, rt, 12 h 2. HCl, THF, rt, 1 h	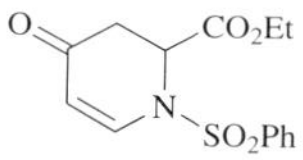 (76)	124

Conditions and yields for the cyclopentadiene reaction (ref. 40):

Base	Solvent	Temp	Time	Yield (%)
Et_3N	CH_2Cl_2	rt	10 min	(22)[b]
Et_3N	toluene	0°	1 h	(80)[b]
i-Pr_2NEt	toluene	0°	1 h	(95)[b]
i-Pr_2NEt[k]	toluene	0°	1 h	(56)[b]
DBU	CH_2Cl_2	0°	1 h	(5)[b]
n-BuLi	toluene	0°	10 min	(12)[b]
n-BuLi	toluene	0°	1 h	(5)[b]
n-BuLi	toluene	–78°	1 h	(5)[b]
MeLi	toluene	0°	30 min	(38)[b]
MeLi	toluene	0°	1 h	(42)[b]
MeLi	CH_2Cl_2	0°	30 min	(38)[b]
t-BuLi	toluene	0°	30 min	(15)[b]
NaH	THF	0°	10 min	(95)[b]
NaH[k]	THF	0°	10 min	(56)[b]
DIBALH	THF	–78°	10 min	(98)[b]
KB(*s*-Bu)$_3$H	THF	–78°	10 min	(97)[b]
NaH	THF	–100°	10 min	(89)[b]

	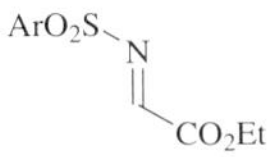 Ar = 4-$O_2NC_6H_4$		1. Base[a], toluene 2. Cyclopentadiene	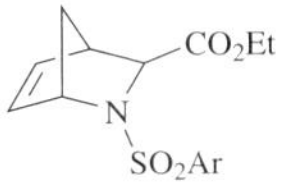	40
			Base / Temp / Time: Et_3N, 0°, 1 h i-Pr_2NEt, 0°, 1 h n-BuLi, 0°, 30 min n-BuLi, 0°, 1 h MeLi, –78°, 30 min MeLi, 0°, 1 h	(72)[b] (96)[b] (20)[b] (17)[b] (10)[b] (10)[b]	
C_{11}	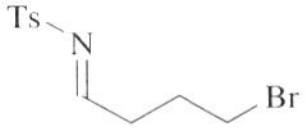	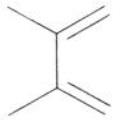	BF_3•Et_2O, $MgSO_4$, toluene, CH_2Cl_2, –30°, 3 h	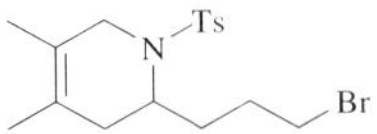 (60)[d]	33
	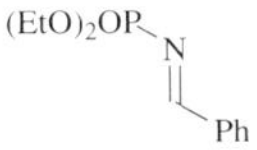	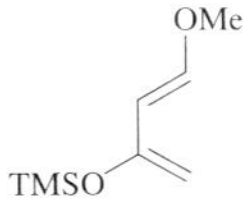	1. $Cu(OTf)_2$, CH_2Cl_2, rt 2. TFA, CH_2Cl_2	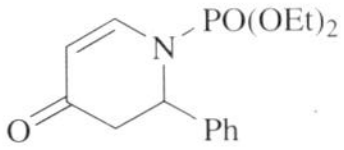 (55)	129
	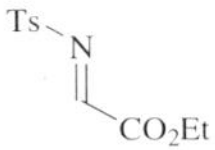			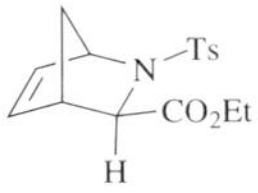	
			Benzene, 0°	(84)	110
			Toluene, 90°, 18 h	(—)	116
		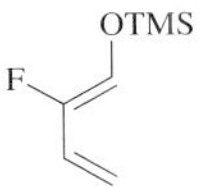	1. Toluene, 80°, 2 h 2. HCl, H_2O, THF, rt, 30 min	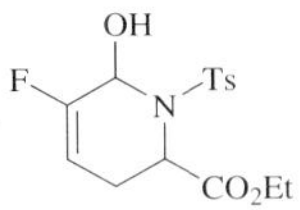 (91)	112

TABLE 4. CYCLOADDITIONS OF *N*-SULFONYLIMINES AND *N*-PHOSPHORYLIMINES (*Continued*)

Imine	Diene	Conditions	Product(s) and Yield(s) (%)	Refs.
C_{11} 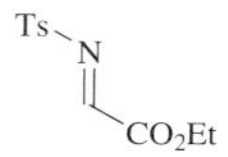	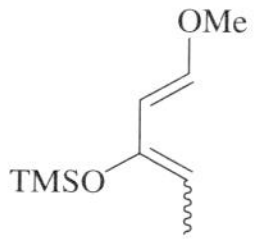	1. Lewis acid, toluene, 3 h 2. HCl, H_2O, rt, 30 min L. A. / Temp $AlCl_3$ / −78° $ZnCl_2$ / −78° — / rt	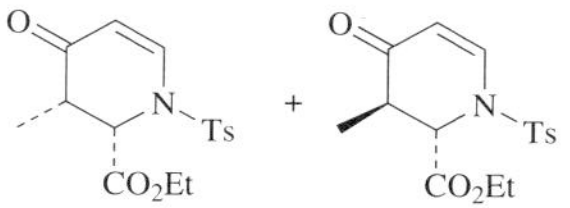 (53) 7:1 (60) 22:1 (51) 4.7:1	114
	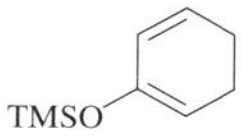	"Thermal conditions"	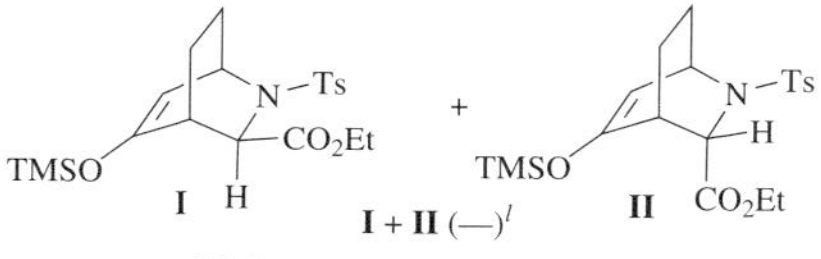 **I** + **II** (—)[l]	252
	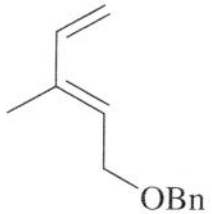	$ZnCl_2$, toluene, rt, 6 d	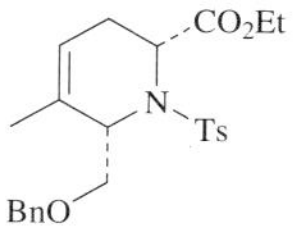 (58)	114
	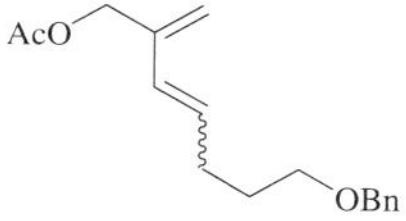	CH_2Cl_2, 70°, 36-48 h, 12 kbar	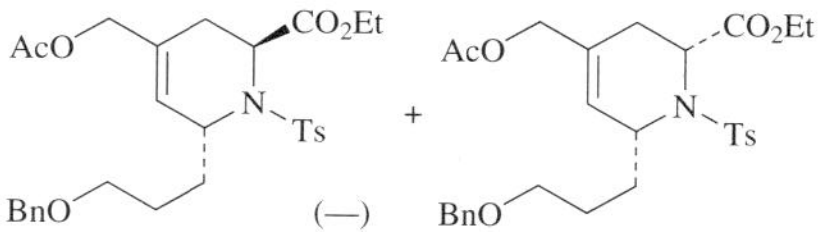 (—)	113
	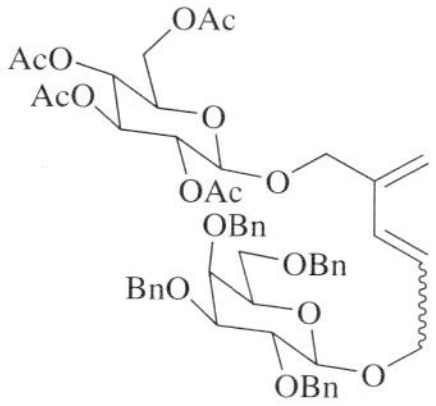	CH_2Cl_2, 70°,40 h, 12 kbar	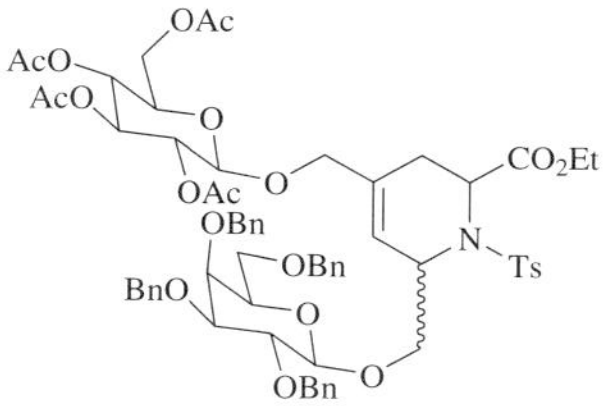 (61)[m]	341

	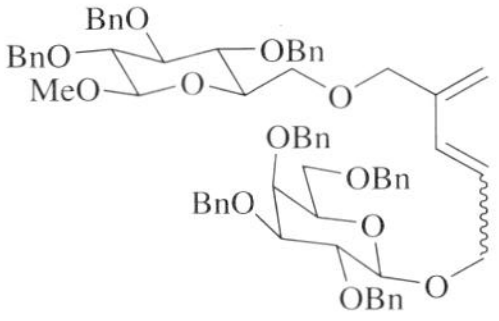BnO BnO MeO O OBn O OBn OBn BnO O OBn O	CH_2Cl_2, 70°,40 h, 12 kbar	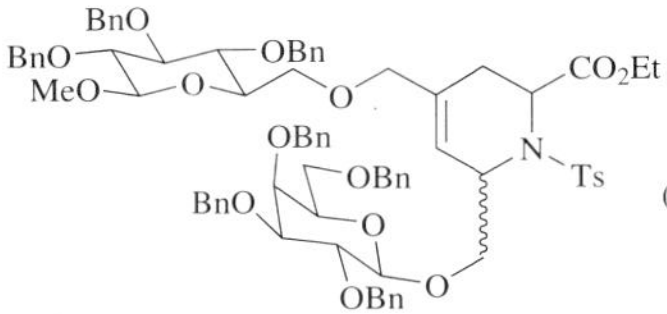BnO BnO MeO O OBn O CO_2Et N Ts OBn OBn BnO O OBn O	(55)[m]	341
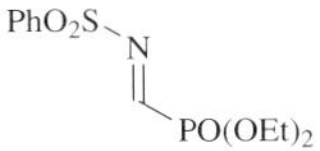PhO_2S N $PO(OEt)_2$		1. THF, –78° to rt, 1 h 2. 40°, 1 h	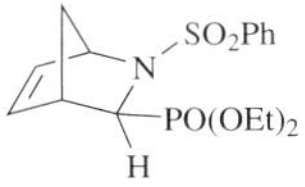N SO_2Ph $PO(OEt)_2$ H	(66)	57
	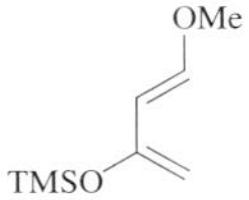OMe TMSO	1. THF, –78° to rt, 13 h 2. Citric acid, H_2O	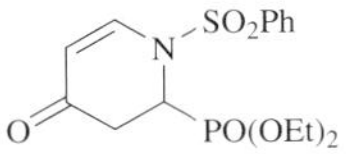N SO_2Ph O $PO(OEt)_2$	(36)	57
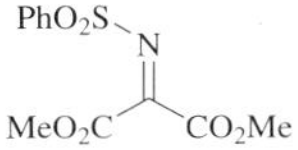PhO_2S N MeO_2C CO_2Me	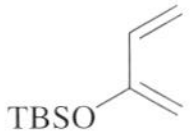TBSO	THF, 100°, 13 h	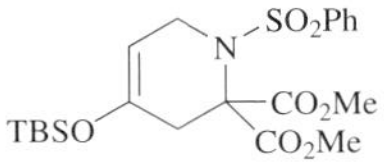N SO_2Ph TBSO CO_2Me CO_2Me	(63)	59
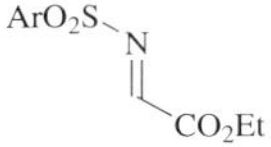ArO_2S N CO_2Et Ar = 4-$MeOC_6H_4$		1. Base[a], toluene 2. Cyclopentadiene	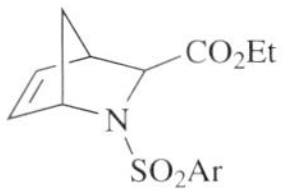CO_2Et N SO_2Ar		40

Base	Temp	Time	
Et_3N	rt	1 h	(65)[b]
i-Pr_2NEt	0°	1 h	(98)[b]
n-BuLi	0°	30 min	(5)[b]
n-BuLi	0°	1 h	(13)[b]
MeLi	0°	30 min	(22)[b]
MeLi	0°	1 h	(28)[b]

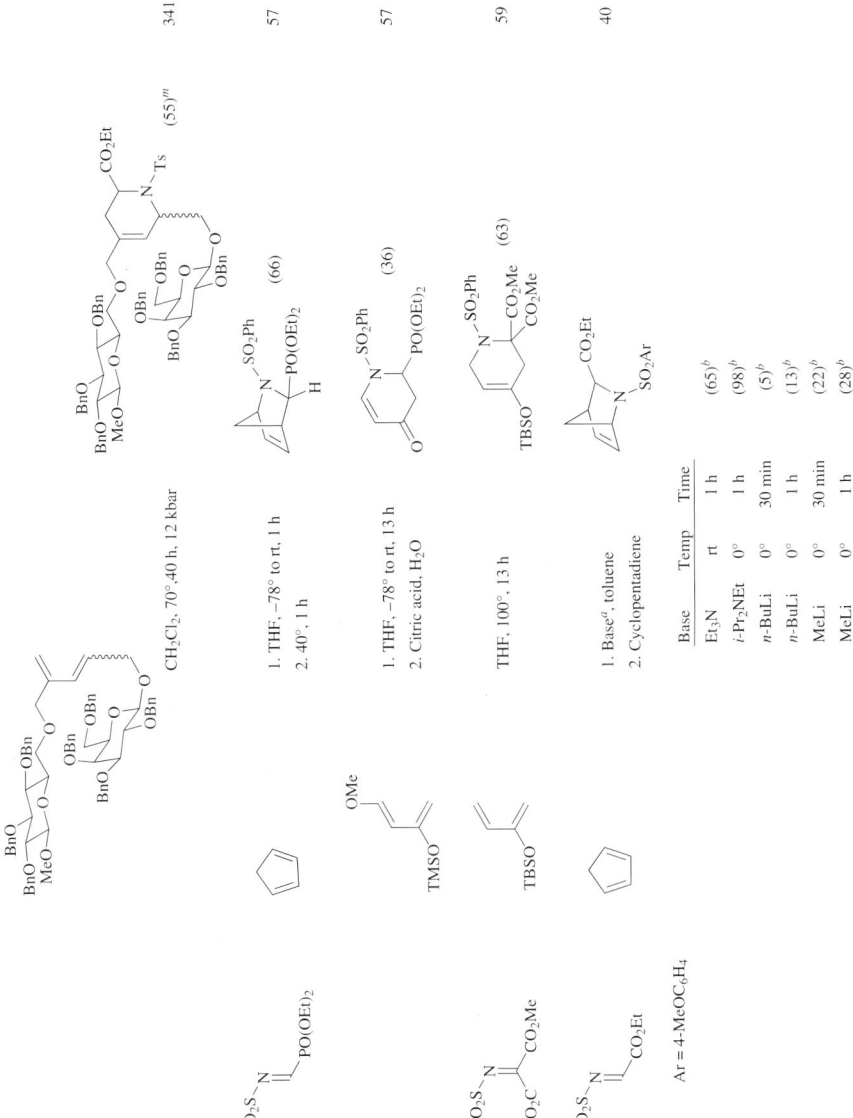

TABLE 4. CYCLOADDITIONS OF *N*-SULFONYLIMINES AND *N*-PHOSPHORYLIMINES (*Continued*)

	Imine	Diene	Conditions	Product(s) and Yield(s) (%)	Refs.
C_{12}	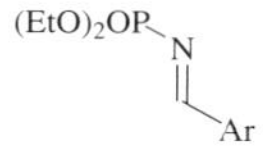 (EtO)2OP N Ar	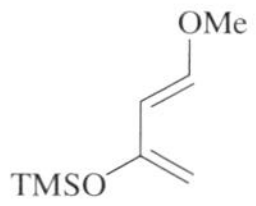 OMe TMSO	1. $Cu(OTf)_2$, CH_2Cl_2, rt 2. TFA, CH_2Cl_2	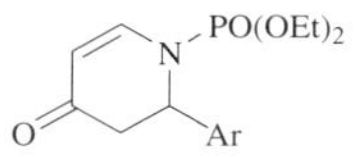 N PO(OEt)2 O Ar	129
	Ar: 4-MeC_6H_4			(57)	
	Ar: 4-$MeOC_6H_4$			(62)	
C_{13}	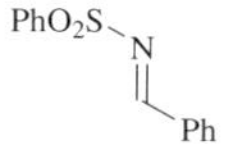 PhO2S N Ph	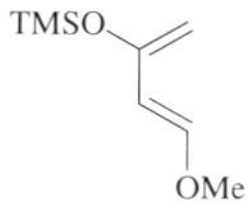 TMSO OMe	Toluene, 110°, 4 h	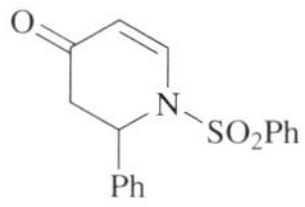 O N SO2Ph Ph (80)	128, 127
		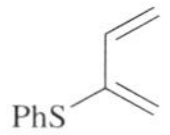 PhS	"Thermal or acidic conditions"	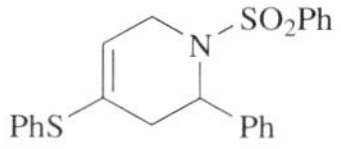 N SO2Ph PhS Ph ("very low")	311
		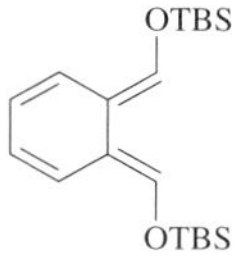 OTBS OTBS	4 h[n]	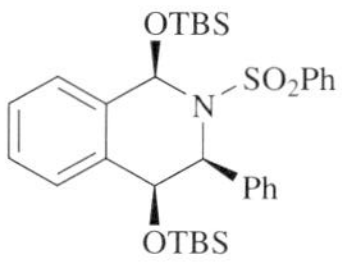 OTBS N SO2Ph Ph OTBS (68)	115
		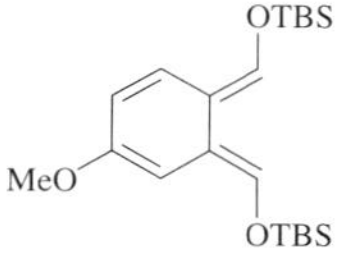 OTBS MeO OTBS	3 h[n]	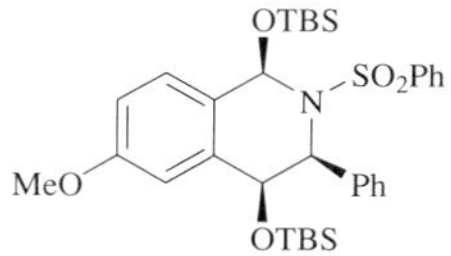 OTBS N SO2Ph MeO Ph OTBS (57)	115
	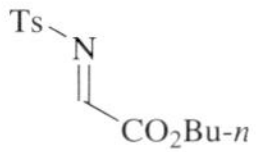 Ts N CO2Bu-n	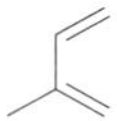	34°, 10 h	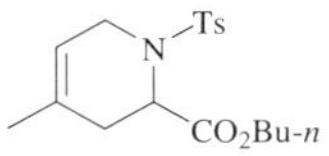 N Ts CO2Bu-n (70)	109

	Benzene, 80°, 15 min	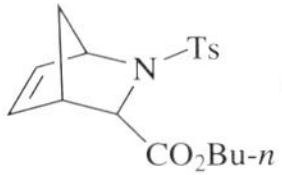 (41)[c]	109
	—[o] Benzene, 0° to rt, 12 h	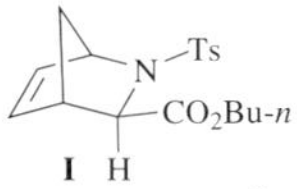 I + 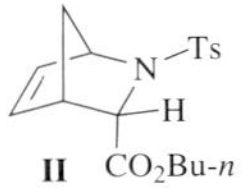 II **I** + **II** (41) **I** (84)	277
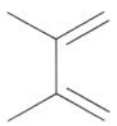	Benzene, 80°, 10 h	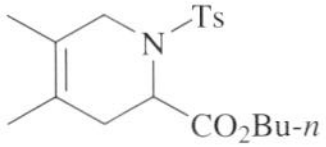 (—)	109
	Benzene, 80°, 12.5 h	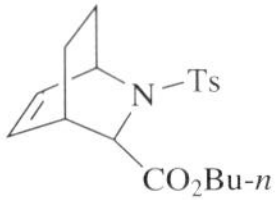 (—)[c]	109
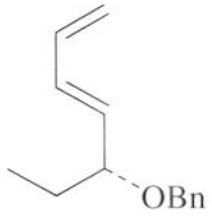	40°, 48 h	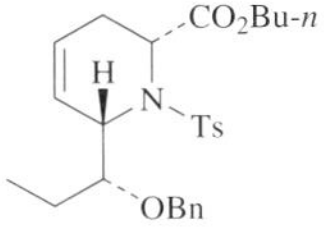 (83)	39

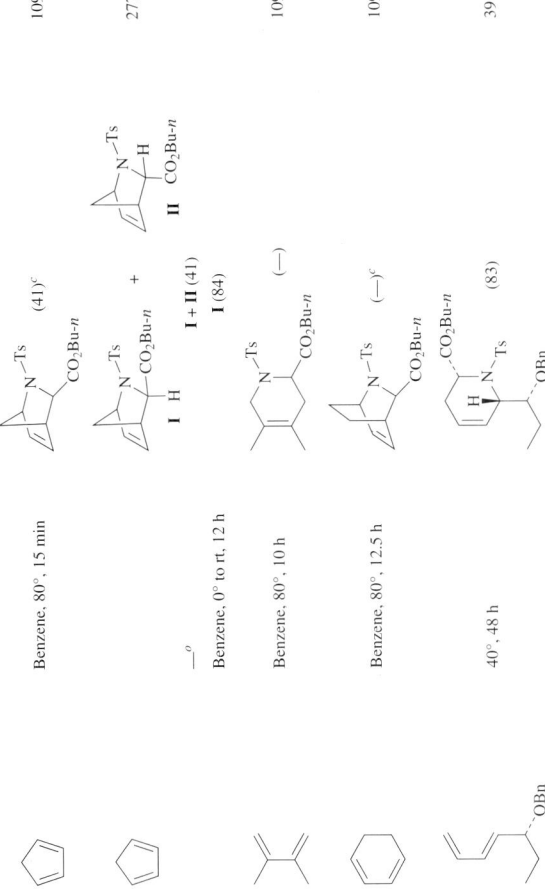

TABLE 4. CYCLOADDITIONS OF *N*-SULFONYLIMINES AND *N*-PHOSPHORYLIMINES (*Continued*)

Imine	Diene	Conditions	Product(s) and Yield(s) (%)	Refs.
C_{13} 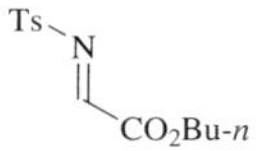	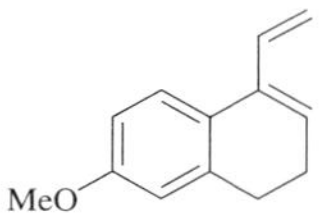		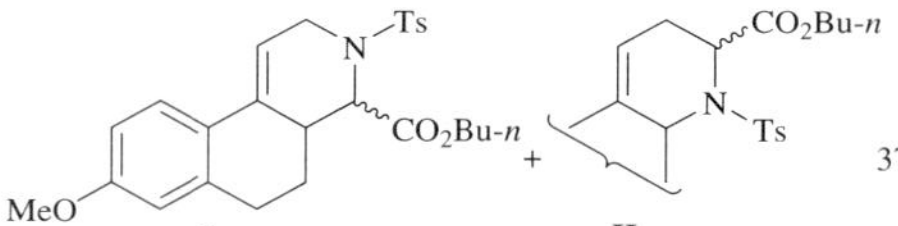 **I** + **II** 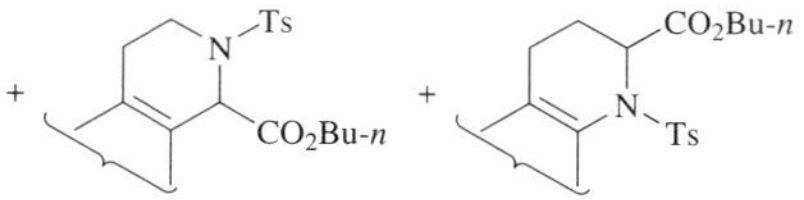 **III** + **IV**	37
		Benzene, 5°, overnight	**I** + **II** ("almost quantitative"), **I**:**II** = 3:1[p]	
		Benzene, rt, overnight	**I** + **II** + **III** + **IV** ("lower")[p]	
	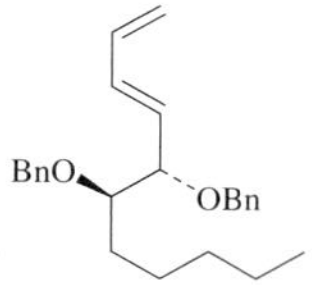	40°, 6 d	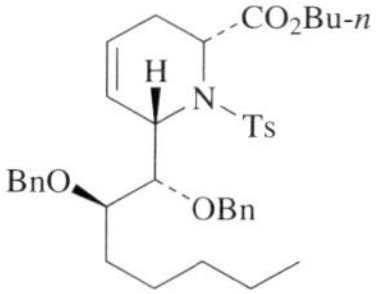 (81)	123
C_{14} 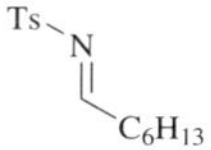	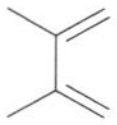		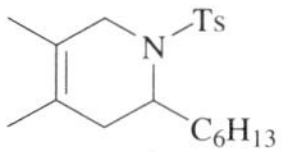	
		BF_3•OEt_2, $MgSO_4$, toluene, CH_2Cl_2, –30°, 3 h	(63)[d]	33
		BF_3•OEt_2, CH_2Cl_2, –20°	(83)	34
			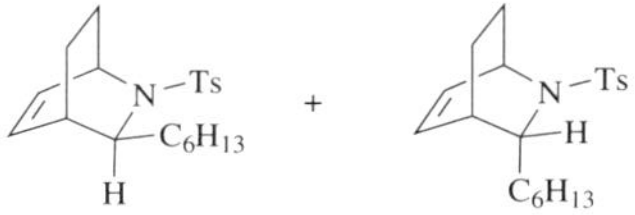	

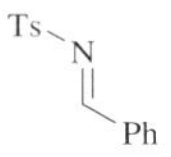	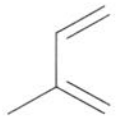	BF_3•OEt_2, $MgSO_4$, toluene, CH_2Cl_2, –30°, 3 h	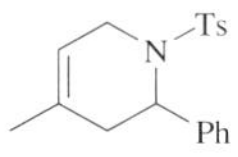 (74)[d] 1.5:1	33
		BF_3•OEt_2, CH_2Cl_2, –20°	(54) 1.7:1	34
		BF_3•OEt_2, $MgSO_4$, toluene, CH_2Cl_2, 0°, 3 h	(60)[d]	33
		BF_3•OEt_2, CH_2Cl_2, –20°	(46)	34
		BF_3•OEt_2, CH_2Cl_2, –20°	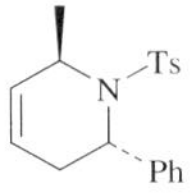 + 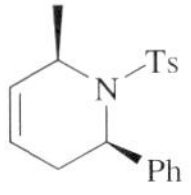 (52) 3:1	34
		BF_3•OEt_2, $MgSO_4$, toluene, CH_2Cl_2, 0°, 3h	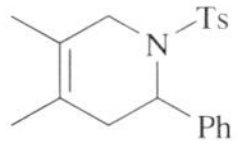 (75)[d]	33
		BF_3•OEt_2, CH_2Cl_2, –20°	(56)	34
		BF_3•OEt_2, CH_2Cl_2, –20°	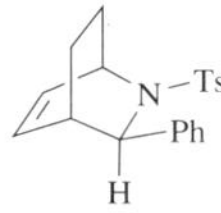 + 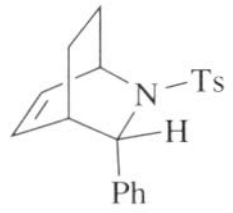 (46) 1.5:1	34
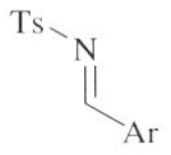 Ar = Ph, 4-$O_2NC_6H_4$	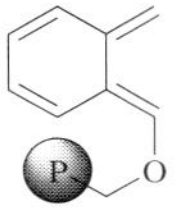	Propionitrile, 105-110°, 14 h[q]	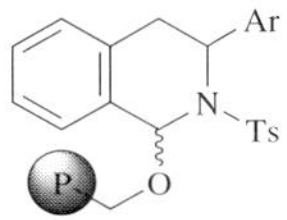 (—)[c]	342

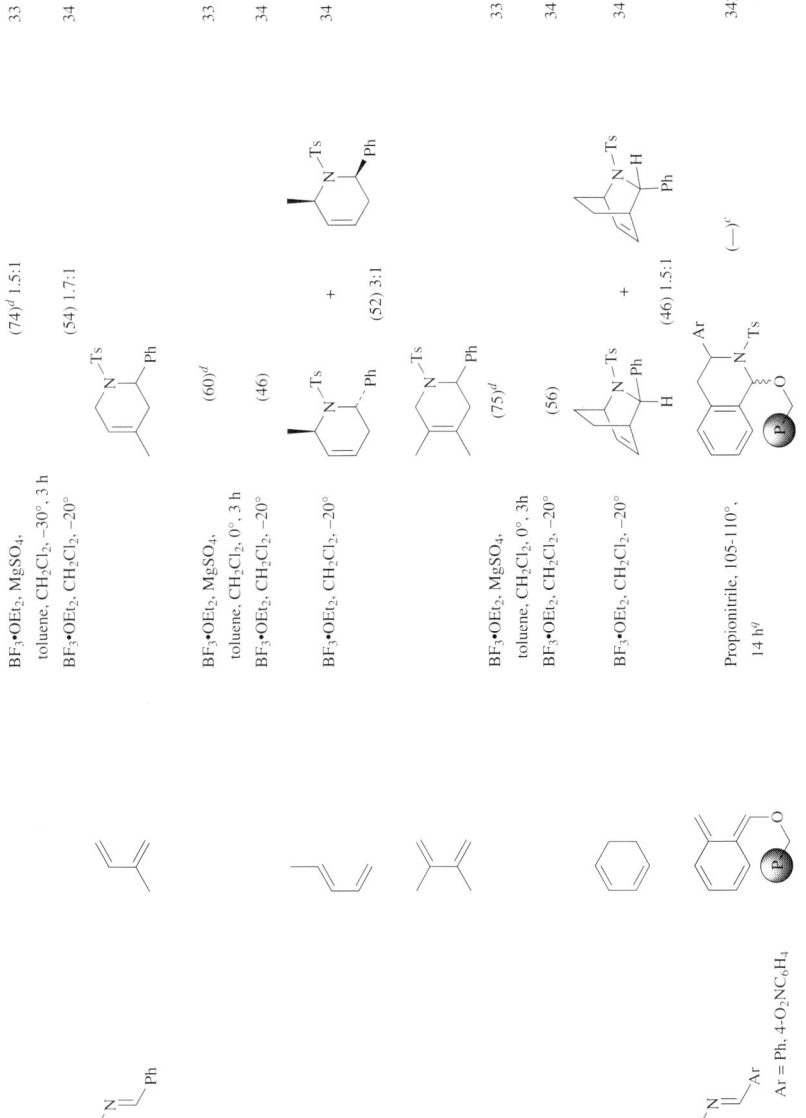

TABLE 4. CYCLOADDITIONS OF *N*-SULFONYLIMINES AND *N*-PHOSPHORYLIMINES (*Continued*)

Imine	Diene	Conditions	Product(s) and Yield(s) (%)	Refs.
C_{14} 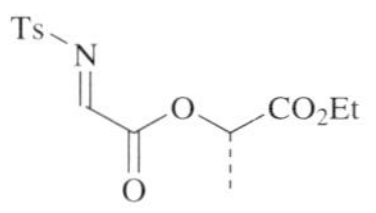			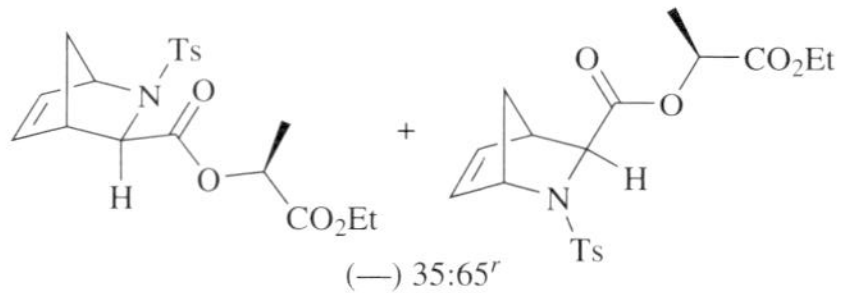	125
		Toluene, rt	(—) 35:65[r]	
		Et_2AlCl, toluene, –78°	(50-60) 12:88[s]	
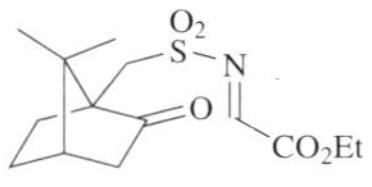	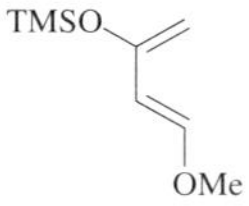	1. Base[a], solvent 2. HCl, THF, rt, 1 h	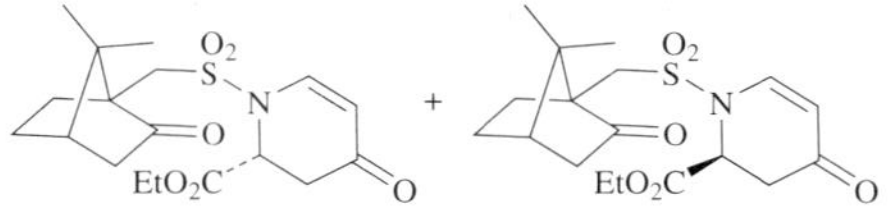 **I** + **II**	
		Base / Solvent / Temp / Time		
		Et_3N / CCl_4 / 20° / 8 h	(58) 65:35	111
		Et_3N / CCl_4 / rt / 12 h	(58) 65:35	124
		Et_3N / Et_2O / 20° / 8 h	(90)[t] 60:40	111
		Et_3N / CH_2Cl_2 / 20° / 8 h	(100)[t] 54:46	111
		Et_3N / MeCN / 20° / 8 h	(100)[t] 58:42	111
		Et_3N / CCl_4 / –15° / 8 h	(100)[t] 67:33	111, 124
		n-BuLi / toluene / 20° / 8 h	(100)[t] 57:43	111
	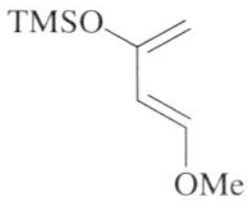	1. *n*-BuLi[a], Lewis acid, toluene 2. HCl, THF, rt, 1 h	**I** + **II**	

L. A.[u] (eq)	Time	Temp		
Et_2AlCl (0.25)	6 h	–75°	(70)[t] 50:50	111
Et_2AlCl (0.50)	6 h	–75°	(45) 41:59	111
Et_2AlCl (0.50)	8 h	–78°	(45) 41:59	124
Et_2AlCl (0.75)	6 h	–75°	(60)[t] 41:59	111
$TiCl_4$ (0.25)	6 h	–75°	(70)[t] 57:43	111
$Ti(OPr\text{-}i)_4$ (0.25)	6 h	–75°	(25)[t] 70:30	111
$Ti(OPr\text{-}i)_4$ (0.25)	4 h	–78°	(25) 2.33:1	124
$Ti(OPr\text{-}i)_4$ (1.00)	6 h	–75°	(25)[t] 67:33	111
$BF_3{\bullet}OEt_2$ (0.25)	6 h	–75°	(60)[t] 64:36	111
$BF_3{\bullet}OEt_2$ (0.50)	6 h	–75°	(50)[t] 56:44	111
Cp_2TiCl_2 (0.10)	6 h	–75°	(40)[t] 65:35	111

C_{15}

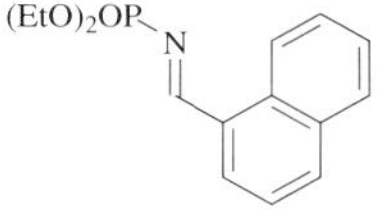

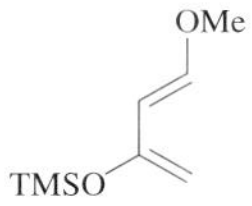

1. $Cu(OTf)_2$, CH_2Cl_2, rt
2. TFA, CH_2Cl_2

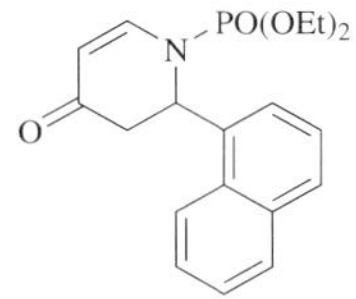

(60) 129

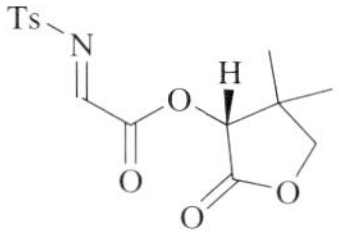

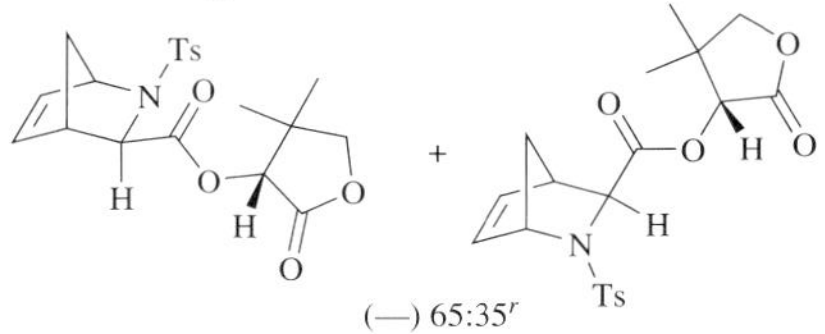

125

Toluene, rt (—) 65:35[r]

Et_2AlCl, toluene, –78° (50-60) 85:15[s]

C_{19}

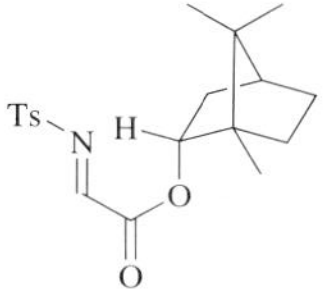

Benzene, 0°

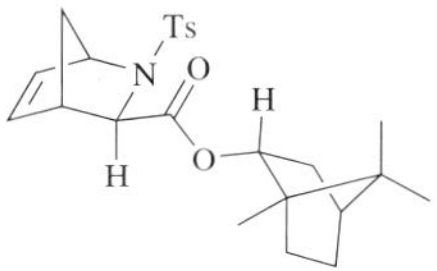

("excellent") 110

53:47 diastereomeric ratio[v]

TABLE 4. CYCLOADDITIONS OF *N*-SULFONYLIMINES AND *N*-PHOSPHORYLIMINES (*Continued*)

Imine	Diene	Conditions	Product(s) and Yield(s) (%)	Refs.
C_{19-25}				
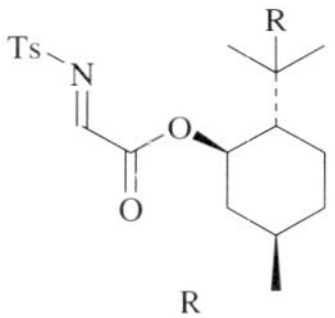 R: H; Ph		Benzene, 0°	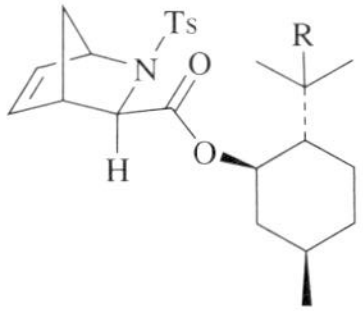 ("excellent") 56:44 diastereomeric ratio[v] (R = H); ("excellent") 60:40 diastereomeric ratio[v] (R = Ph)	110
C_{19}				
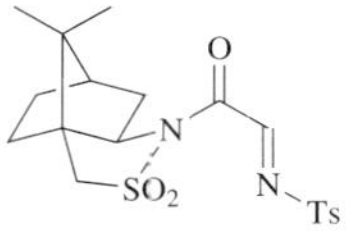		Toluene	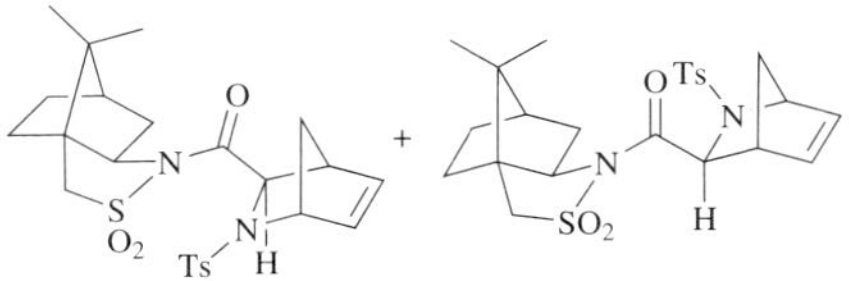	
		Catalyst (mol %) / Temp / Time / Pressure (atm)		
		— / 20° / 48 h / 1	(64) 36:64	126, 92
		— / 20° / 24 h / 15000	(34) 27:73	126
		$Eu(fod)_3$ (2) / 20° / 24 h / 1	(65) 34:66	126
		$Eu(fod)_3$ (2) / 20° / 96 h / 15000	(64) 25:75	126
		$SnCl_4$ (10) / –78° / 2.5 h / 1	(20) 47:53	126
		$SnCl_4$ (50) / –78° / 2.5 h / 1	(14) 53:47	126
		$SnCl_4$ (100) / –78° / 2.5 h / 1	(14) 73:27	126
		$TiCl_4$ (20) / –78° / 2 h / 1	(17) 80:20	126
		$TiCl_4$ (50) / –78° / 2 h / 1	(38) 80:20	126, 92

[a] The base was utilized during the conversion of the corresponding α-bromoglycinate to the imine, which was then reacted with the diene in situ.

[b] The stereochemistry of the cycloadduct was not specified, but it is probably the exo-adduct based on the results of similar imines with cyclopentadiene.[110]

[c] No stereochemical information is provided.

[d] The yields of the reaction are substantially lower without using $MgSO_4$ and slightly lower using molecular sieves instead.

[e] One isomer is formed, but the stereochemistry is not specified.

[f] The yields for both isomers (cis and trans) are shown, but the stereochemistry of the major product is not specified.

[g] The two diastereomers were obtained in ratios varying from 0.8 to 1.5 based on the E/Z ratios of the starting dienes. The stereochemistry of the major products is not specified.

[h] The isolated yields of the pure products are **I** (14) and **II** (27).

[i] The *N*-sulfonylimine was prepared by condensation of the appropriate aldehyde with a sulfonamide, followed by in situ reaction with cyclopentadiene.

[j] The *N*-sulfonylimine was prepared by an aza-Wittig reaction of the corresponding *N*-sulfonylphosphinamide with ethyl glyoxylate, followed by in situ reaction with cyclopentadiene.

[k] Before addition of cyclopentadiene, the precipitate resulting from reaction of the α-bromoglycinate and base was removed from the mixture by filtration.

[l] The exo-adduct **I** is the major product, although the ratio of products is not provided. It is implied that side products resulting from Mannich addition were also observed.

[m] A mixture of the cis and trans isomers was obtained, but no other stereochemical information is provided.

[n] Although the exact reaction conditions are not specified, the reactions were generally performed in benzene or toluene and at slightly elevated temperatures (40-50°). The reactions also took place at room temperature at a slower rate. The precursor to the diene was the corresponding benzocyclobutene.

[o] Although exact conditions are not specified, the conditions of benzene, 80°, 15 min, are referenced.[109]

[p] The stereochemistry and stereochemical ratios of adducts **I** and **II** are not provided.

[q] The diene was formed in situ from the corresponding benzocyclobutene.

[r] The ratio of products could be reversed by conducting the reaction in more polar solvents.

[s] Other Lewis acids, such as $Al(OEt)_3$, $MgBr_2$, $ZnCl_2$, Me_2AlCl, and *i*-Bu_2AlCl, provided diastereomeric ratios between 64:36 and 85:15.

[t] The yield was estimated by ^1H NMR using the weight of the crude product.

[u] $ZnCl_2$ was also used and provided the same major diastereomer as the boron and titanium Lewis acids.

[v] The absolute stereochemistry of the predominant stereoisomer is not specified.

TABLE 5. CYCLOADDITIONS OF *N*-ALKYL- AND *N*-ARYLIMINES

Imine	Diene	Conditions	Product(s) and Yield(s) (%)	Refs.
C_1				
NH=CH_2	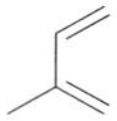	HCl, H_2O, 35°, 96 h	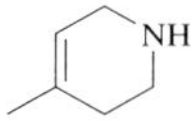 (40)[a,b]	133

		HCl, H_2O, rt, 6 h	(44)[b]	133
		HCl, H_2O, rt	(40-50)	279
		HCl, H_2O, rt, 18 h	(—)	308
	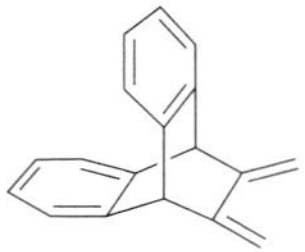	1. HCl, H_2O, EtOH, 80°, 8 h 2. 165-175°, 8 h	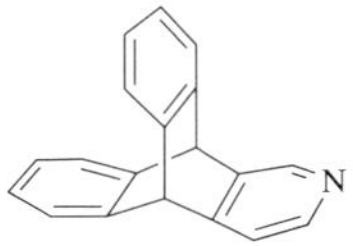 (60)	343
C_2				
Me–N=CH_2			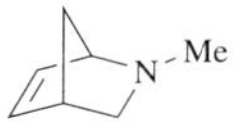	
		HCl, H_2O, rt, 3 h	(82)[b]	133, 234
		HCl, H_2O, rt, 16 h	(55)	284
		HCl, H_2O	(—)	280
		HCl[c]	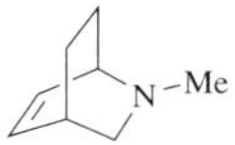 ("modest")	234
	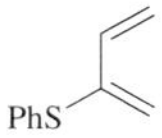	HCl, H_2O, DMF, 60°, 24 h	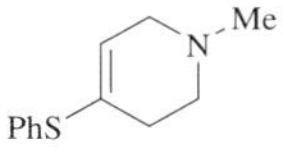 (88)	311

C_3 		HBr, H_2O	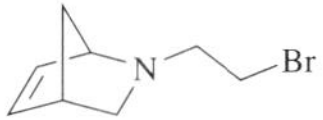 (—)	344
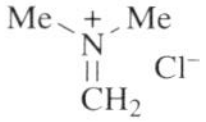		MeCN, 65°, 16 d	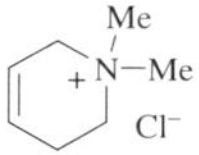 (13)	13
	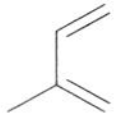	CH_2Cl_2, overnight	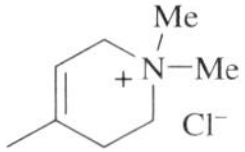 (—)	131
	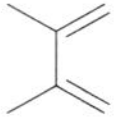	CH_2Cl_2, 40°, 10 h	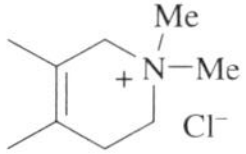 (41)	345
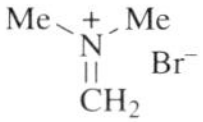	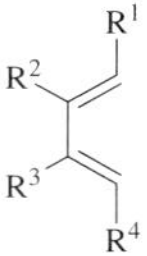	MeCN	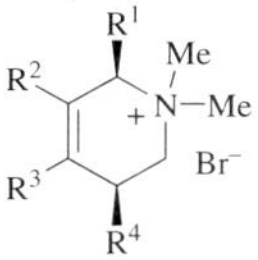	13
	R^1 R^2 R^3 R^4	Temp Time		
	H H H H	40° —	(—)	
	H H Me H	40° 80 d	(66)	
	Me H H H	40° 54 d	(13)	
	H Me Me H	40° 51 d	(73)	
	Me H H Me	rt 80 d	(58)	
	Me H H Me	40° —	(—)	
	Ph H H Ph	82° 1 d	(51)	
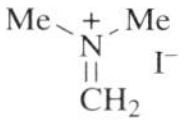	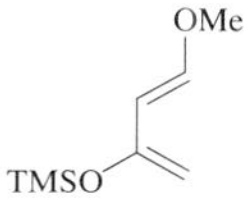	rt	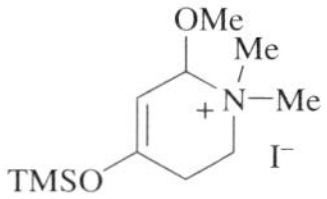 (—)	132

TABLE 5. CYCLOADDITIONS OF *N*-ALKYL- AND *N*-ARYLIMINES (*Continued*)

Imine	Diene	Conditions	Product(s) and Yield(s) (%)	Refs.
C_3				
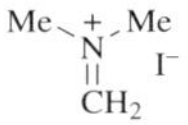	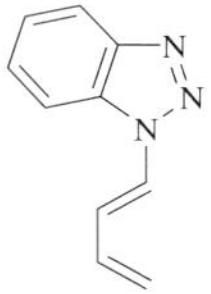	CH_2Cl_2, rt, 2 d	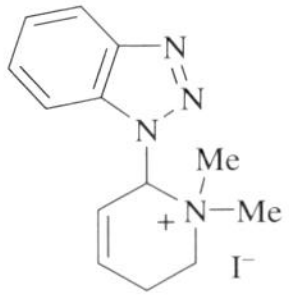 (77)	346
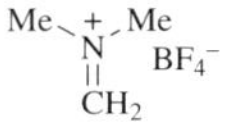	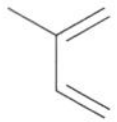	$LiBF_4$, THF, 85°, 3 h	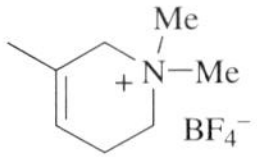 (70)[d,e]	346
	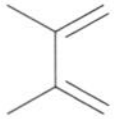	1. $LiBF_4$, THF, 85°, 3 h 2. 0°, 10 h	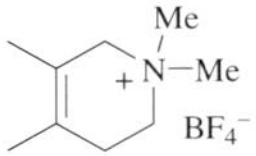 (35)[d]	347
	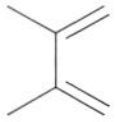	1. $LiBF_4$, THF, 85°, 3 h 2. $NaBPh_4$, H_2O	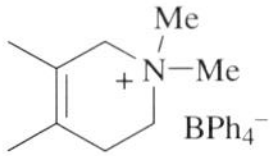 (87)[d]	347
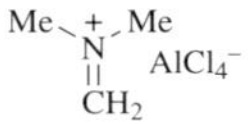	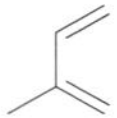	CD_3CN, 20°	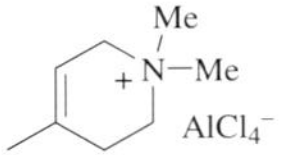 (—)	13
	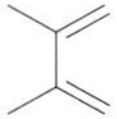	CD_3CN, 20°	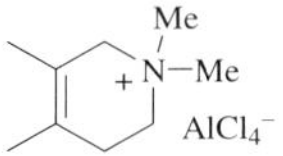 (—)	13
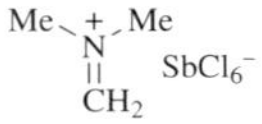	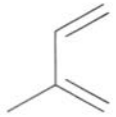		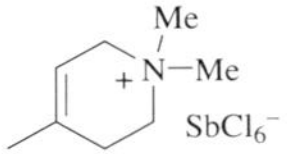	13
		CH_2Cl_2, rt, 27 h	(81)	
		CD_3CN, 20°	(—)	

			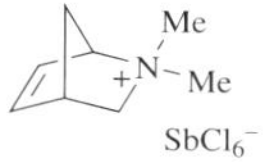Me N Me + SbCl6−		13
		CH_2Cl_2, rt, 10 min		(85)	
		CD_3CN, 20°		(—)	
	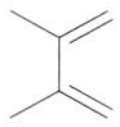	CD_3CN, 20°	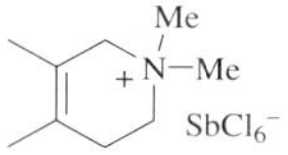Me N−Me + SbCl6−	(—)	13
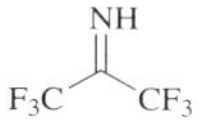NH F3C CF3		1. Hydroquinone, 100°, 13 h 2. 150°, 6 h	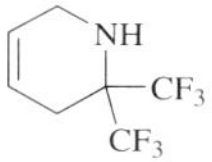NH CF3 CF3	(70)	348
	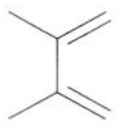	100°, 18 h	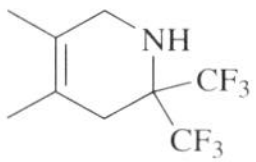NH CF3 CF3	(74)	349
		rt, overnight	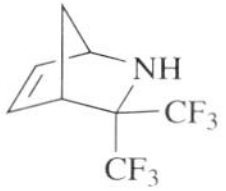NH CF3 CF3	(64)	349
C_4 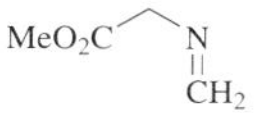MeO2C N CH2		HCl, H_2O	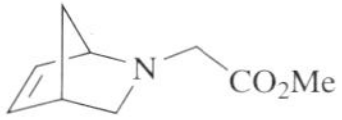N CO2Me	(>90)	350
N	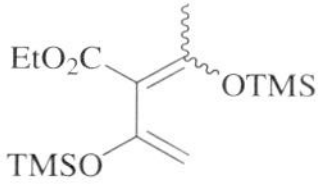EtO2C OTMS TMSO	1. $ZnCl_2$, MeCN, rt, 24 h 2. $NaHCO_3$, H_2O	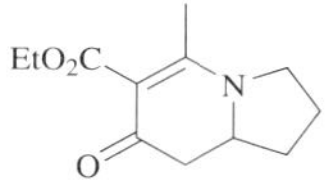EtO2C N O	(67)	351
	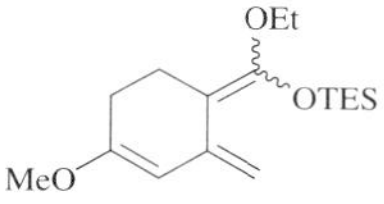OEt OTES MeO	1. CH_2Cl_2, $CHCl_3$, rt, 30 h 2. $NaHCO_3$, H_2O	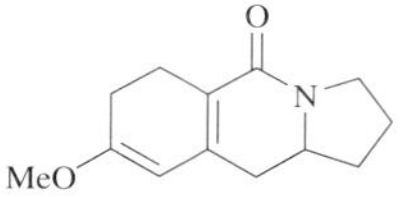O N MeO	(58)	351

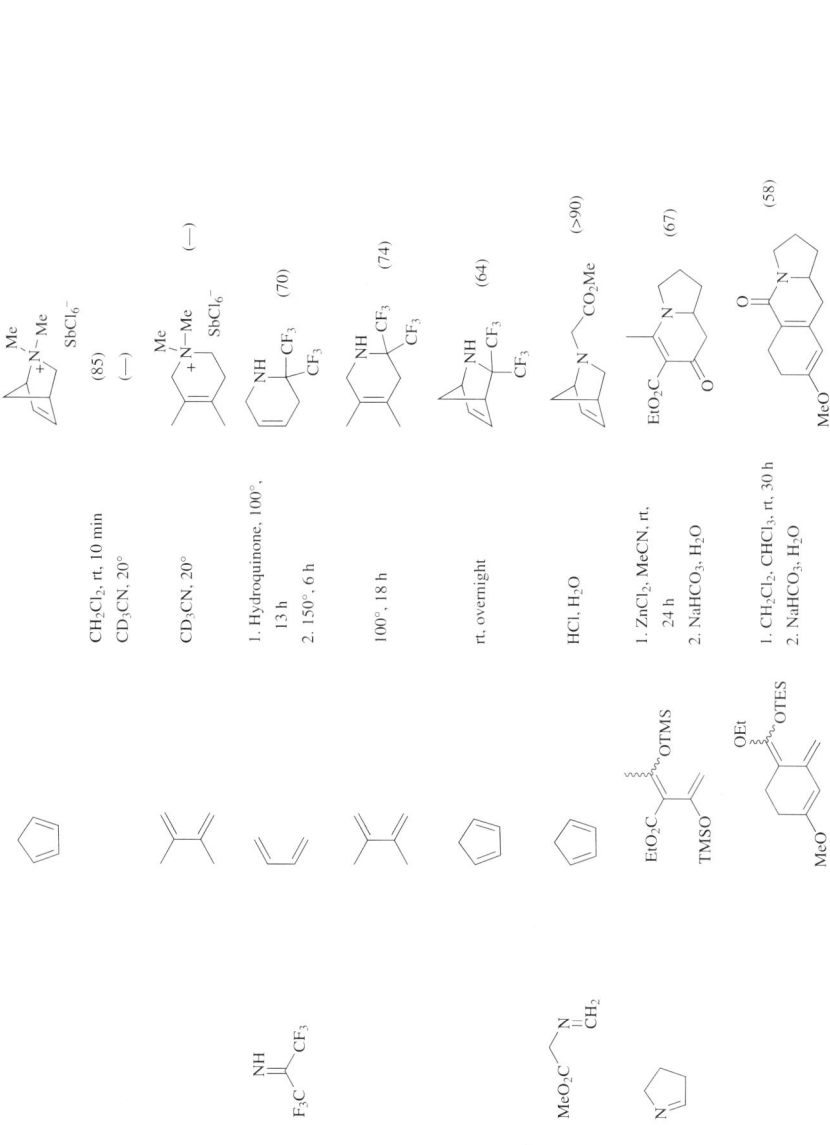

TABLE 5. CYCLOADDITIONS OF *N*-ALKYL- AND *N*-ARYLIMINES (*Continued*)

Imine	Diene	Conditions	Product(s) and Yield(s) (%)	Refs.
C_4				
	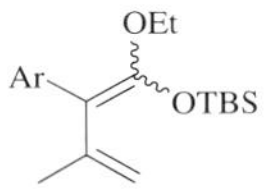 Ar = 4-$MeOC_6H_4$	1. BF_3•OEt_2, CH_2Cl_2, –78° to rt 2. $NaHCO_3$, H_2O	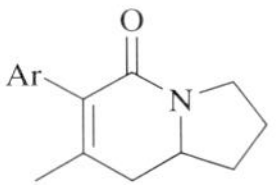 (40-46)	351
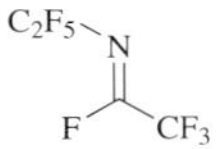		Ether, 160-180°, 10-12 h	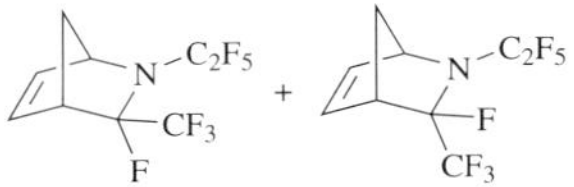 (57.1) 17:5[f]	352
		Ether, 150-160°, 16 h	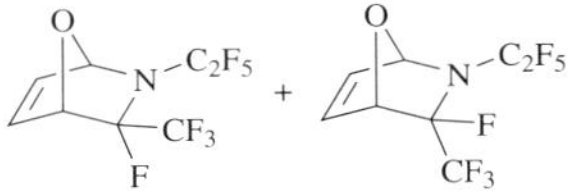 (7) 2.5:1[f]	352
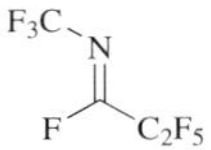		Ether, 160-180°, 10-12 h	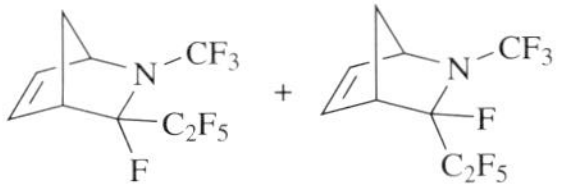 (63.8) 12:5[f]	352
C_5				
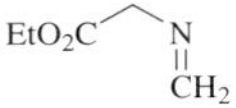		HCl, THF, H_2O, rt, 72 h	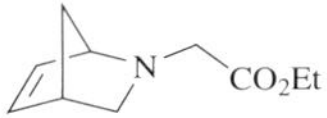 (77)	153
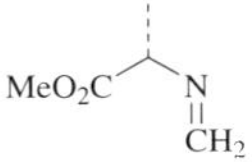		HCl, THF, H_2O, 0°, 72 h	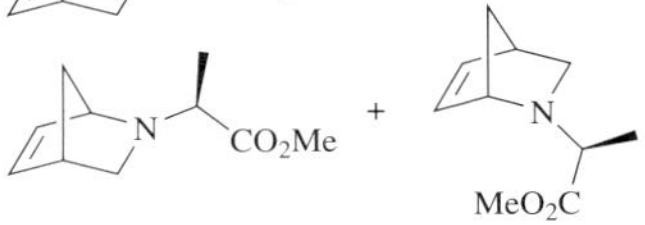 (99) 73:27	153
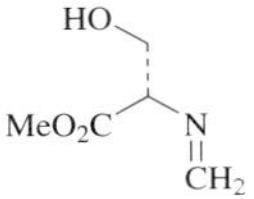			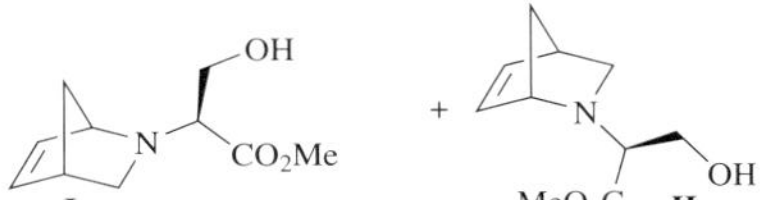 **I** + **II**	
		HCl, H_2O, rt, 1 h	**I** + **II** (98) 3.3:1[g]	354
		HCl, THF, H_2O, 0°, 72 h	**I** + **II** (63), **I**:**II** 87:13	152, 153

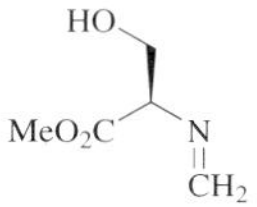	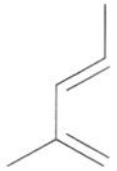	HCl, THF, H_2O, 0°, 72 h	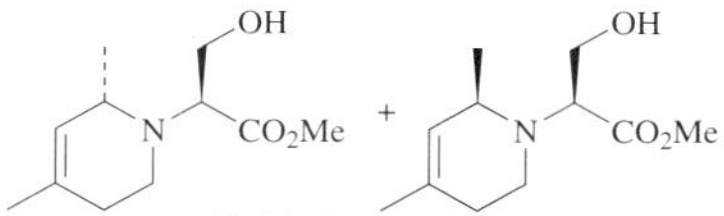 (15) 55:45	153
	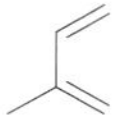	HCl, THF, H_2O, rt, 72 h	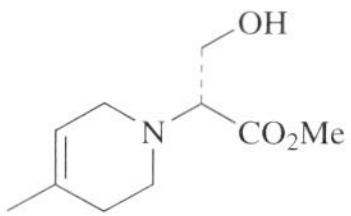 (50)	152, 153
		HCl, THF, H_2O, 0°, 72 h	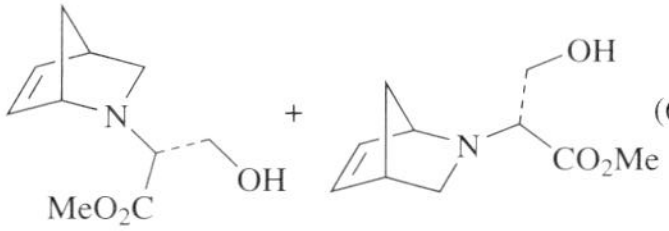 (66) 88:12	152, 153
	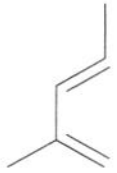	HCl, THF, H_2O, 0°, 72 h	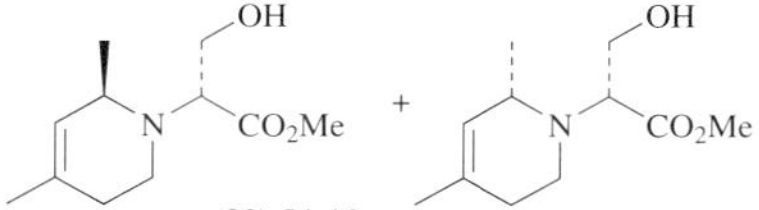 (22) 54:46	153
		HCl, THF, H_2O, rt, 72 h	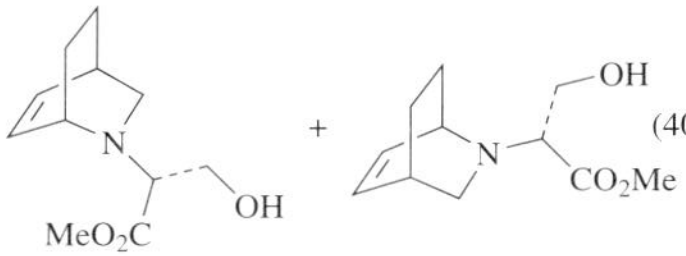 (40) 73:27	152, 153
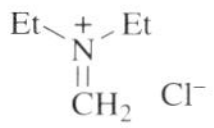	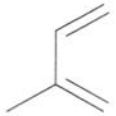	CH_2Cl_2, overnight	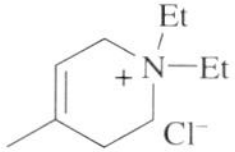 (—)	131
	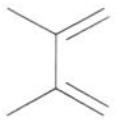	CH_2Cl_2, rt, 48 h	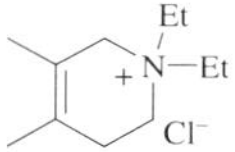 (74)	130

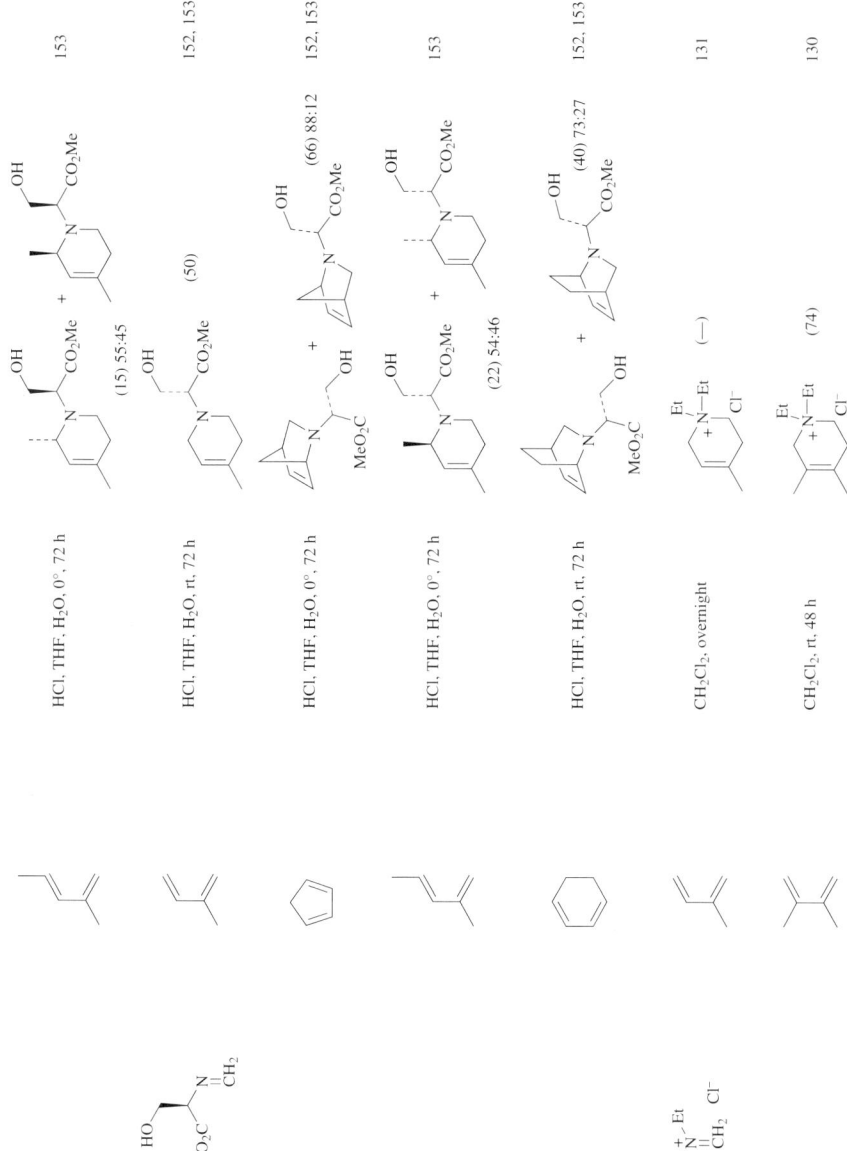

TABLE 5. CYCLOADDITIONS OF *N*-ALKYL- AND *N*-ARYLIMINES (*Continued*)

Imine	Diene	Conditions	Product(s) and Yield(s) (%)	Refs.
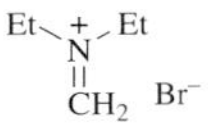	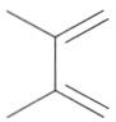	CH_2Cl_2, rt	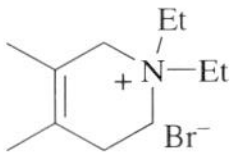 (69)	130
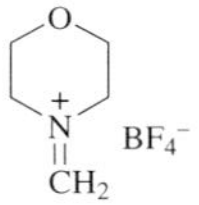	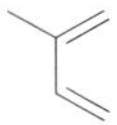	$LiBF_4$, THF, 85°, 3 h	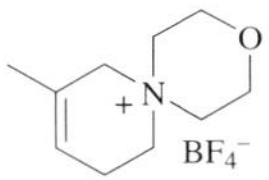 (79)[d,e]	347
		1. $LiBF_4$, THF, 85°, 3 h 2. $NaBPh_4$, H_2O	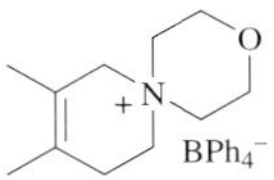 (90)[d]	347
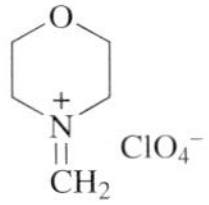	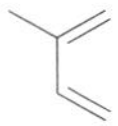	$LiClO_4$, BF_3•OEt_2, THF, 85°, 3 h	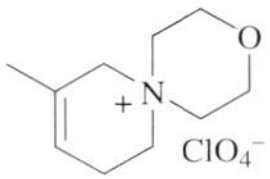 (50)[e]	347
		$KHSO_4$, 150°, 24 h	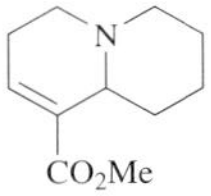 (—)	355
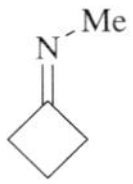	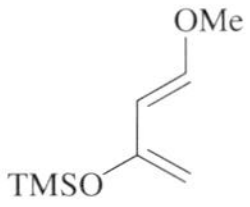	1. $Zn(OTf)_2$, CH_2Cl_2, rt, 12 h 2. NH_4Cl, H_2O, 2 h	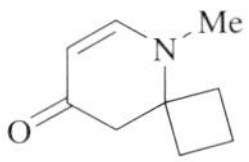 (18)	356
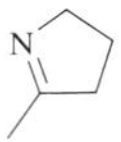	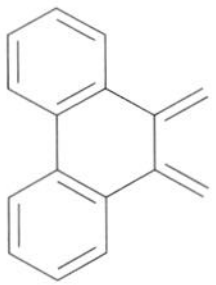	Zn, DMF, 0° to rt, overnight	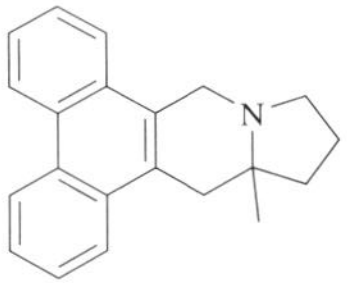 (—)	357

C_6 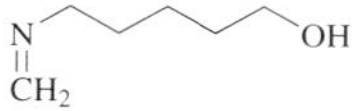		TFA, H_2O, 1 h	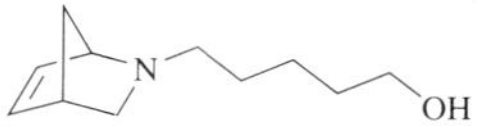	(95)	354
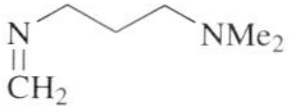		HCl, H_2O, rt, 5 h	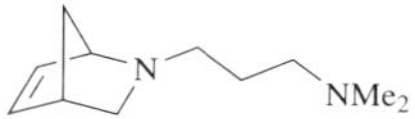	(59)	358
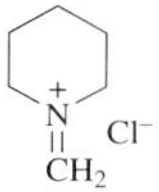	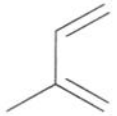	1. CH_2Cl_2, overnight 2. 40°, 5 h	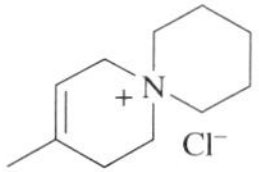	(—)	131
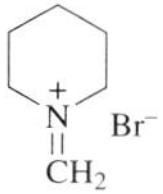	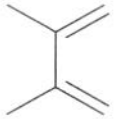	CH_2Cl_2, 40°, 6 h	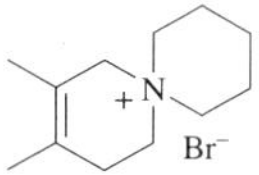	(92)	130
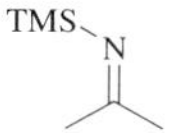	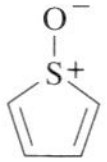	Ether, –78° to rt, 192 h	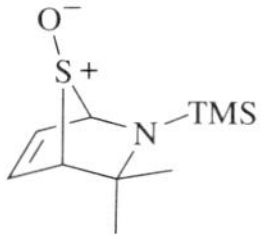	(20)[h]	359, 200
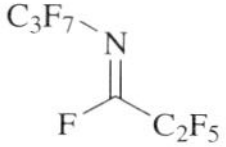		Ether, 160-180°, 10-12 h	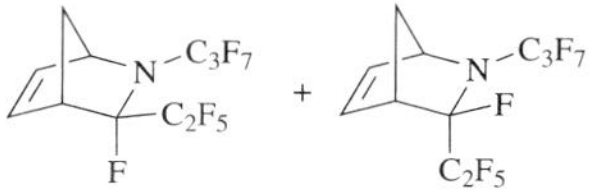	(43.4) 17:10[f]	352
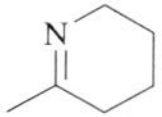		$KHSO_4$, 100°, 24 h	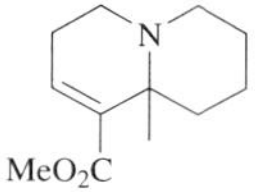	(—)[i]	355

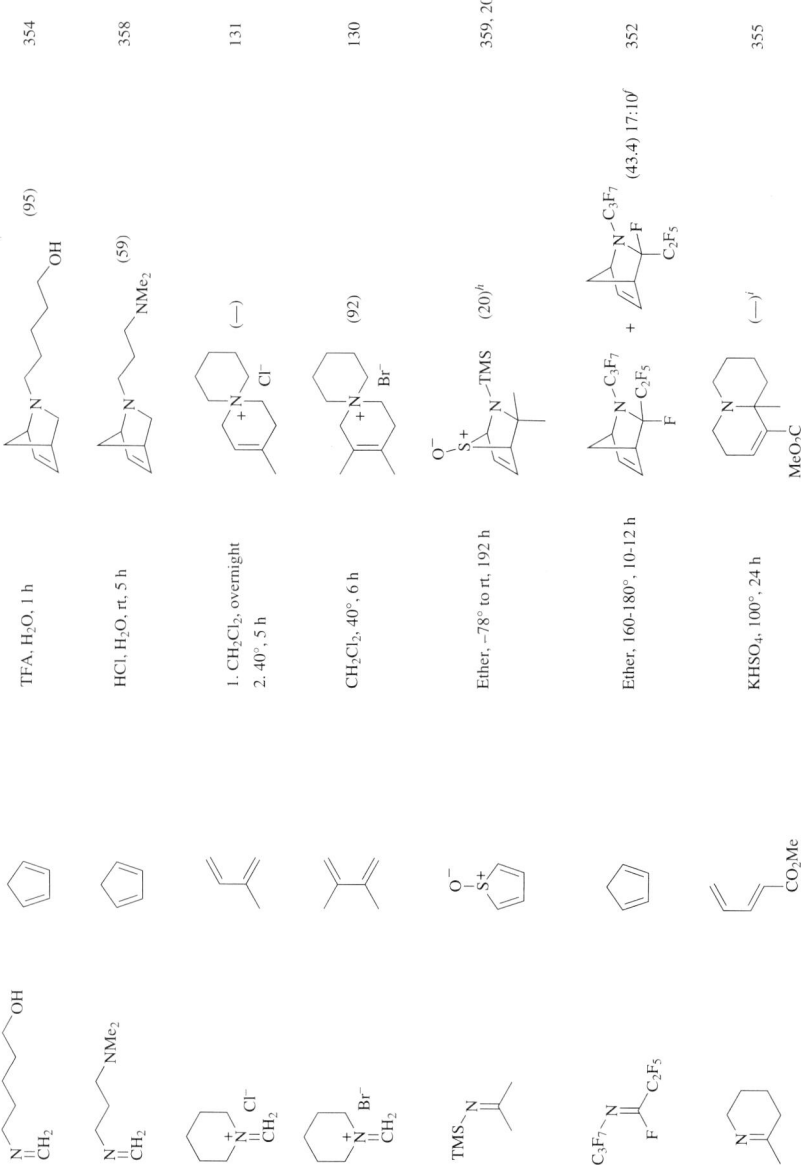

TABLE 5. CYCLOADDITIONS OF *N*-ALKYL- AND *N*-ARYLIMINES (*Continued*)

Imine	Diene	Conditions	Product(s) and Yield(s) (%)	Refs.
C_7 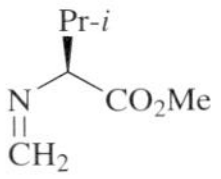			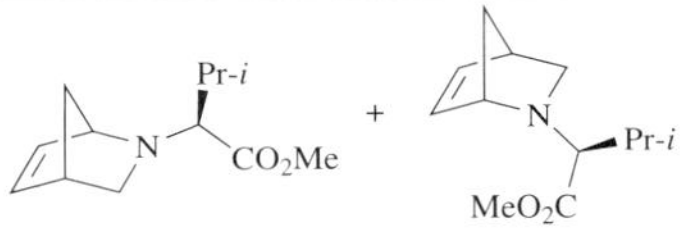 **I** **II**	
		HCl, H_2O, rt, 2 h	**I** + **II** (82) 4.0:1[g]	354
		HCl, THF, H_2O, 0°, 72 h	**I** + **II** (74), **I**:**II** 86:14	152, 153
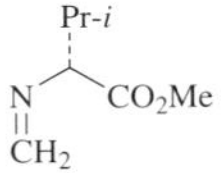		HCl, THF, H_2O, 0°, 72 h	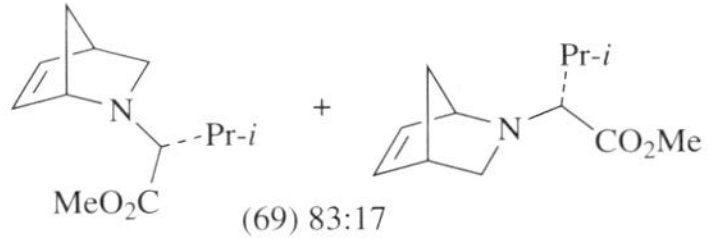 (69) 83:17	152, 153
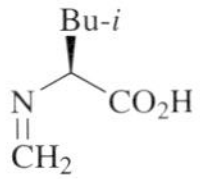		H_2O, rt, 5 h	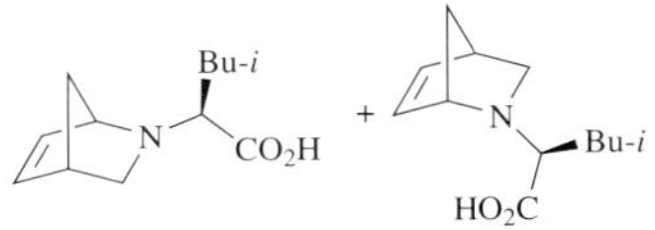 (—)[j]	354
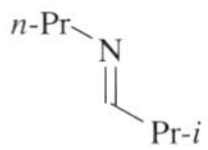	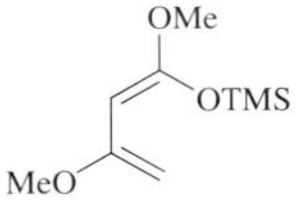	1. Et_2AlCl, CH_2Cl_2, –78° to rt, 2-6 h 2. H_2O, ether	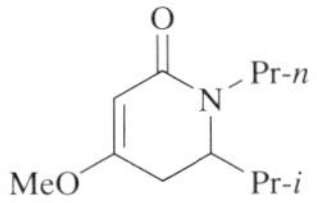 (62)	159
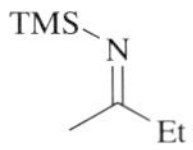	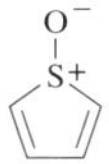	—	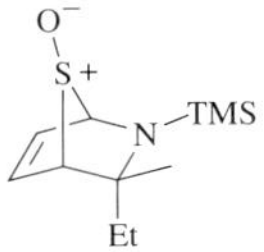 (34.5)[h,j]	200
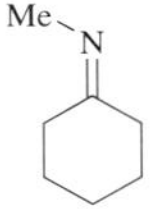	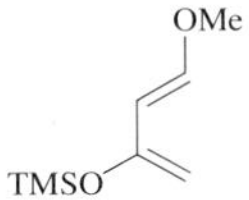	1. $Zn(OTf)_2$, CH_2Cl_2, rt, 12 h 2. NH_4Cl, H_2O, 2 h	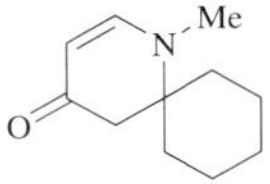 (51-64)	356

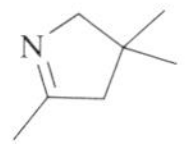	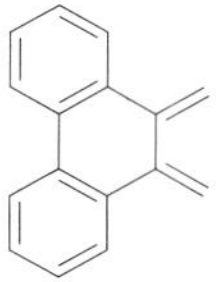	Zn, DMF, 0° to rt, overnight	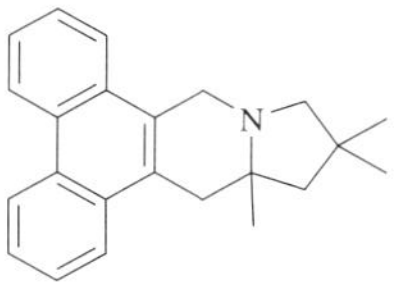 (—)	357
C_8 Bn–N=CH_2	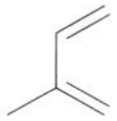		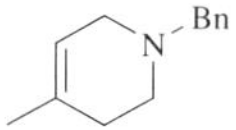	
		HCl, H_2O, 35°, 70 h	(59)[a,b]	133
		H_2O, 7-10 d	(60)[k]	360
		HCl, H_2O	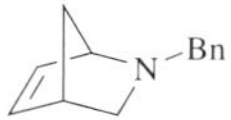	
		Temp / Time		
		rt / 3 h	(99)[b]	133, 235, 361
		rt / 4 h	(91-97)	279
		rt / 16 h	(—)	284
		rt / 16-20 h	(—)	362
		— / —	(—)	363
	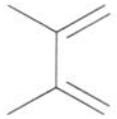		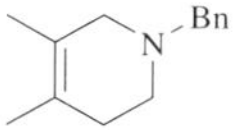	
		HCl, H_2O, 35°, 48 h	(64)[b]	133
		HCl, H_2O[l]	(23)	147
		$Nd(OTf)_3$, HCl, H_2O, rt, 12-20 h[m]	(93)	147
	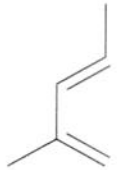	HCl, H_2O, rt, 23 h	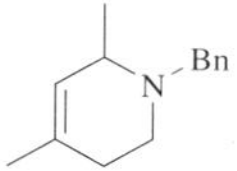 (45)[b]	133

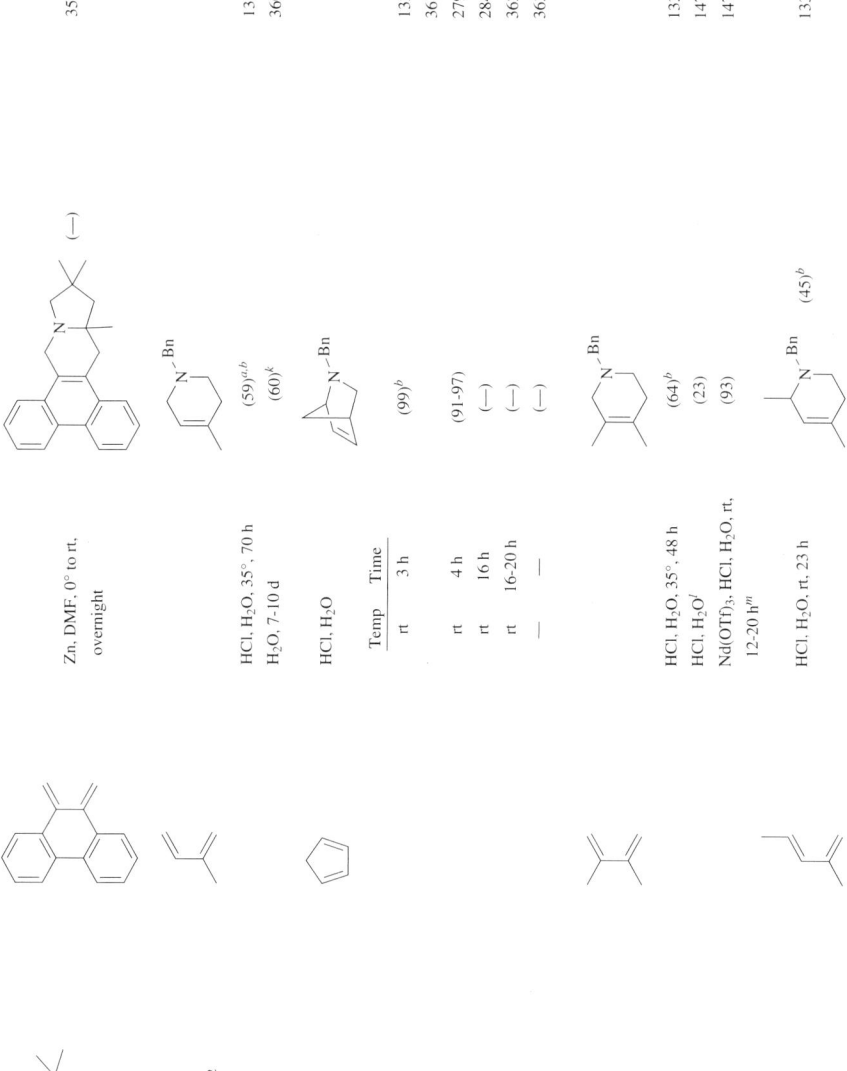

TABLE 5. CYCLOADDITIONS OF *N*-ALKYL- AND *N*-ARYLIMINES (*Continued*)

Imine	Diene	Conditions	Product(s) and Yield(s) (%)	Refs.
C_8 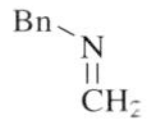	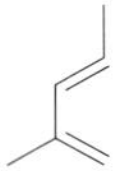		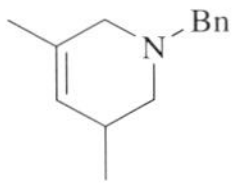	147
		HCl, H_2O[l]	(54)	
		$Yb(OTf)_3$, HCl, H_2O, rt, 12-20 h[m]	(92)	
		HCl, H_2O, 55°, 96 h	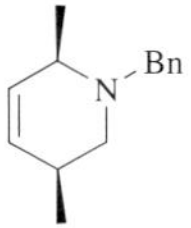 (62)[b]	133
		HCl, H_2O, 55°, 42 h	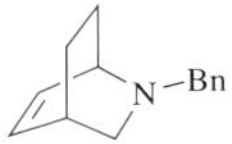 (35)[b]	133, 234
		HCl, H_2O, rt, 8 h	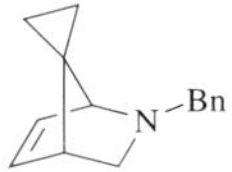 (62)	364
	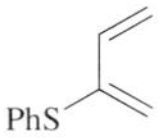	HCl, H_2O, DMF, 60°, 24 h	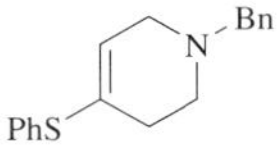 (81)	311
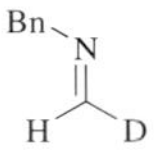		HCl, H_2O, 60°, 42 h	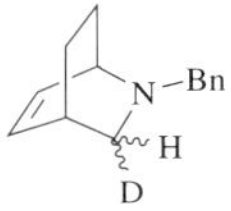 (32)[j]	365
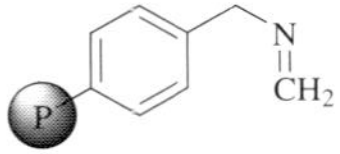	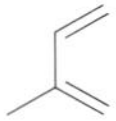	1. $Yb(OTf)_3$, TFA, CH_2Cl_2, rt, 12-48 h 2. $ClCO_2CH(Cl)CH_3$, CH_2Cl_2, rt, 3 h 3. MeOH, 65°, 2 h	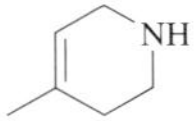 (62)	320

	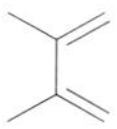	1. $Yb(OTf)_3$, TFA, CH_2Cl_2, rt, 12-48 h 2. $ClCO_2CH(Cl)CH_3$, CH_2Cl_2, rt, 3 h 3. MeOH, 65°, 2 h	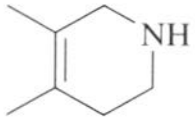 (70)	320
	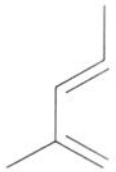	1. $Yb(OTf)_3$, TFA, CH_2Cl_2, rt, 12-48 h 2. $ClCO_2CH(Cl)CH_3$, dichloroethane, 83°, 3 h 3. MeOH, 65°, 2 h	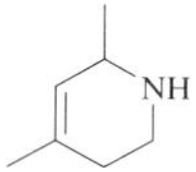 (52)	320
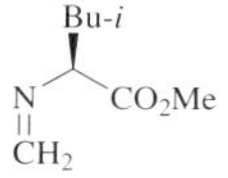			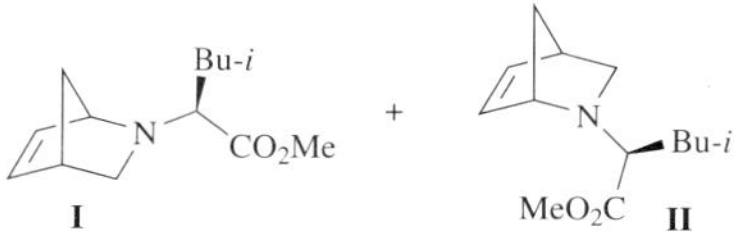	
		HCl, H_2O, rt, 45 min	**I** + **II** (94) 1.7:1[g]	354, 350
		HCl, THF, H_2O, 0°, 72 h	**I** + **II** (85), **I**:**II** 64:36	153
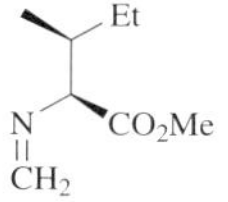	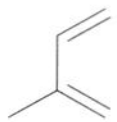	HCl, THF, H_2O, rt, 72 h	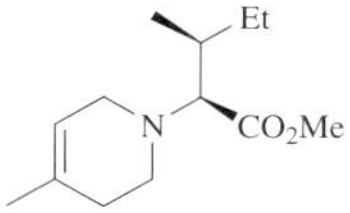 (50)	153
		HCl, THF, H_2O, 0°, 72 h	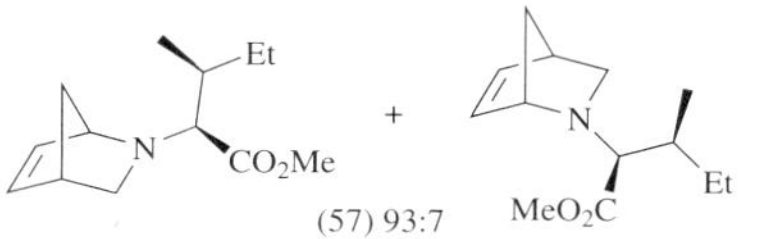 (57) 93:7	152, 153
	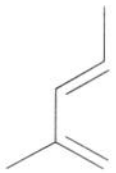	HCl, THF, H_2O, 0°, 72 h	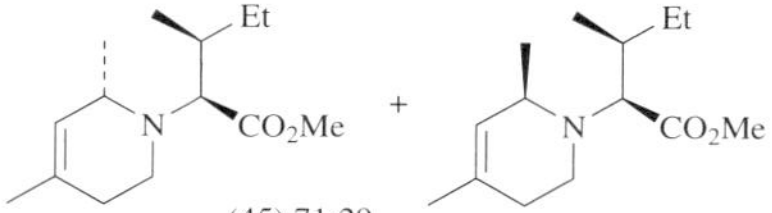 (45) 71:29	152, 153

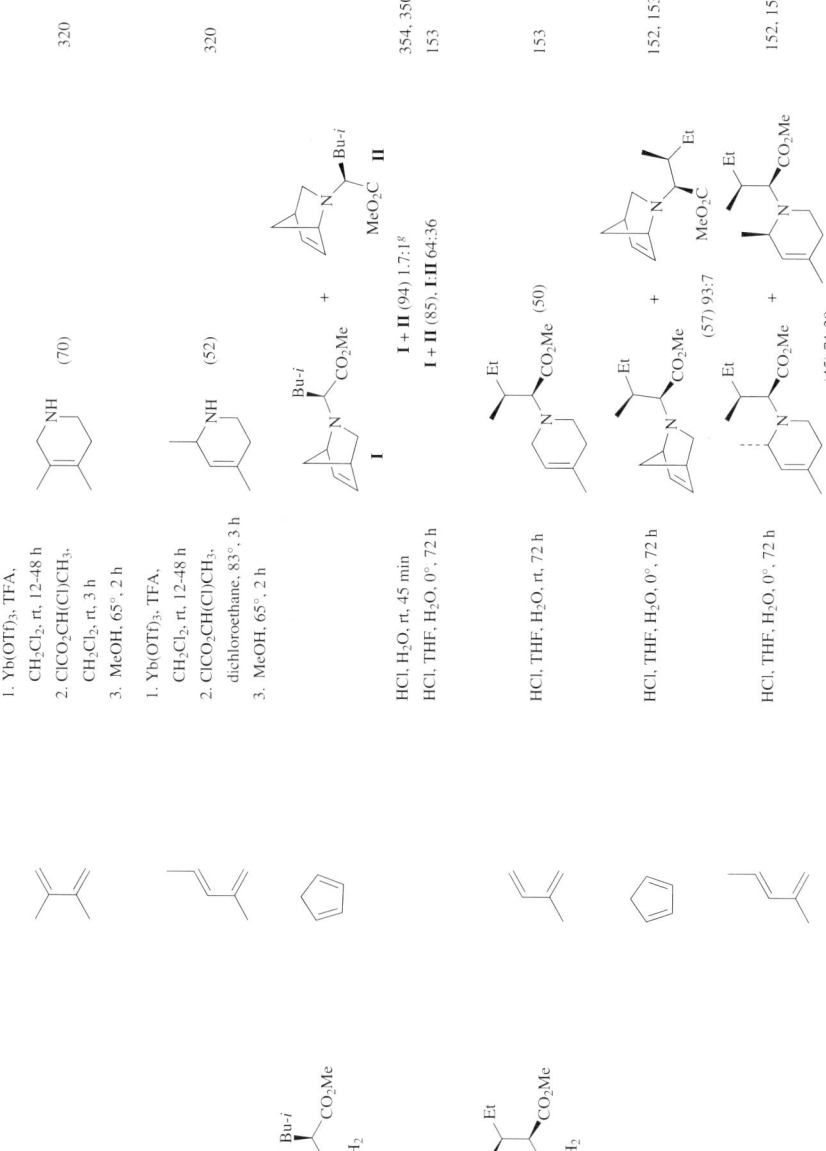

TABLE 5. CYCLOADDITIONS OF *N*-ALKYL- AND *N*-ARYLIMINES (*Continued*)

Imine	Diene	Conditions	Product(s) and Yield(s) (%)	Refs.
C_8 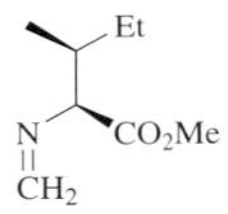		HCl, THF, H_2O, rt, 72 h	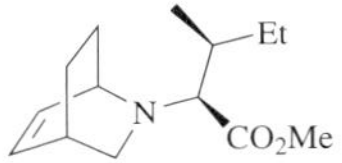 + 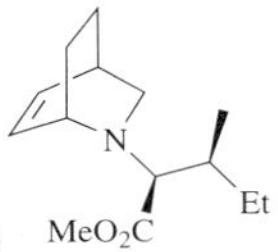 (35) 80:20	152, 153
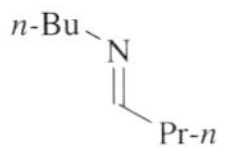	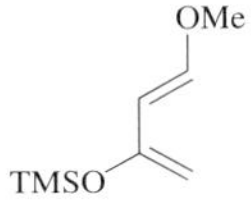	1. $ZnCl_2$, THF, rt, 36-48 h 2. H_2O	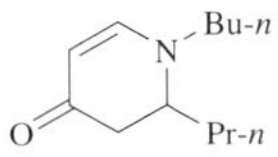 (49)[n], (68)[o]	141
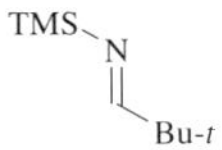	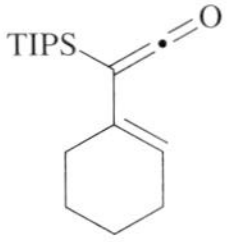	1. MeCN, 81°, 4 h 2. 110-130°, 88 h 3. SiO_2	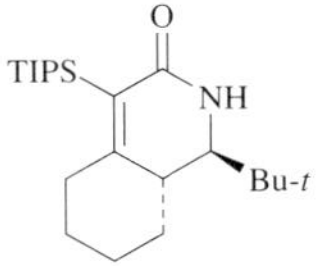 (56)	142
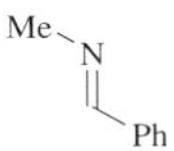	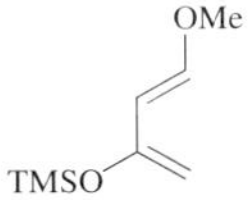	1. Et_2AlCl, CH_2Cl_2, rt, 4 h 2. H_2O, ether	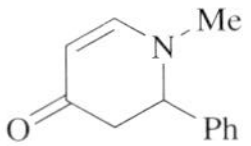 (62)	159
	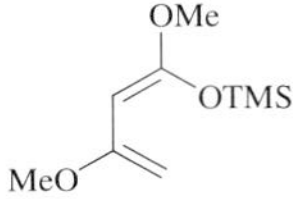	1. Lewis acid 2. H_2O, ether	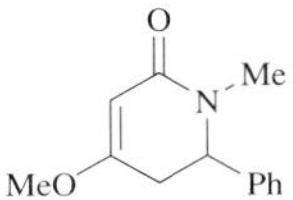	159

L. A.	Solvent	Temp	Time	
$MgBr_2$	—	rt	—	("moderate")
$TiCl_4$	—	—	—	("higher")
BF_3•OEt_2	—	—	—	("higher")
BBr_3	—	—	—	("higher")
Et_2AlCl	CH_2Cl_2	−78° to rt	2 h	(82)

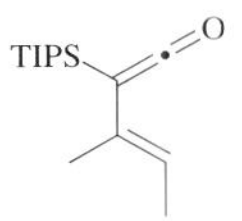	CH_3CN, 120°, 42 h	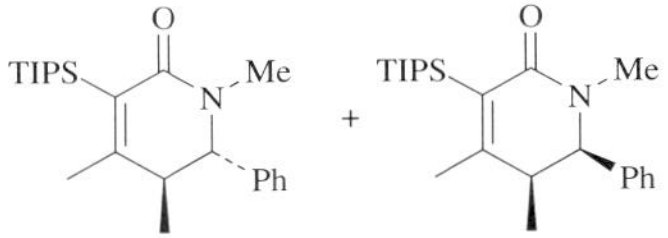 (71) 3:1	142
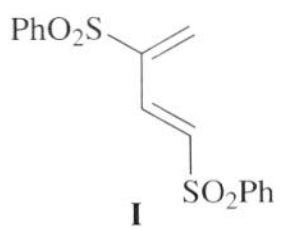 **I**	CH_2Cl_2, rt	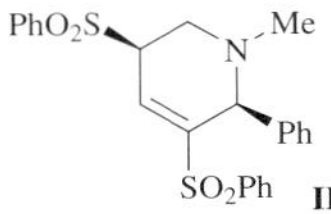 **II**	
	Time		
	24 h[p]	(90)	145
	24 h[p,q]	(81)	151
	2 h[r]	(88)	151, 145
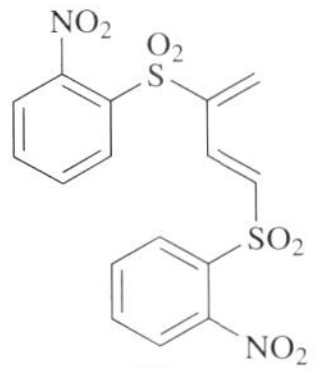 **III**	CH_2Cl_2, rt, 24 h	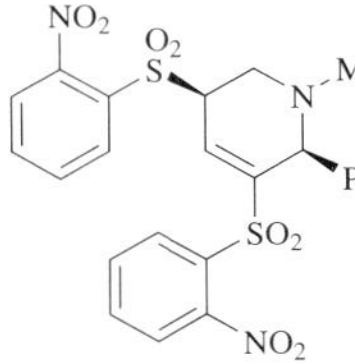 **IV** (73)[p]	151, 145
I + **III** + 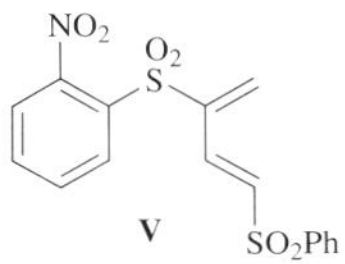 **V**	CH_2Cl_2, rt, 24 h[s]	**II** (—) + **IV** (—) + 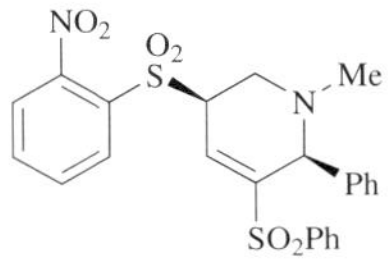 (25)	151, 145
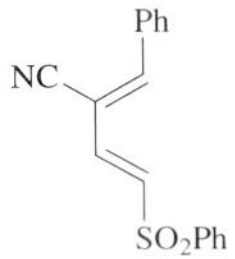	Toluene, 120°, 22 h	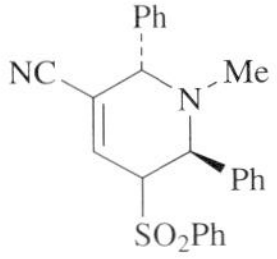 (94)[t]	151

TABLE 5. CYCLOADDITIONS OF *N*-ALKYL- AND *N*-ARYLIMINES (*Continued*)

Imine	Diene	Conditions	Product(s) and Yield(s) (%)	Refs.
C_8 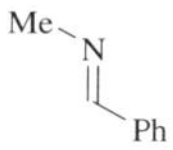	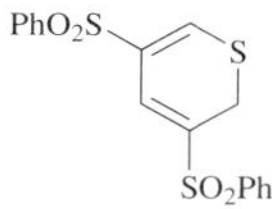	Toluene, 150°, 52 h	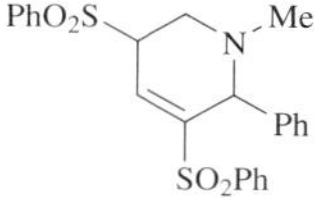 (73)	151
	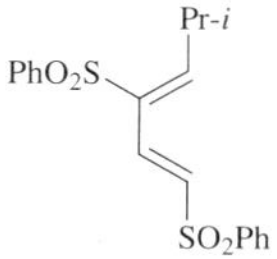	CH_2Cl_2, 50°, 24 h	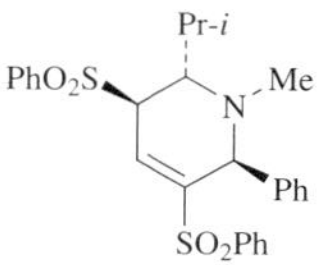 (80)	151
	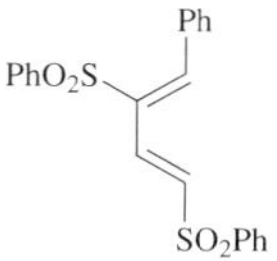	CH_2Cl_2, 50°, 24 h	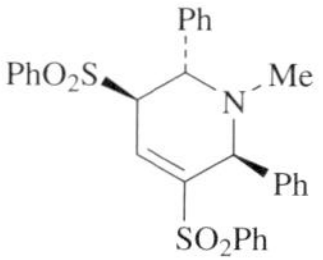 (82)	151
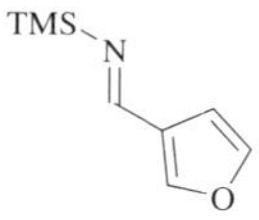	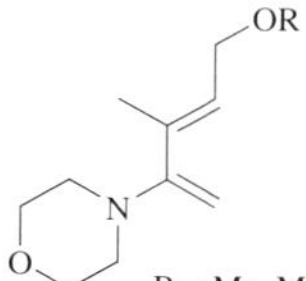 R = Me, MOM, TBS	—	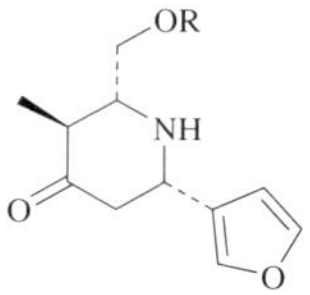 + 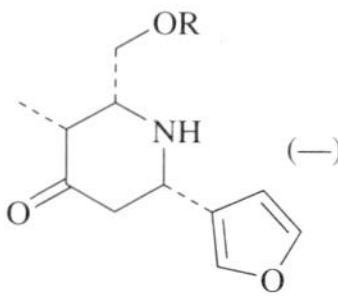 (—)	177
	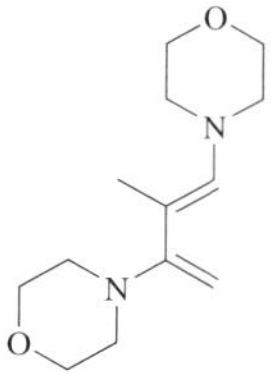	1. $ZnCl_2$, THF, rt, 12 h 2. SiO_2	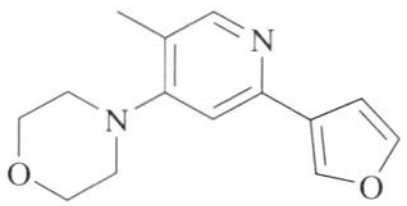 (11)	216

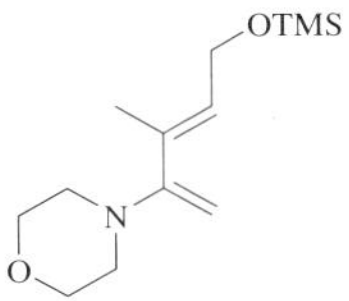

—

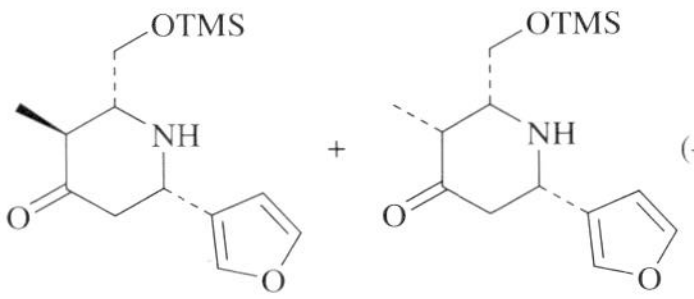

(—) 177

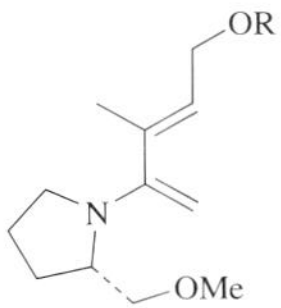

1. $ZnCl_2$, THF, ether, –80° to rt, 16 h
2. $NaHCO_3$, H_2O

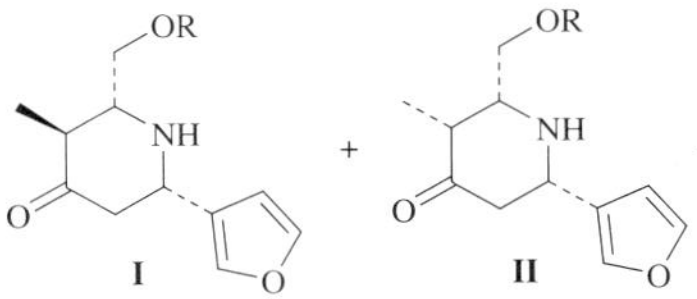

I + II +

177

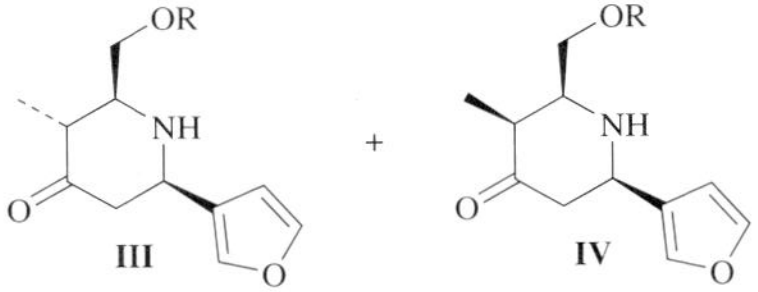

III + IV

R	I + II + III + IV	I:III
Me	(23)	93:7[u]
MOM	(33)	91:9[u]
TBS	(30)	88.5:11.5[u]

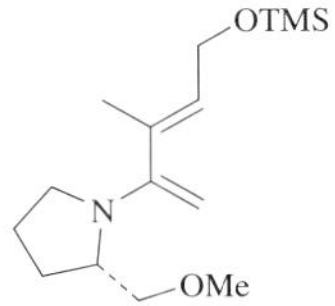

1. $ZnCl_2$, THF, ether, –80° to rt, 16 h
2. $NaHCO_3$, H_2O
3. Na_2CO_3, MeOH, 6 h

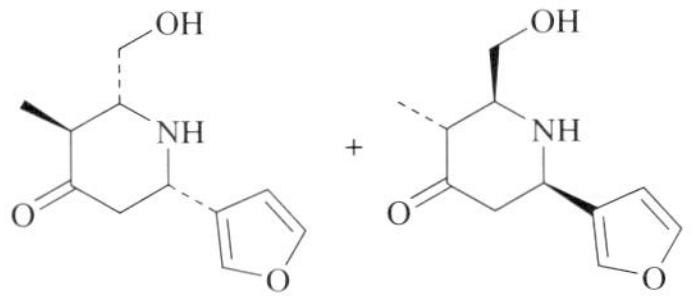

(51) >99:1 — 177, 178

(51) >99.5:0.5 — 179

TABLE 5. CYCLOADDITIONS OF *N*-ALKYL- AND *N*-ARYLIMINES (*Continued*)

Imine	Diene	Conditions	Product(s) and Yield(s) (%)	Refs.
C_8				
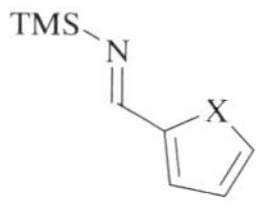TMS N X X: O, O, S	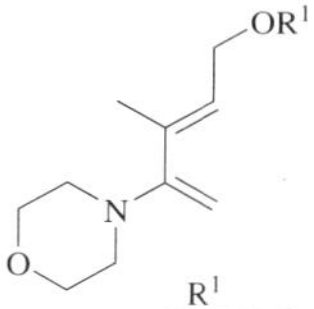OR[1] N O R[1]: TMS, Bn, Me	1. $ZnCl_2$, THF, ether, rt, 10 h 2. SiO_2	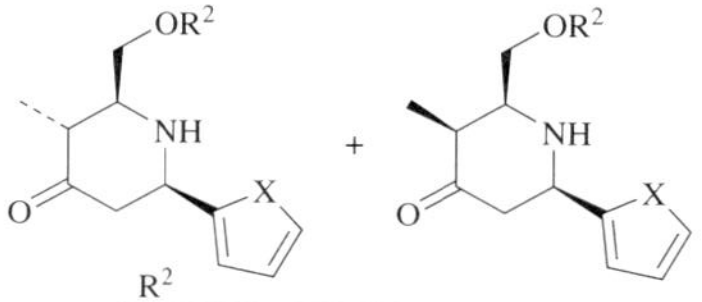OR[2] NH O X + R[2]: H (71) 71:29; Bn (73) 72:28; Me (80) 64:36	157
X: O, S	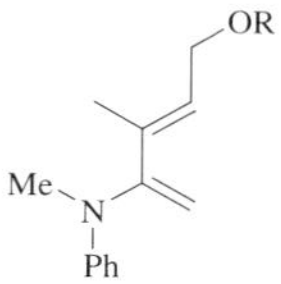OR Me N Ph R: TMS, Me	1. $ZnCl_2$, THF, ether, –40° to rt, 11 h 2. $NaHCO_3$, H_2O	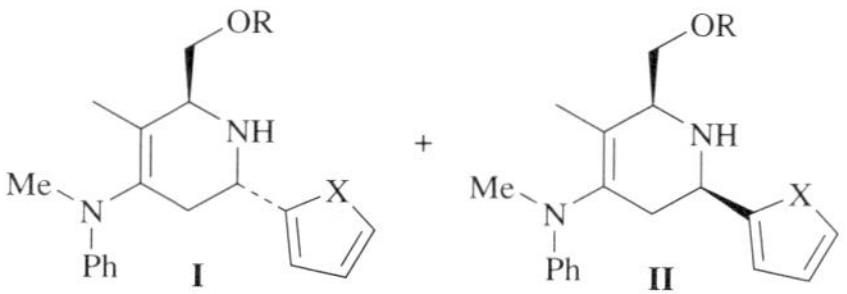OR NH Me N Ph X I + II **I + II**, **I:II**: (89) 12:88[v]; (80) 15:85[w]	157
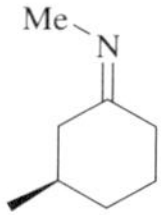Me N	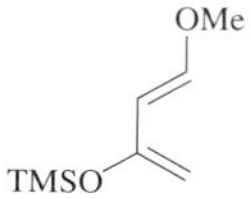OMe TMSO	1. $Zn(OTf)_2$, CH_2Cl_2, rt, 12 h 2. NH_4Cl, H_2O, 2 h	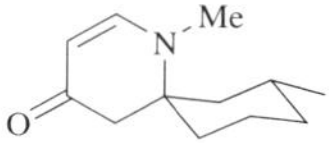N Me O (76)	356
C_9				
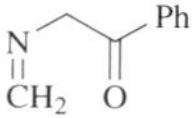N CH2 O Ph		HCl, H_2O	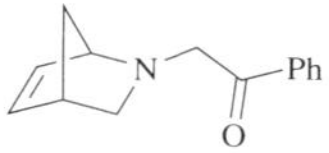N Ph O (—)	350
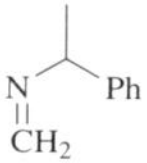N CH2 Ph	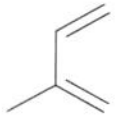	—	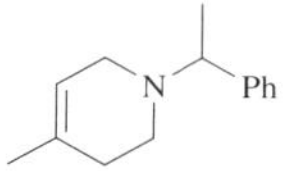N Ph (—)	360

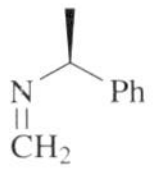 

Acid, H_2O

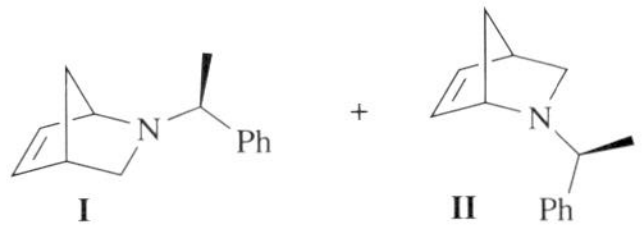

Acid	Temp	Time	Product(s) and Yield(s) (%)	Refs.
HCl	0°	20 h	**I** + **II** (86) 4:1[x]	133
HCl	—	—	**I** (50) + **II** (—)	366, 367
HCl	0°	4 h	**I** (50) + **II** (12)[y]	362
HCl	0°	—	**I** + **II** (—)[z]	368
AcOH	0-8°	20 h	**I** + **II** (81) 79:21	369

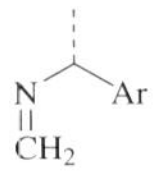 

HCl, H_2O, 0°

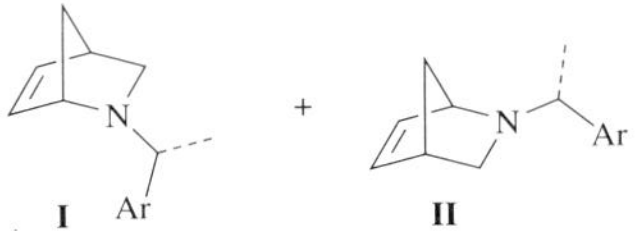

Ar	Time	I	II	I:II	Refs.
Ph	4 h	(60)	(—)	—	367, 363
Ph	overnight	(83)	(—)	—	368
$4-O_2NC_6H_4$	4 h	(—)	(—)	1.2:1	362

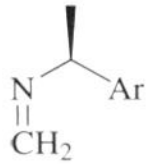 

Ar = $4-BrC_6H_4$

HCl, H_2O

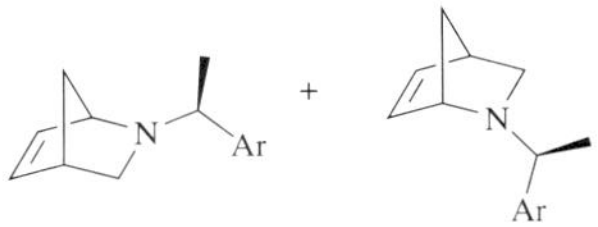

Solvent	Temp	Time	Product(s) and Yield(s) (%)	Refs.
H_2O	0°	4 h	("higher") 1.4:1	362
DMF	—	—	(65) 2.4:1	

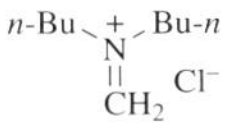 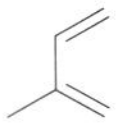

—

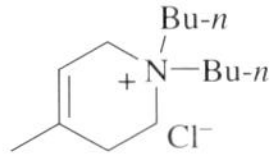 (—) 131

TABLE 5. CYCLOADDITIONS OF *N*-ALKYL- AND *N*-ARYLIMINES (*Continued*)

Imine	Diene	Conditions	Product(s) and Yield(s) (%)	Refs.
C_9 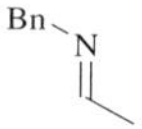		HCl, H_2O, 16 h	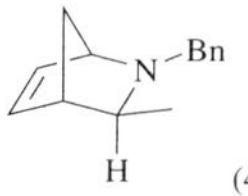 + 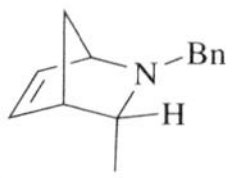 (47) 1.5:1	133
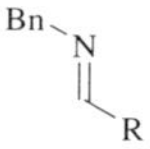	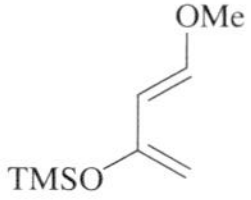		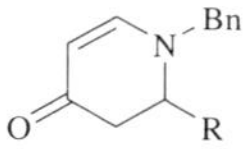	370
R: Me		$ZnCl_2$, THF, 0°, 6 h	(48)	
R: CHF_2		$ZnCl_2$, THF, rt, 4 h	(66)	
R: CF_3		$AlCl_3$, CH_2Cl_2, –78 to 40°	(19)	
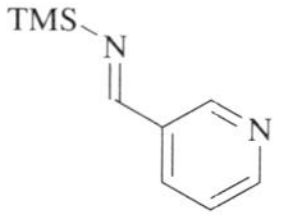	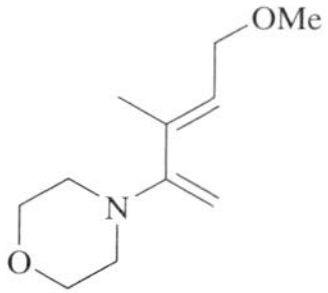	1. $ZnCl_2$, THF, ether, rt, 10 h 2. SiO_2	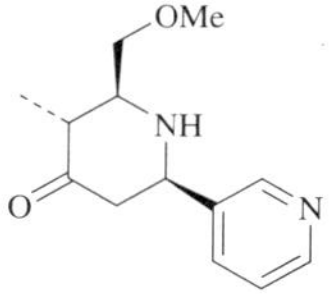 + 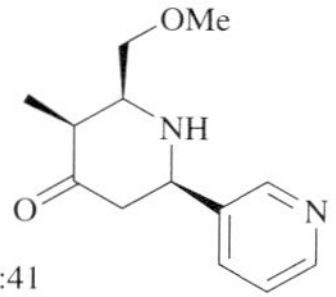 (62) 59:41	157
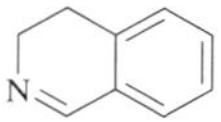		130°, 24 h	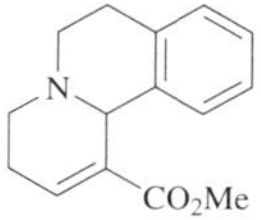 (—)	355
	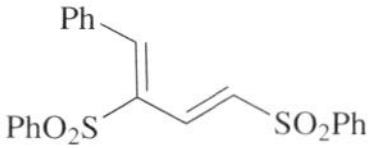	CH_2Cl_2, 50°, 24 h	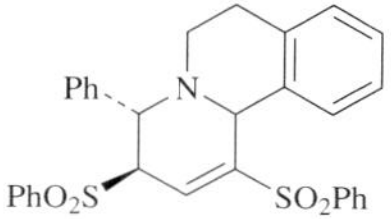 (93)[aa]	151
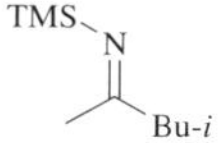	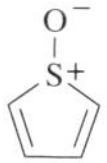	—	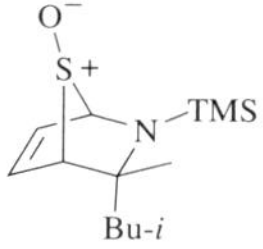 (70.8)[h,j]	200

C$_{10}$ 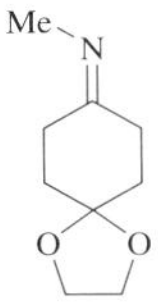	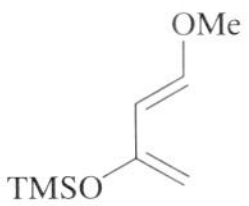	1. Zn(OTf)$_2$, CH$_2$Cl$_2$, rt, 12 h 2. NH$_4$Cl, H$_2$O, 2 h	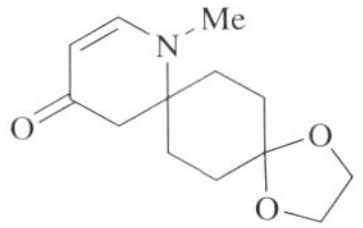	(84)	356
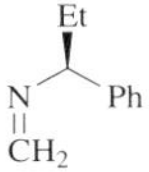		—	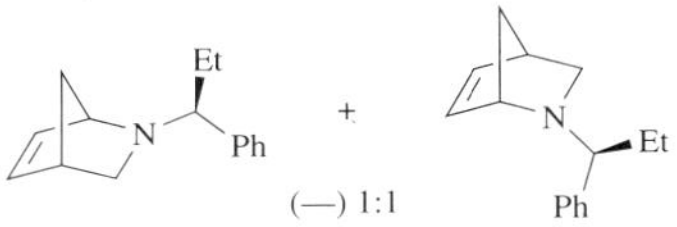	(—) 1:1	362
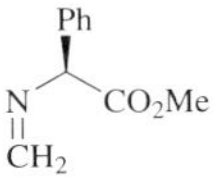	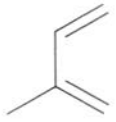	HCl, THF, H$_2$O, rt, 72 h	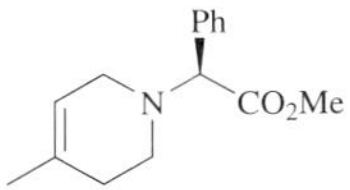	(88)	153
			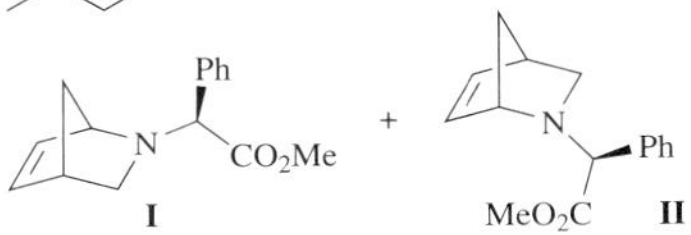 I + II		
		HCl, H$_2$O, rt, 30 min		(94) 1.7:1[g]	354
		HCl, THF, H$_2$O, 0°, 72 h		(97) **I**:**II** 78:22	152, 153
	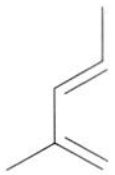	HCl, THF, H$_2$O, 0°, 72 h	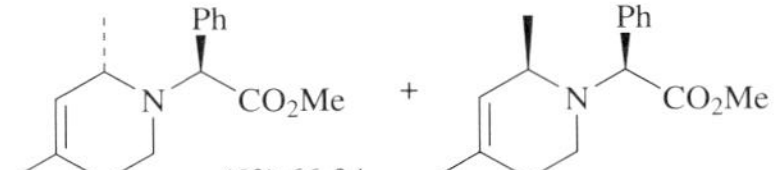	(63) 66:34	152, 153
		HCl, THF, H$_2$O, rt, 72 h	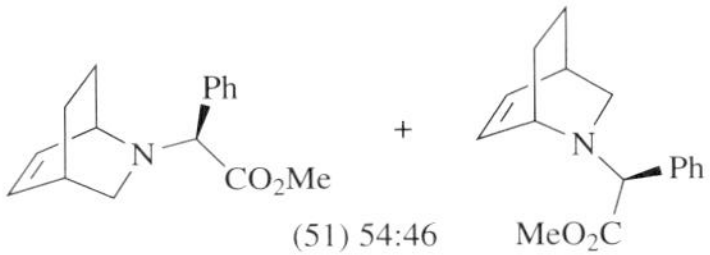	(51) 54:46	153

TABLE 5. CYCLOADDITIONS OF *N*-ALKYL- AND *N*-ARYLIMINES (*Continued*)

Imine	Diene	Conditions	Product(s) and Yield(s) (%)	Refs.
C_{10}				
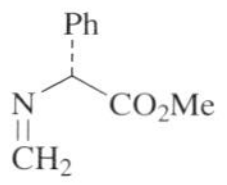		HCl, THF, H_2O, 0°, 72 h	 (90) 80:20	152, 153
	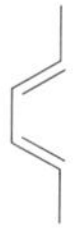	HCl, THF, H_2O, 0°, 72 h	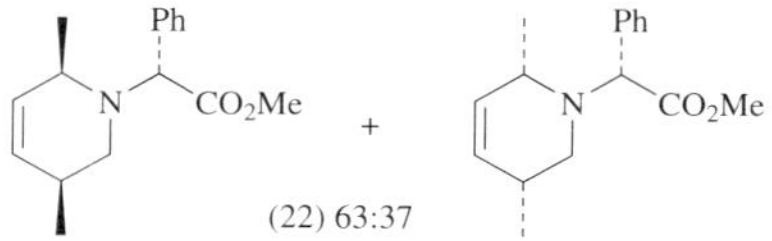 (22) 63:37	152, 153
		HCl, THF, H_2O, 0°, 72 h	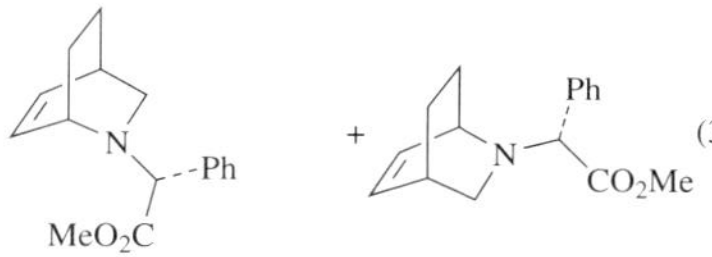 (35) 57:43	153
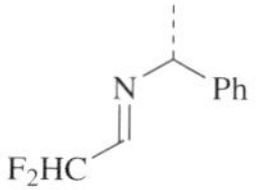	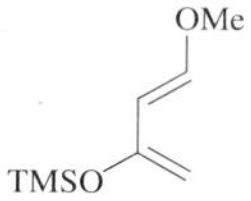	Lewis acid	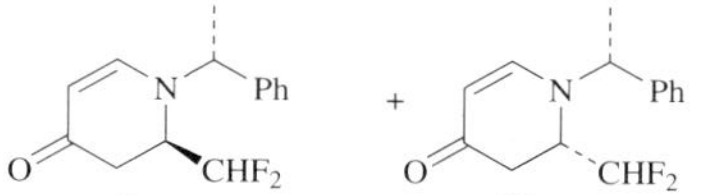 I + II	371

L. A.	Solvent	Temp	Time	**I + II**	**I:II**
$ZnCl_2$	THF	rt	3 h	**I** (43) + **II** (40)	—
$ZnCl_2$	THF	rt	4 h	(70)	48:52
$ZnCl_2$	CH_2Cl_2	–78°	5 h	(50)	46:54
BF_3•OEt_2	CH_2Cl_2	–78°	5 h	(82)	19:81
$TiCl_4$	CH_2Cl_2	–78°	5 h	(61)	41:59
$AlCl_3$	CH_2Cl_2	–78°	5 h	(21)	35:65
$B(OPh)_3$	CH_2Cl_2	–78° to rt	24 h	(65)	25:75
$LiClO_4$	ether	rt	22 h	(85)	53:47

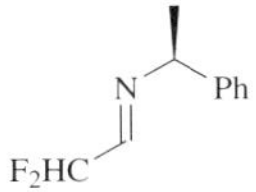	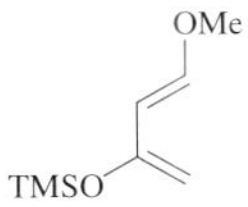	$ZnCl_2$, THF, rt, 3 h	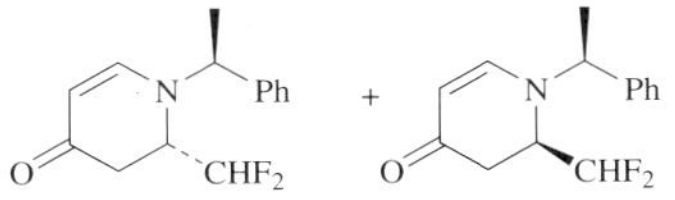 (40)	371
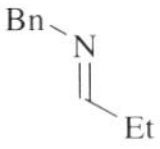		Lewis acid, HCl, H_2O L. A. / Time $Gd(OTf)_3$ 12-20 h $La(OTf)_3$ 12-20 h none —	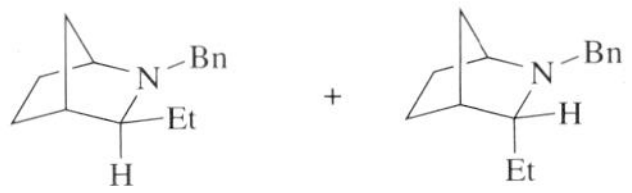 (24) (64) 2.5:1 (4)	147
	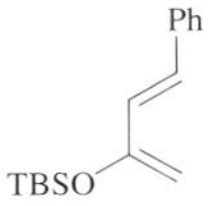	1. TMSOTf, CH_2Cl_2, rt, 2 d 2. TBAF, THF, 10 h	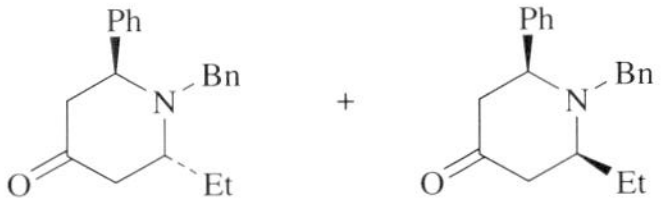 (49) 89:11	149, 150
	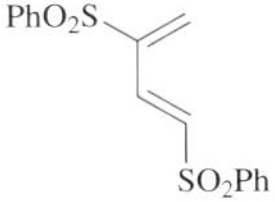	CH_2Cl_2, rt, 12 h	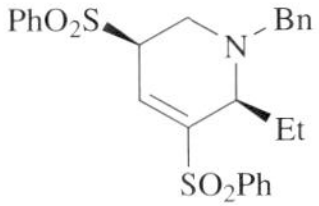 (68)	151
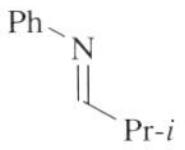	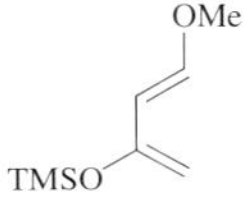		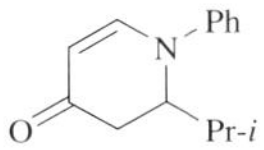	
		HBF_4, H_2O, MeOH, –40°, 30 min	(77)	372
		1. MeOH, rt, 2 h 2. HCl, H_2O	(76)	373

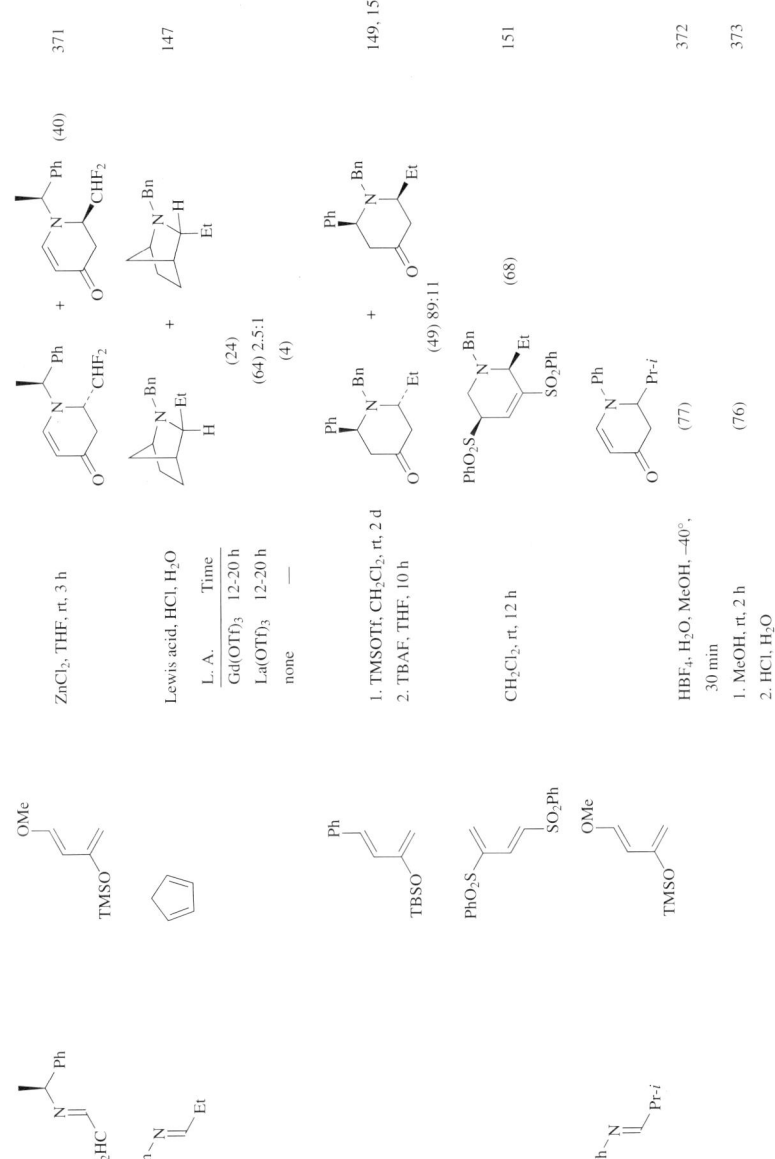

TABLE 5. CYCLOADDITIONS OF *N*-ALKYL- AND *N*-ARYLIMINES (*Continued*)

Imine	Diene	Conditions	Product(s) and Yield(s) (%)	Refs.
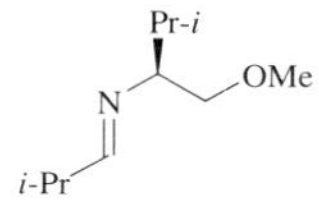	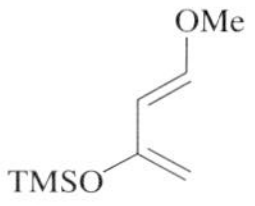	Lewis acid, CH_2Cl_2, rt L. A. BF_3•OEt_2 $Zn(OTf)_2$	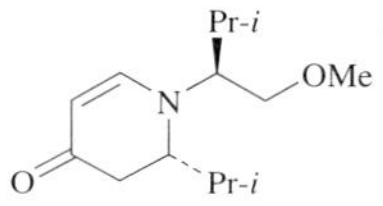 + 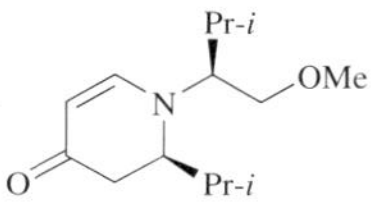 (48) 77:23 (50) 72:28	170
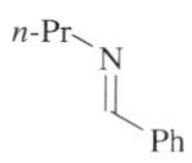	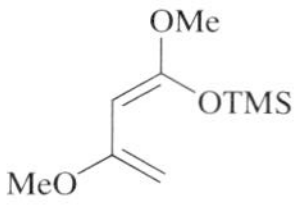	1. Et_2AlCl, CH_2Cl_2, −78° to rt, 6 h 2. H_2O, ether	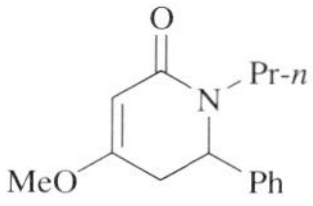 (65)	159
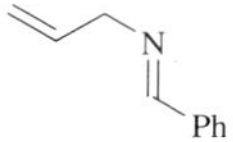	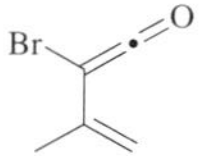	TEA, CH_2Cl_2, 40°	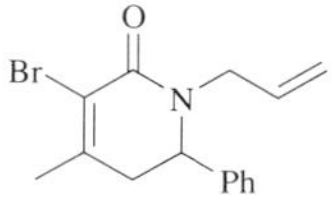 (96)	374
	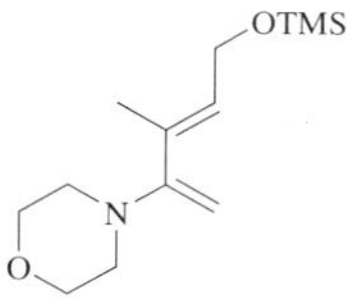	1. $ZnCl_2$ 2. $NaHCO_3$ 3. SiO_2	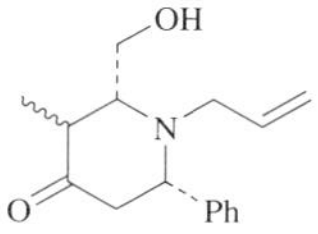 + 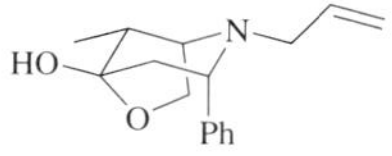 (—)	375
	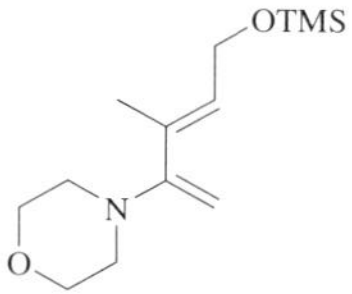	1. $ZnCl_2$ 2. $NaHCO_3$ 3. SiO_2 4. $NaBH_4$ 5. Na_2CO_3, H_2O	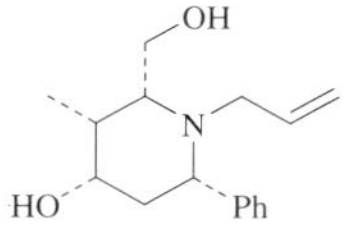 (low)	375

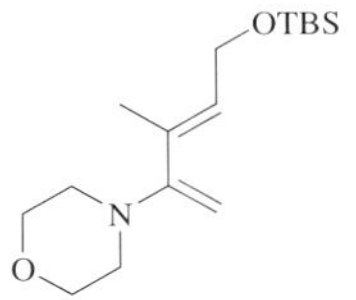

1. $Yb(OTf)_3$, THF, 0°
   to rt, overnight
2. $NaHCO_3$, H_2O
3. SiO_2

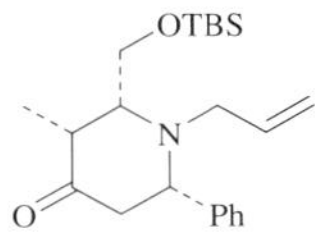

(60) 375

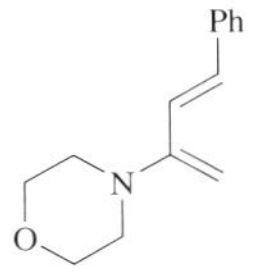

1. Lewis acid, THF
2. $NaHCO_3$, H_2O

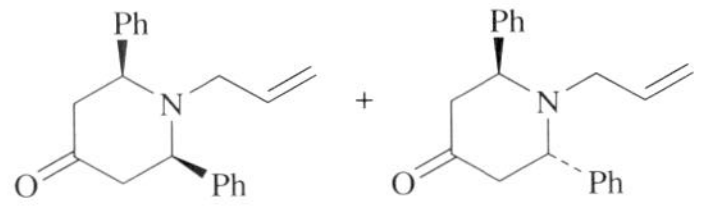

376

L. A.	Temp	
$Cu(OTf)_2$	rt	(33) 96.5:3.5
$CuClO_4$•MeCN	rt	(25) 96:4
$Yb(OTf)_3$	rt	(30) 96:4
$Yb(OTf)_3$	50°	(traces)
$ZnCl_2$	rt	(<10)

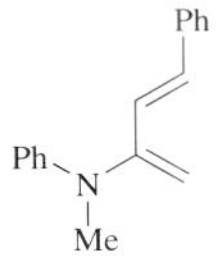

1. Lewis acid, THF
2. $NaHCO_3$, H_2O

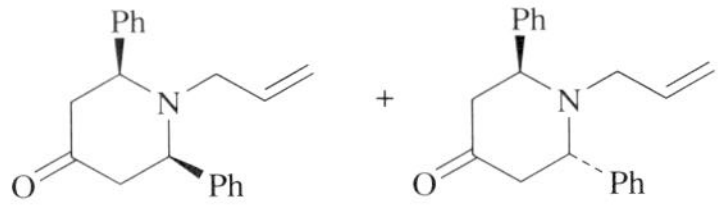

376

L. A.	Temp	Time	
$Cu(OTf)_2$	rt	—	(60) 96.5:3.5
$CuClO_4$•MeCN	rt	—	(46) 96.5:3.5
$Yb(OTf)_3$	rt	—	(53) 95.5:4.5
$Sc(OTf)_3$	rt	—	(39) 95:5
$ZnCl_2$	rt	—	(<10)
$Cu(OTf)_2$	–20°	24 h	(90) 97:3
$Yb(OTf)_3$	–20°	—	(74) 97:3

TABLE 5. CYCLOADDITIONS OF *N*-ALKYL- AND *N*-ARYLIMINES (*Continued*)

Imine	Diene	Conditions	Product(s) and Yield(s) (%)	Refs.
C_{10} 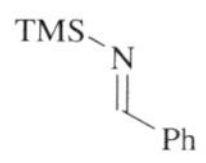	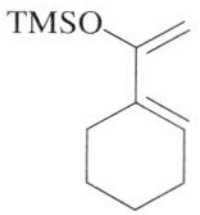	1. Lewis acid, CH_2Cl_2 2. $NaHCO_3$, H_2O	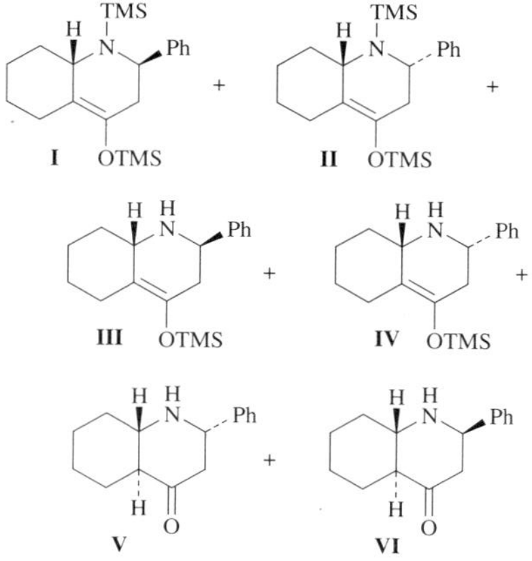 I + II + III + IV + V + VI	155
		L. A. / Temp / Time: $ZrCl_4$ / –10° / 2 min	**I** + **II** + **III** + **IV** (—), **I** + **III**:**II** + **IV** 43:57[bb]	
		$ZrCl_4$ / –10° / 10 min	**I** + **II** + **III** + **IV** + **V** (—), **I** + **III**:**II** + **IV**:**V** 50:40:10[bb]	
		$ZrCl_4$ / 20° / 2 h	**I** + **II** + **III** + **IV** + **V** (—), **I** + **III**:**II** + **IV**:**V** 28:57:15[bb]	
		ZnI_2 / 20° / 2 h	**I** + **II** + **III** + **IV** + **VI** (—), **I** + **III**:**II** + **IV**:**VI** 41:44:15[bb]	
	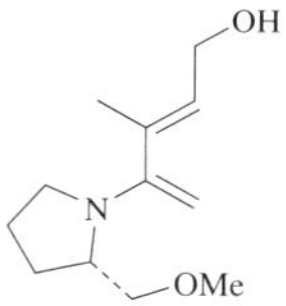	1. $ZnCl_2$, THF, –80° to rt, 10 h 2. $NaHCO_3$, H_2O	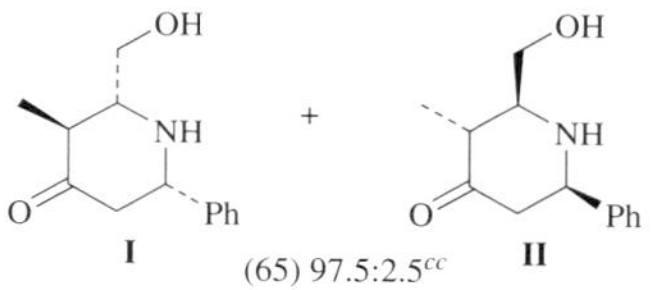 **I** + **II** (65) 97.5:2.5[cc]	176

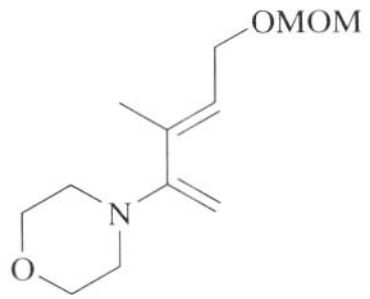	—	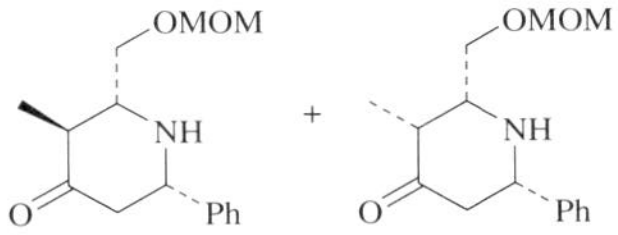	(—)	177
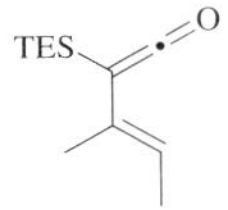	1. rt, 2 h 2. SiO_2	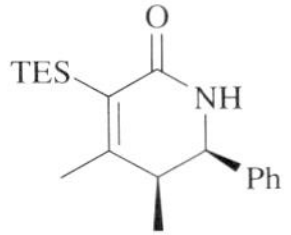	(76)	142
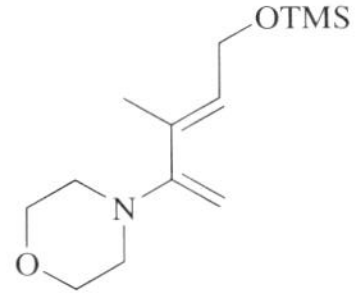	1. $ZnCl_2$, THF, ether, rt, 10 h 2. SiO_2	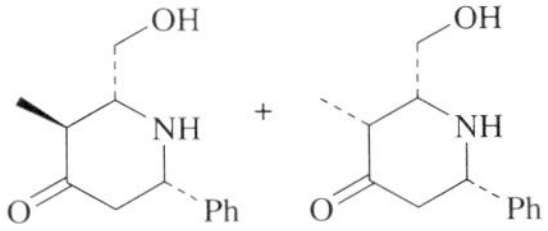	(79) 78:22	157
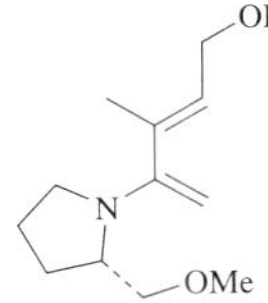	1. $ZnCl_2$, THF, ether, –80° to rt, 16 h 2. $NaHCO_3$, H_2O	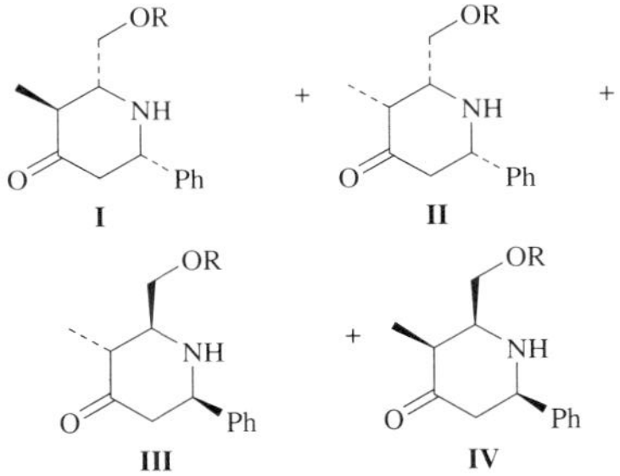		177

R	I + II + III + IV	I:III
MOM	(35)	93.5:6.5[dd]
TBS	(28)	92:8[dd]

TABLE 5. CYCLOADDITIONS OF *N*-ALKYL- AND *N*-ARYLIMINES (*Continued*)

Imine	Diene	Conditions	Product(s) and Yield(s) (%)	Refs.
C_{10} 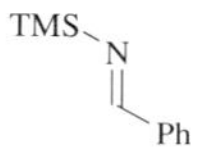	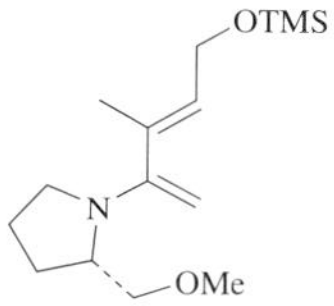	1. $ZnCl_2$, THF, ether, –80° to rt, 16 h 2. $NaHCO_3$, H_2O 3. Na_2CO_3, MeOH, 6 h	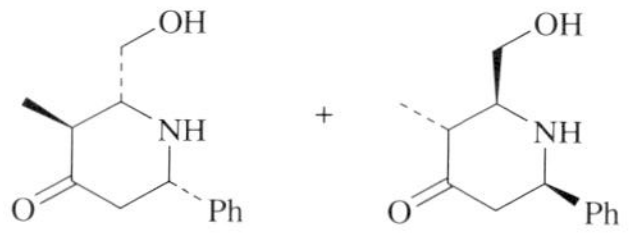 (65) 97.5:2.5	177, 178
	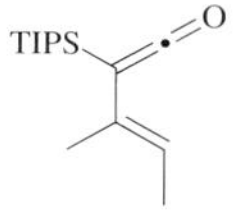	1. MeCN, 81°, 1.5 h 2. SiO_2	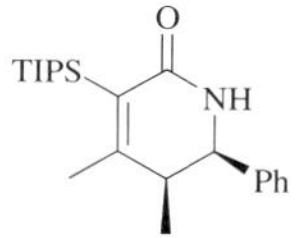 (79-83)	142
	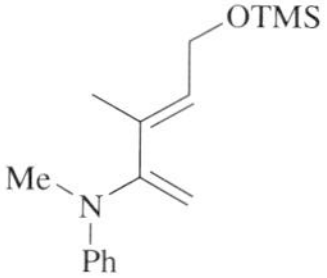	1. $ZnCl_2$, THF, ether, –40° to rt, 11 h 2. $NaHCO_3$, H_2O	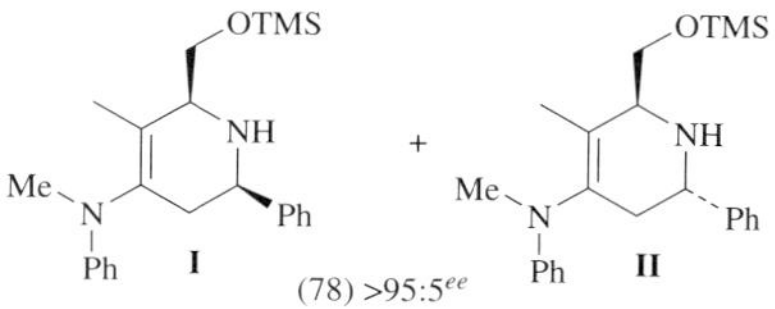 **I** + **II** (78) >95:5[ee]	157
	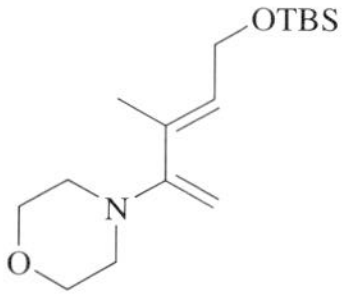	—	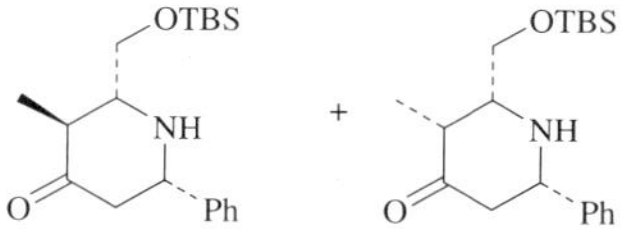 (—)	177
	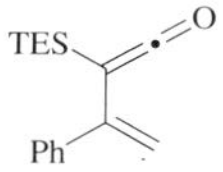	1. MeCN, 81°, 45 min 2. SiO_2	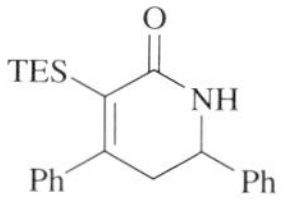 (84)[ff]	142

	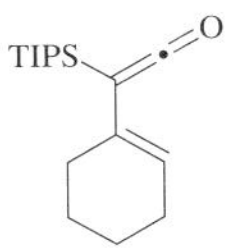	1. MeCN, 81°, 25 h 2. SiO_2	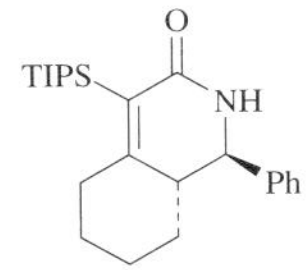 (91)	142
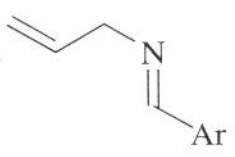 Ar: $4\text{-}FC_6H_4$, $2\text{-}BrC_6H_4$	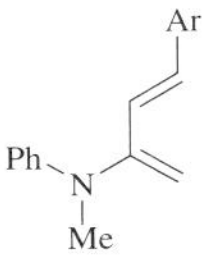	1. $Cu(OTf)_2$, THF, –20°, 24 h 2. $NaHCO_3$, H_2O	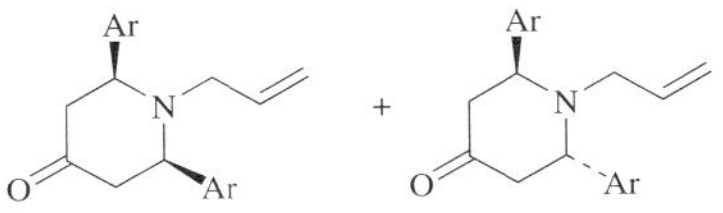 (80) >99.5:0.5 (89) >99.5:0.5	376
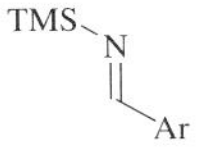 Ar = $2\text{-}BrC_6H_4$	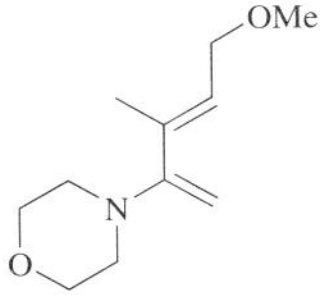	1. $ZnCl_2$, THF, ether, rt, 10 h 2. SiO_2	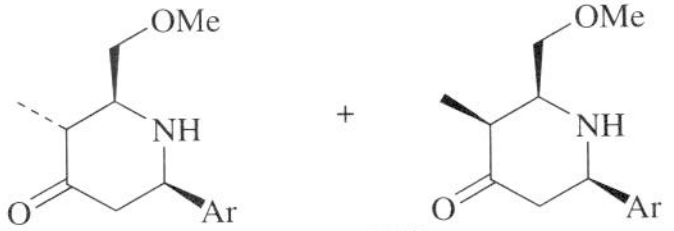 (71) >95:5	157, 177
	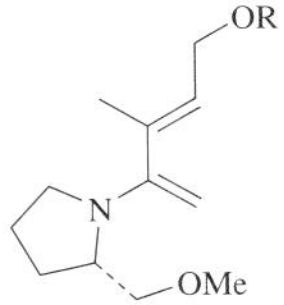	1. $ZnCl_2$, –80° to rt 2. $NaHCO_3$, H_2O	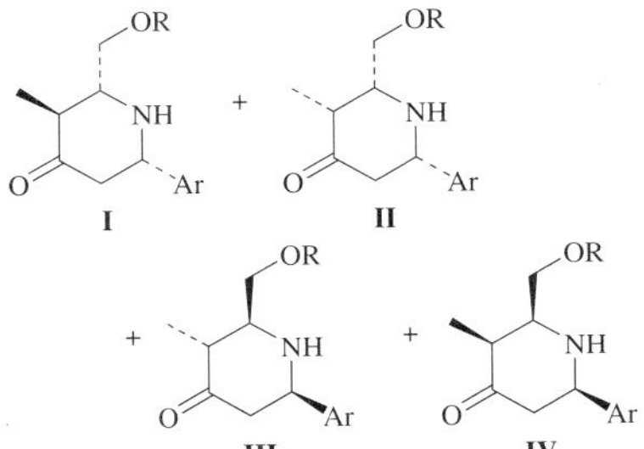 **I** **II** **III** **IV**	

R	Solvent	Time		
Me	THF	10 h	**I** + **III** (63), **I**:**III** 93:7	176
Me	THF, ether	16 h	**I** + **II** + **III** + **IV** (63), **I**:**III** 93:7	177
TBS	THF, ether	16 h	**I** + **II** + **III** + **IV** (32), **I**:**III** 76.5:23.5[gg]	177

TABLE 5. CYCLOADDITIONS OF *N*-ALKYL- AND *N*-ARYLIMINES (*Continued*)

Imine	Diene	Conditions	Product(s) and Yield(s) (%)	Refs.
C_{10}				
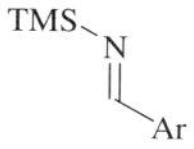 Ar = 2-BrC$_6$H$_4$	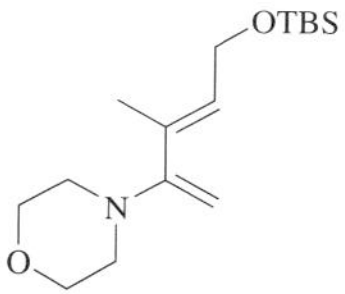	—	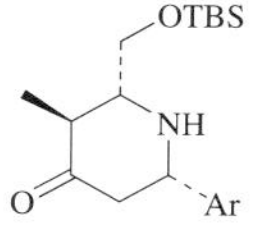 + 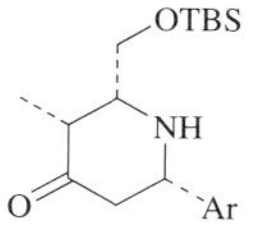 (—)	177
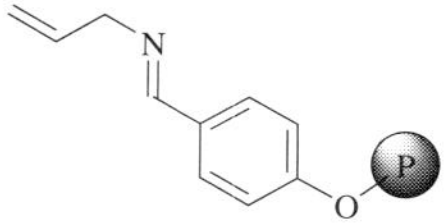	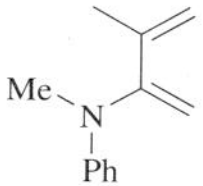	1. Yb(OTf)$_3$, THF, rt, 12 h 2. HCl, H$_2$O, THF, overnight 3. TMSOTf, CH$_2$Cl$_2$, rt, 3 h	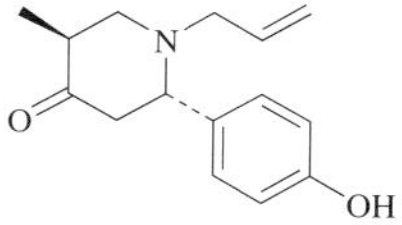 (66)	377
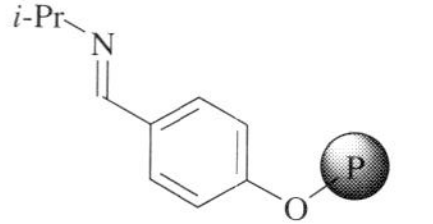	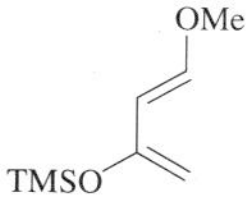	1. Yb(OTf)$_3$, THF, 60°, 3 h 2. H$_2$O 3. TFA, CH$_2$Cl$_2$	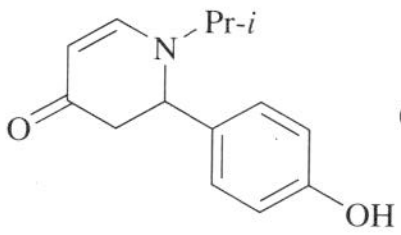 (85)[hh]	378
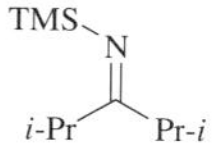	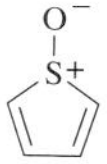	—	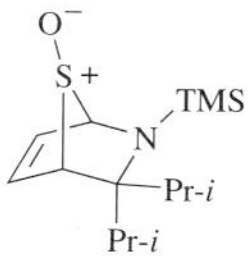 (56.4)[h]	200
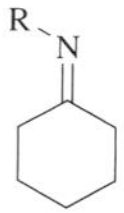 R: *n*-Bu; *s*-Bu	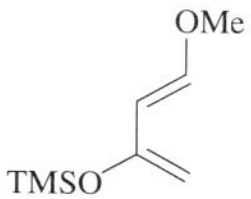	1. Zn(OTf)$_2$, CH$_2$Cl$_2$, r., 12 h 2. NH$_4$Cl, H$_2$O, 2 h	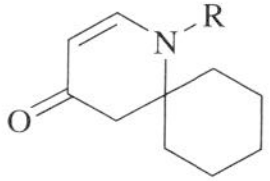 (36); (7)	356

C1 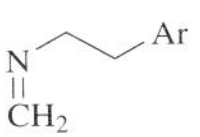 Ar = 3,4-$(MeO)_2C_6H_3$		HCl, H_2O	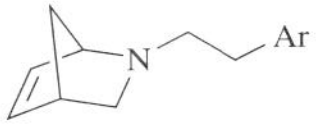 (>90)	350
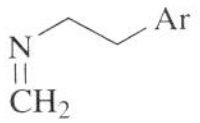 Ar = 2-Br-4,5-$(MeO)_2C_6H_2$	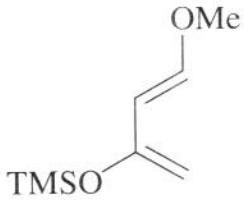	$ZnCl_2$, THF, 0°	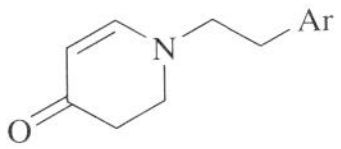 (47)	379
		1. $ZnCl_2$, THF, rt, 24 h 2. HCl, H_2O	(47)	380
	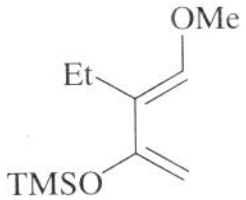	1. $ZnCl_2$, THF, rt, 72 h 2. HCl, H_2O	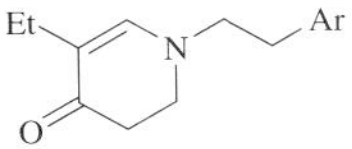 (64)	380 379
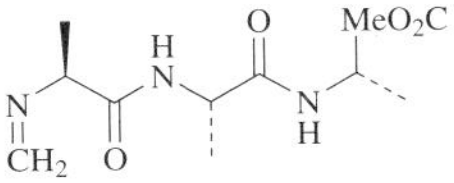		HCl, H_2O, rt, 1 h	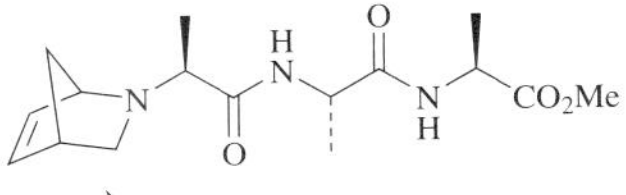 + 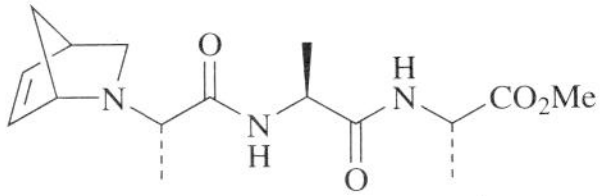 (81) 1:1	354
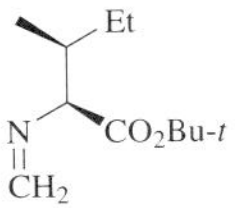		HCl, THF, H_2O	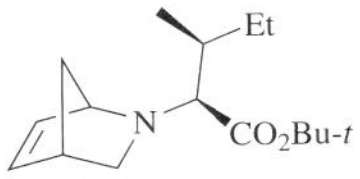 **I** + 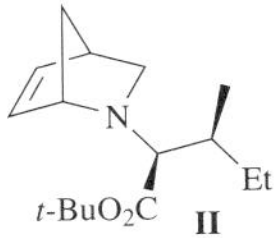 **II** (—)[ii]	153
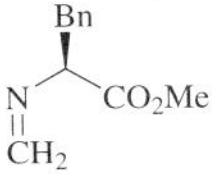	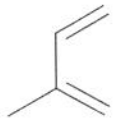	HCl, H_2O[l] $Nd(OTf)_3$, HCl, H_2O, rt, 12-20 h[m]	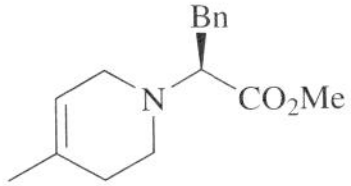 (37) (96)	147

TABLE 5. CYCLOADDITIONS OF *N*-ALKYL- AND *N*-ARYLIMINES (*Continued*)

Imine	Diene	Conditions	Product(s) and Yield(s) (%)	Refs.
C_{11} 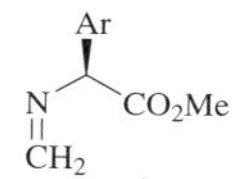			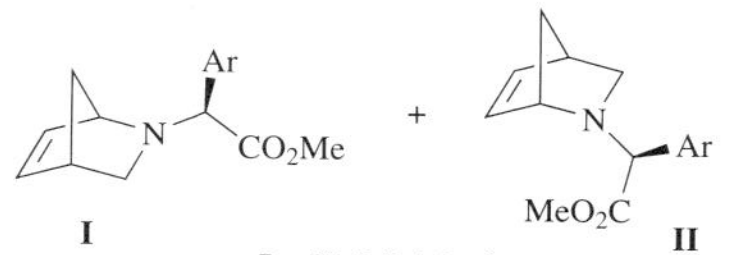	
Ar				
Bn		HCl, H_2O, rt, 45 min	**I** + **II** (84) 1.8:1[g]	354
Bn		HCl, THF, H_2O, 0°, 72 h	**I** + **II** (75), **I**:**II** 68:32	153
$CH_2C_6H_4OH$-4		HCl, H_2O, rt, 1 h	**I** + **II** (87) 2:1[g]	354
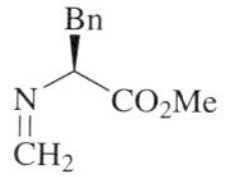	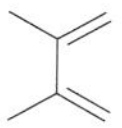		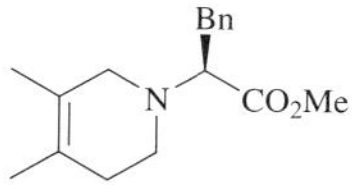	147
		HCl, H_2O[l]	(58)	
		$Nd(OTf)_3$, HCl, H_2O, rt, 12-20 h[m]	(98)	
			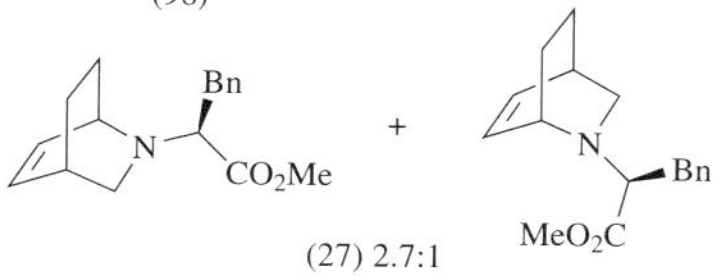	147
		HCl, H_2O[l]	(27) 2.7:1	
		$Nd(OTf)_3$, HCl, H_2O, rt, 12-20 h[m]	(84) 3:1	
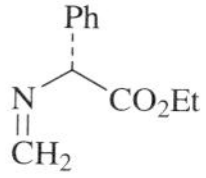		—	 (—) 4:1[f]	362
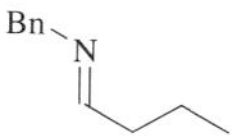	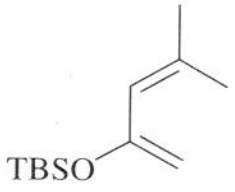	1. TMSOTf, CH_2Cl_2, rt, 2 d 2. TBAF, THF, 10 h	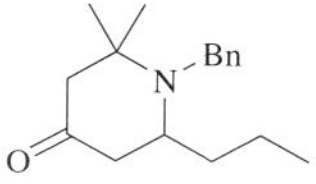 (52)	149

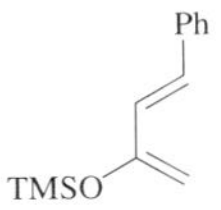

1. Lewis acid, rt, 1 d
2. TBAF, rt

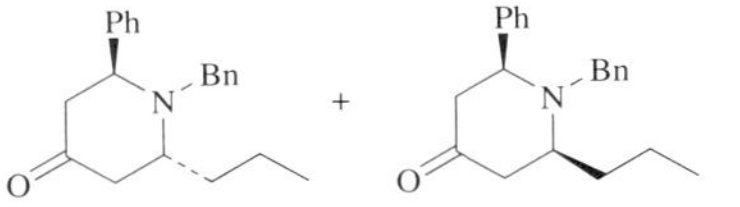

L. A.	Product ratio	Refs.
TMSOTf	(52) 87:13	381, 169
TBSOTf	(58) 88:12	381
TIPSOTf	(75) 90:10	381

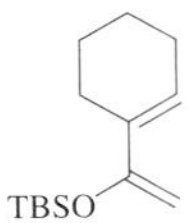

1. TMSOTf, CH_2Cl_2, rt, 2 d
2. TBAF, THF, 10 h

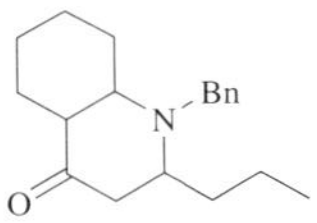

(90)[j]

150

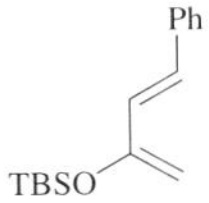

1. TMSOTf, CH_2Cl_2, rt, 2 d
2. TBAF, THF, 10 h

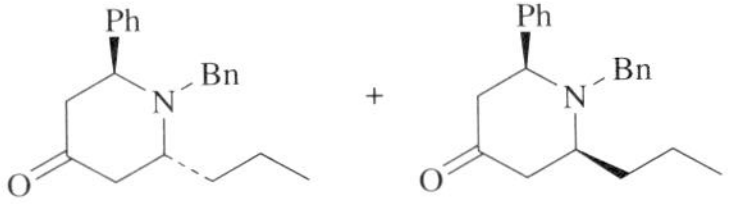

(52) 87:13

149, 150

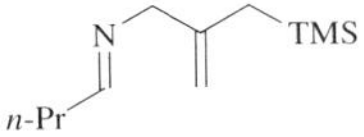

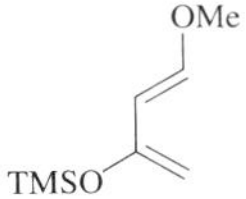

1. $Yb(OTf)_3$, MeCN, rt, 10-15 h
2. H_2O

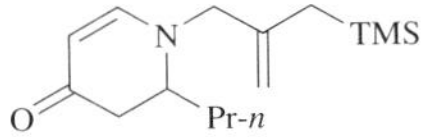

(78)

382

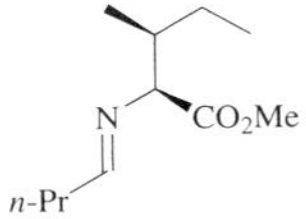

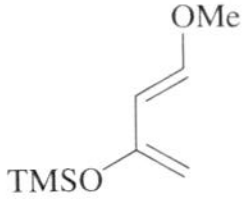

1. Lewis acid
2. H^+, H_2O

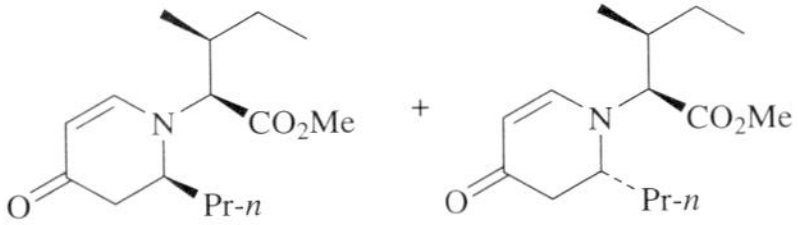

175a

L. A.[jj]	Solvent	Temp	Product ratio
$ZnCl_2$	THF	–15°	(50) 90:10
$EtAlCl_2$	CH_2Cl_2	–78 to 20°	(80) 94:6
$MeAlCl_2$	CH_2Cl_2	–78 to 20°	(74) 93:7

TABLE 5. CYCLOADDITIONS OF *N*-ALKYL- AND *N*-ARYLIMINES (*Continued*)

Imine	Diene	Conditions	Product(s) and Yield(s) (%)	Refs.
C_{11} 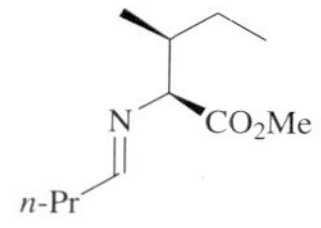	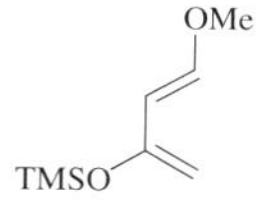	1. Lewis acid 2. $NaHCO_3$, H_2O	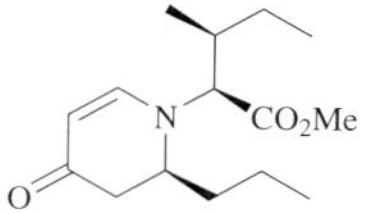 + 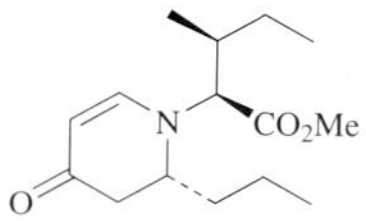	384
		L. A. / Solvent / Temp / Time		
		$ZnCl_2$ / THF / 0° / 7 h	(11) 15:85	
		$ZnCl_2$ / THF / –15° / 7 h	(50) 90:10	
		$TiCl_4$ / CH_2Cl_2 / –78 to –20° / —	(37) 91:9	
		$EtAlCl_2$ / CH_2Cl_2 / –78° to rt / 1 h	(80) 94:6	
		Et_2AlCl / CH_2Cl_2 / –78° to rt / 1 h	(80) 92:8	
		$MeAlCl_2$ / CH_2Cl_2 / –78° to rt / 1 h	(74) 93:7	
		Me_2AlCl / CH_2Cl_2 / –78° to rt / 1 h	(80) 93:7	
		$BF_3•OEt_2$ / CH_2Cl_2 / –78 to –20° / —	(64) 80:20	
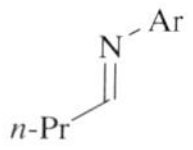 Ar = 4-$MeOC_6H_4$	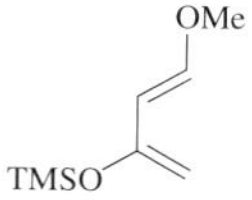	Montmorillonite K10, 1.5 h	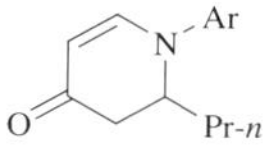	140
		Solvent / Temp[kk]		
		H_2O / 0°	(85)	
		MeCN, H_2O / –10°	(98)	
		MeCN / –10°	(76)	
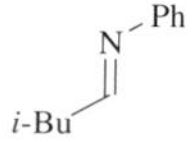	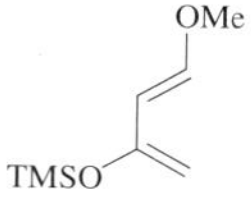	Catalyst, H_2O, rt, 2-3 h	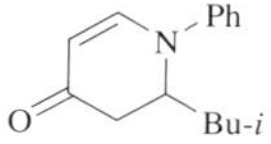	
		L. A.		
		NaOTf	(76)	385
		AgOTf	(72)	386

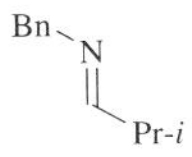	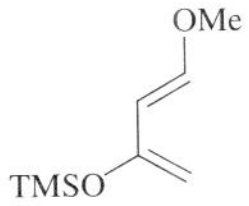	1. $ZnCl_2$, THF, rt, 36-48 h 2. H_2O	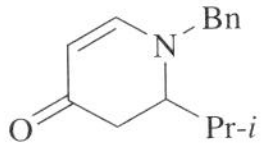 (44)[n], (69)[ll]	141
		$ZnCl_2$, 36 h	(44)	370
	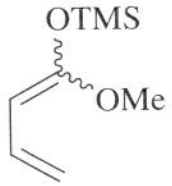	1. $TiCl_4$, CH_2Cl_2, −100° to rt, 3-4 h 2. H_2O	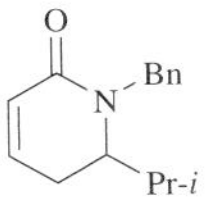 (41)[mm]	134
	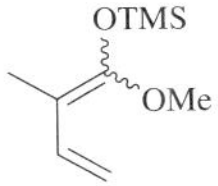	1. $TiCl_4$, CH_2Cl_2, −100° to rt, 3-4 h 2. H_2O	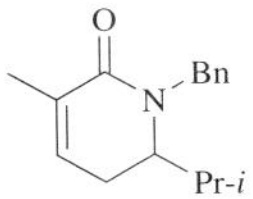 (27)[mm]	134
	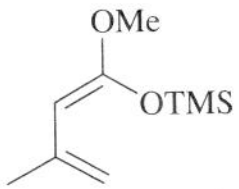	1. $TiCl_4$, CH_2Cl_2, −100° to rt, 3-4 h 2. H_2O	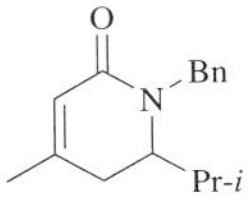 (73)[mm]	134
	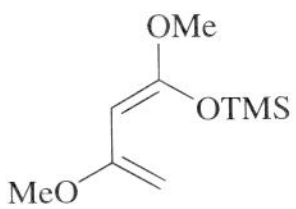	1. Et_2AlCl, CH_2Cl_2, −78° to rt, 2-6 h 2. H_2O, ether	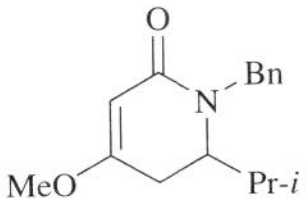 (68)	159

TABLE 5. CYCLOADDITIONS OF *N*-ALKYL- AND *N*-ARYLIMINES (*Continued*)

Imine	Diene	Conditions	Product(s) and Yield(s) (%)	Refs.
C_{11} 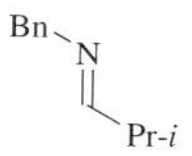	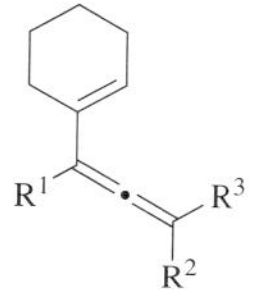	BF_3•OEt_2, CH_2Cl_2	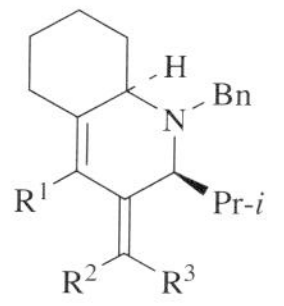	143
	R^1 R^2 R^3	Temp Time		
	Me Me H	rt 1.5 d	(43)[nn]	
	Et Me H	rt 3 d	(39)[nn]	
	Me Me Me	40° 5 d	(37)[nn]	
	Me Ph H	rt 4 d	(61)[nn]	
	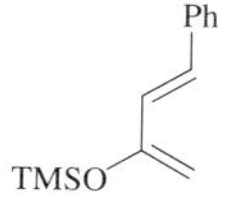	1. TMSOTf, rt 2. TBAF, rt	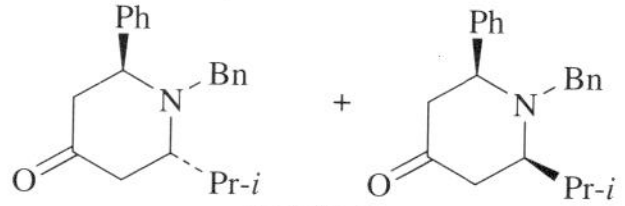 (65) 98:2	169
	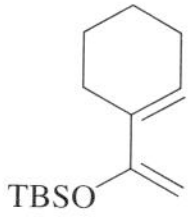		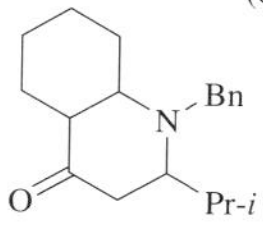	149
		BF_3•OEt_2, –78°, 5 min	(32)[j]	
		$TiCl_4$, rt, 5 min	(32)[j]	
		$AlCl_3$, rt, 15 h	(9)[j]	
	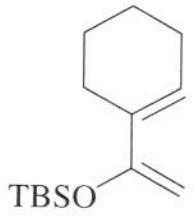	1. TMSOTf, CH_2Cl_2, rt 2. TBAF, THF, 10 h	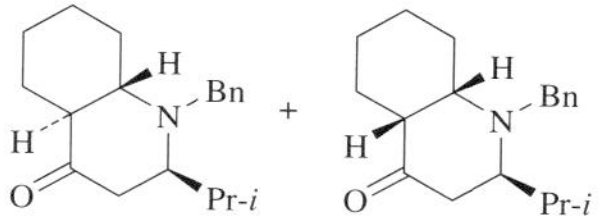	149
		Time		
		2 d	(89) 5:3	
		7 d	(91)	

	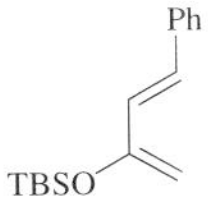	1. TMSOTf, CH_2Cl_2 2. TBAF, THF, 10 h Temp / Time rt / 2 d — / —	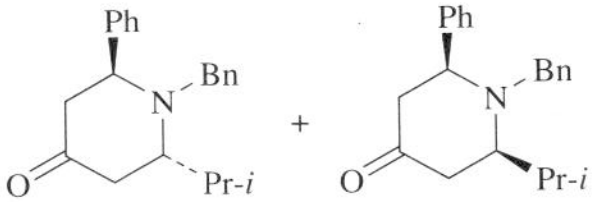 (65) 98:2 (75) 98:2	149 150
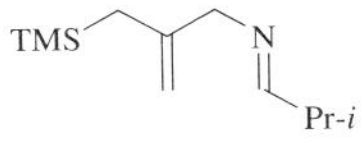	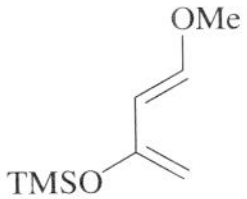	1. $Yb(OTf)_3$, MeCN, rt, 10-15 h 2. H_2O	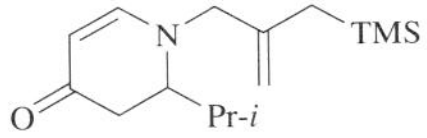 (70)	382
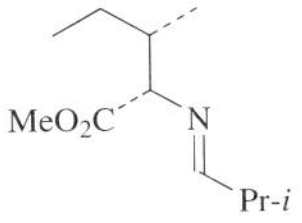	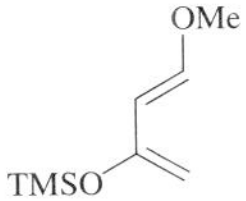	1. $EtAlCl_2$[jj], CH_2Cl_2, −78 to 20° 2. H^+, H_2O	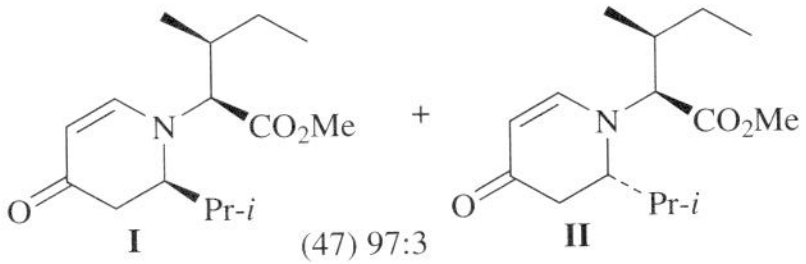 I + II (47) 97:3	175a
	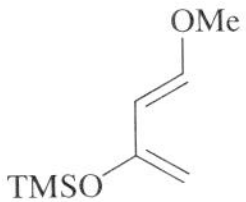	1. Lewis acid, CH_2Cl_2, −78 to 20°, 1 h 2. $NaHCO_3$, H_2O L. A. $EtAlCl_2$ $MeAlCl_2$	**I + II** (48) 97:3 (50) 93:7	384
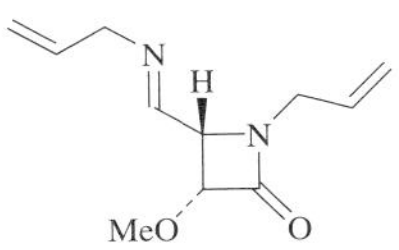	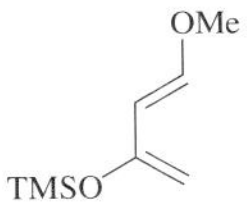	1. ZnI_2, MeCN, −20° 2. $NaHCO_3$, H_2O	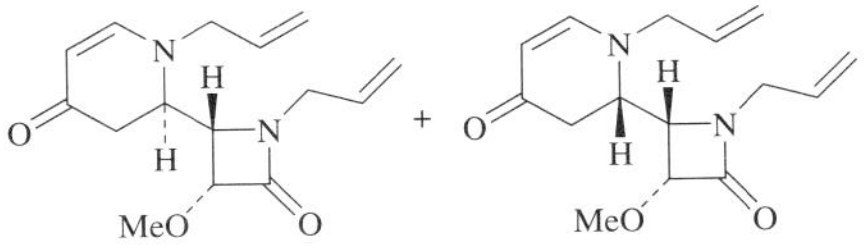 (75) 80:20	442

TABLE 5. CYCLOADDITIONS OF *N*-ALKYL- AND *N*-ARYLIMINES (*Continued*)

Imine	Diene	Conditions	Product(s) and Yield(s) (%)	Refs.
C_{11} 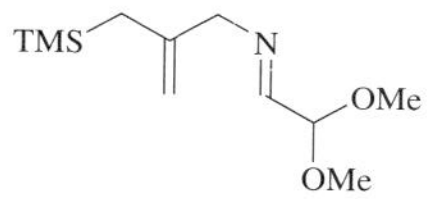	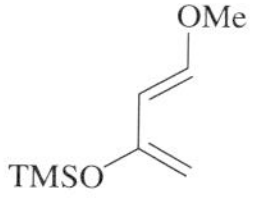	1. $Yb(OTf)_3$, MeCN, H_2O, rt, 10-15 h 2. H_2O	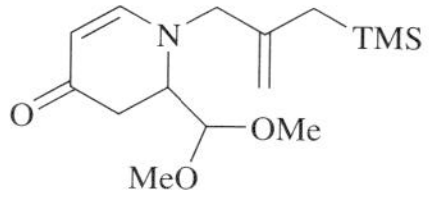 (60)	382
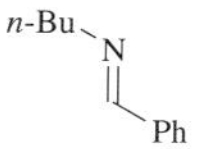	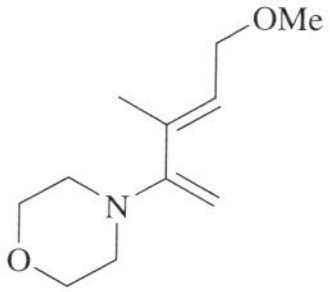	1. $ZnCl_2$, THF, ether, rt, 10 h 2. SiO_2	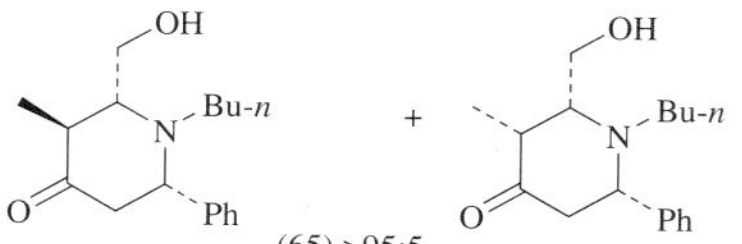 (65) >95:5	157
	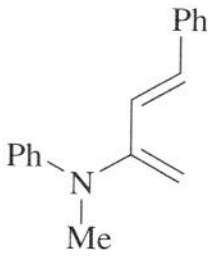	1. $Cu(OTf)_2$, THF, –20°, 24 h 2. $NaHCO_3$, H_2O	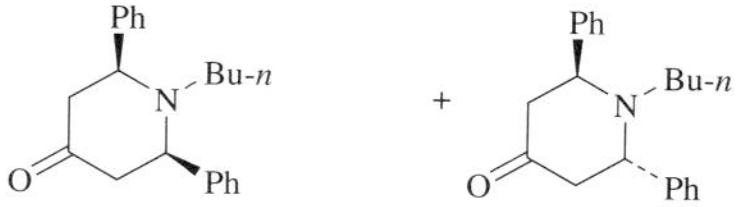 (87) >99.5:0.5	376
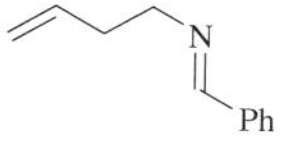	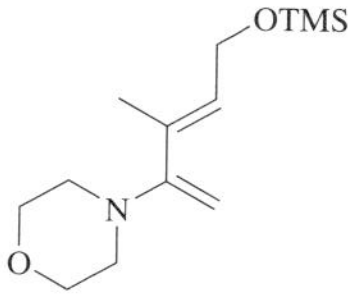	1. $ZnCl_2$ 2. $NaHCO_3$ 3. SiO_2	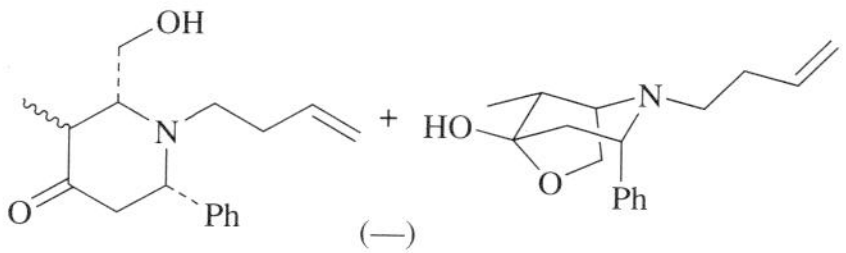 (—)	375
	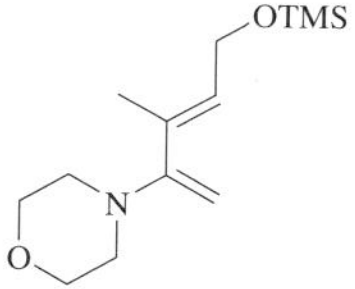	1. $ZnCl_2$ 2. $NaHCO_3$ 3. SiO_2 4. $NaBH_4$ 5. Na_2CO_3, H_2O	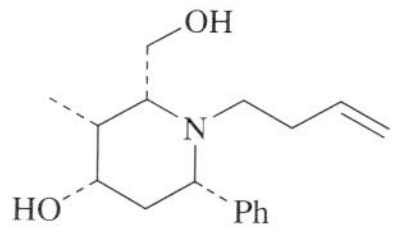 (low)	375

	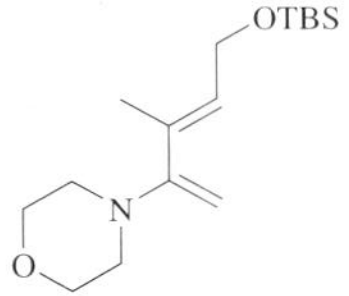	1. $Yb(OTf)_3$, THF, 0° to rt, overnight 2. $NaHCO_3$, H_2O 3. SiO_2	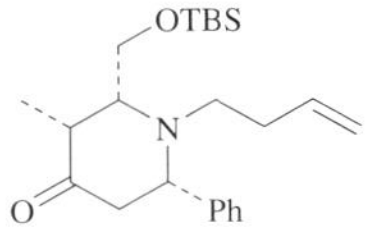 (64)	375
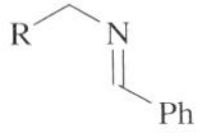 R: CH_2CO_2Me; CO_2Et	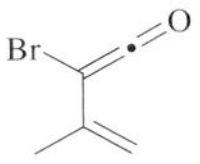	TEA, CH_2Cl_2, 40°	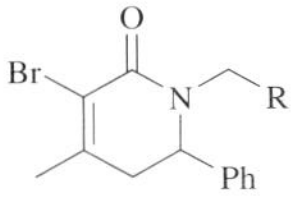 (64) (34)	374
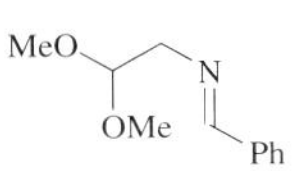	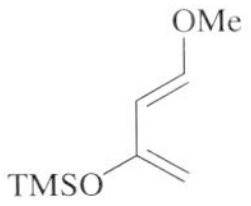	1. $ZnCl_2$, THF, rt, 36-48 h 2. H_2O	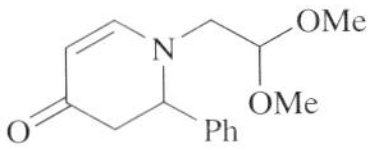 (53)[n], (76)[oo]	141
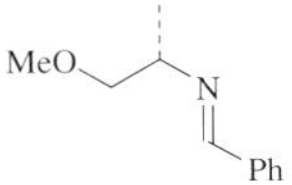	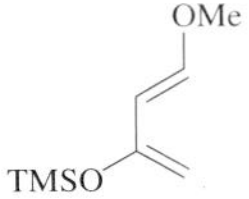	Lewis acid, CH_2Cl_2, rt L. A.: BF_3•OEt_2; $Zn(OTf)_2$	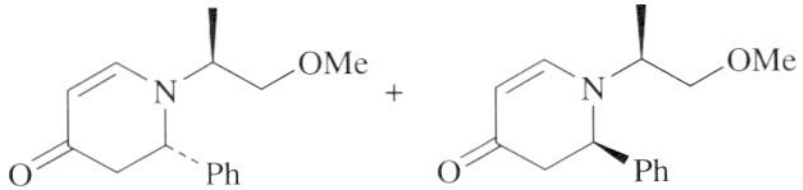 (54) 53:47 (63) 59:41	170
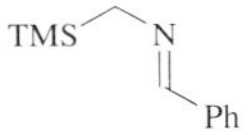	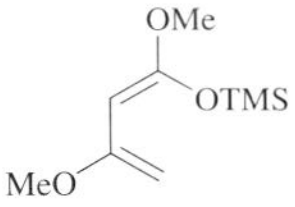	1. Et_2AlCl, CH_2Cl_2, –78° to rt, 2-6 h 2. H_2O, ether	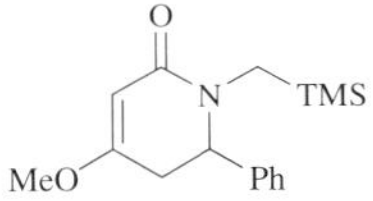 (80)	159
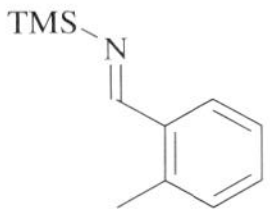	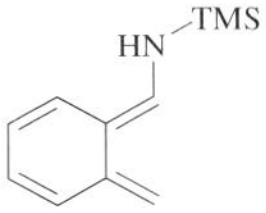	hν[pp]	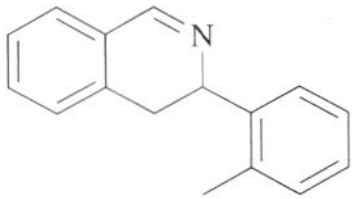 (67)	387

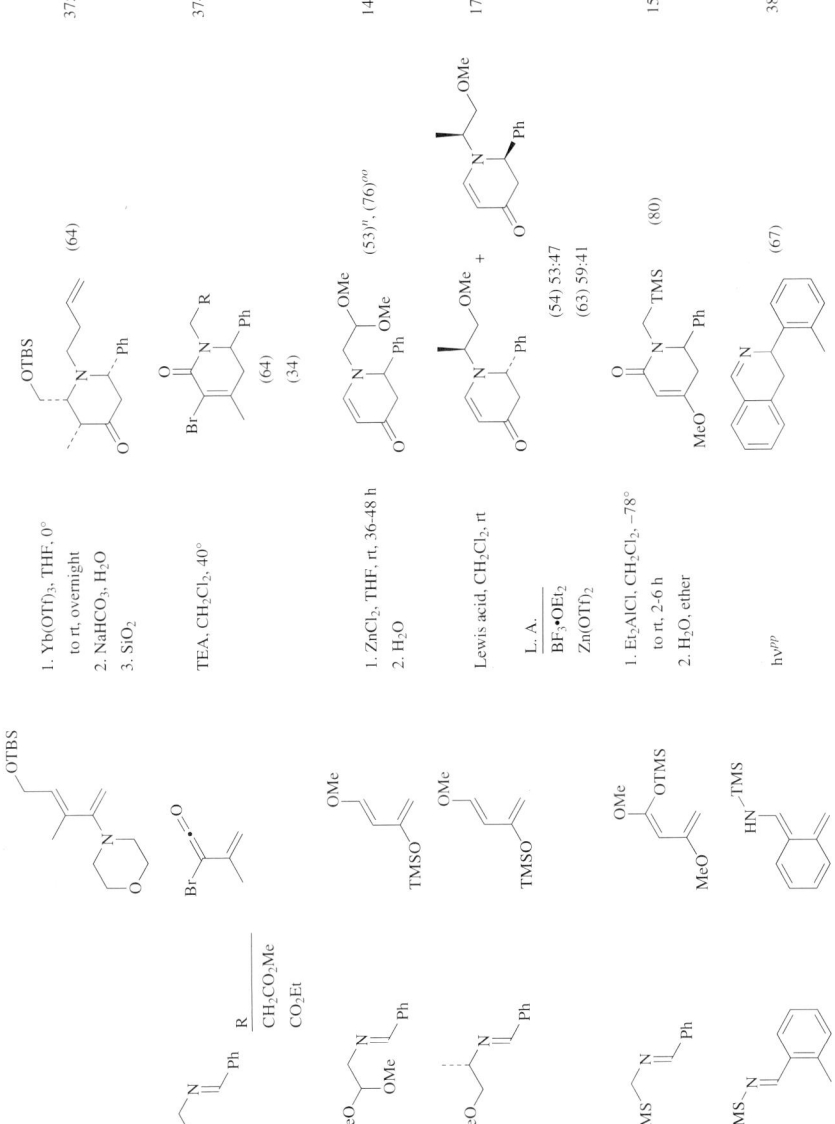

TABLE 5. CYCLOADDITIONS OF *N*-ALKYL- AND *N*-ARYLIMINES (*Continued*)

Imine	Diene	Conditions	Product(s) and Yield(s) (%)	Refs.
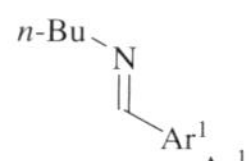 Ar^1 4-FC_6H_4 2-BrC_6H_4	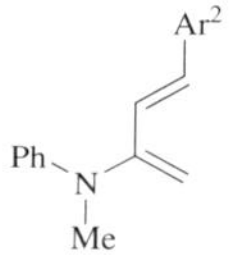 Ar^2 = 4-$MeOC_6H_4$	1. $Cu(OTf)_2$, THF, –20°, 24 h 2. $NaHCO_3$, H_2O	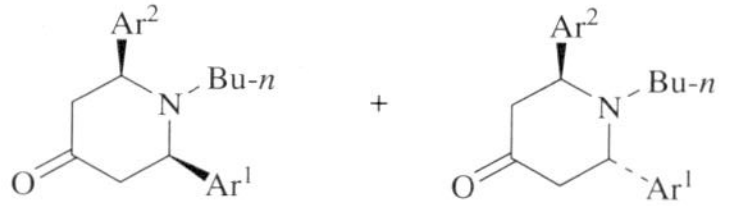 (73) >99.5:0.5 (85) >99.5:0.5	376
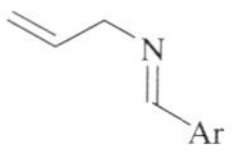 Ar = 4-$MeOC_6H_4$	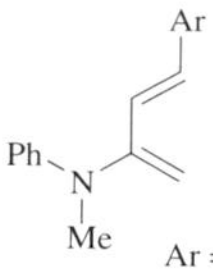 Ar = 4-$MeOC_6H_4$	1. $Cu(OTf)_2$, THF, –20°, 24 h 2. $NaHCO_3$, H_2O	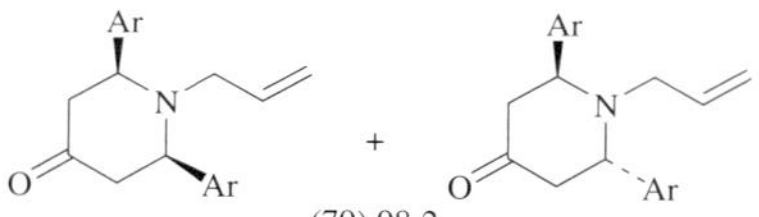 (79) 98:2	376
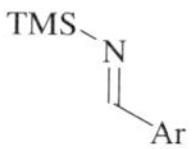 Ar = 4-$MeOC_6H_4$	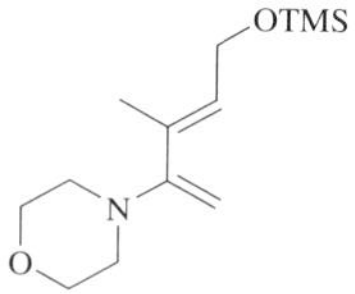	—	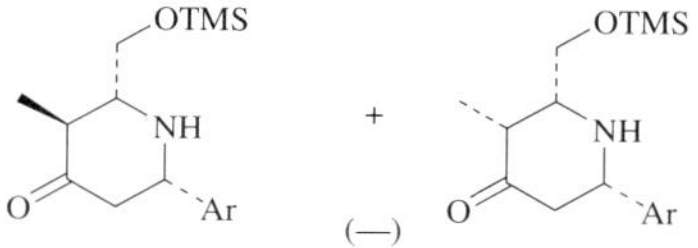 (—)	177
	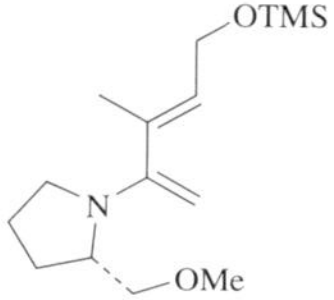	1. $ZnCl_2$, THF, ether, –80° to rt, 16 h 2. $NaHCO_3$, H_2O 3. Na_2CO_3, MeOH, 6 h	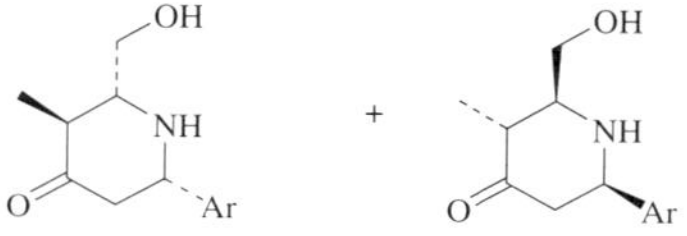 (43) 95:5	177

	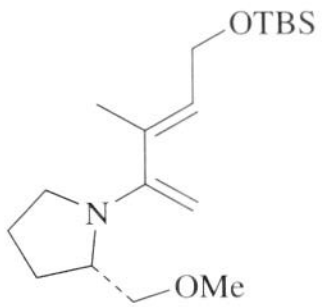	1. $ZnCl_2$, THF, ether, –80° to rt, 16 h 2. $NaHCO_3$, H_2O	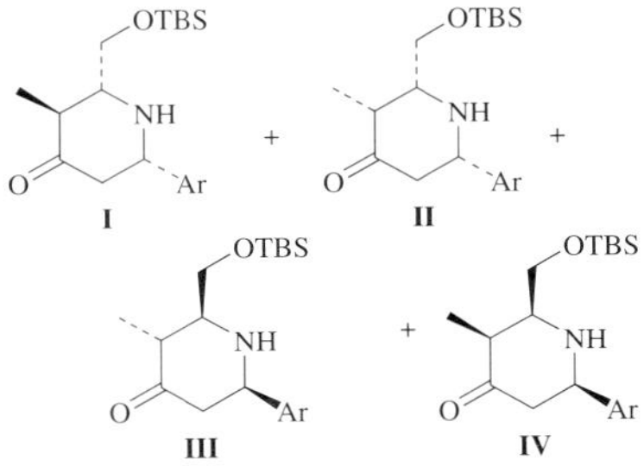 **I + II + III + IV** (29), **I**:**III** 92:8[qq]	177
	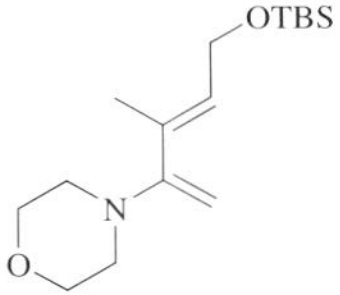	—	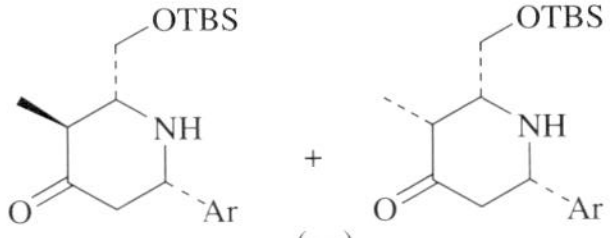 (—)	177
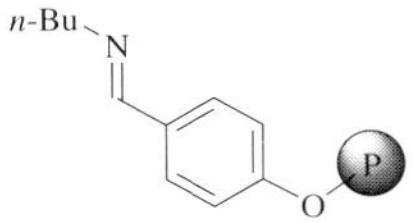	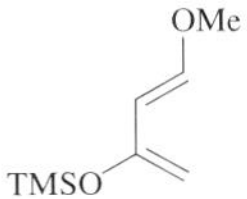	1. $Yb(OTf)_3$, THF, 60°, 3 h 2. H_2O 3. TFA, CH_2Cl_2	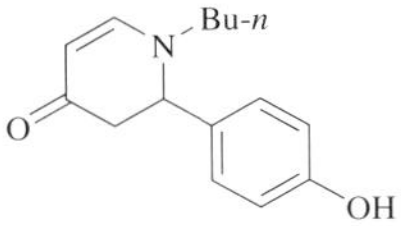 (90)[hh]	378
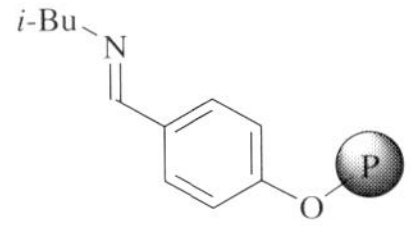	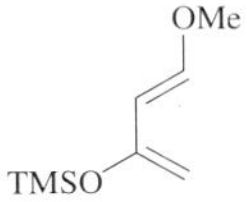	1. $Yb(OTf)_3$, THF, 60°, 3 h 2. H_2O 3. TFA, CH_2Cl_2	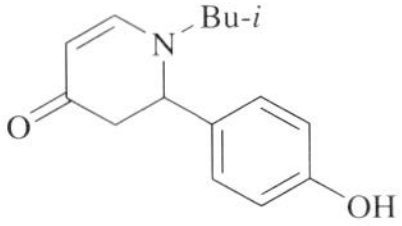 (82)[hh]	378

TABLE 5. CYCLOADDITIONS OF *N*-ALKYL- AND *N*-ARYLIMINES (*Continued*)

Imine	Diene	Conditions	Product(s) and Yield(s) (%)	Refs.
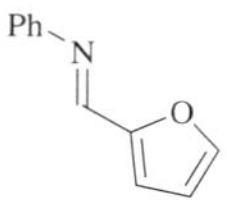	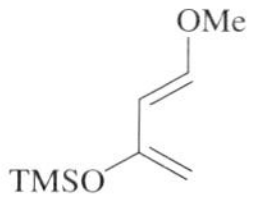		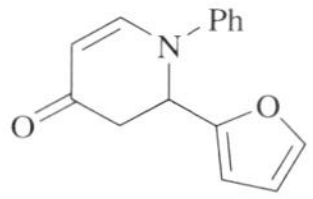	
		$Yb(OTf)_3$, CH_3CN, 0°[c]	(90)	388
		$Sc(OTf)_3$, CH_3CN, 0°	(91)	388
		1. $In(OTf)_3$, $MgSO_4$, MeCN, rt 2. $NaHCO_3$, H_2O	(61)	136
		HBF_4, H_2O, MeOH, –40°, 30 min	(58)	372
		Nafion-H, 10 h	(89)	138
		1. MeOH, rt, 2 h 2. HCl, H_2O	(74)	373
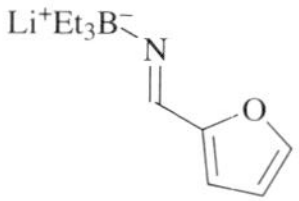	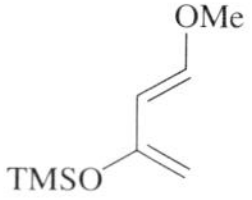	1. THF, rt, 4-5 h 2. $NaHCO_3$, H_2O	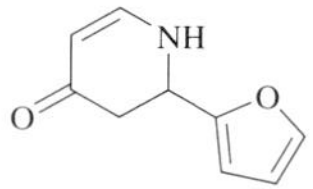 (58)	389
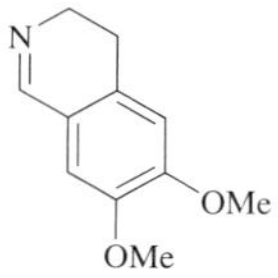	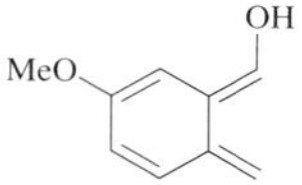	Benzene, 80°, 5 h[rr]	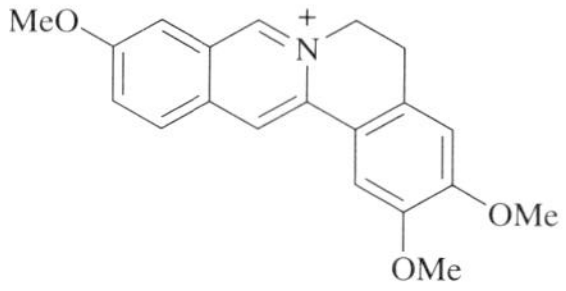 (—)	390
	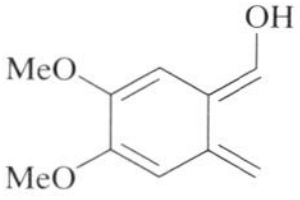	Benzene, 80°, 5 h[rr]	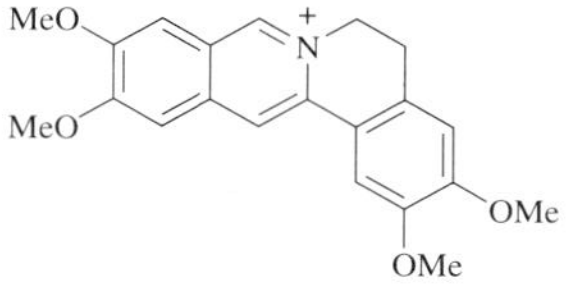 (52)	390

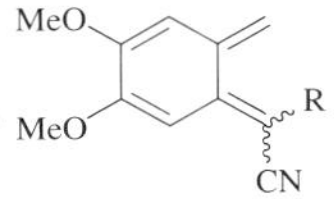

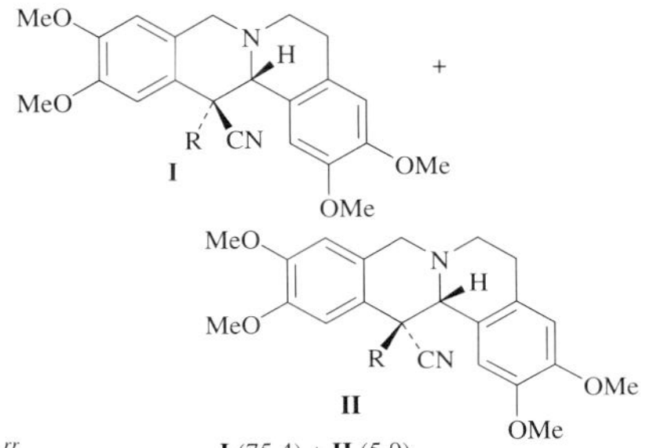

R	Conditions	Product(s) (%)	Refs.
H	Bromobenzene, 150-160°, 3 h[rr]	**I** (75.4) + **II** (5.0)	391
H	150-160°, 45 min[rr]	**II** (88.0)	
Me	Bromobenzene, 150-160°, 3 h[rr]	**I** (32.5) + **II** (41.7)	
Me	150-160°, 1 h[rr]	**I** (18.8) + **II** (24.0)	

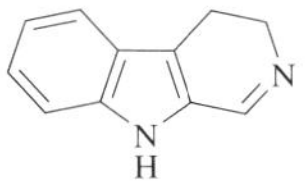

Hydroquinone, chlorobenzene, 120°, 2 h

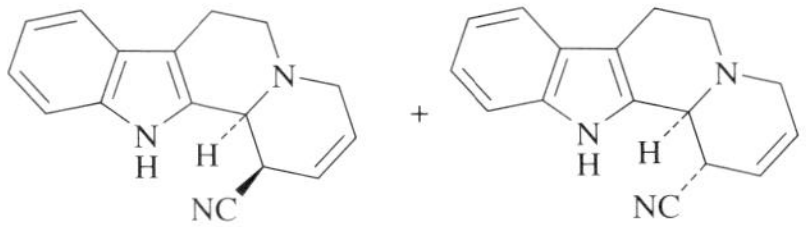 + 392

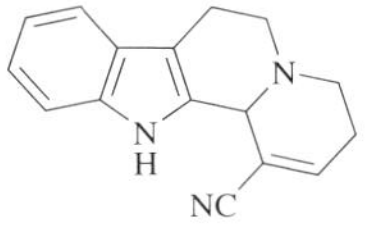

(—) 1.3:1:3.6

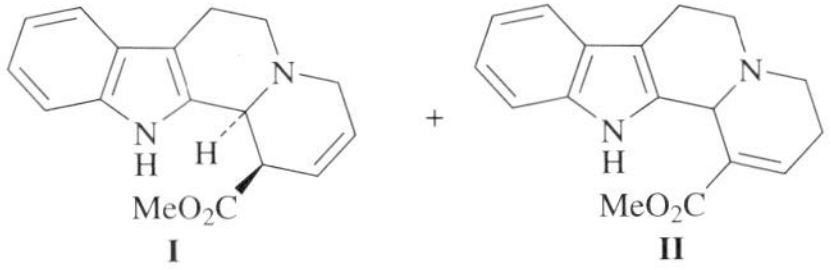

Conditions	Product(s) (%)	Refs.
Chlorobenzene, 120°, 90 min	**I** + **II** (69)	393
—	**I** + **II** (—)[ss]	394

TABLE 5. CYCLOADDITIONS OF *N*-ALKYL- AND *N*-ARYLIMINES (*Continued*)

Imine	Diene	Conditions	Product(s) and Yield(s) (%)	Refs.
C_{11} 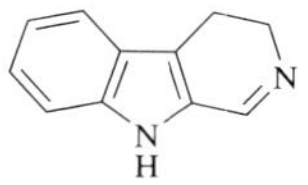	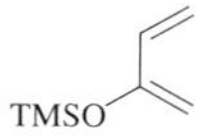	1. $ZnCl_2$, MeCN 2. HCl, H_2O, 10 min	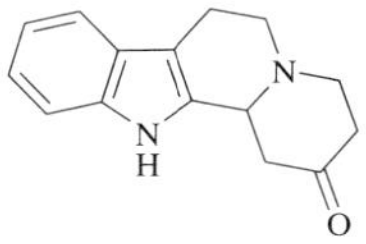	
		Temp / Time		
		50° / 4 h	(55)	395
		81° / several h	(—)[tt]	396
	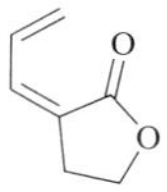	3-(*tert*-Butyl)-4-hydroxy-5-methylphenyl sulfide, chlorobenzene, 160°, 6 h	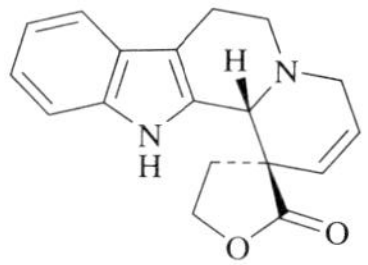 (53)	394
	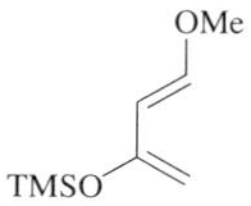	1. $ZnCl_2$ 2. HCl, H_2O, 10 min	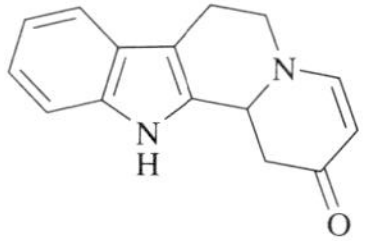	
		Solvent / Temp / Time		
		THF / rt / 5 h	(39)	395
		MeCN / rt / 1 h	(43)	395
		MeCN / 50° / 30 min	(39)	395
		MeCN / 81° / several h	(—)[tt]	396
	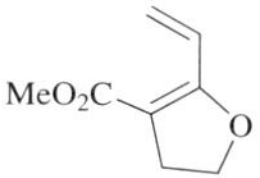	3-(*tert*-Butyl)-4-hydroxy-5-methylphenyl sulfide, MeCN, 120-125°, 6 h	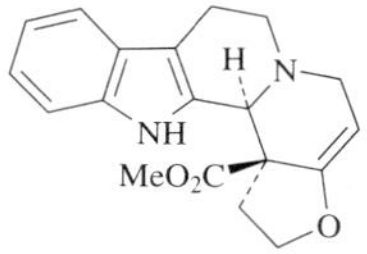 (64)	397

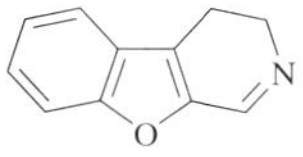	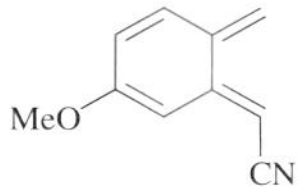	140-150°, 2 h[rr]	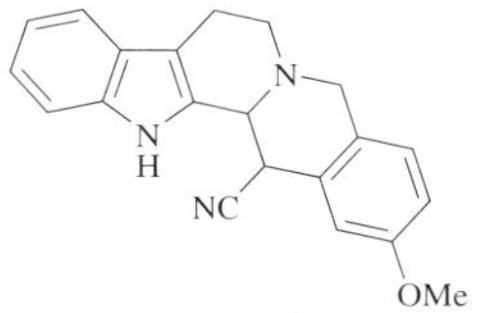	(84.1)[i]	398
	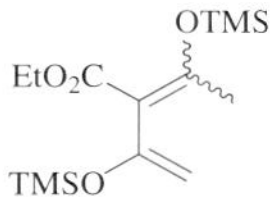	$ZnCl_2$, MeCN, rt	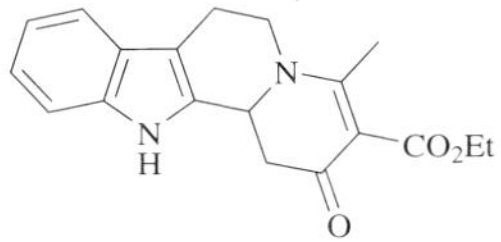	(53)	399
	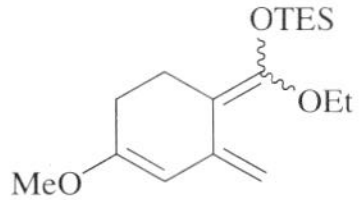	$CHCl_3$, rt	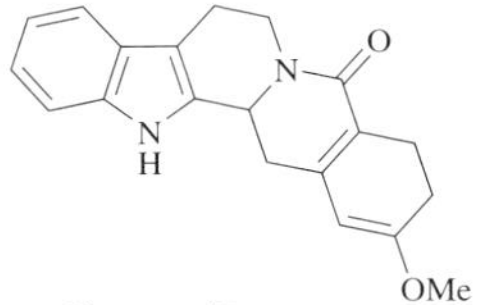	(50)	399
	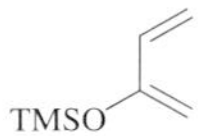		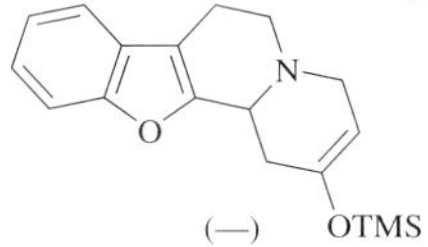		
		BF_3•OEt_2, $CHCl_3$, MeCN, 0°, 5 min	(—)		395, 400
		BF_3•OEt_2[uu]	(—)		401
		TFA, BF_3•OEt_2, CH_2Cl_2, 22-24°, 30 min[uu,vv]	(quantitative)		401
		HBF_4[uu]	(quantitative)		401
	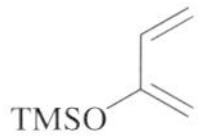	1. $ZnCl_2$, MeCN, 50°, 1 h 2. HCl, H_2O, 10 min	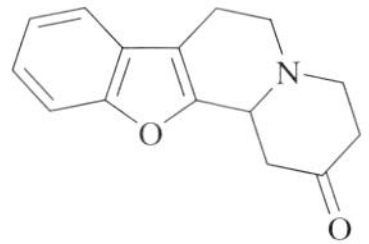	(41)	395

TABLE 5. CYCLOADDITIONS OF *N*-ALKYL- AND *N*-ARYLIMINES (*Continued*)

Imine	Diene	Conditions	Product(s) and Yield(s) (%)	Refs.
C_{11} 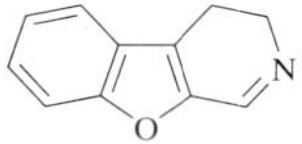	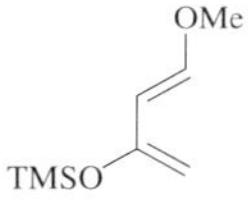	1. $ZnCl_2$, rt 2. HCl, H_2O, 10 min	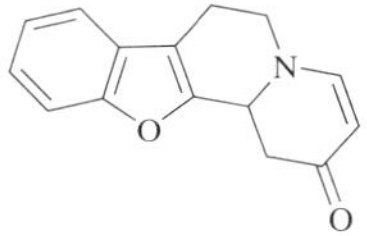	395
		Solvent / Time		
		THF / 1 h	(73)	
		$CHCl_3$ / 1 h	(71)	
		MeCN / 5 min	(65)	
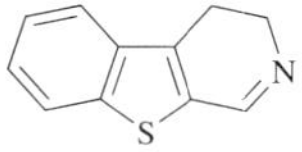	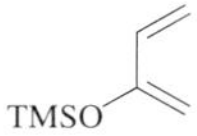	1. $ZnCl_2$, MeCN, 50°, 4 h 2. HCl, H_2O, 10 min	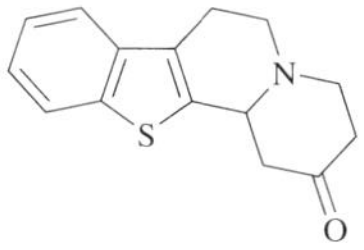 (30)	395
	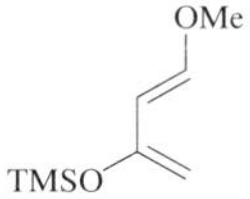	1. $ZnCl_2$, MeCN, rt, 1 h 2. HCl, H_2O, 10 min	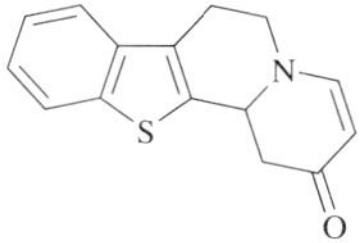 (62)	395
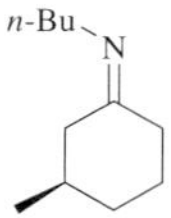	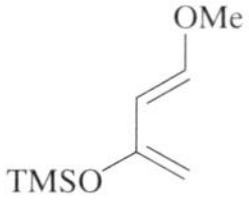	1. $Zn(OTf)_2$, CH_2Cl_2, rt, 12 h 2. NH_4Cl, H_2O, 2 h	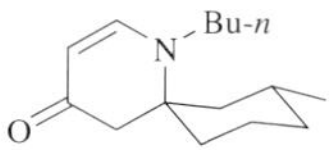 (70)	356
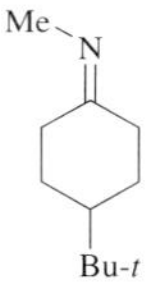	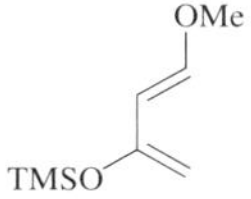	1. $Zn(OTf)_2$, CH_2Cl_2, rt, 12 h 2. NH_4Cl, H_2O, 2 h	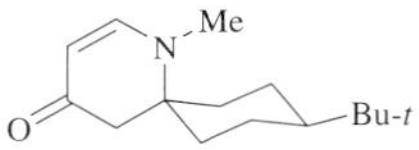 (49)	356

C_{12}	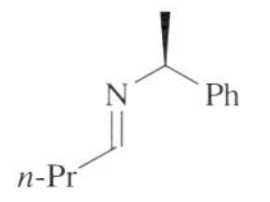	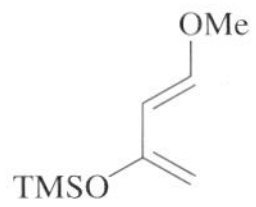	1. $B(OPh)_3$, CH_2Cl_2, –78°, several h 2. H_2O	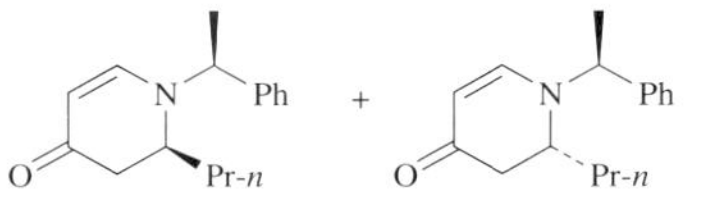 (59) 91:9	243, 242
			—	(—) 75:25[f]	384
		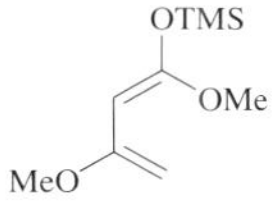	—	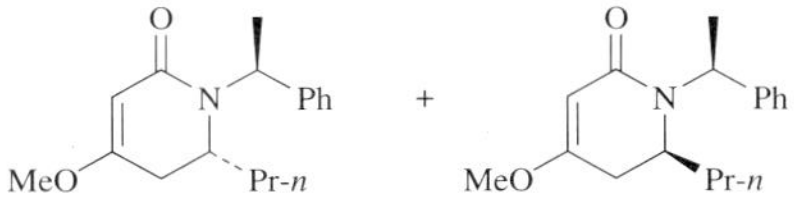 (—) 90:10[f]	402
	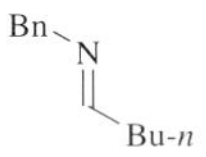	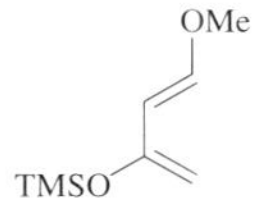	Lewis acid, MeCN, 0° L. A. / $Yb(OTf)_3$ / $Sc(OTf)_3$	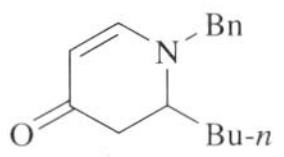 (35) (45)	388
	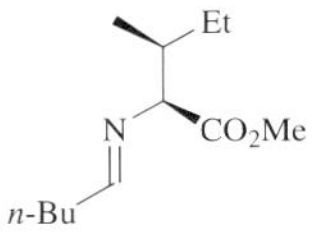	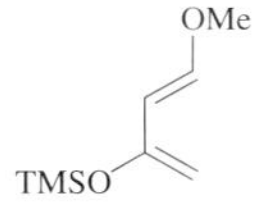	1. Me_2AlCl[jj], CH_2Cl_2, –78 to 20° 2. H^+, H_2O	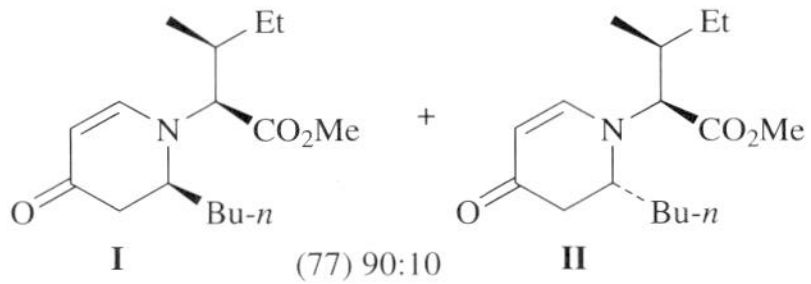 **I** (77) 90:10 **II**	175a
		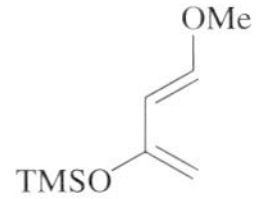	1. Me_2AlCl, CH_2Cl_2, –78 to 20°, 1 h 2. $NaHCO_3$, H_2O	**I** + **II** (77) 90:10	384
	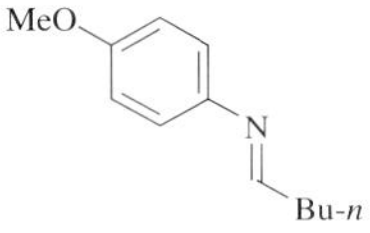	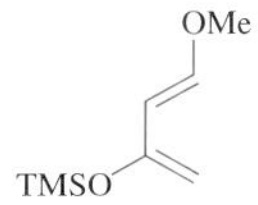	Montmorillonite K10, CH_3CN, H_2O, –10°, 1.5 h	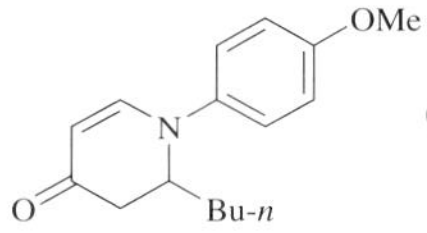 (91)[kk]	140

TABLE 5. CYCLOADDITIONS OF *N*-ALKYL- AND *N*-ARYLIMINES (*Continued*)

Imine	Diene	Conditions	Product(s) and Yield(s) (%)	Refs.
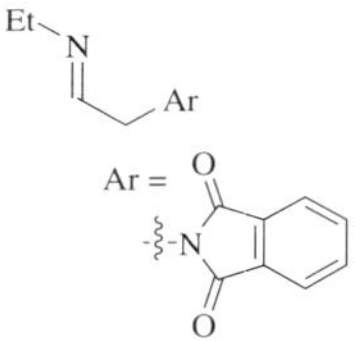		BF_3•OEt_2, TFA, CH_2Cl_2, –78° to rt, overnight	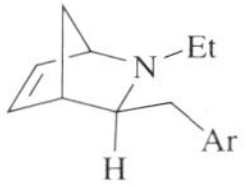 (50) + 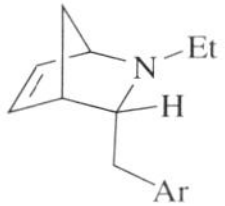 (34)	403
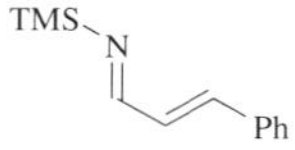	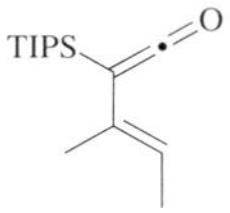	1. rt, 15 min 2. SiO_2	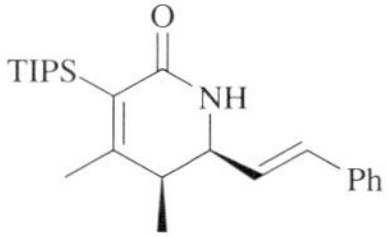 (78)	142
	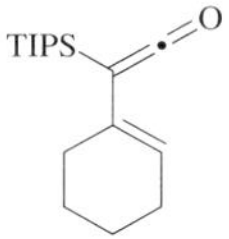	1. rt, 10 min 2. SiO_2	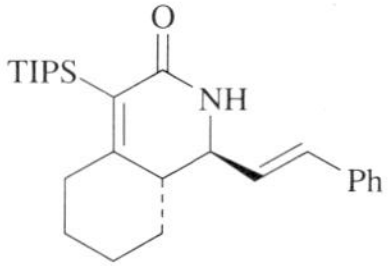 (73)	142
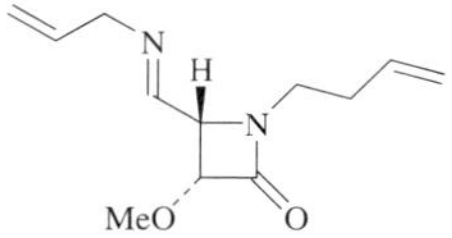	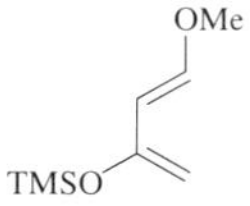	1. ZnI_2, MeCN, –20° 2. $NaHCO_3$, H_2O	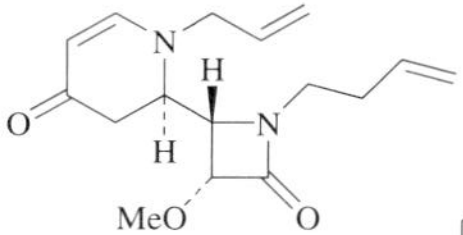 + 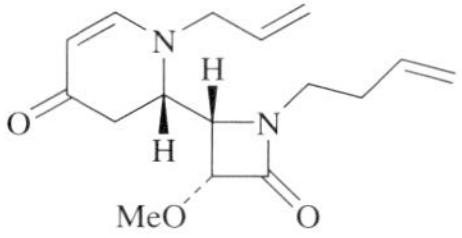 (72) 75:25	442
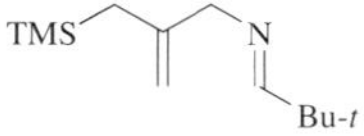	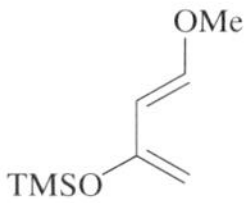	1. $Yb(OTf)_3$, MeCN, rt, 10-15 h 2. H_2O	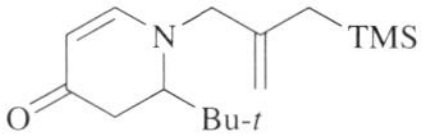 (56)	382

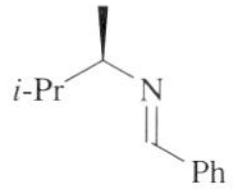	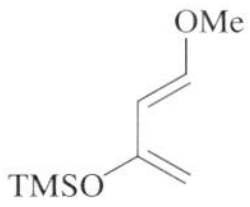	$Zn(OTf)_2$, CH_2Cl_2, rt	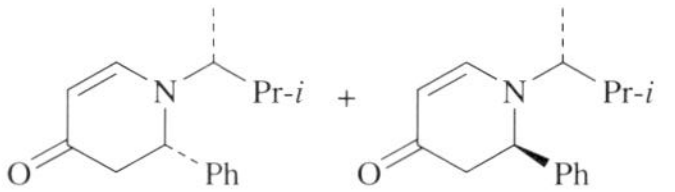 (60) 72:28	170
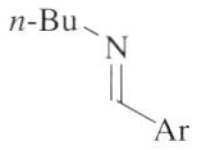 Ar = 4-$MeOC_6H_4$	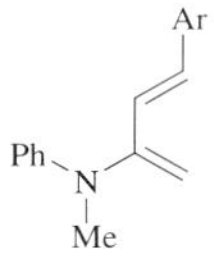	1. $Cu(OTf)_2$, THF, –20°, 24 h 2. $NaHCO_3$, H_2O	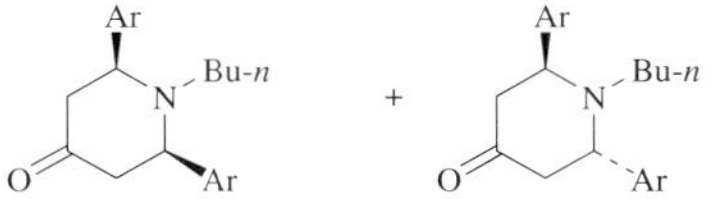 (73) >99.5:0.5	376
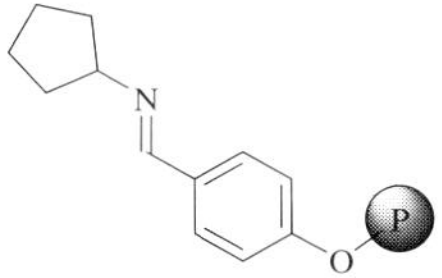	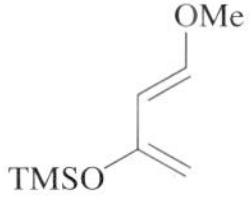	1. $Yb(OTf)_3$, THF, 60°, 3 h 2. H_2O 3. TFA, CH_2Cl_2	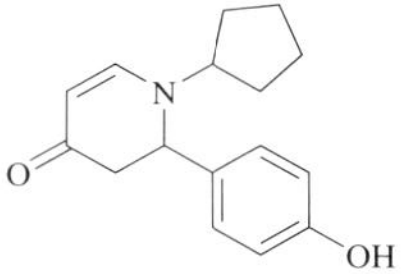 (75)[hh]	378
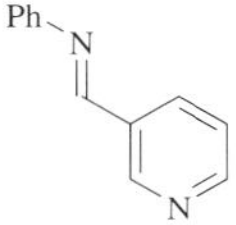	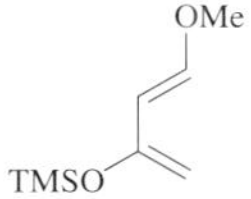		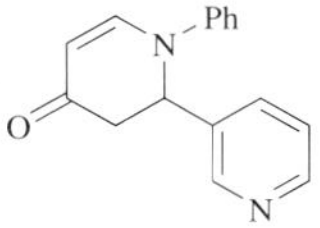	
		Nafion-H, 14 h	(88)	138
		NaOTf, H_2O, rt, 2-3 h	(68)[ww]	385
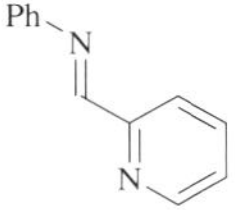	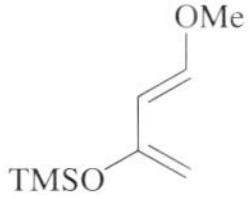		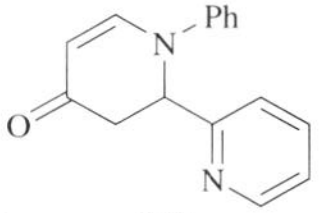	
		1. $In(OTf)_3$, $MgSO_4$, MeCN, rt 2. $NaHCO_3$, H_2O	(95)	136
		$Sc(OSO_2C_8F_{17})_3$, $scCO_2$, 100 atm, 40°, 1 h	(68)[a]	404
		1. MeOH, rt, 2 h 2. HCl, H_2O	(91)	373

TABLE 5. CYCLOADDITIONS OF *N*-ALKYL- AND *N*-ARYLIMINES (*Continued*)

Imine	Diene	Conditions	Product(s) and Yield(s) (%)	Refs.
C_{12}				
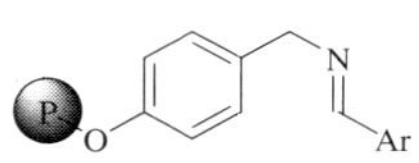 Ar: 2-furyl; 3-furyl	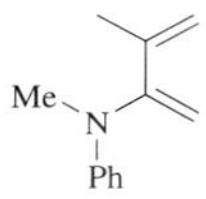	1. $Yb(OTf)_3$, THF, rt, 12 h 2. HCl, H_2O, THF, overnight 3. TMSOTf, CH_2Cl_2, rt, 3 h	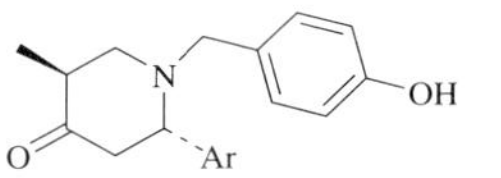 (71)[xx] (76)[xx]	377
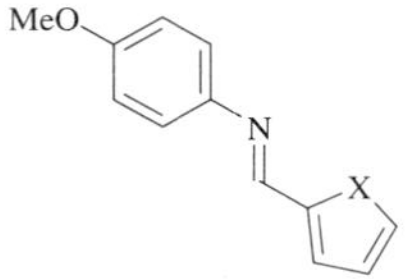	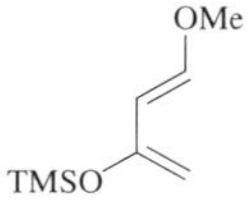	Montmorillonite K10, rt	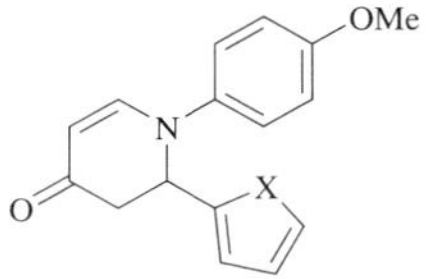	140
X = O		Solvent: H_2O; Time: 2.5 h	(90)	
X = O		Solvent: CH_3CN, H_2O; Time: 1 h	("quant.")	
X = S		Solvent: H_2O; Time: 3 h	(86)	
X = S		Solvent: CH_3CN, H_2O; Time: 3 h	(99)	
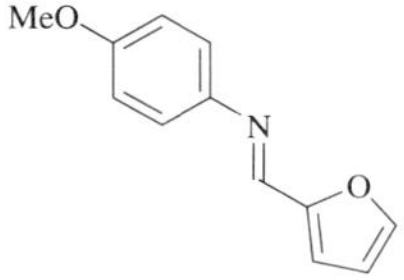	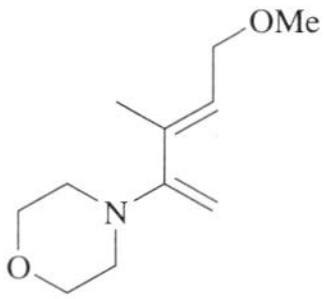	1. $ZnCl_2$, THF, ether, rt, 10 h 2. SiO_2	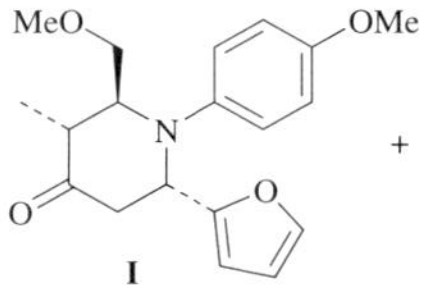 **I** + 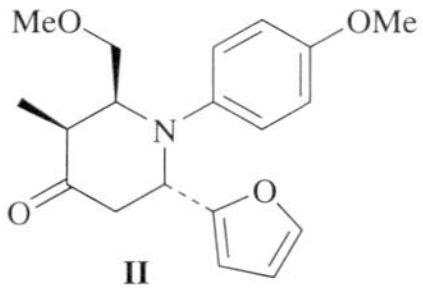 **II** **I** + **II** (79), **I**:**II** 76:21[yy]	157
	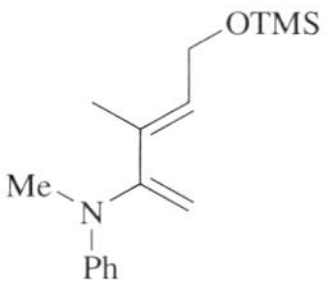	$ZnCl_2$, THF, ether, –40° to rt, 11 h	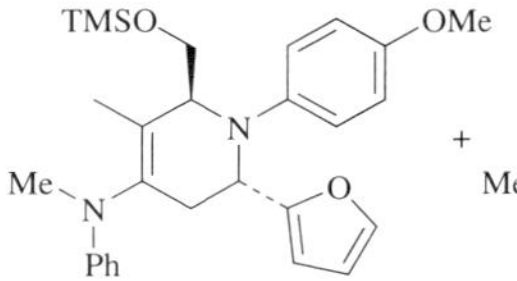 **I** + 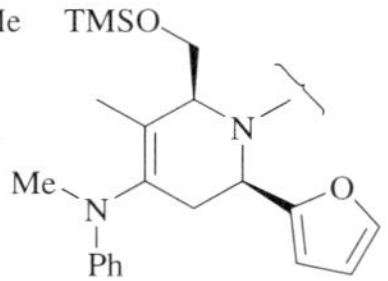 **II** **I** + **II** (81), **I**:**II** 90:10[zz]	157

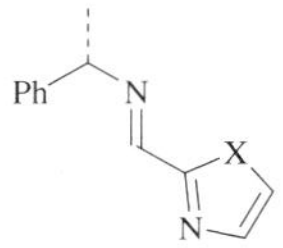 X: NH, S		$MeSO_3H$, CH_2Cl_2, –78° to rt, overnight	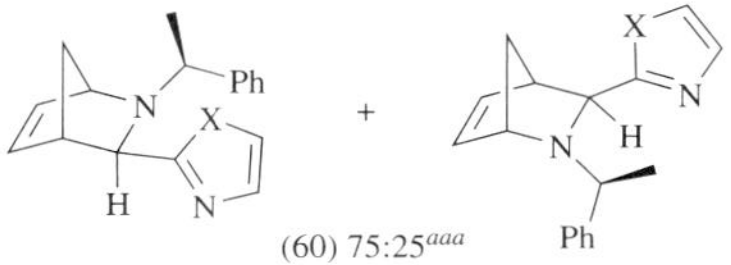 X = NH: (60) 75:25[aaa] X = S: (80) 90:10[aaa]	45, 91
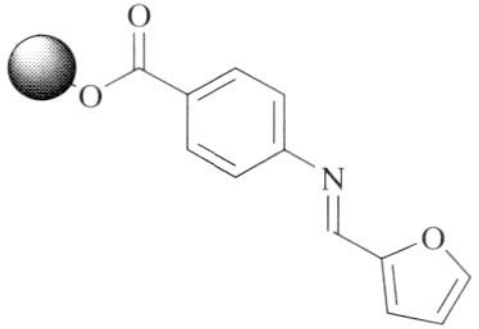	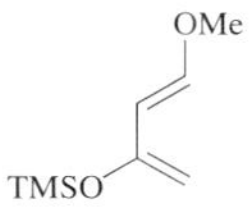	1. $Zn(ClO_4)_2$•$6H_2O$, MeOH, rt, 12 h 2. NaOH, H_2O, rt, 12 h 3. HCl, H_2O	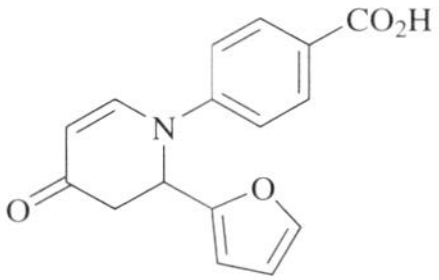 (99)	405
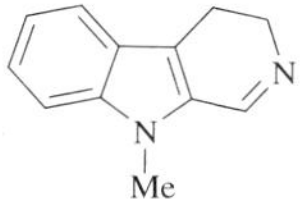		Chlorobenzene, 120°, 4 h	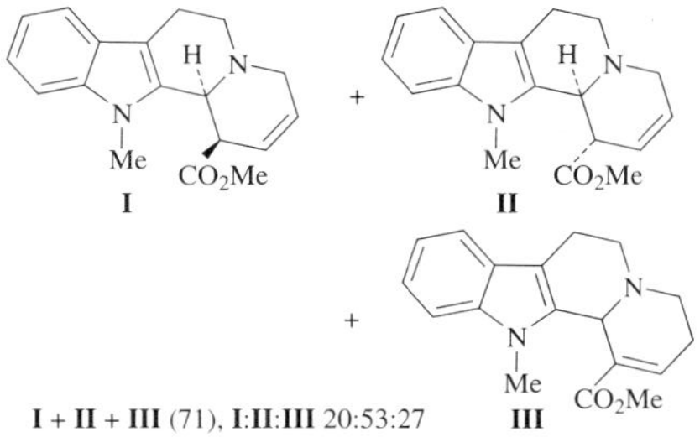 **I + II + III** (71), **I**:**II**:**III** 20:53:27	406, 407
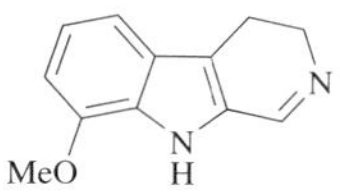		Chlorobenzene, 6 h Temp: 150°	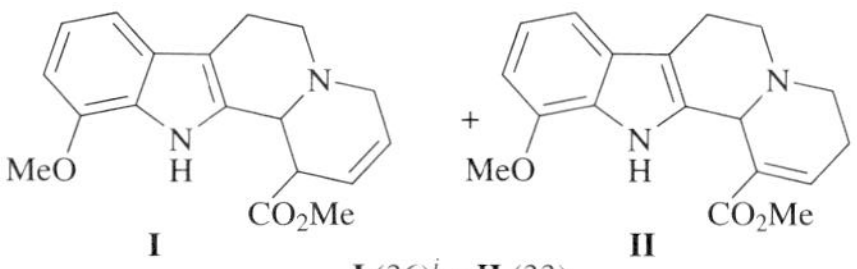 **I** (36)[j] + **II** (33)	394
		Temp: 130°	**I** (31)[j] + **II** (36)	408

TABLE 5. CYCLOADDITIONS OF *N*-ALKYL- AND *N*-ARYLIMINES (*Continued*)

	Imine	Diene	Conditions	Product(s) and Yield(s) (%)	Refs.
C_{12}	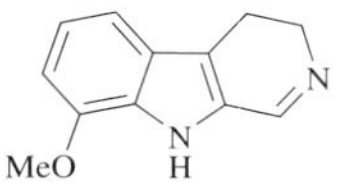	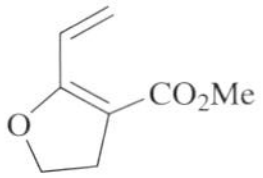	3-(*tert*-Butyl)-4-hydroxy-5-methylphenyl sulfide, MeCN, 120-125°, 6 h	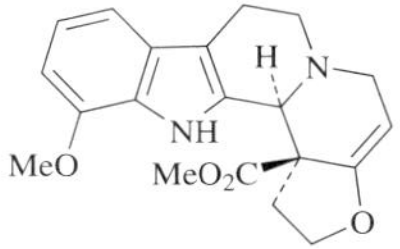 (52)	397
C_{13}	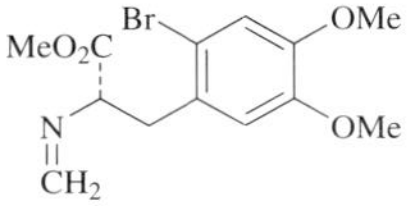	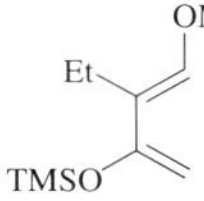	1. $ZnCl_2$, THF, rt 2. HCl, H_2O	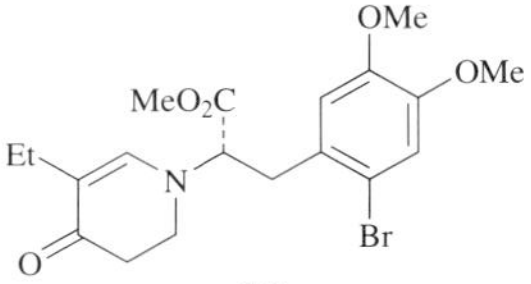	
			$ZnCl_2$ 2 eq, Time 20 h	(48)	380, 379
			$ZnCl_2$ 3 eq, Time 18 h	(34)	380
	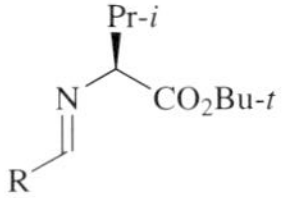	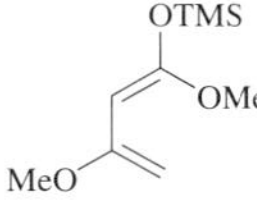	1. Lewis acid, –78 to 20°, CH_2Cl_2, 1-1.5 h 2. $NaHCO_3$, H_2O	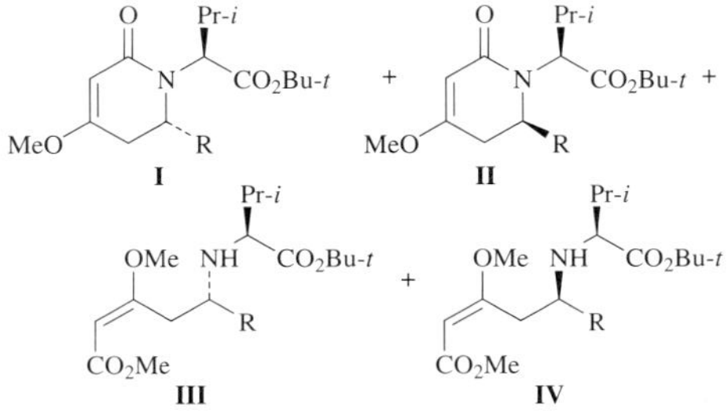	

R	L. A.[bbb]	I + II + III + IV	I + II:III + IV	I + III:II + IV	Refs.
n-Pr	$EtAlCl_2$	(81)	82:18[ccc]	93:7	175a, 402
n-Pr	$SnCl_4$	(33)	78:22[ccc]	93:7	402
i-Pr	$EtAlCl_2$	(60)	46:54[ccc]	96:4	175a, 402

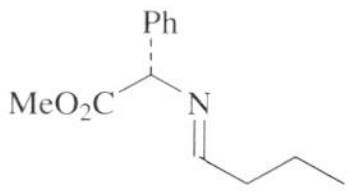

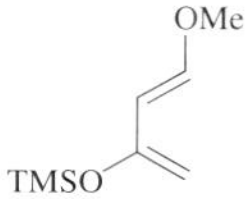

1. $ZnCl_2$, THF, 0°, 7 h
2. $NaHCO_3$, H_2O

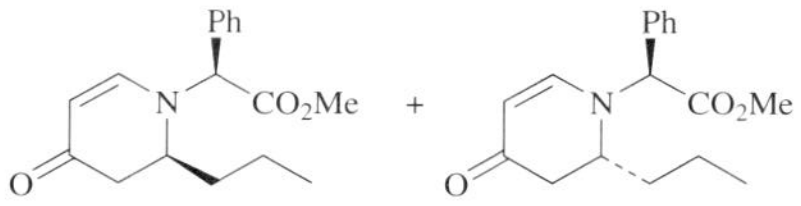

(69) 71:29

384

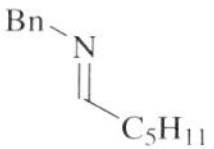

Lewis acid, HCl, H_2O

I II

L. A.	Time	I + II	I:II	Refs.
none	—	(47)	2.7:1	147
$La(OTf)_3$	12-20 h	(47)	2.6:1	147
$Pr(OTf)_3$	12-20 h	(68)	2.9:1	147
$Nd(OTf)_3$	12-20 h	(57)	3.1:1	147
$Gd(OTf)_3$	12-20 h	(19)	2.7:1	147
$Dy(OTf)_3$	12-20 h	(49)	2.8:1	147
$Er(OTf)_3$	12-20 h	(46)	2.6:1	147
$Yb(OTf)_3$	12-20 h	(62)	2.8:1	147
Yb-Amberlyst 15	2 d	**I** (22)	—	409

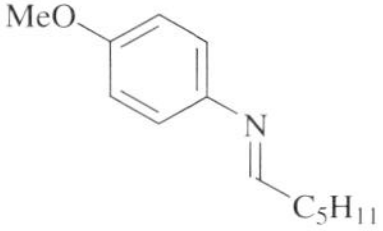

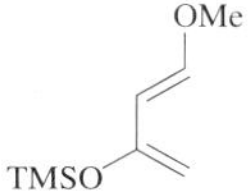

Montmorillonite K10, 1.5 h

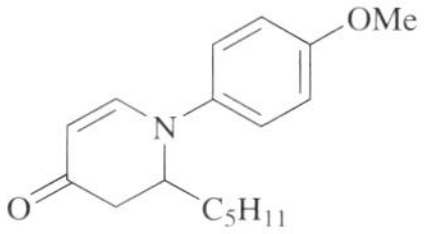

140

Solvent	Temp[kk]	
H_2O	0°	(86)
MeCN, H_2O	–10°	(92)
MeCN	–10°	(61)

TABLE 5. CYCLOADDITIONS OF *N*-ALKYL- AND *N*-ARYLIMINES (*Continued*)

	Imine	Diene	Conditions	Product(s) and Yield(s) (%)	Refs.
C_{13}	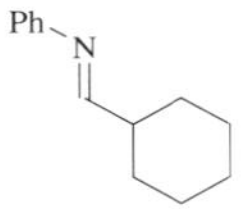	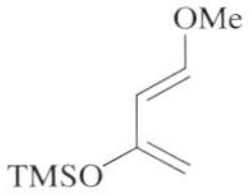		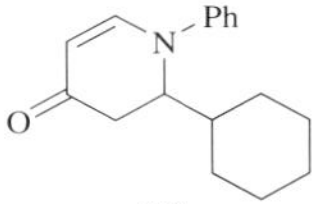	
			HBF_4, H_2O, MeOH, –40°, 30 min	(80)	372
			$Sc(OSO_2C_8F_{17})_3$, $scCO_2$, 100 atm, 40°, 1 h	(60)[a]	404
			NaOTf, H_2O, rt, 2-3 h	(74)	385
			AgOTf, H_2O, rt, 2-3 h	(70)	386
			AgOTf, Triton X-100, H_2O, rt, 2-3 h	(51)	386
	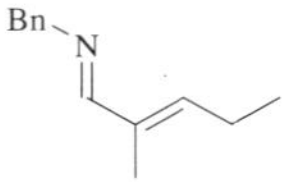	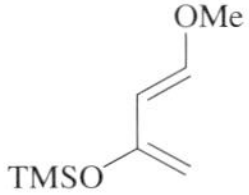	1. $ZnCl_2$, THF, rt, 36-48 h 2. H_2O	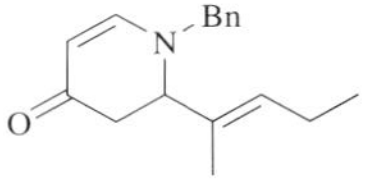 (33)[n], (47)[ddd]	141
	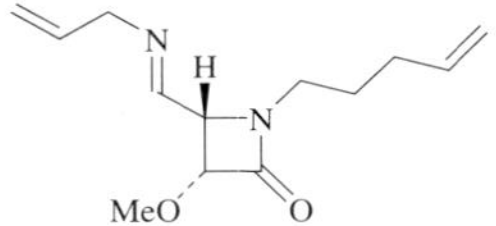	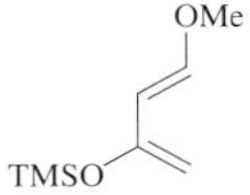	1. ZnI_2, MeCN, –20° 2. $NaHCO_3$, H_2O	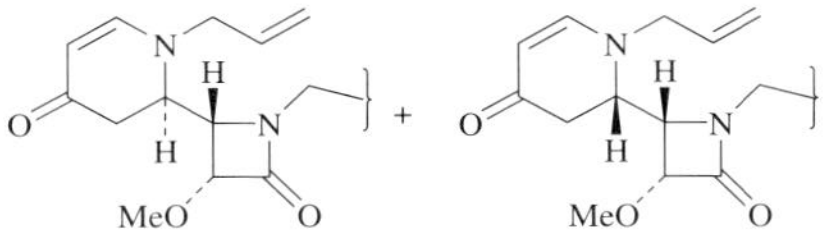 (53) 85:15	442
	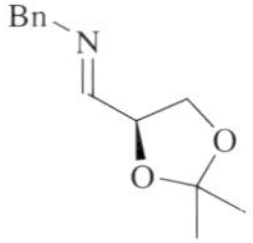		Lewis acid, HCl, H_2O, 48 h	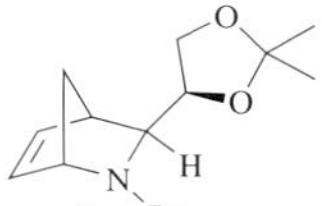 I + 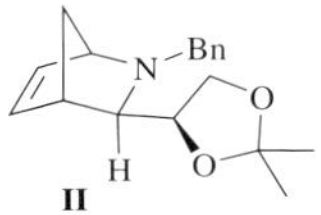 II + 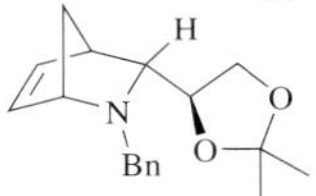 III	363

L. A.	I + II + III	I:II:III
$La(OTf)_3$	(7)	1:8:3
$Pr(OTf)_3$	(14)	1:8:3
$Nd(OTf)_3$[eee]	(44)	1:8:3
$Nd(OTf)_3$[fff]	(72)	—
$Gd(OTf)_3$	(32)	1:8:3
$Dy(OTf)_3$	(10)	1:8:3
$Er(OTf)_3$	(15)	1:8:3
$Yb(OTf)_3$	(8)	1:8:3

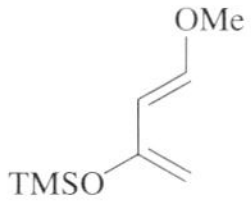

1. ZnI_2, –40°
2. $NaHCO_3$, H_2O

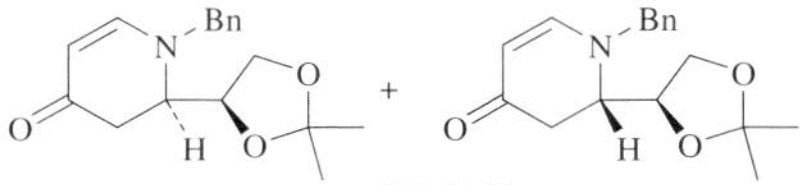

(93) 69:31

163

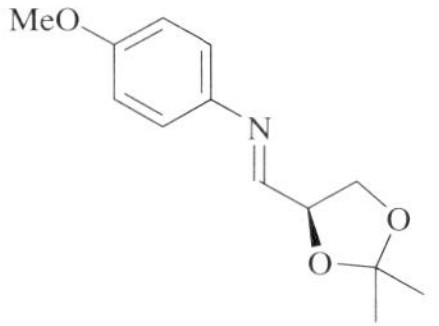

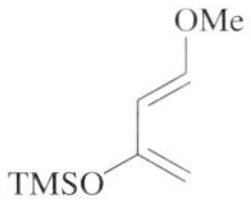

1. MeOH, rt, 2 h
2. HCl, H_2O

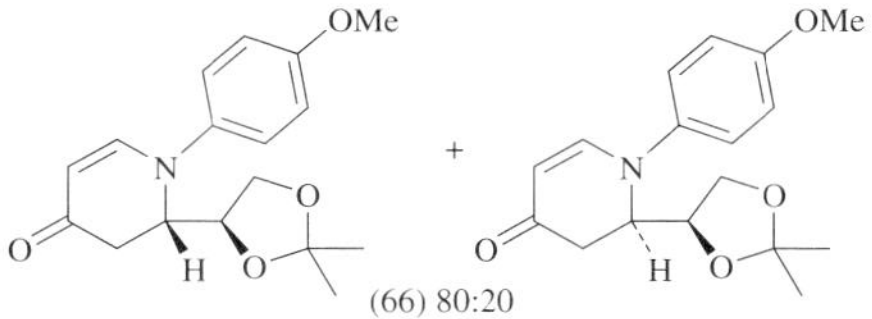

(66) 80:20

373

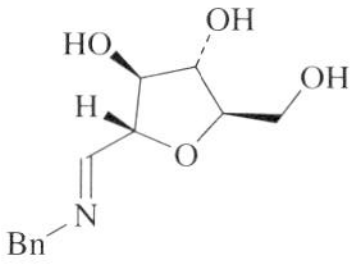

$Nd(OTf)_3$, HCl, H_2O, 60 h

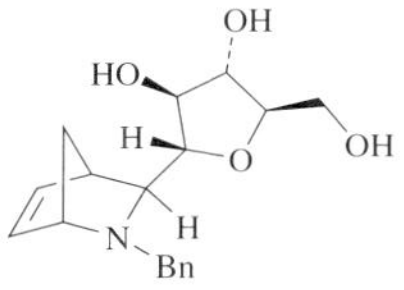

(—)

363

TABLE 5. CYCLOADDITIONS OF *N*-ALKYL- AND *N*-ARYLIMINES (*Continued*)

Imine	Diene	Conditions	Product(s) and Yield(s) (%)	Refs.
C_{13} 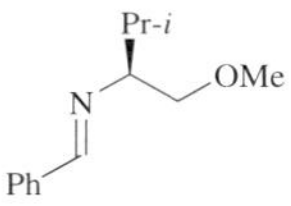	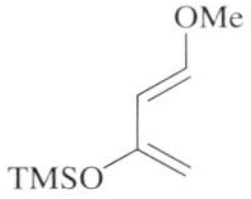	Lewis acid, rt L. A. / Solvent Et_2AlCl / CH_2Cl_2 $BF_3{\bullet}OEt_2$ / CH_2Cl_2 $Zn(OTf)_2$ / CH_2Cl_2 $SnCl_2$ / CH_2Cl_2 $ZnCl_2$ / THF	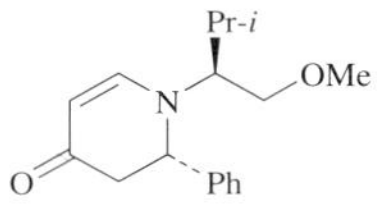 + 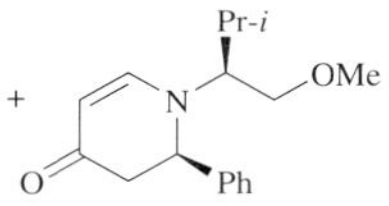 (24) 72:28 (55) 79:21 (67) 87:13 (66) 83:17 (33) 76:24	170
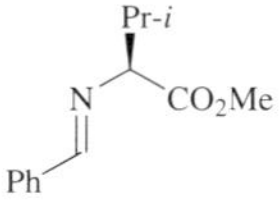	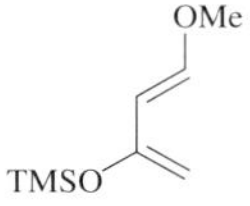	1. $ZnCl_2$[jj], THF, –20° 2. H^+, H_2O	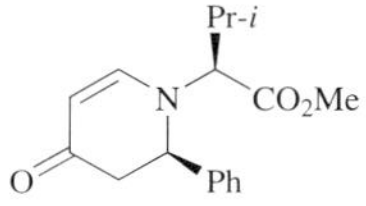 + 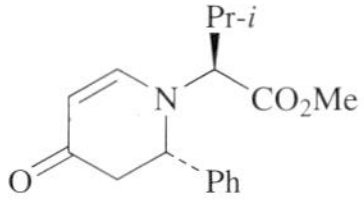 (45) 92:8	175a
		1. $ZnCl_2$, THF, –20°, 7 h 2. $NaHCO_3$, H_2O	(45) 92:8	384
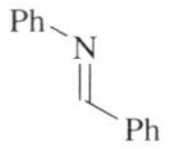	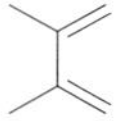	$Sc(OTf)_3$, MeCN, rt, 20 h	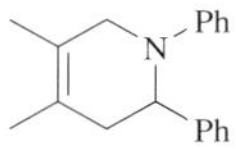 (37)[ggg]	135
	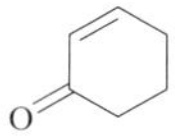	$InCl_3$, MeCN, rt, 24 h[l,hh]	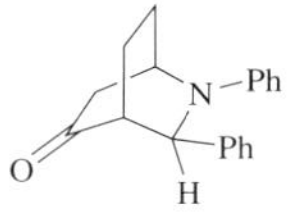 (45) + 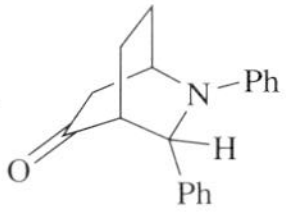 (20)	148
	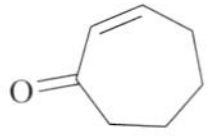	$InCl_3$, MeCN, rt, 24 h[h,hh]	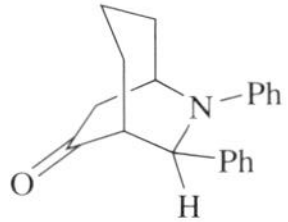 (29) + 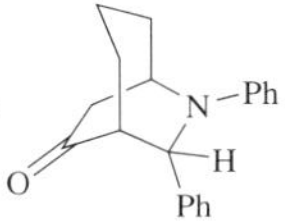 (27)	148

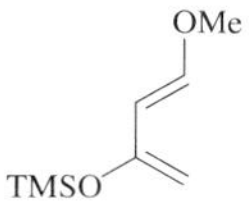
OMe
TMSO

1. $ZnCl_2$, THF, rt, 36-48 h
2. H_2O

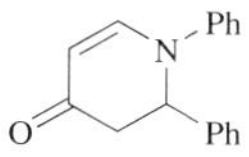
N
Ph
O
Ph

$(48)^{n}$, $(62)^{iii}$

141

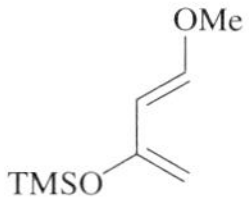
OMe
TMSO

Lewis acid, MeCN, 0°

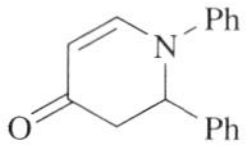
N
Ph
O
Ph

L. A.	
$Yb(OTf)_3$	(93)
$Sc(OTf)_3$	(99)

388

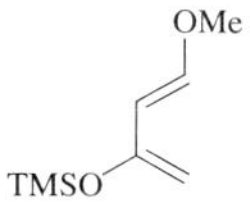
OMe
TMSO

1. Lewis acid, $MgSO_4$,
   MeCN
2. H_2O

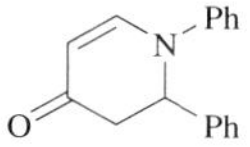
N
Ph
O
Ph

L. A.	Temp	Time	
$Yb(OTf)_3$	0°	20 h	$(80)^{jjj}$
BF_3•OEt_2	rt	—	(23)
$ZnCl_2$	rt	—	(12)
$Sc(OTf)_3$	rt	—	(83)

135, 330

TABLE 5. CYCLOADDITIONS OF *N*-ALKYL- AND *N*-ARYLIMINES (*Continued*)

Imine	Diene	Conditions	Product(s) and Yield(s) (%)	Refs.
C_{13} 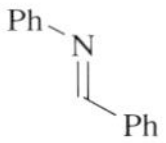	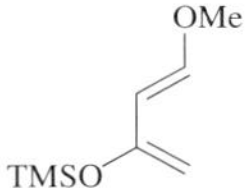	1. Lewis acid, rt 2. $NaHCO_3$, H_2O	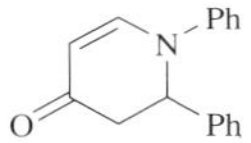	

L. A. (eq)	Solvent	Time	Yield (%)	Refs.
BF_3•OEt_2	CH_2Cl_2	15 min	(60)	155
$Yb(OTf)_3$ (0.1)	MeCN	20 h	(93)	135
$Sc(OTf)_3$ (0.1)	MeCN	20 h	(99)	135
none	MeCN	24 h	(<5)	136
$Sc(OTf)_3$ (0.1)	MeCN	20 h	(83)	136
$In(OTf)_3$ (0.1)	MeCN	30 min	(89)	136
$In(OTf)_3$ (0.005)	MeCN	30 min	(93)	136
$In(OTf)_3$ (0.005)	MeCN	—	(83)[kkk]	136
$BiCl_3$ (0.05)	THF	30 min	(96)	137
$BiCl_3$ (0.05)	MeCN	40 min	(90)	137
$Bi(OTf)_3$ (0.02)	THF	15 min	(95)	137
$Bi(OTf)_3$ (0.02)	MeCN	1.5 h	(91)	137
AgOTf (0.1)	THF, H_2O	16 h	(63)	386
$AgClO_4$ (0.1)	THF, H_2O	16 h	(61)	386
$AgNO_3$ (0.1)	THF, H_2O	16 h	(53)	386
$AgSbF_6$ (0.1)	THF, H_2O	16 h	(42)	386
$AgBF_4$ (0.1)	THF, H_2O	16 h	(35)	386

Imine	Diene	Conditions	Product(s) and Yield(s) (%)	Refs.
	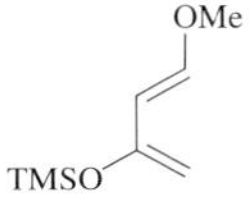	1. Lewis acid, THF, H_2O 2. $NaHCO_3$, H_2O	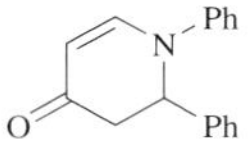	386

L. A.	Yield (%)
$Sc(OTf)_3$	(58)
$InCl_3$	(52)
$BiCl_3$	(41)
$CuCl_2$	(50)

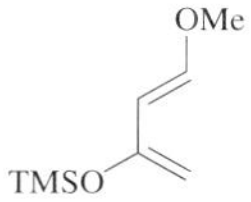

1. Lewis acid, H_2O[lll],
   CH_2Cl_2, rt
2. $NaHCO_3$, H_2O

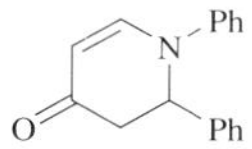

L. A. (eq)	Time		
none	24 h	(<5)	410, 411
$Ag(PPh_3)(CB_{11}H_6Br_6)$ (0.01)	2 h	(>99)[mmm]	411
$Ag(PPh_3)(CB_{11}H_6Br_6)$ (0.01)	1 h	(99)	410
$Ag(PPh_3)(CB_{11}H_6Br_6)$ (0.001)	30 min	(99)	410
$Ag(PPh_3)(CB_{11}H_6Br_6)$ (0.01)	15 min	(70)[kkk]	410, 411
$Ag(PPh_3)_2(CB_{11}H_6Br_6)$ (0.01)	1 h	(85)	410
$Ag(PPh_3)(CB_{11}H_{12})$ (0.01)	2 h	(>99)[mmm]	411
$Ag(PPh_3)(CB_{11}H_{12})$ (0.01)	1 h	(98)	410
$Ag(PPh_3)_2(CB_{11}H_{12})$ (0.01)	1 h	(99)	410
$Ag(PPh_3)(OTf)$ (0.01)	2 h	(70)[mmm]	411
$Ag(PPh_3)(OTf)$ (0.01)	1 h	(70)	410
$Ag(PPh_3)(BF_4)$ (0.01)	1 h	(35)	410
$Ag(PPh_3)(ClO_4)$ (0.01)	1 h	(90)	410

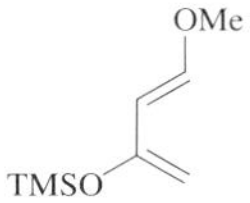

1. $In(OTf)_3$, $MgSO_4$,
   MeCN, rt
2. $NaHCO_3$, H_2O

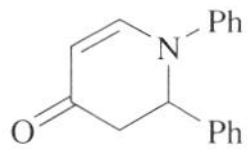

(51) 136

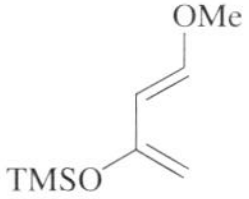

$Sc(OSO_2C_8F_{17})_3$, $scCO_2$,
100 atm, 40°, 1 h

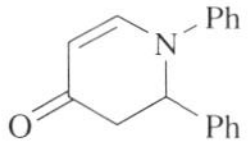

404

Diene (eq)	$Sc(OSO_2C_8F_{17})_3$	
1.5	5 mol %	(72)[a]
2.0	5 mol%	(>99)[a]
2.0	0.5 mol%	(81)[a]

TABLE 5. CYCLOADDITIONS OF *N*-ALKYL- AND *N*-ARYLIMINES (*Continued*)

Imine	Diene	Conditions	Product(s) and Yield(s) (%)	Refs.
C_{13} 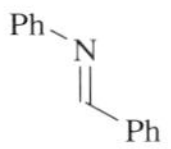	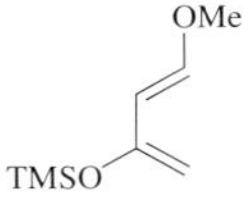	1. $SmI_2(THF)_2$, CH_2Cl_2, rt, 24 h 2. HCl, H_2O	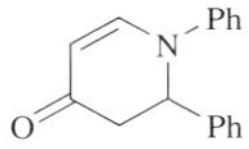 (61)	139
	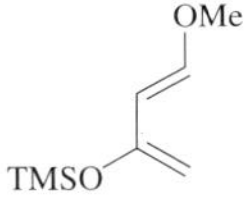	Brønsted acid, H_2O, –40°, 30 min	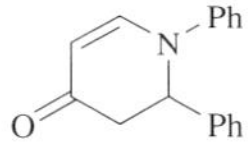	371
		B. A. / Solvent		
		HBF_4 / MeCN	(91)	
		TsOH / MeCN	(76)	
		TFA / MeCN	(89)	
		HBF_4 / MeOH	(98)[nnn]	
		HBF_4 / MeOH	(95)[ooo]	
	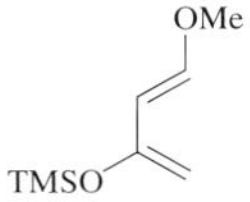	$Sc(SO_3CF_3)_3$, ionic liquid, rt	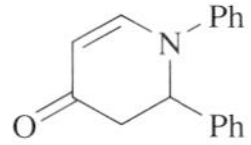	412
		Ionic Liquid / Time		
		8-Ethyl-1,8-diazabicyclo-[5.4.0]-7-undecenium trifluoromethanesulfonate / —	(75)[nnn]	
		1-Ethyl-3-methyl-1*H*-imidazolium trifluoromethanesulfonate / —	(67)[nnn]	
		8-Ethyl-1,8-diazabicyclo-[5.4.0]-7-undecenium trifluoromethanesulfonate / 20 h	(82)[ooo]	
		1-Ethyl-3-methyl-1*H*-imidazolium trifluoromethanesulfonate / 20 h	(80)[ooo]	
	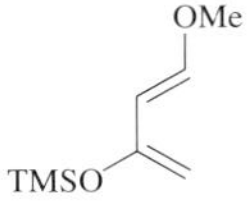	Nafion-H, 10 h	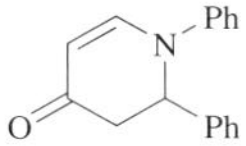 (72)	138

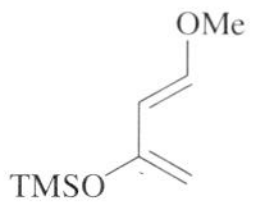

Catalyst, H_2O, rt, 1-2 h

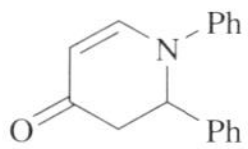

385

Catalyst	
none	(<5)
n-$C_{12}H_{25}OSO_3Na$	(68)
n-$C_{12}H_{25}SO_3Na$	(83)[ppp]
n-$C_8H_{17}SO_3Na$	(80)[ppp]
$PhSO_3Na$	(65)[ppp]
$EtSO_3Na$	(<5)
LiOTf	(89)
NaOTf	(87)
KOTf	(80)
$LiClO_4$	(90)
NaI	(87)[ppp]
n-Bu_4NI	(89)[ppp]
$NaBPh_4$	(90)
NaBr	(<5)
NaCl	(<5)
NaF	(<5)
n-$C_{16}H_{33}NMe_3Br$	(<5)
$PhCO_2Na$	(<5)
CF_3CO_2Na	(44)[ppp]
$MeCO_2Na$	(<5)
n-$C_{11}H_{25}CO_2Na$	(<5)

TABLE 5. CYCLOADDITIONS OF *N*-ALKYL- AND *N*-ARYLIMINES (*Continued*)

Imine	Diene	Conditions	Product(s) and Yield(s) (%)	Refs.
C_{13} 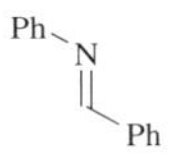	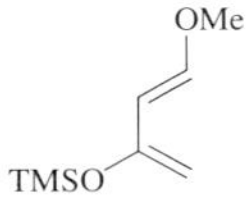	Catalyst, H_2O, rt, 2-3 h	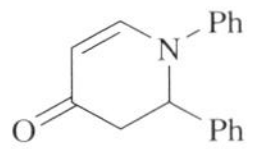	
		Catalyst		
		none	(trace)	386
		NaOTf	(87)[nnn]	385
		NaOTf	(80)[ooo,qqq]	385
		NaOTf	(80)[ooo,rrr]	385
		AgOTf	(83)[nnn]	386
		AgOTf	(63)[ooo,qqq]	386
		AgOTf	(trace)[ooo,rrr]	386
		AgOTf	(80)[ooo,sss]	386
	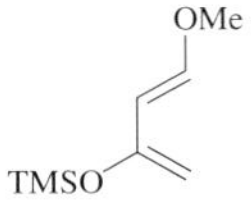	1. rt, 2 h 2. HCl, H_2O	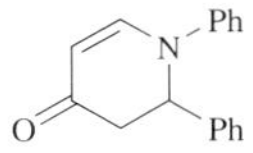	373
		Solvent		
		MeOH	(>99)[nnn]	
		MeOH	(>99)[ooo]	
		MeOH	(100)[nnn,ttt]	
		MeOH/ 2,6-di-*tert*-butyl-4-methylpyridine	(100)[nnn]	
		MeCN	(>99)	
		EtOH	(32)	
		DMF	(54)	
		THF	(47)	
		toluene	(<5)	
		CH_2Cl_2	(<5)	
		$CHCl_3$	(<5)	

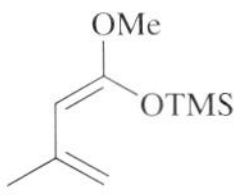	1. $TiCl_4$, CH_2Cl_2, −100° to rt, 3-4 h 2. H_2O	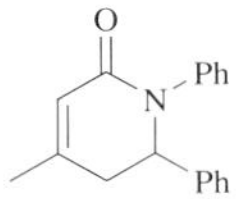 (34)[mm]	134
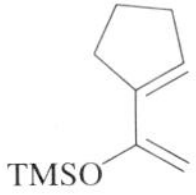	$AlCl_3$, CH_2Cl_2 Temp / Time −40° / 15 min rt / 2 h	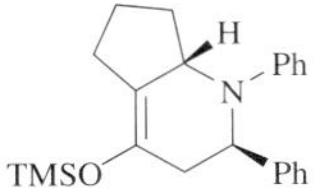 + 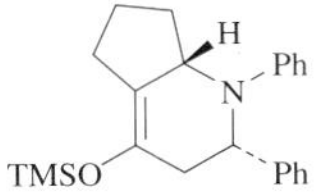 (75) 87:13 (75) 2:98	154
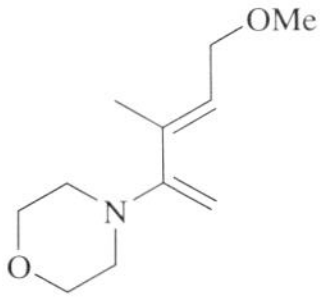	$ZnCl_2$, THF, 0° to rt, 12 h	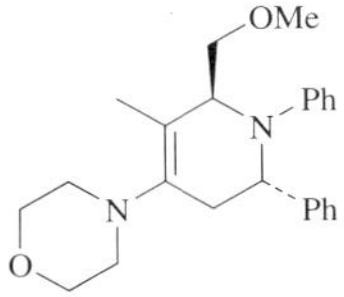 (48-52)	413 157, 414
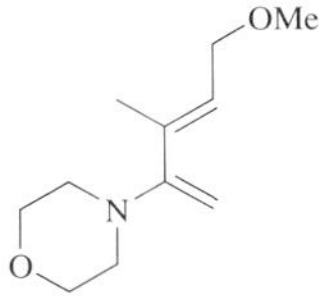	1. $ZnCl_2$, THF, ether, rt, 10 h 2. SiO_2	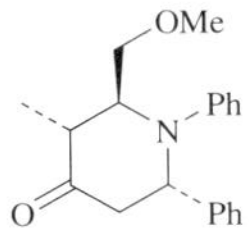 (62) + 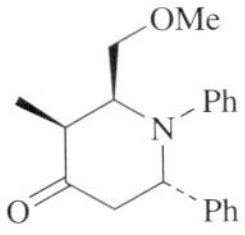 (26)	157
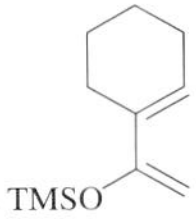	1. Lewis acid, CH_2Cl_2 2. $NaHCO_3$, Et_3N, H_2O L. A. (eq) / Temp / Time BF_3•OEt_2 (1.0) / −78° / 5 min $ZrCl_4$ (1.0) / 20° / 1 h	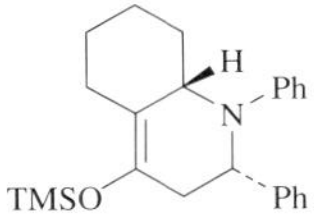 + 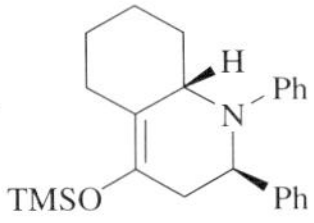 (—) 67:33 (—) 40:60	415

TABLE 5. CYCLOADDITIONS OF *N*-ALKYL- AND *N*-ARYLIMINES (*Continued*)

Imine	Diene	Conditions	Product(s) and Yield(s) (%)	Refs.
C_{13} 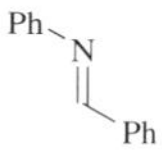	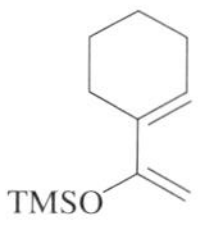	1. Lewis acid, CH_2Cl_2 2. Basic hydrolysis 3. Et_3N, MeOH	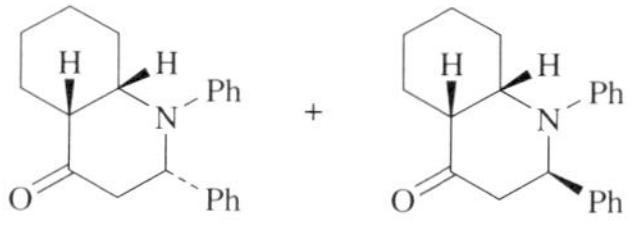 I + II	415
		L. A. (eq) / Temp / Time	I + II / I:II	
		BF_3•OEt_2 (1.0) / 0° / 5 min	(—) / 55:45[uuuu]	
		BF_3•OEt_2 (0.1) / 20° / 1.5 h	(72) / 70:30[uuuu]	
		$AlCl_3$ (1.0) / −10° / 3.5 h	(—) / 70:30	
		$AlCl_3$ (1.0) / 20° / 5 min	(—) / 55:45	
		$AlCl_3$ (1.0) / 20° / 1.5 h	(60) / 10:90	
		Et_2AlCl (1.0) / 20° / 1.5 h	(41) / 70:30	
		$ZrCl_4$ (1.0) / −40° / 5 min	(—) / 70:30	
		$ZrCl_4$ (1.0) / 20° / 4.5 h	(50) / 1:99	
		$TiCl_4$ (1.0) / 20° / 5 min	(—) / 2:98	
	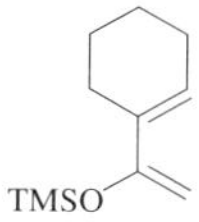	1. Lewis acid, CH_2Cl_2 2. $NaHCO_3$, H_2O	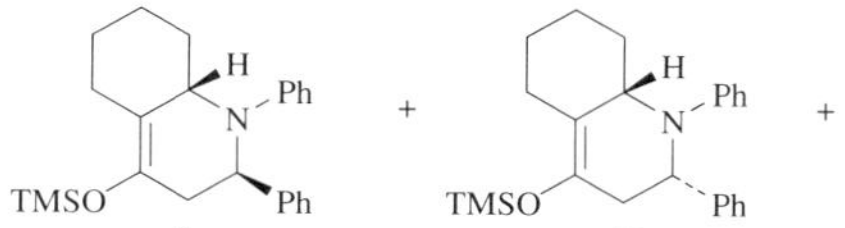 I + II + 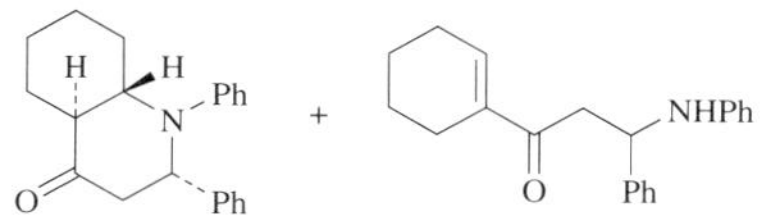 III + IV	

L. A. (eq)	Temp	Time		
$TiCl_4$ (1.0)	—	40 min	**I** + **II** + **III** (—), **I**:**II**:**III** 73:2:25	415
$TiCl_4$ (3.0)	—	1 h	**III** (—)	415
$ZrCl_4$ (3.0)	—	160 min	**III** (—)	415
BF_3•OEt_2 (0.1)	20°	1.5 h	**I** + **II** (—), **I**:**II** 70:30	155
BF_3•OEt_2 (1.0)	–78°	5 min	**I** + **II** (—), **I**:**II** 67:33	155
BF_3•OEt_2 (1.0)	0°	5 min	**I** + **II** (—), **I**:**II** 55:45	155
$AlCl_3$ (1.0)	–10°	3.5 h	**I** + **II** (—), **I**:**II** 70:30	155
$AlCl_3$ (1.0)	20°	5 min	**I** + **II** (—), **I**:**II** 55:45	155
$AlCl_3$ (0.1)	20°	3.5 h	**I** + **II** (—), **I**:**II** 55:45	155
$AlCl_3$ (1.0)	20°	1.5 h	**I** + **II** (—), **I**:**II** 10:90	155
Et_2AlCl (1.0)	20°	1.5 h	**I** + **II** (—), **I**:**II** 70:30	155
$ZrCl_4$ (1.0)	–40°	5 min	**I** + **II** (—), **I**:**II** 70:30	155
$ZrCl_4$ (1.0)	20°	1 h	**I** + **II** (—), **I**:**II** 40:60	155
$ZrCl_4$ (1.0)	20°	4.5 h	**I** + **II** (—), **I**:**II** 1:99	155
$TiCl_4$ (1.0)	20°	5 min	**I** + **II** (—), **I**:**II** <2:98	155
$AlCl_3$ (1.0)	–40°	15 min	**I** + **II** (42)[a], **I**:**II** 84:16	156
$AlCl_3$ (1.0)	20°	1.5 h	**I** + **II** (80)[a], **I**:**II** 5:95	156
TBSOTf (0.1)	20°	1.5 h	**I** + **II** + **IV** (86)[a], **I**:**II**:**IV** 54:28:17	156
TBSOTf (0.1)	20°	15 h	**I** + **II** + **IV** (75)[a], **I**:**II**:**IV** 32:27:41	156

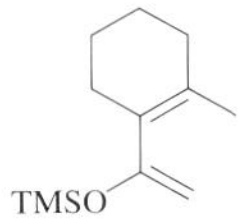
TMSO

1. $TiCl_4$, CH_2Cl_2, 20°, 1 h
2. $NaHCO_3$, H_2O

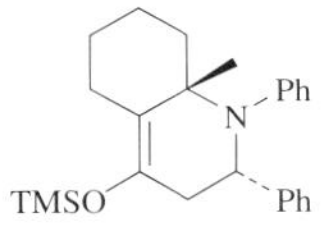
TMSO
N
Ph
Ph

(58) 416

TABLE 5. CYCLOADDITIONS OF *N*-ALKYL- AND *N*-ARYLIMINES (*Continued*)

Imine	Diene	Conditions	Product(s) and Yield(s) (%)	Refs.
C_{13}				
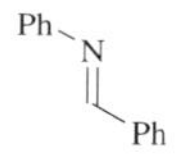	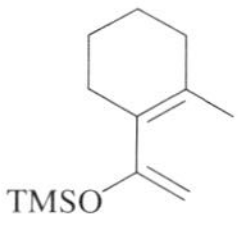	1. Lewis acid, CH_2Cl_2, 20° 2. $NaHCO_3$, H_2O 3. Et_3N, MeOH, overnight	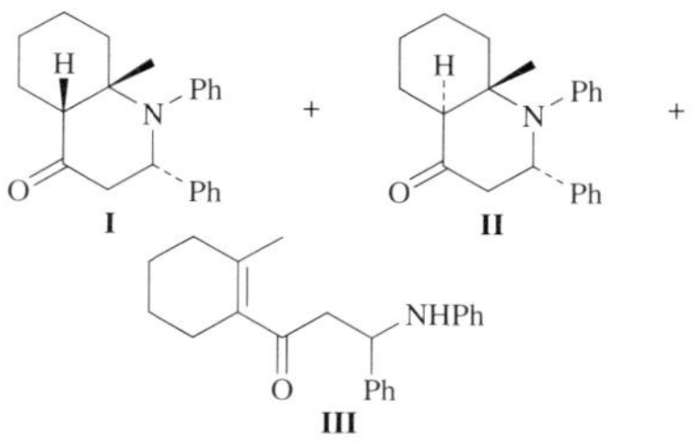 **I** + **II** + **III**	
		L.A. / Time		
		$TiCl_4$ 1 h	**I** + **II** (74), **I**:**II** 1:1.5	416, 417
		$SnCl_4$ 1 h	**I** + **II** (—), **I** + **II**:recovered imine 60:40	416, 417
		$SnCl_4$ 20 h	**I** + **II** (65)	416, 417
		ZnI_2 2 h	**I** + **II** + **III** (—), **I** + **II**:**III**:recovered imine 80:6:12	416, 417
		ZnI_2 7 h	**I** + **II** + **III** (—), **I** + **II**:**III**:recovered imine 82:12:16	416, 417
		Et_2AlCl 20 h	**I** + **II** (—), **I** + **II**:recovered imine 71:29	416
	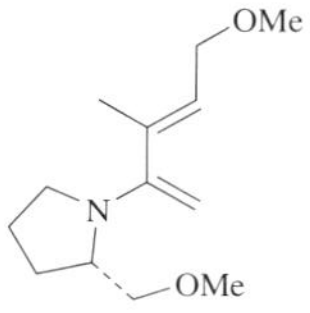	1. $ZnCl_2$, THF, ether, −80° to rt, 16 h 2. $NaHCO_3$, H_2O	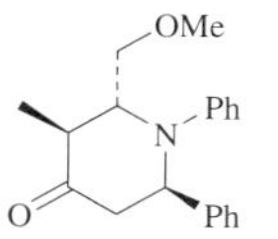 (30) *ee* 35%[vvv] + 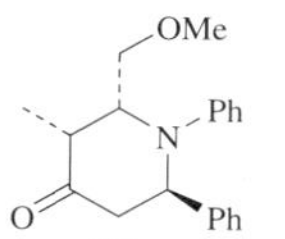 (15)	177

		1. $AlCl_3$ or $BF_3{\bullet}OEt_2$ 2. $NaHCO_3$, H_2O	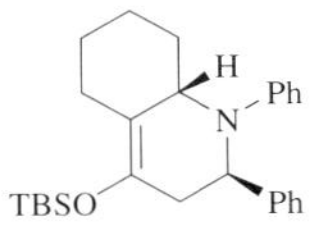 I + 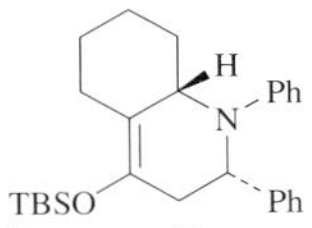 II (—)[www]	155
	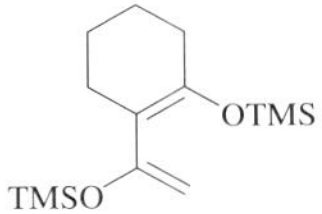	1. $ZnCl_2$, THF, rt, 20 h 2. H_2O	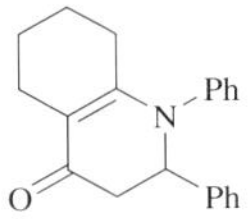 (93)	278
	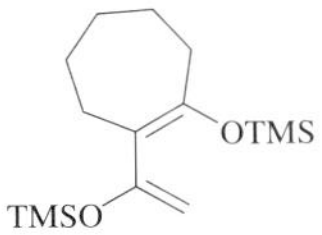	1. $ZnCl_2$, THF, rt, 20 h 2. H_2O	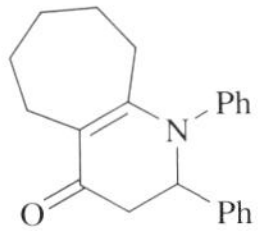 (69)	278
	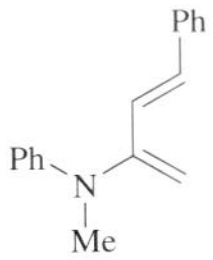	1. $Cu(OTf)_2$, THF, –20°, 24 h 2. $NaHCO_3$, H_2O	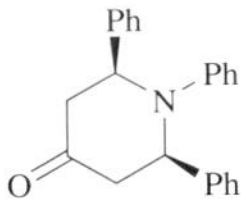 + 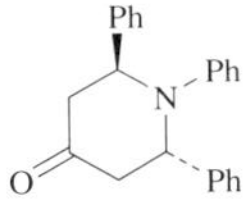 (20) 98:2	376
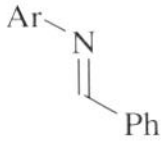 Ar: 4-$O_2NC_6H_4$; 4-ClC_6H_4	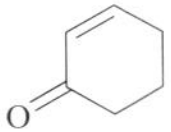	$InCl_3$, MeCN, rt, 24 h[hhh]	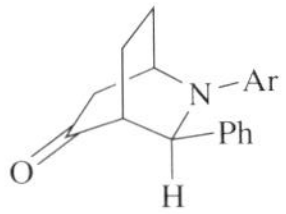 (47), (49) + 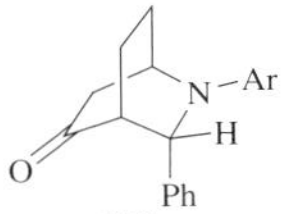 (23), (19)	148
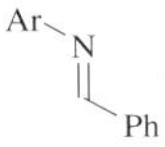 Ar = 4-$O_2NC_6H_4$	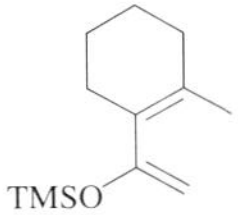	1. ZnI_2, CH_2Cl_2, 20°, 22 h 2. $NaHCO_3$, H_2O	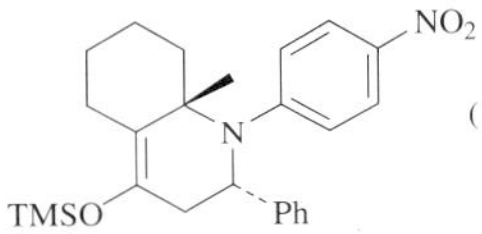 (13)	416

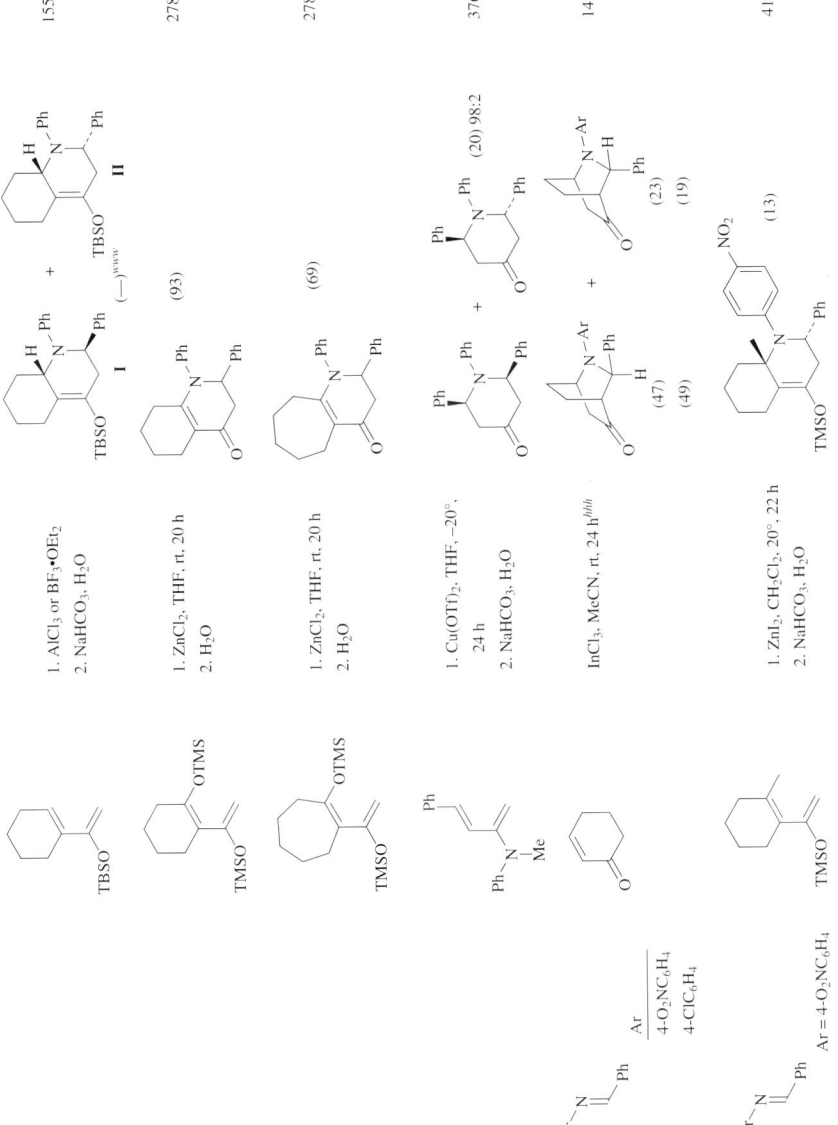

TABLE 5. CYCLOADDITIONS OF *N*-ALKYL- AND *N*-ARYLIMINES (*Continued*)

Imine	Diene	Conditions	Product(s) and Yield(s) (%)	Refs.
C_{13}				
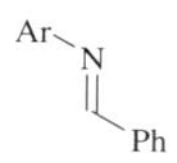 Ar = 4-$O_2NC_6H_4$	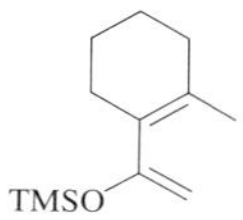	1. ZnI_2, CH_2Cl_2, 20°, 22 h 2. $NaHCO_3$, H_2O 3. TEA, MeOH, overnight	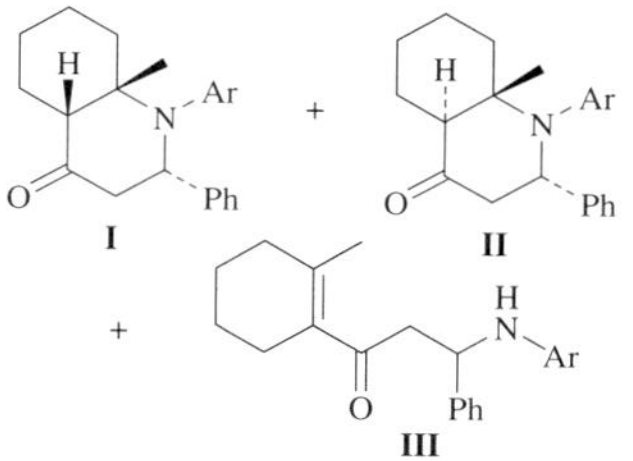 I + II + III (91), I + II:III 70:30 I + II (62), III (29), I:II 1:1[xxx]	416, 417 416
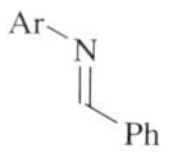	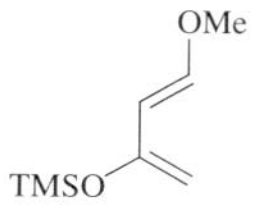	$Sc(SO_3CF_3)_3$, ionic liquid, rt	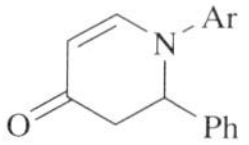	412

Ar	Ionic Liquid	Time	
4-FC_6H_4	8-Ethyl-1,8-diazabicyclo[5.4.0]-7-undecenium trifluoromethanesulfonate	20 h	(75-95)[yyy]
4-FC_6H_4	1-Ethyl-3-methyl-1*H*-imidazolium trifluoromethanesulfonate	20 h	(79)[yyy]
3,4-$F_2C_6H_3$	8-Ethyl-1,8-diazabicyclo[5.4.0]-7-undecenium trifluoromethanesulfonate	—	(74)[nnn]
3,4-$F_2C_6H_3$	8-Ethyl-1,8-diazabicyclo[5.4.0]-7-undecenium trifluoromethanesulfonate	20 h	(95)[ooo,yyy]
3,4-$F_2C_6H_3$	1-Ethyl-3-methyl-1*H*-imidazolium trifluoromethanesulfonate	20 h	(99)[ooo,yyy]

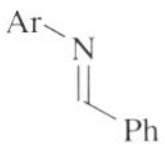 Ar = 4-ClC_6H_4		$Sc(OTf)_3$, MeCN, rt, 20 h	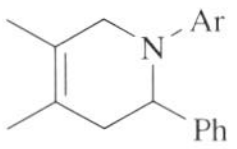 (37)[ggg]	135
	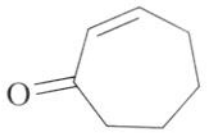	$InCl_3$, MeCN, rt, 20 h[hhh]	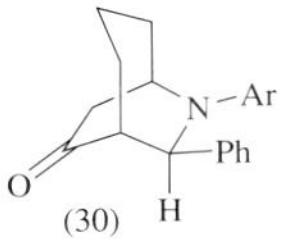 (30) + 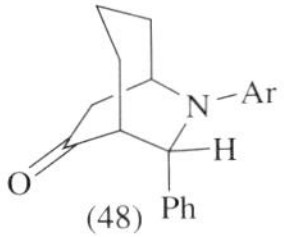 (48)	148
	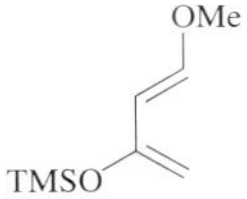 **I**	1. Lewis acid, MeCN, rt 2. $NaHCO_3$, H_2O	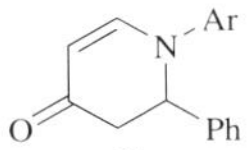 **II**	
		L. A. — Time		
		$Yb(OTf)_3$ — 20 h	(92)	135
		$BiCl_3$ — 30 min	(95)	137
		$Bi(OTf)_3$ — 1.25 h	(90)	137
	I	$Sc(OSO_2C_8F_{17})_3$, $scCO_2$, 100 atm, 40°, 1 h	**II** (73)[a]	404
	I	1. MeOH, rt, 2 h 2. HCl, H_2O	**II** (96)	373
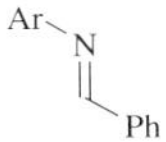 Ar = 4-BrC_6H_4	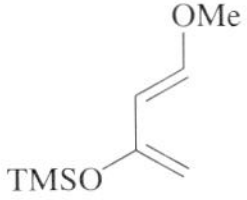	Catalyst, H_2O, rt, 2-3 h	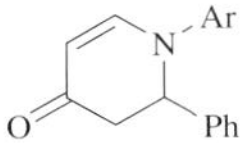	
		Catalyst		
		AgOTf	(57)[nnn,zzz]	386
		AgOTf	(83)[nnn,aaaa]	386
		AgOTf	(90)[ooo,sss]	386
		NaOTf	(96)[nnn]	385
		NaOTf	(83)[ooo]	385

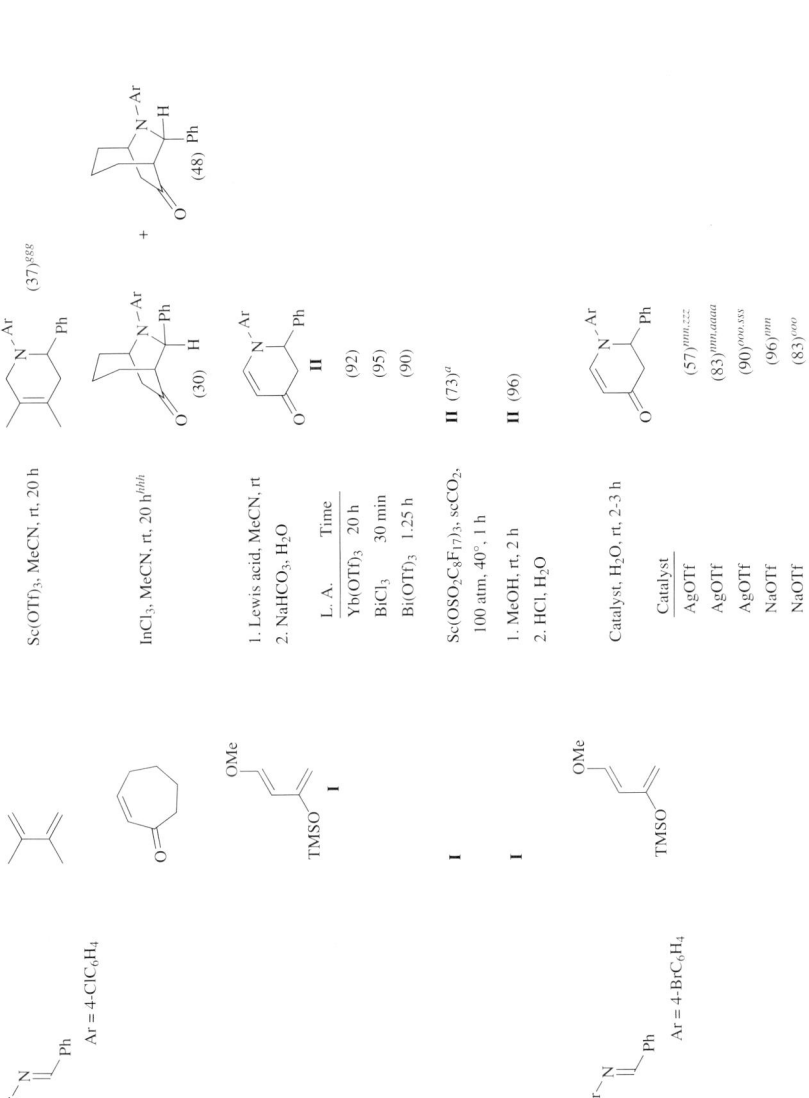

TABLE 5. CYCLOADDITIONS OF *N*-ALKYL- AND *N*-ARYLIMINES (*Continued*)

Imine	Diene	Conditions	Product(s) and Yield(s) (%)	Refs.
C_{13}				
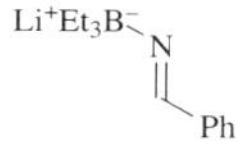	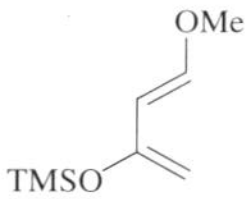	1. THF, rt, 4-5 h 2. $NaHCO_3$, H_2O	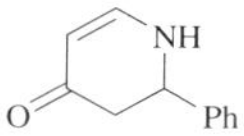 (90)	389
	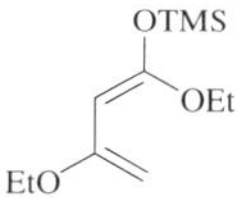	1. THF, rt, 18 h 2. $NaHCO_3$, H_2O	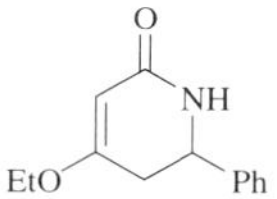 (25)	389
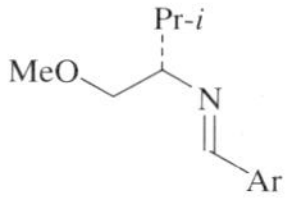 Ar = 4-$O_2NC_6H_4$	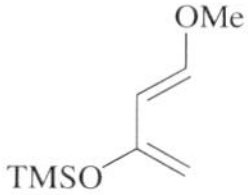	BF_3•OEt_2, CH_2Cl_2, rt	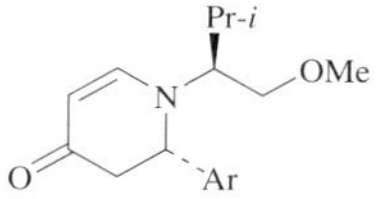 + 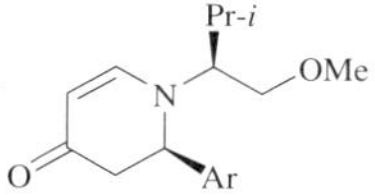 (57) 77:23	170
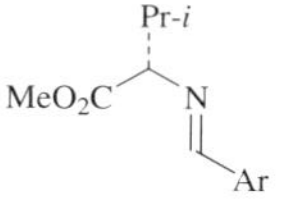 Ar = 4-$O_2NC_6H_4$	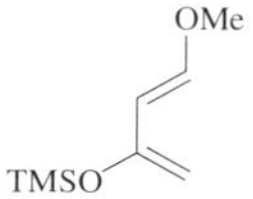		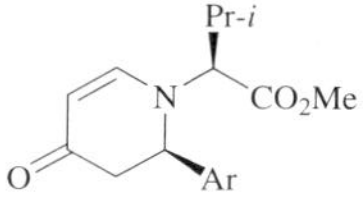 + 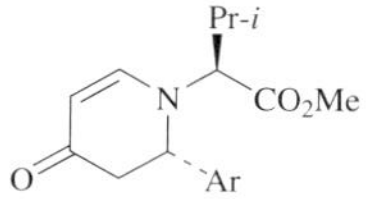	
		1. $ZnCl_2$[jj], THF, 0° 2. H^+, H_2O	(65) 94:6	175a
		1. $ZnCl_2$, THF, 0°, 7 h 2. $NaHCO_3$, H_2O	(65) 94:6	384
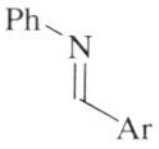 Ar = 4-$O_2NC_6H_4$	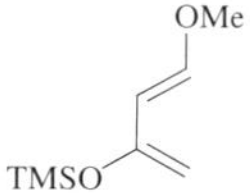		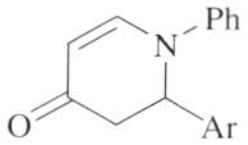	
		HBF_4, H_2O, MeOH, –40°, 30 min	(87)[nnn], (65)[ooo]	372
		1. MeOH, rt, 2 h 2. HCl, H_2O	(63)[nnn], (67)[ooo]	373
		AgOTf, H_2O, rt, 2-3 h	(69)	386
		NaOTf, H_2O, rt, 2-3 h	(94)	385

	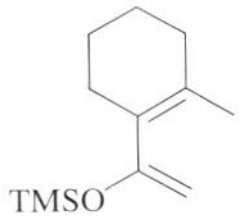	1. ZnI_2, CH_2Cl_2, 20°, 19 h 2. $NaHCO_3$, H_2O	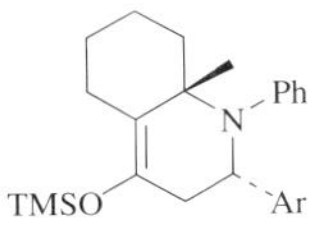	(9)	416
	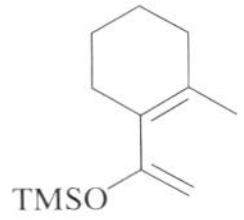	1. $TiCl_4$, CH_2Cl_2, 20°, 1 h 2. $NaHCO_3$, H_2O 3. Et_3N, MeOH	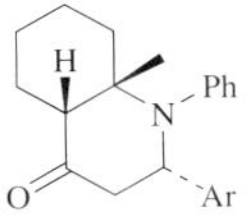 (50) + 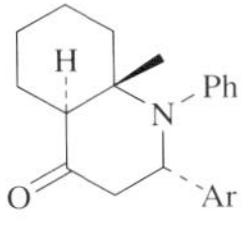		416
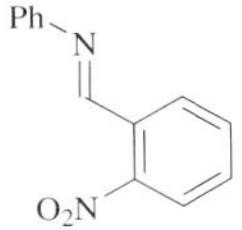	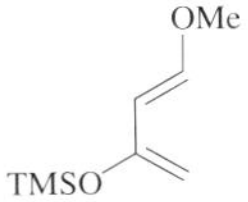	Nafion-H, 15 h	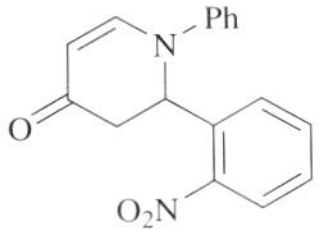	(76)	138
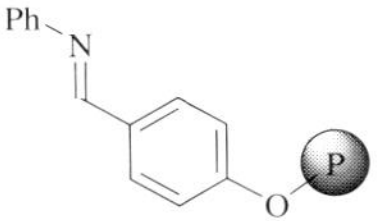	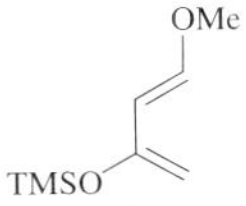	1. $Yb(OTf)_3$, THF, 60°, 3 h 2. H_2O 3. TFA, CH_2Cl_2	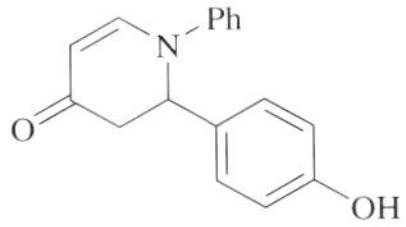	(60)[hh]	378
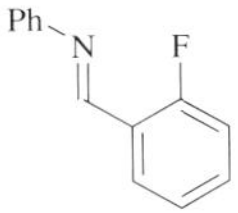	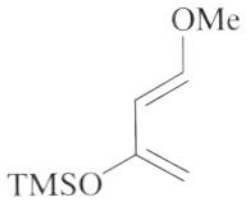	1. $In(OTf)_3$, $MgSO_4$, MeCN, rt 2. $NaHCO_3$, H_2O	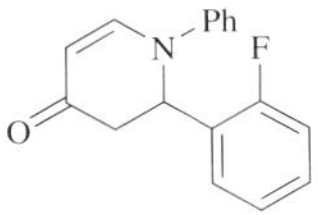	(84)	136
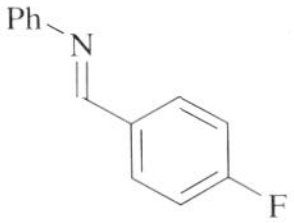	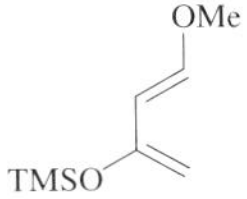	$Sc(SO_3CF_3)_3$, ionic liquid, rt	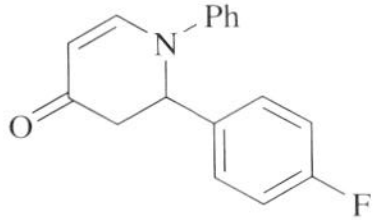		412

Ionic Liquid	
8-Ethyl-1,8-diazabicyclo-[5.4.0]-7-undecenium trifluoromethanesulfonate	(85)[yyy]
1-Ethyl-3-methyl-1*H*-imidazolium trifluoromethanesulfonate	(95)[yyy]

TABLE 5. CYCLOADDITIONS OF *N*-ALKYL- AND *N*-ARYLIMINES (*Continued*)

Imine	Diene	Conditions	Product(s) and Yield(s) (%)	Refs.
C_{13} 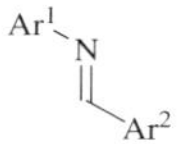	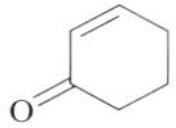	$InCl_3$, MeCN, rt, 24 h[hhh]	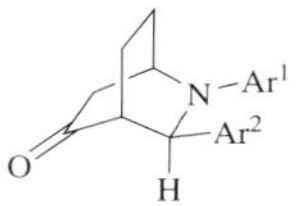 + 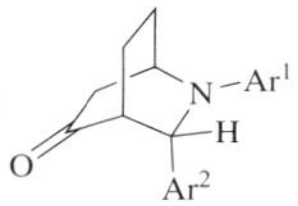	148
Ar¹ / Ar²: Ph / 4-ClC_6H_4			(31) + (34)	
Ar¹ / Ar²: 4-ClC_6H_4 / 4-ClC_6H_4			(34) + (40)	
	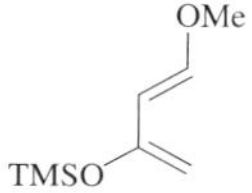	1. MeOH, rt, 2 h 2. HCl, H_2O	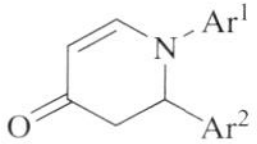 (89)[nnn], (80)[ooo]	373
	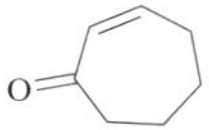	$InCl_3$, MeCN, rt, 21 h[hhh]	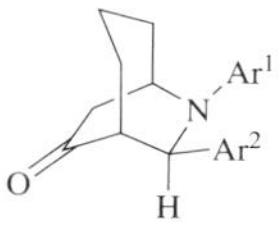 (29) + 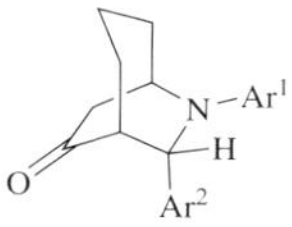 (39)	148
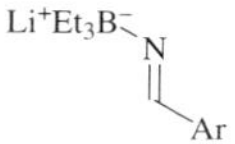	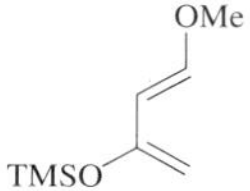	1. THF, rt, 4-5 h 2. $NaHCO_3$, H_2O	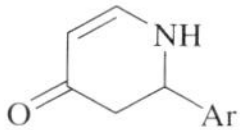	389
Ar = 3-ClC_6H_4			(95)	
Ar = 4-ClC_6H_4			(81)	
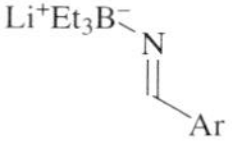 Ar = 3-ClC_6H_4	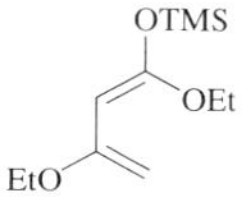	1. THF, rt, 18 h 2. $NaHCO_3$, H_2O	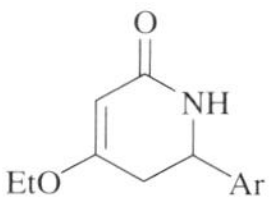 (25)	389
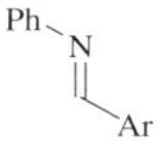 Ar = 4-BrC_6H_4	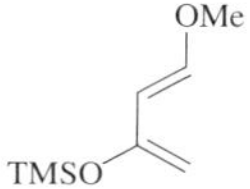	Catalyst, H_2O, rt, 2-3 h	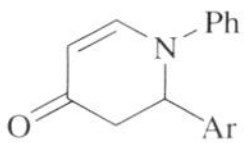	
		Catalyst: AgOTf	(75)[zzz]	386
		Catalyst: AgOTf	(87)[aaaa]	386
		Catalyst: NaOTf	(92)	385

	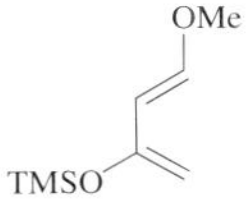	Nafion-H, 13 h	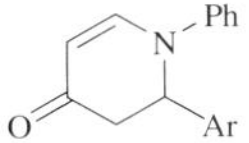 (74)	138
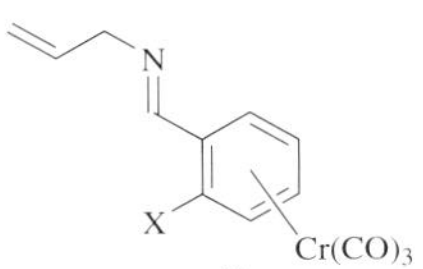	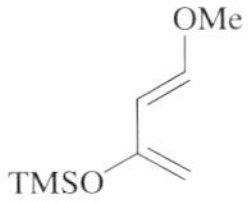	$SnCl_4$,THF	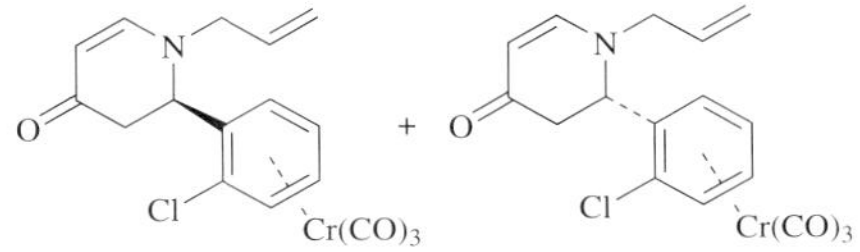	
X: Cl		Temp: −78° to rt; Time: 15 h	(91) >98:2[bbbb]	158
X: Cl		Temp: −78° to rt; Time: —	(95) >98:2	168
X: I		Temp: −78 to 0°; Time: 18 h	(90) >98:2	168
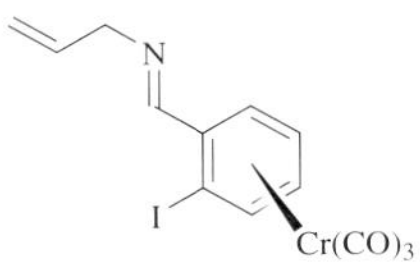	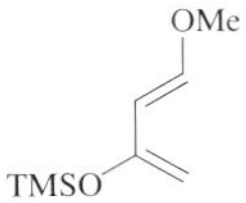	$SnCl_4$,THF, −78° to rt	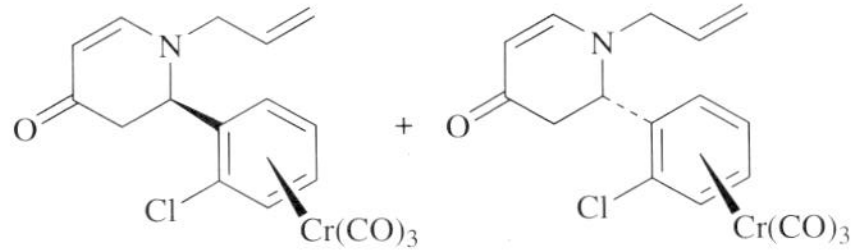 (84) >98:2	168
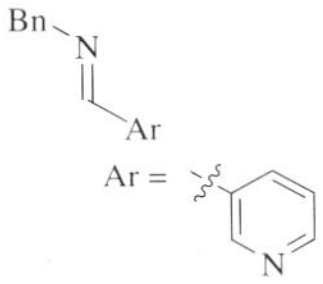			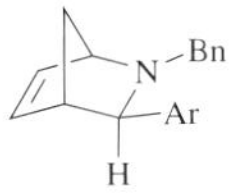	418
		HCl, Yb(OTf)$_3$•3H_2O,)))), H_2O	(5)[cccc,dddd]	
		HCl, $InCl_3$•3H_2O,)))), H_2O	(8)[cccc,dddd]	
		HCl, $AlCl_3$,)))), CH_2Cl_2	(18)[cccc,dddd]	
		HCl, $AlCl_3$, Et_3N,)))), CH_2Cl_2, 0°	(25)[ooo,cccc,dddd]	
		$AlCl_3$, Et_3N,)))), CH_2Cl_2, 0-5°, 2-3 d	(41)[nnn,cccc,dddd,eeee]	

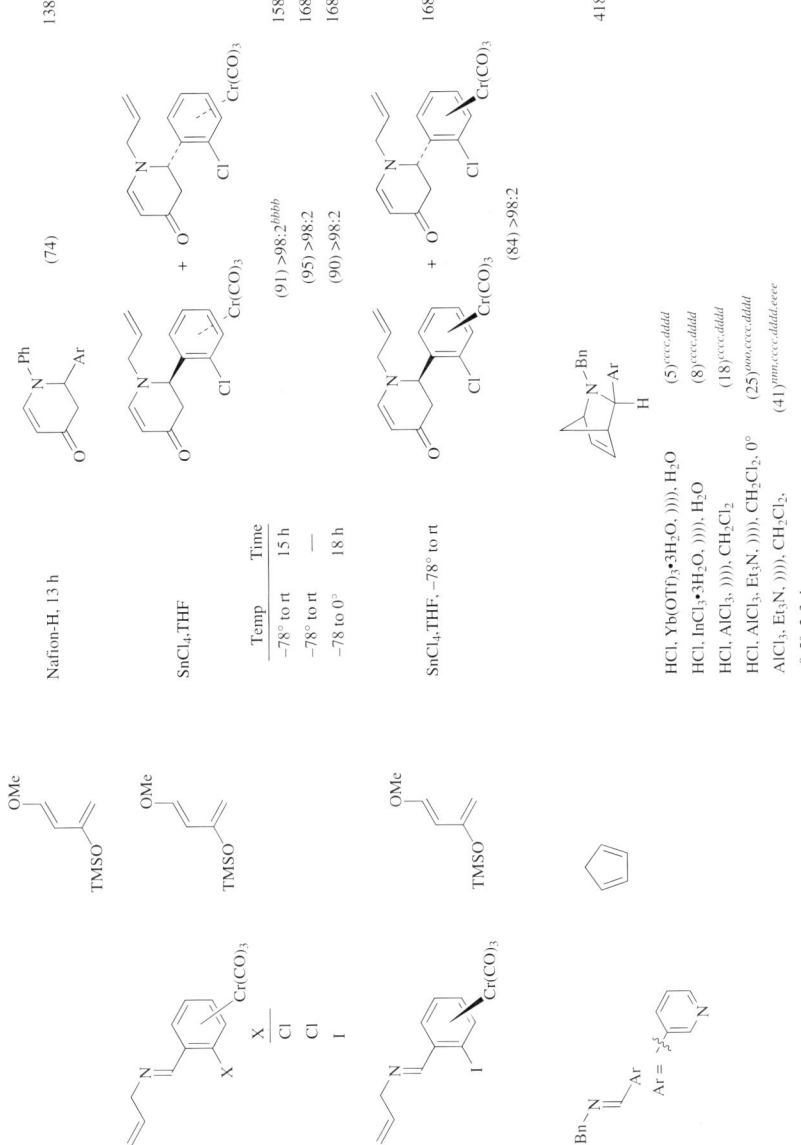

TABLE 5. CYCLOADDITIONS OF *N*-ALKYL- AND *N*-ARYLIMINES (*Continued*)

Imine	Diene	Conditions	Product(s) and Yield(s) (%)	Refs.
C_{13}				
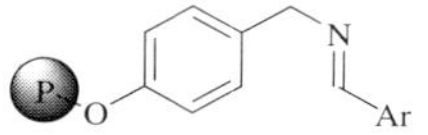 Ar = 3-pyridinyl	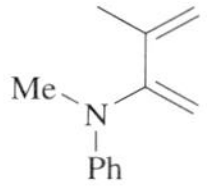	1. $Yb(OTf)_3$, THF, rt, 12 h 2. TFA, CH_2Cl_2, 25 min 3. TMSOTf, CH_2Cl_2, rt, 3 h	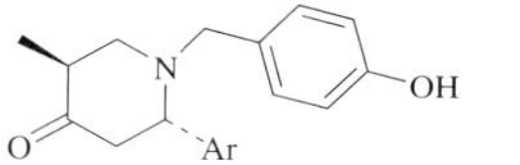 (71)	377
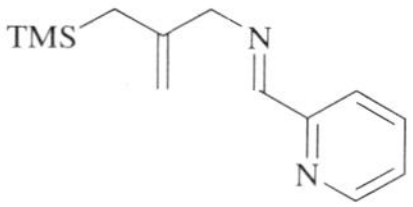	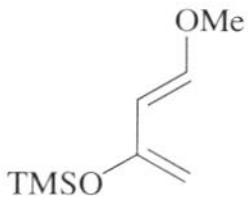	1. $Yb(OTf)_3$, MeCN, rt, 10-15 h 2. H_2O	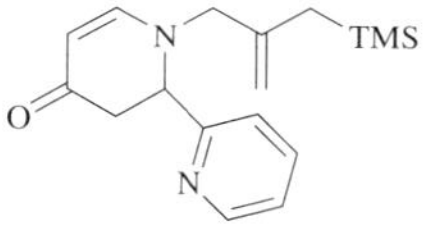 (80)	382
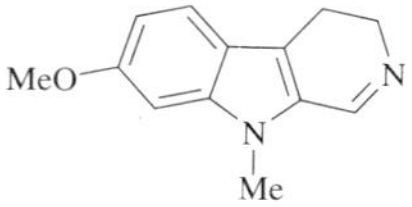		Chlorobenzene, 120°, 4 h	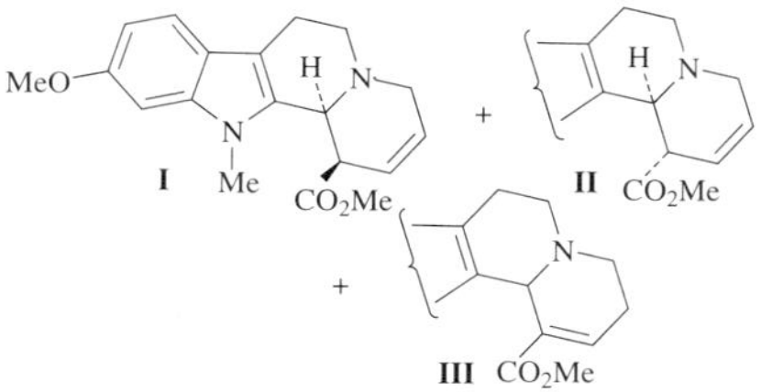 **I + II + III** (87.5), **I**:**II**:**III** 13.3:66:13.3	406
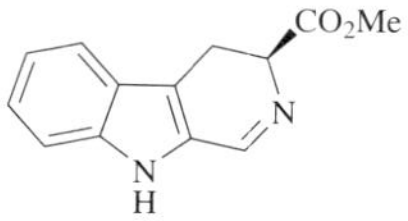	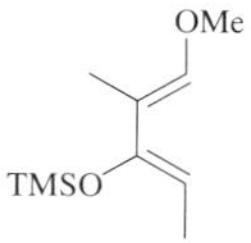	$ZnCl_2$, MeCN, rt	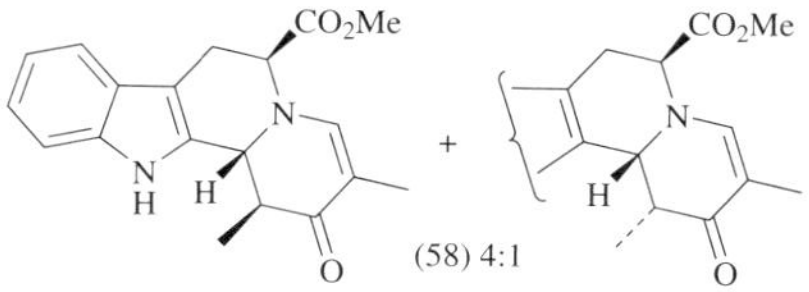 (58) 4:1	399
	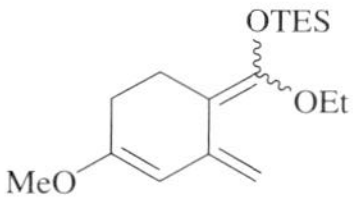	$CHCl_3$, rt	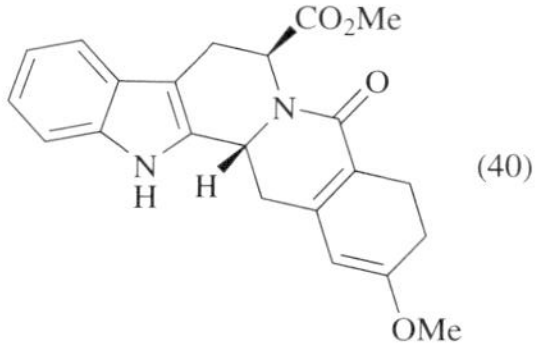 (40)	399

C$_{14}$

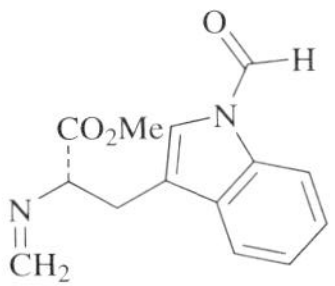	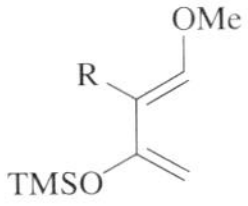		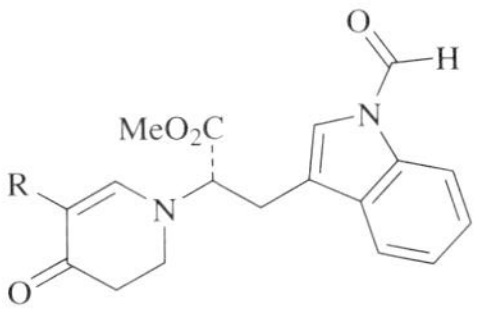	
	R			
	H	$ZnCl_2$	("low")	246
	H	$LiClO_4$, THF, H_2O, rt, 3 d	(45)[ffff]	246, 245
	Et	$ZnCl_2$	("low")	246
	Et	$LiClO_4$, THF, H_2O, rt, 3 d	(62)[ffff]	246, 245
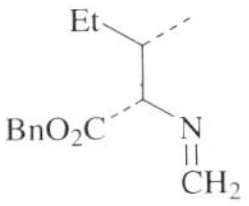		HCl, THF, H_2O	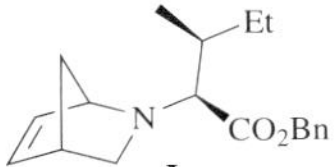 **I** + 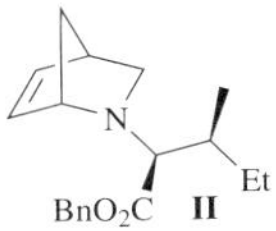 **II** (—)[ii]	153
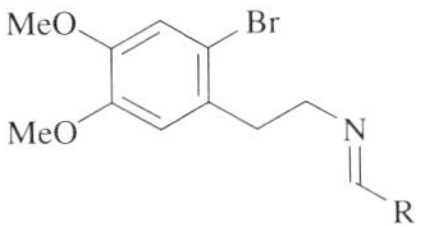	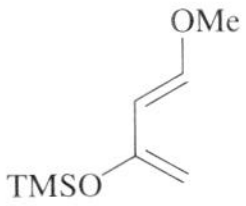	1. $EtAlCl_2$, CH_2Cl_2, −78° to rt, 2. $NaHCO_3$, H_2O	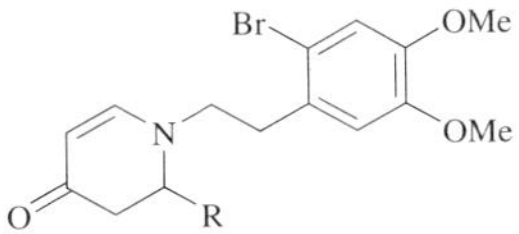	380, 379
R		Time		
n-Pr		20 h	(43)	
i-Pr		3 h	(65)	
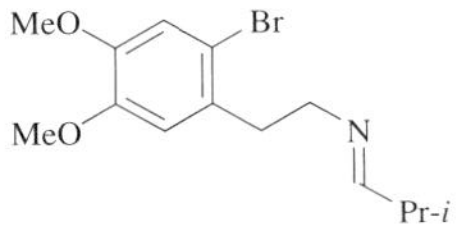	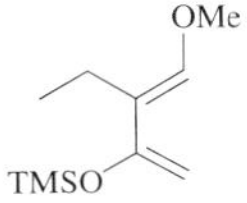	$ZnCl_2$, THF	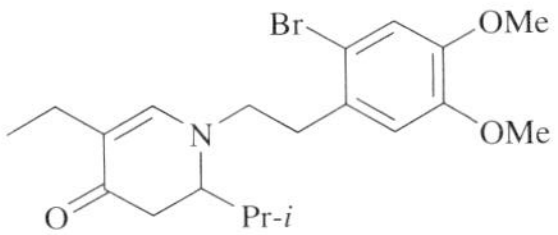 (27)	379

TABLE 5. CYCLOADDITIONS OF *N*-ALKYL- AND *N*-ARYLIMINES (*Continued*)

Imine	Diene	Conditions	Product(s) and Yield(s) (%)	Refs.
C_{14}				
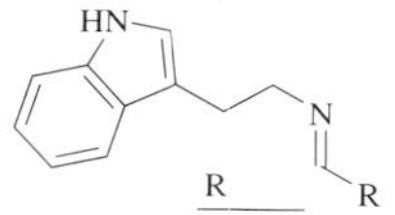 R: *n*-Pr; *i*-Pr	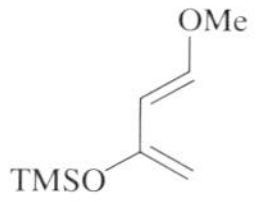	1. $ZnCl_2$[gggg], THF, rt, overnight 2. $NaHCO_3$, H_2O	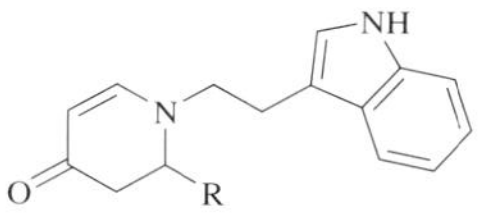 (49) (41)	419
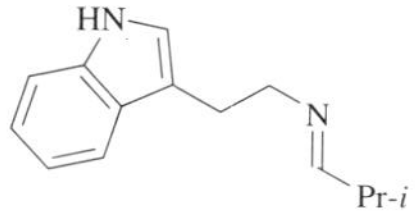	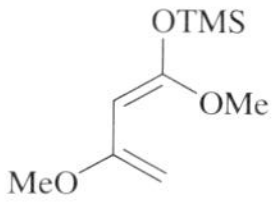	1. $EtAlCl_2$, THF, –60° to rt, 2 h 2. $NaHCO_3$, H_2O	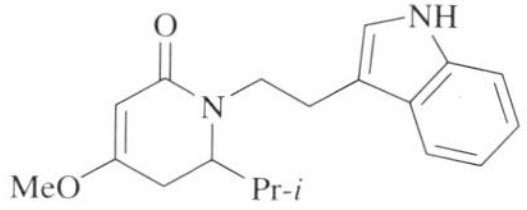 (35)[hhhh]	420, 421
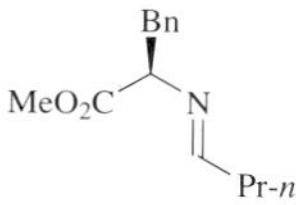	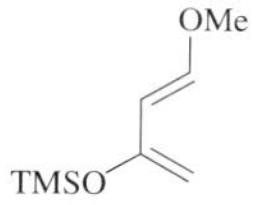	1. $ZnCl_2$, THF, –10°, 7 h 2. $NaHCO_3$, H_2O	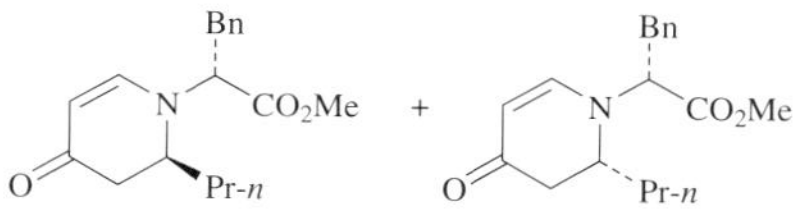 (65) 66:34	384
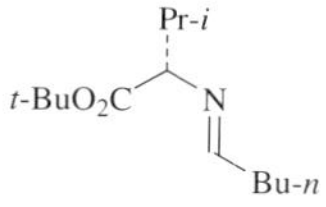	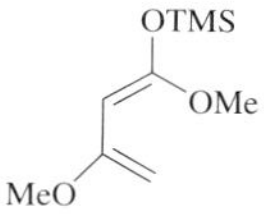	1. $EtAlCl_2$[bbb], –78 to 20°, CH_2Cl_2 2. $NaHCO_3$, H_2O	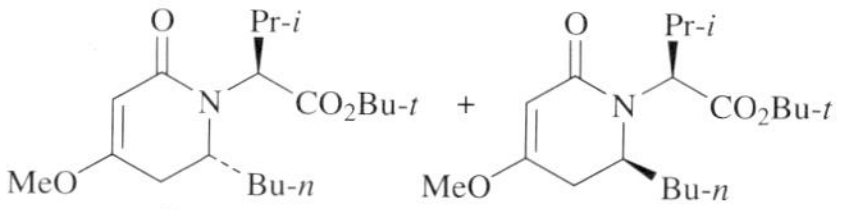 **I** + **II** 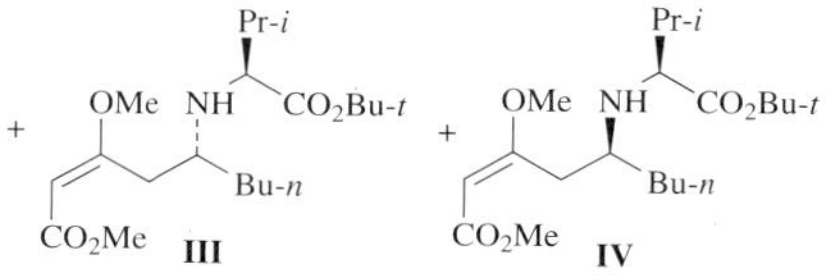 + **III** + **IV** **I** + **II** + **III** + **IV** (57), **I** + **II**:**III** + **IV** 44:56, **I** + **III**:**II** + **IV** 93:7	175a

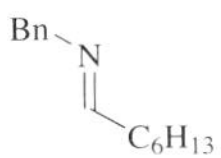

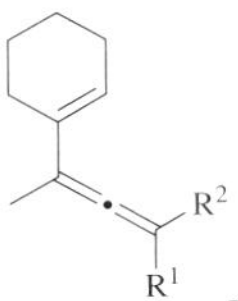

R^1	R^2
Me	Me
Ph	H

$BF_3•OEt_2$, CH_2Cl_2, 40°, 5 d

Temp	Time
40°	5 d
rt	3 d

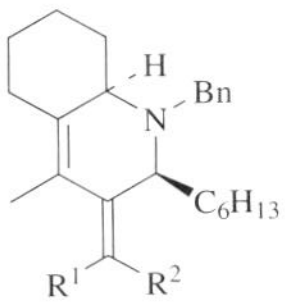

(41)[nn]
(57)[nn]

143

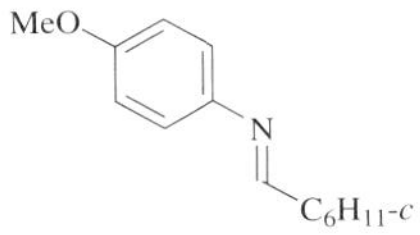

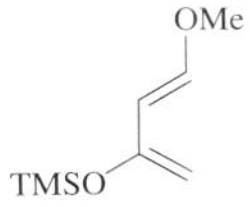

I

HBF_4, SDS, H_2O, rt, 1 h

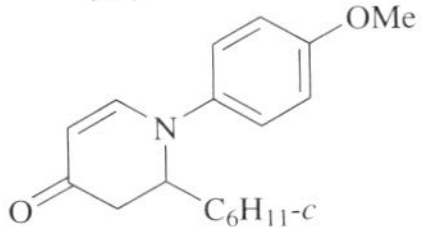

II

(75)

372

I

Montmorillonite K10, 1.5 h **II**

Solvent	Temp[kk]	
H_2O	0°	(85)
MeCN, H_2O	–10°	(86)
MeCN	–10°	(79)

140

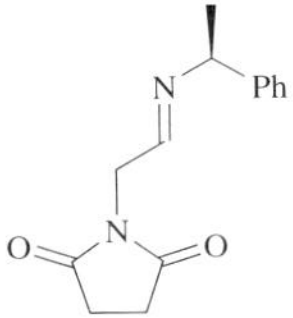

$BF_3•OEt_2$, TFA, CH_2Cl_2, –78° to rt, overnight

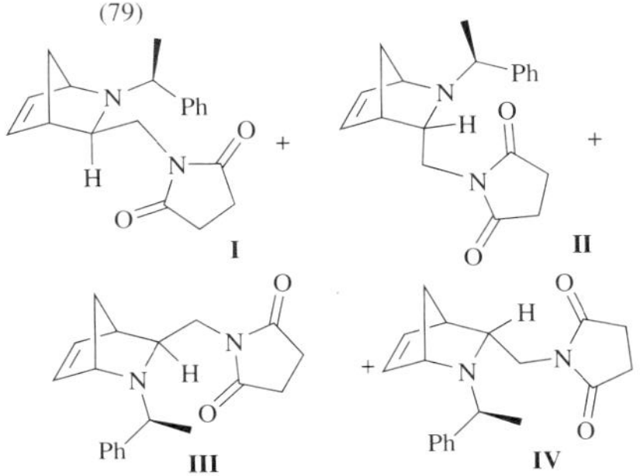

I + **II** + **III** + **IV** (62), **I** + **III**:**II** + **IV** 80:20, **I**:**III** 88:12

403

TABLE 5. CYCLOADDITIONS OF *N*-ALKYL- AND *N*-ARYLIMINES (*Continued*)

Imine	Diene	Conditions	Product(s) and Yield(s) (%)	Refs.
C_{14} 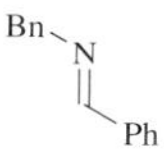		$Yb(OTf)_3$, HCl, H_2O, 12-20 h	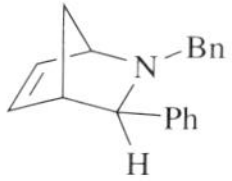 (7)	147
	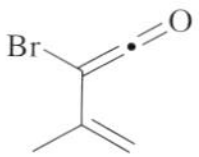	Lewis acid, TEA, CH_2Cl_2	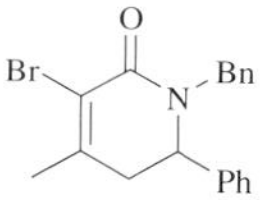	374
		L. A. / Temp		
		$AlMe_2Cl$ / −78° to rt	(54)	
		none / −50° to rt	(65)	
		none / 40°	(92)	
	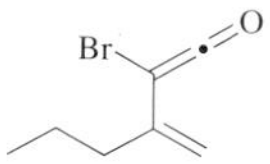	TEA, CH_2Cl_2, 40°	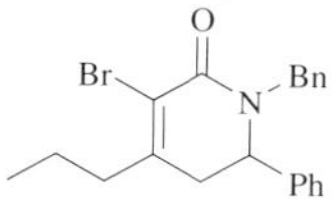 (94)	374
	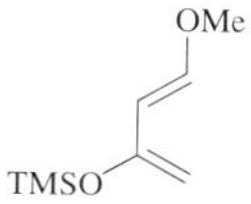	Lewis acid, MeCN, 0°	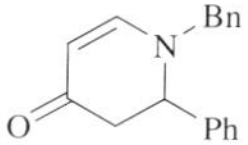	388
		L. A.		
		$Yb(OTf)_3$	(77)	
		$Sc(OTf)_3$	(77)	
	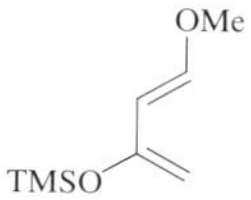	1. $LiClO_4$, CH_2Cl_2, 22°, 20 h 2. $NaHCO_3$, H_2O	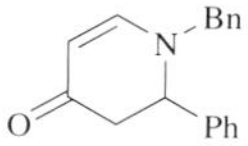 (54)	422
	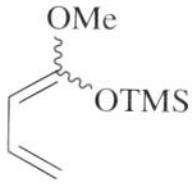	1. $TiCl_4$, CH_2Cl_2, −100° to rt, 3-4 h 2. H_2O	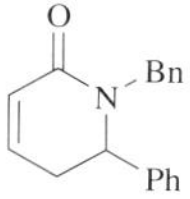 (10)[mm]	134

Reactant	Conditions	Product(s) and Yield(s) (%)	Refs.
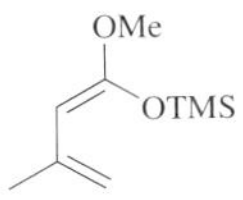	1. $TiCl_4$, CH_2Cl_2, −100° to rt, 3-4 h 2. H_2O	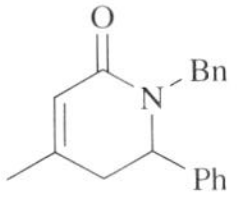 (85)[iiii]	134
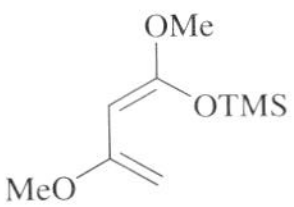	1. Et_2AlCl, CH_2Cl_2, −78° to rt, 4 h 2. H_2O, ether	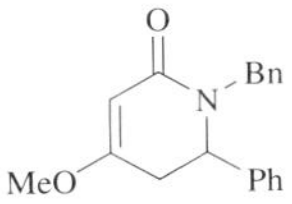 (73)	159
	1. Lewis acid, CH_2Cl_2 2. $NaHCO_3$, Et_3N, H_2O L. A. / Temp / Time $AlCl_3$ / 20° / 10 min $AlCl_3$ / 20° / 1 h BF_3•OEt_2 / 20° / 1 h $ZrCl_4$ / 20° / 1 h	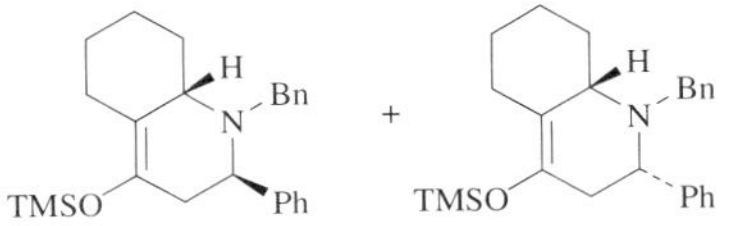 (—) 53:47 (—) 53:47 (—) 53:47 (—) 52:48	155
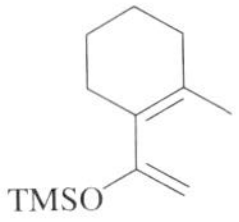	1. Lewis acid, CH_2Cl_2, 20° 2. $NaHCO_3$, H_2O 3. Et_3N, MeOH, overnight L. A. / Time $TiCl_4$ / 1 h $SnCl_4$ / 24 h $SnCl_4$ / 16 h	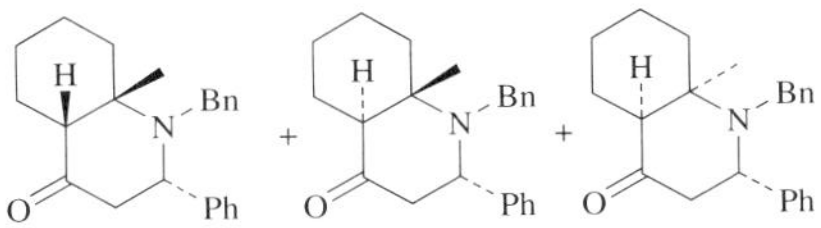 **I** + **II** + **III** **I + II + III** (trace) **I + II + III** (55), **I + II**:**III** 53:47 **I + II** (29), **III** (26), **I**:**II** 1:1.1	416

TABLE 5. CYCLOADDITIONS OF *N*-ALKYL- AND *N*-ARYLIMINES (*Continued*)

Imine	Diene	Conditions	Product(s) and Yield(s) (%)	Refs.
C_{14}				
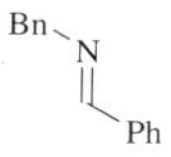	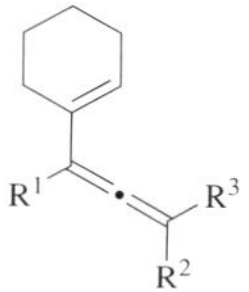	BF_3•OEt_2, CH_2Cl_2	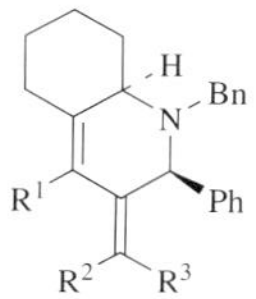	143
	R¹ R² R³	Temp Time		
	Me Me H	rt 1.5 d	(54)[nn]	
	Me Me Me	40° 3 d	(67)[nn]	
	t-Bu Me H	40° 5 d	(34)	
	Me Ph H	rt 4 d	(71)[nn]	
	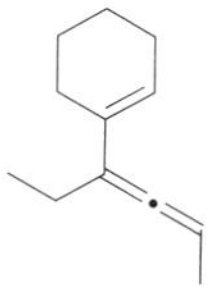	Lewis acid, CH_2Cl_2[jjjj], rt	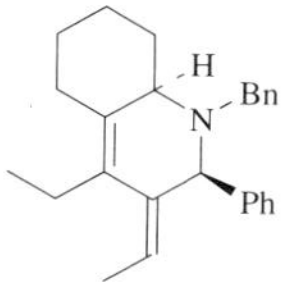	143
		L. A. Time		
		BF_3•OEt_2 4 d	(67)	
		$AlCl_3$ 1.5 d	(55)	
		$TiCl_4$ 3 d	(49)[nn]	
		Et_2AlCl 5 d	(48)[nn]	
	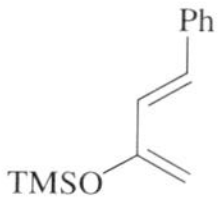	BF_3•OEt_2, H_2O, CH_2Cl_2	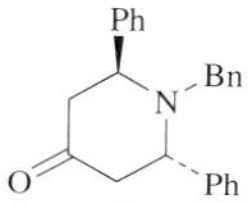 **I** + 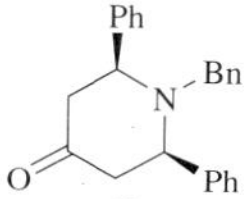 **II** + 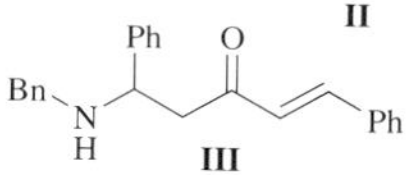 **III** **I** + **II** + **III** (—), **I** + **II**:**III** 16:84	423

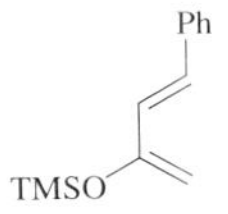

Lewis acid, H_2O, CH_2Cl_2

L. A.	Time
$Zn(OTf)_2$	1 h
$Sc(OTf)_3$	—

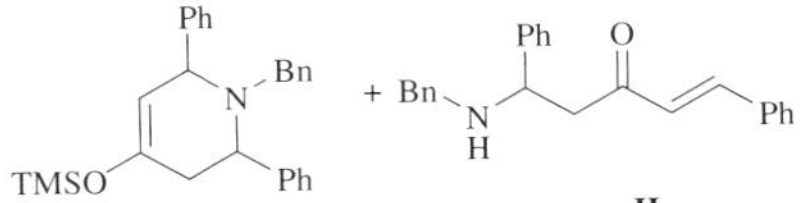

I **II**

I[j] + **II** (—), **I**:**II** 25:75

I[j] + **II** (—)

423

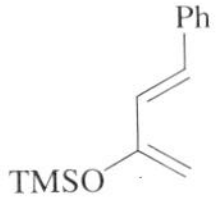

1. Lewis acid, CH_2Cl_2
2. TBAF, THF, rt, 1 h

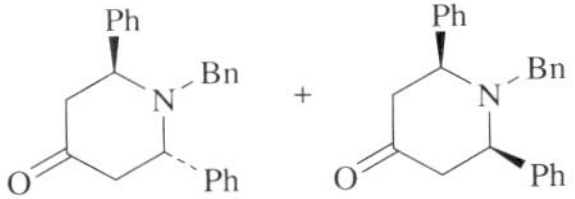

L. A.	Temp	Time	Yield (ratio)	Refs.
TMSOTf	rt	1 h	(97) 81:19	169, 423
TMSOTf	0°	1 h	(95)	423
TMSOTf	–20°	30 min	(20)	423
$TiCl_4$	0°	1 h	(79)	423
$Sn(OTf)_2$	rt	1 h	(59)	423
Et_2AlCl	rt	1 h	(64)	423
BF_3•OEt_2	rt	4 h	(95) 85:15	169, 423
$Zn(OTf)_2$	rt	3 h	(80)	423

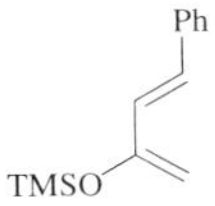

TMSOTf, CH_2Cl_2

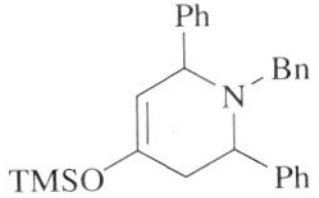

(—)[j]

423

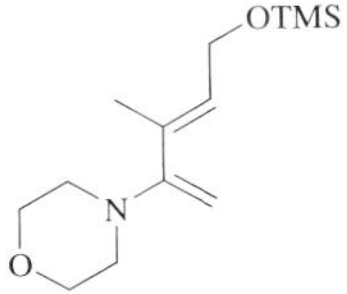

1. $ZnCl_2$, THF, ether, rt, 10 h
2. SiO_2

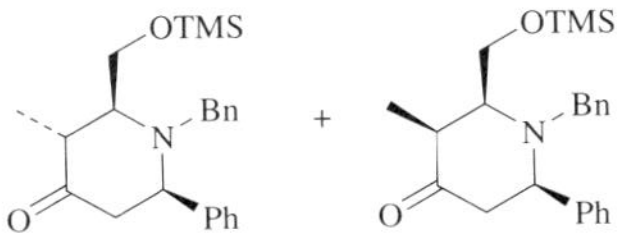

(85) 61:39

157

TABLE 5. CYCLOADDITIONS OF *N*-ALKYL- AND *N*-ARYLIMINES (*Continued*)

Imine	Diene	Conditions	Product(s) and Yield(s) (%)	Refs.
C_{14}				
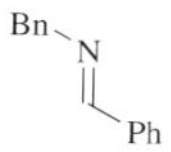	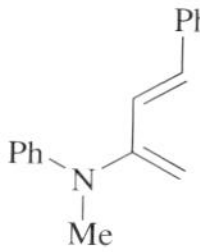	1. $Cu(OTf)_2$, THF, –20°, 24 h 2. $NaHCO_3$, H_2O	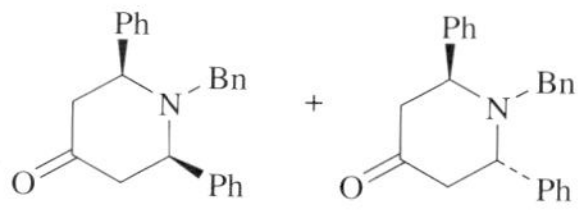 (55) >99.5:0.5	376
	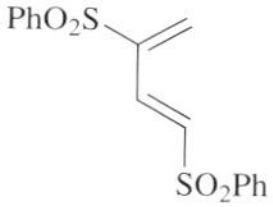	CH_2Cl_2, rt Time 24 h[p,q] 2 h[r]	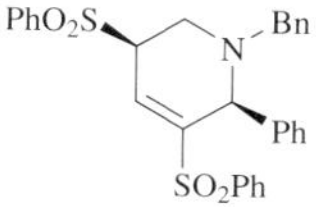 (90) ("excellent")	151, 145
	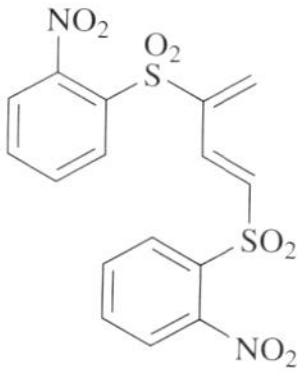	CH_2Cl_2, rt, 44 h	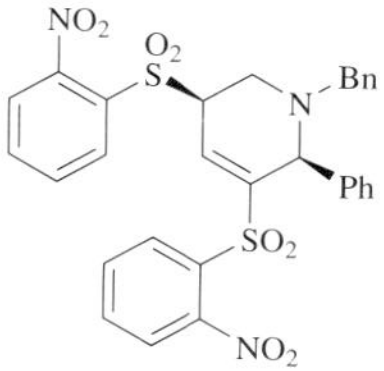 (62)[p]	151, 145
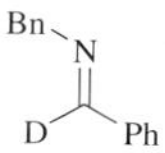	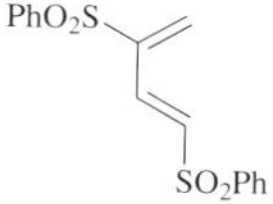	CH_2Cl_2, rt, 20 h	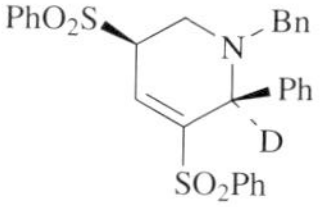 (81)	151
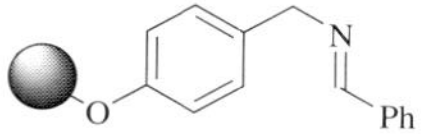	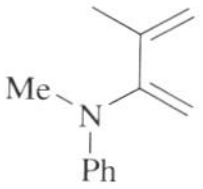	1. $Yb(OTf)_3$, THF, rt, 12 h 2. TFA, CH_2Cl_2, 25 min 3. TMSOTf, CH_2Cl_2, rt, 3 h	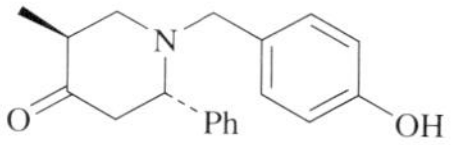 (93)	377

	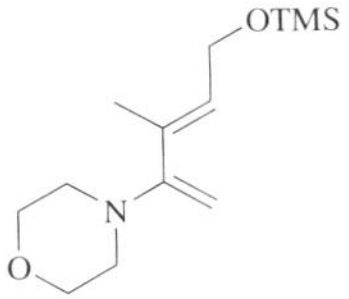	1. Yb(OTf)$_3$, THF, rt, 12 h 2. TFA, CH_2Cl_2, 25 min 3. TMSOTf, CH_2Cl_2, rt, 3 h	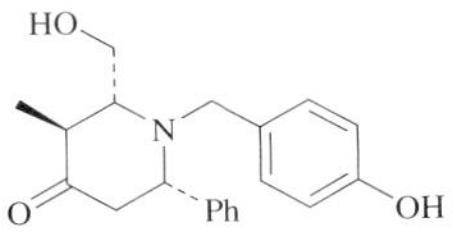 (82)	377
	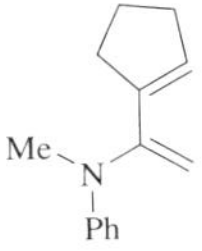	1. Yb(OTf)$_3$[kkkk], THF 2. TFA, CH_2Cl_2, 25 min 3. TMSOTf, CH_2Cl_2, rt, 3 h Temp / Time rt / — –90° / 48 h	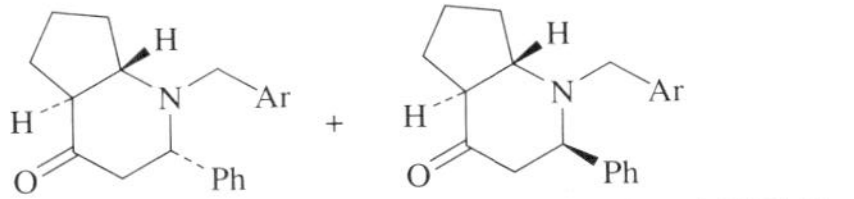 **I** + **II** Ar = 4-HOC_6H_4 **I** + **II** (—), **I**:**II** 1:1 **I** + **II** (78), **I**:**II** 4:1	377
	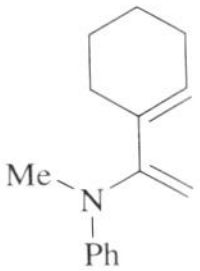	1. Yb(OTf)$_3$, THF 2. TFA, CH_2Cl_2, 25 min 3. TMSOTf, CH_2Cl_2, rt, 3 h Temp / Time rt / — –60° / 48 h	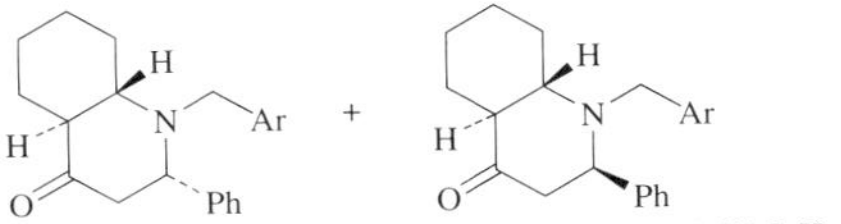 **I** + **II** Ar = 4-HOC_6H_4 **I** + **II** (—), **I**:**II** 1:1 **I** (90)	377
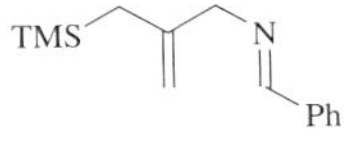	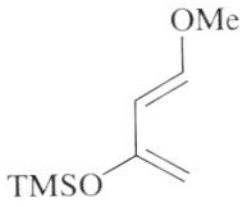	1. Yb(OTf)$_3$, MeCN, rt 2. H_2O Time 10 h[nnn] 10-15 h[ooo]	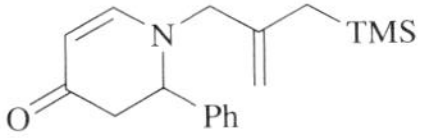 (80) (96)	382

TABLE 5. CYCLOADDITIONS OF *N*-ALKYL- AND *N*-ARYLIMINES (*Continued*)

Imine	Diene	Conditions	Product(s) and Yield(s) (%)	Refs.
C_{14} 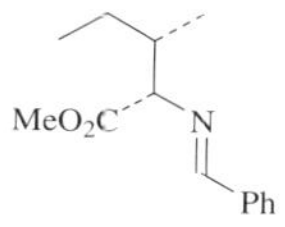	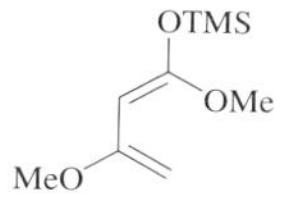	1. Lewis acid, –78° to rt, CH_2Cl_2, 1-1.5 h 2. $NaHCO_3$, H_2O	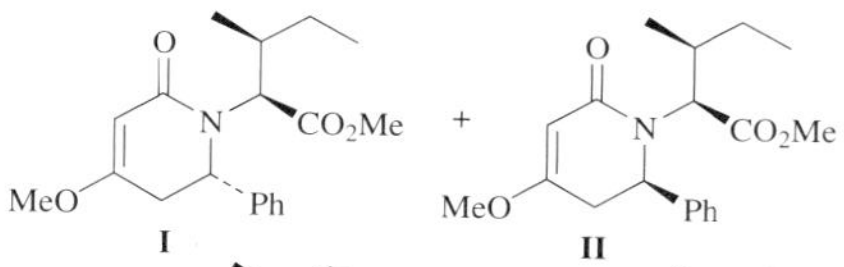 + 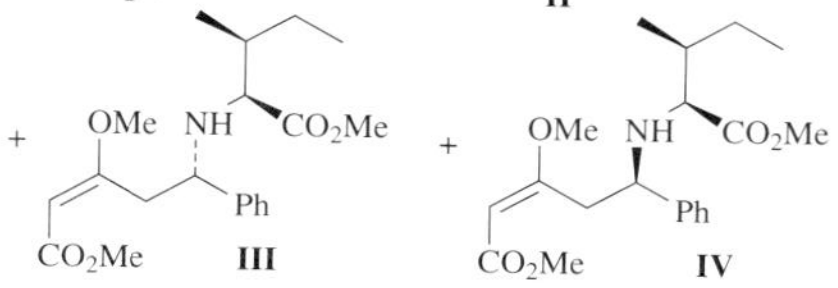	402
		L. A.[bbb]		
		$EtAlCl_2$	**I + II + III + IV** (40), **I + II**:**III + IV** <2:98, **I + III**:**II + IV** 93:7[llll]	
		$MeAlCl_2$	**I + II + III + IV** (34), **I + II**:**III + IV** <2:98, **I + III**:**II + IV** 92:8[llll]	
		$TiCl_4$	**I + II + III + IV** (30), **I + II**:**III + IV** <2:98, **I + III**:**II + IV** 92:7[llll]	
	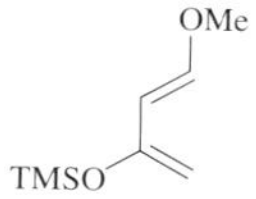	1. Lewis acid 2. $NaHCO_3$, H_2O	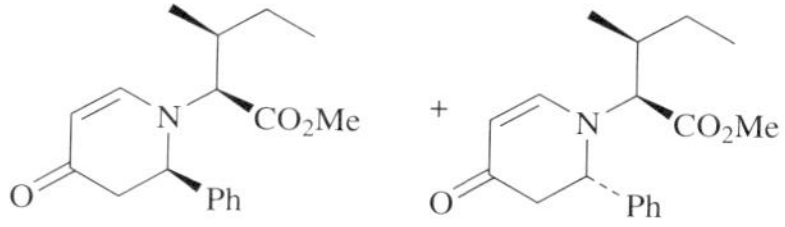	384
		L. A. / Solvent / Temp / Time		
		$ZnCl_2$ / THF / –15° / 7 h	(62) 92:8	
		$EtAlCl_2$ / CH_2Cl_2 / –78° to rt / 1 h	(21) 95:5	
		Et_2AlCl / CH_2Cl_2 / –78° to rt / 1 h	(28) 89:11	
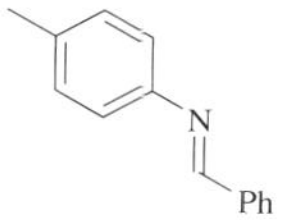	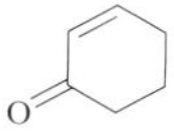	$InCl_3$, MeCN, rt, 24 h[hhh]	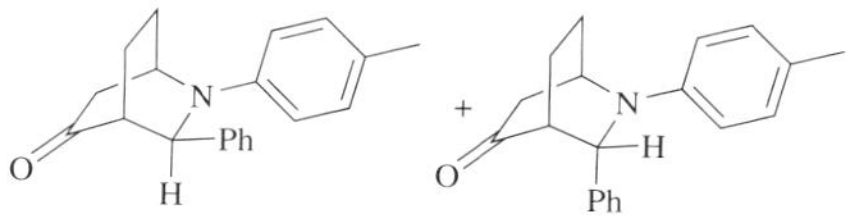 (45) + (17)	148

	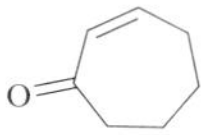	$InCl_3$, MeCN, rt, 22 h[hhh]	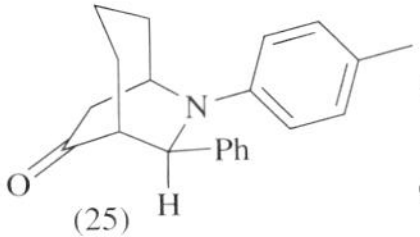 (25) + 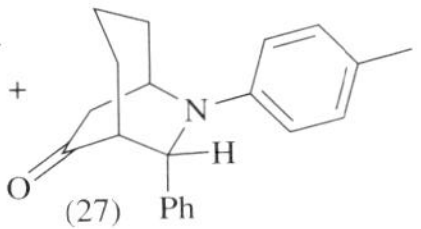 (27)	148
	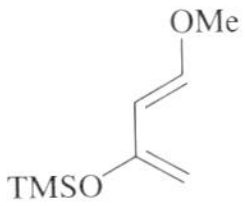	1. $SmI_2(THF)_2$, CH_2Cl_2, rt, 36 h 2. HCl, H_2O	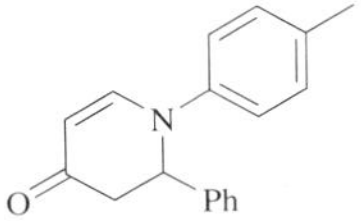 (63)	139
	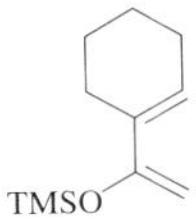	1. $AlCl_3$, CH_2Cl_2 2. $NaHCO_3$, H_2O[mmmm] Temp / Time –40° / 15 min 20° / 1.5 h	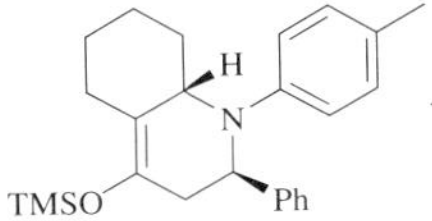 + 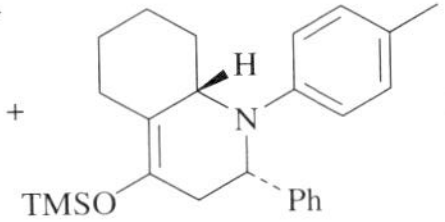 (—)70:30 (—) <2:98	155, 424
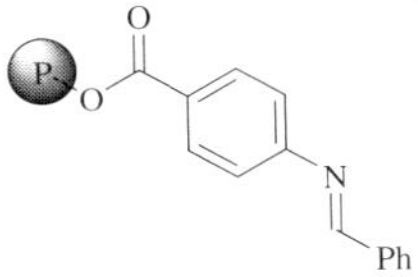	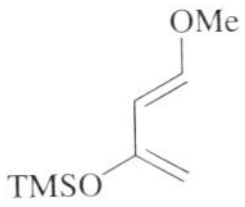	1. $Zn(ClO_4)_2$•$6H_2O$, MeOH, rt, 12 h 2. NaOH, H_2O, rt, 12 h 3. HCl, H_2O	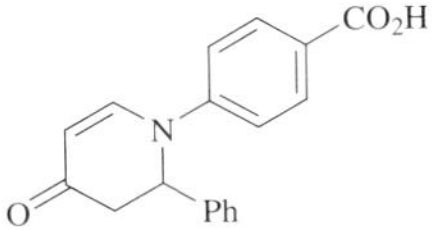 (83)	405
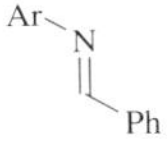 Ar = 4-$MeOC_6H_4$	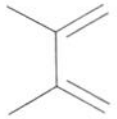	$Sc(OTf)_3$, MeCN, rt, 20 h	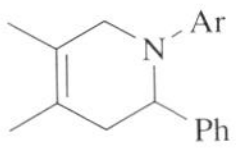 (8)[ggg]	135
	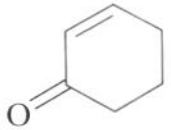	$InCl_3$, MeCN, rt, 24 h[hhh]	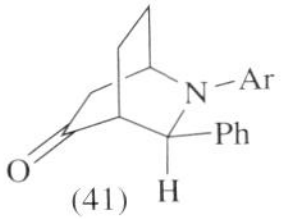 (41) + 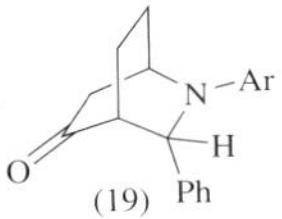 (19)	148

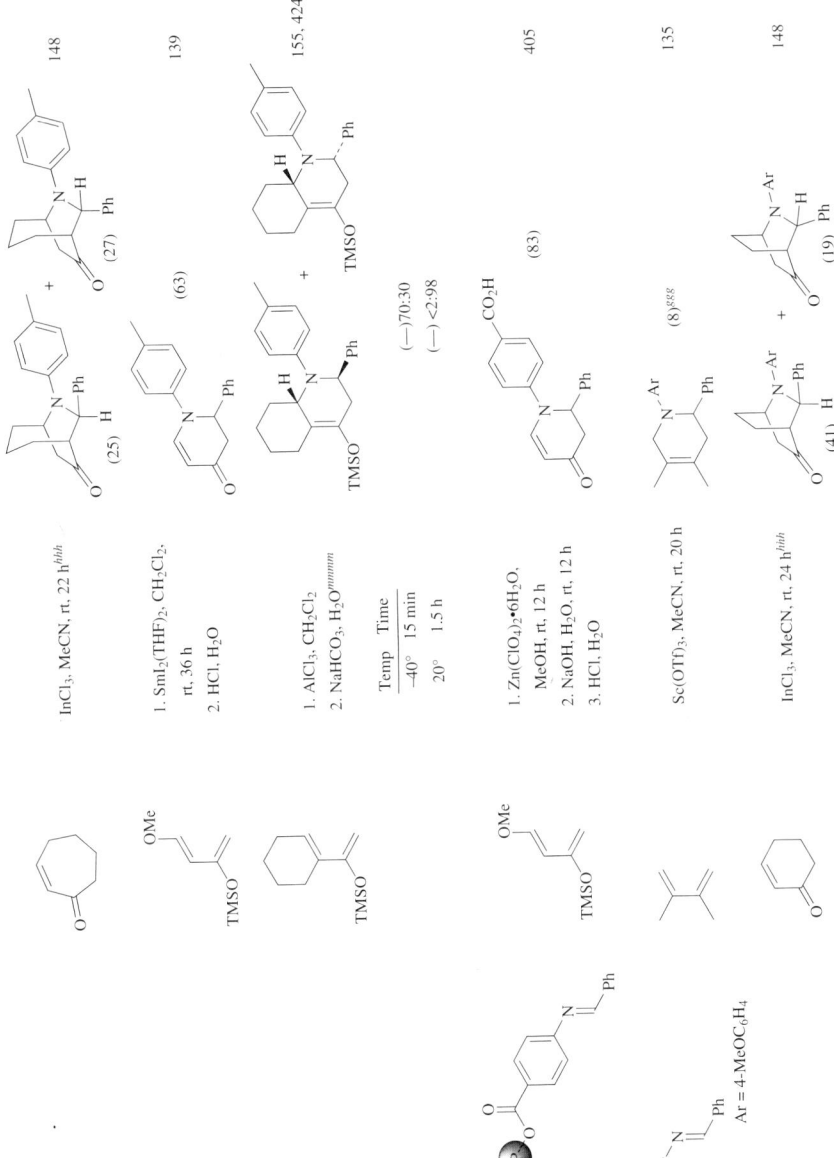

TABLE 5. CYCLOADDITIONS OF *N*-ALKYL- AND *N*-ARYLIMINES (*Continued*)

Imine	Diene	Conditions	Product(s) and Yield(s) (%)	Refs.
C_{14} 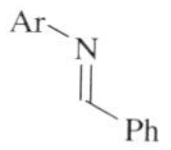 Ar = 4-$MeOC_6H_4$	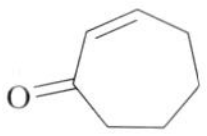	$InCl_3$, MeCN, rt, 26 h[hhh]	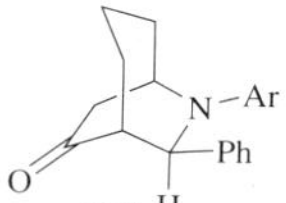 (31) + 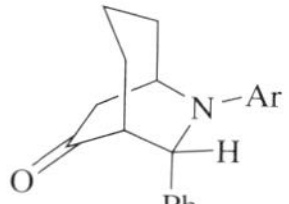 (17)	148
	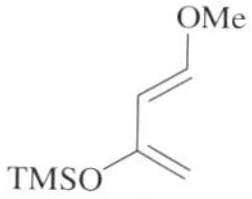 **I**	1. Lewis acid, MeCN, rt 2. $NaHCO_3$, H_2O	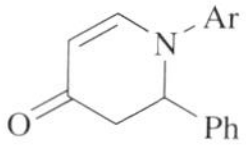 **II**	
		L. A. — Time		
		$Yb(OTf)_3$ — 20 h	(82)	135
		$BiCl_3$ — 30 min	(95)	137
		$Bi(OTf)_3$ — 1.5 h	(90)	137
	I	1. $Yb(OTf)_3$, $MgSO_4$, MeCN, rt 2. H_2O	**II** (83)[jjj]	330
	I		**II**	372
		HBF_4, H_2O, MeOH, –40°, 30 min	(90)[nnn], (88)[ooo]	
		HBF_4, SDS, H_2O, rt, 1 h	(88)	
	I	1. 4-$C_{12}H_{25}C_6H_4SO_3H$, H_2O, 0°, 1 h 2. $NaHCO_3$, NaCl, H_2O	**II** (61)	425
	I	1. $SmI_2(THF)_2$, CH_2Cl_2, rt, 36 h 2. HCl, H_2O	**II** (86)	139
	I	$Sc(OSO_2C_8F_{17})_3$, $scCO_2$, 100 atm, 40°, 1 h	**II** (86)[a]	404

I	Montmorillonite K10, rt	**II**	140
	Solvent / Time: H_2O, 1 h	(98)	
	$MeCN, H_2O$, 20 min	(97)	
	MeCN, 20 min	("quant.")	
I	NaOTf, H_2O, rt, 2-3 h	**II** (83)	385
I	1. MeOH, rt, 2 h 2. HCl, H_2O	**II** (95)[nnn], (91)[ooo]	373
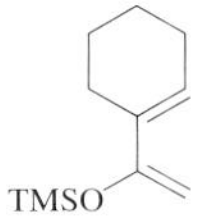	1. $AlCl_3$, CH_2Cl_2 2. $NaHCO_3$, H_2O[mmmm]	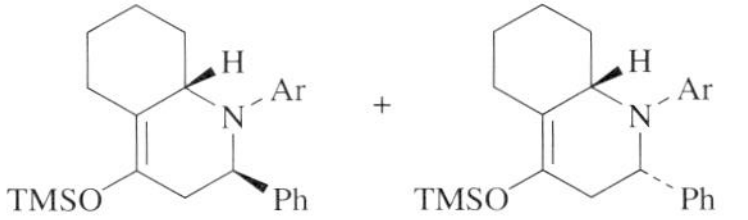	155, 424
	Temp / Time: –40°, 15 min	(—) 70:30	
	20°, 1.5 h	(—) <2:98	
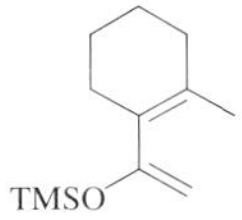	1. Lewis acid, CH_2Cl_2, 20° 2. $NaHCO_3$, H_2O	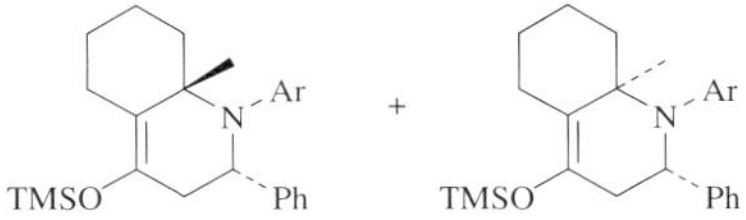 **I** + **II**	416
	L. A. / Time: $SnCl_4$, 22 h	**I** (26)	
	$TiCl_4$, 1 h	**I** + **II** (10), **I**:**II** 4:1	

TABLE 5. CYCLOADDITIONS OF *N*-ALKYL- AND *N*-ARYLIMINES (*Continued*)

Imine	Diene	Conditions	Product(s) and Yield(s) (%)	Refs.
C_{14} 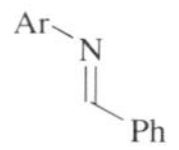 Ar = 4-MeOC$_6$H$_4$	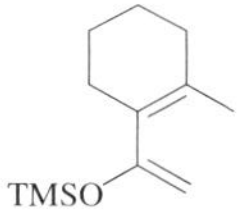	1. Lewis acid, CH_2Cl_2, 20° 2. $NaHCO_3$, H_2O 3. TEA, MeOH, overnight	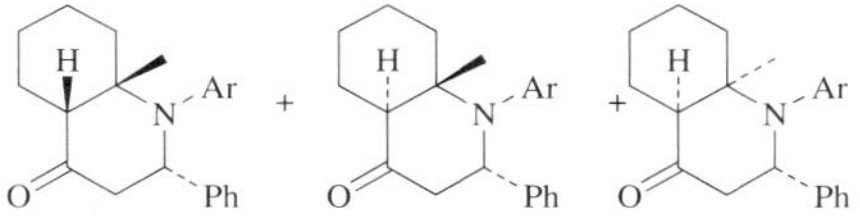 I II III	
		L. A. Time		
		$TiCl_4$ 1 h	**I** + **II** (49), **III** (26), **I**:**II** 1:2.5	416
		$TiCl_4$ 2 h	**I** + **II** + **III** (75), **I** + **II**:**III** 65:35	416, 417
		$SnCl_4$ 20 h	**I** + **II** + **III** (—), **I** + **II**:**III**:recovered imine 84:8:8	416, 417
	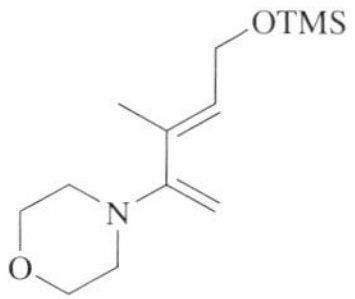	1. $ZnCl_2$, THF, ether, rt, 10 h 2. SiO_2	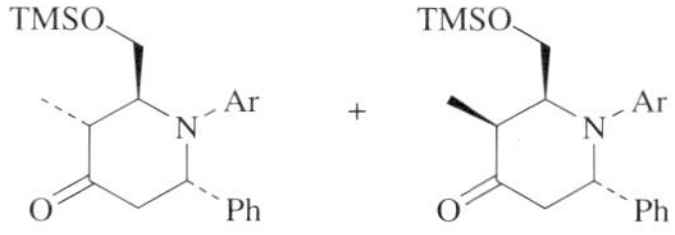 I II **I** + **II** (87), **I**:**II** 11:89[nnnn]	157
	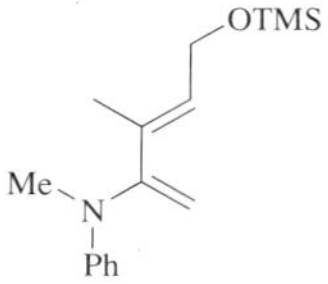	$ZnCl_2$, THF, ether, –40° to rt, 11 h	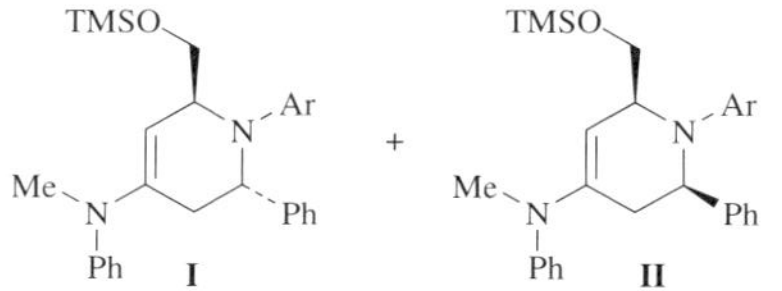 **I** + **II** (91), **I**:**II** >95:5[oooo]	157
	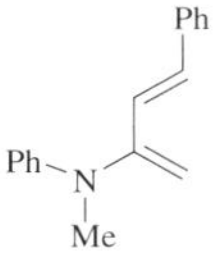	1. $Cu(OTf)_2$, THF, –20°, 24 h 2. $NaHCO_3$, H_2O	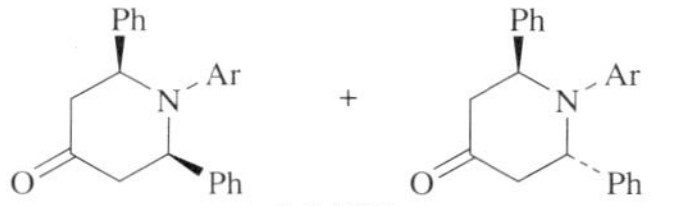 (15) 98:2	376

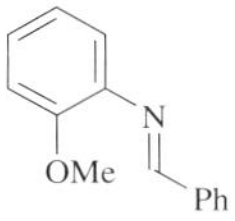	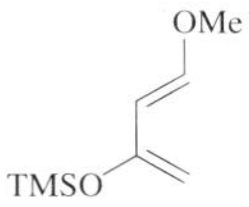	1. $SmI_2(THF)_2$, CH_2Cl_2, rt, 16 h 2. HCl, H_2O	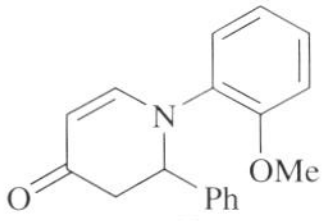 **I** (86)	139
	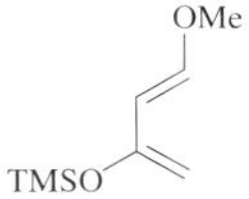	Catalyst, H_2O, rt, 2-3 h	**I**	
		Catalyst	**I**	
		NaOTf	(74)	385
		AgOTf	(56)[sss]	386
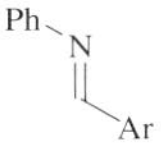	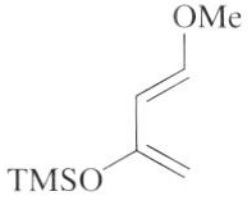		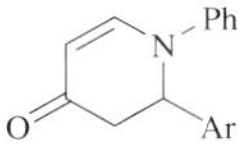	
Ar				
3-MeC_6H_4		Nafion-H, 16 h	(81)	138
4-MeC_6H_4		Nafion-H, 4 h	(73)	138
4-MeC_6H_4		HBF_4, H_2O, MeOH, –40°, 30 min	(95)[nnn], (79)[ooo]	372
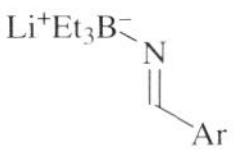	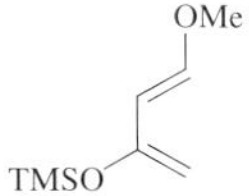	1. THF, rt, 4-5 h 2. $NaHCO_3$, H_2O	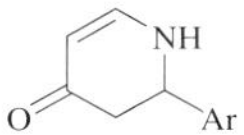	389
Ar: 4-MeC_6H_4			(38)	
Ar: 4-$MeOC_6H_4$			(62)	
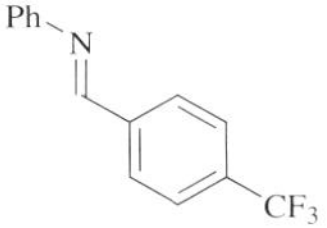	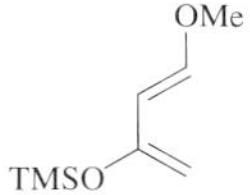	$Sc(SO_3CF_3)_3$, ionic liquid, rt, 20 h	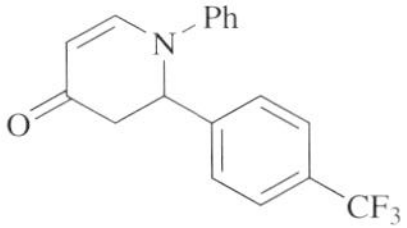	412
		Ionic Liquid		
		8-Ethyl-1,8-diazabicyclo-[5.4.0]-7-undecenium trifluoromethanesulfonate	(85)[yyy]	
		1-Ethyl-3-methyl-1*H*-imidazolium trifluoromethanesulfonate	(88)[yyy]	

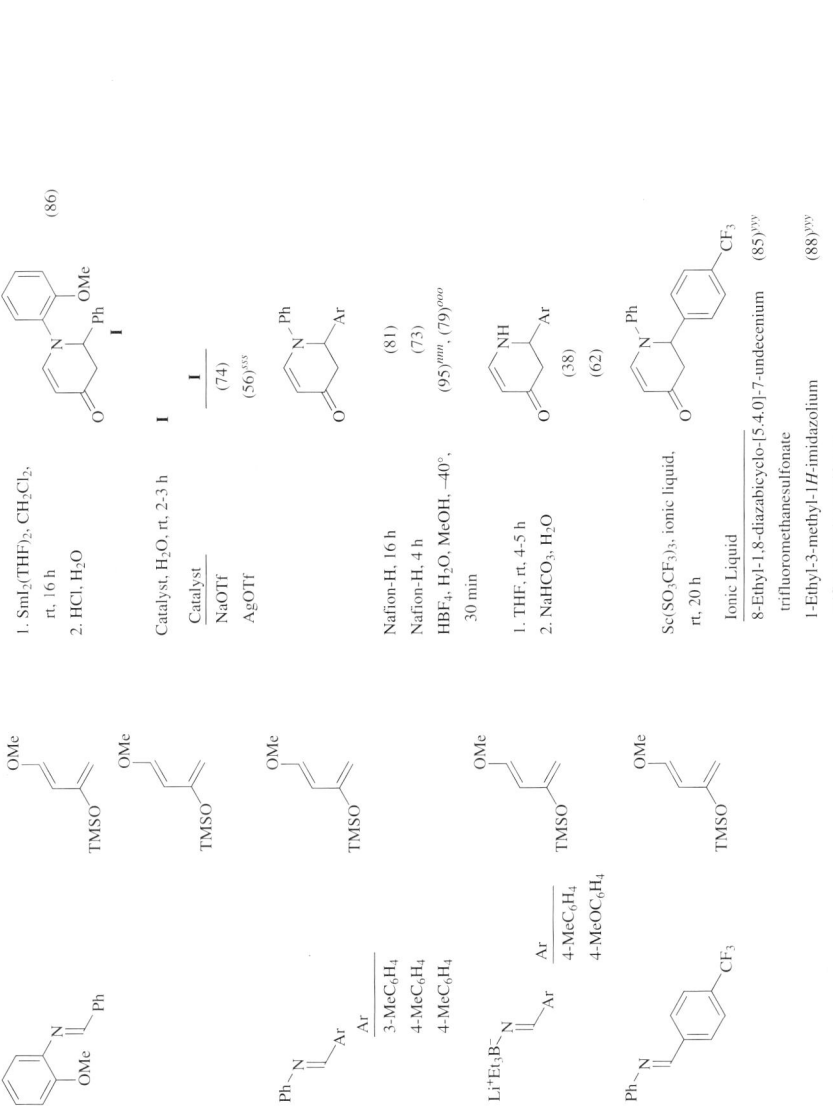

TABLE 5. CYCLOADDITIONS OF *N*-ALKYL- AND *N*-ARYLIMINES (*Continued*)

	Imine	Diene	Conditions	Product(s) and Yield(s) (%)	Refs.
C_{14}	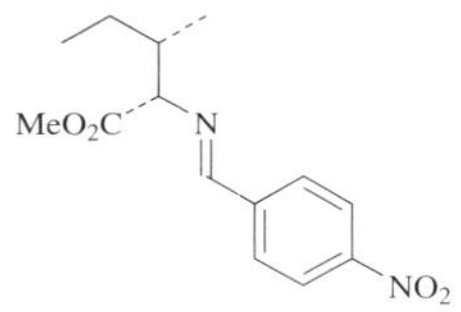	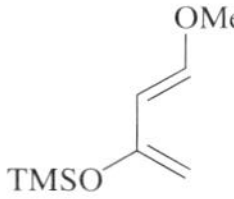	1. $ZnCl_2$, THF, –10°, 7 h 2. $NaHCO_3$, H_2O	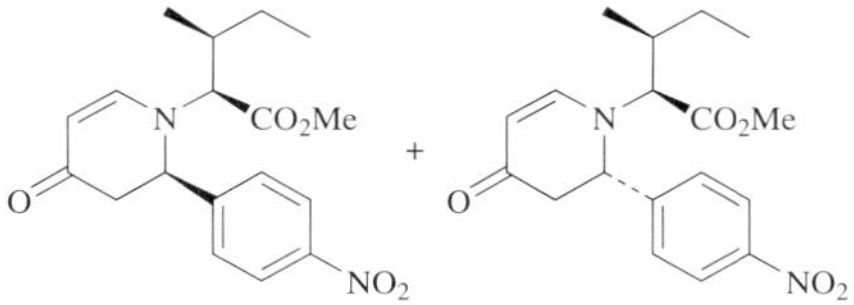 (57) 93:7	384
	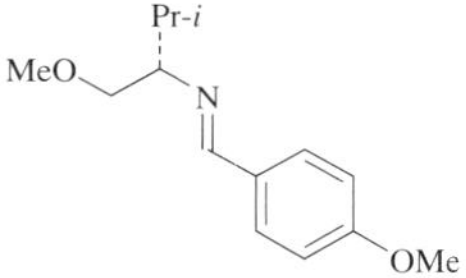	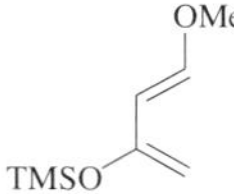	$Zn(OTf)_2$, CH_2Cl_2, rt	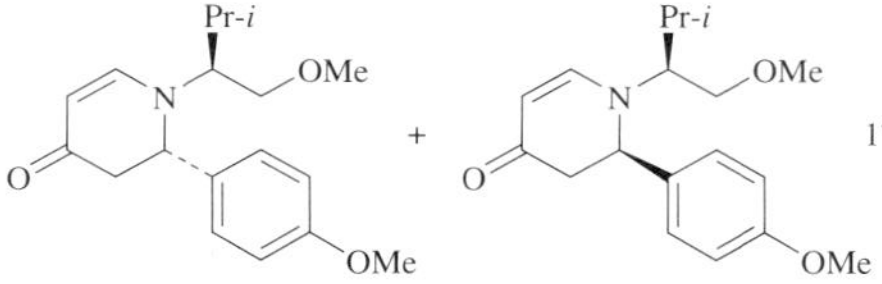 (79) 82:18	170
	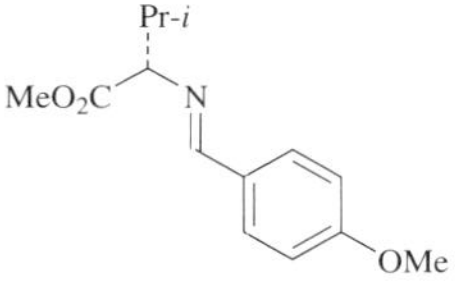	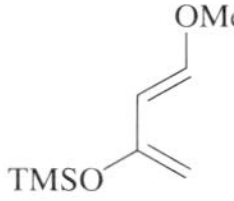		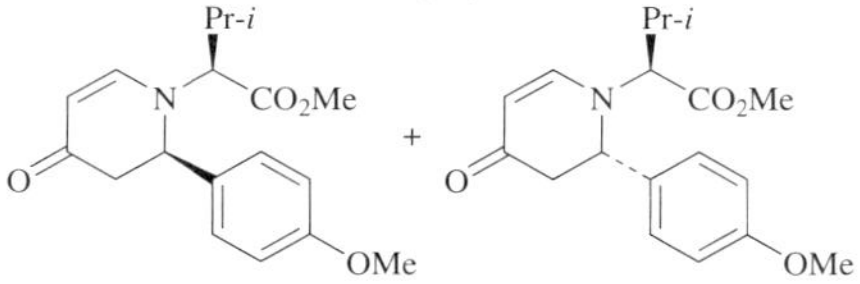	
			1. $ZnCl_2$[jj], THF, 0° 2. H^+, H_2O	(54) 92:8	175a
			1. $ZnCl_2$, THF, 0°, 7 h 2. $NaHCO_3$, H_2O	(54) 92:8	384
	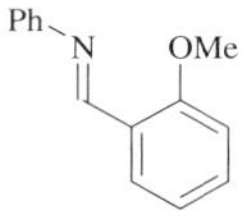	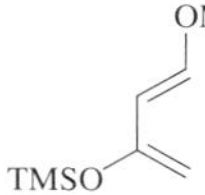	1. Lewis acid, CH_2Cl_2, rt, 5 h 2. HCl, H_2O	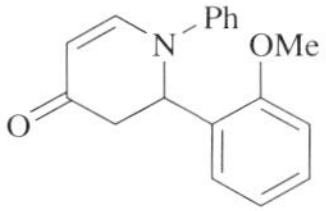	139
			L. A.		
			$SmI_2(THF)_2$	(72)	
			$SmI_3(THF)_3$	(71)	

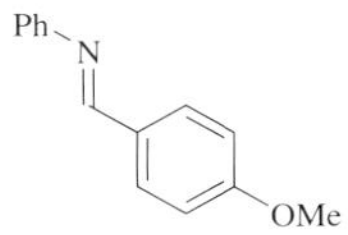 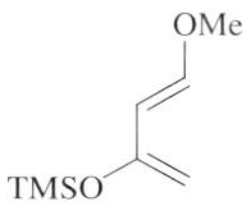

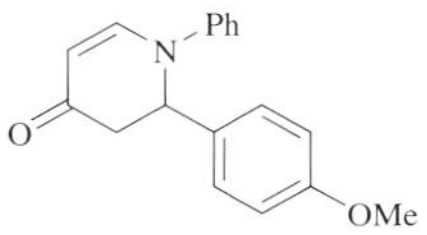

Conditions	Product(s) and Yield(s) (%)	Refs.
Nafion-H, 12 h	(77)	138
1. $SmI_2(THF)_2$, CH_2Cl_2, rt, 18 h 2. HCl, H_2O	(82)	139
1. MeOH, rt, 2 h 2. HCl, H_2O	(95)[nnn], (91)[ooo]	373
AgOTf, H_2O, rt, 2-3 h	(77)	386

1. Lewis acid, CH_2Cl_2
2. $NaHCO_3$, H_2O

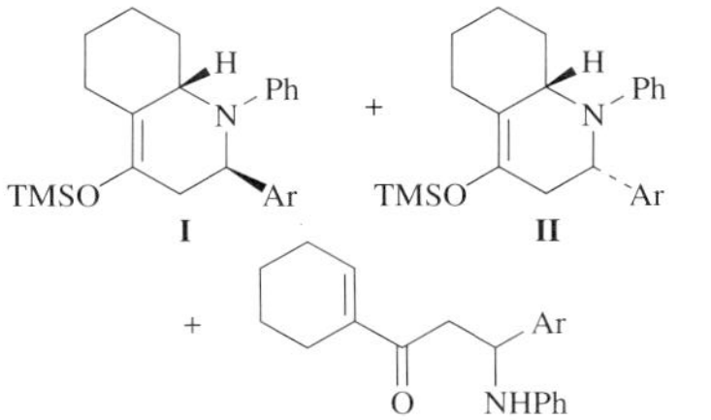

I + II + III

Ar = 4-$MeOC_6H_4$

156

L. A. (eq)	Temp	Time	
$AlCl_3$ (1.0)	–40°	15 min	**I** + **II** (52)[a], **I**:**II** 60:40
$AlCl_3$ (1.0)	20°	1.5 h	**I** + **II** (71)[a], **I**:**II** <2:98
TBSOTf (0.1)	20°	1.5 h	**I** + **II** (84)[a], **I**:**II** >98:2
TBSOTf (0.1)	20°	15 h	**I** + **II** + **III** (96)[a], **I**:**II**:**III** 41:45:14
TBSOTf (1.0)	20°	0.5 h	**I** + **II** (79)[a], **I**:**II** 70:30
TBSOTf (1.0)	20°	1.5 h	**I** + **II** (93)[a], **I**:**II** 62:38
TMSOTf (0.1)	–40°	15 min	**I** + **II** (19)[a], **I**:**II** 70:30
TMSOTf (0.1)	20°	1.5 h	**I** + **II** (70)[a], **I**:**II** 70:30
$Eu(fod)_3$ (0.1)	20°	72 h	**I** + **II** + **III** (33)[a], **I**:**II**:**III** 35:55:10

TABLE 5. CYCLOADDITIONS OF *N*-ALKYL- AND *N*-ARYLIMINES (*Continued*)

Imine	Diene	Conditions	Product(s) and Yield(s) (%)	Refs.
C_{14} 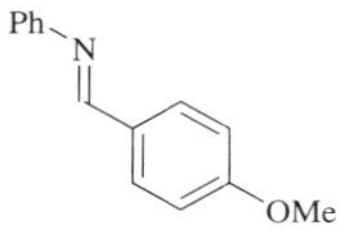	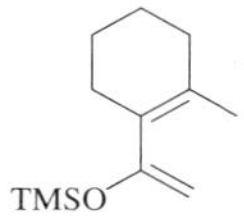	1. $TiCl_4$, CH_2Cl_2, 20°, 2 h 2. $NaHCO_3$, H_2O	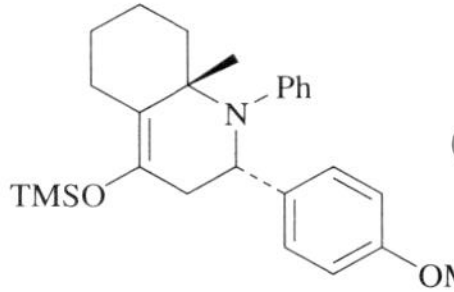 (27)	416
	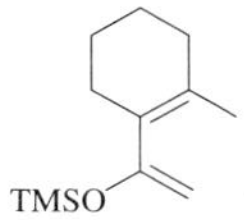	1. $TiCl_4$, CH_2Cl_2, 20°, 4 h 2. $NaHCO_3$, H_2O 3. Et_3N, MeOH	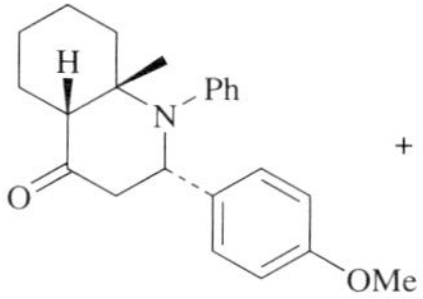 + 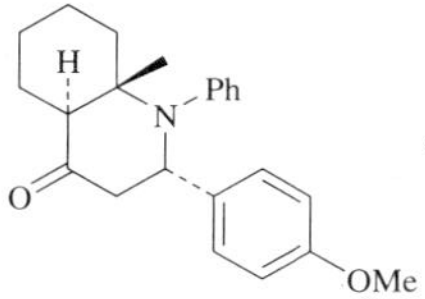 (50) 1:2	416
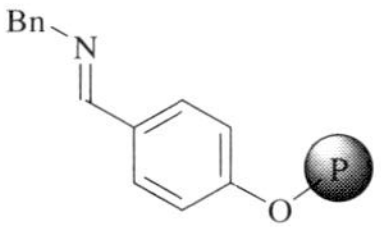	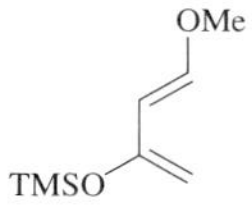	1. $Yb(OTf)_3$, THF, 60°, 3 h 2. H_2O 3. TFA, CH_2Cl_2	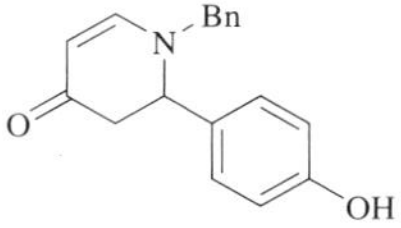 (80)[hh]	378
	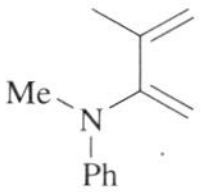	1. $Yb(OTf)_3$, THF, rt, 12 h 2. TFA, CH_2Cl_2, 25 min 3. TMSOTf, CH_2Cl_2, rt, 3 h	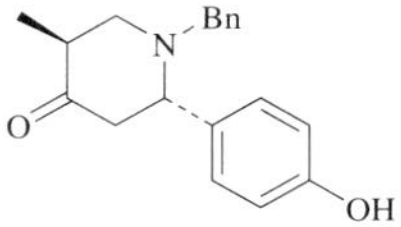 (54)	377
	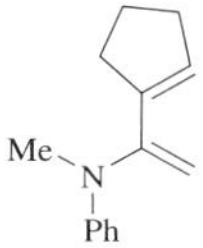	1. $Yb(OTf)_3$[kkkk], THF 2. TFA, CH_2Cl_2, 25 min 3. TMSOTf, CH_2Cl_2, rt, 3 h Temp / Time rt / — –90° / 48 h	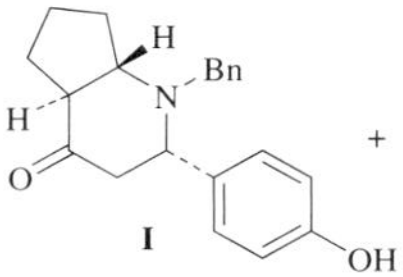 I + 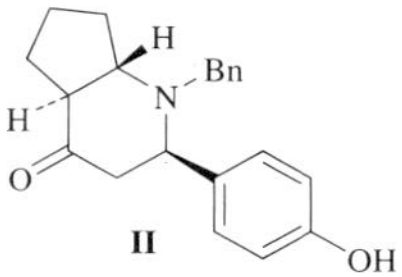 II **I** + **II** (—), **I**:**II** 1:1 **I** + **II** (42), **I**:**II** 4:1	377

	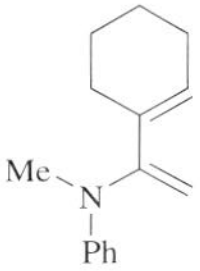	1. Yb(OTf)$_3$, THF 2. TFA, CH_2Cl_2, 25 min 3. TMSOTf, CH_2Cl_2, rt, 3 h Temp / Time rt / — –60° / 48 h	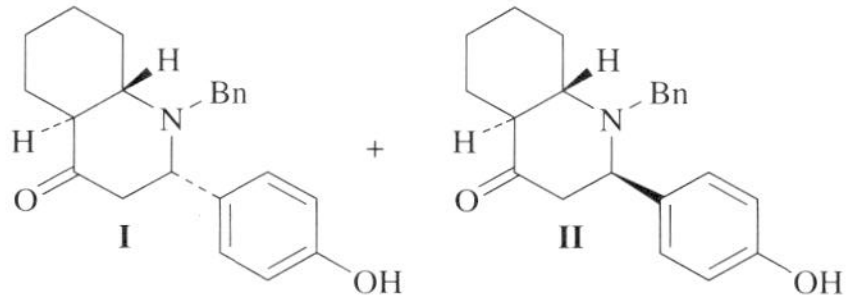 **I** + **II** (—), **I**:**II** 1:1 **I** (59)	377
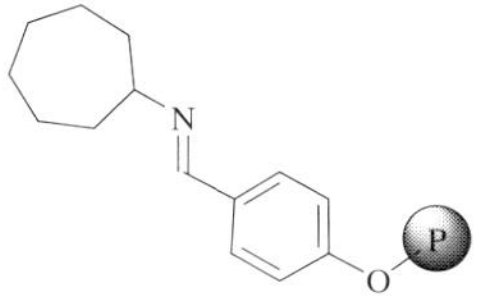	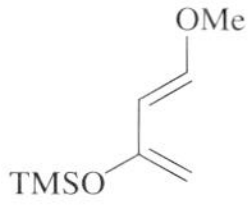	1. Yb(OTf)$_3$, THF, 60°, 3 h 2. H_2O 3. TFA, CH_2Cl_2	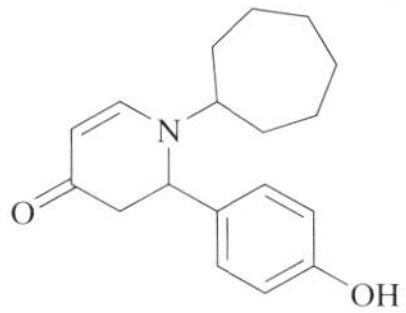 (63)[hh]	378
	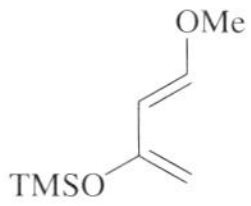	1. In(OTf)$_3$, $MgSO_4$, MeCN, rt 2. $NaHCO_3$, H_2O	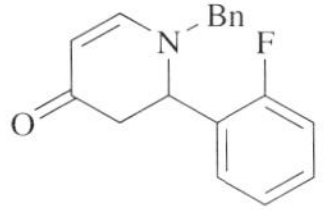 (50)	136
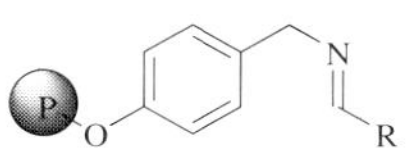 R 2-BrC_6H_4 3-BrC_6H_4 4-BrC_6H_4	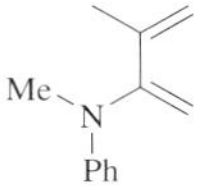	1. Yb(OTf)$_3$, THF, rt, 12 h 2. TFA, CH_2Cl_2, 25 min 3. TMSOTf, CH_2Cl_2, rt, 3 h	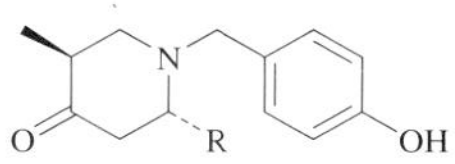 (83) (91) (67)	377

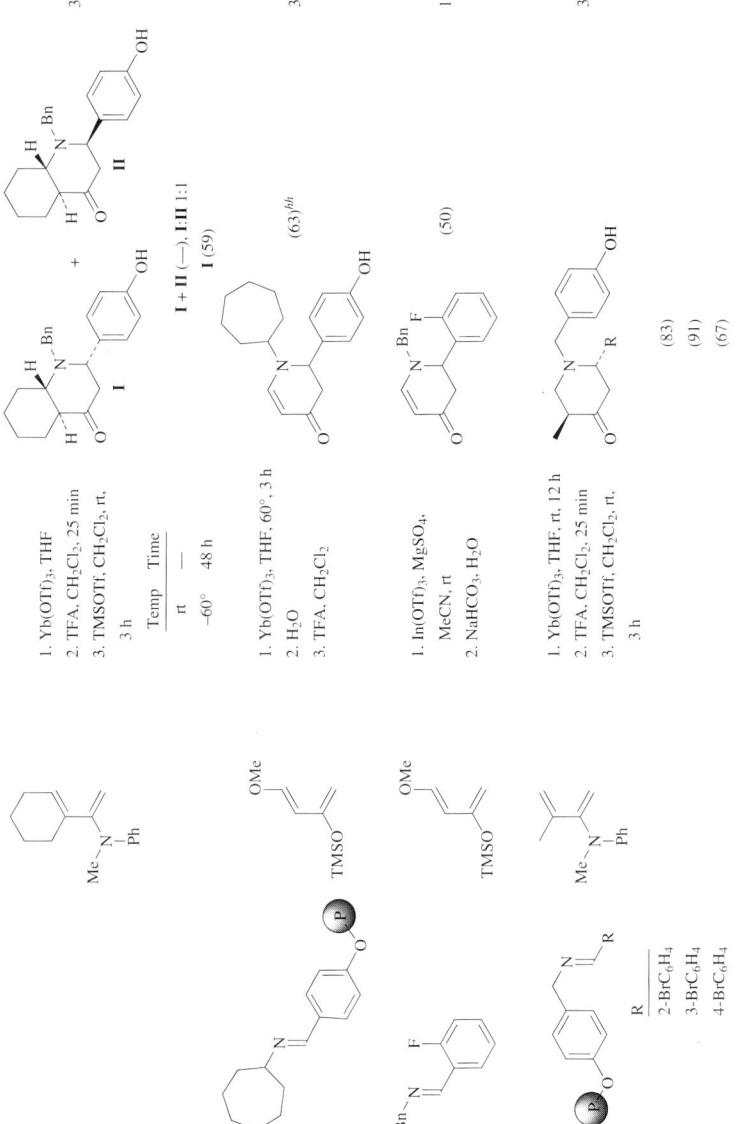

TABLE 5. CYCLOADDITIONS OF *N*-ALKYL- AND *N*-ARYLIMINES (*Continued*)

Imine	Diene	Conditions	Product(s) and Yield(s) (%)	Refs.
C_{14}				
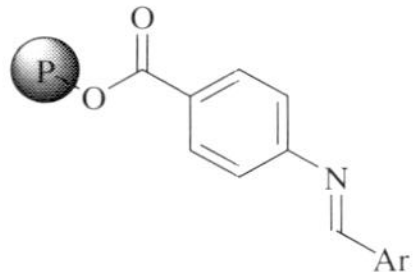	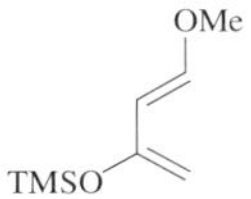	1. $Zn(ClO_4)_2 \cdot 6H_2O$, MeOH, rt 2. NaOH, H_2O, rt, 12 h 3. HCl, H_2O	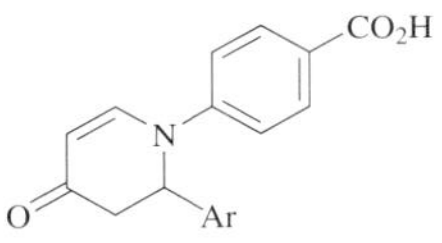	405
Ar		Time		
3-FC_6H_4		12 h	(92)	
3,5-$F_2C_6H_3$		12 h	(84)	
4-ClC_6H_4		12 h	(44)	
		30 min	(90)	
3-ClC_6H_4		12 h	(99)	
3,5-$Cl_2C_6H_3$		12 h	(58)	
		30 min	(78)	
4-BrC_6H_4		12 h	(95)	
3-BrC_6H_4		12 h	(31)	
		30 min	(83)	
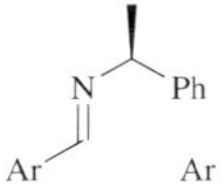		$MeSO_3H$, TFA, CH_2Cl_2, –78° to rt, overnight		393, 91
Ar				
4-pyridinyl			(60) 80:20[aaa]	
2-pyridinyl			(80) 87:13[aaa]	
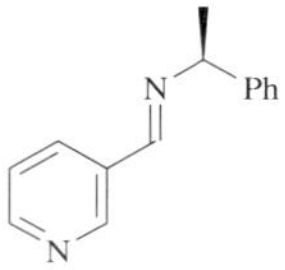	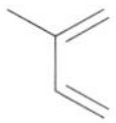	$ZnCl_2$, CH_2Cl_2, ether, rt, 96 h	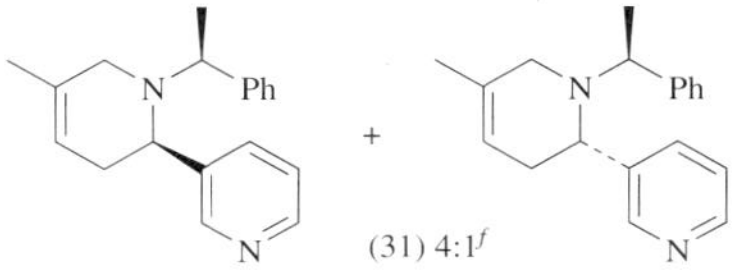 (31) 4:1[f]	393

	Imine	Diene	Conditions	Product(s) and Yield(s) (%)	Refs.
	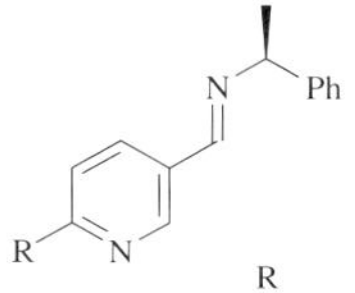Ph N R N R: H, Cl		$AlCl_3$, Et_3N,)))), CH_2Cl_2, 0-5°, 2-3 d	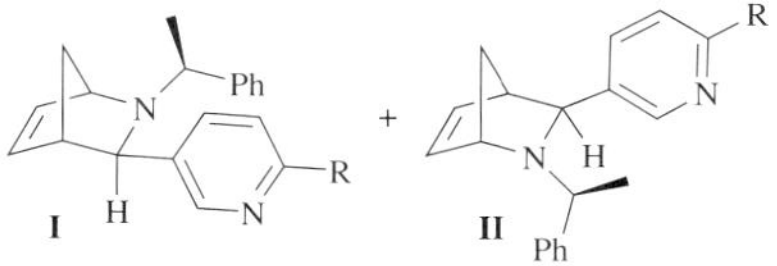I II Ph R N H I + II: R = H (53)[cccc,pppp]; R = Cl (28) 98:2	418
	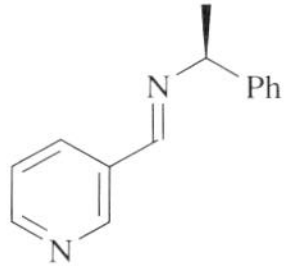N Ph	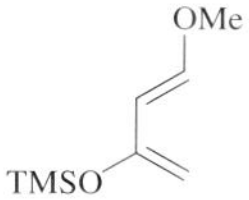OMe TMSO	1. $B(OPh)_3$, CH_2Cl_2, –78°, several h 2. H_2O	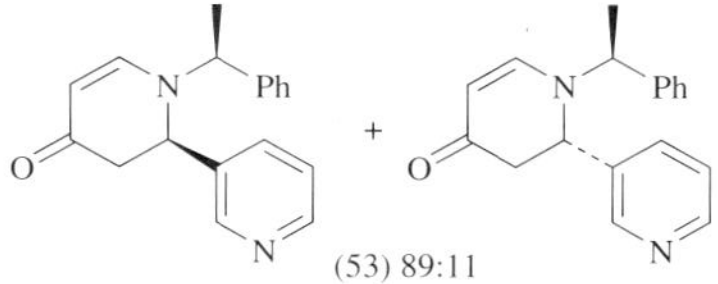O N Ph (53) 89:11	243, 242
	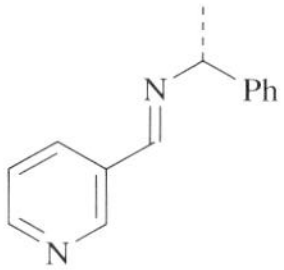N Ph		$AlCl_3$, Et_3N,)))), CH_2Cl_2, 0-5°, 2-3 d	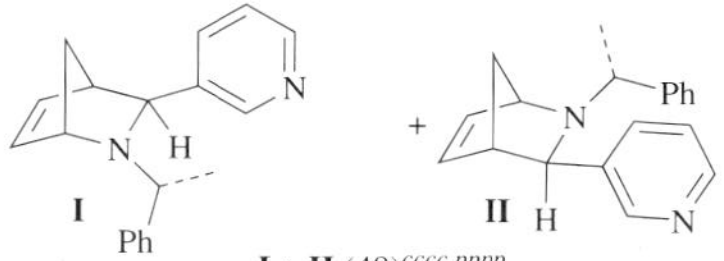I II N H Ph **I** + **II** (48)[cccc,pppp]	418
C_{15}	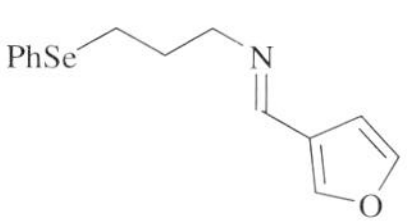PhSe N O	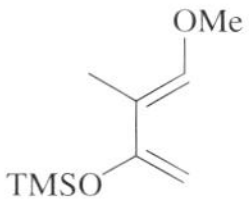OMe TMSO	$ZnCl_2$, THF, rt, 40 h	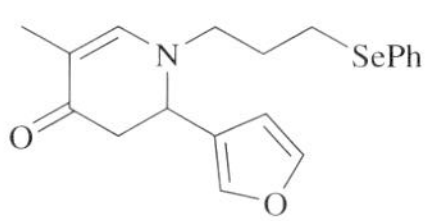O N SePh (72)	426
	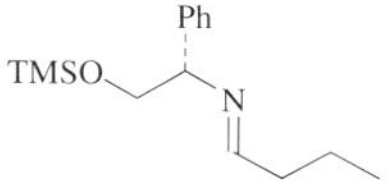TMSO Ph N	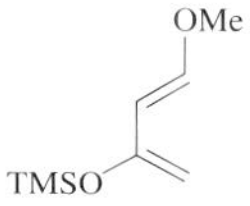OMe TMSO	$Zn(OTf)_2$, CH_2Cl_2, rt	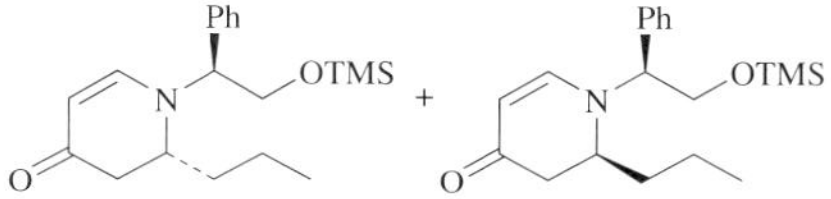Ph OTMS O N (45) 58:42	170

TABLE 5. CYCLOADDITIONS OF *N*-ALKYL- AND *N*-ARYLIMINES (*Continued*)

	Imine	Diene	Conditions	Product(s) and Yield(s) (%)	Refs.
C_{15}	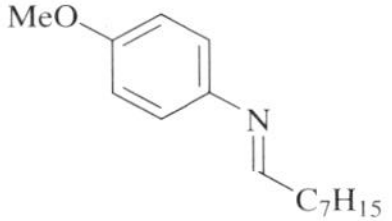	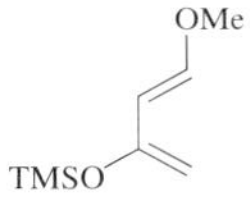	Montmorillonite K10, MeCN, H_2O, –10°, 1.5 h	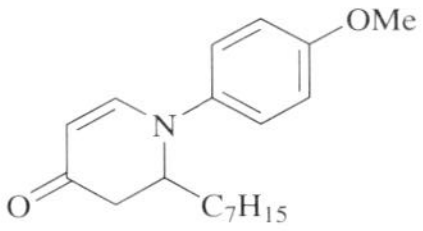 (91)[kk]	140
	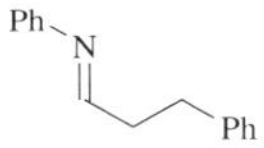	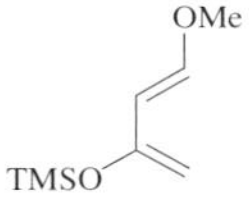		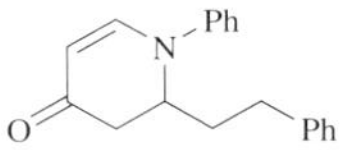	
			HBF_4, H_2O, MeOH, –40°, 30 min	(75)	372
			NaOTf, H_2O, rt, 2-3 h	(70)	385
			AgOTf, H_2O, rt, 2-3 h	(53)	386
			1. MeOH, rt, 2 h 2. HCl, H_2O	(89)	373
	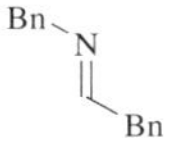		Lewis acid, HCl, H_2O	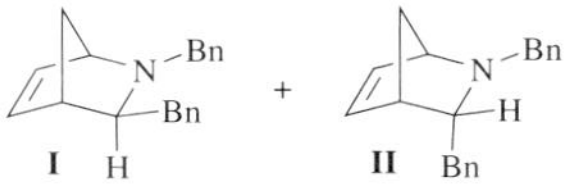	147
			L. A. — Time		
			$Yb(OTf)_3$ — 12-20 h	**I** + **II** (72), **I**:**II** 4:1[qqqq]	
			none — —	**I** + **II** (3)	
	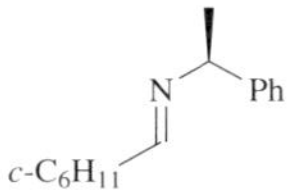	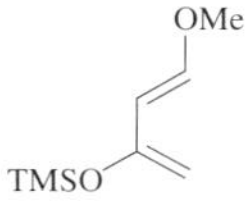	1. $B(OPh)_3$, CH_2Cl_2, –73°, several h 2. H_2O	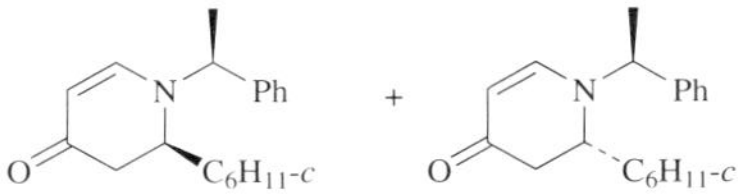 (40) 90:10	243, 242
	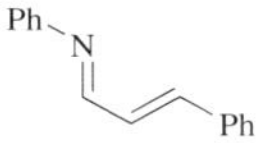	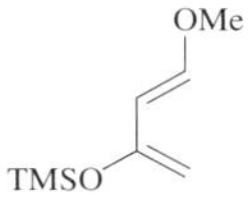 **I**		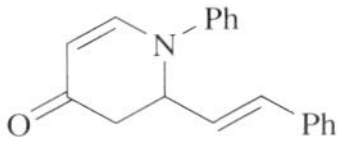 **II**	

Reactant	Conditions	Product(s) and Yield(s) (%)	Refs.
	1. $ZnCl_2$, THF, rt, 36-48 h 2. H_2O	(41)[rrrr]	141, 427
	1. $ZnCl_2$, THF, rt, 36-48 h 2. H_2O	(72)[ssss]	141
	HBF_4, H_2O, MeOH, –40°, 30 min	(89)	372
I	1. Lewis acid, THF, rt 2. $NaHCO_3$, H_2O	**II**	137
	L. A. / Time	**II**	
	$BiCl_3$ / 25 min	(95)	
	$Bi(OTf)_3$ / 40 min	(90)	
I	Catalyst, H_2O, rt, 2-3 h	**II**	
	Catalyst	**II**	
	NaOTf	(72)[nnn]	385
	NaOTf	(81)[ooo]	385
	AgOTf	(63)[zzz]	386
	AgOTf	(92)[aaaa]	386
I	1. MeOH, rt, 2 h 2. HCl, H_2O	**II** (97)[nnn], (93)[ooo]	373

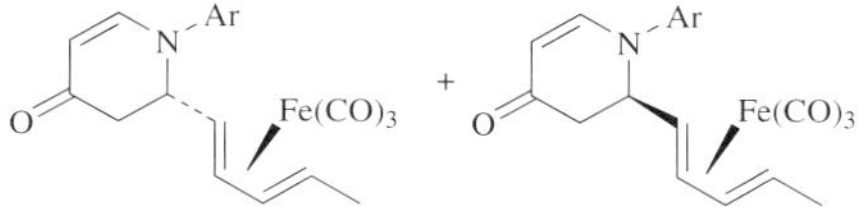

Ar	Conditions	Yield (ratio)	Refs.
Ph	$AlCl_3$, CH_2Cl_2, –78 to 30°, 6 h	(70) 65:35	428
Ph	$LiClO_4$, CH_2Cl_2, rt, 2 h	(92) 90.5:9.5	
4-ClC_6H_4	$LiClO_4$, CH_2Cl_2, rt, 30 min	(82) 89.5:10.5	

TABLE 5. CYCLOADDITIONS OF *N*-ALKYL- AND *N*-ARYLIMINES (*Continued*)

Imine	Diene	Conditions	Product(s) and Yield(s) (%)	Refs.
C_{15} 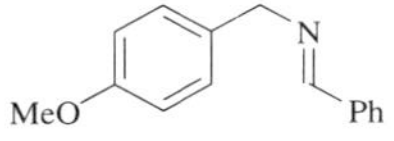	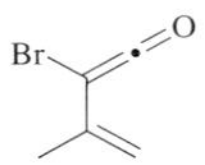	TEA, CH_2Cl_2, 40°	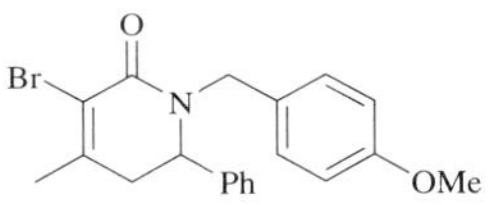 (92)	374
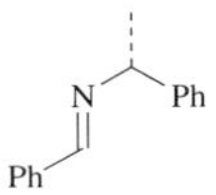	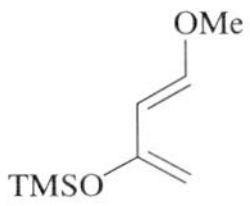	$Zn(OTf)_2$, CH_2Cl_2, rt	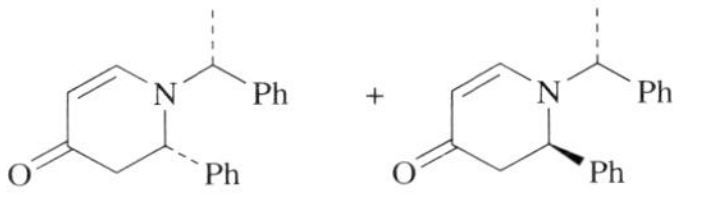 (69) 80:20	170
		Lewis acid, CH_2Cl_2	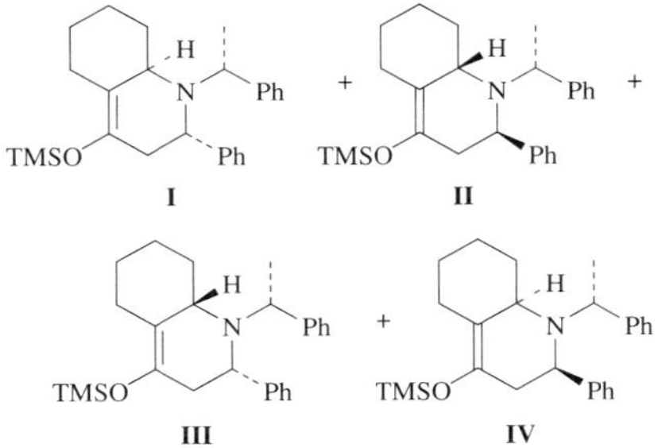 I + II + III + IV	429

L. A.	Temp	Time	Product(s) and Yield(s) (%)
$AlCl_3$	–40°	5 h	**I + II + III + IV** (20)[a]
$AlCl_3$	rt	2 h	**I + II + III + IV** (84)[a], **I + II:III + IV** 33:67[tttt]
$AlCl_3$	rt	48 h	**I + II + III + IV** (55)[a], **I + II:III + IV** 50:50[tttt]
TBSOTf	rt	3 h	**I + II + III + IV** (72)[a], **I + II:III + IV** 57:43[tttt]
TMSOTf	rt	2 h	**I + II + III + IV** (72)[a], **I + II:III + IV** 46:54[tttt]
TMSOTf	rt	48 h	**I + II + III + IV** (95)[a], **I + II:III + IV** 38:62, **I:II** 78:22, **III + IV** 82:18

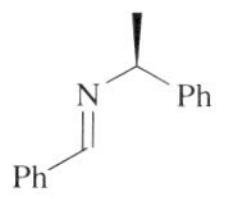

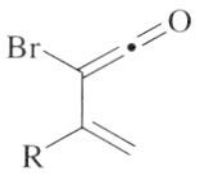

R
Me
n-Pr

TEA, CH_2Cl_2, 40°

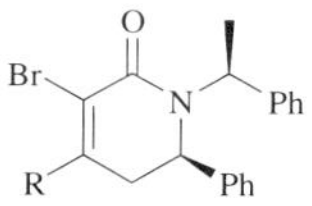 + 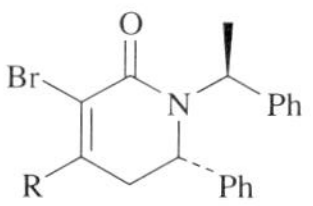 374

(61)	(37)
(37)	(18)

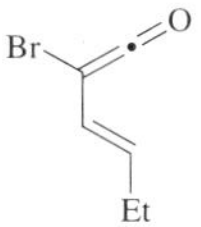

$AlMe_2Cl$, TEA, CH_2Cl_2, −78°

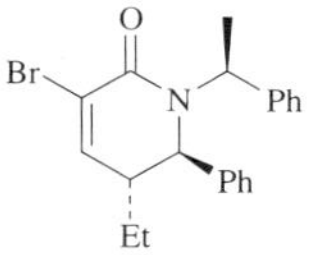 (59) + 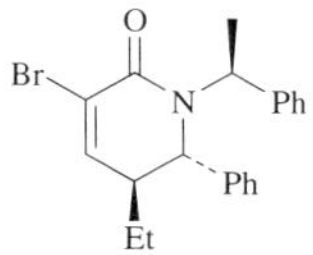 (17) 374

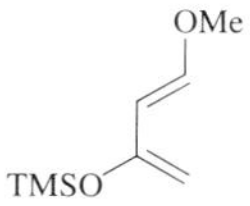

1. Lewis acid, CH_2Cl_2, −78°
2. H_2O

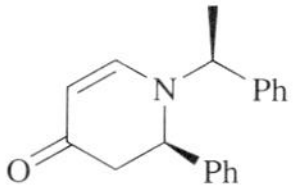 + 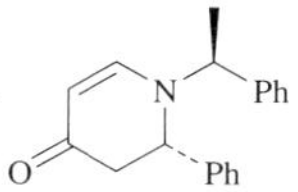

L. A. (eq)	Time		
BF_3•OEt_2 (1.0)	8 h	(63) 85:15	243, 242
BF_3•OEt_2 (1.5)	several h	(41) 92:8	242
BF_3•OEt_2 (1.5)	8 h	(41) 94:6	243
$B(OPh)_3$ (1.0)	several h	(57) 96:4	243, 242, 244
$B(OPh)_3$ (1.5)	8 h	(61) 96:4	243, 242
$EtAlCl_2$ (1.5)	8 h	(16) 77:23	243, 242
$MeAl(OPh)_2$ (1.5)	8 h	(23) 90:10	243, 242
$ZnCl_2$ (1.5)	8 h	(75) 96:4	243, 242

TABLE 5. CYCLOADDITIONS OF *N*-ALKYL- AND *N*-ARYLIMINES (*Continued*)

Imine	Diene	Conditions	Product(s) and Yield(s) (%)	Refs.
C_{15} 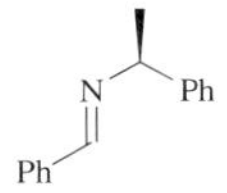	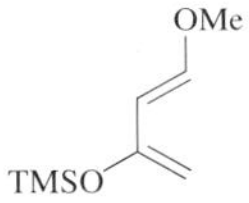	1. Lewis acid, CH_2Cl_2, –78° 2. H_2O	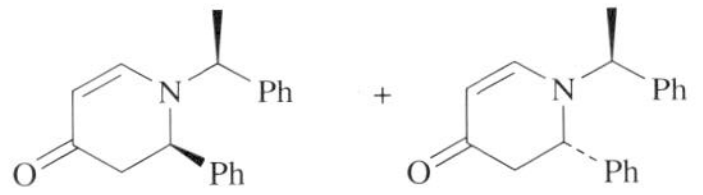	
		L. A. (eq) — Time		
		$TiCl_4$ (1.0) — 8 h	(7) 87:13	243, 242
		$TiCl_2(OPr\text{-}i)_2$ (1.5) — 8 h	(56) 95:5	243, 242
		$B(OPh)_3$/ catechol[uuuu] — several h	(53) 94:6	243, 242
		$B(OPh)_3$/ biphenol[uuuu] — several h	(55) 98:2	243, 242
		$B(OPh)_3$/ 2,2':6',2''-trihydroxyterphenol[uuuu] — several h	(69) 98:2	243, 242
		$B(OPh)_3$/ binaphthol[uuuu] — several h	(58) 99:1	242
		BH_3/ biphenol[uuuu] — several h	(38) 90:10	243
		$B(OMe)_3$/ biphenol[uuuu] — several h	(38) 87:13	243
		$B[O(2\text{-}MeC_6H_4)]_3$/ biphenol[uuuu] — several h	(62) 98:2	243
		$B[O(2,3,5\text{-}Me_3C_6H_2)]_3$/ biphenol[uuuu] — several h	(50) 92:8	243
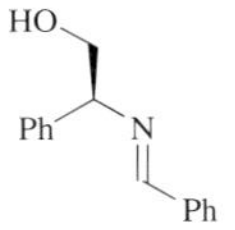	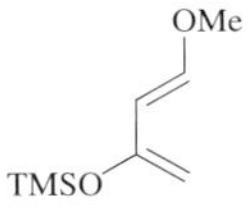	BF_3•OEt_2, CH_2Cl_2, rt	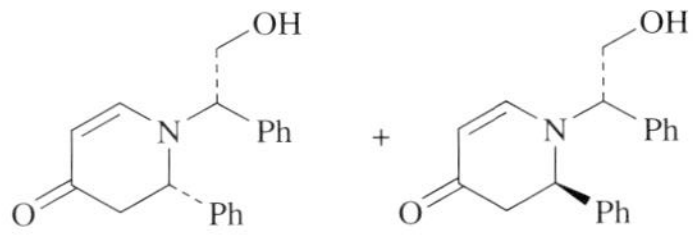 (60) >95:5[bbbb]	170

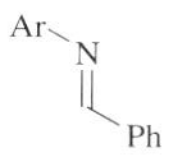 Ar = 4-$Me_2NC_6H_4$	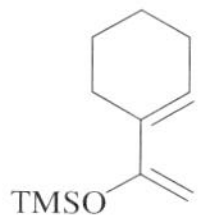	1. $AlCl_3$, CH_2Cl_2 2. $NaHCO_3$, H_2O Temp / Time –40° / 15 min 20° / 1.5 h 20° / 4 h	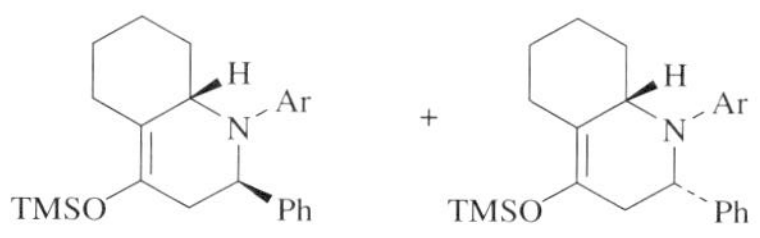 (—) 70:30 (—) 50:50 (—) 50:50	155, 424
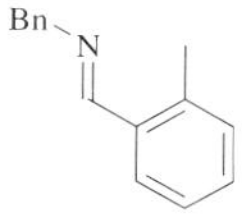	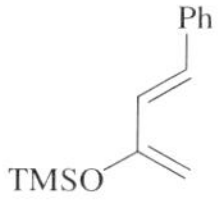	1. Lewis acid 2. TBAF	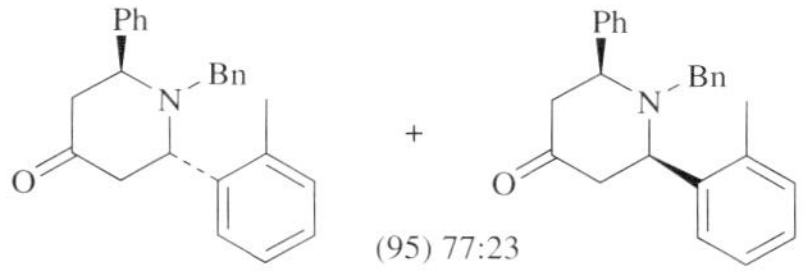 (95) 77:23	381
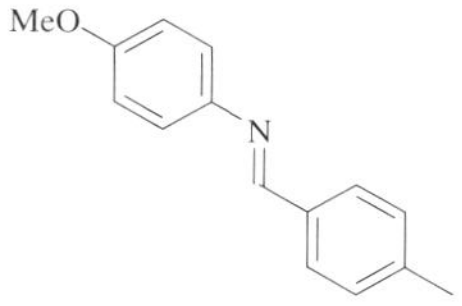	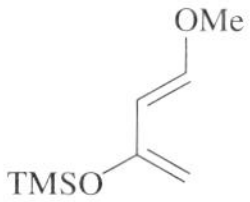	HBF_4, SDS, H_2O, rt, 1 h	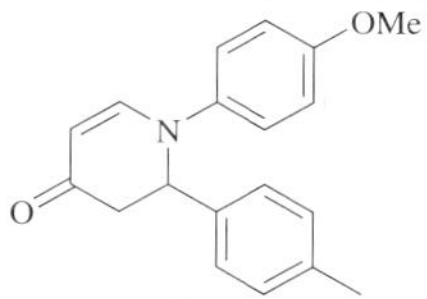 **I** (86)	372
	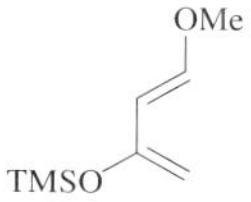	Montmorillonite K10, rt, 3 h Solvent H_2O MeCN, H_2O	**I** (89) (90)	140
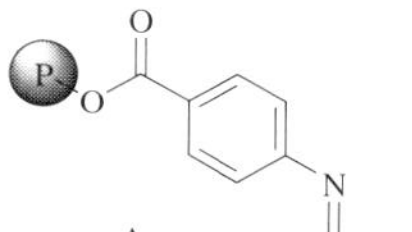 Ar 4-MeC_6H_4 3-MeC_6H_4 4-NCC_6H_4	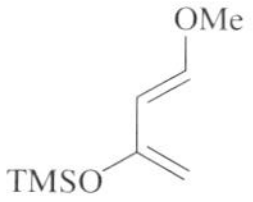	1. $Zn(ClO_4)_2$•$6H_2O$, MeOH, rt, 12 h 2. NaOH, H_2O, rt, 12 h 3. HCl, H_2O	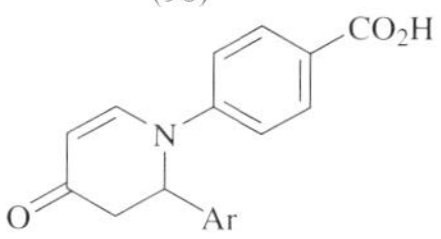 (99) (87) (80)	405

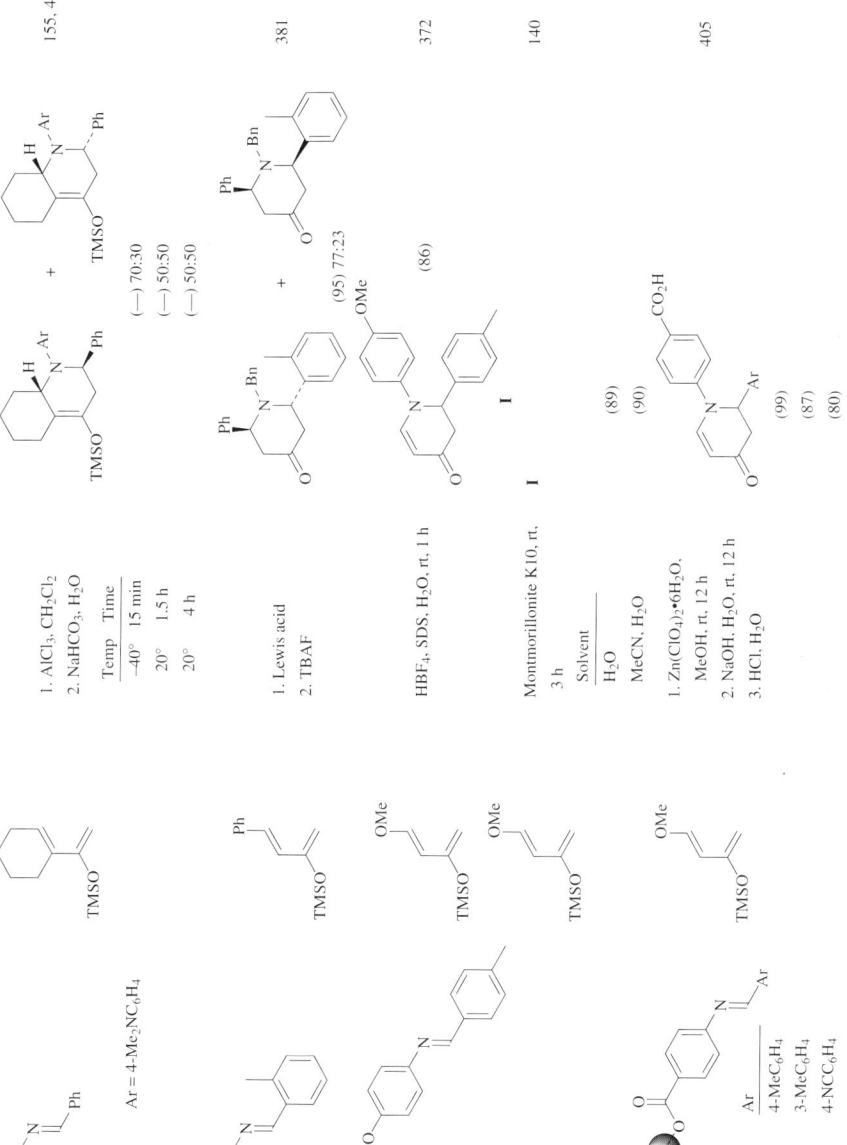

TABLE 5. CYCLOADDITIONS OF *N*-ALKYL- AND *N*-ARYLIMINES (*Continued*)

	Imine	Diene	Conditions	Product(s) and Yield(s) (%)	Refs.
C_{15}	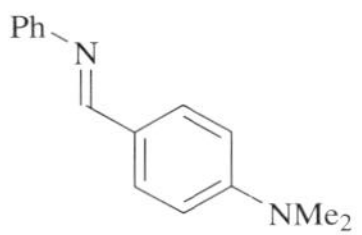	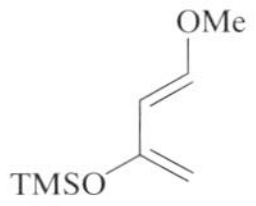	NaOTf, H_2O, rt, 2-3 h	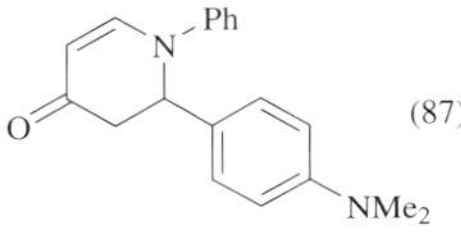 (87)	385
	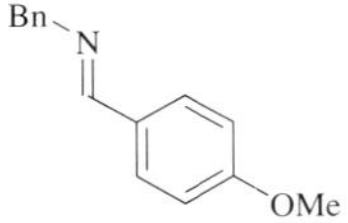	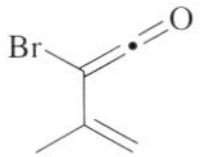	TEA, CH_2Cl_2, 40°	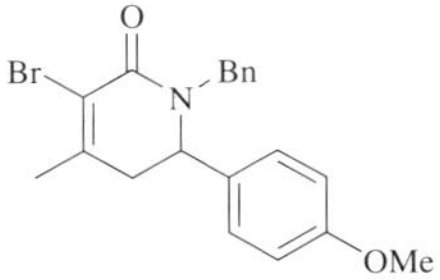 (94)	374
		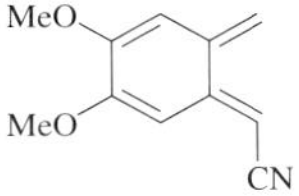	150-160°, 1 h[rr]	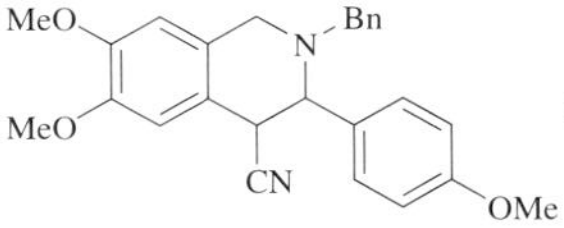 (41.8)[vvvv]	144
	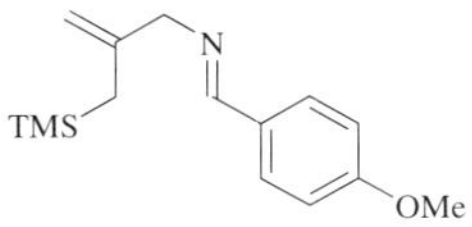	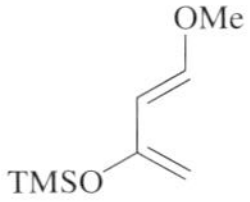	1. $Yb(OTf)_3$, MeCN, rt, 10-15 h 2. H_2O	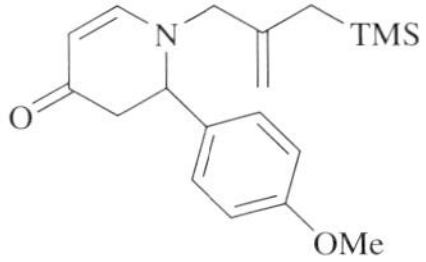 (85)	382
	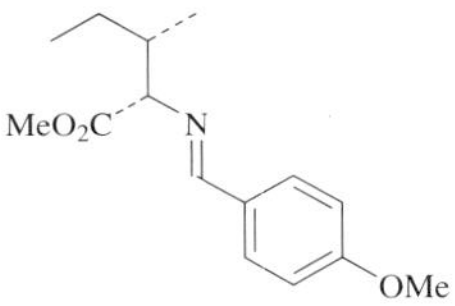	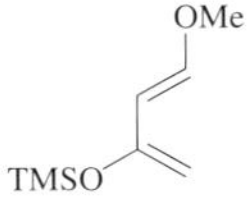	1. $ZnCl_2$, THF, −10°, 7 h 2. $NaHCO_3$, H_2O	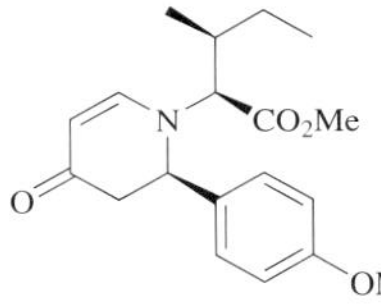 + 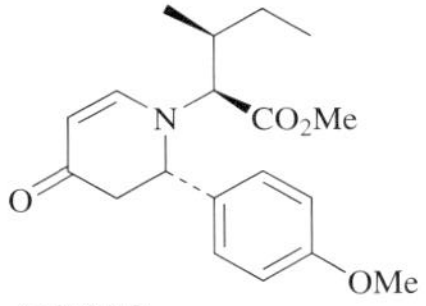 (56) 91:9	384

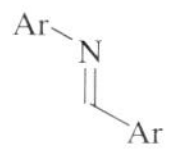 Ar = 4-MeOC_6H_4	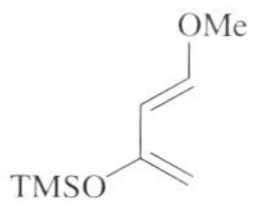	1. $SmI_2(THF)_2$, CH_2Cl_2, rt, 36 h 2. HCl, H_2O	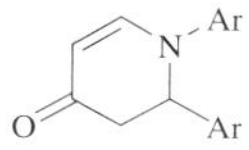 (87)	139
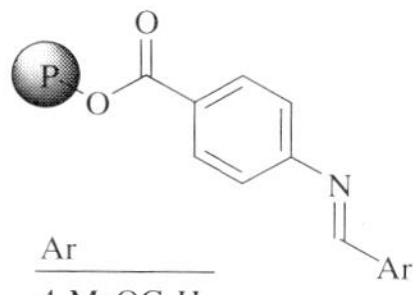 Ar 4-MeOC_6H_4 3-MeOC_6H_4	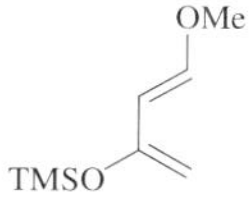	1. $Zn(ClO_4)_2$•$6H_2O$, MeOH, rt, 12 h 2. NaOH, H_2O, rt, 12 h 3. HCl, H_2O	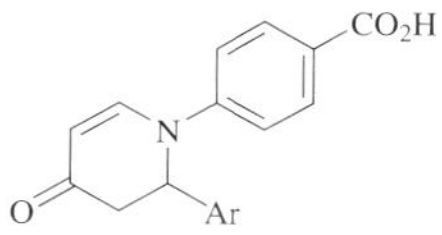 (96) (89)	405
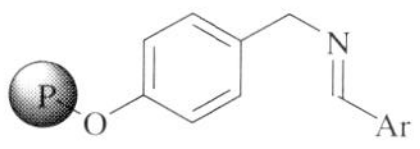 Ar = 3-MeOC_6H_4	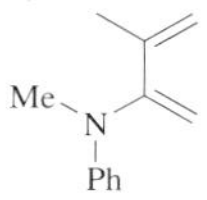	1. $Yb(OTf)_3$, THF, rt, 12 h 2. TFA, CH_2Cl_2, 25 min 3. TMSOTf, CH_2Cl_2, rt, 3 h	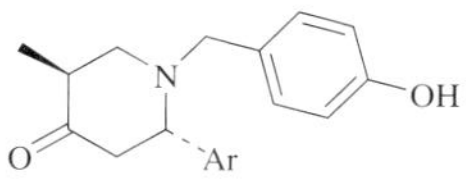 (80)	377
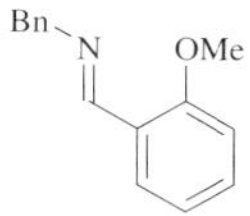	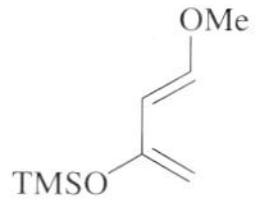	1. $SmI_2(THF)_2$, CH_2Cl_2, rt, 6 h 2. HCl, H_2O	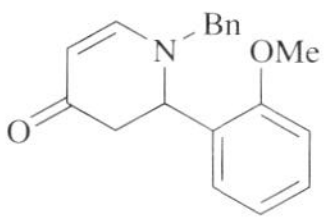 (79)	139
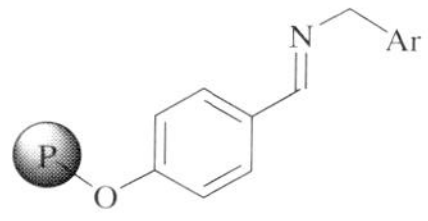 Ar = 4-MeOC_6H_4	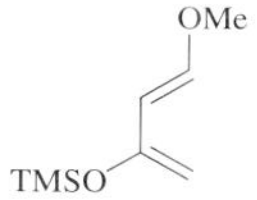	1. $Yb(OTf)_3$, THF, 60°, 3 h 2. H_2O 3. TFA, CH_2Cl_2	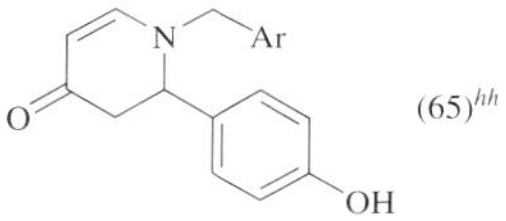 (65)[hh]	378

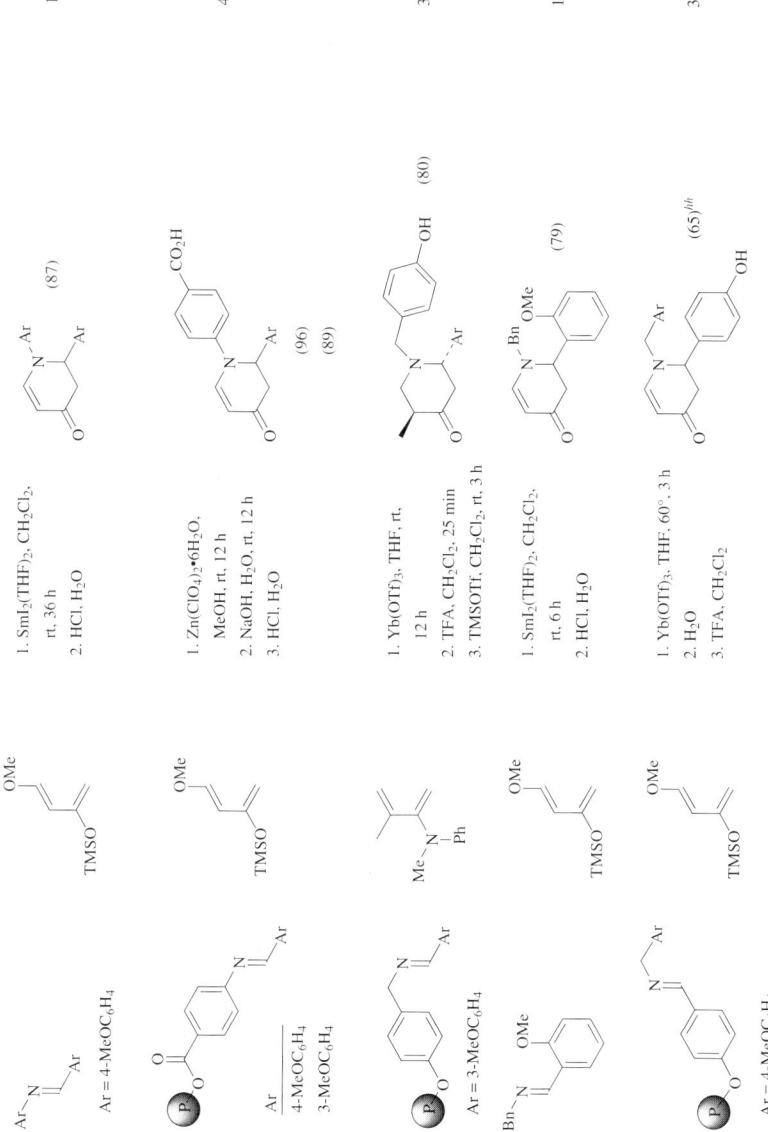

TABLE 5. CYCLOADDITIONS OF *N*-ALKYL- AND *N*-ARYLIMINES (*Continued*)

	Imine	Diene	Conditions	Product(s) and Yield(s) (%)	Refs.
C_{15}	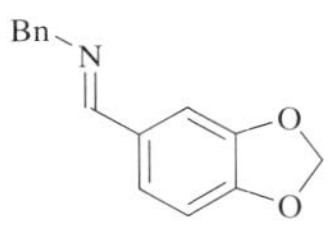	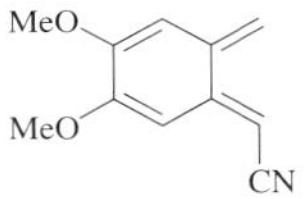	150-160°, 2 h[rr]	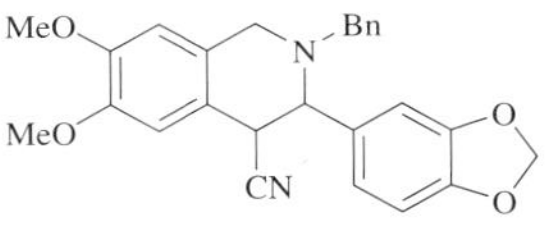 (25.0)[vvvv]	144
	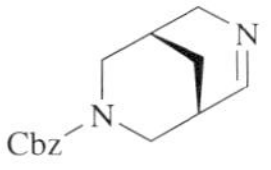	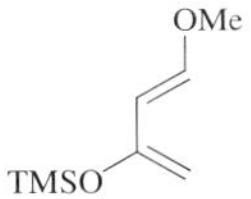	$TiCl_4$	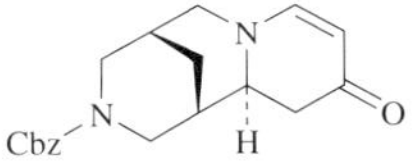 ("traces")[wwww]	305
	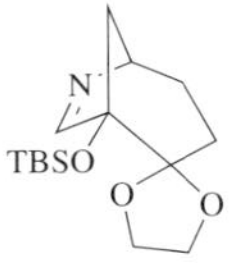	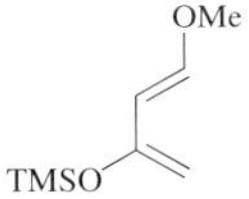		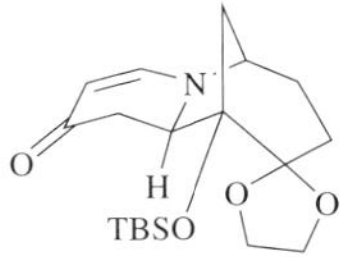	256, 257
			$Yb(OTf)_3$, MeCN, 0° to rt, overnight	(84)[xxxx]	
			CH_2Cl_2, 12 kbar, rt, 48 h	(71)	
	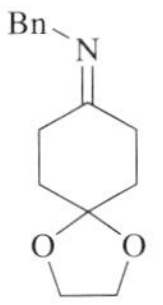	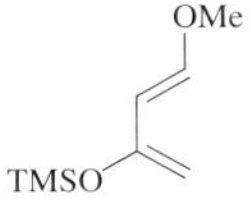	1. $Zn(OTf)_2$, CH_2Cl_2, rt, 12 h 2. NH_4Cl, H_2O, 2 h	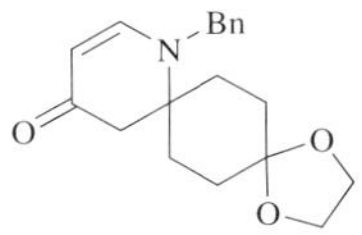 (40)	356
C_{16}	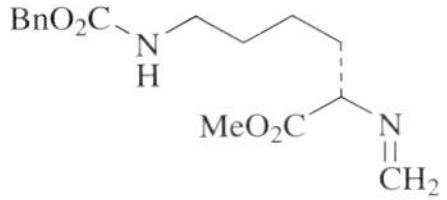		HCl, H_2O, rt, 45 min	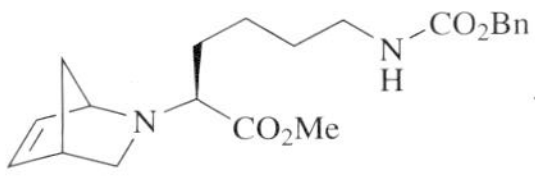 I + 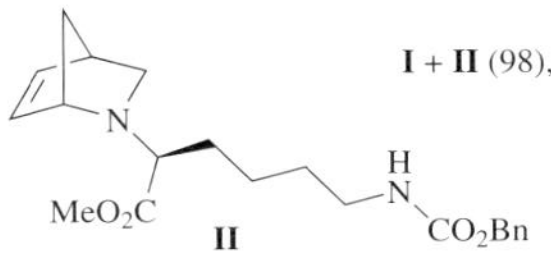 II **I** + **II** (98), 1.1:1[f]	354

Imine	Diene	Conditions	Product(s) and Yield(s) (%)	Refs.
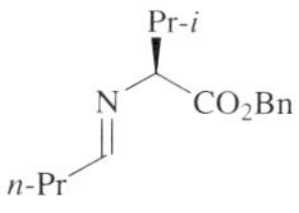	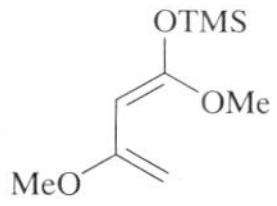	1. $EtAlCl_2$[bbb], –78° to rt, CH_2Cl_2, 1-1.5 h 2. $NaHCO_3$, H_2O	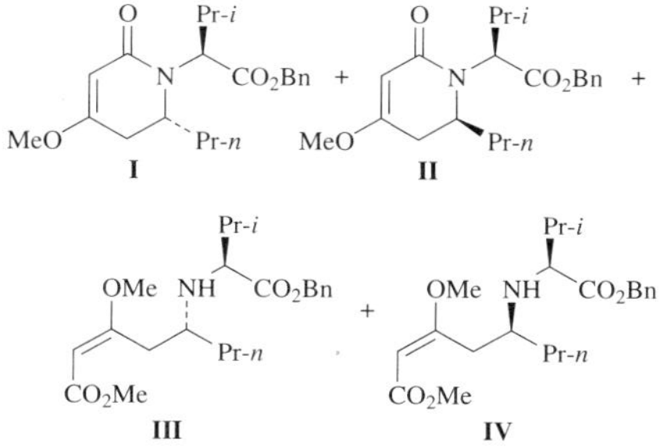 **I** + **II** + **III** + **IV** (35), **I** + **II**:**III** + **IV** 80:20, **I** + **III**:**II** + **IV** 92:8[ccc]	402, 175a
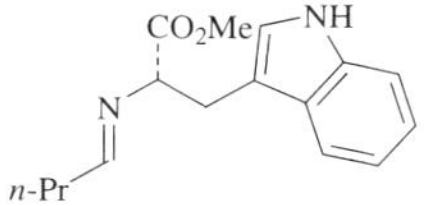	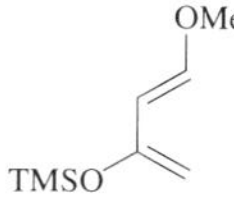	1. $ZnCl_2$[gggg], THF 2. $NaHCO_3$, H_2O	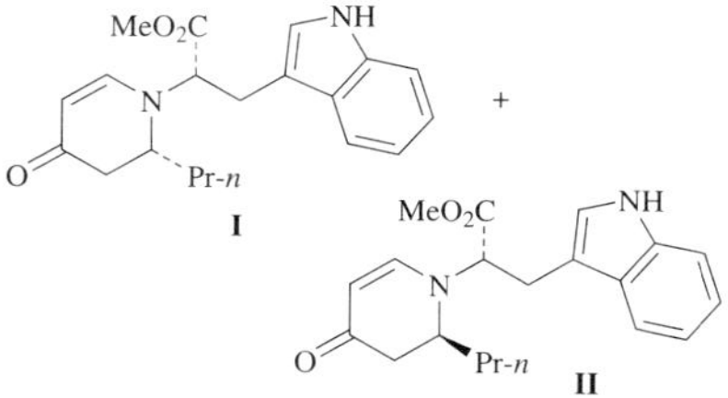	
		Temp: –20 to 0°, Time: 12 h	**I** + **II** (55), **I**:**II** 83:17	430
		Temp: rt, Time: overnight	**I** + **II** (55), **I**:**II** 83:17	419

TABLE 5. CYCLOADDITIONS OF *N*-ALKYL- AND *N*-ARYLIMINES (*Continued*)

Imine	Diene	Conditions	Product(s) and Yield(s) (%)	Refs.
C_{16}				
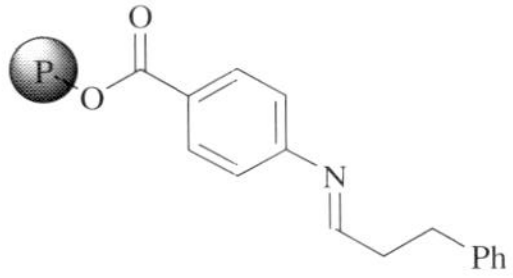	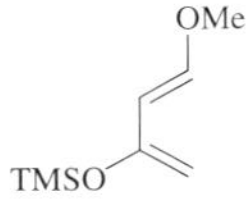	$Zn(ClO_4)_2 \cdot 6H_2O$, MeOH, rt	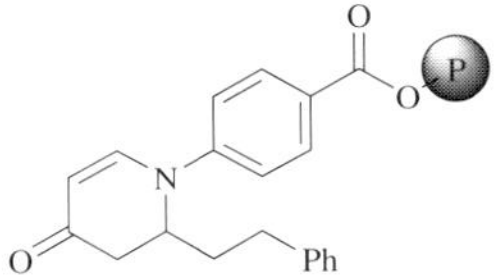 (<50)	405
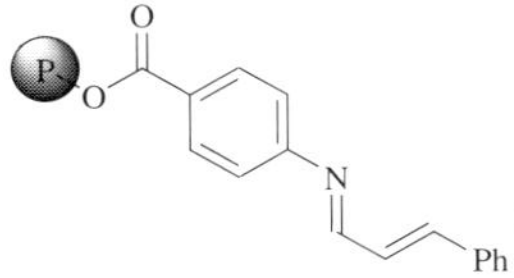	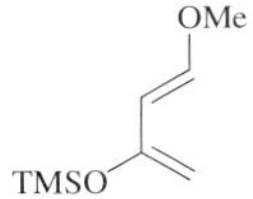	$Zn(ClO_4)_2 \cdot 6H_2O$, MeOH, rt	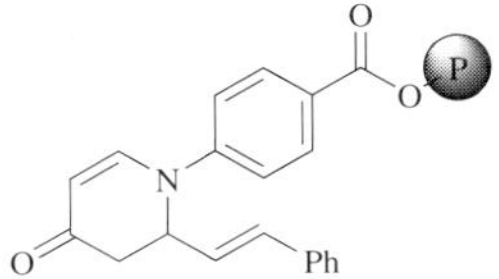 (<50)	405
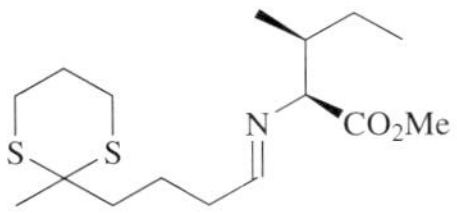	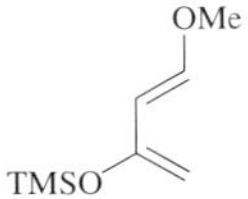		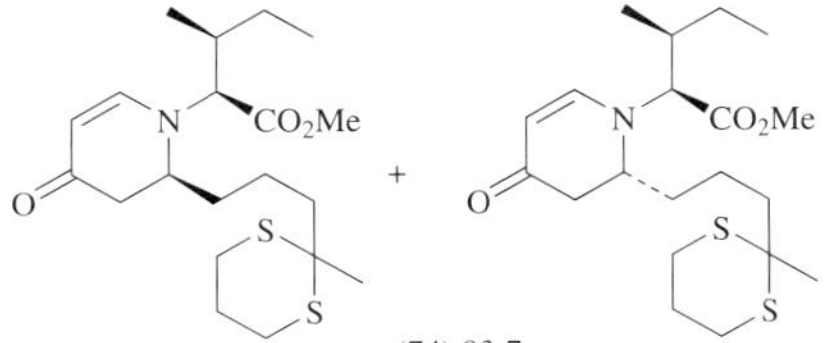	
		1. $EtAlCl_2$[jj], CH_2Cl_2, –78 to 20° 2. H^+, H_2O	(74) 93:7	175a
		1. $EtAlCl_2$, CH_2Cl_2, –78 to rt, 1 h 2. $NaHCO_3$, H_2O	(75) 93:7	384
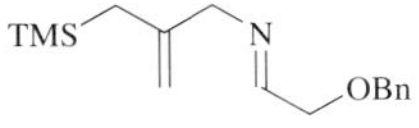	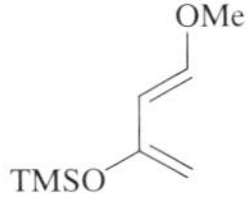	1. $Yb(OTf)_3$, MeCN, rt, 10-15 h 2. H_2O	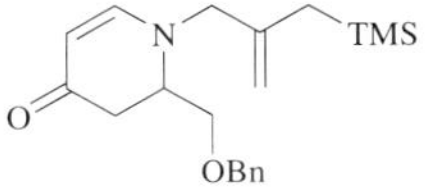 (70)	382

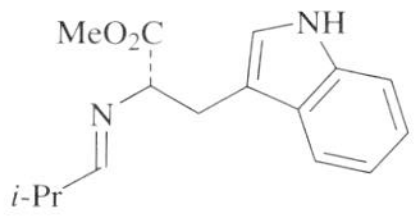

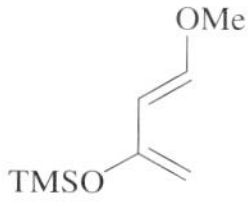

1. $ZnCl_2$[gggg], THF
2. $NaHCO_3$, H_2O

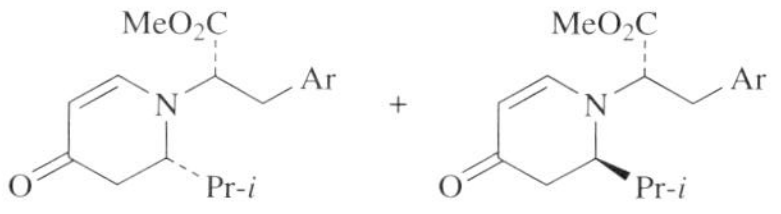

I Ar = 3-indolyl II

Temp	Time	Product(s) and Yield(s) (%)	Refs.
–20 to 0°	12 h	I + II (55), I:II 85:15	430
rt	overnight	I + II (55), I:II 85:15	419

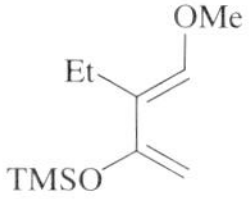

1. $B(OPh)_3$, 4 Å MS, CH_2Cl_2, –78° to rt, overnight
2. HCl, H_2O

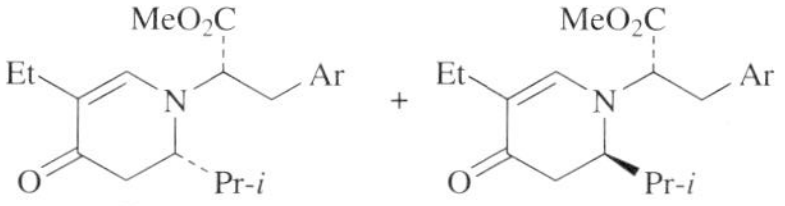

I Ar = 3-indolyl II

I + **II** (29), **I**:**II** >98:2[yyyy] — 246, 245

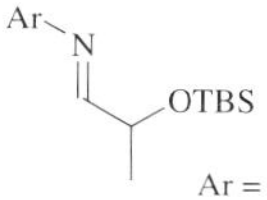

Ar = 4-$MeOC_6H_4$

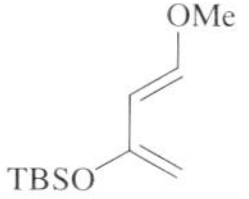

Montmorillonite K10, H_2O, 0°, 4 h

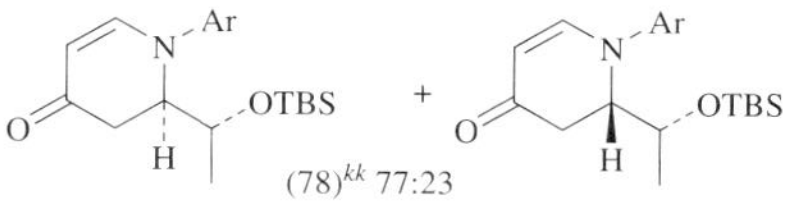

(78)[kk] 77:23 — 140

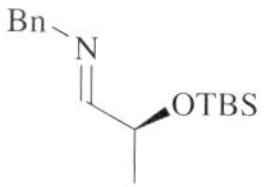

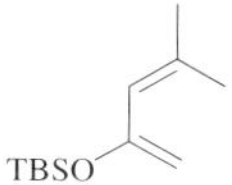

1. Lewis acid
2. $NaHCO_3$, H_2O
3. TBAF, THF, 10 h

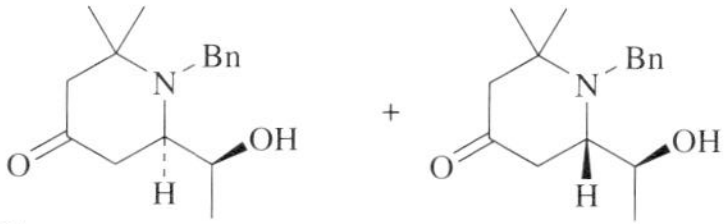

L. A.	Solvent	Temp	Time	Product(s) and Yield(s) (%)	Refs.
$TiCl_4$	*i*-PrCN	–30°	1 d	(65) 9:91	150, 161
$Zn(OTf)_2$	*n*-hexane	4°	2 d	(75) 43:57	150, 161
TMSOTf	CH_2Cl_2	4°	1 d	(80) 43:57	150, 161
TMSOTf	—	0°	15 min	(63)	150

TABLE 5. CYCLOADDITIONS OF *N*-ALKYL- AND *N*-ARYLIMINES (*Continued*)

Imine	Diene	Conditions	Product(s) and Yield(s) (%)	Refs.
C_{16}				
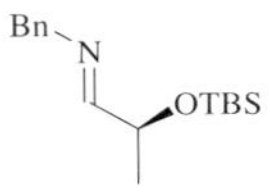	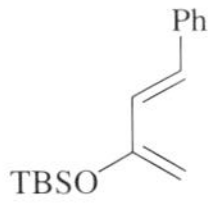	1. Lewis acid 2. $NaHCO_3$, H_2O 3. TBAF, THF, 10 h	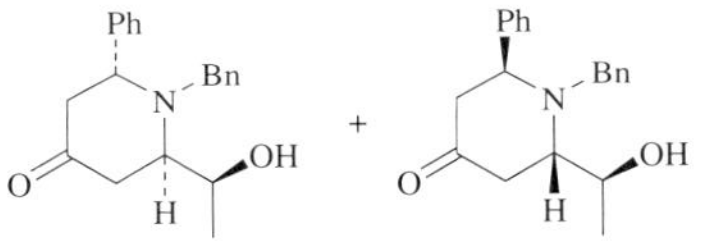	

L. A.	Solvent	Diene eq	Temp	Time	Product(s) and Yield(s) (%)	Refs.
TMSOTf	CH_2Cl_2	2	4°	1-7 d	(90) 77:23	150, 161
TMSOTf	toluene	—	4°	3 d	(97) 78:22	150, 161
TMSOTf	CH_2Cl_2	—	rt	3 d	(76) 66:34	150
TMSOTf	CH_2Cl_2	—	–78°	7 d	(39) 51:49	150
TMSOTf	—	—	0°	1 min	(20)	150
H_3SiOTf	CH_2Cl_2	2	4°	1-7 d	(91) 55:45	150
TIPSOTf	CH_2Cl_2	2	4°	1-7 d	(64) 56:44	150
$Zn(OTf)_2$	CH_2Cl_2	2	4°	1-7 d	(34) 92:8	150, 161
$Zn(OTf)_2$	CH_2Cl_2	4	4°	4 d	(98) 77:23	150, 161
$Zn(OTf)_2$	*n*-hexane	4	4°	2 d	(93) 89:11	150, 161
$Zn(OTf)_2$	*n*-hexane	—	rt	1 d	(53) 94:6	150, 161
$Zn(OTf)_2$	*n*-hexane	—	—	—	(92) 62:38[zzzz]	150, 161
$BF_3•OEt_2$	CH_2Cl_2	2	4°	1-7 d	(45) 45:55	150, 161
$ZnCl_2$	CH_2Cl_2	2	4°	1-7 d	(78) 37:63	150, 161
$ZnBr_2$	CH_2Cl_2	2	4°	1-7 d	(76) 47:53	150, 161
$SnCl_4$	CH_2Cl_2	2	4°	1-7 d	(24) 37:63	150, 161
$TiCl_4$	CH_2Cl_2	2	4°	1-7 d	(50) 19:81	150, 161
$TiCl_4$	CH_2Cl_2	4	4°	1 d	(86) 30:70	150, 161
$TiCl_4$	CH_2Cl_2/ sulfolane	—	4°	1 d	(65) 24:76	150

$TiCl_4$	*i*-PrCN	—	4°	1 d	(67) 24:76	150, 161
$TiCl_4$	*i*-PrCN	4	−30°	1 d	(87) <2:98	150, 161
$TiCl_4$	*t*-BuCN	—	4°	1 d	(60)18:82	150, 161
$TiCl_4$	ether	—	4°	1 d	(96) 53:47	150, 161
$TiCl_4$	toluene	—	4°	1 d	(90) 68:32	150, 161
$TiCl_4$	CCl_4	—	4°	1 d	(71) 49:51	150
$TiCl_4$	*n*-hexane	—	4°	1 d	(12) >98:2	150

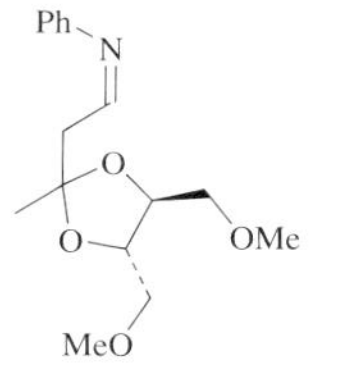
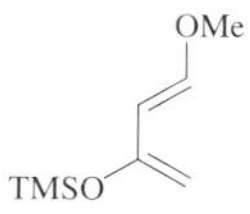

ZnI_2

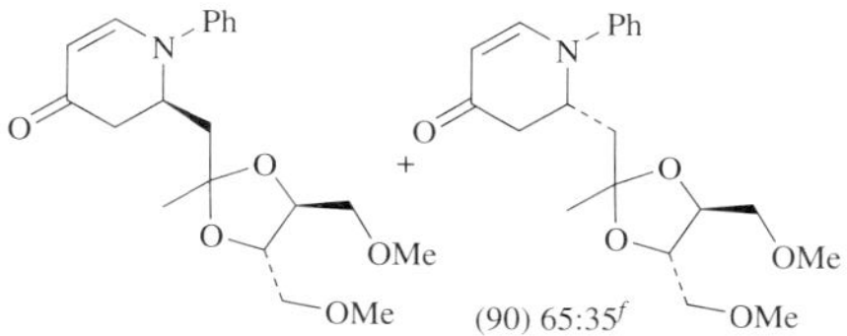

(90) 65:35[f]

138

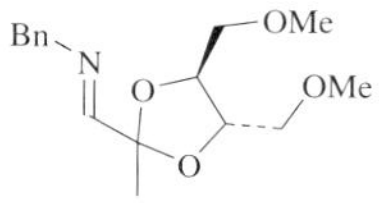
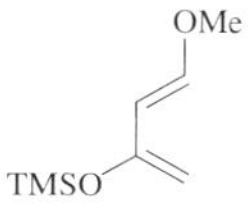

Lewis acid

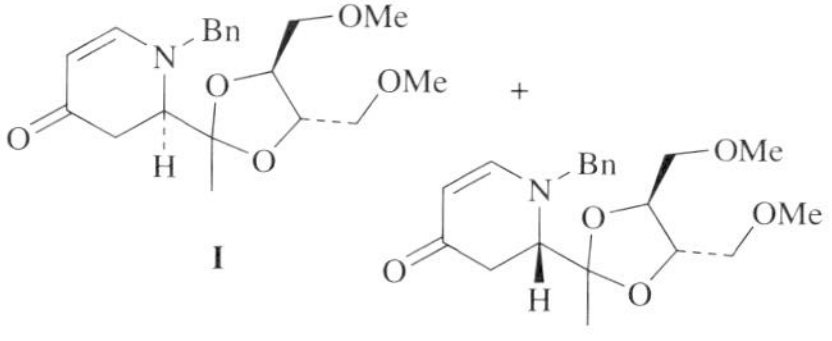

I + II

431

L. A. (eq)	Solvent	Temp	Time	**I + II**	**I:II**
$ZnCl_2$ (1.1)	THF	−78° to rt	26 h	(20)	63:37
$EtAlCl_2$ (1.1)	CH_2Cl_2	−78°	5 min	(13)	82:18
$BF_3•OEt_2$ (1.1)	CH_2Cl_2	−78°	5 min	(27)	87:13
$BF_3•OEt_2$ (2.2)	CH_2Cl_2	−78°	5 min	(27)	>99:1
$BF_3•OEt_2$ (3.3)	CH_2Cl_2	−78°	5 min	(32)	>99:1

TABLE 5. CYCLOADDITIONS OF *N*-ALKYL- AND *N*-ARYLIMINES (*Continued*)

Imine	Diene	Conditions	Product(s) and Yield(s) (%)	Refs.
C_{16}				
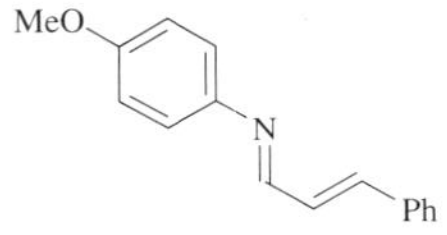	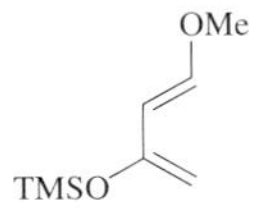	HBF_4, SDS, H_2O, rt, 1 h	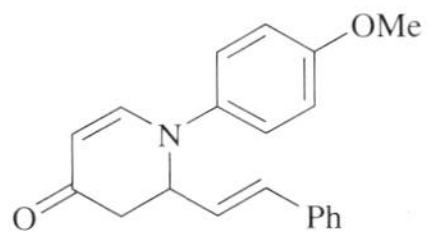 (86)	372
	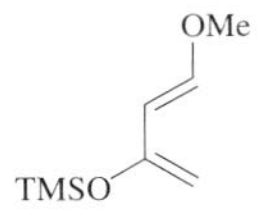	Montmorillonite K10, rt	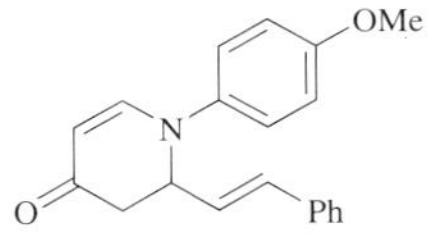	140
		Solvent / Time		
		H_2O / 4 h	(<5)	
		CH_3CN, H_2O / 1 h	(95)	
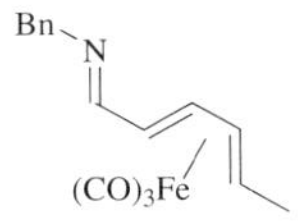	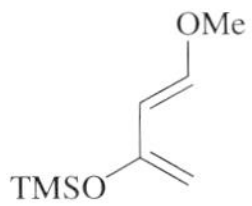	Lewis acid	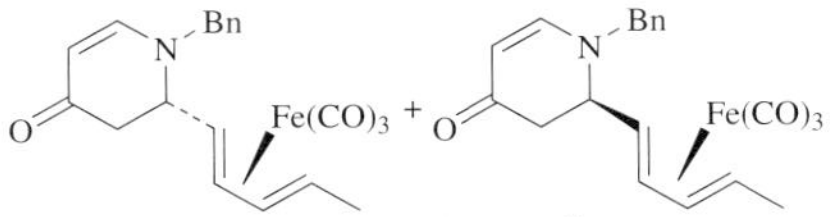 ("small amount")	428

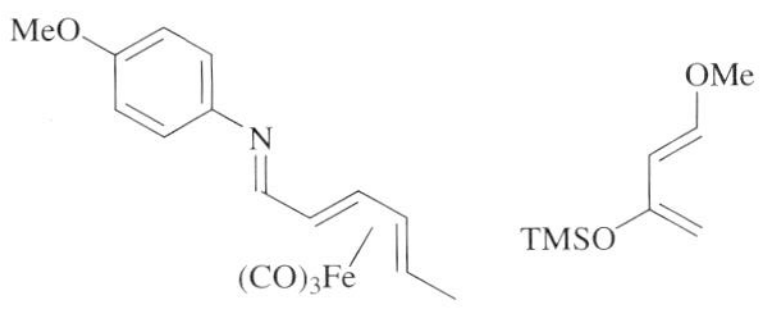

Lewis acid

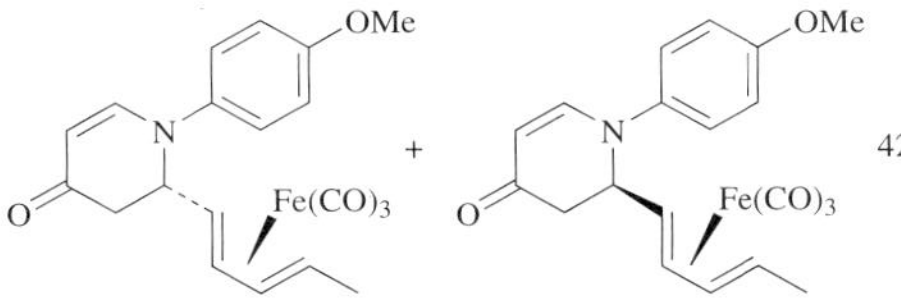

428

L. A. (eq)	Solvent	Temp	Time	
$AlCl_3$ (1.1)	CH_2Cl_2	–78 to –30°	6 h	(59) 81.5:18.5
$AlCl_3$ (0.2)	CH_2Cl_2	–78 to –30°	6 h	(28) 72.5:27.5
TMSOTf (1.1)	CH_2Cl_2	–78 to –30°	8 h	(77) 87:13
$LiClO_4$ (5 M)	ether	rt	1 h	(89) 93:7
$LiClO_4$ (2.0)	CH_2Cl_2	rt	3 h	(87) 90.5:9.5
$LiClO_4$ (1.1)	CH_2Cl_2	rt	3 h	(80) >97.5:2.5
$LiClO_4$ (0.2)	CH_2Cl_2	rt	8 h	(93) >97.5:2.5
$LiClO_4$ (0.2)	THF	rt	8 h	(73) 88.5:11.5

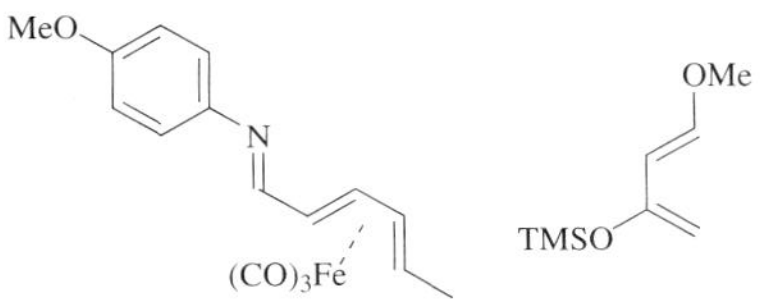

$LiClO_4$, CH_2Cl_2

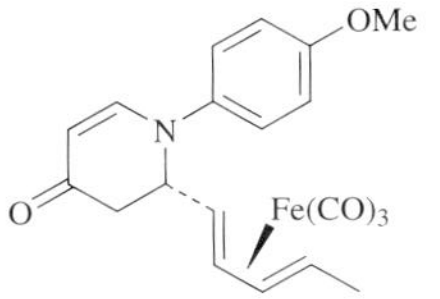

(93)

428

$Zn(OTf)_2$, CH_2Cl_2, rt

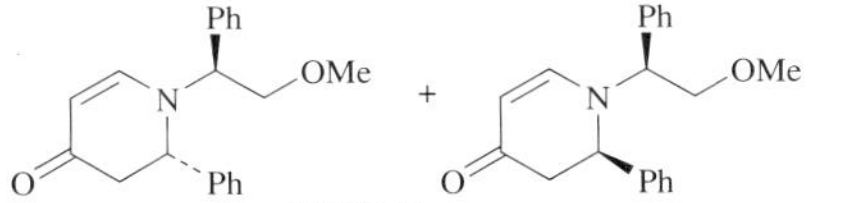

(51) 87:13

170

TABLE 5. CYCLOADDITIONS OF *N*-ALKYL- AND *N*-ARYLIMINES (*Continued*)

Imine	Diene	Conditions	Product(s) and Yield(s) (%)	Refs.
C_{16} 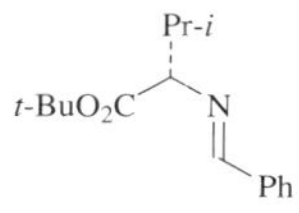	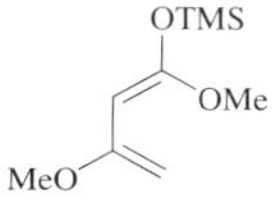	1. Lewis acid[bbb], –78 to 20°, CH_2Cl_2, 1-1.5 h 2. $NaHCO_3$, H_2O	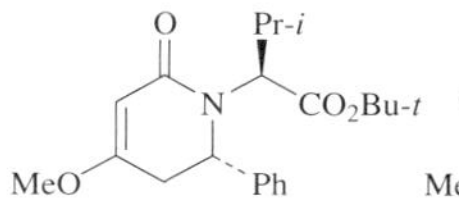 + 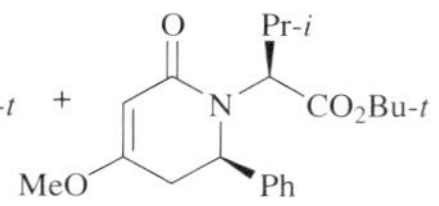 + **I** **II** 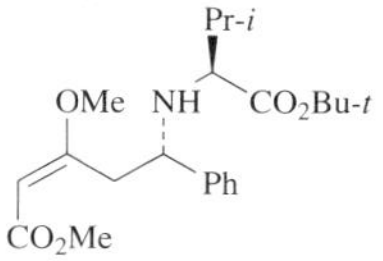 + 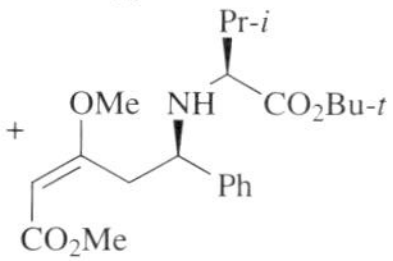 **III** **IV**	
		L. A.		
		$EtAlCl_2$	**I** + **II** + **III** + **IV** (84), **I** + **II**:**III** + **IV** <2:98, **I** + **III**:**II** + **IV** 97.5:2.5[llll]	402, 175a
		$TiCl_4$	**I** + **II** + **III** + **IV** (30), **I** + **II**:**III** + **IV** <2:98, **I** + **III**:**II** + **IV** 94:6[llll]	402, 175a
		BF_3•OEt_2	**I** + **II** + **III** + **IV** (28), **I** + **II**:**III** + **IV** <2:98, **I** + **III**:**II** + **IV** 92:8[llll]	402
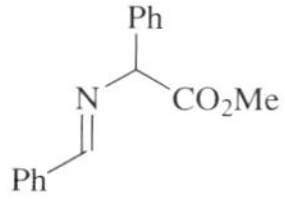	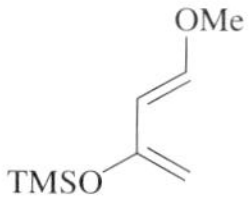	$Zn(OTf)_2$, CH_2Cl_2, rt	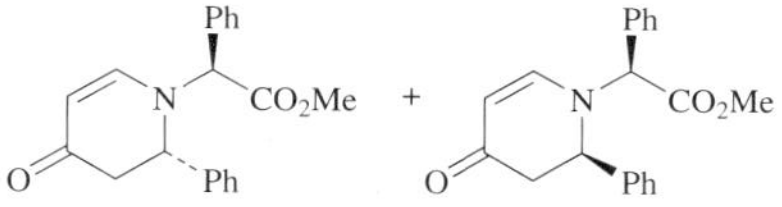 ("small amount")	170
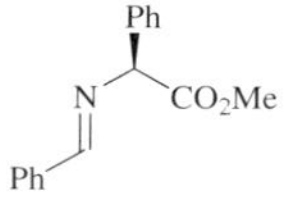	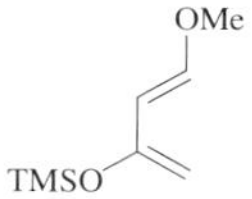	1. $ZnCl_2$, THF, 0°, 7 h 2. $NaHCO_3$, H_2O	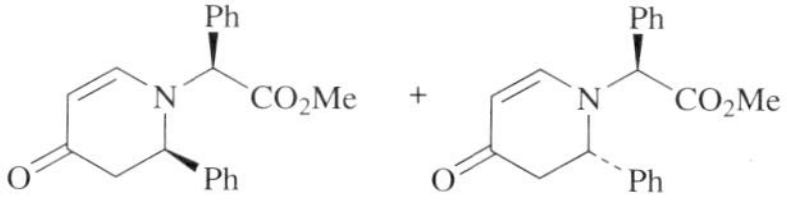 (25) 60:40	384

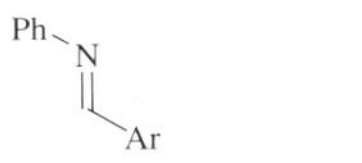

Ar = 2,4,6-$Me_3C_6H_2$

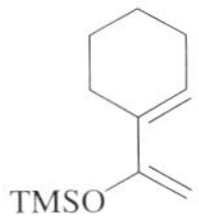

1. TBSOTf, CH_2Cl_2, 20°
2. $NaHCO_3$, H_2O

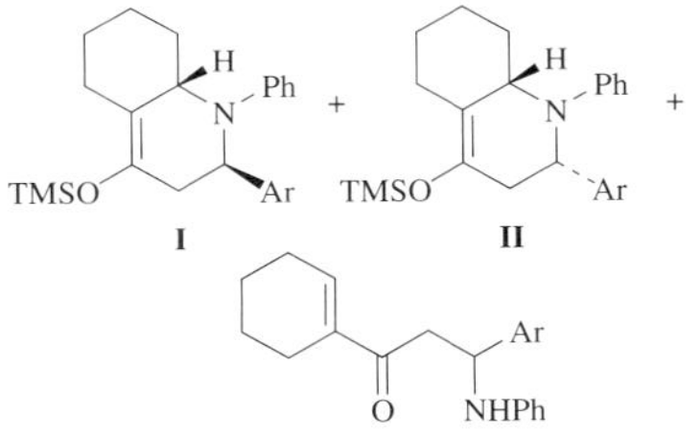

I
II
III

TBSOTf eq	Time	
0.1	1.5 h	**I** + **II** (43)[a], **I**:**II** >98:2
0.1	15 h	**I** + **II** + **III** (70)[a], **I**:**II**:**III** 18:56:25
1.0	0.5 h	**I** + **II** + **III** (71)[a], **I**:**II**:**III** 70:15:10
1.0	1.5 h	**I** + **II** + **III** (80)[a], **I**:**II**:**III** 75:16:9

156

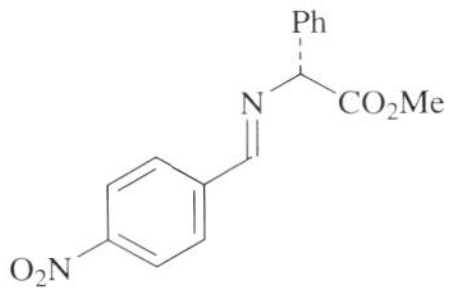

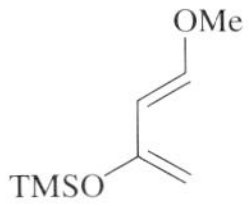

1. $ZnCl_2$, THF, 0°, 7 h
2. $NaHCO_3$, H_2O

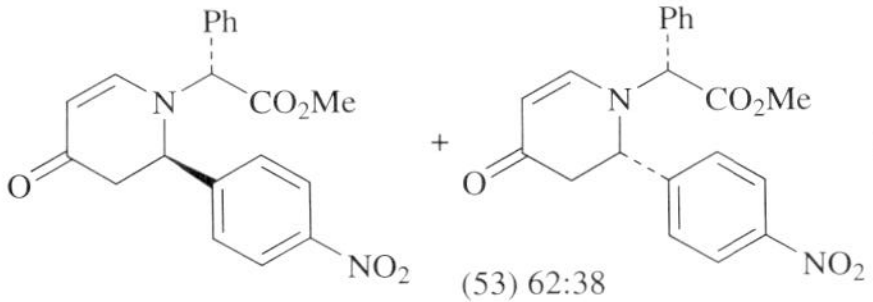

(53) 62:38

384

	Imine	Diene	Conditions	Product(s) and Yield(s) (%)	Refs.
C_{16}	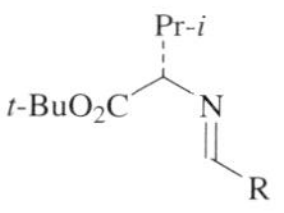	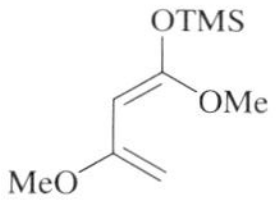	1. $EtAlCl_2$[bbb], –78 to 20°, CH_2Cl_2, 1-1.5 h 2. $NaHCO_3$, H_2O	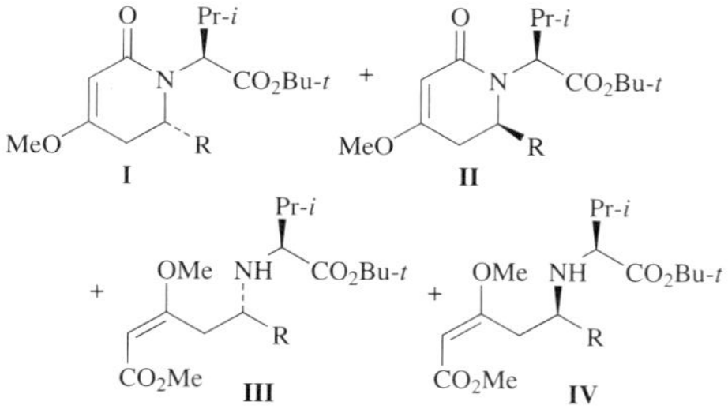	
	R				
	3-ClC_6H_4			**I** + **II** + **III** + **IV** (64), **I** + **II**:**III** + **IV** <2:98, **I** + **III**:**II** + **IV** 95:5[llll]	402, 175a
	4-ClC_6H_4			**I** + **II** + **III** + **IV** (67), **I** + **II**:**III** + **IV** <2:98, **I** + **III**:**II** + **IV** 97:3[llll]	402, 175a
	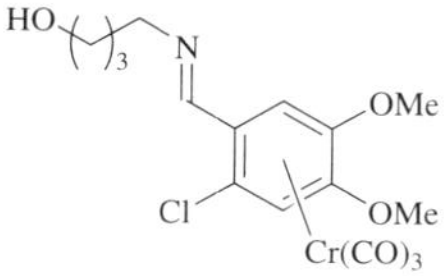	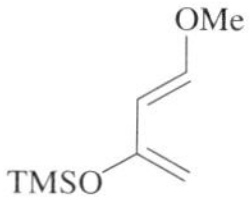	$SnCl_4$,THF, –78° to rt 16 h	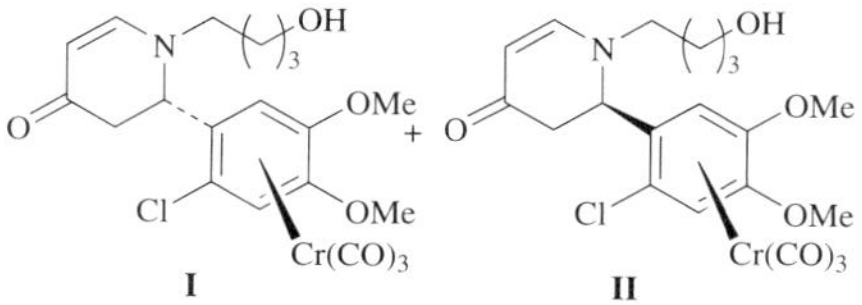 **I** + **II** (70), **I**:**II** >98:2[bbbb]	167
	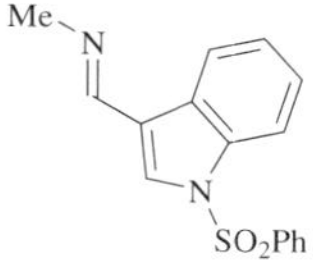	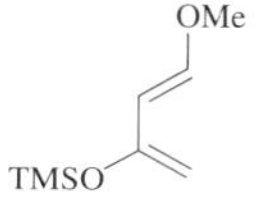	$Zn(OTf)_2$, H_2O, CH_2Cl_2, rt, 30 h	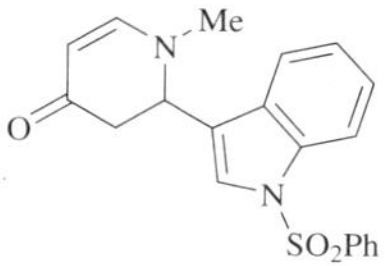 (75)	432

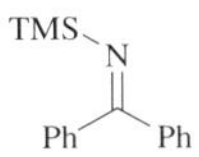TMS N Ph Ph	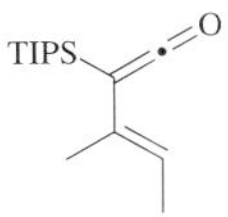TIPS O	1. MeCN, 81°, 45 min 2. SiO_2	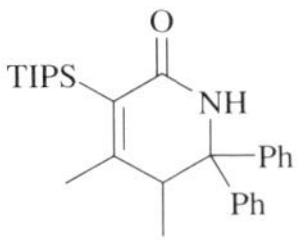TIPS O NH Ph Ph (79)	142
	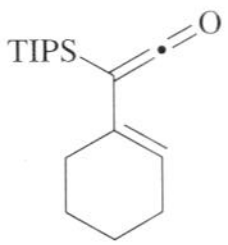TIPS O	1. MeCN, 81°, 24 h 2. SiO_2	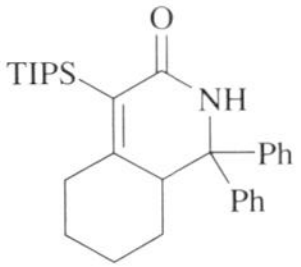TIPS O NH Ph Ph (66)	142
C_{17} 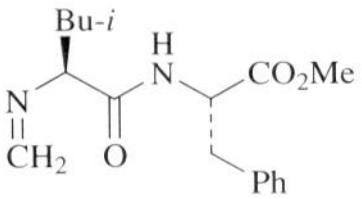Bu-*i* H N CO2Me N CH2 O Ph		HCl, H_2O, rt, 2 h	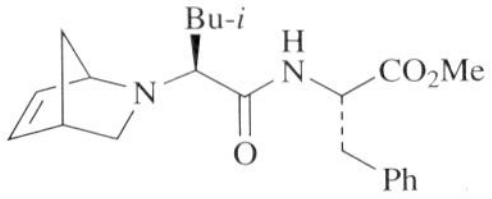Bu-*i* H N CO2Me N O Ph + 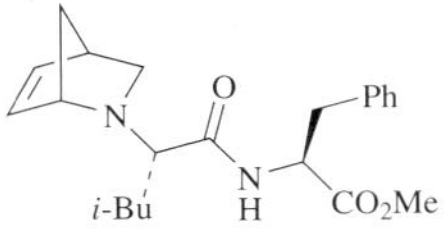O Ph N *i*-Bu N H CO2Me (87) 1:1	354
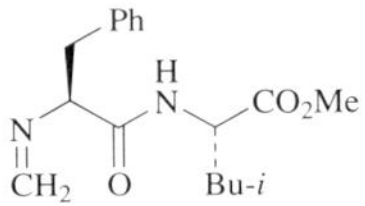Ph H N CO2Me N CH2 O Bu-*i*		HCl, H_2O, rt, 2 h	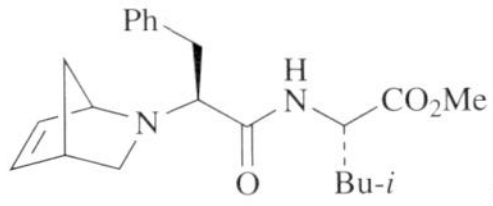Ph H N CO2Me N O Bu-*i* + 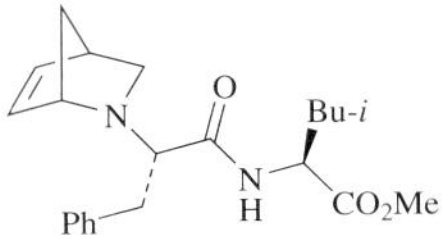O Bu-*i* N N H CO2Me Ph (95) 1:1	354, 350

TABLE 5. CYCLOADDITIONS OF *N*-ALKYL- AND *N*-ARYLIMINES (*Continued*)

Imine	Diene	Conditions	Product(s) and Yield(s) (%)	Refs.
C_{17} 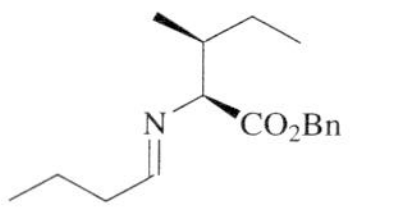	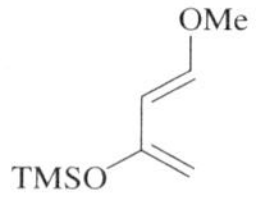		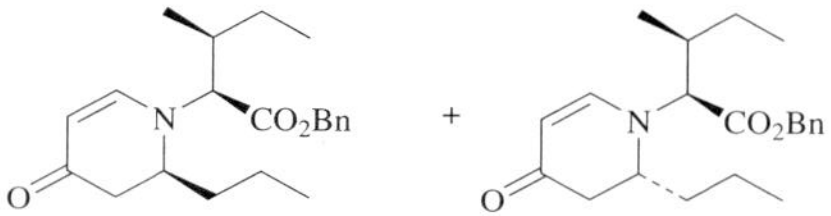	
		1. $ZnCl_2$[jj], THF, –13° 2. H^+, H_2O	(65) 90:10	175a
		1. $ZnCl_2$, THF, 0°, 7 h 2. $NaHCO_3$, H_2O	(65) 90:10	384
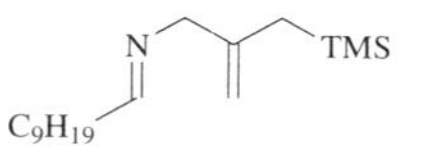	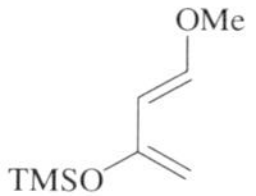	1. $Yb(OTf)_3$, MeCN, rt, 10-15 h 2. H_2O	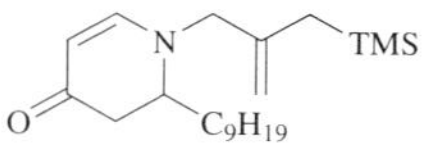 (68)	382
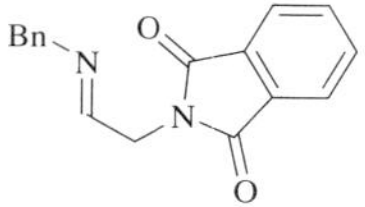		BF_3•OEt_2, TFA, CH_2Cl_2, –78° to rt, overnight	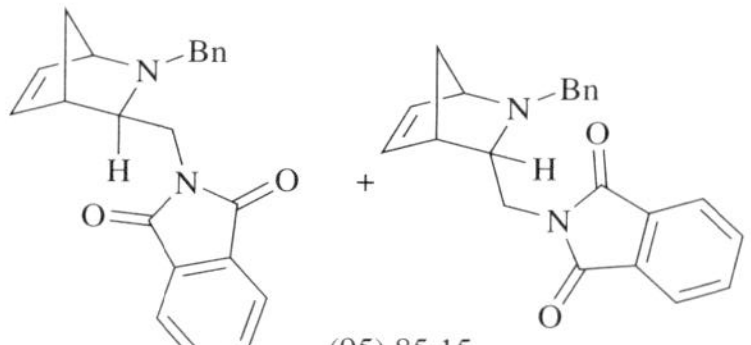 (95) 85:15	403
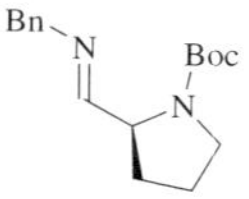		BF_3•OEt_2, TFA, CH_2Cl_2, –78° to rt, overnight	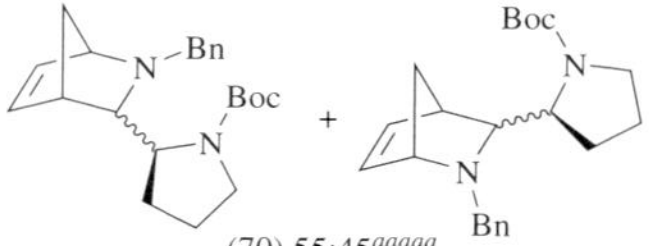 (70) 55:45[aaaaa]	403
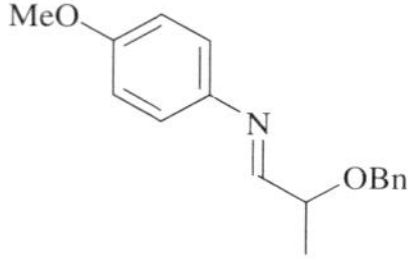	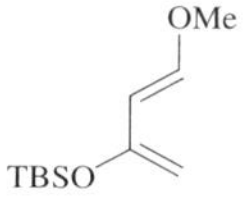	Montmorillonite K10, H_2O, 0°, 2.5 h	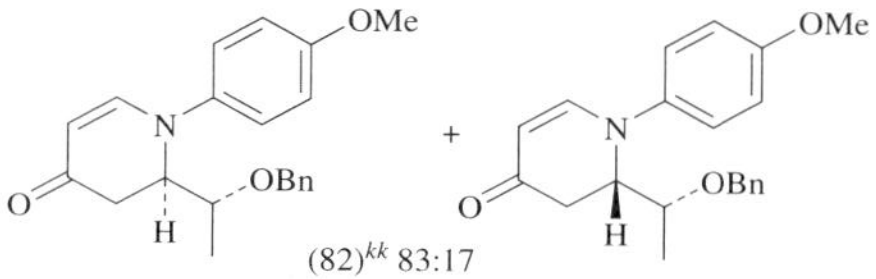 (82)[kk] 83:17	140

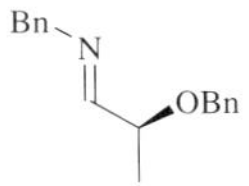	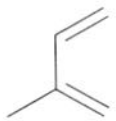	$ZnCl_2$•OEt_2, 7-10 d	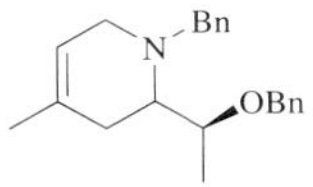 (20)[j,bbbbb]	403
		BF_3•OEt_2, TFA, CH_2Cl_2, –78° to rt, overnight	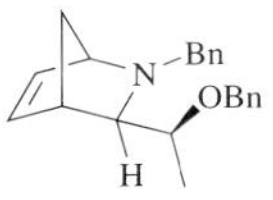 (23) + 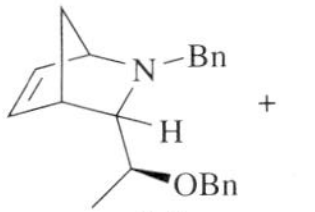 (14) + 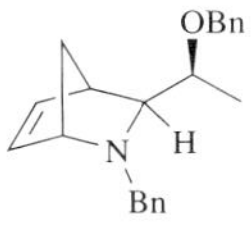 (15) + 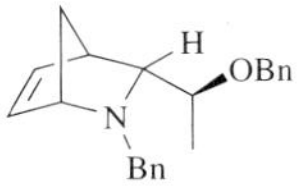 (18)	403
	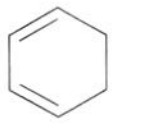	7-10 d	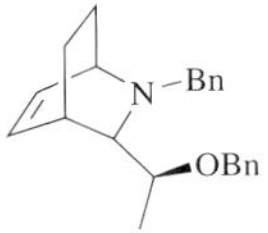 + 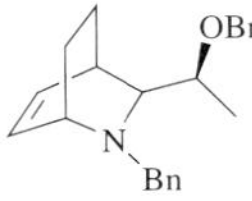 (<10)[j,bbbbb]	403
	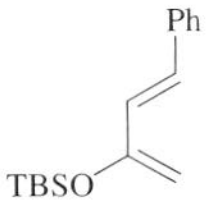	1. Lewis acid, CH_2Cl_2, 4°, 1 d 2. $NaHCO_3$, H_2O	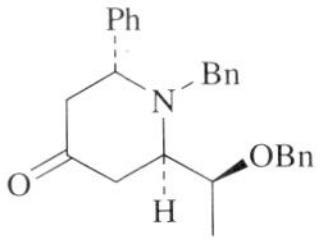 + 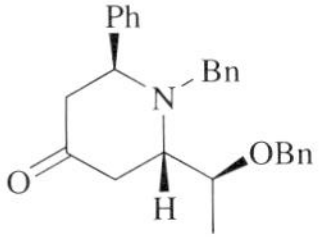	150
		L. A.		
		$ZnCl_2$	(25) 84:16	
		$Zn(OTf)_2$	(51) 70:30	
		TMSOTf	(95) 78:22	

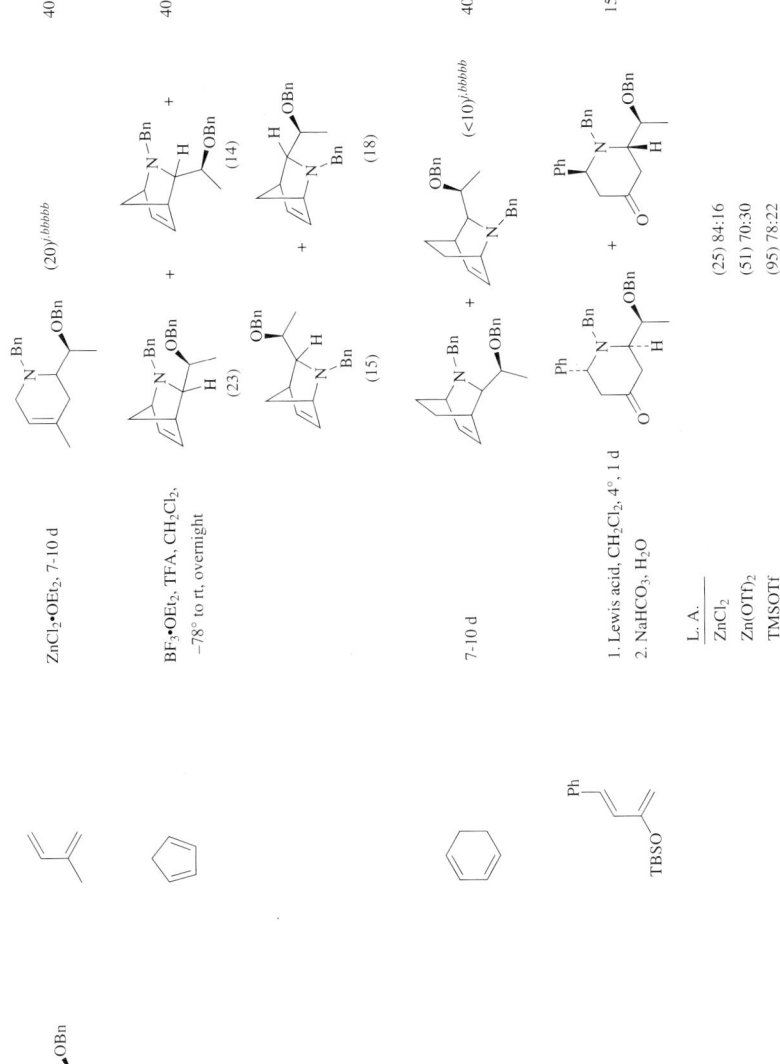

TABLE 5. CYCLOADDITIONS OF *N*-ALKYL- AND *N*-ARYLIMINES (*Continued*)

	Imine	Diene	Conditions	Product(s) and Yield(s) (%)	Refs.
C_{17}	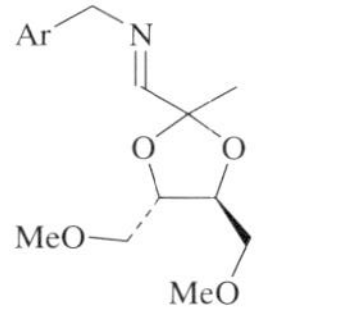 Ar = 4-$MeOC_6H_4$	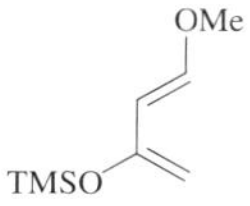	Lewis acid	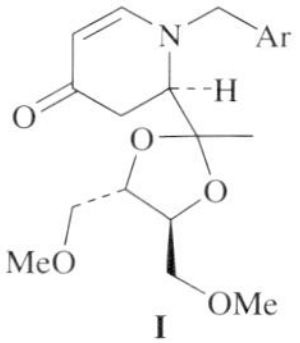 **I** + 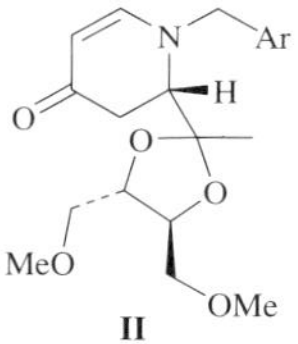 **II**	431

L. A. (eq)	Solvent	Temp	Time	I + II	I:II
$ZnCl_2$ (1.1)	CH_2Cl_2	–78 to 0°	13 h	(57)	>99:1
$ZnCl_2$ (1.1)	THF	–78 to 0°	23 h	(95)	>99:1
BF_3•OEt_2 (1.1)	CH_2Cl_2	–78°	5 min	(100)	>99:1
BF_3•OEt_2 (2.2)	CH_2Cl_2	–78°	30 min	(66)	>99:1
BF_3•OEt_2 (3.3)	CH_2Cl_2	–78°	20 min	(52)	>99:1
$TiCl_2(OPr\text{-}i)_2$ (1.1)	THF	–78°	3.5 h	(52)	83:17
TMSOTf (1.1)	CH_2Cl_2	–78°	10 min	(73)	94:6
$EtAlCl_2$ (1.1)	CH_2Cl_2	–78°	40 min	(30)	77:23

	Imine	Diene	Conditions	Product(s) and Yield(s) (%)	Refs.
	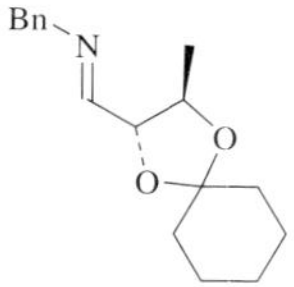	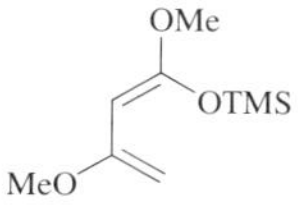	1. Et_2AlCl, CH_2Cl_2, –78° to rt, 4 h 2. H_3O^+	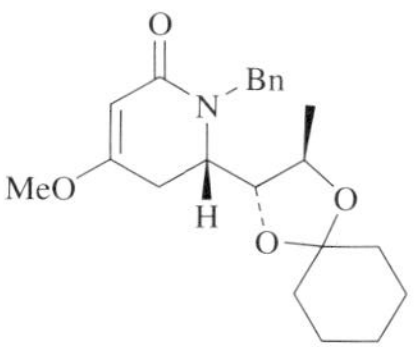 (82)	159
	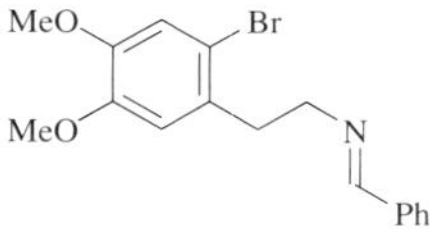	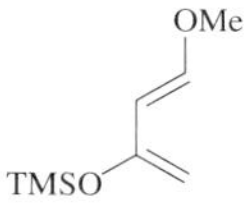	$ZnCl_2$, THF, 0°	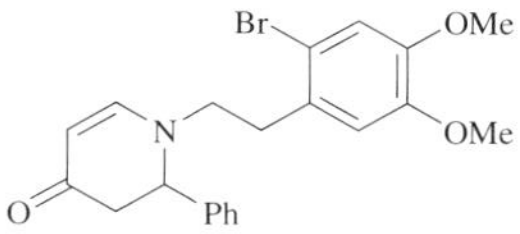 (43)	379

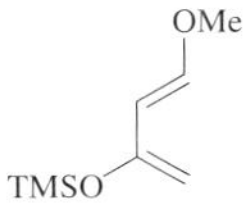

1. Lewis acid
2. HCl, H_2O

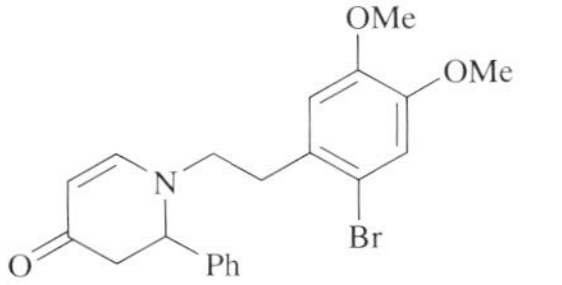

380

L. A.	Solvent	Temp	Time	
$ZnCl_2$	THF	rt	40 h	(47)
$ZnCl_2$	THF	rt	18 h	(58)
$B(OPh)_3$	CH_2Cl_2	–78° to rt	5 h	(14)

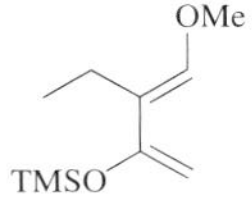

1. $ZnCl_2$, THF, rt, 20 h
2. HCl, H_2O

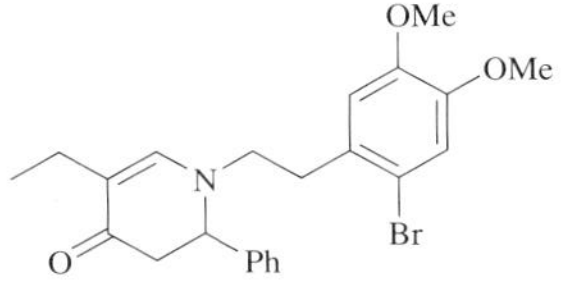

(45) 380, 379

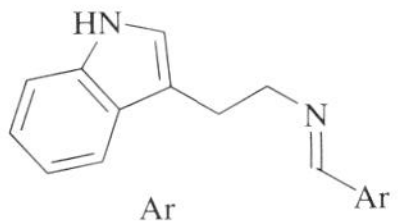

Ar	
Ph	(78)
$4\text{-}O_2NC_6H_4$	(43)
$4\text{-}ClC_6H_4$	(79)

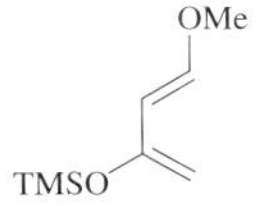

1. $ZnCl_2$[gggg], THF, rt, overnight
2. $NaHCO_3$, H_2O

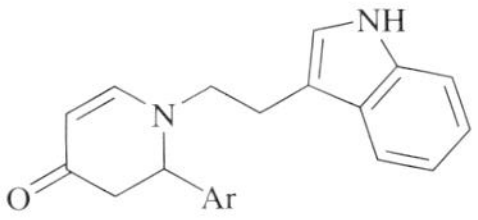

419

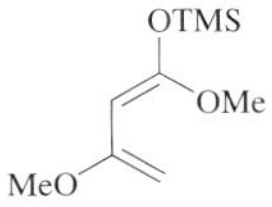

1. $EtAlCl_2$, CH_2Cl_2, ether, –60° to rt, 2 h
2. $NaHCO_3$, H_2O

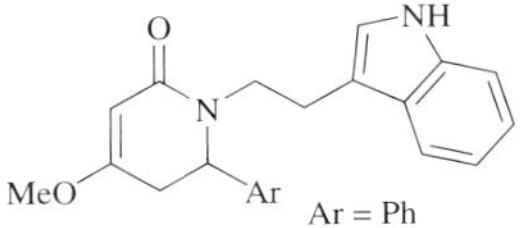

Ar = Ph

(68)[hhhh] 420, 421

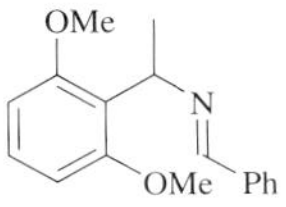

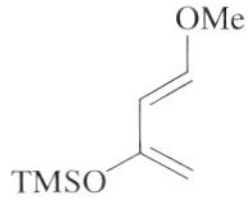

1. $ZnCl_2$, CH_2Cl_2, –78°, 24 h
2. $NaHCO_3$, H_2O

(63) 79:21 Ar = 2,6-$(MeO)_2C_6H_3$

433

TABLE 5. CYCLOADDITIONS OF *N*-ALKYL- AND *N*-ARYLIMINES (*Continued*)

	Imine	Diene	Conditions	Product(s) and Yield(s) (%)	Refs.
C_{17}	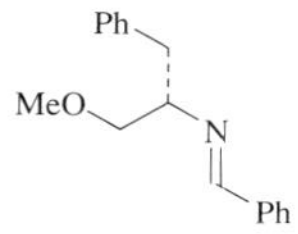	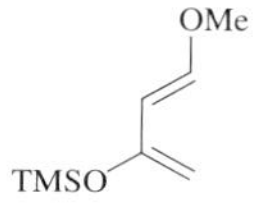	Lewis acid, CH_2Cl_2, rt	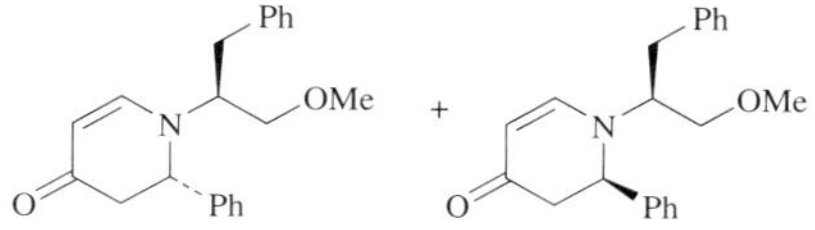	170
			L. A.		
			$BF_3{\bullet}OEt_2$	(53) 73:27	
			$Zn(OTf)_2$	(47) 73:27	
	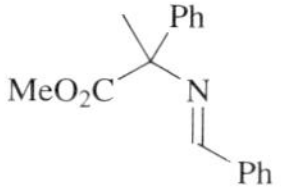	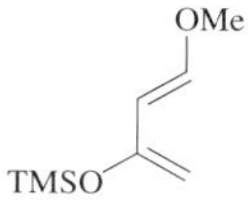	$Zn(OTf)_2$, CH_2Cl_2, rt	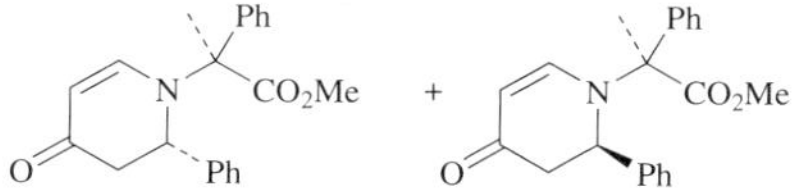 (60) >95:5[vvvv]	170
	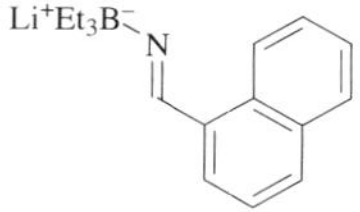	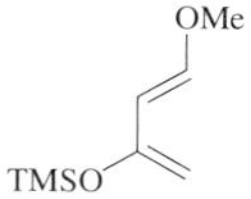	1. THF, rt, 4-5 h 2. $NaHCO_3$, H_2O	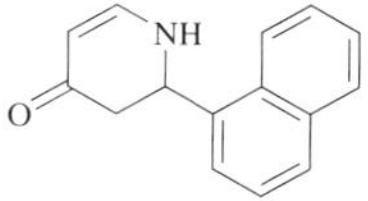 (67)	389
	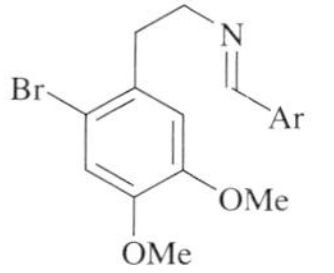	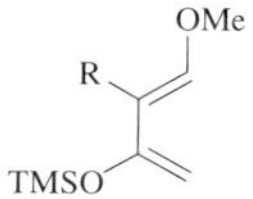		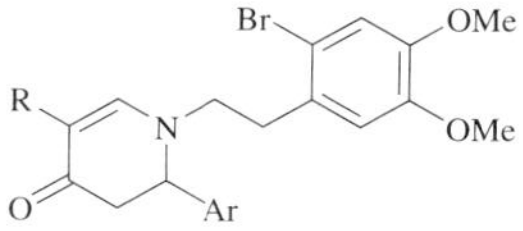	
	Ar	R			
	2-$O_2NC_6H_4$	H	$ZnCl_2$, THF, 0°	(20)	379
	2-$O_2NC_6H_4$	H	1. $ZnCl_2$, THF, rt, 20 h 2. HCl, H_2O	(33)	380
	4-$O_2NC_6H_4$	Et	$ZnCl_2$, THF	(73)	379
	4-$O_2NC_6H_4$	Et	1. $ZnCl_2$, THF, rt, 72 h 2. HCl, H_2O	(77)	380

Imine	Diene	Conditions	Product(s) and Yield(s) (%)	Refs.
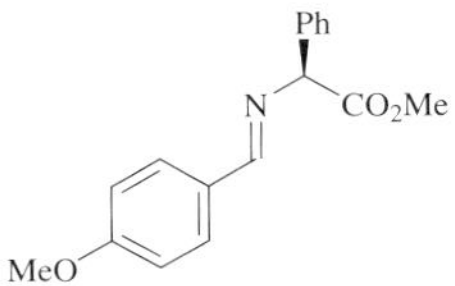	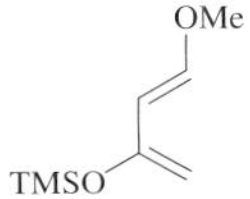	1. $ZnCl_2$, THF, –20°, 7 h 2. $NaHCO_3$, H_2O	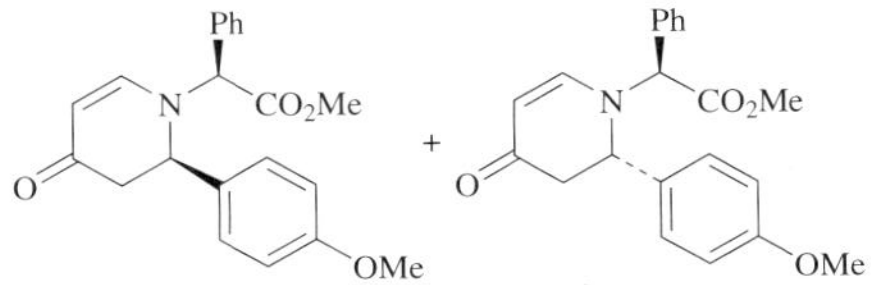 (63) 67:33	384
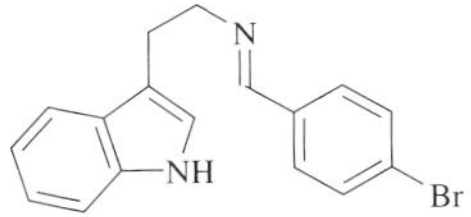	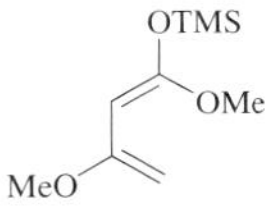	1. $EtAlCl_2$, THF, –60° to rt, 2 h 2. $NaHCO_3$, H_2O	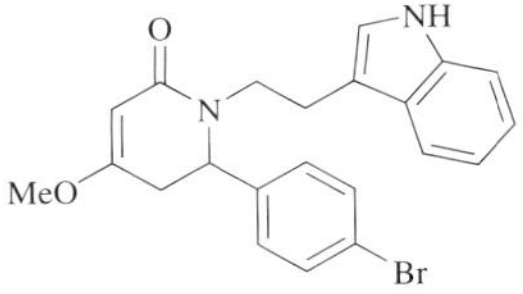 (43)[hhhh]	420, 421
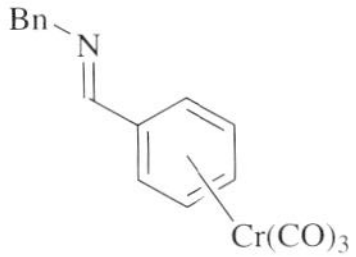	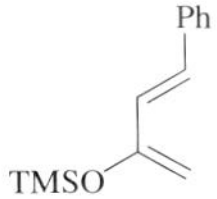	1. Lewis acid, CH_2Cl_2 2. hν, O_2, 3 h 3. TBAF, THF, rt, 1 h	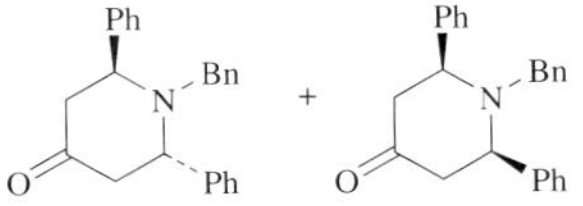	169
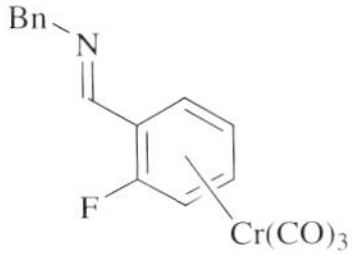	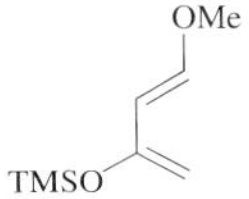	$SnCl_4$,THF, –78° to rt, 15 h	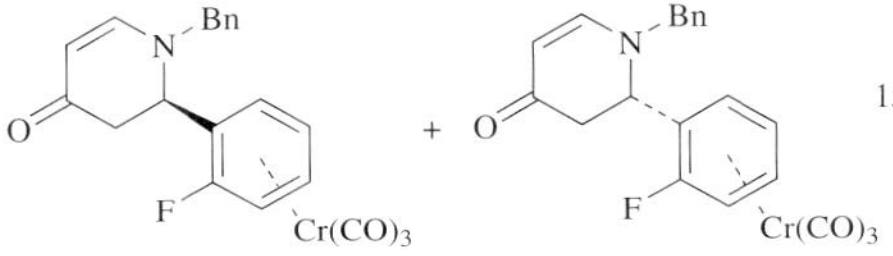 (89) 98:2	158

L. A. (eq)	Temp	Time	
TMSOTf (1)	rt	3 h	(84) 90:10
TMSOTf (1)	0°	12 h	(70) 88:12
TMSOTf (1)	40°	1 h	(66) 90:10
TMSOTf (0.1)	rt	12 h	(64) 83:17
BF_3•OEt_2	rt	12 h	(<30)
$TiCl_4$	–78°	30 min	(30) 85:15

TABLE 5. CYCLOADDITIONS OF *N*-ALKYL- AND *N*-ARYLIMINES (*Continued*)

	Imine	Diene	Conditions	Product(s) and Yield(s) (%)	Refs.
C_{17}	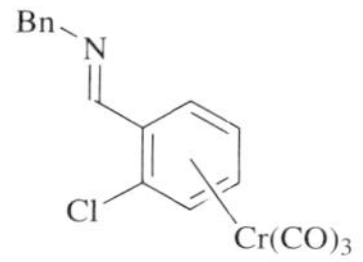	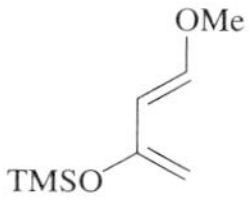	Lewis acid, THF, −78° to rt, 15 h L. A. $ZnCl_2$ $SnCl_4$	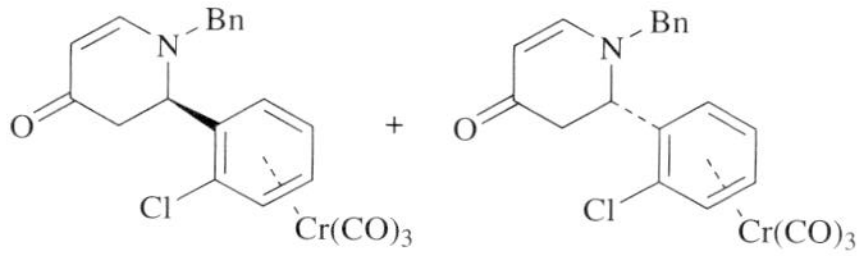 (88) 89:11 (98) >98:2[bbbb]	158
	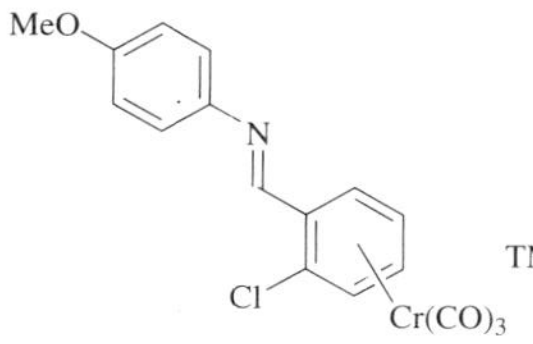	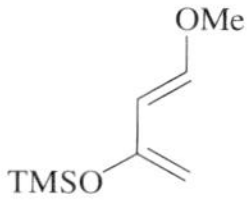	$ZnCl_2$, CH_2Cl_2, rt, 24 h	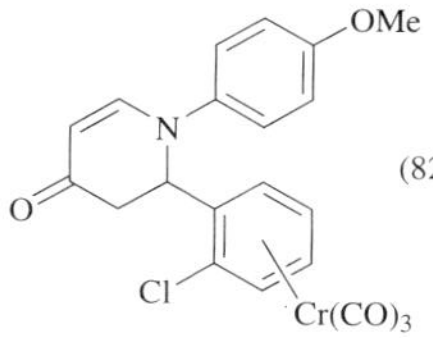 (82)[ccccc]	166
	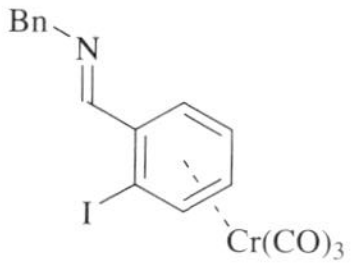	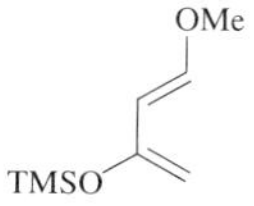	$SnCl_4$, THF, −78° to rt	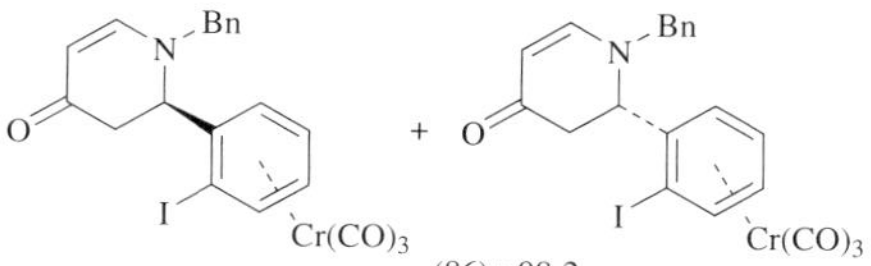 (86) >98:2	168
	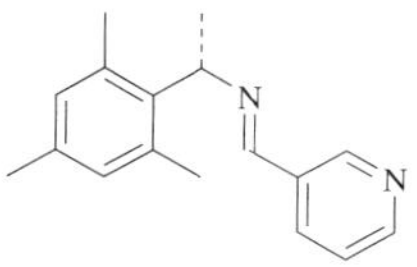	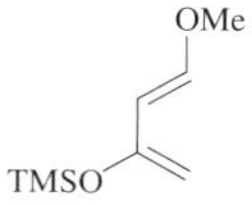	1. BF_3•OEt_2, CH_2Cl_2, −78 to 0°, 24 h 2. $NaHCO_3$, H_2O	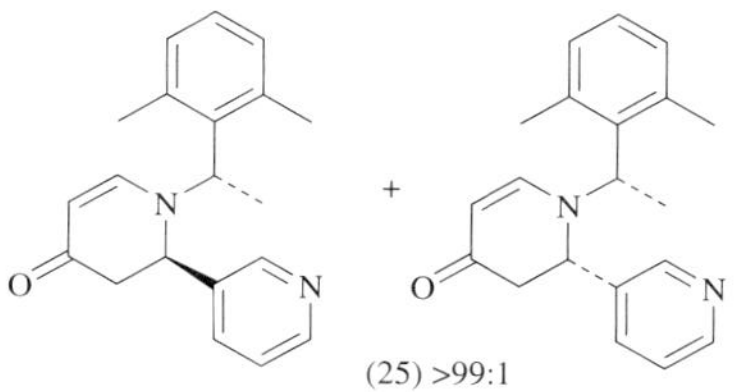 (25) >99:1	433

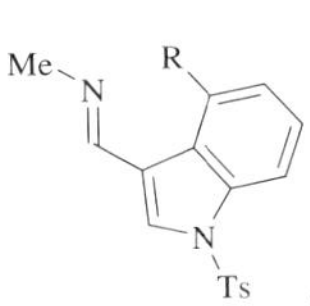 R: H, I	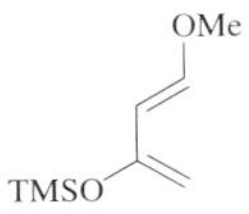	$Zn(OTf)_2$, H_2O, CH_2Cl_2, rt, 29 h	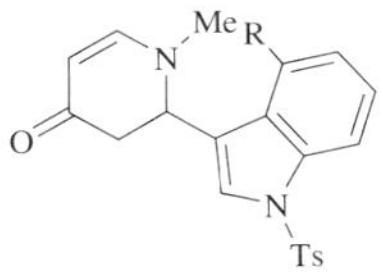 (92) (94)	432
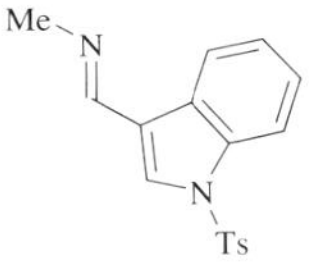	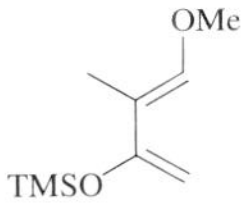	$Zn(OTf)_2$, $MgSO_4$, rt, 14 h	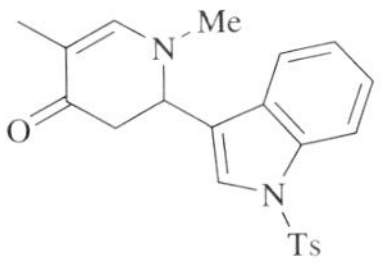 (86)	432
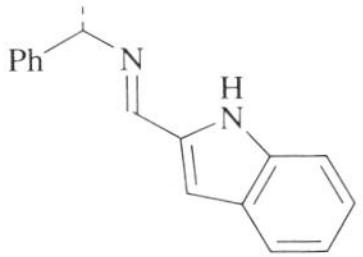	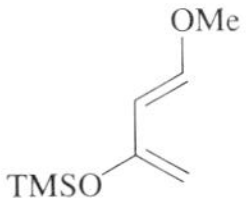	1. $ZnCl_2$, CH_2Cl_2, –20°, 2 h 2. HCl, H_2O, –20° to rt	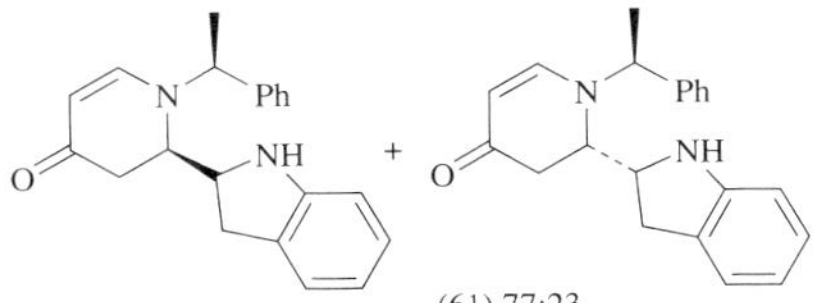 (61) 77:23	434
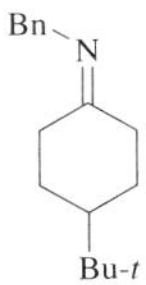	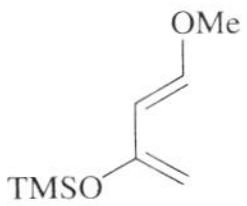	1. $Zn(OTf)_2$, CH_2Cl_2, rt, 12 h 2. NH_4Cl, H_2O, 2 h	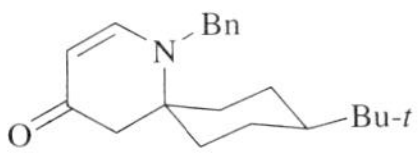 (29)	356
C_{18}				
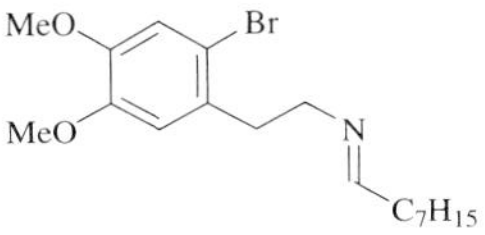	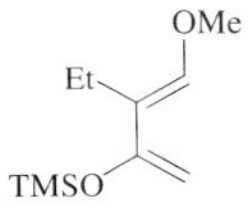	1. $ZnCl_2$, THF, rt, 96 h 2. HCl, H_2O	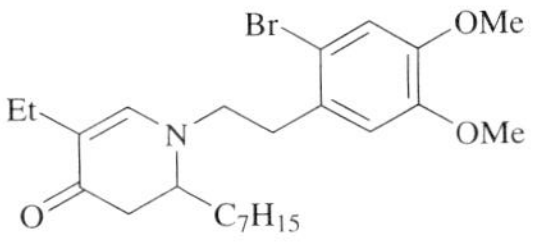 (22)	380, 379

TABLE 5. CYCLOADDITIONS OF *N*-ALKYL- AND *N*-ARYLIMINES (*Continued*)

Imine	Diene	Conditions	Product(s) and Yield(s) (%)	Refs.
C_{18} 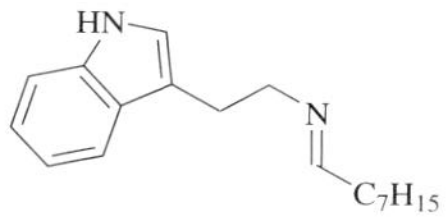	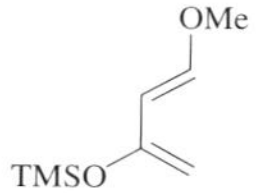	1. $ZnCl_2$[gggg], THF, rt, overnight 2. $NaHCO_3$, H_2O	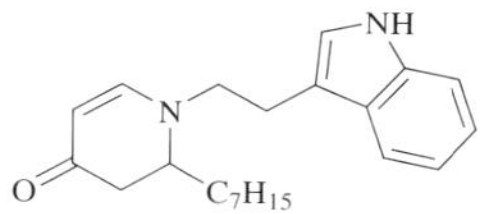 (40)	419
	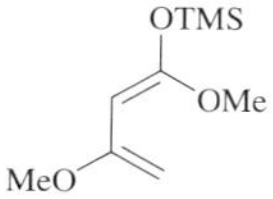	1. $EtAlCl_2$, CH_2Cl_2, –60° to rt, 2 h 2. $NaHCO_3$, H_2O	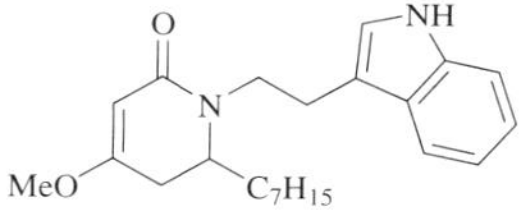 (35)[hhhh]	420, 421
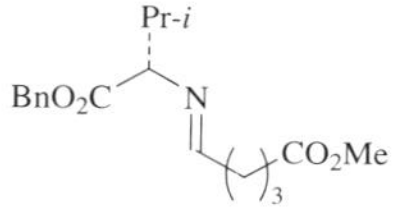	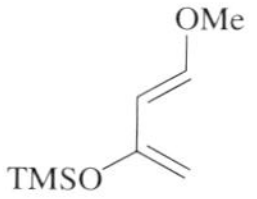	1. $EtAlCl_2$, CH_2Cl_2, –78° to rt, 1 h 2. $NaHCO_3$, H_2O	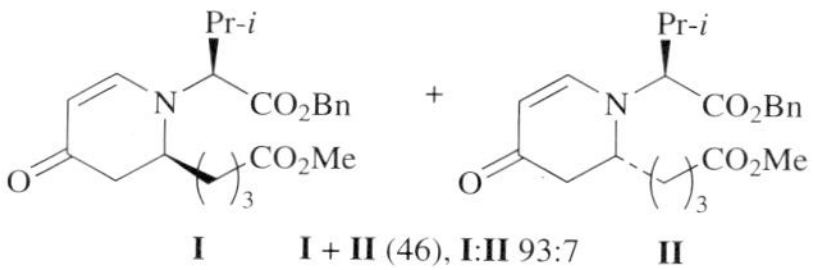 **I** + **II** (46), **I**:**II** 93:7	384
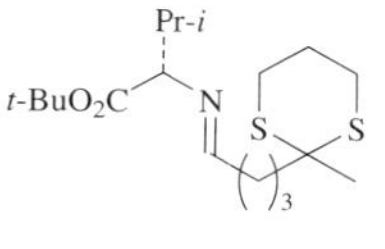	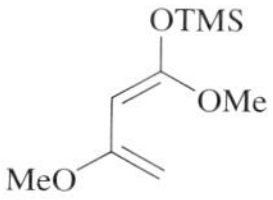	1. $EtAlCl_2$[bbb], –78 to 20°, CH_2Cl_2, 1-1.5 h 2. $NaHCO_3$, H_2O	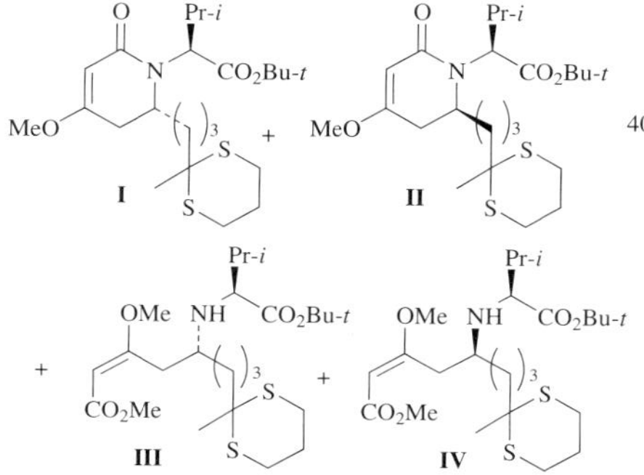 **I** + **II** + **III** + **IV** (40), **I** + **II**:**III** + **IV** <2:98, **I** + **III**:**II** + **IV** 96:4	402, 175a

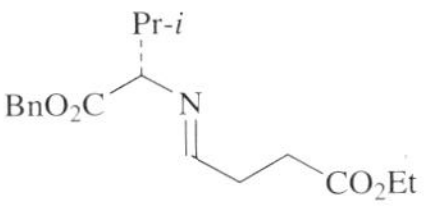	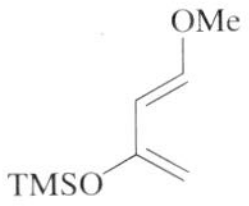	1. $EtAlCl_2$[jj], CH_2Cl_2, –78 to 20° 2. H^+, H_2O	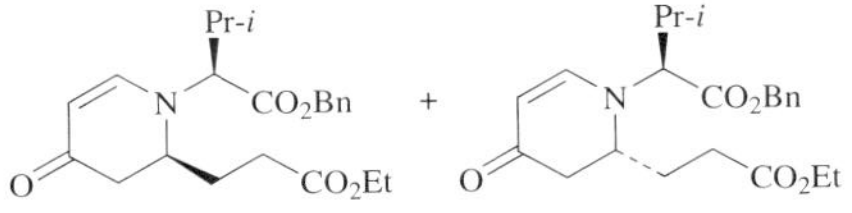 (50) 93:7	175a
	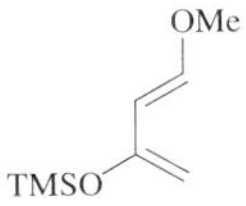	1. $EtAlCl_2$, CH_2Cl_2, –78° to rt, 1 h 2. $NaHCO_3$, H_2O	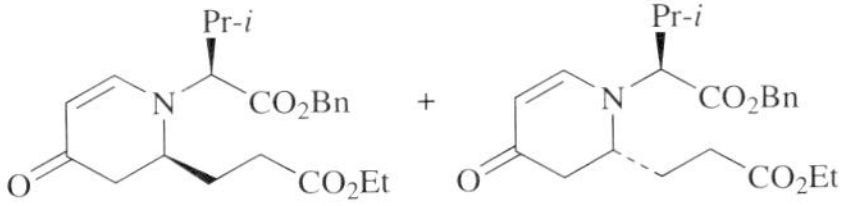 (50) 93:7	384
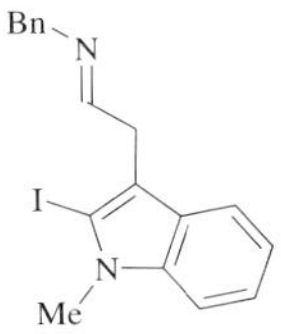	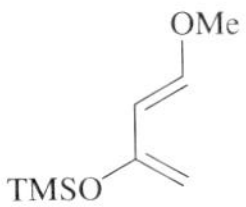	1. $Zn(OTf)_2$, CH_2Cl_2, rt 2. HCl, H_2O	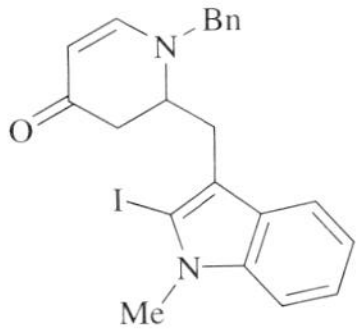 (70)	435
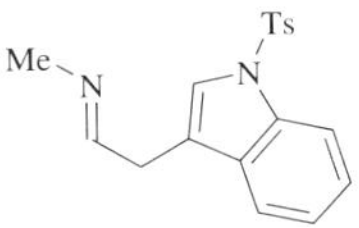	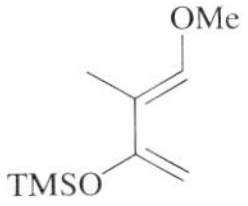	$Zn(OTf)_2$, H_2O, CH_2Cl_2, 0°, 10 h	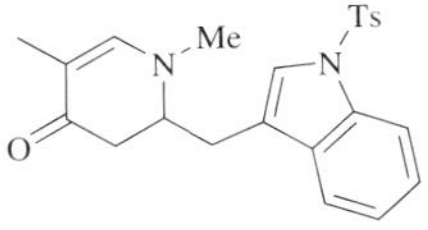 (55)	432
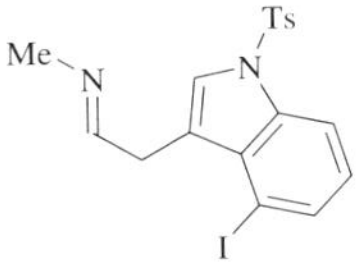	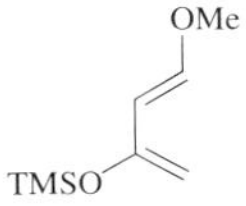	$Zn(OTf)_2$, H_2O, CH_2Cl_2, rt, 2 h	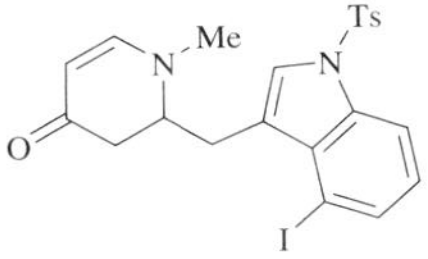 (85)[ddddd,eeeee]	432

TABLE 5. CYCLOADDITIONS OF *N*-ALKYL- AND *N*-ARYLIMINES (*Continued*)

Imine	Diene	Conditions	Product(s) and Yield(s) (%)	Refs.
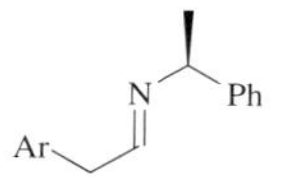 Ar = *N*-phthaloyl		BF_3•OEt_2, TFA, CH_2Cl_2, −78° to rt, overnight	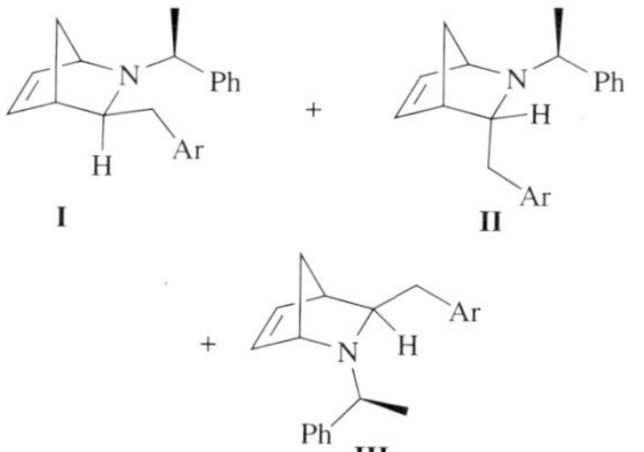 I + II + III **I** + **II** + **III** + **IV** (87), **I**:**II** 85:15, **I**:**III** 99:1	403
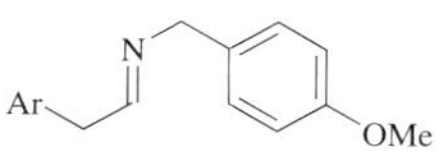 Ar = *N*-phthaloyl	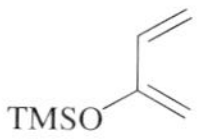		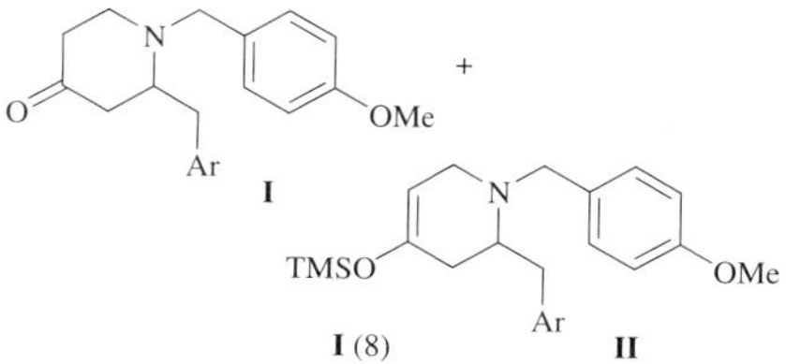 I + II	436
		1. BF_3•OEt_2, CH_2Cl_2, 0° 2. HCl, H_2O	**I** (8)	
		1. $AlCl_3$, CH_2Cl_2, −78° to rt, 18 h 2. HCl, H_2O, 0°, 2 h	**I** (33-38) + **II** (11)	

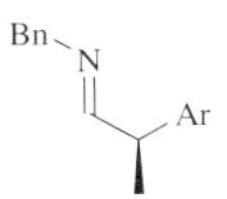

Ar = *N*-phthaloyl

Acid, CH_2Cl_2, –78° to rt, overnight

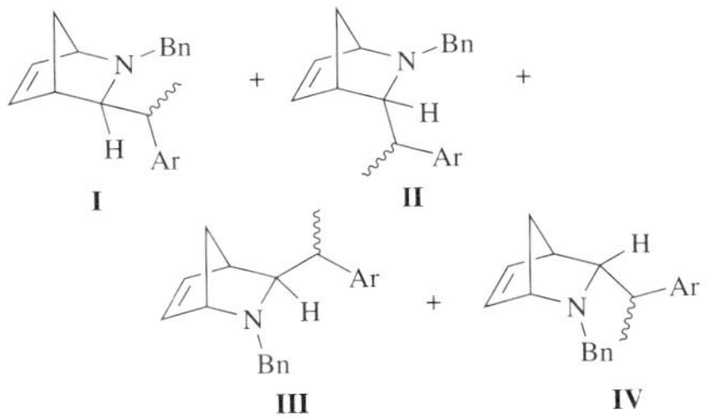

I + II + III + IV

Acid	
BF_3•OEt_2, TFA	**I + II + III + IV** (89), **I+III:II+IV** 72:28, **I:II** 72:28[fffff]
$TiCl_4$	**I + III + IV** (40), **III:IV** 45:55, **III:I** 99:1[fffff]

Ref. 403

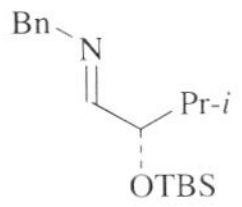

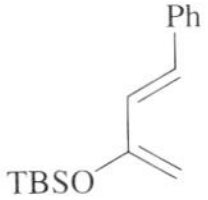

1. Lewis acid
2. $NaHCO_3$, H_2O
3. TBAF, THF, 10 h

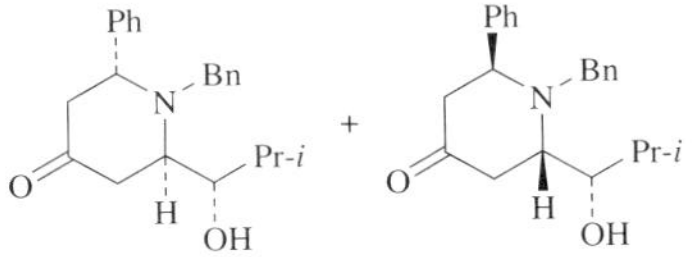

L. A.	Solvent	Temp	Time	
TMSOTf	CH_2Cl_2	4°	7 d	(91) 43:57
$Zn(OTf)_2$	CH_2Cl_2	4°	7 d	(59) 49:51
$TiCl_4$	CH_2Cl_2	4°	11 d	(91) 30:70
$TiCl_4$-AgOTf	*i*-PrCN	–30°	1 d	(49) <2:98

Ref. 150

TABLE 5. CYCLOADDITIONS OF *N*-ALKYL- AND *N*-ARYLIMINES (*Continued*)

	Imine	Diene	Conditions	Product(s) and Yield(s) (%)	Refs.
C_{18}	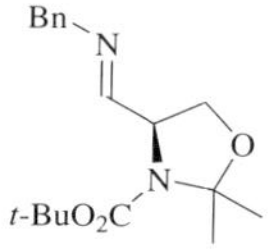	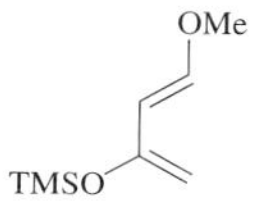	1. Lewis acid, CH_2Cl_2 2. HCl, H_2O	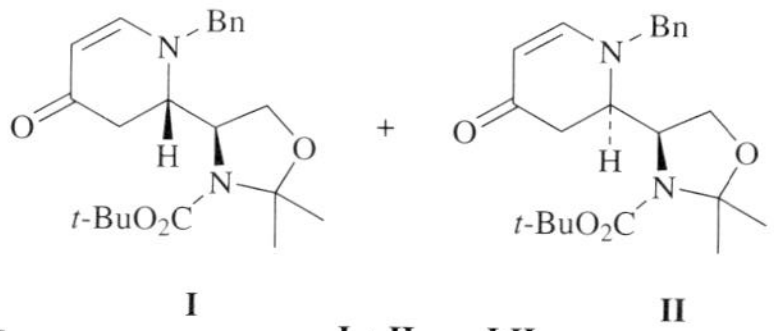 I + II	437

L. A. (eq)	Temp	Time	I + II	I:II
$ZnCl_2$ (0.7)	0°	6 h	(28)	38:62
$ZnCl_2$ (1.0)	–50°	28 h	(20)	36:64
$ZnCl_2$ (1.2)	–40°	48 h	(29)	21:79
Et_2AlCl (1.0)	0°	72 h	(38)	41:59
Et_2AlCl (1.0)	–20°	24 h	(41)	26:74
Et_2AlCl (1.0)	–40°	26 h	(57)	25:75
Et_2AlCl (1.5)	–40°	72 h	(34)	24:76
Et_2AlCl (2.0)	–40°	17 h	**I** (13) + **II** (52)	
$EtAlCl_2$ (1.0)	–40°	48 h	(55)	19:81
$EtAlCl_2$ (2.0)	–40°	15 h	(57)	20:80

	Imine	Diene	Conditions	Product(s) and Yield(s) (%)	Refs.
	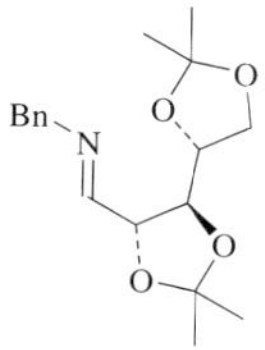	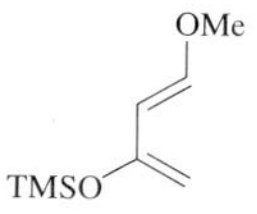	1. $ZnCl_2$ 2. $NaHCO_3$, H_2O	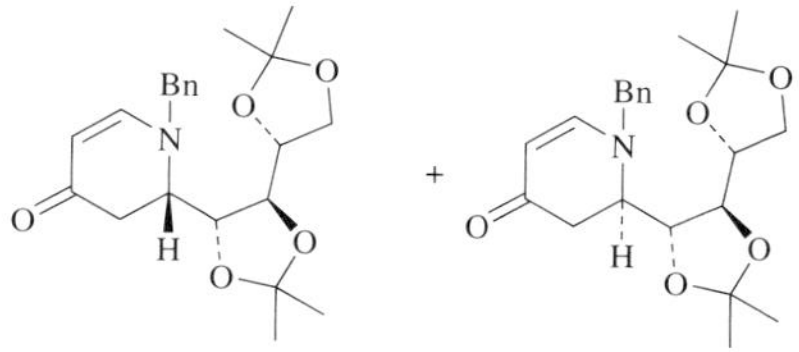	

Solvent	Temp	Time	Product(s) and Yield(s) (%)	Refs.
dioxane	rt	1 h	(79) 4.7:1	438
dioxane	—	2-3 h	(79) 83:17	439, 440
THF	—	2-3 h	(79) 94:6	439, 440

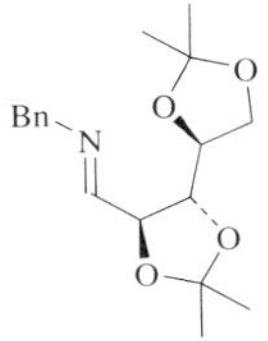	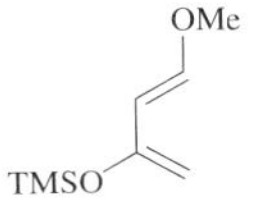	1. $ZnCl_2$, THF, 2-3 h 2. $NaHCO_3$, H_2O	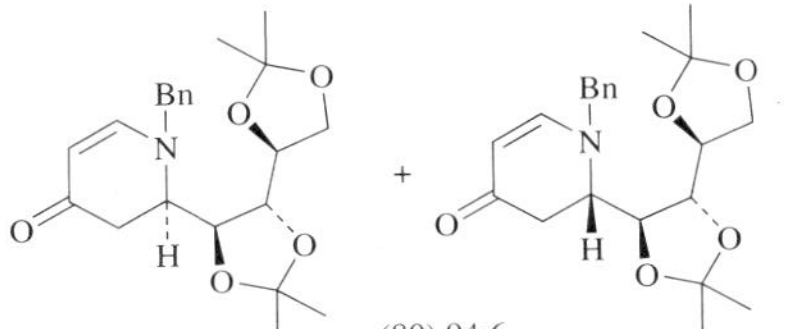 (80) 94:6	439, 440, 438
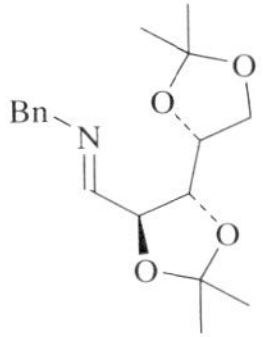	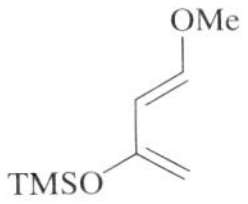	1. $ZnCl_2$, dioxane, 2-3 h 2. $NaHCO_3$, H_2O	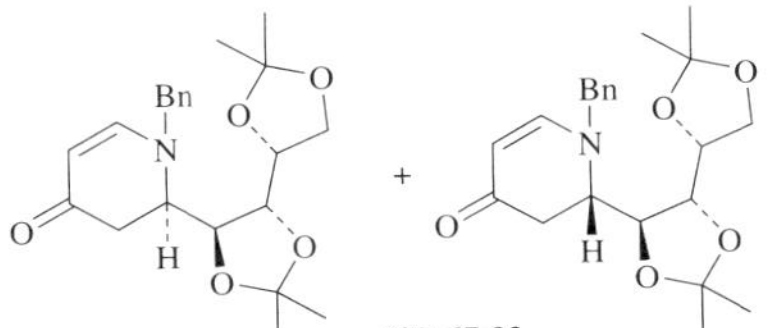 (65) 67:33	439, 440
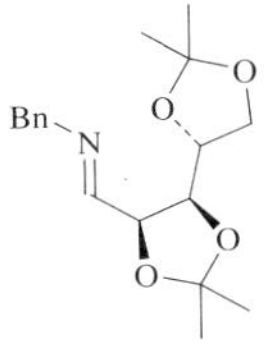	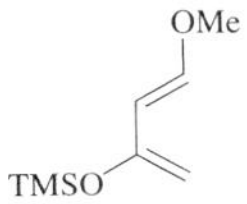	1. $ZnCl_2$, THF, 2-3 h 2. $NaHCO_3$, H_2O	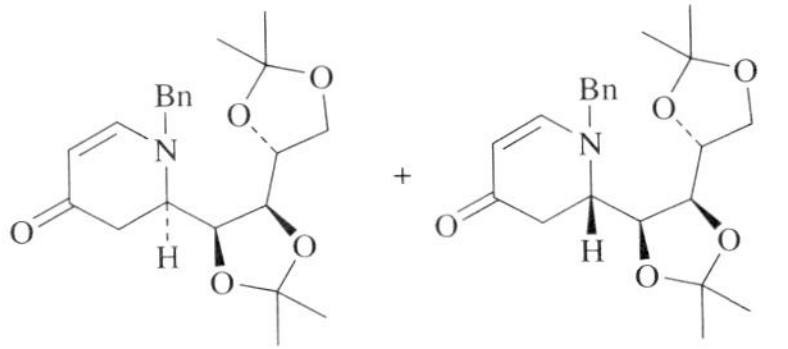 (61) 90:10	439, 440
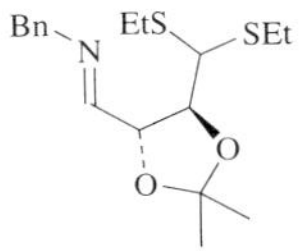	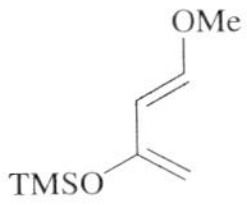	1. $ZnCl_2$, dioxane, 2-3 h 2. $NaHCO_3$, H_2O	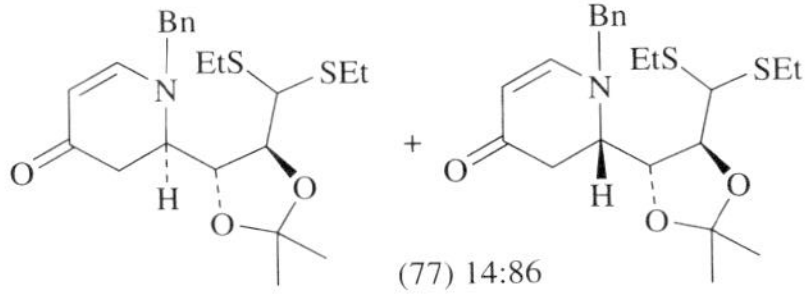 (77) 14:86	439, 440

TABLE 5. CYCLOADDITIONS OF *N*-ALKYL- AND *N*-ARYLIMINES (*Continued*)

Imine	Diene	Conditions	Product(s) and Yield(s) (%)	Refs.
C_{18}				
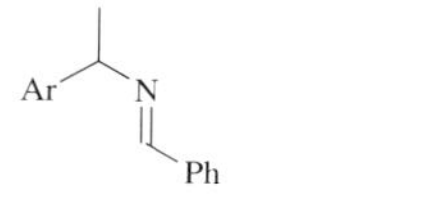 Ar = 2,4,6-$Me_3C_6H_2$	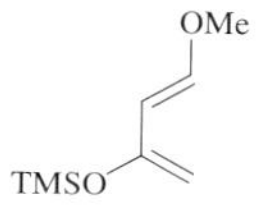	1. Lewis acid, CH_2Cl_2, 24 h 2. $NaHCO_3$, H_2O L. A. — Temp $ZnCl_2$ — –78° $ZnBr_2$ — –78° $Zn(OTf)_2$ — –78° $BF_3•OEt_2$ — –78 to 0°	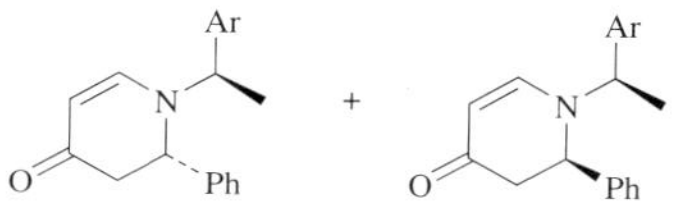 (52) 99:1 (49) 95:5 (25) >99:1 (60) 99:1	433
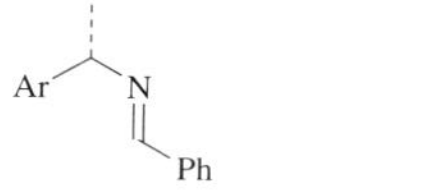 Ar = 2,4,6-$Me_3C_6H_2$	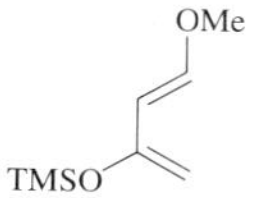	1. $BF_3•OEt_2$, CH_2Cl_2, –78 to 0°, 24 h 2. $NaHCO_3$, H_2O	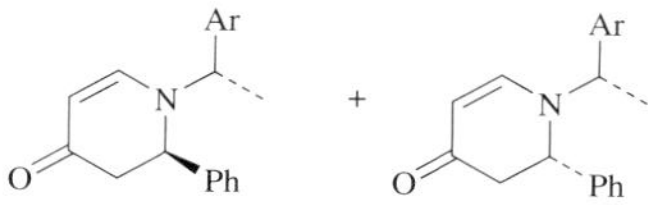 (59) 99:1	433
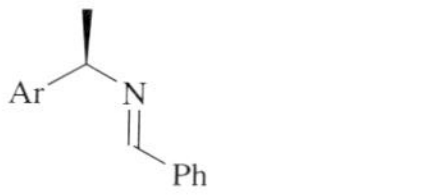 Ar = 2,4,6-$Me_3C_6H_2$	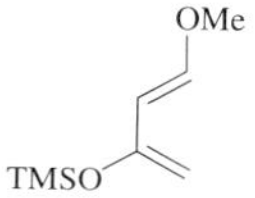	1. $BF_3•OEt_2$, CH_2Cl_2, –78 to 0°, 24 h 2. $NaHCO_3$, H_2O	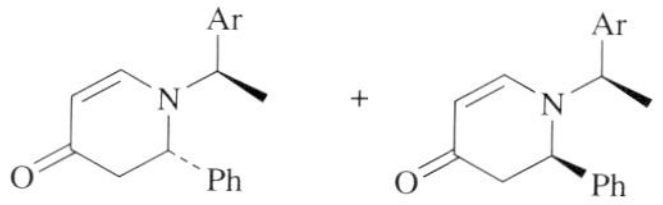 (53) 99:1	433
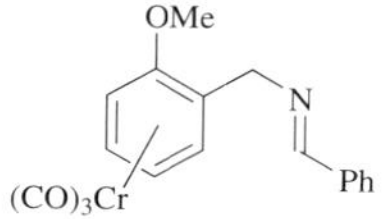	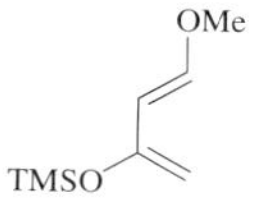	$ZnCl_2$, CH_2Cl_2, –78°	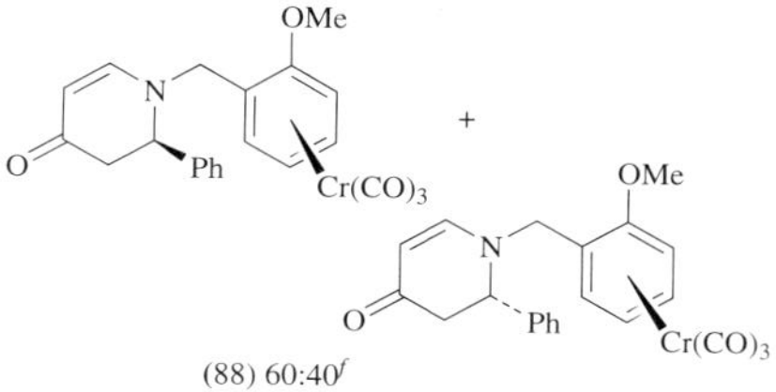 (88) 60:40[f]	166

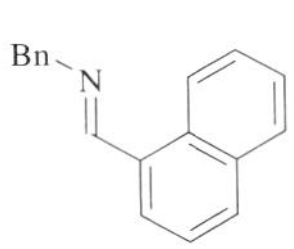	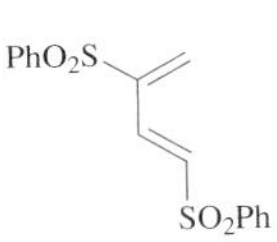	CH_2Cl_2, rt Time 24 h[p,q] 2 h[r]	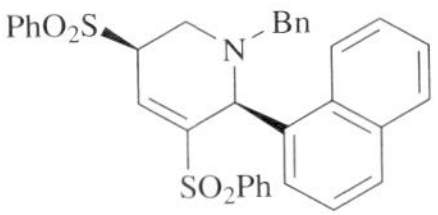 (75) ("excellent")	151, 145
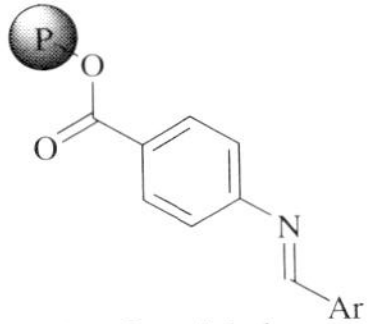 Ar = 2-naphthyl	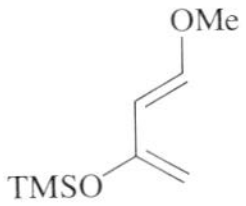	1. $Zn(ClO_4)_2{\bullet}6H_2O$, MeOH, rt, 12 h 2. NaOH, H_2O, rt, 12 h 3. HCl, H_2O	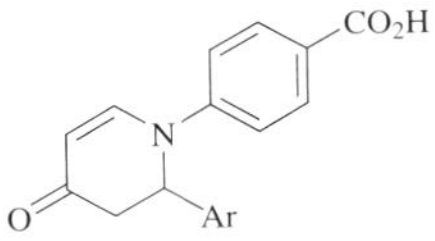 (84)	405
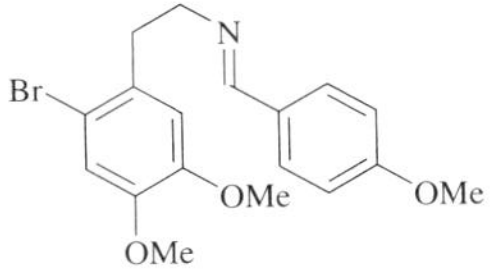	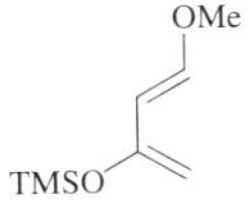	$ZnCl_2$, THF, 0° 1. $ZnCl_2$, THF, rt, 18 h 2. HCl, H_2O	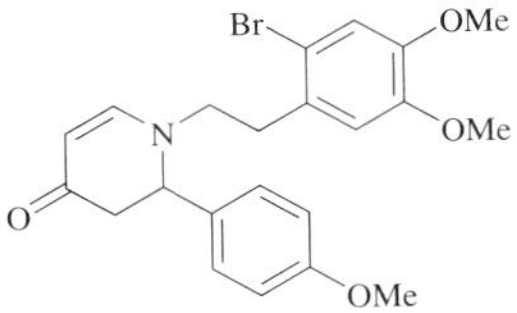 (66) (66)	379 380
	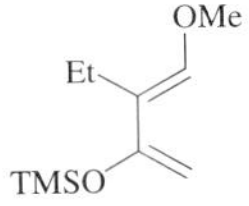	$ZnCl_2$, THF 1. $ZnCl_2$, THF, rt, 24 h 2. HCl, H_2O	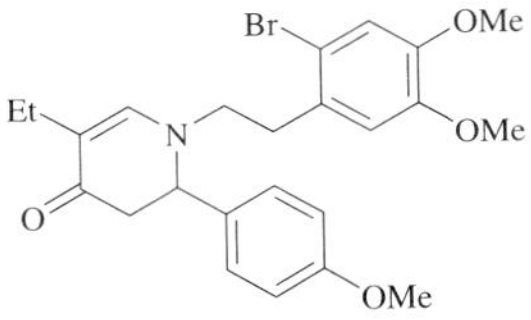 (52) (54)	379 380

TABLE 5. CYCLOADDITIONS OF *N*-ALKYL- AND *N*-ARYLIMINES (*Continued*)

	Imine	Diene	Conditions	Product(s) and Yield(s) (%)	Refs.
C_{18}	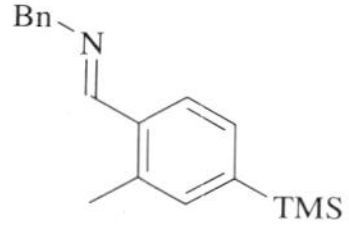	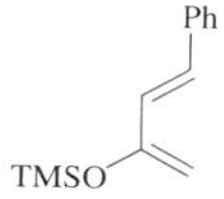	1. TMSOTf, CH_2Cl_2, rt, 12 h 3. TBAF, THF, rt, 1 h	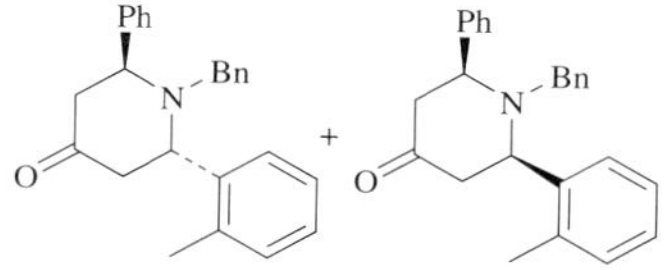 (71) >97:3[bbbb]	381
	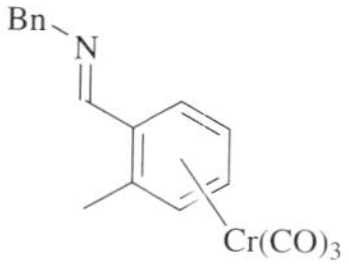	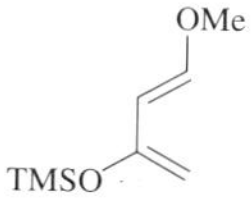	Et_2AlCl, THF, −78° to rt, 15 h	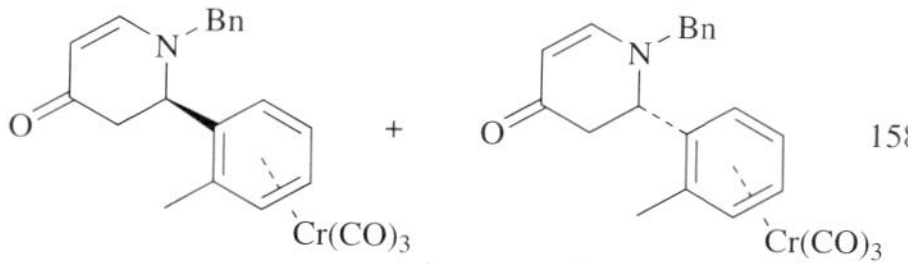 (82) >98:2[bbbb]	158
	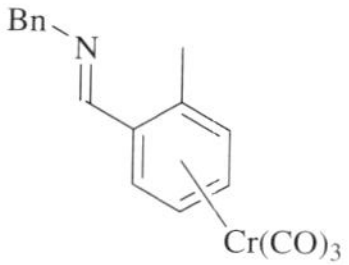	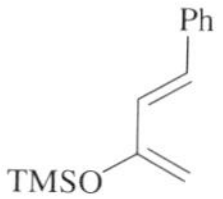	1. Lewis acid 2. hν, O_2 3. TBAF	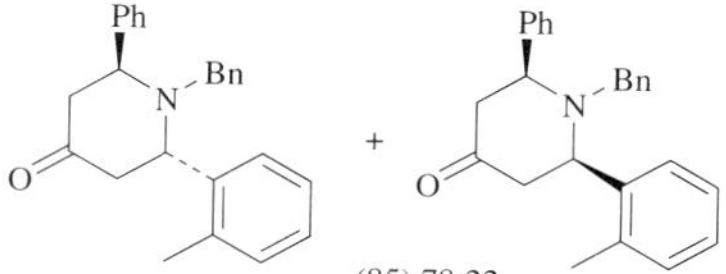 (85) 78:22	381
	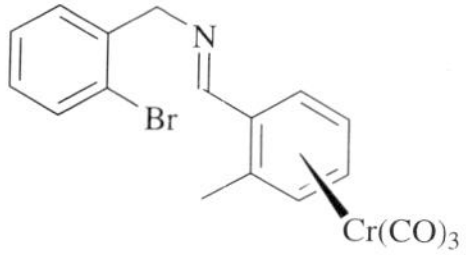	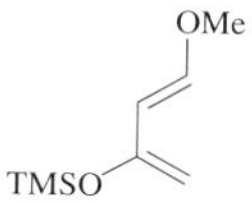	$SnCl_4$,THF, −78° to rt, 15 h	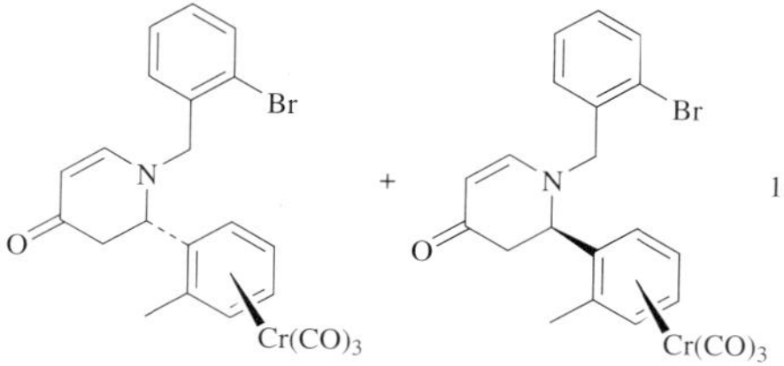 (92) >98:2[bbbb]	158

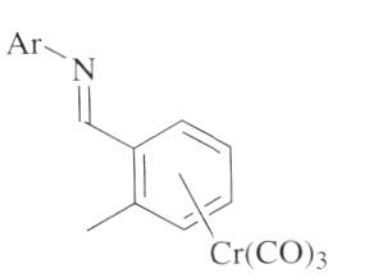 Ar = 4-MeOC$_6$H$_4$	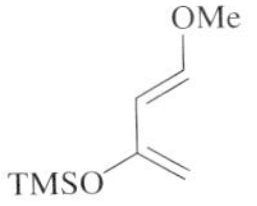	$ZnCl_2$, CH_2Cl_2, rt, 24 h	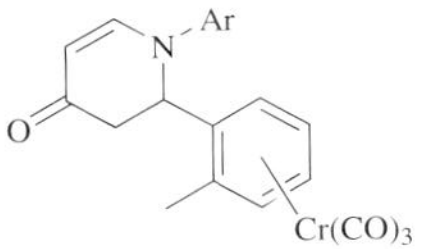	(80)[ccccc]	166
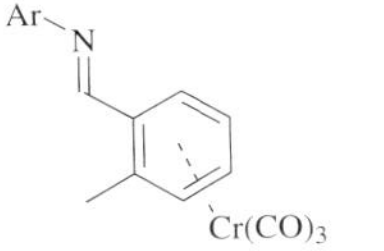 Ar = 4-MeOC$_6$H$_4$	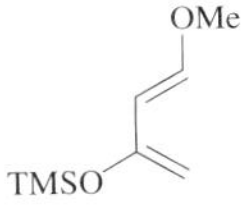	$ZrCl_2$, CH_2Cl_2, rt, 24 h	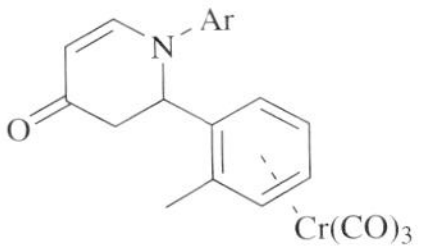	(76)[ggggg]	166
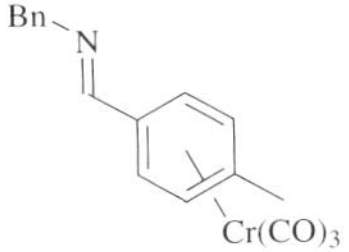	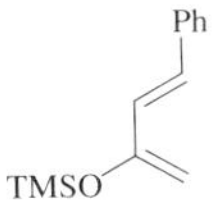	1 TMSOTf, CH_2Cl_2, rt, 12 h 2. hν, O_2, 3 h 3. TBAF, THF, rt, 1 h	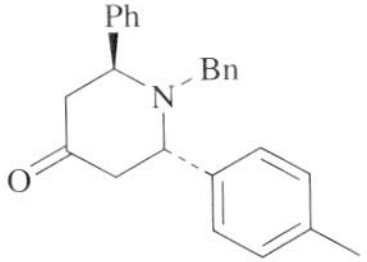 + 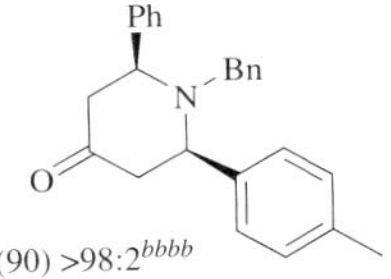	(90) >98:2[bbbb]	169
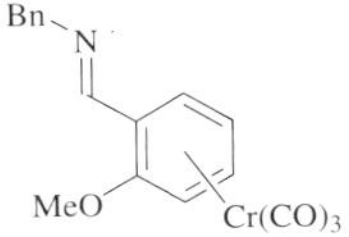	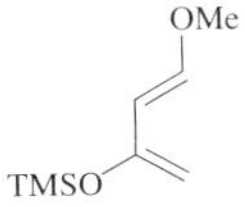	Lewis acid, –78° to rt, 15 h	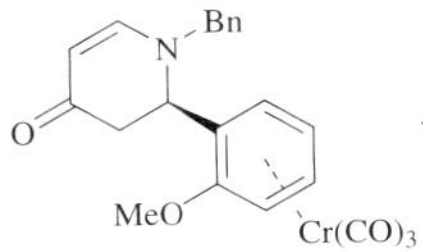 + 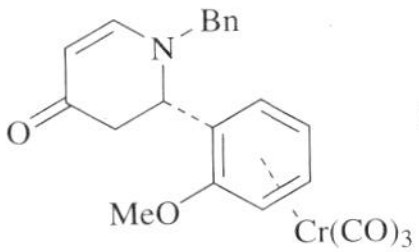		158

L. A.	Solvent	
$ZnCl_2$	THF	(82) 93:17
$SnCl_4$	THF	(82) 86:14
$TiCl_4$	THF	(57) 87:13
$MgBr_2$	THF	(24) 87:13
Et_2AlCl	THF	(88) 85:15
Et_2AlCl	CH_2Cl_2	(72) 60:40

TABLE 5. CYCLOADDITIONS OF *N*-ALKYL- AND *N*-ARYLIMINES (*Continued*)

Imine	Diene	Conditions	Product(s) and Yield(s) (%)	Refs.
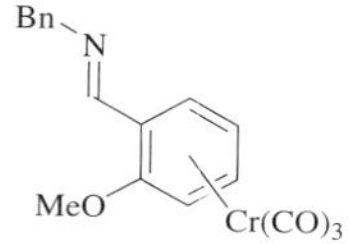	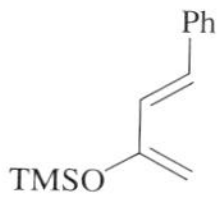	1. TMSOTf, CH_2Cl_2, rt, 12 h 2. hν, O_2, 3 h 3. TBAF, THF, rt, 1 h	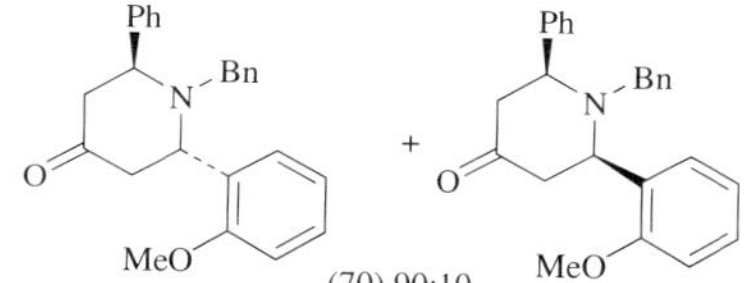 (70) 90:10	169
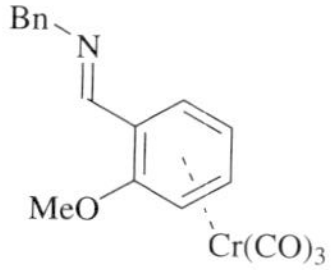	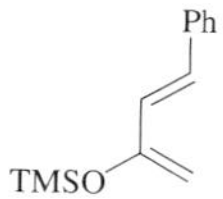	1. TMSOTf, CH_2Cl_2, rt, 12 h 2. hν, O_2, 3 h 3. TBAF, THF, rt, 1 h	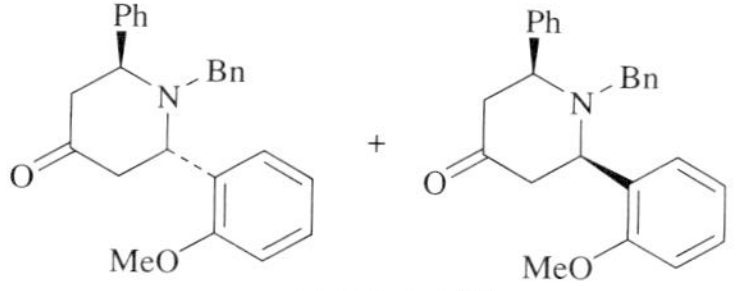 (70) 90:10[hhhhh]	169
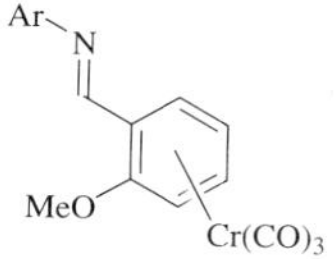 Ar = 4-$MeOC_6H_4$	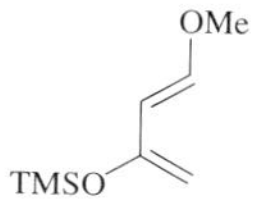	$ZnCl_2$, CH_2Cl_2, rt, 24 h	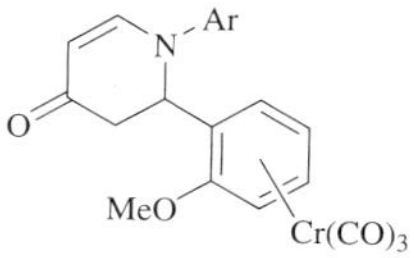 (75)[ccccc]	166
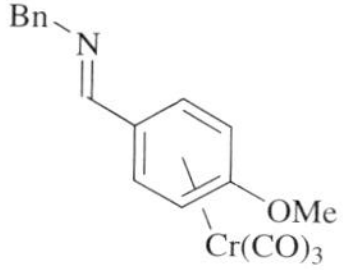	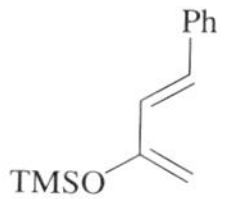	1. TMSOTf, CH_2Cl_2, rt, 12 h 2. hν, O_2, 3 h 3. TBAF, THF, rt, 1 h	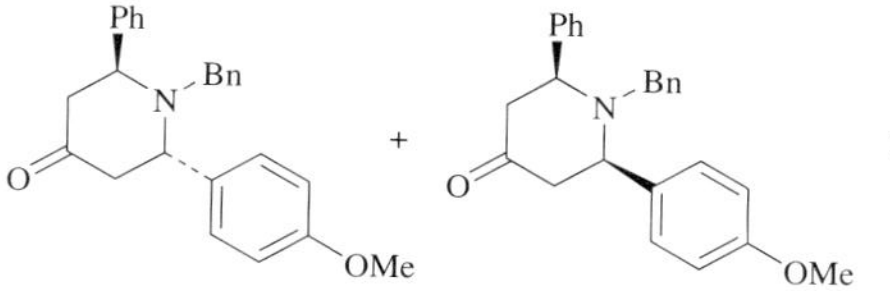 (78) 90:10	169

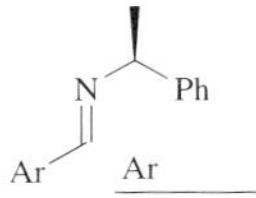 Ar: 2-quinolinyl 4-quinolinyl		$MeSO_3H$, CH_2Cl_2, –78° to rt, overnight	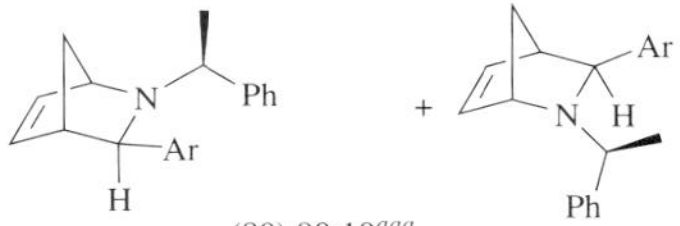 (80) 90:10[aaa] (79) 90:10[aaa]	45, 91
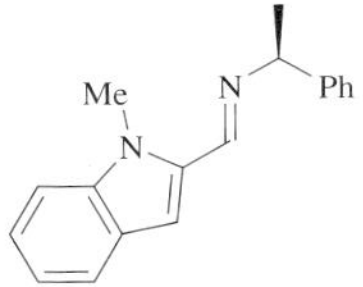	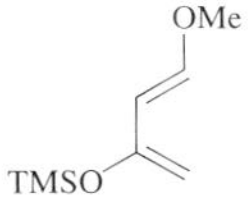	1. $ZnCl_2$, CH_2Cl_2, –20°, 2 h 2. HCl, H_2O, –20° to rt	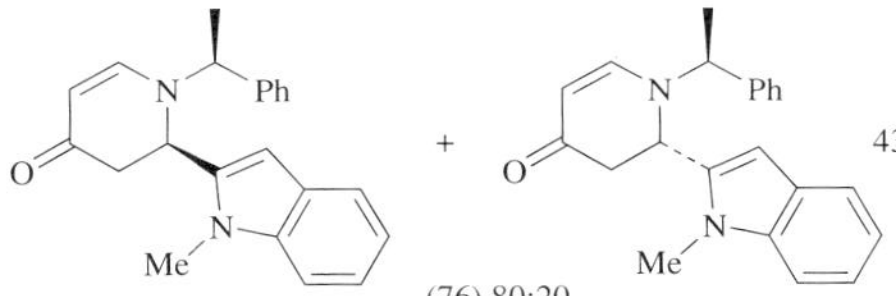 (76) 80:20	434
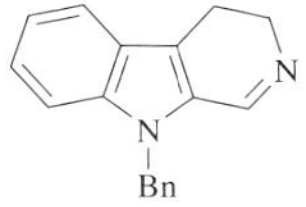		Hydroquinone, chlorobenzene, 120°, 22 h	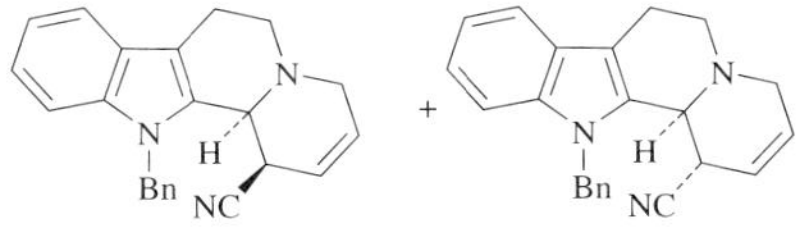 I + II + 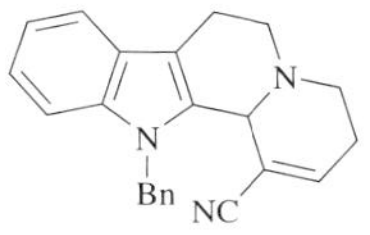 III **I** + **II** + **III** (86)	392
C_{19}				
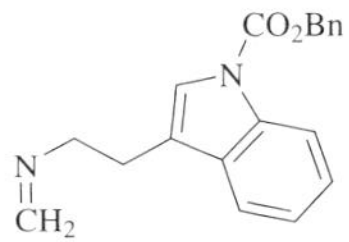	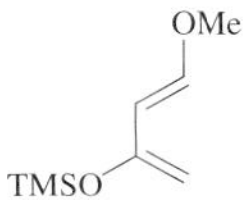	1. $ZnCl_2$, THF, rt 2. $NaHCO_3$, H_2O	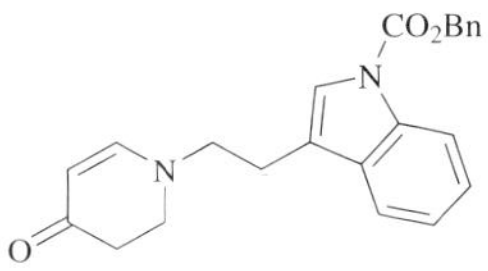 (65)	430

TABLE 5. CYCLOADDITIONS OF *N*-ALKYL- AND *N*-ARYLIMINES (*Continued*)

Imine	Diene	Conditions	Product(s) and Yield(s) (%)	Refs.
C_{19} 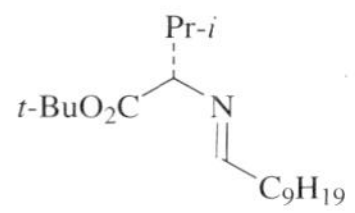	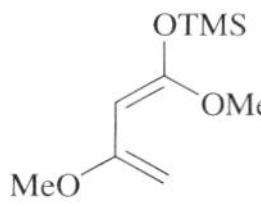	1. $EtAlCl_2$[bbb], –78° to rt, CH_2Cl_2, 1-1.5 h 2. $NaHCO_3$, H_2O	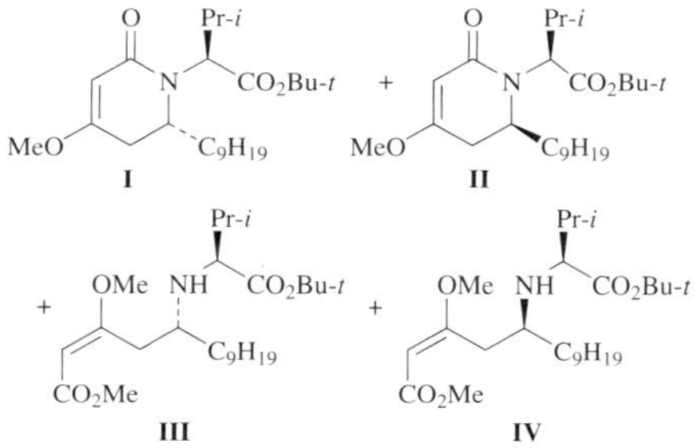 **I** + **II** + **III** + **IV** (57), **I** + **II**:**III** + **IV** 44:56, **I** + **III**:**II** + **IV** 93:7[ccc]	402
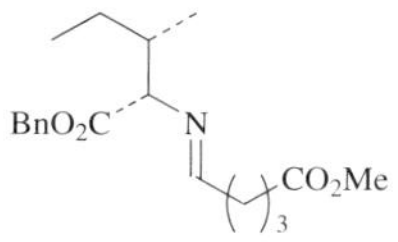	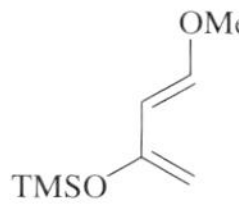	1. $EtAlCl_2$[jj], CH_2Cl_2, –78 to 20° 2. H^+, H_2O	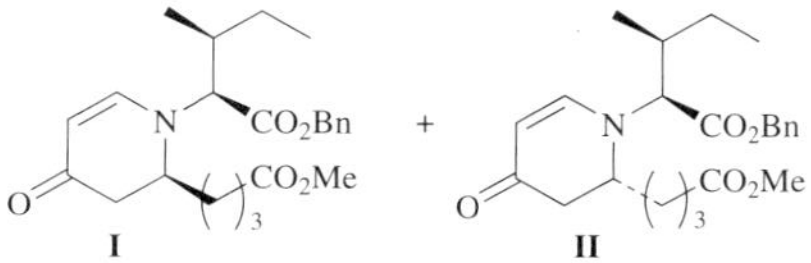 **I** + **II** (46), **I**:**II** 93:7	175a
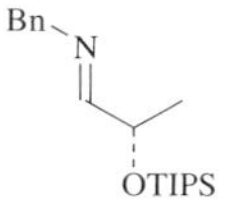	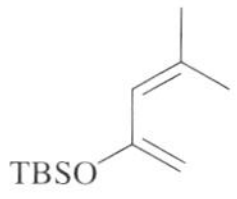	1. TMSOTf, CH_2Cl_2, 4° 2. $NaHCO_3$, H_2O 3. TBAF, THF, 10 h	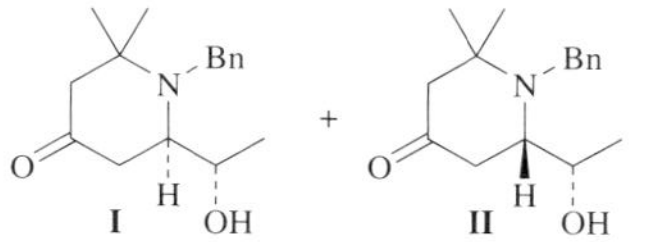 **I** + **II** (—)[iiiii]	150
	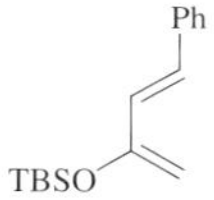	1. $TiCl_4$, CH_2Cl_2, 4° 2. $NaHCO_3$, H_2O 3. TBAF, THF, 10 h	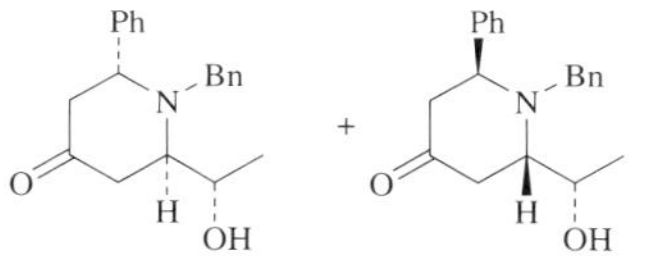 (75) 40:60	150

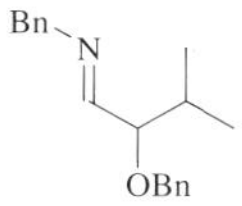 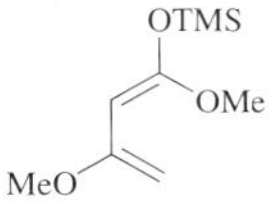

Lewis acid, CH_2Cl_2, −78° to rt

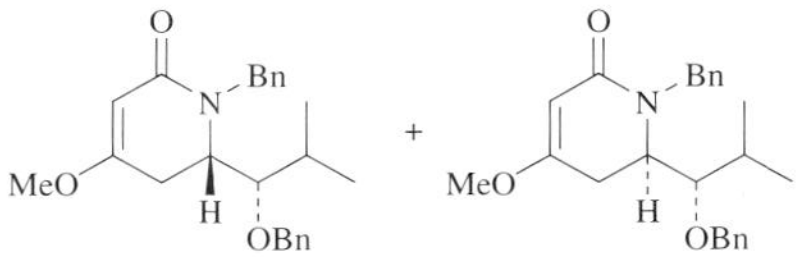

160

L. A. (eq)	Time	
$SnCl_4$	16-24 h	(20) 91:9
Et_2AlCl (0.9)	20-22 h	(64) 40:60
Et_2AlCl (2.0-2.1)	20-22 h	(58) 60:40

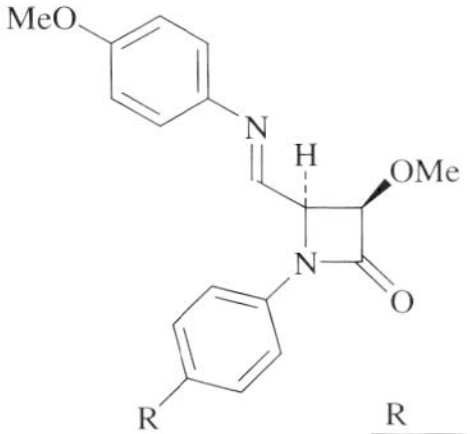 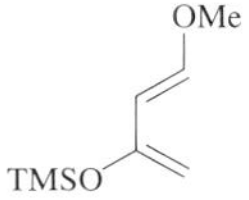

1. ZnI_2, MeCN, −20°, 6 h
2. $NaHCO_3$, H_2O

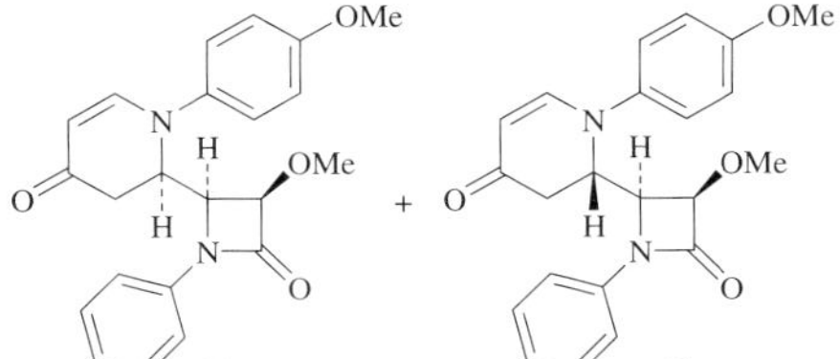

I + II

R		
Me	**I** + **II** (75), **I**:**II** 75:25[jjjjj,kkkkk]	441, 165
MeO	**I** (57) + **II** (25)	165

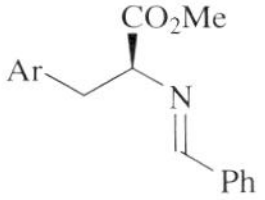

Ar = 3-indolyl

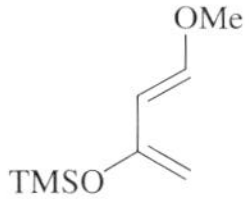

1. $ZnCl_2$[gggg], THF
2. $NaHCO_3$, H_2O

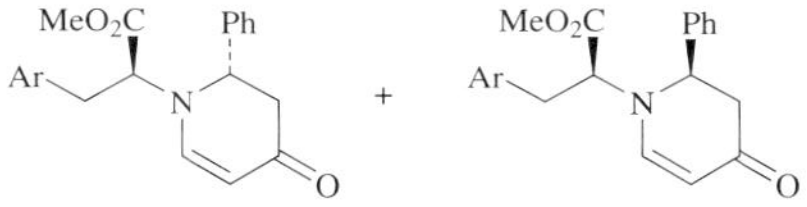

I + II

Temp	Time		
−20 to 0°	12 h	**I** + **II** (68), **I**:**II** 70:30	430
rt	overnight	**I** + **II** (68), **I**:**II** 70:30	419

Imine	Diene	Conditions	Product(s) and Yield(s) (%)	Refs.
C_{19}				
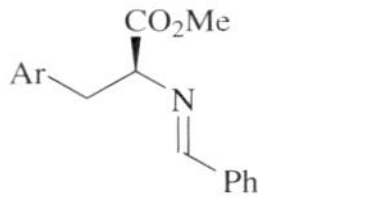 Ar = 3-indolyl	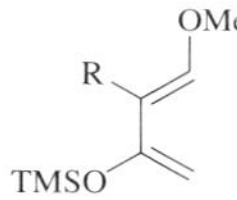 R: H; Et	1. $B(OPh)_3$, 4 Å MS, CH_2Cl_2, –78° to rt, overnight 2. HCl, H_2O	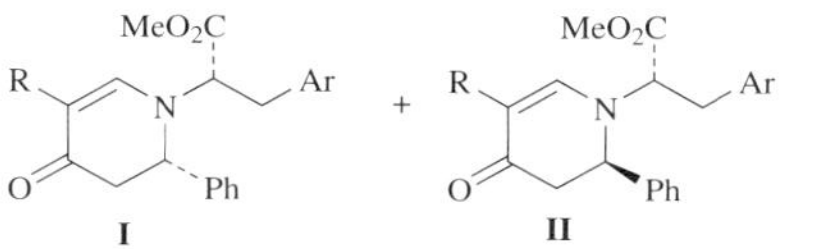 I + II R = H: **I** + **II** (51), **I**:**II** 88:12 R = Et: **I** + **II** (61), **I**:**II** 93:7[yyyy]	246, 245
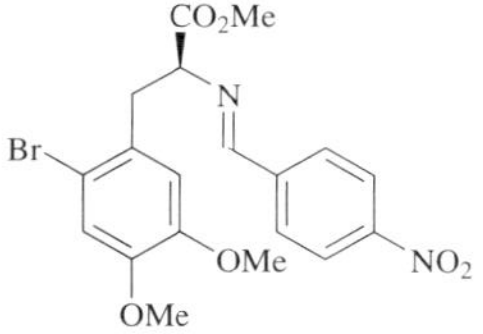	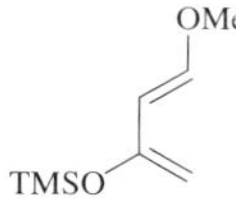	1. $B(OPh)_3$, CH_2Cl_2, –78° to rt, 20 h 2. HCl, H_2O	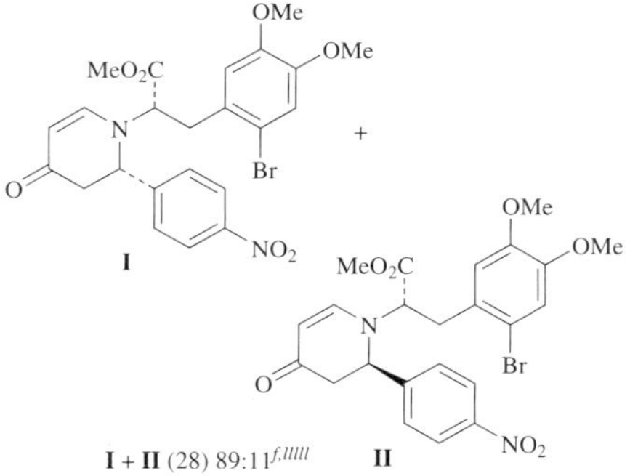 I + II **I** + **II** (28) 89:11[f,llllll]	380
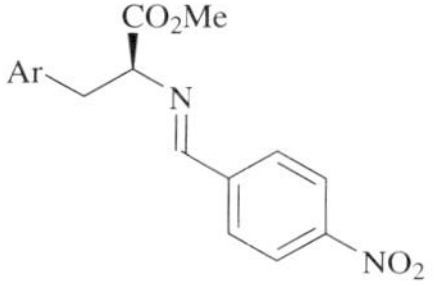 Ar = 3-indolyl	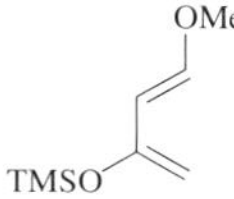	1. $ZnCl_2$[gggg], THF 2. $NaHCO_3$, H_2O	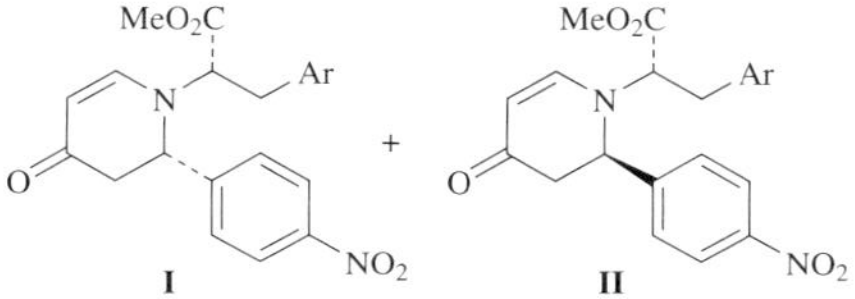 I + II	
		Temp –20 to 0°, Time 12 h	**I** + **II** (46), **I**:**II** 66:34	430
		Temp rt, Time overnight	**I** + **II** (46), **I**:**II** 66:34	419

	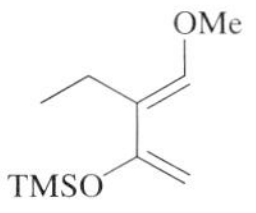	1. $B(OPh)_3$, 4 Å MS, CH_2Cl_2, –78° to rt, overnight 2. HCl, H_2O	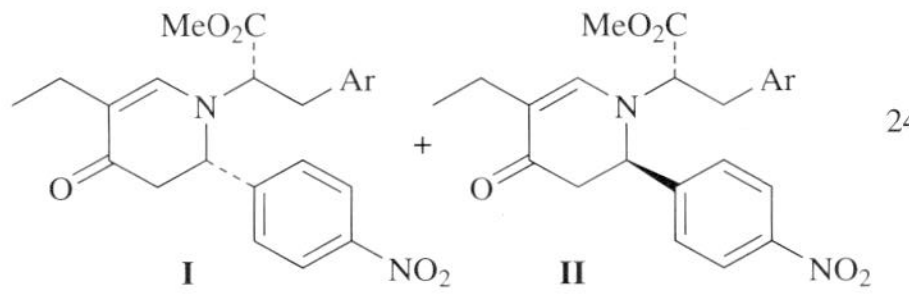 **I** + **II** (35), **I**:**II** 94:6[yyyy]	246, 245
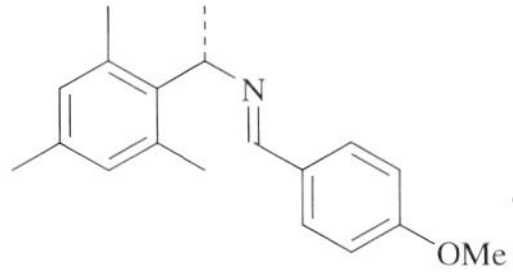	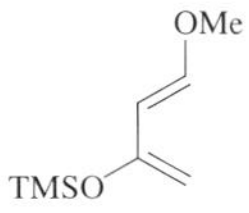	1. BF_3•OEt_2, CH_2Cl_2, –78 to 0°, 24 h 2. $NaHCO_3$, H_2O	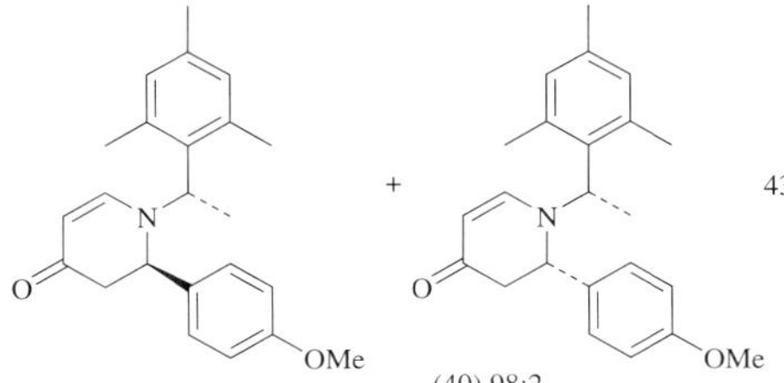 (40) 98:2	433
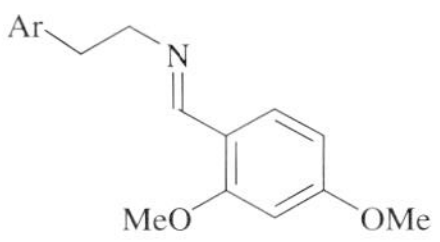 Ar = 3-indolyl	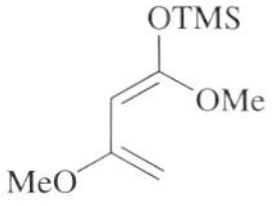	1. $EtAlCl_2$, THF, –60° to rt, 2 h 2. $NaHCO_3$, H_2O	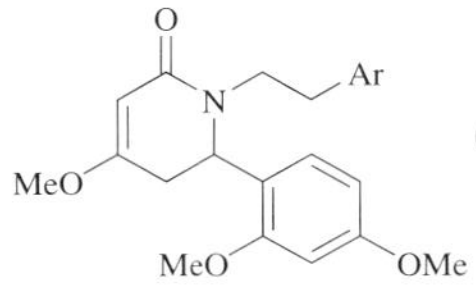 (40)[hhhh]	420, 421
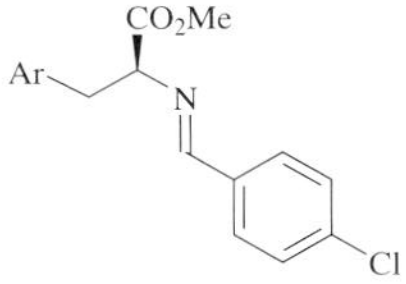 Ar = 3-indolyl	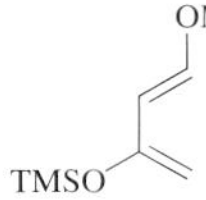	1. $ZnCl_2$[gggg], THF 2. $NaHCO_3$, H_2O	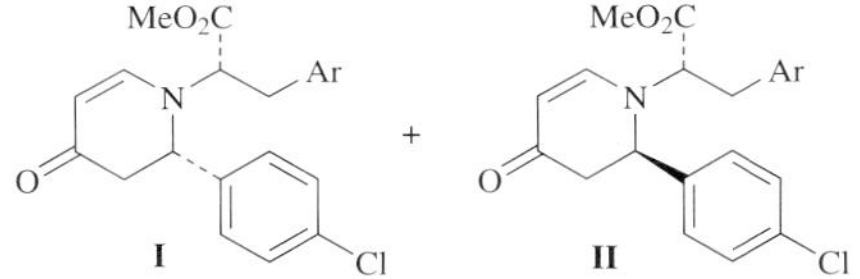	
		Temp / Time		
		–20 to 0° / 12 h	**I** + **II** (73), **I**:**II** 66:34	430
		rt / overnight	**I** + **II** (73), **I**:**II** 66:34	419

TABLE 5. CYCLOADDITIONS OF *N*-ALKYL- AND *N*-ARYLIMINES (*Continued*)

	Imine	Diene	Conditions	Product(s) and Yield(s) (%)	Refs.
C_{19}	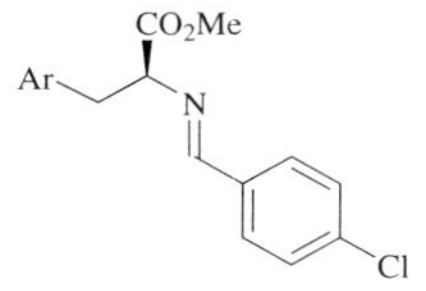 Ar = 3-indolyl	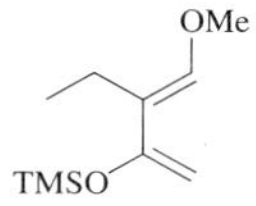	1. $B(OPh)_3$, 4 Å MS, CH_2Cl_2, –78° to rt, overnight 2. HCl, H_2O	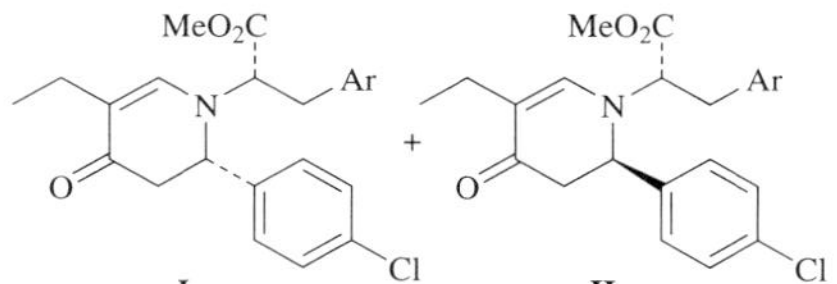 **I** **II** **I** + **II** (38), **I**:**II** 91:9[yyyy]	246, 245
	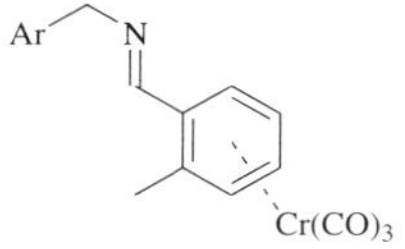 Ar = $CH_2C_6H_4Br$-2	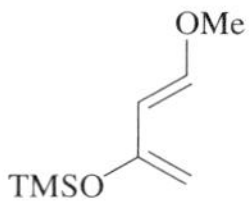	$SnCl_4$, THF, –78° to rt, 15 h	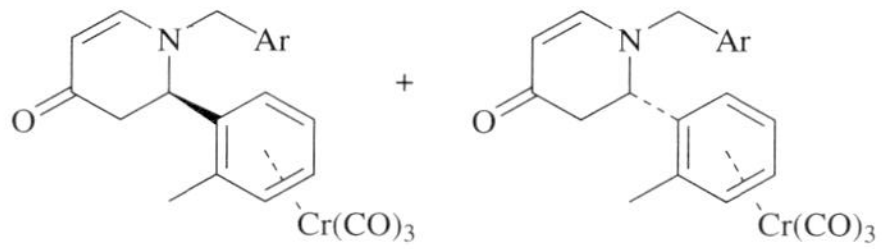 (63) >98:2[bbbb]	158
	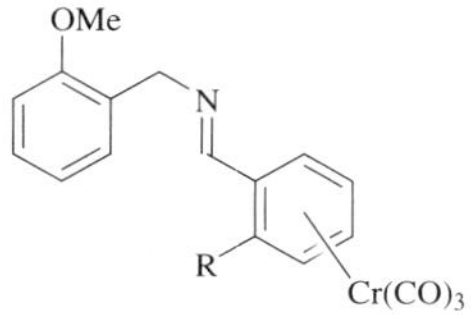	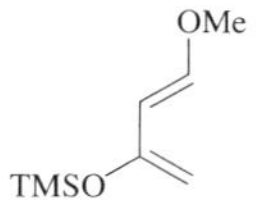	Et_2AlCl, –78° to rt, 15 h	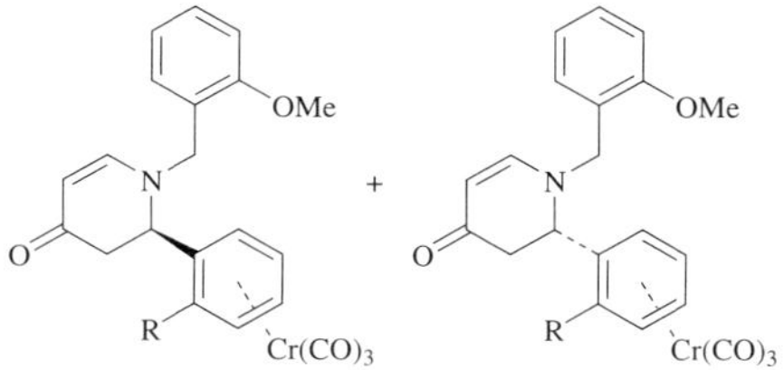	158
	R		Solvent		
	Me		THF	(77) >98:2[bbbb]	
	MeO		THF	(83) 85:15	
	MeO		CH_2Cl_2	(74) 64:36	

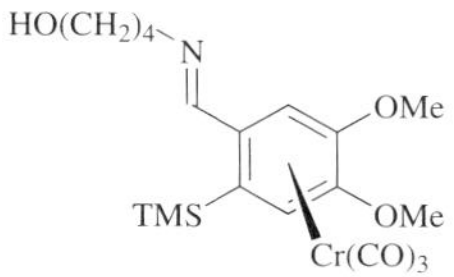	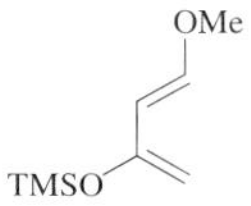	$SnCl_4$, THF, –78 to 0°, 18 h $SnCl_4$ eq: 1.0 / 1.6 / 2.0	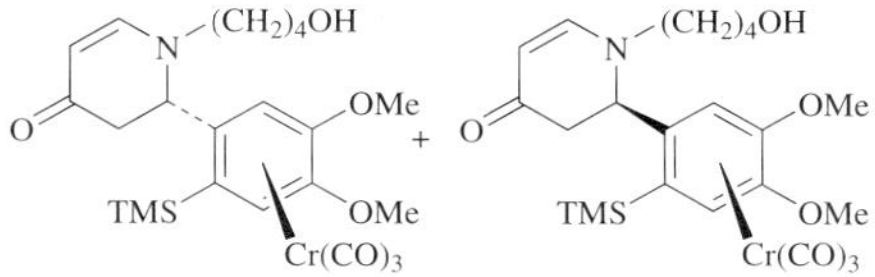 **I** + **II** **I + II** (15), **I:II** >98:2[bbbb] **I + II** (48), **I:II** >98:2[bbbb] **I + II** (35), **I:II** >98:2[bbbb]	167
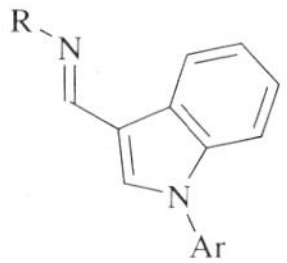 R / Ar: *n*-Bu / SO_2Ph; *i*-Pr / Ts	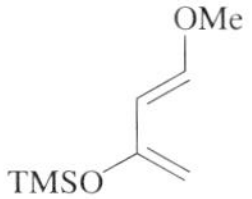	$Zn(OTf)_2$, H_2O, CH_2Cl_2, rt, 31.5 h Temp / Time: rt / 31.5 h; 0° / 13 h	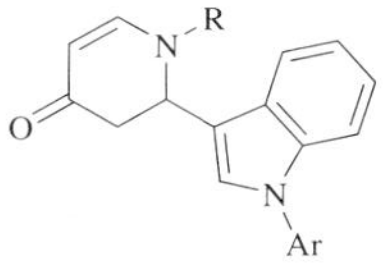 (81) (49)	432
C_{20}				
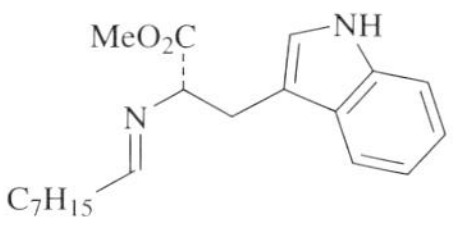	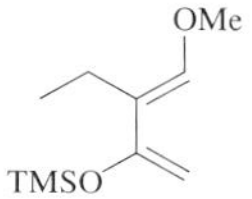	1. $B(OPh)_3$, 4 Å MS, CH_2Cl_2, –78° to rt, overnight 2. HCl, H_2O	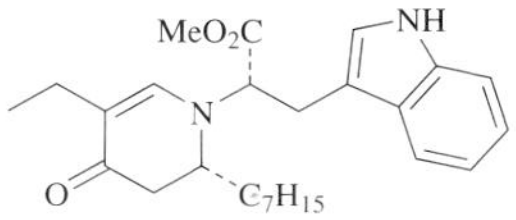 **I** + 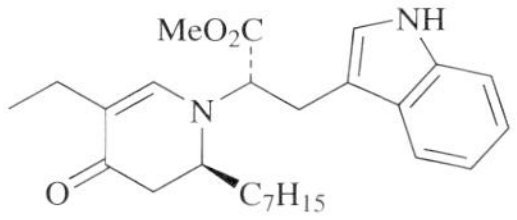 **II** **I + II** (31), **I:II** 90:10[yyyy]	246, 245

TABLE 5. CYCLOADDITIONS OF *N*-ALKYL- AND *N*-ARYLIMINES (*Continued*)

Imine	Diene	Conditions	Product(s) and Yield(s) (%)	Refs.
C_{20}				
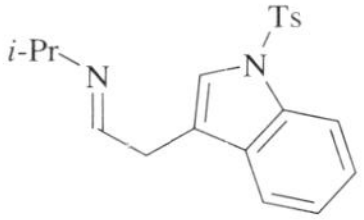	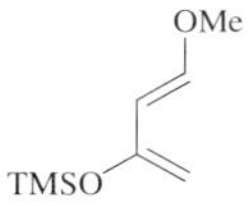	$Zn(OTf)_2$, H_2O, CH_2Cl_2, 0°, 13 h	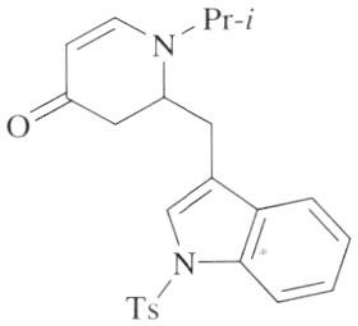 (49-54)[eeeee]	432
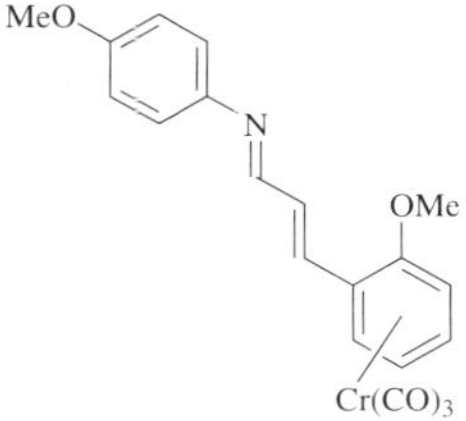	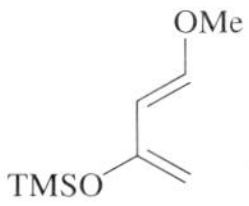	$ZnCl_2$, CH_2Cl_2, –78°	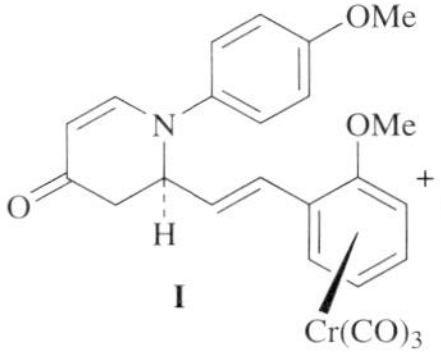 **I** + 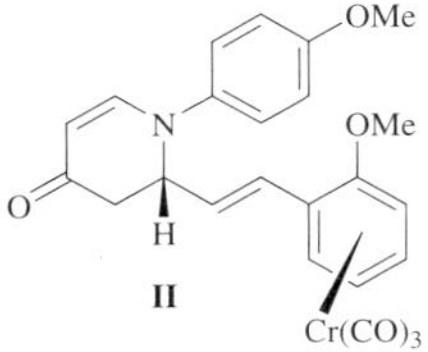 **II** **I** + **II** (75), 56.5:43.5[f]	166
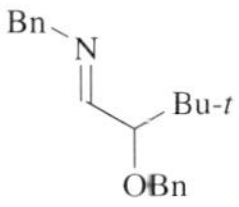	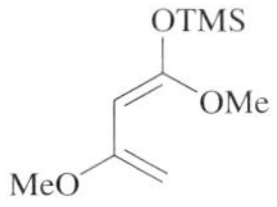	Lewis acid, CH_2Cl_2, –78° to rt	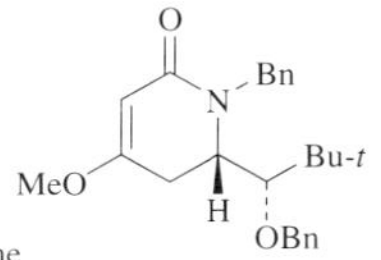 + 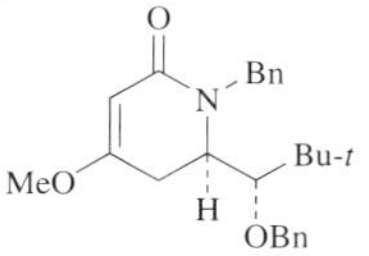	160
		L. A. (eq) / Time		
		$SnCl_4$ / 16-24 h	(10) >99:1[bbbb]	
		Et_2AlCl (0.9) / 20-22 h	(68) 91:9	
		Et_2AlCl (2.0-2.1) / 20-22 h	(75) >99:1[bbbb]	
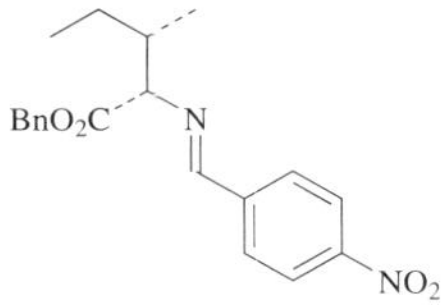	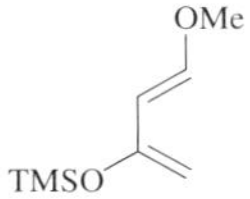	1. $ZnCl_2$, THF, –10°, 7 h 2. $NaHCO_3$, H_2O	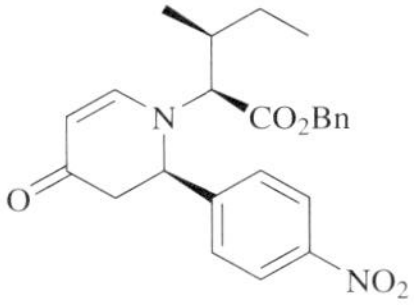 + 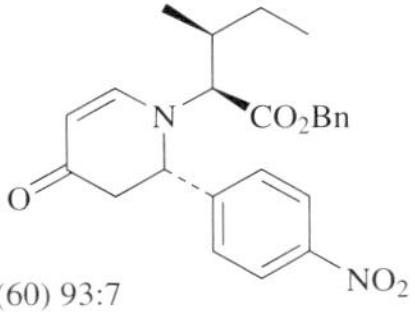 (60) 93:7	384

	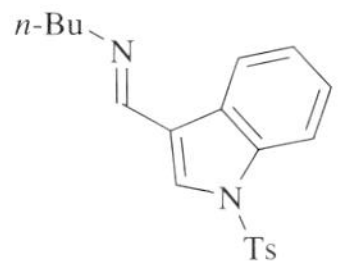	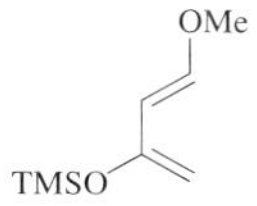	$Zn(OTf)_2$, H_2O, CH_2Cl_2, 0°, 13 h	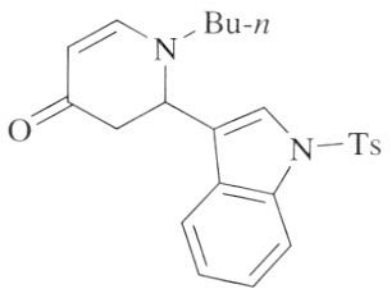 (71)	432
C_{21}	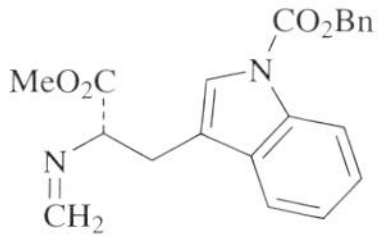	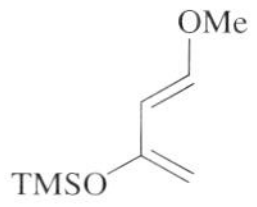	1. $ZnCl_2$, THF, rt, overnight 2. $NaHCO_3$, H_2O	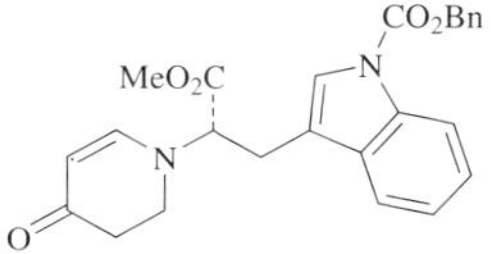 (63)	430, 419
	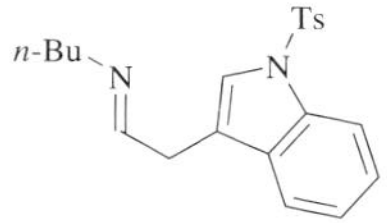	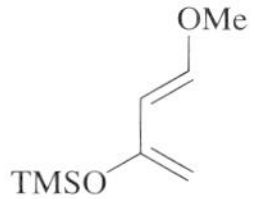	Lewis acid[ddddd], H_2O, CH_2Cl_2, 0°	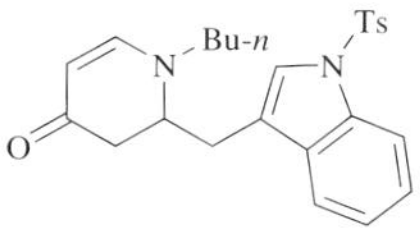	432
			L. A. — Time		
			$Zn(OTf)_2$ — 13 h	(71)	
			$Zn(OTf)_2$[mmmmm] — 5 h	(39)	
			TMSOTf — 9 h	(27)	
	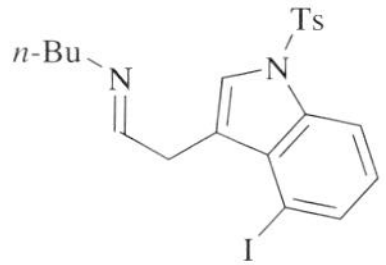	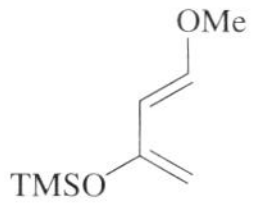	$Zn(OTf)_2$, H_2O, CH_2Cl_2, 0°, 8 h	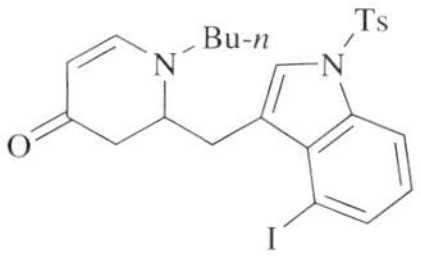 (57-62)[eeeee]	432
	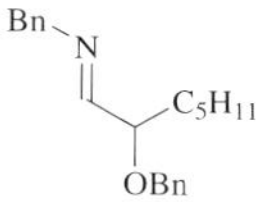	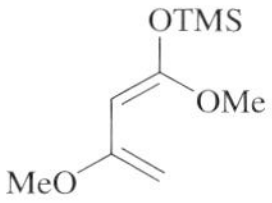	Lewis acid, CH_2Cl_2, –78° to rt	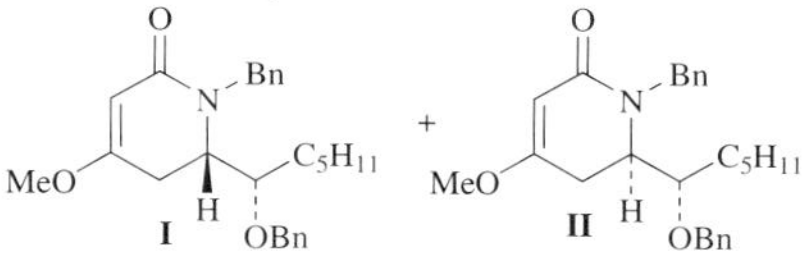	160

L. A. (eq)	Time	I + II	I:II
$SnCl_4$	16-24 h	(15)	83:17
Et_2AlCl (0.9)	20-22 h	(60)	50:50
Et_2AlCl (2.0-2.1)	20-22 h	(55)	90:10

TABLE 5. CYCLOADDITIONS OF *N*-ALKYL- AND *N*-ARYLIMINES (*Continued*)

Imine	Diene	Conditions	Product(s) and Yield(s) (%)	Refs.
C_{22}				
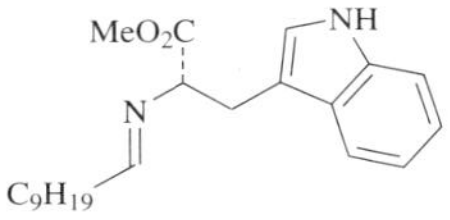	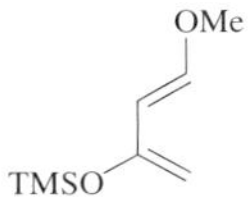	1. $ZnCl_2$[gggg], THF 2. $NaHCO_3$, H_2O	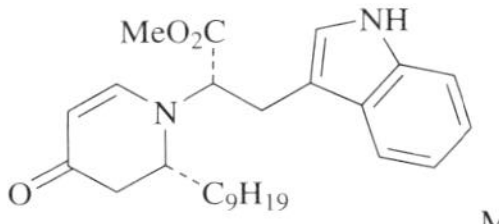 I + 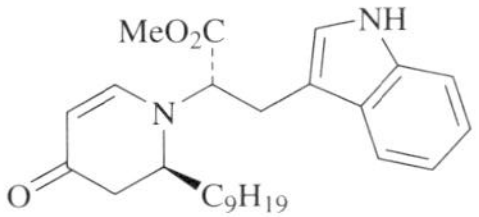 II	
		Temp / Time	I + II / I:II	
		–20 to 0° / 12 h	(45) / 82:18	430
		rt / overnight	(45) / 82:18	419
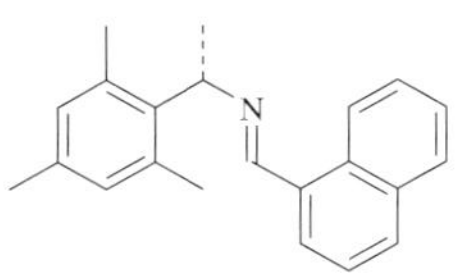	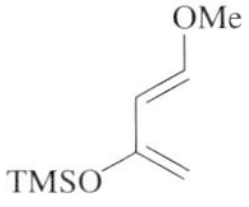	1. BF_3•OEt_2, CH_2Cl_2, –78 to 0°, 24 h 2. $NaHCO_3$, H_2O	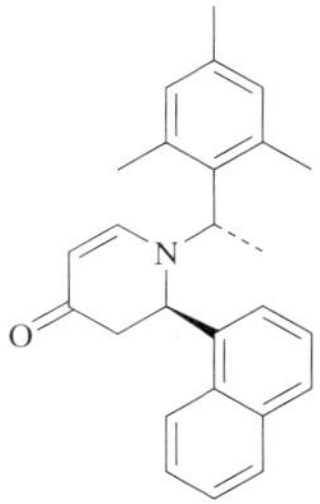 + 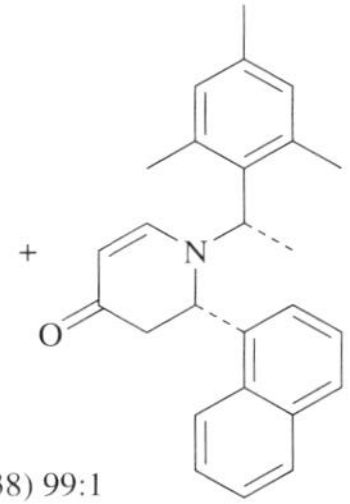 (38) 99:1	433
C_{23}				
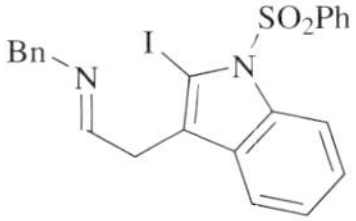	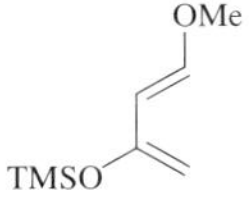	1. $Zn(OTf)_2$, CH_2Cl_2, rt, 3 h 2. HCl, H_2O	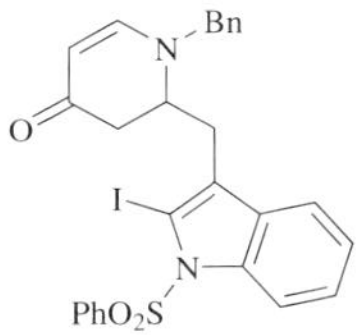 (74)	435

	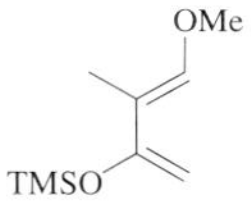	Zn(OTf)$_2$, CH_2Cl_2, rt	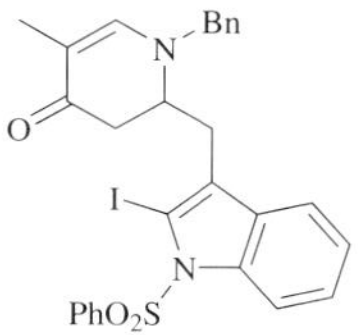 (52)	435
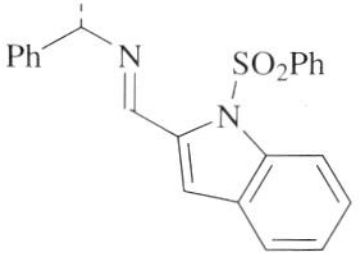	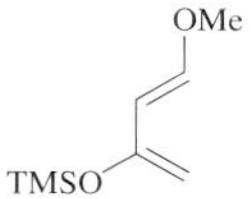	1. $ZnCl_2$, CH_2Cl_2, –20°, 2 h 2. HCl, H_2O, –20° to rt	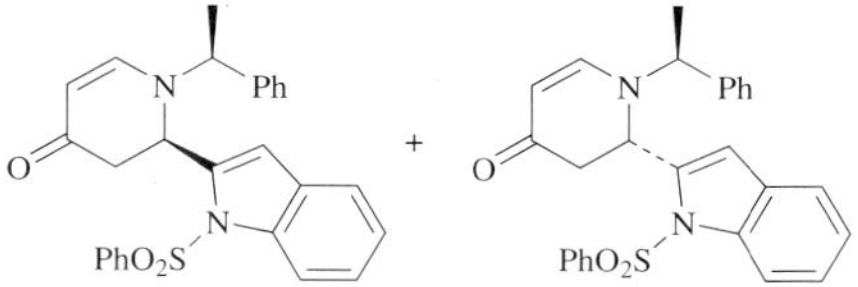 (96) 97:3	434
C_{24} 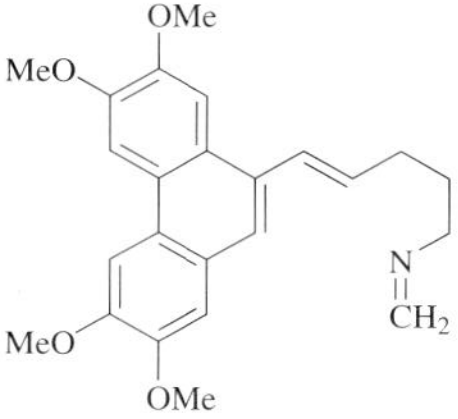		—	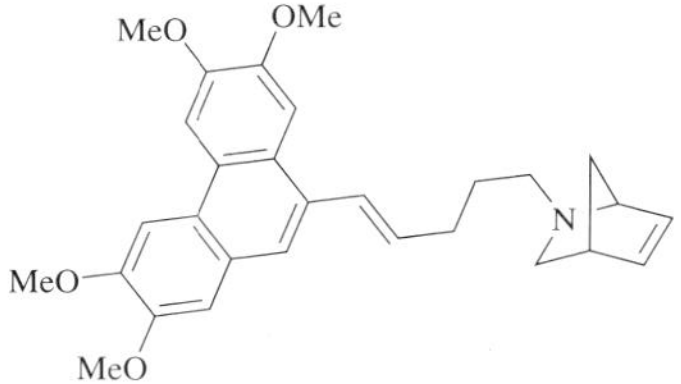 (—)	233
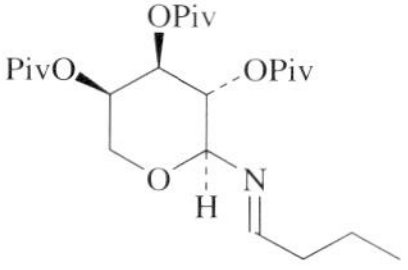	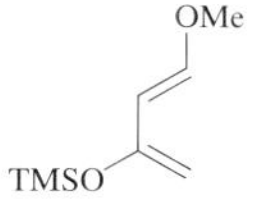	$ZnCl_2$•OEt_2, THF, –78 to –20°, 48 h	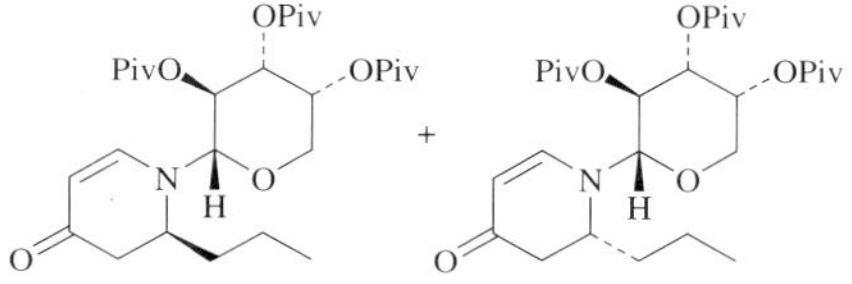 (76) 91:9	175

TABLE 5. CYCLOADDITIONS OF *N*-ALKYL- AND *N*-ARYLIMINES (*Continued*)

Imine	Diene	Conditions	Product(s) and Yield(s) (%)	Refs.
C_{24}				
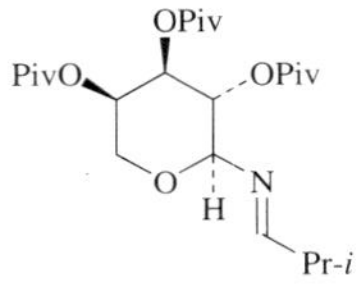	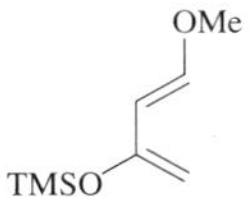	$ZnCl_2•OEt_2$, THF, –78 to –20°, 48 h	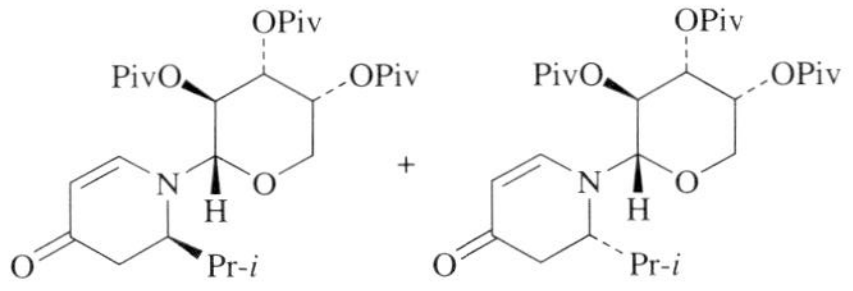 (73) 95:5	175
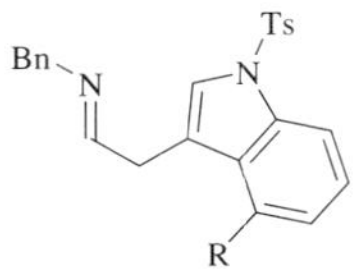 R: H; I	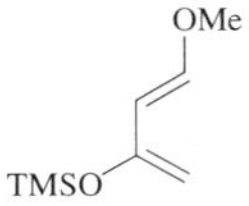	$Zn(OTf)_2$, H_2O, CH_2Cl_2 Temp / Time: 0° / 10 h; 0° to rt / 8 h	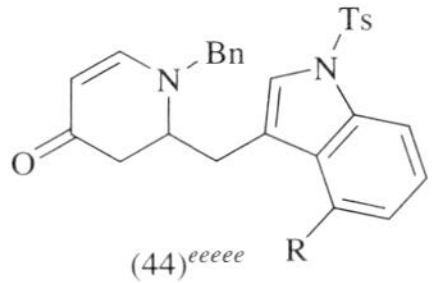 (44)[eeeee]; (68)[eeeee]	432
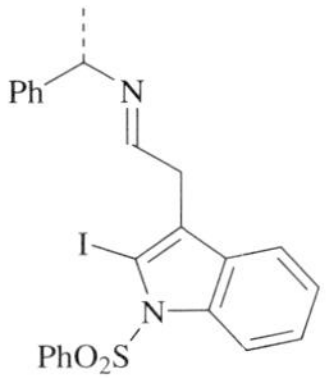	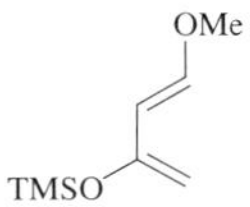	—	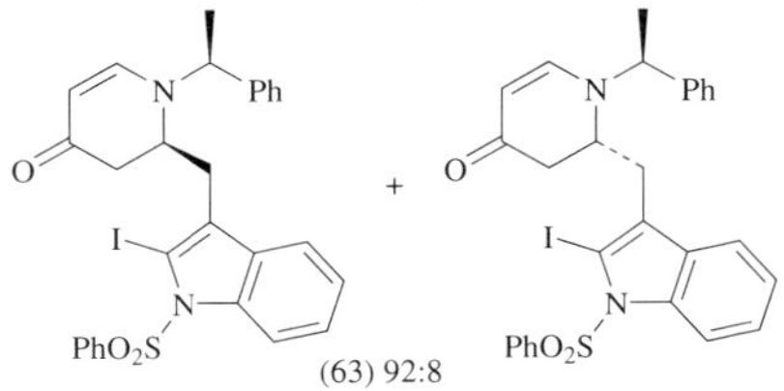 (63) 92:8	435
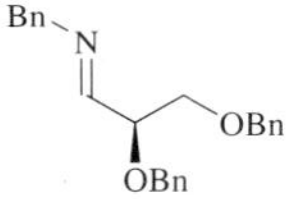	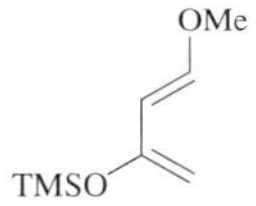	1. Lewis acid 2. $NaHCO_3$, H_2O	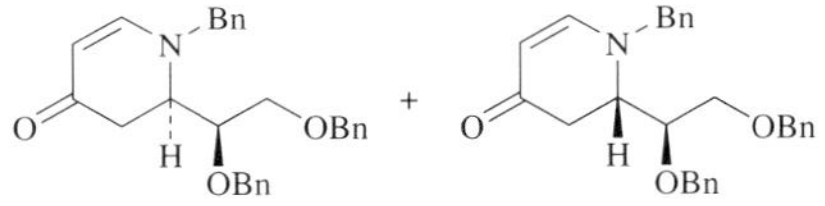	

L. A.	Solvent	Temp	Time		
Et_2AlCl	CH_2Cl_2	rt	15 min	(16) 82:18	163, 162
Et_2AlCl	CH_2Cl_2	–40°	4 h	(53) 84:16	163, 162
Et_2AlCl	CH_2Cl_2	–78°	10 h	(52) 82:18	163
BF_3•OEt_2	CH_2Cl_2	rt	15 min	(15) 80:20	163, 162
BF_3•OEt_2	CH_2Cl_2	–40°	4 h	(35) 86:14	163, 162
BF_3•OEt_2	CH_2Cl_2	–78°	10 h	(34) 85:15	163
$ZnCl_2$	CH_2Cl_2	–40°	—	(65) 90:10	162
ZnI_2	CH_2Cl_2	rt	15 min	(16) 85:15	163, 162
ZnI_2	CH_2Cl_2	–40°	4 h	(67) 91:9	163, 162
ZnI_2	CH_2Cl_2	–78°	10 h	(67) 90:10	163
ZnI_2	toluene	–40°	4 h	(65) 75:25	163
ZnI_2	ether	–40°	4 h	(30) 76:24	163
ZnI_2	THF	–40°	4 h	(63) 80:20	163
ZnI_2 (1.1 eq)	nitroethane	–40°	4 h	(65) 95:5	163, 162
ZnI_2 (0.2 eq)	nitroethane	–40°	4 h	(59) 95:5	163, 162
ZnI_2 (1.1 eq)	MeCN	–40°	—	(65) 95:5	162
ZnI_2 (1.1 eq)	MeCN	–40°	4 h	(88) 95:5	163
ZnI_2 (0.2 eq)	MeCN	–40°	4 h	(70) 95:5	163, 162

TABLE 5. CYCLOADDITIONS OF *N*-ALKYL- AND *N*-ARYLIMINES (*Continued*)

	Imine	Diene	Conditions	Product(s) and Yield(s) (%)	Refs.
C_{24}	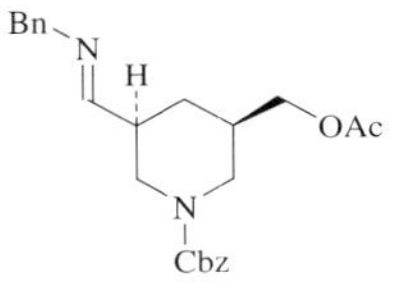	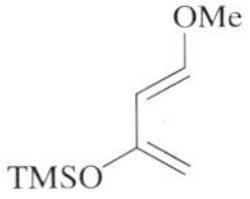		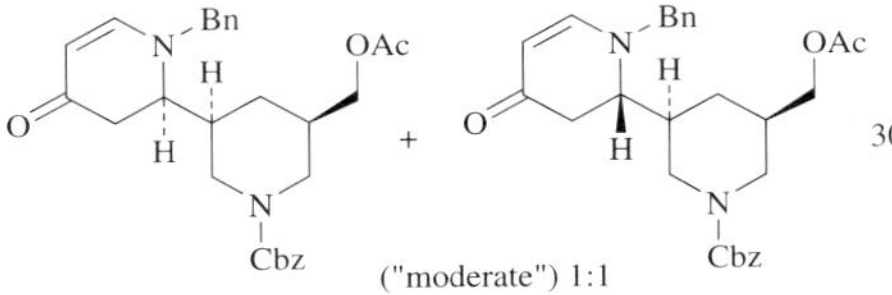	305
			1. $ZnCl_2$, THF, –80° to rt, 16 h 2. H_2O	("moderate") 1:1	
			$Sc(OTf)_3$, $MgSO_4$, MeCN, rt	(89) 1:1	
	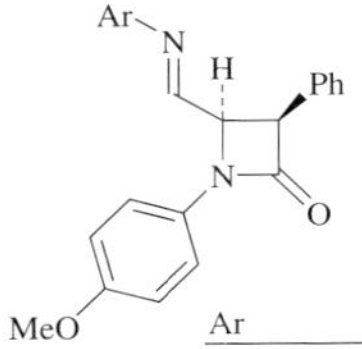	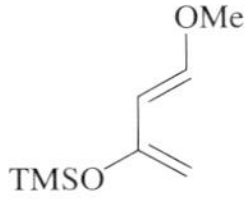		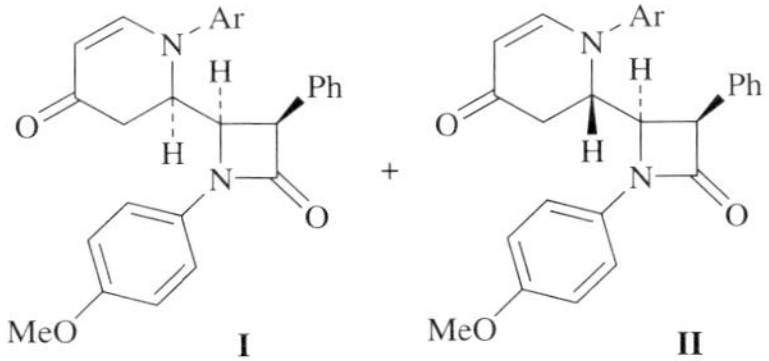 **I** **II**	
	Ar: Bn		1. ZnI_2, MeCN, –20°, 6 h 2. $NaHCO_2$, H_2O	**I** (64)	165
	Ar: 4-$MeOC_6H_4$		ZnI_2, MeCN, –20°	**I** + **II** (95), **I**:**II** 66:34[jjjjj,nnnnn]	441
	Ar: 4-$MeOC_6H_4$		1. ZnI_2, MeCN, –20°, 6 h 2. $NaHCO_2$, H_2O	**I** (60) + **II** (25)	165
	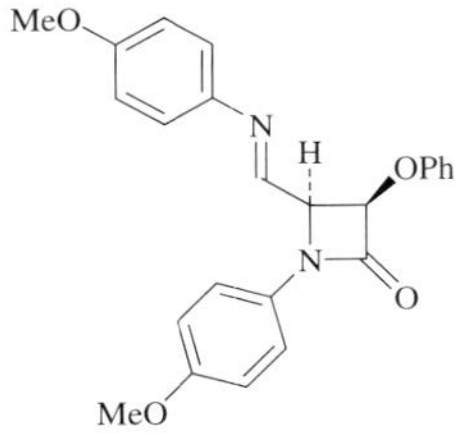	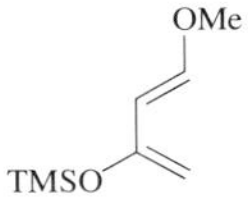	ZnI_2, MeCN, –20°	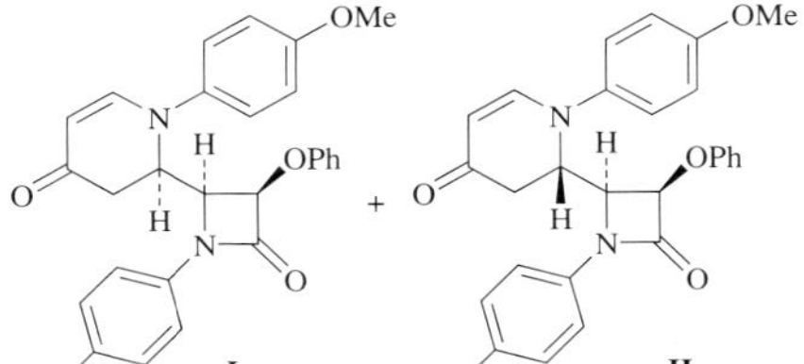 **I** **II** **I** + **II** (95), **I**:**II** 69:31[jjjjj,ooooo]	441

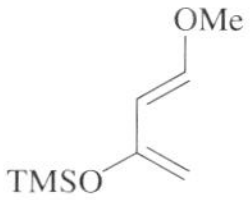

1. Lewis acid, MeCN[ppppp], −20°, 6 h
2. $NaHCO_3$, H_2O

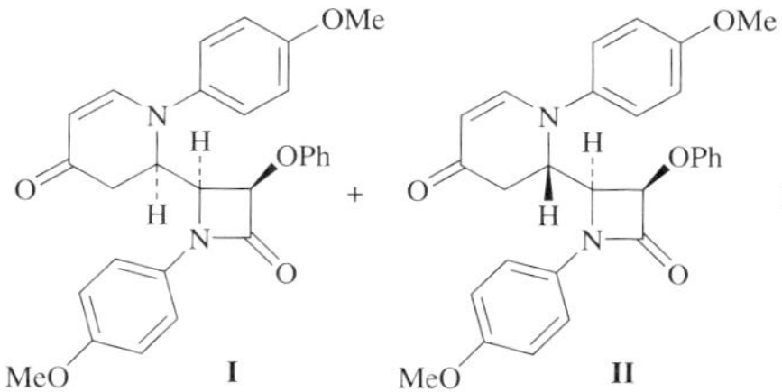

165

L. A. (eq)	Time	I	II
ZnI_2 (1.0)	6 h	(65)	(22)
ZnI_2 (0.2)	6 h	(61)	(24)
$InCl_3$ (1.0)	2 h	(62)	(20)
$InCl_3$ (0.2)	2 h	(63)	(18)
$In(OTf)_3$ (1.0)	2 h	(33)	(8)
$In(OTf)_3$ (0.2)	2 h	(33)	(7)
$HfCl_4$ (1.0)	3.5 h	(49)	(17)
$HfCl_4$ (0.2)	3.5 h	(49)	(16)

C_{25}

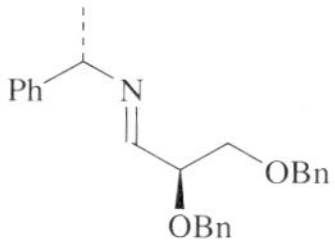

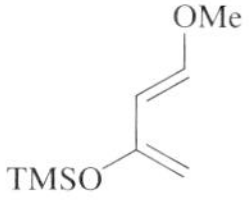

1. ZnI_2, MeCN, −20°, 4 h
2. $NaHCO_3$, H_2O

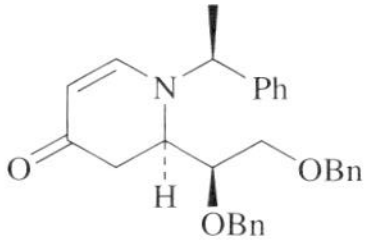

(75) 163

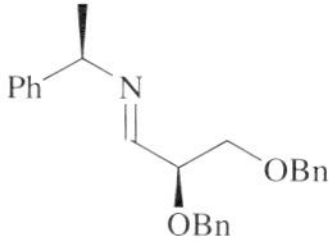

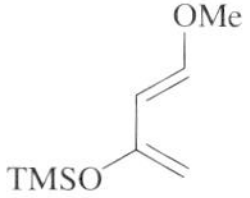

1. ZnI_2, MeCN, −20°, 4 h
2. $NaHCO_3$

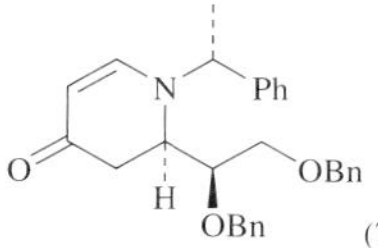

+

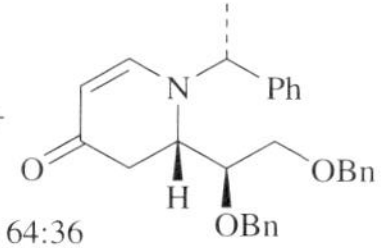

(73) 64:36 163

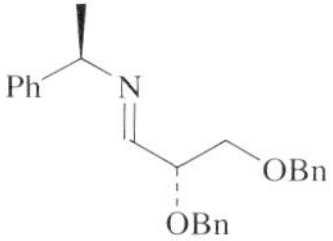

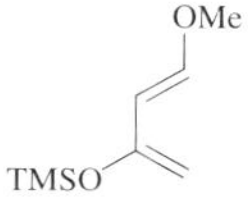

1. ZnI_2, MeCN, −20°, overnight
2. $NaHCO_3$, H_2O

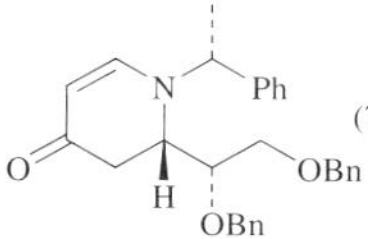

(78) 164

TABLE 5. CYCLOADDITIONS OF *N*-ALKYL- AND *N*-ARYLIMINES (*Continued*)

	Imine	Diene	Conditions	Product(s) and Yield(s) (%)	Refs.
C_{25}	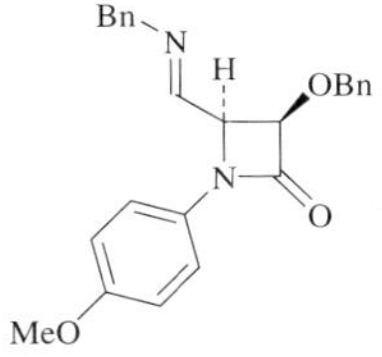	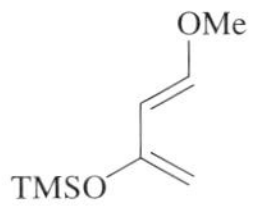	1. ZnI_2, MeCN, –20°, 6 h 2. $NaHCO_3$, H_2O	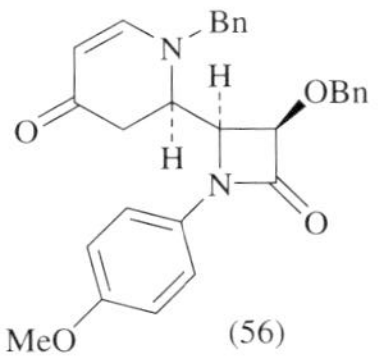 (56) + 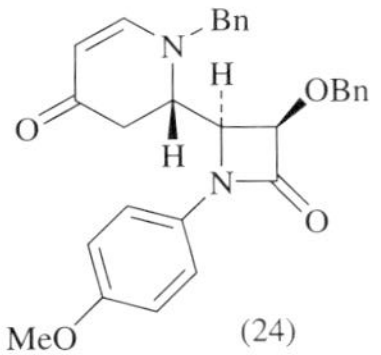 (24)	165
	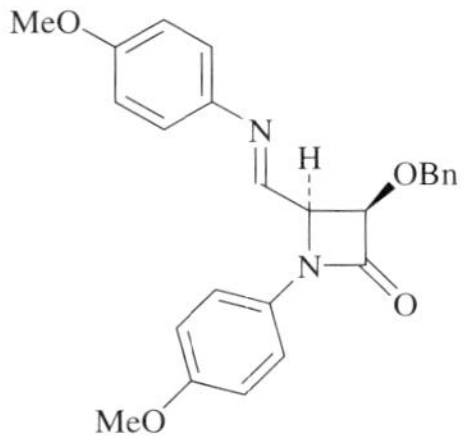	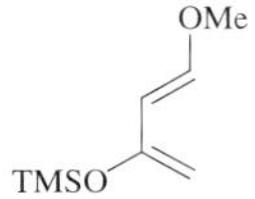	1. ZnI_2, MeCN, –20°, 5 h 2. $NaHCO_3$, H_2O	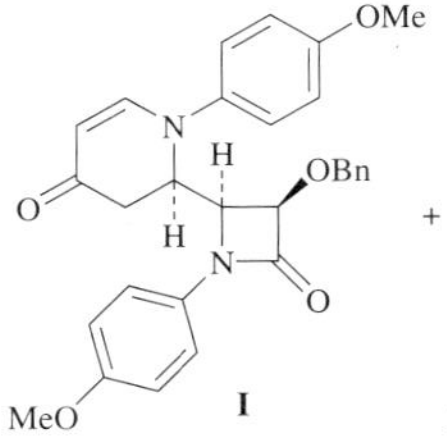 **I** + 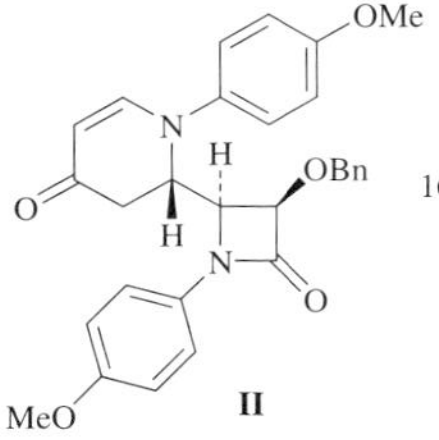 **II** **I** + **II** (90), **I**:**II** 60:40[jjjjj,qqqqq]	165, 441
C_{26}	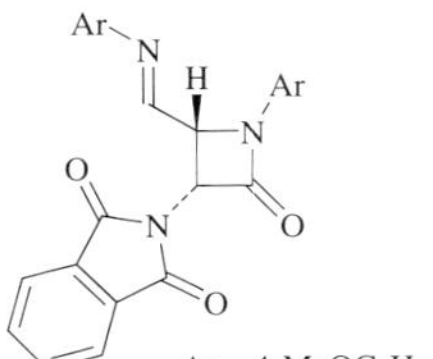 Ar = 4-$MeOC_6H_4$	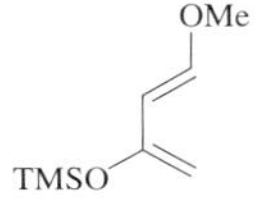	1. ZnI_2, MeCN, –20°, 6 h 2. $NaHCO_3$, H_2O	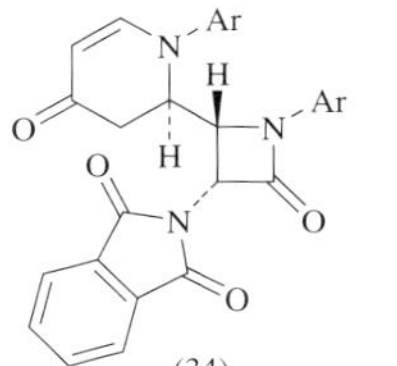 (34) + 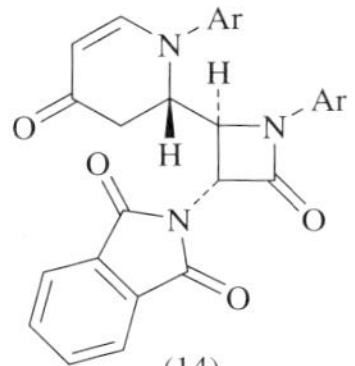 (14)	165

	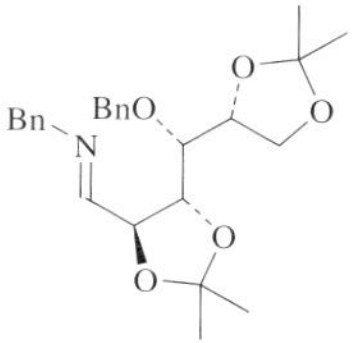	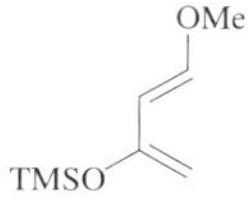		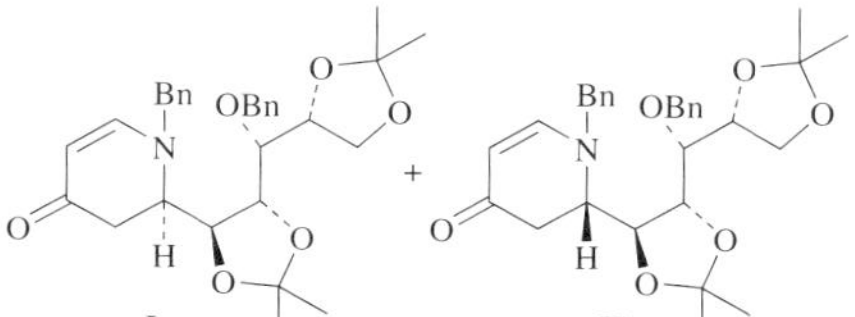 **I** **II**	
			1. $ZnCl_2$, dioxane, 3 h 2. $NaHCO_3$, H_2O	**I** (65.7) + **II** (7.3)	443
			$ZnCl_2$, THF	**I** + **II** (72), **I**:**II** 92:8	440
C_{27}	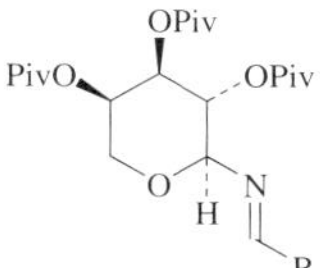 R Ph 4-$NO_2C_6H_4$ 4-ClC_6H_4	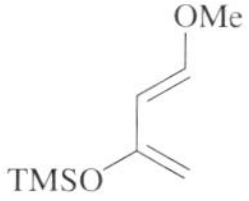	$ZnCl_2$•OEt_2, THF, –78 to –20°, 48 h	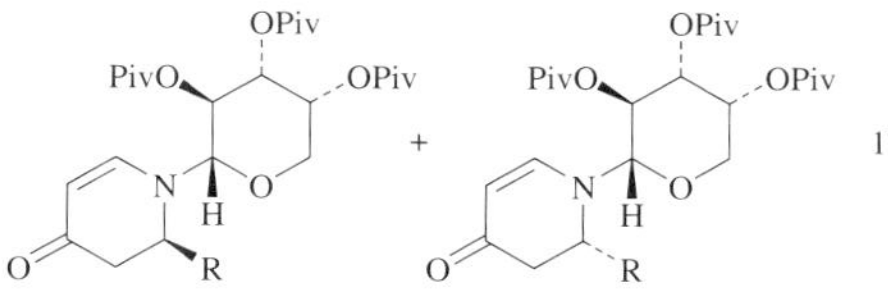 (84) 92:8 (74) 86:14 (96) 97:3	175
C_{29}	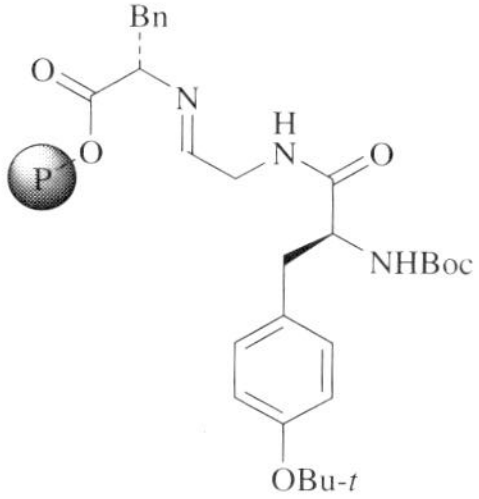	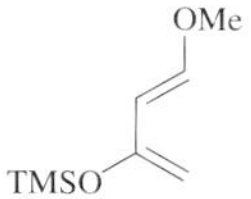	1. $ZnCl_2$[rrrrr], THF, 4°, 24 h 2. TFA, CH_2Cl_2	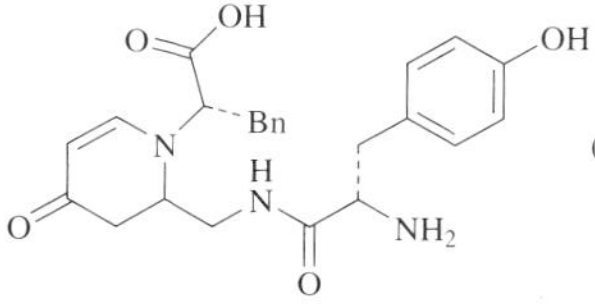 (55) >95% de[f]	444

TABLE 5. CYCLOADDITIONS OF *N*-ALKYL- AND *N*-ARYLIMINES (*Continued*)

	Imine	Diene	Conditions	Product(s) and Yield(s) (%)	Refs.
C_{29}	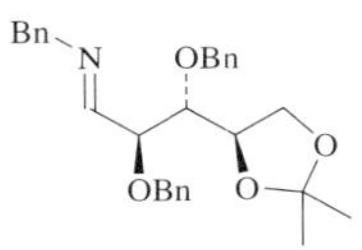	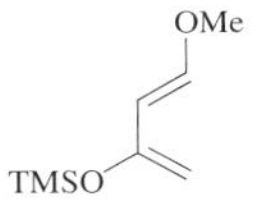	1. $ZnCl_2$, 2-3 h 2. $NaHCO_3$, H_2O Solvent THF MeCN	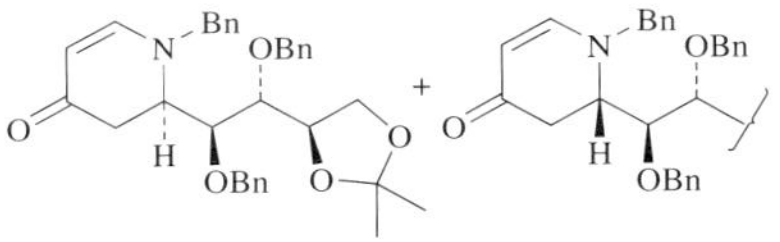 (62) 65:35 (76) 95:5	439, 440
C_{30}	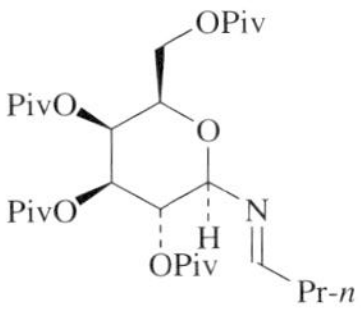	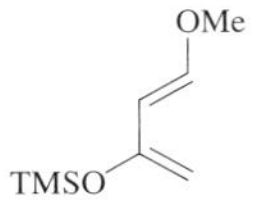		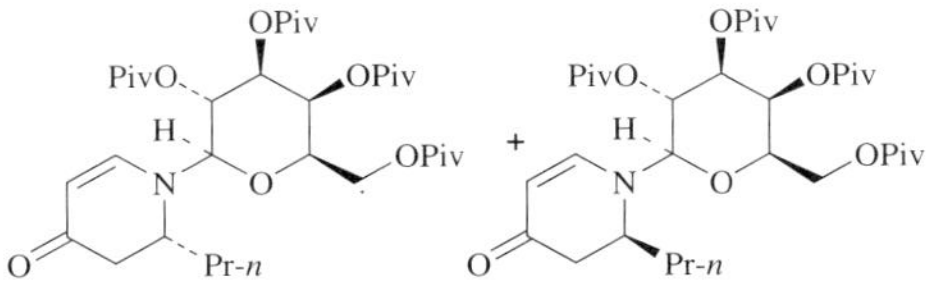 I + II	
			1. $ZnCl_2{\bullet}OEt_2$, THF, –20° 2. HCl, H_2O	**I** + **II** (96), **I**:**II** 97.5:2.5[sssss,ttttt]	171
			1. $ZnCl_2{\bullet}OEt_2$, THF, –20° 2. NH_4Cl, H_2O	**I** + **II** (90), **I**:**II** 97.5:2.5[ttttt]	171
			1. $ZnCl_2$, THF, CH_2Cl_2, –20°, 24-48 h 2. HCl, H_2O	**I** + **II** (96), **I**:**II** 38:1[uuuuu]	173
			$ZnCl_2{\bullet}OEt_2$, THF, –20°	**I** + **II** (68), **I**:**II** >35:1	445
	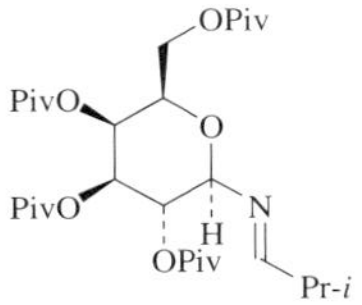	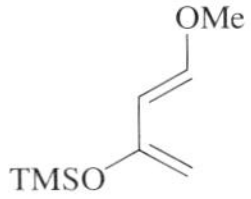	1. $ZnCl_2$, THF, CH_2Cl_2, –20°, 24-48 h 2. HCl, H_2O	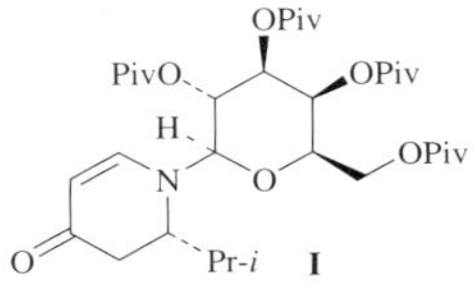 I + 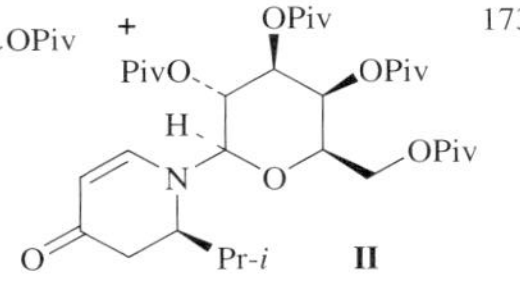 II **I** + **II** (58), **I**:**II** 33.5:1	173

C_{30-31}

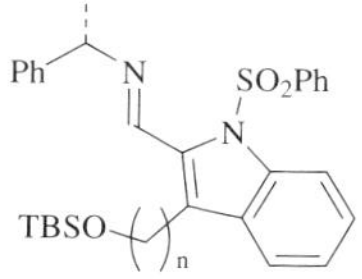

n
1
2

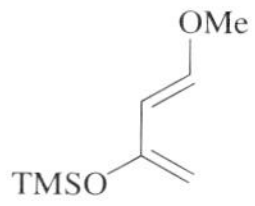

1. $ZnCl_2$, CH_2Cl_2, –20°, 2 h
2. HCl, H_2O, –20° to rt

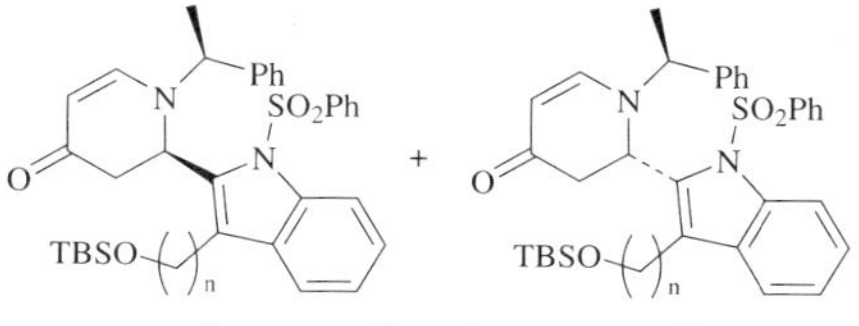

I + **II** 434

I	I:II
(75)	85:15
(73)	86:14

C_{31}

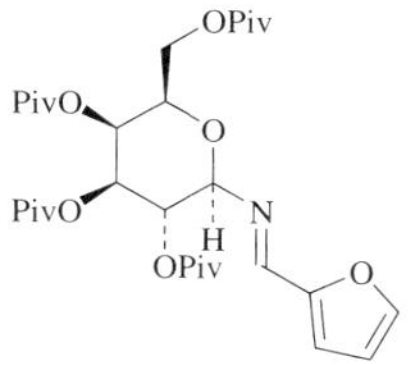

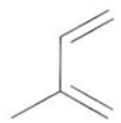

$ZnCl_2$•OEt_2, CH_2Cl_2, 20°, 96 h

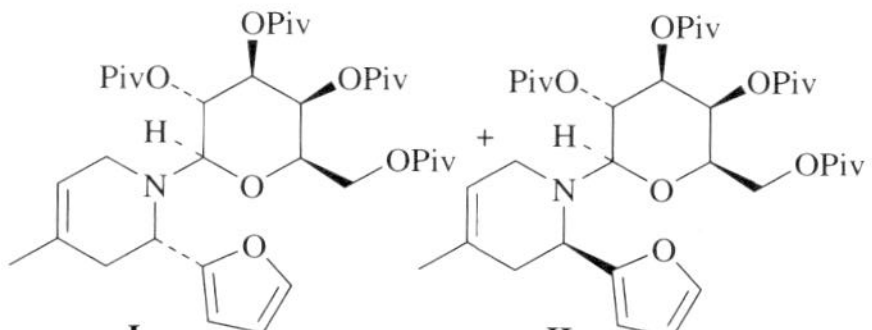

I + **II** 172

I + II (98), **I:II** 79:21

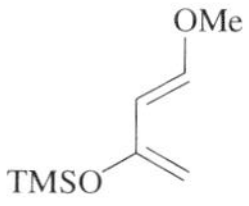

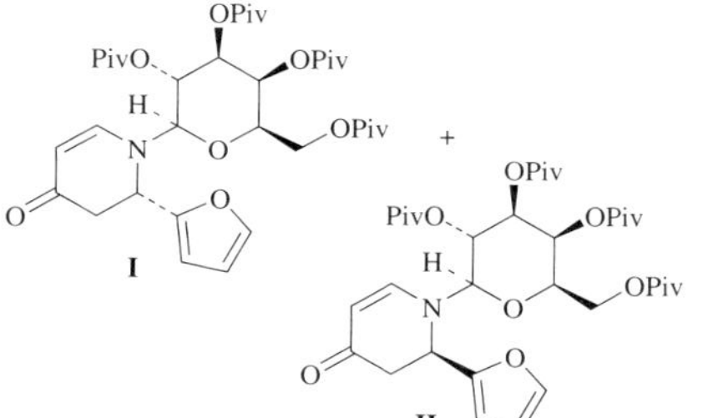

I + **II** 171

1. $ZnCl_2$•OEt_2, THF, –20°
2. HCl, H_2O

I + II (95), **I:II** 96:4[ttttt,vvvvv]

1. $ZnCl_2$•OEt_2, THF, –20°
2. NH_4Cl, H_2O

I + II (60), **I:II** 96:4[lllll,ttttt]

TABLE 5. CYCLOADDITIONS OF *N*-ALKYL- AND *N*-ARYLIMINES (*Continued*)

Imine	Diene	Conditions	Product(s) and Yield(s) (%)	Refs.
C_{31} 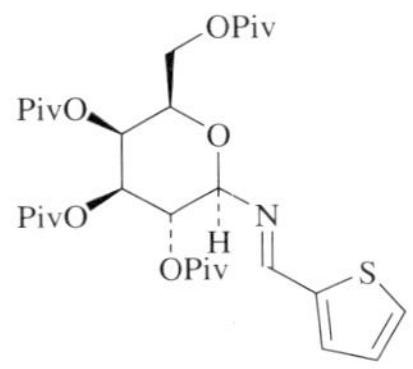	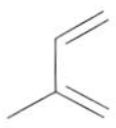	$ZnCl_2{\bullet}OEt_2$, CH_2Cl_2, 20°, 96 h	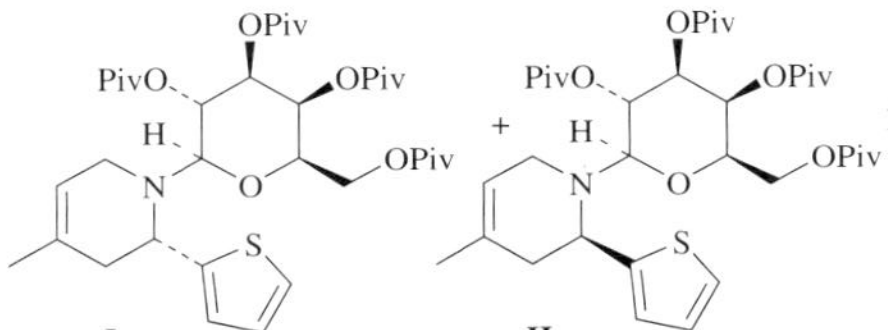 **I** + **II** 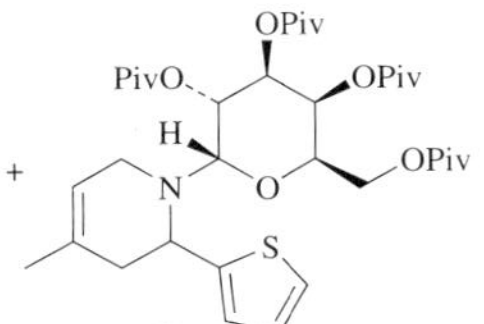 + **III** **I + II** (95), **I:II** 85:15, **III** (8)[wwwww]	172
C_{32} 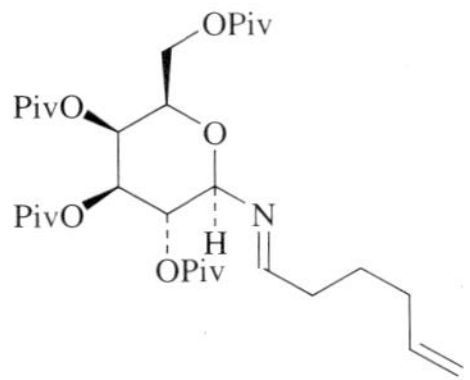	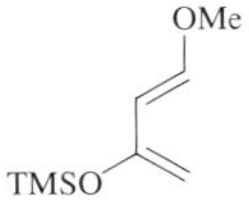	1. $ZnCl_2{\bullet}OEt_2$, THF, –20°, 24-48 h 2. HCl, H_2O	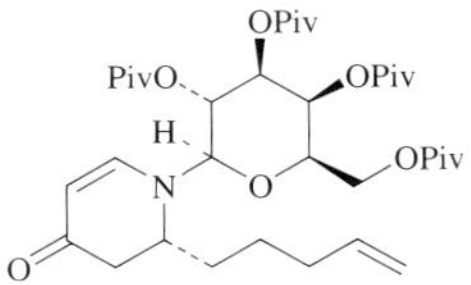 **I** + 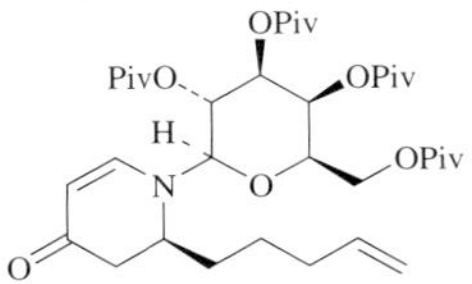 **II** **I + II** (68), **I:II** >60:1	445, 174

Reactant	Diene	Conditions	Product(s) and Yield(s) (%)	Refs.
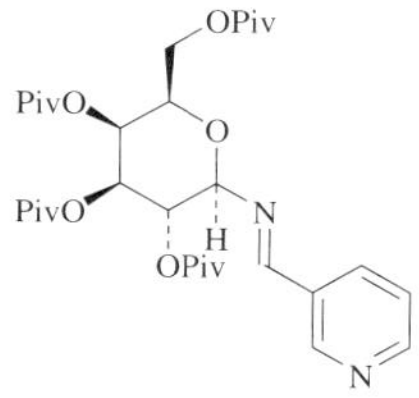	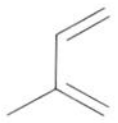	$ZnCl_2•OEt_2$, CH_2Cl_2, 20°, 96 h	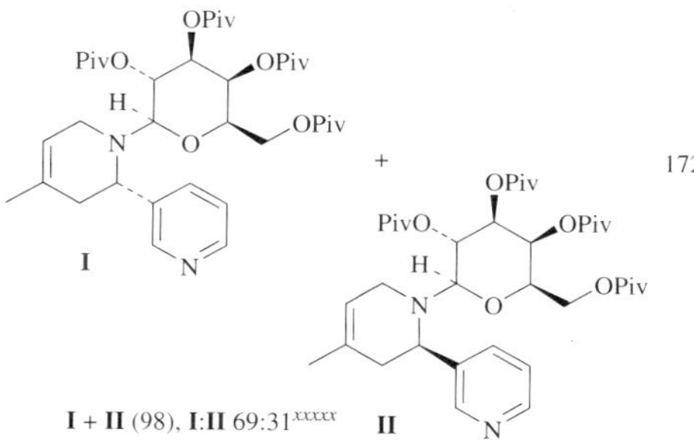 I + II (98), I:II 69:31[xxxxx]	172
	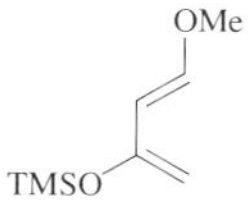		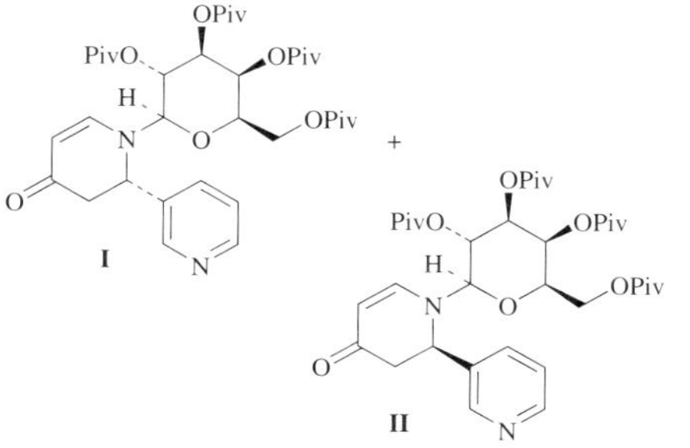	
		1. $ZnCl_2•OEt_2$, THF, –20°, 12 h 2. HCl, H_2O	I + II (92), I:II >20:1[yyyyy]	172
		1. $ZnCl_2•OEt_2$, THF, –20° 2. HCl, H_2O	I + II (90), I:II 96:4[yyyyy]	171
		1. $ZnCl_2$, THF, CH_2Cl_2, –20°, 24-48 h 2. HCl, H_2O	I + II (90), I:II 34:1	173

Imine	Diene	Conditions	Product(s) and Yield(s) (%)	Refs.
C_{33} 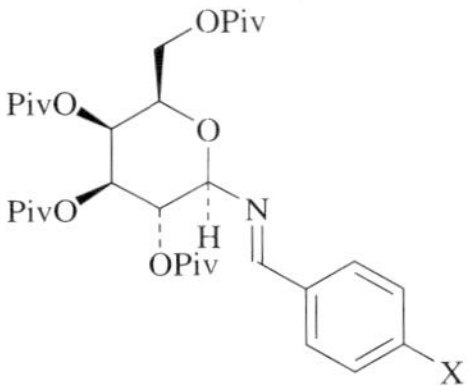	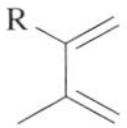	$ZnCl_2{\bullet}OEt_2$, CH_2Cl_2, 4°, 12 h	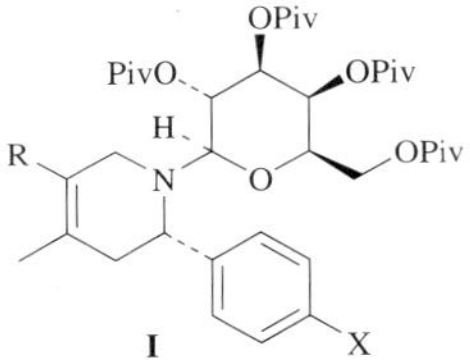 **I** + 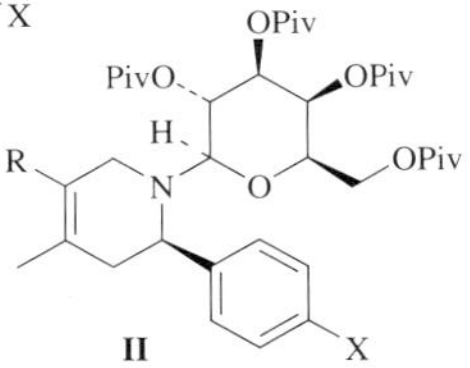 **II** + 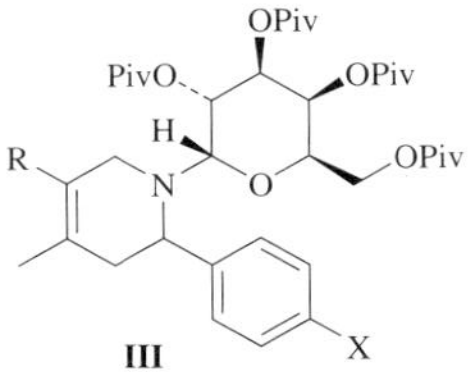 **III**	172

X	R	I + II	I:II	III
F	H	(90)	90:10	(15)[wwwww,zzzzz]
Cl	H	(95)	85:15	(12)[vvvvv,wwwww]
Cl	Me	(96)	87:13	(5)[wwwww,aaaaaa]

C_{34}

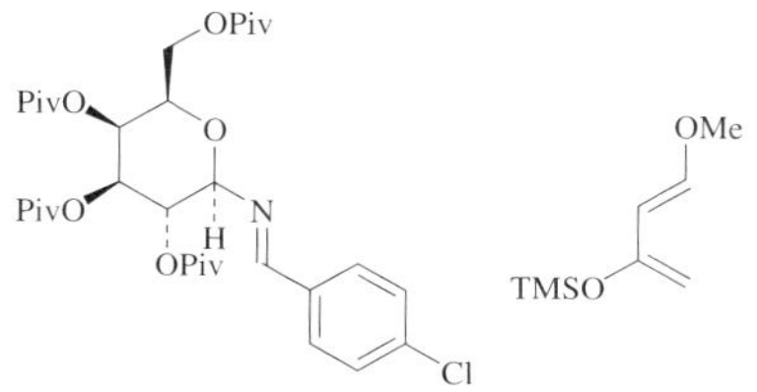	1. $ZnCl_2{\bullet}OEt_2$, THF, –20° 2. HCl, H_2O	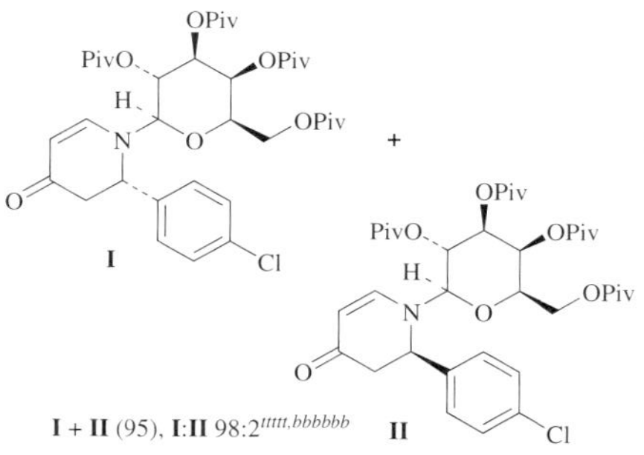 **I** + **II** (95), **I**:**II** 98:2[ttttt,bbbbbb]	171
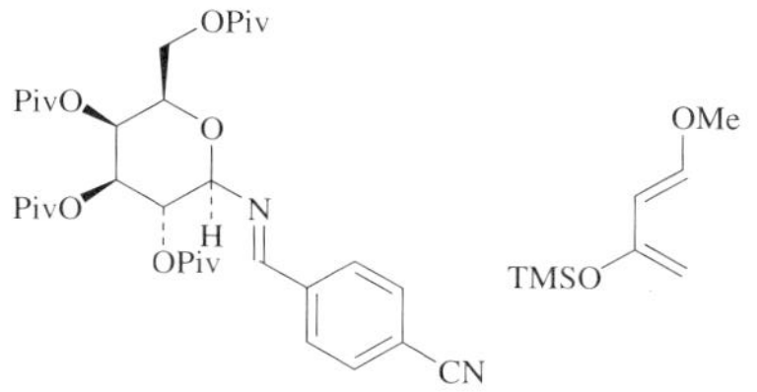	1. $ZnCl_2$, THF, CH_2Cl_2, –20°, 24-48 h 2. HCl, H_2O	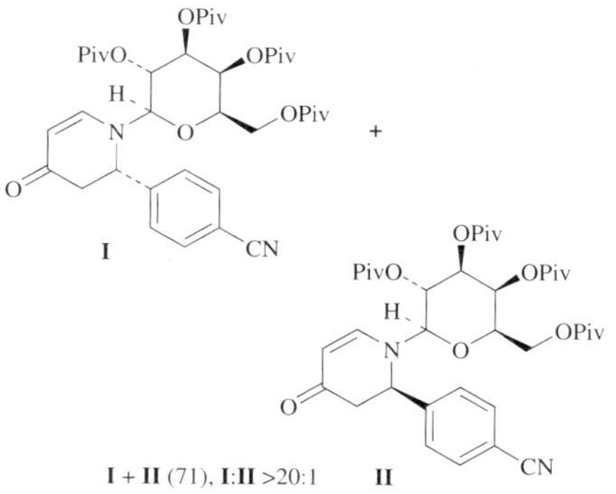 **I** + **II** (71), **I**:**II** >20:1	173

TABLE 5. CYCLOADDITIONS OF *N*-ALKYL- AND *N*-ARYLIMINES (*Continued*)

	Imine	Diene	Conditions	Product(s) and Yield(s) (%)	Refs.
C_{35}	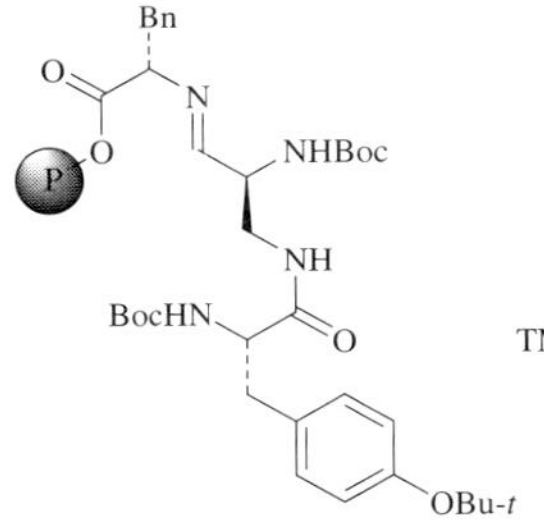	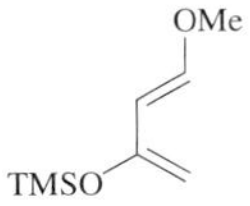	1. $ZnCl_2$[rrrrr], THF, 4°, 16 h 2. AcOH, MeOH 3. TFA, CH_2Cl_2, 24 h	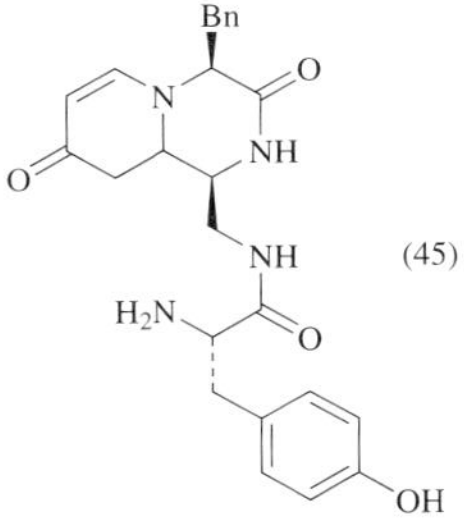 (45) >95% de[f]	444
C_{37}	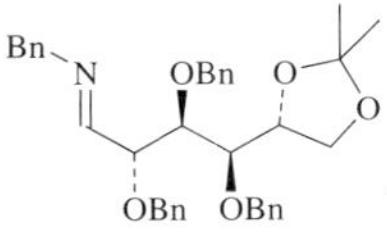	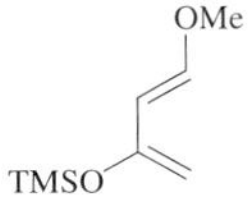		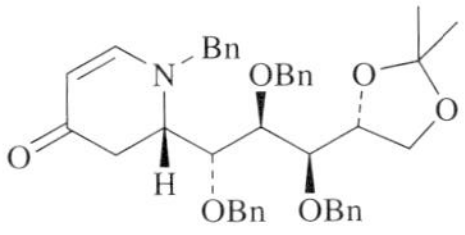	
			1. $ZnCl_2$, MeCN, 3 h 2. $NaHCO_3$, H_2O	(86)	443
			$ZnCl_2$, THF	(86)	440
	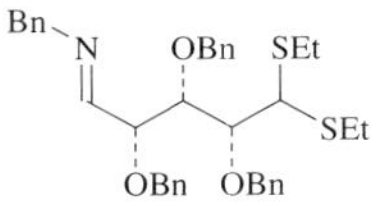	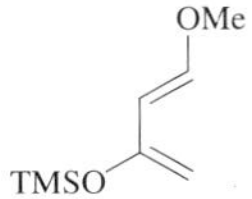	1. $ZnCl_2$, THF, 2-3 h 2. $NaHCO_3$, H_2O	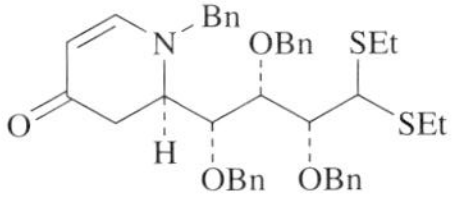 I + 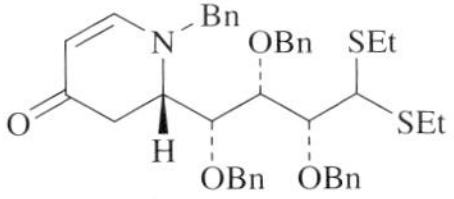 II **I** + **II** (65), **I**:**II** 20:80	439, 440

C_{44}	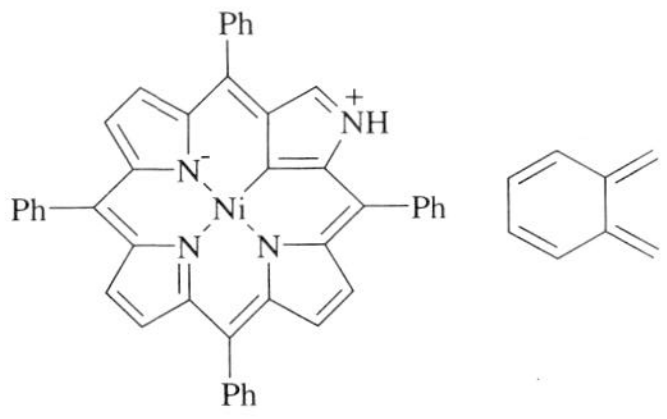	Benzene, 80°, 2 d	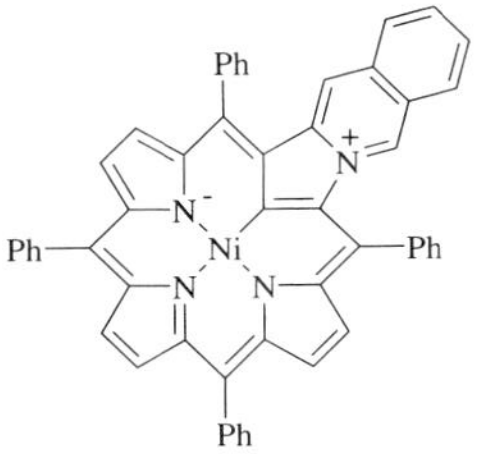 (23)[cccccc]	446
C_{48}	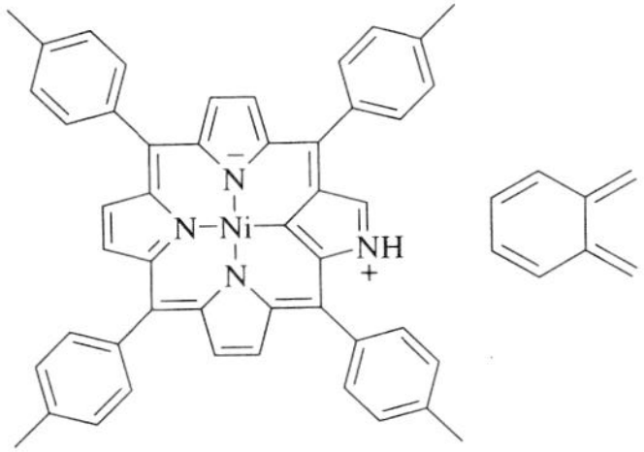	—	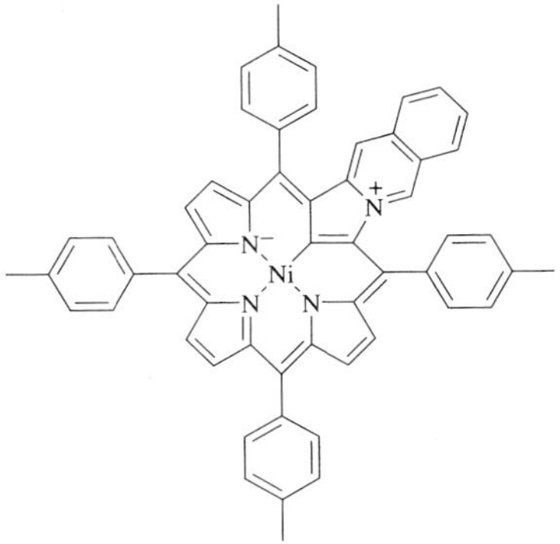 (—)[cccccc]	446

[a] The yield was determined by ^{1}H NMR.

[b] The use of aqueous tetrahydrofuran mixtures as solvent did not affect the reaction rate or yield. However, the reaction rate decreased with alcoholic solvents.

[c] Although the exact reaction conditions are not specified, products are only obtained after long reaction times and elevated temperatures.

[d] This reaction also takes place at 25°, but requires 15-20 days to reach completion.

[e] Although the 3-methyl regioisomer is shown as the product, the authors mention that the usual regioselectivity is observed. Thus, the regiochemistry of this product is unclear.

Imine	Diene	Conditions	Product(s) and Yield(s) (%)	Refs.

[f] Although the diastereoselectivity is given, the stereochemistry of the major isomer is not indicated.

[g] The major diastereomer is not specified. Based on other results[152, 153] the major isomer is probably compound **I**.

[h] The reactive *N*-silylimine and thiophene oxide were formed in situ from the thiophene and the *O*-silyloxime.

[i] *cis*-8a-Methyl-8-methoxycarbonyl-1,2,3,4,4a,5,6,8a-octahydroquinoline was obtained as a side product.

[j] No stereochemical information is provided.

[k] Lewis acids such as $Nb(OTf)_3$ did not improve the rate or yield of the cycloaddition.

[l] Specific reaction conditions are not provided, although prolonged reaction times and higher temperatures were required in the absence of lanthanide catalysis.

[m] The reactions were also performed with other lanthanide triflates, but only the best results were provided.

[n] The reaction was performed with 1.1 equiv. of diene.

[o] The reaction was performed with 3.1 equiv. of diene.

[p] The reactive diene was obtained in situ from the precursor 2,3-bis(arylsulfonyl)-1,3-butadiene.

[q] When a small amount of sodium benzenesulfinate is added, a shorter reaction time (approx. 2 h) was required for the reaction to reach completion.

[r] The reactive diene, 1,3-bis(phenylsulfonyl)-1,3-butadiene was independently synthesized prior to the reaction.

[s] The reaction was run with the precursor 1,3-bis(arylsulfonyl)-1,3-butadienes of diene **I** and **III** used in previous reactions. Diene **V** was formed in situ from the precursors to **I** and **III**.

[t] The relative stereochemistry of the 5-phenylsulfonyl group is not specified.

[u] The combined yields of the purified products **I** and **III** are 18% (R = Me), 26% (R = MOM), and 18% (R = TBS).

[v] The major isomer **II** was isolated in 68% yield after chromatography.

[w] The major isomer **II** was isolated in 62% yield after chromatography.

[x] The stereochemistry of the major product is not specified. However, based on other results[366, 362] the major isomer is probably adduct **I**.

[y] The ratio of isomers before purification is 3:1 (**I**:**II**).

[z] No yield or isomeric ratio is provided, but it is implied that the major product is adduct **I**.

[aa] The relative stereochemistry between the phenyl and phenylsulfonyl substituents and the proton adjacent to the nitrogen is not specified.

[bb] The ratios of the silylamines **I** and **II** to the deprotected amines **III** and **IV** are not provided. The compounds were characterized as the crude mixtures.

[cc] The absolute stereochemistry of the major enantiomer is not provided. However, based on later results with similar dienes,[177] the major enantiomer is probably adduct **I**.

[dd] The combined yields of the purified products **I** and **III** are 27% (R = MOM) and 22% (R = TBS).

[ee] The major isomer **I** was isolated in 72% yield after chromatography.

[ff] The vinylketene was formed in situ from the corresponding cyclobutenone.

[gg] The combined yield of the purified products **I** and **III** is 22%.

[hh] Other Lewis acids, including $ZnCl_2$, $AlCl_3$, Et_2AlCl, $TiCl_4$, and BF_3•OEt_2, gave lower yields of product.

[ii] The *tert*-butyl and benzyl esters of isoleucine did not give higher enantiomeric excesses than the methyl ester of isoleucine. The enantiomeric excess obtained with the methyl ester was **I**:**II** 93:7.

[jj] Different Lewis acids ($TiCl_4$, BF_3•OEt_2, $MeAlCl_2$, $EtAlCl_2$, Me_2AlCl, Et_2AlCl, and $ZnCl_2$) were used in this reaction, but not all of the results are provided.

[kk] Lower yields were obtained at room temperature.

[ll] The reaction was performed with 3.8 equiv. of diene.

[mm] 5-Amino-2-alkenoates, resulting from addition of the terminal end of the silyl enol ether to the imine, were also observed.

[nn] The ene reaction products were also obtained in 3-12% yields.

[oo] The reaction was performed with 4.2 equiv. of diene.

[pp] The diene was formed in situ by irradiation of the starting imine.

[qq] The combined yield of the purified products **I** and **II** is 23%.

[rr] The reactive diene was formed in situ from the corresponding benzocyclobutene.

[ss] The stereochemistry of adduct **I** is not specified, but based on other results[393] it is probably the isomer shown.

[tt] The reaction was also successfully performed using $Eu(fod)_3$ as the catalyst.

[uu] The imine was formed in situ from the benzofuropyridine trimer.

[vv] Other combinations of acid and Lewis acid were less effective. The reaction could also be performed with chloroform as the solvent.

[ww] The Mannich product resulting from silyl enol ether addition to the imine was obtained in 26% yield (detected by ^1H NMR).

[xx] Hydrolysis using TFA in CH_2Cl_2 led to a mixture of epimers at C-3.

[yy] The major isomer **I** was isolated in 52% yield after chromatography.

[zz] The major isomer **I** was isolated in 61% yield after chromatography.

[aaa] The authors report an exo/endo ratio of >99:1, but only the two exo isomers were observed.

[bbb] Different Lewis acids ($TiCl_4$, BF_3•OEt_2, $EtAlCl_2$, $SnCl_4$, and $ZnCl_2$) were used in this reaction, but not all of the results are provided. The best results were obtained with $EtAlCl_2$.

[ccc] The ratio of cyclized to open products (**I+II**:**III+IV**) is lowered if the reaction is quenched with water.

[ddd] The reaction was performed with 2.3 equiv. of diene.

[eee] The reaction was performed with 0.1 equiv. of $Nd(OTf)_3$.

[fff] The reaction was performed with 0.4 equiv. of $Nd(OTf)_3$.

[ggg] The cycloadduct resulting from the 2,3-dimethylbutadiene acting as a dienophile was also observed.

[hhh] The reactive diene is the dienolate ion formed in situ by coordination of indium trichloride with the enone.

[iii] The reaction was performed with 4.3 equiv. of diene.

[jjj] Good yields were also obtained with other lanthanide triflates.

TABLE 5. CYCLOADDITIONS OF *N*-ALKYL- AND *N*-ARYLIMINES (*Continued*)

Imine	Diene	Conditions	Product(s) and Yield(s) (%)	Refs.

[kkk] The reaction was performed in the presence of benzaldehyde as a competition experiment.

[lll] Replacement of water with methanol results in nearly identical yields and reaction times.

[mmm] The reaction was also performed with the Lewis acid catalysts coordinated to polymer-bound triphenylphosphine, which provided yields in excess of 95% after 1 hour at room temperature.

[nnn] The imine was isolated before performing the cycloaddition.

[ooo] The imine was formed in situ from the aldehyde and amine.

[ppp] The Mannich product resulting from silyl enol ether addition to the imine was observed in the crude mixture. The mixture was treated with aqueous HCl in CH_2Cl_2 to convert this material to the cycloadduct, which is included in the yield shown.

[qqq] The diene was added slowly.

[rrr] The diene was added in a single portion.

[sss] The reaction was performed in the presence of Triton X-100.

[ttt] The methanol was freshly distilled over calcium hydride.

[uuu] Traces of 1-cyclohex-1-enyl-3-phenyl-3-(phenylamino)propan-1-one, resulting from silyl enol ether addition to the imine, were obtained as a byproduct.

[vvv] The absolute stereochemistry of the major enantiomer is not specified.

[www] The two products were obtained in ratios varying from 70:30 to 2:98 depending on the specific conditions employed.

[xxx] After purification, a 15% yield of **II** was obtained.

[yyy] The yield was determined by ^{19}F NMR.

[zzz] The reaction was performed with 1.5 equiv. of diene.

[aaaa] The reaction was performed with 3.0 equiv. of diene.

[bbbb] Although a diastereoselectivity is given, only one isomer was observed.

[cccc] Although the exo/endo selectivity is not specified, it is implied that only the exo isomer(s) was obtained.

[dddd] The use of $Yb(OTf)_3$, $La(OTf)_3$, $Zn(OTf)_2$, $Sn(OTf)_2$, $ZnCl_2$, or $ZnCl_2/ZnO$ under aqueous or anhydrous conditions gave the desired product in low yield.

[eeee] In the absence of sonication or with $EtAlCl_2$ instead of $AlCl_3$, lower yields were observed.

[ffff] Attempts to catalyze the reaction in an aqueous medium with lanthanide triflates were less successful.

[gggg] Alkylaluminum halide Lewis acids in CH_2Cl_2 at lower temperatures gave inferior results.

[hhhh] With other Lewis acids ($TiCl_4$, BF_3•OEt_2, or $ZnCl_2$) the products are obtained in lower yields or not observed.

[iiii] The corresponding 5-amino-alkenoate, resulting from addition of the terminal end of the silyl enol ether to the imine, was obtained selectively (cycloadduct:acyclic product 1:4) if the reaction was quenched below –50°.

[jjjj] Using diethyl ether as a solvent gave similar results, but the reaction was considerably slower.

[kkkk] Lower yields without improving the selectivity were observed when the reactions were performed with $Cu(OTf)_2$, $CuClO_4$, $Sc(OTf)_3$, or $ZnCl_2$.

[llll] Although a ratio for the cyclized and uncyclized products is given, only the uncyclized products **III** and **IV** were observed.

[mmmm] The same ratio is observed when the reaction mixture is treated with aqueous NH_4Cl or aqueous HCl for 2 minutes.

[nnnn] The major isomer **II** was isolated in 73% yield after chromatography.

[oooo] The major isomer **I** was isolated in 85% yield after chromatography.

[pppp] No enantiomeric ratio is given, but adduct **I** is the major product.

[qqqq] The other lanthanide triflates used ($La(OTf)_3$, $Pr(OTf)_3$, $Nd(OTf)_3$, $Gd(OTf)_3$, $Dy(OTf)_3$, and $Er(OTf)_3$) under the same reaction conditions afforded the cycloadducts in the same ratio (**I**:**II** 4:1) with combined yields from 63-72%.

[rrrr] The reaction was performed with 1.2 equiv. of diene.

[ssss] The reaction was performed with 4.6 equiv. of diene.

[tttt] Although the exact ratios are not provided, poor diastereoselectivities (approximately **I**:**II** 78:22 and **III**:**IV** 82:18) were observed for the endo and exo isomers.

[uuuu] The reaction was performed in the presence of 4 Å molecular sieves.

[vvvv] Only one stereoisomer was obtained, but its stereochemistry was not determined.

[wwww] A variety of Lewis acids, including lanthanide triflates, did not improve the yield.

[xxxx] Low yields of the desired cycloadduct were obtained with a variety of common Lewis acid catalysts, including $SnCl_4$ and $TiCl_4$.

[yyyy] When the reaction was performed with $ZnCl_2$, low stereoselectivities were observed.

[zzzz] Instead of adding the diene immediately, the imine and $Zn(OTf)_2$ were stirred for 5-30 minutes before the addition of diene.

[aaaaa] The exo/endo selectivity was not determined. Although the diastereoselectivity is given, the major diastereomers are not indicated.

[bbbbb] Other Lewis acids, such as BF_3•OEt_2/TFA, $LiClO_4$, $ZnCl_2$, MgI_2, MgI_2•OEt_2, Et_3Al, and $Nd(OTf)_3$, did not improve the yield.

[ccccc] The product was obtained with complete diastereoselectivity (d.e. of approximately 99:1). The stereochemistry of the product is not specified.

[ddddd] The use of BF_3•OEt_2, Et_2AlCl, or $ZnCl_2$ as the Lewis acid gave 0-20% yields of product.

[eeeee] Lower yields were observed using TMSOTf.

[fffff] During the reaction, racemization of the chiral center next to the phthalimide moiety occurred. The ee of adduct **I** after using BF_3•OEt_2 was 0-80%. After utilization of $TiCl_4$, the ee of adduct **IV** was 90% (little or no racemization occurred under these conditions).

[ggggg] The product was obtained with complete enantioselectivity, but the absolute stereochemistry is not specified.

[hhhhh] The cycloaddition yielded a single trans diastereomer, but the absolute stereochemistry was not determined. The enantiomeric excess for the cis isomer is not specified.

[iiiii] The use of the TIPS-protected aldimine did not improve the ratio of products (**I**:**II** 43:57) obtained with the TBS-protected imine.

[jjjjj] Other Lewis acids tested, including hafnium chloride and BF_3•OEt_2, were less effective.

[kkkkk] The isolated yields of the purified products were 43% (**I**) and 14% (**II**).

TABLE 5. CYCLOADDITIONS OF *N*-ALKYL- AND *N*-ARYLIMINES (*Continued*)

Imine	Diene	Conditions	Product(s) and Yield(s) (%)	Refs.

[lllll] The product of silyl enol ether addition to the imine was also observed.

[mmmmm] The reaction was run in the presence of 4 Å molecular sieves to remove the water.

[nnnnn] The isolated yields of the purified products were 49% (**I**) and 24% (**II**).

[ooooo] The isolated yields of the purified products were 36% (**I**) and 16% (**II**).

[ppppp] Using THF, Et_2O, or CH_2Cl_2 as the solvent had no significant effect on the diastereoselectivity of the reaction.

[qqqqq] The isolated yields of the purified products were 45% (**I**) and 30% (**II**).

[rrrrr] A variety of Lewis acids were used. $ZnCl_2$ was a better catalyst than $Yb(OTf)_3$.

[sssss] The major product was isolated in 81% yield.

[ttttt] The incorrect stereochemistry of the major product was reported in Ref. 171, but is corrected in Ref. 173.

[uuuuu] The major product **I** was isolated in 72% yield.

[vvvvv] The major product **I** was isolated in 60% yield.

[wwwww] The relative stereochemistry of the α-anomer is not specified.

[xxxxx] The major product **I** was isolated in 48% yield.

[yyyyy] The major product **I** was isolated in 86% yield.

[zzzzz] The major product **I** was isolated in 52% yield.

[aaaaaa] The major product **I** was isolated in 71% yield.

[bbbbbb] The major product **I** was isolated in 90% yield.

[cccccc] The reactive diene was obtained from 1,4-dihydro-2,3-benzoxathiin-3-oxide.

TABLE 6. CYCLOADDITIONS OF AZIRINES

Imine	Diene	Conditions	Product(s) and Yield(s) (%)	Refs.
C_5 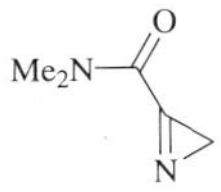		Toluene, rt, 24 h	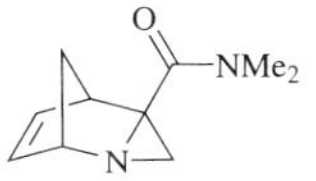 (42)	191
	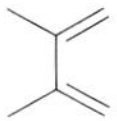	Toluene, rt, 6 d	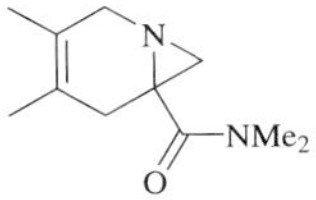 (31)	191
	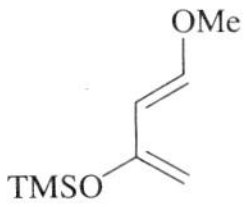	Toluene, rt, 4 d	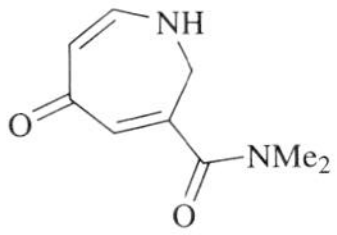 (20)[a]	191
C_6 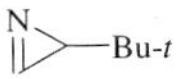	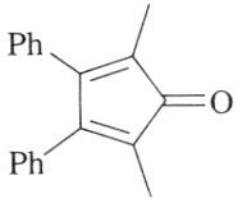	Toluene, 110°, 5 h	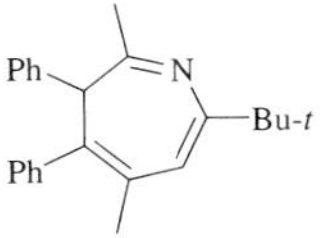 (71)[b,c]	184
	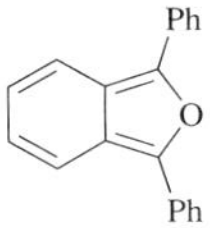	Toluene, 110°, 16 h	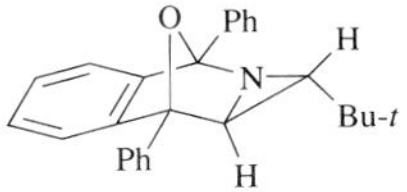 (82)[c]	186
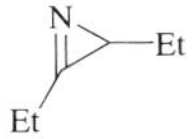	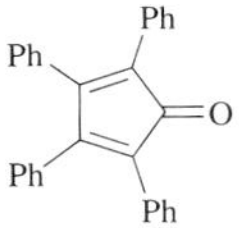	Toluene, 110°, 10 d Xylene, 140°, 10 d	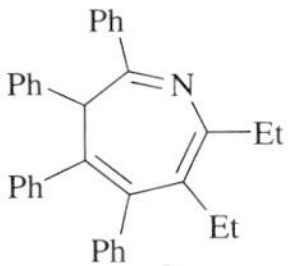 I + 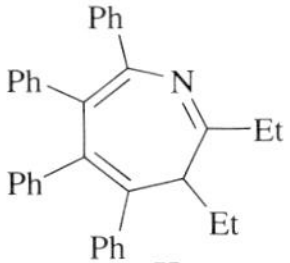 II (85)[b,d] (—) 3:8[b]	182, 183 182

TABLE 6. CYCLOADDITIONS OF AZIRINES (*Continued*)

Imine	Diene	Conditions	Product(s) and Yield(s) (%)	Refs.
	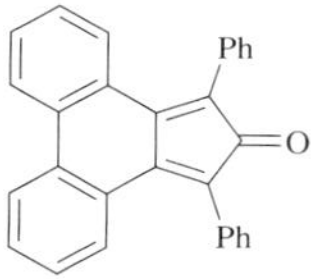	—[e]	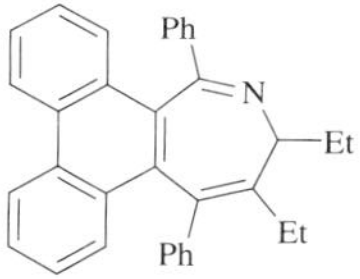 (50-80)[b]	181
		Toluene, 110°, 2-3 d	(90)[b]	182
C_7 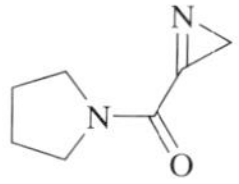		Toluene, rt, 2 d	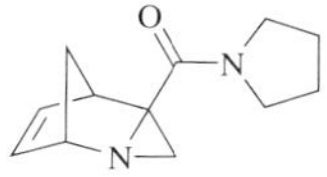 (38)	191
	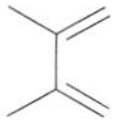	Toluene, rt, 1-several d	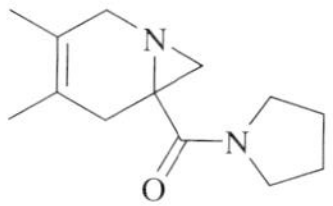 (—)	191
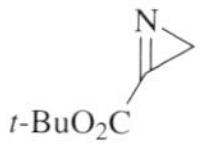		Heptane, rt	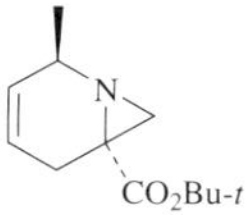 (43)	194
	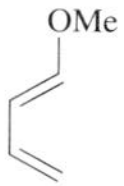	Heptane, rt	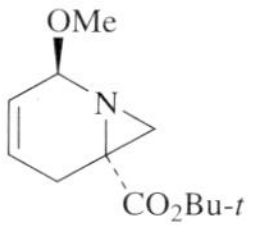 (67)	194
		Heptane, rt	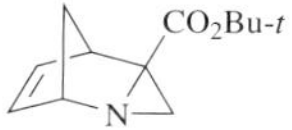 (50)	194

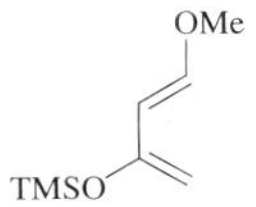	Heptane, rt	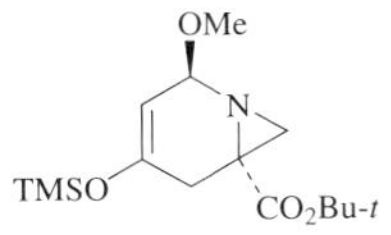 (30)	194
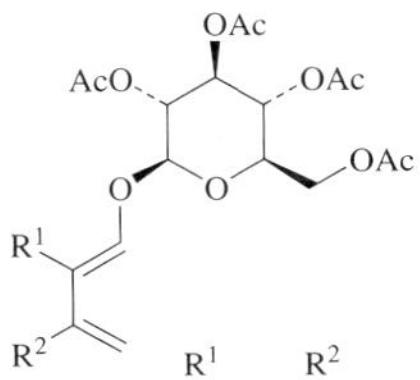	Toluene, 50-55°, 1.5 h	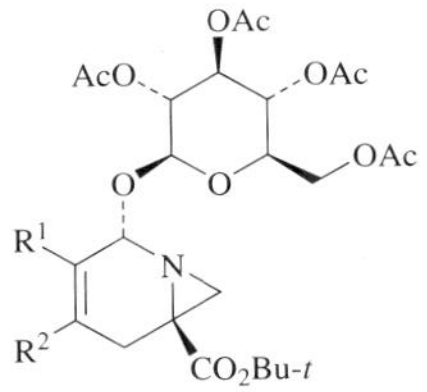	195
R^1 R^2			
H H		(57)	
H Me		(42)	
Me Me		(89)	
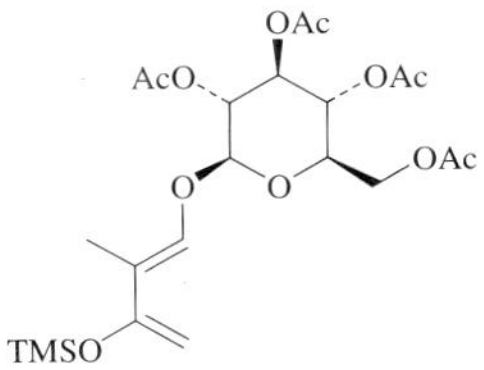	Heptane, heat, 1 h	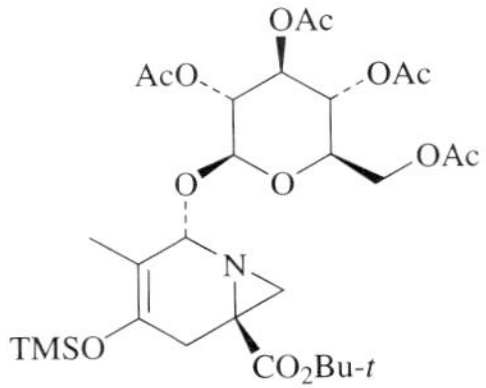 (84)	194

TABLE 6. CYCLOADDITIONS OF AZIRINES (*Continued*)

Imine	Diene	Conditions	Product(s) and Yield(s) (%)	Refs.
C_8				
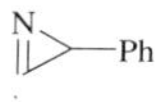	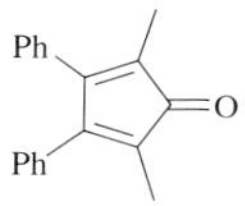	Toluene, 110°, 2 h	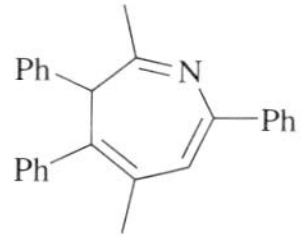 (69)[b,c]	184
	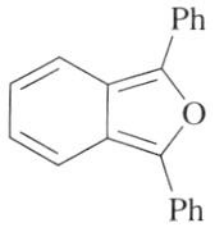	Toluene, 110°, 1 5 h	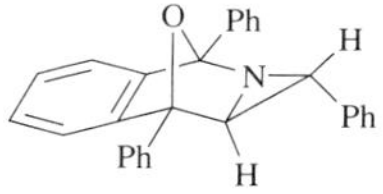 (65)[c]	186
	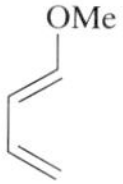	$ZnCl_2$, toluene, 75°	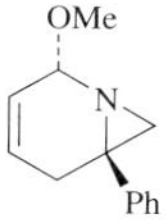	
		Time		
		6 h	(50)	188
		8 h	(50)	189
		BF_3•OEt_2, CH_2Cl_2, −78°, 2 h	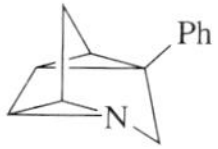 (25)[f]	189
	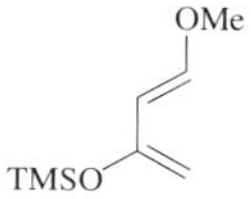	Lewis acid	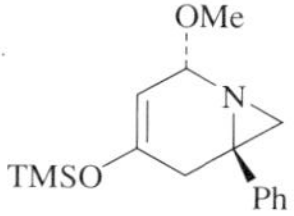	

L. A.	Solvent	Temp	Time		
$ZnCl_2$	toluene	75°	8 h	(40)	188
$ZnCl_2$	toluene	75°	12 h	(40)	189
$YbCl_3$	toluene	75°	6 h	(55)	188
$YbCl_3$	toluene	75°	12 h	(55)	189
$ScCl_3$	toluene	75°	12 h	(30)	189, 188
$CuCl_2$	toluene	75°	12 h	(35)	189, 188
$BF_3 \cdot OEt_2$	CH_2Cl_2	–70°	0.3 h	(42)	188
$BF_3 \cdot OEt_2$	CH_2Cl_2	–78°	0.3 h	(45)	189

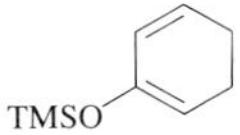
TMSO

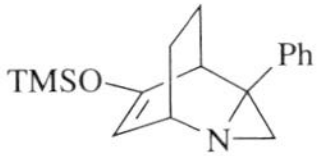
TMSO
Ph
N

189

$YbCl_3$, toluene, 90°, 6 h	(50)	
Toluene, 90°, 12 h	(48)	

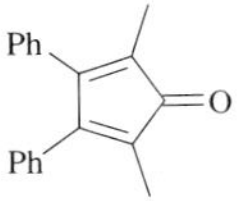
Ph
Ph
O

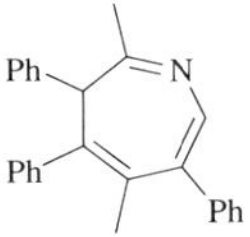
Ph
N
Ph
Ph

Toluene, 110°, 13 h	(86)[b,g]	182, 183
Toluene, 110°, 12 h	(85)[b,h]	182
Benzene, 80°, 4 d	(65)[b]	447

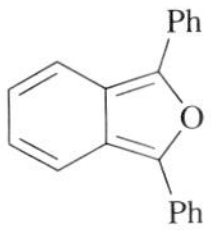
Ph
O
Ph

Toluene, 110°, 17 h

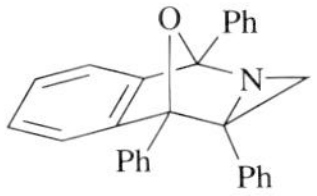
O
Ph
N
Ph
Ph

(84) 186

TABLE 6. CYCLOADDITIONS OF AZIRINES (*Continued*)

Imine	Diene	Conditions	Product(s) and Yield(s) (%)	Refs.
	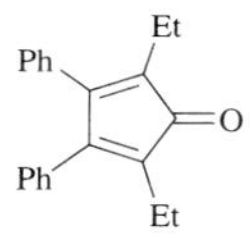	Toluene, 110°, 17 h	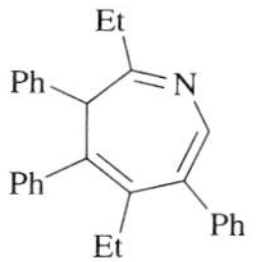 (90)[b]	182
	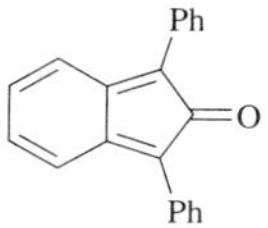	Xylene, ~140°, 3 h[i]	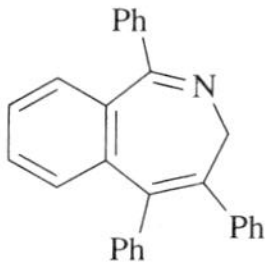 (65)[b]	182, 181
	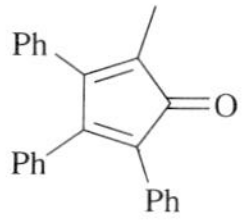	Toluene, 110°, 30 h	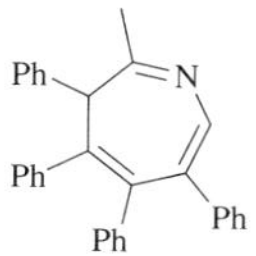 **I** (59)[b,j] + 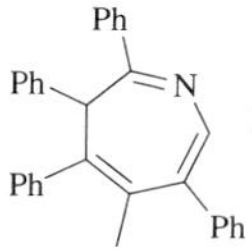 **II** (28)[b,j]	182
	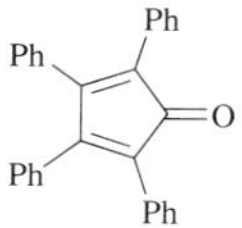		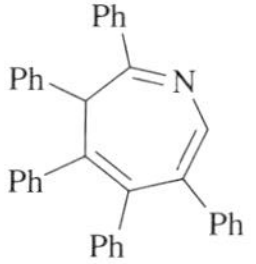	
		Mesitylene, 162°	(90)[b]	447
		Toluene, 110°, 4 d	(87)[b]	182, 183
	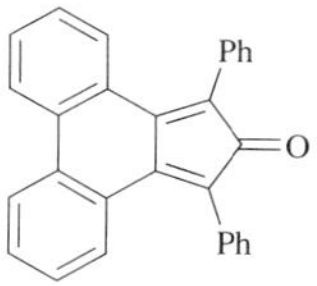	Toluene, 110°, 2-3 d	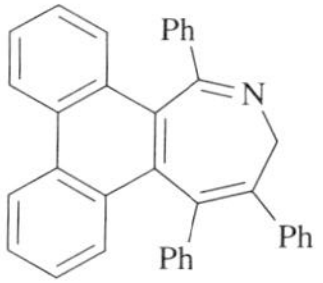 (69)[b]	182, 181

C_9 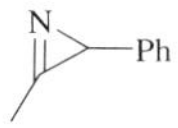	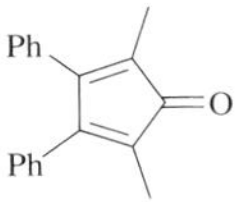	Toluene, 110°, 2 d	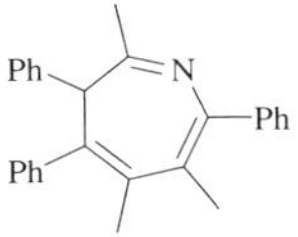 (6)[b,k]	182
	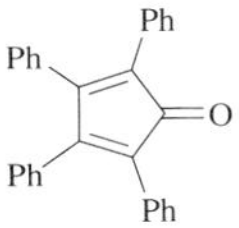	Toluene, 110°, 3 d	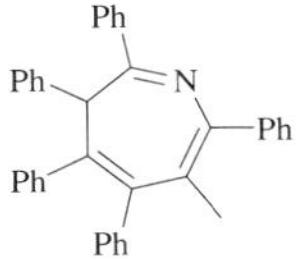 (5)[b,k]	182
	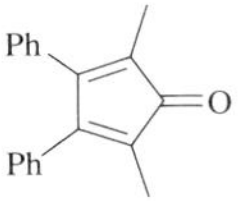	Toluene, 110°, 2 d	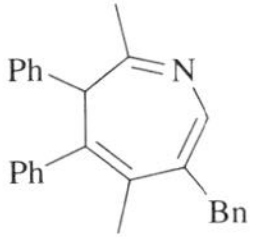 (51)[b,l]	182
	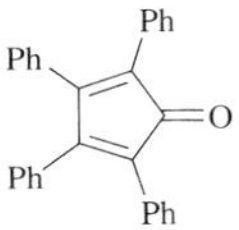	Toluene, 110°	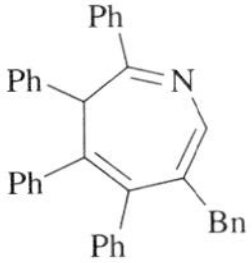	
		Time		
		—	(57)[b]	183
		3 d	(51)[b,l]	182
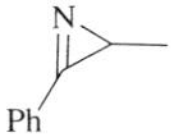	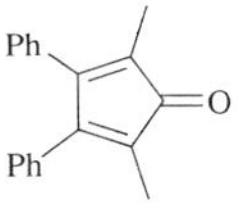		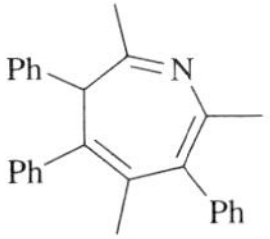	
		Toluene, 110°	(83)[b,g]	182, 183
		Toluene, 110°, 12 h	(88)[b,h]	182
		—	(69)[b]	447

TABLE 6. CYCLOADDITIONS OF AZIRINES (*Continued*)

Imine	Diene	Conditions	Product(s) and Yield(s) (%)	Refs.
C_9 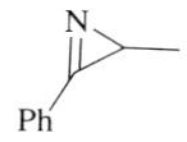	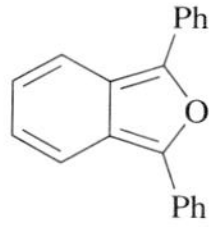	Toluene, 110°	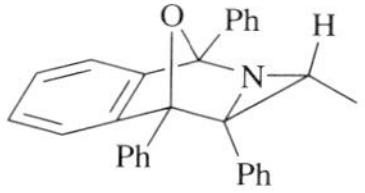	
		Time		
		18 h	(73)	187
		24 h	(90)	186, 185
	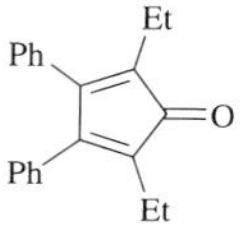	Toluene, 110°	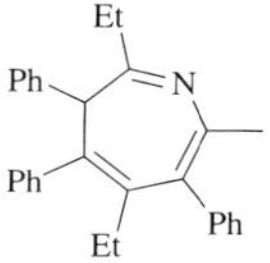 (66)[b]	182
	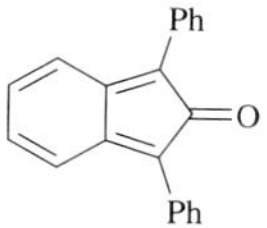	Xylene, ~140°, 3 h[m]	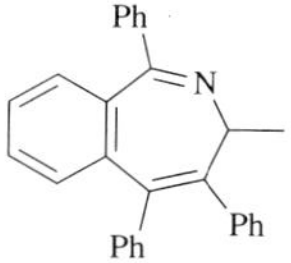 (68)[b]	182, 181
	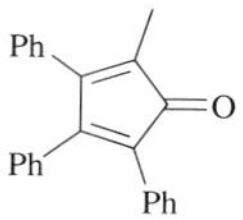	Toluene, 110°, 4 d	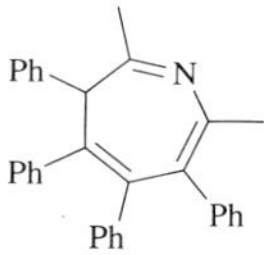 **I** + 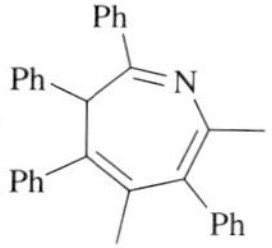 **II** **I** + **II** (—), **I**:**II** 1:1.4[b,n]	182
	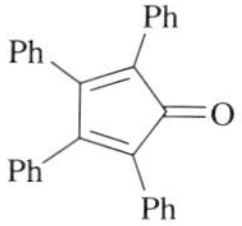		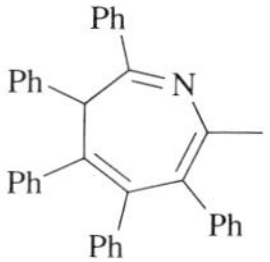	
		—	(84)[b]	447
		Toluene, 110°, 6 d	(65)[b]	182, 185, 183

	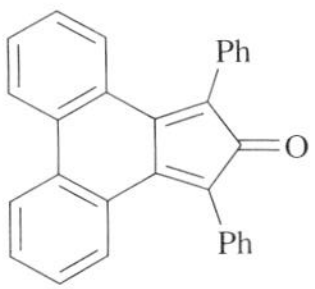	Toluene, 110°, 2-3 d	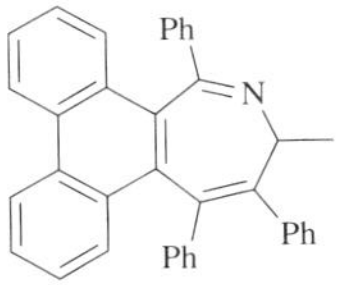 (71)[b]	182, 181
	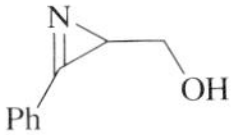 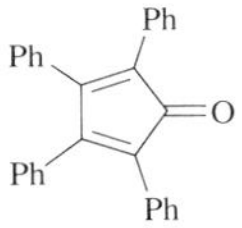	Toluene, 110°, 6 d	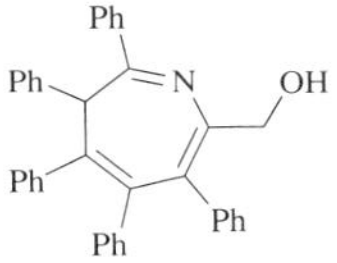 (55)[b]	182
	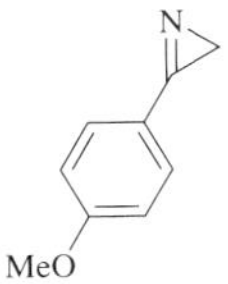 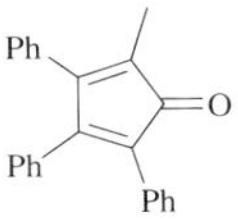	Toluene, 110°, 30 h[c]	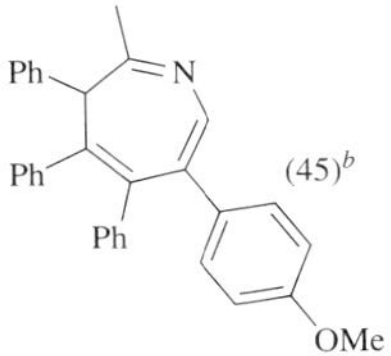 (45)[b] + 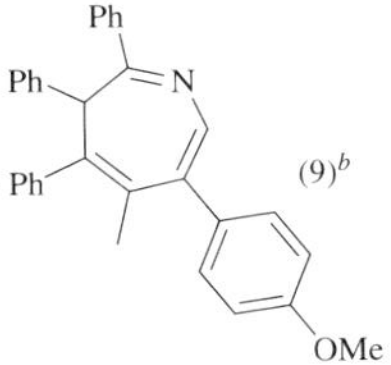 (9)[b]	182
C_{10}	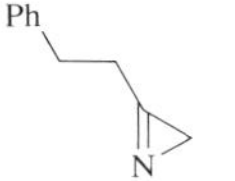 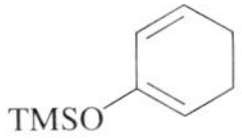		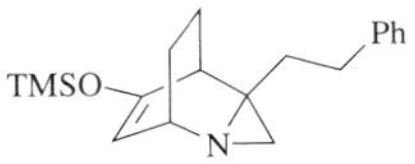	
		$YbCl_3$, toluene, 90°, 72 h	(36)	188
		$YbCl_3$, toluene, 90°, 12 h	(34)	189
		Toluene, 90°, 72 h	(32)	189
	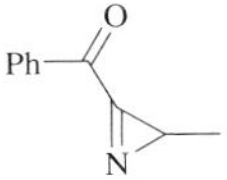 	0° to rt, 12 h	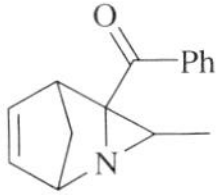 (67)[o]	448

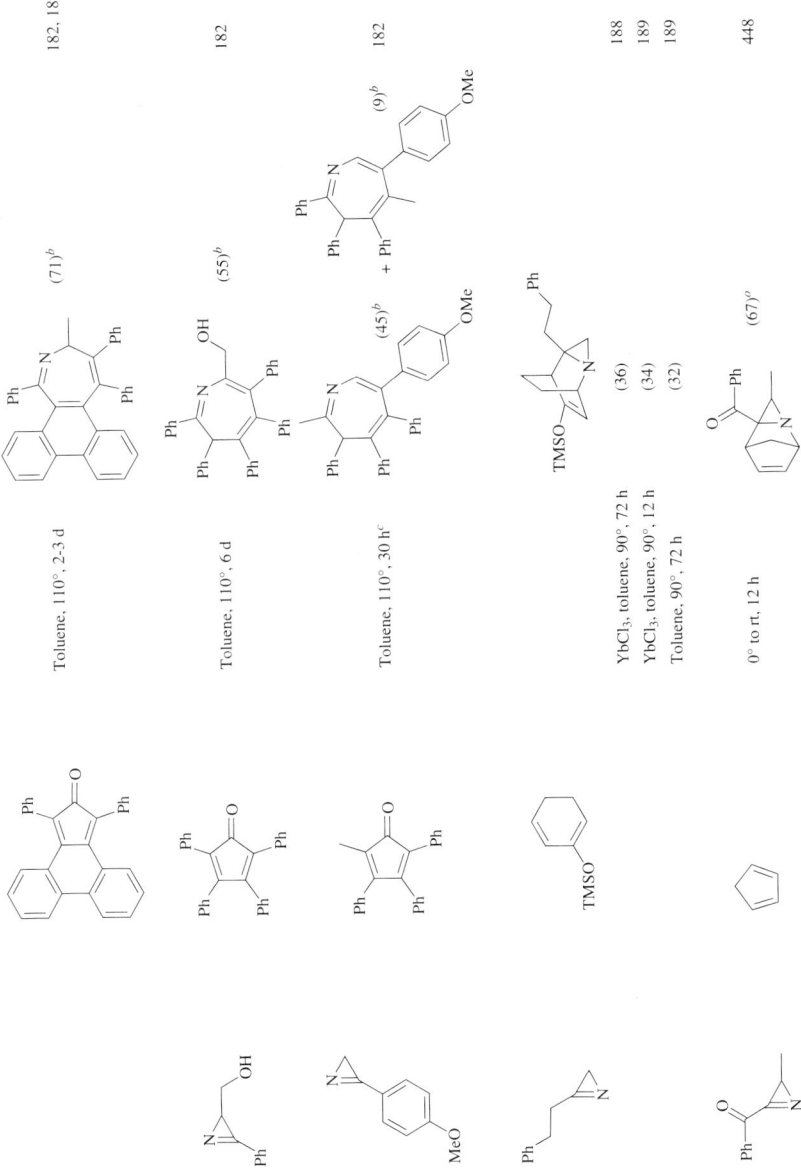

TABLE 6. CYCLOADDITIONS OF AZIRINES (*Continued*)

Imine	Diene	Conditions	Product(s) and Yield(s) (%)	Refs.
C_{10} 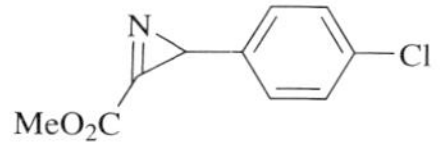		rt, 18 h	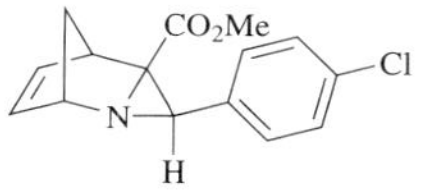 (40)	449
	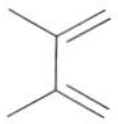	rt, 18 h	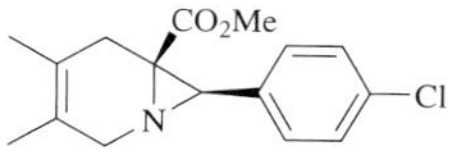 (58)[p]	449
		rt, 18 h	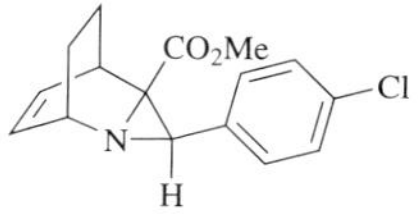 (48)[p]	449
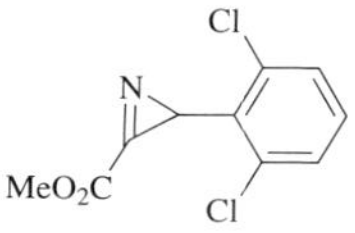		Ether, rt, 7 d	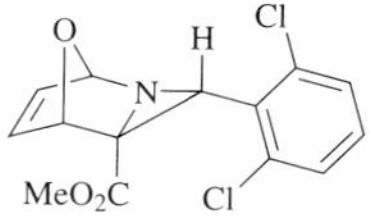 (100)	190
	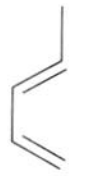	THF or ether, rt, 5 d	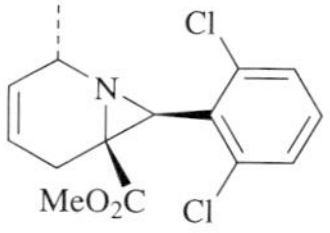 (78)	190
	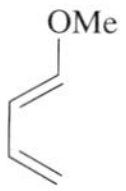	THF or ether, rt, 24 h	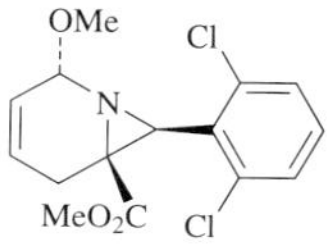 (60)	190
		THF or ether, rt, 2 h	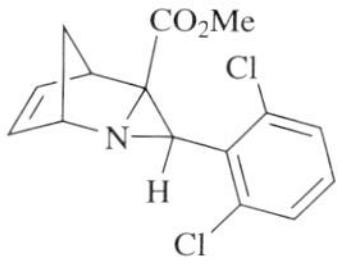 (65)	190

	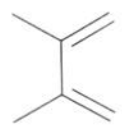	THF or ether, rt, 24 h	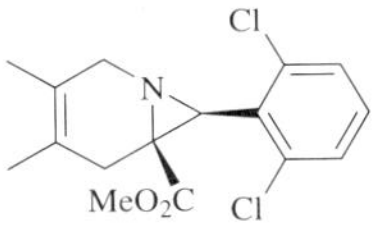 (40)	190
	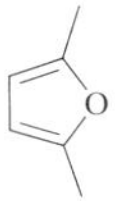	rt, 24 h	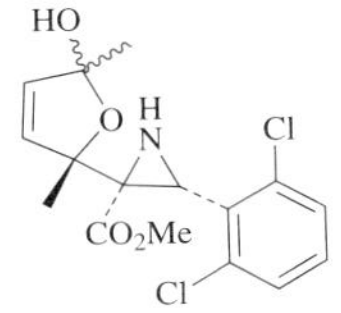 (39)[q]	192
	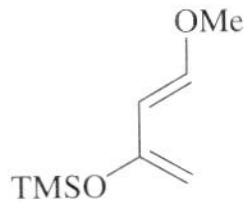	THF or ether, rt, 48 h	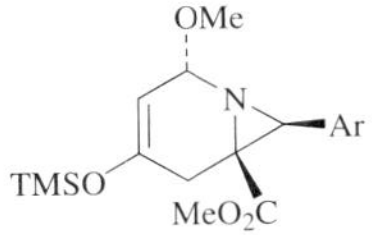 (61) + 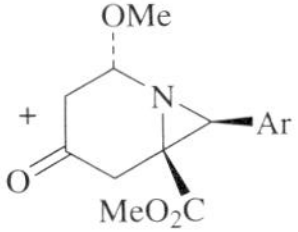 (17) Ar = 2,5-$Cl_2C_6H_3$	190
	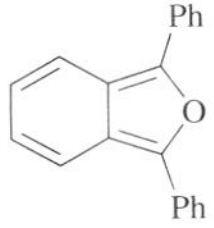	rt	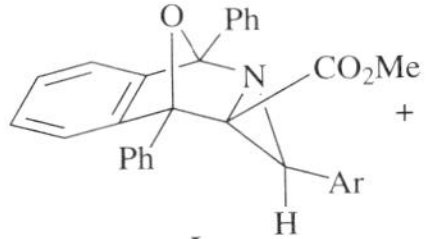 **I** + 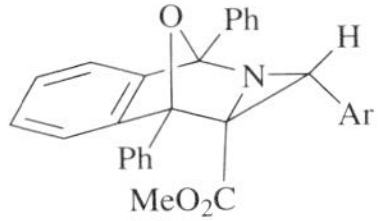 **II** Ar = 2,5-$Cl_2C_6H_3$	192
		Solvent / Time THF 24 h ether 2 h ether 20 h	 **I** + **II** (67), **I**:**II** 1:1 **I** (48)[r] **I** + **II** (48)[r]	
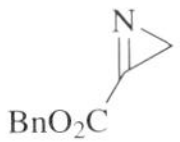		Toluene, rt, 7 d	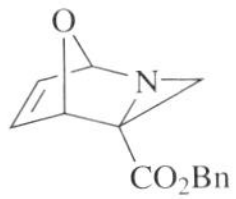 (100)	192

TABLE 6. CYCLOADDITIONS OF AZIRINES (*Continued*)

Imine	Diene	Conditions	Product(s) and Yield(s) (%)	Refs.
C_{10} 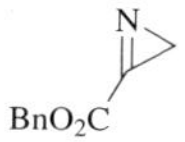			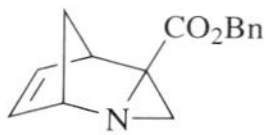	
		$ZnCl_2$, ether, –20°, 12 h	(35)	188
		$ZnCl_2$, ether, –20°, 12 h	(32)	189
		$Rh_2(OAc)_4$, CH_2Cl_2, –45°, 24 h	(35)	188
		Ether, rt, 24 h	(45)	188
		Ether, rt, 24 h	(31)	189
		Toluene, rt, 10 h	(60)	450
		CH_2Cl_2, –40°, 1.5 h	(75)	199
		rt, 15 min	(—)	199
		–78°	(—)	199
	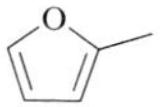	Toluene, rt, 3 d	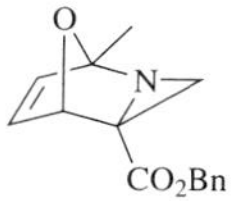 (55)	192
	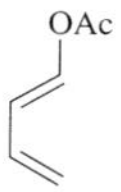	Toluene, rt, 3 d	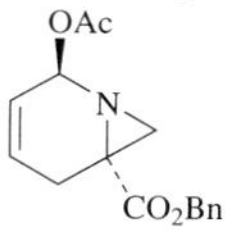 (28)	450
		Toluene, rt, 12 h	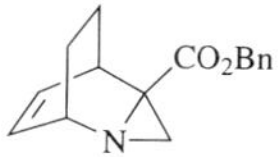 (60)	450
	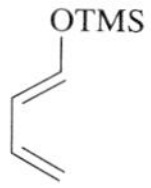		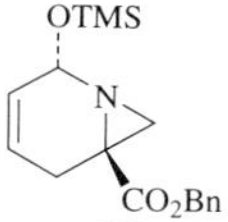	189
		$YbCl_3$, ether, –20°, 4 h	(52)	
		Ether, rt, 18 h	(30)	

	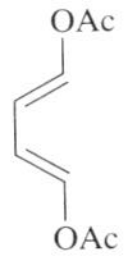	Toluene, rt, 7 d	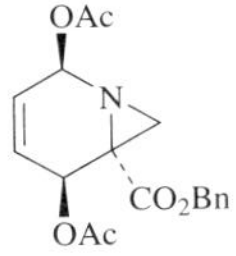	(55)	450
	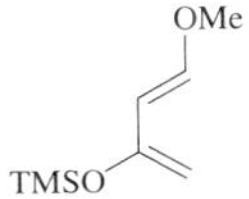	YbCl$_3$, ether, –20°, 3 h Ether, rt, 18 h	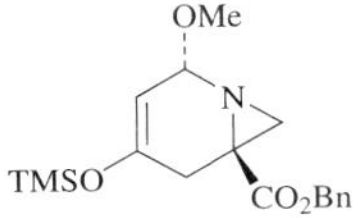	(65) (61)	189
		Toluene, rt, 7 d	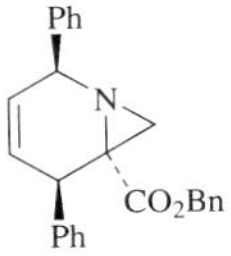	(24)	450
	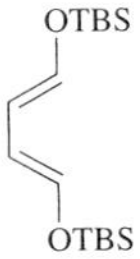	Toluene, rt, 7 d	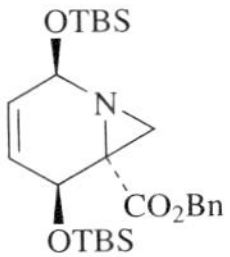	(42)	192
	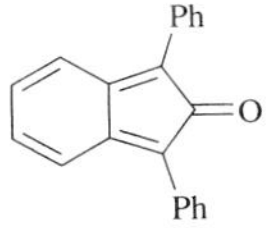	Toluene, rt, 48 h	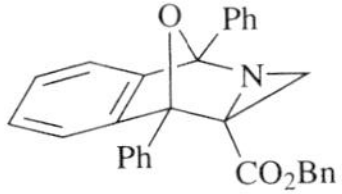 + 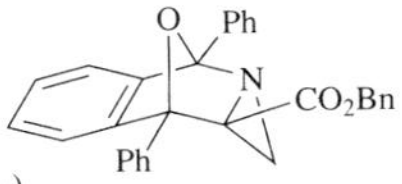	(—)	192
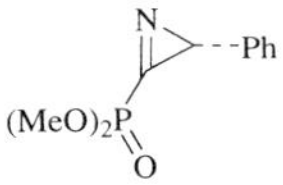	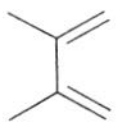	rt, 2-4 d	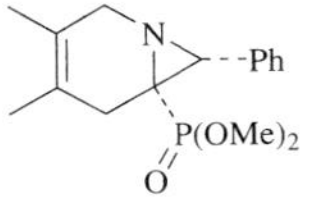	(98)	193

TABLE 6. CYCLOADDITIONS OF AZIRINES (*Continued*)

Imine	Diene	Conditions	Product(s) and Yield(s) (%)	Refs.
C_{10}				
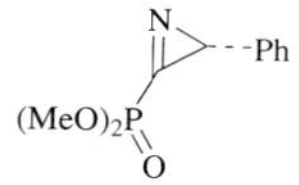	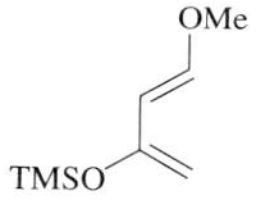	Toluene, rt, 8 h	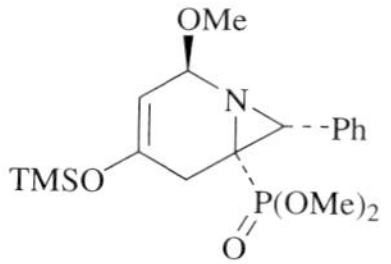 (99)	193
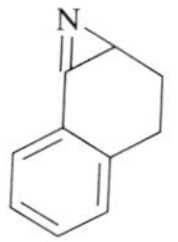	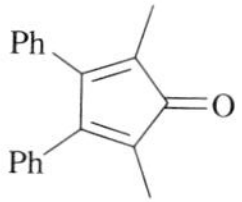	Toluene, 110°, 8 h	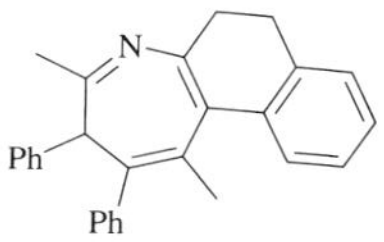 (32)[c,s], (45)[c,t]	184
C_{11}				
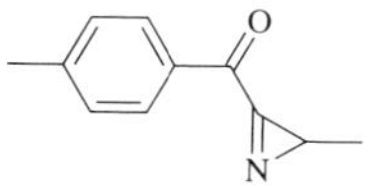		0° to rt, 12 h	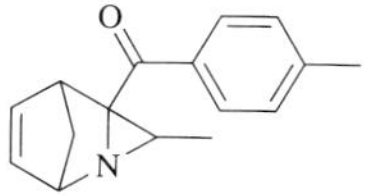 (57)[o]	448
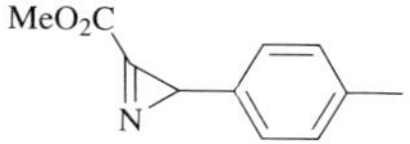		rt, 18 h	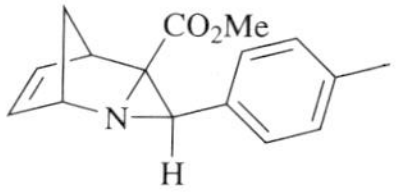 (51)	449
		THF or ether, rt, 72 h	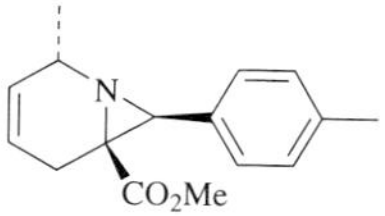 (46)	190
	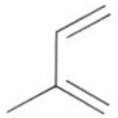	THF or ether, rt, 72 h	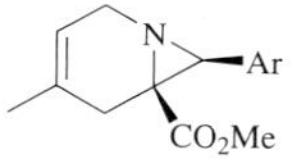 + 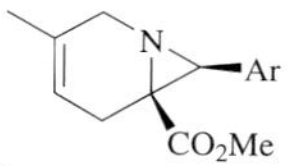 (69) 2.7:1 Ar = 4-MeC_6H_4	190

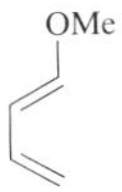	THF or ether, rt, 19 h	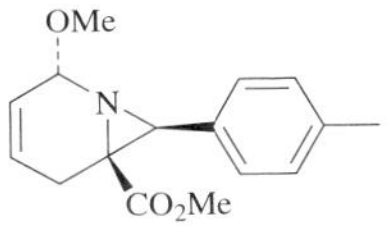 (63)	190
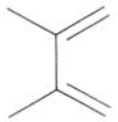	rt, 18 h	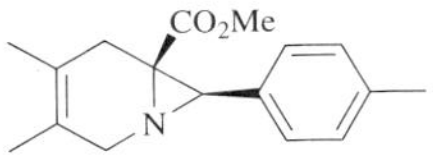 (56)[p]	449
	rt, 18 h	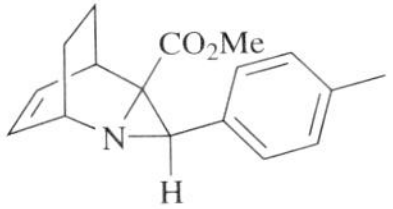 (19)[p]	449
	THF or ether, rt, 60 h	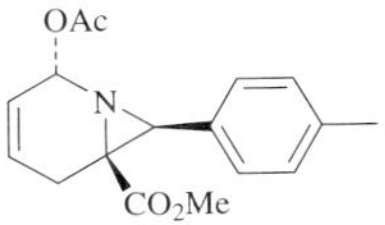 (48)	190
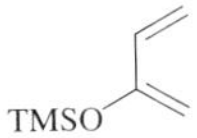	THF or ether, rt, 48 h	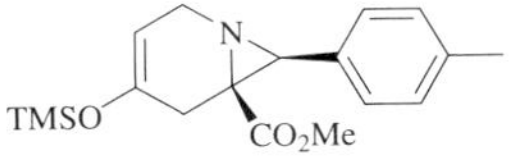 (42)	190
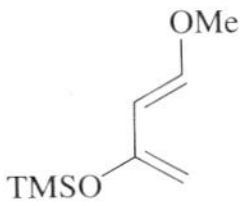	THF or ether, rt, 15 min	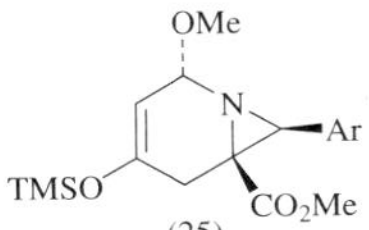 (25) + 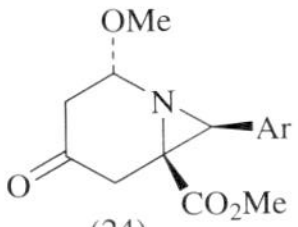 (24) Ar = 4-MeC_6H_4	190
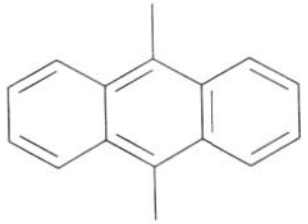	THF or ether, rt, 7 d	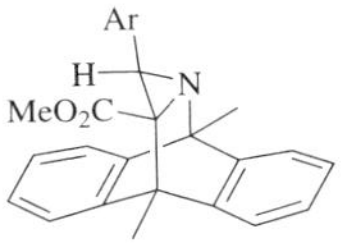 (60) Ar = 4-MeC_6H_4	190

TABLE 6. CYCLOADDITIONS OF AZIRINES (*Continued*)

Imine	Diene	Conditions	Product(s) and Yield(s) (%)	Refs.
C_{11}				
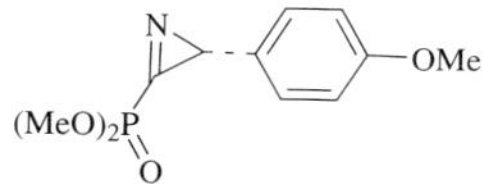N OMe (MeO)2P O		Toluene, rt, 16 h	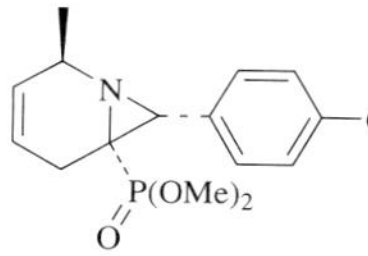N OMe P(OMe)2 O (89)	193
	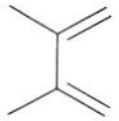	Toluene, rt, 4 d	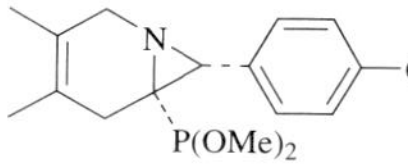N OMe P(OMe)2 (97)	193
C_{12}				
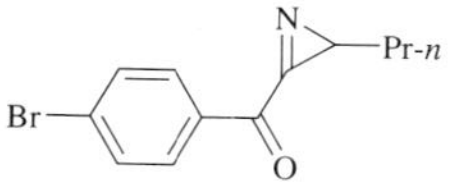N Pr-n Br O		—	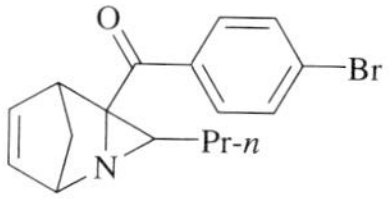O Br Pr-n N (42)[o]	448
C_{13}				
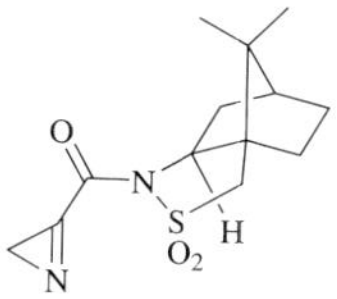O N S H O2 N	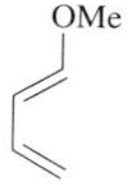OMe	Lewis acid, CH_2Cl_2	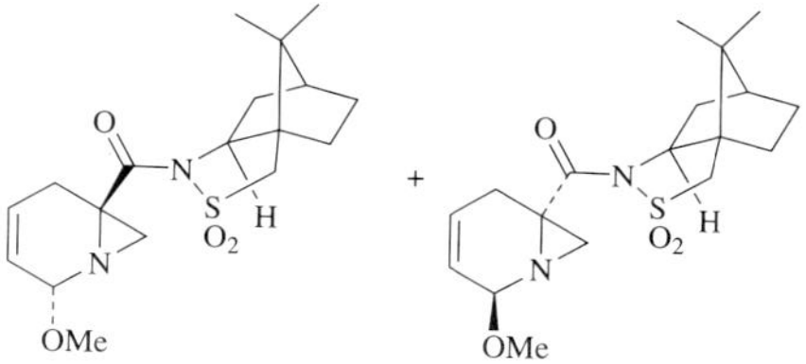O N S H O2 N OMe + O N S H O2 N OMe	
		L. A. / Temp		
		none / rt	(95) 60.5:39.5[u]	198
		$MgBr_2$•OEt_2 / –78°	(38) 70:30[u]	198, 199
		$ZnCl_2$•OEt_2 / –78°	(69) 36.5:63.5[u]	198
		$YbCl_3$ / –78 to –60°	(100) 62.5:37.5[u]	198

C_{14} 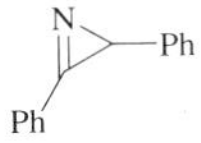	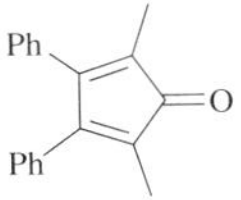		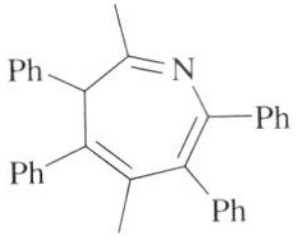	
		Toluene, 110°	(12)[b]	183
		Toluene, 110°, 3.5 d	(63)[b]	182
		—	(58)[b]	447
	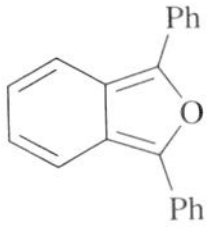	Heat, 44 h	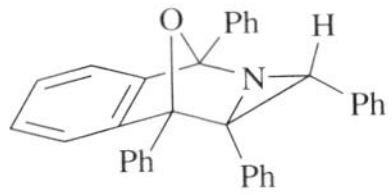 (70.5)	187
	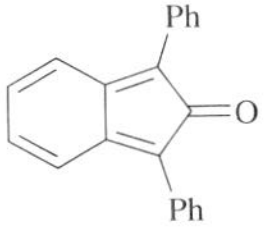	Xylene, ~140°, 10 h[i]	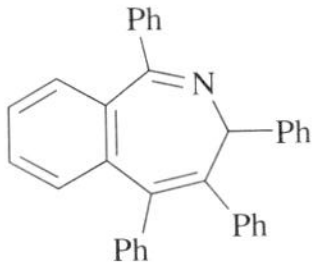 (63)[b]	182, 181
	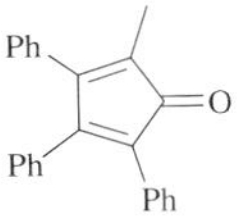	Toluene, 110°, 7 d	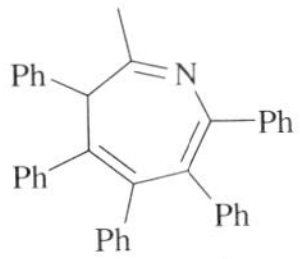 **I** (8)[b,v] + 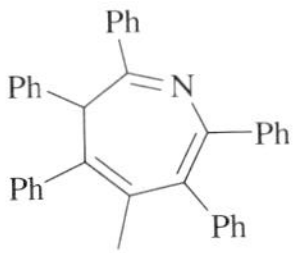 **II** (51)[b,v]	182
	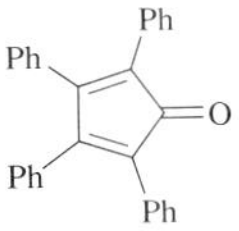		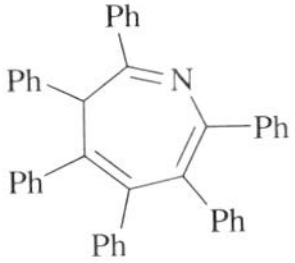	
		—	(91)[b]	447
		Toluene, 110°, 12 d	(61)[b]	182, 183

TABLE 6. CYCLOADDITIONS OF AZIRINES (*Continued*)

Imine	Diene	Conditions	Product(s) and Yield(s) (%)	Refs.
C_{14} 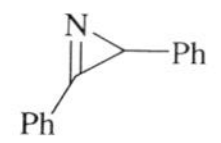	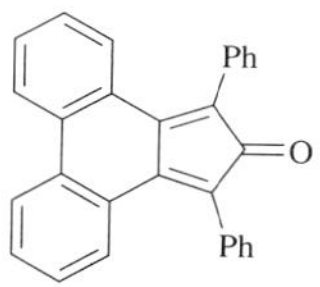	—[e]	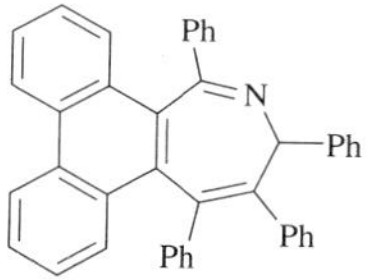 (50-80)[b]	181
		Toluene, 110°, 2-3 d	(84)[b]	182
C_{16} 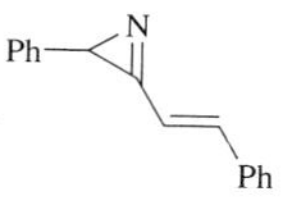	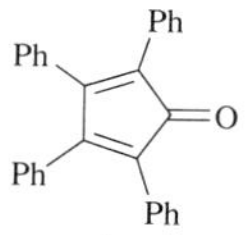	Toluene, 110°, 4 d	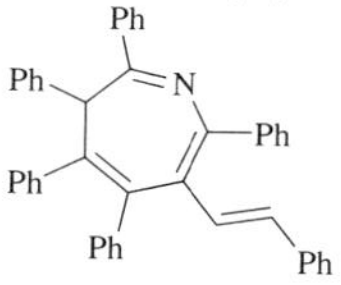 (38)[b]	451
C_{17} 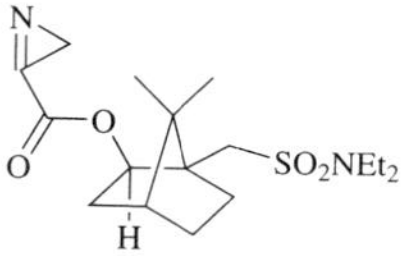		Toluene	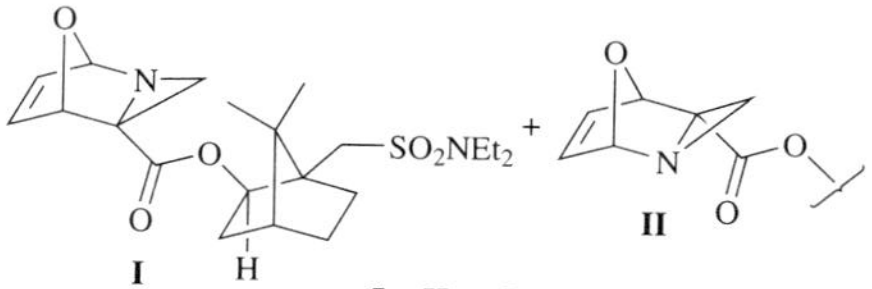 I + II	197
		Temp / Time: 5° / —	I + II: (—); I:II: 1:1	
		Temp / Time: rt / 3 d	I + II: (—); I:II: 1:1	
	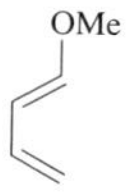	Toluene, rt, 5 d	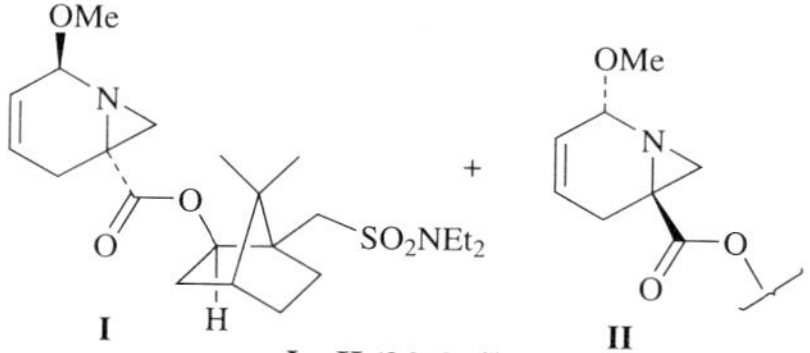 I + II (36) 2:1[w]	197

		Toluene, rt, 19 h	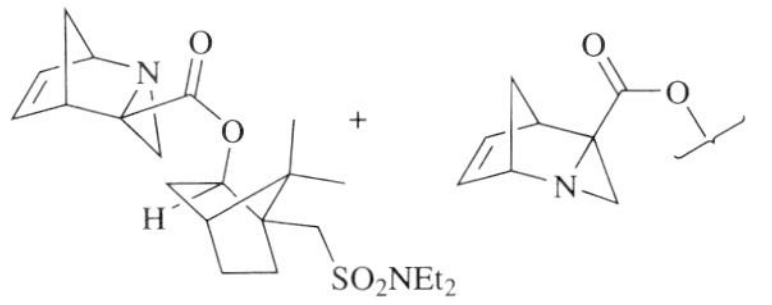 (73) 2:1	197, 196
	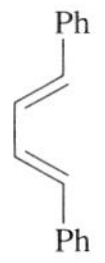	Toluene, 110°, 7 h	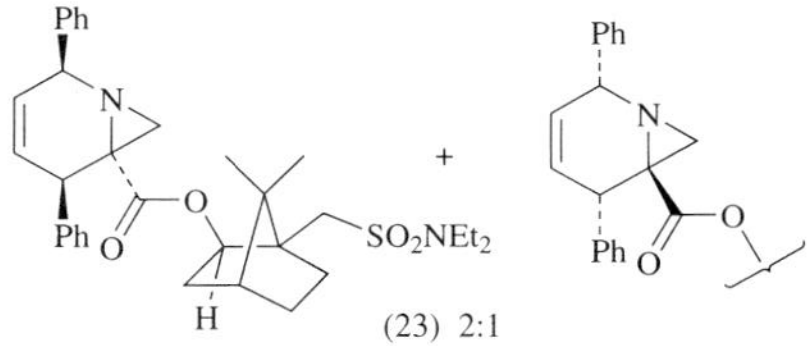 (23) 2:1	197
C_{19} 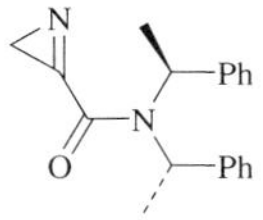		Toluene, rt, 24 h	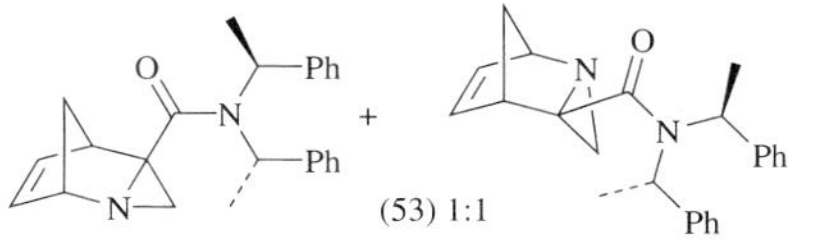 (53) 1:1	191

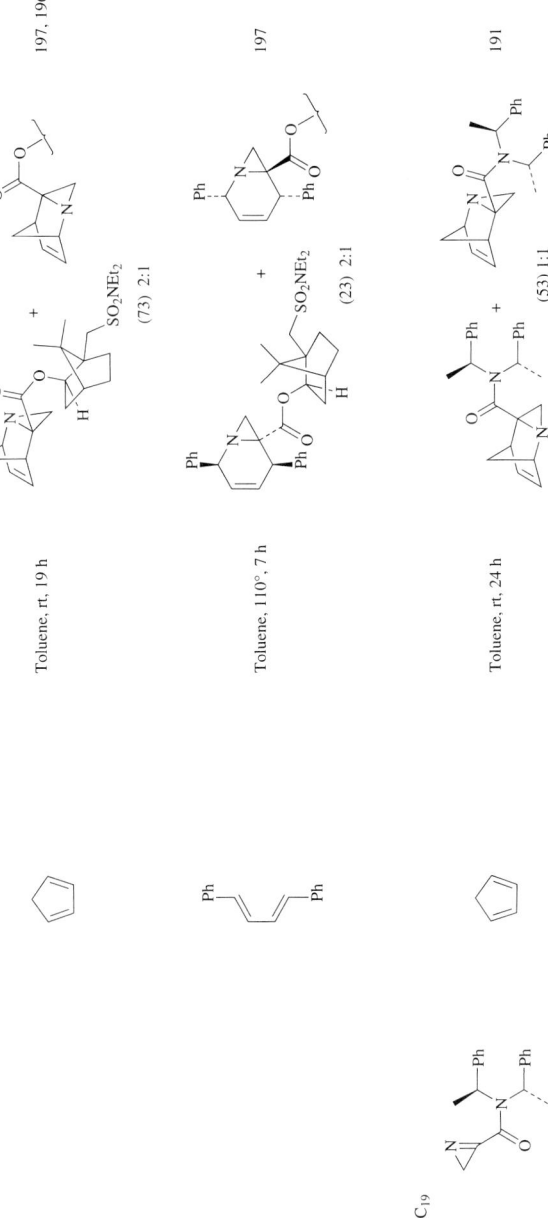

TABLE 6. CYCLOADDITIONS OF AZIRINES (*Continued*)

Imine	Diene	Conditions	Product(s) and Yield(s) (%)	Refs.
C_{19} 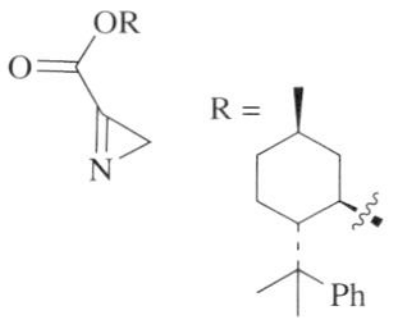		Lewis acid, CH_2Cl_2	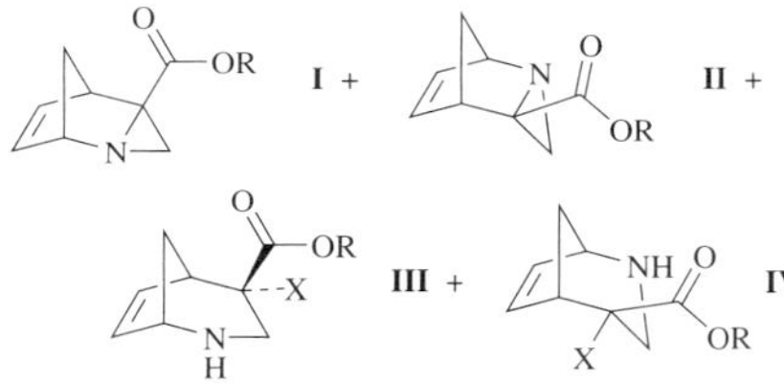 **I** + **II** + **III** + **IV**	

L. A.[x]	Temp	Time	Product(s) and Yield(s) (%)	Refs.
none	–78 to –40°	several d	**I** + **II** (99), **I**:**II** 54:46	198, 199
$MgBr_2 \cdot OEt_2$	–100°	—	**I** + **II** (88), **I**:**II** 92.5:7.5	198
$MgBr_2 \cdot OEt_2$	–100°	<10 min	**I** + **II** + **III** + **IV** (88)[y], **I**+**II**:**III**+**IV** 92.5:7.5	199
$ZnCl_2 \cdot OEt_2$	–100°	<10 min	**I** + **II** (99), **I**:**II** 79:21	199, 198
$MgI_2 \cdot (OEt_2)_x$	–78 to –40°	—	**III** + **IV** (100)[z], **III**:**IV** 89:11	199
$MgBr_2$	–100 to –72°	—	**I** + **II** + **III** + **IV** (100)[y], **I**+**II**:**III**+**IV** 55:45	199
$MgCl_2$	–78 to –40°	—	**I** + **II** + **III** + **IV** (100)[aa], **I**+**II**:**III**+**IV** 57.5:42.5	199
$YbCl_3$	–73°	—	**I** + **II** + **III** + **IV** (100)[aa], **I**+**II**:**III**+**IV** 65.5:34.5	199
$Mg(OTf)_2$	–100 to –72°	—	**I** + **II** (100), **I**:**II** 41.5:58.5	199
$ZnCl_2 \cdot OEt_2$[bb]	–77°	4.5 h	**I** + **II** (54), **I**:**II** 59.5:40.5	199

Imine	Diene	Conditions	Product(s) and Yield(s) (%)	Refs.
	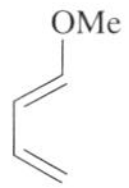	Lewis acid, CH_2Cl_2	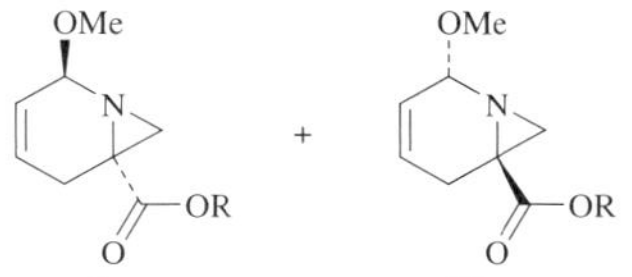 **I** + **II**	

L. A.	Temp	Time		
none	rt	>24 h	**I + II** (100), **I:II** 50:50	198
$MgBr_2•OEt_2$	–100 to –75°	—	**I + II** (82), **I:II** 93.5:6.5	198
$MgBr_2•OEt_2$	–78°	—	**I + II** (82), **I:II** 93.5:6.5	199
$ZnCl_2•OEt_2$	–100°	<10 min	**I + II** (62), **I:II** 90:10	198
$ZnCl_2•OEt_2$	–100°	10 min	**I** (56) + **II** (6)	199
$YbCl_3$	–78 to –20°	—	**I + II** (43), **I:II** 63.5:36.5	198

Lewis acid, CH_2Cl_2

L. A.	Temp		
none	rt	(100) 60:40	198
none	rt	(100)[cc] 60:40	199
$ZnCl_2•OEt_2$	–78°	(99) 90:10	199, 198

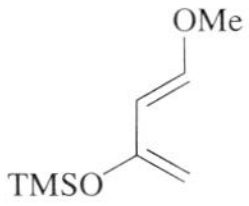

Lewis acid, CH_2Cl_2

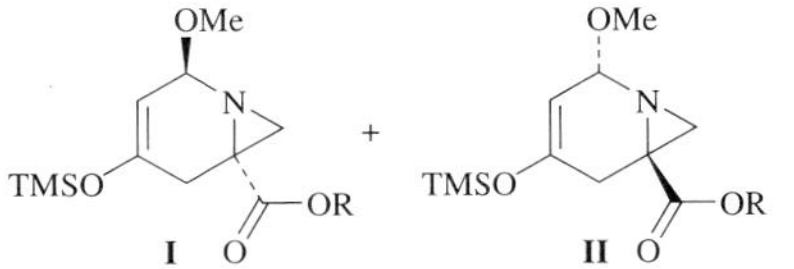

I + **II** 199, 198

L. A.	Temp	Time	**I + II**	**I:II**
none	–75 to –40°	several d	(90)	65:35
$MgBr_2•OEt_2$	–100°	<10 min	(56)	98:2
$ZnCl_2•OEt_2$	–100 to –90°	—	(31)	93.5:6.5

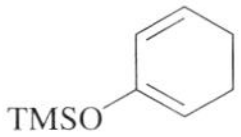

Lewis acid, CH_2Cl_2

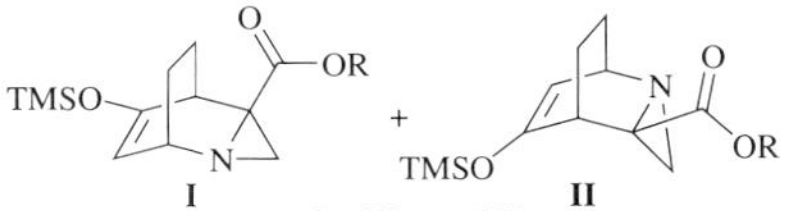

I + **II** 199, 198

L. A.	Temp	**I + II**	**I:II**
none	–75 to –40°	(80)	65:35
$MgBr_2•OEt_2$[dd]	–75°	(99)	98.5:1.5
$ZnCl_2•OEt_2$[dd]	–77°	(99)	67:33

TABLE 6. CYCLOADDITIONS OF AZIRINES (*Continued*)

Imine	Diene	Conditions	Product(s) and Yield(s) (%)	Refs.
C_{25}				
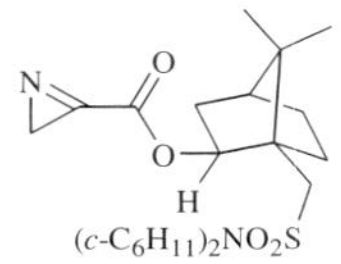		Toluene, rt, 3 d	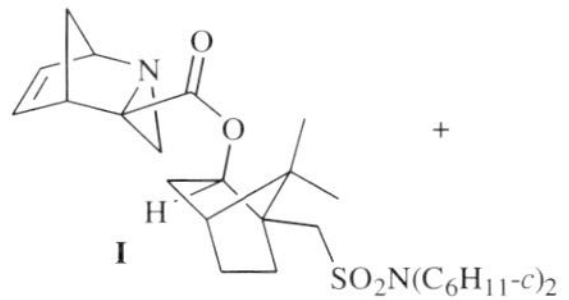 **I** + 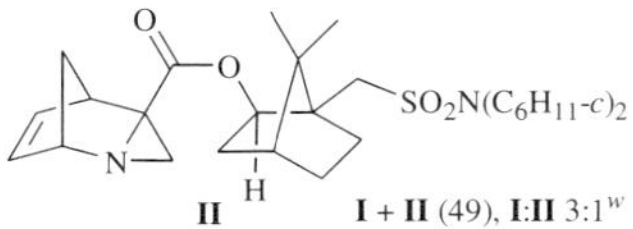 **II** **I** + **II** (49), **I**:**II** 3:1[w]	197, 196

[a] The azepinone is formed via the intermediate Diels-Alder adduct.

[b] The azepine is formed from the intermediate Diels-Alder adduct via loss of carbon monoxide and rearrangement.

[c] The azirine was formed in situ by heating the corresponding vinyl azide.

[d] A 28% yield of **I** and a 13% yield of **II** were obtained by fractional crystallization.

[e] The reaction was either carried out in refluxing toluene or refluxing xylene, but no other details are provided.

[f] The observed product resulted from Lewis acid mediated rearrangement of the initial Diels-Alder adduct.

[g] The azirine was formed and isolated prior to the cycloaddition.

[h] The azirine was formed by decomposition of the vinyl azide immediately before addition of the dienone to the same flask.

[i] The reaction was slower in refluxing toluene.

[j] The crude product was a 2:1 mixture of **I**:**II**.

[k] 2-Methyl-3-phenyl-1-azirine was formed as a side product during the preparation of 2-benzyl-1-azirine. The cycloadducts shown were obtained as minor products during the cycloadditions of 2-benzyl-1-azirine with tetracyclone or dimethylcyclone.

[l] 2-Methyl-3-phenyl-1-azirine was formed as a side product during the preparation of 2-benzyl-1-azirine, and its cycloadducts with tetracyclone or dimethylcyclone were obtained as minor products.

[m] In refluxing toluene the reaction required 7 d to reach completion.

[n] A 53% yield of adduct **II** was obtained after purification.

[o] Only one isomer was obtained, but its stereochemical configuration was not determined.

[p] The reaction may have been heated to 50° for 2 h after stirring at room temperature in order to reach completion.

[q] The product resulted from hydrolysis of the intermediate cycloadduct.

[r] After 2 hours, a 48% yield of **I** was obtained by filtration. The remaining mixture was then stirred for 18 hours and yielded 48% of a mixture of **I** and **II**.

[s] The reaction was performed with 1 equiv. of the starting azide.

[t] The reaction was performed with 3 equiv. of the starting azide.

[u] Although the diastereoselectivities are given, the stereochemistry of the major isomer is not specified. In the case of $ZnCl_2$•OEt_2, the oppposite isomer is the major product.

[v] The crude product was a 1:6 mixture of **I**:**II**.

[w] Although the diastereoselectivity is given, the sterochemistry of the major product is not specified.

[x] A stoichiometric amount of Lewis acid was utilized. $Yb(OTf)_3$, Ti(OPr-*i*)$_4$, and $SnCl_4$ gave lower selectivities than those observed with MgI_2•$(OEt_2)_x$ or $MgBr_2$•OEt_2.

[y] The ring-opened products (X = Br) were also obtained.

[z] Only the ring-opened products (X = I) were obtained.

[aa] The ring-opened products (X = Cl) were also obtained.

[bb] Instead of a stoichiometric amount, 10 mol% of Lewis acid was used.

[cc] The yield is based on unreacted azirine.

[dd] Although reaction times are not provided, the reaction rates were increased in the presence of Lewis acid.

Table 7. Cycloadditions of Oximino Compounds

	Imine	Diene	Conditions	Product(s) and Yield(s) (%)	Refs.
C_4	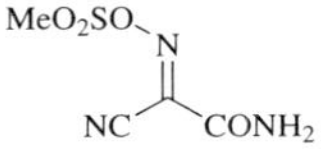		Dioxane, 70°, 36 h	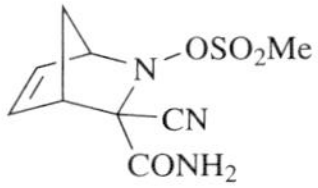 (39)	203
C_5	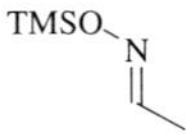		Ether, –78 to 20°	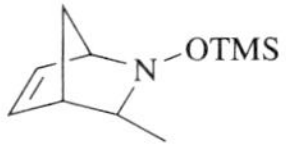 (54.3)[a]	200
	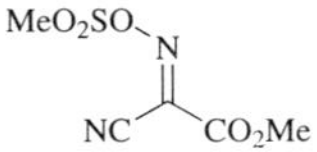		Benzene, 80°, 4 h	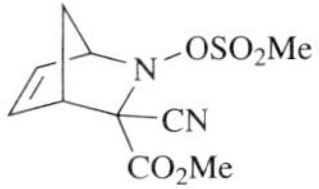 (50)	203
C_6	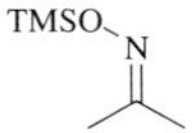		Ether, –78 to 20°	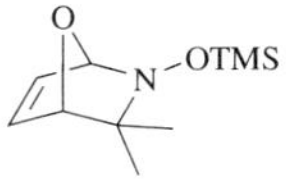	
			Time		
			190 h	(32.2)	201
			—	(40.0)	200
			Ether, –78 to 20°, 190 h	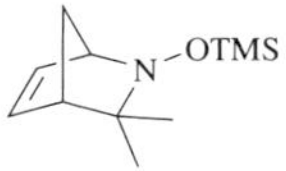 (50.0)	200, 201
	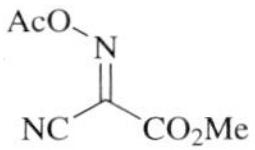		rt	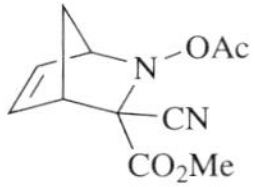 **I** + 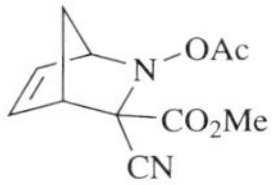 **II**	206

Catalyst	Solvent	Time	
none	none	18 h	**I** (22) + **II** (3)
none	ether	18 h	**I** + **II** (12), **I**:**II** 6.4:1
$LiClO_4$	ether	5 h	**I** (46) + **II** (5)
none[b]	toluene	40 h	**I** + **II** (99), **I**:**II** 7.6:1

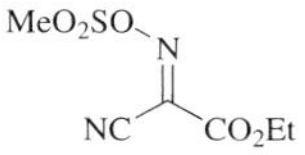

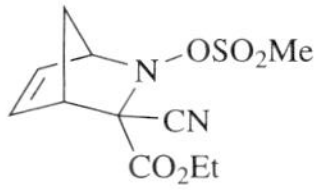

Ether, rt	(73)	202
Ether, 20°, 12 d	(65)	203
Benzene, 80°, 3 h	(51)	203

C_7

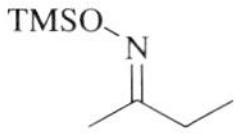

Ether, –78 to 20°

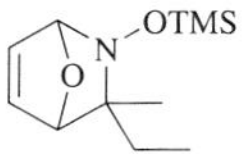

(78.3)[a] 200

Ether, –78 to 20°

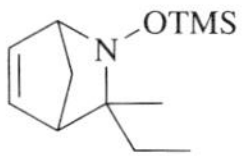

(55.4)[a] 200

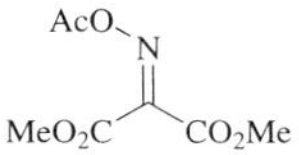

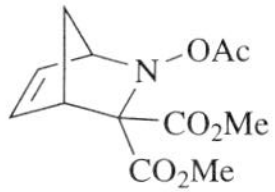

Catalyst (conc.)	Solvent	Temp	Time		
$LiClO_4$ (1 M)	ether	rt	24 h	(32)	206
$LiClO_4$ (3 M)	ether	rt	24 h	(41)	206
$LiClO_4$ (5 M)	ether	rt	24 h	(49)	206, 207
none[b]	toluene	rt	48 h	(17)	206
none[c]	toluene	60°	48 h	(25)	207

TABLE 7. CYCLOADDITIONS OF OXIMINO COMPOUNDS (*Continued*)

Imine	Diene	Conditions	Product(s) and Yield(s) (%)	Refs.
C_8 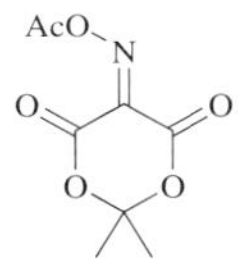			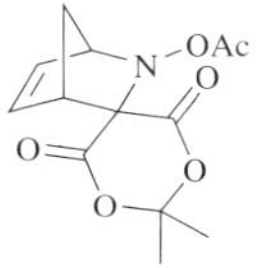	208
		rt, 24 h	(99)	
		Benzene, rt, 5 d	(58)[d]	
	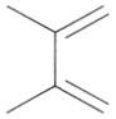	Toluene, rt, 18 h, 8 kbar	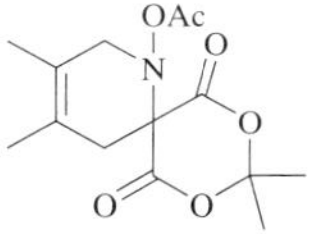 (68)	208
	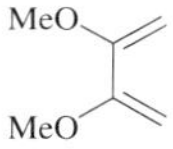	Toluene, rt, 18 h, 8 kbar	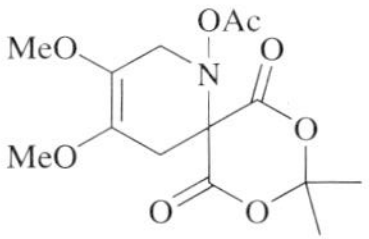 (62)	208
C_9 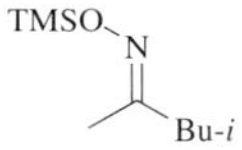		Ether, 20°, 190 h	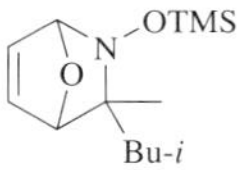 (—)[a]	201
		Ether, –78 to 20°	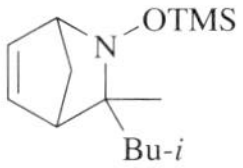	
		Time		
		190 h	(76.7)[a]	201
		—	(74.0)[a]	200
C_{10} 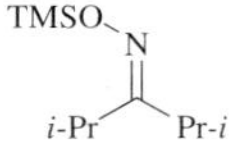		Ether, –78 to 20°	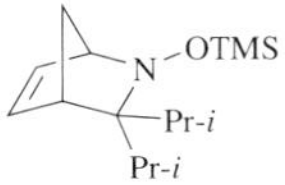 (62.0)	200

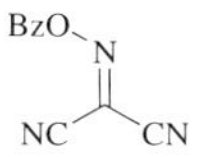		Ether, 20°, 48 h	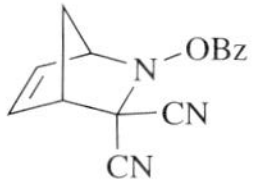 (80)	203, 202
		Benzene, 80°, 20 h	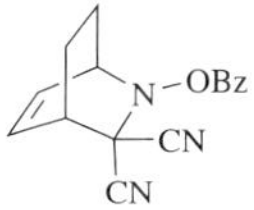 (26)	204
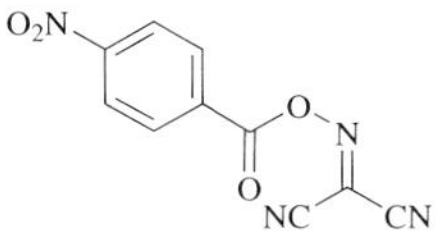		Ether, 20°, 24 h	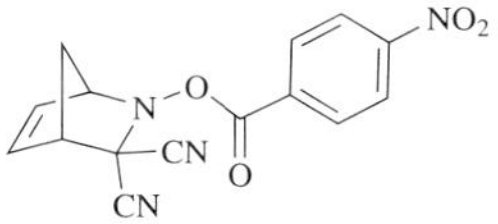 (90)	203, 202
		Benzene, 80°, 20 h	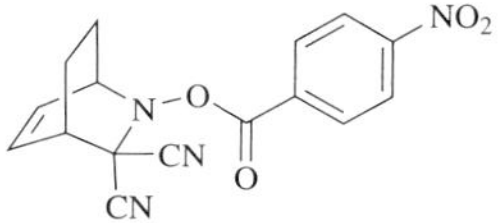 (52)	204
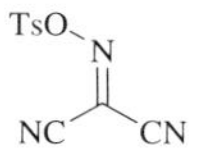			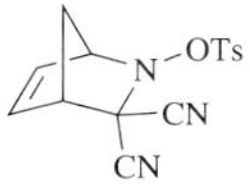	
		Ether, 20°, 12 h	(88)	203, 202
		$CHCl_3$, 20°	(—)	452
		—	(—)	65, 64
	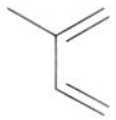	Benzene, 80°, 28 h	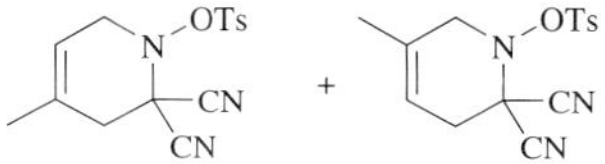 (77) 42:58	204

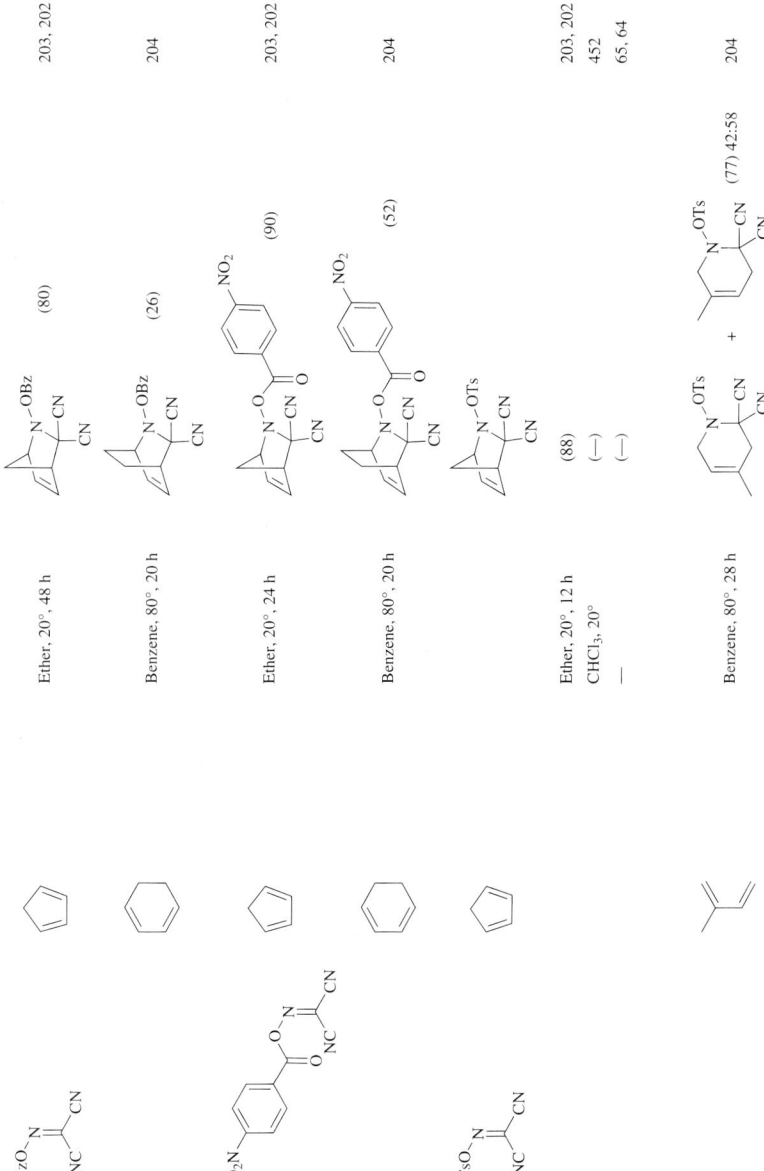

TABLE 7. CYCLOADDITIONS OF OXIMINO COMPOUNDS (*Continued*)

Imine	Diene					Conditions	Product(s) and Yield(s) (%)	Refs.
C_{10} 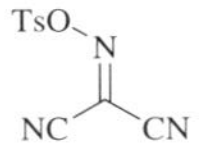	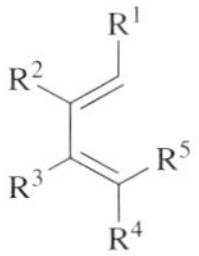						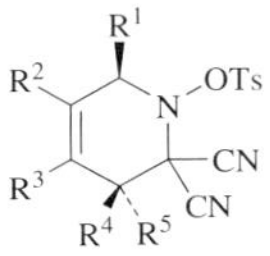	
	R^1	R^2	R^3	R^4	R^5			
	H	H	H	Me	H	Benzene, 80°, 14 h	(30)	204, 452
	H	Me	Me	H	H	$AlCl_3$	(—)	453
	H	Me	Me	H	H	Benzene, 80°, 14 h	(70)	204
	Me	H	H	Me	H	Benzene, 80°, 14 h	(70)	204, 452
	H	Me	H	Me	H	Benzene, 80°, 14 h	(70)	204
	H	Me	H	Me	Me	Benzene, 80°, 14 h	(70)	204
	OEt	Me	H	H	H	THF, H_2O, 0 to 20°, 3 d	(79)	205
	OEt	Et	H	H	H	THF, H_2O, 0 to 20°, 3 d	(77)	205
	OEt	*n*-Pr	H	H	H	THF, H_2O, 0 to 20°, 3 d	(66)	205
	OEt	*i*-Pr	H	H	H	THF, H_2O, 0 to 20°, 3 d	(69)	205
	H	H	H	Ph	H	Benzene, 80°, 14 h	(70)	204
	OEt	*n*-Bu	H	H	H	THF, H_2O, 0 to 20°, 3 d	(59)	205
	Me	H	H	Ph	H	Benzene, 80°, 14 h	(70)	204
	OEt	*n*-C_5H_{11}	H	H	H	THF, H_2O, 0 to 20°, 3 d	(47)	205
							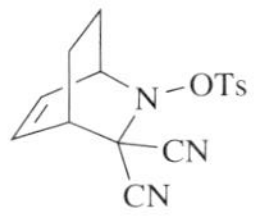	
						$AlCl_3$	(—)	453
						$CHCl_3$, 20°, 8 d	(35)	204

C$_{10-12}$

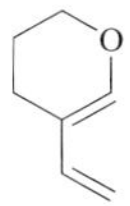

THF, H_2O, 0 to 20°, 3 d

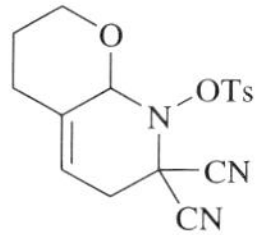

(20) 205

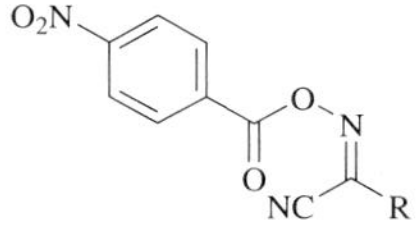

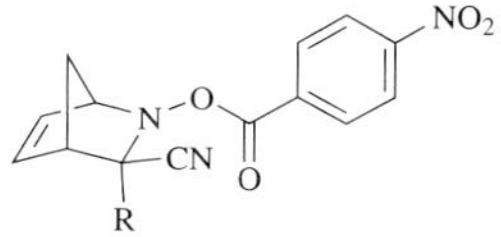

203

R	Conditions	Yield (%)
$CONH_2$	Dioxane, 70°, 48 h	(20)
CO_2Me	Benzene, 80°, 4 h	(62)
CO_2Et	Benzene, 80°, 4 h	(65)

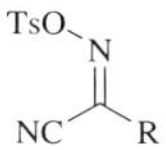

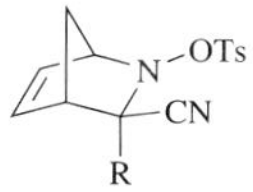

R	Conditions	Yield (%)	Refs.
$CONH_2$	Ether, rt	(42)	202
$CONH_2$	Acetone, 20°, 15 d	(45)	203
$CONH_2$	Acetone, 60°, 20 h	(42)	203
CO_2Me	Benzene, 80°, 4 h	(57)	203
CO_2Et	Ether, rt	(55)	202, 453
CO_2Et	Benzene, 80°, 3 h	(55)	203
CO_2Et	Ether, 20°, 10 d	(61)	203
CO_2Et	—	(55)	454, 455

TABLE 7. CYCLOADDITIONS OF OXIMINO COMPOUNDS (*Continued*)

Imine	Diene	Conditions	Product(s) and Yield(s) (%)	Refs.
C_{11}				
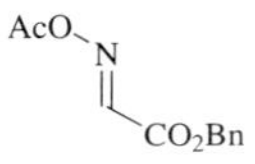		$LiClO_4$, ether, rt, 2 h	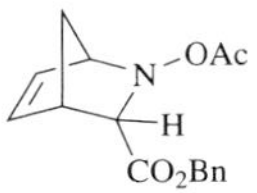 (38)	208
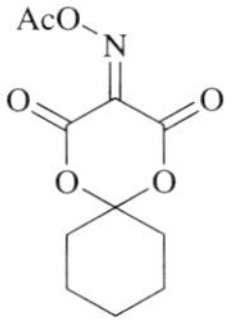		CH_2Cl_2, rt, 18 h	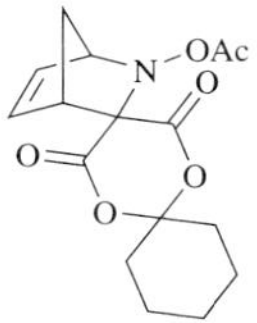 (72)[e]	208
C_{13}				
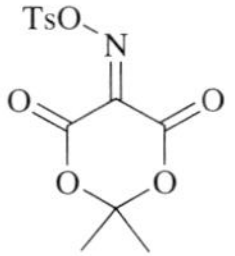		Benzene, 80°	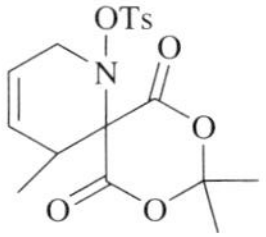 (32)	209

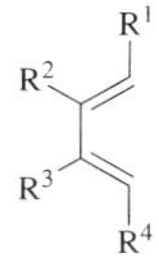

Me_2AlCl, CH_2Cl_2, –78°

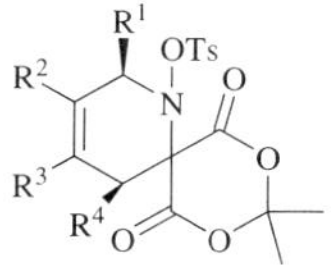

R[1]	R[2]	R[3]	R[4]	Time		
H	H	H	Me	4 h	(76-78)[f]	209
H	H	H	Me	3 h	(—)	209, 210
H	Me	H	H	4 h	(84-91)[f]	209
H	Me	H	H	3 h	(—)	210, 209
H	Me	H	Me	1 h	(—)	209
H	Me	H	Me	2-4 h	(—)	210
H	H	Me	Me	2.5 h	(—)	209, 210
Me	H	H	Me	1.5 h	(—)	209
Me	H	H	Me	2-4 h	(—)	210
H	*n*-Bu	H	H	4 h	(—)	209, 210
H	*t*-Bu	H	H	3.5 h	(—)	209, 210

TABLE 7. CYCLOADDITIONS OF OXIMINO COMPOUNDS (*Continued*)

Imine	Diene	Conditions	Product(s) and Yield(s) (%)	Refs.
C_{13} 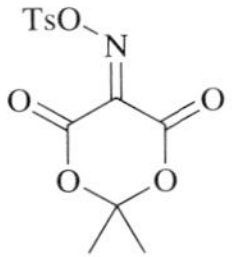		Toluene, 60°, 2 h	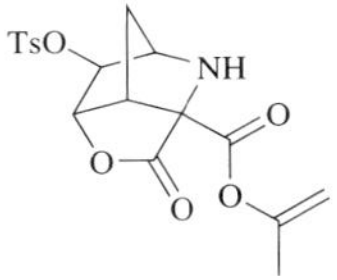 (38)[g] + 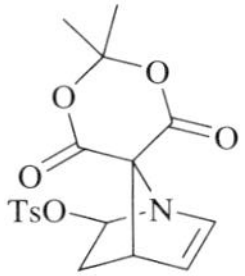 (9)[g]	209
	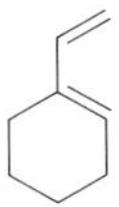	Me_2AlCl, CH_2Cl_2, –78°, 4 h	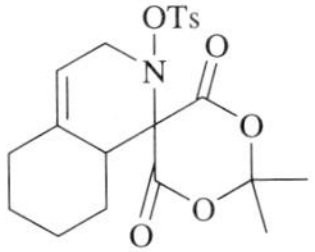 + 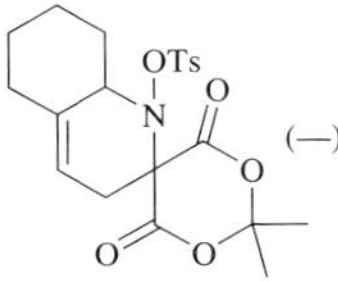 (—) 83:17	209, 210
	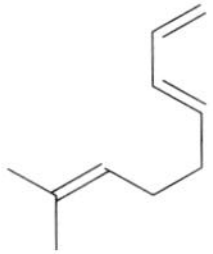	Me_2AlCl, CH_2Cl_2, –78°, 3 h	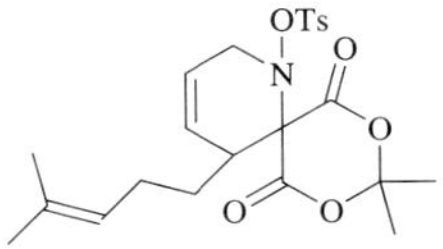 (—)	209, 210
	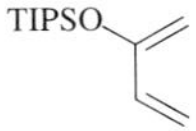	Benzene, 55-65°, 24 h	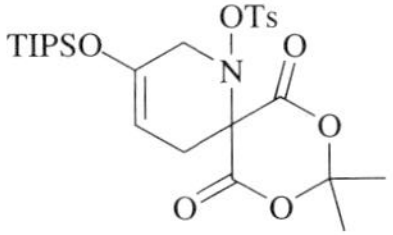 (46-56)	209
	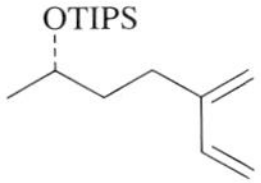	Me_2AlCl, CH_2Cl_2, –78°, 3 h	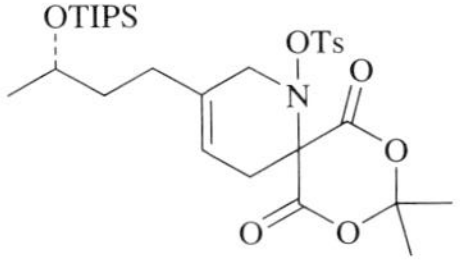 (—)	209

C_{15}	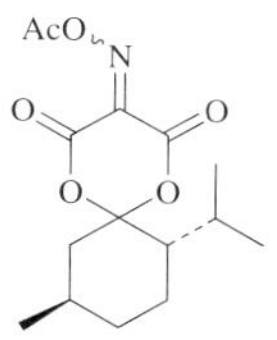		rt, 1 d	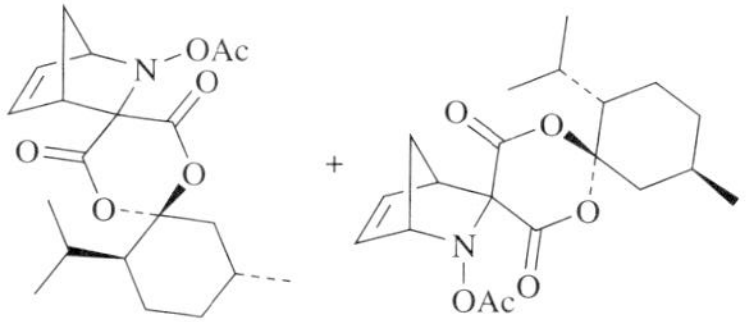 (—) 1:1	208
C_{21}	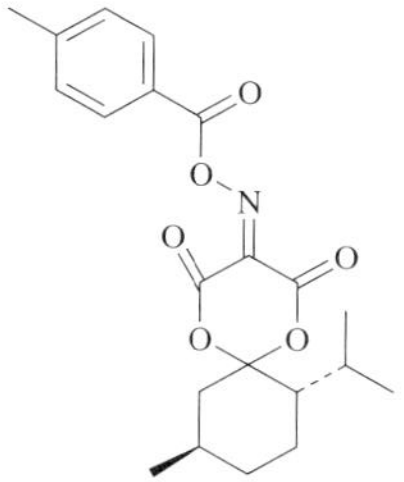		rt, 1 d	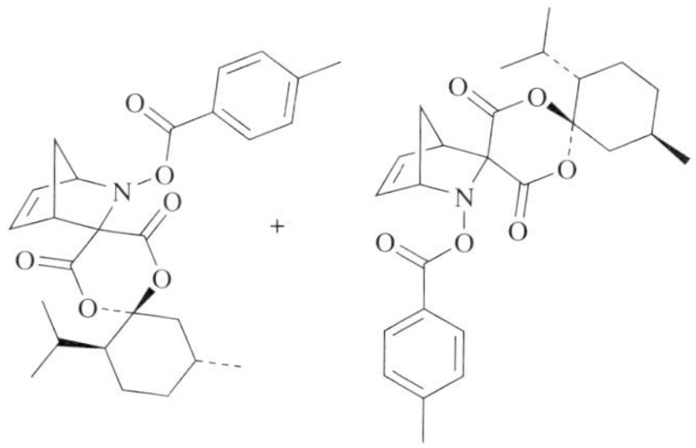 (—) 5:3	208

[a] No stereochemical information is provided.

[b] The reaction was performed at 10 kbar of pressure.

[c] The reaction was performed at 11 kbar of pressure.

[d] The tricyclic adduct resulting from the oxime acting as the diene component was obtained in 10% yield.

[e] The tricyclic adduct resulting from the oxime acting as the diene component was obtained in 18% yield as determined by ^{1}H NMR.

[f] Other reagents such as $TiCl_4$, $ZnCl_2$, $AlCl_3$, $MeSO_3H$, and camphorsulfonic acid were less effective.

[g] The two products shown result from rearrangements of the initially formed Diels-Alder cycloadduct.

TABLE 8. CYCLOADDITIONS OF ELECTRON-RICH IMINES

Imine	Diene	Conditions	Product(s) and Yield(s) (%)	Refs.
C_1 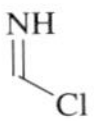	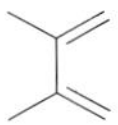	$POCl_3$[a], nitrobenzene, 170-180°, 2-3 h	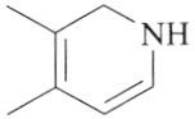 ("low")	456, 457
	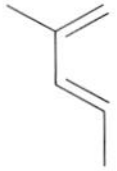	$POCl_3$[a], nitrobenzene, 170-180°, 2-3 h	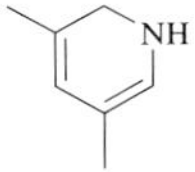 + 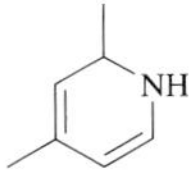 ("low")	456, 457
	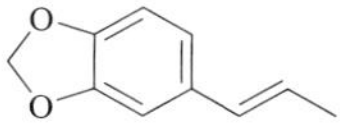	$POCl_3$[a], toluene, 120-130°, 1-2 h	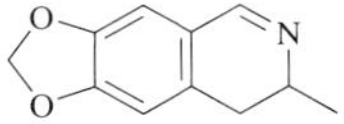 (30-60)	211
	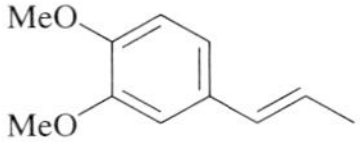	$POCl_3$[a], toluene, 120-130°, 1-2 h	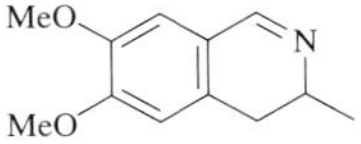 (30-60)	211
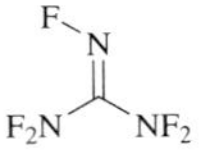		rt, 2 h	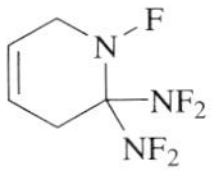 (—)	215
	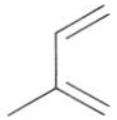	rt, 2 h	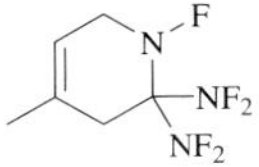 (—)	215
		rt, 2 h	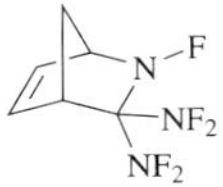 (—)	215
	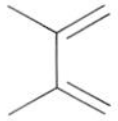	rt, 2 h	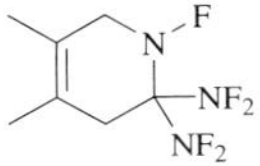 (—)	215

	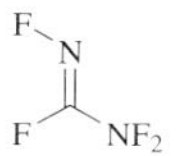	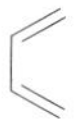	rt, 2 h	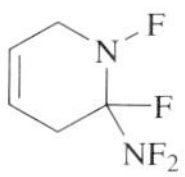 (—)	215
			rt, 2 h	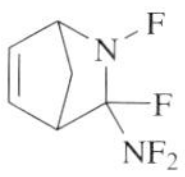 (—)[b]	215
C_2					
	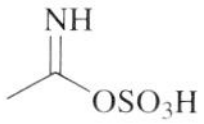	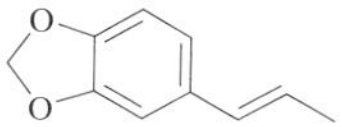	1. H_2SO_4[c], rt, 10-12 h 2. 100°, 2 h	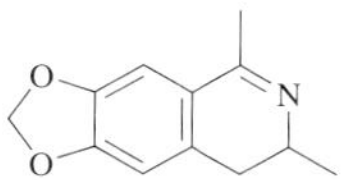 ("low")	214, 457
		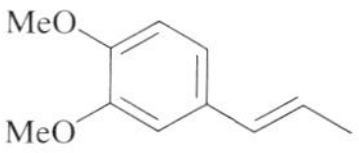	1. H_2SO_4[c], rt, 10-12 h 2. 100°, 2 h	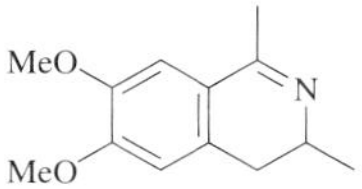 ("low")	214, 457
	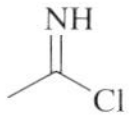		$POCl_3$[a], 80-100°, 3 h	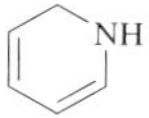 (—)	456
		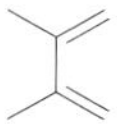	$POCl_3$[d], nitrobenzene, 170-180°, 2-3 h	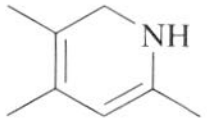 ("low")	456, 457
		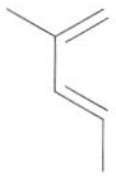	$POCl_3$[a], nitrobenzene, 170-180°, 2-3 h	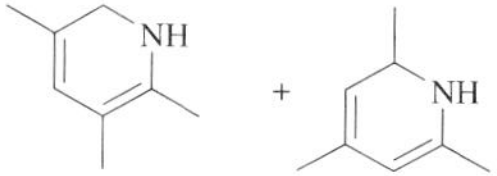 ("low")	456, 457

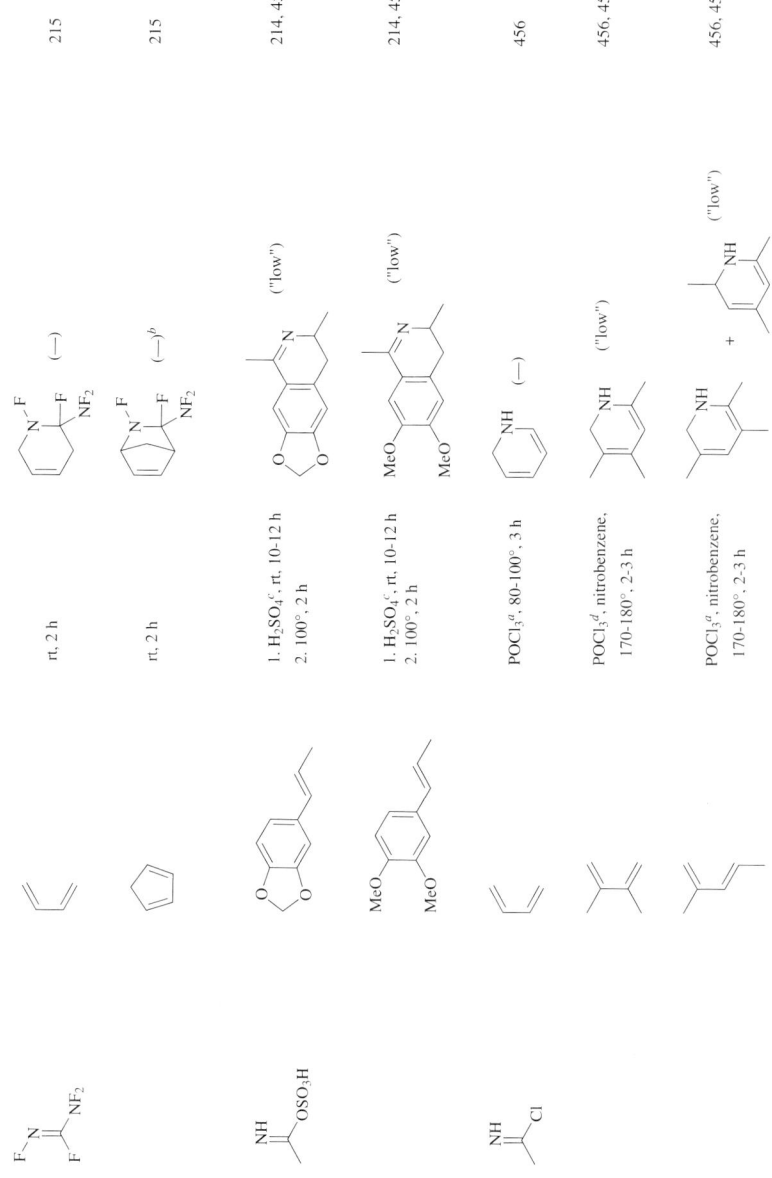

TABLE 8. CYCLOADDITIONS OF ELECTRON-RICH IMINES (*Continued*)

Imine	Diene	Conditions	Product(s) and Yield(s) (%)	Refs.
C_2 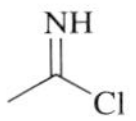	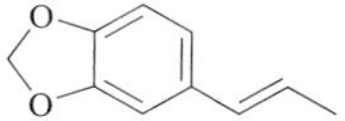	$POCl_3$, toluene, 1-2 h	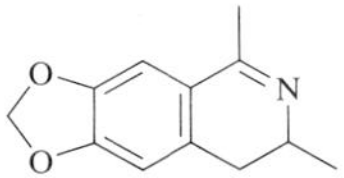	
		Temp		
		120-130°	(30-60)[a]	211
		100-120°	(—)[e]	212
	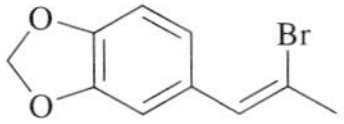	—[a]	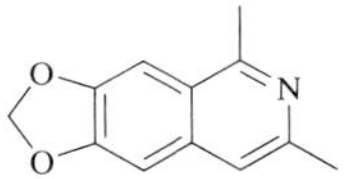 (—)	213
	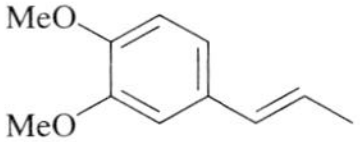	$POCl_3$, toluene, 1-2 h	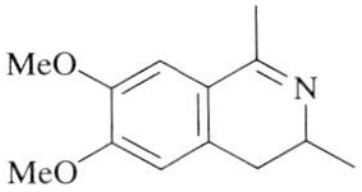	
		Temp		
		120-130°	(30-60)[a]	211
		100-120°	(—)[e]	212
C_7 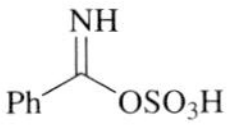	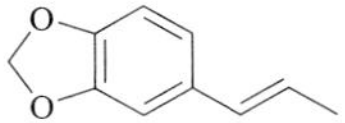	1. H_2SO_4[c], rt, 10-12 h 2. 100°, 2 h	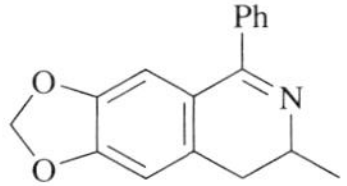 ("low")	214, 457
	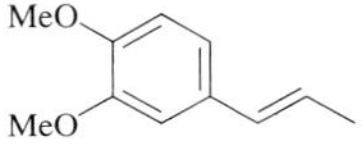	1. H_2SO_4[c], rt, 10-12 h 2. 100°, 2 h	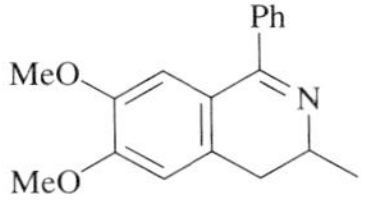 ("low")	214, 457
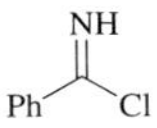	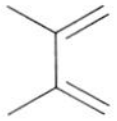	$POCl_3$[d], nitrobenzene, 170-180°, 2-3 h	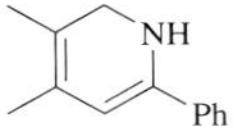 ("low")	456, 457

	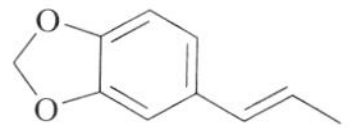	$POCl_3$, toluene, 1-2 h	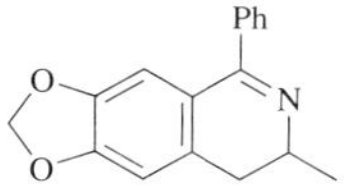		
		Temp			
		120-130°	(30-60)[a]		211
		100-120°	(—)[e]		212
	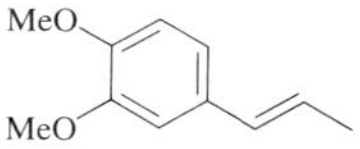	$POCl_3$, toluene, 1-2 h	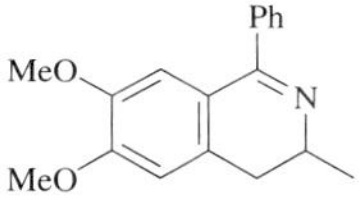		
		Temp			
		120-130°	(30-60)[a]		211
		100-120°	(—)[e]		212
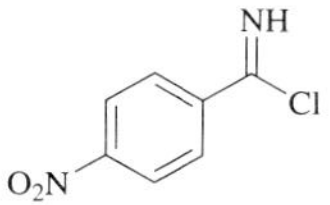	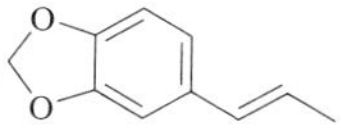	—[a]	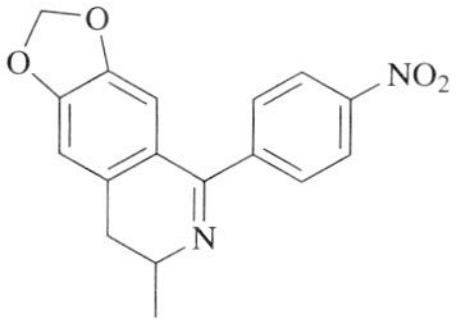	("poor")	213
C_8					
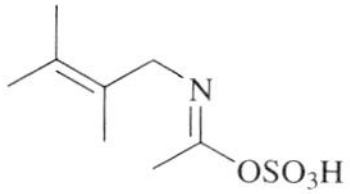	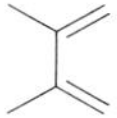	1. MeCN, H_2SO_4[f], rt, overnight 2. 100°, 1 h	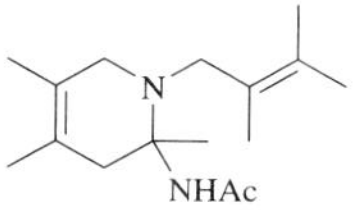	(—)	214
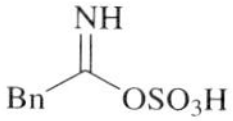	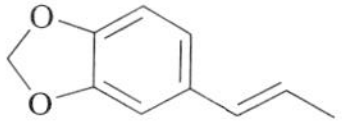	1. H_2SO_4[c], rt, 10-12 h 2. 100°, 2 h	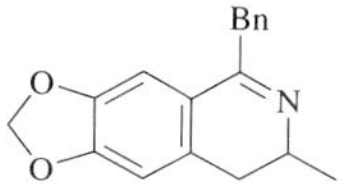	("low")	214, 457

TABLE 8. CYCLOADDITIONS OF ELECTRON-RICH IMINES (*Continued*)

Imine	Diene	Conditions	Product(s) and Yield(s) (%)	Refs.
C_8				
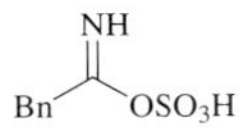	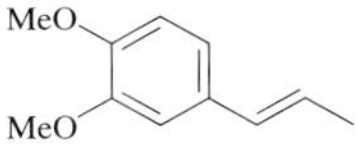	1. H_2SO_4[c], rt, 10-12 h 2. 100°, 2 h	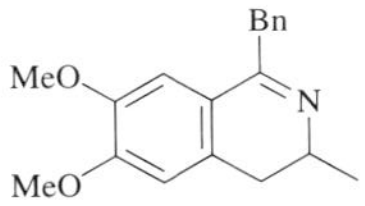 ("low")	214, 457
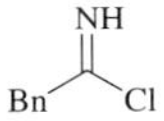	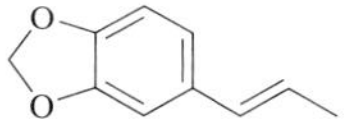	$POCl_3$[a], toluene, 120-130°, 1-2 h	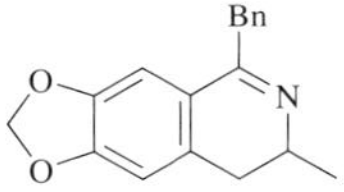 (30-60)	211
	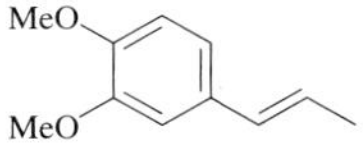	$POCl_3$[a], toluene, 120-130°, 1-2 h	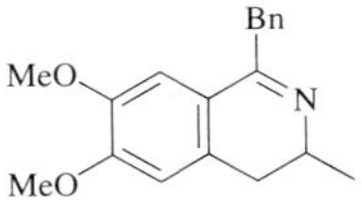 (30-60)	211
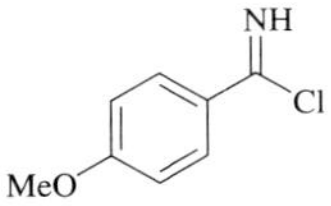	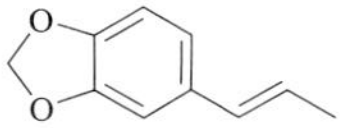	—[d]	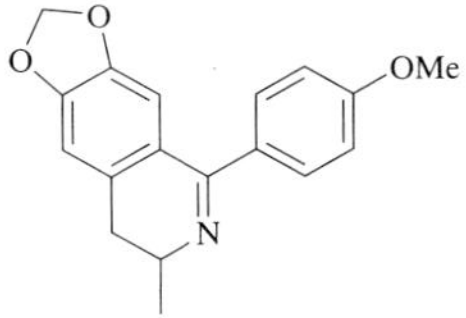 (—)	213
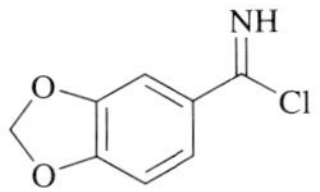	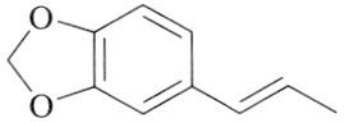	$POCl_3$, 1-2 h	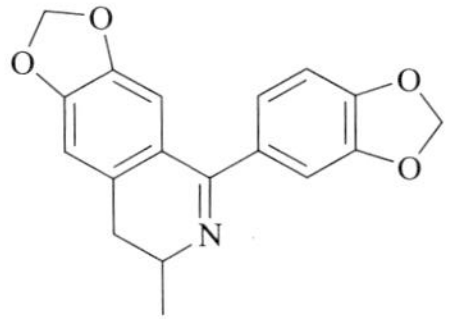	
		Solvent / Temp		
		toluene / 100-120°	(—)[e]	212
		benzene or toluene / 90-100°	(—)[a]	458

	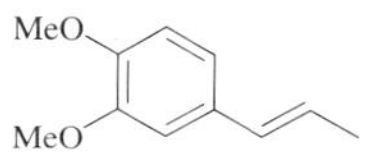MeO MeO	$POCl_3$, toluene, 100-120°, 1-2 h	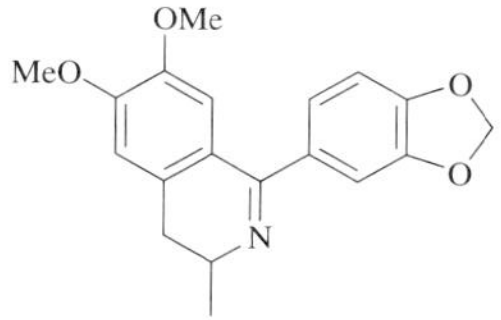OMe MeO O O N	(—)[e]	212
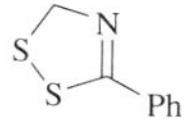N S S Ph	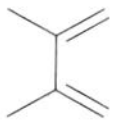	1. $HClO_4$, H_2O, $MeNO_2$ 2. $MeNO_2$, 3 h 3. Et_3N, ether	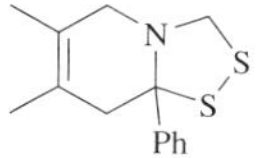N S S Ph	(91)	459, 460
C_9					
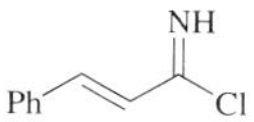NH Ph Cl	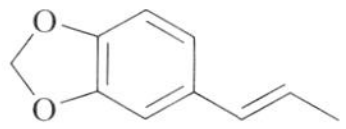O O	$POCl_3$, toluene, 100-120°, 1-2 h	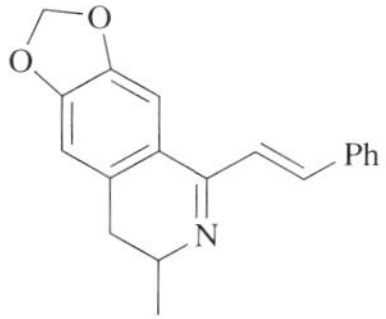O O N Ph	(—)[e]	212
	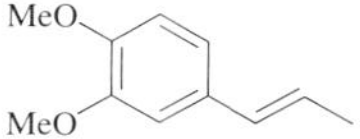MeO MeO	$POCl_3$, toluene, 100-120°, 1-2 h	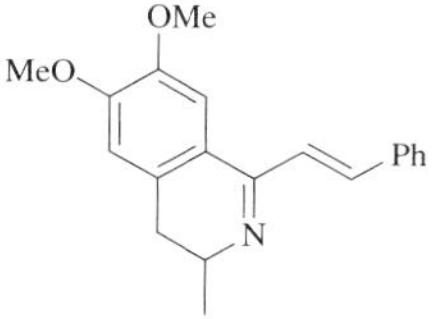OMe MeO N Ph	(—)[e]	212
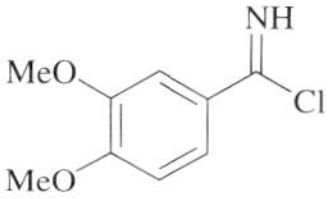NH MeO Cl MeO	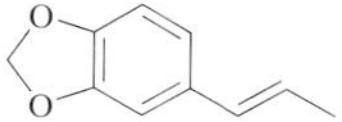O O	$POCl_3$, toluene, 100-120°, 1-2 h	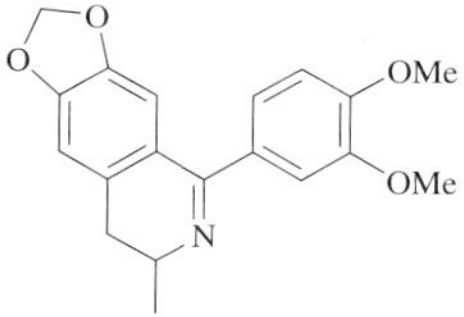O O OMe OMe N	(—)[e]	212

TABLE 8. CYCLOADDITIONS OF ELECTRON-RICH IMINES (*Continued*)

Imine	Diene	Conditions	Product(s) and Yield(s) (%)	Refs.
C_9 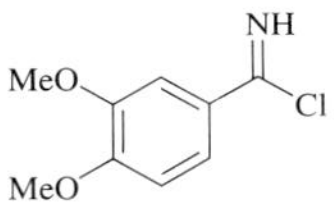	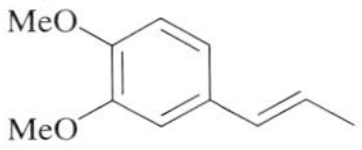	$POCl_3$, 1-2 h	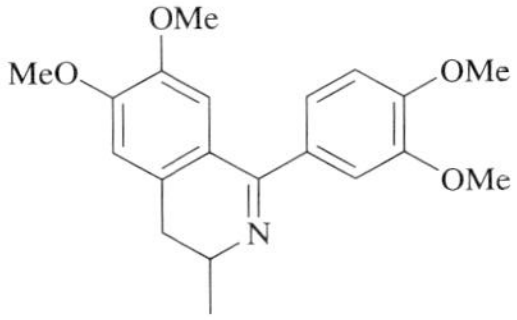	
		Solvent / Temp: toluene, 100-120°	(—)[e]	212
		benzene or toluene, 90-100°	(—)[a]	458
C_{10} 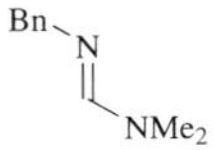	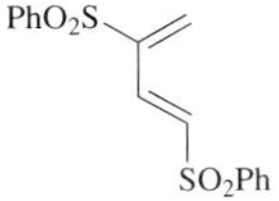		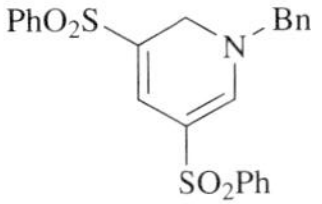	146
		Benzene, 80°, 12 h	(52)[g]	
		Et_3N, benzene, 80°, 12 h	(87)[h]	
	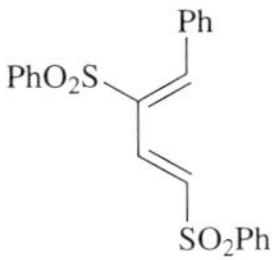	Benzene, 80°, 7 d	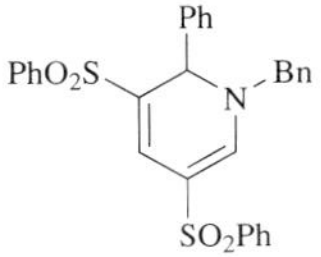 (83)	146
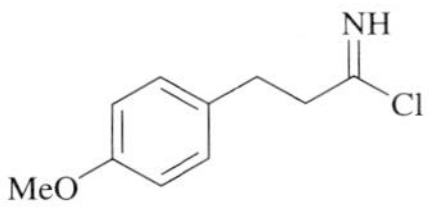	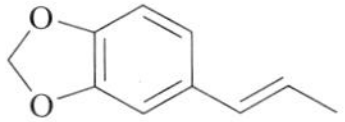	$POCl_3$, benzene or toluene, 90-100°, 1-2 h	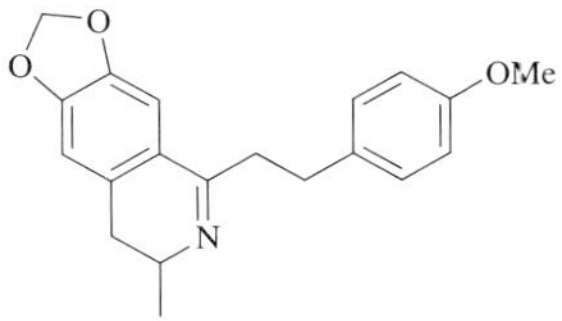 (—)[a]	458

	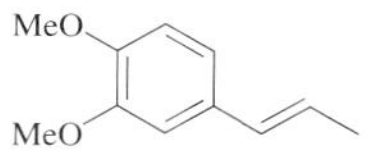	$POCl_3$, benzene or toluene, 90-100°, 1-2 h	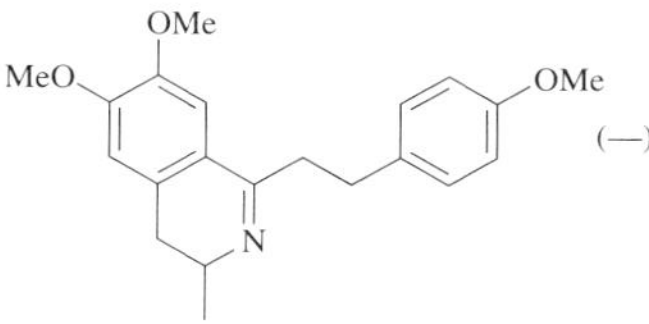 (—)[a]	458
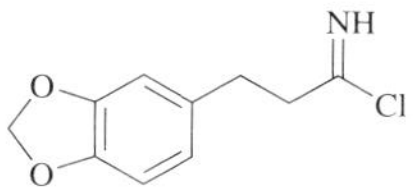	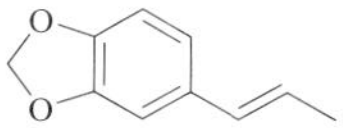	$POCl_3$, benzene or toluene, 90-100°, 1-2 h	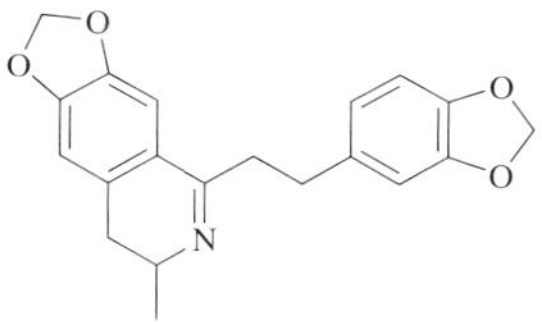 (—)[a]	458
C_{11}				
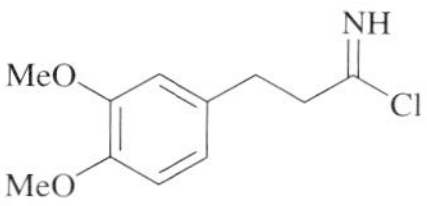	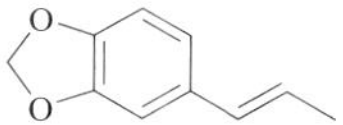	$POCl_3$, benzene or toluene, 90-100°, 1-2 h	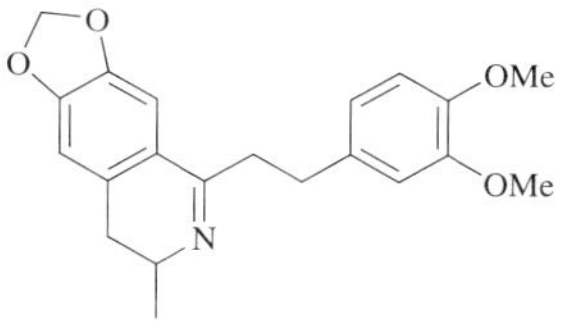 (—)[a]	458

[a] The imine was produced in situ by treating the corresponding amide with $POCl_3$.

[b] No stereochemical information is provided.

[c] The imine was formed in situ by treating the corresponding nitrile with sulfuric acid.

[d] The imine was produced in situ by treating the corresponding amide or oxime with $POCl_3$.

[e] The imine was produced in situ by treating the corresponding oxime with $POCl_3$.

[f] The reactive imine was produced in situ from 2,3-dimethylbutadiene, acetonitrile, and sulfuric acid.

[g] The reactive diene was formed in situ by rearrangement of 2,3-bis(phenylsulfonyl)-1,3-butadiene.

[h] The reactive diene was formed in situ by the reaction of 1,2,4-tris(phenylsulfonyl)-2-butene with triethylamine.

TABLE 9. CYCLOADDITIONS OF ISOCYANATES AND ISOTHIOCYANATES

	Imine	Diene	Conditions	Product(s) and Yield(s) (%)	Refs.
C_3	MeOC 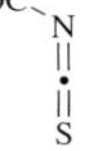	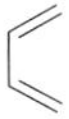	Hydroquinone, 18-20°, 6 months	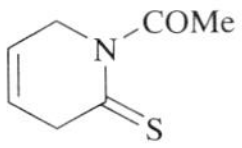 (40.7)	219
		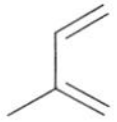	Hydroquinone, 18-20°, 7 months	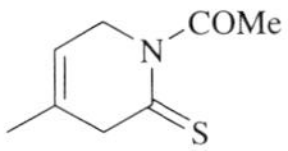 or 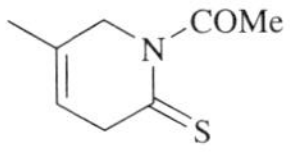 (39.2)[a]	219
			Hydroquinone, 18-20°, 6 months	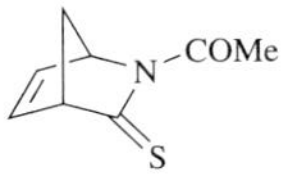 (49.8)	219
		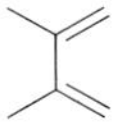	Hydroquinone, 18-20°, 6.5 months	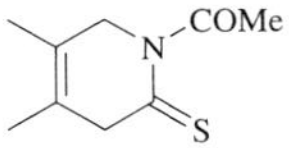 (42.5)	219
C_7	PhO_2S 	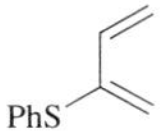	1. Hydroquinone, 110°[b] 2. SiO_2[c]	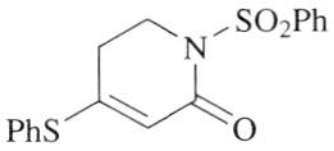	218, 217
			Solvent / Time		
			toluene 4 h	(30)	
			THF[d] 4.5 h	(23)	
			toluene[d] 4.5 h	(50)	
		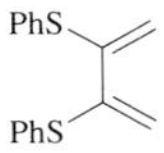	1. Hydroquinone, toluene, 130°, 8 h[b] 2. SiO_2[c]	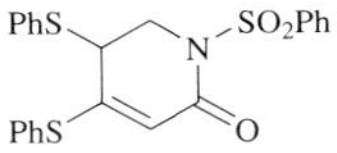 (73)	218, 217

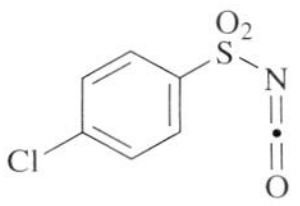	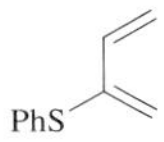	1. Hydroquinone, toluene, 110°, 4 h[b] 2. SiO_2[c]	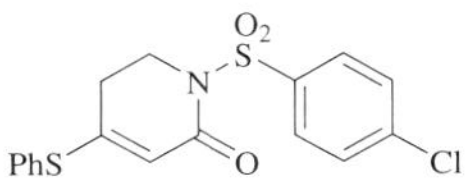 (38)	218, 217
	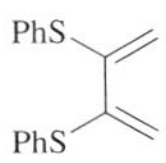	1. Hydroquinone, toluene, 130°, 8 h[b] 2. SiO_2[c]	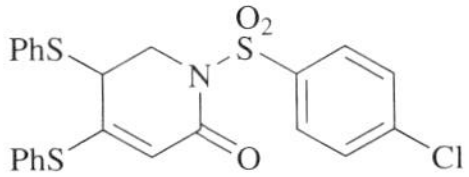 (47)	218, 217
C_8 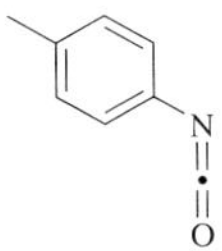	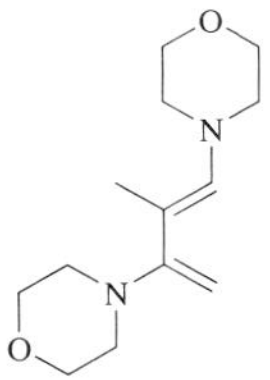	1. THF, 12 h 2. H^+, H_2O	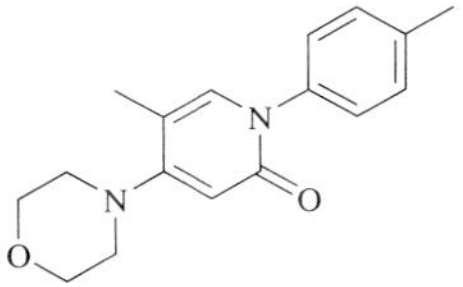 (65)	216
	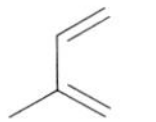	—	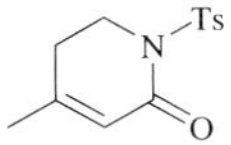 (5-7)	218, 217
	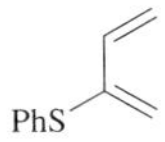	1. Hydroquinone, 110°, 4.5 h[b] 2. SiO_2[c]	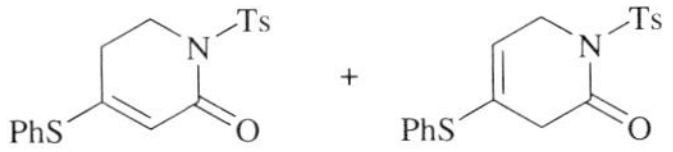 **I** + **II**	218, 217
		Solvent		
		toluene[d,e]	**I** (6) + **II** (51)	
		toluene[d]	**I** (71)	
		toluene	**I** (46)	
		THF[d]	**I** (17)	

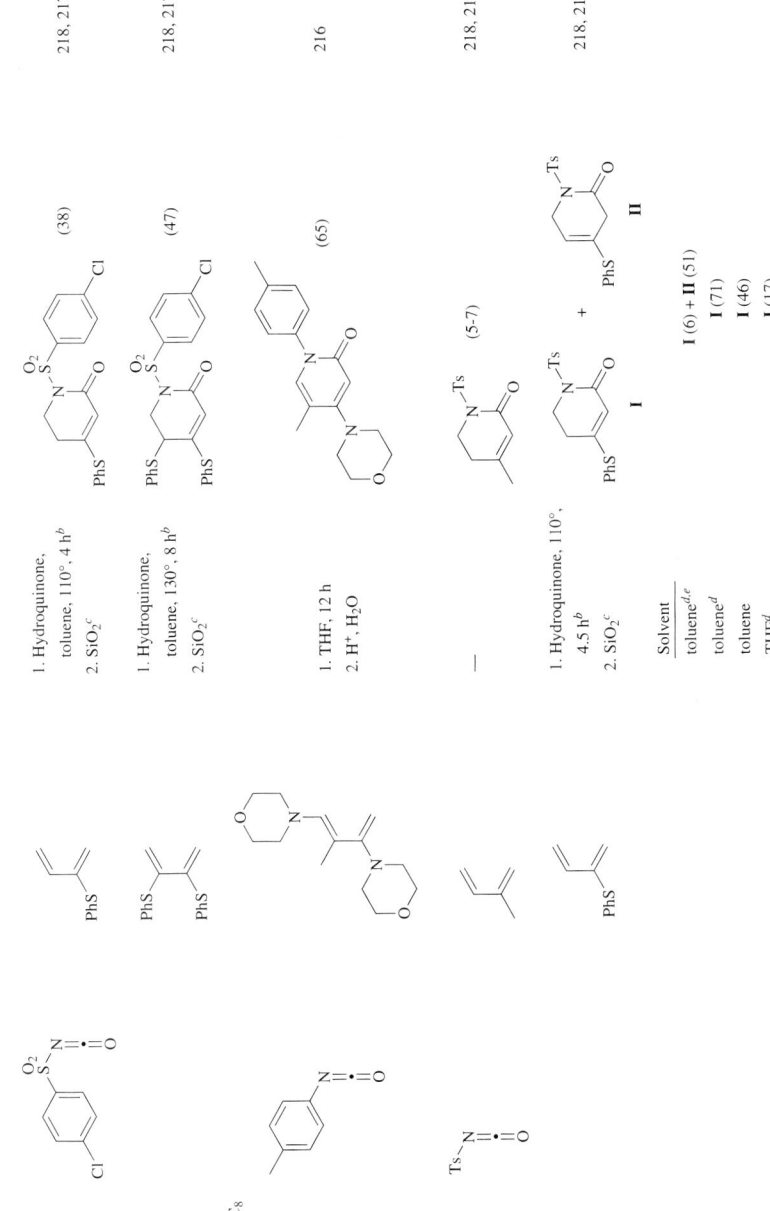

TABLE 9. CYCLOADDITIONS OF ISOCYANATES AND ISOTHIOCYANATES (*Continued*)

Imine	Diene	Conditions	Product(s) and Yield(s) (%)	Refs.
C_8 	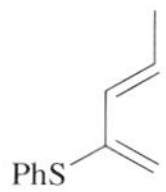	1. Hydroquinone, toluene, 110°, 4.5 h[b] 2. SiO_2[c]	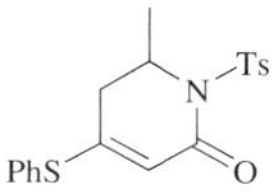	
			(62)	218, 217
			(81)[d]	218
	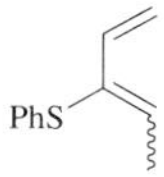	1. Hydroquinone, toluene, 150° 2. SiO_2[c]	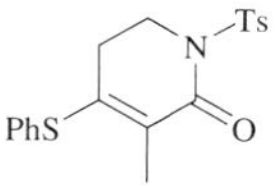 (53)	218
	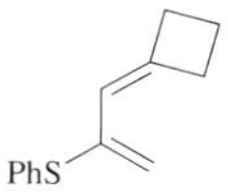	1. $NaHCO_3$, toluene, 110°[b] 2. SiO_2[c]	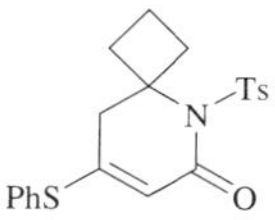 (58)	218
	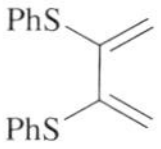	1. Hydroquinone, toluene[b] 2. SiO_2[c]	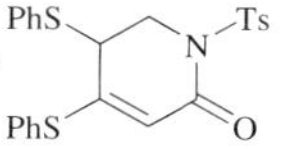 I + 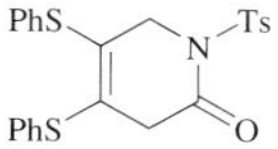 II	
		Temp / Time		
		130°[d,e,f] / 8 h	**I** (34) + **II** (21)	217
		130°[g] / 8 h	**I** (60)	217
		150°[g] / 5 h	**I** (42)	217
		130°[g] / 12 h	**I** (55)	217
		130°[f] / 8 h	**I** (72)	218, 217
		130°[d,f] / 8 h	**I** (67)	218

	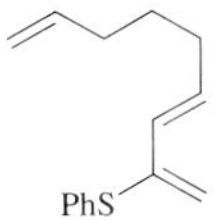	1. Hydroquinone, $NaHCO_3$, toluene, 110°, 4.5 h[b] 2. SiO_2[c]	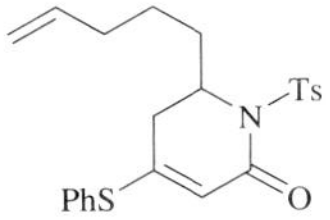	(70)	218, 217
		Hydroquinone, 18-20°, 6 months	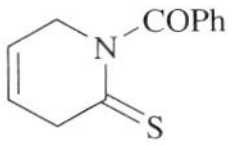	(45.3)	219
	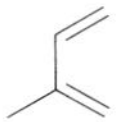	Hydroquinone, 18-20°, 7 months	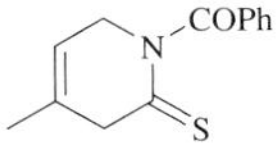 or 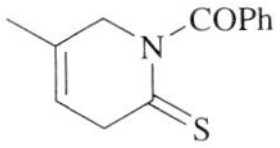	(44.5)[a]	219
		Hydroquinone, 18-20°, 6 months	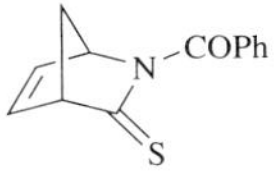	(50.6)	219
	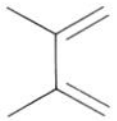	Hydroquinone, 18-20°, 6 months	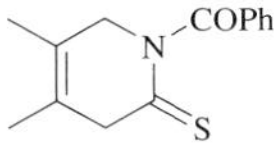	(46)	219

[a] The authors imply that only one regioisomer was obtained, but its identity is not specified.

[b] The diene was generated in situ from the corresponding 3-sulfolene.

[c] The crude product was purified by silica gel chromatography eluting with hexane/ethyl acetate/triethylamine unless otherwise noted.

[d] One equivalent of $NaHCO_3$ was added to the reaction mixture.

[e] Triethylamine was not added to the eluent for the chromatography.

[f] Five equivalents of toluenesulfonyl isocyanate were used.

[g] Three equivalents of toluenesulfonyl isocyanate were used.

TABLE 10. CYCLOADDITIONS OF KETENIMINES AND 2-AZAALLENES

Imine	Diene	Conditions	Product(s) and Yield(s) (%)	Refs.
C_6 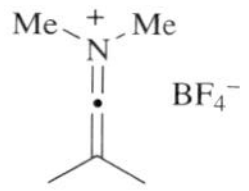		CH_2Cl_2, rt	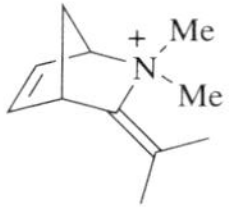 BF_4^- (85)	220
	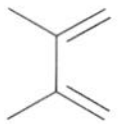	1. CH_2Cl_2, 20° 2. H_2O, 100°, 8 h	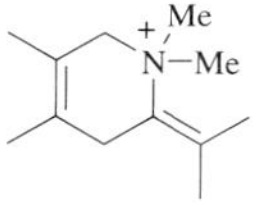 BF_4^- (46)[a]	220
		CH_2Cl_2, rt	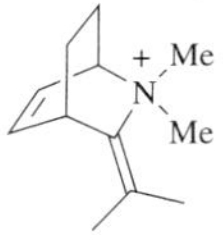 BF_4^- (75)	220
C_{14} 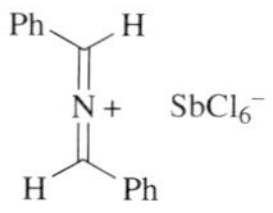	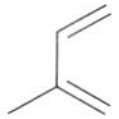	–78° to rt	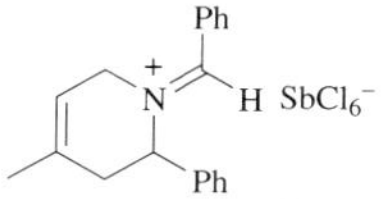 + 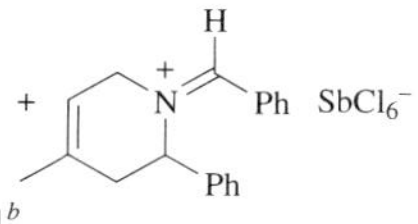 (11) 7.5:1[b]	221
	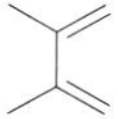	–78° to rt	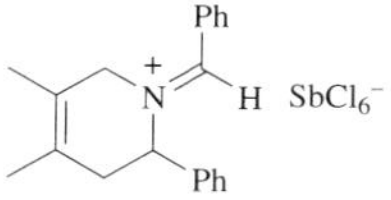 (8)[c]	221
	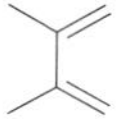	1. –78° to rt 2. $NaHCO_3$, H_2O	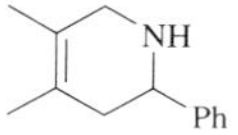 + 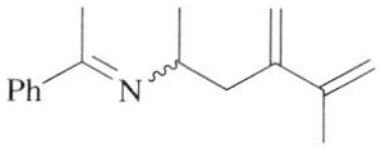 (16) 1:1[d]	221
		–78° to rt	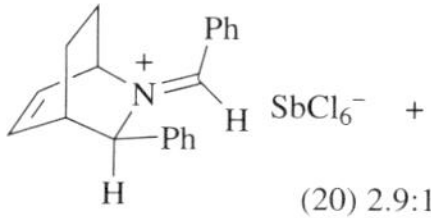 + 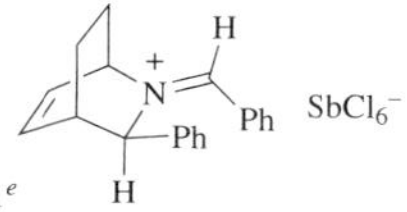 (20) 2.9:1[e]	221

		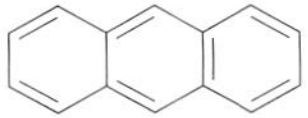	−78° to rt	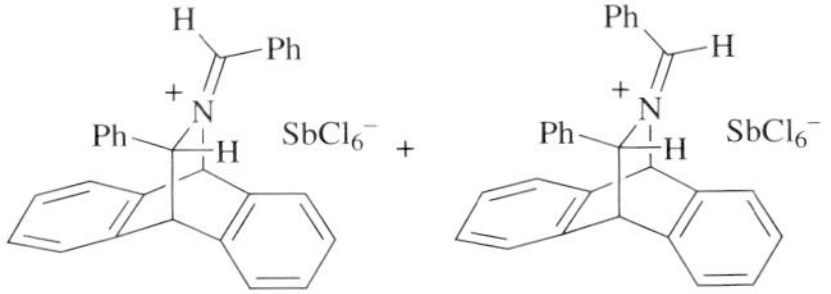 (22) 1.3:1[e,f]	221
C_{16}	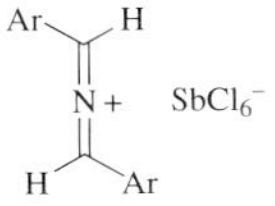 Ar = 4-MeC_6H_4	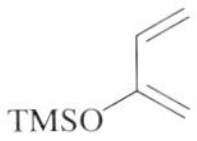	−78° to rt	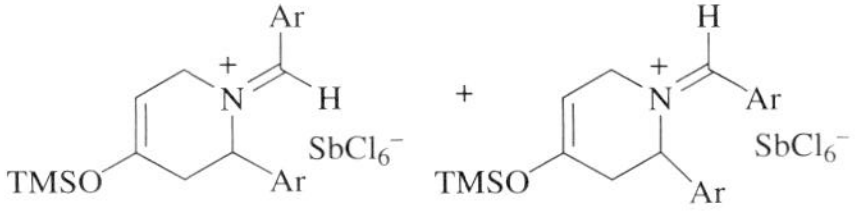 (—) 9:1	221

[a] The cyclobutanone resulting from hydrolysis of the intermediate [2+2]-cycloadduct was obtained in 20% yield.

[b] The crude yield of the cycloadducts was 49%.

[c] The crude yield of the cycloadduct was 33%.

[d] The crude yield of the mixture was 42%.

[e] Although a ratio is provided, the structure of the major isomer is not specified.

[f] The crude yield of the cycloadducts was 51%.

Table 11. Intramolecular Cycloadditions

	Imine	Conditions	Product(s) and Yield(s) (%)	Refs.
C_8	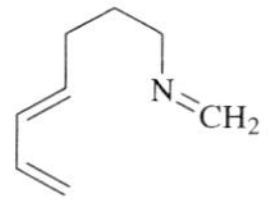	HCl, H_2O, 50°, 48 h	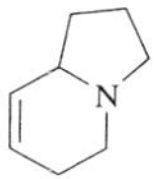 (95)	133
	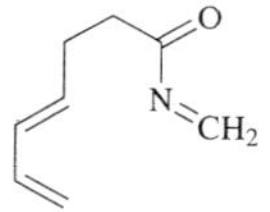		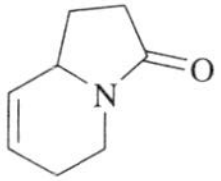	
		Toluene, 370-390°, 2.75 h	(73)	223, 222
		450°, 10^{-5} torr	(—)[a]	461
C_9	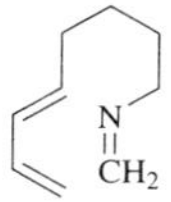	HCl, H_2O, 50°, 48 h	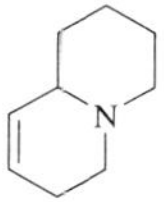 (65)	133, 259
	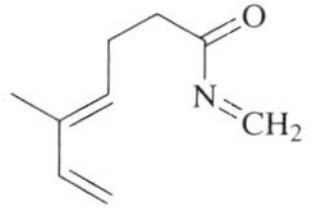	800°	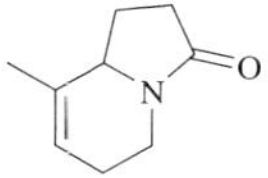 (<20)[b]	462
	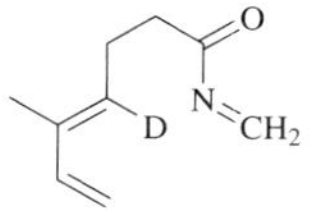	800°	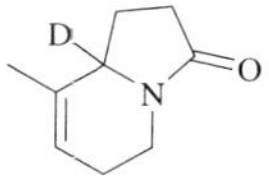 (—)[b]	462
	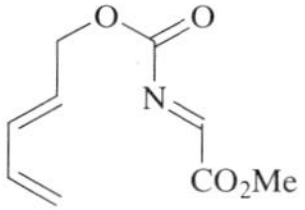	Toluene, 215°, 2 h	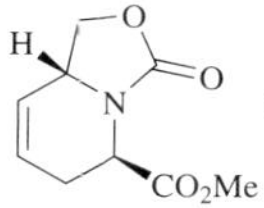 (30)	229, 231

C_{10} 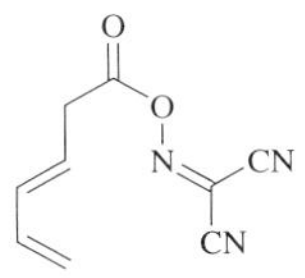	Toluene, 110°, 48 h	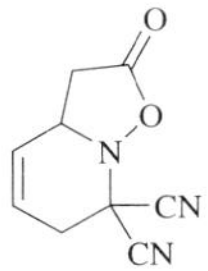 (67)	239
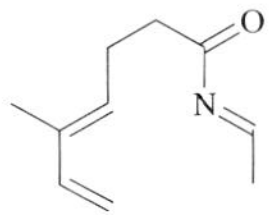	800°	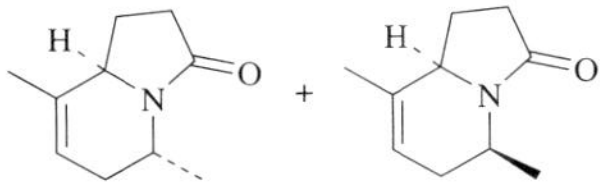 (<20) 2:1[b]	462
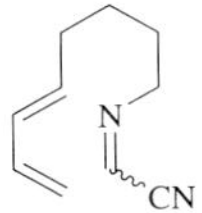	BHT, toluene, 120°, 18 h	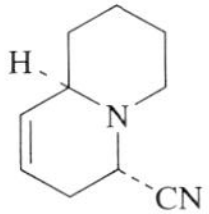 (67-70)	232
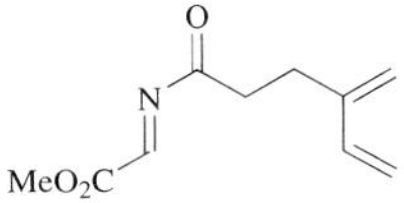	Xylenes, 2.5 min	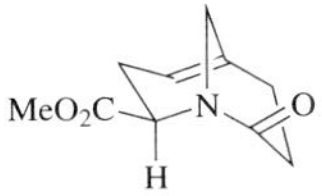	
	Temp		
	250°	(35)	225
	252°	(29)	226
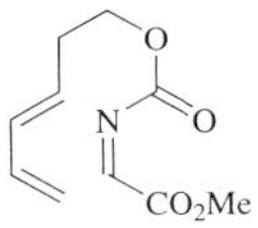	Toluene, 210°, 2 h	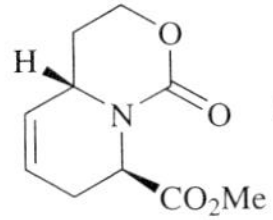 (80)	229
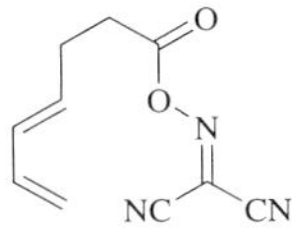		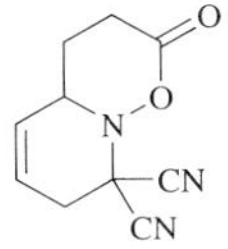	239
	Toluene, 110°, 24 h	(68)	
	CH_2Cl_2, rt, 48 h, 12 kbar	(65)	

TABLE 11. INTRAMOLECULAR CYCLOADDITIONS (*Continued*)

Imine	Conditions	Product(s) and Yield(s) (%)	Refs.
C_{10} 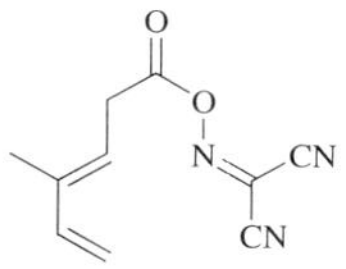	Toluene, 110°, 48 h	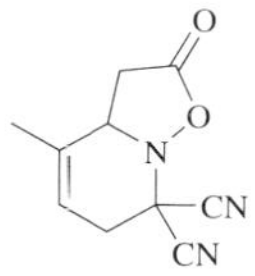 (62)	239
C_{11} 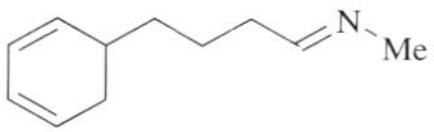	HCl, H_2O, EtOH, 70°, 39 h	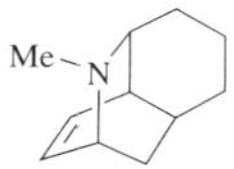 (66)[c]	234, 235
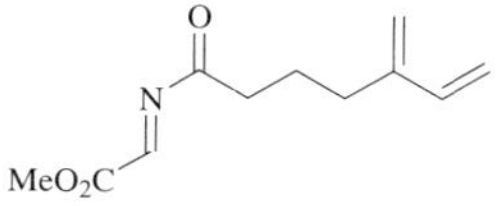	Xylenes, 200°, 2 h	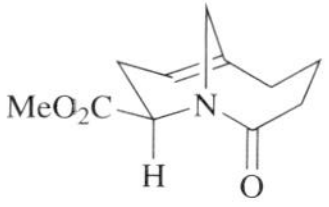 (82)	226
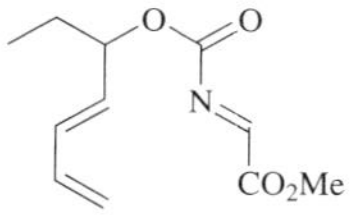		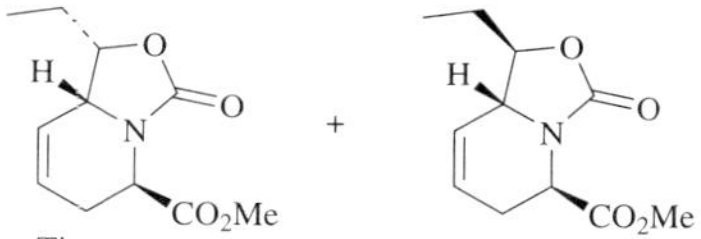	

Solvent	Temp	Time	Product(s) and Yield(s) (%)	Refs.
1,2-dichlorobenzene	550°	—	(trace)[d]	229
xylene	500°	—	(trace)[d]	229
bromobenzene	156°	12 h	(trace)[e]	229
bromobenzene	500°	—	(21) 55:45[e]	229
bromobenzene	260°	—	(40) 55:45[e]	229
bromobenzene	230-240°	2.5 h	(46) 55:45[e]	229, 231
bromobenzene	230-240°	2.5 h	(59)[f]	231
toluene	210°	2 h	(59) 55:45[f]	229
toluene	210°	22.5 h	(18) 55:45[g]	229

C_{12}	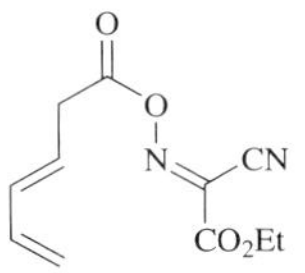	Toluene, 110°, 48 h	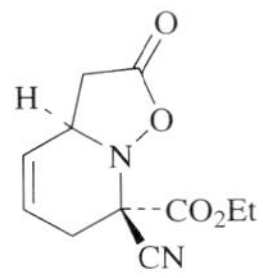 (63)	239
	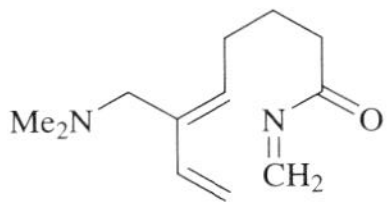	$NaHCO_3$[h], toluene, 200°, 2 h	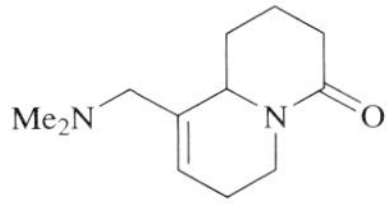 (80)	463
	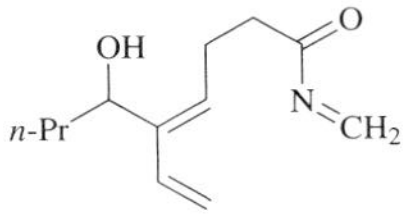	—[i]	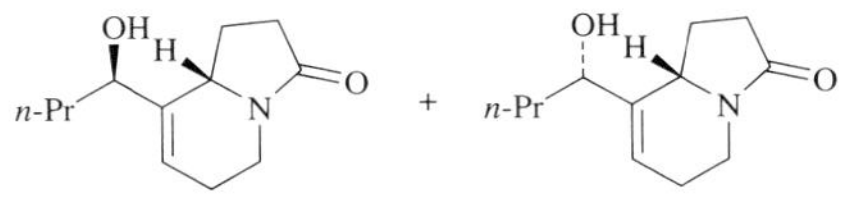 ("low")	258
	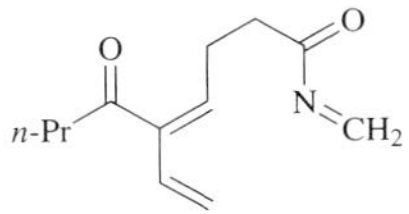	—[i]	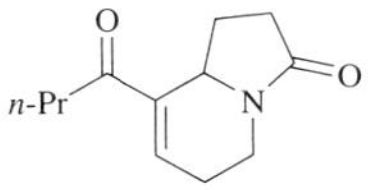 ("poor")	258, 223
	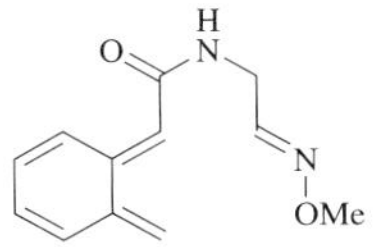	Bromobenzene, 155°, 20 h[j]	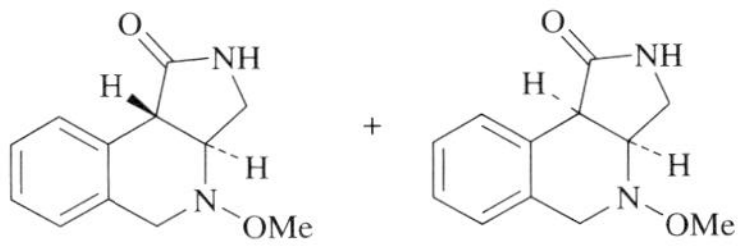 (—)	238
	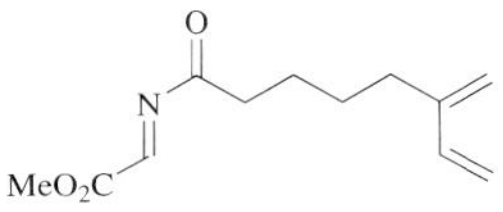	Xylenes, 215°, 2 h	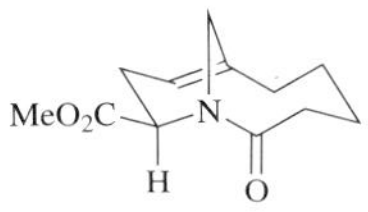 (76)	226

TABLE 11. INTRAMOLECULAR CYCLOADDITIONS (*Continued*)

Imine	Conditions	Product(s) and Yield(s) (%)	Refs.
C_{12} 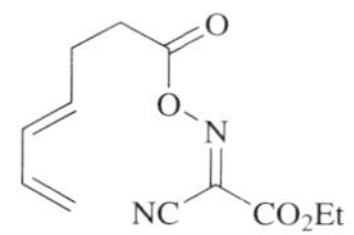		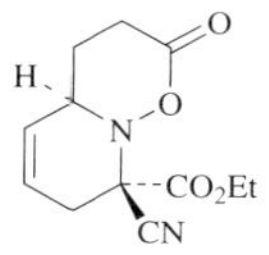	239
	Toluene, 110°, 24 h	(70)	
	CH_2Cl_2, rt, 48 h, 12 kbar	(76)	
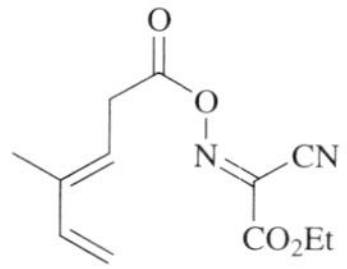	Toluene, 110°, 48 h	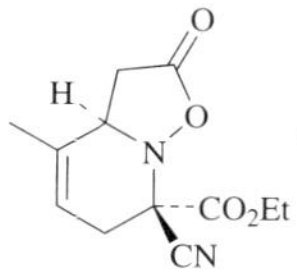 (67)	239
C_{13} 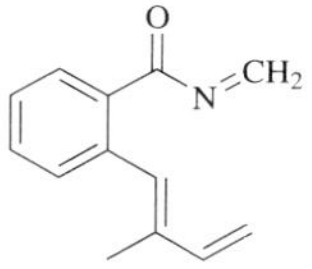	800°	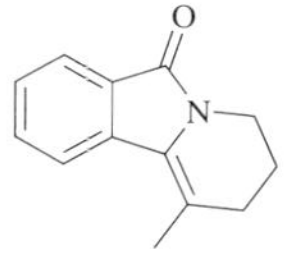 (<20)[b]	462
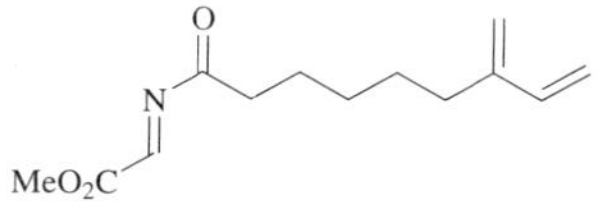	Xylenes	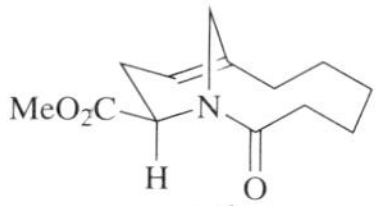	226
	Temp Time		
	307° 5 min	(9)[k]	
	200° 1-2 h	(9)[k]	
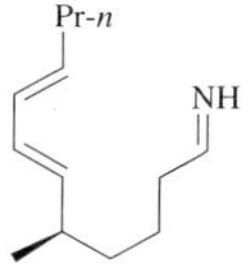	HCl, H_2O, EtOH, 75°, 48 h	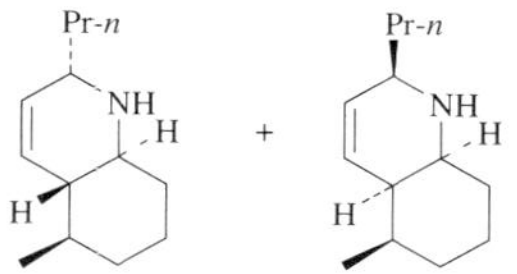 (55) 2.2:1	235

C14

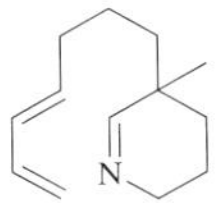	1. TFA, ether, –78° to rt, 1 h 2. H_2O, 95°, 38 h	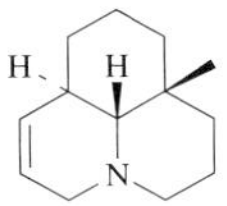 (55)	236
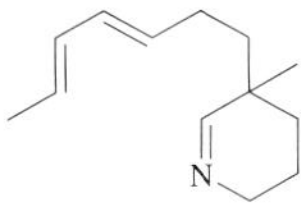	TFA, $LiClO_4$, ether, rt, 74 h	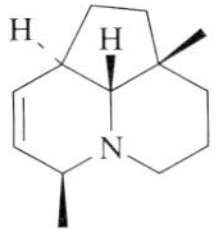 (26)[l]	236
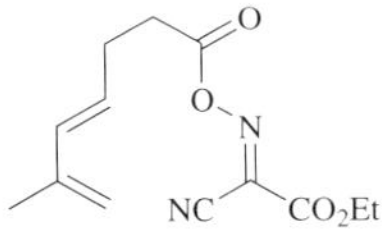	Toluene, 110°, 24 h	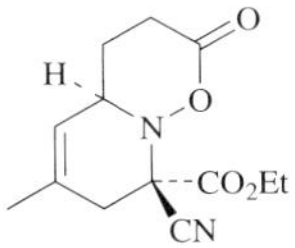 (35)	239
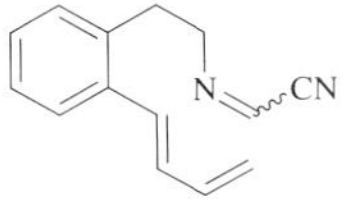	BHT, toluene, 120°, 36 h	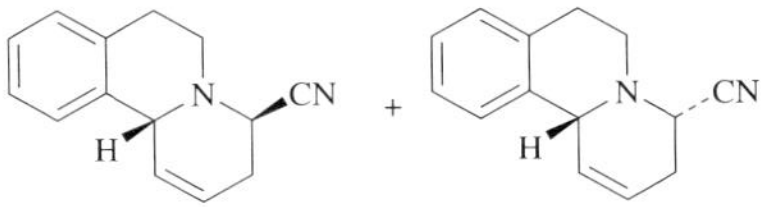 (44-45) 79:21	232
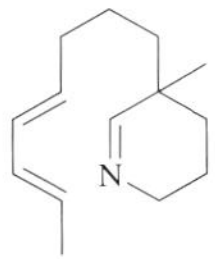	1. TFA, ether, –78°, 30 min 2. H_2O, 70°, 48 h	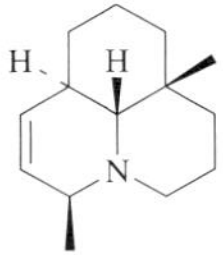 (80)	236, 261
	1. TFA, ether, –78°, 30 min 2. $LiClO_4$, ether, rt, 66 h	(13)	236

TABLE 11. INTRAMOLECULAR CYCLOADDITIONS (*Continued*)

	Imine	Conditions	Product(s) and Yield(s) (%)	Refs.
C_{15}	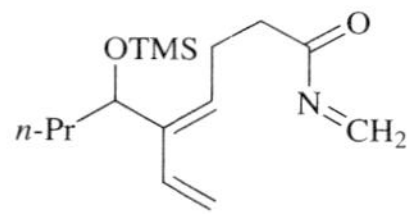	Toluene[h] Temp / Time 370-390° / — 370° / 2 h	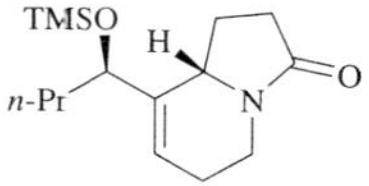 + 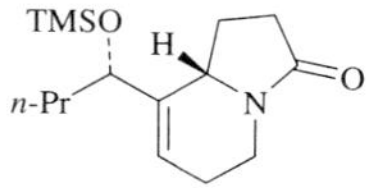 (68) (54) 5:4[m]	 258 223
	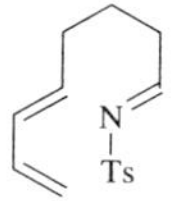	BF_3•OEt_2, benzene, 5°	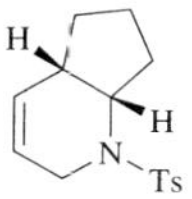 (71)	33
	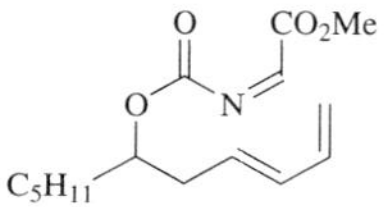	*i*-Pr_2NEt, toluene, 215°, 3 h	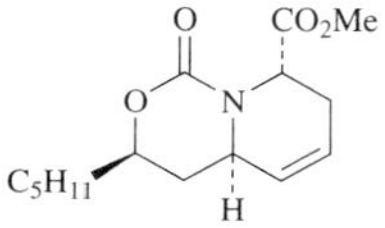 (83)	464
C_{16}	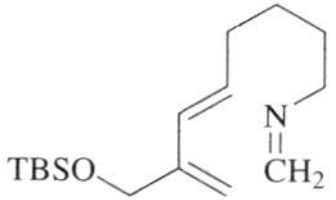	HCl, H_2O, 65°, 24 h	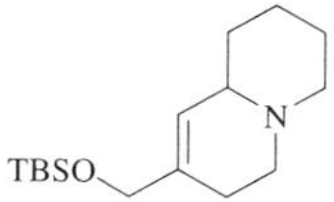 (63-66)	232
	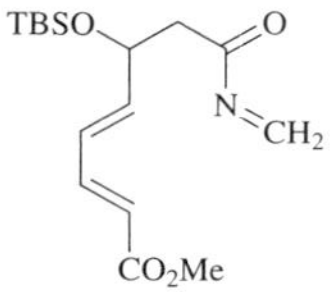	1,2-Dichlorobenzene, 180°, 4 h	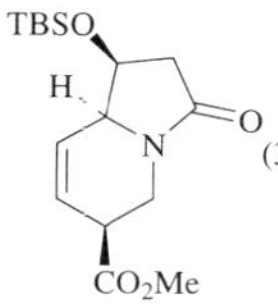 (33) + 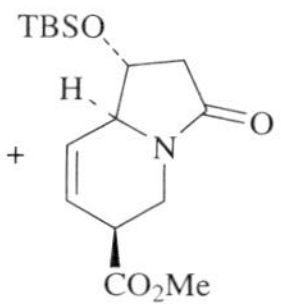 (57)	230
	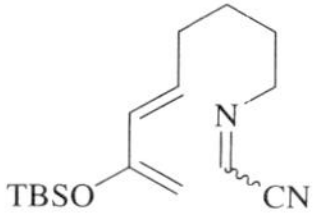	BHT, toluene, 120°, 25 h	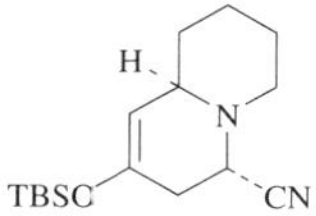 (71-78)	232

C_{17} 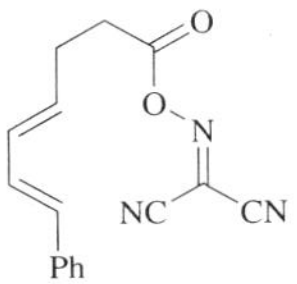	Benzene, 80°, 20 h	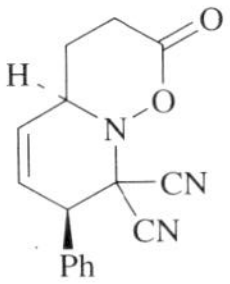 (52)	239
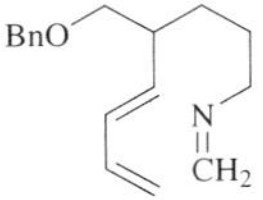	HCl, H_2O, 65°, 28 h	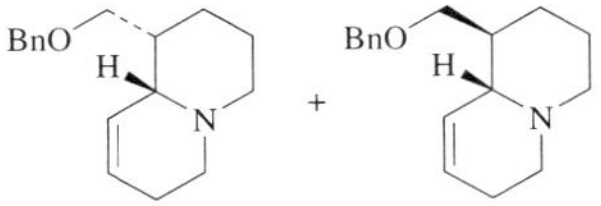 (82) 1:1.6	259
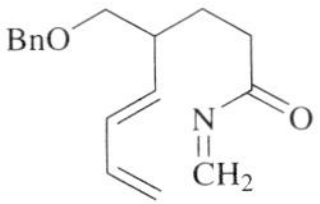	1,2-Dichlorobenzene,180°	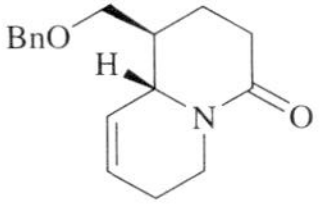	
	Time		
	6 h	(93)	228
	2 h	(93)	227
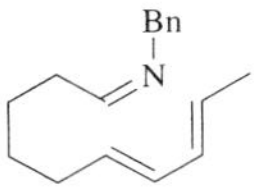	HCl, H_2O, EtOH, 70°, 20 h	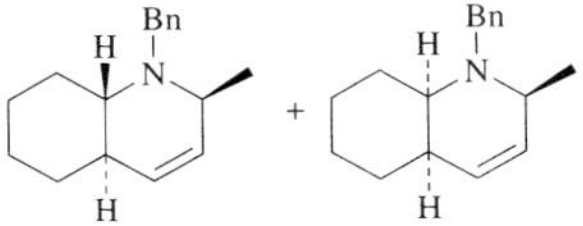 (63) 2.5:1	133, 235
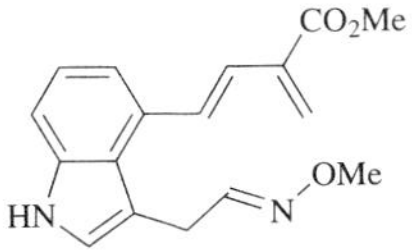	1,2,4-Trichlorobenzene, 200°, 5 h[n]	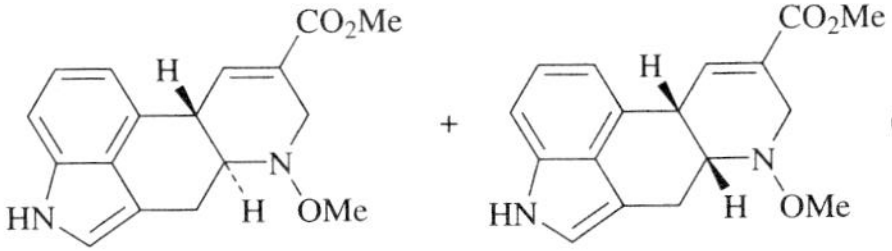 (67) 3:2[o]	262
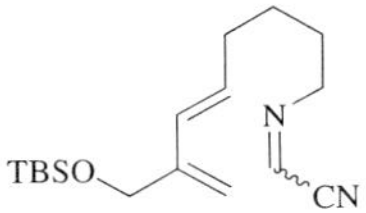	BHT, toluene, 120°, 14 h	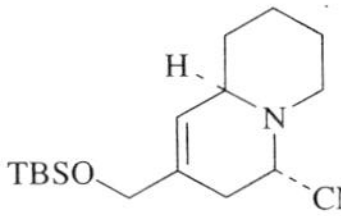 (79-87)	232

TABLE 11. INTRAMOLECULAR CYCLOADDITIONS (*Continued*)

	Imine	Conditions	Product(s) and Yield(s) (%)	Refs.
	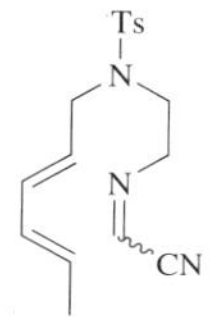	BHT, CH_3CN, 81°, 18 h	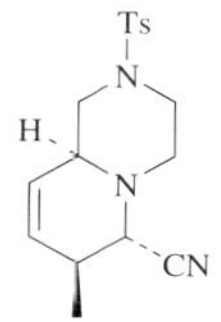 (61-64)	232
C_{18}	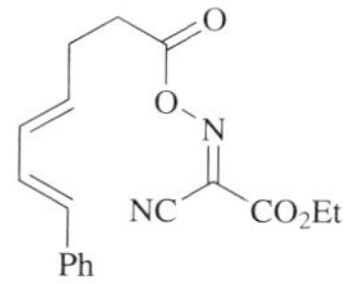	Toluene, 110°, 38 h CH_2Cl_2, rt, 40 h, 12 kbar	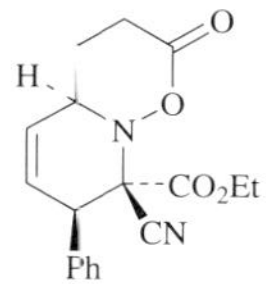 (74) (63)	239
C_{19}	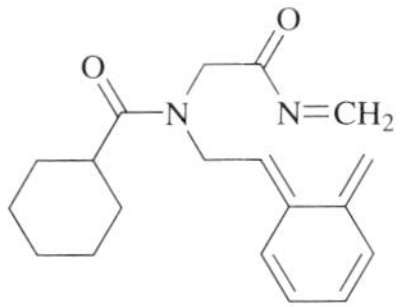	400-425°, 8 h[j]	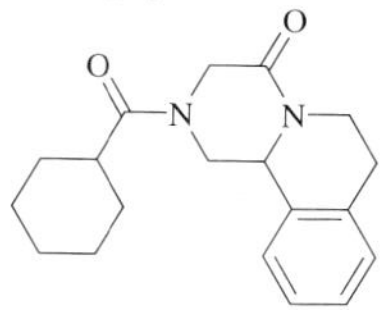 (49)	224
	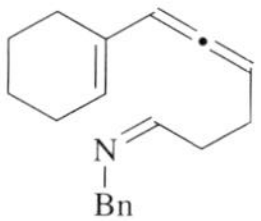	$BF_3{\bullet}OEt_2$, CH_2Cl_2, –78° to rt, 8 h	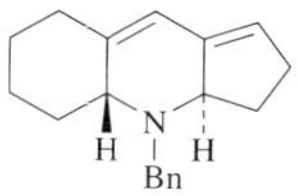 (41)	237
	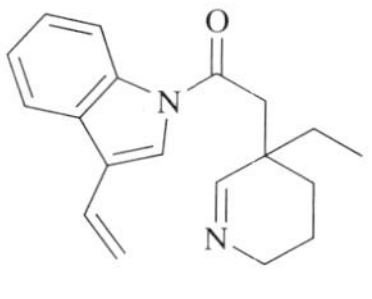	Acid, benzene	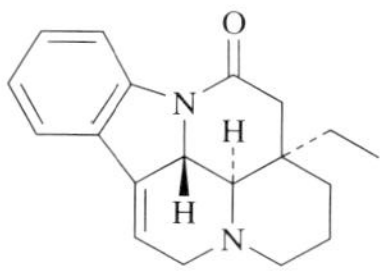 **I** + 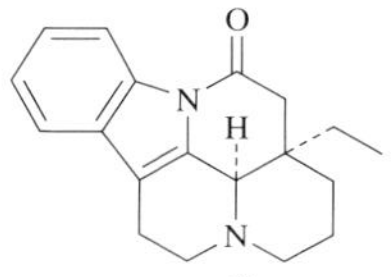 **II**	

Acid (eq)	Temp	Time	I	II	Refs.
TFA (1.0)	50°	46 h	(60)	(0)	261
TFA (1.1)	80°	3 h	(70)	(5)	261, 260
CSA (1.0)	80°	5 h	(65)	(7)	261
TFA (1.1)	80°	40 h	(20)	(60)	261
TFA (0.1)	80°	48 h	(45)	(23)	261
CSA (0.1)	80°	48 h	(56)	(22)	261

Conditions	Product(s) and Yield(s) (%)	Refs.
1,2-Dichlorobenzene, 180°, 22 h	**I** (32)	261, 260
Florisil, EtOAc, 50°, 9 h	**I** (82)	261, 260
Florisil, EtOAc, 40 h	**I** (47)	261
$LiClO_4$, CSA, ether, rt, 60 h	**I** (96)	261, 260
$LiClO_4$, ether, rt, 40 h	**I** (70)	261
$LiClO_4$, CH_2Cl_2, 14 d	**I** (—)	261
Lithium salt, benzene, 80°	**I**	261

Lithium salt	Time	I
$LiClO_4$	9 h	(81)
$LiPF_6$	5 h	(62)
$LiBF_4$	64 h	(60)
$LiCo(B_9C_2H_{11})_2$	2 h	(80)

C_{20}

N
Bn

CH_2Cl_2, 0° to rt, 12 h

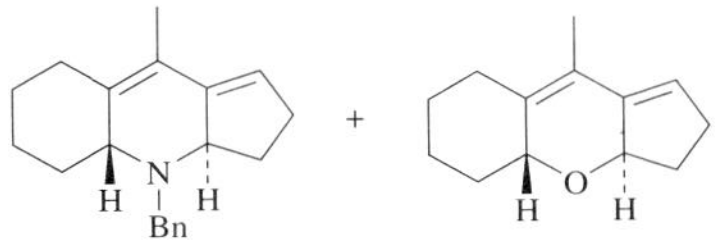

H
N
H
Bn
+
H
O
H

I

(70) 1:1.5[p] — 237

CH_2Cl_2, 0° to rt, 12 h — **I** (69)[q] — 237

TABLE 11. INTRAMOLECULAR CYCLOADDITIONS (*Continued*)

	Imine	Conditions	Product(s) and Yield(s) (%)	Refs.
C_{21}	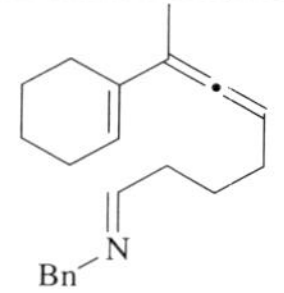	$BF_3{\bullet}OEt_2$, CH_2Cl_2, –78° to rt, 8 h	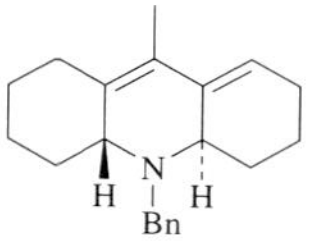 (54)	237
C_{23}	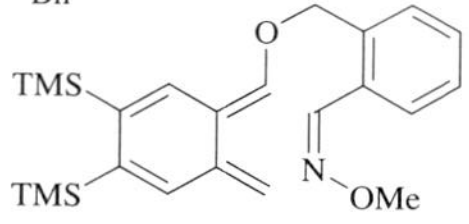	$CpCo(CO)_2$, bis(trimethylsilyl)-acetylene, 136°, 117 h[r]	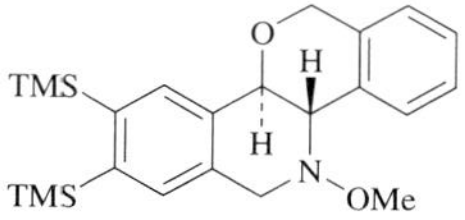 (45)	465, 466
C_{24}	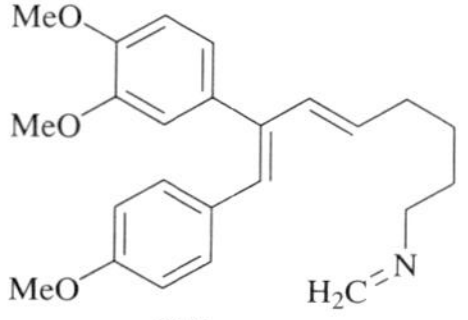	HCl, H_2O, EtOH, 110°, 23 h	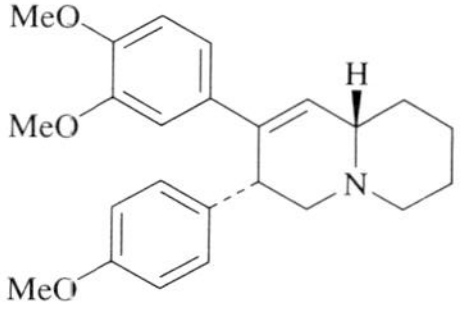 (60)	259
	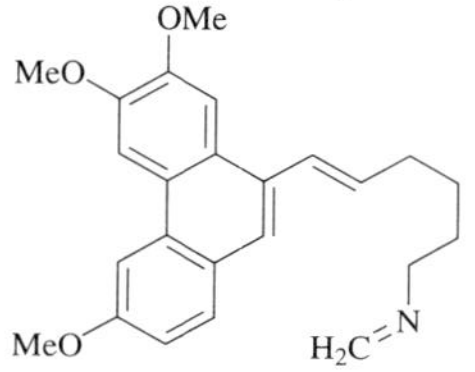	HCl, H_2O, EtOH, 180°, 10 h	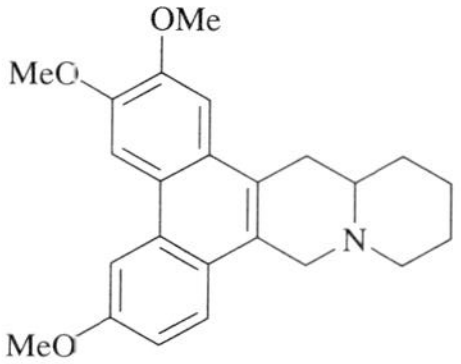 (84)	259
	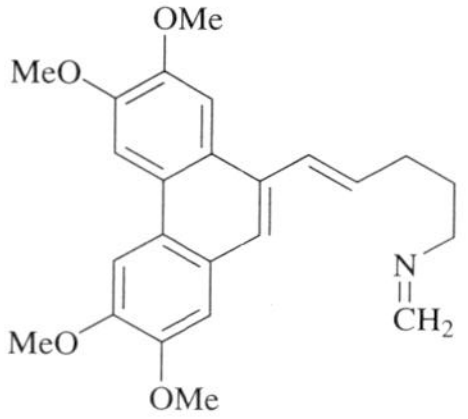	Camphorsulfonic acid, 1,2-dichlorobenzene, 155-160°, 5 h[s]	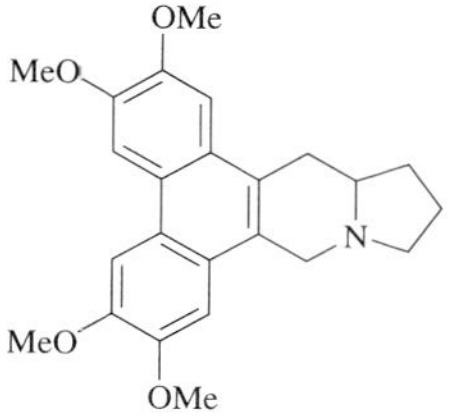 (35)	233

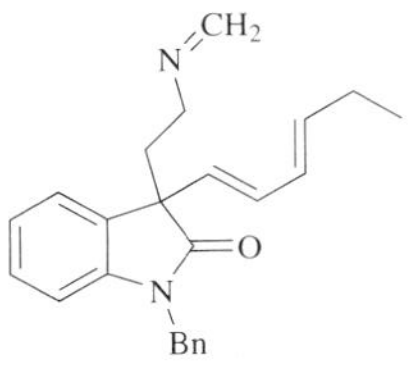	BF_3•OEt_2, toluene, 100°, 2 h[s]	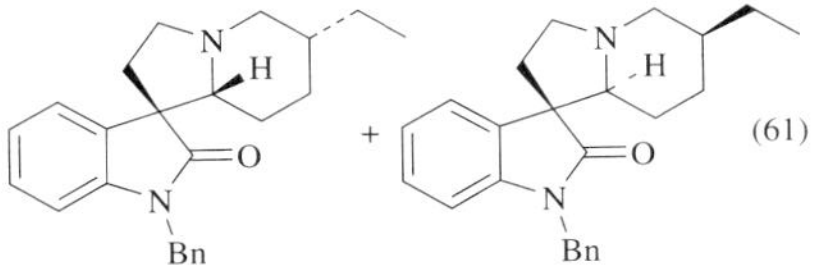 (61) 1.5:1	467
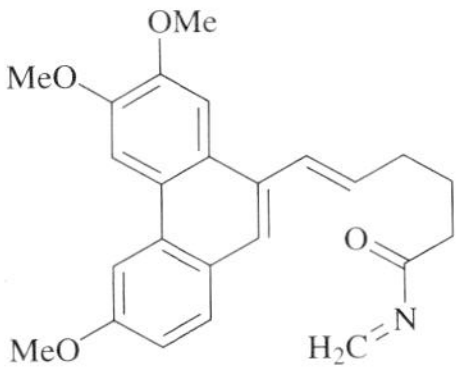	1,2-Dichlorobenzene, 210-212°, 70 min	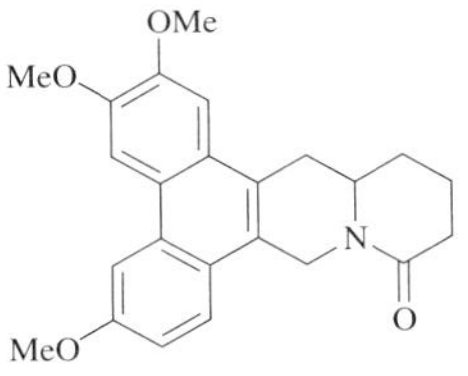 (66)	227
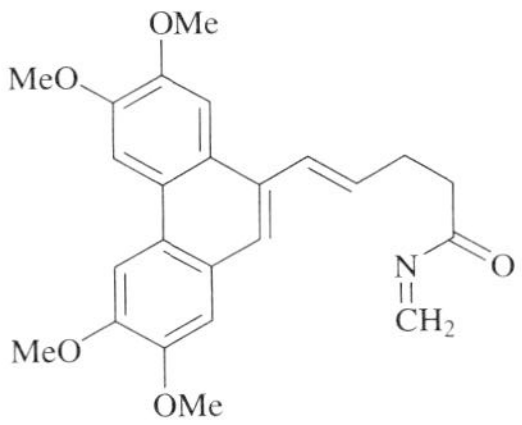	Bromobenzene, 219-220°, 5 h	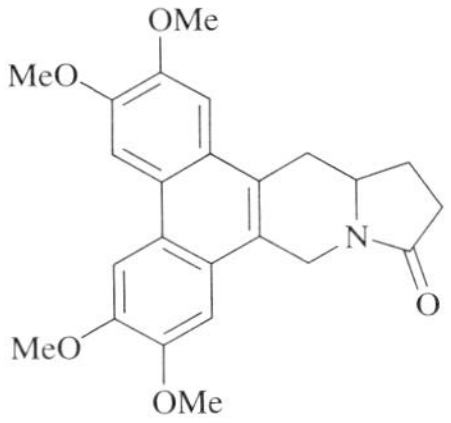 (50)	223, 222
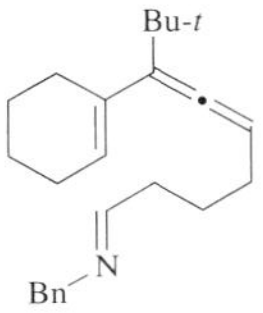	BF_3•OEt_2, CH_2Cl_2, −78° to rt, 8 h	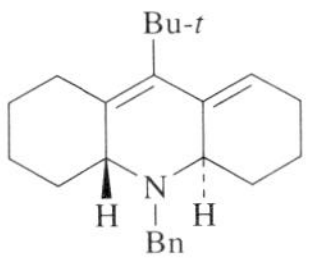 (88)	237

TABLE 11. INTRAMOLECULAR CYCLOADDITIONS (*Continued*)

Imine	Conditions	Product(s) and Yield(s) (%)	Refs.

[a] The reactive imine was generated by the retro-imino-Diels-Alder reaction of the corresponding 2-azabicyclo[2.2.1]hept-5-ene.

[b] The reactive imine was generated in situ by the retro-ene reaction of the corresponding δ-lactam.

[c] Rapid addition of the aldehyde to the methylamine hydrochloride in water or aqueous ethanol led to poor yields. The yield shown was obtained by adding the aldehyde over 29 hours.

[d] The acylimine was generated in situ by elimination of methanol from the methyl 2-methoxy-2-aminoacetate.

[e] The acylimine was generated in situ by elimination of acetic acid from the methyl 2-acetoxy-2-aminoacetate.

[f] The acylimine was generated in situ from the methyl 2-phenylcarbamoyloxy-2-aminoacetate.

[g] The acylimine was generated in situ from the methyl 2-methylcarbamoyloxy-2-aminoacetate.

[h] The diene was generated in situ from the corresponding sulfolene.

[i] The diene was generated in situ by pyrolysis of the corresponding sulfolene, but the conditions are not specified.

[j] The diene was generated in situ from the corresponding benzocyclobutene.

[k] A 10-membered macrocyclic lactam that resulted from an intramolecular ene reaction was isolated in 25% yield.

[l] The reactive diene was formed in situ by acid-catalyzed rearrangement of 5-(hepta-4(*E*),6-dienyl)-5-methyl-2,3,4,5-tetrahydropyridine.

[m] Although a ratio of isomers is given, the stereochemistry of the major isomer is not specified.

[n] The reactive diene was formed in situ via the retro-Diels-Alder reaction of the cyclopentadiene adduct.

[o] Although a ratio of isomers is given, the stereochemistry of the major isomer is not specified. In addition to the two products shown, compounds resulting from isomerization of the double bond and dimers of the Diels-Alder precursor were obtained.

[p] The imine was formed in situ from the corresponding aldehyde, which was obtained via Swern oxidation. The aldehyde was used without purification, but after quenching the oxidation. The side product from Diels-Alder cyclization of the aldehyde is readily formed and was obtained along with the desired product.

[q] Formation of the oxa-Diels-Alder product was avoided by forming the imine directly after completion of the Swern oxidation (addition of triethylamine), without work-up or purification of the aldehyde precursor.

[r] The reactive diene was obtained via an intermediate benzocyclobutene, which was formed in situ from bis(trimethylsilyl)acetylene and the corresponding substituted 1,5-hexadiyne.

[s] The imine was formed in situ via retro-Diels-Alder reaction of the corresponding azanorbornene.

TABLE 12. CYCLOADDITIONS WITH ASYMMETRIC CATALYSIS

Refer to Chart 1 at the beginning of the Tabular Survey (p. 224) for catalyst (bold numbers) structures.

Imine	Diene	Conditions	Product(s) and Yield(s) (%)	Refs.
C_7 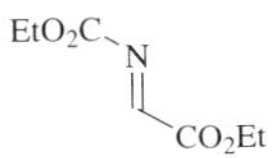	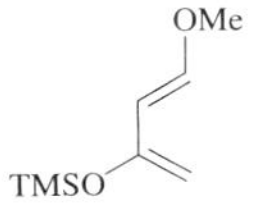	1. **1b**, $CuClO_4$•4MeCN, –78° 2. TFA, CH_2Cl_2, –78° to rt, 30 min Solvent / Time THF / 20 h CH_2Cl_2 / 18 h	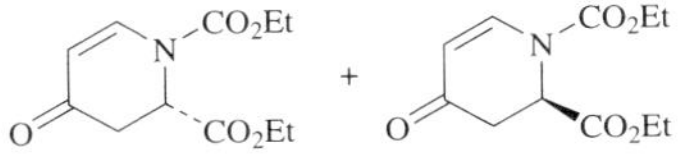 (10) 89.5:10.5[a] (23) 88.5:11.5[b]	252
C_9 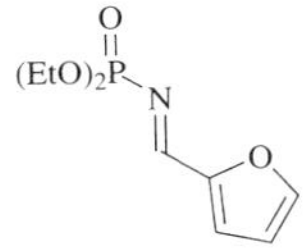	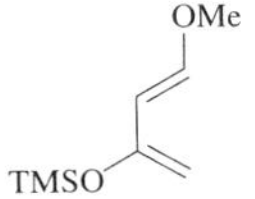	1. Et_2Zn, (*S*)-1,1'-bi-2-naphthol, CH_2Cl_2, 36 h, rt 2. TFA, CH_2Cl_2	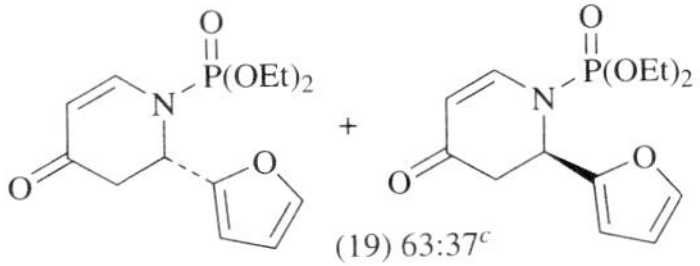 (19) 63:37[c]	129
C_{10} 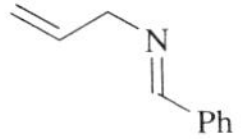	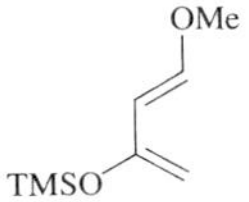	1. $B(OPh)_3$, (*R*)-1,1'-bi-2-naphthol, 4 Å MS, CH_2Cl_2, –78°, 5 h 2. H_2O	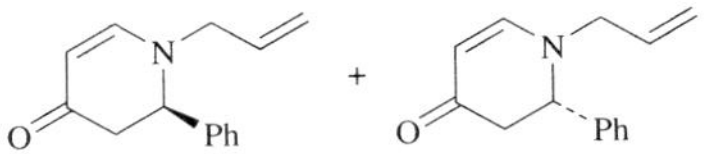 (97) 85:15	243
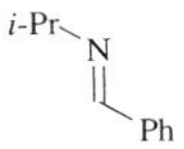	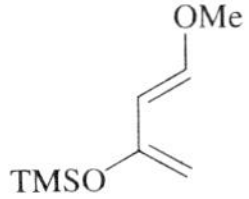	1. $B(OPh)_3$, (*R*)-1,1'-bi-2-naphthol, 4 Å MS, CH_2Cl_2, –78°, 5 h 2. H_2O	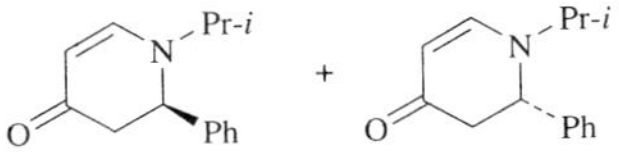 (13) 52:48	243
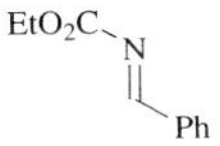	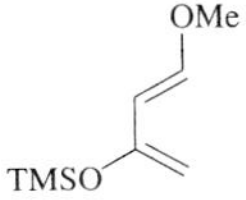	—[d]	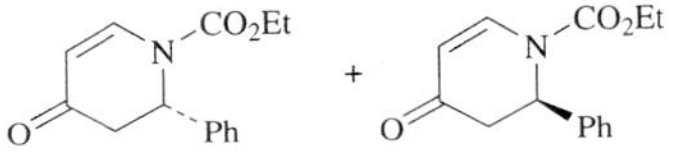 ("good") 1:1	251

TABLE 12. CYCLOADDITIONS WITH ASYMMETRIC CATALYSIS (*Continued*)

	Imine	Diene	Conditions	Product(s) and Yield(s) (%)	Refs.
C_{10}	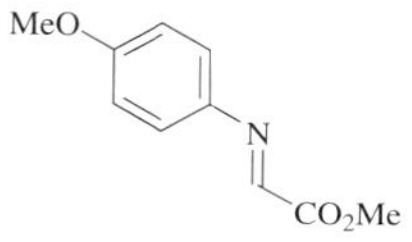	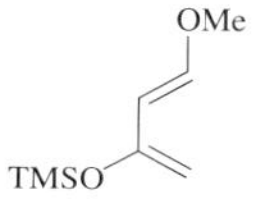	1. (1*R*, 2*R*)-1,2-Diphenylethylenediamine, Lewis acid, 18 h[e] 2. SiO_2	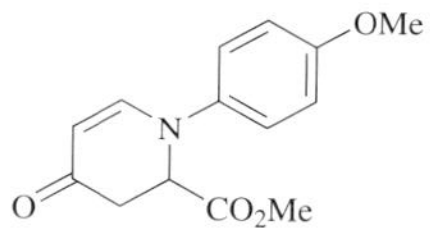	80

L. A.	Solvent	Additive	Product(s) and Yield(s) (%)
$Yb(OTf)_3$	MeCN	2,6-lutidine	(—) 77% ee[f]
$Yb(OTf)_3$	CH_2Cl_2	4 Å MS	(—) 72% ee[f]
$Yb(OTf)_3$	CH_2Cl_2	2,6-lutidine	(—) 81% ee[f]
$Yb(OTf)_3$	toluene	4 Å MS	(—) 64% ee[f]
$Yb(OTf)_3$	toluene	2,6-lutidine	(60) 87% ee[f,g]
$Cu(OTf)_2$	MeCN	4 Å MS	(—) 51% ee[f]
$Cu(OTf)_2$	MeCN	2,6-lutidine	(—) 64% ee[f]
$Cu(OTf)_2$	MeCN	—	(—) 88% ee[f,h]
$Cu(OTf)_2$	MeCN	—	(58) 86% ee[f,g,i]
$Cu(OTf)_2$	CH_2Cl_2	—	(—) 72% ee[f]
$Cu(OTf)_2$	toluene	4 Å MS	(—) 60% ee[f]
$Cu(OTf)_2$	toluene	—	(—) 56% ee[f]
MgI_2	MeCN	4 Å MS	(—) 79% ee[f]
MgI_2	MeCN	2,6-lutidine	(64) 97% ee[f,j]
MgI_2	MeCN	—	(—) 83% ee[f]
MgI_2	CH_2Cl_2	4 Å MS	(—) 65% ee[f]
MgI_2	CH_2Cl_2	2,6-lutidine	(—) 54% ee[f]
MgI_2	CH_2Cl_2	—	(—) 94% ee[f]
MgI_2	toluene	4 Å MS	(—) 90% ee[f]
MgI_2	toluene	2,6-lutidine	(—) 91% ee[f]

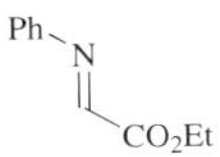

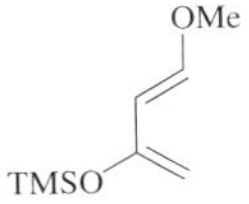

1. **2**, $FeCl_3$, 18 h[e]
2. SiO_2

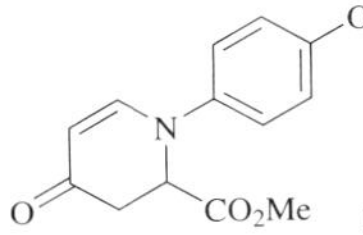

I

Ref. 80

Solvent	Additive	
MeCN	—	(—) 57% ee[f]
CH_2Cl_2	4 Å MS	(67) 92% ee[f,g]
CH_2Cl_2	2,6-lutidine	(—) 70% ee[f]

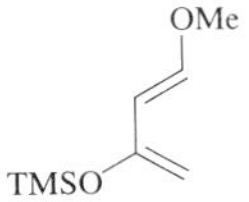

1. (*R*)-1,1'-Bi-2-naphthol, Lewis acid, 18 h[d]
2. SiO_2

I

Ref. 80

L. A.	Solvent	Additive	I
$Yb(OTf)_3$	MeCN	2,6-lutidine	(—) 41% ee[f]
MgI_2	toluene	—	(—) 64% ee[f]

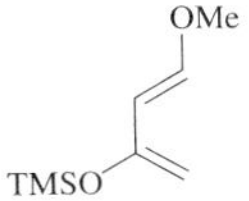

1. **1b**, $CuClO_4$•4MeCN, CH_2Cl_2, rt, 20 h
2. TFA, CH_2Cl_2, –78° to rt, 20 min

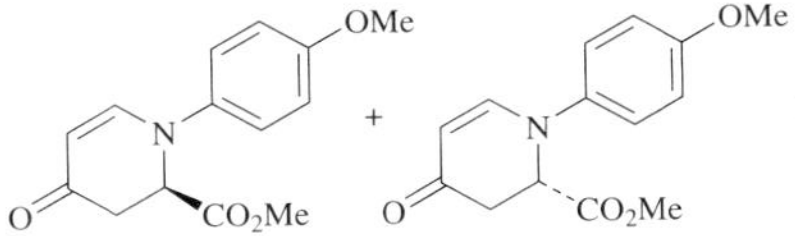

(72) 93.5:6.5

Ref. 253

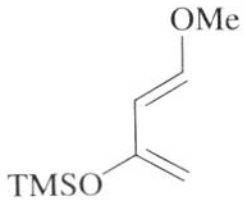

1. **1b**, $CuClO_4$•4MeCN, rt
2. TFA, CH_2Cl_2, –78° to rt, 20 min

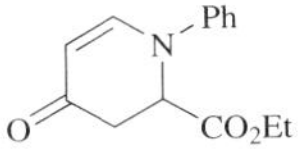

Ref. 252

Solvent	Time	
THF	—	(61) 86% ee[k]
CH_2Cl_2	20 h	(78) 91% ee[k]

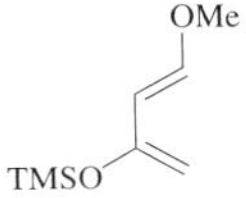

—[d]

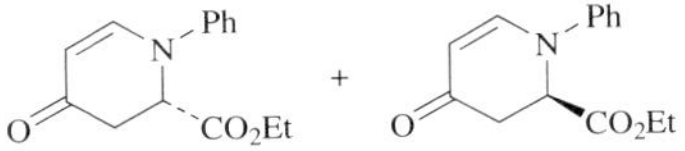

("good") 1:1

Ref. 251

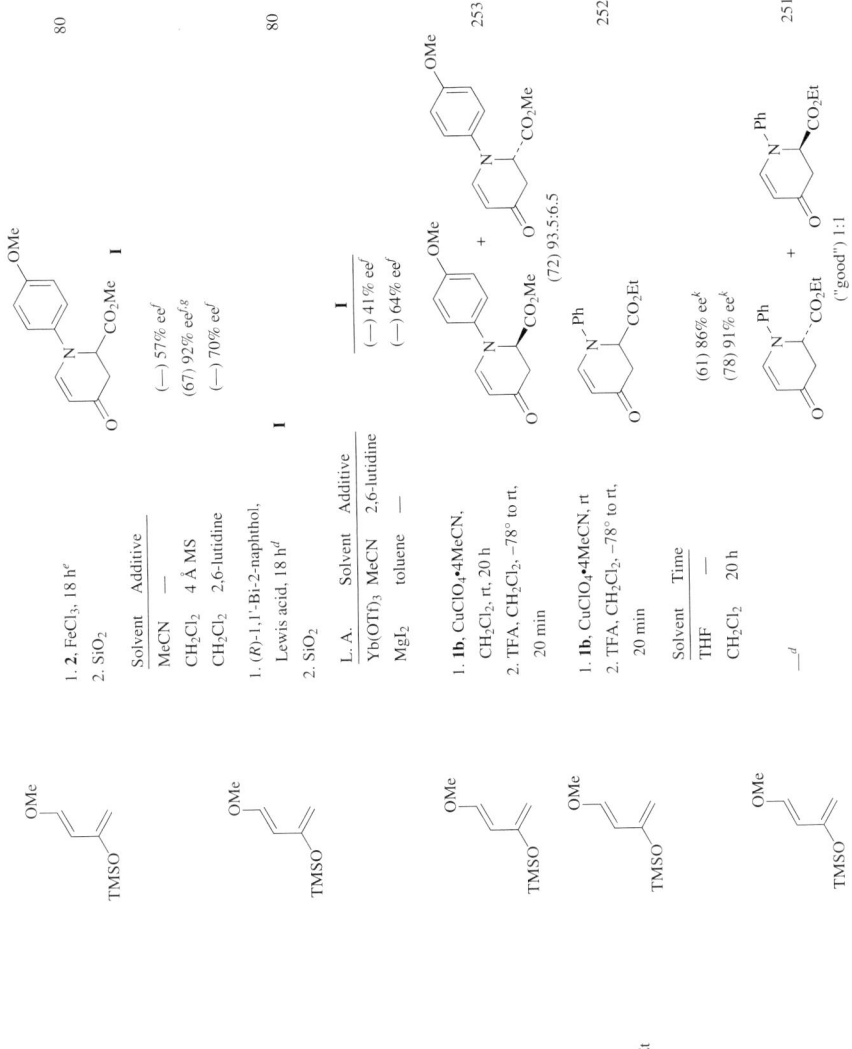

TABLE 12. CYCLOADDITIONS WITH ASYMMETRIC CATALYSIS (*Continued*)

Imine	Diene	Conditions	Product(s) and Yield(s) (%)	Refs.
C_{10} 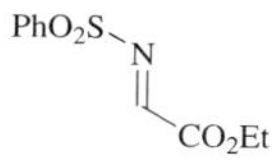		1. DIBALH[l], (*R*)-1,1'-bi-2-naphthol, CH_2Cl_2, –78°, 15 min 2. Cyclopentadiene, –78°, 10 min	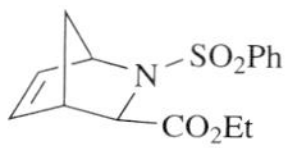 (96)[m] 0% ee	40
		1. NaH[l], (*R*)-1,1'-bi-2-naphthol, Lewis acid, –100°, 30 min 2. Cyclopentadiene, –100°, 15 min	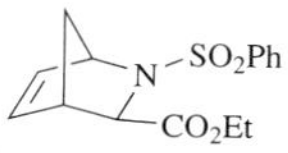	40
		L. A.		
		$TiCl_4$	(86)[m] 0% ee	
		Ti(OPr-*i*)$_2$Cl$_2$	(92)[m] 0% ee	
		Et_2AlCl	(94)[m] 0% ee	
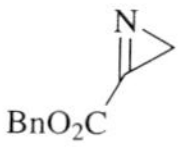			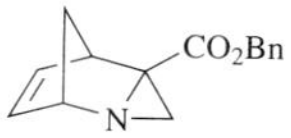	199
		$AlMe_3$, (*S*)-1,1'-bi-2-naphthol, CH_2Cl_2, –35°, 24 h	(41) 51% ee[k]	
		3, $AlMe_3$, CH_2Cl_2, –40°	(27) 35% ee[k]	
		2, $Mg(ClO_4)_2$, 4Å MS, CH_2Cl_2, –40°, 40 min	(22) 32% ee[k]	
		2, $Mg(ClO_4)_2$, 4Å MS, CH_2Cl_2, –60°, 4 h	(25) 52% ee[k]	
		4, $AlMe_3$, CH_2Cl_2, –60°	(22) 12% ee[k]	
		5, $AlMe_3$, CH_2Cl_2, –60°	(20) 19% ee[k]	

C$_{11}$

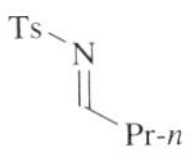	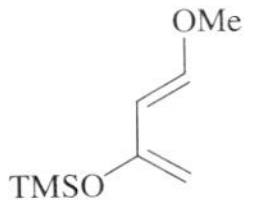	1. **6h**, $AgClO_4$, CH_2Cl_2 2. TFA, CH_2Cl_2, rt, 30-60 min Temp / Time –78° / 5 h –20° / 30 min	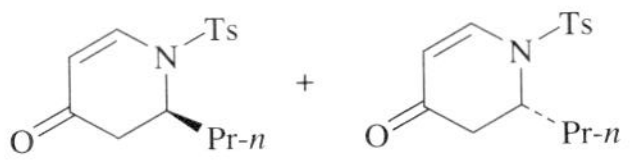 (51) 91:9[n] (65) 86.5:13.5[n]	2
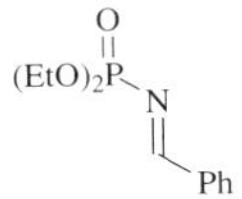	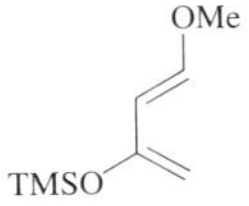	1. Et_2Zn, (*S*)-1,1'-bi-2-naphthol, CH_2Cl_2, 36 h, rt 2. TFA, CH_2Cl_2	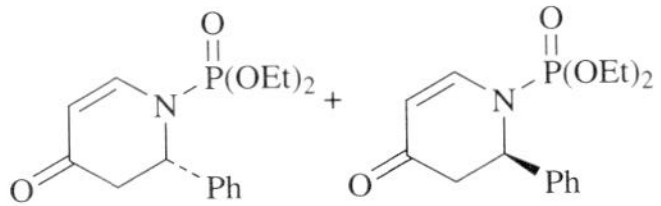 (20) 88.5:11.5[c]	129
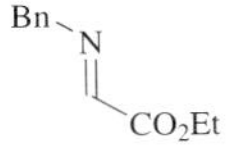	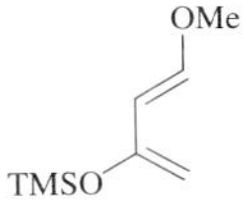	—[d]	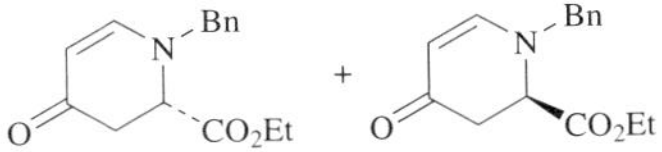 ("good") 1:1	251
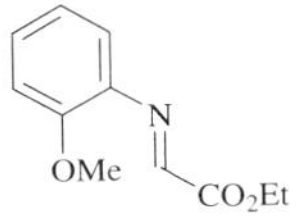	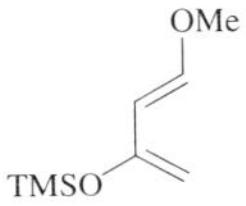	1. **1b**, $CuClO_4$•4MeCN, rt 2. SiO_2 Solvent / Time THF / — CH_2Cl_2 / 40 h	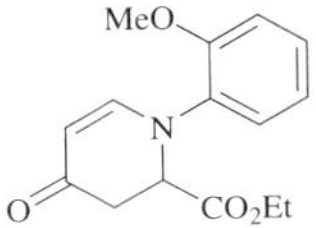 (25) 60% ee[k] (25) 55% ee[k]	252

TABLE 12. CYCLOADDITIONS WITH ASYMMETRIC CATALYSIS (*Continued*)

Imine	Diene	Conditions	Product(s) and Yield(s) (%)	Refs.
C_{11} 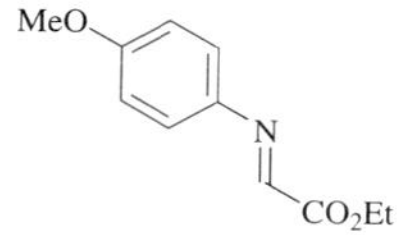	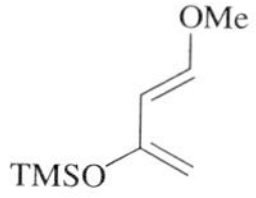	1. (*S*)-1,1'-Bi-2-naphthol, Lewis acid 2. HCl, H_2O	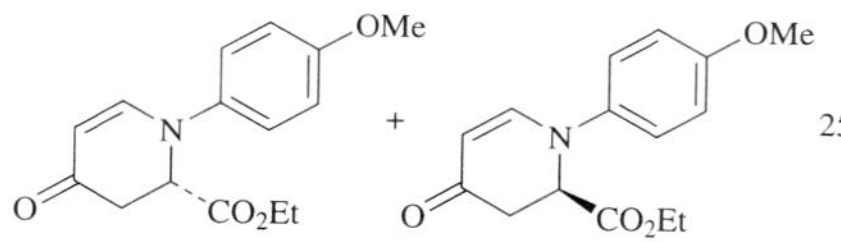	253a

L. A. (eq)	Solvent	Temp	Time	Yield (%) ratio
Et_3Al (0.1)	CH_2Cl_2	rt	—	(65) 57.5:42.5
Et_3B (0.1)	CH_2Cl_2	rt	—	(53) 1:1
(*i*-PrO)$_3$B (0.1)	CH_2Cl_2	rt	—	(56) 1:1
(*i*-PrO)$_4$Ti (0.1)	CH_2Cl_2	rt	—	(<5) 1:1
Et_2Zn (0.1)	CH_2Cl_2	rt	—	(57) 68:32
Et_2Zn (1.0)	CH_2Cl_2	rt	2.5 h	(78) 96.5:3.5
Et_2Zn (0.1)	CH_2Cl_2	rt	15 h	(62) 70:30
Et_2Zn (0.1)	THF	rt	15 h	(61) 56:44
Et_2Zn (0.1)	MeCN	rt	15 h	(67) 62:38
Et_2Zn (0.1)	toluene	rt	15 h	(52) 92:8
Et_2Zn (0.1)	toluene	rt	36 h	(35) 71:29
Et_2Zn (0.1)	toluene	rt	60 h	(56) 58.5:41.5
Et_2Zn (0.1)	toluene	0°	15 h	(47) 84:16
Et_2Zn (0.1)	toluene	–40°	15 h	(45) 61.5:38.5
Et_2Zn (0.1)	toluene	–78°	15 h	(38) 56:44
Et_2Zn (0.1)	CH_2Cl_2	0°	15 h	(50) 68.5:31.5
Et_2Zn (0.1)	CH_2Cl_2	–78°	15 h	(38) 53.5:46.5
Et_2Zn (1.0)	CH_2Cl_2	0°	2.5 h	(72) 96:4
Et_2Zn (1.0)	CH_2Cl_2	–40°	2.5 h	(66) 94:6
Et_2Zn (1.0)	CH_2Cl_2	–78°	2.5 h	(63) 86:14

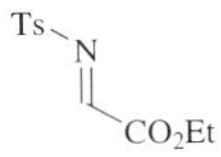

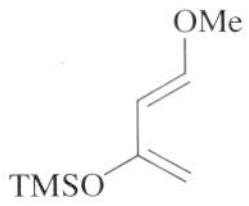

1. **1b**, $CuClO_4$•4MeCN
2. TFA, CH_2Cl_2, –78° to rt, 20 min

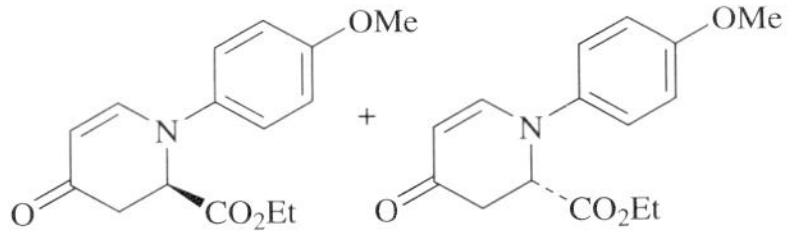

Solvent	Temp	Time	Product(s) and Yield(s) (%)	Refs.
THF	rt	—	(93) 57.5:42.5	252
CH_2Cl_2	rt	20 h	(89) 86:14	252
THF	–78° to rt	—	(82) 58:42	252
CH_2Cl_2	–78° to rt	—	(75) 89:11	252
CH_2Cl_2	rt	20 h	(66) 91.5:8.5	253

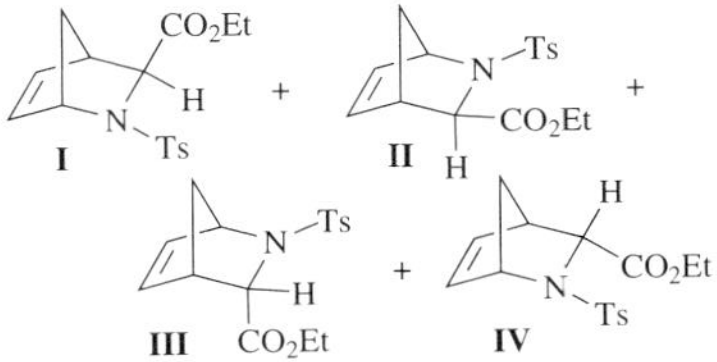

I + II + III + IV (252)

1b, $CuClO_4$•4MeCN

Solvent	Temp[o]	Time	Product(s) and Yield(s) (%)
THF	—	—	**I** + **II** (88), **I**:**II** 80:20, **III** + **IV** (9), **III**:**IV** 99.5:0.5
CH_2Cl_2	–20°[p]	30 min	**I** + **II** (85), **I**:**II** 91.5:8.5, **III** + **IV** (8), **III**:**IV** 91.5:8.5

7b, $CuClO_4$•4MeCN, THF — **I** + **II** (81), **I**:**II** 52:48, **III** + **IV** (9)

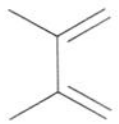

1b, $CuPF_6$

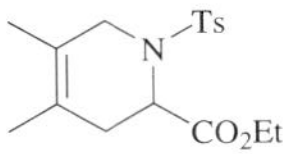

(—) 65% ee[k,q] — 468

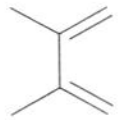

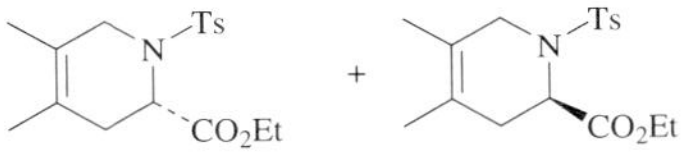

252

1b, $CuClO_4$•4MeCN, rt

Solvent	Time	Product(s) and Yield(s) (%)
THF	20 h	(64) 82.5:17.5[q]
CH_2Cl_2	24 h	(63) 80:20[q]

7b, $CuClO_4$•4MeCN, THF, rt, 20 h — (65) 1:1

8b, $CuClO_4$•4MeCN, THF, rt, 20 h — (71) 1:1

TABLE 12. CYCLOADDITIONS WITH ASYMMETRIC CATALYSIS (*Continued*)

Imine	Diene	Conditions	Product(s) and Yield(s) (%)	Refs.
C_{11} 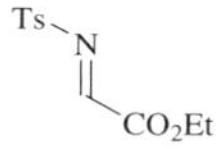	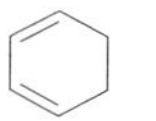		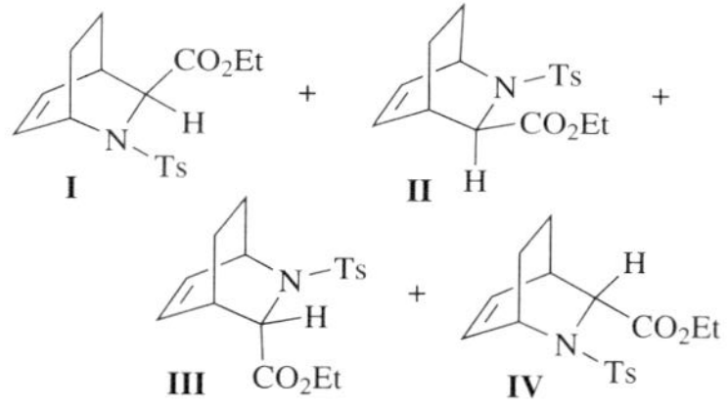	252
		1b, $CuClO_4$•4MeCN, rt		
		Solvent: THF; Time: —	**I** + **II** (31), **I**:**II** 95.5:4.5, **III** + **IV** (6), **III**:**IV** 70:30	
		Solvent: CH_2Cl_2; Time: 50 h	**I** + **II** (52), **I**:**II** 97.5:2.5, **III** + **IV** (7), **III**:**IV** 68.5:31.5	
		7b, $CuClO_4$•4MeCN, THF, rt	**I** + **II** (26), **I**:**II** 56.5:43.5, **III** + **IV** (10), **III**:**IV** 55:45	
	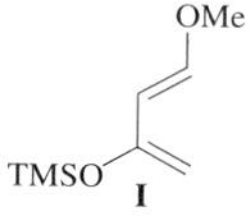	Catalyst, Lewis acid, –78°	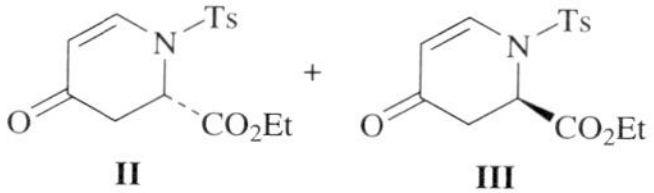	251

Cat.	L. A.	Solvent	Product(s) and Yield(s) (%)
1a	$CuClO_4$•4MeCN	THF	(65) 82:18
1a	$CuClO_4$•4MeCN	THF	(78) 83.5:16.5[r]
1a	2CuOTf•C_6H_6	THF	(60) 80.5:19.5
1a	$AgSbF_6$	THF	(75) 66.5:33.5
1b	$CuClO_4$•4MeCN	THF	(68) 90:10
1b	$Cu(OTf)_2$	THF	(42) 88.5:11.5
1b	$Cu(OTf)_2$	CH_2Cl_2	(72) 70:30
1b	AgOTf	THF	(85) 65:35
1b	$AgClO_4$	THF	(90) 67:33
1b	$Pd(SbF_6)_2$	THF	(76) 65:35
1b	$Pd(ClO_4)_2$	THF	(68) 55.5:44.5
1b	$Pd(OTf)_2$	THF	(88) 55.5:44.5
1b	$RuSbF_6$	THF	(70) 50:50

Reactant	Conditions	Product(s) and Yield(s) (%)			Refs.
I		II + III			251
			II + III	II:III	
	2, 2 $CuOTf{\bullet}C_6H_6$, THF, –78°		(74)	56:44	
	2, $Cu(OTf)_2$, THF, –78°		(60)	55:45	
	9a, $Zn(OTf)_2$, THF, –78°		(74)	58.5:41.5	
	9b, $Zn(OTf)_2$, THF, –78°		(60)	54:46	
I	1. Catalyst, $CuClO_4{\bullet}4MeCN^{s}$, –78°, 15-20 h 2. TFA, CH_2Cl_2, –78° to rt, 20 min	II + III			252

Cat.	Solvent[f]	Time	II + III	II:III
1a	THF	15-20 h	(78)	83.5:16.5
1a	CH_2Cl_2	15-20 h	(85)	55:45
1b	THF	15-20 h	(80)	89.5:10.5
1b	CH_2Cl_2	15-20 h	(89)	63:37
10	THF	15-20 h	(93)	36.5:63.5
10	CH_2Cl_2	15-20 h	(90)	35.5:64.5
7a	THF	15-20 h	(97)	88.5:11.5
7a	CH_2Cl_2	15-20 h	(73)	85.5:14.5
7b	THF	20 h	(82)	93.5:6.5
7b	CH_2Cl_2	15-20 h	(96)	88.5:11.5
7c	THF	15-20 h	(74)	93.5:6.5
7c	CH_2Cl_2	15-20 h	(91)	85.5:14.5
11a	THF	15-20 h	(94)	19.5:80.5
11a	CH_2Cl_2	15-20 h	(81)	19:81
11b	THF	15-20 h	(97)	19:81
11b	CH_2Cl_2	15-20 h	(93)	19:81
11c	THF	15-20 h	(97)	10.5:89.5
11c	CH_2Cl_2	15-20 h	(94)	23.5:76.5
8a	THF	15-20 h	(49)	10:90
8a	CH_2Cl_2	15-20 h	(92)	12.5:87.5
8b	THF	15-20 h	(66)	9.5:90.5
8b	CH_2Cl_2	15-20 h	(77)	7:93
8c	THF	15-20 h	(90)	17:83
8c	CH_2Cl_2	15-20 h	(70)	27.5:72.5

TABLE 12. CYCLOADDITIONS WITH ASYMMETRIC CATALYSIS (*Continued*)

Imine	Diene	Conditions	Product(s) and Yield(s) (%)	Refs.
C_{11} 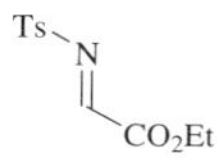	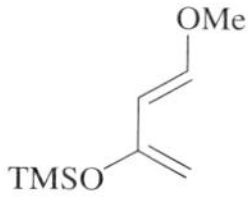	1. **12**, $CuClO_4$•4MeCN, –78° 2. MeOH, CH_2Cl_2, 10 min	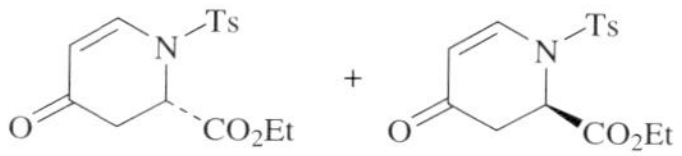	252
		Solvent: THF, Time: 18 h	(55) 59:41[u]	
		Solvent: CH_2Cl_2, Time: 20 h	(61) 52:48[u]	
	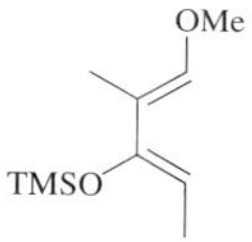	**1b**, $CuClO_4$•4MeCN	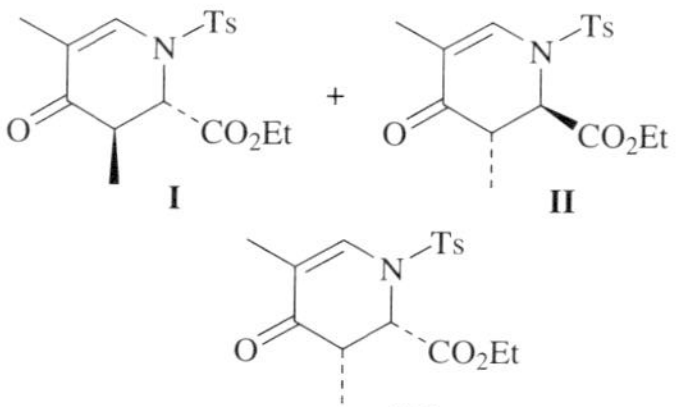 **I** + **II** + **III**	251

Cat. eq	Solvent	Temp	Time	Product(s) and Yield(s) (%)
0.1	THF	–78°	overnight	**I** + **II** + **III** (67), **I**:**II** 97:3 **I+II**:**III** 10:1[v,w]
0.05	THF	–78°	—	**I** + **II** + **III** (70), **I**:**II** 97:3 **I+II**:**III** 10:1[v]
0.01[x]	THF	–78°	—	**I** + **II** + **III** (70), **I**:**II** 98:2 **I+II**:**III** 10:1[v]
0.01[y]	THF	–78°	—	**I** + **II** + **III** (90), **I**:**II** 96.5:3.5 **I+II**:**III** 4:1[v]
0.1	THF	20°	—	**I** + **II** + **III** (70), **I**:**II** 90.5:9.5[v,z]
—	ether	–78°	—	**I** + **II** + **III** (—), **I**:**II** 89.5:10.5[v,z]
—	CH_2Cl_2	–78°	—	**I** + **II** + **III** (—), **I**:**II** 90:10[v,z]
—	toluene	–78°	—	**I** + **II** + **III** (—), **I**:**II** 82.5:17.5[v,z]
—	DMF	–78°	—	**I** + **II** + **III** (—), **I**:**II** 50:50[v,z]

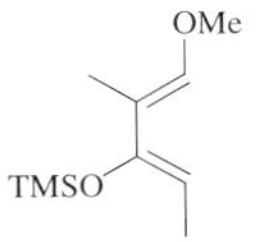

1. Catalyst, $CuClO_4$•4MeCN, THF, –78°
2. TFA, CH_2Cl_2, –78° to rt, 20 min

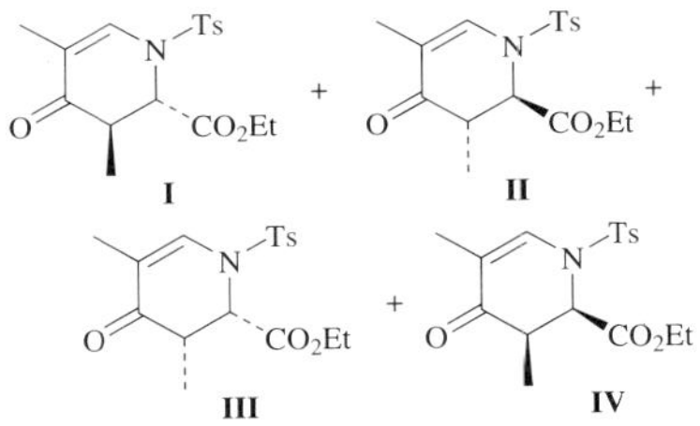

Cat. (eq)	Solvent	Time	I + II	I:II	III + IV	III:IV
1b (0.1)	THF	20 h	(83)	97:3	(8)	50:50
1b (0.01)	THF	15-20 h	(70)	98:2	(6)	50:50
8a (—)	THF	15-20 h	(64)	89.5:10.5	(32)	57.5:42.5
8b (—)	CH_2Cl_2	15-20 h	(48)	82:18	(44)	52.5:47.5
7b (—)	THF	15-20 h	(71)	83:17	(23)	57.5:42.5

252

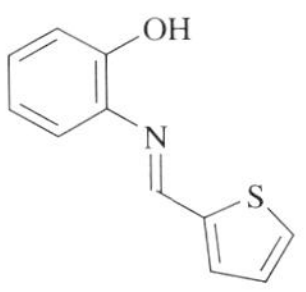

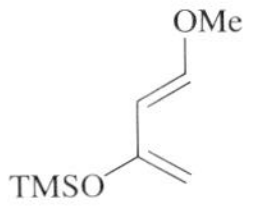

1. **13a**, Zr(OBu-*t*)$_4$, 1-methylimidazole, toluene, –45°
2. HCl, H_2O, THF, 0°, 30 min

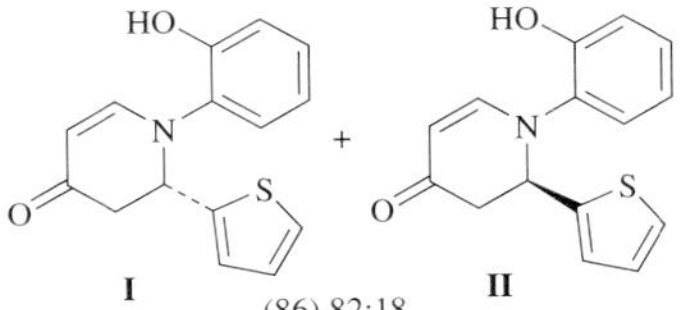

(86) 82:18

247

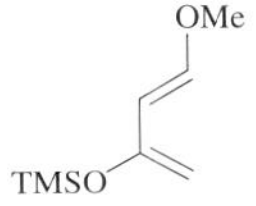

1. **13f,** Zr(OBu-*t*)$_4$, 1-methylimidazole, TMSCN, 3 Å MS, benzene, rt, 2 h
2. HCl, H_2O, THF, 0°, 30 min

I + **II** (61), **II:I** 91.5:8.5

249

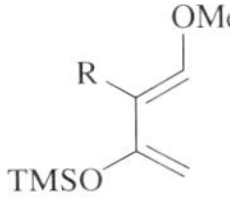

R
H
Me

1. **13h**, Zr(OBu-*t*)$_4$[aa], 1-methylimidazole, 3 Å MS, benzene, rt, 48 h
2. HCl, H_2O, THF, 0°, 30 min

I + **II**

I + II	II:I
(74)	93:7
(70)	93:7

248

TABLE 12. CYCLOADDITIONS WITH ASYMMETRIC CATALYSIS (*Continued*)

Imine	Diene	Conditions	Product(s) and Yield(s) (%)	Refs.
C_{12} 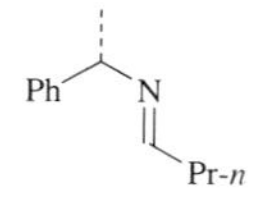	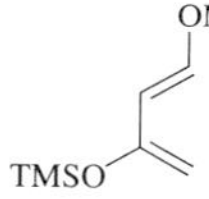	1. $B(OPh)_3$, (*R*) or (*S*)-1,1'-bi-2-naphthol, 4 Å MS, CH_2Cl_2, –78°, several h 2. H_2O Cat. config. R S	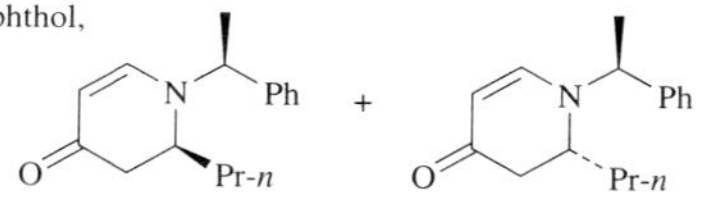 (41-49) 95:5 (31) 91:9	243, 242
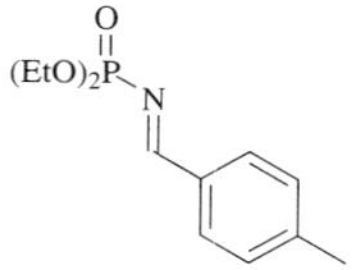	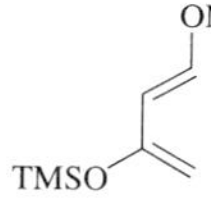	1. Et_2Zn, (*S*)-1,1'-bi-2-naphthol, CH_2Cl_2, 36 h, rt 2. TFA, CH_2Cl_2	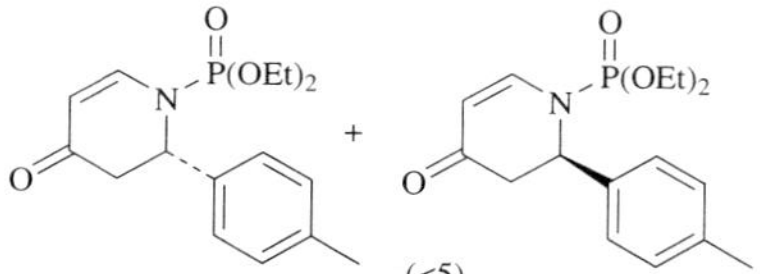 (<5)	129
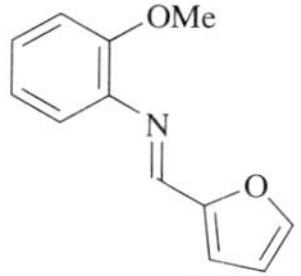	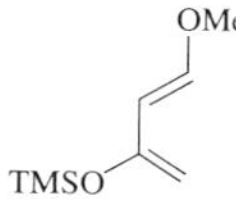	1. **15**, AgOAc, *i*-PrOH[bb], THF, 4°, 12 h 2. HCl, H_2O, 5 min	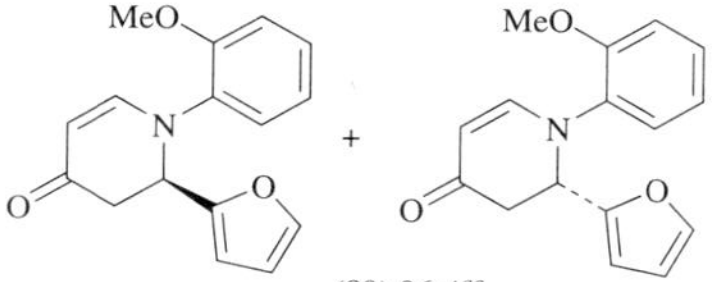 (89) 96:4[cc]	250
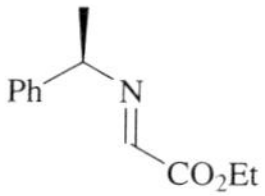	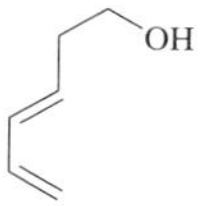	Aza-BSA-3 (polyclonal antibody), PBS (pH 7.0), 37°	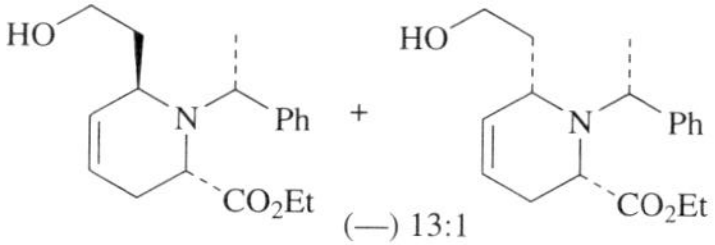 (—) 13:1	326
C_{13} 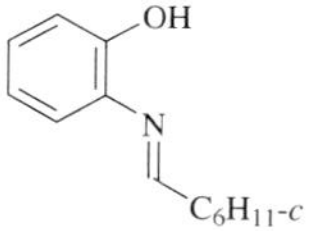	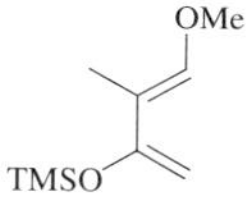	1. **13a**, $Zr(OBu\text{-}t)_4$, 1-methylimidazole, toluene, –45° 2. HCl, H_2O, THF, 0°, 30 min	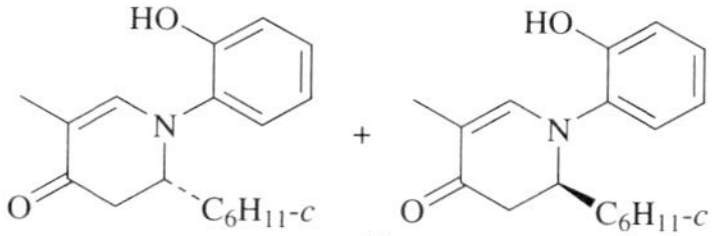 (—)[dd]	247

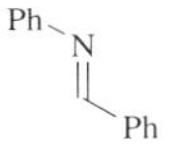	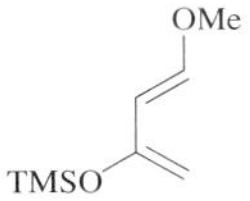	(*S*)-6-Methoxy-α-methyl-2-naphthaleneacetic acid, MeOH	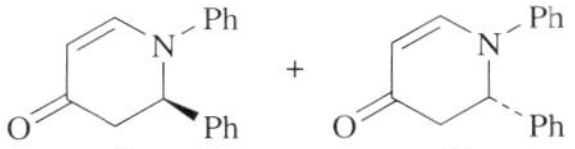 **I** + **II** (—) 1:1	373
	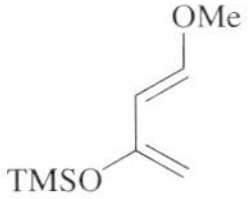	1. $B(OPh)_3$, (*R*)-1,1'-bi-2-naphthol, 4 Å MS, CH_2Cl_2, –78°, 5 h 2. H_2O	**I** + **II** (77) 62:38	243
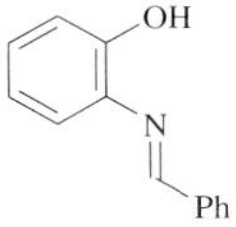	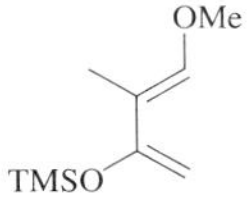	1. **13a**, Zr(OBu-*t*)$_4$, 1-methylimidazole, toluene, –45° 2. HCl, H_2O, THF, 0°, 30 min	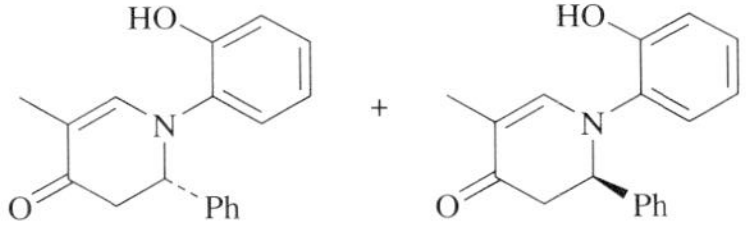 **I** + **II** (83) 82.5:17.5	247
	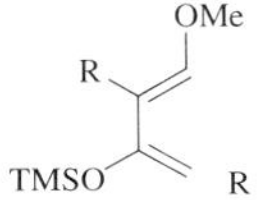 R: H, Me	1. **13b**, Zr(OBu-*t*)$_4$[aa], 1-methylimidazole, 3 Å MS, benzene, rt, 48 h 2. HCl, H_2O, THF, 0°, 30 min	**I** + **II** **I + II** / **II:I** R = H: (94) 91:9 R = Me: (quant.) 90:10	248
	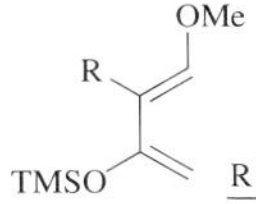 R: H, Me	1. **13f**, Zr(OBu-*t*)$_4$, 1-methylimidazole, TMSCN, 3 Å MS, benzene, rt, 2 h 2. HCl, H_2O, THF, 0°, 30 min	**I** + **II** **I + II** / **II:I** R = H: (76) 96:4 R = Me: (81) 95.5:4.5	249
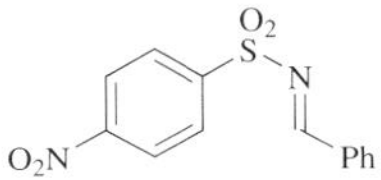	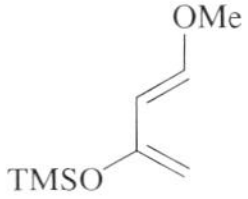	1. **6h**, $AgClO_4$, CH_2Cl_2, rt, 3 h 2. TFA, CH_2Cl_2, rt, 30-60 min	 **I** + **II** **I + II** (58), **I**:**II** 95:5[n] Ar = 4-$O_2NC_6H_4$	2

TABLE 12. CYCLOADDITIONS WITH ASYMMETRIC CATALYSIS (*Continued*)

Imine	Diene	Conditions	Product(s) and Yield(s) (%)	Refs.
C_{13} 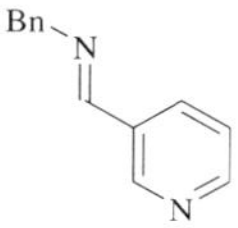	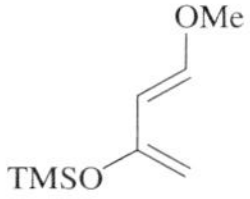	1. $B(OPh)_3$, (*R*) or (*S*)-1,1'-bi-2-naphthol, 4 Å MS, CH_2Cl_2, –78°, 5 h 2. H_2O	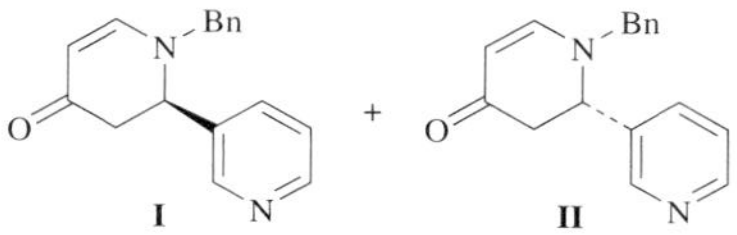	
		Cat. config.		
		R	**I** + **II** (71), **I**:**II** 95:5	243, 241
		S	**I** + **II** (68), **I**:**II** 5:95	241
		S	**I** (55)	243
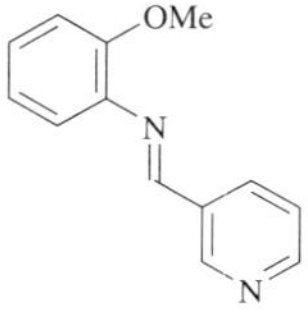	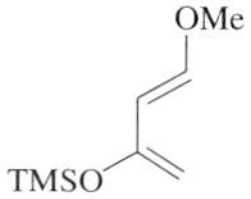	1. **15**, AgOAc, *i*-PrOH[bb], THF, 4°, 12 h 2. HCl, H_2O, 5 min	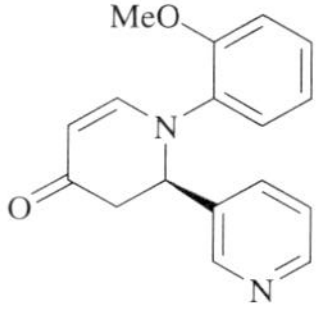 + 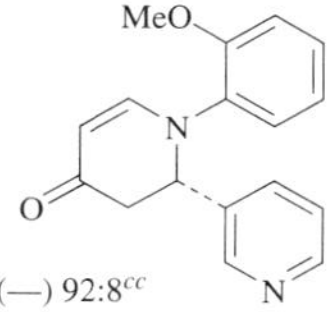 (—) 92:8[cc]	250
C_{14} 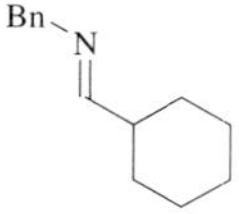	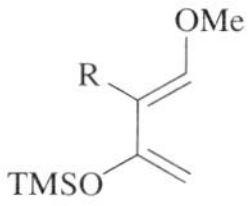	1. $B(OAr)_3$, (*R*)-1,1'-bi-2-naphthol, 4 Å MS, CH_2Cl_2, –78°, 5 h 2. H_2O	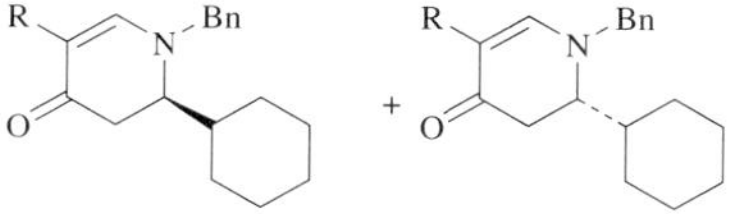	
	R	Ar		
	H	Ph	(45) 88:12	243, 241
	H	3,5-$Me_2C_6H_3$	(49) 86:14	241
	Me	Ph	(35) 89:11	243, 241
	Me	2-MeC_6H_4	(31) 90.5:9.5	241

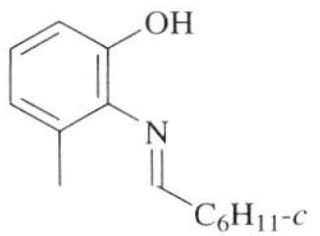

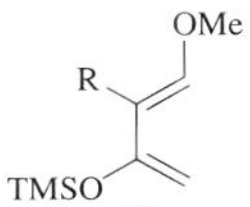

R
H
Me

1. **13b**, Zr(OBu-*t*)$_4$[aa],
   1-methylimidazole,
   3 Å MS, benzene, rt, 48 h
2. HCl, H_2O, THF, 0°, 30 min

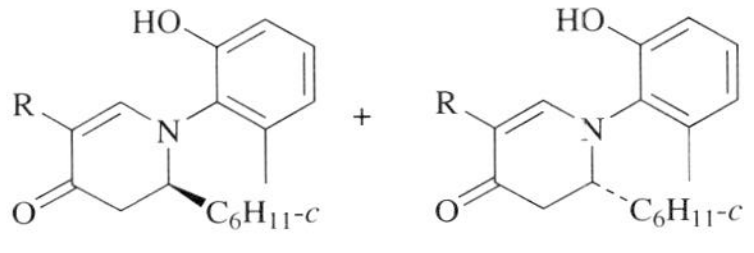

I + **II**

(64) 90.5:9.5
(67) 90:10

248

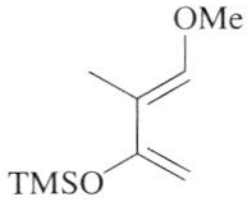

1. **13a**, Zr(OBu-*t*)$_4$,
   1-methylimidazole,
   toluene, –45°
2. HCl, H_2O, THF, 0°, 30 min

I + **II**

247

Cat. eq	I + II	II:I
0.1	(47)	89:11
0.2	(51)	93:7

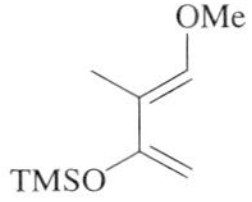

1. **13f**, Zr(OBu-*t*)$_4$,
   1-methylimidazole, TMSCN
   3 Å MS, benzene, rt, 2 h
2. HCl, H_2O, THF, 0°, 30 min

I + **II** (75), **I**:**II** 92:8

249

TABLE 12. CYCLOADDITIONS WITH ASYMMETRIC CATALYSIS (*Continued*)

Imine	Diene	Conditions	Product(s) and Yield(s) (%)	Refs.
C_{14} 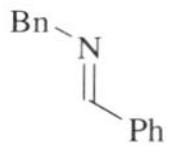	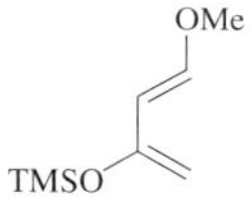	1. Lewis acid, (*R*) or (*S*)-1,1'-bi-2-naphthol, 4 Å MS 2. H_2O	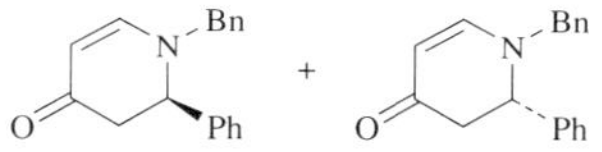	

L. A.	Cat. config.	Solvent	Temp	Time		Refs.
$B(OPh)_3$	R	CH_2Cl_2	–78°	5 h	(75) 91:9	243, 241, 244
$B(OPh)_3$	R	CH_2Cl_2	–100°	5 h	(70) 92.5:7.5	241
$B(OPh)_3$	R	THF	–78°	5 h	(22) 61:39	243, 241
$B(OPh)_3$	R	EtCN	–78°	5 h	(38) 61:39	243, 241
$B(OPh)_3$	R	toluene	–78°	5 h	(15) 87:13	243
$B[O(2\text{-}MeC_6H_4)]_3$	R	CH_2Cl_2	–78°	5 h	(76) 92:8	243, 241
$B[O(3,5\text{-}Me_2C_6H_3)]_3$	R	CH_2Cl_2	–78°	5 h	(75) 93:7	243, 241
$B(OPh)_3$	S	CH_2Cl_2	–78°	5 h	(72) 9:91	243
$B[O(3,5\text{-}Me_2C_6H_3)]_3$	S	CH_2Cl_2	–78°	5 h	(82) 7:93	241
BH_3	R	CH_2Cl_2	–78°	5 h	(62) 86:14	243
$PhB(OH)_2$	R	CH_2Cl_2	–78°	5 h	(15) 65:35	243
$B(OMe)_3$	R	CH_2Cl_2	–78°	12 h	(78) 93:7[ee]	244
$B(OMe)_3$	R	CH_2Cl_2	–78°	5 h	(42) 86:14[ff]	243
Me_3Al	R	CH_2Cl_2	–78°	5 h	(15) 56:44	243
$TiCl_2(OPr\text{-}i)_2$	R	CH_2Cl_2	–78°	5 h	(20) 58.5:41.5	243

	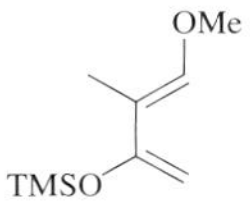	1. $B(OAr)_3$, (*R*)-1,1'-bi-2-naphthol, 4 Å MS, CH_2Cl_2, –78°, 5 h 2. H_2O	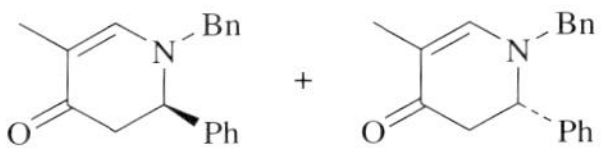	
		Ar		
		Ph	(72) 90.5:9.5	243, 241
		3,5-$Me_2C_6H_3$	(66) 93.5:6.5	241
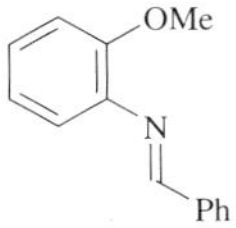	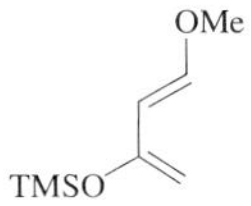	**14**, CD_3CN, –40°, 2 h	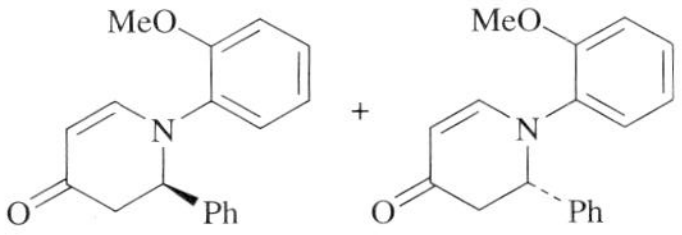 (74) 1:1	469
	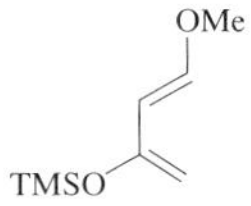	1. **15**, AgOAc, *i*-PrOH[bb], THF, 4°, 12 h 2. HCl, H_2O, 5 min	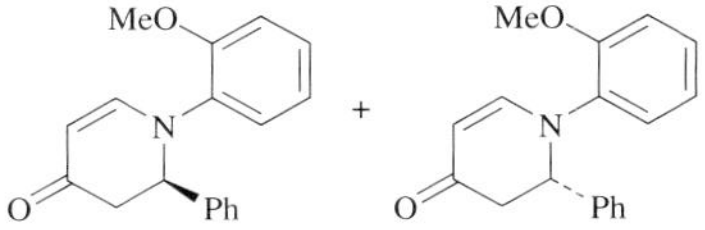	250
		Cat. eq		
		0.01	(94) 96.5:3.5[cc]	
		0.005	(92) 96:4[cc]	
		0.001	(78) 94:6[cc]	
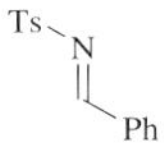	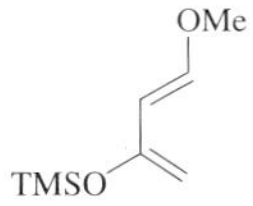	1. **1b**, $CuClO_4$•4MeCN, rt 2. TFA, CH_2Cl_2, –78° to rt, 20 min	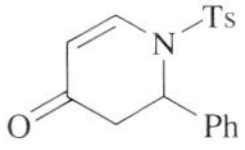	252
		Solvent / Time		
		THF / —	(65) 46% *ee*[k,u]	
		CH_2Cl_2 / 60 h	(63) 48% *ee*[k]	

TABLE 12. CYCLOADDITIONS WITH ASYMMETRIC CATALYSIS (*Continued*)

Imine	Diene	Conditions	Product(s) and Yield(s) (%)	Refs.
C_{14} 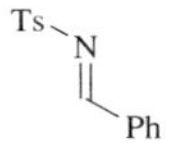	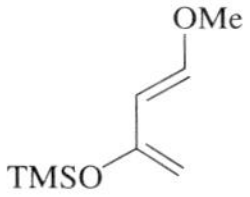	1. **16a**, copper salt, CH_2Cl_2, rt, 5-7 h 2. TFA, CH_2Cl_2, rt, 30-60 min	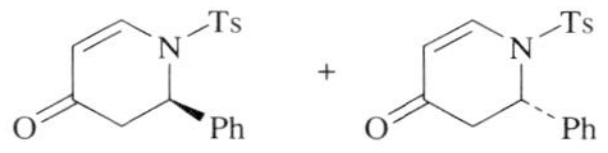	2
		Cu salt		
		$Cu(MeCN)_4ClO_4$	(83) 87:13[n]	
		$Cu(MeCN)_4PF_6$	(72) 88.5:11.5[n]	
		CuOTf	(80) 88:12[n]	
	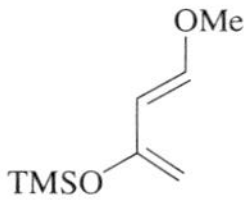	1. Catalyst, $AgClO_4$, CH_2Cl_2, rt, 24 h 2. TFA, CH_2Cl_2, rt, 30-60 min	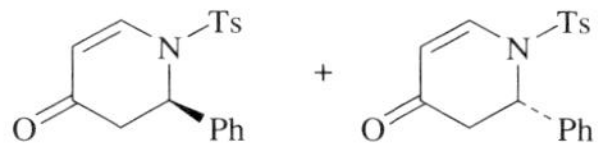	2
		Cat.		
		16a	(67-68) 77.5:22.5[n]	
		16b	(67-68) 86.5:13.5[n]	
	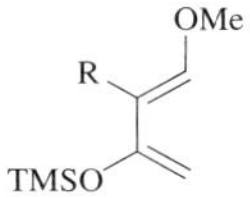	1. Catalyst, $AgClO_4$, CH_2Cl_2 2. TFA, CH_2Cl_2, rt, 30-60 min	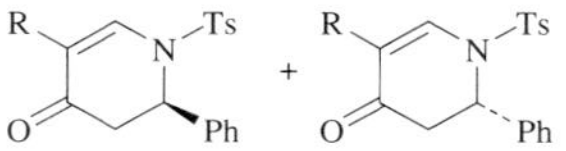	2

R	Cat.	Temp	Time	Product(s) and Yield(s) (%)
H	**6a**	rt	6 h	(79) 90:10[n]
H	**6b**	rt	20 h	(70) 85.5:14.5[n]
H	**6c**	rt	20 h	(58) 85.5:14.5[n]
H	**6d**	rt	20 h	(60) 77.5:22.5[n]
H	**6e**	rt	20 h	(66) 78.5:21.5[n]
H	**6f**	rt	3 h	(96) 90:10[n]
H	**6g**	rt	2 h	(89) 90:10[n]
H	**6h**	rt	1 h	(90) 96.5:3.5[n]
H	**6h**	−20°	1 h	(87) 98.5:1.5[n]
Me	**6h**	rt	1-5 h	(64) 93.5:6.5[n]
Me	**6h**	0°	5 h	(47) 96:4[n]

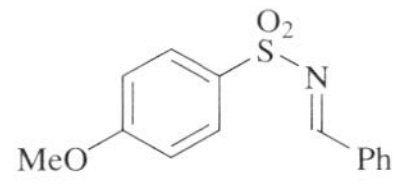

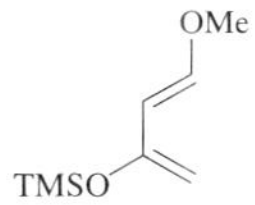

1. **6h,** AgClO$_4$, CH$_2$Cl$_2$, rt, 3 h
2. TFA, CH$_2$Cl$_2$, rt, 30-60 min

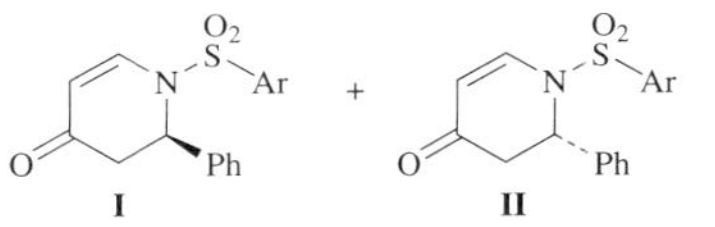

I + **II** (61), **I**:**II** 97:3[n]   Ar = 4-MeOC$_6$H$_4$ 2

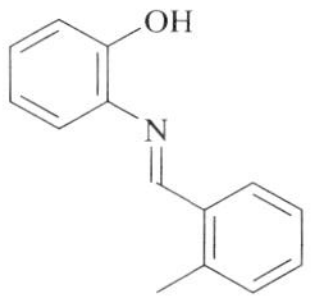

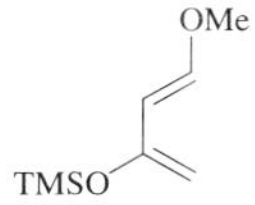

1. Catalyst, Zr(OBu-*t*)$_4$, 3Å MS, benzene, rt
2. HCl, H$_2$O, THF, 0°, 30 min

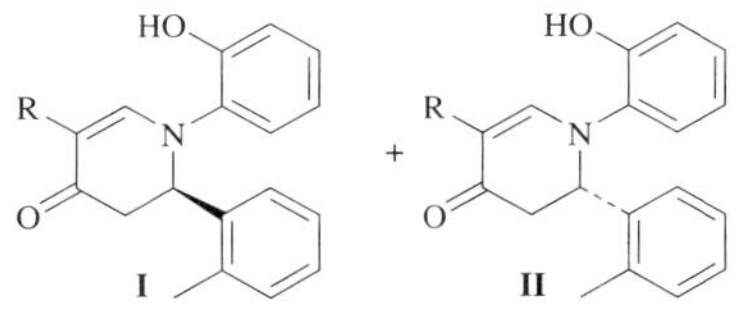

R = H   249

Cat. (eq)	
13a (0.2)	(92) 88.5:11.5
13b (0.2)	(93) 95.5:4.5
13b (0.1)	(81) 88.5:11.5
13b (0.05)	(75) 76:24

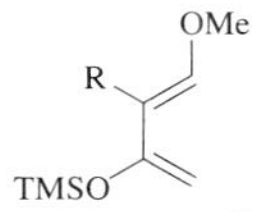

1. **13a**, Zr(OBu-*t*)$_4$, 1-methylimidazole, toluene, –45°
2. HCl, H$_2$O, THF, 0°, 30 min

I + **II**

R	Cat. eq	Time	I + II	II:I		
H	0.1	—	(81)	88:12		247
H	0.2	35 h	(83)	91:9		247, 248
Me	0.2	—	(97)	88.5:11.5	R = Me	247

TABLE 12. CYCLOADDITIONS WITH ASYMMETRIC CATALYSIS (*Continued*)

Imine	Diene	Conditions	Product(s) and Yield(s) (%)	Refs.
C_{14} 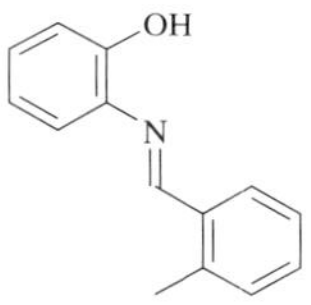	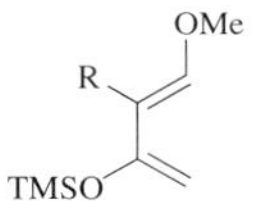	1. Catalyst, $Zr(OBu\text{-}t)_4$, 1-methylimidazole, TMSCN, 3Å MS, benzene, rt, 2 h 2. HCl, H_2O, THF, 0°, 30 min	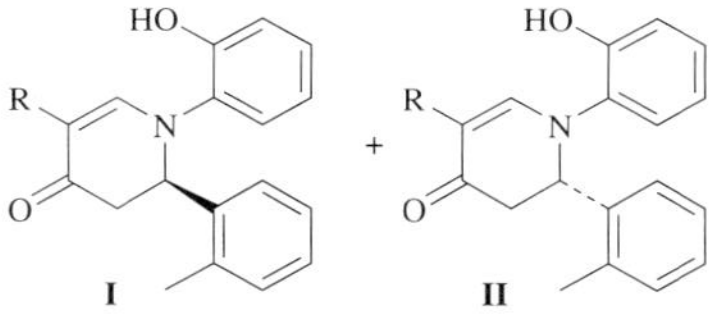	249

R	Cat. (eq)	I + II	I:II
H	**13b** (0.2)	(94)	97:3
H	**13b** (0.1)	(90)	94:6
H	**13b** (0.05)	(84)	84:16
H	**13b** (0.05)	(92)	96:4[gg]
H	**13b** (0.02)	(77)	73.5:26.5
H	**13b** (0.02)	(83)	90:10[gg]
H	**13d** (0.02)	(59)	86.5:13.5
H	**13e** (0.02)	(70)	88.5:11.5
H	**13e** (0.02)	(70)	92.5:7.5[gg]
H	**13f** (0.05)	(93)	95.5:4.5[gg]
H	**13f** (0.02)	(73)	93.5:6.5
H	**13f** (0.02)	(68)	97:3[gg]
H	**13f** (0.01)	(64)	91.5:8.5[gg]
H	**13g** (0.02)	(74)	94:6
Me	**13f** (0.02)	(72)	94:6

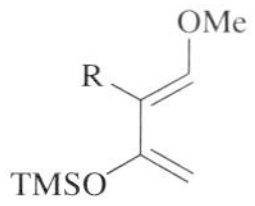
OMe
R
TMSO

Catalyst, $Zr(OBu\text{-}t)_4$,
3 Å MS , benzene, rt, 24 h

I + **II**

249

R	Cat.	I + II	I:II
H	**17a**	(61)	88.5:11.5
H	**17b**	(80)	86.5:13.5
H	**17c**	(80)	86:14
H	**17d**	(58)	74:26
H	**17e**	(70)	91:9
H	**17f**	(59)	90:10
H	**17g**	(74)	87:13
H	**17h**	(61)	85:15
H	**17i**	(80)	91.5:8.5
H	**17j**	(82)	85.5:14.5
H	**17k**	(87)	90:10
H	**17l**	(92)	80:20
H	**17m**	(75)	88:12
H	**17n**	(75)	70.5:29.5
H	**17o**	(82)	80:20
H	**17p**	(63)	79.5:20.5
H	**17q**	(61)	72:28
Me	**17i**	(78)	94:6
Me	**17k**	(>99)	95.5:4.5[hh]

TABLE 12. CYCLOADDITIONS WITH ASYMMETRIC CATALYSIS (*Continued*)

Imine	Diene	Conditions	Product(s) and Yield(s) (%)	Refs.
C_{14} 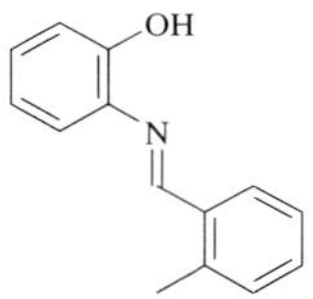	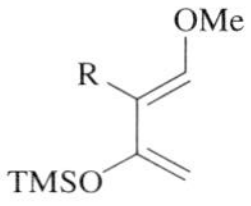	1. **13b**, Zr(OBu-*t*)$_4$[aa], 1-methylimidazole 2. HCl, H_2O, THF, 0°, 30 min	**I** + **II**	248

R	Cat. eq	Solvent	Additive	Temp	Time	I + II	I:II
H	0.2	toluene	—	–45°	—	(66)	92:8
H	0.2	toluene	—	0°	—	(45)	78.5:21.5
H	0.2	toluene	3 Å MS	0°	—	(80)	95:5
H	0.2	toluene	4 Å MS	0°	—	(76)	94.5:5.5
H	0.2	toluene	5 Å MS	0°	—	(77)	94.5:5.5
H	0.2	toluene	3 Å MS	–45°	—	(54)	88.5:11.5
H	0.2	toluene	3 Å MS	rt	48 h	(96)	94:6
H	0.2	benzene	3 Å MS	rt	48 h	(93)	95.5:4.5
H	0.1	benzene	3 Å MS	rt	48 h	(81)	88.5:11.5
H	0.2[ii]	benzene	3 Å MS	rt	48 h	(>98)	94.5:5.5
Me	0.2	toluene	3 Å MS	rt	48 h	(98)	94.5:5.5

Imine	Diene	Conditions	Product(s) and Yield(s) (%)	Refs.
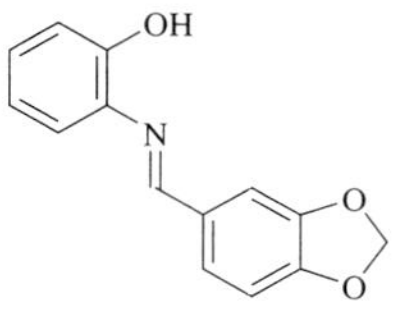	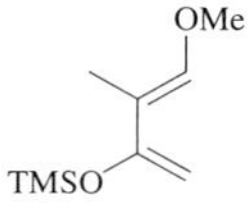	1. **13a**, Zr(OBu-*t*)$_4$, 1-methylimidazole, toluene, –45° 2. HCl, H_2O, THF, 0°, 30 min	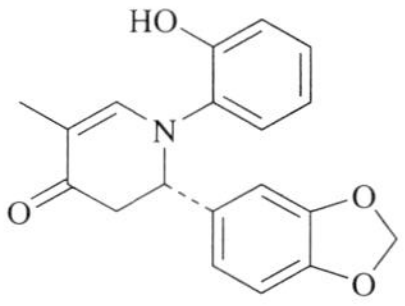 + 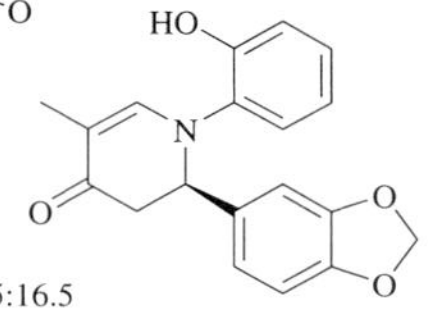 (80) 83.5:16.5	247

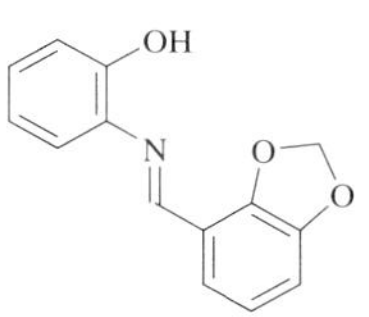

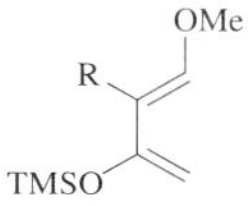

1. **13b**, Zr(OBu-*t*)$_4$[aa],
   1-methylimidazole,
   3 Å MS, benzene, rt, 48 h
2. HCl, H_2O, THF, 0°, 30 min

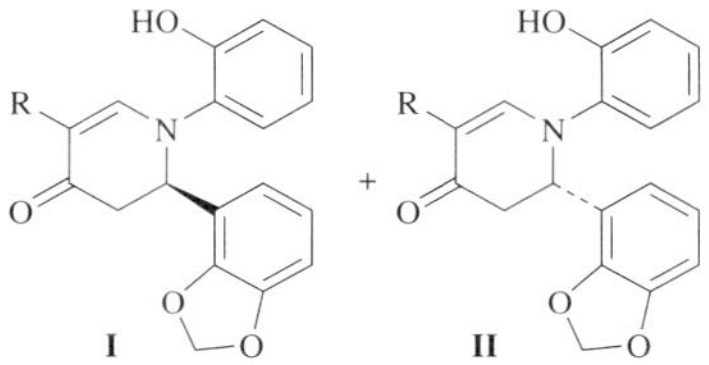

R	(%) I:II
H	(90) 95:5
Me	(87) 94:6

248

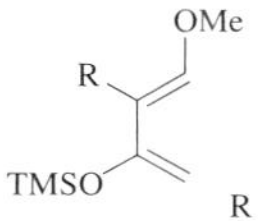

1. **13f**, Zr(OBu-*t*)$_4$,
   1-methylimidazole, TMSCN,
   3 Å MS, benzene, rt, 2 h
2. HCl, H_2O, THF, 0°, 30 min

I + **II**

R	I + II	I:II
H	(75)	95:5
Me	(68)	95:5

249

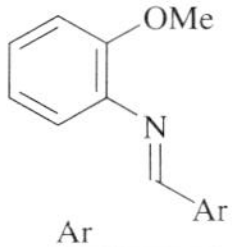

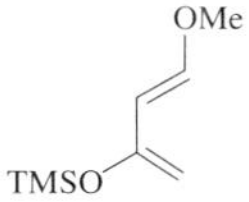

1. Catalyst, AgOAc, 4°, 12 h
2. HCl, H_2O, 5 min

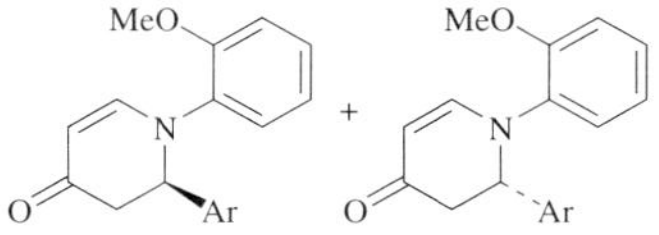

250

Ar	Cat.	Additive[bb]	Solvent	
3-$O_2NC_6H_4$	**18a**	*i*-PrOH	THF	(92) 95.5:4.5[cc]
3-$O_2NC_6H_4$	**18a**	*i*-PrOH	none	(88) 90:10
4-$O_2NC_6H_4$	**18a**	*i*-PrOH	THF	(>98) 96:4[cc]
4-ClC_6H_4	**18a**	*i*-PrOH	THF	(98) 95:5[cc,jj]
4-ClC_6H_4	**18a**	H_2O	THF	(81) 95:5[kk]
4-ClC_6H_4	**18a**	*i*-PrOH	THF	(>98) 96:4
4-ClC_6H_4	**18b**	*i*-PrOH	THF	(75) 94:6
4-ClC_6H_4	**18c**	*i*-PrOH	THF	(53) 64:36
4-ClC_6H_4	**18d**	*i*-PrOH	THF	(>98) 90:10
4-ClC_6H_4	**18e**	*i*-PrOH	THF	(>98) 90:10
4-ClC_6H_4	**18f**	*i*-PrOH	THF	(52) 60:40
2-BrC_6H_4	**18a**	*i*-PrOH	THF	(91) 94.5:5.5[cc]

TABLE 12. CYCLOADDITIONS WITH ASYMMETRIC CATALYSIS (*Continued*)

Imine	Diene	Conditions	Product(s) and Yield(s) (%)	Refs.
C_{14} 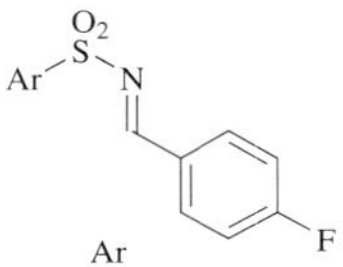	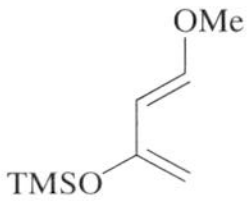	1. **6h**, $AgClO_4$, CH_2Cl_2 2. TFA, CH_2Cl_2, rt, 30-60 min	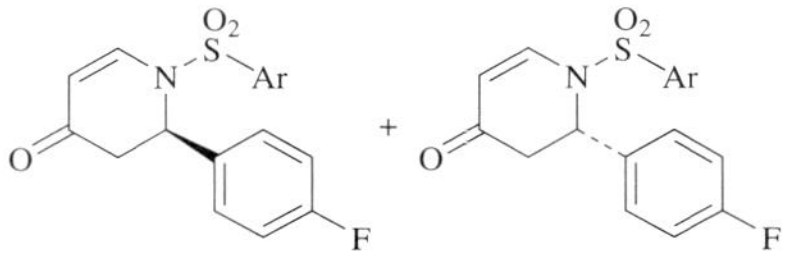	2
Ar		Temp / Time		
$4\text{-}MeC_6H_4$		rt / 1-5 h	(78) 94:6[n]	
$4\text{-}MeC_6H_4$		–20° / 3 h	(50) 96.5:3.5[n]	
$4\text{-}MeOC_6H_4$		rt / 3 h	(78) 95:5[n]	
$C_{14\text{-}15}$ 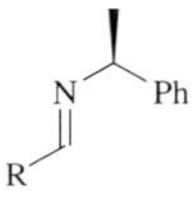	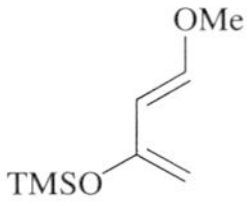	1. $B(OPh)_3$, (*R*) or (*S*)-1,1'-bi-2-naphthol, 4 Å MS, CH_2Cl_2, –73°, several h 2. H_2O	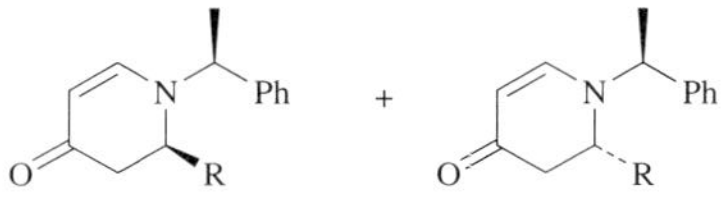	243, 242
R		Cat. config.		
3-pyridinyl		R	(63) 99:1	
3-pyridinyl		S	(35) 86:14	
$c\text{-}C_6H_{11}$		R	(31) 99:1	
$c\text{-}C_6H_{11}$		S	(20) 89:11	
C_{15} 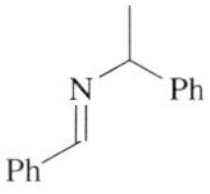	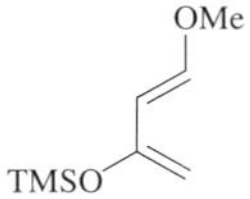	1. $B(OPh)_3$, (*R*)-1,1'-bi-2-naphthol, 4 Å MS, CH_2Cl_2, –78°, several h 2. H_2O	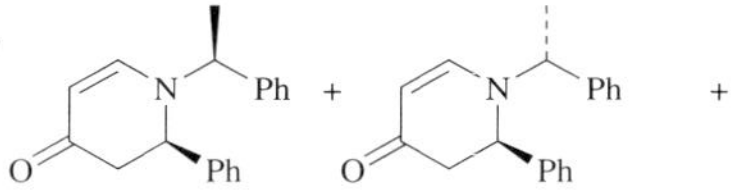 **I** + **II** + 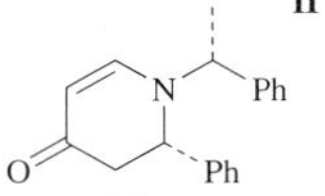 **III** **I** + **II** + **III** (50), **I**:**II**:**III** 72:3:25	243

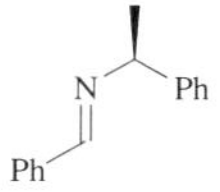

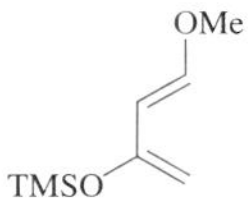

1. $B(OR)_3$,
(*R*) or (*S*)-1,1'-bi-2-naphthol,
4 Å MS, CH_2Cl_2, –78°
2. H_2O

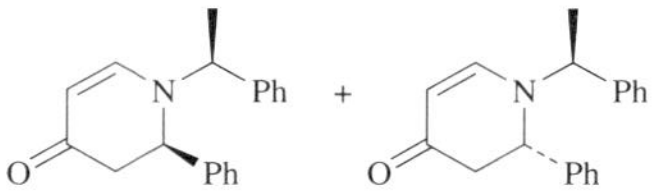

R	Cat. config.	Time		
Ph	R	several h	(61) 99:1	243, 242, 244
Ph	S	several h	(30) 93:7	243, 242, 244
Ph	R	12 h	(64) >99.5:0.5[ll]	244
Ph	S	12 h	(49) 92:8[ll]	244
Me	R	12 h	(64) 99.5:0.5	244
Me	S	12 h	(49) 91.5:8.5	244

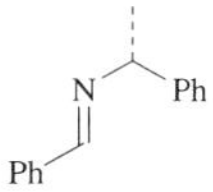

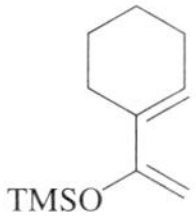

$TiCl_2(OPr\text{-}i)_2$,
(*R*) or (*S*)-1,1'-bi-2-naphthol, rt

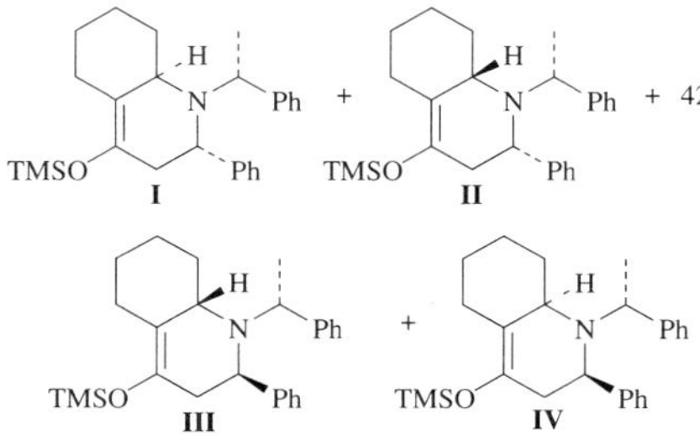
I
II
III
IV

429

Cat. config.	Time	
R	0.25 h	**I + II + III + IV** (trace)
R	0.5 h	**I + II + III + IV** (52)[mm], **I + II**:**III + IV** 38:62, **I**:**II** 84:16, **III**:**IV** 83:17
R	1 h	**I + II + III + IV** (trace)
S	0.5 h	**I + II + III + IV** (31)[mm], **I + II**:**III + IV** 40:60, **I**:**II** 70:30, **III**:**IV** >99:1

TABLE 12. CYCLOADDITIONS WITH ASYMMETRIC CATALYSIS (*Continued*)

Imine	Diene	Conditions	Product(s) and Yield(s) (%)	Refs.
C_{15}				
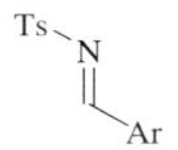 Ar 2-MeC_6H_4 4-$MeOC_6H_4$	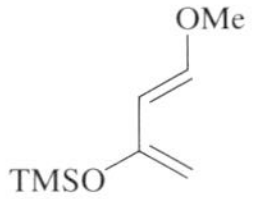	1. **6h**, $AgClO_4$, CH_2Cl_2, rt, 3 h 2. TFA, CH_2Cl_2, rt, 30-60 min	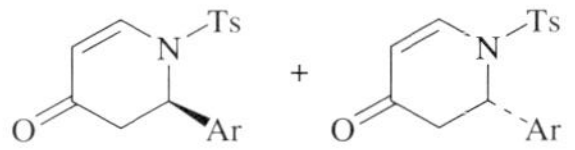 (82) 96.5:3.5[n] (76) 95.5:4.5[n]	2
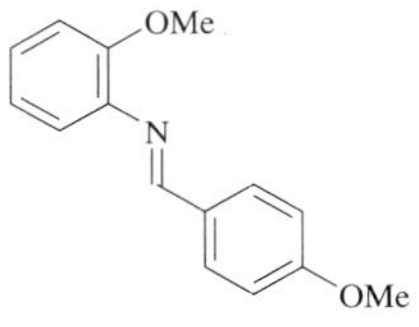	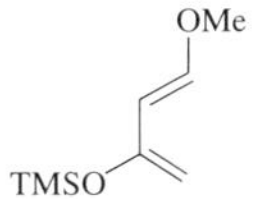	1. **15**, AgOAc, *i*-PrOH[bb], 4°, 12 h 2. HCl, H_2O, 5 min Solvent THF none	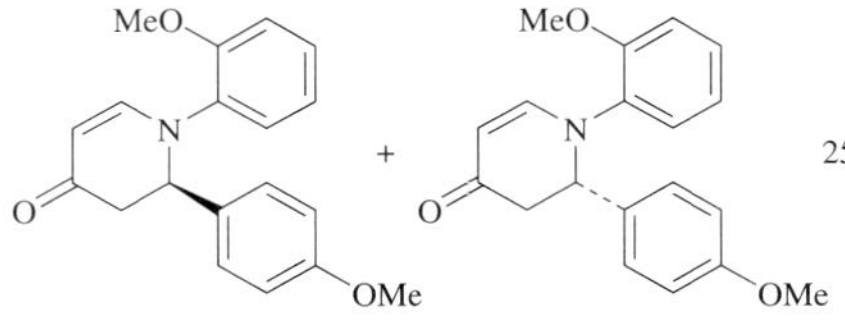 (86) 95.5:4.5[cc] (66) 95:5	250
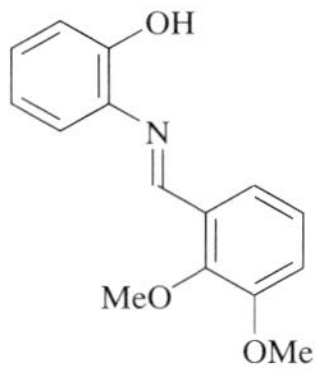	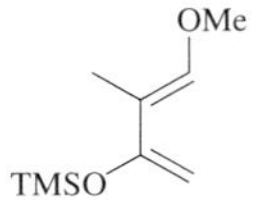	1. **13a**, Zr(OBu-*t*)$_4$, 1-methylimidazole, toluene, –45° 2. HCl, H_2O, THF, 0°, 30 min	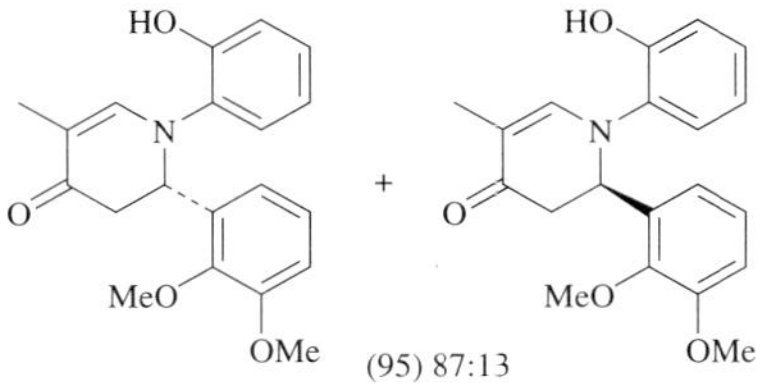 (95) 87:13	247
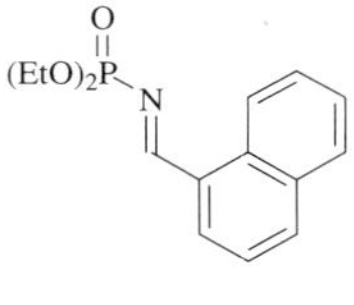	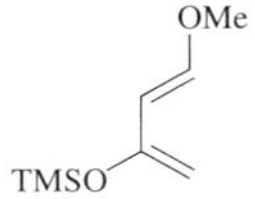	1. Et_2Zn, (*S*)-1,1'-bi-2-naphthol, CH_2Cl_2, 36 h, rt 2. TFA, CH_2Cl_2	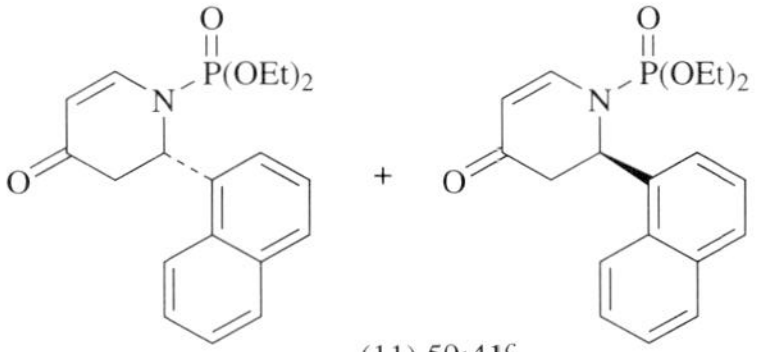 (11) 59:41[c]	129

C$_{16}$

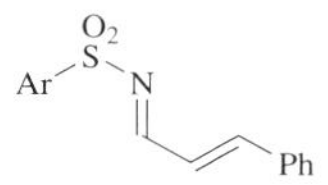	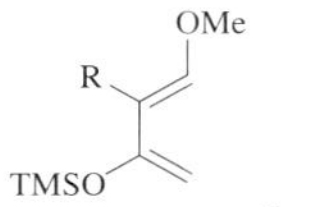	1. **6h**, $AgClO_4$, CH_2Cl_2 2. TFA, CH_2Cl_2, rt, 30-60 min	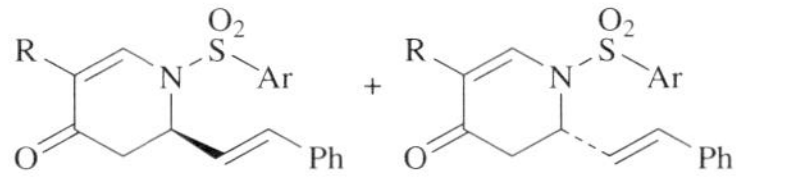	2
Ar: 4-MeC_6H_4	R: H	Temp: rt, Time: 1-5 h	(66) 91.5:8.5^n	
Ar: 4-MeC_6H_4	R: H	Temp: –20°, Time: 1.5 h	(82) 98:2^n	
Ar: 4-MeC_6H_4	R: Me	Temp: –20°, Time: 5 h	(57) 94:6^n	
Ar: 4-$MeOC_6H_4$	R: H	Temp: rt, Time: 3 h	(58) 88:12^n	
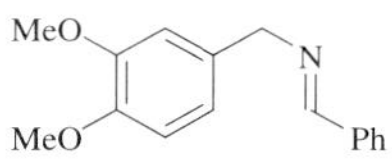	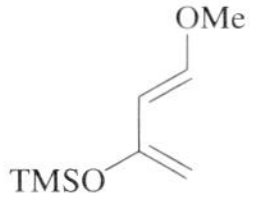	1. $B(OPh)_3$, (*R*)-1,1'-bi-2-naphthol, 4 Å MS, CH_2Cl_2, –78°, 5 h 2. H_2O	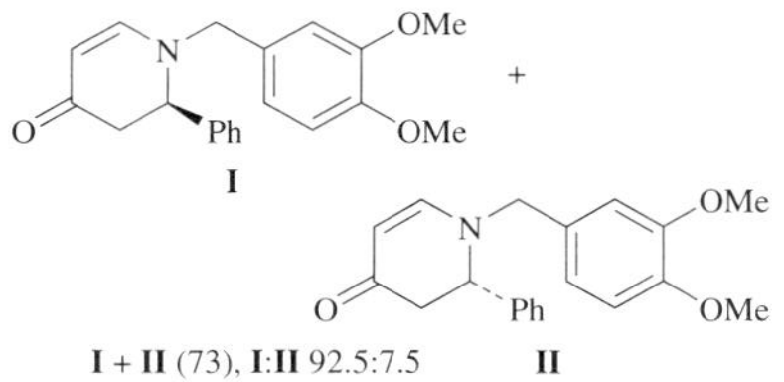 **I** + **II** (73), **I**:**II** 92.5:7.5	243, 241
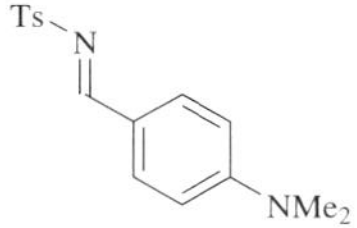	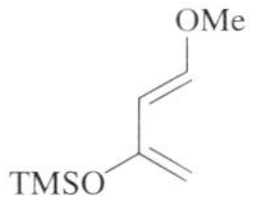	1. **6h**, $AgClO_4$, CH_2Cl_2, rt, 3 h 2. TFA, CH_2Cl_2, rt, 30-60 min	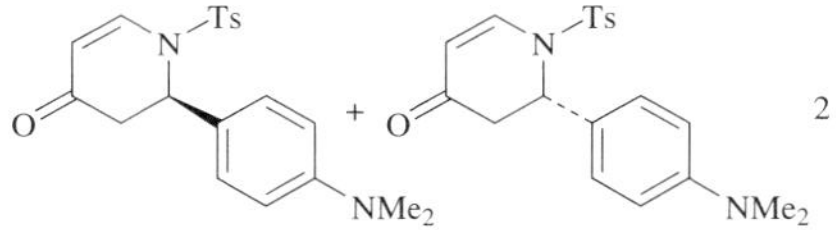 (39) 96.5:3.5^n	2
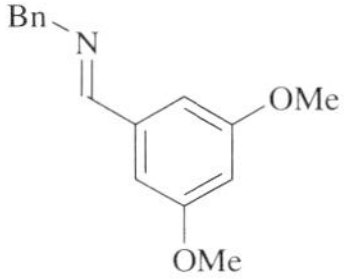	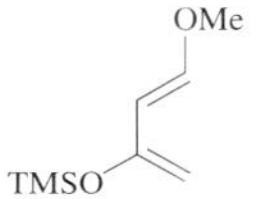	1. $B(OPh)_3$, (*R*)-1,1'-bi-2-naphthol, 4 Å MS, CH_2Cl_2, –78°, 5 h 2. H_2O	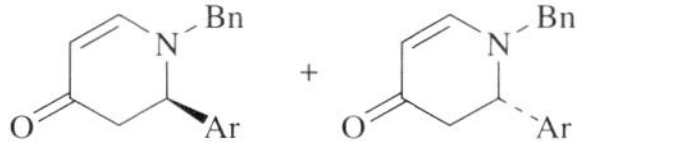 (89) 87:13 Ar = 3,5-$(MeO)_2C_6H_3$	243, 241

TABLE 12. CYCLOADDITIONS WITH ASYMMETRIC CATALYSIS (*Continued*)

Imine	Diene	Conditions	Product(s) and Yield(s) (%)	Refs.
C_{17} 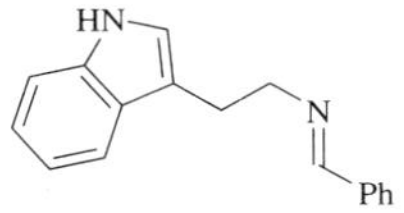	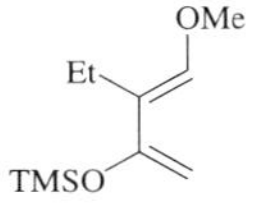	1. $B(OPh)_3$, (*R*)-1,1'-bi-2-naphthol, 4 Å MS, CH_2Cl_2, –60° to rt, overnight 2. HCl, H_2O	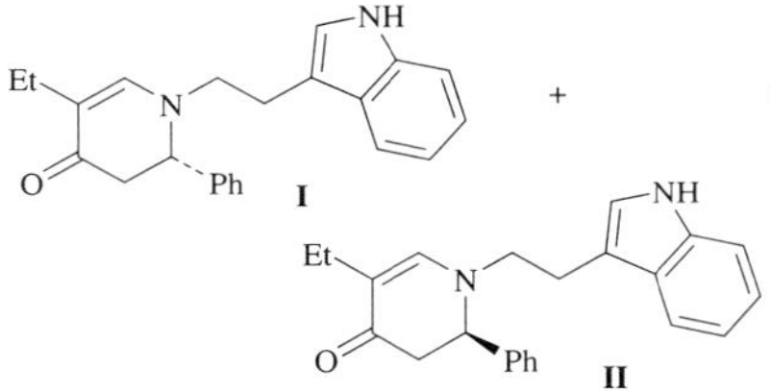 I + II **I + II** (62), **I**:**II** 63:37	246, 245
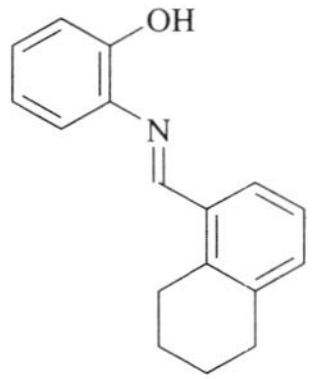	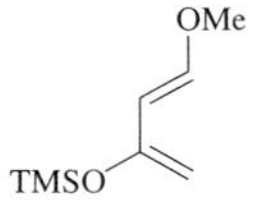	1. **13a**, Zr(OBu-*t*)$_4$, 1-methylimidazole, toluene, –45° 2. HCl, H_2O, THF, 0°, 30 min	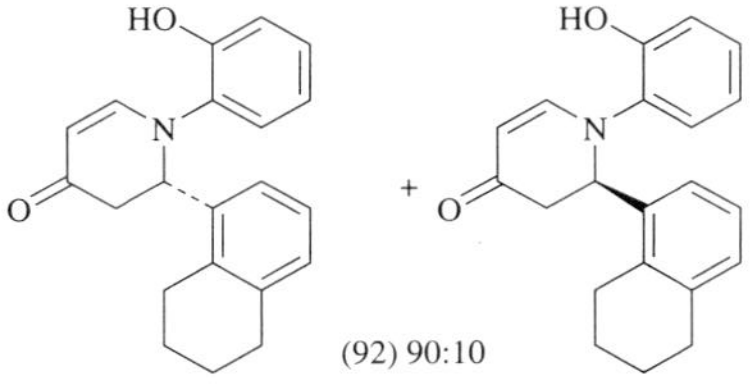 **I** + **II** (92) 90:10	247
	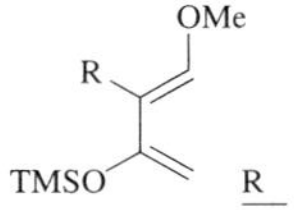 R: H, Me	1. **13b**, Zr(OBu-*t*)$_4$[aa], 1-methylimidazole, 3 Å MS, benzene, rt, 48 h 2. HCl, H_2O, THF, 0°, 30 min	**I** + **II** R = H: **I + II** (67), **II**:**I** 92.5:7.5 R = Me: **I + II** (78), **II**:**I** 93.5:6.5	248
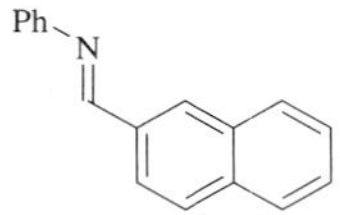	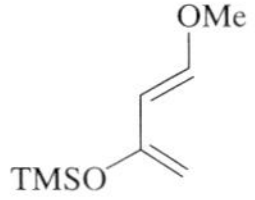	1. **15**, AgOAc, *i*-PrOH, THF, 4°, 12 h 2. HCl, H_2O, 5 min	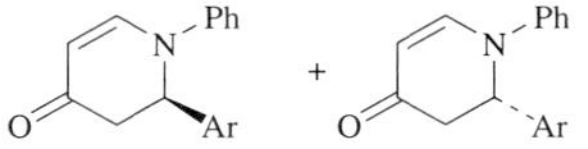 (>90) 67.5-70:30-32.5 Ar = 2-naphthyl	250

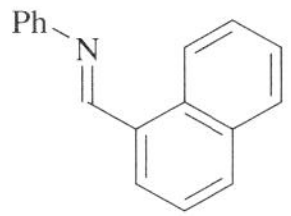

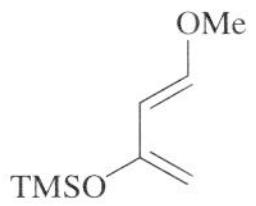

1. **13a**, $Zr(OBu\text{-}t)_4$, 1-methylimidazole, toluene, –45°
2. HCl, H_2O, THF, 0°, 30 min

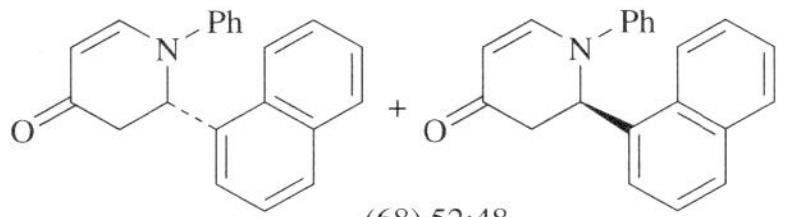

(68) 52:48

247

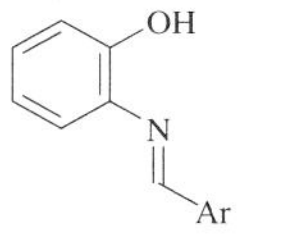

Ar = 1-naphthyl

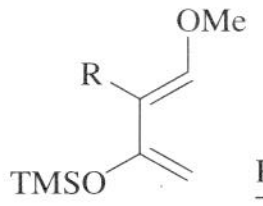

I

1. **13b**, $Zr(OBu\text{-}t)_4$[aa], 1-methylimidazole, 3 Å MS, benzene, rt, 48 h
2. HCl, H_2O, THF, 0°, 30 min

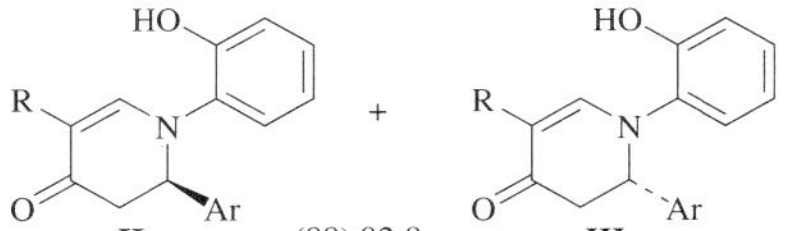

II + **III**

R	
H	(88) 92:8
Me	(78) 90:10

248

1. **13a**, Lewis acid, ligand, –45°
2. HCl, H_2O, THF, 0°, 30 min

II + **III**

247

R	L. A.	Ligand	Cat. eq	Solvent	**II + III**	**III:II**
H	$Zr(OBu\text{-}t)_4$	1-methylimidazole	0.1	CH_2Cl_2	(74)	70:30
H	$Zr(OBu\text{-}t)_4$	1-methylimidazole	0.1[nn]	CH_2Cl_2	(81)	80.5:19.5
H	$Zr(OBu\text{-}t)_4$	1-methylimidazole	0.1[oo]	CH_2Cl_2	(81)	85.5:14.5
H	$Zr(OBu\text{-}t)_4$	1-methylimidazole	0.1	toluene	(86)	91:9
H	$Zr(OBu\text{-}t)_4$	1,2-dimethylimidazole	0.1	toluene	(76)	79.5:20.5
H	$Zr(OBu\text{-}t)_4$	2-methylimidazole	0.1	toluene	(28)	62:38
H	$Zr(OBu\text{-}t)_4$	4,5-dihydrooxazole	0.1	toluene	(86)	75:25
H	$Zr(OBu\text{-}t)_4$	2-methyl-4,5-dihydrooxazole	0.1	toluene	(81)	73:27
H	$Zr(OBu\text{-}t)_4$	none	0.1	toluene	(65)	45:55
H	$Zr(OBu\text{-}t)_4$	1-methylimidazole	0.05	toluene	(72)	83.5:16.5
H	$Zr(OBu\text{-}t)_4$	1-methylimidazole	0.2	toluene	(96)	94:6
H	$Zr(OBu\text{-}t)_4$	1-methylimidazole	0.3	toluene	(98)	94.5:5.5
H	$Zr(OBu\text{-}t)_4$	1-methylimidazole	0.5	toluene	(88)	95:5
H	$Hf(OBu\text{-}t)_4$	1-methylimidazole	0.1	toluene	(89)	86.5:13.5
H	$Hf(OBu\text{-}t)_4$	1-methylimidazole	0.2	toluene	(96)	92:8
H	$Ti(OBu\text{-}t)_4$	1-methylimidazole	0.1	toluene	(68)	69.5:30.5
H	$Ti(OBu\text{-}t)_4$	1-methylimidazole	0.2	toluene	(70)	81:19
Me	$Zr(OBu\text{-}t)_4$	1-methylimidazole	0.1	toluene	(79)	94.5:5.5
Me	$Zr(OBu\text{-}t)_4$	1-methylimidazole	0.2	toluene	(93)	96.5:3.5

TABLE 12. CYCLOADDITIONS WITH ASYMMETRIC CATALYSIS (*Continued*)

Imine	Diene	Conditions	Product(s) and Yield(s) (%)	Refs.
C_{17} 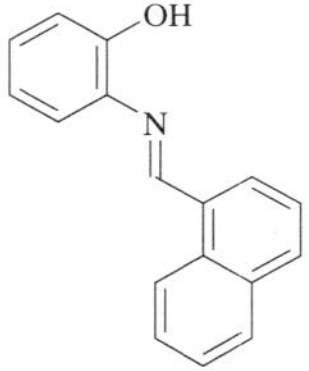	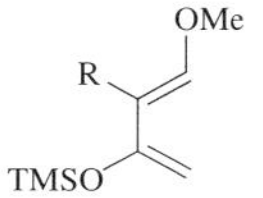	1. **13f**, Zr(OBu-*t*)$_4$, 1-methylimidazole, TMSCN, 3 Å MS, benzene, rt, 2 h 2. HCl, H_2O, THF, 0°, 30 min	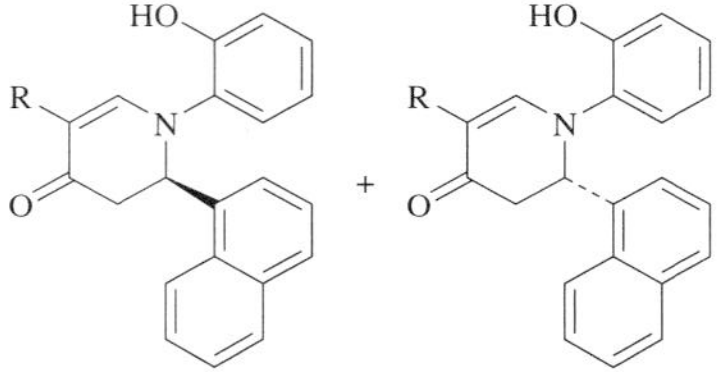	249
	R	Cat. eq		
	H	0.05	(80) 96:4	
	H	0.02	(67) 93:7	
	Me	0.02	(71) 92:8	
	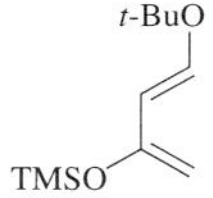	1. **13a**, Zr(OBu-*t*)$_4$, 1-methylimidazole[nn], CH_2Cl_2, –45° 2. HCl, H_2O, THF, 0°, 30 min	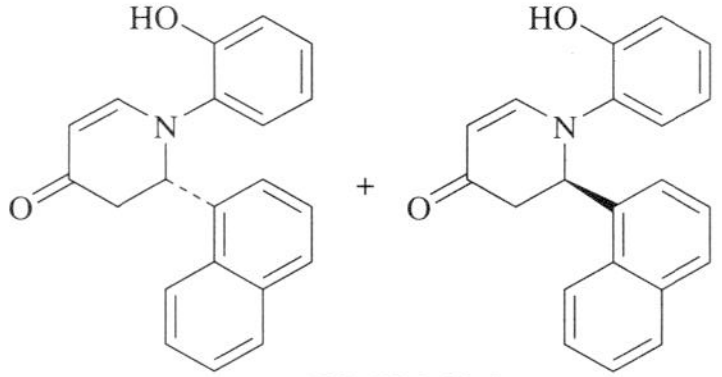 (83) 70.5:29.5	247
C_{18} 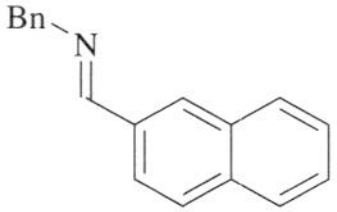	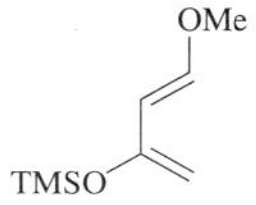	1. B(OPh)$_3$, (*R*)-1,1'-bi-2-naphthol, 4 Å MS, CH_2Cl_2, –78°, 5 h 2. H_2O	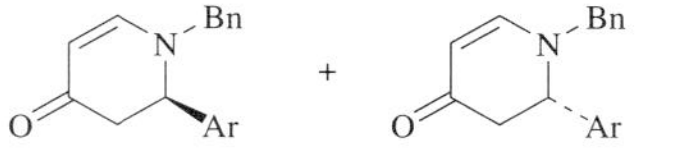 (83) 92:8 Ar = 2-naphthyl	243, 241
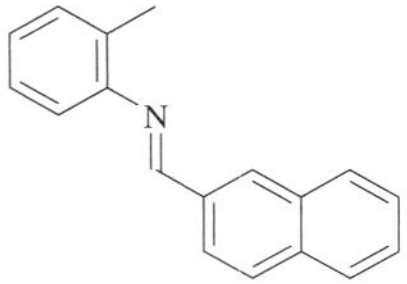	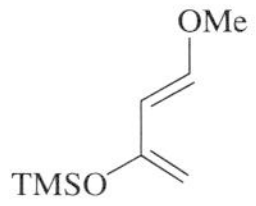	1. **15**, AgOAc, *i*-PrOH, THF, 4°, 12 h 2. HCl, H_2O, 5 min	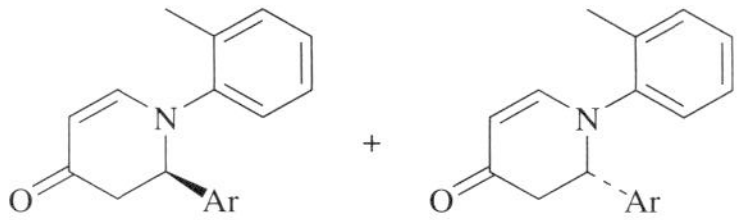 (>90) 67.5-70:30-32.5 Ar = 2-naphthyl	250

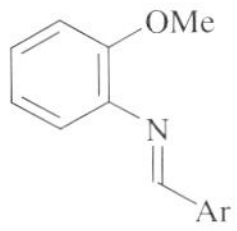

Ar = 2-naphthyl

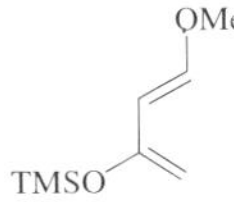

1. **15**, AgOAc, 4°
2. HCl, H_2O, 5 min

Additive[bb]	Solvent	Time
i-PrOH	THF	12 h
—	THF	24 h
—	H_2O, THF	12 h
H_2O	THF	12 h

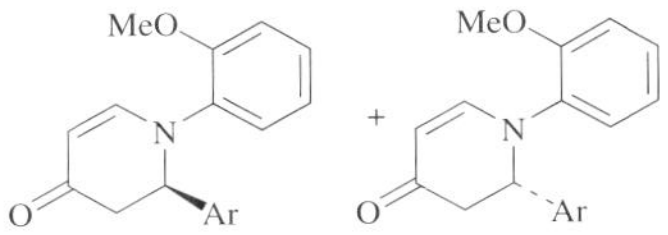

I + **II**

(>98) 97.5:2.5[cc,jj]
(30) 71.5:28.5
(>98) 95.5:4.5
(82) 97:3[kk]

250

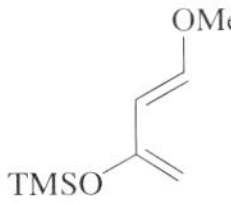

1. **19**, AgOAc, *i*-PrOH, THF, 4°, 12 h
2. HCl, H_2O

I + **II** (96) 93:7[pp]

250

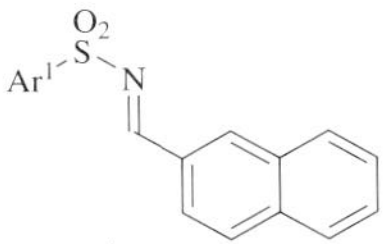

Ar[1]
4-MeC_6H_4
4-MeC_6H_4
4-$MeOC_6H_4$

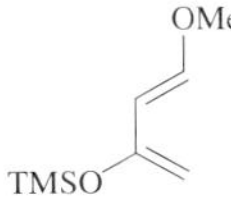

1. **6h**, $AgClO_4$, CH_2Cl_2
2. TFA, CH_2Cl_2, rt, 30-60 min

Temp	Time
rt	1-5 h
–20°	3 h
rt	3 h

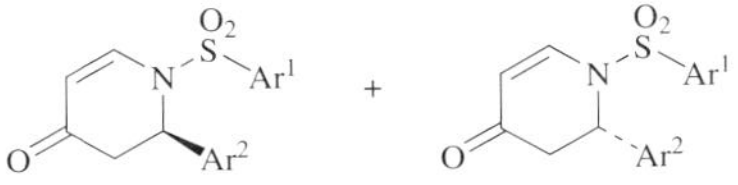

Ar = 2-naphthyl

(85) 93:7[n]
(58) 96.5:3.5[n]
(56) 91:9[n]

2

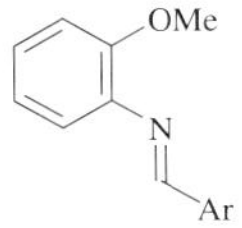

Ar = 1-naphthyl

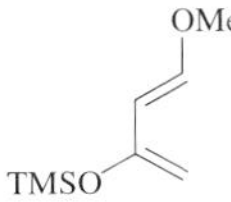

1. **13a**, $Zr(OBu\text{-}t)_4$, 1-methylimidazole, toluene, –45°
2. HCl, H_2O, THF, 0°, 30 min

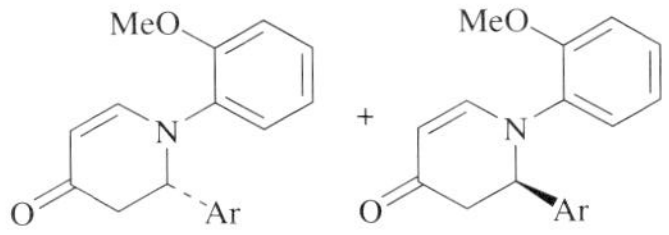

I + **II**

I + II (74) 62.5:37.5

247

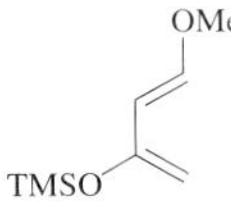

1. **15**, AgOAc, *i*-PrOH[bb], THF, 4°, 12 h
2. HCl, H_2O, 5 min

II + I (94) 95:5[cc]

250

TABLE 12. CYCLOADDITIONS WITH ASYMMETRIC CATALYSIS (*Continued*)

Imine	Diene	Conditions	Product(s) and Yield(s) (%)	Refs.
C_{19}				
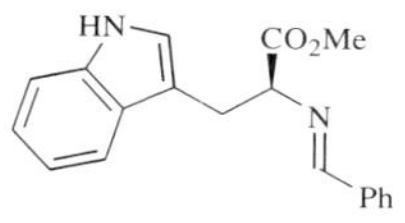	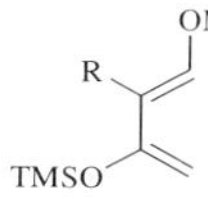	1. $B(OPh)_3$, (*R*) or (*S*)-1,1'-bi-2-naphthol, 4 Å MS, CH_2Cl_2, –40° to rt, overnight 2. HCl, H_2O	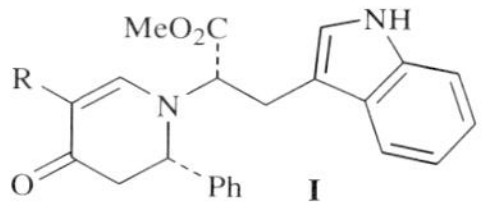 I + 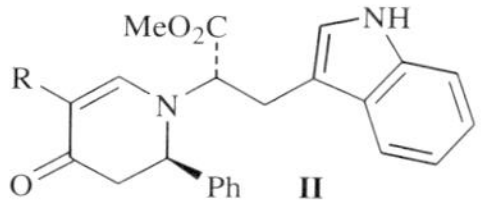 II	246, 245

R	Cat. config.	I + II	I:II
H	R	(34)	95:5
H	S	(24)	91:9
Et	R	(53)	96:4
Et	S	(40)	94:6

[a] The major product resulted from Mannich addition of the silyl enol ether to the imine and was obtained in 56% yield.

[b] The major product resulted from Mannich addition, and some of the Mannich product with a trimethylsilylamine was also observed.

[c] Using different Lewis acids and chiral ligands, such as Cu(I), Cu(II), Yb(III), Sn(II), Zn(II), Mg(II), and Ag(I) salts with chiral diols, diamines, diphosphines, amino alcohols, and bis-oxazolines, gave no detectable enantioselectivity.

[d] The reaction was run in the presence of various complexes of chiral BINAP and bisoxazoline ligands and Lewis acids such as $Zn(OTf)_2$, $Cu(OTf)_2$, CuOTf, $CuClO_4$, $AgSbF_6$, AgOTf, $AgClO_4$, $Pd(SbF_6)_2$, $Pd(ClO_4)_2$, $Pd(OTf)_2$, and $RuSbF_6$. No enantioselectivity was observed.

[e] Four Lewis acids ($Yb(OTf)_3$, MgI_2, $Cu(OTf)_2$, and $FeCl_3$), three ligands, two additives, and three solvents were tested using parallel screening. Only the results with greater than 40% ee were reported.

[f] The absolute stereochemistry of the major enantiomer was not determined. However, the same enantiomer was the major product under all of the conditions attempted.

[g] In a corrigendum[253] the authors state that after repeated efforts to duplicate the results, the yields are reproducible, but no asymmetric induction has been observed.

[h] The reaction was run on small scale as part of a parallel screen.

[i] The reaction was run on preparative scale (1 mmol).

[j] In a corrigendum[253] the authors state that after repeated efforts to duplicate the results, the yields are reproducible and typical enantiomeric excesses have varied from 0 to 55%.

[k] The absolute stereochemistry of the major enantiomer was not determined.

[l] The base was utilized during the conversion of the corresponding α-bromoglycinate to the imine, which was then reacted with the diene in situ.

[m] The stereochemistry of the cycloadduct was not specified, but it is probably the exo-adduct based on the results of similar imines with cyclopentadiene (see Ref. 110).

[n] The acyclic Mannich products are formed initially, but cyclize to the final products upon treatment with TFA.

[o] The reaction was performed at several temperatures (–78°, –40°, –20°, 0°, and rt). In all cases, adduct **I** is the major product.

[p] The reaction is very slow when conducted at –78°, and at rt the enantiomeric excess of the reaction is reduced significantly.

[q] The ene reaction products were also obtained as minor products.

[r] The diene was added slowly to the reaction.

[s] Other Lewis acids, such as $Zn(OTf)_2$, $Cu(OTf)_2$, $AgSbF_6$, AgOTf, $AgClO_4$, $Pd(SbF_6)_2$, $Pd(ClO_4)_2$, $Pd(OTf)_2$, and $RuSbF_6$, also catalyze the reaction. Good yields are obtained, but the enantioselectivity is significantly lower than with $CuClO_4$•MeCN. Use of 2CuOTf•C_6H_6 or $CuPF_6$•4MeCN leads to slight differences in enantioselectivity.

[t] Lower enantioselectivities are observed when diethyl ether, *t*-butyl methyl ether, acetonitrile, DMF, toluene, or benzotrifluoride are used as the solvents.

[u] A minor amount of the product resulting from silyl enol ether addition to the imine was also formed.

[v] The absolute stereochemistry of the cis-adduct **III** is not provided. The enantioselective excess for adduct **III** was less than 30%.

[w] The reaction mixture was treated with TFA in CH_2Cl_2 for 30 minutes at 0° before purification.

[x] The reaction was performed on 0.2 mmol scale.

[y] The reaction was performed on gram scale.

[z] The diastereoselectivity between the trans products **I** and **II** and the cis products **III** decreased in solvents other than THF.

[aa] Replacing the *tert*-butoxy groups with *n*-propoxy groups slightly decreased the selectivity of the reaction.

[bb] Other proton sources such as water, *tert*-butyl alcohol, or methanol could also be used.

[cc] Reactions with dipeptide phosphine catalysts were significantly less selective and efficient.

[dd] The cycloadduct was obtained with low enantiomeric excess.

[ee] A 2:1 ratio of (*R*)-binaphthol to trimethyl borate was utilized.

[ff] A 1:1 ratio of (*R*)-binaphthol to trimethyl borate was utilized.

[gg] A mixture of the imine and diene was slowly added to the catalyst over 1 hour.

[hh] Subsequent recycling of the catalyst gave 97% yields with a selectivity of 95:5.

[ii] The $Zr(OBu\text{-}t)_4$, 1-methylimidazole, 3 Å MS, and the binaphthol were premixed in benzene at 80° for 2.5 hours.

TABLE 12. CYCLOADDITIONS WITH ASYMMETRIC CATALYSIS (*Continued*)

Imine	Diene	Conditions	Product(s) and Yield(s) (%)	Refs.

[jj] The reaction was performed under a nitrogen atmosphere.

[kk] The reaction was performed in air.

[ll] The catalyst was premixed with the imine before addition of the diene.

[mm] The yields were determined by ^{1}H NMR.

[nn] The catalyst was prepared in benzene prior to performing the cycloaddition.

[oo] The catalyst was prepared in toluene prior to performing the cycloaddition.

[pp] The catalyst could be recycled four times. The results are as follows: cycle 2 (98) 91:9, cycle 3 (91) 88:12, cycle 4 (78) 86:14, and cycle 5 (77) 86:14.

REFERENCES

[1] Bur, S. K.; Martin, S. F. *Tetrahedron* **2001**, *57*, 3221.
[2] Mancheno, O. G.; Arrayas, R. G.; Carretero, J. C. *J. Am. Chem. Soc.* **2004**, *126*, 456.
[3] Weinreb, S. M.; Levin, J. I. *Heterocycles* **1979**, *12*, 949.
[4] Boger, D. L.; Weinreb, S. M. *Hetero Diels-Alder Methodology in Organic Synthesis*; Academic Press: San Diego, 1987.
[5] Weinreb, S. M. In *Comprehensive Organic Synthesis*; Trost, B. M., Fleming, I., Eds.; Pergamon Press: Oxford, 1991; Vol. 5, pp 401–449.
[6] Waldmann, H. *Synthesis* **1994**, 535.
[7] Jorgensen, K. A. *Angew. Chem., Int. Ed. Engl.* **2000**, *39*, 3558.
[8] Buonora, P.; Olsen, J.-C.; Oh, T. *Tetrahedron* **2001**, *57*, 6099.
[9] McCarrick, M. A.; Wu, Y.-D.; Houk, K. N. *J. Am. Chem. Soc.* **1992**, *114*, 1499.
[10] McCarrick, M. A.; Wu, Y.-D.; Houk, K. N. *J. Org. Chem.* **1993**, *58*, 3330.
[11] Whiting, A.; Windsor, C. M. *Tetrahedron* **1998**, *54*, 6035.
[12] Park, Y. S.; Lee, B.-S.; Lee, I. *New J. Chem.* **1999**, *23*, 707.
[13] Mayr, H.; Ofial, A. R.; Sauer, J.; Schmied, B. *Eur. J. Org. Chem.* **2000**, 2013.
[14] Domingo, L. R. *J. Org. Chem.* **2001**, *66*, 3211.
[15] Domingo, L. R.; Oliva, M.; Andres, J. *J. Mol. Struct. (Theochem)* **2001**, *544*, 79.
[16] Domingo, L. R.; Oliva, M.; Andres, J. *J. Org. Chem.* **2001**, *66*, 6151.
[17] Fleming, I. *Frontier Orbitals and Organic Chemical Reactions*; John Wiley and Sons: New York, 1976.
[18] Krow, G.; Rodebaugh, R.; Carmosin, R.; Figures, W.; Pannella, H.; DeVicaris, G.; Grippi, M. *J. Am. Chem. Soc.* **1973**, *95*, 5273.
[19] Lucchini, V.; Prato, M.; Scorrano, G.; Tecilla, P. *J. Org. Chem.* **1988**, *53*, 2251.
[20] Raasch, M. S. *J. Org. Chem.* **1975**, *40*, 161.
[21] Kasper, F.; Böttger, H. East German Patent DD 234676 A1 (1986); *Chem. Abstr.* **1987**, *106*, 50075.
[22] Krow, G. R.; Johnson, C. A.; Guare, J. P.; Kubrak, D.; Henz, K. J.; Shaw, D. A.; Szczepanski, S. W.; Carey, J. T. *J. Org. Chem.* **1982**, *47*, 5239.
[23] Krow, G. R.; Henz, K. J.; Szczepanski, S. W. *J. Org. Chem.* **1985**, *50*, 1888.
[24] Imagawa, T.; Sisido, K.; Kawanisi, M. *Bull. Chem. Soc. Jpn.* **1973**, *46*, 2922.
[25] Jankowski, K. *Tetrahedron Lett.* **1976**, 3309.
[26] Szczepanski, S. W.; Anouna, K. G. *Tetrahedron Lett.* **1996**, *37*, 8841.
[27] Krow, G.; Rodebaugh, R. *Org. Magn. Res.* **1973**, *5*, 73.
[28] Krow, G. R.; Damodaran, K. M.; Fan, D. M.; Rodebaugh, R.; Gaspari, A.; Nadir, U. K. *J. Org. Chem.* **1977**, *42*, 2486.
[29] Krow, G. R.; Johnson, C. A.; Boyle, M. *Tetrahedron Lett.* **1978**, 1971.
[30] Corey, E. J.; Yuen, P.-W. *Tetrahedron Lett.* **1989**, *30*, 5825.
[31] Krow, G. R.; Pyun, C.; Rodebaugh, R.; Marakowski, J. *Tetrahedron* **1974**, *30*, 2977.
[32] Krow, G.; Rodebaugh, R.; Marakowski, J.; Ramey, K. C. *Tetrahedron Lett.* **1973**, 1899.
[33] Sisko, J.; Weinreb, S. M. *Tetrahedron Lett.* **1989**, *30*, 3037.
[34] Fujii, T.; Kimura, T.; Furukawa, N. *Tetrahedron Lett.* **1995**, *36*, 4813.
[35] Hamley, P.; Holmes, A. B.; Kee, A.; Ladduwahetty, T.; Smith, D. F. *Synlett* **1991**, 29.
[36] Loven, R. P.; Zunnebeld, W. A.; Speckamp, W. N. *Tetrahedron* **1975**, *31*, 1723.
[37] Zunnebeld, W. A.; Speckamp, W. N. *Tetrahedron* **1975**, *31*, 1717.
[38] Rijsenbrij, P. P. M.; Loven, R.; Wijnberg, J. B. P. A.; Speckamp, W. N.; Huisman, H. O. *Tetrahedron Lett.* **1972**, 1425.
[39] Hamada, T.; Sato, H.; Hikota, M.; Yonemitsu, O. *Tetrahedron Lett.* **1989**, *30*, 6405.
[40] Morgan, P. E.; McCague, R.; Whiting, A. *J. Chem. Soc., Perkin Trans. 1* **2000**, 515.
[41] Jung, M. E.; Shishido, K.; Light, L.; Davis, L. *Tetrahedron Lett.* **1981**, *22*, 4607.
[42] Yamazaki, H.; Horikawa, H.; Nishitani, T.; Iwasaki, T.; Nosaka, K.; Tamaki, H. *Chem. Pharm. Bull.* **1992**, *40*, 102.
[43] Gaitanopoulos, D. E.; Weinstock, J. *J. Heterocycl. Chem.* **1985**, *22*, 957.
[44] Jäger, M.; Polborn, K.; Steglich, W. *Tetrahedron Lett.* **1995**, *36*, 861.
[45] Hedberg, C.; Pinho, P.; Roth, P.; Andersson, P. G. *J. Org. Chem.* **2000**, *65*, 2810.

[46] Hermitage, S.; Jay, D. A.; Whiting, A. *Tetrahedron Lett.* **2002**, *43*, 9633.
[47] Merten, R.; Müller, G. *Angew. Chem.* **1962**, *74*, 866.
[48] Cava, M. P.; Wilkins, C. K., Jr.; Dalton, D. R.; Bessho, K. *J. Org. Chem.* **1965**, *30*, 3772.
[49] Krow, G.; Rodebaugh, R.; Grippi, M.; Carmosin, R. *Synth. Commun.* **1972**, *2*, 211.
[50] Borne, R. F.; Clark, C. R.; Holbrook, J. M. *J. Med. Chem.* **1973**, *16*, 853.
[51] Baldwin, J. E.; Forrest, A. K.; Monaco, S.; Young, R. J. *J. Chem. Soc., Chem. Commun.* **1985**, 1586.
[52] Speckamp, W. N.; Barends, R. J. P.; de Gee, A. J.; Huisman, H. O. *Tetrahedron Lett.* **1970**, 383.
[53] Huang, Y.; Rawal, V. H. *Org. Lett.* **2000**, *2*, 3321.
[54] Quan, P. M.; Karns, T. K. B.; Quin, L. D. *J. Org. Chem.* **1965**, *30*, 2769.
[55] Merten, R.; Müller, G. *Chem. Ber.* **1964**, *97*, 682.
[56] Hartmann, P.; Obrecht, J.-P. *Synth. Commun.* **1988**, *18*, 553.
[57] Schrader, T.; Steglich, W. *Synthesis* **1990**, 1153.
[58] Kryukov, L. N.; Kryukova, L. Y.; Kolomiets, A. F. *J. Org. Chem. USSR* **1982**, *18*, 1638.
[59] Herdeis, C.; Engel, W. *Arch. Pharm. (Weinheim, Ger.)* **1992**, *325*, 411.
[60] Vor der Brück, D.; Bühler, R.; Plieninger, H. *Tetrahedron* **1972**, *28*, 791.
[61] Krow, G. R.; Rodebaugh, R.; Grippi, M.; DeVicaris, G.; Hyndman, C.; Marakowski, J. *J. Org. Chem.* **1973**, *38*, 3094.
[62] Baxter, A. J. G.; Holmes, A. B. *J. Chem. Soc., Perkin Trans. 1* **1977**, 2343.
[63] Fischer, G.; Fritz, H.; Prinzbach, H. *Tetrahedron Lett.* **1986**, *27*, 1269.
[64] Hall, H. K., Jr.; Ramezanian, M.; Saeva, F. D. *Tetrahedron Lett.* **1988**, *29*, 1235.
[65] Ramezanian, M.; Padias, A. B.; Saeva, F. D.; Hall, H. K., Jr. *J. Org. Chem.* **1990**, *55*, 1768.
[66] Ben-Ishai, D.; Goldstein, E. *Tetrahedron* **1971**, *27*, 3119.
[67] Goldstein, E.; Ben-Ishai, D. *Tetrahedron Lett.* **1969**, 2631.
[68] Tokita, S.; Hiruta, K.; Yaginuma, Y.; Ishikawa, S.; Nishi, H. *Synthesis* **1984**, 270.
[69] Kim, D.; Weinreb, S. M. *J. Org. Chem.* **1978**, *43*, 121.
[70] Weinreb, S. M.; Basha, F. Z.; Hibino, S.; Khatri, N. A.; Kim, D.; Pye, W. E.; Wu, T.-T. *J. Am. Chem. Soc.* **1982**, *104*, 536.
[71] Evnin, A. B.; Lam, A.; Blyskal, J. *J. Org. Chem.* **1970**, *35*, 3097.
[72] Gavina, F.; Costero, A. M.; Andreu, M. R.; Ayet, M. D. *J. Org. Chem.* **1991**, *56*, 5417.
[73] Occhiato, E. G.; Scarpi, D.; Machetti, F.; Guarna, A. *Tetrahedron* **1998**, *54*, 11589.
[74] Ben-Ishai, D.; Warshawsky, A. *J. Heterocycl. Chem.* **1971**, *8*, 865.
[75] Ben-Ishai, D.; Gillon, I.; Warshansky, A. *J. Heterocycl. Chem.* **1973**, *10*, 149.
[76] Pearce, D. S.; Locke, M. J.; Moore, H. W. *J. Am. Chem. Soc.* **1975**, *97*, 6181.
[77] Ueda, Y.; Maynard, S. C. *Tetrahedron Lett.* **1985**, *26*, 6309.
[78] Meyers, A. I.; Sowin, T. J.; Scholz, S.; Ueda, Y. *Tetrahedron Lett.* **1987**, *28*, 5103.
[79] Sowin, T. J.; Meyers, A. I. *J. Org. Chem.* **1988**, *53*, 4154.
[80] Bromidge, S.; Wilson, P. C.; Whiting, A. *Tetrahedron Lett.* **1998**, *39*, 8905.
[81] Bailey, P. D.; Smith, P. D.; Pederson, F.; Clegg, W.; Rosair, G. M.; Teat, S. J. *Tetrahedron Lett.* **2002**, *43*, 1067.
[82] Abraham, H.; Stella, L. *Tetrahedron* **1992**, *48*, 9707.
[83] Bailey, P. D.; Wilson, R. D.; Brown, G. R. *J. Chem. Soc., Perkin Trans. 1* **1991**, 1337.
[84] Bailey, P. D.; Brown, G. R.; Korber, F.; Reed, A.; Wilson, R. D. *Tetrahedron: Asymmetry* **1991**, *2*, 1263.
[85] Bailey, P. D.; Smith, P. D.; Morgan, K. M.; Rosair, G. M. *Tetrahedron Lett.* **2002**, *43*, 1071.
[86] Grieco, P. A.; Larsen, S. D.; Fobare, W. F. *Tetrahedron Lett.* **1986**, *27*, 1975.
[87] Mellor, J. M.; Richards, N. G. J.; Sargood, K. J.; Anderson, D. W.; Chamberlin, S. G.; Davies, D. E. *Tetrahedron Lett.* **1995**, *36*, 6765.
[88] Ward, S. E.; Holmes, A. B.; McCague, R. *Chem. Commun.* **1997**, 2085.
[89] Abraham, H.; Theus, E.; Stella, L. *Bull. Soc. Chim. Belg.* **1994**, *103*, 361.
[90] Tararov, V. I.; Kadyrov, R.; Kadyrova, Z.; Dubrovina, N.; Borner, A. *Tetrahedron: Asymmetry* **2002**, *13*, 25.
[91] Bertilsson, S. K.; Ekegren, J. K.; Modin, S. A.; Andersson, P. G. *Tetrahedron* **2001**, *57*, 6399.
[92] Szymanski, S.; Chapuis, C.; Jurczak, J. *Tetrahedron: Asymmetry* **2001**, *12*, 1939.
[93] Modin, S. A.; Andersson, P. G. *J. Org. Chem.* **2000**, *65*, 6736.

[94] Alonso, D. A.; Bertilsson, S. K.; Johnsson, S. Y.; Nordin, S. J. M.; Södergren, M. J.; Andersson, P. G. *J. Org. Chem.* **1999**, *64*, 2276.
[95] Alonso, D. A.; Guijarro, D.; Pinho, P.; Temme, O.; Andersson, P. G. *J. Org. Chem.* **1998**, *63*, 2749.
[96] Guijarro, D.; Pinho, P.; Andersson, P. G. *J. Org. Chem.* **1998**, *63*, 2530.
[97] Pinho, P.; Guijarro, D.; Andersson, P. G. *Tetrahedron* **1998**, *54*, 7897.
[98] Blanco, J. M.; Caamano, O.; Fernandez, F.; Garcia-Mera, X.; Lopez, C.; Rodriguez, G.; Rodriguez-Borges, J. E.; Rodriguez-Hergueta, A. *Tetrahedron Lett.* **1998**, *39*, 5663.
[99] Waldmann, H.; Braun, M. *Liebigs Ann. Chem.* **1991**, 1045.
[100] Bailey, P. D.; Londesbrough, D. J.; Hancox, T. C.; Heffernan, J. D.; Holmes, A. B. *J. Chem. Soc., Chem. Commun.* **1994**, 2543.
[101] Alonso-Silva, I. J.; Pardo, M.; Rabasco, D.; Soto, J. L. *J. Chem. Res. (S)* **1988**, 390.
[102] Alonso-Silva, I. J.; Pardo, M.; Rabasco, D.; Soto, J. L. *J. Chem. Res. (M)* **1988**, 2972.
[103] Ch'ng, H. S.; Hooper, M. *Tetrahedron Lett.* **1969**, 1527.
[104] Feineis, E.; Schwarz, H.; Hegmann, J.; Christl, M.; Peters, E.-M.; Peters, K.; Von Schnering, H. G. *Chem. Ber.* **1993**, *126*, 1743.
[105] Ager, D.; Cooper, N.; Cox, G. G.; Garro-Helion, F.; Harwood, L. M. *Tetrahedron: Asymmetry* **1996**, *7*, 2563.
[106] Fagan, P. J.; Nye, M. J. *J. Chem. Soc., Chem. Commun.* **1971**, 537.
[107] Fagan, P. J.; Neidert, E. E.; Nye, M. J.; O'Hare, M. J.; Tang, W.-P. *Can. J. Chem.* **1979**, *57*, 904.
[108] Trost, B. M.; Whitman, P. J. *J. Am. Chem. Soc.* **1974**, *96*, 7421.
[109] Albrecht, R.; Kresze, G. *Chem. Ber.* **1965**, *98*, 1431.
[110] Maggini, M.; Prato, M.; Scorrano, G. *Tetrahedron Lett.* **1990**, *31*, 6243.
[111] McFarlane, A. K.; Thomas, G.; Whiting, A. *Tetrahedron Lett.* **1993**, *34*, 2379.
[112] Shi, G.-Q.; Schlosser, M. *Tetrahedron* **1993**, *49*, 1445.
[113] Schürer, S. C.; Blechert, S. *Tetrahedron Lett.* **1999**, *40*, 1877.
[114] Heintzelman, G. R.; Weinreb, S. M.; Parvez, M. *J. Org. Chem.* **1996**, *61*, 4594.
[115] Hentemann, M. F.; Allen, J. G.; Danishefsky, S. J. *Angew. Chem., Int. Ed. Engl.* **2000**, *39*, 1937.
[116] Edwards, M. L.; Matt, J. E., Jr.; Wenstrup, D. L.; Kemper, C. A.; Perishetti, R. A.; Margolin, A. L. *Org. Prep. Proced. Int.* **1996**, *28*, 193.
[117] Holmes, A. B.; Raithby, P. R.; Thompson, J.; Baxter, A. J. G.; Dixon, J. *J. Chem. Soc., Chem. Commun.* **1983**, 1490.
[118] Holmes, A. B.; Kee, A.; Ladduwahetty, T.; Smith, D. F. *J. Chem. Soc., Chem. Commun.* **1990**, 1412.
[119] Holmes, A. B.; Thompson, J.; Baxter, A. J. G.; Dixon, J. *J. Chem. Soc., Chem. Commun.* **1985**, 37.
[120] Birkinshaw, T. N.; Holmes, A. B. *Tetrahedron Lett.* **1987**, *28*, 813.
[121] Birkinshaw, T. N.; Tabor, A. B.; Holmes, A. B.; Kaye, P.; Mayne, P. M.; Raithby, P. R. *J. Chem. Soc., Chem. Commun.* **1988**, 1599.
[122] Birkinshaw, T. N.; Tabor, A. B.; Holmes, A. B.; Raithby, P. R. *J. Chem. Soc., Chem. Commun.* **1988**, 1601.
[123] Hamada, T.; Zenkoh, T.; Sato, H.; Yonemitsu, O. *Tetrahedron Lett.* **1991**, *32*, 1649.
[124] McFarlane, A. K.; Thomas, G.; Whiting, A. *J. Chem. Soc., Perkin Trans. 1* **1995**, 2803.
[125] Hamley, P.; Helmchen, G.; Holmes, A. B.; Marshall, D. R.; MacKinnon, J. W. M.; Smith, D. F.; Ziller, J. W. *J. Chem. Soc., Chem. Commun.* **1992**, 786.
[126] Bauer, T.; Szymanski, S.; Jezewski, A.; Gluzinski, P.; Jurczak, J. *Tetrahedron: Asymmetry* **1997**, *8*, 2619.
[127] Abramovitch, R. A.; Stowers, J. R. *Heterocycles* **1984**, *22*, 671.
[128] Abramovitch, R. A.; Shinkai, I.; Mavunkel, B. J.; More, K. M.; O'Connor, S.; Ooi, G. H.; Pennington, W. T.; Srinivasan, P. C.; Stowers, J. R. *Tetrahedron* **1996**, *52*, 3339.
[129] DiBari, L.; Guillarme, S.; Hermitage, S.; Howard, J. A. K.; Jay, D. A.; Pescitelli, G.; Whiting, A.; Yufit, D. S. *Synlett* **2004**, 708.
[130] Böhme, H.; Hartke, K.; Muller, A. *Chem. Ber.* **1963**, *96*, 607.
[131] Babayan, A. T.; Martirosyan, G. T.; Grigoryan, D. V. *J. Org. Chem., USSR* **1968**, *4*, 955.
[132] Danishefsky, S.; Kitahara, T.; McKee, R.; Schuda, P. F. *J. Am. Chem. Soc.* **1976**, *98*, 6715.
[133] Larsen, S. D.; Grieco, P. A. *J. Am. Chem. Soc.* **1985**, *107*, 1768.
[134] Brandstadter, S. M.; Ojima, I. *Tetrahedron Lett.* **1987**, *28*, 613.

[135] Kobayashi, S.; Ishitani, H.; Nagayama, S. *Synthesis* **1995**, 1195.
[136] Ali, T.; Chauhan, K. K.; Frost, C. G. *Tetrahedron Lett.* **1999**, *40*, 5621.
[137] Laurent-Robert, H.; Garrigues, B.; Dubac, J. *Synlett* **2000**, 1160.
[138] Kumareswaran, R.; Reddy, B. G.; Vankar, Y. D. *Tetrahedron Lett.* **2001**, *42*, 7493.
[139] Collin, J.; Jaber, N.; Lannou, M. I. *Tetrahedron Lett.* **2001**, *42*, 7405.
[140] Akiyama, T.; Matsuda, K.; Fuchibe, K. *Synlett* **2002**, 1898.
[141] Kerwin, J. F., Jr.; Danishefsky, S. *Tetrahedron Lett.* **1982**, *23*, 3739.
[142] Bennett, D. M.; Okamoto, I.; Danheiser, R. L. *Org. Lett.* **1999**, *1*, 641.
[143] Regas, D.; Afonso, M. M.; Rodriguez, M. L.; Palenzuela, J. A. *J. Org. Chem.* **2003**, *68*, 7845.
[144] Kametani, T.; Takahashi, T.; Ogasawara, K.; Fukumoto, K. *Tetrahedron* **1974**, *30*, 1047.
[145] Padwa, A.; Harrison, B.; Norman, B. H. *Tetrahedron Lett.* **1989**, *30*, 3259.
[146] Padwa, A.; Gareau, Y.; Harrison, B.; Rodriguez, A. *J. Org. Chem.* **1992**, *57*, 3540.
[147] Yu, L.; Chen, D.; Wang, P. G. *Tetrahedron Lett.* **1996**, *37*, 2169.
[148] Babu, G.; Perumal, P. T. *Tetrahedron* **1998**, *54*, 1627.
[149] Akiba, K.-Y.; Motoshima, T.; Ishimaru, K.; Yabuta, K.; Hirota, H.; Yamamoto, Y. *Synlett* **1993**, 657.
[150] Ishimaru, K.; Yamamoto, Y.; Akiba, K.-Y. *Tetrahedron* **1997**, *53*, 5423.
[151] Padwa, A.; Gareau, Y.; Harrison, B.; Norman, B. H. *J. Org. Chem.* **1991**, *56*, 2713.
[152] Waldmann, H. *Angew. Chem., Int. Ed. Engl.* **1988**, *27*, 274.
[153] Waldmann, H. *Liebigs Ann. Chem.* **1989**, 231.
[154] Paugam, R.; Wartski, L. *Tetrahedron Lett.* **1991**, *32*, 491.
[155] Le Coz, L.; Veyrat-Martin, C.; Wartski, L.; Seyden-Penne, J.; Bois, C.; Philoche-Levisalles, M. *J. Org. Chem.* **1990**, *55*, 4870.
[156] Nogue, D.; Paugam, R.; Wartski, L. *Tetrahedron Lett.* **1992**, *33*, 1265.
[157] Barluenga, J.; Aznar, F.; Valdes, C.; Cabal, M.-P. *J. Org. Chem.* **1993**, *58*, 3391.
[158] Kündig, E. P.; Xu, L. H.; Romanens, P.; Bernardinelli, G. *Synlett* **1996**, 270.
[159] Midland, M. M.; McLoughlin, J. I. *Tetrahedron Lett.* **1988**, *29*, 4653.
[160] Midland, M. M.; Koops, R. W. *J. Org. Chem.* **1992**, *57*, 1158.
[161] Ishimaru, K.; Watanabe, K.; Yamamoto, Y.; Akiba, K.-Y. *Synlett* **1994**, 495.
[162] Badorrey, R.; Cativiela, C.; Diaz-de-Villegas, M. D.; Galvez, J. A. *Tetrahedron Lett.* **1997**, *38*, 2547.
[163] Badorrey, R.; Cativiela, C.; Diaz-de-Villegas, M. D.; Galvez, J. A. *Tetrahedron* **1999**, *55*, 7601.
[164] Badorrey, R.; Cativiela, C.; Diaz-de-Villegas, M. D.; Galvez, J. A. *Tetrahedron* **2002**, *58*, 341.
[165] Alcaide, B.; Almendros, P.; Alonso, J. M.; Aly, M. F. *Chem. Eur. J.* **2003**, *9*, 3415.
[166] Baldoli, C.; Del Buttero, P.; Di Ciolo, M.; Maiorana, S.; Papagni, A. *Synlett* **1996**, 258.
[167] Ratni, H.; Kündig, E. P. *Org. Lett.* **1999**, *1*, 1997.
[168] Ratni, H.; Crousse, B.; Kündig, E. P. *Synlett* **1999**, 626.
[169] Ishimaru, K.; Kojima, T. *J. Chem. Soc., Perkin Trans. 1* **2000**, 2105.
[170] Devine, P. N.; Reilly, M.; Oh, T. *Tetrahedron Lett.* **1993**, *34*, 5827.
[171] Kunz, H.; Pfrengle, W. *Angew. Chem., Int. Ed. Engl.* **1989**, *28*, 1067.
[172] Pfrengle, W.; Kunz, H. *J. Org. Chem.* **1989**, *54*, 4261.
[173] Weymann, M.; Pfrengle, W.; Schollmeyer, D.; Kunz, H. *Synthesis* **1997**, 1151.
[174] Weymann, M.; Schultz-Kukula, M.; Kunz, H. *Tetrahedron Lett.* **1998**, *39*, 7835.
[175] Kranke, B.; Hebrault, D.; Schultz-Kukula, M.; Kunz, H. *Synlett* **2004**, 671.
[175a] Waldmann, H.; Braun, M.; Drager, M. *Angew. Chem., Int. Ed. Engl.* **1990**, *29*, 1468.
[176] Barluenga, J.; Aznar, F.; Valdes, C.; Martin, A.; Garcia-Granda, S.; Martin, E. *J. Am. Chem. Soc.* **1993**, *115*, 4403.
[177] Barluenga, J.; Aznar, F.; Ribas, C.; Valdes, C.; Fernandez, M.; Cabal, M.-P.; Trujillo, J. *Chem. Eur. J.* **1996**, *2*, 805.
[178] Barluenga, J.; Aznar, F.; Valdes, C.; Ribas, C. *J. Org. Chem.* **1998**, *63*, 3918.
[179] Barluenga, J.; Aznar, F.; Ribas, C.; Valdes, C. *J. Org. Chem.* **1999**, *64*, 3736.
[180] Gilchrist, T. L. *Aldrichimica Acta* **2001**, *34*, 51.
[181] Hassner, A.; Anderson, D. J. *J. Am. Chem. Soc.* **1972**, *94*, 8255.
[182] Hassner, A.; Anderson, D. J. *J. Org. Chem.* **1974**, *39*, 3070.
[183] Anderson, D. J.; Hassner, A. *J. Am. Chem. Soc.* **1971**, *93*, 4339.
[184] Anderson, D. J.; Hassner, A. *J. Org. Chem.* **1973**, *38*, 2565.

[185] Anderson, D. J.; Hassner, A. *Synthesis* **1975**, 483.
[186] Hassner, A.; Anderson, D. J. *J. Org. Chem.* **1974**, *39*, 2031.
[187] Nair, V. *J. Org. Chem.* **1972**, *37*, 2508.
[188] Ray, C. A.; Risberg, E.; Somfai, P. *Tetrahedron Lett.* **2001**, *42*, 9289.
[189] Ray, C. A.; Risberg, E.; Somfai, P. *Tetrahedron* **2002**, *58*, 5983.
[190] Alves, M. J.; Gilchrist, T. L. *J. Chem. Soc., Perkin Trans. 1* **1998**, 299.
[191] Gilchrist, T. L.; Mendonca, R. *Arkivoc* **2000**, *1*, 769.
[192] Alves, M. J.; Azoia, N. G.; Bickley, J. F.; Fortes, A. G.; Gilchrist, T. L.; Mendonca, R. *J. Chem. Soc., Perkin Trans. 1* **2001**, 2969.
[193] Davis, F. A.; Wu, Y.; Yan, H.; Prasad, K. R.; McCoull, W. *Org. Lett.* **2002**, *4*, 655.
[194] Alves, M. J.; Gilchrist, T. L. *Tetrahedron Lett.* **1998**, *39*, 7579.
[195] Alves, M. J.; Almeida, I. G.; Fortes, A. G.; Freitas, A. P. *Tetrahedron Lett.* **2003**, *44*, 6561.
[196] Alves, M. J.; Buckley, J. F.; Gilchrist, T. L. *J. Chem. Soc., Perkin Trans. 1* **1999**, 1399.
[197] Alvares, Y. S. P.; Alves, M. J.; Azoia, N. G.; Bickley, J. F.; Gilchrist, T. L. *J. Chem. Soc., Perkin Trans. 1* **2002**, 1911.
[198] Timen, A. S.; Fischer, A.; Somfai, P. *Chem. Commun.* **2003**, 1150.
[199] Timen, A. S.; Somfai, P. *J. Org. Chem.* **2003**, *68*, 9958.
[200] Krolevets, A. A.; Popov, A. G.; Adamov, A. V.; Martynov, I. V. *Dokl. Chem. (Engl. Transl.)* **1988**, *303*, 353.
[201] Krolevets, A. A.; Adamov, A. V.; Popov, A. G.; Martynov, I. V. *Bull. Acad. Sci. USSR, Div. Chem. Sci.* **1988**, 1737.
[202] Biehler, J.-M.; Fleury, J.-P.; Perchais, J.; Regent, A. *Tetrahedron Lett.* **1968**, 4227.
[203] Biehler, J.-M.; Fleury, J.-P. *J. Heterocycl. Chem.* **1971**, *8*, 431.
[204] Fleury, J.-P.; Desbois, M.; See, J. *Bull. Soc. Chim. Fr.* **1978**, II-147.
[205] Dormagen, W.; Rotscheidt, K.; Breitmaier, E. *Synthesis* **1988**, 636.
[206] Katagiri, N.; Kurimoto, A.; Kaneko, C. *Chem. Pharm. Bull.* **1992**, *40*, 1737.
[207] Katagiri, N.; Kurimoto, A.; Kitano, K.; Nochi, H.; Sato, H.; Kaneko, C. In *Ninteenth Symposium on Nucleic Acids Chemistry*; IRL Press: Oxford: Fukuoka, Japan, 1992, p 83-84.
[208] Katagiri, N.; Nochi, H.; Kurimoto, A.; Sato, H.; Kaneko, C. *Chem. Pharm. Bull.* **1994**, *42*, 1251.
[209] Renslo, A. R.; Danheiser, R. L. *J. Org. Chem.* **1998**, *63*, 7840.
[210] Danheiser, R. L.; Renslo, A. R.; Amos, D. T.; Wright, G. T. *Org. Synth.* **2003**, *80*, 133.
[211] Madronero Pelaez, R.; Alvarez, E. F.; Lora-Tamayo, M. *Anales real soc. espan. fis y quim.* **1955**, *51B*, 276.
[212] Madronero, R.; Alvarez, E. F.; Lora-Tamayo, M. *Anales real soc. espan. fis y quim.* **1955**, *51B*, 465.
[213] Lora-Tamayo, M.; Lopez Aparicio, T.; Madronero, R. *Anales real soc. espan. fis y quim.* **1958**, *54B*, 567.
[214] Lora-Tamayo, M.; Garcia Munoz, G.; Madronero, R. *Bull. Soc. Chim. Fr.* **1958**, 1334.
[215] Brownlee, T. H. US Patent 3,594,385 (1971); *Chem. Abstr.* **1971**, *75*, 88486.
[216] Barluenga, J.; Aznar, F.; Fernandez, M. *Tetrahedron Lett.* **1995**, *36*, 6551.
[217] Chou, S.-S. P.; Hung, C.-C. *Tetrahedron Lett.* **2000**, *41*, 8323.
[218] Chou, S.-S. P.; Hung, C.-C. *Synthesis* **2001**, 2450.
[219] Arbuzov, B. A.; Zobova, N. N. *Dokl. Chem. (Engl. Transl.)* **1966**, *167*, 379.
[220] Marchand-Brynaert, J.; Ghosez, L. *Tetrahedron Lett.* **1974**, 377.
[221] Geisler, A.; Würthwein, E.-U. *Tetrahedron Lett.* **1994**, *35*, 77.
[222] Weinreb, S. M.; Khatri, N. A.; Shringarpure, J. *J. Am. Chem. Soc.* **1979**, *101*, 5073.
[223] Khatri, N. A.; Schmitthenner, H. F.; Shringarpure, J.; Weinreb, S. M. *J. Am. Chem. Soc.* **1981**, *103*, 6387.
[224] Berkowitz, W. F.; John, T. V. *J. Org. Chem.* **1984**, *49*, 5269.
[225] Shea, K. J.; Lease, T. G.; Ziller, J. W. *J. Am. Chem. Soc.* **1990**, *112*, 8627.
[226] Lease, T. G.; Shea, K. J. *J. Am. Chem. Soc.* **1993**, *115*, 2248.
[227] Bremmer, M. L.; Khatri, N. A.; Weinreb, S. M. *J. Org. Chem.* **1983**, *48*, 3661.
[228] Bremmer, M. L.; Weinreb, S. M. *Tetrahedron Lett.* **1983**, *24*, 261.
[229] Nader, B.; Bailey, T. R.; Franck, R. W.; Weinreb, S. M. *J. Am. Chem. Soc.* **1981**, *103*, 7573.
[230] Gobao, R. A.; Bremmer, M. L.; Weinreb, S. M. *J. Am. Chem. Soc.* **1982**, *104*, 7065.

[231] Nader, B.; Franck, R. W.; Weinreb, S. M. *J. Am. Chem. Soc.* **1980**, *102*, 1153.
[232] Amos, D. T.; Renslo, A. R.; Danheiser, R. L. *J. Am. Chem. Soc.* **2003**, *125*, 4970.
[233] Parker, D. T. In *Organic Synthesis in Water*; Grieco, P. A., Ed.; Blackie Academic & Professional: London, 1998; pp 47–81.
[234] Grieco, P. A.; Larsen, S. D. *J. Org. Chem.* **1986**, *51*, 3553.
[235] Grieco, P. A.; Parker, D. T. *J. Org. Chem.* **1988**, *53*, 3658.
[236] Grieco, P. A.; Kaufman, M. D. *J. Org. Chem.* **1999**, *64*, 6041.
[237] Regas, D.; Afonso, M. M.; Palenzuela, J. A. *Synthesis* **2004**, 757.
[238] Oppolzer, W. *Angew. Chem., Int. Ed. Engl.* **1972**, *11*, 1031.
[239] Bland, D. C.; Raudenbush, B. C.; Weinreb, S. M. *Org. Lett.* **2000**, *2*, 4007.
[240] Jorgensen, K. A. *Eur. J. Org. Chem.* **2004**, 2093.
[241] Hattori, K.; Yamamoto, H. *J. Org. Chem.* **1992**, *57*, 3264.
[242] Hattori, K.; Yamamoto, H. *Synlett* **1993**, 129.
[243] Hattori, K.; Yamamoto, H. *Tetrahedron* **1993**, *49*, 1749.
[244] Ishihara, K.; Miyata, M.; Hattori, K.; Tada, T.; Yamamoto, H. *J. Am. Chem. Soc.* **1994**, *116*, 10520.
[245] Lock, R.; Waldmann, H. *Tetrahedron Lett.* **1996**, *37*, 2753.
[246] Lock, R.; Waldmann, H. *Chem. Eur. J.* **1997**, *3*, 143.
[247] Kobayashi, S.; Komiyama, S.; Ishitani, H. *Angew. Chem., Int. Ed. Engl.* **1998**, *37*, 979.
[248] Kobayashi, S.; Kusakabe, K.; Komiyama, S.; Ishitani, H. *J. Org. Chem.* **1999**, *64*, 4220.
[249] Kobayashi, S.; Kusakabe, K.; Ishitani, H. *Org. Lett.* **2000**, *2*, 1225.
[250] Josephsohn, N. S.; Snapper, M. L.; Hoveyda, A. H. *J. Am. Chem. Soc.* **2003**, *125*, 4018.
[251] Yao, S.; Johannsen, M.; Hazell, R. G.; Jorgensen, K. A. *Angew. Chem., Int. Ed. Engl.* **1998**, *37*, 3121.
[252] Yao, S.; Saaby, S.; Hazell, R. G.; Jorgensen, K. A. *Chem. Eur. J.* **2000**, *6*, 2435.
[253] Bundu, A.; Guillarme, S.; Hannan, J.; Wan, H.; Whiting, A. *Tetrahedron Lett.* **2003**, *44*, 7849.
[253a] Guillarme, S.; Whiting, A. *Synlett* **2004**, 711.
[254] Kametani, T.; Hibino, S. *Adv. Heterocycl. Chem.* **1987**, *42*, 245.
[255] Basha, F. Z.; Hibino, S.; Kim, D.; Pye, W. E.; Wu, T.-T.; Weinreb, S. M. *J. Am. Chem. Soc.* **1980**, *102*, 3962.
[256] Han, G.; LaPorte, M. G.; Folmer, J. J.; Werner, K. M.; Weinreb, S. M. *J. Org. Chem.* **2000**, *65*, 6293.
[257] Han, G.; LaPorte, M. G.; Folmer, J. J.; Werner, K. M.; Weinreb, S. M. *Angew. Chem., Int. Ed. Engl.* **2000**, *39*, 237.
[258] Schmitthenner, H. F.; Weinreb, S. M. *J. Org. Chem.* **1980**, *45*, 3372.
[259] Grieco, P. A.; Parker, D. T. *J. Org. Chem.* **1988**, *53*, 3325.
[260] Kaufman, M. D.; Grieco, P. A. *J. Org. Chem.* **1994**, *59*, 7197.
[261] Grieco, P. A.; Kaufman, M. D. *J. Org. Chem.* **1999**, *64*, 7586.
[262] Oppolzer, W.; Francotte, E.; Bättig, K. *Helv. Chim. Acta* **1981**, *64*, 478.
[263] Laschat, S.; Dickner, T. *Synthesis* **2000**, 1781.
[264] Felpin, F.-X.; Lebreton, J. *Current Organic Synthesis* **2004**, *1*, 83.
[265] Thyagarajan, G.; May, E. L. *J. Heterocycl. Chem.* **1971**, *8*, 465.
[266] Guilloteau-Bertin, B.; Compere, D.; Gil, L.; Marazano, C.; Das, B. C. *Eur. J. Org. Chem.* **2000**, 1391.
[267] Angle, S. R.; Henry, R. M. *J. Org. Chem.* **1998**, *63*, 7490.
[268] Walters, M. A. *Prog. Heterocycl. Chem.* **2003**, *15*, 1.
[269] Deiters, A.; Martin, S. F. *Chem. Rev.* **2004**, *104*, 2199.
[270] Agami, C.; Couty, F.; Rabasso, N. *Tetrahedron* **2001**, *57*, 5393.
[271] Ahman, J.; Somfai, P. *Tetrahedron* **1995**, *51*, 9747.
[272] Wuts, P. G. M.; Jung, Y.-W. *J. Org. Chem.* **1988**, *53*, 1957.
[273] Blumenkopf, T.; Overman, L. E. *Chem. Rev.* **1986**, *86*, 857.
[274] Franciotti, M.; Mann, A.; Mordini, A.; Taddei, M. *Tetrahedron Lett.* **1993**, *34*, 1355.
[275] Comins, D. L. *J. Heterocycl. Chem.* **1999**, *36*, 1491.
[276] Heintzelman, G. R.; Fang, W.-K.; Keen, S. P.; Wallace, G. A.; Weinreb, S. M. *J. Am. Chem. Soc.* **2002**, *124*, 3939.
[277] Barco, A.; Benetti, S.; Baraldi, P. G.; Moroder, F.; Pollini, G. P.; Simoni, D. *Liebigs Ann. Chem.* **1982**, 960.
[278] Sestelo, J. P.; del Mar Real, M.; Sarandeses, L. A. *J. Org. Chem.* **2001**, *66*, 1395.
[279] Grieco, P. A.; Larsen, S. D. *Org. Synth.* **1990**, *68*, 206.

[280] Fennhoff, G.; Heesing, A. *Chem. Ber.* **1989**, *122*, 1153.
[281] Merten, R. Belgian Patent 608,904 (1962); *Chem. Abstr.* **1963**, *59*, 2781d.
[282] Merten, R. Belgian Patent 611,643 (1962); *Chem. Abstr.* **1962**, *57*, 16573b.
[283] Cava, M. P.; Wilkins, C. K., Jr. *Chem. Ind. (London)* **1964**, 1422.
[284] Fray, A. H.; Augeri, D. J.; Kleinman, E. F. *J. Org. Chem.* **1988**, *53*, 896.
[285] Hobson, J. D.; Riddell, W. D. *J. Chem. Soc., Chem. Commun.* **1968**, 1180.
[286] Skvarchenko, V. R.; Koshkina, N. P. *J. Org. Chem. USSR* **1979**, *15*, 2142.
[287] Kasper, F.; Bottger, H. *Z. Chem.* **1987**, *27*, 70.
[288] Arbuzov, Y. A.; Klimova, E. I.; Antonova, N. D.; Tomilov, Y. V. *J. Org. Chem. USSR* **1974**, *10*, 1178.
[289] Ben-Ishai, D.; Hirsch, S. *Tetrahedron Lett.* **1983**, *24*, 955.
[290] Harter, H. P.; Liisberg, S. *Acta Chem. Scand.* **1968**, *22*, 2685.
[291] Quan, P. M.; Karns, T. K. B.; Quin, L. D. *Chem. Ind. (London)* **1964**, 1553.
[292] Bunch, L.; Liljefors, T.; Greenwood, J. R.; Frydenvang, K.; Bräuner-Osborne, H.; Krogsgaard-Larsen, P.; Madsen, U. *J. Org. Chem.* **2003**, *68*, 1489.
[293] Heine, H. W.; Schairer, W. C.; Suriano, J. A.; Williams, E. A. *Tetrahedron* **1988**, *44*, 3181.
[294] Heine, H. W.; Barchiesi, B. J.; Williams, E. A. *J. Org. Chem.* **1984**, *49*, 2560.
[295] Edwards, O. E.; Greaves, A. M.; Sy, W.-W. *Can. J. Chem.* **1988**, *66*, 1163.
[296] Gavina, F.; Costero, A. M.; Andreu, M. R. *J. Org. Chem.* **1990**, *55*, 434.
[297] Gavina, F.; Costero, A. M.; Andreu, M. R.; Carda, M.; Luis, S. V. *J. Am. Chem. Soc.* **1988**, *110*, 4017.
[298] Warshawsky, A.; Ben-Ishai, D. *J. Heterocycl. Chem.* **1970**, *7*, 917.
[299] Pilli, R. A.; Dias, L. C.; Maldaner, A. O. *Tetrahedron Lett.* **1993**, *34*, 2729.
[300] Pilli, R. A.; Dias, L. C.; Maldaner, A. O. *J. Org. Chem.* **1995**, *60*, 717.
[301] Ben-Ishai, D.; Inbal, Z.; Warshawsky, A. *J. Heterocycl. Chem.* **1970**, *7*, 615.
[302] Alonso-Silva, I. J.; Pardo, M.; Soto, J. L. *Heterocycles* **1988**, *27*, 357.
[303] Warshawsky, A.; Ben-Ishai, D. *J. Heterocycl. Chem.* **1969**, *6*, 681.
[304] Guarna, A.; Occhiato, E. G.; Scarpi, D.; Tsai, R.; Danza, G.; Comerci, A.; Mancina, R.; Serio, M. *Bioorg. Med. Chem. Lett.* **1998**, *8*, 2871.
[305] Danieli, B.; Lesma, G.; Passarella, D.; Piacenti, P.; Sacchetti, A.; Silvani, A.; Virdis, A. *Tetrahedron Lett.* **2002**, *43*, 7155.
[306] Guarna, A.; Occhiato, E. G.; Machetti, F.; Scarpi, D. *J. Org. Chem.* **1998**, *63*, 4111.
[307] Guarna, A.; Occhiato, E. G.; Machetti, F.; Giacomelli, V. *J. Org. Chem.* **1999**, *64*, 4985.
[308] Kobayashi, T.; Ono, K.; Kato, H. *Bull. Chem. Soc. Jpn.* **1992**, *65*, 61.
[309] Hursthouse, M. B.; Malik, K. M. A.; Hibbs, D. E.; Roberts, S. M.; Seago, A. J. H.; Sik, V.; Storer, R. *J. Chem. Soc., Perkin Trans. 1* **1995**, 2419.
[310] Chou, S.-S. P.; Hung, C.-C. *Synth. Commun.* **2002**, *32*, 3119.
[311] Chou, S.-S. P.; Hung, C.-C. *Synth. Commun.* **2001**, *31*, 1097.
[312] Bourgeois-Cury, A.; Doan, D.; Gore, J. *Tetrahedron Lett.* **1992**, *33*, 1277.
[313] Perrin, V.; Riveron, V.; Traversa, C.; Balme, G.; Gore, J. *J. Chem. Res. (M)* **1998**, 3121.
[314] Brandt, P.; Hedberg, C.; Lawonn, K.; Pinho, P.; Andersson, P. G. *Chem. Eur. J.* **1999**, *5*, 1692.
[315] Stella, L.; Abraham, H.; Feneau-Dupont, J.; Tinant, B.; Declercq, J. P. *Tetrahedron Lett.* **1990**, *31*, 2603.
[316] Bertilsson, S. K.; Andersson, P. G. *J. Organomet. Chem.* **2000**, *603*, 13.
[317] Ekegren, J. K.; Modin, S. A.; Alonso, D. A.; Andersson, P. G. *Tetrahedron: Asymmetry* **2002**, *13*, 447.
[318] Bailey, P. D.; Wilson, R. D.; Brown, G. R. *Tetrahedron Lett.* **1989**, *30*, 6781.
[319] Barluenga, J.; Fernandez, M. A.; Aznar, F.; Valdes, C. *Tetrahedron Lett.* **2002**, *43*, 8159.
[320] Zhang, W.; Xie, W.; Fang, J.; Wang, P. G. *Tetrahedron Lett.* **1999**, *40*, 7929.
[321] Nakano, H.; Kumagai, N.; Kabuto, C.; Matsuzaki, H.; Hongo, H. *Tetrahedron: Asymmetry* **1995**, *6*, 1233.
[322] Nakano, H.; Kumagai, N.; Matsuzaki, H.; Kabuto, C.; Hongo, H. *Tetrahedron: Asymmetry* **1997**, *8*, 1391.
[323] Nakano, H.; Iwasa, K.; Hongo, H. *Heterocycles* **1997**, *44*, 435.
[324] Maison, W.; Adiwidjaja, G. *Tetrahedron Lett.* **2002**, *43*, 5957.
[325] Maison, W.; Kuntzer, D.; Grohs, D. *Synlett* **2002**, 1795.
[326] Shi, Z.-D.; Yang, B.-H.; Wu, Y.-L.; Pan, Y.-J.; Ji, Y.-Y.; Yeh, M. *Bioorg. Med. Chem. Lett.* **2002**, *12*, 2321.

[327] Trova, M. P.; McGee, K. F., Jr. *Tetrahedron* **1995**, *51*, 5951.
[328] McKay, W. R.; Proctor, G. R. *J. Chem. Soc., Perkin Trans. 1* **1981**, 2443.
[329] Prato, M.; Quintily, U.; Scorrano, G. *Gazz. Chim. Ital.* **1984**, *114*, 405.
[330] Kobayashi, S.; Ishitani, H.; Nagayama, S. *Chem. Lett.* **1995**, 423.
[331] Kobayashi, S.; Akiyama, R.; Kitagawa, H. *J. Comb. Chem.* **2001**, *3*, 196.
[332] Kobayashi, S.; Akiyama, R.; Kitagawa, H. *J. Comb. Chem.* **2000**, *2*, 438.
[333] Södergren, M. J.; Andersson, P. G. *Tetrahedron Lett.* **1996**, *37*, 7577.
[334] Genov, M.; Scherer, G.; Studer, M.; Pfaltz, A. *Synthesis* **2002**, 2037.
[335] Fernandez, F.; Garcia-Mera, X.; Rodriguez-Borges, J. E.; Vale, M. L. C. *Tetrahedron Lett.* **2003**, *44*, 431.
[336] Il'in, G. F.; Kolomiets, A. F. *Zh. Vses. Khim. O-va.* **1980**, *25*, 705.
[337] Kasper, F.; Dathe, S. *J. Prakt. Chem.* **1985**, *327*, 1041.
[338] Kresze, G.; Albrecht, R. *Chem. Ber.* **1964**, *97*, 490.
[339] Kresze, G.; Wagner, U. *Liebigs Ann. Chem.* **1972**, *762*, 106.
[340] Kresze, G.; Albrecht, R. *Angew. Chem., Int. Ed. Engl.* **1962**, *1*, 595.
[341] Schürer, S. C.; Blechert, S. *Chem. Commun.* **1999**, 1203.
[342] Craig, D.; Robson, M. J.; Shaw, S. J. *Synlett* **1998**, 1381.
[343] Skvarchenko, V. R.; Lapteva, V. L.; Gorbunova, M. A. *J. Org. Chem., USSR* **1990**, *26*, 2244.
[344] Grieco, P. A.; Carroll, W. A. *Tetrahedron Lett.* **1992**, *33*, 4401.
[345] Möhrle, H.; Dwuletzki, H. *Arch. Pharm. (Weinheim, Ger.)* **1987**, *320*, 298.
[346] Katritzky, A. R.; Gupta, V.; Gordeev, M. *J. Heterocycl. Chem.* **1993**, *30*, 1073.
[347] Katritzky, A. R.; Gordeev, M. F. *J. Org. Chem.* **1993**, *58*, 4049.
[348] Zeifman, Y. V.; Gambaryan, N. P.; Knunyants, I. L. *Bull. Acad. Sci. USSR, Div. Chem. Sci.* **1965**, 1431.
[349] Middleton, W. J.; Krespan, C. G. *J. Org. Chem.* **1965**, *30*, 1398.
[350] Grieco, P. A.; Parker, D. T.; Fobare, W. F.; Ruckle, R. *J. Am. Chem. Soc.* **1987**, *109*, 5859.
[351] Danishefsky, S. J.; Vogel, C. *J. Org. Chem.* **1986**, *51*, 3915.
[352] Al'bekov, V. A.; Benda, A. F.; Gontar, A. F.; Sokol'skii, G. A.; Knunyants, I. L. *Bull. Acad. Sci. USSR, Div. Chem. Sci.* **1986**, 1305.
[353] Al'bekov, V. A.; Benda, A. F.; Gontar, A. F.; Sokol'skii, G. A.; Knunyants, I. L. *Bull. Acad. Sci. USSR, Div. Chem. Sci.* **1988**, 777.
[354] Grieco, P. A.; Bahsas, A. *J. Org. Chem.* **1987**, *52*, 5746.
[355] Bohlmann, F.; Habeck, D.; Poetsch, E.; Schumann, D. *Chem. Ber.* **1967**, *100*, 2742.
[356] Huang, P.; Isayan, K.; Sarkissian, A.; Oh, T. *J. Org. Chem.* **1998**, *63*, 4500.
[357] Dannhardt, G.; Wiegrebe, W. *Arch. Pharm. (Weinheim, Ger.)* **1977**, *310*, 802.
[358] Staninets, V. I.; Mironova, D. F.; Iksanova, S. V.; Sinitsa, A. D. *Ukr. Khim. Zh. (Russian Ed.)* **1990**, *56*, 1321.
[359] Krolevets, A. A.; Adamov, A. V.; Popov, A. G.; Martynov, I. V. *J. Gen. Chem. USSR (Engl. Transl.)* **1988**, *58*, 2331.
[360] Brown Ripin, D. H.; Abele, S.; Cai, W.; Blumenkopf, T.; Casavant, J.-M.; Doty, J. L.; Flanagan, M.; Koecher, C.; Laue, K. W.; McCarthy, K.; Meltz, C.; Munchhoff, M.; Pouwer, K.; Shah, B.; Sun, J.; Teixeira, J.; Vries, T.; Whipple, D. A.; Wilcox, G. *Org. Process Res. Dev.* **2003**, *7*, 115.
[361] Grieco, P. A.; Bahsas, A. *Tetrahedron Lett.* **1988**, *29*, 5855.
[362] Pombo-Villar, E.; Boelsterli, J.; Cid, M. M.; France, J.; Fuchs, B.; Walkinshaw, M.; Weber, H.-P. *Helv. Chim. Acta* **1993**, *76*, 1203.
[363] Yu, L.; Li, J.; Ramirez, J.; Chen, D.; Wang, P. G. *J. Org. Chem.* **1997**, *62*, 903.
[364] Roberts, S.; Smith, C.; Thomas, R. J. *J. Chem. Soc., Perkin Trans. 1* **1990**, 1493.
[365] Hanley, J. A.; Forsyth, D. A. *J. Labeled Comp. Radiopharm.* **1990**, *28*, 307.
[366] Cid, M. M.; Eggnauer, U.; Weber, H. P.; Pombo-Villar, E. *Tetrahedron Lett.* **1991**, *32*, 7233.
[367] Pombo-Villar, E.; Weber, H.P.; Boddeke, H.W.G.M. *Bioorg. Med. Chem. Lett.* **1992**, *2*, 501.
[368] Boumendjel, A.; Roberts, J. C.; Hu, E.; Pallai, P. V.; Rebek, J., Jr. *J. Org. Chem.* **1996**, *61*, 4434.
[369] Chiu, C. K.-F. *Synth. Commun.* **1996**, *26*, 577.
[370] Murata, K.; Kitazume, T. *Isr. J. Chem.* **1999**, *39*, 163.
[371] Kitazume, T.; Murata, K.; Okabe, A.; Takahashi, Y.; Yamazaki, T. *Tetrahedron: Asymmetry* **1994**, *5*, 1029.

[372] Akiyama, T.; Takaya, J.; Kagoshima, H. *Tetrahedron Lett.* **1999**, *40*, 7831.
[373] Yuan, Y.; Li, X.; Ding, K. *Org. Lett.* **2002**, *4*, 3309.
[374] Cardillo, G.; Fabbroni, S.; Gentilucci, L.; Perciaccante, R.; Piccinelli, F.; Tolomelli, A. *Tetrahedron* **2004**, *60*, 5031.
[375] Barluenga, J.; Mateos, C.; Aznar, F.; Valdes, C. *Org. Lett.* **2002**, *4*, 1971.
[376] Garcia, A.-B.; Valdes, C.; Cabal, M. P. *Tetrahedron Lett.* **2004**, *45*, 4357.
[377] Barluenga, J.; Mateos, C.; Aznar, F.; Valdes, C. *Org. Lett.* **2002**, *4*, 3667.
[378] Wang, Y.; Wilson, S. R. *Tetrahedron Lett.* **1997**, *38*, 4021.
[379] Kirschbaum, S.; Waldmann, H. *Tetrahedron Lett.* **1997**, *38*, 2829.
[380] Kirschbaum, S.; Waldmann, H. *J. Org. Chem.* **1998**, *63*, 4936.
[381] Ishimaru, K.; Kojima, T. *Heterocycles* **2001**, *55*, 1591.
[382] Furman, B.; Dziedzic, M. *Tetrahedron Lett.* **2003**, *44*, 6629.
[383] Deleted reference.
[384] Waldmann, H.; Braun, M. *J. Org. Chem.* **1992**, *57*, 4444.
[385] Loncaric, C.; Manabe, K.; Kobayashi, S. *Chem. Commun.* **2003**, 574.
[386] Loncaric, C.; Manabe, K.; Kobayashi, S. *Adv. Synth. Catal.* **2003**, *345*, 475.
[387] Kessar, S. V.; Singh, T.; Mankotia, A. K. S. *J. Chem. Soc., Chem. Commun.* **1989**, 1692.
[388] Kobayashi, S.; Araki, M.; Ishitani, H.; Nagayama, S.; Hachiya, I. *Synlett* **1995**, 233.
[389] Kawecki, R. *Synthesis* **2001**, 828.
[390] Kametani, T.; Katoh, Y.; Fukumoto, K. *J. Chem. Soc., Perkin Trans. 1* **1974**, 1712.
[391] Kametani, T.; Takahashi, T.; Honda, T.; Ogasawara, K.; Fukumoto, K. *J. Org. Chem.* **1974**, *39*, 447.
[392] Langlois, Y.; Pouilhes, A.; Genin, D.; Andriamialisoa, R. Z.; Langlois, N. *Tetrahedron* **1983**, *39*, 3755.
[393] Genin, D.; Andriamialisoa, R. Z.; Langlois, N.; Langlois, Y. *J. Org. Chem.* **1987**, *52*, 353.
[394] Palmisano, G.; Danieli, B.; Lesma, G.; Passarella, D. *Tetrahedron* **1989**, *45*, 3583.
[395] Vacca, J. P. *Tetrahedron Lett.* **1985**, *26*, 1277.
[396] Lu, D. W.; Jankowski, K. *Spectroscopy* **1993**, *11*, 59.
[397] Palmisano, G.; Danieli, B.; Lesma, G.; Passarella, D.; Toma, L. *J. Org. Chem.* **1991**, *56*, 2380.
[398] Kametani, T.; Kajiwara, M.; Takahashi, T.; Fukumoto, K. *J. Chem. Soc., Perkin Trans. 1* **1975**, 737.
[399] Danishefsky, S.; Langer, M. E.; Vogel, C. *Tetrahedron Lett.* **1985**, *26*, 5983.
[400] Huff, J. R.; Anderson, P. S.; Baldwin, J. J.; Clineschmidt, B. V.; Guare, J. P.; Lotti, V. J.; Pettibone, D. J.; Randall, W. C.; Vacca, J. P. *J. Med. Chem.* **1985**, *28*, 1756.
[401] Ryan, K. M.; Reamer, R. A.; Volante, R. P.; Shinkai, I. *Tetrahedron Lett.* **1987**, *28*, 2103.
[402] Waldmann, H.; Braun, M.; Drager, M. *Tetrahedron: Asymmetry* **1991**, *2*, 1231.
[403] Trifonova, A.; Andersson, P. G. *Tetrahedron: Asymmetry* **2004**, *15*, 445.
[404] Matsuo, J.; Tsuchiya, T.; Odashima, K.; Kobayashi, S. *Chem. Lett.* **2000**, 178.
[405] Guo, H.; Ding, K. *Tetrahedron Lett.* **2003**, *44*, 7103.
[406] Andriamialisoa, R. Z.; Langlois, N.; Langlois, Y. *J. Org. Chem.* **1985**, *50*, 961.
[407] Andriamialisoa, R. Z.; Langlois, N.; Langlois, Y. *J. Chem. Soc., Chem. Commun.* **1982**, 1118.
[408] Ortuno, J.-C.; Langlois, Y. *Tetrahedron Lett.* **1991**, *32*, 4491.
[409] Yu, L.; Chen, D.; Li, J.; Wang, P. G. *J. Org. Chem.* **1997**, *62*, 3575.
[410] Patmore, N. J.; Hague, C.; Cotgreave, J. H.; Mahon, M. F.; Frost, C. G.; Weller, A. S. *Chem. Eur. J.* **2002**, *8*, 2088.
[411] Hague, C.; Patmore, N. J.; Frost, C. G.; Mahon, M. F.; Weller, A. S. *Chem. Commun.* **2001**, 2286.
[412] Zulfiqar, F.; Kitazume, T. *Green Chem.* **2000**, *2*, 137.
[413] Barluenga, J.; Aznar, F.; Cabal, M.-P.; Valdes, C. *J. Chem. Soc., Perkin Trans. 1* **1990**, 633.
[414] Barluenga, J.; Aznar, F.; Cabal, M.-P.; Cano, F. H.; Foces-Foces, M. C. *J. Chem. Soc., Chem. Commun.* **1988**, 1247.
[415] Veyrat, C.; Wartski, L.; Seyden-Penne, J. *Tetrahedron Lett.* **1986**, *27*, 2981.
[416] Stanetty, P.; Mihovilovic, M. D.; Mereiter, K.; Vollenkle, H.; Renz, F. *Tetrahedron* **1998**, *54*, 875.
[417] Stanetty, P.; Mihovilovic, M. D. *Chem. Lett.* **1997**, 849.
[418] Loh, T.-P.; Koh, K. S.-V.; Sim, K.-Y.; Leong, W.-K. *Tetrahedron Lett.* **1999**, *40*, 8447.
[419] Waldmann, H.; Braun, M.; Weymann, M.; Gewehr, M. *Tetrahedron* **1992**, *49*, 397.
[420] Lock, R.; Waldmann, H. *Liebigs Ann. Chem.* **1994**, 511.
[421] Lock, R.; Waldmann, H. *Nat. Prod. Lett.* **1993**, *2*, 49.

[422] Reetz, M. T.; Gansauer, A. *Tetrahedron* **1993**, *49*, 6025.
[423] Ishimaru, K.; Kojima, T. *J. Org. Chem.* **2000**, *65*, 8395.
[424] Le Coz, L.; Wartski, L.; Seyden-Penne, J.; Charpin, P.; Nierlich, M. *Tetrahedron Lett.* **1989**, *30*, 2795.
[425] Manabe, K.; Mori, Y.; Kobayashi, S. *Tetrahedron* **2001**, *57*, 2537.
[426] Clive, D. L. J.; Bergstra, R. J. *J. Org. Chem.* **1991**, *56*, 4976.
[427] Danishefsky, S.; Kerwin, J. F., Jr. *J. Org. Chem.* **1982**, *47*, 3183.
[428] Takemoto, Y.; Ueda, S.; Takeuchi, J.; Nakamoto, T.; Iwata, C. *Tetrahedron Lett.* **1994**, *35*, 8821.
[429] Paugam, R.; Valenciennes, E.; Le Coz-Bardol, L.; Garde, J.-C.; Wartski, L.; Lance, M.; Nierlich, M. *Tetrahedron: Asymmetry* **2000**, *11*, 2509.
[430] Waldmann, H.; Braun, M.; Weymann, M.; Gewehr, M. *Synlett* **1991**, 881.
[431] Shimizu, M.; Arai, A.; Fujisawa, T. *Heterocycles* **2000**, *52*, 137.
[432] Lin, Y. M.; Oh, T. *Tetrahedron Lett.* **1997**, *38*, 727.
[433] Kohara, T.; Hashimoto, Y.; Shioya, R.; Saigo, K. *Tetrahedron: Asymmetry* **1999**, *10*, 4831.
[434] Kuethe, J. T.; Davies, I. W.; Dormer, P. G.; Reamer, R. A.; Mathre, D. J.; Reider, P. J. *Tetrahedron Lett.* **2002**, *43*, 29.
[435] Kuethe, J. T.; Wong, A.; Davies, I. W.; Reider, P. J. *Tetrahedron Lett.* **2002**, *43*, 3871.
[436] Holsworth, D. D.; Bakir, F.; Uyeda, R. T.; Ge, Y.; Lau, J.; Hansen, T. K.; Andersen, H. S. *Synth. Commun.* **2003**, *33*, 1789.
[437] Avenoza, A.; Busto, J. H.; Cativiela, C.; Corzana, F.; Peregrina, J. M.; Zurbano, M. M. *J. Org. Chem.* **2002**, *67*, 598.
[438] Herczegh, P.; Kovacs, I.; Szilagyi, L.; Zsely, M.; Sztaricskai, F.; Berecibar, A.; Olesker, A.; Lukacs, G. *Tetrahedron Lett.* **1992**, *33*, 3133.
[439] Herczegh, P.; Kovacs, I.; Szilagyi, L.; Sztaricskai, F.; Berecibar, A.; Riche, C.; Chiaroni, A.; Olesker, A.; Lukacs, G. *Tetrahedron* **1994**, *50*, 13671.
[440] Herczegh, P.; Kovacs, I.; Erdosi, G.; Varga, T.; Agocs, A.; Szilagyi, L.; Sztaricskai, F.; Berecibar, A.; Lukacs, G.; Olesker, A. *Pure Appl. Chem.* **1997**, *69*, 519.
[441] Alcaide, B.; Almendros, P.; Alonso, J. M.; Aly, M. F.; Torres, M. R. *Synlett* **2001**, 1531.
[442] Alcaide, B.; Almendros, P.; Alonso, J. M.; Redondo, M. C. *J. Org. Chem.* **2003**, *68*, 1426.
[443] Herczegh, P.; Kovacs, I.; Szilagyi, L.; Sztaricskai, F.; Berecibar, A.; Riche, C.; Chiaroni, A.; Olesker, A.; Lukacs, G. *Tetrahedron* **1995**, *51*, 2969.
[444] Creighton, C. J.; Zapf, C. W.; Bu, J. H.; Goodman, M. *Org. Lett.* **1999**, *1*, 1407.
[445] Weymann, M.; Schulz-Kukula, M.; Knauer, S.; Kunz, H. *Monatsh. Chem.* **2002**, *133*, 571.
[446] Xiao, Z.; Patrick, B. O.; Dolphin, D. *Chem. Commun.* **2002**, 1816.
[447] Nair, V. *J. Org. Chem.* **1972**, *37*, 802.
[448] Hemetsberger, H.; Knittel, D. *Monatsh. Chem.* **1972**, *103*, 205.
[449] Bhullar, P.; Gilchrist, T. L.; Maddocks, P. *Synthesis* **1997**, 271.
[450] Bickley, J. F.; Gilchrist, T. L.; Mendonca, R. *Arkivoc* **2002**, 192.
[451] Padwa, A.; Smolanoff, J.; Tremper, A. *J. Org. Chem.* **1976**, *41*, 543.
[452] Fleury, J.-P. *Chimia* **1977**, *31*, 143.
[453] Fleury, J.-P.; Biehler, J.-M.; Desbois, M. *Tetrahedron Lett.* **1969**, 4091.
[454] Blondet, D.; Morin, C. *J. Chem. Soc., Perkin Trans. 1* **1984**, 1085.
[455] Blondet, D.; Morin, C. *Tetrahedron Lett.* **1982**, *23*, 3681.
[456] Lora-Tamayo, M.; Garcia Munoz, G.; Madronero, R. *Bull. Soc. Chim. Fr.* **1958**, 1331.
[457] Lora-Tamayo, M.; Madronero, R. In *1, 4-Cycloaddition Reactions: The Diels-Alder Reaction in Heterocyclic Syntheses*; Hamer, J., Ed.; Academic Press: New York, 1967; pp 127–142.
[458] Aparicio, T. L.; Lora-Tamayo, M.; Madronero, R.; Marzal, J. M. *Publ. Inst. Quim. "Alonso Barba"* **1961**, *15*, 41.
[459] Böhme, H.; Ahrens, K. H. *Arch. Pharm. (Weinheim, Ger.)* **1974**, *307*, 828.
[460] Böhme, H.; Ahrens, K. H. *Tetrahedron Lett.* **1971**, 149.
[461] Lasne, M.-C.; Ripoll, J.-L.; Thuillier, A. *J. Chem. Res. (S)* **1982**, 214.
[462] Earl, R. A.; Vollhardt, K. P. C. *Heterocycles* **1982**, *19*, 265.
[463] Nomoto, T.; Takayama, H. *Heterocycles* **1985**, *23*, 2913.
[464] Bailey, T. R.; Garigipati, R. S.; Morton, J. A.; Weinreb, S. M. *J. Am. Chem. Soc.* **1984**, *106*, 3240.

[465] Funk, R. L.; Vollhardt, K. P. C. *J. Am. Chem. Soc.* **1980**, *102*, 5245.
[466] Funk, R. L.; Vollhardt, K. P. C. *J. Am. Chem. Soc.* **1976**, *98*, 6755.
[467] Carroll, W. A.; Grieco, P. A. *J. Am. Chem. Soc.* **1993**, *115*, 1164.
[468] Yao, S.; Fang, X.; Jorgensen, K. A. *Chem. Commun.* **1998**, 2547.
[469] Johannsen, M.; Jorgensen, K. A.; Helmchen, G. *J. Am. Chem. Soc.* **1998**, *120*, 7637.

CUMULATIVE CHAPTER TITLES BY VOLUME

Volume 1 (1942)

1. **The Reformatsky Reaction**: Ralph L. Shriner
2. **The Arndt-Eistert Reaction**: W. E. Bachmann and W. S. Struve
3. **Chloromethylation of Aromatic Compounds**: Reynold C. Fuson and C. H. McKeever
4. **The Amination of Heterocyclic Bases by Alkali Amides**: Marlin T. Leffler
5. **The Bucherer Reaction**: Nathan L. Drake
6. **The Elbs Reaction**: Louis F. Fieser
7. **The Clemmensen Reduction**: Elmore L. Martin
8. **The Perkin Reaction and Related Reactions**: John R. Johnson
9. **The Acetoacetic Ester Condensation and Certain Related Reactions**: Charles R. Hauser and Boyd E. Hudson, Jr.
10. **The Mannich Reaction**: F. F. Blicke
11. **The Fries Reaction**: A. H. Blatt
12. **The Jacobson Reaction**: Lee Irvin Smith

Volume 2 (1944)

1. **The Claisen Rearrangement**: D. Stanley Tarbell
2. **The Preparation of Aliphatic Fluorine Compounds**: Albert L. Henne
3. **The Cannizzaro Reaction**: T. A. Geissman
4. **The Formation of Cyclic Ketones by Intramolecular Acylation**: William S. Johnson
5. **Reduction with Aluminum Alkoxides (The Meerwein-Ponndorf-Verley Reduction)**: A. L. Wilds

6. **The Preparation of Unsymmetrical Biaryls by the Diazo Reaction and the Nitrosoacetylamine Reaction**: Werner E. Bachmann and Roger A. Hoffman

7. **Replacement of the Aromatic Primary Amino Group by Hydrogen**: Nathan Kornblum

8. **Periodic Acid Oxidation**: Ernest L. Jackson

9. **The Resolution of Alcohols**: A. W. Ingersoll

10. **The Preparation of Aromatic Arsonic and Arsinic Acids by the Bart, Béchamp, and Rosenmund Reactions**: Cliff S. Hamilton and Jack F. Morgan

Volume 3 (1946)

1. **The Alkylation of Aromatic Compounds by the Friedel-Crafts Method**: Charles C. Price

2. **The Willgerodt Reaction**: Marvin Carmack and M. A. Spielman

3. **Preparation of Ketenes and Ketene Dimers**: W. E. Hanford and John C. Sauer

4. **Direct Sulfonation of Aromatic Hydrocarbons and Their Halogen Derivatives**: C. M. Suter and Arthur W. Weston

5. **Azlactones**: H. E. Carter

6. **Substitution and Addition Reactions of Thiocyanogen**: John L. Wood

7. **The Hofmann Reaction**: Everett L. Wallis and John F. Lane

8. **The Schmidt Reaction**: Hans Wolff

9. **The Curtius Reaction**: Peter A. S. Smith

Volume 4 (1948)

1. **The Diels-Alder Reaction with Maleic Anhydride**: Milton C. Kloetzel

2. **The Diels-Alder Reaction: Ethylenic and Acetylenic Dienophiles**: H. L. Holmes

3. **The Preparation of Amines by Reductive Alkylation**: William S. Emerson

4. **The Acyloins**: S. M. McElvain

5. **The Synthesis of Benzoins**: Walter S. Ide and Johannes S. Buck

6. **Synthesis of Benzoquinones by Oxidation**: James Cason

7. **The Rosenmund Reduction of Acid Chlorides to Aldehydes**: Erich Mosettig and Ralph Mozingo

8. **The Wolff-Kishner Reduction**: David Todd

Volume 5 (1949)

1. **The Synthesis of Acetylenes**: Thomas L. Jacobs
2. **Cyanoethylation**: Herman L. Bruson
3. **The Diels-Alder Reaction: Quinones and Other Cyclenones**: Lewis L. Butz and Anton W. Rytina
4. **Preparation of Aromatic Fluorine Compounds from Diazonium Fluoborates: The Schiemann Reaction**: Arthur Roe
5. **The Friedel and Crafts Reaction with Aliphatic Dibasic Acid Anhydrides**: Ernst Berliner
6. **The Gattermann-Koch Reaction**: Nathan N. Crounse
7. **The Leuckart Reaction**: Maurice L. Moore
8. **Selenium Dioxide Oxidation**: Norman Rabjohn
9. **The Hoesch Synthesis**: Paul E. Spoerri and Adrien S. DuBois
10. **The Darzens Glycidic Ester Condensation**: Melvin S. Newman and Barney J. Magerlein

Volume 6 (1951)

1. **The Stobbe Condensation**: William S. Johnson and Guido H. Daub
2. **The Preparation of 3,4-Dihydroisoquinolines and Related Compounds by the Bischler-Napieralski Reaction**: Wilson M. Whaley and Tutucorin R. Govindachari
3. **The Pictet-Spengler Synthesis of Tetrahydroisoquinolines and Related Compounds**: Wilson M. Whaley and Tutucorin R. Govindachari
4. **The Synthesis of Isoquinolines by the Pomeranz-Fritsch Reaction**: Walter J. Gensler
5. **The Oppenauer Oxidation**: Carl Djerassi
6. **The Synthesis of Phosphonic and Phosphinic Acids**: Gennady M. Kosolapoff
7. **The Halogen-Metal Interconversion Reaction with Organolithium Compounds**: Reuben G. Jones and Henry Gilman
8. **The Preparation of Thiazoles**: Richard H. Wiley, D. C. England, and Lyell C. Behr
9. **The Preparation of Thiophenes and Tetrahydrothiophenes**: Donald E. Wolf and Karl Folkers
10. **Reductions by Lithium Aluminum Hydride**: Weldon G. Brown

Volume 7 (1953)

1. **The Pechmann Reaction**: Suresh Sethna and Ragini Phadke
2. **The Skraup Synthesis of Quinolines**: R. H. F. Manske and Marshall Kulka
3. **Carbon-Carbon Alkylations with Amines and Ammonium Salts**: James H. Brewster and Ernest L. Eliel
4. **The von Braun Cyanogen Bromide Reaction**: Howard A. Hageman
5. **Hydrogenolysis of Benzyl Groups Attached to Oxygen, Nitrogen, or Sulfur**: Walter H. Hartung and Robert Simonoff
6. **The Nitrosation of Aliphatic Carbon Atoms**: Oscar Touster
7. **Epoxidation and Hydroxylation of Ethylenic Compounds with Organic Peracids**: Daniel Swern

Volume 8 (1954)

1. **Catalytic Hydrogenation of Esters to Alcohols**: Homer Adkins
2. **The Synthesis of Ketones from Acid Halides and Organometallic Compounds of Magnesium, Zinc, and Cadmium**: David A. Shirley
3. **The Acylation of Ketones to Form β-Diketones or β-Keto Aldehydes**: Charles R. Hauser, Frederic W. Swamer, and Joe T. Adams
4. **The Sommelet Reaction**: S. J. Angyal
5. **The Synthesis of Aldehydes from Carboxylic Acids**: Erich Mosettig
6. **The Metalation Reaction with Organolithium Compounds**: Henry Gilman and John W. Morton, Jr.
7. **β-Lactones**: Harold E. Zaugg
8. **The Reaction of Diazomethane and Its Derivatives with Aldehydes and Ketones**: C. David Gutsche

Volume 9 (1957)

1. **The Cleavage of Non-enolizable Ketones with Sodium Amide**: K. E. Hamlin and Arthur W. Weston
2. **The Gattermann Synthesis of Aldehydes**: William E. Truce
3. **The Baeyer-Villiger Oxidation of Aldehydes and Ketones**: C. H. Hassall
4. **The Alkylation of Esters and Nitriles**: Arthur C. Cope, H. L. Holmes, and Herbert O. House

2. **Halocyclopropanes from Halocarbenes**: William E. Parham and Edward E. Schweizer
3. **Free Radical Addition to Olefins to Form Carbon-Carbon Bonds**: Cheves Walling and Earl S. Huyser
4. **Formation of Carbon-Heteroatom Bonds by Free Radical Chain Additions to Carbon-Carbon Multiple Bonds**: F. W. Stacey and J. F. Harris, Jr.

Volume 14 (1965)

1. **The Chapman Rearrangement**: J. W. Schulenberg and S. Archer
2. **α-Amidoalkylations at Carbon**: Harold E. Zaugg and William B. Martin
3. **The Wittig Reaction**: Adalbert Maercker

Volume 15 (1967)

1. **The Dieckmann Condensation**: John P. Schaefer and Jordan J. Bloomfield
2. **The Knoevenagel Condensation**: G. Jones

Volume 16 (1968)

1. **The Aldol Condensation**: Arnold T. Nielsen and William J. Houlihan

Volume 17 (1969)

1. **The Synthesis of Substituted Ferrocenes and Other π-Cyclopentadienyl-Transition Metal Compounds**: Donald E. Bublitz and Kenneth L. Rinehart, Jr.
2. **The γ-Alkylation and γ-Arylation of Dianions of β-Dicarbonyl Compounds**: Thomas M. Harris and Constance M. Harris
3. **The Ritter Reaction**: L. I. Krimen and Donald J. Cota

Volume 18 (1970)

1. **Preparation of Ketones from the Reaction of Organolithium Reagents with Carboxylic Acids**: Margaret J. Jorgenson
2. **The Smiles and Related Rearrangements of Aromatic Systems**: W. E. Truce, Eunice M. Kreider, and William W. Brand
3. **The Reactions of Diazoacetic Esters with Alkenes, Alkynes, Heterocyclic, and Aromatic Compounds**: Vinod Dave and E. W. Warnhoff
4. **The Base-Promoted Rearrangements of Quaternary Ammonium Salts**: Stanley H. Pine

Volume 19 (1972)

1. **Conjugate Addition Reactions of Organocopper Reagents**: Gary H. Posner
2. **Formation of Carbon-Carbon Bonds via π-Allylnickel Compounds**: Martin F. Semmelhack
3. **The Thiele-Winter Acetoxylation of Quinones**: J. F. W. McOmie and J. M. Blatchly
4. **Oxidative Decarboxylation of Acids by Lead Tetraacetate**: Roger A. Sheldon and Jay K. Kochi

Volume 20 (1973)

1. **Cyclopropanes from Unsaturated Compounds, Methylene Iodide, and Zinc-Copper Couple**: H. E. Simmons, T. L. Cairns, Susan A. Vladuchick, and Connie M. Hoiness
2. **Sensitized Photooxygenation of Olefins**: R. W. Denny and A. Nickon
3. **The Synthesis of 5-Hydroxyindoles by the Nenitzescu Reaction**: George R. Allen, Jr.
4. **The Zinin Reaction of Nitroarenes**: H. K. Porter

Volume 21 (1974)

1. **Fluorination with Sulfur Tetrafluoride**: G. A. Boswell, Jr., W. C. Ripka, R. M. Scribner, and C. W. Tullock
2. **Modern Methods to Prepare Monofluoroaliphatic Compounds**: Clay M. Sharts and William A. Sheppard

Volume 22 (1975)

1. **The Claisen and Cope Rearrangements**: Sara Jane Rhoads and N. Rebecca Raulins
2. **Substitution Reactions Using Organocopper Reagents**: Gary H. Posner
3. **Clemmensen Reduction of Ketones in Anhydrous Organic Solvents**: E. Vedejs
4. **The Reformatsky Reaction**: Michael W. Rathke

Volume 23 (1976)

1. **Reduction and Related Reactions of α,β-Unsaturated Compounds with Metals in Liquid Ammonia**: Drury Caine
2. **The Acyloin Condensation**: Jordan J. Bloomfield, Dennis C. Owsley, and Janice M. Nelke
3. **Alkenes from Tosylhydrazones**: Robert H. Shapiro

Volume 24 (1976)

1. **Homogeneous Hydrogenation Catalysts in Organic Solvents**: Arthur J. Birch and David H. Williamson

2. **Ester Cleavages via S_N2-Type Dealkylation**: John E. McMurry

3. **Arylation of Unsaturated Compounds by Diazonium Salts (The Meerwein Arylation Reaction)**: Christian S. Rondestvedt, Jr.

4. **Selenium Dioxide Oxidation**: Norman Rabjohn

Volume 25 (1977)

1. **The Ramberg-Bäcklund Rearrangement**: Leo A. Paquette

2. **Synthetic Applications of Phosphoryl-Stabilized Anions**: William S. Wadsworth, Jr.

3. **Hydrocyanation of Conjugated Carbonyl Compounds**: Wataru Nagata and Mitsuru Yoshioka

Volume 26 (1979)

1. **Heteroatom-Facilitated Lithiations**: Heinz W. Gschwend and Herman R. Rodriguez

2. **Intramolecular Reactions of Diazocarbonyl Compounds**: Steven D. Burke and Paul A. Grieco

Volume 27 (1982)

1. **Allylic and Benzylic Carbanions Substituted by Heteroatoms**: Jean-François Biellmann and Jean-Bernard Ducep

2. **Palladium-Catalyzed Vinylation of Organic Halides**: Richard F. Heck

Volume 28 (1982)

1. **The Reimer-Tiemann Reaction**: Hans Wynberg and Egbert W. Meijer

2. **The Friedländer Synthesis of Quinolines**: Chia-Chung Cheng and Shou-Jen Yan

3. **The Directed Aldol Reaction**: Teruaki Mukaiyama

Volume 29 (1983)

1. **Replacement of Alcoholic Hydroxy Groups by Halogens and Other Nucleophiles via Oxyphosphonium Intermediates**: Bertrand R. Castro

Volume 36 (1988)

1. **The [3 + 2] Nitrone-Olefin Cycloaddition Reaction**: Pat N. Confalone and Edward M. Huie
2. **Phosphorus Addition at sp^2 Carbon**: Robert Engel
3. **Reduction by Metal Alkoxyaluminum Hydrides. Part II. Carboxylic Acids and Derivatives, Nitrogen Compounds, and Sulfur Compounds**: Jaroslav Málek

Volume 37 (1989)

1. **Chiral Synthons by Ester Hydrolysis Catalyzed by Pig Liver Esterase**: Masaji Ohno and Masami Otsuka
2. **The Electrophilic Substitution of Allylsilanes and Vinylsilanes**: Ian Fleming, Jacques Dunoguès, and Roger Smithers

Volume 38 (1990)

1. **The Peterson Olefination Reaction**: David J. Ager
2. **Tandem Vicinal Difunctionalization: β-Addition to α,β-Unsaturated Carbonyl Substrates Followed by α-Functionalization**: Marc J. Chapdelaine and Martin Hulce
3. **The Nef Reaction**: Harold W. Pinnick

Volume 39 (1990)

1. **Lithioalkenes from Arenesulfonylhydrazones**: A. Richard Chamberlin and Steven H. Bloom
2. **The Polonovski Reaction**: David Grierson
3. **Oxidation of Alcohols to Carbonyl Compounds via Alkoxysulfonium Ylides: The Moffatt, Swern, and Related Oxidations**: Thomas T. Tidwell

Volume 40 (1991)

1. **The Pauson-Khand Cycloaddition Reaction for Synthesis of Cyclopentenones**: Neil E. Schore
2. **Reduction with Diimide**: Daniel J. Pasto and Richard T. Taylor
3. **The Pummerer Reaction of Sulfinyl Compounds**: Ottorino DeLucchi, Umberto Miotti, and Giorgio Modena
4. **The Catalyzed Nucleophilic Addition of Aldehydes to Electrophilic Double Bonds**: Hermann Stetter and Heinrich Kuhlmann

Volume 41 (1992)

1. **Divinylcyclopropane-Cycloheptadiene Rearrangement**: Tomáš Hudlický, Rulin Fan, Josephine W. Reed, and Kumar G. Gadamasetti

2. **Organocopper Reagents: Substitution, Conjugate Addition, Carbo/Metallocupration, and Other Reactions**: Bruce H. Lipshutz and Saumitra Sengupta

Volume 42 (1992)

1. **The Birch Reduction of Aromatic Compounds**: Peter W. Rabideau and Zbigniew Marcinow

2. **The Mitsunobu Reaction**: David L. Hughes

Volume 43 (1993)

1. **Carbonyl Methylenation and Alkylidenation Using Titanium-Based Reagents**: Stanley H. Pine

2. **Anion-Assisted Sigmatropic Rearrangements**: Stephen R. Wilson

3. **The Baeyer-Villiger Oxidation of Ketones and Aldehydes**: Grant R. Krow

Volume 44 (1993)

1. **Preparation of α,β-Unsaturated Carbonyl Compounds and Nitriles by Selenoxide Elimination**: Hans J. Reich and Susan Wollowitz

2. **Enone Olefin [2 + 2] Photochemical Cyclizations**: Michael T. Crimmins and Tracy L. Reinhold

Volume 45 (1994)

1. **The Nazarov Cyclization**: Karl L. Habermas, Scott E. Denmark, and Todd K. Jones

2. **Ketene Cycloadditions**: John A. Hyatt and Peter W. Raynolds

Volume 46 (1994)

1. **Tin(II) Enolates in the Aldol, Michael, and Related Reactions**: Teruaki Mukaiyama and Shū Kobayashi

2. **The [2,3]-Wittig Reaction**: Takeshi Nakai and Koichi Mikami

3. **Reductions with Samarium(II) Iodide**: Gary A. Molander

Volume 47 (1995)

1. **Lateral Lithiation Reactions Promoted by Heteroatomic Substituents**: Robin D. Clark and Alam Jahangir

2. **The Intramolecular Michael Reaction**: R. Daniel Little, Mohammad R. Masjedizadeh, Olof Wallquist (in part), and Jim I. McLoughlin (in part)

Volume 48 (1996)

1. **Asymmetric Epoxidation of Allylic Alcohols: The Katsuki–Sharpless Epoxidation Reaction**: Tsutomu Katsuki and Victor S. Martin

2. **Radical Cyclization Reactions**: B. Giese, B. Kopping, T. Göbel, J. Dickhaut, G. Thoma, K. J. Kulicke, and F. Trach

Volume 49 (1997)

1. **The Vilsmeier Reaction of Fully Conjugated Carbocycles and Heterocycles**: Gurnos Jones and Stephen P. Stanforth

2. **[6 + 4] Cycloaddition Reactions**: James H. Rigby

3. **Carbon–Carbon Bond-Forming Reactions Promoted by Trivalent Manganese**: Gagik G. Melikyan

Volume 50 (1997)

1. **The Stille Reaction**: Vittorio Farina, Venkat Krishnamurthy, and William J. Scott

Volume 51 (1997)

1. **Asymmetric Aldol Reactions Using Boron Enolates**: Cameron J. Cowden and Ian Paterson

2. **The Catalyzed α-Hydroxylation and α-Aminoalkylation of Activated Olefins (The Morita–Baylis–Hillman Reaction)**: Engelbert Ciganek

3. **[4 + 3] Cycloaddition Reactions**: James H. Rigby and F. Christopher Pigge

Volume 52 (1998)

1. **The Retro–Diels–Alder Reaction. Part I. C—C Dienophiles**: Bruce Rickborn

2. **Enantioselective Reduction of Ketones**: Shinichi Itsuno

Volume 53 (1998)

1. **The Oxidation of Alcohols by Modified Oxochromium(VI)-Amine Reagents**: Frederick A. Luzzio

2. **The Retro–Diels–Alder Reaction. Part II. Dienophiles with One or More Heteroatoms**: Bruce Rickborn

Volume 54 (1999)

1. **Aromatic Substitution by the $S_{RN}1$ Reaction**: Roberto Rossi, Adriana B. Pierini, and Ana N. Santiago

2. **Oxidation of Carbonyl Compounds with Organohypervalent Iodine Reagents**: Robert M. Moriarty and Om Prakash

AUTHOR INDEX, VOLUMES 1–65

Volume number only is designated in this index.

CHAPTER AND TOPIC INDEX, VOLUMES 1–65

Many chapters contain brief discussions of reactions and comparisons of alternative synthetic methods related to the reaction that is the subject of the chapter. These related reactions and alternative methods are not usually listed in this index. In this index, the volume number is in **boldface**, the chapter number is in ordinary type.